Code of Federal Regulations

21

Parts 800 to 1299

Revised as of April 1, 2006

Food and Drugs

Containing a codification of documents of general applicability and future effect

As of April 1, 2006

With Ancillaries

Published by
Office of the Federal Register
National Archives and Records Administration

A Special Edition of the Federal Register

U.S. GOVERNMENT OFFICIAL EDITION NOTICE

Legal Status and Use of Seals and Logos

The seal of the National Archives and Records Administration (NARA) authenticates the Code of Federal Regulations (CFR) as the official codification of Federal regulations established under the Federal Register Act. Under the provisions of 44 U.S.C. 1507, the contents of the CFR, a special edition of the Federal Register, shall be judicially noticed. The CFR is prima facie evidence of the original documents published in the Federal Register (44 U.S.C. 1510).

It is prohibited to use NARA's official seal and the stylized Code of Federal Regulations logo on any republication of this material without the express, written permission of the Archivist of the United States or the Archivist's designee. Any person using NARA's official seals and logos in a manner inconsistent with the provisions of 36 CFR part 1200 is subject to the penalties specified in 18 U.S.C. 506, 701, and 1017.

Use of ISBN Prefix

This is the Official U.S. Government edition of this publication and is herein identified to certify its authenticity. Use of the 0–16 ISBN prefix is for U.S. Government Printing Office Official Editions only. The Superintendent of Documents of the U.S. Government Printing Office requests that any reprinted edition clearly be labeled as a copy of the authentic work with a new ISBN.

U.S. GOVERNMENT PRINTING OFFICE

U.S. Superintendent of Documents • Washington, DC 20402–0001

http://bookstore.gpo.gov

Phone: toll-free (866) 512-1800; DC area (202) 512-1800

Table of Contents

Cite this Code: CFR

To cite the regulations in this volume use title, part and section number. Thus, 21 CFR 800.10 *refers to title 21, part 800, section 10.*

Explanation

The Code of Federal Regulations is a codification of the general and permanent rules published in the Federal Register by the Executive departments and agencies of the Federal Government. The Code is divided into 50 titles which represent broad areas subject to Federal regulation. Each title is divided into chapters which usually bear the name of the issuing agency. Each chapter is further subdivided into parts covering specific regulatory areas.

Each volume of the Code is revised at least once each calendar year and issued on a quarterly basis approximately as follows:

Title 1 through Title 16..........as of January 1
Title 17 through Title 27..........as of April 1
Title 28 through Title 41..........as of July 1
Title 42 through Title 50..........as of October 1

The appropriate revision date is printed on the cover of each volume.

LEGAL STATUS

The contents of the Federal Register are required to be judicially noticed (44 U.S.C. 1507). The Code of Federal Regulations is prima facie evidence of the text of the original documents (44 U.S.C. 1510).

HOW TO USE THE CODE OF FEDERAL REGULATIONS

The Code of Federal Regulations is kept up to date by the individual issues of the Federal Register. These two publications must be used together to determine the latest version of any given rule.

To determine whether a Code volume has been amended since its revision date (in this case, April 1, 2006), consult the "List of CFR Sections Affected (LSA)," which is issued monthly, and the "Cumulative List of Parts Affected," which appears in the Reader Aids section of the daily Federal Register. These two lists will identify the Federal Register page number of the latest amendment of any given rule.

EFFECTIVE AND EXPIRATION DATES

Each volume of the Code contains amendments published in the Federal Register since the last revision of that volume of the Code. Source citations for the regulations are referred to by volume number and page number of the Federal Register and date of publication. Publication dates and effective dates are usually not the same and care must be exercised by the user in determining the actual effective date. In instances where the effective date is beyond the cut-off date for the Code a note has been inserted to reflect the future effective date. In those instances where a regulation published in the Federal Register states a date certain for expiration, an appropriate note will be inserted following the text.

OMB CONTROL NUMBERS

The Paperwork Reduction Act of 1980 (Pub. L. 96–511) requires Federal agencies to display an OMB control number with their information collection request.

Many agencies have begun publishing numerous OMB control numbers as amendments to existing regulations in the CFR. These OMB numbers are placed as close as possible to the applicable recordkeeping or reporting requirements.

OBSOLETE PROVISIONS

Provisions that become obsolete before the revision date stated on the cover of each volume are not carried. Code users may find the text of provisions in effect on a given date in the past by using the appropriate numerical list of sections affected. For the period before January 1, 2001, consult either the List of CFR Sections Affected, 1949–1963, 1964–1972, 1973–1985, or 1986–2000, published in 11 separate volumes. For the period beginning January 1, 2001, a ''List of CFR Sections Affected'' is published at the end of each CFR volume.

INCORPORATION BY REFERENCE

What is incorporation by reference? Incorporation by reference was established by statute and allows Federal agencies to meet the requirement to publish regulations in the Federal Register by referring to materials already published elsewhere. For an incorporation to be valid, the Director of the Federal Register must approve it. The legal effect of incorporation by reference is that the material is treated as if it were published in full in the Federal Register (5 U.S.C. 552(a)). This material, like any other properly issued regulation, has the force of law.

What is a proper incorporation by reference? The Director of the Federal Register will approve an incorporation by reference only when the requirements of 1 CFR part 51 are met. Some of the elements on which approval is based are:

(a) The incorporation will substantially reduce the volume of material published in the Federal Register.

(b) The matter incorporated is in fact available to the extent necessary to afford fairness and uniformity in the administrative process.

(c) The incorporating document is drafted and submitted for publication in accordance with 1 CFR part 51.

Properly approved incorporations by reference in this volume are listed in the Finding Aids at the end of this volume.

What if the material incorporated by reference cannot be found? If you have any problem locating or obtaining a copy of material listed in the Finding Aids of this volume as an approved incorporation by reference, please contact the agency that issued the regulation containing that incorporation. If, after contacting the agency, you find the material is not available, please notify the Director of the Federal Register, National Archives and Records Administration, Washington DC 20408, or call 202-741-6010.

CFR INDEXES AND TABULAR GUIDES

A subject index to the Code of Federal Regulations is contained in a separate volume, revised annually as of January 1, entitled CFR INDEX AND FINDING AIDS. This volume contains the Parallel Table of Statutory Authorities and Agency Rules (Table I). A list of CFR titles, chapters, and parts and an alphabetical list of agencies publishing in the CFR are also included in this volume.

An index to the text of ''Title 3—The President'' is carried within that volume.

The Federal Register Index is issued monthly in cumulative form. This index is based on a consolidation of the ''Contents'' entries in the daily Federal Register.

A List of CFR Sections Affected (LSA) is published monthly, keyed to the revision dates of the 50 CFR titles.

REPUBLICATION OF MATERIAL

There are no restrictions on the republication of textual material appearing in the Code of Federal Regulations.

INQUIRIES

For a legal interpretation or explanation of any regulation in this volume, contact the issuing agency. The issuing agency's name appears at the top of odd-numbered pages.

For inquiries concerning CFR reference assistance, call 202-741-6000 or write to the Director, Office of the Federal Register, National Archives and Records Administration, Washington, DC 20408 or e-mail fedreg.info@nara.gov.

SALES

The Government Printing Office (GPO) processes all sales and distribution of the CFR. For payment by credit card, call toll-free, 866-512-1800 or DC area, 202-512-1800, M-F, 8 a.m. to 4 p.m. e.s.t. or fax your order to 202-512-2250, 24 hours a day. For payment by check, write to the Superintendent of Documents, Attn: New Orders, P.O. Box 371954, Pittsburgh, PA 15250-7954. For GPO Customer Service call 202-512-1803.

ELECTRONIC SERVICES

The full text of the Code of Federal Regulations, the LSA (List of CFR Sections Affected), The United States Government Manual, the Federal Register, Public Laws, Public Papers, Weekly Compilation of Presidential Documents and the Privacy Act Compilation are available in electronic format at *www.gpoaccess.gov/nara* ("GPO Access"). For more information, contact Electronic Information Dissemination Services, U.S. Government Printing Office. Phone 202-512-1530, or 888-293-6498 (toll-free). E-mail, *gpoaccess@gpo.gov*.

The Office of the Federal Register also offers a free service on the National Archives and Records Administration's (NARA) World Wide Web site for public law numbers, Federal Register finding aids, and related information. Connect to NARA's web site at *www.archives.gov/federal-register*. The NARA site also contains links to GPO Access.

RAYMOND A. MOSLEY,
Director,
Office of the Federal Register.

April 1, 2006.

THIS TITLE

Title 21—FOOD AND DRUGS is composed of nine volumes. The parts in these volumes are arranged in the following order: Parts 1–99, 100–169, 170–199, 200–299, 300–499, 500–599, 600–799, 800–1299 and 1300–end. The first eight volumes, containing parts 1–1299, comprise Chapter I—Food and Drug Administration, Department of Health and Human Services. The ninth volume, containing part 1300 to end, includes Chapter II—Drug Enforcement Administration, Department of Justice, and Chapter III—Office of National Drug Control Policy. The contents of these volumes represent all current regulations codified under this title of the CFR as of April 1, 2006.

For this volume, Robert J. Sheehan was Chief Editor. The Code of Federal Regulations publication program is under the direction of Frances D. McDonald, assisted by Alomha S. Morris.

Title 21—Food and Drugs

(This book contains Parts 800 to 1299)

CHAPTER I—FOOD AND DRUG ADMINISTRATION, DEPARTMENT OF HEALTH AND HUMAN SERVICES (CONTINUED)

EDITORIAL NOTES: 1. For nomenclature changes to chapter I see 59 FR 14366, Mar. 28, 1994.

2. For nomenclature changes to chapter I see 68 FR 24879, May 9, 2003.

3. For nomenclature changes to chapter I see 69 FR 13717, Mar. 24, 2004.

SUBCHAPTER H—MEDICAL DEVICES

SUBCHAPTER H—MEDICAL DEVICES

PART 800—GENERAL

AUTHORITY: 21 U.S.C. 321, 334, 351, 352, 355, 360e, 360i, 360k, 361, 362, 371.

Subpart A [Reserved]

Subpart B—Requirements for Specific Medical Devices

§ 800.10 Contact lens solutions; sterility.

(a)(1) Informed medical opinion is in agreement that all preparations offered or intended for ophthalmic use, including contact lens solutions, should be sterile. It is further evident that such preparations purport to be of such purity and quality as to be suitable for safe use in the eye.

(2) The Food and Drug Administration concludes that all such preparations, if they are not sterile, fall below their professed standard of purity or quality and may be unsafe. In a statement of policy issued on September 1, 1964, the Food and Drug Administration ruled that liquid preparations offered or intended for ophthalmic use that are not sterile may be regarded as adulterated within the meaning of section 501(c) of the Federal Food, Drug, and Cosmetic Act (the act), and, further, may be deemed misbranded within the meaning of section 502(j) of the act. By this regulation, this ruling is applicable to all preparations for ophthalmic use that are regulated as medical devices, i.e., contact lens solutions. By the regulation in § 200.50 of this chapter, this ruling is applicable to ophthalmic preparations that are regulated as drugs.

(3) The containers shall be sterile at the time of filling and closing, and the container or individual carton shall be so sealed that the contents cannot be used without destroying the seal. The packaging and labeling of these solutions shall also comply with § 800.12 on tamper-resistant packaging requirements.

(b) Liquid ophthalmic preparations packed in multiple-dose containers should:

(1) Contain one or more suitable and harmless substances that will inhibit the growth of microorganisms; or

(2) Be so packaged as to volume and type of container and so labeled as to duration of use and with such necessary warnings as to afford adequate protection and minimize the hazard of injury resulting from contamination during use.

(c) Eye cups, eye droppers, and other dispensers intended for ophthalmic use should be sterile, and may be regarded as falling below their professed standard of purity or quality if they are not sterile. These articles, which are regulated as medical devices unless packaged with the drugs with which they are to be used, should be packaged so as to maintain sterility until the package is opened and be labeled, on or within the retail package, so as to afford adequate directions and necessary warnings to minimize the hazard of injury resulting from contamination during use.

[47 FR 50455, Nov. 5, 1982]

§ 800.12 Contact lens solutions and tablets; tamper-resistant packaging.

(a) *General.* Unless contact lens solutions used, for example, to clean, disinfect, wet, lubricate, rinse, soak, or store contact lenses and salt tablets or other dosage forms to be used to make any such solutions are packaged in tamper-resistant retail packages, there is the opportunity for the malicious adulteration of these products with risks both to individuals who unknowingly purchase adulterated products and with loss of consumer confidence in the security of the packages of over-the-counter (OTC) health care products. The Food and Drug Administration has the authority and responsibility under the Federal Food, Drug, and Cosmetic Act (the act) to establish a uniform national standard for tamper-resistant packaging of those OTC products vulnerable to malicious adulteration that will improve the security of OTC packaging and help assure the safety and effectiveness of the products contained therein. A contact lens solution or tablet or other dosage form to be used to make such a solution for retail sale that is not packaged in a tamper-resistant package and labeled in accordance with this section is adulterated under section 501 of the act or misbranded under section 502 of the act, or both.

(b) *Requirement for tamper-resistant package.* Each manufacturer and packer who packages for retail sale a product regulated as a medical device that is a solution intended for use with contact lenses, e.g., for cleaning, disinfecting, wetting, lubricating, rinsing, soaking, or storing contact lenses or tablets or other dosage forms to be used to make any such solution shall package the product in a tamper-resistant package, if this product is accessible to the public while held for sale. A tamper-resistant package is one having an indicator or barrier to entry which, if breached or missing, can reasonably be expected to provide visible evidence to consumers that tampering has occurred. To reduce the likelihood of substitution of a tamper-resistant feature after tampering, the indicator or barrier to entry is required to be distinctive by design or by the use of an identifying characteristic (e.g., a pattern, name, registered trademark, logo, or picture). For purposes of this section, the term "distinctive by design" means the package cannot be duplicated with commonly available material or through commonly available processes. A tamper-resistant package may involve an immediate-container and closure system or secondary-container or carton system or any combination of systems intended to provide a visual indication of package integrity. The tamper-resistant feature shall be designed to and shall remain intact when handled in a reasonable manner during manufacture, distribution, and retail display.

(c) *Labeling.* Each retail package of a product covered by this section is required to bear a statement that is prominently placed so that consumers are alerted to the tamper-resistant feature of the package. The labeling statement is also required to be so placed that it will be unaffected if the tamper-resistant feature of the package is breached or missing. If the tamper-resistant feature chosen to meet the requirement in paragraph (b) of this section is one that uses an identifying characteristic, that characteristic is required to be referred to in the labeling statement. For example, the labeling statement on a bottle with a shrink band could say "For your protection, this bottle has an imprinted seal around the neck."

(d) *Requests for exemptions from packaging and labeling requirements.* A manufacturer or packer may request an exemption from the packaging and labeling requirements of this section. A request for an exemption is required to be submitted in the form of a citizen petition under § 10.30 of this chapter and should be clearly identified on the envelope as a "Request for Exemption from Tamper-resistant Rule." A petition for an exemption from a requirement of this section is required to contain the same kind of information about the product as is specified for OTC drugs in § 211.132(d) of this chapter. This information collection requirement has been approved by the Office of Management and Budget under number 0910–0150.

(e) *Products subject to approved premarket approval applications.* Holders of approved premarket approval applications for products subject to this section are required to submit supplements to provide for changes in packaging to comply with the requirement of paragraph (b) of this section unless these changes do not affect the composition of the container, the torque (tightness) of the container, or the composition of the closure component in contact with the contents (cap liner or innerseal) as these features are described in the approved premarket approval application. Any supplemental premarket approval application under this paragraph is required to include data sufficient to show that these changes do not adversely affect the product.

(f) *Effective date.* Each product subject to this section is required to comply with the requirements of this section on the dates listed below except to the extent that a product's manufacturer or packer has obtained an exemption from a packaging or labeling requirement:

(1) *Initial effective date for packaging requirements.* (i) The packaging requirement in paragraph (b) of this section is effective on February 7, 1983 for each contact lens solution packaged for retail sale on or after that date, except for the requirement in paragraph (b) of this section for a distinctive indicator or barrier to entry.

(ii) The packaging requirement in paragraph (b) of this section is effective on May 5, 1983 for each tablet that is to be used to make a contact lens solution and that is packaged for retail sale on or after that date.

(2) *Initial effective date for labeling requirements.* The requirement in paragraph (b) of this section that the indicator or barrier to entry be distinctive by design and the requirement in paragraph (c) of this section for a labeling statement are effective on May 5, 1983 for each product subject to this section packaged for retail sale on or after that date, except that the requirement for a specific label reference to any identifying characteristic is effective on February 6, 1984 for each affected product subject to this section packaged for retail sale on or after that date.

(3) *Retail level effective date.* The tamper-resistant packaging requirement of paragraph (b) of this section is effective on February 6, 1984 for each product subject to this section that is held for sale at retail level on or after that date that was packaged for retail sale before May 5, 1983. This does not include the requirement in paragraph (b) of this section that the indicator or barrier to entry be distinctive by design. Products packaged for retail sale after May 5, 1983, are required to be in compliance with all aspects of the regulations without regard to the retail level effective date.

[47 FR 50455, Nov. 5, 1982; 48 FR 1706, Jan. 14, 1983, as amended at 48 FR 16666, Apr. 19, 1983; 48 FR 37625, Aug. 19, 1983; 53 FR 11252, Apr. 6, 1988]

EFFECTIVE DATE NOTE: A document published at 48 FR 41579, Sept. 16, 1983, stayed the effective date of § 800.12(f)(3) until further notice.

§ 800.20 Patient examination gloves and surgeons' gloves; sample plans and test method for leakage defects; adulteration.

(a) *Purpose.* The prevalence of human immunodeficiency virus (HIV), which causes acquired immune deficiency syndrome (AIDS), and its risk of transmission in the health care context, have caused the Food and Drug Administration (FDA) to look more closely at the quality control of barrier devices, such as surgeons' gloves and patient examination gloves (collectively known as medical gloves) to reduce the risk of transmission of HIV and other blood-borne infectious diseases. The Centers for Disease Control (CDC) recommend that health care workers wear medical gloves to reduce the risk of transmission of HIV and other blood-borne infectious deseases. The CDC recommends that health care workers wear medical gloves when touching blood or other body fluids, mucous membranes, or nonintact skin of all patients; when handling items or surfaces soiled with blood or other body fluids; and when performing venipuncture and other vascular access procedures. Among other things, CDC's recommendation that health care providers wear medical gloves demonstrates the proposition that devices labeled as medical gloves purport to be and are represented to be effective barriers against the transmission of blood- and fluid-borne pathogens. Therefore, FDA, through this regulation, is defining adulteration for patient examination and surgeons' gloves as a means of assuring safe and effective devices.

(1) For a description of a patient examination glove, see § 880.6250. Finger cots, however, are excluded from the

test method and sample plans in paragraphs (b) and (c) of this section.

(2) For a description of a surgeons' glove, see § 878.4460 of this chapter.

(b) *Test method.* For the purposes of this regulation, FDA's analysis of gloves for leaks will be conducted by a water leak method, using 1,000 milliliters (mL) of water. Each medical glove will be analyzed independently. When packaged as pairs, each glove is considered separately, and both gloves will be analyzed. A defect on one of the gloves is counted as one defect; a defect in both gloves is counted as two defects. Defects are defined as leaks, tears, mold, embedded foreigh objects, etc. A leak is defined as the appearance of water on the outside of the glove. This emergence of water from the glove constitutes a watertight barrier failure. Leaks within 1 and ½ inches of the cuff are to be disregarded.

(1) The following materials are required for testing: A 2⅜-inch by 15-inch (clear) plastic cylinder with a hook on one end and a mark scored 1½ inches from the other end (a cylinder of another size may be used if it accommodates both cuff diameter and any water above the glove capacity); elastic strapping with velcro or other fastening material; automatic water-dispensing apparatus or manual device capable of delivering 1,000 mL of water; a stand with horizontal rod for hanging the hook end of the plastic tube. The support rod must be capable of holding the weight of the total number of gloves that will be suspended at any one time, e.g., five gloves suspended will weigh about 11 pounds.

(2) The following methodology is used: Examine the sample and identify code/ lot number, size, and brand as appropriate. Examine gloves for defects as follows: carefully remove the glove from the wrapper, box, etc., visually examining each glove for defects. Visual defects in the top 1½ inches of a glove will not be counted as a defect for the purposes of this rule. Visually defective gloves do not require further testing but are to be included in the total number of defective gloves counted for the sample. Attach the glove to the plastic fill tube by bringing the cuff end to the 1½-inch mark and fastening with elastic strapping to make a watertight seal. Add 1,000 mL of room temperature water (i.e., 20 °C to 30 °C) into the open end of the fill tube. The water shall pass freely into the glove. (With some larger sizes of long-cuffed surgeons' gloves, the water level may reach only the base of the thumb. With some smaller gloves, the water level may extend several inches up the fill tube.)

(3) Immediately after adding the water, examine the glove for water leaks. Do not squeeze the glove; use only minimal manipulation to spread the fingers to check for leaks. Water drops may be blotted to confirm leaking. If the glove does not leak immediately, keep the glove/filling tube assembly upright and hang the assembly vertically from the horizontal rod, using the wire hook on the open end of the fill tube (do not support the filled glove while transferring). Make a second observation for leaks 2 minutes after addition of the water to the glove. Use only minimal manipulation of the fingers to check for leaks. Record the number of defective gloves.

(c) *Sample plans.* FDA will collect samples from lots of gloves to perform the test for defects described in paragraph (b) of this section in accordance with FDA's sampling inspection plans which are based on the tables of MIL-STD-105E (the military sampling standard, "Sampling Procedures and Tables for Inspection by Attributes," May 10, 1989). Based on the acceptable quality levels found in this standard, FDA has defined adulteration as follows: 2.5 or higher for surgeons' gloves and 4.0 or higher for patient examination gloves at a general inspection level II. FDA will use single normal sampling for lots of 1,200 gloves or less and multiple normal sampling for all larger lots. For convenience, the sample plans (sample size and accept/reject numbers) are shown in the following tables:

ADULTERATION LEVEL AT 2.5 FOR SURGEONS' GLOVES

Lot size	Sample	Sample size	Number examined	Number defective	
				Accept	Reject
35,001 and above	First	125	125	2	9
	Second	125	250	7	14
	Third	125	375	13	19
	Fourth	125	500	19	25
	Fifth	125	625	25	29
	Sixth	125	750	31	33
	Seventh	125	875	37	38
35,000 to 10,001	First	80	80	1	7
	Second	80	160	4	10
	Third	80	240	8	13
	Fourth	80	320	12	17
	Fifth	80	400	17	20
	Sixth	80	480	21	23
	Seventh	80	560	25	26
10,000 to 3,201	First	50	50	0	5
	Second	50	100	3	8
	Third	50	150	6	10
	Fourth	50	200	8	13
	Fifth	50	250	11	15
	Sixth	50	300	14	17
	Seventh	50	350	18	19
3,200 to 1,201	First	32	32	0	4
	Second	32	64	1	6
	Third	32	96	3	8
	Fourth	32	128	5	10
	Fifth	32	160	7	11
	Sixth	32	192	10	12
	Seventh	32	224	13	14
1,200 to 501	Single sample		80	5	6
500 to 281	Single sample		50	3	4
280 to 151	Single sample		32	2	3
150 to 51	Single sample		20	1	2
50 to 0	Single sample		5	0	1

ADULTERATION LEVEL AT 4.0 FOR PATIENT EXAMINATION GLOVES

Lot size	Sample	Sample size	Number examined	Number defective	
				Accept	Reject
10,001 and above	First	80	80	2	9
	Second	80	160	7	14
	Third	80	240	13	19
	Fourth	80	320	19	25
	Fifth	80	400	25	29
	Sixth	80	480	31	33
	Seventh	80	560	37	38
10,000 to 3,201	First	50	50	1	7
	Second	50	100	4	10
	Third	50	150	8	13
	Fourth	50	200	12	17
	Fifth	50	250	17	20
	Sixth	50	300	21	23
	Seventh	50	350	25	26
3,200 to 1,201	First	32	32	0	5
	Second	32	64	3	8
	Third	32	96	6	10
	Fourth	32	128	8	13
	Fifth	32	160	11	15
	Sixth	32	192	14	17
	Seventh	32	224	18	19
1,200 to 501	Single sample		80	7	8
500 to 281	Single sample		50	5	6
280 to 151	Single sample		32	3	4
150 to 91	Single sample		20	2	3
90 to 26	Single sample		13	1	2
25 to 0	Single sample		3	0	1

(d) Lots of gloves which are tested and rejected using the test method according to paragraph (b) of this section, are adulterated within the meaning of section 501(c) of the Federal Food, Drug, and Cosmetic Act, and are subject to regulatory action, such as detention of imported products and seizure of domestic products.

[55 FR 51256, Dec. 12, 1990]

Subpart C—Administrative Practices and Procedures

§ 800.55 Administrative detention.

(a) *General.* This section sets forth the procedures for detention of medical devices intended for human use believed to be adulterated or misbranded. Administrative detention is intended to protect the public by preventing distribution or use of devices encountered during inspections that may be adulterated or misbranded, until the Food and Drug Administration (FDA) has had time to consider what action it should take concerning the devices, and to initiate legal action, if appropriate. Devices that FDA orders detained may not be used, moved, altered, or tampered with in any manner by any person during the detention period, except as authorized under paragraph (h) of this section, until FDA terminates the detention order under paragraph (j) of this section, or the detention period expires, whichever occurs first.

(b) *Criteria for ordering detention.* Administrative detention of devices may be ordered in accordance with this section when an authorized FDA representative, during an inspection under section 704 of the Federal Food, Drug, and Cosmetic Act (the act), has reason to believe that a device, as defined in section 201(h) of the act, is adulterated or misbranded.

(c) *Detention period.* The detention is to be for a reasonable period that may not exceed 20 calendar days after the detention order is issued, unless the FDA District Director in whose district the devices are located determines that a greater period is required to seize the devices, to institute injuction proceedings, or to evaluate the need for legal action, in which case the District Director may authorize detention for 10 additional calendar days. The additional 10-calendar-day detention period may be ordered at the time the detention order is issued or at any time thereafter. The entire detention period may not exceed 30 calendar days, except when the detention period is extended under paragraph (g)(6) of this section. An authorized FDA representative may, in accordance with paragraph (j) of this section, terminate a detention before the expiration of the detention period.

(d) *Issuance of detention order.* (1) The detention order shall be issued in writing, in the form of a detention notice, signed by the authorized FDA representative who has reason to believe that the devices are adulterated or misbranded, and issued to the owner, operator, or agent in charge of the place where the devices are located. If the owner or the user of the devices is different from the owner, operator, or agent in charge of the place where the devices are detained, a copy of the detention order shall be provided to the owner or user of the devices if the owner's or user's identity can be readily determined.

(2) If detention of devices in a vehicle or other carrier is ordered, a copy of the detention order shall be provided to the shipper of record and the owner of the vehicle or other carrier, if their identities can be readily determined.

(3) The detention order shall include the following information: (i) A statement that the devices identified in the order are detained for the period shown; (ii) a brief, general statement of the reasons for the detention; (iii) the location of the devices; (iv) a statement that these devices are not to be used, moved, altered, or tampered with in any manner during that period, except as permitted under paragraph (h) of this section, without the written permission of an authorized FDA representative; (v) identification of the detained devices; (vi) the detention order number; (vii) the date and hour of the detention order; (viii) the period of the detention; (ix) the text of section 304(g) of the act and paragraph (g) (1) and (2) of this section; (x) a statement that any informal hearing on an appeal of a detention order shall be conducted as a

regulatory hearing under part 16 of this chapter, with certain exceptions described in paragraph (g)(3) of this section; and (xi) the location and telephone number of the FDA district office and the name of the FDA District Director.

(e) *Approval of detention order.* A detention order, before issuance, shall be approved by the FDA District Director in whose district the devices are located. If prior written approval is not feasible, prior oral approval shall be obtained and confirmed by written memorandum within FDA as soon as possible.

(f) *Labeling or marking a detained device.* An FDA representative issuing a detention order under paragraph (d) of this section shall label or mark the devices with official FDA tags that include the following information:

(1) A statement that the devices are detained by the United States Government in accordance with section 304(g) of the Federal Food, Drug, and Cosmetic Act (21 U.S.C. 334(g)).

(2) A statement that the devices shall not be used, moved, altered, or tampered with in any manner for the period shown, without the written permission of an authorized FDA representative, except as authorized in paragraph (h) of this section.

(3) A statement that the violation of a detention order or the removal or alteration of the tag is punishable by fine or imprisonment or both (section 303 of the act, 21 U.S.C. 333).

(4) The detention order number, the date and hour of the detention order, the detention period, and the name of the FDA representative who issued the detention order.

(g) *Appeal of a detention order.* (1) A person who would be entitled to claim the devices, if seized, may appeal a detention order. Any appeal shall be submitted in writing to the FDA District Director in whose district the devices are located within 5 working days of receipt of a detention order. If the appeal includes a request for an informal hearing, as defined in section 201(y) of the act, the appellant shall request either that a hearing be held within 5 working days after the appeal is filed or that the hearing be held at a later date, which shall not be later than 20 calendar days after receipt of a detention order.

(2) The appellant of a detention order shall state the ownership or proprietary interest the appellant has in the detained devices. If the detained devices are located at a place other than an establishment owned or operated by the appellant, the appellant shall include documents showing that the appellant would have legitimate authority to claim the devices if seized.

(3) Any informal hearing on an appeal of a detention order shall be conducted as a regulatory hearing pursuant to regulation in accordance with part 16 of this chapter, except that:

(i) The detention order under paragraph (d) of this section, rather than the notice under § 16.22(a) of this chapter, provides notice of opportunity for a hearing under this section and is part of the administrative record of the regulatory hearing under § 16.80(a) of this chapter.

(ii) A request for a hearing under this section should be addressed to the FDA District Director.

(iii) The last sentence of § 16.24(e) of this chapter, stating that a hearing may not be required to be held at a time less than 2 working days after receipt of the request for a hearing, does not apply to a hearing under this section.

(iv) Paragraph (g)(4) of this section, rather than § 16.42(a) of this chapter, describes the FDA employees, i.e., regional food and drug directors, who preside at hearings under this section.

(4) The presiding officer of a regulatory hearing on an appeal of a detention order, who also shall decide the appeal, shall be a regional food and drug director (i.e., a director of an FDA regional office listed in part 5, subpart M of this chapter) who is permitted by § 16.42(a) of this chapter to preside over the hearing.

(5) If the appellant requests a regulatory hearing and requests that the hearing be held within 5 working days after the appeal is filed, the presiding officer shall, within 5 working days, hold the hearing and render a decision affirming or revoking the detention.

(6) If the appellant requests a regulatory hearing and requests that the hearing be held at a date later than

within 5 working days after the appeal is filed, but not later than 20 calendar days after receipt of a detention order, the presiding officer shall hold the hearing at a date agreed upon by FDA and the appellant. The presiding officer shall decide whether to affirm or revoke the detention within 5 working days after the conclusion of the hearing. The detention period extends to the date of the decision even if the 5-working-day period for making the decision extends beyond the otherwise applicable 20-calendar-day or 30-calendar-day detention period.

(7) If the appellant appeals the detention order but does not request a regulatory hearing, the presiding officer shall render a decision on the appeal affirming or revoking the detention within 5 working days after the filing of the appeal.

(8) If the presiding officer affirms a detention order, the devices continue to be detained until FDA terminates the detention under paragraph (j) of this section or the detention period expires, whichever occurs first.

(9) If the presiding officer revokes a detention order, FDA shall terminate the detention under paragraph (j) of this section.

(h)(1) *Movement of detained devices.* Except as provided in this paragraph, no person shall move detained devices within or from the place where they have been ordered detained until FDA terminates the detention under paragraph (j) of this section or the detention period expires, whichever occurs first.

(2) If detained devices are not in final form for shipment, the manufacturer may move them within the establishment where they are detained to complete the work needed to put them in final form. As soon as the devices are moved for this purpose, the individual responsible for their movement shall orally notify the FDA representative who issued the detention order, or another responsible district office official, of the movement of the devices. As soon as the devices are put in final form, they shall be segregated from other devices, and the individual responsible for their movement shall orally notify the FDA representative who issued the detention order, or another responsible district office official, of their new location. The devices put in final form shall not be moved further without FDA approval.

(3) The FDA representative who issued the detention order, or another responsible district office official, may approve, in writing, the movement of detained devices for any of the following purposes:

(i) To prevent interference with an establishment's operations or harm to the devices.

(ii) To destroy the devices.

(iii) To bring the devices into compliance.

(iv) For any other purpose that the FDA representative who issued the detention order, or other responsible district office official, believes is appropriate in the case.

(4) If an FDA representative approves the movement of detained devices under paragraph (h)(3) of this section, the detained devices shall remain segregated from other devices and the person responsible for their movement shall immediately orally notify the official who approved the movement of the devices, or another responsible FDA district office official, of the new location of the detained devices.

(5) Unless otherwise permitted by the FDA representative who is notified of, or who approves, the movement of devices under this paragraph, the required tags shall accompany the devices during and after movement and shall remain with the devices until FDA terminates the detention or the detention period expires, whichever occurs first.

(i) *Actions involving adulterated or misbranded devices.* If FDA determines that the detained devices, including any that have been put in final form, are adulterated or misbranded, or both, it may initiate legal action against the devices or the responsible individuals, or both, or request that the devices be destroyed or otherwise brought into compliance with the act under FDA's supervision.

(j) *Detention termination.* If FDA decides to terminate a detention or when the detention period expires, whichever occurs first, an FDA representative authorized to terminate a detention will issue a detention termination notice releasing the devices to any person who received the original detention order or that person's representative and will remove, or authorize in writing the removal of, the required labels or tags.

(k) *Recordkeeping requirements.* (1) After issuance of a detention order under paragraph (d) of this section, the owner, operator, or agent is charge of any factory, warehouse, other establishment, or consulting laboratory where detained devices are manufactured, processed, packed, or held shall have, or establish, and maintain adequate records relating to how the detained devices may have become adulterated or misbranded, records on any distribution of the devices before and after the detention period, records on the correlation of any in-process detained devices that are put in final form under paragraph (h) of this section to the completed devices, records of any changes in, or processing of, the devices permitted under the detention order, and records of any other movement under paragraph (h) of this section. Records required under this paragraph shall be provided to the FDA on request for review and copying. Any FDA request for access to records required under this paragraph shall be made at a reasonable time, shall state the reason or purpose for the request, and shall identify to the fullest extent practicable the information or type of information sought in the records to which access is requested.

(2) Records required under this paragraph shall be maintained for a maximum period of 2 years after the issuance of the detention order or for such other shorter period as FDA directs. When FDA terminates the detention or when the detention period expires, whichever occurs first, FDA will advise all persons required under this paragraph to keep records concerning that detention whether further recordkeeping is required for the remainder of the 2-year, or shorter, period. FDA ordinarily will not require further recordkeeping if the agency determines that the devices are not adulterated or misbranded or that recordkeeping is not necessary to protect the public health, unless the records are required under other regulations in this chapter (e.g., the good manufacturing practice regulation in part 820 of this chapter).

[44 FR 13239, Mar. 9, 1979, as amended at 49 FR 3174, Jan. 26, 1984; 69 FR 17292, Apr. 2, 2004]

PART 801—LABELING

Subpart A—General Labeling Provisions

Sec.

Subpart B [Reserved]

Subpart C—Labeling Requirements for Over-the-Counter Devices

Subpart D—Exemptions From Adequate Directions for Use

Subpart E—Other Exemptions

Subparts F–G [Reserved]

Subpart H—Special Requirements for Specific Devices

AUTHORITY: 21 U.S.C. 321, 331, 351, 352, 360i, 360j, 371, 374.

SOURCE: 41 FR 6896, Feb. 13, 1976, unless otherwise noted.

Subpart A—General Labeling Provisions

§ 801.1 Medical devices; name and place of business of manufacturer, packer or distributor.

(a) The label of a device in package form shall specify conspicuously the name and place of business of the manufacturer, packer, or distributor.

(b) The requirement for declaration of the name of the manufacturer, packer, or distributor shall be deemed to be satisfied, in the case of a corporation, only by the actual corporate name which may be preceded or followed by the name of the particular division of the corporation. Abbreviations for "Company," "Incorporated," etc., may be used and "The" may be omitted. In the case of an individual, partnership, or association, the name under which the business is conducted shall be used.

(c) Where a device is not manufactured by the person whose name appears on the label, the name shall be qualified by a phrase that reveals the connection such person has with such device; such as, "Manufactured for _____", "Distributed by _____", or any other wording that expresses the facts.

(d) The statement of the place of business shall include the street address, city, State, and Zip Code; however, the street address may be omitted if it is shown in a current city directory or telephone directory. The requirement for inclusion of the ZIP Code shall apply only to consumer commodity labels developed or revised after the effective date of this section. In the case of nonconsumer packages, the ZIP Code shall appear on either the label or the labeling (including the invoice).

(e) If a person manufactures, packs, or distributes a device at a place other than his principal place of business, the label may state the principal place of business in lieu of the actual place where such device was manufactured or packed or is to be distributed, unless such statement would be misleading.

§ 801.4 Meaning of *intended uses*.

The words *intended uses* or words of similar import in §§ 801.5, 801.119, and 801.122 refer to the objective intent of the persons legally responsible for the labeling of devices. The intent is determined by such persons' expressions or may be shown by the circumstances surrounding the distribution of the article. This objective intent may, for example, be shown by labeling claims, advertising matter, or oral or written statements by such persons or their representatives. It may be shown by the circumstances that the article is, with the knowledge of such persons or their representatives, offered and used for a purpose for which it is neither labeled nor advertised. The intended uses of an article may change after it has been introduced into interstate commerce by its manufacturer. If, for example, a packer, distributor, or seller intends an article for different uses than those intended by the person from whom he received the devices, such packer, distributor, or seller is required to supply adequate labeling in accordance with the new intended uses. But if a manufacturer knows, or has knowledge of facts that would give him notice that a device introduced into interstate commerce by him is to be used for conditions, purposes, or uses other than the ones for which he offers it, he is required to provide adequate labeling for such a device which accords with such other uses to which the article is to be put.

§ 801.5 Medical devices; adequate directions for use.

Adequate directions for use means directions under which the layman can use a device safely and for the purposes for which it is intended. Section 801.4 defines *intended use*. Directions for use may be inadequate because, among other reasons, of omission, in whole or in part, or incorrect specification of:

(a) Statements of all conditions, purposes, or uses for which such device is intended, including conditions, purposes, or uses for which it is prescribed, recommended, or suggested in its oral, written, printed, or graphic advertising, and conditions, purposes, or uses for which the device is commonly used; except that such statements shall not refer to conditions, uses, or purposes for which the device can be safely used only under the supervision of a practitioner licensed by law and for which it is advertised solely to such practitioner.

(b) Quantity of dose, including usual quantities for each of the uses for which it is intended and usual quantities for persons of different ages and different physical conditions.

(c) Frequency of administration or application.

(d) Duration of administration or application.

(e) Time of administration or application, in relation to time of meals, time of onset of symptoms, or other time factors.

(f) Route or method of administration or application.

(g) Preparation for use, i.e., adjustment of temperature, or other manipulation or process.

§ 801.6 Medical devices; misleading statements.

Among representations in the labeling of a device which render such device misbranded is a false or misleading representation with respect to another device or a drug or food or cosmetic.

§ 801.15 Medical devices; prominence of required label statements.

(a) A word, statement, or other information required by or under authority of the act to appear on the label may lack that prominence and conspicuousness required by section 502(c) of the act by reason, among other reasons, of:

(1) The failure of such word, statement, or information to appear on the part or panel of the label which is presented or displayed under customary conditions of purchase;

(2) The failure of such word, statement, or information to appear on two or more parts or panels of the label, each of which has sufficient space therefor, and each of which is so designed as to render it likely to be, under customary conditions of purchase, the part or panel displayed;

(3) The failure of the label to extend over the area of the container or package available for such extension, so as to provide sufficient label space for the prominent placing of such word, statement, or information;

(4) Insufficiency of label space for the prominent placing of such word, statement, or information, resulting from the use of label space for any word, statement, design, or device which is not required by or under authority of the act to appear on the label;

(5) Insufficiency of label space for the placing of such word, statement, or information, resulting from the use of label space to give materially greater conspicuousness to any other word, statement, or information, or to any design or device; or

(6) Smallness or style of type in which such word, statement, or information appears, insufficient background contrast, obscuring designs or vignettes, or crowding with other written, printed, or graphic matter.

(b) No exemption depending on insufficiency of label space, as prescribed in regulations promulgated under section 502(b) of the act, shall apply if such insufficiency is caused by:

(1) The use of label space for any word, statement, design, or device which is not required by or under authority of the act to appear on the label;

(2) The use of label space to give greater conspicuousness to any word,

statement, or other information than is required by section 502(c) of the act; or

(3) The use of label space for any representation in a foreign language.

(c)(1) All words, statements, and other information required by or under authority of the act to appear on the label or labeling shall appear thereon in the English language: *Provided, however,* That in the case of articles distributed solely in the Commonwealth of Puerto Rico or in a Territory where the predominant language is one other than English, the predominant language may be substituted for English.

(2) If the label contains any representation in a foreign language, all words, statements, and other information required by or under authority of the act to appear on the label shall appear thereon in the foreign language.

(3) If the labeling contains any representation in a foreign language, all words, statements, and other information required by or under authority of the act to appear on the label or labeling shall appear on the labeling in the foreign language.

§ 801.16 Medical devices; Spanish-language version of certain required statements.

If devices restricted to prescription use only are labeled solely in Spanish for distribution in the Commonwealth of Puerto Rico where Spanish is the predominant language, such labeling is authorized under § 801.15(c).

Subpart B [Reserved]

Subpart C—Labeling Requirements for Over-the-Counter Devices

§ 801.60 Principal display panel.

The term *principal display panel,* as it applies to over-the-counter devices in package form and as used in this part, means the part of a label that is most likely to be displayed, presented, shown, or examined under customary conditions of display for retail sale. The principal display panel shall be large enough to accommodate all the mandatory label information required to be placed thereon by this part with clarity and conspicuousness and without obscuring designs, vignettes, or crowding. Where packages bear alternate principal display panels, information required to be placed on the principal display panel shall be duplicated on each principal display panel. For the purpose of obtaining uniform type size in declaring the quantity of contents for all packages of substantially the same size, the term *area of the principal display panel* means the area of the side or surface that bears the principal display panel, which area shall be:

(a) In the case of a rectangular package where one entire side properly can be considered to be the principal display panel side, the product of the height times the width of that side;

(b) In the case of a cylindrical or nearly cylindrical container, 40 percent of the product of the height of the container times the circumference; and

(c) In the case of any other shape of container, 40 percent of the total surface of the container: *Provided, however,* That where such container presents an obvious "principal display panel" such as the top of a triangular or circular package, the area shall consist of the entire top surface.

In determining the area of the principal display panel, exclude tops, bottoms, flanges at the tops and bottoms of cans, and shoulders and necks of bottles or jars. In the case of cylindrical or nearly cylindrical containers, information required by this part to appear on the principal display panel shall appear within that 40 percent of the circumference which is most likely to be displayed, presented, shown, or examined under customary conditions of display for retail sale.

§ 801.61 Statement of identity.

(a) The principal display panel of an over-the-counter device in package form shall bear as one of its principal features a statement of the identity of the commodity.

(b) Such statement of identity shall be in terms of the common name of the device followed by an accurate statement of the principal intended action(s) of the device. Such statement shall be placed in direct conjunction with the most prominent display of the name and shall employ terms descriptive of the principal intended action(s). The indications for use shall be included in the directions for use of the device, as required by section 502(f)(1) of the act and by the regulations in this part.

(c) The statement of identity shall be presented in bold face type on the principal display panel, shall be in a size reasonably related to the most prominent printed matter on such panel, and shall be in lines generally parallel to the base on which the package rests as it is designed to be displayed.

§ 801.62 Declaration of net quantity of contents.

(a) The label of an over-the-counter device in package form shall bear a declaration of the net quantity of contents. This shall be expressed in the terms of weight, measure, numerical count, or a combination of numerical count and weight, measure, or size: *Provided*, That:

(1) In the case of a firmly established general consumer usage and trade custom of declaring the quantity of a device in terms of linear measure or measure of area, such respective term may be used. Such term shall be augmented when necessary for accuracy of information by a statement of the weight, measure, or size of the individual units or of the entire device.

(2) If the declaration of contents for a device by numerical count does not give accurate information as to the quantity of the device in the package, it shall be augmented by such statement of weight, measure, or size of the individual units or of the total weight, measure, or size of the device as will give such information; for example, "100 tongue depressors, adult size", "1 rectal syringe, adult size", etc. Whenever the Commissioner determines for a specific packaged device that an existing practice of declaring net quantity of contents by weight, measure, numerical count, or a combination of these does not facilitate value comparisions by consumers, he shall by regulation designate the appropriate term or terms to be used for such article.

(b) Statements of weight of the contents shall be expressed in terms of avoirdupois pound and ounce. A statement of liquid measure of the contents shall be expressed in terms of the U.S. gallon of 231 cubic inches and quart, pint, and fluid-ounce subdivisions thereof, and shall express the volume at 68 °F (20 °C). See also paragraph (p) of this section.

(c) The declaration may contain common or decimal fractions. A common fraction shall be in terms of halves, quarters, eighths, sixteenths, or thirty-seconds; except that if there exists a firmly established, general consumer usage and trade custom of employing different common fractions in the net quantity declaration of a particular commodity, they may be employed. A common fraction shall be reduced to its lowest terms; a decimal fraction shall not be carried out to more than two places. A statement that includes small fractions of an ounce shall be deemed to permit smaller variations than one which does not include such fractions.

(d) The declaration shall be located on the principal display panel of the label, and with respect to packages bearing alternate principal panels it shall be duplicated on each principal display panel.

(e) The declaration shall appear as a distinct item on the principal display panel, shall be separated, by at least a space equal to the height of the lettering used in the declaration, from other printed label information appearing above or below the declaration and, by at least a space equal to twice the width of the letter "N" of the style of type used in the quantity of contents statement, from other printed label information appearing to the left or right of the declaration. It shall not include any term qualifying a unit of weight, measure, or count, such as "giant pint" and "full quart", that tends to exaggerate. It shall be placed on the principal display panel within the bottom 30 percent of the area of the label panel in lines generally parallel to the base on which the package rests as it is designed to be displayed: *Provided*, That:

(1) On packages having a principal display panel of 5 square inches or less the requirement for placement within the bottom 30 percent of the area of the label panel shall not apply when the

declaration of net quantity of contents meets the other requirements of this part; and

(2) In the case of a device that is marketed with both outer and inner retail containers bearing the mandatory label information required by this part and the inner container is not intended to be sold separately, the net quantity of contents placement requirement of this section applicable to such inner container is waived.

(3) The principal display panel of a device marketed on a display card to which the immediate container is affixed may be considered to be the display panel of the card, and the type size of the net quantity of contents statement is governed by the dimensions of the display card.

(f) The declaration shall accurately reveal the quantity of device in the package exclusive of wrappers and other material packed therewith.

(g) The declaration shall appear in conspicuous and easily legible boldface print or type in distinct contrast (by typography, layout, color, embossing, or molding) to other matter on the package; except that a declaration of net quantity blown, embossed, or molded on a glass or plastic surface is permissible when all label information is so formed on the surface. Requirements of conspicuousness and legibility shall include the specifications that:

(1) The ratio of height to width of the letter shall not exceed a differential of 3 units to 1 unit, i.e., no more than 3 times as high as it is wide.

(2) Letter heights pertain to upper case or capital letters. When upper and lower case or all lower case letters are used, it is the lower case letter "o" or its equivalent that shall meet the minimum standards.

(3) When fractions are used, each component numeral shall meet one-half the minimum height standards.

(h) The declaration shall be in letters and numerals in a type size established in relationship to the area of the principal display panel of the package and shall be uniform for all packages of substantially the same size by complying with the following type specifications:

(1) Not less than one-sixteenth inch in height on packages the principal display panel of which has an area of 5 square inches or less.

(2) Not less than one-eighth inch in height on packages the principal display panel of which has an area of more than 5 but not more than 25 square inches.

(3) Not less than three-sixteenths inch in height on packages the principal display panel of which has an area of more than 25 but not more than 100 square inches.

(4) Not less than one-fourth inch in height on packages the principal display panel of which has an area of more than 100 square inches, except not less than one-half inch in height if the area is more than 400 square inches.

Where the declaration is blown, embossed, or molded on a glass or plastic surface rather than by printing, typing, or coloring, the lettering sizes specified in paragraphs (h)(1) through (4) of this section shall be increased by one-sixteenth of an inch.

(i) On packages containing less than 4 pounds or 1 gallon and labeled in terms of weight or fluid measure:

(1) The declaration shall be expressed both in ounces, with identification by weight or by liquid measure and, if applicable (1 pound or 1 pint or more) followed in parentheses by a declaration in pounds for weight units, with any remainder in terms of ounces or common or decimal fractions of the pound (see examples set forth in paragraphs (k) (1) and (2) of this section), or in the case of liquid measure, in the largest whole units (quarts, quarts and pints, or pints, as appropriate) with any remainder in terms of fluid ounces or common or decimal fractions of the pint or quart (see examples set forth in paragraphs (k) (3) and (4) of this section). If the net weight of the package is less than 1 ounce avoirdupois or the net fluid measure is less than 1 fluid ounce, the declaration shall be in terms of common or decimal fractions of the respective ounce and not in terms of drams.

(2) The declaration may appear in more than one line. The term "net weight" shall be used when stating the net quantity of contents in terms of weight. Use of the terms "net" or "net contents" in terms of fluid measure or

numerical count is optional. It is sufficient to distinguish avoirdupois ounce from fluid ounce through association of terms; for example, "Net wt. 6 oz" or "6 oz net wt.," and "6 fl oz" or "net contents 6 fl oz."

(j) On packages containing 4 pounds or 1 gallon or more and labeled in terms of weight or fluid measure, the declaration shall be expressed in pounds for weight units with any remainder in terms of ounces or common or decimal fractions of the pound; in the case of fluid measure, it shall be expressed in the largest whole unit, i.e., gallons, followed by common or decimal fractions of a gallon or by the next smaller whole unit or units (quarts or quarts and pints), with any remainder in terms of fluid ounces or common or decimal fractions of the pint or quart; see paragraph (k)(5) of this section.

(k) *Examples:* (1) A declaration of 1½ pounds weight shall be expressed as "net wt. 24 oz (1 lb 8 oz)," or "Net wt. 24 oz (1½ lb)" or "Net wt. 24 oz (1.5 lb)."

(2) A declaration of three-fourths pound avoirdupois weight shall be expressed as "Net wt. 12 oz.".

(3) A declaration of 1 quart liquid measure shall be expressed as "Net contents 32 fl oz (1 qt)" or "32 fl oz (1 qt)."

(4) A declaration of 1¾ quarts liquid measure shall be expressed as, "Net contents 56 fl oz (1 qt 1 pt 8 oz)" or "Net contents 56 fl oz (1 qt 1.5 pt)," but not in terms of quart and ounce such as "Net contents 56 fl oz (1 qt 24 oz)."

(5) A declaration of 2½ gallons liquid measure shall be expressed as "Net contents 2 gal 2 qt", "Net contents 2.5 gallons," or "Net contents 2½ gal" but not as "2 gal 4 pt".

(l) For quantities, the following abbreviations and none other may be employed. Periods and plural forms are optional:

gallon gal	liter l
milliliter ml	cubic centimeter cc
quart qt	yard yd
pint pt	feet or foot ft
ounce oz	inch in
pound lb	meter m
grain gr	centimeter cm
kilogram kg	millimeter mm
gram g	fluid fl
milligram mg	square sq
microgram mcg	weight wt

(m) On packages labeled in terms of linear measure, the declaration shall be expressed both in terms of inches and, if applicable (1 foot or more), the largest whole units (yards, yards and feet, feet). The declaration in terms of the largest whole units shall be in parentheses following the declaration in terms of inches and any remainder shall be in terms of inches or common or decimal fractions of the foot or yard; if applicable, as in the case of adhesive tape, the initial declaration in linear inches shall be preceded by a statement of the width. Examples of linear measure are "86 inches (2 yd 1 ft 2 in)", "90 inches (2½ yd)", "30 inches (2.5 ft)", "¾ inch by 36 in (1 yd)", etc.

(n) On packages labeled in terms of area measure, the declaration shall be expressed both in terms of square inches and, if applicable (1 square foot or more), the largest whole square unit (square yards, square yards and square feet, square feet). The declaration in terms of the largest whole units shall be in parentheses following the declaration in terms of square inches and any remainder shall be in terms of square inches or common or decimal fractions of the square foot or square yard; for example, "158 sq inches (1 sq ft 14 sq in)".

(o) Nothing in this section shall prohibit supplemental statements at locations other than the principal display panel(s) describing in nondeceptive terms the net quantity of contents, provided that such supplemental statements of net quantity of contents shall not include any term qualifying a unit of weight, measure, or count that tends to exaggerate the amount of the device contained in the package; for example, "giant pint" and "full quart". Dual or combination declarations of net quantity of contents as provided for in paragraphs (a) and (i) of this section are not regarded as supplemental net quantity statements and shall be located on the principal display panel.

(p) A separate statement of net quantity of contents in terms of the metric system of weight or measure is not regarded as a supplemental statement and an accurate statement of the net quantity of contents in terms of the metric system of weight or measure may also appear on the principal display panel or on other panels.

(q) The declaration of net quantity of contents shall express an accurate statement of the quantity of contents of the package. Reasonable variations caused by loss or gain of moisture during the course of good distribution practice or by unavoidable deviations in good manufacturing practice will be recognized. Variations from stated quantity of contents shall not be unreasonably large.

§ 801.63 Medical devices; warning statements for devices containing or manufactured with chlorofluorocarbons and other class I ozone-depleting substances.

(a) All over-the-counter devices containing or manufactured with chlorofluorocarbons, halons, carbon tetrachloride, methyl chloride, or any other class I substance designated by the Environmental Protection Agency (EPA) shall carry one of the following warnings:

(1) The EPA warning statement:

WARNING: Contains [or Manufactured with, if applicable] [*insert name of substance*], a substance which harms public health and environment by destroying ozone in the upper atmosphere.

(2) The alternative statement:

NOTE: The indented statement below is required by the Federal government's Clean Air Act for all products containing or manufactured with chlorofluorocarbons (CFC's) [or other class I substance, if applicable]:

WARNING: Contains [or Manufactured with, if applicable] [*insert name of substance*], a substance which harms public health and environment by destroying ozone in the upper atmosphere.

CONSULT WITH YOUR PHYSICIAN, HEALTH PROFESSIONAL, OR SUPPLIER IF YOU HAVE ANY QUESTION ABOUT THE USE OF THIS PRODUCT.

(b) The warning statement shall be clearly legible and conspicuous on the product, its immediate container, its outer packaging, or other labeling in accordance with the requirements of 40 CFR part 82 and appear with such prominence and conspicuousness as to render it likely to be read and understood by consumers under normal conditions of purchase. This provision does not replace or relieve a person from any requirements imposed under 40 CFR part 82.

[61 FR 20101, May 3, 1996]

Subpart D—Exemptions From Adequate Directions for Use

§ 801.109 Prescription devices.

A device which, because of any potentiality for harmful effect, or the method of its use, or the collateral measures necessary to its use is not safe except under the supervision of a practitioner licensed by law to direct the use of such device, and hence for which "adequate directions for use" cannot be prepared, shall be exempt from section 502(f)(1) of the act if all the following conditions are met:

(a) The device is:

(1)(i) In the possession of a person, or his agents or employees, regularly and lawfully engaged in the manufacture, transportation, storage, or wholesale or retail distribution of such device; or

(ii) In the possession of a practitioner, such as physicians, dentists, and veterinarians, licensed by law to use or order the use of such device; and

(2) Is to be sold only to or on the prescription or other order of such practitioner for use in the course of his professional practice.

(b) The label of the device, other than surgical instruments, bears:

(1) The statement "Caution: Federal law restricts this device to sale by or on the order of a ______", the blank to be filled with the word "physician", "dentist", "veterinarian", or with the descriptive designation of any other practitioner licensed by the law of the State in which he practices to use or order the use of the device; and

(2) The method of its application or use.

(c) Labeling on or within the package from which the device is to be dispensed bears information for use, including indications, effects, routes, methods, and frequency and duration of administration, and any relevant hazards, contraindications, side effects, and precautions under which practitioners licensed by law to administer the device can use the device safely and for the purpose for which it is intended, including all purposes for which it is advertised or represented: *Provided, however*, That such information may be omitted from the dispensing package if, but only if, the article is a device for which directions, hazards, warnings, and other information are commonly known to practitioners licensed by law to use the device. Upon written request, stating reasonable grounds therefor, the Commissioner will offer an opinion on a proposal to omit such information from the dispensing package under this proviso.

(d) Any labeling, as defined in section 201(m) of the act, whether or not it is on or within a package from which the device is to be dispensed, distributed by or on behalf of the manufacturer, packer, or distributor of the device, that furnishes or purports to furnish information for use of the device contains adequate information for such use, including indications, effects, routes, methods, and frequency and duration of administration and any relevant hazards, contraindications, side effects, and precautions, under which practitioners licensed by law to employ the device can use the device safely and for the purposes for which it is intended, including all purposes for which it is advertised or represented. This information will not be required on so-called reminder—piece labeling which calls attention to the name of the device but does not include indications or other use information.

(e) All labeling, except labels and cartons, bearing information for use of the device also bears the date of the issuance or the date of the latest revision of such labeling.

§ 801.110 Retail exemption for prescription devices.

A device subject to § 801.109 shall be exempt at the time of delivery to the ultimate purchaser or user from section 502(f)(1) of the act if it is delivered by a licensed practitioner in the course of his professional practice or upon a prescription or other order lawfully issued in the course of his professional practice, with labeling bearing the name and address of such licensed practitioner and the directions for use and cautionary statements, if any, contained in such order.

§ 801.116 Medical devices having commonly known directions.

A device shall be exempt from section 502(f)(1) of the act insofar as adequate directions for common uses thereof are known to the ordinary individual.

§ 801.119 In vitro diagnostic products.

A product intended for use in the diagnosis of disease and which is an in vitro diagnostic product as defined in § 809.3(a) of this chapter shall be deemed to be in compliance with the requirements of this section and section 502(f)(1) of the act if it meets the requirements of § 809.10 of this chapter.

§ 801.122 Medical devices for processing, repacking, or manufacturing.

A device intended for processing, repacking, or use in the manufacture of another drug or device shall be exempt from section 502(f)(1) of the act if its label bears the statement "Caution: For manufacturing, processing, or repacking".

§ 801.125 Medical devices for use in teaching, law enforcement, research, and analysis.

A device subject to § 801.109 shall be exempt from section 502(f)(1) of this act if shipped or sold to, or in the possession of, persons regularly and lawfully engaged in instruction in pharmacy, chemistry, or medicine not involving clinical use, or engaged in law enforcement, or in research not involving clinical use, or in chemical analysis, or physical testing, and is to be used only for such instruction, law enforcement, research, analysis, or testing.

§ 801.127 Medical devices; expiration of exemptions.

(a) If a shipment or delivery, or any part thereof, of a device which is exempt under the regulations in this section is made to a person in whose possession the article is not exempt, or is made for any purpose other than those specified, such exemption shall expire, with respect to such shipment or delivery or part thereof, at the beginning of that shipment or delivery. The causing of an exemption to expire shall be considered an act which results in such device being misbranded unless it is disposed of under circumstances in which it ceases to be a drug or device.

(b) The exemptions conferred by §§ 801.119, 801.122, and 801.125 shall continue until the devices are used for the purposes for which they are exempted, or until they are relabeled to comply with section 502(f)(1) of the act. If, however, the device is converted, or manufactured into a form limited to prescription dispensing, no exemption shall thereafter apply to the article unless the device is labeled as required by § 801.109.

Subpart E—Other Exemptions

§ 801.150 Medical devices; processing, labeling, or repacking.

(a) Except as provided by paragraphs (b) and (c) of this section, a shipment or other delivery of a device which is, in accordance with the practice of the trade, to be processed, labeled, or repacked, in substantial quantity at an establishment other than that where originally processed or packed, shall be exempt, during the time of introduction into and movement in interstate commerce and the time of holding in such establishment, from compliance with the labeling and packaging requirements of section 502(b) and (f) of the act if:

(1) The person who introduced such shipment or delivery into interstate commerce is the operator of the establishment where such device is to be processed, labeled, or repacked; or

(2) In case such person is not such operator, such shipment or delivery is made to such establishment under a written agreement, signed by and containing the post office addresses of such person and such operator, and containing such specifications for the processing, labeling, or repacking, as the case may be, of such device in such establishment as will insure, if such specifications are followed, that such device will not be adulterated or misbranded within the meaning of the act upon completion of such processing, labeling, or repacking. Such person and such operator shall each keep a copy of such agreement until 2 years after the final shipment or delivery of such device from such establishment, and shall make such copies available for inspection at any reasonable hour to any officer or employee of the Department who requests them.

(b) An exemption of a shipment or other delivery of a device under paragraph (a)(1) of this section shall, at the beginning of the act of removing such shipment or delivery, or any part thereof, from such establishment, become void ab initio if the device comprising such shipment, delivery, or part is adulterated or misbranded within the meaning of the act when so removed.

(c) An exemption of a shipment or other delivery of a device under paragraph (a)(2) of this section shall become void ab initio with respect to the person who introduced such shipment or delivery into interstate commerce upon refusal by such person to make available for inspection a copy of the agreement, as required by such paragraph (a)(2).

(d) An exemption of a shipment or other delivery of a device under paragraph (a)(2) of this section shall expire:

(1) At the beginning of the act of removing such shipment or delivery, or any part thereof, from such establishment if the device comprising such shipment, delivery, or part is adulterated or misbranded within the meaning of the act when so removed; or

(2) Upon refusal by the operator of the establishment where such device is to be processed, labeled, or repacked, to make available for inspection a copy of the agreement, as required by such clause.

(e) As it is a common industry practice to manufacture and/or assemble, package, and fully label a device as sterile at one establishment and then ship such device in interstate commerce to another establishment or to a contract sterilizer for sterilization, the Food and Drug Administration will initiate no regulatory action against the device as misbranded or adulterated when the nonsterile device is labeled sterile, provided all the following conditions are met:

(1) There is in effect a written agreement which:

(i) Contains the names and post office addresses of the firms involved and is signed by the person authorizing such shipment and the operator or person in charge of the establishment receiving the devices for sterilization.

(ii) Provides instructions for maintaining proper records or otherwise accounting for the number of units in each shipment to insure that the number of units shipped is the same as the number received and sterilized.

(iii) Acknowledges that the device is nonsterile and is being shipped for further processing, and

(iv) States in detail the sterilization process, the gaseous mixture or other media, the equipment, and the testing method or quality controls to be used by the contract sterilizer to assure that the device will be brought into full compliance with the Federal Food, Drug, and Cosmetic Act.

(2) Each pallet, carton, or other designated unit is conspicuously marked to show its nonsterile nature when it is introduced into and is moving in interstate commerce, and while it is being held prior to sterilization. Following sterilization, and until such time as it is established that the device is sterile and can be released from quarantine, each pallet, carton, or other designated unit is conspicuously marked to show that it has not been released from quarantine, e.g., "sterilized—awaiting test results" or an equivalent designation.

Subparts F–G [Reserved]

Subpart H—Special Requirements for Specific Devices

§ 801.405 Labeling of articles intended for lay use in the repairing and/or refitting of dentures.

(a) The American Dental Association and leading dental authorities have advised the Food and Drug Administration of their concern regarding the safety of denture reliners, repair kits, pads, cushions, and other articles marketed and labeled for lay use in the repairing, refitting, or cushioning of ill-fitting, broken, or irritating dentures. It is the opinion of dental authorities and the Food and Drug Administration that to properly repair and properly refit dentures a person must have professional knowledge and specialized technical skill. Laymen cannot be expected to maintain the original vertical dimension of occlusion and the centric relation essential in the proper repairing or refitting of dentures. The continued wearing of improperly repaired or refitted dentures may cause acceleration of bone resorption, soft tissue hyperplasia, and other irreparable damage to the oral cavity. Such articles designed for lay use should be limited to emergency or temporary situations pending the services of a licensed dentist.

(b) The Food and Drug Administration therefore regards such articles as unsafe and misbranded under the Federal Food, Drug, and Cosmetic Act, unless the labeling:

(1)(i) Limits directions for use for denture repair kits to emergency repairing pending unavoidable delay in obtaining professional reconstruction of the denture;

(ii) Limits directions for use for denture reliners, pads, and cushions to temporary refitting pending unavoidable delay in obtaining professional reconstruction of the denture;

(2) Contains in a conspicuous manner the word "emergency" preceding and modifying each indication-for-use statement for denture repair kits and the word "temporary" preceding and modifying each indication-for-use statement for reliners, pads, and cushions; and

(3) Includes a conspicuous warning statement to the effect:

(i) For denture repair kits: "*Warning—For emergency repairs only.* Long term use of home-repaired dentures may cause faster bone loss, continuing irritation, sores, and tumors. This kit for emergency use only. See Dentist Without Delay."

(ii) For denture reliners, pads, and cushions: "*Warning—For temporary use only*. Longterm use of this product may lead to faster bone loss, continuing irritation, sores, and tumors. For Use Only Until a Dentist Can Be Seen."

(c) Adequate directions for use require full information of the temporary and emergency use recommended in order for the layman to understand the limitations of usefulness, the reasons

208-072 D-2

therefor, and the importance of adhering to the warnings. Accordingly, the labeling should contain substantially the following information:

(1) For denture repair kits: Special training and tools are needed to repair dentures to fit properly. Home-repaired dentures may cause irritation to the gums and discomfort and tiredness while eating. Long term use may lead to more troubles, even permanent changes in bones, teeth, and gums, which may make it impossible to wear dentures in the future. For these reasons, dentures repaired with this kit should be used only in an emergency until a dentist can be seen. Dentures that don't fit properly cause irritation and injury to the gums and faster bone loss, which is permanent. Dentures that don't fit properly cause gum changes that may require surgery for correction. Continuing irritation and injury may lead to cancer in the mouth. You must see your dentist as soon as possible.

(2) For denture reliners, pads, and cushions: Use of these preparations or devices may temporarily decrease the discomfort; however, their use will not make the denture fit properly. Special training and tools are needed to repair a denture to fit properly. Dentures that do not fit properly cause irritation and injury to the gums and faster bone loss, which is permanent and may require a completely new denture. Changes in the gums caused by dentures that do not fit properly may require surgery for correction. Continuing irritation and injury may lead to cancer in the mouth. You must see your dentist as soon as possible.

(3) If the denture relining or repairing material forms a permanent bond with the denture, a warning statement to the following effect should be included: "This reliner becomes fixed to the denture and a completely new denture may be required because of its use."

(d) Labeling claims exaggerating the usefulness or the safety of the material or failing to disclose all facts relevant to the claims of usefulness will be regarded as false and misleading under sections 201(n) and 502(a) of the Federal Food, Drug, and Cosmetic Act.

(e) Regulatory action may be initiated with respect to any article found within the jurisdiction of the act contrary to the provisions of this policy statement after 90 days following the date of publication of this section in the FEDERAL REGISTER.

§ 801.410 Use of impact-resistant lenses in eyeglasses and sunglasses.

(a) Examination of data available on the frequency of eye injuries resulting from the shattering of ordinary crown glass lenses indicates that the use of such lenses constitutes an avoidable hazard to the eye of the wearer.

(b) The consensus of the ophthalmic community is that the number of eye injuries would be substantially reduced by the use in eyeglasses and sunglasses of impact-resistant lenses.

(c)(1) To protect the public more adequately from potential eye injury, eyeglasses and sunglasses must be fitted with impact-resistant lenses, except in those cases where the physician or optometrist finds that such lenses will not fulfill the visual requirements of the particular patient, directs in writing the use of other lenses, and gives written notification thereof to the patient.

(2) The physician or optometrist shall have the option of ordering glass lenses, plastic lenses, or laminated glass lenses made impact resistant by any method; however, all such lenses shall be capable of withstanding the impact test described in paragraph (d)(2) of this section.

(3) Each finished impact-resistant glass lens for prescription use shall be individually tested for impact resistance and shall be capable of withstanding the impact test described in paragraph (d)(2) of this section. Raised multifocal lenses shall be impact resistant but need not be tested beyond initial design testing. Prism segment multifocal, slab-off prism, lenticular cataract, iseikonic, depressed segment one-piece multifocal, bioconcave, myodisc and minus lenticular, custom laminate and cemented assembly lenses shall be impact resistant but need not be subjected to impact testing. To demonstrate that all other types of impact-resistant lenses, including impact-resistant laminated

glass lenses (i.e., lenses other than those described in the three preceding sentences of this paragraph (c)(3)), are capable of withstanding the impact test described in this regulation, the manufacturer of these lenses shall subject to an impact test a statistically significant sampling of lenses from each production batch, and the lenses so tested shall be representative of the finished forms as worn by the wearer, including finished forms that are of minimal lens thickness and have been subjected to any treatment used to impart impact resistance. All nonprescription lenses and plastic prescription lenses tested on the basis of statistical significance shall be tested in uncut-finished or finished form.

(d)(1) For the purpose of this regulation, the impact test described in paragraph (d)(2) of this section shall be the "referee test," defined as "one which will be utilized to determine compliance with a regulation." The referee test provides the Food and Drug Administration with the means of examining a medical device for performance and does not inhibit the manufacturer from using equal or superior test methods. A lens manufacturer shall conduct tests of lenses using the impact test described in paragraph (d)(2) of this section or any equal or superior test. Whatever test is used, the lenses shall be capable of withstanding the impact test described in paragraph (d)(2) of this section if the Food and Drug Administration examines them for performance.

(2) In the impact test, a ⅝-inch steel ball weighing approximately 0.56 ounce is dropped from a height of 50 inches upon the horizontal upper surface of the lens. The ball shall strike within a ⅝-inch diameter circle located at the geometric center of the lens. The ball may be guided but not restricted in its fall by being dropped through a tube extending to within approximately 4 inches of the lens. To pass the test, the lens must not fracture; for the purpose of this section, a lens will be considered to have fractured if it cracks through its entire thickness, including a laminar layer, if any, and across a complete diameter into two or more separate pieces, or if any lens material visible to the naked eyes becomes detached from the ocular surface. The test shall be conducted with the lens supported by a tube (1-inch inside diameter, 1¼-inch outside diameter, and approximately 1-inch high) affixed to a rigid iron or steel base plate. The total weight of the base plate and its rigidly attached fixtures shall be not less than 27 pounds. For lenses of small minimum diameter, a support tube having an outside diameter of less than 1¼ inches may be used. The support tube shall be made of rigid acrylic plastic, steel, or other suitable substance and shall have securely bonded on the top edge a ⅛- by ⅛-inch neoprene gasket having a hardness of 40 ±5, as determined by ASTM Method D 1415–88, "Standard Test Method for Rubber Property—International Hardness" a minimum tensile strength of 1,200 pounds, as determined by ASTM Method D 412–98A, 'Standard Test Methods for Vulcanized Rubber and Thermoplastic Elastomers—Tension, and a minimum ultimate elongation of 400 percent, as determined by ASTM Method D 412–68 (Both methods are incorporated by reference and are available from the American Society for Testing Materials, 100 Barr Harbor Dr., West Conshohocken, Philadelphia, PA 19428, or available for inspection at the Center for Devices and Radiological Health's Library, 9200 Corporate Blvd., Rockville, MD 20850, or at the National Archives and Records Administration (NARA). For information on the availability of this material at NARA, call 202–741–6030, or go to: *http://www.archives.gov/federal_register/code_of_federal_regulations/ibr_locations.html.* The diameter or contour of the lens support may be modified as necessary so that the ⅛- by ⅛-inch neoprene gasket supports the lens at its periphery.

(e) Copies of invoice(s), shipping document(s), and records of sale or distribution of all impact resistant lenses, including finished eyeglasses and sunglasses, shall be kept and maintained for a period of 3 years; however, the names and addresses of individuals purchasing nonprescription eyeglasses and sunglasses at the retail level need not be kept and maintained by the retailer. The records kept in compliance with this paragraph shall be made available upon request at all reasonable hours by any officer or employee of the Food and Drug Administration or by any other officer or employee acting on behalf of the Secretary of Health and Human Services and such officer or employee shall be permitted to inspect and copy such records, to make such inventories of stock as he deems necessary, and otherwise to check the correctness of such inventories.

(f) In addition, those persons conducting tests in accordance with paragraph (d) of this section shall maintain the results thereof and a description of the test method and of the test apparatus for a period of 3 years. These records shall be made available upon request at any reasonable hour by any officer or employee acting on behalf of the Secretary of Health and Human Services. The persons conducting tests shall permit the officer or employee to inspect and copy the records, to make such inventories of stock as the officer or employee deems necessary, and otherwise to check the correctness of the inventories.

(g) For the purpose of this section, the term "manufacturer" includes an importer for resale. Such importer may have the tests required by paragraph (d) of this section conducted in the country of origin but must make the results thereof available, upon request, to the Food and Drug Administration, as soon as practicable.

(h) All lenses must be impact-resistant except when the physician or optometrist finds that impact-resistant lenses will not fulfill the visual requirements for a particular patient.

(i) This statement of policy does not apply to contact lenses.

[41 FR 6896, Feb. 13, 1976, as amended at 44 FR 20678, Apr. 6, 1979; 47 FR 9397, Mar. 5, 1982; 65 FR 3586, Jan. 24, 2000; 65 FR 44436, July 18, 2000; 69 FR 18803, Apr. 9, 2004]

§801.415 Maximum acceptable level of ozone.

(a) Ozone is a toxic gas with no known useful medical application in specific, adjunctive, or preventive therapy. In order for ozone to be effective as a germicide, it must be present in a concentration far greater than that which can be safely tolerated by man and animals.

(b) Although undesirable physiological effects on the central nervous system, heart, and vision have been reported, the predominant physiological effect of ozone is primary irritation of the mucous membranes. Inhalation of ozone can cause sufficient irritation to the lungs to result in pulmonary edema. The onset of pulmonary edema is usually delayed for some hours after exposure; thus, symptomatic response is not a reliable warning of exposure to toxic concentrations of ozone. Since olfactory fatigue develops readily, the odor of ozone is not a reliable index of atmospheric ozone concentration.

(c) A number of devices currently on the market generate ozone by design or as a byproduct. Since exposure to ozone above a certain concentration can be injurious to health, any such device will be considered adulterated and/or misbranded within the meaning of sections 501 and 502 of the act if it is used or intended for use under the following conditions:

(1) In such a manner that it generates ozone at a level in excess of 0.05 part per million by volume of air circulating through the device or causes an accumulation of ozone in excess of 0.05 part per million by volume of air (when measured under standard conditions at 25 °C (77 °F) and 760 millimeters of mercury) in the atmosphere of enclosed space intended to be occupied by people for extended periods of time, e.g., houses, apartments, hospitals, and offices. This applies to any such device, whether portable or permanent or part of any system, which generates ozone by design or as an inadvertent or incidental product.

(2) To generate ozone and release it into the atmosphere in hospitals or other establishments occupied by the ill or infirm.

(3) To generate ozone and release it into the atmosphere and does not indicate in its labeling the maximum acceptable concentration of ozone which may be generated (not to exceed 0.05 part per million by volume of air circulating through the device) as established herein and the smallest area in which such device can be used so as not to produce an ozone accumulation in excess of 0.05 part per million.

(4) In any medical condition for which there is no proof of safety and effectiveness.

(5) To generate ozone at a level less than 0.05 part per million by volume of air circulating through the device and it is labeled for use as a germicide or deodorizer.

(d) This section does not affect the present threshold limit value of 0.10 part per million (0.2 milligram per cubic meter) of ozone exposure for an 8-hour-day exposure of industrial workers as recommended by the American Conference of Governmental Industrial Hygienists.

(e) The method and apparatus specified in 40 CFR part 50, or any other equally sensitive and accurate method, may be employed in measuring ozone pursuant to this section.

§801.417 Chlorofluorocarbon propellants.

The use of chlorofluorocarbon in devices as propellants in self-pressurized containers is generally prohibited except as provided in §2.125 of this chapter.

[43 FR 11318, Mar. 17, 1978]

§801.420 Hearing aid devices; professional and patient labeling.

(a) *Definitions for the purposes of this section and §801.421.* (1) *Hearing aid* means any wearable instrument or device designed for, offered for the purpose of, or represented as aiding persons with or compensating for, impaired hearing.

(2) *Ear specialist* means any licensed physician who specializes in diseases of the ear and is medically trained to identify the symptoms of deafness in the context of the total health of the patient, and is qualified by special training to diagnose and treat hearing loss. Such physicians are also known as otolaryngologists, otologists, and otorhinolaryngologists.

(3) *Dispenser* means any person, partnership, corporation, or association engaged in the sale, lease, or rental of hearing aids to any member of the consuming public or any employee, agent, sales person, and/or representative of such a person, partnership, corporation, or association.

(4) *Audiologist* means any person qualified by training and experience to specialize in the evaluation and rehabilitation of individuals whose communication disorders center in whole or in part in the hearing function. In some states audiologists must satisfy specific requirements for licensure.

(5) *Sale* or *purchase* includes any lease or rental of a hearing aid to a member of the consuming public who is a user or prospective user of a hearing aid.

(6) *Used hearing aid* means any hearing aid that has been worn for any period of time by a user. However, a hearing aid shall not be considered "used" merely because it has been worn by a prospective user as a part of a bona fide hearing aid evaluation conducted to determine whether to select that particular hearing aid for that prospective user, if such evaluation has been conducted in the presence of the dispenser or a hearing aid health professional selected by the dispenser to assist the buyer in making such a determination.

(b) *Label requirements for hearing aids.* Hearing aids shall be clearly and permanently marked with:

(1) The name of the manufacturer or distributor, the model name or number, the serial number, and the year of manufacture.

(2) A "+" symbol to indicate the positive connection for battery insertion, unless it is physically impossible to insert the battery in the reversed position.

(c) *Labeling requirements for hearing aids*—(1) *General.* All labeling information required by this paragraph shall be included in a User Instructional Brochure that shall be developed by the manufacturer or distributor, shall accompany the hearing aid, and shall be provided to the prospective user by the dispenser of the hearing aid in accordance with §801.421(c). The User Instructional Brochure accompanying each hearing aid shall contain the following information and instructions for use, to the extent applicable to the particular requirements and characteristics of the hearing aid:

(i) An illustration(s) of the hearing aid, indicating operating controls, user adjustments, and battery compartment.

(ii) Information on the function of all controls intended for user adjustment.

(iii) A description of any accessory that may accompany the hearing aid, e.g., accessories for use with a television or telephone.

(iv) Specific instructions for:

(*a*) Use of the hearing aid.

(*b*) Maintenance and care of the hearing aid, including the procedure to follow in washing the earmold, when replacing tubing on those hearing aids that use tubing, and in storing the hearing aid when it will not be used for an extended period of time.

(*c*) Replacing or recharging the batteries, including a generic designation of replacement batteries.

(v) Information on how and where to obtain repair service, including at least one specific address where the user can go, or send the hearing aid to, to obtain such repair service.

(vi) A description of commonly occurring avoidable conditions that could adversely affect or damage the hearing aid, such as dropping, immersing, or exposing the hearing aid to excessive heat.

(vii) Identification of any known side effects associated with the use of a hearing aid that may warrant consultation with a physician, e.g., skin irritation and accelerated accumulation of cerumen (ear wax).

(viii) A statement that a hearing aid will not restore normal hearing and will not prevent or improve a hearing impairment resulting from organic conditions.

(ix) A statement that in most cases infrequent use of a hearing aid does not permit a user to attain full benefit from it.

(x) A statement that the use of a hearing aid is only part of hearing habilitation and may need to be supplemented by auditory training and instruction in lipreading.

(xi) The warning statement required by paragraph (c)(2) of this section.

(xii) The notice for prospective hearing aid users required by paragraph (c)(3) of this section.

(xiii) The technical data required by paragraph (c)(4) of this section, unless such data is provided in separate labeling accompanying the device.

(2) *Warning statement.* The User Instructional Brochure shall contain the following warning statement:

WARNING TO HEARING AID DISPENSERS

A hearing aid dispenser should advise a prospective hearing aid user to consult promptly with a licensed physician (preferably an ear specialist) before dispensing a hearing aid if the hearing aid dispenser determines through inquiry, actual observation, or review of any other available information concerning the prospective user, that the prospective user has any of the following conditions:

(i) Visible congenital or traumatic deformity of the ear.

(ii) History of active drainage from the ear within the previous 90 days.

(iii) History of sudden or rapidly progressive hearing loss within the previous 90 days.

(iv) Acute or chronic dizziness.

(v) Unilateral hearing loss of sudden or recent onset within the previous 90 days.

(vi) Audiometric air-bone gap equal to or greater than 15 decibels at 500 hertz (Hz), 1,000 Hz, and 2,000 Hz.

(vii) Visible evidence of significant cerumen accumulation or a foreign body in the ear canal.

(viii) Pain or discomfort in the ear.

Special care should be exercised in selecting and fitting a hearing aid whose maximum sound pressure level exceeds 132 decibels because there may be risk of impairing the remaining hearing of the hearing aid user. (This provision is required only for those hearing aids with a maximum sound pressure capability greater than 132 decibels (dB).)

(3) *Notice for prospective hearing aid users.* The User Instructional Brochure shall contain the following notice:

IMPORTANT NOTICE FOR PROSPECTIVE HEARING AID USERS

Good health practice requires that a person with a hearing loss have a medical evaluation by a licensed physician (preferably a physician who specializes in diseases of the ear) before purchasing a hearing aid. Licensed physicians who specialize in diseases of the ear are often referred to as otolaryngologists, otologists or otorhinolaryngologists. The purpose of medical evaluation is to assure that all medically treatable conditions that may affect hearing are identified and treated before the hearing aid is purchased.

Following the medical evaluation, the physician will give you a written statement that states that your hearing loss has been medically evaluated and that you may be considered a candidate for a hearing aid. The physician will refer you to an audiologist or a hearing aid dispenser, as appropriate, for a hearing aid evaluation.

The audiologist or hearing aid dispenser will conduct a hearing aid evaluation to assess your ability to hear with and without a hearing aid. The hearing aid evaluation will enable the audiologist or dispenser to select and fit a hearing aid to your individual needs.

If you have reservations about your ability to adapt to amplification, you should inquire about the availability of a trial-rental or purchase-option program. Many hearing aid dispensers now offer programs that permit you to wear a hearing aid for a period of time for a nominal fee after which you may decide if you want to purchase the hearing aid.

Federal law restricts the sale of hearing aids to those individuals who have obtained a medical evaluation from a licensed physician. Federal law permits a fully informed adult to sign a waiver statement declining the medical evaluation for religious or personal beliefs that preclude consultation with a physician. The exercise of such a waiver is not in your best health interest and its use is strongly discouraged.

CHILDREN WITH HEARING LOSS

In addition to seeing a physician for a medical evaluation, a child with a hearing loss should be directed to an audiologist for evaluation and rehabilitation since hearing loss may cause problems in language development and the educational and social growth of a child. An audiologist is qualified by training and experience to assist in the evaluation and rehabilitation of a child with a hearing loss.

(4) *Technical data.* Technical data useful in selecting, fitting, and checking the performance of a hearing aid shall be provided in the User Instructional Brochure or in separate labeling that accompanies the device. The determination of technical data values for the hearing aid labeling shall be conducted in accordance with the test procedures of the American National Standard "Specification of Hearing Aid Characteristics," ANSI S3.22–1996 (ASA 70–1996) (Revision of ANSI S3.22–1987), which is incorporated by reference in accordance with 5 U.S.C. 552(a) and 1 CFR part 51. Copies are available from the Standards Secretariat of the Acoustical Society of America, 120 Wall St., New York, NY 10005–3993, or are available for inspection at the Regulations Staff, CDRH (HFZ–215), FDA, 1350 Piccard Dr., rm. 240, Rockville, MD 20850, and or at the National Archives and Records Administration (NARA). For information on the availability of this material at NARA, call 202–741–6030, or go to: *http://www.archives.gov/federal_register/code_of_federal_regulations/ibr_locations.html.* As a minimum, the User Instructional Brochure or such other labeling shall include the appropriate values or information for the following technical data elements as these elements are defined or used in such standard:

(i) Saturation output curve (SSPL 90 curve).

(ii) Frequency response curve.

(iii) Average saturation output (HF-Average SSPL 90).

(iv) Average full-on gain (HF-Average full-on gain).

(v) Reference test gain.

(vi) Frequency range.

(vii) Total harmonic distortion.

(viii) Equivalent input noise.

(ix) Battery current drain.

(x) Induction coil sensitivity (telephone coil aids only).

(xi) Input-output curve (ACG aids only).

(xii) Attack and release times (ACG aids only).

(5) *Statement if hearing aid is used or rebuilt.* If a hearing aid has been used or rebuilt, this fact shall be declared on the container in which the hearing aid is packaged and on a tag that is physically attached to such hearing aid. Such fact may also be stated in the User Instructional Brochure.

(6) *Statements in User Instructional Brochure other than those required.* A User Instructional Brochure may contain statements or illustrations in addition to those required by paragraph (c) of this section if the additional statements:

(i) Are not false or misleading in any particular, e.g., diminishing the impact of the required statements; and

(ii) Are not prohibited by this chapter or by regulations of the Federal Trade Commission.

(d) *Submission of all labeling for each type of hearing aid.* Any manufacturer of a hearing aid described in paragraph (a) of this section shall submit to the Food and Drug Administration, Bureau of Medical Devices and Diagnostic Products, Division of Compliance,

HFK–116, 8757 Georgia Ave., Silver Spring, MD 20910, a copy of the User Instructional Brochure described in paragraph (c) of this section and all other labeling for each type of hearing aid on or before August 15, 1977.

[42 FR 9294, Feb. 15, 1977, as amended at 47 FR 9398, Mar. 5, 1982; 50 FR 30154, July 24, 1985; 54 FR 52396, Dec. 21, 1989; 64 FR 59620, Nov. 3, 1999; 69 FR 18803, Apr. 9, 2004]

§ 801.421 Hearing aid devices; conditions for sale.

(a) *Medical evaluation requirements*—(1) *General.* Except as provided in paragraph (a)(2) of this section, a hearing aid dispenser shall not sell a hearing aid unless the prospective user has presented to the hearing aid dispenser a written statement signed by a licensed physician that states that the patient's hearing loss has been medically evaluated and the patient may be considered a candidate for a hearing aid. The medical evaluation must have taken place within the preceding 6 months.

(2) *Waiver to the medical evaluation requirements.* If the prospective hearing aid user is 18 years of age or older, the hearing aid dispenser may afford the prospective user an opportunity to waive the medical evaluation requirement of paragraph (a)(1) of this section provided that the hearing aid dispenser:

(i) Informs the prospective user that the exercise of the waiver is not in the user's best health interest;

(ii) Does not in any way actively encourage the prospective user to waive such a medical evaluation; and

(iii) Affords the prospective user the opportunity to sign the following statement:

I have been advised by ______ ______ (Hearing aid dispenser's name) that the Food and Drug Administration has determined that my best health interest would be served if I had a medical evaluation by a licensed physician (preferably a physician who specializes in diseases of the ear) before purchasing a hearing aid. I do not wish a medical evaluation before purchasing a hearing aid.

(b) *Opportunity to review User Instructional Brochure.* Before signing any statement under paragraph (a)(2)(iii) of this section and before the sale of a hearing aid to a prospective user, the hearing aid dispenser shall:

(1) Provide the prospective user a copy of the User Instructional Brochure for a hearing aid that has been, or may be selected for the prospective user;

(2) Review the content of the User Instructional Brochure with the prospective user orally, or in the predominate method of communication used during the sale;

(3) Afford the prospective user an opportunity to read the User Instructional Brochure.

(c) *Availability of User Instructional Brochure.* (1) Upon request by an individual who is considering purchase of a hearing aid, a dispenser shall, with respect to any hearing aid that he dispenses, provide a copy of the User Instructional Brochure for the hearing aid or the name and address of the manufacturer or distributor from whom a User Instructional Brochure for the hearing aid may be obtained.

(2) In addition to assuring that a User Instructional Brochure accompanies each hearing aid, a manufacturer or distributor shall with respect to any hearing aid that he manufactures or distributes:

(i) Provide sufficient copies of the User Instructional Brochure to sellers for distribution to users and prospective users;

(ii) Provide a copy of the User Instructional Brochure to any hearing aid professional, user, or prospective user who requests a copy in writing.

(d) *Recordkeeping.* The dispenser shall retain for 3 years after the dispensing of a hearing aid a copy of any written statement from a physician required under paragraph (a)(1) of this section or any written statement waiving medical evaluation required under paragraph (a)(2)(iii) of this section.

(e) *Exemption for group auditory trainers.* Group auditory trainers, defined as a group amplification system purchased by a qualified school or institution for the purpose of communicating with and educating individuals with hearing impairments, are exempt from the requirements of this section.

[42 FR 9296, Feb. 15, 1977]

§ 801.430 User labeling for menstrual tampons.

(a) This section applies to scented or scented deodorized menstrual tampons as identified in § 884.5460 and unscented menstrual tampons as identified in § 884.5470 of this chapter.

(b) Data show that toxic shock syndrome (TSS), a rare but serious and sometimes fatal disease, is associated with the use of menstrual tampons. To protect the public and to minimize the serious adverse effects of TSS, menstrual tampons shall be labeled as set forth in paragraphs (c), (d), and (e) of this section and tested for absorbency as set forth in paragraph (f) of this section.

(c) If the information specified in paragraph (d) of this section is to be included as a package insert, the following alert statement shall appear prominently and legibly on the package label:

ATTENTION: Tampons are associated with Toxic Shock Syndrome (TSS). TSS is a rare but serious disease that may cause death. Read and save the enclosed information.

(d) The labeling of menstrual tampons shall contain the following consumer information prominently and legibly, in such terms as to render the information likely to be read and understood by the ordinary individual under customary conditions of purchase and use:

(1)(i) Warning signs of TSS, e.g., sudden fever (usually 102° or more) and vomiting, diarrhea, fainting or near fainting when standing up, dizziness, or a rash that looks like a sunburn;

(ii) What to do if these or other signs of TSS appear, including the need to remove the tampon at once and seek medical attention immediately;

(2) The risk of TSS to all women using tampons during their menstrual period, especially the reported higher risks to women under 30 years of age and teenage girls, the estimated incidence of TSS of 1 to 17 per 100,000 menstruating women and girls per year, and the risk of death from contracting TSS;

(3) The advisability of using tampons with the minimum absorbency needed to control menstrual flow in order to reduce the risk of contracting TSS;

(4) Avoiding the risk of getting tampon-associated TSS by not using tampons, and reducing the risk of getting TSS by alternating tampon use with sanitary napkin use during menstrual periods; and

(5) The need to seek medical attention before again using tampons if TSS warning signs have occurred in the past, or if women have any questions about TSS or tampon use.

(e) The statements required by paragraph (e) of this section shall be prominently and legibly placed on the package label of menstrual tampons in conformance with section 502(c) of the Federal Food, Drug, and Cosmetic Act (the act) (unless the menstrual tampons are exempt under paragraph (g) of this section).

(1) Menstrual tampon package labels shall bear one of the following absorbency terms representing the absorbency of the production run, lot, or batch as measured by the test described in paragraph (f)(2) of this section;

Ranges of absorbency in grams[1]	Corresponding term of absorbency
6 and under	Light absorbency
6 to 9	Regular absorbency
9 to 12	Super absorbency
12 to 15	Super plus absorbency
15 to 18	Ultra absorbency
Above 18	No term

[1]These ranges are defined, respectively, as follows: Less than or equal to 6 grams (g); greater than 6 g up to and including 9 g; greater than 9 g up to and including 12 g; greater than 12 g up to and including 15 g; greater than 15 g up to and including 18 g; and greater than 18 g.

(2) The package label shall include an explanation of the ranges of absorbency and a description of how consumers can use a range of absorbency, and its corresponding absorbency term, to make comparisons of absorbency of tampons to allow selection of the tampons with the minimum absorbency needed to control menstrual flow in order to reduce the risk of contracting TSS.

(f) A manufacturer shall measure the absorbency of individual tampons using the test method specified in paragraph (f)(2) of this section and calculate the mean absorbency of a production run, lot, or batch by rounding to the nearest 0.1 gram.

(1) A manufacturer shall design and implement a sampling plan that includes collection of probability samples of adequate size to yield consistent tolerance intervals such that the probability is 90 percent that at least 90 percent of the absorbencies of individual tampons within a brand and type are within the range of absorbency stated on the package label.

(2) In the absorbency test, an unlubricated condom, with tensile strength between 17 Mega Pascals (MPa) and 30 MPa, as measured according to the procedure in the American Society for Testing and Materials (ASTM) D 3492–97, ''Standard Specification for Rubber Contraceptives (Male Condoms)''[1] for determining tensile strength, which is incorporated by reference in accordance with 5 U.S.C. 552(a), is attached to the large end of a glass chamber (or a chamber made from hard transparent plastic) with a rubber band (see figure 1) and pushed through the small end of the chamber using a smooth, finished rod. The condom is pulled through until all slack is removed. The tip of the condom is cut off and the remaining end of the condom is stretched over the end of the tube and secured with a rubber band. A preweighed (to the nearest 0.01 gram) tampon is placed within the condom membrane so that the center of gravity of the tampon is at the center of the chamber. An infusion needle (14 gauge) is inserted through the septum created by the condom tip until it contacts the end of the tampon. The outer chamber is filled with water pumped from a temperature-controlled waterbath to maintain the average temperature at 27±1 °C. The water returns to the waterbath as shown in figure 2. Syngyna fluid (10 grams sodium chloride, 0.5 gram Certified Reagent Acid Fushsin, 1,000 milliliters distilled water) is then pumped through the infusion needle at a rate of 50 milliliters per hour. The test shall be terminated when the tampon is saturated and the first drop of fluid exits the apparatus. (The test result shall be discarded if fluid is detected in the folds of the condom before the tampon is saturated). The water is then drained and the tampon is removed and immediately weighed to the nearest 0.01 gram. The absorbency of the tampon is determined by subtracting its dry weight from this value. The condom shall be replaced after 10 tests or at the end of the day during which the condom is used in testing, whichever occurs first.

[1] Copies of the standard are available from the American Society for Testing and Materials, 100 Barr Harbor Dr., West Conshohocken, PA 19428, or available for inspection at the Center for Devices and Radiological Health's Library, 9200 Corporate Blvd., Rockville, MD 20850, or at the National Archives and Records Administration (NARA). For information on the availability of this material at NARA, call 202–741–6030, or go to: *http://www.archives.gov/federal_register/code_of_federal_regulations/ibr_locations.html.*

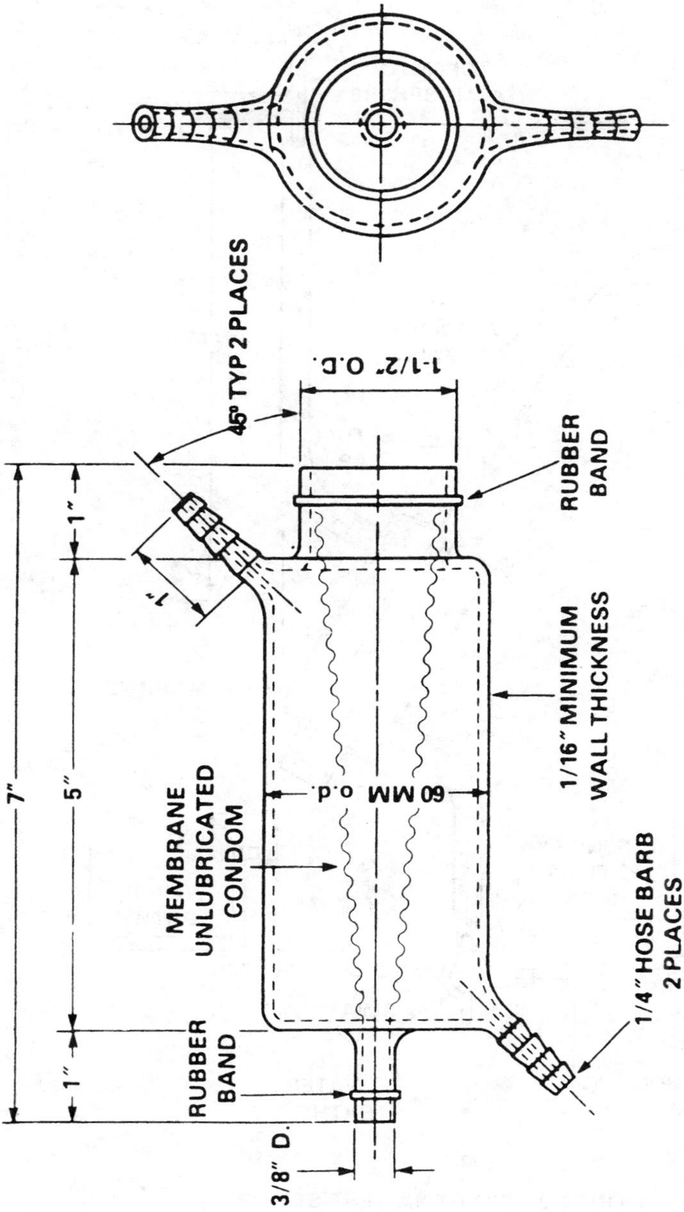

FIGURE 1 — SYNGYNA TEST CHAMBER

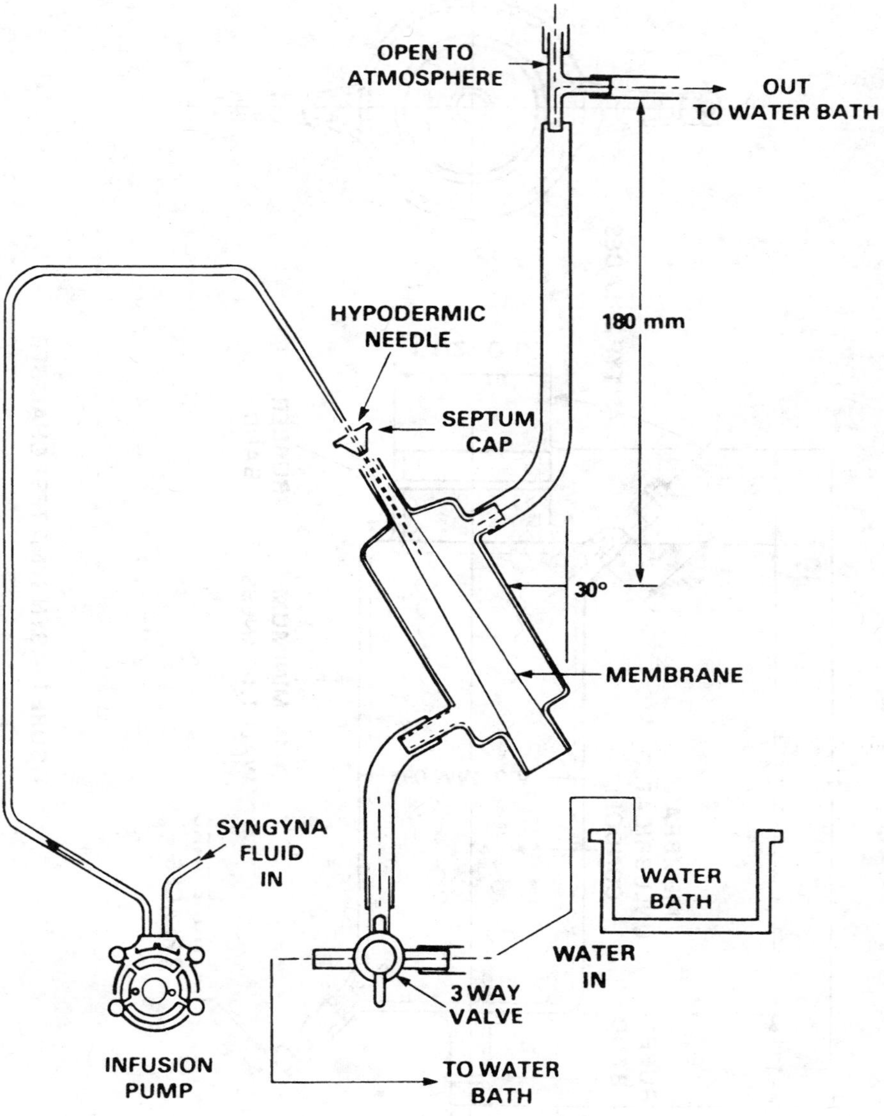

FIGURE 2—SYNGYNA TEST SET-UP

(3) The Food and Drug Administration may permit the use of an absorbency test method different from the test method specified in this section if each of the following conditions is met:

(i) The manufacturer presents evidence, in the form of a citizen petition submitted in accordance with the requirements of § 10.30 of this chapter, demonstrating that the alternative test method will yield results that are equivalent to the results yielded by the test method specified in this section; and

(ii) FDA approves the method and has published notice of its approval of the alternative test method in the FEDERAL REGISTER.

(g) Any menstrual tampon intended to be dispensed by a vending machine is exempt from the requirements of this section.

(h) Any menstrual tampon that is not labeled as required by paragraphs (c), (d), and (e) of this section and that is initially introduced or initially delivered for introduction into commerce after March 1, 1990, is misbranded under sections 201(n), 502 (a) and (f) of the act.

(Information collection requirements contained in paragraphs (e) and (f) were approved by the Office of Management and Budget under control number 0910–0257)

[47 FR 26989, June 22, 1982, as amended at 54 FR 43771, Oct. 26, 1989; 55 FR 17600, Apr. 26, 1990; 65 FR 3586, Jan. 24, 2000; 65 FR 44436, July 18, 2000; 65 FR 62284, Oct. 18, 2000; 69 FR 18803, Apr. 9, 2004; 69 FR 52171, Aug. 25, 2004]

§ 801.433 Warning statements for prescription and restricted device products containing or manufactured with chlorofluorocarbons or other ozone-depleting substances.

(a)(1) All prescription and restricted device products containing or manufactured with chlorofluorocarbons, halons, carbon tetrachloride, methyl chloride, or any other class I substance designated by the Environmental Protection Agency (EPA) shall, except as provided in paragraph (b) of this section, bear the following warning statement:

WARNING: Contains [or Manufactured with, if applicable] [*insert name of substance*], a substance which harms public health and environment by destroying ozone in the upper atmosphere.

(2) The warning statement shall be clearly legible and conspicuous on the product, its immediate container, its outer packaging, or other labeling in accordance with the requirements of 40 CFR part 82 and appear with such prominence and conspicuousness as to render it likely to be read and understood by consumers under normal conditions of purchase.

(b)(1) For prescription and restricted device products, the following alternative warning statement may be used:

NOTE: The indented statement below is required by the Federal government's Clean Air Act for all products containing or manufactured with chlorofluorocarbons (CFC's) [or name of other class I substance, if applicable]:

This product contains [or is manufactured with, if applicable] [*insert name of substance*], a substance which harms the environment by destroying ozone in the upper atmosphere.

Your physician has determined that this product is likely to help your personal health. USE THIS PRODUCT AS DIRECTED, UNLESS INSTRUCTED TO DO OTHERWISE BY YOUR PHYSICIAN. If you have any questions about alternatives, consult with your physician.

(2) The warning statement shall be clearly legible and conspicuous on the product, its immediate container, its outer packaging, or other labeling in accordance with the requirements of 40 CFR part 82 and appear with such prominence and conspicuousness as to render it likely to be read and understood by consumers under normal conditions of purchase.

(3) If the warning statement in paragraph (b)(1) of this section is used, the following warning statement must be placed on the package labeling intended to be read by the physician (physician package insert) after the "How supplied" section, which describes special handling and storage conditions on the physician labeling:

NOTE: The indented statement below is required by the Federal government's Clean Air Act for all products containing or manufactured with chlorofluorocarbons (CFC's) [or name of other class I substance, if applicable]:

WARNING: Contains [or Manufactured with, if applicable] [*insert name of substance*], a substance which harms public health and environment by destroying ozone in the upper atmosphere.

A notice similar to the above WARNING has been placed in the information for the patient [or patient information leaflet, if applicable] of this product under Environmental Protection Agency (EPA) regulations. The patient's warning states that the patient should consult his or her physician if there are questions about alternatives.

(c) This section does not replace or relieve a person from any requirements imposed under 40 CFR part 82.

[61 FR 20101, May 3, 1996]

§ 801.435 User labeling for latex condoms.

(a) This section applies to the subset of condoms as identified in § 884.5300 of this chapter, and condoms with spermicidal lubricant as identified in § 884.5310 of this chapter, which products are formed from latex films.

(b) Data show that the material integrity of latex condoms degrade over time. To protect the public health and minimize the risk of device failure, latex condoms must bear an expiration date which is supported by testing as described in paragraphs (d) and (h) of this section.

(c) The expiration date, as demonstrated by testing procedures required by paragraphs (d) and (h) of this section, must be displayed prominently and legibly on the primary packaging (i.e., individual package), and higher levels of packaging (e.g., boxes of condoms), in order to ensure visibility of the expiration date by consumers.

(d) Except as provided under paragraph (f) of this section, the expiration date must be supported by data demonstrating physical and mechanical integrity of the product after three discrete and representative lots of the product have been subjected to each of the following conditions:

(1) Storage of unpackaged bulk product for the maximum amount of time the manufacturer allows the product to remain unpackaged, followed by storage of the packaged product at 70 °C (plus or minus 2 °C) for 7 days;

(2) Storage of unpackaged bulk product for the maximum amount of time the manufacturer allows the product to remain unpackaged, followed by storage of the packaged product at a selected temperature between 40 and 50 °C (plus or minus 2 °C) for 90 days; and

(3) Storage of unpackaged bulk product for the maximum amount of time the manufacturer allows the product to remain unpackaged, followed by storage of the packaged product at a monitored or controlled temperature between 15 and 30 °C for the lifetime of the product (real time storage).

(e) If a product fails the physical and mechanical integrity tests commonly used by industry after the completion of the accelerated storage tests described in paragraphs (d)(1) and (d)(2) of this section, the product expiration date must be demonstrated by real time storage conditions described in paragraph (d)(3) of this section. If all of the products tested after storage at temperatures as described in paragraphs (d)(1) and (d)(2) of this section pass the manufacturer's physical and mechanical integrity tests, the manufacturer may label the product with an expiration date of up to 5 years from the date of product packaging. If the extrapolated expiration date under paragraphs (d)(1) and (d)(2) of this section is used, the labeled expiration date must be confirmed by physical and mechanical integrity tests performed at the end of the stated expiration period as described in paragraph (d)(3) of this section. If the data from tests following real time storage described in paragraph (d)(3) of this section fails to confirm the extrapolated expiration date, the manufacturer must, at that time, relabel the product to reflect the actual shelf life.

(f) Products that already have established shelf life data based upon real time storage and testing and have such storage and testing data available for inspection are not required to confirm such data using accelerated and intermediate aging data described in paragraphs (d)(1) and (d)(2) of this section. If, however, such real time expiration dates were based upon testing of products that were not first left unpackaged for the maximum amount of time as described in paragraph (d)(3) of this section, the real time testing must be confirmed by testing products consistent with the requirements of paragraph (d)(3) of this section. This testing shall be initiated no later than the effective date of this regulation. Until the confirmation testing in accordance with paragraph (d)(3) of this section is completed, the product may remain on the market labeled with the expiration date based upon previous real time testing.

(g) If a manufacturer uses testing data from one product to support expiration dating on any variation of that product, the manufacturer must document and provide, upon request, an appropriate justification for the application of the testing data to the variation of the tested product.

(h) If a latex condom contains a spermicide, and the expiration date based on spermicidal stability testing is different from the expiration date based upon latex integrity testing, the product shall bear only the earlier expiration date.

(i) The time period upon which the expiration date is based shall start with the date of packaging.

(j) As provided in part 820 of this chapter, all testing data must be retained in each company's files, and shall be made available upon request for inspection by the Food and Drug Administration.

(k) Any latex condom not labeled with an expiration date as required by paragraph (c) of this section, and initially delivered for introduction into interstate commerce after the effective date of this regulation is misbranded under sections 201(n) and 502(a) and (f) of Federal Food, Drug, and Cosmetic Act (21 U.S.C. 321(n) and 352(a) and (f)).

[62 FR 50501, Sept. 26, 1997]

§ 801.437 User labeling for devices that contain natural rubber.

(a) Data in the Medical Device Reporting System and the scientific literature indicate that some individuals are at risk of severe anaphylactic reactions to natural latex proteins. This labeling regulation is intended to minimize the risk to individuals sensitive to natural latex proteins and protect the public health.

(b) This section applies to all devices composed of or containing, or having packaging or components that are composed of, or contain, natural rubber that contacts humans. The term "natural rubber" includes natural rubber latex, dry natural rubber, and synthetic latex or synthetic rubber that contains natural rubber in its formulation.

(1) The term "natural rubber latex" means rubber that is produced by the natural rubber latex process that involves the use of natural latex in a concentrated colloidal suspension. Products are formed from natural rubber latex by dipping, extruding, or coating.

(2) The term "dry natural rubber" means rubber that is produced by the dry natural rubber process that involves the use of coagulated natural latex in the form of dried or milled sheets. Products are formed from dry natural rubber by compression molding, extrusion, or by converting the sheets into a solution for dipping.

(3) The term "contacts humans" means that the natural rubber contained in a device is intended to contact or is likely to contact the user or patient. This includes contact when the device that contains natural rubber is connected to the patient by a liquid path or an enclosed gas path; or the device containing the natural rubber is fully or partially coated with a powder, and such powder may carry natural rubber proteins that may contaminate the environment of the user or patient.

(c) Devices containing natural rubber shall be labeled as set forth in paragraphs (d) through (h) of this section. Each required labeling statement shall be prominently and legibly displayed in conformance with section 502(c) of the Federal Food, Drug, and Cosmetic Act (the act) (21 U.S.C. 352(c)).

(d) Devices containing natural rubber latex that contacts humans, as described in paragraph (b) of this section, shall bear the following statement in bold print on the device labeling:

"Caution: This Product Contains Natural Rubber Latex Which May Cause Allergic Reactions."

This statement shall appear on all device labels, and other labeling, and shall appear on the principal display panel of the device packaging, the outside package, container or wrapper, and the immediate device package, container, or wrapper.

(e) Devices containing dry natural rubber that contacts humans, as described in paragraph (b) of this section, that are not already subject to paragraph (d) of this section, shall bear the following statement in bold print on the device labeling:

"This Product Contains Dry Natural Rubber."

This statement shall appear on all device labels, and other labeling, and shall appear on the principal display panel of the device packaging, the outside package, container or wrapper, and the immediate device package, container, or wrapper.

(f) Devices that have packaging containing natural rubber latex that contacts humans, as described in paragraph (b) of this section, shall bear the following statement in bold print on the device labeling:

"Caution: The Packaging of This Product Contains Natural Rubber Latex Which May Cause Allergic Reactions."

This statement shall appear on the packaging that contains the natural rubber, and the outside package, container, or wrapper.

(g) Devices that have packaging containing dry natural rubber that contacts humans, as described in paragraph (b) of this section, shall bear the following statement in bold print on the device labeling:

"The Packaging of This Product Contains Dry Natural Rubber."

This statement shall appear on the packaging that contains the natural rubber, and the outside package, container, or wrapper.–

(h) Devices that contain natural rubber that contacts humans, as described in paragraph (b) of this section, shall not contain the term "hypoallergenic" on their labeling.

(i) Any affected person may request an exemption or variance from the requirements of this section by submitting a citizen petition in accordance with § 10.30 of this chapter.

(j) Any device subject to this section that is not labeled in accordance with paragraphs (d) through (h) of this section and that is initially introduced or initially delivered for introduction into interstate commerce after the effective date of this regulation is misbranded under sections 201(n) and 502(a), (c), and (f) of the act (21 U.S.C. 321(n) and 352(a), (c), and (f)).

NOTE TO § 801.437: Paragraphs (f) and (g) are stayed until June 27, 1999, as those regulations relate to device packaging that uses "cold seal" adhesives.

[62 FR 51029, Sept. 30, 1997, as amended at 63 FR 46175, Aug. 31, 1998]

PART 803—MEDICAL DEVICE REPORTING

Subpart A—General Provisions

Sec.

Subpart B—Generally Applicable Requirements for Individual Adverse Event Reports

Subpart C—User Facility Reporting Requirements

Subpart D—Importer Reporting Requirements

Subpart E—Manufacturer Reporting Requirements

AUTHORITY: 21 U.S.C. 352, 360, 360i, 360j, 371, 374.

SOURCE: 70 FR 9519, July 13, 2005, unless otherwise noted

Subpart A—General Provisions

§ 803.1 What does this part cover?

(a) This part establishes the requirements for medical device reporting for device user facilities, manufacturers, importers, and distributors. If you are a device user facility, you must report deaths and serious injuries that a device has or may have caused or contributed to, establish and maintain adverse event files, and submit summary annual reports. If you are a manufacturer or importer, you must report deaths and serious injuries that your device has or may have caused or contributed to, you must report certain device malfunctions, and you must establish and maintain adverse event files. If you are a manufacturer, you must also submit specified followup and baseline reports. These reports help us to protect the public health by helping to ensure that devices are not adulterated or misbranded and are safe and effective for their intended use. If you are a medical device distributor, you must maintain records (files) of incidents, but you are not required to report these incidents.

(b) This part supplements and does not supersede other provisions of this chapter, including the provisions of part 820 of this chapter.

(c) References in this part to regulatory sections of the Code of Federal Regulations are to chapter I of title 21, unless otherwise noted.

§ 803.3 How does FDA define the terms used in this part?

Some of the terms we use in this part are specific to medical device reporting and reflect the language used in the statute (law). Other terms are more general and reflect our interpretation of the law. This section defines the following terms as used in this part:

Act means the Federal Food, Drug, and Cosmetic Act, 21 U.S.C. 301 *et seq.*, as amended.

Ambulatory surgical facility (ASF) means a distinct entity that operates for the primary purpose of furnishing same day outpatient surgical services to patients. An ASF may be either an independent entity (i.e., not a part of a provider of services or any other facility) or operated by another medical entity (e.g., under the common ownership, licensure, or control of an entity). An ASF is subject to this regulation regardless of whether it is licensed by a Federal, State, municipal, or local government or regardless of whether it is accredited by a recognized accreditation organization. If an adverse event meets the criteria for reporting, the ASF must report that event regardless of the nature or location of the medical service provided by the ASF.

Become aware means that an employee of the entity required to report has acquired information that reasonably suggests a reportable adverse event has occurred.

(1) If you are a device user facility, you are considered to have "become aware" when medical personnel, as defined in this section, who are employed by or otherwise formally affiliated with your facility, obtain information about a reportable event.

(2) If you are a manufacturer, you are considered to have become aware of an

event when any of your employees becomes aware of a reportable event that is required to be reported within 30 calendar days or that is required to be reported within 5 work days because we had requested reports in accordance with § 803.53(b). You are also considered to have become aware of an event when any of your employees with management or supervisory responsibilities over persons with regulatory, scientific, or technical responsibilities, or whose duties relate to the collection and reporting of adverse events, becomes aware, from any information, including any trend analysis, that a reportable MDR event or events necessitates remedial action to prevent an unreasonable risk of substantial harm to the public health.

(3) If you are an importer, you are considered to have become aware of an event when any of your employees becomes aware of a reportable event that is required to be reported by you within 30 days.

Caused or contributed means that a death or serious injury was or may have been attributed to a medical device, or that a medical device was or may have been a factor in a death or serious injury, including events occurring as a result of:

(1) Failure;

(2) Malfunction;

(3) Improper or inadequate design;

(4) Manufacture;

(5) Labeling; or

(6) User error.

Device family. (1) Device family means a group of one or more devices manufactured by or for the same manufacturer and having the same:

(i) Basic design and performance characteristics related to device safety and effectiveness,

(ii) Intended use and function, and

(iii) Device classification and product code.

(2) You may consider devices that differ only in minor ways not related to safety or effectiveness to be in the same device family. When grouping products in device families, you may consider factors such as brand name and common name of the device and whether the devices were introduced into commercial distribution under the same 510(k) or premarket approval application (PMA).

Device user facility means a hospital, ambulatory surgical facility, nursing home, outpatient diagnostic facility, or outpatient treatment facility as defined in this section, which is not a physician's office, as defined in this section. School nurse offices and employee health units are not device user facilities.

Distributor means any person (other than the manufacturer or importer) who furthers the marketing of a device from the original place of manufacture to the person who makes final delivery or sale to the ultimate user, but who does not repackage or otherwise change the container, wrapper, or labeling of the device or device package. If you repackage or otherwise change the container, wrapper, or labeling, you are considered a manufacturer as defined in this section.

Expected life of a device means the time that a device is expected to remain functional after it is placed into use. Certain implanted devices have specified "end of life" (EOL) dates. Other devices are not labeled as to their respective EOL, but are expected to remain operational through activities such as maintenance, repairs, or upgrades, for an estimated period of time.

FDA, we, or us means the Food and Drug Administration.

Five-day report means a medical device report that must be submitted by a manufacturer to us under § 803.53, on FDA Form 3500A or an electronic equivalent approved under § 803.14, within 5 work days.

Hospital means a distinct entity that operates for the primary purpose of providing diagnostic, therapeutic (such as medical, occupational, speech, physical), surgical, and other patient services for specific and general medical conditions. Hospitals include general, chronic disease, rehabilitative, psychiatric, and other special-purpose facilities. A hospital may be either independent (e.g., not a part of a provider of services or any other facility) or may be operated by another medical entity (e.g., under the common ownership, licensure, or control of another entity). A hospital is covered by this regulation regardless of whether it is licensed by a Federal, State, municipal or local government or whether it is accredited by a recognized accreditation organization. If an adverse event meets the criteria for reporting, the hospital must report that event regardless of the nature or location of the medical service provided by the hospital.

Importer means any person who imports a device into the United States and who furthers the marketing of a device from the original place of manufacture to the person who makes final delivery or sale to the ultimate user, but who does not repackage or otherwise change the container, wrapper, or labeling of the device or device package. If you repackage or otherwise change the container, wrapper, or labeling, you are considered a manufacturer as defined in this section.

Malfunction means the failure of a device to meet its performance specifications or otherwise perform as intended. Performance specifications include all claims made in the labeling for the device. The intended performance of a device refers to the intended use for which the device is labeled or marketed, as defined in § 801.4 of this chapter.

Manufacturer means any person who manufactures, prepares, propagates, compounds, assembles, or processes a device by chemical, physical, biological, or other procedure. The term includes any person who either:

(1) Repackages or otherwise changes the container, wrapper, or labeling of a device in furtherance of the distribution of the device from the original place of manufacture;

(2) Initiates specifications for devices that are manufactured by a second party for subsequent distribution by the person initiating the specifications;

(3) Manufactures components or accessories that are devices that are ready to be used and are intended to be commercially distributed and intended to be used as is, or are processed by a licensed practitioner or other qualified person to meet the needs of a particular patient; or

(4) Is the U.S. agent of a foreign manufacturer.

Manufacturer or importer report number. Manufacturer or importer report number means the number that uniquely identifies each individual adverse event report submitted by a manufacturer or importer. This number consists of the following three parts:

(1) The FDA registration number for the manufacturing site of the reported device, or the registration number for the importer. If the manufacturing site or the importer does not have an establishment registration number, we will assign a temporary MDR reporting number until the site is registered in accordance with part 807 of this chapter. We will inform the manufacturer or importer of the temporary MDR reporting number;

(2) The four-digit calendar year in which the report is submitted; and

(3) The five-digit sequence number of the reports submitted during the year, starting with 00001. (For example, the complete number will appear as follows: 1234567–1995–00001.)

MDR means medical device report.

MDR reportable event (or reportable event) means:

(1) An event that user facilities become aware of that reasonably suggests that a device has or may have caused or contributed to a death or serious injury; or

(2) An event that manufacturers or importers become aware of that reasonably suggests that one of their marketed devices:

(i) May have caused or contributed to a death or serious injury, or

(ii) Has malfunctioned and that the device or a similar device marketed by the manufacturer or importer would be likely to cause or contribute to a death or serious injury if the malfunction were to recur.

Medical personnel means an individual who:

(1) Is licensed, registered, or certified by a State, territory, or other governing body, to administer health care;

(2) Has received a diploma or a degree in a professional or scientific discipline;

(3) Is an employee responsible for receiving medical complaints or adverse event reports; or

(4) Is a supervisor of these persons.

Nursing home means:

(1) An independent entity (i.e., not a part of a provider of services or any other facility) or one operated by another medical entity (e.g., under the common ownership, licensure, or control of an entity) that operates for the primary purpose of providing:

(i) Skilled nursing care and related services for persons who require medical or nursing care;

(ii) Hospice care to the terminally ill; or

(iii) Services for the rehabilitation of the injured, disabled, or sick.

(2) A nursing home is subject to this regulation regardless of whether it is licensed by a Federal, State, municipal, or local government or whether it is accredited by a recognized accreditation organization. If an adverse event meets the criteria for reporting, the nursing home must report that event regardless of the nature or location of the medical service provided by the nursing home.

Outpatient diagnostic facility. (1) Outpatient diagnostic facility means a distinct entity that:

(i) Operates for the primary purpose of conducting medical diagnostic tests on patients,

(ii) Does not assume ongoing responsibility for patient care, and

(iii) Provides its services for use by other medical personnel.

(2) Outpatient diagnostic facilities include outpatient facilities providing radiography, mammography, ultrasonography, electrocardiography, magnetic resonance imaging, computerized axial tomography, and in vitro testing. An outpatient diagnostic facility may be either independent (i.e., not a part of a provider of services or any other facility) or operated by another medical entity (e.g., under the common ownership, licensure, or control of an entity). An outpatient diagnostic facility is covered by this regulation regardless of whether it is licensed by a Federal, State, municipal, or local government or whether it is accredited by a recognized accreditation organization. If an adverse event meets the criteria for reporting, the outpatient diagnostic facility must report that event regardless of the nature or location of the medical service provided by the outpatient diagnostic facility.

Outpatient treatment facility means a distinct entity that operates for the primary purpose of providing nonsurgical therapeutic (medical, occupational, or physical) care on an outpatient basis or in a home health care setting. Outpatient treatment facilities include ambulance providers, rescue services, and home health care groups. Examples of services provided by outpatient treatment facilities include the following: Cardiac defibrillation, chemotherapy, radiotherapy, pain control, dialysis, speech or physical therapy, and treatment for substance abuse. An outpatient treatment facility may be either independent (i.e., not a part of a provider of services or any other facility) or operated by another medical entity (e.g., under the common ownership, licensure, or control of an entity). An outpatient treatment facility is covered by this regulation regardless of whether it is licensed by a Federal, State, municipal, or local government or whether it is accredited by a recognized accreditation organization. If an adverse event meets the criteria for reporting, the outpatient treatment facility must report that event regardless of the nature or location of the medical service provided by the outpatient treatment facility.

Patient of the facility means any individual who is being diagnosed or treated and/or receiving medical care at or under the control or authority of the facility. This includes employees of the facility or individuals affiliated with the facility who, in the course of their duties, suffer a device-related death or serious injury that has or may have been caused or contributed to by a device used at the facility.

Physician's office means a facility that operates as the office of a physician or other health care professional for the primary purpose of examination, evaluation, and treatment or referral of patients. Examples of physician offices include dentist offices, chiropractor offices, optometrist offices, nurse practitioner offices, school nurse offices, school clinics, employee health clinics, or freestanding care units. A physician's office may be independent, a group practice, or part of a Health Maintenance Organization.

Remedial action means any action other than routine maintenance or servicing of a device where such action is necessary to prevent recurrence of a reportable event.

Serious injury means an injury or illness that:

(1) Is life-threatening,

(2) Results in permanent impairment of a body function or permanent damage to a body structure, or

(3) Necessitates medical or surgical intervention to preclude permanent impairment of a body function or permanent damage to a body structure.

Permanent means irreversible impairment or damage to a body structure or function, excluding trivial impairment or damage.

Shelf life means the maximum time a device will remain functional from the date of manufacture until it is used in patient care. Some devices have an expiration date on their labeling indicating the maximum time they can be stored before losing their ability to perform their intended function.

User facility report number means the number that uniquely identifies each report submitted by a user facility to manufacturers and to us. This number consists of the following three parts:

(1) The user facility's 10-digit Centers for Medicare and Medicaid Services (CMS) number (if the CMS number has fewer than 10 digits, fill the remaining spaces with zeros);

(2) The four-digit calendar year in which the report is submitted; and

(3) The four-digit sequence number of the reports submitted for the year, starting with 0001. (For example, a complete user facility report number will appear as follows: 1234560000–2004–0001. If a user facility has more than one CMS number, it must select one that will be used for all of its MDR reports. If a user facility has no CMS number, it should use all zeros in the appropriate space in its initial report (e.g., 0000000000–2004–0001). We will assign a number for future use and send that number to the user facility. This number is used in our record of the initial report, in subsequent reports, and in any correspondence with the user facility. If a facility has multiple sites, the primary site may submit reports for all sites and use one reporting number for all sites if the primary site provides the name, address, and CMS number for each respective site.)

Work day means Monday through Friday, except Federal holidays.

§ 803.9 What information from the reports do we disclose to the public?

(a) We may disclose to the public any report, including any FDA record of a telephone report, submitted under this part. Our disclosures are governed by part 20 of this chapter.

(b) Before we disclose a report to the public, we will delete the following:

(1) Any information that constitutes trade secret or confidential commercial or financial information under § 20.61 of this chapter;

(2) Any personal, medical, and similar information, including the serial number of implanted devices, which would constitute an invasion of personal privacy under § 20.63 of this chapter. However, if a patient requests a report, we will disclose to that patient all the information in the report concerning that patient, as provided in § 20.61 of this chapter; and

(3) Any names and other identifying information of a third party that voluntarily submitted an adverse event report.

(c) We may not disclose the identity of a device user facility that makes a report under this part except in connection with:

(1) An action brought to enforce section 301(q) of the act, including the failure or refusal to furnish material or information required by section 519 of the act;

(2) A communication to a manufacturer of a device that is the subject of a report required to be submitted by a user facility under § 803.30; or

(3) A disclosure to employees of the Department of Health and Human Services, to the Department of Justice, or to the duly authorized committees and subcommittees of the Congress.

§ 803.10 Generally, what are the reporting requirements that apply to me?

(a) If you are a device user facility, you must submit reports (described in subpart C of this part), as follows:

(1) Submit reports of individual adverse events no later than 10 work days after the day that you become aware of a reportable event:

(i) Submit reports of device-related deaths to us and to the manufacturer, if known; or

(ii) Submit reports of device-related serious injuries to the manufacturers or, if the manufacturer is unknown, submit reports to us.

(2) Submit annual reports (described in § 803.33) to us.

(b) If you are an importer, you must submit reports (described in subpart D of this part), as follows:

(1) Submit reports of individual adverse events no later than 30 calendar days after the day that you become aware of a reportable event:

(i) Submit reports of device-related deaths or serious injuries to us and to the manufacturer; or

(ii) Submit reports of device-related malfunctions to the manufacturer.

(2) [Reserved]

(c) If you are a manufacturer, you must submit reports (described in subpart E of this part) to us, as follows:

(1) Submit reports of individual adverse events no later than 30 calendar days after the day that you become aware of a reportable death, serious injury, or malfunction.

(2) Submit reports of individual adverse events no later than 5 work days after the day that you become aware of:

(i) A reportable event that requires remedial action to prevent an unreasonable risk of substantial harm to the public health, or

(ii) A reportable event for which we made a written request.

(3) Submit annual baseline reports.

(4) Submit supplemental reports if you obtain information that you did not submit in an initial report.

§ 803.11 What form should I use to submit reports of individual adverse events and where do I obtain these forms?

If you are a user facility, importer, or manufacturer, you must submit all reports of individual adverse events on FDA MEDWATCH Form 3500A or in an electronic equivalent as approved under § 803.14. You may obtain this form and all other forms referenced in this section from any of the following:

(1) The Consolidated Forms and Publications Office, Beltsville Service Center, 6351 Ammendale Rd., Landover, MD 20705;

(2) FDA, MEDWATCH (HF–2), 5600 Fishers Lane, Rockville, MD 20857, 301–827–7240;

(3) Division of Small Manufacturers, International, and Consumer Assistance, Office of Communication, Education, and Radiation Programs, Center for Devices and Radiological Health (CDRH) (HFZ–220), 1350 Piccard Dr. Rockville, MD 20850, by e-mail: *DSMICA@CDRH.FDA.GOV*, or FAX: 301–443–8818; or

(4) On the Internet at *http://www.fda.gov/cdrh/mdr/mdr-forms.html.*

§ 803.12 Where and how do I submit reports and additional information?

(a) You must submit any written report or additional information required under this part to FDA, CDRH, Medical Device Reporting, P.O. Box 3002, Rockville, MD 20847–3002.

(b) You must specifically identify each report (e.g., "User Facility Report," "Annual Report," "Importer Report," "Manufacturer Report," "10-Day Report").

(c) If an entity is confronted with a public health emergency, this can be brought to FDA's attention by contacting the FDA Office of Emergency Operations (HFA–615), Office of Crisis Management, Office of the Commissioner, at 301–443–1240, followed by the submission of an e-mail to *emergency.operations@fda.hhs.gov* or a fax report to 301–827–3333.

(d) You may submit a voluntary telephone report to the MEDWATCH office at 800–FDA–1088. You may also obtain information regarding voluntary reporting from the MEDWATCH office at 800–FDA–1088. You may also find the voluntary MEDWATCH 3500 form and instructions to complete it at *http://www.fda.gov/medwatch/getforms.htm.*

[70 FR 9519, July 13, 2005, as amended at 71 FR 1488, Jan. 10, 2006]

§ 803.13 Do I need to submit reports in English?

(a) Yes. You must submit all written or electronic equivalent reports required by this part in English.

(b) If you submit any reports required by this part in an electronic medium, that submission must be done in accordance with § 803.14.

§ 803.14 How do I submit a report electronically?

(a) You may electronically submit any report required by this part if you have our prior written consent. We may revoke this consent at anytime. Electronic report submissions include alternative reporting media (magnetic tape, disc, etc.) and computer-to-computer communication.

(b) If your electronic report meets electronic reporting standards, guidance documents, or other MDR reporting procedures that we have developed, you may submit the report electronically without receiving our prior written consent.

§ 803.15 How will I know if you require more information about my medical device report?

(a) We will notify you in writing if we require additional information and will tell you what information we need. We will require additional information if we determine that protection of the public health requires additional or clarifying information for medical device reports submitted to us and in cases when the additional information is beyond the scope of FDA reporting forms or is not readily accessible to us.

(b) In any request under this section, we will state the reason or purpose for the information request, specify the due date for submitting the information, and clearly identify the reported event(s) related to our request. If we verbally request additional information, we will confirm the request in writing.

§ 803.16 When I submit a report, does the information in my report constitute an admission that the device caused or contributed to the reportable event?

No. A report or other information submitted by you, and our release of that report or information, is not necessarily an admission that the device, or you or your employees, caused or contributed to the reportable event. You do not have to admit and may deny that the report or information submitted under this part constitutes an admission that the device, you, or your employees, caused or contributed to a reportable event.

§ 803.17 What are the requirements for developing, maintaining, and implementing written MDR procedures that apply to me?

If you are a user facility, importer, or manufacturer, you must develop, maintain, and implement written MDR procedures for the following:

(a) Internal systems that provide for:

(1) Timely and effective identification, communication, and evaluation of events that may be subject to MDR requirements;

(2) A standardized review process or procedure for determining when an event meets the criteria for reporting under this part; and

(3) Timely transmission of complete medical device reports to manufacturers or to us, or to both if required.

(b) Documentation and recordkeeping requirements for:

(1) Information that was evaluated to determine if an event was reportable;

(2) All medical device reports and information submitted to manufacturers and/or us;

(3) Any information that was evaluated for the purpose of preparing the submission of annual reports; and

(4) Systems that ensure access to information that facilitates timely followup and inspection by us.

§ 803.18 What are the requirements for establishing and maintaining MDR files or records that apply to me?

(a) If you are a user facility, importer, or manufacturer, you must establish and maintain MDR event files. You must clearly identify all MDR event files and maintain them to facilitate timely access.

(b)(1) For purposes of this part, "MDR event files" are written or electronic files maintained by user facilities, importers, and manufacturers. MDR event files may incorporate references to other information (e.g., medical records, patient files, engineering reports), in lieu of copying and maintaining duplicates in this file. Your MDR event files must contain:

(i) Information in your possession or references to information related to the adverse event, including all documentation of your deliberations and decisionmaking processes used to determine if a device-related death, serious injury, or malfunction was or was not reportable under this part; and

(ii) Copies of all MDR forms, as required by this part, and other information related to the event that you submitted to us and other entities such as an importer, distributor, or manufacturer.

(2) If you are a user facility, importer, or manufacturer, you must permit any authorized FDA employee, at all reasonable times, to access, to copy, and to verify the records required by this part.

(c) If you are a user facility, you must retain an MDR event file relating to an adverse event for a period of 2 years from the date of the event. If you are a manufacturer or importer, you must retain an MDR event file relating to an adverse event for a period of 2 years from the date of the event or a period of time equivalent to the expected life of the device, whichever is greater. If the device is no longer distributed, you still must maintain MDR event files for the time periods described in this paragraph.

(d)(1) If you are a device distributor, you must establish and maintain device complaint records (files). Your records must contain any incident information, including any written, electronic, or oral communication, either received or generated by you, that alleges deficiencies related to the identity (e.g., labeling), quality, durability, reliability, safety, effectiveness, or performance of a device. You must also maintain information about your evaluation of the allegations, if any, in the incident record. You must clearly identify the records as device incident records and file these records by device name. You may maintain these records in written or electronic format. You must back up any file maintained in electronic format.

(2) You must retain copies of the required device incident records for a period of 2 years from the date of inclusion of the record in the file or for a period of time equivalent to the expected life of the device, whichever is greater. You must maintain copies of these records for this period even if you no longer distribute the device.

(3) You must maintain the device complaint files established under this section at your principal business establishment. If you are also a manufacturer, you may maintain the file at the same location as you maintain your complaint file under part 820 of this chapter. You must permit any authorized FDA employee, at all reasonable times, to access, to copy, and to verify the records required by this part.

(e) If you are a manufacturer, you may maintain MDR event files as part of your complaint file, under part 820 of this chapter, if you prominently identify these records as MDR reportable events. We will not consider your submitted MDR report to comply with this part unless you evaluate an event in accordance with the quality system requirements described in part 820 of this chapter. You must document and maintain in your MDR event files an explanation of why you did not submit or could not obtain any information required by this part, as well as the results of your evaluation of each event.

§ 803.19 Are there exemptions, variances, or alternative forms of adverse event reporting requirements?

(a) We exempt the following persons from the adverse event reporting requirements in this part:

(1) A licensed practitioner who prescribes or administers devices intended for use in humans and manufactures or imports devices solely for use in diagnosing and treating persons with whom the practitioner has a ‘‘physician-patient’’ relationship;

(2) An individual who manufactures devices intended for use in humans solely for this person’s use in research or teaching and not for sale. This includes any person who is subject to alternative reporting requirements under the investigational device exemption regulations (described in part 812 of

this chapter), which require reporting of all adverse device effects; and

(3) Dental laboratories or optical laboratories.

(b) If you are a manufacturer, importer, or user facility, you may request an exemption or variance from any or all of the reporting requirements in this part. You must submit the request to us in writing. Your request must include information necessary to identify you and the device; a complete statement of the request for exemption, variance, or alternative reporting; and an explanation why your request is justified.

(c) If you are a manufacturer, importer, or user facility, we may grant in writing an exemption or variance from, or alternative to, any or all of the reporting requirements in this part and may change the frequency of reporting to quarterly, semiannually, annually or other appropriate time period. We may grant these modifications in response to your request, as described in paragraph (b) of this section, or at our discretion. When we grant modifications to the reporting requirements, we may impose other reporting requirements to ensure the protection of public health.

(d) We may revoke or modify in writing an exemption, variance, or alternative reporting requirement if we determine that revocation or modification is necessary to protect the public health.

(e) If we grant your request for a reporting modification, you must submit any reports or information required in our approval of the modification. The conditions of the approval will replace and supersede the regular reporting requirement specified in this part until such time that we revoke or modify the alternative reporting requirements in accordance with paragraph (d) of this section.

Subpart B—Generally Applicable Requirements for Individual Adverse Event Reports

§ 803.20 How do I complete and submit an individual adverse event report?

(a) What form must I complete and submit? There are two versions of the MEDWATCH form for individual reports of adverse events. If you are a health professional or consumer, you may use the FDA Form 3500 to submit voluntary reports regarding FDA-regulated products. If you are a user facility, importer, or manufacturer, you must use the FDA Form 3500A to submit mandatory reports about FDA-regulated products.

(1) If you are a user facility, importer, or manufacturer, you must complete the applicable blocks on the front of FDA Form 3500A. The front of the form is used to submit information about the patient, the event, the device, and the "initial reporter" (i.e., the first person or entity who reported the information to you).

(2) If you are a user facility, importer, or manufacturer, you must complete the applicable blocks on the back of the form. If you are a user facility or importer, you must complete block F. If you are a manufacturer, you must complete blocks G and H. If you are a manufacturer, you do not have to recopy information that you received on a Form 3500A unless you are copying the information onto an electronic medium. If you are a manufacturer and you are correcting or supplying information that is missing from another reporter's Form 3500A, you must attach a copy of that form to your report form. If you are a manufacturer and the information from another reporter's Form 3500A is complete and correct, you may fill in the remaining information on the same form and submit it to us.

(b) To whom must I submit reports and when?

(1) If you are a user facility, you must submit MDR reports to:

(i) The manufacturer and to us no later than 10 work days after the day that you become aware of information that reasonably suggests that a device has or may have caused or contributed to a death; or

(ii) The manufacturer no later than 10 work days after the day that you become aware of information that reasonably suggests that a device has or may have caused or contributed to a serious injury. If the manufacturer is not known, you must submit this report to us.

(2) If you are an importer, you must submit MDR reports to:

(i) The manufacturer and to us, no later than 30 calendar days after the day that you become aware of information that reasonably suggests that a device has or may have caused or contributed to a death or serious injury; or

(ii) The manufacturer, no later than 30 days calendar after receiving information that a device you market has malfunctioned and that this device or a similar device that you market would be likely to cause or contribute to a death or serious injury if the malfunction were to recur.

(3) If you are a manufacturer, you must submit MDR reports to us:

(i) No later than 30 calendar days after the day that you become aware of information that reasonably suggests that a device may have caused or contributed to a death or serious injury; or

(ii) No later than 30 calendar days after the day that you become aware of information that reasonably suggests a device has malfunctioned and that this device or a similar device that you market would be likely to cause or contribute to a death or serious injury if the malfunction were to recur; or

(iii) Within 5 work days if required by § 803.53.

(c) What kind of information reasonably suggests that a reportable event has occurred?

(1) Any information, including professional, scientific, or medical facts, observations, or opinions, may reasonably suggest that a device has caused or may have caused or contributed to an MDR reportable event. An MDR reportable event is a death, a serious injury, or, if you are a manufacturer or importer, a malfunction that would be likely to cause or contribute to a death or serious injury if the malfunction were to recur.

(2) If you are a user facility, importer, or manufacturer, you do not have to report an adverse event if you have information that would lead a person who is qualified to make a medical judgment reasonably to conclude that a device did not cause or contribute to a death or serious injury, or that a malfunction would not be likely to cause or contribute to a death or serious injury if it were to recur. Persons qualified to make a medical judgment include physicians, nurses, risk managers, and biomedical engineers. You must keep in your MDR event files (described in § 803.18) the information that the qualified person used to determine whether or not a device-related event was reportable.

§ 803.21 Where can I find the reporting codes for adverse events that I use with medical device reports?

(a) The MEDWATCH Medical Device Reporting Code Instruction Manual contains adverse event codes for use with FDA Form 3500A. You may obtain the coding manual from CDRH's Web site at *http://www.fda.gov/cdrh/mdr/373.html*; and from the Division of Small Manufacturers, International, and Consumer Assistance, Center for Devices and Radiological Health, 1350 Piccard Dr., Rockville, MD 20850, FAX: 301–443–8818, or e-mail to *DSMICA@CDRH.FDA.GOV*.

(b) We may sometimes use additional coding of information on the reporting forms or modify the existing codes. If we do make modifications, we will ensure that we make the new coding information available to all reporters.

§ 803.22 What are the circumstances in which I am not required to file a report?

(a) If you become aware of information from multiple sources regarding the same patient and same reportable event, you may submit one medical device report.

(b) You are not required to submit a medical device report if:

(1) You are a user facility, importer, or manufacturer, and you determine that the information received is erroneous in that a device-related adverse event did not occur. You must retain documentation of these reports in your MDR files for the time periods specified in § 803.18.

(2) You are a manufacturer or importer and you did not manufacture or import the device about which you have adverse event information. When you receive reportable event information in error, you must forward this information to us with a cover letter explaining that you did not manufacture or import the device in question.

Subpart C—User Facility Reporting Requirements

§ 803.30 If I am a user facility, what reporting requirements apply to me?

(a) You must submit reports to the manufacturer or to us, or both, as specified below:

(1) *Reports of death.* You must submit a report to us as soon as practicable but no more than 10 work days after the day that you become aware of information, from any source, that reasonably suggests that a device has or may have caused or contributed to the death of a patient of your facility. You must also submit the report to the device manufacturer, if known. You must report information required by § 803.32 on FDA Form 3500A or an electronic equivalent approved under § 803.14.

(2) *Reports of serious injury.* You must submit a report to the manufacturer of the device no later than 10 work days after the day that you become aware of information, from any source, that reasonably suggests that a device has or may have caused or contributed to a serious injury to a patient of your facility. If the manufacturer is not known, you must submit the report to us. You must report information required by § 803.32 on FDA Form 3500A or an electronic equivalent approved under § 803.14.

(b) What information does FDA consider "reasonably known" to me? You must submit all information required in this subpart C that is reasonably known to you. This information includes information found in documents that you possess and any information that becomes available as a result of reasonable followup within your facility. You are not required to evaluate or investigate the event by obtaining or evaluating information that you do not reasonably know.

§ 803.32 If I am a user facility, what information must I submit in my individual adverse event reports?

You must include the following information in your report, if reasonably known to you, as described in § 803.30(b). These types of information correspond generally to the elements of FDA Form 3500A:

(a) Patient information (Form 3500A, Block A). You must submit the following:

(1) Patient name or other identifier;

(2) Patient age at the time of event, or date of birth;

(3) Patient gender; and

(4) Patient weight.

(b) Adverse event or product problem (Form 3500A, Block B). You must submit the following:

(1) Identification of adverse event or product problem;

(2) Outcomes attributed to the adverse event (e.g., death or serious injury). An outcome is considered a serious injury if it is:

(i) Life-threatening injury or illness;

(ii) Disability resulting in permanent impairment of a body function or permanent damage to a body structure; or

(iii) Injury or illness that requires intervention to prevent permanent impairment of a body structure or function;

(3) Date of event;

(4) Date of report by the initial reporter;

(5) Description of event or problem, including a discussion of how the device was involved, nature of the problem, patient followup or required treatment, and any environmental conditions that may have influenced the event;

(6) Description of relevant tests, including dates and laboratory data; and

(7) Description of other relevant history, including preexisting medical conditions.

(c) Device information (Form 3500A, Block D). You must submit the following:

(1) Brand name;

(2) Type of device;

(3) Manufacturer name and address;

(4) Operator of the device (health professional, patient, lay user, other);

(5) Expiration date;

(6) Model number, catalog number, serial number, lot number, or other identifying number;

(7) Date of device implantation (month, day, year);

(8) Date of device explantation (month, day, year);

(9) Whether the device was available for evaluation and whether the device was returned to the manufacturer; if

so, the date it was returned to the manufacturer; and

(10) Concomitant medical products and therapy dates. (Do not report products that were used to treat the event.)

(d) Initial reporter information (Form 3500A, Block E). You must submit the following:

(1) Name, address, and telephone number of the reporter who initially provided information to you, or to the manufacturer or distributor;

(2) Whether the initial reporter is a health professional;

(3) Occupation; and

(4) Whether the initial reporter also sent a copy of the report to us, if known.

(e) User facility information (Form 3500A, Block F). You must submit the following:

(1) An indication that this is a user facility report (by marking the user facility box on the form);

(2) Your user facility number;

(3) Your address;

(4) Your contact person;

(5) Your contact person's telephone number;

(6) Date that you became aware of the event (month, day, year);

(7) Type of report (initial or followup); if it is a followup, you must include the report number of the initial report;

(8) Date of your report (month, day, year);

(9) Approximate age of device;

(10) Event problem codes—patient code and device code (refer to the "MEDWATCH Medical Device Reporting Code Instructions");

(11) Whether a report was sent to us and the date it was sent (month, day, year);

(12) Location where the event occurred;

(13) Whether the report was sent to the manufacturer and the date it was sent (month, day, year); and

(14) Manufacturer name and address, if available.

§ 803.33 If I am a user facility, what must I include when I submit an annual report?

(a) You must submit to us an annual report on FDA Form 3419, or electronic equivalent as approved by us under § 803.14. You must submit an annual report by January 1, of each year. You must clearly identify your annual report as such. Your annual report must include:

(1) Your CMS provider number used for medical device reports, or the number assigned by us for reporting purposes in accordance with § 803.3;

(2) Reporting year;

(3) Your name and complete address;

(4) Total number of reports attached or summarized;

(5) Date of the annual report and report numbers identifying the range of medical device reports that you submitted during the report period (e.g., 1234567890–2004–0001 through 1000);

(6) Name, position title, and complete address of the individual designated as your contact person responsible for reporting to us and whether that person is a new contact for you; and

(7) Information for each reportable event that occurred during the annual reporting period including:

(i) Report number;

(ii) Name and address of the device manufacturer;

(iii) Device brand name and common name;

(iv) Product model, catalog, serial and lot number;

(v) A brief description of the event reported to the manufacturer and/or us; and

(vi) Where the report was submitted, i.e., to the manufacturer, importer, or us.

(b) In lieu of submitting the information in paragraph (a)(7) of this section, you may submit a copy of FDA Form 3500A, or an electronic equivalent approved under § 803.14, for each medical device report that you submitted to the manufacturers and/or to us during the reporting period.

(c) If you did not submit any medical device reports to manufacturers or us during the time period, you do not need to submit an annual report.

Subpart D—Importer Reporting Requirements

§ 803.40 If I am an importer, what kinds of individual adverse event reports must I submit, when must I submit them, and to whom must I submit them?

(a) *Reports of deaths or serious injuries.* You must submit a report to us, and a copy of this report to the manufacturer, as soon as practicable but no later than 30 calendar days after the day that you receive or otherwise become aware of information from any source, including user facilities, individuals, or medical or scientific literature, whether published or unpublished, that reasonably suggests that one of your marketed devices may have caused or contributed to a death or serious injury. This report must contain the information required by § 803.42, on FDA form 3500A or an electronic equivalent approved under § 803.14.

(b) *Reports of malfunctions.* You must submit a report to the manufacturer as soon as practicable but no later than 30 calendar days after the day that you receive or otherwise become aware of information from any source, including user facilities, individuals, or through your own research, testing, evaluation, servicing, or maintenance of one of your devices, that reasonably suggests that one of your devices has malfunctioned and that this device or a similar device that you market would be likely to cause or contribute to a death or serious injury if the malfunction were to recur. This report must contain information required by § 803.42, on FDA form 3500A or an electronic equivalent approved under § 803.14.

§ 803.42 If I am an importer, what information must I submit in my individual adverse event reports?

You must include the following information in your report, if the information is known or should be known to you, as described in § 803.40. These types of information correspond generally to the format of FDA Form 3500A:

(a) Patient information (Form 3500A, Block A). You must submit the following:

(1) Patient name or other identifier;

(2) Patient age at the time of event, or date of birth;

(3) Patient gender; and

(4) Patient weight.

(b) Adverse event or product problem (Form 3500A, Block B). You must submit the following:

(1) Identification of adverse event or product problem;

(2) Outcomes attributed to the adverse event (e.g., death or serious injury). An outcome is considered a serious injury if it is:

(i) Life-threatening injury or illness;

(ii) Disability resulting in permanent impairment of a body function or permanent damage to a body structure; or

(iii) Injury or illness that requires intervention to prevent permanent impairment of a body structure or function;

(3) Date of event;

(4) Date of report by the initial reporter;

(5) Description of the event or problem, including a discussion of how the device was involved, nature of the problem, patient followup or required treatment, and any environmental conditions that may have influenced the event;

(6) Description of relevant tests, including dates and laboratory data; and

(7) Description of other relevant patient history, including preexisting medical conditions.

(c) Device information (Form 3500A, Block D). You must submit the following:

(1) Brand name;

(2) Type of device;

(3) Manufacturer name and address;

(4) Operator of the device (health professional, patient, lay user, other);

(5) Expiration date;

(6) Model number, catalog number, serial number, lot number, or other identifying number;

(7) Date of device implantation (month, day, year);

(8) Date of device explanation (month, day, year);

(9) Whether the device was available for evaluation, and whether the device was returned to the manufacturer, and if so, the date it was returned to the manufacturer; and

(10) Concomitant medical products and therapy dates. (Do not report products that were used to treat the event.)

(d) Initial reporter information (Form 3500A, Block E). You must submit the following:

(1) Name, address, and telephone number of the reporter who initially provided information to the manufacturer, user facility, or distributor;

(2) Whether the initial reporter is a health professional;

(3) Occupation; and

(4) Whether the initial reporter also sent a copy of the report to us, if known.

(e) Importer information (Form 3500A, Block F). You must submit the following:

(1) An indication that this is an importer report (by marking the importer box on the form);

(2) Your importer report number;

(3) Your address;

(4) Your contact person;

(5) Your contact person's telephone number;

(6) Date that you became aware of the event (month, day, year);

(7) Type of report (initial or followup). If it is a followup report, you must include the report number of your initial report;

(8) Date of your report (month, day, year);

(9) Approximate age of device;

(10) Event problem codes—patient code and device code (refer to FDA MEDWATCH Medical Device Reporting Code Instructions);

(11) Whether a report was sent to us and the date it was sent (month, day, year);

(12) Location where event occurred;

(13) Whether a report was sent to the manufacturer and the date it was sent (month, day, year); and

(14) Manufacturer name and address, if available.

Subpart E—Manufacturer Reporting Requirements

§ 803.50 If I am a manufacturer, what reporting requirements apply to me?

(a) If you are a manufacturer, you must report to us no later than 30 calendar days after the day that you receive or otherwise become aware of information, from any source, that reasonably suggests that a device that you market:

(1) May have caused or contributed to a death or serious injury; or

(2) Has malfunctioned and this device or a similar device that you market would be likely to cause or contribute to a death or serious injury, if the malfunction were to recur.

(b) What information does FDA consider "reasonably known" to me?

(1) You must submit all information required in this subpart E that is reasonably known to you. We consider the following information to be reasonably known to you:

(i) Any information that you can obtain by contacting a user facility, importer, or other initial reporter;

(ii) Any information in your possession; or

(iii) Any information that you can obtain by analysis, testing, or other evaluation of the device.

(2) You are responsible for obtaining and submitting to us information that is incomplete or missing from reports submitted by user facilities, importers, and other initial reporters.

(3) You are also responsible for conducting an investigation of each event and evaluating the cause of the event. If you cannot submit complete information on a report, you must provide a statement explaining why this information was incomplete and the steps you took to obtain the information. If you later obtain any required information that was not available at the time you filed your initial report, you must submit this information in a supplemental report under § 803.56.

§ 803.52 If I am a manufacturer, what information must I submit in my individual adverse event reports?

You must include the following information in your reports, if known or reasonably known to you, as described in § 803.50(b). These types of information correspond generally to the format of FDA Form 3500A:

(a) Patient information (Form 3500A, Block A). You must submit the following:

(1) Patient name or other identifier;

(2) Patient age at the time of event, or date of birth;

(3) Patient gender; and

(4) Patient weight.

(b) Adverse event or product problem (Form 3500A, Block B). You must submit the following:

(1) Identification of adverse event or product problem;

(2) Outcomes attributed to the adverse event (e.g., death or serious injury). An outcome is considered a serious injury if it is:

(i) Life-threatening injury or illness;

(ii) Disability resulting in permanent impairment of a body function or permanent damage to a body structure; or

(iii) Injury or illness that requires intervention to prevent permanent impairment of a body structure or function;

(3) Date of event;

(4) Date of report by the initial reporter;

(5) Description of the event or problem, including a discussion of how the device was involved, nature of the problem, patient followup or required treatment, and any environmental conditions that may have influenced the event;

(6) Description of relevant tests, including dates and laboratory data; and

(7) Other relevant patient history including preexisting medical conditions.

(c) Device information (Form 3500A, Block D). You must submit the following:

(1) Brand name;

(2) Type of device;

(3) Your name and address;

(4) Operator of the device (health professional, patient, lay user, other);

(5) Expiration date;

(6) Model number, catalog number, serial number, lot number, or other identifying number;

(7) Date of device implantation (month, day, year);

(8) Date of device explantation (month, day, year);

(9) Whether the device was available for evaluation, and whether the device was returned to you, and if so, the date it was returned to you; and

(10) Concomitant medical products and therapy dates. (Do not report products that were used to treat the event.)

(d) Initial reporter information (Form 3500A, Block E). You must submit the following:

(1) Name, address, and phone number of the reporter who initially provided information to you, or to the user facility or importer;

(2) Whether the initial reporter is a health professional;

(3) Occupation; and

(4) Whether the initial reporter also sent a copy of the report to us, if known.

(e) Reporting information for all manufacturers (Form 3500A, Block G). You must submit the following:

(1) Your reporting office's contact name and address and device manufacturing site;

(2) Your telephone number;

(3) Your report sources;

(4) Date received by you (month, day, year);

(5) Type of report being submitted (e.g., 5-day, initial, followup); and

(6) Your report number.

(f) Device manufacturer information (Form 3500A, Block H). You must submit the following:

(1) Type of reportable event (death, serious injury, malfunction, etc.);

(2) Type of followup report, if applicable (e.g., correction, response to FDA request, etc);

(3) If the device was returned to you and evaluated by you, you must include a summary of the evaluation. If you did not perform an evaluation, you must explain why you did not perform an evaluation;

(4) Device manufacture date (month, day, year);

(5) Whether the device was labeled for single use;

(6) Evaluation codes (including event codes, method of evaluation, result, and conclusion codes) (refer to FDA MEDWATCH Medical Device Reporting Code Instructions);

(7) Whether remedial action was taken and the type of action;

(8) Whether the use of the device was initial, reuse, or unknown;

(9) Whether remedial action was reported as a removal or correction under section 519(f) of the act, and if it was, provide the correction/removal report number; and

(10) Your additional narrative; and/or

(11) Corrected data, including:

(i) Any information missing on the user facility report or importer report, including any event codes that were not reported, or information corrected on these forms after your verification;

(ii) For each event code provided by the user facility under § 803.32(e)(10) or the importer under 803.42(e)(10), you must include a statement of whether the type of the event represented by the code is addressed in the device labeling; and

(iii) If your report omits any required information, you must explain why this information was not provided and the steps taken to obtain this information.

§ 803.53 If I am a manufacturer, in which circumstances must I submit a 5-day report?

You must submit a 5-day report to us, on Form 3500A or an electronic equivalent approved under § 803.14, no later than 5 work days after the day that you become aware that:

(a) An MDR reportable event necessitates remedial action to prevent an unreasonable risk of substantial harm to the public health. You may become aware of the need for remedial action from any information, including any trend analysis; or

(b) We have made a written request for the submission of a 5-day report. If you receive such a written request from us, you must submit, without further requests, a 5-day report for all subsequent events of the same nature that involve substantially similar devices for the time period specified in the written request. We may extend the time period stated in the original written request if we determine it is in the interest of the public health.

§ 803.55 If I am a manufacturer, in what circumstances must I submit a baseline report, and what are the requirements for such a report?

(a) You must submit a baseline report for a device when you submit the first report under § 803.50 involving that device model. Submit this report on FDA Form 3417 or an electronic equivalent approved under § 803.14.

(b) You must update each baseline report annually on the anniversary month of the initial submission, after the initial baseline report is submitted. Report changes to baseline information in the manner described in § 803.56 (i.e., include only the new, changed, or corrected information in the appropriate portion(s) of the report form). In each baseline report, you must include the following information:

(1) Name, complete address, and establishment registration number of your reporting site. If your reporting site is not registered under part 807, we will assign a temporary number for use in MDR reporting until you register your reporting site in accordance with part 807. We will inform you of the temporary MDR reporting number;

(2) FDA registration number of each site where you manufacture the device;

(3) Name, complete address, and telephone number of the individual who you have designated as your MDR contact, and the date of the report. For foreign manufacturers, we require a confirmation that the individual submitting the report is the agent of the manufacturer designated under § 803.58(a);

(4) Product identification, including device family, brand name, generic name, model number, catalog number, product code, and any other product identification number or designation;

(5) Identification of any device that you previously reported in a baseline report that is substantially similar (e.g., same device with a different model number, or same device except for cosmetic differences in color or shape) to the device being reported. This includes additional identification of the previously reported device by model number, catalog number, or other product identification, and the date of the baseline report for the previously reported device;

(6) Basis for marketing, including your 510(k) premarket notification number or PMA number, if applicable, and whether the device is currently the subject of an approved postmarket study under section 522 of the act;

(7) Date that you initially marketed the device and, if applicable, the date on which you stopped marketing the device;

(8) Shelf life of the device, if applicable, and expected life of the device;

(9) The number of devices manufactured and distributed in the last 12 months and an estimate of the number of devices in current use; and

(10) Brief description of any methods that you used to estimate the number of devices distributed and the number of devices in current use. If this information was provided in a previous baseline report, in lieu of resubmitting the information, it may be referenced by providing the date and product identification for the previous baseline report.

EFFECTIVE DATE NOTE: At 61 FR 39869, July 31, 1996, in §803.55, paragraphs (b)(9) and (10) were stayed indefinitely.

§803.56 If I am a manufacturer, in what circumstances must I submit a supplemental or followup report and what are the requirements for such reports?

If you are a manufacturer, when you obtain information required under this part that you did not provide because it was not known or was not available when you submitted the initial report, you must submit the supplemental information to us within 1 month of the day that you receive this information. On a supplemental or followup report, you must:

(a) Indicate on the envelope and in the report that the report being submitted is a supplemental or followup report. If you are using FDA form 3500A, indicate this in Block Item H–2;

(b) Submit the appropriate identification numbers of the report that you are updating with the supplemental information (e.g., your original manufacturer report number and the user facility or importer report number of any report on which your report was based), if applicable; and

(c) Include only the new, changed, or corrected information in the appropriate portion(s) of the respective form(s) for reports that cross reference previous reports.

§803.58 Foreign manufacturers.

(a) Every foreign manufacturer whose devices are distributed in the United States shall designate a U.S. agent to be responsible for reporting in accordance with §807.40 of this chapter. The U.S. designated agent accepts responsibility for the duties that such designation entails. Upon the effective date of this regulation, foreign manufacturers shall inform FDA, by letter, of the name and address of the U.S. agent designated under this section and §807.40 of this chapter, and shall update this information as necessary. Such updated information shall be submitted to FDA, within 5 days of a change in the designated agent information.

(b) U.S.-designated agents of foreign manufacturers are required to:

(1) Report to FDA in accordance with §§803.50, 803.52, 803.53, 803.55, and 803.56;

(2) Conduct, or obtain from the foreign manufacturer the necessary information regarding, the investigation and evaluation of the event to comport with the requirements of §803.50;

(3) Forward MDR complaints to the foreign manufacturer and maintain documentation of this requirement;

(4) Maintain complaint files in accordance with §803.18; and

(5) Register, list, and submit premarket notifications in accordance with part 807 of this chapter.

EFFECTIVE DATE NOTE: At 61 FR 38347, July 23, 1996, §803.58 was stayed indefinitely.

PART 806—MEDICAL DEVICES; REPORTS OF CORRECTIONS AND REMOVALS

Subpart A—General Provisions

Subpart B—Reports and Records

AUTHORITY: 21 U.S.C. 352, 360, 360i, 360j, 371, 374.

SOURCE: 62 FR 27191, May 19, 1997, unless otherwise noted.

Subpart A—General Provisions

§806.1 Scope.

(a) This part implements the provisions of section 519(f) of the Federal Food, Drug, and Cosmetic Act (the act) requiring device manufacturers and importers to report promptly to the Food and Drug Administration (FDA) certain actions concerning device corrections and removals, and to maintain records of all corrections and removals regardless of whether such corrections and removals are required to be reported to FDA.

(b) The following actions are exempt from the reporting requirements of this part:

(1) Actions taken by device manufacturers or importers to improve the performance or quality of a device but that do not reduce a risk to health posed by the device or remedy a violation of the act caused by the device.

(2) Market withdrawals as defined in § 806.2(h).

(3) Routine servicing as defined in § 806.2(k).

(4) Stock recoveries as defined in § 806.2(l).

[62 FR 27191, May 19, 1997, as amended at 63 FR 42232, Aug. 7, 1998]

§ 806.2 Definitions.

As used in this part:

(a) *Act* means the Federal Food, Drug, and Cosmetic Act.

(b) *Agency* or *FDA* means the Food and Drug Administration.

(c) *Consignee* means any person or firm that has received, purchased, or used a device subject to correction or removal.

(d) *Correction* means the repair, modification, adjustment, relabeling, destruction, or inspection (including patient monitoring) of a device without its physical removal from its point of use to some other location.

(e) *Correction or removal report number* means the number that uniquely identifies each report submitted.

(f) *Importer* means, for the purposes of this part, any person who imports a device into the United States.

(g) *Manufacturer* means any person who manufactures, prepares, propagates, compounds, assembles, or processes a device by chemical, physical, biological, or other procedures. The term includes any person who:

(1) Repackages or otherwise changes the container, wrapper, or labeling of a device in furtherance of the distribution of the device from the original place of manufacture to the person who makes final delivery or sale to the ultimate user or consumer;

(2) Initiates specifications for devices that are manufactured by a second party for subsequent distribution by the person initiating the specifications; or

(3) Manufactures components or accessories which are devices that are ready to be used and are intended to be commercially distributed and are intended to be used as is, or are processed by a licensed practitioner or other qualified person to meet the needs of a particular patient.

(h) *Market withdrawal* means a correction or removal of a distributed device that involves a minor violation of the act that would not be subject to legal action by FDA or that involves no violation of the act, e.g., normal stock rotation practices.

(i) *Removal* means the physical removal of a device from its point of use to some other location for repair, modification, adjustment, relabeling, destruction, or inspection.

(j) *Risk to health* means

(1) A reasonable probability that use of, or exposure to, the product will cause serious adverse health consequences or death; or

(2) That use of, or exposure to, the product may cause temporary or medically reversible adverse health consequences, or an outcome where the probability of serious adverse health consequences is remote.

(k) *Routine servicing* means any regularly scheduled maintenance of a device, including the replacement of parts at the end of their normal life expectancy, e.g., calibration, replacement of batteries, and responses to normal wear and tear. Repairs of an unexpected nature, replacement of parts earlier than their normal life expectancy, or identical repairs or replacements of multiple units of a device are not routine servicing.

(l) *Stock recovery* means the correction or removal of a device that has not been marketed or that has not left the direct control of the manufacturer, i.e., the device is located on the premises owned, or under the control of, the manufacturer, and no portion of the lot, model, code, or other relevant unit involved in the corrective or removal action has been released for sale or use.

[62 FR 27191, May 19, 1997, as amended at 63 FR 42232, Aug. 7, 1998]

Subpart B—Reports and Records

§ 806.10 Reports of corrections and removals.

(a) Each device manufacturer or importer shall submit a written report to FDA of any correction or removal of a device initiated by such manufacturer or importer if the correction or removal was initiated:

(1) To reduce a risk to health posed by the device; or

(2) To remedy a violation of the act caused by the device which may present a risk to health unless the information has already been provided as set forth in paragraph (f) of this section or the corrective or removal action is exempt from the reporting requirements under § 806.1(b).

(b) The manufacturer or importer shall submit any report required by paragraph (a) of this section within 10-working days of initiating such correction or removal.

(c) The manufacturer or importer shall include the following information in the report:

(1) The seven digit registration number of the entity responsible for submission of the report of corrective or removal action (if applicable), the month, day, and year that the report is made, and a sequence number (i.e., 001 for the first report, 002 for the second report, 003 etc.), and the report type designation "C" or "R". For example, the complete number for the first correction report submitted on June 1, 1997, will appear as follows for a firm with the registration number 1234567: 1234567–6/1/97–001–C. The second correction report number submitted by the same firm on July 1, 1997, would be 1234567–7/1/97–002–C etc. For removals, the number will appear as follows: 1234567–6/1/97–001–R and 1234567–7/1/97–002–R, etc. Firms that do not have a seven digit registration number may use seven zeros followed by the month, date, year, and sequence number (i.e. 0000000–6/1/97–001–C for corrections and 0000000–7/1/97–001–R for removals). Reports received without a seven digit registration number will be assigned a seven digit central file number by the district office reviewing the reports.

(2) The name, address, and telephone number of the manufacturer or importer, and the name, title, address, and telephone number of the manufacturer or importer representative responsible for conducting the device correction or removal.

(3) The brand name and the common name, classification name, or usual name of the device and the intended use of the device.

(4) Marketing status of the device, i.e., any applicable premarket notification number, premarket approval number, or indication that the device is a preamendments device, and the device listing number. A manufacturer or importer that does not have an FDA establishment registration number shall indicate in the report whether it has ever registered with FDA.

(5) The model, catalog, or code number of the device and the manufacturing lot or serial number of the device or other identification number.

(6) The manufacturer's name, address, telephone number, and contact person if different from that of the person submitting the report.

(7) A description of the event(s) giving rise to the information reported and the corrective or removal actions that have been, and are expected to be taken.

(8) Any illness or injuries that have occurred with use of the device. If applicable, include the medical device report numbers.

(9) The total number of devices manufactured or distributed subject to the correction or removal and the number in the same batch, lot, or equivalent unit of production subject to the correction or removal.

(10) The date of manufacture or distribution and the device's expiration date or expected life.

(11) The names, addresses, and telephone numbers of all domestic and foreign consignees of the device and the dates and number of devices distributed to each such consignee.

(12) A copy of all communications regarding the correction or removal and

the names and addresses of all recipients of the communications not provided in accordance with paragraph (c)(11) of this section.

(13) If any required information is not immediately available, a statement as to why it is not available and when it will be submitted.

(d) If, after submitting a report under this part, a manufacturer or importer determines that the same correction or removal should be extended to additional lots or batches of the same device, the manufacturer or importer shall within 10-working days of initiating the extension of the correction or removal, amend the report by submitting an amendment citing the original report number assigned according to paragraph (c)(1) of this section, all of the information required by paragraph (c)(2), and any information required by paragraphs (c)(3) through (c)(12) of this section that is different from the information submitted in the original report. The manufacturer or importer shall also provide a statement in accordance with paragraph (c)(13) of this section for any required information that is not readily available.

(e) A report submitted by a manufacturer or importer under this section (and any release by FDA of that report or information) does not necessarily reflect a conclusion by the manufacturer, importer, or FDA that the report or information constitutes an admission that the device caused or contributed to a death or serious injury. A manufacturer or importer need not admit, and may deny, that the report or information submitted under this section constitutes an admission that the device caused or contributed to a death or serious injury.

(f) No report of correction or removal is required under this part, if a report of the correction or removal is required and has been submitted under parts 803 or 1004 of this chapter.

[62 FR 27191, May 19, 1997, as amended at 63 FR 42232, Aug. 7, 1998; 69 FR 11311, Mar. 10, 2004]

§ 806.20 Records of corrections and removals not required to be reported.

(a) Each device manufacturer or importer who initiates a correction or removal of a device that is not required to be reported to FDA under § 806.10 shall keep a record of such correction or removal.

(b) Records of corrections and removals not required to be reported to FDA under § 806.10 shall contain the following information:

(1) The brand name, common or usual name, classification, name and product code if known, and the intended use of the device.

(2) The model, catalog, or code number of the device and the manufacturing lot or serial number of the device or other identification number.

(3) A description of the event(s) giving rise to the information reported and the corrective or removal action that has been, and is expected to be taken.

(4) Justification for not reporting the correction or removal action to FDA, which shall contain conclusions and any followups, and be reviewed and evaluated by a designated person.

(5) A copy of all communications regarding the correction or removal.

(c) The manufacturer or importer shall retain records required under this section for a period of 2 years beyond the expected life of the device, even if the manufacturer or importer has ceased to manufacture or import the device. Records required to be maintained under paragraph (b) of this section must be transferred to the new manufacturer or importer of the device and maintained for the required period of time.

[62 FR 27191, May 19, 1997, as amended at 63 FR 42233, Aug. 7, 1998]

§ 806.30 FDA access to records.

Each device manufacturer or importer required under this part to maintain records and every person who is in charge or custody of such records shall, upon request of an officer or employee designated by FDA and under section 704(e) of the act, permit such officer or employee at all reasonable times to have access to, and to copy and verify, such records and reports.

[63 FR 42233, Aug. 7, 1998]

§ 806.40 Public availability of reports.

(a) Any report submitted under this part is available for public disclosure in accordance with part 20 of this chapter.

(b) Before public disclosure of a report, FDA will delete from the report:

(1) Any information that constitutes trade secret or confidential commercial or financial information under § 20.61 of this chapter; and

(2) Any personnel, medical, or similar information, including the serial numbers of implanted devices, which would constitute a clearly unwarranted invasion of personal privacy under § 20.63 of this chapter or 5 U.S.C. 552(b)(6); provided, that except for the information under § 20.61 of this chapter or 5 U.S.C. 552(b)(4), FDA will disclose to a patient who requests a report all the information in the report concerning that patient.

PART 807—ESTABLISHMENT REGISTRATION AND DEVICE LISTING FOR MANUFACTURERS AND INITIAL IMPORTERS OF DEVICES

Subpart A—General Provisions

Sec.

Subpart B—Procedures for Device Establishments

Subpart C—Registration Procedures for Foreign Device Establishments

Subpart D—Exemptions

Subpart E—Premarket Notification Procedures

AUTHORITY: 21 U.S.C. 321, 331, 351, 352, 360, 360c, 360e, 360i, 360j, 371, 374, 381, 393; 42 U.S.C. 264, 271.

SOURCE: 42 FR 42526, Aug. 23, 1977, unless otherwise noted.

Subpart A—General Provisions

§ 807.3 Definitions.

(a) *Act* means the Federal Food, Drug, and Cosmetic Act.

(b) *Commercial distribution* means any distribution of a device intended for human use which is held or offered for sale but does not include the following:

(1) Internal or interplant transfer of a device between establishments within the same parent, subsidiary, and/or affiliate company;

(2) Any distribution of a device intended for human use which has in effect an approved exemption for investigational use under section 520(g) of the act and part 812 of this chapter;

(3) Any distribution of a device, before the effective date of part 812 of this chapter, that was not introduced or delivered for introduction into interstate commerce for commercial distribution before May 28, 1976, and that is classified into class III under section 513(f) of the act: *Provided*, That the device is intended solely for investigational use, and under section 501(f)(2)(A) of the act the device is not

required to have an approved premarket approval application as provided in section 515 of the act; or

(4) For foreign establishments, the distribution of any device that is neither imported nor offered for import into the United States.

(c) *Establishment* means a place of business under one management at one general physical location at which a device is manufactured, assembled, or otherwise processed.

(d) *Manufacture, preparation, propagation, compounding, assembly, or processing* of a device means the making by chemical, physical, biological, or other procedures of any article that meets the definition of device in section 201(h) of the act. These terms include the following activities:

(1) Repackaging or otherwise changing the container, wrapper, or labeling of any device package in furtherance of the distribution of the device from the original place of manufacture to the person who makes final delivery or sale to the ultimate consumer;

(2) Initial importation of devices manufactured in foreign establishments; or

(3) Initiation of specifications for devices that are manufactured by a second party for subsequent commercial distribution by the person initiating specifications.

(e) *Official correspondent* means the person designated by the owner or operator of an establishment as responsible for the following:

(1) The annual registration of the establishment;

(2) Contact with the Food and Drug Administration for device listing;

(3) Maintenance and submission of a current list of officers and directors to the Food and Drug Administration upon the request of the Commissioner;

(4) The receipt of pertinent correspondence from the Food and Drug Administration directed to and involving the owner or operator and/or any of the firm's establishments; and

(5) The annual certification of medical device reports required by § 804.30 of this chapter or forwarding the certification form to the person designated by the firm as responsible for the certification.

(f) *Owner or operator* means the corporation, subsidiary, affiliated company, partnership, or proprietor directly responsible for the activities of the registering establishment.

(g) *Initial importer* means any importer who furthers the marketing of a device from a foreign manufacturer to the person who makes the final delivery or sale of the device to the ultimate consumer or user, but does not repackage, or otherwise change the container, wrapper, or labeling of the device or device package.

(h) Any term defined in section 201 of the act shall have that meaning.

(i) *Restricted device* means a device for which the Commissioner, by regulation under § 801.109 of this chapter or otherwise under section 520(e) of the act, has restricted sale, distribution, or use only upon the written or oral authorization of a practitioner licensed by law to administer or use the device or upon such other conditions as the Commissioner may prescribe.

(j) *Classification name* means the term used by the Food and Drug Administration and its classification panels to describe a device or class of devices for purposes of classifying devices under section 513 of the act.

(k) *Representative sampling of advertisements* means typical advertising material that gives the promotional claims made for the device.

(l) *Representative sampling of any other labeling* means typical labeling material (excluding labels and package inserts) that gives the promotional claims made for the device.

(m) *Material change* includes any change or modification in the labeling or advertisements that affects the identity or safety and effectiveness of the device. These changes may include, but are not limited to, changes in the common or usual or proprietary name, declared ingredients or components, intended use, contraindications, warnings, or instructions for use. Changes that are not material may include graphic layouts, grammar, or correction of typographical errors which do not change the content of the labeling, changes in lot number, and, for devices where the biological activity or known composition differs with each lot produced, the labeling containing the actual values for each lot.

(n) *510(k) summary* (summary of any information respecting safety and effectiveness) means a summary, submitted under section 513(i) of the act, of the safety and effectiveness information contained in a premarket notification submission upon which a determination of substantial equivalence can be based. Safety and effectiveness information refers to safety and effectiveness data and information supporting a finding of substantial equivalence, including all adverse safety and effectiveness information.

(o) *510(k) statement* means a statement, made under section 513(i) of the act, asserting that all information in a premarket notification submission regarding safety and effectiveness will be made available within 30 days of request by any person if the device described in the premarket notification submission is determined to be substantially equivalent. The information to be made available will be a duplicate of the premarket notification submission, including any adverse safety and effectiveness information, but excluding all patient identifiers, and trade secret or confidential commercial information, as defined in § 20.61 of this chapter.

(p) *Class III certification* means a certification that the submitter of the 510(k) has conducted a reasonable search of all known information about the class III device and other similar, legally marketed devices.

(q) *Class III summary* means a summary of the types of safety and effectiveness problems associated with the type of device being compared and a citation to the information upon which the summary is based. The summary must be comprehensive and describe the problems to which the type of device is susceptible and the causes of such problems.

(r) *United States* agent means a person residing or maintaining a place of business in the United States whom a foreign establishment designates as its agent. This definition excludes mailboxes, answering machines or services, or other places where an individual acting as the foreign establishment's agent is not physically present.

(s) *Wholesale distributor* means any person (other than the manufacturer or the initial importer) who distributes a device from the original place of manufacture to the person who makes the final delivery or sale of the device to the ultimate consumer or user.

[42 FR 42526, Aug. 23, 1977, as amended at 43 FR 37997, Aug. 25, 1978; 57 FR 18066, Apr. 28, 1992; 58 FR 46522, Sept. 1, 1993; 59 FR 64295, Dec. 14, 1994; 60 FR 63606, Dec. 11, 1995; 63 FR 51826, Sept. 29, 1998; 66 FR 59159, Nov. 27, 2001]

Subpart B—Procedures for Device Establishments

§ 807.20 Who must register and submit a device list?

(a) An owner or operator of an establishment not exempt under section 510(g) of the act or subpart D of this part who is engaged in the manufacture, preparation, propagation, compounding, assembly, or processing of a device intended for human use shall register and submit listing information for those devices in commercial distribution, except that registration and listing information may be submitted by the parent, subsidiary, or affiliate company for all the domestic or foreign establishments under the control of one of these organizations when operations are conducted at more than one establishment and there exists joint ownership and control among all the establishments. The term "device" includes all in vitro diagnostic products and in vitro diagnostic biological products not subject to licensing under section 351 of the Public Health Service Act. An owner or operator of an establishment located in any State as defined in section 201(a)(1) of the act shall register its name, places of business, and all establishments and list the devices whether or not the output of the establishments or any particular device so listed enters interstate commerce. The registration and listing requirements shall pertain to any person who:

(1) Initiates or develops specifications for a device that is to be manufactured by a second party for commercial distribution by the person initiating specifications;

(2) Manufactures for commercial distribution a device either for itself or for another person. However, a person who only manufactures devices according to another person's specifications, for commercial distribution by the person initiating specifications, is not required to list those devices.

(3) Repackages or relabels a device;

(4) Acts as an initial importer; or

(5) Manufactures components or accessories which are ready to be used for any intended health-related purpose and are packaged or labeled for commercial distribution for such health-related purpose, e.g., blood filters, hemodialysis tubing, or devices which of necessity must be further processed by a licensed practitioner or other qualified person to meet the needs of a particular patient, e.g., a manufacturer of ophthalmic lens blanks.

(b) No registration or listing fee is required. Registration or listing does not constitute an admission or agreement or determination that a product is a device within the meaning of section 201(h) of the act.

(c) Registration and listing requirements shall not pertain to any person who:

(1) Manufacturers devices for another party who both initiated the specifications and commercially distributes the device;

(2) Sterilizes devices on a contract basis for other registered facilities who commercially distribute the devices.

(3) Acts as a wholesale distributor, as defined in § 807.3(s), and who does not manufacture, repackage, process, or relabel a device.

(d) Owners and operators of establishments or persons engaged in the recovery, screening, testing, processing, storage, or distribution of human cells, tissues, and cellular and tissue-based products, as defined in § 1271.3(d) of this chapter, that are regulated under the Federal Food, Drug, and Cosmetic Act must register and list those human cells, tissues, and cellular and tissue-based products with the Center for Biologics Evaluation and Research on Form FDA 3356 following the procedures set out in subpart B of part 1271 of this chapter, instead of the procedures for registration and listing contained in this part, except that the additional listing information requirements of § 807.31 remain applicable.

[42 FR 42526, Aug. 23, 1977, as amended at 43 FR 37997, Aug. 25, 1978; 58 FR 46522, Sept. 1, 1993; 60 FR 63606, Dec. 11, 1995; 63 FR 51826, Sept. 29, 1998; 66 FR 5466, Jan. 19, 2001; 66 FR 59160, Nov. 27, 2001]

§ 807.21 Times for establishment registration and device listing.

(a) An owner or operator of an establishment who has not previously entered into an operation defined in § 807.20 shall register within 30 days after entering into such an operation and submit device listing information at that time. An owner or operator of an establishment shall update its registration information annually within 30 days after receiving registration forms from FDA. FDA will mail form FDA-2891a to the owners or operators of registered establishments according to a schedule based on the first letter of the name of the owner or operator. The schedule is as follows:

First letter of owner or operator name	Date FDA will mail forms
A, B, C, D, E	March.
F, G, H, I, J, K, L, M	June.
N, O, P, Q, R	August.
S, T, U, V, W, X, Y, Z	November.

(b) Owners or operators of all registered establishments shall update their device listing information every June and December or, at their discretion, at the time the change occurs.

[58 FR 46522, Sept. 1, 1993]

§ 807.22 How and where to register establishments and list devices.

(a) The first registration of a device establishment shall be on Form FDA–2891 (Initial Registration of Device Establishment). Forms are available upon request from the Office of Compliance, Center for Devices and Radiological Health (HFZ–308), Food and Drug Administration, 9200 Corporate Blvd., Rockville, MD 20850–4015, or from Food and Drug Administration district offices. Subsequent annual registration shall be accomplished on Form FDA–2891a (Annual Registration of Device Establishment), which will be furnished by FDA to establishments whose registration for that year was validated under § 807.35(a). The forms will be mailed to the owner or operators of all establishments via the official correspondent in accordance with the schedule as described in § 807.21(a). The completed form shall be mailed to the address designated in this paragraph 30 days after receipt from FDA.

(b) The initial listing of devices and subsequent June and December updatings shall be on form FDA–2892 (Medical Device Listing). Forms are obtainable upon request as described in paragraph (a) of this section. A separate form FDA–2892 shall be submitted for each device or device class listed with the Food and Drug Administration. Devices having variations in physical characteristics such as size, package, shape, color, or composition should be considered to be one device: *Provided,* The variation does not change the function or intended use of the device. In lieu of form FDA–2892, tapes for computer input or hard copy computer output may by submitted if equivalent in all elements of information as specified in form FDA–2892. All formats proposed for use in lieu of form FDA–2892 require initial review and approval by the Food and Drug Administration.''

(c) The listing obligations of the initial importer are satisfied as follows:

(1) The initial importer is not required to submit a form FDA-2892 for those devices for which such initial importer did not initiate or develop the specifications for the device or repackage or relabel the device. However, the initial importer shall submit, for each device, the name and address of the manufacturer. Initial importers shall also be prepared to submit, when requested by FDA, the proprietary name, if any, and the common or usual name of each device for which they are the initial importers; and

(2) The initial importer shall update the information required by paragraphs (c)(1) of this section at the intervals specified in § 807.30.

[43 FR 37997, Aug. 25, 1978, as amended at 58 FR 46522, Sept. 1, 1993; 60 FR 63606, Dec. 11, 1995; 63 FR 51826, Sept. 29, 1998; 69 FR 11311, Mar. 10, 2004; 69 FR 18473, Apr. 8, 2004; 69 FR 25489, May 7, 2004]

§ 807.25 Information required or requested for establishment registration and device listing.

(a) Form FDA–2891 and Form FDA–2891(a) are the approved forms for initially providing the information required by the act and for providing annual registration, respectively. The required information includes the name and street address of the device establishment, including post office code, all trade names used by the establishment, and the business trading name of the owner or operator of such establishment.

(b) The owner or operator shall identify the device activities of the establishment such as manufacturing, repackaging, or distributing devices.

(c) Each owner or operator is required to maintain a listing of all officers, directors, and partners for each establishment he registers and to furnish this information to the Food and Drug Administration upon request.

(d) Each owner or operator shall provide the name of an official correspondent who will serve as a point of contact between the Food and Drug Administration and the establishment for matters relating to the registration of device establishments and the listing of device products. All future correspondence relating to registration, including requests for the names of partners, officers, and directors, will be directed to this official correspondent. In the event no person is designated by the owner or operator, the owner or operator of the establishment will be the official correspondent.

(e) The designation of an official correspondent does not in any manner affect the liability of the owner or operator of the establishment or any other individual under section 301(p) or any other provision of the act.

(f) Form FD–2892 is the approved form for providing the device listing information required by the act. This required information includes the following:

(1) The identification by classification name and number, proprietary name, and common or usual name of each device being manufactured, prepared, propagated, compounded, or processed for commercial distribution that has not been included in any list

of devices previously submitted on form FDA–2892.

(2) The Code of Federal Regulations citation for any applicable standard for the device under section 514 of the act or section 358 of the Public Health Service Act.

(3) The assigned Food and Drug Administration number of the approved application for each device listed that is subject to section 505 or 515 of the act.

(4) The name, registration number, and establishment type of every domestic or foreign device establishment under joint ownership and control of the owner or operator at which the device is manufactured, repackaged, or relabeled.

(5) Whether the device, as labeled, is intended for distribution to and use by the general public.

(6) Other general information requested on form FDA–2892, i.e.,

(i) If the submission refers to a previously listed device, as in the case of an update, the document number from the initial listing document for the device,

(ii) The reason for submission,

(iii) The date on which the reason for submission occurred,

(iv) The date that the form FDA–2892 was completed,

(v) The owner's or operator's name and identification number.

(7) Labeling or other descriptive information (e.g., specification sheets or catalogs) adequate to describe the intended use of a device when the owner or operator is unable to find an appropriate FDA classification name for the device.

[42 FR 42526, Aug. 23, 1977, as amended at 43 FR 37998, Aug. 25, 1978; 58 FR 46523, Sept. 1, 1993; 64 FR 404, Jan. 5, 1999; 66 FR 59160, Nov. 27, 2001; 69 FR 11312, Mar. 10, 2004]

§ 807.26 Amendments to establishment registration.

Changes in individual ownership, corporate or partnership structure, or location of an operation defined in § 807.3(c) shall be submitted on Form FDA–2891(a) at the time of annual registration, or by letter if the changes occur at other times. This information shall be submitted within 30 days of such changes. Changes in the names of officers and/or directors of the corporation(s) shall be filed with the establishment's official correspondent and shall be provided to the Food and Drug Administration upon receipt of a written request for this information.

[69 FR 11312, Mar. 10, 2004]

§ 807.30 Updating device listing information.

(a) Form FDA–2892 shall be used to update device listing information. The preprinted original document number of each form FDA–2892 on which the device was initially listed shall appear on the form subsequently used to update the listing information for the device and on any correspondence related to the device.

(b) An owner or operator shall update the device listing information during each June and December or, at its discretion, at the time the change occurs. Conditions that require updating and information to be submitted for each of these updates are as follows:

(1) If an owner or operator introduces into commercial distribution a device identified with a classification name not currently listed by the owner or operator, then the owner or operator must submit form FDA–2892 containing all the information required by § 807.25(f).

(2) If an owner or operator discontinues commercial distribution of all devices in the same device class, i.e., with the same classification name, the owner or operator must submit form FDA–2892 containing the original document number of the form FDA–2892 on which the device class was initially listed, the reason for submission, the date of discontinuance, the owner or operator's name and identification number, the classification name and number, the proprietary name, and the common or usual name of the discontinued device.

(3) If commercial distribution of a discontinued device identified on a form FDA–2892 filed under paragraph (b)(2) of this section is resumed, the owner or operator must submit on form FDA–2892 a notice of resumption containing: the original document number of the form initially used to list that device class, the reason for submission,

date of resumption, and all other information required by § 807.25(f).

(4) If one or more classification names for a previously listed device with multiple classification names has been added or deleted, the owner or operator must supply the original document number from the form FDA–2892 on which the device was initially listed and a supplemental sheet identifying the names of any new or deleted classification names.

(5) Other changes to information on form FDA–2892 will be updated as follows:

(i) Whenever a change occurs only in the owner or operator name or number, e.g., whenever one company's device line is purchased by another owner or operator, it will not be necessary to supply a separate form FDA–2892 for each device. In such cases, the new owner or operator must follow the procedures in § 807.26 and submit a letter informing the Food and Drug Administration of the original document number from form FDA–2892 on which each device was initially listed for those devices affected by the change in ownership.

(ii) The owner or operator must also submit update information whenever establishment registration numbers, establishment names, and/or activities are added to or deleted from form FDA 2892. The owner or operator must supply the original document number from the form FDA–2892 on which the device was initially listed, the reason for submission, and all other information required by § 807.25(f).

(6) Updating is not required if the above information has not changed since the previously submitted list. Also, updating is not required if changes occur in proprietary names, in common or usual names, or to supplemental lists of unclassified components or accessories.

[69 FR 11312, Mar. 10, 2004]

§ 807.31 Additional listing information.

(a) Each owner or operator shall maintain a historical file containing the labeling and advertisements in use on the date of initial listing, and in use after October 10, 1978, but before the date of initial listing, as follows:

(1) For each device subject to section 514 or 515 of the act that is not a restricted device, a copy of all labeling for the device;

(2) For each restricted device, a copy of all labeling and advertisements for the device;

(3) For each device that is neither restricted nor subject to section 514 or 515 of the act, a copy of all labels, package inserts, and a representative sampling of any other labeling.

(b) In addition to the requirements set forth in paragraph (a) of this section, each owner or operator shall maintain in the historical file any labeling or advertisements in which a material change has been made anytime after initial listing.

(c) Each owner or operator may discard labeling and advertisements from the historical file 3 years after the date of the last shipment of a discontinued device by an owner or operator.

(d) Location of the file:

(1) Currently existing systems for maintenance of labeling and advertising may be used for the purpose of maintaining the historical file as long as the information included in the systems fulfills the requirements of this section, but only if the labeling and advertisements are retrievable in a timely manner.

(2) The contents of the historical file may be physically located in more than one place in the establishment or in more than one establishment provided there exists joint ownership and control among all the establishments maintaining the historical file. If no joint ownership and control exists, the registered establishment must provide the Food and Drug Administration with a letter authorizing the establishment outside its control to maintain the historical file.

(3) A copy of the certification and disclosure statements as required by part 54 of this chapter shall be retained and physically located at the establishment maintaining the historical file.

(e) Each owner or operator shall be prepared to submit to the Food and Drug Administration, only upon specific request, the following information:

(1) For a device subject to section 514 or 515 of the act that is not a restricted

device, a copy of all labeling for the device.

(2) For a device that is a restricted device, a copy of all labeling for the device, a representative sampling of advertisements for the device, and for good cause, a copy of all advertisements for a particular device. A request for all advertisements will, where feasible, be accompanied by an explanation of the basis for such request.

(3) For a device that is neither a restricted device, nor subject to section 514 of 515 of the act, the label and package insert for the device and a representative sampling of any other labeling for the device.

(4) For a particular device, a statement of the basis upon which the registrant has determined that the device is not subject to section 514 or 515 of the act.

(5) For a particular device, a statement of the basis upon which the registrant has determined the device is not a restricted device.

(6) For a particular device, a statement of the basis for determining that the product is a device rather than a drug.

(7) For a device that the owner or operator has manufactured for distribution under a label other than its own, the names of all distributors for whom it has been manufactured.

[43 FR 37999, Aug. 25, 1978, as amended at 51 FR 33033, Sept. 18, 1986; 63 FR 5253, Feb. 2, 1998]

§ 807.35 Notification of registrant.

(a) The Commissioner will provide to the official correspondent, at the address listed on the form, a validated copy of Form FDA–2891 or Form FDA–2891(a) (whichever is applicable) as evidence of registration. A permanent registration number will be assigned to each device establishment registered in accordance with these regulations.

(b) Owners and operators of device establishments who also manufacture or process blood or drug products at the same establishment shall also register with the Center for Biologics Evaluation and Research and Center for Drug Evaluation and Research, as appropriate. Blood products shall be listed with the Center for Biologics Evaluation and Research, Food and Drug Administration, pursuant to part 607 of this chapter; drug products shall be listed with the Center for Drug Evaluation and Research, Food and Drug Administration, pursuant to part 207 of this chapter.

(c) Although establishment registration and device listing are required to engage in the device activities described in § 807.20, validation of registration and the assignment of a device listing number in itself does not establish that the holder of the registration is legally qualified to deal in such devices and does not represent a determination by the Food and Drug Administration as to the status of any device.

[69 FR 11312, Mar. 10, 2004]

§ 807.37 Inspection of establishment registration and device listings.

(a) A copy of the forms FDA–2891 and FDA–2891a filed by the registrant will be available for inspection in accordance with section 510(f) of the act, at the Center for Devices and Radiological Health (HFZ–308), Food and Drug Administration, Department of Health and Human Services, 9200 Corporate Blvd., Rockville, MD 20850–4015. In addition, there will be available for inspection at each of the Food and Drug Administration district offices the same information for firms within the geographical area of such district office. Upon request, verification of registration number or location of a registered establishment will be provided.

(b)(1) The following information filed under the device listing requirements will be available for public disclosure:

(i) Each form FDA–2892 submitted;

(ii) All labels submitted;

(iii) All labeling submitted;

(iv) All advertisements submitted;

(v) All data or information that has already become a matter of public knowledge.

(2) Requests for device listing information identified in paragraph (b)(1) of this section should be directed to the Center for Devices and Radiological Health (HFZ–308), Food and Drug Administration, Department of Health and Human Services, 9200 Corporate Blvd., Rockville, MD 20850–4015.

(3) Requests for device listing information not identified in paragraph (b)(1) of this section shall be submitted and handled in accordance with part 20 of this chapter.

[69 FR 11313, Mar. 10, 2004]

§807.39 Misbranding by reference to establishment registration or to registration number.

Registration of a device establishment or assignment of a registration number does not in any way denote approval of the establishment or its products. Any representation that creates an impression of official approval because of registration or possession of a registration number is misleading and constitutes misbranding.

Subpart C—Registration Procedures for Foreign Device Establishments

§807.40 Establishment registration and device listing for foreign establishments importing or offering for import devices into the United States.

(a) Any establishment within any foreign country engaged in the manufacture, preparation, propagation, compounding, or processing of a device that is imported or offered for import into the United States shall register and list such devices in conformance with the requirements in subpart B of this part unless the device enters a foreign trade zone and is re-exported from that foreign trade zone without having entered U. S. commerce. The official correspondent for the foreign establishment shall facilitate communication between the foreign establishment's management and representatives of the Food and Drug Administration for matters relating to the registration of device establishments and the listing of device products.

(b) Each foreign establishment required to register under paragraph (a) of this section shall submit the name, address, and phone number of its United States agent as part of its initial and updated registration information in accordance with subpart B of this part. Each foreign establishment shall designate only one United States agent and may designate the United States agent to act as its official correspondent.

(1) The United States agent shall reside or maintain a place of business in the United States.

(2) Upon request from FDA, the United States agent shall assist FDA in communications with the foreign establishment, respond to questions concerning the foreign establishment's products that are imported or offered for import into the United States, and assist FDA in scheduling inspections of the foreign establishment. If the agency is unable to contact the foreign establishment directly or expeditiously, FDA may provide information or documents to the United States agent, and such an action shall be considered to be equivalent to providing the same information or documents to the foreign establishment.

(3) The foreign establishment or the United States agent shall report changes in the United States agent's name, address, or phone number to FDA within 10-business days of the change.

(c) No device may be imported or offered for import into the United States unless it is the subject of a device listing as required under subpart B of this part and is manufactured, prepared, propagated, compounded, or processed at a registered foreign establishment; however, this restriction does not apply to devices imported or offered for import under the investigational use provisions of part 812 of this chapter or to a component, part, or accessory of a device or other article of a device imported under section 801(d)(3) of the act. The establishment registration and device listing information shall be in the English language.

[66 FR 59160, Nov. 27, 2001]

Subpart D—Exemptions

§807.65 Exemptions for device establishments.

The following classes of persons are exempt from registration in accordance with §807.20 under the provisions of section 510(g)(1), (g)(2), and (g)(3) of the act, or because the Commissioner of Food and Drugs has found, under section 510(g)(5) of the act, that such registration is not necessary for the protection of the public health. The exemptions in paragraphs (d), (e), (f), and (i) of this section are limited to those classes of persons located in any State as defined in section 201(a)(1) of the act.

(a) A manufacturer of raw materials or components to be used in the manufacture or assembly of a device who would otherwise not be required to register under the provisions of this part.

(b) A manufacturer of devices to be used solely for veterinary purposes.

(c) A manufacturer of general purpose articles such as chemical reagents or laboratory equipment whose uses are generally known by persons trained in their use and which are not labeled or promoted for medical uses.

(d) Licensed practitioners, including physicians, dentists, and optometrists, who manufacture or otherwise alter devices solely for use in their practice.

(e) Pharmacies, surgical supply outlets, or other similar retail establishments making final delivery or sale to the ultimate user. This exemption also applies to a pharmacy or other similar retail establishment that purchases a device for subsequent distribution under its own name, e.g., a properly labeled health aid such as an elastic bandage or crutch, indicating "distributed by" or "manufactured for" followed by the name of the pharmacy.

(f) Persons who manufacture, prepare, propagate, compound, or process devices solely for use in research, teaching, or analysis and do not introduce such devices into commercial distribution.

(g) [Reserved]

(h) Carriers by reason of their receipt, carriage, holding or delivery of devices in the usual course of business as carriers.

(i) Persons who dispense devices to the ultimate consumer or whose major responsibility is to render a service necessary to provide the consumer (i.e., patient, physician, layman, etc.) with a device or the benefits to be derived from the use of a device; for example, a hearing aid dispenser, optician, clinical laboratory, assembler of diagnostic x-ray systems, and personnel from a hospital, clinic, dental laboratory, orthotic or prosthetic retail facility, whose primary responsibility to the ultimate consumer is to dispense or provide a service through the use of a previously manufactured device.

[42 FR 42526, Aug. 23, 1977, as amended at 58 FR 46523, Sept. 1, 1993; 61 FR 44615, Aug. 28, 1996; 65 FR 17136, Mar. 31, 2000; 66 FR 59160, Nov. 27, 2001]

Subpart E—Premarket Notification Procedures

§ 807.81 When a premarket notification submission is required.

(a) Except as provided in paragraph (b) of this section, each person who is required to register his establishment pursuant to § 807.20 must submit a premarket notification submission to the Food and Drug Administration at least 90 days before he proposes to begin the introduction or delivery for introduction into interstate commerce for commercial distribution of a device intended for human use which meets any of the following criteria:

(1) The device is being introduced into commercial distribution for the first time; that is, the device is not of the same type as, or is not substantially equivalent to, (i) a device in commercial distribution before May 28, 1976, or (ii) a device introduced for commercial distribution after May 28, 1976, that has subsequently been reclassified into class I or II.

(2) The device is being introduced into commercial distribution for the first time by a person required to register, whether or not the device meets the criteria in paragraph (a)(1) of this section.

(3) The device is one that the person currently has in commercial distribution or is reintroducing into commercial distribution, but that is about to be significantly changed or modified in design, components, method of manufacture, or intended use. The following constitute significant changes or modifications that require a premarket notification:

(i) A change or modification in the device that could significantly affect the safety or effectiveness of the device, e.g., a significant change or modification in design, material, chemical composition, energy source, or manufacturing process.

(ii) A major change or modification in the intended use of the device.

(b) A premarket notification under this subpart is not required for a device for which a premarket approval application under section 515 of the act, or for which a petition to reclassify under section 513(f)(2) of the act, is pending before the Food and Drug Administration.

(c) In addition to complying with the requirements of this part, owners or operators of device establishments that manufacture radiation-emitting electronic products, as defined in § 1000.3 of this chapter, shall comply with the reporting requirements of part 1002 of this chapter.

§ 807.85 Exemption from premarket notification.

(a) A device is exempt from the premarket notification requirements of this subpart if the device intended for introduction into commercial distribution is not generally available in finished form for purchase and is not offered through labeling or advertising by the manufacturer, importer, or distributor thereof for commercial distribution, and the device meets one of the following conditions:

(1) It is intended for use by a patient named in the order of the physician or dentist (or other specially qualified person); or

(2) It is intended solely for use by a physician or dentist (or other specially qualified person) and is not generally available to, or generally used by, other physicians or dentists (or other specially qualified persons).

(b) A distributor who places a device into commercial distribution for the first time under his own name and a repackager who places his own name on a device and does not change any other labeling or otherwise affect the device shall be exempted from the premarket notification requirements of this subpart if:

(1) The device was in commercial distribution before May 28, 1976; or

(2) A premarket notification submission was filed by another person.

§ 807.87 Information required in a premarket notification submission.

Each premarket notification submission shall contain the following information:

(a) The device name, including both the trade or proprietary name and the common or usual name or classification name of the device.

(b) The establishment registration number, if applicable, of the owner or operator submitting the premarket notification submission.

(c) The class in which the device has been put under section 513 of the act and, if known, its appropriate panel; or, if the owner or operator determines that the device has not been classified under such section, a statement of that determination and the basis for the person's determination that the device is not so classified.

(d) Action taken by the person required to register to comply with the requirements of the act under section 514 for performance standards.

(e) Proposed labels, labeling, and advertisements sufficient to describe the device, its intended use, and the directions for its use. Where applicable, photographs or engineering drawings should be supplied.

(f) A statement indicating the device is similar to and/or different from other products of comparable type in commercial distribution, accompanied by data to support the statement. This information may include an identification of similar products, materials, design considerations, energy expected to be used or delivered by the device, and a description of the operational principles of the device.

(g) Where a person required to register intends to introduce into commercial distribution a device that has undergone a significant change or modification that could significantly affect the safety or effectiveness of the device, or the device is to be marketed for a new or different indication for use, the premarket notification submission must include appropriate supporting data to show that the manufacturer has considered what consequences and effects the change or modification or new use might have on the safety and effectiveness of the device.

(h) A 510(k) summary as described in § 807.92 or a 510(k) statement as described in § 807.93.

(i) A financial certification or disclosure statement or both, as required by part 54 of this chapter.

(j) For submissions claiming substantial equivalence to a device which has been classified into class III under section 513(b) of the act:

(1) Which was introduced or delivered for introduction into interstate commerce for commercial distribution before December 1, 1990; and

(2) For which no final regulation requiring premarket approval has been issued under section 515(b) of the act, a summary of the types of safety and effectiveness problems associated with the type of devices being compared and a citation to the information upon which the summary is based (class III summary). The 510(k) submitter shall also certify that a reasonable search of all information known or otherwise available about the class III device and other similar legally marketed devices has been conducted (class III certification), as described in § 807.94. This information does not refer to information that already has been submitted to the Food and Drug Administration (FDA) under section 519 of the act. FDA may require the submission of the adverse safety and effectiveness data described in the class III summary or citation.

(k) A statement that the submitter believes, to the best of his or her knowledge, that all data and information submitted in the premarket notification are truthful and accurate and that no material fact has been omitted.

(l) Any additional information regarding the device requested by the Commissioner that is necessary for the Commissioner to make a finding as to whether or not the device is substantially equivalent to a device in commercial distribution. A request for additional information will advise the owner or operator that there is insufficient information contained in the original premarket notification submission for the Commissioner to make this determination and that the owner or operator may either submit the requested data or a new premarket notification containing the requested information at least 90 days before the owner or operator intends to market the device, or submit a premarket approval application in accordance with section 515 of the act. If the additional information is not submitted within 30 days following the date of the request, the Commissioner will consider the premarket notification to be withdrawn.

(Information collection requirements in this section were approved by the Office of Management and Budget (OMB) and assigned OMB control number 0910–0281)

[42 FR 42526, Aug 23, 1977, as amended at 57 FR 18066, Apr. 28, 1992; 59 FR 64295, Dec. 14, 1994; 63 FR 5253, Feb. 2, 1998]

§ 807.90 Format of a premarket notification submission.

Each premarket notification submission pursuant to this part shall be submitted in accordance with this section. Each submission shall:

(a)(1) For devices regulated by the Center for Devices and Radiological Health, be addressed to the Food and Drug Administration, Center for Devices and Radiological Health (HFZ–401), 9200 Corporate Blvd., Rockville, MD 20850.

(2) For devices regulated by the Center for Biologics Evaluation and Research, be addressed to the Document Control Center (HFM–99), Center for Biologics Evaluation and Research, Food and Drug Administration, 1401 Rockville Pike, suite 200N, Rockville, MD 20852–1448; or for devices regulated by the Center for Drug Evaluation and Research, be addressed to the Central Document Room, Center for Drug Evaluation and Research, Food and Drug Administration, 5901–B Ammendale Rd., Beltsville, MD 20705–1266. Information about devices regulated by the Center for Biologics Evaluation and Research is available at *http://www.fda.gov/cber/dap/devlst.htm* on the Internet.

(3) All inquiries regarding a premarket notification submission should be in writing and sent to one of the addresses above.

(b) Be bound into a volume or volumes, where necessary.

(c) Be submitted in duplicate on standard size paper, including the original and two copies of the cover letter.

(d) Be submitted separately for each product the manufacturer intends to market.

(e) Designated "510(k) Notification" in the cover letter.

[42 FR 42526, Aug. 23, 1977, as amended at 53 FR 11252, Apr. 6, 1988; 55 FR 11169, Mar. 27, 1990; 65 FR 17137, Mar. 31, 2000; 70 FR 14986, Mar. 24, 2005]

§ 807.92 Content and format of a 510(k) summary.

(a) A 510(k) summary shall be in sufficient detail to provide an understanding of the basis for a determination of substantial equivalence. FDA will accept summaries as well as amendments thereto until such time as FDA issues a determination of substantial equivalence. All 510(k) summaries shall contain the following information:

(1) The submitter's name, address, telephone number, a contact person, and the date the summary was prepared;

(2) The name of the device, including the trade or proprietary name if applicable, the common or usual name, and the classification name, if known;

(3) An identification of the legally marketed device to which the submitter claims equivalence. A legally marketed device to which a new device may be compared for a determination regarding substantial equivalence is a device that was legally marketed prior to May 28, 1976, or a device which has been reclassified from class III to class II or I (the predicate), or a device which has been found to be substantially equivalent through the 510(k) premarket notification process;

(4) A description of the device that is the subject of the premarket notification submission, such as might be found in the labeling or promotional material for the device, including an explanation of how the device functions, the scientific concepts that form the basis for the device, and the significant physical and performance characteristics of the device, such as device design, material used, and physical properties;

(5) A statement of the intended use of the device that is the subject of the premarket notification submission, including a general description of the diseases or conditions that the device will diagnose, treat, prevent, cure, or mitigate, including a description, where appropriate, of the patient population for which the device is intended. If the indication statements are different from those of the legally marketed device identified in paragraph (a)(3) of this section, the 510(k) summary shall contain an explanation as to why the differences are not critical to the intended therapeutic, diagnostic, prosthetic, or surgical use of the device, and why the differences do not affect the safety and effectiveness of the device when used as labeled; and

(6) If the device has the same technological characteristics (i.e., design, material, chemical composition, energy source) as the predicate device identified in paragraph (a)(3) of this section, a summary of the technological characteristics of the new device in comparison to those of the predicate device. If the device has different technological characteristics from the predicate device, a summary of how the technological characteristics of the device compare to a legally marketed device identified in paragraph (a)(3) of this section.

(b) 510(k) summaries for those premarket submissions in which a determination of substantial equivalence is also based on an assessment of performance data shall contain the following information:

(1) A brief discussion of the nonclinical tests submitted, referenced, or relied on in the premarket notification submission for a determination of substantial equivalence;

(2) A brief discussion of the clinical tests submitted, referenced, or relied on in the premarket notification submission for a determination of substantial equivalence. This discussion shall include, where applicable, a description of the subjects upon whom the device was tested, a discussion of the safety or effectiveness data obtained from the testing, with specific reference to adverse effects and complications, and any other information from the clinical testing relevant to a determination of substantial equivalence; and

(3) The conclusions drawn from the nonclinical and clinical tests that demonstrate that the device is as safe, as effective, and performs as well as or better than the legally marketed device identified in paragraph (a)(3) of this section.

(c) The summary should be in a separate section of the submission, beginning on a new page and ending on a page not shared with any other section of the premarket notification submission, and should be clearly identified as a "510(k) summary."

(d) Any other information reasonably deemed necessary by the agency.

[57 FR 18066, Apr. 28, 1992, as amended at 59 FR 64295, Dec. 14, 1994]

§ 807.93 Content and format of a 510(k) statement.

(a)(1) A 510(k) statement submitted as part of a premarket notification shall state as follows:

I certify that, in my capacity as (the position held in company by person required to submit the premarket notification, preferably the official correspondent in the firm), of (company name), I will make available all information included in this premarket notification on safety and effectiveness within 30 days of request by any person if the device described in the premarket notification submission is determined to be substantially equivalent. The information I agree to make available will be a duplicate of the premarket notification submission, including any adverse safety and effectiveness information, but excluding all patient identifiers, and trade secret and confidential commercial information, as defined in 21 CFR 20.61.

(2) The statement in paragraph (a)(1) of this section should be signed by the certifier, made on a separate page of the premarket notification submission, and clearly identified as "510(k) statement."

(b) All requests for information included in paragraph (a) of this section shall be made in writing to the certifier, whose name will be published by FDA on the list of premarket notification submissions for which substantial equivalence determinations have been made.

(c) The information provided to requestors will be a duplicate of the premarket notification submission, including any adverse information, but excluding all patient identifiers, and trade secret and confidential commercial information as defined in § 20.61 of this chapter.

[59 FR 64295, Dec. 14, 1994]

§ 807.94 Format of a class III certification.

(a) A class III certification submitted as part of a premarket notification shall state as follows:

I certify, in my capacity as (position held in company), of (company name), that I have conducted a reasonable search of all information known or otherwise available about the types and causes of safety or effectiveness problems that have been reported for the (type of device). I further certify that I am aware of the types of problems to which the (type of device) is susceptible and that, to the best of my knowledge, the following summary of the types and causes of safety or effectiveness problems about the (type of device) is complete and accurate.

(b) The statement in paragraph (a) of this section should be signed by the certifier, clearly identified as "class III certification," and included at the beginning of the section of the premarket notification submission that sets forth the class III summary.

[59 FR 64296, Dec. 14, 1994]

§ 807.95 Confidentiality of information.

(a) The Food and Drug Administration will disclose publicly whether there exists a premarket notification submission under this part:

(1) Where the device is on the market, i.e., introduced or delivered for introduction into interstate commerce for commercial distribution;

(2) Where the person submitting the premarket notification submission has disclosed, through advertising or any other manner, his intent to market the device to scientists, market analysts, exporters, or other individuals who are not employees of, or paid consultants to, the establishment and who are not in an advertising or law firm pursuant to commercial arrangements with appropriate safeguards for secrecy; or

(3) Where the device is not on the market and the intent to market the device has not been so disclosed, except where the submission is subject to an exception under paragraph (b) or (c) of this section.

(b) The Food and Drug Administration will not disclose publicly the existence of a premarket notification submission for a device that is not on the market and where the intent to market the device has not been disclosed for 90 days from the date of receipt of the submission, if:

(1) The person submitting the premarket notification submission requests in the submission that the Food and Drug Administration hold as confidential commercial information the intent to market the device and submits a written certification to the Commissioner:

(i) That the person considers his intent to market the device to be confidential commercial information;

(ii) That neither the person nor, to the best of his knowledge, anyone else, has disclosed through advertising or any other manner, his intent to market the device to scientists, market analysts, exporters, or other individuals, except employees of, or paid consultants to, the establishment or individuals in an advertising or law firm pursuant to commercial arrangements with appropriate safeguards for secrecy;

(iii) That the person will immediately notify the Food and Drug Administration if he discloses the intent to market the device to anyone, except employees of, or paid consultants to, the establishment or individuals in an advertising or law firm pursuant to commercial arrangements with appropriate safeguards for secrecy;

(iv) That the person has taken precautions to protect the confidentiality of the intent to market the device; and

(v) That the person understands that the submission to the government of false information is prohibited by 18 U.S.C. 1001 and 21 U.S.C. 331(q); and

(2) The Commissioner agrees that the intent to market the device is confidential commercial information.

(c) Where the Commissioner determines that the person has complied with the procedures described in paragraph (b) of this section with respect to a device that is not on the market and where the intent to market the device has not been disclosed, and the Commissioner agrees that the intent to market the device is confidential commercial information, the Commissioner will not disclose the existence of the submission for 90 days from the date of its receipt by the agency. In addition, the Commissioner will continue not to disclose the existence of such a submission for the device for an additional time when any of the following occurs:

(1) The Commissioner requests in writing additional information regarding the device pursuant to § 807.87(h), in which case the Commissioner will not disclose the existence of the submission until 90 days after the Food and Drug Administration's receipt of a complete premarket notification submission;

(2) The Commissioner determines that the device intended to be introduced is a class III device and cannot be marketed without premarket approval or reclassification, in which case the Commissioner will not disclose the existence of the submission unless a petition for reclassification is submitted under section 513(f)(2) of the act and its existence can be disclosed under § 860.5(d) of this chapter; or

(d) FDA will make a 510(k) summary of the safety and effectiveness data available to the public within 30 days of the issuance of a determination that the device is substantially equivalent to another device. Accordingly, even when a 510(k) submitter has complied with the conditions set forth in paragraphs (b) and (c) of this section, confidentiality for a premarket notification submission cannot be granted beyond 30 days after FDA issues a determination of equivalency.

(e) Data or information submitted with, or incorporated by reference in, a premarket notification submission (other than safety and effectiveness data that have not been disclosed to the public) shall be available for disclosure by the Food and Drug Administration when the intent to market the device is no longer confidential in accordance with this section, unless exempt from public disclosure in accordance with part 20 of this chapter. Upon final classification, data and information relating to safety and effectiveness of a device classified in class I (general controls) or class II (performance standards) shall be available for public disclosure. Data and information relating to safety and effectiveness of a device classified in class III (premarket approval) that have not been released to the public shall be retained as confidential unless such data and information become available for release to the public under § 860.5(d) or other provisions of this chapter.

[42 FR 42526, Aug. 23, 1977, as amended at 53 FR 11252, Apr. 6, 1988; 57 FR 18067, Apr. 28, 1992; 59 FR 64296, Dec. 14, 1994]

§ 807.97 Misbranding by reference to premarket notification.

Submission of a premarket notification in accordance with this subpart, and a subsequent determination by the Commissioner that the device intended for introduction into commercial distribution is substantially equivalent to a device in commercial distribution before May 28, 1976, or is substantially equivalent to a device introduced into commercial distribution after May 28, 1976, that has subsequently been reclassified into class I or II, does not in any way denote official approval of the device. Any representation that creates an impression of official approval of a device because of complying with the premarket notification regulations is misleading and constitutes misbranding.

§ 807.100 FDA action on a premarket notification.

(a) After review of a premarket notification, FDA will:

(1) Issue an order declaring the device to be substantially equivalent to a legally marketed predicate device;

(2) Issue an order declaring the device to be not substantially equivalent to any legally marketed predicate device;

(3) Request additional information; or

(4) Withhold the decision until a certification or disclosure statement is submitted to FDA under part 54 of this chapter.

(5) Advise the applicant that the premarket notification is not required. Until the applicant receives an order declaring a device substantially equivalent, the applicant may not proceed to market the device.

(b) FDA will determine that a device is substantially equivalent to a predicate device using the following criteria:

(1) The device has the same intended use as the predicate device; and

(2) The device:

(i) Has the same technological characteristics as the predicate device; or

(ii)(A) Has different technological characteristics, such as a significant change in the materials, design, energy source, or other features of the device from those of the predicate device;

(B) The data submitted establishes that the device is substantially equivalent to the predicate device and contains information, including clinical data if deemed necessary by the Commissioner, that demonstrates that the device is as safe and as effective as a legally marketed device; and

(C) Does not raise different questions of safety and effectiveness than the predicate device.

(3) The predicate device has not been removed from the market at the initiative of the Commissioner of Food and Drugs or has not been determined to be misbranded or adulterated by a judicial order.

[57 FR 58403, Dec. 10, 1992, as amended at 63 FR 5253, Feb. 2, 1998]

PART 808—EXEMPTIONS FROM FEDERAL PREEMPTION OF STATE AND LOCAL MEDICAL DEVICE REQUIREMENTS

Subpart A—General Provisions

Sec.

Subpart B—Exemption Procedures

Subpart C—Listing of Specific State and Local Exemptions

808.73 Minnesota.
808.74 Mississippi.
808.77 Nebraska.
808.80 New Jersey.
808.81 New Mexico.
808.82 New York.
808.85 Ohio.
808.87 Oregon.
808.88 Pennsylvania.
808.89 Rhode Island.
808.93 Texas.
808.97 Washington.
808.98 West Virginia.
808.101 District of Columbia.

AUTHORITY: 21 U.S.C. 360j, 360k, 371.

SOURCE: 43 FR 18665, May 2, 1978, unless otherwise noted.

Subpart A—General Provisions

§ 808.1 Scope.

(a) This part prescribes procedures for the submission, review, and approval of applications for exemption from Federal preemption of State and local requirements applicable to medical devices under section 521 of the act.

(b) Section 521(a) of the act contains special provisions governing the regulation of devices by States and localities. That section prescribes a general rule that after May 28, 1976, no State or political subdivision of a State may establish or continue in effect any requirement with respect to a medical device intended for human use having the force and effect of law (whether established by statute, ordinance, regulation, or court decision), which is different from, or in addition to, any requirement applicable to such device under any provision of the act and which relates to the safety or effectiveness of the device or to any other matter included in a requirement applicable to the device under the act.

(c) Section 521(b) of the act contains a provision whereby the Commissioner of Food and Drugs may, upon application by a State or political subdivision, allow imposition of a requirement which is different from, or in addition to, any requirement applicable under the act to the device (and which is thereby preempted) by promulgating a regulation in accordance with this part exempting the State or local requirement from preemption. The granting of an exemption does not affect the applicability to the device of any requirements under the act. The Commissioner may promulgate an exemption regulation for the preempted requirement if he makes either of the following findings:

(1) That the requirement is more stringent than a requirement under the act applicable to the device; or

(2) That the requirement is required by compelling local conditions and compliance with the requirement would not cause the device to be in violation of any applicable requirement under the act.

(d) State or local requirements are preempted only when the Food and Drug Administration has established specific counterpart regulations or there are other specific requirements applicable to a particular device under the act, thereby making any existing divergent State or local requirements applicable to the device different from, or in addition to, the specific Food and Drug Administration requirements. There are other State or local requirements that affect devices that are not preempted by section 521(a) of the act because they are not "requirements applicable to a device" within the meaning of section 521(a) of the act. The following are examples of State or local requirements that are not regarded as preempted by section 521 of the act:

(1) Section 521(a) does not preempt State or local requirements of general applicability where the purpose of the requirement relates either to other products in addition to devices (e.g., requirements such as general electrical codes, and the Uniform Commercial Code (warranty of fitness)), or to unfair trade practices in which the requirements are not limited to devices.

(2) Section 521(a) does not preempt State or local requirements that are equal to, or substantially identical to, requirements imposed by or under the act.

(3) Section 521(a) does not preempt State or local permits, licensing, registration, certification, or other requirements relating to the approval or sanction of the practice of medicine, dentistry, optometry, pharmacy, nursing, podiatry, or any other of the healing arts or allied medical sciences or related professions or occupations that

administer, dispense, or sell devices. However, regulations issued under section 520(e) or (g) of the act may impose restrictions on the sale, distribution, or use of a device beyond those prescribed in State or local requirements. If there is a conflict between such restrictions and State or local requirements, the Federal regulations shall prevail.

(4) Section 521(a) does not preempt specifications in contracts entered into by States or localities for procurement of devices.

(5) Section 521(a) does not preempt criteria for payment of State or local obligations under Medicaid and similar Federal, State or local health-care programs.

(6)(i) Section 521(a) does not preempt State or local requirements respecting general enforcement, e.g., requirements that State inspection be permitted of factory records concerning all devices, registration, and licensing requirements for manufacturers and others, and prohibition of manufacture of devices in unlicensed establishments. However, Federal regulations issued under sections 519 and 520(f) of the act may impose requirements for records and reports and good manufacturing practices beyond those prescribed in State or local requirements. If there is a conflict between such regulations and State or local requirements, the Federal regulations shall prevail.

(ii) Generally, section 521(a) does not preempt a State or local requirement prohibiting the manufacture of adulterated or misbranded devices. Where, however, such a prohibition has the effect of establishing a substantive requirement for a specific device, e.g., a specific labeling requirement, then the prohibition will be preempted if the requirement is different from, or in addition to, a Federal requirement established under the act. In determining whether such a requirement is preempted, the determinative factor is how the requirement is interpreted and enforced by the State or local government and not the literal language of the statute, which may be identical to a provision in the act.

(7) Section 521(a) does not preempt State or local provisions respecting delegations of authority and related administrative matters relating to devices.

(8) Section 521(a) does not preempt a State or local requirement whose sole purpose is raising revenue or charging fees for services, registration, or regulatory programs.

(9) Section 521(a) does not preempt State or local requirements of the types that have been developed under the Atomic Energy act of 1954 (42 U.S.C. 2011 note), as amended, the Radiation Control for Health and Safety Act of 1968 (Pub. L. 90–602 (42 U.S.C. 263b *et seq.*)) and other Federal statutes, until such time as the Food and Drug Administration issues specific requirements under the Federal Food, Drug, and Cosmetic Act applicable to these types of devices.

(10) Part 820 of this chapter (21 CFR part 820) (CGMP requirements) does not preempt remedies created by States or Territories of the United States, the District of Columbia, or the Commonwealth of Puerto Rico.

(e) It is the responsibility of the Food and Drug Administration, subject to review by Federal courts, to determine whether a State or local requirement is equal to, or substantially identical to, requirements imposed by or under the act, or is different from, or in addition to, such requirements, in accordance with the procedures provided by this part. However, it is the responsibility of States and political subdivisions to determine initially whether to seek exemptions from preemption. Any State or political subdivision whose requirements relating to devices are preempted by section 521(a) may petition the Commissioner of Food and Drugs for exemption from preemption, in accordance with the procedures provided by this part.

(f) The Federal requirement with respect to a device applies whether or not a corresponding State or local requirement is preempted or exempted from preemption. As a result, if a State or local requirement that the Food and Drug Administration has exempted from preemption is not as broad in its application as the Federal requirement, the Federal requirement applies to all circumstances not covered by the State or local requirement.

[43 FR 18665, May 2, 1978, as amended at 45 FR 67336, Oct. 10, 1980; 61 FR 52654, Oct. 7, 1996]

§ 808.3 Definitions.

(a) *Act* means the Federal Food, Drug, and Cosmetic Act.

(b) *Compelling local conditions* includes any factors, considerations, or circumstances prevailing in, or characteristic of, the geographic area or population of the State or political subdivision that justify exemption from preemption.

(c) *More stringent* refers to a requirement of greater restrictiveness or one that is expected to afford to those who may be exposed to a risk of injury from a device a higher degree of protection than is afforded by a requirement applicable to the device under the act.

(d) *Political subdivision* or *locality* means any lawfully established local governmental unit within a State which unit has the authority to establish or continue in effect any requirement having the force and effect of law with respect to a device intended for human use.

(e) *State* means a State, American Samoa, the Canal Zone, the Commonwealth of Puerto Rico, the District of Columbia, Guam, Johnston Island, Kingman Reef, Midway Island, the Trust Territory of the Pacific Islands, the Virgin Islands, and Wake Island.

(f) *Substantially identical to* refers to the fact that a State or local requirement does not significantly differ in effect from a Federal requirement.

§ 808.5 Advisory opinions.

(a) Any State, political subdivision, or other interested person may request an advisory opinion from the Commissioner with respect to any general matter concerning preemption of State or local device requirements or with respect to whether the Food and Drug Administration regards particular State or local requirements, or proposed requirements, as preempted.

(1) Such an advisory opinion may be requested and may be granted in accordance with § 10.85 of this chapter.

(2) The Food and Drug Administration, in its discretion and after consultation with the State or political subdivision, may treat a request by a State or political subdivision for an advisory opinion as an application for exemption from preemption under § 808.20.

(b) The Commissioner may issue an advisory opinion relating to a State or local requirement on his own initiative when he makes one of the following determinations:

(1) A requirement with respect to a device for which an application for exemption from preemption has been submitted under § 808.20 is not preempted by section 521(a) of the act because it is: (i) Equal to or substantially identical to a requirement under the act applicable to the device, or (ii) is not a requirement within the meaning of section 521 of the act and therefore is not preempted;

(2) A proposed State or local requirement with respect to a device is not eligible for exemption from preemption because the State or local requirement has not been issued in final form. In such a case, the advisory opinion may indicate whether the proposed requirement would be preempted and, if it would be preempted, whether the Food and Drug Administration would propose to grant an exemption from preemption;

(3) Issuance of such an advisory opinion is in the public interest.

Subpart B—Exemption Procedures

§ 808.20 Application.

(a) Any State or political subdivision may apply to the Food and Drug Administration for an exemption from preemption for any requirement that it has enacted and that is preempted. An exemption may only be granted for a requirement that has been enacted, promulgated, or issued in final form by the authorized body or official of the State or political subdivision so as to have the force and effect of law. However, an application for exemption may be submitted before the effective date of the requirement.

(b) An application for exemption shall be in the form of a letter to the Commissioner of Food and Drugs and shall be signed by an individual who is authorized to request the exemption on behalf of the State or political subdivision. An original and two copies of the letter and any accompanying material, as well as any subsequent reports or correspondence concerning an application, shall be submitted to the Division of Dockets Management (HFA–305), Food and Drug Administration, 5630 Fishers Lane, rm. 1061, Rockville, MD 20852. The outside wrapper of any application, report, or correspondence should indicate that it concerns an application for exemption from preemption of device requirements.

(c) For each requirement for which an exemption is sought, the application shall include the following information to the fullest extent possible, or an explanation of why such information has not been included:

(1) Identification and a current copy of any statute, rule, regulation, or ordinance of the State or political subdivision considered by the State or political subdivision to be a requirement which is preempted, with a reference to the date of enactment, promulgation, or issuance in final form. The application shall also include, where available, copies of any legislative history or background materials pertinent to enactment, promulgation, or issuance of the requirement, including hearing reports or studies concerning development or consideration of the requirement. If the requirement has been subject to any judicial or administrative interpretations, the State or political subdivision shall furnish copies of such judicial or administrative interpretations.

(2) A comparison of the requirement of the State or political subdivision and any applicable Federal requirements to show similarities and differences.

(3) Information on the nature of the problem addressed by the requirement of the State or political subdivision.

(4) Identification of which (or both) of the following bases is relied upon for seeking an exemption from preemption:

(i) The requirement is more stringent than a requirement applicable to a device under the act. If the State or political subdivision relies upon this basis for exemption from preemption, the application shall include information, data, or material showing how and why the requirement of the State or political subdivision is more stringent than requirements under the act.

(ii) The requirement is required by compelling local conditions, and compliance with the requirement would not cause the device to be in violation of any applicable requirement under the act. If the State or political subdivision relies upon this basis for exemption from preemption, the application shall include information, data, or material showing why compliance with the requirement of the State or political subdivision would not cause a device to be in violation of any applicable requirement under the act and why the requirement is required by compelling local conditions. The application shall also explain in detail the compelling local conditions that justify the requirement.

(5) The title of the chief administrative or legal officers of that State or local agency that has primary responsibility for administration of the requirement.

(6) When requested by the Food and Drug Administration, any records concerning administration of any requirement which is the subject of an exemption or an application for an exemption from preemption.

(7) Information on how the public health may be benefitted and how interstate commerce may be affected, if an exemption is granted.

(8) Any other pertinent information respecting the requirement voluntarily submitted by the applicant.

(d) If litigation regarding applicability of the requirement is pending, the State or political subdivision may so indicate in its application and request expedited action on such application.

[43 FR 18665, May 2, 1978; 43 FR 22010, May 23, 1978, as amended at 49 FR 3646, Jan. 30, 1984; 59 FR 14365, Mar. 28, 1994]

§ 808.25 Procedures for processing an application.

(a) Upon receipt of an application for an exemption from preemption submitted in accordance with § 808.20, the Commissioner shall notify the State or political subdivision of the date of such receipt.

(b) If the Commissioner finds that an application does not meet the requirements of § 808.20, he shall notify the State or political subdivision of the deficiencies in the application and of the opportunity to correct such deficiencies. A deficient application may be corrected at any time.

(c) After receipt of an application meeting the requirements of § 808.20, the Commissioner shall review such application and determine whether to grant or deny an exemption from preemption for each requirement which is the subject of the application. The Commissioner shall then issue in the FEDERAL REGISTER a proposed regulation either to grant or to deny an exemption from preemption. The Commissioner shall also issue in the FEDERAL REGISTER a notice of opportunity to request an oral hearing before the Commissioner or the Commissioner's designee.

(d) A request for an oral hearing may be made by the State or political subdivision or any other interested person. Such request shall be submitted to the Division of Dockets Management within the period of time prescribed in the notice and shall include an explanation of why an oral hearing, rather than submission of written comments only, is essential to the presentation of views on the application for exemption from preemption and the proposed regulation.

(e) If a timely request for an oral hearing is made, the Commissioner shall review such a request and may grant a legislative-type informal oral hearing pursuant to part 15 of this chapter by publishing in the FEDERAL REGISTER a notice of the hearing in accordance with § 15.20 of this chapter. The scope of the oral hearing shall be limited to matters relevant to the application for exemption from preemption and the proposed regulation. Oral or written presentations at the oral hearing which are not relevant to the application shall be excluded from the administrative record of the hearing.

(f) If a request for hearing is not timely made or a notice of appearance is not filed pursuant to § 15.21 of this chapter, the Commissioner shall consider all written comments submitted and publish a final rule in accordance with paragraph (g) of this section.

(g)(1) The Commissioner shall review all written comments submitted on the proposed rule and the administrative record of the oral hearing, if an oral hearing has been granted, and shall publish in the FEDERAL REGISTER a final rule in subpart C of this part identifying any requirement in the application for which exemption from preemption is granted, or conditionally granted, and any requirement in the application for which exemption from preemption is not granted.

(2) The Commissioner may issue a regulation granting or conditionally granting an application for an exemption from preemption for any requirement if the Commissioner makes either of the following findings:

(i) The requirement is more stringent than a requirement applicable to the device under the act;

(ii) The requirement is required by compelling local conditions, and compliance with the requirement would not cause the device to be in violation of any requirement applicable to the device under the act.

(3) The Commissioner may not grant an application for an exemption from preemption for any requirement with respect to a device if the Commissioner determines that the granting of an exemption would not be in the best interest of public health, taking into account the potential burden on interstate commerce.

(h) An advisory opinion pursuant to § 808.5 or a regulation pursuant to paragraph (g) of this section constitutes final agency action.

§ 808.35 Revocation of an exemption.

(a) An exemption from preemption pursuant to a regulation under this part shall remain effective until the Commissioner revokes such exemption.

(b) The Commissioner may by regulation, in accordance with § 808.25, revoke an exemption from preemption for any of the following reasons:

(1) An exemption may be revoked upon the effective date of a newly established requirement under the act which, in the Commissioner's view, addresses the objectives of an exempt requirement and which is described,

when issued, as preempting a previously exempt State or local requirement.

(2) An exemption may be revoked upon a finding that there has occurred a change in the bases listed in § 808.20(c)(4) upon which the exemption was granted.

(3) An exemption may be revoked if it is determined that a condition placed on the exemption by the regulation under which the exemption was granted has not been met or is no longer being met.

(4) An exemption may be revoked if a State or local jurisdiction fails to submit records as provided in § 808.20(c)(6).

(5) An exemption may be revoked if a State or local jurisdiction to whom the exemption was originally granted requests revocation.

(6) An exemption may be revoked if it is determined that it is no longer in the best interests of the public health to continue the exemption.

(c) An exemption that has been revoked may be reinstated, upon request from the State or political subdivision, if the Commissioner, in accordance with the procedures in § 808.25, determines that the grounds for revocation are no longer applicable except that the Commissioner may permit abbreviated submissions of the documents and materials normally required for an application for exemption under § 808.20.

Subpart C—Listing of Specific State and Local Exemptions

§ 808.53 Arizona.

The following Arizona medical device requirements are preempted by section 521(a) of the act, and the Food and Drug Administration has denied them exemptions from preemption under section 521(b) of the act:

(a) Arizona Revised Statutes, Chapter 17, sections 36–1901.7(s) and 36–1901.7(t).

(b) Arizona Code of Revised Regulations, Title 9, Article 3, sections R9–16–303 and R9–16–304.

[45 FR 67336, Oct. 10, 1980]

§ 808.55 California.

(a) The following California medical device requirements are enforceable notwithstanding section 521 of the act because the Food and Drug Administration exempted them from preemption under section 521(b) of the act: Business and Professions Code sections 3365 and 3365.6.

(b) The following California medical device requirements are preempted by section 521 of the act, and FDA has denied them an exemption from preemption:

(1) Sherman Food, Drug, and Cosmetic Law (Division 21 of the California Health and Safety Code), sections 26207, 26607, 26614, 26615, 26618, 26631, 26640, and 26641, to the extent that they apply to devices.

(2) Sherman Food, Drug, and Cosmetic Law, section 26463(m) to the extent that it applies to hearing aids.

(3) Business and Professions Code section 2541.3, to the extent that it requires adoption of American National Standards Institute standards Z–80.1 and Z–80.2.

[45 FR 67324, Oct. 10, 1980]

§ 808.57 Connecticut.

The following Connecticut medical device requirements are enforceable notwithstanding section 521(a) of the act because the Food and Drug Administration has exempted them from preemption under section 521(b) of the act: Connecticut General Statutes, sections 20–403 and 20–404.

[45 FR 67336, Oct. 10, 1980]

§ 808.59 Florida.

The following Florida medical device requirements are preempted by section 521(a) of the act, and the Food and Drug Administration has denied them an exemption from preemption under section 521(b) of the act:

(a) Florida Statutes, section 468.135(5).

(b) Florida Administrative Code, section 10D–48.25(26).

[45 FR 67336, Oct. 10, 1980]

§ 808.61 Hawaii.

(a) The following Hawaii medical device requirements are enforceable notwithstanding section 521 of the act, because the Food and Drug Administration has exempted them from preemption under section 521(b) of the act: Hawaii Revised Statutes, chapter 451A, § 14.1, subsection (a) with respect to medical examination of a child 10 years of age or under, and subsection (c).

(b) The following Hawaii medical device requirements are preempted by section 521(a) of the act, and the Food and Drug Administration has denied them exemption from preemption: Hawaii Revised Statutes, chapter 451A, § 14.1, subsection (a) to the extent that it requires a written authorization by a physician and does not allow adults to waive this requirement for personal, as well as religious reasons, and subsection (b).

[50 FR 30699, July 29, 1985; 50 FR 32694, Aug. 14, 1985]

§ 808.67 Kentucky.

The following Kentucky medical device requirement is preempted by section 521(a) of the act, and the Food and Drug Administration has denied it an exemption from preemption under section 521(b) of the act: Kentucky Revised Statutes, section 334.200(1).

[45 FR 67336, Oct. 10, 1980]

§ 808.69 Maine.

(a) The following Maine medical device requirement is enforceable notwithstanding section 521(a) of the act because the Food and Drug Administration has exempted it from preemption under section 521(b) of the act: Maine Revised Statutes Annotated, Title 32, section 1658–C, on the condition that, in enforcing this requirement, Maine apply the definition of "used hearing aid" in § 801.420(a)(6) of this chapter.

(b) The following Maine medical device requirement is preempted by section 521(a) of the act, and the Food and Drug Administration has denied it an exemption from preemption under section 521(b) of the act: Maine Revised Statutes Annotated, Title 32, section 1658–D and the last sentence of section 1658–E.

[45 FR 67336, Oct. 10, 1980]

§ 808.71 Massachusetts.

(a) The following Massachusetts medical device requirements are enforceable notwithstanding section 521 of the act because the Food and Drug Administration has exempted them from preemption under section 521(b) of the act:

(1) Massachusetts General Laws, Chapter 93, Section 72, to the extent that it requires a hearing test evaluation for a child under the age of 18.

(2) Massachusetts General Laws, Chapter 93, Section 74, except as provided in paragraph (6) of the Section, on the condition that, in enforcing this requirement, Massachusetts apply the definition of "used hearing aid" in § 801.420(a)(6) of this chapter.

(b) The following Massachusetts medical device requirements are preempted by section 521(a) of the act, and the Food and Drug Administration has denied them exemptions from preemption under section 521(b) of the act.

(1) Massachusetts General Laws, Chapter 93, Section 72, except as provided in paragraph (a) of this section.

(2) Massachusetts General Laws, Chapter 93, Section 74, to the extent that it requires that the sales receipt contain a statement that State law requires a medical examination and a hearing test evaluation before the sale of a hearing aid.

[45 FR 67326, Oct. 10, 1980]

§ 808.73 Minnesota.

The following Minnesota medical device requirements are preempted by section 521(a) of the act, and the Food and Drug Administration has denied them an exemption from preemption under section 521(b) of the act: Minnesota Statutes, sections 145.43 and 145.44.

[45 FR 67336, Oct. 10, 1980]

§ 808.74 Mississippi.

The following Mississippi medical device requirement is preempted by section 521(a) of the act, and the Food and Drug Administration has denied it an exemption from preemption under section 521(b) of the act: Mississippi Code, section 73–14–3(g)(9).

[45 FR 67336, Oct. 10, 1980]

§ 808.77 Nebraska.

(a) The following Nebraska medical device requirement is enforceable notwithstanding section 521(a) of the act because the Food and Drug Administration has exempted it from preemption under section 521(b) of the act: Nebraska Revised Statutes, section 71–4712(2)(c)(vi).

(b) The following Nebraska medical device requirement is preempted by section 521(a) of the act, and the Food and Drug Administration has denied it an exemption from preemption under section 521(b) of the act: Nebraska Revised Statutes, section 71–4712(2)(c)(vii).

[45 FR 67336, Oct. 10, 1980]

§ 808.80 New Jersey.

(a) The following New Jersey medical device requirements are enforceable notwithstanding section 521(a) of the act because the Food and Drug Administration has exempted them from preemption under section 521(b) of the act:

(1) New Jersey Statutes Annotated, section 45:9A–23 on the condition that, in enforcing this requirement, New Jersey apply the definition of "used hearing aid" in § 801.420(a)(6) of this chapter;

(2) New Jersey Statutes Annotated, sections 45:9A–24 and 45:9A–25;

(3) Chapter 3, Section 5 of the Rules and Regulations adopted pursuant to New Jersey Statutes Annotated 45:9A–1 *et seq.* except as provided in paragraph (b) of this section.

(b) The following New Jersey medical device requirement is preempted by section 521(a) of the act, and the Food and Drug Administration has denied it an exemption from preemption under section 521(b) of the act: Chapter 3, Section 5 of the Rules and Regulations adopted pursuant to New Jersey Statutes Annotated 45:9A–1 *et seq.* to the extent that it requires testing to be conducted in an environment which meets or exceeds the American National Standards Institute S3.1 Standard.

[45 FR 67337, Oct. 10, 1980]

§ 808.81 New Mexico.

The following New Mexico medical device requirement is enforceable notwithstanding section 521(a) of the act because the Food and Drug Administration has exempted it from preemption under section 521(b) of the act: New Mexico Statutes Annotated, section 67–36–16(F).

[45 FR 67337, Oct. 10, 1980]

§ 808.82 New York.

(a) The following New York medical device requirements are enforceable notwithstanding section 521(a) of the act because the Food and Drug Administration has exempted them from preemption under section 521(b) of the act:

(1) General Business Law, Article 37, sections 784(3) and (4).

(2) Official Compilation of Codes, Rules and Regulations of the State of New York, Chapter V, Title 19, Subchapter G, section 191.10 and section 191.11(a) on the condition that, in enforcing these requirements, New York apply the definition of "used hearing aid" in § 801.420(a)(6) of this chapter and section 191.11(b), (c), (d), and (e).

(b) The following New York medical device requirements are preempted by section 521(a) of the act, and the Food and Drug Administration has denied them an exemptions from preemption under section 521(b) of the act:

(1) General Business Law, Article 37, section 784.1.

(2) Official Compilation of Codes, Rules and Regulations of the State of New York, Chapter V, Title 19, Subchapter G, sections 191.6, 191.7, 191.8, and 191.9.

[45 FR 67337, Oct. 10, 1980]

§ 808.85 Ohio.

(a) The following Ohio medical device requirement is enforceable notwithstanding section 521(a) of the act because the Food and Drug Administration has exempted it from preemption under section 521(b) of the act: Ohio Revised Code, section 4747.09, the first two sentences with respect to disclosure of information to purchasers on the condition that, in enforcing these requirements, Ohio apply the definition of "used hearing aid" in § 801.420(a)(6) of this chapter.

(b) The following Ohio medical device requirement is preempted by section 521(a) of the act, and the Food and Drug Administration has denied it an exemption from preemption under section 521(b) of the act: Ohio Revised Code, section 4747.09, the last two sentences with respect to medical examination of children.

[45 FR 67337, Oct. 10, 1980]

§ 808.87 Oregon.

(a) The following Oregon medical device requriements are enforceable notwithstanding section 521(a) of the act because the Food and Drug Administration has exempted them from preemption under section 521(b) of the act: Oregon Revised Statutes, section 694.036 on the condition that, in enforcing this requirement, Oregon apply the definition of "used hearing aid" in § 801.420(a)(6) of this chapter.

(b) The following Oregon medical device requirements are preempted by section 521(a) of the act, and the Food and Drug Administration has denied them exemptions from preemption under section 521(b) of the act: Oregon Revised Statutes, sections 694.136(6) and (7).

[45 FR 67337, Oct. 10, 1980, as amended at 53 FR 11252, Apr. 6, 1988]

§ 808.88 Pennsylvania.

(a) The following Pennsylvania medical device requirements are enforceable notwithstanding section 521(a) of the act because the Food and Drug Administration has exempted them from preemption under section 521(b) of the act: 35 Purdon's Statutes 6700, section 504(4) on the condition that, in enforcing this requirement, Pennsylvania apply the definition of "used hearing aid" in § 801.420(a)(6) of this chapter; section 506; and, section 507(2).

(b) The following Pennsylvania medical device requirement is preempted by section 521(a) of the act and the Food and Drug Administration has denied it an exemption from preemption under section 521(b) of the act: 35 Purdon's Statutes 6700, section 402.

[45 FR 67326, Oct. 10, 1980]

§ 808.89 Rhode Island.

The following Rhode Island medical device requirements are preempted by section 521(a) of the act, and the Food and Drug Administration has denied them an exemption from preemption under section 521(b) of the act: Rhode Island General Laws, Section 5–49–2.1, and Section 2.2, to the extent that Section 2.2 requires hearing aid dispensers to keep copies of the certificates of need.

[45 FR 67337, Oct. 10, 1980]

§ 808.93 Texas.

(a) The following Texas medical device requirement is enforceable notwithstanding section 521(a) of the act because the Food and Drug Administration has exempted it from preemption under section 521(b) of the act: Vernon's Civil Statutes, Article 4566, section 14(b) on the condition that, in enforcing this requirement, Texas apply the definition of "used hearing aid" in § 801.420(a)(6) of this chapter.

(b) The following Texas medical device requirement is preempted by section 521(a) of the act, and the Food and Drug Administration has denied it an exemption from preemption under section 521(b) of the act: Vernon's Civil Statutes, Article 4566, section 14(d).

[45 FR 67337, Oct. 10, 1980]

§ 808.97 Washington.

(a) The following Washington medical device requirement is enforceable notwithstanding section 521(a) of the act because the Food and Drug Administration has exempted it from preemption under section 521(b) of the act: Revised Code of Washington 18.35.110(2)(e) (i) and (iii) on the condition that it is enforced in addition to the applicable requirements of this chapter.

(b) The following Washington medical device requirements are preempted by section 521(a) of the act, and the Food and Drug Administration has denied them an exemption from preemption under section 521(b) of the act: Revised Code of Washington 18.35.110(2)(e)(ii).

[45 FR 67337, Oct. 10, 1980]

§ 808.98 West Virginia.

(a) The following West Virginia medical device requirements are enforceable notwithstanding section 521(a) of the act because the Food and Drug Administration has exempted them from preemption: West Virginia Code, sections 30–26–14 (b) and (c) and section 30–26–15(a) on the condition that in enforcing section 30–26–15(a) West Virginia apply the definition of "used hearing aid" in § 801.420(a)(6) of this chapter.

(b) The following West Virginia medical device requirement is preempted by section 521(a) of the act, and the Food and Drug Administration has denied it an exemption from preemption under section 521(b) of the act: West Virginia Code, section 30–26–14(a).

[45 FR 67337, Oct. 10, 1980, as amended at 53 FR 35314, Sept. 13, 1988]

§ 808.101 District of Columbia.

(a) The following District of Columbia medical device requirements are enforceable, notwithstanding section 521 of the act, because the Food and Drug Administration has exempted them from preemption under section 521(b) of the act:

(1) Act 2–79, section 5, to the extent that it requires an audiological evaluation for children under the age of 18.

(2) Act 2–79, section 6, on the condition that in enforcing section 6(a)(5), the District of Columbia apply the definition of "used hearing aid" in § 801.420(a)(6) of this chapter.

(b) The following District of Columbia medical device requirement is preempted by section 521(a) of the act, and the Food and Drug Administration has denied it an exemption from preemption under section 521(b) of the act: Act 2–79, section 5, except as provided in paragraph (a) of this section.

[46 FR 59236, Dec. 4, 1981]

PART 809—IN VITRO DIAGNOSTIC PRODUCTS FOR HUMAN USE

Subpart A—General Provisions

Sec.

Subpart B—Labeling

Subpart C—Requirements for Manufacturers and Producers

AUTHORITY: 21 U.S.C. 331, 351, 352, 355, 360b, 360c, 360d, 360h, 360i, 360j, 371, 372, 374, 381.

Subpart A—General Provisions

§ 809.3 Definitions.

(a) *In vitro diagnostic products* are those reagents, instruments, and systems intended for use in the diagnosis of disease or other conditions, including a determination of the state of health, in order to cure, mitigate, treat, or prevent disease or its sequelae. Such products are intended for use in the collection, preparation, and examination of specimens taken from the human body. These products are devices as defined in section 201(h) of the Federal Food, Drug, and Cosmetic Act (the act), and may also be biological products subject to section 351 of the Public Health Service Act.

(b) A *product class* is all those products intended for use for a particular determination or for a related group of determinations or products with common or related characteristics or those intended for common or related uses. A class may be further divided into subclasses when appropriate.

(c) [Reserved]

(d) *Act* means the Federal Food, Drug, and Cosmetic Act.

[41 FR 6903, Feb. 13, 1976, as amended at 45 FR 7484, Feb. 1, 1980]

§ 809.4 Confidentiality of submitted information.

Data and information submitted under § 809.10(c) that are shown to fall within the exemption established in § 20.61 of this chapter shall be treated as confidential by the Food and Drug Administration and any person to whom the data and information are referred. The Food and Drug Administration will determine whether information submitted will be treated as confidential in accordance with the provisions of part 20 of this chapter.

[45 FR 7484, Feb. 1, 1980]

Subpart B—Labeling

§ 809.10 Labeling for in vitro diagnostic products.

(a) The label for an in vitro diagnostic product shall state the following information, except where such information is not applicable, or as otherwise specified in a standard for a particular product class or as provided in paragraph (e) of this section. Section 201(k) of the act provides that "a requirement made by or under authority of this act that any word, statement, or other information appear on the label shall not be considered to be complied with unless such word, statement, or other information also appears on the outside container or wrapper, if any there be, of the retail package of such article, or is easily legible through the outside container or wrapper."

(1) The proprietary name and established name (common or usual name), if any.

(2) The intended use or uses of the product.

(3) For a reagent, a declaration of the established name (common or usual name), if any, and quantity, proportion or concentration of each reactive ingredient; and for a reagent derived from biological material, the source and a measure of its activity. The quantity, proportion, concentration, or activity shall be stated in the system generally used and recognized by the intended user, e.g., metric, international units, etc.

(4) A statement of warnings or precautions for users as established in the regulations contained in 16 CFR part 1500 and any other warnings appropriate to the hazard presented by the product; and a statement "For In Vitro Diagnostic Use" and any other limiting statements appropriate to the intended use of the product.

(5) For a reagent, appropriate storage instructions adequate to protect the stability of the product. When applicable, these instructions shall include such information as conditions of temperature, light, humidity, and other pertinent factors. For products requiring manipulation, such as reconstitution and/or mixing before use, appropriate storage instructions shall be provided for the reconstituted or mixed product which is to be stored in the original container. The basis for such instructions shall be determined by reliable, meaningful, and specific test methods such as those described in § 211.166 of this chapter.

(6) For a reagent, a means by which the user may be assured that the product meets appropriate standards of identity, strength, quality and purity at the time of use. This shall be provided, both for the product as provided and for any resultant reconstituted or mixed product, by including on the label one or more of the following:

(i) An expiration date based upon the stated storage instructions.

(ii) A statement of an observable indication of an alteration of the product, e.g., turbidity, color change, precipitate, beyond its appropriate standards.

(iii) Instructions for a simple method by which the user can reasonably determine that the product meets its appropriate standards.

(7) For a reagent, a declaration of the net quantity of contents, expressed in terms of weight or volume, numerical count, or any combination of these or other terms which accurately reflect the contents of the package. The use of metric designations is encouraged, wherever appropriate. If more than a single determination may be performed using the product, any statement of the number of tests shall be consistent with instructions for use and amount of material provided.

(8) Name and place of business of manufacturer, packer, or distributor.

(9) A lot or control number, identified as such, from which it is possible to determine the complete manufacturing history of the product.

(i) If it is a multiple unit product, the lot or control number shall permit tracing the identity of the individual units.

(ii) For an instrument, the lot or control number shall permit tracing the identity of all functional subassemblies.

(iii) For multiple unit products which require the use of included units together as a system, all units should bear the same lot or control number, if appropriate, or other suitable uniform identification should be used.

(10) Except that for items in paragraphs (a) (1) through (9) of this section: (i) In the case of immediate containers too small or otherwise unable to accommodate a label with sufficient space to bear all such information and which are packaged within an outer container from which they are removed for use, the information required by paragraphs (a) (2), (3), (4), (5), (6) (ii), (iii) and (7) of this section may appear in the outer container labeling only.

(ii) In any case in which the presence of this information on the immediate container will interfere with the test, the information may appear on the outside container or wrapper rather than on the immediate container label.

(b) Labeling accompanying each product, e.g., a package insert, shall state in one place the following information in the format and order specified below, except where such information is not applicable, or as specified in a standard for a particular product class. The labeling for a multiple-purpose instrument used for diagnostic purposes, and not committed to specific diagnostic procedures or systems, may bear only the information indicated in paragraphs (b) (1), (2), (6), (14), and (15) of this section. The labeling for a reagent intended for use as a replacement in a diagnostic system may be limited to that information necessary to identify the reagent adequately and to describe its proper use in the system.

(1) The proprietary name and established name, i.e., common or usual name, if any.

(2) The intended use or uses of the product and the type of procedure, e.g., qualitative or quantitative.

(3) Summary and explanation of the test. Include a short history of the methodology, with pertinent references and a balanced statement of the special merits and limitations of this method or product. If the product labeling refers to any other procedure, appropriate literature citations shall be included and the labeling shall explain the nature of any differences from the original and their effect on the results.

(4) The chemical, physical, physiological, or biological principles of the procedure. Explain concisely, with chemical reactions and techniques involved, if applicable.

(5) Reagents:

(i) A declaration of the established name (common or usual name), if any, and quantity, proportion or concentration or each reactive ingredient; and for biological material, the source and a measure of its activity. The quantity, proportion, concentration or activity shall be stated in the system generally used and recognized by the intended user, e.g., metric, international units, etc. A statement indicating the presence of and characterizing any catalytic or nonreactive ingredients, e.g., buffers, preservatives, stabilizers.

(ii) A statement of warnings or precautions for users as established in the regulations contained in 16 CFR part 1500 and any other warnings appropriate to the hazard presented by the product; and a statement "For In Vitro Diagnostic Use" and any other limiting statements appropriate to the intended use of the product.

(iii) Adequate instructions for reconstitution, mixing, dilution, etc.

(iv) Appropriate storage instructions adequate to protect the stability of the product. When applicable, these instructions shall include such information as conditions of temperature, light, humidity, and other pertinent factors. For products requiring manipulation, such as reconstitution and/or mixing before use, appropriate storage instructions shall be provided for the reconstituted or mixed product. The basis for such instructions shall be determined by reliable, meaningful, and specific test methods such as those described in § 211.166 of this chapter.

(v) A statement of any purification or treatment required for use.

(vi) Physical, biological, or chemical indications of instability or deterioration.

(6) Instruments:

(i) Use or function.

(ii) Installation procedures and special requirements.

(iii) Principles of operation.

(iv) Performance characteristics and specifications.

(v) Operating instructions.

(vi) Calibration procedures including materials and/or equipment to be used.

(vii) Operational precautions and limitations.

(viii) Hazards.

(ix) Service and maintenance information.

(7) Specimen collection and preparation for analysis, including a description of:

(i) Special precautions regarding specimen collection including special preparation of the patient as it bears on the validity of the test.

(ii) Additives, preservatives, etc., necessary to maintain the integrity of the specimen.

(iii) Known interfering substances.

(iv) Recommended storage, handling or shipping instructions for the protection and maintenance of stability of the specimen.

(8) Procedure: A step-by-step outline of recommended procedures from reception of the specimen to obtaining results. List any points that may be useful in improving precision and accuracy.

(i) A list of all materials provided, e.g., reagents, instruments and equipment, with instructions for their use.

(ii) A list of all materials required but not provided. Include such details as sizes, numbers, types, and quality.

(iii) A description of the amounts of reagents necessary, times required for specific steps, proper temperatures, wavelengths, etc.

(iv) A statement describing the stability of the final reaction material to be measured and the time within which it shall be measured to assure accurate results.

(v) Details of calibration: Identify reference material. Describe preparation of reference sample(s), use of blanks, preparation of the standard curve, etc. The description of the range of calibration should include the highest and the lowest values measurable by the procedure.

(vi) Details of kinds of quality control procedures and materials required. If there is need for both positive and negative controls, this should be stated. State what are considered satisfactory limits of performance.

(9) Results: Explain the procedure for calculating the value of the unknown. Give an explanation for each component of the formula used for the calculation of the unknown. Include a sample calculation, step-by-step, explaining the answer. The values shall be expressed to the appropriate number of significant figures. If the test provides other than quantitative results, provide an adequate description of expected results.

(10) Limitation of the procedure: Include a statement of limitations of the procedure. State known extrinsic factors or interfering substances affecting results. If further testing, either more specific or more sensitive, is indicated in all cases where certain results are obtained, the need for the additional test shall be stated.

(11) Expected values: State the range(s) of expected values as obtained with the product from studies of various populations. Indicate how the range(s) was established and identify the population(s) on which it was established.

(12) Specific performance characteristics: Include, as appropriate, information describing such things as accuracy, precision, specificity, and sensitivity. These shall be related to a generally accepted method using biological specimens from normal and abnormal populations. Include a statement summarizing the data upon which the specific performance characteristics are based.

(13) Bibliography: Include pertinent references keyed to the text.

(14) Name and place of business of manufacturer, packer, or distributor.

(15) Date of issuance of the last revision of the labeling identified as such.

(c) A shipment or other delivery of an in vitro diagnostic product shall be exempt from the requirements of paragraphs (a) and (b) of this section and from a standard promulgated under part 861 provided that the following conditions are met:

(1) In the case of a shipment or delivery for an investigation subject to part

812, if there has been compliance with part 812; or

(2) In the case of a shipment or delivery for an investigation that is not subject to part 812 (see § 812.2(c)), if the following conditions are met:

(i) For a product in the laboratory research phase of development, and not represented as an effective in vitro diagnostic product, all labeling bears the statement, prominently placed: "For Research Use Only. Not for use in diagnostic procedures."

(ii) For a product being shipped or delivered for product testing prior to full commercial marketing (for example, for use on specimens derived from humans to compare the usefulness of the product with other products or procedures which are in current use or recognized as useful), all labeling bears the statement, prominently placed: "For Investigational Use Only. The performance characteristics of this product have not been established."

(d) The labeling of general purpose laboratory reagents (e.g., hydrochloric acid) and equipment (e.g., test tubes and pipettes) whose uses are generally known by persons trained in their use need not bear the directions for use required by § 809.10(a) and (b), if their labeling meets the requirements of this paragraph.

(1) The label of a reagent shall bear the following information:

(i) The proprietary name and established name (common or usual name), if any, of the reagent.

(ii) A declaration of the established name (common or usual name), if any, and quantity, proportion or concentration of the reagent ingredient (e.g., hydrochloric acid: Formula weight 36.46, assay 37.9 percent, specific gravity 1.192 at 60 °F); and for a reagent derived from biological material, the source and where applicable a measure of its activity. The quantity, proportion, concentration or activity shall be stated in the system generally used and recognized by the intended user, e.g., metric, international units, etc.

(iii) A statement of the purity and quality of the reagent, including a quantitative declaration of any impurities present. The requirement for this information may be met by a statement of conformity with a generally recognized and generally available standard which contains the same information, e.g., those established by the American Chemical Society, U.S. Pharmacopeia, National Formulary, National Research Council.

(iv) A statement of warnings or precautions for users as established in the regulations contained in 16 CFR part 1500 and any other warnings appropriate to the hazard presented by the product; and a statement "For Laboratory Use."

(v) Appropriate storage instructions adequate to protect the stability of the product. When applicable, these instructions shall include such information as conditions of temperature, light, humidity, and other pertinent factors. The basis for such information shall be determined by reliable, meaningful, and specific test methods such as those described in § 211.166 of this chapter.

(vi) A declaration of the net quantity of contents, expressed in terms of weight or volume, numerical count, or any combination of these or other terms which accurately reflect the contents of the package. The use of metric designations is encouraged, wherever appropriate.

(vii) Name and place of business of manufacturer, packer, or distributor.

(viii) A lot or control number, identified as such, from which it is possible to determine the complete manufacturing history of the product.

(ix) In the case of immediate containers too small or otherwise unable to accommodate a label with sufficient space to bear all such information, and which are packaged within an outer container from which they are removed for use, the information required by paragraphs (d)(1)(ii), (iii), (iv), (v), and (vi) of this section may appear in the outer container labeling only.

(2) The label of general purpose laboratory equipment, e.g., a beaker or a pipette, shall bear a statement adequately describing the product, its composition, and physical characteristics if necessary for its proper use.

(e)(1) The labeling for analyte specific reagents (e.g., monoclonal antibodies, deoxyribonucleic acid (DNA) probes, viral antigens, ligands) shall bear the following information:

(i) The proprietary name and established name (common or usual name), if any, of the reagent;

(ii) A declaration of the established name (common or usual name), if any;

(iii) The quantity, proportion, or concentration of the reagent ingredient; and for a reagent derived from biological material, the source and where applicable, a measure of its activity. The quantity, proportion, concentration, or activity shall be stated in the system generally used and recognized by the intended user, e.g., metric, international units, etc.;

(iv) A statement of the purity and quality of the reagent, including a quantitative declaration of any impurities present and method of analysis or characterization. The requirement for this information may be met by a statement of conformity with a generally recognized and generally available standard that contains the same information, e.g., those established by the American Chemical Society, U.S. Pharmacopeia, National Formulary, and National Research Council. The labeling may also include information concerning chemical/molecular composition, nucleic acid sequence, binding affinity, cross-reactivities, and interaction with substances of known clinical significance;

(v) A statement of warnings or precautions for users as established in the regulations contained in 16 CFR part 1500 and any other warnings appropriate to the hazard presented by the product;

(vi) The date of manufacture and appropriate storage instructions adequate to protect the stability of the product. When applicable, these instructions shall include such information as conditions of temperature, light, humidity, date of expiration, and other pertinent factors. The basis for such instructions shall be determined by reliable, meaningful, and specific test methods, such as those described in § 211.166 of this chapter;

(vii) A declaration of the net quantity of contents, expressed in terms of weight or volume, numerical count, or any combination of these or other terms that accurately reflect the contents of the package. The use of metric designations is encouraged, wherever appropriate;

(viii) The name and place of business of manufacturer, packer, or distributor;

(ix) A lot or control number, identified as such, from which it is possible to determine the complete manufacturing history of the product;

(x) For class I exempt ASR's, the statement: "Analyte Specific Reagent. Analytical and performance characteristics are not established"; and

(xi) For class II and III ASR's, the statement: "Analyte Specific Reagent. Except as a component of the approved/cleared test (Name of approved/cleared test), analytical and performance characteristics of this ASR are not established."

(2) In the case of immediate containers too small or otherwise unable to accommodate a label with sufficient space to bear all such information, and which are packaged within an outer container from which they are removed for use, the information required by paragraphs (e)(1) through (e)(6) of this section may appear in the outer container labeling only.

(f) The labeling for over-the-counter (OTC) test sample collection systems for drugs of abuse testing shall bear the following information in language appropriate for the intended users:

(1) Adequate instructions for specimen collection and handling, and for preparation and mailing of the specimen to the laboratory for testing.

(2) An identification system to ensure that specimens are not mixed up or otherwise misidentified at the laboratory, and that user anonymity is maintained.

(3) The intended use or uses of the product, including what drugs are to be identified in the specimen, a quantitative description of the performance characteristics for those drugs (e.g., sensitivity and specificity) in terms understandable to lay users, and the detection period.

(4) A statement that confirmatory testing will be conducted on all samples that initially test positive.

(5) A statement of warnings or precautions for users as established in the regulations contained in 16 CFR part

1500 and any other warnings appropriate to the hazard presented by the product.

(6) Adequate instructions on how to obtain test results from a person who can explain their meaning, including the probability of false positive and false negative results, as well as how to contact a trained health professional if additional information on interpretation of test results from the laboratory or followup counseling is desired.

(7) Name and place of business of the manufacturer, packer, or distributor.

[41 FR 6903, Feb. 13, 1976, as amended at 45 FR 3750, Jan. 18, 1980; 45 FR 7484, Feb. 1, 1980; 47 FR 41107, Sept. 17, 1982; 47 FR 51109, Nov. 12, 1982; 48 FR 34470, July 29, 1983; 62 FR 62259, Nov. 21, 1997; 65 FR 18234, Apr. 7, 2000]

Subpart C—Requirements for Manufacturers and Producers

§ 809.20 General requirements for manufacturers and producers of in vitro diagnostic products.

(a) [Reserved]

(b) *Compliance with good manufacturing practices.* In vitro diagnostic products shall be manufactured in accordance with the good manufacturing practices requirements found in part 820 of this chapter and, if applicable, with § 610.44 of this chapter.

[41 FR 6903, Feb. 13, 1976, as amended at 42 FR 42530, Aug. 23, 1977; 43 FR 31527, July 21, 1978; 66 FR 31165, June 11, 2001]

§ 809.30 Restrictions on the sale, distribution and use of analyte specific reagents.

(a) Analyte specific reagents (ASR's) (§ 864.4020 of this chapter) are restricted devices under section 520(e) of the Federal Food, Drugs, and Cosmetic Act (the act) subject to the restrictions set forth in this section.

(b) ASR's may only be sold to:

(1) In vitro diagnostic manufacturers;

(2) Clinical laboratories regulated under the Clinical Laboratory Improvement Amendments of 1988 (CLIA), as qualified to perform high complexity testing under 42 CFR part 493 or clinical laboratories regulated under VHA Directive 1106 (available from Department of Veterans Affairs, Veterans Health Administration, Washington, DC 20420); and

(3) Organizations that use the reagents to make tests for purposes other than providing diagnostic information to patients and practitioners, e.g., forensic, academic, research, and other nonclinical laboratories.

(c) ASR's must be labeled in accordance with § 809.10(e).

(d) Advertising and promotional materials for ASR's:

(1) Shall include the identity and purity (including source and method of acquisition) of the analyte specific reagent and the identity of the analyte;

(2) Shall include the statement for class I exempt ASR's: "Analyte Specific Reagent. Analytical and performance characteristics are not established";

(3) Shall include the statement for class II or III ASR's: "Analyte Specific Reagent. Except as a component of the approved/cleared test (name of approved/cleared test), analytical and performance characteristics are not established"; and

(4) Shall not make any statement regarding analytical or clinical performance.

(e) The laboratory that develops an in-house test using the ASR shall inform the ordering person of the test result by appending to the test report the statement: "This test was developed and its performance characteristics determined by (Laboratory Name). It has not been cleared or approved by the U.S. Food and Drug Administration." This statement would not be applicable or required when test results are generated using the test that was cleared or approved in conjunction with review of the class II or III ASR.

(f) Ordering in-house tests that are developed using analyte specific reagents is limited under section 520(e) of the act to physicians and other persons authorized by applicable State law to order such tests.

(g) The restrictions in paragraphs (c) through (f) of this section do not apply when reagents that otherwise meet the analyte specific reagent definition are sold to:

(1) In vitro diagnostic manufacturers; or

(2) Organizations that use the reagents to make tests for purposes other than providing diagnostic information

to patients and practitioners, e.g., forensic, academic, research, and other nonclinical laboratories.

[62 FR 62259, Nov. 21, 1997]

§ 809.40 Restrictions on the sale, distribution, and use of OTC test sample collection systems for drugs of abuse testing.

(a) Over-the-counter (OTC) test sample collection systems for drugs of abuse testing (§ 864.3260 of this chapter) are restricted devices under section 520(e) of the Act subject to the restrictions set forth in this section.

(b) Sample testing shall be performed in a laboratory using screening tests that have been approved, cleared, or otherwise recognized by the Food and Drug Administration as accurate and reliable for the testing of such specimens for identifying drugs of abuse or their metabolites.

(c) The laboratory performing the test(s) shall have, and shall be recognized as having, adequate capability to reliably perform the necessary screening and confirmatory tests, including adequate capability to perform integrity checks of the biological specimens for possible adulteration.

(d) All OTC test sample collection systems for drugs of abuse testing shall be labeled in accordance with § 809.10(f) and shall provide an adequate system to communicate the proper interpretation of test results from the laboratory to the lay purchaser.

[65 FR 18234, Apr. 7, 2000]

PART 810—MEDICAL DEVICE RECALL AUTHORITY

Subpart A—General Provisions

Sec.

Subpart B—Mandatory Medical Device Recall Procedures

AUTHORITY: 21 U.S.C. 321, 331, 332, 333, 334, 351, 352, 360h, 371, 374, 375.

SOURCE: 61 FR 59018, Nov. 20, 1996, unless otherwise noted.

Subpart A—General Provisions

§ 810.1 Scope.

Part 810 describes the procedures that the Food and Drug Administration will follow in exercising its medical device recall authority under section 518(e) of the Federal Food, Drug, and Cosmetic Act.

§ 810.2 Definitions.

As used in this part:

(a) *Act* means the Federal Food, Drug, and Cosmetic Act.

(b) *Agency* or *FDA* means the Food and Drug Administration.

(c) *Cease distribution and notification strategy* or *mandatory recall strategy* means a planned, specific course of action to be taken by the person named in a cease distribution and notification order or in a mandatory recall order, which addresses the extent of the notification or recall, the need for public warnings, and the extent of effectiveness checks to be conducted.

(d) *Consignee* means any person or firm that has received, purchased, or used a device that is subject to a cease distribution and notification order or a mandatory recall order. Consignee does not mean lay individuals or patients, i.e., nonhealth professionals.

(e) *Correction* means repair, modification, adjustment, relabeling, destruction, or inspection (including patient monitoring) of a device, without its physical removal from its point of use to some other location.

(f) *Device user facility* means a hospital, ambulatory surgical facility, nursing home, or outpatient treatment or diagnostic facility that is not a physician's office.

(g) *Health professionals* means practitioners, including physicians, nurses, pharmacists, dentists, respiratory therapists, physical therapists, technologists, or any other practitioners or allied health professionals that have a role in using a device for human use.

(h) *Reasonable probability* means that it is more likely than not that an event will occur.

(i) *Serious, adverse health consequence* means any significant adverse experience, including those that may be either life-threatening or involve permanent or long-term injuries, but excluding injuries that are nonlife-threatening and that are temporary and reasonably reversible.

(j) *Recall* means the correction or removal of a device for human use where FDA finds that there is a reasonable probability that the device would cause serious, adverse health consequences or death.

(k) *Removal* means the physical removal of a device from its point of use to some other location for repair, modification, adjustment, relabeling, destruction, or inspection.

§ 810.3 Computation of time.

In computing any period of time prescribed or allowed by this part, the day of the act or event from which the designated period of time begins to run shall not be included. The computation of time is based only on working days.

§ 810.4 Service of orders.

Orders issued under this part will be served in person by a designated employee of FDA, or by certified or registered mail or similar mail delivery service with a return receipt record reflecting receipt, to the named person or designated agent at the named person's or designated agent's last known address in FDA's records.

Subpart B—Mandatory Medical Device Recall Procedures

§ 810.10 Cease distribution and notification order.

(a) If, after providing the appropriate person with an opportunity to consult with the agency, FDA finds that there is a reasonable probability that a device intended for human use would cause serious, adverse health consequences or death, the agency may issue a cease distribution and notification order requiring the person named in the order to immediately:

(1) Cease distribution of the device;

(2) Notify health professionals and device user facilities of the order; and

(3) Instruct these professionals and device user facilities to cease use of the device.

(b) FDA will include the following information in the order:

(1) The requirements of the order relating to cessation of distribution and notification of health professionals and device user facilities;

(2) Pertinent descriptive information to enable accurate and immediate identification of the device subject to the order, including, where known:

(i) The brand name of the device;

(ii) The common name, classification name, or usual name of the device;

(iii) The model, catalog, or product code numbers of the device; and

(iv) The manufacturing lot numbers or serial numbers of the device or other identification numbers; and

(3) A statement of the grounds for FDA's finding that there is a reasonable probability that the device would cause serious, adverse health consequences or death.

(c) FDA may also include in the order a model letter for notifying health professionals and device user facilities of the order and a requirement that notification of health professionals and device user facilities be completed within a specified timeframe. The model letter will include the key elements of information that the agency in its discretion has determined, based on the circumstances surrounding the issuance of each order, are necessary to inform health professionals and device user facilities about the order.

(d) FDA may also require that the person named in the cease distribution and notification order submit any or all of the following information to the agency by a time specified in the order:

(1) The total number of units of the device produced and the timespan of the production;

(2) The total number of units of the device estimated to be in distribution channels;

(3) The total number of units of the device estimated to be distributed to health professionals and device user facilities;

(4) The total number of units of the device estimated to be in the hands of home users;

(5) Distribution information, including the names and addresses of all consignees;

(6) A copy of any written communication used by the person named in the order to notify health professionals and device user facilities;

(7) A proposed strategy for complying with the cease distribution and notification order;

(8) Progress reports to be made at specified intervals, showing the names and addresses of health professionals and device user facilities that have been notified, names of specific individuals contacted within device user facilities, and the dates of such contacts; and

(9) The name, address, and telephone number of the person who should be contacted concerning implementation of the order.

(e) FDA will provide the person named in a cease distribution and notification order with an opportunity for a regulatory hearing on the actions required by the cease distribution and notification order and on whether the order should be modified, or vacated, or amended to require a mandatory recall of the device.

(f) FDA will also provide the person named in the cease distribution and notification order with an opportunity, in lieu of a regulatory hearing, to submit a written request to FDA asking that the order be modified, or vacated, or amended.

(g) FDA will include in the cease distribution and notification order the name, address, and telephone number of an agency employee to whom any request for a regulatory hearing or agency review is to be addressed.

§810.11 Regulatory hearing.

(a) Any request for a regulatory hearing shall be submitted in writing to the agency employee identified in the order within the timeframe specified by FDA. Under §16.22(b) of this chapter, this timeframe ordinarily will not be fewer than 3 working days after receipt of the cease distribution and notification order. However, as provided in §16.60(h) of this chapter, the Commissioner of Food and Drugs or presiding officer may waive, suspend, or modify any provision of part 16 under §10.19 of this chapter, including those pertaining to the timing of the hearing. As provided in §16.26(a), the Commissioner or presiding officer may deny a request for a hearing, in whole or in part, if he or she determines that no genuine and substantial issue of fact is raised by the material submitted in the request.

(b) If a request for a regulatory hearing is granted, the regulatory hearing shall be limited to:

(1) Reviewing the actions required by the cease distribution and notification order, determining if FDA should affirm, modify, or vacate the order, and addressing an appropriate cease distribution and notification strategy; and

(2) Determining whether FDA should amend the cease distribution and notification order to require a recall of the device that was the subject of the order. The hearing may also address the actions that might be required by a recall order, including an appropriate recall strategy, if FDA later orders a recall.

(c) If a request by the person named in a cease distribution and notification order for a regulatory hearing is granted, the regulatory hearing will be conducted in accordance with the procedures set out in section 201(x) of the act (21 U.S.C. 321(x)) and part 16 of this chapter, except that the order issued under §810.10, rather than a notice under §16.22(a) of this chapter, provides the notice of opportunity for a hearing and is part of the administrative record of the regulatory hearing under §16.80(a) of this chapter. As provided in §16.60(h) of this chapter, the Commissioner of Food and Drugs or presiding officer may waive, suspend, or modify any provision of part 16 under §10.19 of this chapter. As provided in §16.26(b), after the hearing commences, the presiding officer may issue a summary decision on any issue if the presiding officer determines that there is no genuine and substantial issue of fact respecting that issue.

(d) If the person named in the cease distribution and notification order does not request a regulatory hearing within the timeframe specified by FDA in the cease distribution and notification order, that person will be deemed to have waived his or her right to request a hearing.

(e) The presiding officer will ordinarily hold any regulatory hearing requested under paragraph (a) of this section no fewer than 2 working days after receipt of the request for a hearing, under § 16.24(e) of this chapter, and no later than 10 working days after the date of issuance of the cease distribution and notification order. However, FDA and the person named in the order may agree to a later date or the presiding officer may determine that the hearing should be held in fewer than 2 days. Moreover, as provided for in § 16.60(h) of this chapter, the Commissioner of Food and Drugs or presiding officer may waive, suspend, or modify any provision of part 16 under § 10.19 of this chapter, including those pertaining to the timing of the hearing. After the presiding officer prepares a written report of the hearing and the agency issues a final decision based on the report, the presiding officer shall provide the requestor written notification of the final decision to affirm, modify, or vacate the order or to amend the order to require a recall of the device within 15 working days of conducting a regulatory hearing.

§ 810.12 Written request for review of cease distribution and notification order.

(a) In lieu of requesting a regulatory hearing under § 810.11, the person named in a cease distribution and notification order may submit a written request to FDA asking that the order be modified or vacated. Such person shall address the written request to the agency employee identified in the order and shall submit the request within the timeframe specified in the order, unless FDA and the person named in the order agree to a later date.

(b) A written request for review of a cease distribution and notification order shall identify each ground upon which the requestor relies in asking that the order be modified or vacated, as well as addressing an appropriate cease distribution and notification strategy, and shall address whether the order should be amended to require a recall of the device that was the subject of the order and the actions required by such a recall order, including an appropriate recall strategy.

(c) The agency official who issued the cease distribution and notification order shall provide the requestor written notification of the agency's decision to affirm, modify, or vacate the order or amend the order to require a recall of the device within 15 working days of receipt of the written request. The agency official shall include in this written notification:

(1) A statement of the grounds for the decision to affirm, modify, vacate, or amend the order; and

(2) The requirements of any modified or amended order.

§ 810.13 Mandatory recall order.

(a) If the person named in a cease distribution and notification order does not request a regulatory hearing or submit a request for agency review of the order, or, if the Commissioner of Food and Drugs or the presiding officer denies a request for a hearing, or, if after conducting a regulatory hearing under § 810.11 or completing agency review of a cease distribution and notification order under § 810.12, FDA determines that the order should be amended to require a recall of the device with respect to which the order was issued, FDA shall amend the order to require such a recall. FDA shall amend the order to require such a recall within 15 working days of issuance of a cease distribution and notification order if a regulatory hearing or agency review of the order is not requested, or within 15 working days of denying a request for a hearing, or within 15 working days of completing a regulatory hearing under § 810.11, or within 15 working days of receipt of a written request for review of a cease distribution and notification order under § 810.12.

(b) In a mandatory recall order, FDA may:

(1) Specify that the recall is to extend to the wholesale, retail, or user level;

(2) Specify a timetable in accordance with which the recall is to begin and be completed;

(3) Require the person named in the order to submit to the agency a proposed recall strategy, as described in § 810.14, and periodic reports describing the progress of the mandatory recall, as described in § 810.16; and

(4) Provide the person named in the order with a model recall notification letter that includes the key elements of information that FDA has determined are necessary to inform health professionals and device user facilities.

(c) FDA will not include in a mandatory recall order a requirement for:

(1) Recall of a device from individuals; or

(2) Recall of a device from device user facilities, if FDA determines that the risk of recalling the device from the facilities presents a greater health risk than the health risk of not recalling the device from use, unless the device can be replaced immediately with an equivalent device.

(d) FDA will include in a mandatory recall order provisions for notification to individuals subject to the risks associated with use of the device. If a significant number of such individuals cannot be identified, FDA may notify such individuals under section 705(b) of the act.

§ 810.14 Cease distribution and notification or mandatory recall strategy.

(a) *General.* The person named in a cease distribution and notification order issued under § 810.10 shall comply with the order, which FDA will fashion as appropriate for the individual circumstances of the case. The person named in a cease distribution and notification order modified under § 810.11(e) or § 810.12(c) or a mandatory recall order issued under § 810.13 shall develop a strategy for complying with the order that is appropriate for the individual circumstances and that takes into account the following factors:

(1) The nature of the serious, adverse health consequences related to the device;

(2) The ease of identifying the device;

(3) The extent to which the risk presented by the device is obvious to a health professional or device user facility; and

(4) The extent to which the device is used by health professionals and device user facilities.

(b) *Submission and review.* (1) The person named in the cease distribution and notification order modified under § 810.11(e) or § 810.12(c) or mandatory recall order shall submit a copy of the proposed strategy to the agency within the timeframe specified in the order.

(2) The agency will review the proposed strategy and make any changes to the strategy that it deems necessary within 7 working days of receipt of the proposed strategy. The person named in the order shall act in accordance with a strategy determined by FDA to be appropriate.

(c) *Elements of the strategy.* A proposed strategy shall meet all of the following requirements:

(1)(i) The person named in the order shall specify the level in the chain of distribution to which the cease distribution and notification order or mandatory recall order is to extend as follows:

(A) Consumer or user level, e.g., health professionals, consignee, or device user facility level, including any intermediate wholesale or retail level; or

(B) Retail level, to the level immediately preceding the consumer or user level, and including any intermediate level; or

(C) Wholesale level.

(ii) The person named in the order shall not recall a device from individuals; and

(iii) The person named in the order shall not recall a device from device user facilities if FDA notifies the person not to do so because of a risk determination under § 810.13(c)(2).

(2) The person named in a recall order shall ensure that the strategy provides for notice to individuals subject to the risks associated with use of the recalled device. The notice may be provided through the individuals' health professionals if FDA determines that such consultation is appropriate and would be the most effective method of notifying patients.

(3) Effectiveness checks by the person named in the order are required to verify that all health professionals, device user facilities, consignees, and individuals, as appropriate, have been notified of the cease distribution and notification order or mandatory recall order and of the need to take appropriate action. The person named in the cease distribution and notification order or the mandatory recall order shall specify in the strategy the method(s) to be used in addition to written communications as required by § 810.15, i.e., personal visits, telephone calls, or a combination thereof to contact all health professionals, device user facilities, consignees, and individuals, as appropriate. The agency may conduct additional audit checks where appropriate.

§ 810.15 Communications concerning a cease distribution and notification or mandatory recall order.

(a) *General.* The person named in a cease distribution and notification order issued under § 810.10 or a mandatory recall order issued under § 810.13 is responsible for promptly notifying each health professional, device user facility, consignee, or individual, as appropriate, of the order. In accordance with § 810.10(c) or § 810.13(b)(4), FDA may provide the person named in the cease distribution and notification or mandatory recall order with a model letter for notifying each health professional, device user facility, consignee, or individual, as appropriate, of the order. However, if FDA does not provide the person named in the cease distribution and notification or mandatory recall order with a model letter, the person named in a cease distribution and notification order issued under § 810.10, or a mandatory recall order issued under § 810.13, is responsible for providing such notification. The purpose of the communication is to convey:

(1) That FDA has found that there is a reasonable probability that use of the device would cause a serious, adverse health consequence or death;

(2) That the person named in the order has ceased distribution of the device;

(3) That health professionals and device user facilities should cease use of the device immediately;

(4) Where appropriate, that the device is subject to a mandatory recall order; and

(5) Specific instructions on what should be done with the device.

(b) *Implementation.* The person named in a cease distribution and notification order, or a mandatory recall order, shall notify the appropriate person(s) of the order by verified written communication, e.g., telegram, mailgram, or fax. The written communication and any envelope in which it is sent or enclosed shall be conspicuously marked, preferably in bold red ink: "URGENT—[DEVICE CEASE DISTRIBUTION AND NOTIFICATION ORDER] or [MANDATORY DEVICE RECALL ORDER]." Telephone calls or other personal contacts may be made in addition to, but not as a substitute for, the verified written communication, and shall be documented in an appropriate manner.

(c) *Contents.* The person named in the order shall ensure that the notice of a cease distribution and notification order or mandatory recall order:

(1) Is brief and to the point;

(2) Identifies clearly the device, size, lot number(s), code(s), or serial number(s), and any other pertinent descriptive information to facilitate accurate and immediate identification of the device;

(3) Explains concisely the serious, adverse health consequences that may occur if use of the device were continued;

(4) Provides specific instructions on what should be done with the device;

(5) Provides a ready means for the recipient of the communication to confirm receipt of the communication and to notify the person named in the order of the actions taken in response to the communication. Such means may include, but are not limited to, the return of a postage-paid, self-addressed post card or a toll-free call to the person named in the order; and

(6) Does not contain irrelevant qualifications, promotional materials, or any other statement that may detract from the message.

(d) *Followup communications.* The person named in the cease distribution and notification order or mandatory recall order shall ensure that followup communications are sent to all who fail to respond to the initial communication.

(e) *Responsibility of the recipient.* Health professionals, device user facilities, and consignees who receive a communication concerning a cease distribution and notification order or a mandatory recall order should immediately follow the instructions set forth in the communication. Where appropriate, these recipients should immediately notify their consignees of the order in accordance with paragraphs (b) and (c) of this section.

§ 810.16 Cease distribution and notification or mandatory recall order status reports.

(a) The person named in a cease distribution and notification order issued under § 810.10 or a mandatory recall order issued under § 810.13 shall submit periodic status reports to FDA to enable the agency to assess the person's progress in complying with the order. The frequency of such reports and the agency official to whom such reports shall be submitted will be specified in the order.

(b) Unless otherwise specified in the order, each status report shall contain the following information:

(1) The number and type of health professionals, device user facilities, consignees, or individuals notified about the order and the date and method of notification;

(2) The number and type of health professionals, device user facilities, consignees, or individuals who have responded to the communication and the quantity of the device on hand at these locations at the time they received the communication;

(3) The number and type of health professionals, device user facilities, consignees, or individuals who have not responded to the communication;

(4) The number of devices returned or corrected by each health professional, device user facility, consignee, or individual contacted, and the quantity of products accounted for;

(5) The number and results of effectiveness checks that have been made; and

(6) Estimated timeframes for completion of the requirements of the cease distribution and notification order or mandatory recall order.

(c) The person named in the cease distribution and notification order or recall order may discontinue the submission of status reports when the agency terminates the order in accordance with § 810.17.

§ 810.17 Termination of a cease distribution and notification or mandatory recall order.

(a) The person named in a cease distribution and notification order issued under § 810.10 or a mandatory recall order issued under § 810.13 may request termination of the order by submitting a written request to FDA. The person submitting a request shall certify that he or she has complied in full with all of the requirements of the order and shall include a copy of the most current status report submitted to the agency under § 810.16. A request for termination of a recall order shall include a description of the disposition of the recalled device.

(b) FDA may terminate a cease distribution and notification order issued under § 810.10 or a mandatory recall order issued under § 810.13 when the agency determines that the person named in the order:

(1) Has taken all reasonable efforts to ensure and to verify that all health professionals, device user facilities, consignees, and, where appropriate, individuals have been notified of the cease distribution and notification order, and to verify that they have been instructed to cease use of the device and to take other appropriate action; or

(2) Has removed the device from the market or has corrected the device so that use of the device would not cause serious, adverse health consequences or death.

(c) FDA will provide written notification to the person named in the order when a request for termination of a cease distribution and notification order or a mandatory recall order has been granted or denied. FDA will respond to a written request for termination of a cease distribution and notification or recall order within 30 working days of its receipt.

§810.18 Public notice.

The agency will make available to the public in the weekly FDA Enforcement Report a descriptive listing of each new mandatory recall issued under §810.13. The agency will delay public notification of orders when the agency determines that such notification may cause unnecessary and harmful anxiety in individuals and that initial consultation between individuals and their health professionals is essential.

PART 812—INVESTIGATIONAL DEVICE EXEMPTIONS

Subpart A—General Provisions

Subpart B—Application and Administrative Action

Subpart C—Responsibilities of Sponsors

Subpart D—IRB Review and Approval

Subpart E—Responsibilities of Investigators

Subpart F [Reserved]

Subpart G—Records and Reports

AUTHORITY: 21 U.S.C. 331, 351, 352, 353, 355, 360, 360c–360f, 360h–360j, 371, 372, 374, 379e, 381, 382, 383; 42 U.S.C. 216, 241, 262, 263b–263n.

SOURCE: 45 FR 3751, Jan. 18, 1980, unless otherwise noted.

Subpart A—General Provisions

§812.1 Scope.

(a) The purpose of this part is to encourage, to the extent consistent with the protection of public health and safety and with ethical standards, the discovery and development of useful devices intended for human use, and to that end to maintain optimum freedom for scientific investigators in their pursuit of this purpose. This part provides procedures for the conduct of clinical investigations of devices. An approved investigational device exemption (IDE) permits a device that otherwise would be required to comply with a performance standard or to have premarket approval to be shipped lawfully for the purpose of conducting investigations of that device. An IDE approved under §812.30 or considered approved under §812.2(b) exempts a device from the requirements of the following sections of the Federal Food, Drug, and Cosmetic Act (the act) and regulations issued thereunder: Misbranding under section 502 of the act, registration, listing, and premarket notification under section 510, performance standards under section 514, premarket approval under section 515, a banned device regulation under section 516, records and reports under section 519, restricted device requirements under section 520(e), good manufacturing practice requirements under section 520(f) except for the requirements found in § 820.30, if applicable (unless the sponsor states an intention to comply with these requirements under § 812.20(b)(3) or § 812.140(b)(4)(v)) and color additive requirements under section 721.

(b) References in this part to regulatory sections of the Code of Federal Regulations are to chapter I of title 21, unless otherwise noted.

[45 FR 3751, Jan. 18, 1980, as amended at 59 FR 14366, Mar. 28, 1994; 61 FR 52654, Oct. 7, 1996]

§ 812.2 Applicability.

(a) *General.* This part applies to all clinical investigations of devices to determine safety and effectiveness, except as provided in paragraph (c) of this section.

(b) *Abbreviated requirements.* The following categories of investigations are considered to have approved applications for IDE's, unless FDA has notified a sponsor under § 812.20(a) that approval of an application is required:

(1) An investigation of a device other than a significant risk device, if the device is not a banned device and the sponsor:

(i) Labels the device in accordance with § 812.5;

(ii) Obtains IRB approval of the investigation after presenting the reviewing IRB with a brief explanation of why the device is not a significant risk device, and maintains such approval;

(iii) Ensures that each investigator participating in an investigation of the device obtains from each subject under the investigator's care, informed consent under part 50 and documents it, unless documentation is waived by an IRB under § 56.109(c).

(iv) Complies with the requirements of § 812.46 with respect to monitoring investigations;

(v) Maintains the records required under § 812.140(b) (4) and (5) and makes the reports required under § 812.150(b) (1) through (3) and (5) through (10);

(vi) Ensures that participating investigators maintain the records required by § 812.140(a)(3)(i) and make the reports required under § 812.150(a) (1), (2), (5), and (7); and

(vii) Complies with the prohibitions in § 812.7 against promotion and other practices.

(2) An investigation of a device other than one subject to paragraph (e) of this section, if the investigation was begun on or before July 16, 1980, and to be completed, and is completed, on or before January 19, 1981.

(c) *Exempted investigations.* This part, with the exception of § 812.119, does not apply to investigations of the following categories of devices:

(1) A device, other than a transitional device, in commercial distribution immediately before May 28, 1976, when used or investigated in accordance with the indications in labeling in effect at that time.

(2) A device, other than a transitional device, introduced into commercial distribution on or after May 28, 1976, that FDA has determined to be substantially equivalent to a device in commercial distribution immediately before May 28, 1976, and that is used or investigated in accordance with the indications in the labeling FDA reviewed under subpart E of part 807 in determining substantial equivalence.

(3) A diagnostic device, if the sponsor complies with applicable requirements in § 809.10(c) and if the testing:

(i) Is noninvasive,

(ii) Does not require an invasive sampling procedure that presents significant risk,

(iii) Does not by design or intention introduce energy into a subject, and

(iv) Is not used as a diagnostic procedure without confirmation of the diagnosis by another, medically established diagnostic product or procedure.

(4) A device undergoing consumer preference testing, testing of a modification, or testing of a combination of two or more devices in commercial distribution, if the testing is not for the purpose of determining safety or effectiveness and does not put subjects at risk.

(5) A device intended solely for veterinary use.

(6) A device shipped solely for research on or with laboratory animals and labeled in accordance with § 812.5(c).

(7) A custom device as defined in §812.3(b), unless the device is being used to determine safety or effectiveness for commercial distribution.

(d) *Limit on certain exemptions.* In the case of class II or class III device described in paragraph (c)(1) or (2) of this section, this part applies beginning on the date stipulated in an FDA regulation or order that calls for the submission of premarket approval applications for an unapproved class III device, or establishes a performance standard for a class II device.

(e) *Investigations subject to IND's.* A sponsor that, on July 16, 1980, has an effective investigational new drug application (IND) for an investigation of a device shall continue to comply with the requirements of part 312 until 90 days after that date. To continue the investigation after that date, a sponsor shall comply with paragraph (b)(1) of this section, if the device is not a significant risk device, or shall have obtained FDA approval under §812.30 of an IDE application for the investigation of the device.

[45 FR 3751, Jan. 18, 1980, as amended at 46 FR 8956, Jan. 27, 1981; 46 FR 14340, Feb. 27, 1981; 53 FR 11252, Apr. 6, 1988; 62 FR 4165, Jan, 29, 1997; 62 FR 12096, Mar. 14, 1997]

§812.3 Definitions.

(a) *Act* means the Federal Food, Drug, and Cosmetic Act (sections 201–901, 52 Stat. 1040 *et seq.*, as amended (21 U.S.C. 301–392)).

(b) *Custom device* means a device that:

(1) Necessarily deviates from devices generally available or from an applicable performance standard or premarket approval requirement in order to comply with the order of an individual physician or dentist;

(2) Is not generally available to, or generally used by, other physicians or dentists;

(3) Is not generally available in finished form for purchase or for dispensing upon prescription;

(4) Is not offered for commercial distribution through labeling or advertising; and

(5) Is intended for use by an individual patient named in the order of a physician or dentist, and is to be made in a specific form for that patient, or is intended to meet the special needs of the physician or dentist in the course of professional practice.

(c) *FDA* means the Food and Drug Administration.

(d) *Implant* means a device that is placed into a surgically or naturally formed cavity of the human body if it is intended to remain there for a period of 30 days or more. FDA may, in order to protect public health, determine that devices placed in subjects for shorter periods are also "implants" for purposes of this part.

(e) *Institution* means a person, other than an individual, who engages in the conduct of research on subjects or in the delivery of medical services to individuals as a primary activity or as an adjunct to providing residential or custodial care to humans. The term includes, for example, a hospital, retirement home, confinement facility, academic establishment, and device manufacturer. The term has the same meaning as "facility" in section 520(g) of the act.

(f) *Institutional review board* (IRB) means any board, committee, or other group formally designated by an institution to review biomedical research involving subjects and established, operated, and functioning in conformance with part 56. The term has the same meaning as "institutional review committee" in section 520(g) of the act.

(g) *Investigational device* means a device, including a transitional device, that is the object of an investigation.

(h) *Investigation* means a clinical investigation or research involving one or more subjects to determine the safety or effectiveness of a device.

(i) *Investigator* means an individual who actually conducts a clinical investigation, i.e., under whose immediate direction the test article is administered or dispensed to, or used involving, a subject, or, in the event of an investigation conducted by a team of individuals, is the responsible leader of that team.

(j) *Monitor*, when used as a noun, means an individual designated by a sponsor or contract research organization to oversee the progress of an investigation. The monitor may be an employee of a sponsor or a consultant to the sponsor, or an employee of or consultant to a contract research organization. *Monitor*, when used as a verb, means to oversee an investigation.

(k) *Noninvasive*, when applied to a diagnostic device or procedure, means one that does not by design or intention: (1) Penetrate or pierce the skin or mucous membranes of the body, the ocular cavity, or the urethra, or (2) enter the ear beyond the external auditory canal, the nose beyond the nares, the mouth beyond the pharynx, the anal canal beyond the rectum, or the vagina beyond the cervical os. For purposes of this part, blood sampling that involves simple venipuncture is considered noninvasive, and the use of surplus samples of body fluids or tissues that are left over from samples taken for noninvestigational purposes is also considered noninvasive.

(l) *Person* includes any individual, partnership, corporation, association, scientific or academic establishment, Government agency or organizational unit of a Government agency, and any other legal entity.

(m) *Significant risk device* means an investigational device that:

(1) Is intended as an implant and presents a potential for serious risk to the health, safety, or welfare of a subject;

(2) Is purported or represented to be for a use in supporting or sustaining human life and presents a potential for serious risk to the health, safety, or welfare of a subject;

(3) Is for a use of substantial importance in diagnosing, curing, mitigating, or treating disease, or otherwise preventing impairment of human health and presents a potential for serious risk to the health, safety, or welfare of a subject; or

(4) Otherwise presents a potential for serious risk to the health, safety, or welfare of a subject.

(n) *Sponsor* means a person who initiates, but who does not actually conduct, the investigation, that is, the investigational device is administered, dispensed, or used under the immediate direction of another individual. A person other than an individual that uses one or more of its own employees to conduct an investigation that it has initiated is a sponsor, not a sponsor-investigator, and the employees are investigators.

(o) *Sponsor-investigator* means an individual who both initiates and actually conducts, alone or with others, an investigation, that is, under whose immediate direction the investigational device is administered, dispensed, or used. The term does not include any person other than an individual. The obligations of a sponsor-investigator under this part include those of an investigator and those of a sponsor.

(p) *Subject* means a human who participates in an investigation, either as an individual on whom or on whose specimen an investigational device is used or as a control. A subject may be in normal health or may have a medical condition or disease.

(q) *Termination* means a discontinuance, by sponsor or by withdrawal of IRB or FDA approval, of an investigation before completion.

(r) *Transitional device* means a device subject to section 520(l) of the act, that is, a device that FDA considered to be a new drug or an antibiotic drug before May 28, 1976.

(s) *Unanticipated adverse device effect* means any serious adverse effect on health or safety or any life-threatening problem or death caused by, or associated with, a device, if that effect, problem, or death was not previously identified in nature, severity, or degree of incidence in the investigational plan or application (including a supplementary plan or application), or any other unanticipated serious problem associated with a device that relates to the rights, safety, or welfare of subjects.

[45 FR 3751, Jan. 18, 1980, as amended at 46 FR 8956, Jan. 27, 1981; 48 FR 15622, Apr. 12, 1983]

§812.5 Labeling of investigational devices.

(a) *Contents*. An investigational device or its immediate package shall bear a label with the following information: the name and place of business of the manufacturer, packer, or distributor (in accordance with §801.1), the quantity of contents, if appropriate, and the following statement: "CAUTION—Investigational device. Limited by Federal (or United States) law to investigational use." The label or other labeling shall describe all relevant contraindications, hazards, adverse effects, interfering substances or devices, warnings, and precautions.

(b) *Prohibitions.* The labeling of an investigational device shall not bear any statement that is false or misleading in any particular and shall not represent that the device is safe or effective for the purposes for which it is being investigated.

(c) *Animal research.* An investigational device shipped solely for research on or with laboratory animals shall bear on its label the following statement: "CAUTION—Device for investigational use in laboratory animals or other tests that do not involve human subjects."

[45 FR 3751, Jan. 18, 1980, as amended at 45 FR 58842, Sept. 5, 1980]

§ 812.7 Prohibition of promotion and other practices.

A sponsor, investigator, or any person acting for or on behalf of a sponsor or investigator shall not:

(a) Promote or test market an investigational device, until after FDA has approved the device for commercial distribution.

(b) Commercialize an investigational device by charging the subjects or investigators for a device a price larger than that necessary to recover costs of manufacture, research, development, and handling.

(c) Unduly prolong an investigation. If data developed by the investigation indicate in the case of a class III device that premarket approval cannot be justified or in the case of a class II device that it will not comply with an applicable performance standard or an amendment to that standard, the sponsor shall promptly terminate the investigation.

(d) Represent that an investigational device is safe or effective for the purposes for which it is being investigated.

§ 812.10 Waivers.

(a) *Request.* A sponsor may request FDA to waive any requirement of this part. A waiver request, with supporting documentation, may be submitted separately or as part of an application to the address in § 812.19.

(b) *FDA action.* FDA may by letter grant a waiver of any requirement that FDA finds is not required by the act and is unnecessary to protect the rights, safety, or welfare of human subjects.

(c) *Effect of request.* Any requirement shall continue to apply unless and until FDA waives it.

§ 812.18 Import and export requirements.

(a) *Imports.* In addition to complying with other requirements of this part, a person who imports or offers for importation an investigational device subject to this part shall be the agent of the foreign exporter with respect to investigations of the device and shall act as the sponsor of the clinical investigation, or ensure that another person acts as the agent of the foreign exporter and the sponsor of the investigation.

(b) *Exports.* A person exporting an investigational device subject to this part shall obtain FDA's prior approval, as required by section 801(e) of the act or comply with section 802 of the act.

[45 FR 3751, Jan. 18, 1980, as amended at 62 FR 26229, May 13, 1997]

§ 812.19 Address for IDE correspondence.

If you are sending an application, supplemental application, report, request for waiver, request for import or export approval, or other correspondence relating to matters covered by this part, you must address it to the Center for Devices and Radiological Health, Document Mail Center (HFZ–401), Food and Drug Administration, 9200 Corporate Blvd., Rockville, MD 20850. You must state on the outside wrapper of each submission what the submission is, for example, an "IDE application," a "supplemental IDE application," or a "correspondence concerning an IDE (or an IDE application)."

[65 FR 17137, Mar. 31, 2000]

Subpart B—Application and Administrative Action

§812.20 Application.

(a) *Submission.* (1) A sponsor shall submit an application to FDA if the sponsor intends to use a significant risk device in an investigation, intends to conduct an investigation that involves an exception from informed consent under §50.24 of this chapter, or if FDA notifies the sponsor that an application is required for an investigation.

(2) A sponsor shall not begin an investigation for which FDA's approval of an application is required until FDA has approved the application.

(3) A sponsor shall submit three copies of a signed "Application for an Investigational Device Exemption" (IDE application), together with accompanying materials, by registered mail or by hand to the address in §812.19. Subsequent correspondence concerning an application or a supplemental application shall be submitted by registered mail or by hand.

(4)(i) A sponsor shall submit a separate IDE for any clinical investigation involving an exception from informed consent under §50.24 of this chapter. Such a clinical investigation is not permitted to proceed without the prior written authorization of FDA. FDA shall provide a written determination 30 days after FDA receives the IDE or earlier.

(ii) If the investigation involves an exception from informed consent under §50.24 of this chapter, the sponsor shall prominently identify on the cover sheet that the investigation is subject to the requirements in §50.24 of this chapter.

(b) *Contents.* An IDE application shall include, in the following order:

(1) The name and address of the sponsor.

(2) A complete report of prior investigations of the device and an accurate summary of those sections of the investigational plan described in §812.25(a) through (e) or, in lieu of the summary, the complete plan. The sponsor shall submit to FDA a complete investigational plan and a complete report of prior investigations of the device if no IRB has reviewed them, if FDA has found an IRB's review inadequate, or if FDA requests them.

(3) A description of the methods, facilities, and controls used for the manufacture, processing, packing, storage, and, where appropriate, installation of the device, in sufficient detail so that a person generally familiar with good manufacturing practices can make a knowledgeable judgment about the quality control used in the manufacture of the device.

(4) An example of the agreements to be entered into by all investigators to comply with investigator obligations under this part, and a list of the names and addresses of all investigators who have signed the agreement.

(5) A certification that all investigators who will participate in the investigation have signed the agreement, that the list of investigators includes all the investigators participating in the investigation, and that no investigators will be added to the investigation until they have signed the agreement.

(6) A list of the name, address, and chairperson of each IRB that has been or will be asked to review the investigation and a certification of the action concerning the investigation taken by each such IRB.

(7) The name and address of any institution at which a part of the investigation may be conducted that has not been identified in accordance with paragraph (b)(6) of this section.

(8) If the device is to be sold, the amount to be charged and an explanation of why sale does not constitute commercialization of the device.

(9) A claim for categorical exclusion under §25.30 or 25.34 or an environmental assessment under §25.40.

(10) Copies of all labeling for the device.

(11) Copies of all forms and informational materials to be provided to subjects to obtain informed consent.

(12) Any other relevant information FDA requests for review of the application.

(c) *Additional information.* FDA may request additional information concerning an investigation or revision in the investigational plan. The sponsor may treat such a request as a disapproval of the application for purposes of requesting a hearing under part 16.

(d) *Information previously submitted.* Information previously submitted to the Center for Devices and Radiological Health in accordance with this chapter ordinarily need not be resubmitted, but may be incorporated by reference.

[45 FR 3751, Jan. 18, 1980, as amended at 46 FR 8956, Jan. 27, 1981; 50 FR 16669, Apr. 26, 1985; 53 FR 11252, Apr. 6, 1988; 61 FR 51530, Oct. 2, 1996; 62 FR 40600, July 29, 1997; 64 FR 10942, Mar. 8, 1999]

§ 812.25 Investigational plan.

The investigational plan shall include, in the following order:

(a) *Purpose.* The name and intended use of the device and the objectives and duration of the investigation.

(b) *Protocol.* A written protocol describing the methodology to be used and an analysis of the protocol demonstrating that the investigation is scientifically sound.

(c) *Risk analysis.* A description and analysis of all increased risks to which subjects will be exposed by the investigation; the manner in which these risks will be minimized; a justification for the investigation; and a description of the patient population, including the number, age, sex, and condition.

(d) *Description of device.* A description of each important component, ingredient, property, and principle of operation of the device and of each anticipated change in the device during the course of the investigation.

(e) *Monitoring procedures.* The sponsor's written procedures for monitoring the investigation and the name and address of any monitor.

(f) *Labeling.* Copies of all labeling for the device.

(g) *Consent materials.* Copies of all forms and informational materials to be provided to subjects to obtain informed consent.

(h) *IRB information.* A list of the names, locations, and chairpersons of all IRB's that have been or will be asked to review the investigation, and a certification of any action taken by any of those IRB's with respect to the investigation.

(i) *Other institutions.* The name and address of each institution at which a part of the investigation may be conducted that has not been identified in paragraph (h) of this section.

(j) *Additional records and reports.* A description of records and reports that will be maintained on the investigation in addition to those prescribed in subpart G.

§ 812.27 Report of prior investigations.

(a) *General.* The report of prior investigations shall include reports of all prior clinical, animal, and laboratory testing of the device and shall be comprehensive and adequate to justify the proposed investigation.

(b) *Specific contents.* The report also shall include:

(1) A bibliography of all publications, whether adverse or supportive, that are relevant to an evaluation of the safety or effectiveness of the device, copies of all published and unpublished adverse information, and, if requested by an IRB or FDA, copies of other significant publications.

(2) A summary of all other unpublished information (whether adverse or supportive) in the possession of, or reasonably obtainable by, the sponsor that is relevant to an evaluation of the safety or effectiveness of the device.

(3) If information on nonclinical laboratory studies is provided, a statement that all such studies have been conducted in compliance with applicable requirements in the good laboratory practice regulations in part 58, or if any such study was not conducted in compliance with such regulations, a brief statement of the reason for the noncompliance. Failure or inability to comply with this requirement does not justify failure to provide information on a relevant nonclinical test study.

[45 FR 3751, Jan. 18, 1980, as amended at 50 FR 7518, Feb. 22, 1985]

§ 812.30 FDA action on applications.

(a) *Approval or disapproval.* FDA will notify the sponsor in writing of the date it receives an application. FDA may approve an investigation as proposed, approve it with modifications, or disapprove it. An investigation may not begin until:

(1) Thirty days after FDA receives the application at the address in § 812.19 for the investigation of a device other than a banned device, unless FDA notifies the sponsor that the investigation may not begin; or

(2) FDA approves, by order, an IDE for the investigation.

(b) *Grounds for disapproval or withdrawal.* FDA may disapprove or withdraw approval of an application if FDA finds that:

(1) There has been a failure to comply with any requirement of this part or the act, any other applicable regulation or statute, or any condition of approval imposed by an IRB or FDA.

(2) The application or a report contains an untrue statement of a material fact, or omits material information required by this part.

(3) The sponsor fails to respond to a request for additional information within the time prescribed by FDA.

(4) There is reason to believe that the risks to the subjects are not outweighed by the anticipated benefits to the subjects and the importance of the knowledge to be gained, or informed consent is inadquate, or the investigation is scientifically unsound, or there is reason to believe that the device as used is ineffective.

(5) It is otherwise unreasonable to begin or to continue the investigation owing to the way in which the device is used or the inadequacy of:

(i) The report of prior investigations or the investigational plan;

(ii) The methods, facilities, and controls used for the manufacturing, processing, packaging, storage, and, where appropriate, installation of the device; or

(iii) Monitoring and review of the investigation.

(c) *Notice of disapproval or withdrawal.* If FDA disapproves an application or proposes to withdraw approval of an application, FDA will notify the sponsor in writing.

(1) A disapproval order will contain a complete statement of the reasons for disapproval and a statement that the sponsor has an opportunity to request a hearing under part 16.

(2) A notice of a proposed withdrawal of approval will contain a complete statement of the reasons for withdrawal and a statement that the sponsor has an opportunity to request a hearing under part 16. FDA will provide the opportunity for hearing before withdrawal of approval, unless FDA determines in the notice that continuation of testing under the exemption will result in an unreasonble risk to the public health and orders withdrawal of approval before any hearing.

[45 FR 3751, Jan. 18, 1980, as amended at 45 FR 58842, Sept. 5, 1980]

§ 812.35 Supplemental applications.

(a) *Changes in investigational plan*—(1) *Changes requiring prior approval.* Except as described in paragraphs (a)(2) through (a)(4) of this section, a sponsor must obtain approval of a supplemental application under § 812.30(a), and IRB approval when appropriate (see §§ 56.110 and 56.111 of this chapter), prior to implementing a change to an investigational plan. If a sponsor intends to conduct an investigation that involves an exception to informed consent under § 50.24 of this chapter, the sponsor shall submit a separate investigational device exemption (IDE) application in accordance with § 812.20(a).

(2) *Changes effected for emergency use.* The requirements of paragraph (a)(1) of this section regarding FDA approval of a supplement do not apply in the case of a deviation from the investigational plan to protect the life or physical well-being of a subject in an emergency. Such deviation shall be reported to FDA within 5-working days after the sponsor learns of it (see § 812.150(a)(4)).

(3) *Changes effected with notice to FDA within 5 days.* A sponsor may make certain changes without prior approval of a supplemental application under paragraph (a)(1) of this section if the sponsor determines that these changes meet the criteria described in paragraphs (a)(3)(i) and (a)(3)(ii) of this section, on the basis of credible information defined in paragraph (a)(3)(iii) of this section, and the sponsor provides notice to FDA within 5-working days of making these changes.

(i) *Developmental changes.* The requirements in paragraph (a)(1) of this section regarding FDA approval of a supplement do not apply to developmental changes in the device (including manufacturing changes) that do not constitute a significant change in design or basic principles of operation and that are made in response to information gathered during the course of an investigation.

(ii) *Changes to clinical protocol.* The requirements in paragraph (a)(1) of this section regarding FDA approval of a supplement do not apply to changes to clinical protocols that do not affect:

(A) The validity of the data or information resulting from the completion of the approved protocol, or the relationship of likely patient risk to benefit relied upon to approve the protocol;

(B) The scientific soundness of the investigational plan; or

(C) The rights, safety, or welfare of the human subjects involved in the investigation.

(iii) *Definition of credible information.* (A) Credible information to support developmental changes in the device (including manufacturing changes) includes data generated under the design control procedures of § 820.30, preclinical/animal testing, peer reviewed published literature, or other reliable information such as clinical information gathered during a trial or marketing.

(B) Credible information to support changes to clinical protocols is defined as the sponsor's documentation supporting the conclusion that a change does not have a significant impact on the study design or planned statistical analysis, and that the change does not affect the rights, safety, or welfare of the subjects. Documentation shall include information such as peer reviewed published literature, the recommendation of the clinical investigator(s), and/or the data gathered during the clinical trial or marketing.

(iv) *Notice of IDE change.* Changes meeting the criteria in paragraphs (a)(3)(i) and (a)(3)(ii) of this section that are supported by credible information as defined in paragraph (a)(3)(iii) of this section may be made without prior FDA approval if the sponsor submits a notice of the change to the IDE not later than 5-working days after making the change. Changes to devices are deemed to occur on the date the device, manufactured incorporating the design or manufacturing change, is distributed to the investigator(s). Changes to a clinical protocol are deemed to occur when a clinical investigator is notified by the sponsor that the change should be implemented in the protocol or, for sponsor-investigator studies, when a sponsor-investigator incorporates the change in the protocol. Such notices shall be identified as a "notice of IDE change."

(A) For a developmental or manufacturing change to the device, the notice shall include a summary of the relevant information gathered during the course of the investigation upon which the change was based; a description of the change to the device or manufacturing process (cross-referenced to the appropriate sections of the original device description or manufacturing process); and, if design controls were used to assess the change, a statement that no new risks were identified by appropriate risk analysis and that the verification and validation testing, as appropriate, demonstrated that the design outputs met the design input requirements. If another method of assessment was used, the notice shall include a summary of the information which served as the credible information supporting the change.

(B) For a protocol change, the notice shall include a description of the change (cross-referenced to the appropriate sections of the original protocol); an assessment supporting the conclusion that the change does not have a significant impact on the study design or planned statistical analysis; and a summary of the information that served as the credible information supporting the sponsor's determination that the change does not affect the rights, safety, or welfare of the subjects.

(4) *Changes submitted in annual report.* The requirements of paragraph (a)(1) of this section do not apply to minor changes to the purpose of the study, risk analysis, monitoring procedures, labeling, informed consent materials, and IRB information that do not affect:

(i) The validity of the data or information resulting from the completion of the approved protocol, or the relationship of likely patient risk to benefit relied upon to approve the protocol;

(ii) The scientific soundness of the investigational plan; or

(iii) The rights, safety, or welfare of the human subjects involved in the investigation. Such changes shall be reported in the annual progress report for the IDE, under §812.150(b)(5).

(b) *IRB approval for new facilities.* A sponsor shall submit to FDA a certification of any IRB approval of an investigation or a part of an investigation not included in the IDE application. If the investigation is otherwise unchanged, the supplemental application shall consist of an updating of the information required by §812.20(b) and (c) and a description of any modifications in the investigational plan required by the IRB as a condition of approval. A certification of IRB approval need not be included in the initial submission of the supplemental application, and such certification is not a precondition for agency consideration of the application. Nevertheless, a sponsor may not begin a part of an investigation at a facility until the IRB has approved the investigation, FDA has received the certification of IRB approval, and FDA, under §812.30(a), has approved the supplemental application relating to that part of the investigation (see §56.103(a)).

[50 FR 25909, June 24, 1985; 50 FR 28932, July 17, 1985, as amended at 61 FR 51531, Oct. 2, 1996; 63 FR 64625, Nov. 23, 1998]

§812.36 Treatment use of an investigational device.

(a) *General.* A device that is not approved for marketing may be under clinical investigation for a serious or immediately life-threatening disease or condition in patients for whom no comparable or satisfactory alternative device or other therapy is available. During the clinical trial or prior to final action on the marketing application, it may be appropriate to use the device in the treatment of patients not in the trial under the provisions of a treatment investigational device exemption (IDE). The purpose of this section is to facilitate the availability of promising new devices to desperately ill patients as early in the device development process as possible, before general marketing begins, and to obtain additional data on the device's safety and effectiveness. In the case of a serious disease, a device ordinarily may be made available for treatment use under this section after all clinical trials have been completed. In the case of an immediately life-threatening disease, a device may be made available for treatment use under this section prior to the completion of all clinical trials. For the purpose of this section, an "immediately life-threatening" disease means a stage of a disease in which there is a reasonable likelihood that death will occur within a matter of months or in which premature death is likely without early treatment. For purposes of this section, "treatment use"of a device includes the use of a device for diagnostic purposes.

(b) *Criteria.* FDA shall consider the use of an investigational device under a treatment IDE if:

(1) The device is intended to treat or diagnose a serious or immediately life-threatening disease or condition;

(2) There is no comparable or satisfactory alternative device or other therapy available to treat or diagnose that stage of the disease or condition in the intended patient population;

(3) The device is under investigation in a controlled clinical trial for the same use under an approved IDE, or such clinical trials have been completed; and

(4) The sponsor of the investigation is actively pursuing marketing approval/clearance of the investigational device with due diligence.

(c) *Applications for treatment use.* (1) A treatment IDE application shall include, in the following order:

(i) The name, address, and telephone number of the sponsor of the treatment IDE;

(ii) The intended use of the device, the criteria for patient selection, and a written protocol describing the treatment use;

(iii) An explanation of the rationale for use of the device, including, as appropriate, either a list of the available regimens that ordinarily should be tried before using the investigational device or an explanation of why the use of the investigational device is preferable to the use of available marketed treatments;

(iv) A description of clinical procedures, laboratory tests, or other measures that will be used to evaluate the effects of the device and to minimize risk;

(v) Written procedures for monitoring the treatment use and the name and address of the monitor;

(vi) Instructions for use for the device and all other labeling as required under § 812.5(a) and (b);

(vii) Information that is relevant to the safety and effectiveness of the device for the intended treatment use. Information from other IDE's may be incorporated by reference to support the treatment use;

(viii) A statement of the sponsor's commitment to meet all applicable responsibilities under this part and part 56 of this chapter and to ensure compliance of all participating investigators with the informed consent requirements of part 50 of this chapter;

(ix) An example of the agreement to be signed by all investigators participating in the treatment IDE and certification that no investigator will be added to the treatment IDE before the agreement is signed; and

(x) If the device is to be sold, the price to be charged and a statement indicating that the price is based on manufacturing and handling costs only.

(2) A licensed practitioner who receives an investigational device for treatment use under a treatment IDE is an "investigator" under the IDE and is responsible for meeting all applicable investigator responsibilities under this part and parts 50 and 56 of this chapter.

(d) *FDA action on treatment IDE applications*—(1) *Approval of treatment IDE's.* Treatment use may begin 30 days after FDA receives the treatment IDE submission at the address specified in § 812.19, unless FDA notifies the sponsor in writing earlier than the 30 days that the treatment use may or may not begin. FDA may approve the treatment use as proposed or approve it with modifications.

(2) *Disapproval or withdrawal of approval of treatment IDE's.* FDA may disapprove or withdraw approval of a treatment IDE if:

(i) The criteria specified in § 812.36(b) are not met or the treatment IDE does not contain the information required in § 812.36(c);

(ii) FDA determines that any of the grounds for disapproval or withdrawal of approval listed in § 812.30(b)(1) through (b)(5) apply;

(iii) The device is intended for a serious disease or condition and there is insufficient evidence of safety and effectiveness to support such use;

(iv) The device is intended for an immediately life-threatening disease or condition and the available scientific evidence, taken as a whole, fails to provide a reasonable basis for concluding that the device:

(A) May be effective for its intended use in its intended population; or

(B) Would not expose the patients to whom the device is to be administered to an unreasonable and significant additional risk of illness or injury;

(v) There is reasonable evidence that the treatment use is impeding enrollment in, or otherwise interfering with the conduct or completion of, a controlled investigation of the same or another investigational device;

(vi) The device has received marketing approval/clearance or a comparable device or therapy becomes available to treat or diagnose the same indication in the same patient population for which the investigational device is being used;

(vii) The sponsor of the controlled clinical trial is not pursuing marketing approval/clearance with due diligence;

(viii) Approval of the IDE for the controlled clinical investigation of the device has been withdrawn; or

(ix) The clinical investigator(s) named in the treatment IDE are not qualified by reason of their scientific training and/or experience to use the investigational device for the intended treatment use.

(3) *Notice of disapproval or withdrawal.* If FDA disapproves or proposes to withdraw approval of a treatment IDE, FDA will follow the procedures set forth in § 812.30(c).

(e) *Safeguards.* Treatment use of an investigational device is conditioned upon the sponsor and investigators complying with the safeguards of the

IDE process and the regulations governing informed consent (part 50 of this chapter) and institutional review boards (part 56 of this chapter).

(f) *Reporting requirements.* The sponsor of a treatment IDE shall submit progress reports on a semi-annual basis to all reviewing IRB's and FDA until the filing of a marketing application. These reports shall be based on the period of time since initial approval of the treatment IDE and shall include the number of patients treated with the device under the treatment IDE, the names of the investigators participating in the treatment IDE, and a brief description of the sponsor's efforts to pursue marketing approval/clearance of the device. Upon filing of a marketing application, progress reports shall be submitted annually in accordance with § 812.150(b)(5). The sponsor of a treatment IDE is responsible for submitting all other reports required under § 812.150.

[62 FR 48947, Sept. 18, 1997]

§ 812.38 Confidentiality of data and information.

(a) *Existence of IDE.* FDA will not disclose the existence of an IDE unless its existence has previously been publicly disclosed or acknowledged, until FDA approves an application for premarket approval of the device subject to the IDE; or a notice of completion of a product development protocol for the device has become effective.

(b) *Availability of summaries or data.* (1) FDA will make publicly available, upon request, a detailed summary of information concerning the safety and effectiveness of the device that was the basis for an order approving, disapproving, or withdrawing approval of an application for an IDE for a banned device. The summary shall include information on any adverse effect on health caused by the device.

(2) If a device is a banned device or if the existence of an IDE has been publicly disclosed or acknowledged, data or information contained in the file is not available for public disclosure before approval of an application for premarket approval or the effective date of a notice of completion of a product development protocol except as provided in this section. FDA may, in its discretion, disclose a summary of selected portions of the safety and effectiveness data, that is, clinical, animal, or laboratory studies and tests of the device, for public consideration of a specific pending issue.

(3) If the existence of an IDE file has not been publicly disclosed or acknowledged, no data or information in the file are available for public disclosure except as provided in paragraphs (b)(1) and (c) of this section.

(4) Notwithstanding paragraph (b)(2) of this section, FDA will make available to the public, upon request, the information in the IDE that was required to be filed in Docket Number 95S–0158 in the Division of Dockets Management (HFA–305), Food and Drug Administration, 5630 Fishers Lane, rm. 1061, Rockville, MD 20852, for investigations involving an exception from informed consent under § 50.24 of this chapter. Persons wishing to request this information shall submit a request under the Freedom of Information Act.

(c) *Reports of adverse effects.* Upon request or on its own initiative, FDA shall disclose to an individual on whom an investigational device has been used a copy of a report of adverse device effects relating to that use.

(d) *Other rules.* Except as otherwise provided in this section, the availability for public disclosure of data and information in an IDE file shall be handled in accordance with § 814.9.

[45 FR 3751, Jan. 18, 1980, as amended at 53 FR 11253, Apr. 6, 1988; 61 FR 51531, Oct. 2, 1996]

Subpart C—Responsibilities of Sponsors

§ 812.40 General responsibilities of sponsors.

Sponsors are responsible for selecting qualified investigators and providing them with the information they need to conduct the investigation properly, ensuring proper monitoring of the investigation, ensuring that IRB review and approval are obtained, submitting an IDE application to FDA, and ensuring that any reviewing IRB and FDA are promptly informed of significant new information about an investigation. Additional responsibilities of sponsors are described in subparts B and G.

§ 812.42 FDA and IRB approval.

A sponsor shall not begin an investigation or part of an investigation until an IRB and FDA have both approved the application or supplemental application relating to the investigation or part of an investigation.

[46 FR 8957, Jan. 27, 1981]

§ 812.43 Selecting investigators and monitors.

(a) *Selecting investigators.* A sponsor shall select investigators qualified by training and experience to investigate the device.

(b) *Control of device.* A sponsor shall ship investigational devices only to qualified investigators participating in the investigation.

(c) *Obtaining agreements.* A sponsor shall obtain from each participating investigator a signed agreement that includes:

(1) The investigator's curriculum vitae.

(2) Where applicable, a statement of the investigator's relevant experience, including the dates, location, extent, and type of experience.

(3) If the investigator was involved in an investigation or other research that was terminated, an explanation of the circumstances that led to termination.

(4) A statement of the investigator's commitment to:

(i) Conduct the investigation in accordance with the agreement, the investigational plan, this part and other applicable FDA regulations, and conditions of approval imposed by the reviewing IRB or FDA;

(ii) Supervise all testing of the device involving human subjects; and

(iii) Ensure that the requirements for obtaining informed consent are met.

(5) Sufficient accurate financial disclosure information to allow the sponsor to submit a complete and accurate certification or disclosure statement as required under part 54 of this chapter. The sponsor shall obtain a commitment from the clinical investigator to promptly update this information if any relevant changes occur during the course of the investigation and for 1 year following completion of the study. This information shall not be submitted in an investigational device exemption application, but shall be submitted in any marketing application involving the device.

(d) *Selecting monitors.* A sponsor shall select monitors qualified by training and experience to monitor the investigational study in accordance with this part and other applicable FDA regulations.

[45 FR 3751, Jan. 18, 1980, as amended at 63 FR 5253, Feb. 2, 1998]

§ 812.45 Informing investigators.

A sponsor shall supply all investigators participating in the investigation with copies of the investigational plan and the report of prior investigations of the device.

§ 812.46 Monitoring investigations.

(a) *Securing compliance.* A sponsor who discovers that an investigator is not complying with the signed agreement, the investigational plan, the requirements of this part or other applicable FDA regulations, or any conditions of approval imposed by the reviewing IRB or FDA shall promptly either secure compliance, or discontinue shipments of the device to the investigator and terminate the investigator's participation in the investigation. A sponsor shall also require such an investigator to dispose of or return the device, unless this action would jeopardize the rights, safety, or welfare of a subject.

(b) *Unanticipated adverse device effects.* (1) A sponsor shall immediately conduct an evaluation of any unanticipated adverse device effect.

(2) A sponsor who determines that an unanticipated adverse device effect presents an unreasonable risk to subjects shall terminate all investigations or parts of investigations presenting that risk as soon as possible. Termination shall occur not later than 5 working days after the sponsor makes this determination and not later than 15 working days after the sponsor first received notice of the effect.

(c) *Resumption of terminated studies.* If the device is a significant risk device, a sponsor may not resume a terminated investigation without IRB and FDA approval. If the device is not a significant risk device, a sponsor may not resume a terminated investigation without IRB approval and, if the investigation was terminated under paragraph (b)(2) of this section, FDA approval.

§812.47 Emergency research under §50.24 of this chapter.

(a) The sponsor shall monitor the progress of all investigations involving an exception from informed consent under §50.24 of this chapter. When the sponsor receives from the IRB information concerning the public disclosures under §50.24(a)(7)(ii) and (a)(7)(iii) of this chapter, the sponsor shall promptly submit to the IDE file and to Docket Number 95S–0158 in the Division of Dockets Management (HFA–305), Food and Drug Administration, 5630 Fishers Lane, rm. 1061, Rockville, MD 20852, copies of the information that was disclosed, identified by the IDE number.

(b) The sponsor also shall monitor such investigations to determine when an IRB determines that it cannot approve the research because it does not meet the criteria in the exception in §50.24(a) of this chapter or because of other relevant ethical concerns. The sponsor promptly shall provide this information in writing to FDA, investigators who are asked to participate in this or a substantially equivalent clinical investigation, and other IRB's that are asked to review this or a substantially equivalent investigation.

[61 FR 51531, Oct. 2, 1996, as amended at 64 FR 10943, Mar. 8, 1999]

Subpart D—IRB Review and Approval

§812.60 IRB composition, duties, and functions.

An IRB reviewing and approving investigations under this part shall comply with the requirements of part 56 in all respects, including its composition, duties, and functions.

[46 FR 8957, Jan. 27, 1981]

§812.62 IRB approval.

(a) An IRB shall review and have authority to approve, require modifications in (to secure approval), or disapprove all investigations covered by this part.

(b) If no IRB exists or if FDA finds that an IRB's review is inadequate, a sponsor may submit an application to FDA.

[46 FR 8957, Jan. 27, 1981]

§812.64 IRB's continuing review.

The IRB shall conduct its continuing review of an investigation in accordance with part 56.

[46 FR 8957, Jan. 27, 1981]

§812.65 [Reserved]

§812.66 Significant risk device determinations.

If an IRB determines that an investigation, presented for approval under §812.2(b)(1)(ii), involves a significant risk device, it shall so notify the investigator and, where appropriate, the sponsor. A sponsor may not begin the investigation except as provided in §812.30(a).

[46 FR 8957, Jan. 27, 1981]

Subpart E—Responsibilities of Investigators

§812.100 General responsibilities of investigators.

An investigator is responsible for ensuring that an investigation is conducted according to the signed agreement, the investigational plan and applicable FDA regulations, for protecting the rights, safety, and welfare of subjects under the investigator's care, and for the control of devices under investigation. An investigator also is responsible for ensuring that informed consent is obtained in accordance with part 50 of this chapter. Additional responsibilities of investigators are described in subpart G.

[45 FR 3751, Jan. 18, 1980, as amended at 46 FR 8957, Jan. 27, 1981]

§812.110 Specific responsibilities of investigators.

(a) *Awaiting approval.* An investigator may determine whether potential subjects would be interested in participating in an investigation, but shall not request the written informed consent of any subject to participate, and shall not allow any subject to participate before obtaining IRB and FDA approval.

(b) *Compliance.* An investigator shall conduct an investigation in accordance with the signed agreement with the sponsor, the investigational plan, this part and other applicable FDA regulations, and any conditions of approval imposed by an IRB or FDA.

(c) *Supervising device use.* An investigator shall permit an investigational device to be used only with subjects under the investigator's supervision. An investigator shall not supply an investigational device to any person not authorized under this part to receive it.

(d) *Financial disclosure.* A clinical investigator shall disclose to the sponsor sufficient accurate financial information to allow the applicant to submit complete and accurate certification or disclosure statements required under part 54 of this chapter. The investigator shall promptly update this information if any relevant changes occur during the course of the investigation and for 1 year following completion of the study.

(e) *Disposing of device.* Upon completion or termination of a clinical investigation or the investigator's part of an investigation, or at the sponsor's request, an investigator shall return to the sponsor any remaining supply of the device or otherwise dispose of the device as the sponsor directs.

[45 FR 3751, Jan. 18, 1980, as amended at 63 FR 5253, Feb. 2, 1998]

§812.119 Disqualification of a clinical investigator.

(a) If FDA has information indicating that an investigator has repeatedly or deliberately failed to comply with the requirements of this part, part 50, or part 56 of this chapter, or has repeatedly or deliberately submitted false information either to the sponsor of the investigation or in any required report, the Center for Devices and Radiological Health will furnish the investigator written notice of the matter under complaint and offer the investigator an opportunity to explain the matter in writing, or, at the option of the investigator, in an informal conference. If an explanation is offered and accepted by the Center for Devices and Radiological Health, the disqualification process will be terminated. If an explanation is offered but not accepted by the Center for Devices and Radiological Health, the investigator will be given an opportunity for a regulatory hearing under part 16 of this chapter on the question of whether the investigator is entitled to receive investigational devices.

(b) After evaluating all available information, including any explanation presented by the investigator, if the Commissioner determines that the investigator has repeatedly or deliberately failed to comply with the requirements of this part, part 50, or part 56 of this chapter, or has deliberately or repeatedly submitted false information either to the sponsor of the investigation or in any required report, the Commissioner will notify the investigator, the sponsor of any investigation in which the investigator has been named as a participant, and the reviewing IRB that the investigator is not entitled to receive investigational devices. The notification will provide a statement of basis for such determination.

(c) Each investigational device exemption (IDE) and each cleared or approved application submitted under this part, subpart E of part 807 of this chapter, or part 814 of this chapter containing data reported by an investigator who has been determined to be ineligible to receive investigational devices will be examined to determine whether the investigator has submitted unreliable data that are essential to the continuation of the investigation or essential to the approval or clearance of any marketing application.

(d) If the Commissioner determines, after the unreliable data submitted by the investigator are eliminated from consideration, that the data remaining are inadequate to support a conclusion that it is reasonably safe to continue the investigation, the Commissioner will notify the sponsor who shall have an opportunity for a regulatory hearing under part 16 of this chapter. If a danger to the public health exists, however, the Commissioner shall terminate the IDE immediately and notify the sponsor and the reviewing IRB of the determination. In such case, the sponsor shall have an opportunity for a regulatory hearing before FDA under part 16 of this chapter on the question of whether the IDE should be reinstated.

(e) If the Commissioner determines, after the unreliable data submitted by the investigator are eliminated from consideration, that the continued clearance or approval of the marketing application for which the data were submitted cannot be justified, the Commissioner will proceed to withdraw approval or rescind clearance of the medical device in accordance with the applicable provisions of the act.

(f) An investigator who has been determined to be ineligible to receive investigational devices may be reinstated as eligible when the Commissioner determines that the investigator has presented adequate assurances that the investigator will employ investigational devices solely in compliance with the provisions of this part and of parts 50 and 56 of this chapter.

[62 FR 12096, Mar. 14, 1997]

Subpart F [Reserved]

Subpart G—Records and Reports

§812.140 Records.

(a) *Investigator records.* A participating investigator shall maintain the following accurate, complete, and current records relating to the investigator's participation in an investigation:

(1) All correspondence with another investigator, an IRB, the sponsor, a monitor, or FDA, including required reports.

(2) Records of receipt, use or disposition of a device that relate to:

(i) The type and quantity of the device, the dates of its receipt, and the batch number or code mark.

(ii) The names of all persons who received, used, or disposed of each device.

(iii) Why and how many units of the device have been returned to the sponsor, repaired, or otherwise disposed of.

(3) Records of each subject's case history and exposure to the device. Case histories include the case report forms and supporting data including, for example, signed and dated consent forms and medical records including, for example, progress notes of the physician, the individual's hospital chart(s), and the nurses' notes. Such records shall include:

(i) Documents evidencing informed consent and, for any use of a device by the investigator without informed consent, any written concurrence of a licensed physician and a brief description of the circumstances justifying the failure to obtain informed consent. The case history for each individual shall document that informed consent was obtained prior to participation in the study.

(ii) All relevant observations, including records concerning adverse device effects (whether anticipated or unanticipated), information and data on the condition of each subject upon entering, and during the course of, the investigation, including information about relevant previous medical history and the results of all diagnostic tests.

(iii) A record of the exposure of each subject to the investigational device, including the date and time of each use, and any other therapy.

(4) The protocol, with documents showing the dates of and reasons for each deviation from the protocol.

(5) Any other records that FDA requires to be maintained by regulation or by specific requirement for a category of investigations or a particular investigation.

(b) *Sponsor records.* A sponsor shall maintain the following accurate, complete, and current records relating to an investigation:

(1) All correspondence with another sponsor, a monitor, an investigator, an IRB, or FDA, including required reports.

(2) Records of shipment and disposition. Records of shipment shall include the name and address of the consignee, type and quantity of device, date of shipment, and batch number or code mark. Records of disposition shall describe the batch number or code marks of any devices returned to the sponsor,

repaired, or disposed of in other ways by the investigator or another person, and the reasons for and method of disposal.

(3) Signed investigator agreements including the financial disclosure information required to be collected under §812.43(c)(5) in accordance with part 54 of this chapter.

(4) For each investigation subject to §812.2(b)(1) of a device other than a significant risk device, the records described in paragraph (b)(5) of this section and the following records, consolidated in one location and available for FDA inspection and copying:

(i) The name and intended use of the device and the objectives of the investigation;

(ii) A brief explanation of why the device is not a significant risk device:

(iii) The name and address of each investigator:

(iv) The name and address of each IRB that has reviewed the investigation:

(v) A statement of the extent to which the good manufacturing practice regulation in part 820 will be followed in manufacturing the device; and

(vi) Any other information required by FDA.

(5) Records concerning adverse device effects (whether anticipated or unanticipated) and complaints and

(6) Any other records that FDA requires to be maintained by regulation or by specific requirement for a category of investigation or a particular investigation.

(c) *IRB records.* An IRB shall maintain records in accordance with part 56 of this chapter.

(d) *Retention period.* An investigator or sponsor shall maintain the records required by this subpart during the investigation and for a period of 2 years after the latter of the following two dates: The date on which the investigation is terminated or completed, or the date that the records are no longer required for purposes of supporting a premarket approval application or a notice of completion of a product development protocol.

(e) *Records custody.* An investigator or sponsor may withdraw from the responsibility to maintain records for the period required in paragraph (d) of this section and transfer custody of the records to any other person who will accept responsibility for them under this part, including the requirements of §812.145. Notice of a transfer shall be given to FDA not later than 10 working days after transfer occurs.

[45 FR 3751, Jan. 18, 1980, as amended at 45 FR 58843, Sept. 5, 1980; 46 FR 8957, Jan. 27, 1981; 61 FR 57280, Nov. 5, 1996; 63 FR 5253, Feb. 2, 1998]

§812.145 Inspections.

(a) *Entry and inspection.* A sponsor or an investigator who has authority to grant access shall permit authorized FDA employees, at reasonable times and in a reasonable manner, to enter and inspect any establishment where devices are held (including any establishment where devices are manufactured, processed, packed, installed, used, or implanted or where records of results from use of devices are kept).

(b) *Records inspection.* A sponsor, IRB, or investigator, or any other person acting on behalf of such a person with respect to an investigation, shall permit authorized FDA employees, at reasonable times and in a reasonable manner, to inspect and copy all records relating to an investigation.

(c) *Records identifying subjects.* An investigator shall permit authorized FDA employees to inspect and copy records that identify subjects, upon notice that FDA has reason to suspect that adequate informed consent was not obtained, or that reports required to be submitted by the investigator to the sponsor or IRB have not been submitted or are incomplete, inaccurate, false, or misleading.

§812.150 Reports.

(a) *Investigator reports.* An investigator shall prepare and submit the following complete, accurate, and timely reports:

(1) *Unanticipated adverse device effects.* An investigator shall submit to the sponsor and to the reviewing IRB a report of any unanticipated adverse device effect occurring during an investigation as soon as possible, but in no event later than 10 working days after the investigator first learns of the effect.

(2) *Withdrawal of IRB approval.* An investigator shall report to the sponsor, within 5 working days, a withdrawal of approval by the reviewing IRB of the investigator's part of an investigation.

(3) *Progress.* An investigator shall submit progress reports on the investigation to the sponsor, the monitor, and the reviewing IRB at regular intervals, but in no event less often than yearly.

(4) *Deviations from the investigational plan.* An investigator shall notify the sponsor and the reviewing IRB (see § 56.108(a) (3) and (4)) of any deviation from the investigational plan to protect the life or physical well-being of a subject in an emergency. Such notice shall be given as soon as possible, but in no event later than 5 working days after the emergency occurred. Except in such an emergency, prior approval by the sponsor is required for changes in or deviations from a plan, and if these changes or deviations may affect the scientific soundness of the plan or the rights, safety, or welfare of human subjects, FDA and IRB in accordance with § 812.35(a) also is required.

(5) *Informed consent.* If an investigator uses a device without obtaining informed consent, the investigator shall report such use to the sponsor and the reviewing IRB within 5 working days after the use occurs.

(6) *Final report.* An investigator shall, within 3 months after termination or completion of the investigation or the investigator's part of the investigation, submit a final report to the sponsor and the reviewing IRB.

(7) *Other.* An investigator shall, upon request by a reviewing IRB or FDA, provide accurate, complete, and current information about any aspect of the investigation.

(b) *Sponsor reports.* A sponsor shall prepare and submit the following complete, accurate, and timely reports:

(1) *Unanticipated adverse device effects.* A sponsor who conducts an evaluation of an unanticipated adverse device effect under § 812.46(b) shall report the results of such evaluation to FDA and to all reviewing IRB's and participating investigators within 10 working days after the sponsor first receives notice of the effect. Thereafter the sponsor shall submit such additional reports concerning the effect as FDA requests.

(2) *Withdrawal of IRB approval.* A sponsor shall notify FDA and all reviewing IRB's and participating investigators of any withdrawal of approval of an investigation or a part of an investigation by a reviewing IRB within 5 working days after receipt of the withdrawal of approval.

(3) *Withdrawal of FDA approval.* A sponsor shall notify all reviewing IRB's and participating investigators of any withdrawal of FDA approval of the investigation, and shall do so within 5 working days after receipt of notice of the withdrawal of approval.

(4) *Current investigator list.* A sponsor shall submit to FDA, at 6-month intervals, a current list of the names and addresses of all investigators participating in the investigation. The sponsor shall submit the first such list 6 months after FDA approval.

(5) *Progress reports.* At regular intervals, and at least yearly, a sponsor shall submit progress reports to all reviewing IRB's. In the case of a significant risk device, a sponsor shall also submit progress reports to FDA. A sponsor of a treatment IDE shall submit semi-annual progress reports to all reviewing IRB's and FDA in accordance with § 812.36(f) and annual reports in accordance with this section.

(6) *Recall and device disposition.* A sponsor shall notify FDA and all reviewing IRB's of any request that an investigator return, repair, or otherwise dispose of any units of a device. Such notice shall occur within 30 working days after the request is made and shall state why the request was made.

(7) *Final report.* In the case of a significant risk device, the sponsor shall notify FDA within 30 working days of the completion or termination of the investigation and shall submit a final report to FDA and all reviewing the IRB's and participating investigators within 6 months after completion or termination. In the case of a device that is not a significant risk device, the sponsor shall submit a final report to all reviewing IRB's within 6 months after termination or completion.

(8) *Informed consent.* A sponsor shall submit to FDA a copy of any report by an investigator under paragraph (a)(5)

of this section of use of a device without obtaining informed consent, within 5 working days of receipt of notice of such use.

(9) *Significant risk device determinations.* If an IRB determines that a device is a significant risk device, and the sponsor had proposed that the IRB consider the device not to be a significant risk device, the sponsor shall submit to FDA a report of the IRB's determination within 5 working days after the sponsor first learns of the IRB's determination.

(10) *Other.* A sponsor shall, upon request by a reviewing IRB or FDA, provide accurate, complete, and current information about any aspect of the investigation.

[45 FR 3751, Jan. 18, 1980, as amended at 45 FR 58843, Sept. 5, 1980; 48 FR 15622, Apr. 12, 1983; 62 FR 48948, Sept. 18, 1997]

PART 813 [RESERVED]

PART 814—PREMARKET APPROVAL OF MEDICAL DEVICES

Subpart A—General

Sec.

Subpart B—Premarket Approval Application (PMA)

Subpart C—FDA Action on a PMA

Subpart D—Administrative Review [Reserved]

Subpart E—Postapproval Requirements

Subparts F-G [Reserved]

Subpart H—Humanitarian Use Devices

AUTHORITY: 21 U.S.C. 351, 352, 353, 360, 360c–360j, 371, 372, 373, 374, 375, 379, 379e, 381.

SOURCE: 51 FR 26364, July 22, 1986, unless otherwise noted.

Subpart A—General

§ 814.1 Scope.

(a) This part implements section 515 of the act by providing procedures for the premarket approval of medical devices intended for human use.

(b) References in this part to regulatory sections of the Code of Federal Regulations are to chapter I of title 21, unless otherwise noted.

(c) This part applies to any class III medical device, unless exempt under section 520(g) of the act, that:

(1) Was not on the market (introduced or delivered for introduction into commerce for commercial distribution) before May 28, 1976, and is not substantially equivalent to a device on the market before May 28, 1976, or to a device first marketed on, or after that date, which has been classified into class I or class II; or

(2) Is required to have an approved premarket approval application (PMA) or a declared completed product development protocol under a regulation

issued under section 515(b) of the act; or

(3) Was regulated by FDA as a new drug or antibiotic drug before May 28, 1976, and therefore is governed by section 520(1) of the act.

(d) This part amends the conditions to approval for any PMA approved before the effective date of this part. Any condition to approval for an approved PMA that is inconsistent with this part is revoked. Any condition to approval for an approved PMA that is consistent with this part remains in effect.

§814.2 Purpose.

The purpose of this part is to establish an efficient and thorough device review process—

(a) To facilitate the approval of PMA's for devices that have been shown to be safe and effective and that otherwise meet the statutory criteria for approval; and

(b) To ensure the disapproval of PMA's for devices that have not been shown to be safe and effective or that do not otherwise meet the statutory criteria for approval. This part shall be construed in light of these objectives.

§814.3 Definitions.

For the purposes of this part:

(a) *Act* means the Federal Food, Drug, and Cosmetic Act (sections 201–902, 52 Stat. 1040 *et seq.*, as amended (21 U.S.C. 321–392)).

(b) *FDA* means the Food and Drug Administration.

(c) *IDE* means an approved or considered approved investigational device exemption under section 520(g) of the act and parts 812 and 813.

(d) *Master file* means a reference source that a person submits to FDA. A master file may contain detailed information on a specific manufacturing facility, process, methodology, or component used in the manufacture, processing, or packaging of a medical device.

(e) *PMA* means any premarket approval application for a class III medical device, including all information submitted with or incorporated by reference therein. "PMA" includes a new drug application for a device under section 520(1) of the act.

(f) *PMA amendment* means information an applicant submits to FDA to modify a pending PMA or a pending PMA supplement.

(g) *PMA supplement* means a supplemental application to an approved PMA for approval of a change or modification in a class III medical device, including all information submitted with or incorporated by reference therein.

(h) *Person* includes any individual, partnership, corporation, association, scientific or academic establishment, Government agency, or organizational unit thereof, or any other legal entity.

(i) *Statement of material fact* means a representation that tends to show that the safety or effectiveness of a device is more probable than it would be in the absence of such a representation. A false affirmation or silence or an omission that would lead a reasonable person to draw a particular conclusion as to the safety or effectiveness of a device also may be a false statement of material fact, even if the statement was not intended by the person making it to be misleading or to have any probative effect.

(j) *30-day PMA supplement* means a supplemental application to an approved PMA in accordance with §814.39(e).

(k) *Reasonable probability* means that it is more likely than not that an event will occur.

(l) *Serious, adverse health consequences* means any significant adverse experience, including those which may be either life-threatening or involve permanent or long term injuries, but excluding injuries that are nonlife-threatening and that are temporary and reasonably reversible.

(m) *HDE* means a premarket approval application submitted pursuant to this subpart seeking a humanitarian device exemption from the effectiveness requirements of sections 514 and 515 of the act as authorized by section 520(m)(2) of the act.

(n) *HUD (humanitarian use device)* means a medical device intended to benefit patients in the treatment or diagnosis of a disease or condition that affects or is manifested in fewer than

4,000 individuals in the United States per year.

[51 FR 26364, July 22, 1986, as amended at 61 FR 15190, Apr. 5, 1996; 61 FR 33244, June 26, 1996]

§ 814.9 Confidentiality of data and information in a premarket approval application (PMA) file.

(a) A "PMA file" includes all data and information submitted with or incorporated by reference in the PMA, any IDE incorporated into the PMA, any PMA supplement, any report under § 814.82, any master file, or any other related submission. Any record in the PMA file will be available for public disclosure in accordance with the provisions of this section and part 20. The confidentiality of information in a color additive petition submitted as part of a PMA is governed by § 71.15.

(b) The existence of a PMA file may not be disclosed by FDA before an approval order is issued to the applicant unless it previously has been publicly disclosed or acknowledged.

(c) If the existence of a PMA file has not been publicly disclosed or acknowledged, data or information in the PMA file are not available for public disclosure.

(d)(1) If the existence of a PMA file has been publicly disclosed or acknowledged before an order approving, or an order denying approval of the PMA is issued, data or information contained in the file are not available for public disclosure before such order issues. FDA may, however, disclose a summary of portions of the safety and effectiveness data before an approval order or an order denying approval of the PMA issues if disclosure is relevant to public consideration of a specific pending issue.

(2) Notwithstanding paragraph (d)(1) of this section, FDA will make available to the public upon request the information in the IDE that was required to be filed in Docket Number 95S–0158 in the Division of Dockets Management (HFA–305), Food and Drug Administration, 12420 Parklawn Dr., rm. 1–23, Rockville, MD 20857, for investigations involving an exception from informed consent under § 50.24 of this chapter. Persons wishing to request this information shall submit a request under the Freedom of Information Act.

(e) Upon issuance of an order approving, or an order denying approval of any PMA, FDA will make available to the public the fact of the existence of the PMA and a detailed summary of information submitted to FDA respecting the safety and effectiveness of the device that is the subject of the PMA and that is the basis for the order.

(f) After FDA issues an order approving, or an order denying approval of any PMA, the following data and information in the PMA file are immediately available for public disclosure:

(1) All safety and effectiveness data and information previously disclosed to the public, as such disclosure is defined in § 20.81.

(2) Any protocol for a test or study unless the protocol is shown to constitute trade secret or confidential commercial or financial information under § 20.61.

(3) Any adverse reaction report, product experience report, consumer complaint, and other similar data and information, after deletion of:

(i) Any information that constitutes trade secret or confidential commercial or financial information under § 20.61; and

(ii) Any personnel, medical, and similar information disclosure of which would constitute a clearly unwarranted invasion of personal privacy under § 20.63; provided, however, that except for the information that constitutes trade secret or confidential commercial or financial information under § 20.61, FDA will disclose to a patient who requests a report all the information in the report concerning that patient.

(4) A list of components previously disclosed to the public, as such disclosure is defined in § 20.81.

(5) An assay method or other analytical method, unless it does not serve any regulatory purpose and is shown to fall within the exemption in § 20.61 for trade secret or confidential commercial or financial information.

(6) All correspondence and written summaries of oral discussions relating to the PMA file, in accordance with the provisions of §§ 20.103 and 20.104.

(g) All safety and effectiveness data and other information not previously disclosed to the public are available for public disclosure if any one of the following events occurs and the data and information do not constitute trade secret or confidential commercial or financial information under §20.61:

(1) The PMA has been abandoned. FDA will consider a PMA abandoned if:

(i)(A) The applicant fails to respond to a request for additional information within 180 days after the date FDA issues the request or

(B) Other circumstances indicate that further work is not being undertaken with respect to it, and

(ii) The applicant fails to communicate with FDA within 7 days after the date on which FDA notifies the applicant that the PMA appears to have been abandoned.

(2) An order denying approval of the PMA has issued, and all legal appeals have been exhausted.

(3) An order withdrawing approval of the PMA has issued, and all legal appeals have been exhausted.

(4) The device has been reclassified.

(5) The device has been found to be substantially equivalent to a class I or class II device.

(6) The PMA is considered voluntarily withdrawn under §814.44(g).

(h) The following data and information in a PMA file are not available for public disclosure unless they have been previously disclosed to the public, as such disclosure is defined in §20.81, or they relate to a device for which a PMA has been abandoned and they no longer represent a trade secret or confidential commercial or financial information as defined in §20.61:

(1) Manufacturing methods or processes, including quality control procedures.

(2) Production, sales, distribution, and similar data and information, except that any compilation of such data and information aggregated and prepared in a way that does not reveal data or information which are not available for public disclosure under this provision is available for public disclosure.

(3) Quantitative or semiquantitative formulas.

[51 FR 26364, July 22, 1986, as amended at 61 FR 51531, Oct. 2, 1996]

§814.15 Research conducted outside the United States.

(a) A study conducted outside the United States submitted in support of a PMA and conducted under an IDE shall comply with part 812. A study conducted outside the United States submitted in support of a PMA and not conducted under an IDE shall comply with the provisions in paragraph (b) or (c) of this section, as applicable.

(b) *Research begun on or after effective date.* FDA will accept studies submitted in support of a PMA which have been conducted outside the United States and begun on or after November 19, 1986, if the data are valid and the investigator has conducted the studies in conformance with the "Declaration of Helsinki" or the laws and regulations of the country in which the research is conducted, whichever accords greater protection to the human subjects. If the standards of the country are used, the applicant shall state in detail any differences between those standards and the "Declaration of Helsinki" and explain why they offer greater protection to the human subjects.

(c) Research begun before effective date. FDA will accept studies submitted in support of a PMA which have been conducted outside the United States and begun before November 19, 1986, if FDA is satisfied that the data are scientifically valid and that the rights, safety, and welfare of human subjects have not been violated.

(d) *As sole basis for marketing approval.* A PMA based solely on foreign clinical data and otherwise meeting the criteria for approval under this part may be approved if:

(1) The foreign data are applicable to the U.S. population and U.S. medical practice;

(2) The studies have been performed by clinical investigators of recognized competence; and

(3) The data may be considered valid without the need for an on-site inspection by FDA or, if FDA considers such an inspection to be necessary, FDA can

validate the data through an on-site inspection or other appropriate means.

(e) *Consultation between FDA and applicants.* Applicants are encouraged to meet with FDA officials in a "presubmission" meeting when approval based solely on foreign data will be sought.

(Approved by the Office of Management and Budget under control number 0910–0231)

[51 FR 26364, July 22, 1986; 51 FR 40415, Nov. 7, 1986, as amended at 51 FR 43344, Dec. 2, 1986]

§814.17 Service of orders.

Orders issued under this part will be served in person by a designated officer or employee of FDA on, or by registered mail to, the applicant or the designated agent at the applicant's or designated agent's last known address in FDA's records.

§814.19 Product development protocol (PDP).

A class III device for which a product development protocol has been declared completed by FDA under this chapter will be considered to have an approved PMA.

Subpart B—Premarket Approval Application (PMA)

§814.20 Application.

(a) The applicant or an authorized representative shall sign the PMA. If the applicant does not reside or have a place of business within the United States, the PMA shall be countersigned by an authorized representative residing or maintaining a place of business in the United States and shall identify the representative's name and address.

(b) Unless the applicant justifies an omission in accordance with paragraph (d) of this section, a PMA shall include:

(1) The name and address of the applicant.

(2) A table of contents that specifies the volume and page number for each item referred to in the table. A PMA shall include separate sections on nonclinical laboratory studies and on clinical investigations involving human subjects. A PMA shall be submitted in six copies each bound in one or more numbered volumes of reasonable size. The applicant shall include information that it believes to be trade secret or confidential commercial or financial information in all copies of the PMA and identify in at least one copy the information that it believes to be trade secret or confidential commercial or financial information.

(3) A summary in sufficient detail that the reader may gain a general understanding of the data and information in the application. The summary shall contain the following information:

(i) *Indications for use.* A general description of the disease or condition the device will diagnose, treat, prevent, cure, or mitigate, including a description of the patient population for which the device is intended.

(ii) *Device description.* An explanation of how the device functions, the basic scientific concepts that form the basis for the device, and the significant physical and performance characteristics of the device. A brief description of the manufacturing process should be included if it will significantly enhance the reader's understanding of the device. The generic name of the device as well as any proprietary name or trade name should be included.

(iii) *Alternative practices and procedures.* A description of existing alternative practices or procedures for diagnosing, treating, preventing, curing, or mitigating the disease or condition for which the device is intended.

(iv) *Marketing history.* A brief description of the foreign and U.S. marketing history, if any, of the device, including a list of all countries in which the device has been marketed and a list of all countries in which the device has been withdrawn from marketing for any reason related to the safety or effectiveness of the device. The description shall include the history of the marketing of the device by the applicant and, if known, the history of the marketing of the device by any other person.

(v) *Summary of studies.* An abstract of any information or report described in the PMA under paragraph (b)(8)(ii) of this section and a summary of the results of technical data submitted under paragraph (b)(6) of this section. Such summary shall include a description of the objective of the study, a description of the experimental design of the study, a brief description of how the data were collected and analyzed, and a brief description of the results, whether positive, negative, or inconclusive. This section shall include the following:

(A) A summary of the nonclinical laboratory studies submitted in the application;

(B) A summary of the clinical investigations involving human subjects submitted in the application including a discussion of subject selection and exclusion criteria, study population, study period, safety and effectiveness data, adverse reactions and complications, patient discontinuation, patient complaints, device failures and replacements, results of statistical analyses of the clinical investigations, contraindications and precautions for use of the device, and other information from the clinical investigations as appropriate (any investigation conducted under an IDE shall be identified as such).

(vi) *Conclusions drawn from the studies.* A discussion demonstrating that the data and information in the application constitute valid scientific evidence within the meaning of §860.7 and provide reasonable assurance that the device is safe and effective for its intended use. A concluding discussion shall present benefit and risk considerations related to the device including a discussion of any adverse effects of the device on health and any proposed additional studies or surveillance the applicant intends to conduct following approval of the PMA.

(4) A complete description of:

(i) The device, including pictorial representations;

(ii) Each of the functional components or ingredients of the device if the device consists of more than one physical component or ingredient;

(iii) The properties of the device relevant to the diagnosis, treatment, prevention, cure, or mitigation of a disease or condition;

(iv) The principles of operation of the device; and

(v) The methods used in, and the facilities and controls used for, the manufacture, processing, packing, storage, and, where appropriate, installation of the device, in sufficient detail so that a person generally familiar with current good manufacturing practice can make a knowledgeable judgment about the quality control used in the manufacture of the device.

(5) Reference to any performance standard under section 514 of the act or the Radiation Control for Health and Safety Act of 1968 (42 U.S.C. 263b *et seq.*) in effect or proposed at the time of the submission and to any voluntary standard that is relevant to any aspect of the safety or effectiveness of the device and that is known to or that should reasonably be known to the applicant. The applicant shall—

(i) Provide adequate information to demonstrate how the device meets, or justify any deviation from, any performance standard established under section 514 of the act or under the Radiation Control for Health and Safety Act, and

(ii) Explain any deviation from a voluntary standard.

(6) The following technical sections which shall contain data and information in sufficient detail to permit FDA to determine whether to approve or deny approval of the application:

(i) A section containing results of the nonclinical laboratory studies with the device including microbiological, toxicological, immunological, biocompatibility, stress, wear, shelf life, and other laboratory or animal tests as appropriate. Information on nonclinical laboratory studies shall include a statement that each such study was conducted in compliance with part 58, or, if the study was not conducted in compliance with such regulations, a brief statement of the reason for the noncompliance.

(ii) A section containing results of the clinical investigations involving human subjects with the device including clinical protocols, number of investigators and subjects per investigator, subject selection and exclusion criteria, study population, study period, safety and effectiveness data, adverse reactions and complications, patient discontinuation, patient complaints, device failures and replacements, tabulations of data from all individual subject report forms and copies of such forms for each subject who died during a clinical investigation or who did not complete the investigation, results of statistical analyses of the clinical investigations, device failures and replacements, contraindications and precautions for use of the device, and any other appropriate information from the clinical investigations. Any investigation conducted under an IDE shall be identified as such. Information on clinical investigations involving human subjects shall include the following:

(A) A statement with respect to each study that it either was conducted in compliance with the institutional review board regulations in part 56, or was not subject to the regulations under § 56.104 or § 56.105, and that it was conducted in compliance with the informed consent regulations in part 50; or if the study was not conducted in compliance with those regulations, a brief statement of the reason for the noncompliance.

(B) A statement that each study was conducted in compliance with part 812 or part 813 concerning sponsors of clinical investigations and clinical investigators, or if the study was not conducted in compliance with those regulations, a brief statement of the reason for the noncompliance.

(7) For a PMA supported solely by data from one investigation, a justification showing that data and other information from a single investigator are sufficient to demonstrate the safety and effectiveness of the device and to ensure reproducibility of test results.

(8)(i) A bibliography of all published reports not submitted under paragraph (b)(6) of this section, whether adverse or supportive, known to or that should reasonably be known to the applicant and that concern the safety or effectiveness of the device.

(ii) An identification, discussion, and analysis of any other data, information, or report relevant to an evaluation of the safety and effectiveness of the device known to or that should reasonably be known to the applicant from any source, foreign or domestic, including information derived from investigations other than those proposed in the application and from commercial marketing experience.

(iii) Copies of such published reports or unpublished information in the possession of or reasonably obtainable by the applicant if an FDA advisory committee or FDA requests.

(9) One or more samples of the device and its components, if requested by FDA. If it is impractical to submit a requested sample of the device, the applicant shall name the location at which FDA may examine and test one or more devices.

(10) Copies of all proposed labeling for the device. Such labeling may include, e.g., instructions for installation and any information, literature, or advertising that constitutes labeling under section 201(m) of the act.

(11) An environmental assessment under § 25.20(n) prepared in the applicable format in § 25.40, unless the action qualifies for exclusion under § 25.30 or § 25.34. If the applicant believes that the action qualifies for exclusion, the PMA shall under § 25.15(a) and (d) provide information that establishes to FDA's satisfaction that the action requested is included within the excluded category and meets the criteria for the applicable exclusion.

(12) A financial certification or disclosure statement or both as required by part 54 of this chapter.

(13) Such other information as FDA may request. If necessary, FDA will obtain the concurrence of the appropriate FDA advisory committee before requesting additional information.

(c) Pertinent information in FDA files specifically referred to by an applicant may be incorporated into a PMA by reference. Information in a master file or other information submitted to FDA by a person other than the applicant will not be considered part of a PMA unless such reference is authorized in writing by the person who submitted the information or the master file. If a master file is not referenced within 5 years after the date that it is submitted to FDA, FDA will return the master file to the person who submitted it.

(d) If the applicant believes that certain information required under paragraph (b) of this section to be in a PMA is not applicable to the device that is the subject of the PMA, and omits any such information from its PMA, the applicant shall submit a statement that identifies the omitted information and justifies the omission. The statement shall be submitted as a separate section in the PMA and identified in the table of contents. If the justification for the omission is not accepted by the agency, FDA will so notify the applicant.

(e) The applicant shall periodically update its pending application with new safety and effectiveness information learned about the device from ongoing or completed studies that may reasonably affect an evaluation of the safety or effectiveness of the device or that may reasonably affect the statement of contraindications, warnings, precautions, and adverse reactions in the draft labeling. The update report shall be consistent with the data reporting provisions of the protocol. The applicant shall submit three copies of any update report and shall include in the report the number assigned by FDA to the PMA. These updates are considered to be amendments to the PMA. The time frame for review of a PMA will not be extended due to the submission of an update report unless the update is a major amendment under §814.37(c)(1). The applicant shall submit these reports—

(1) 3 months after the filing date,

(2) Following receipt of an approvable letter, and

(3) At any other time as requested by FDA.

(f) If a color additive subject to section 706 of the act is used in or on the device and has not previously been listed for such use, then, in lieu of submitting a color additive petition under part 71, at the option of the applicant, the information required to be submitted under part 71 may be submitted as part of the PMA. When submitted as part of the PMA, the information shall be submitted in three copies each bound in one or more numbered volumes of reasonable size. A PMA for a device that contains a color additive that is subject to section 706 of the act will not be approved until the color additive is listed for use in or on the device.

(g) FDA has issued a PMA guidance document to assist the applicant in the arrangement and content of a PMA. This guidance document is available on the Internet at *http://www.fda.gov/cdrh/dsma/pmaman/front.html*. This guidance document is also available upon request from the Center for Devices and Radiological Health, Division of Small Manufacturers Assistance (HFZ–220), 1350 Piccard Dr., Rockville, MD 20850, FAX 301–443–8818.

(h) If you are sending a PMA, PMA amendment, PMA supplement, or correspondence with respect to a PMA, you must send it to the Document Mail Center (HFZ–401), Center for Devices and Radiological Health, Food and Drug Administration, 9200 Corporate Blvd., Rockville, MD 20850.

[51 FR 26364, July 22, 1986; 51 FR 40415, Nov. 7, 1986, as amended at 51 FR 43344, Dec. 2, 1986; 55 FR 11169, Mar. 27, 1990; 62 FR 40600, July 29, 1997; 63 FR 5253, Feb. 2, 1998; 65 FR 17137, Mar. 31, 2000; 65 FR 56480, Sept. 19, 2000; 67 FR 9587, Mar. 4, 2002]

§814.37 PMA amendments and resubmitted PMA's.

(a) An applicant may amend a pending PMA or PMA supplement to revise existing information or provide additional information.

(b) FDA may request the applicant to amend a PMA or PMA supplement with any information regarding the device that is necessary for FDA or the appropriate advisory committee to complete the review of the PMA or PMA supplement.

(c) A PMA amendment submitted to FDA shall include the PMA or PMA supplement number assigned to the original submission and, if submitted on the applicant's own initiative, the reason for submitting the amendment. FDA may extend the time required for its review of the PMA, or PMA supplement, as follows:

(1) If the applicant on its own initiative or at FDA's request submits a major PMA amendment (e.g., an amendment that contains significant new data from a previously unreported study, significant updated data from a previously reported study, detailed new analyses of previously submitted data, or significant required information previously omitted), the review period may be extended up to 180 days.

(2) If an applicant declines to submit a major amendment requested by FDA, the review period may be extended for the number of days that elapse between the date of such request and the date that FDA receives the written response declining to submit the requested amendment.

(d) An applicant may on its own initiative withdraw a PMA or PMA supplement. If FDA requests an applicant to submit a PMA amendment and a written response to FDA's request is not received within 180 days of the date of the request, FDA will consider the pending PMA or PMA supplement to be withdrawn voluntarily by the applicant.

(e) An applicant may resubmit a PMA or PMA supplement after withdrawing it or after it is considered withdrawn under paragraph (d) of this section, or after FDA has refused to accept it for filing, or has denied approval of the PMA or PMA supplement. A resubmitted PMA or PMA supplement shall comply with the requirements of § 814.20 or § 814.39, respectively, and shall include the PMA number assigned to the original submission and the applicant's reasons for resubmission of the PMA or PMA supplement.

§ 814.39 PMA supplements.

(a) After FDA's approval of a PMA, an applicant shall submit a PMA supplement for review and approval by FDA before making a change affecting the safety or effectiveness of the device for which the applicant has an approved PMA, unless the change is of a type for which FDA, under paragraph (e) of this section, has advised that an alternate submission is permitted or is of a type which, under section 515(d)(6)(A) of the act and paragraph (f) of this section, does not require a PMA supplement under this paragraph. While the burden for determining whether a supplement is required is primarily on the PMA holder, changes for which an applicant shall submit a PMA supplement include, but are not limited to, the following types of changes if they affect the safety or effectiveness of the device:

(1) New indications for use of the device.

(2) Labeling changes.

(3) The use of a different facility or establishment to manufacture, process, or package the device.

(4) Changes in sterilization procedures.

(5) Changes in packaging.

(6) Changes in the performance or design specifications, circuits, components, ingredients, principle of operation, or physical layout of the device.

(7) Extension of the expiration date of the device based on data obtained under a new or revised stability or sterility testing protocol that has not been approved by FDA. If the protocol has been approved, the change shall be reported to FDA under paragraph (b) of this section.

(b) An applicant may make a change in a device after FDA's approval of a PMA for the device without submitting a PMA supplement if the change does not affect the device's safety or effectiveness and the change is reported to FDA in postapproval periodic reports required as a condition to approval of the device, e.g., an editorial change in labeling which does not affect the safety or effectiveness of the device.

(c) All procedures and actions that apply to an application under § 814.20 also apply to PMA supplements except that the information required in a supplement is limited to that needed to support the change. A summary under § 814.20(b)(3) is required for only a supplement submitted for new indications for use of the device, significant changes in the performance or design specifications, circuits, components, ingredients, principles of operation, or physical layout of the device, or when otherwise required by FDA. The applicant shall submit three copies of a PMA supplement and shall include information relevant to the proposed changes in the device. A PMA supplement shall include a separate section that identifies each change for which approval is being requested and explains the reason for each such change. The applicant shall submit additional copies and additional information if requested by FDA. The time frames for review of, and FDA action on, a PMA supplement are the same as those provided in § 814.40 for a PMA.

(d)(1) After FDA approves a PMA, any change described in paragraph (d)(2) of this section that enhances the safety of the device or the safety in the use of the device may be placed into effect by the applicant prior to the receipt under § 814.17 of a written FDA order approving the PMA supplement provided that:

(i) The PMA supplement and its mailing cover are plainly marked "Special PMA Supplement—Changes Being Effected";

(ii) The PMA supplement provides a full explanation of the basis for the changes;

(iii) The applicant has received acknowledgement from FDA of receipt of the supplement; and

(iv) The PMA supplement specifically identifies the date that such changes are being effected.

(2) The following changes are permitted by paragraph (d)(1) of this section:

(i) Labeling changes that add or strengthen a contraindication, warning, precaution, or information about an adverse reaction.

(ii) Labeling changes that add or strengthen an instruction that is intended to enhance the safe use of the device.

(iii) Labeling changes that delete misleading, false, or unsupported indications.

(iv) Changes in quality controls or manufacturing process that add a new specification or test method, or otherwise provide additional assurance of purity, identity, strength, or reliability of the device.

(e)(1) FDA will identify a change to a device for which an applicant has an approved PMA and for which a PMA supplement under paragraph (a) is not required. FDA will identify such a change in an advisory opinion under § 10.85, if the change applies to a generic type of device, or in correspondence to the applicant, if the change applies only to the applicant's device. FDA will require that a change for which a PMA supplement under paragraph (a) is not required be reported to FDA in:

(i) A periodic report under § 814.84 or

(ii) A 30-day PMA supplement under this paragraph.

(2) FDA will identify, in the advisory opinion or correspondence, the type of information that is to be included in the report or 30-day PMA supplement. If the change is required to be reported to FDA in a periodic report, the change may be made before it is reported to FDA. If the change is required to be reported in a 30-day PMA supplement, the change may be made 30 days after FDA files the 30-day PMA supplement unless FDA requires the PMA holder to provide additional information, informs the PMA holder that the supplement is not approvable, or disapproves the supplement. The 30-day PMA supplement shall follow the instructions in the correspondence or advisory opinion. Any 30-day PMA supplement that does not meet the requirements of the correspondence or advisory opinion will not be filed and, therefore, will not be deemed approved 30 days after receipt.

(f) Under section 515(d) of the act, modifications to manufacturing procedures or methods of manufacture that affect the safety and effectiveness of a device subject to an approved PMA do not require submission of a PMA supplement under paragraph (a) of this section and are eligible to be the subject of a 30-day notice. A 30-day notice shall describe in detail the change, summarize the data or information supporting the change, and state that the change has been made in accordance with the requirements of part 820 of this chapter. The manufacturer may distribute the device 30 days after the date on which FDA receives the 30-day notice, unless FDA notifies the applicant within 30 days from receipt of the notice that the notice is not adequate. If the notice is not adequate, FDA shall inform the applicant in writing that a 135-day PMA supplement is needed and shall describe what further information or action is required for acceptance of such change. The number of days under review as a 30-day notice shall be deducted from the 135-day PMA supplement review period if the notice meets appropriate content requirements for a PMA supplement.

[51 FR 26364, July 22, 1986, as amended at 51 FR 43344, Dec. 2, 1986; 63 FR 54044, Oct. 8, 1998; 67 FR 9587, Mar. 4, 2002; 69 FR 11313, Mar. 10, 2004]

Subpart C—FDA Action on a PMA

§ 814.40 Time frames for reviewing a PMA.

Within 180 days after receipt of an application that is accepted for filing and to which the applicant does not submit a major amendment, FDA will review the PMA and, after receiving the report and recommendation of the appropriate FDA advisory committee, send the applicant an approval order under § 814.44(d), an approvable letter under § 814.44(e), a not approvable letter under § 814.44(f), or an order denying approval under § 814.45. The approvable letter and the not approvable letter will provide an opportunity for the applicant to amend or withdraw the application, or to consider the letter to be a denial of approval of the PMA under § 814.45 and to request administrative review under section 515 (d)(3) and (g) of the act.

§ 814.42 Filing a PMA.

(a) The filing of an application means that FDA has made a threshold determination that the application is sufficiently complete to permit a substantive review. Within 45 days after a PMA is received by FDA, the agency will notify the applicant whether the application has been filed.

(b) If FDA does not find that any of the reasons in paragraph (e) of this section for refusing to file the PMA applies, the agency will file the PMA and will notify the applicant in writing of the filing. The notice will include the PMA reference number and the date FDA filed the PMA. The date of filing is the date that a PMA accepted for filing was received by the agency. The 180-day period for review of a PMA starts on the date of filing.

(c) If FDA refuses to file a PMA, the agency will notify the applicant of the reasons for the refusal. This notice will identify the deficiencies in the application that prevent filing and will include the PMA reference number.

(d) If FDA refuses to file the PMA, the applicant may:

(1) Resubmit the PMA with additional information necessary to comply with the requirements of section 515(c)(1) (A)–(G) of the act and § 814.20. A resubmitted PMA shall include the PMA reference number of the original submission. If the resubmitted PMA is accepted for filing, the date of filing is the date FDA receives the resubmission;

(2) Request in writing within 10 working days of the date of receipt of the notice refusing to file the PMA, an informal conference with the Director of the Office of Device Evaluation to review FDA's decision not to file the PMA. FDA will hold the informal conference within 10 working days of its receipt of the request and will render its decision on filing within 5 working days after the informal conference. If, after the informal conference, FDA accepts the PMA for filing, the date of filing will be the date of the decision to accept the PMA for filing. If FDA does not reverse its decision not to file the PMA, the applicant may request reconsideration of the decision from the Director of the Center for Devices and Radiological Health. The Director's decision will constitute final administrative action for the purpose of judicial review.

(e) FDA may refuse to file a PMA if any of the following applies:

(1) The application is incomplete because it does not on its face contain all the information required under section 515(c)(1) (A)–(G) of the act;

(2) The PMA does not contain each of the items required under § 814.20 and justification for omission of any item is inadequate;

(3) The applicant has a pending premarket notification under section 510(k) of the act with respect to the same device, and FDA has not determined whether the device falls within the scope of § 814.1(c).

(4) The PMA contains a false statement of material fact.

(5) The PMA is not accompanied by a statement of either certification or disclosure as required by part 54 of this chapter.

[51 FR 26364, July 22, 1986, as amended at 63 FR 5254, Feb. 2, 1998]

§ 814.44 Procedures for review of a PMA.

(a) FDA will begin substantive review of a PMA after the PMA is accepted for filing under § 814.42. FDA may refer the PMA to a panel on its own initiative, and will do so upon request of an applicant, unless FDA determines that the application substantially duplicates information previously reviewed by a panel. If FDA refers an application to a panel, FDA will forward the PMA, or relevant portions thereof, to each member of the appropriate FDA panel for review. During the review process, FDA may communicate with the applicant as set forth under § 814.37(b), or with a panel to respond to questions that may be posed by panel members or to provide additional information to the panel. FDA will maintain a record of all communications with the applicant and with the panel.

(b) The advisory committee shall submit a report to FDA which includes the committee's recommendation and the basis for such recommendation on the PMA. Before submission of this report, the committee shall hold a public meeting to review the PMA in accordance with part 14. This meeting may be held by a telephone conference under § 14.22(g). The advisory committee report and recommendation may be in the form of a meeting transcript signed by the chairperson of the committee.

(c) FDA will complete its review of the PMA and the advisory committee report and recommendation and, within the later of 180 days from the date of filing of the PMA under § 814.42 or the number of days after the date of filing as determined under § 814.37(c), issue an approval order under paragraph (d) of this section, an approvable letter under paragraph (e) of this section, a not approvable letter under paragraph (f) of this section, or an order denying approval of the application under § 814.45(a).

(d)(1) FDA will issue to the applicant an order approving a PMA if none of the reasons in § 814.45 for denying approval of the application applies. FDA will approve an application on the basis of draft final labeling if the only deficiencies in the application concern editorial or similar minor deficiencies in the draft final labeling. Such approval will be conditioned upon the applicant incorporating the specified labeling changes exactly as directed and upon the applicant submitting to FDA a copy of the final printed labeling before marketing. FDA will also give the public notice of the order, including notice of and opportunity for any interested persons to request review under section 515(d)(3) of the act. The notice of approval will be placed on FDA's home page on the Internet (*http://www.fda.gov*), and it will state that a detailed summary of information respecting the safety and effectiveness of the device, which was the basis for the order approving the PMA, including information about any adverse effects of the device on health, is available on the Internet and has been placed on public display, and that copies are available upon request. FDA will publish in the FEDERAL REGISTER after each quarter a list of the approvals announced in that quarter. When a notice of approval is published, data and information in the PMA file will be available for public disclosure in accordance with § 814.9.

(2) A request for copies of the current PMA approvals and denials document and for copies of summaries of safety and effectiveness shall be sent in writing to the Division of Dockets Management (HFA–305), Food and Drug Administration, 5630 Fishers Lane, rm. 1061, Rockville, MD 20852.

(e) FDA will send the applicant an approvable letter if the application substantially meets the requirements of this part and the agency believes it can approve the application if specific additional information is submitted or specific conditions are agreed to by the applicant.

(1) The approvable letter will describe the information FDA requires to be provided by the applicant or the conditions the applicant is required to meet to obtain approval. For example, FDA may require, as a condition to approval:

(i) The submission of certain information identified in the approvable letter, e.g., final labeling;

(ii) An FDA inspection that finds the manufacturing facilities, methods, and controls in compliance with part 820 and, if applicable, that verifies records pertinent to the PMA;

(iii) Restrictions imposed on the device under section 515(d)(1)(B)(ii) or 520(e) of the act;

(iv) Postapproval requirements as described in subpart E of this part.

(2) In response to an approvable letter the applicant may:

(i) Amend the PMA as requested in the approvable letter; or

(ii) Consider the approvable letter to be a denial of approval of the PMA under § 814.45 and request administrative review under section 515(d)(3) of the act by filing a petition in the form of a petition for reconsideration under § 10.33; or

(iii) Withdraw the PMA.

(f) FDA will send the applicant a not approvable letter if the agency believes that the application may not be approved for one or more of the reasons given in § 814.45(a). The not approvable letter will describe the deficiencies in the application, including each applicable ground for denial under section 515(d)(2) (A)–(E) of the act, and, where practical, will identify measures required to place the PMA in approvable form. In response to a not approvable letter, the applicant may:

(1) Amend the PMA as requested in the not approvable letter (such an amendment will be considered a major amendment under § 814.37(c)(1)); or

(2) Consider the not approvable letter to be a denial of approval of the PMA under § 814.45 and request administrative review under section 515(d)(3) of the act by filing a petition in the form of a petition for reconsideration under § 10.33; or

(3) Withdraw the PMA.

(g) FDA will consider a PMA to have been withdrawn voluntarily if:

(1) The applicant fails to respond in writing to a written request for an amendment within 180 days after the date FDA issues such request;

(2) The applicant fails to respond in writing to an approvable or not approvable letter within 180 days after the date FDA issues such letter; or

(3) The applicant submits a written notice to FDA that the PMA has been withdrawn.

[51 FR 26364, July 22, 1986, as amended at 57 FR 58403, Dec. 10, 1992; 63 FR 4572, Jan. 30, 1998]

§ 814.45 Denial of approval of a PMA.

(a) FDA may issue an order denying approval of a PMA if the applicant fails to follow the requirements of this part or if, upon the basis of the information submitted in the PMA or any other information before the agency, FDA determines that any of the grounds for denying approval of a PMA specified in section 515(d)(2) (A)–(E) of the act applies. In addition, FDA may deny approval of a PMA for any of the following reasons:

(1) The PMA contains a false statement of material fact;

(2) The device's proposed labeling does not comply with the requirements in part 801 or part 809;

(3) The applicant does not permit an authorized FDA employee an opportunity to inspect at a reasonable time and in a reasonable manner the facilities, controls, and to have access to and to copy and verify all records pertinent to the application;

(4) A nonclinical laboratory study that is described in the PMA and that is essential to show that the device is safe for use under the conditions prescribed, recommended, or suggested in its proposed labeling, was not conducted in compliance with the good laboratory practice regulations in part 58 and no reason for the noncompliance is provided or, if it is, the differences between the practices used in conducting the study and the good laboratory practice regulations do not support the validity of the study; or

(5) Any clinical investigation involving human subjects described in the PMA, subject to the institutional review board regulations in part 56 or informed consent regulations in part 50, was not conducted in compliance with those regulations such that the rights or safety of human subjects were not adequately protected.

(b) FDA will issue any order denying approval of the PMA in accordance with § 814.17. The order will inform the applicant of the deficiencies in the PMA, including each applicable ground for denial under section 515(d)(2) of the act and the regulations under this part, and, where practical, will identify measures required to place the PMA in approvable form. The order will include a notice of an opportunity to request review under section 515(d)(3) of the act.

(c) FDA will use the criteria specified in § 860.7 to determine the safety and effectiveness of a device in deciding whether to approve or deny approval of a PMA. FDA may use information other than that submitted by the applicant in making such determination.

(d)(1) FDA will give the public notice of an order denying approval of the PMA. The notice will be placed on the FDA's home page on the Internet (*http://www.fda.gov*), and it will state that a detailed summary of information respecting the safety and effectiveness of the device, including information about any adverse effects of the device on health, is available on the Internet and has been placed on public display and that copies are available upon request. FDA will publish in the FEDERAL REGISTER after each quarter a list of the denials announced in that quarter. When a notice of denial of approval is made publicly available, data and information in the PMA file will be available for public disclosure in accordance with § 814.9.

(2) A request for copies of the current PMA approvals and denials document and copies of summaries of safety and effectiveness shall be sent in writing to the Freedom of Information Staff (HFI–35), Food and Drug Administration, 5600 Fishers Lane, Rockville, MD 20857.

(e) FDA will issue an order denying approval of a PMA after an approvable or not approvable letter has been sent and the applicant:

(1) Submits a requested amendment but any ground for denying approval of the application under section 515(d)(2) of the act still applies; or

(2) Notifies FDA in writing that the requested amendment will not be submitted; or

(3) Petitions for review under section 515(d)(3) of the act by filing a petition in the form of a petition for reconsideration under § 10.33.

[51 FR 26364, July 22, 1986, as amended at 63 FR 4572, Jan. 30, 1998]

§ 814.46 Withdrawal of approval of a PMA.

(a) FDA may issue an order withdrawing approval of a PMA if, from any information available to the agency, FDA determines that:

(1) Any of the grounds under section 515(e)(1) (A)–(G) of the act applies.

(2) Any postapproval requirement imposed by the PMA approval order or by regulation has not been met.

(3) A nonclinical laboratory study that is described in the PMA and that is essential to show that the device is safe for use under the conditions prescribed, recommended, or suggested in its proposed labeling, was not conducted in compliance with the good laboratory practice regulations in part 58 and no reason for the noncompliance is provided or, if it is, the differences between the practices used in conducting the study and the good laboratory practice regulations do not support the validity of the study.

(4) Any clinical investigation involving human subjects described in the PMA, subject to the institutional review board regulations in part 56 or informed consent regulations in part 50, was not conducted in compliance with those regulations such that the rights or safety of human subjects were not adequately protected.

(b)(1) FDA may seek advice on scientific matters from any appropriate FDA advisory committee in deciding whether to withdraw approval of a PMA.

(2) FDA may use information other than that submitted by the applicant in deciding whether to withdraw approval of a PMA.

(c) Before issuing an order withdrawing approval of a PMA, FDA will issue the holder of the approved application a notice of opportunity for an informal hearing under part 16.

(d) If the applicant does not request a hearing or if after the part 16 hearing is held the agency decides to proceed with the withdrawal, FDA will issue to the holder of the approved application an order withdrawing approval of the application. The order will be issued under § 814.17, will state each ground for withdrawing approval, and will include a notice of an opportunity for administrative review under section 515(e)(2) of the act.

(e) FDA will give the public notice of an order withdrawing approval of a PMA. The notice will be published in the FEDERAL REGISTER and will state that a detailed summary of information respecting the safety and effectiveness of the device, including information about any adverse effects of the device on health, has been placed on public display and that copies are available upon request. When a notice of withdrawal of approval is published, data and information in the PMA file will be available for public disclosure in accordance with § 814.9.

§ 814.47 Temporary suspension of approval of a PMA.

(a) *Scope.* (1) This section describes the procedures that FDA will follow in exercising its authority under section 515(e)(3) of the act (21 U.S.C. 360e(e)(3)). This authority applies to the original PMA, as well as any PMA supplement(s), for a medical device.

(2) FDA will issue an order temporarily suspending approval of a PMA if FDA determines that there is a reasonable probability that continued distribution of the device would cause serious, adverse health consequences or death.

(b) *Regulatory hearing.* (1) If FDA believes that there is a reasonable probability that the continued distribution of a device subject to an approved PMA would cause serious, adverse health consequences or death, FDA may initiate and conduct a regulatory hearing to determine whether to issue an order temporarily suspending approval of the PMA.

(2) Any regulatory hearing to determine whether to issue an order temporarily suspending approval of a PMA shall be initiated and conducted by FDA pursuant to part 16 of this chapter. If FDA believes that immediate action to remove a dangerous device from the market is necessary to protect the public health, the agency may, in accordance with § 16.60(h) of this chapter, waive, suspend, or modify any part 16 procedure pursuant to § 10.19 of this chapter.

(3) FDA shall deem the PMA holder's failure to request a hearing within the timeframe specified by FDA in the notice of opportunity for hearing to be a waiver.

(c) *Temporary suspension order.* If the PMA holder does not request a regulatory hearing or if, after the hearing, and after consideration of the administrative record of the hearing, FDA determines that there is a reasonable probability that the continued distribution of a device under an approved PMA would cause serious, adverse health consequences or death, the agency shall, under the authority of section 515(e)(3) of the act, issue an order to the PMA holder temporarily suspending approval of the PMA.

(d) *Permanent withdrawal of approval of the PMA.* If FDA issues an order temporarily suspending approval of a PMA, the agency shall proceed expeditiously, but within 60 days, to hold a hearing on whether to permanently withdraw approval of the PMA in accordance with section 515(e)(1) of the act and the procedures set out in § 814.46.

[61 FR 15190, Apr. 5, 1996]

Subpart D—Administrative Review [Reserved]

Subpart E—Postapproval Requirements

§ 814.80 General.

A device may not be manufactured, packaged, stored, labeled, distributed, or advertised in a manner that is inconsistent with any conditions to approval specified in the PMA approval order for the device.

§ 814.82 Postapproval requirements.

(a) FDA may impose postapproval requirements in a PMA approval order or by regulation at the time of approval of the PMA or by regulation subsequent to approval. Postapproval requirements may include as a condition to approval of the device:

(1) Restriction of the sale, distribution, or use of the device as provided by section 515(d)(1)(B)(ii) or 520(e) of the act.

(2) Continuing evaluation and periodic reporting on the safety, effectiveness, and reliability of the device for its intended use. FDA will state in the PMA approval order the reason or purpose for such requirement and the number of patients to be evaluated and the reports required to be submitted.

(3) Prominent display in the labeling of a device and in the advertising of

any restricted device of warnings, hazards, or precautions important for the device's safe and effective use, including patient information, e.g., information provided to the patient on alternative modes of therapy and on risks and benefits associated with the use of the device.

(4) Inclusion of identification codes on the device or its labeling, or in the case of an implant, on cards given to patients if necessary to protect the public health.

(5) Maintenance of records that will enable the applicant to submit to FDA information needed to trace patients if such information is necessary to protect the public health. Under section 519(a)(4) of the act, FDA will require that the identity of any patient be disclosed in records maintained under this paragraph only to the extent required for the medical welfare of the individual, to determine the safety or effectiveness of the device, or to verify a record, report, or information submitted to the agency.

(6) Maintenance of records for specified periods of time and organization and indexing of records into identifiable files to enable FDA to determine whether there is reasonable assurance of the continued safety and effectiveness of the device.

(7) Submission to FDA at intervals specified in the approval order of periodic reports containing the information required by §814.84(b).

(8) Batch testing of the device.

(9) Such other requirements as FDA determines are necessary to provide reasonable assurance, or continued reasonable assurance, of the safety and effectiveness of the device.

(b) An applicant shall grant to FDA access to any records and reports required under the provisions of this part, and shall permit authorized FDA employees to copy and verify such records and reports and to inspect at a reasonable time and in a reasonable manner all manufacturing facilities to verify that the device is being manufactured, stored, labeled, and shipped under approved conditions.

(c) Failure to comply with any post-approval requirement constitutes a ground for withdrawal of approval of a PMA.

(Approved by the Office of Management and Budget under control number 0910–0231)

[51 FR 26364, July 22, 1986, as amended at 51 FR 43344, Dec. 2, 1986]

§814.84 Reports.

(a) The holder of an approved PMA shall comply with the requirements of part 803 and with any other requirements applicable to the device by other regulations in this subchapter or by order approving the device.

(b) Unless FDA specifies otherwise, any periodic report shall:

(1) Identify changes described in §814.39(a) and changes required to be reported to FDA under §814.39(b).

(2) Contain a summary and bibliography of the following information not previously submitted as part of the PMA:

(i) Unpublished reports of data from any clinical investigations or nonclinical laboratory studies involving the device or related devices and known to or that reasonably should be known to the applicant.

(ii) Reports in the scientific literature concerning the device and known to or that reasonably should be known to the applicant. If, after reviewing the summary and bibliography, FDA concludes that the agency needs a copy of the unpublished or published reports, FDA will notify the applicant that copies of such reports shall be submitted.

[51 FR 26364, July 22, 1986, as amended at 51 FR 43344, Dec. 2, 1986; 67 FR 9587, Mar. 4, 2002]

Subparts F–G [Reserved]

Subpart H—Humanitarian Use Devices

SOURCE: 61 FR 33244, June 26, 1996, unless otherwise noted.

§814.100 Purpose and scope.

(a) This subpart H implements section 520(m) of the act. The purpose of section 520(m) is, to the extent consistent with the protection of the public health and safety and with ethical standards, to encourage the discovery and use of devices intended to benefit patients in the treatment or diagnosis of diseases or conditions that affect or are manifested in fewer than 4,000 individuals in the United States per year. This subpart provides procedures for obtaining:

(1) HUD designation of a medical device; and

(2) Marketing approval for the HUD notwithstanding the absence of reasonable assurance of effectiveness that would otherwise be required under sections 514 and 515 of the act.

(b) Although a HUD may also have uses that differ from the humanitarian use, applicants seeking approval of any non-HUD use shall submit a PMA as required under § 814.20, or a premarket notification as required under part 807 of this chapter.

(c) Obtaining marketing approval for a HUD involves two steps:

(1) Obtaining designation of the device as a HUD from FDA's Office of Orphan Products Development, and

(2) Submitting an HDE to the Office of Device Evaluation (ODE), Center for Devices and Radiological Health (CDRH).

(d) A person granted an exemption under section 520(m) of the act shall submit periodic reports as described in § 814.126(b).

(e) FDA may suspend or withdraw approval of an HDE after providing notice and an opportunity for an informal hearing.

[61 FR 33244, June 26, 1996, as amended at 63 FR 59220, Nov. 3, 1998]

§ 814.102 Designation of HUD status.

(a) *Request for designation.* Prior to submitting an HDE application, the applicant shall submit a request for HUD designation to FDA's Office of Orphan Products Development. The request shall contain the following:

(1) A statement that the applicant requests HUD designation for a rare disease or condition or a valid subset of a disease or condition which shall be identified with specificity;

(2) The name and address of the applicant, the name of the applicant's primary contact person and/or resident agent, including title, address, and telephone number;

(3) A description of the rare disease or condition for which the device is to be used, the proposed indication or indications for use of the device, and the reasons why such therapy is needed. If the device is proposed for an indication that represents a subset of a common disease or condition, a demonstration that the subset is medically plausible should be included;

(4) A description of the device and a discussion of the scientific rationale for the use of the device for the rare disease or condition; and

(5) Documentation, with appended authoritative references, to demonstrate that the device is designed to treat or diagnose a disease or condition that affects or is manifested in fewer than 4,000 people in the United States per year. If the device is for diagnostic purposes, the documentation must demonstrate that fewer than 4,000 patients per year would be subjected to diagnosis by the device in the United States. Authoritative references include literature citations in specialized medical journals, textbooks, specialized medical society proceedings, or governmental statistics publications. When no such studies or literature citations exist, the applicant may be able to demonstrate the prevalence of the disease or condition in the United States by providing credible conclusions from appropriate research or surveys.

(b) *FDA action.* Within 45 days of receipt of a request for HUD designation, FDA will take one of the following actions:

(1) Approve the request and notify the applicant that the device has been designated as a HUD based on the information submitted;

(2) Return the request to the applicant pending further review upon submission of additional information. This action will ensue if the request is incomplete because it does not on its face contain all of the information required under § 814.102(a). Upon receipt of this additional information, the review period may be extended up to 45 days; or

(3) Disapprove the request for HUD designation based on a substantive review of the information submitted. FDA may disapprove a request for HUD designation if:

(i) There is insufficient evidence to support the estimate that the disease or condition for which the device is designed to treat or diagnose affects or is manifested in fewer than 4,000 people in the United States per year;

(ii) FDA determines that, for a diagnostic device, 4,000 or more patients in the United States would be subjected to diagnosis using the device per year; or

(iii) FDA determines that the patient population defined in the request is not a medically plausible subset of a larger population.

(c) *Revocation of designation.* FDA may revoke a HUD designation if the agency finds that:

(1) The request for designation contained an untrue statement of material fact or omitted material information; or

(2) Based on the evidence available, the device is not eligible for HUD designation.

(d) *Submission.* The applicant shall submit two copies of a completed, dated, and signed request for HUD designation to: Office of Orphan Products Development (HF–35), Food and Drug Administration, 5600 Fishers Lane, Rockville, MD 20857.

§ 814.104 Original applications.

(a) *United States applicant or representative.* The applicant or an authorized representative shall sign the HDE. If the applicant does not reside or have a place of business within the United States, the HDE shall be countersigned by an authorized representative residing or maintaining a place of business in the United States and shall identify the representative's name and address.

(b) *Contents.* Unless the applicant justifies an omission in accordance with paragraph (d) of this section, an HDE shall include:

(1) A copy of or reference to the determination made by FDA's Office of Orphan Products Development (in accordance with § 814.102) that the device qualifies as a HUD;

(2) An explanation of why the device would not be available unless an HDE were granted and a statement that no comparable device (other than another HUD approved under this subpart or a device under an approved IDE) is available to treat or diagnose the disease or condition. The application also shall contain a discussion of the risks and benefits of currently available devices or alternative forms of treatment in the United States;

(3) An explanation of why the probable benefit to health from the use of the device outweighs the risk of injury or illness from its use, taking into account the probable risks and benefits of currently available devices or alternative forms of treatment. Such explanation shall include a description, explanation, or theory of the underlying disease process or condition, and known or postulated mechanism(s) of action of the device in relation to the disease process or condition;

(4) All of the information required to be submitted under § 814.20(b), except that:

(i) In lieu of the summaries, conclusions, and results from clinical investigations required under §§ 814.20(b)(3)(v)(B), (b)(3)(vi), and (b)(6)(ii), the applicant shall include the summaries, conclusions, and results of all clinical experience or investigations (whether adverse or supportive) reasonably obtainable by the applicant that are relevant to an assessment of the risks and probable benefits of the device; and

(ii) In addition to the proposed labeling requirement set forth in § 814.20(b)(10), the labeling shall bear the following statement: Humanitarian Device. Authorized by Federal law for use in the [treatment or diagnosis] of [specify disease or condition]. The effectiveness of this device for this use has not been demonstrated; and

(5) The amount to be charged for the device and, if the amount is more than $250, a report by an independent certified public accountant, made in accordance with the Statement on Standards for Attestation established by the American Institute of Certified Public Accountants, or in lieu of such a report, an attestation by a responsible individual of the organization, verifying that the amount charged does not exceed the costs of the device's research, development, fabrication, and distribution. If the amount charged is $250 or less, the requirement for a report by an independent certified public

accountant or an attestation by a responsible individual of the organization is waived.

(c) *Omission of information.* If the applicant believes that certain information required under paragraph (b) of this section is not applicable to the device that is the subject of the HDE, and omits any such information from its HDE, the applicant shall submit a statement that identifies and justifies the omission. The statement shall be submitted as a separate section in the HDE and identified in the table of contents. If the justification for the omission is not accepted by the agency, FDA will so notify the applicant.

(d) *Address for submissions and correspondence.* Copies of all original HDE's, amendments and supplements, as well as any correspondence relating to an HDE, shall be sent or delivered to the Document Mail Center (HFZ–401), Office of Device Evaluation, Center for Devices and Radiological Health, Food and Drug Administration, 9200 Corporate Blvd., Rockville, MD 20850.

[61 FR 33244, June 26, 1996, as amended at 63 FR 59220, Nov. 3, 1998]

§ 814.106 HDE amendments and resubmitted HDE's.

An HDE or HDE supplement may be amended or resubmitted upon an applicant's own initiative, or at the request of FDA, for the same reasons and in the same manner as prescribed for PMA's in § 814.37, except that the timeframes set forth in § 814.37(c)(1) and (d) do not apply. If FDA requests an HDE applicant to submit an HDE amendment, and a written response to FDA's request is not received within 75 days of the date of the request, FDA will consider the pending HDE or HDE supplement to be withdrawn voluntarily by the applicant. Furthermore, if the HDE applicant, on its own initiative or at FDA's request, submits a major amendment as described in § 814.37(c)(1), the review period may be extended up to 75 days.

[63 FR 59220, Nov. 3, 1998]

§ 814.108 Supplemental applications.

After FDA approval of an original HDE, an applicant shall submit supplements in accordance with the requirements for PMA's under § 814.39, except that a request for a new indication for use of a HUD shall comply with requirements set forth in § 814.110. The timeframes for review of, and FDA action on, an HDE supplement are the same as those provided in § 814.114 for an HDE.

[63 FR 59220, Nov. 3, 1998]

§ 814.110 New indications for use.

(a) An applicant seeking a new indication for use of a HUD approved under this subpart H shall obtain a new designation of HUD status in accordance with § 814.102 and shall submit an original HDE in accordance with § 814.104.

(b) An application for a new indication for use made under § 814.104 may incorporate by reference any information or data previously submitted to the agency under an HDE.

§ 814.112 Filing an HDE.

(a) The filing of an HDE means that FDA has made a threshold determination that the application is sufficiently complete to permit substantive review. Within 30 days from the date an HDE is received by FDA, the agency will notify the applicant whether the application has been filed. FDA may refuse to file an HDE if any of the following applies:

(1) The application is incomplete because it does not on its face contain all the information required under § 814.104(b);

(2) FDA determines that there is a comparable device available (other than another HUD approved under this subpart or a device under an approved IDE) to treat or diagnose the disease or condition for which approval of the HUD is being sought; or

(3) The application contains an untrue statement of material fact or omits material information.

(4) The HDE is not accompanied by a statement of either certification or disclosure, or both, as required by part 54 of this chapter.

(b) The provisions contained in § 814.42(b), (c), and (d) regarding notification of filing decisions, filing dates, the start of the 75-day review period, and applicant's options in response to

FDA refuse to file decisions shall apply to HDE's.

[61 FR 33244, June 26, 1996, as amended at 63 FR 5254, Feb. 2, 1998; 63 FR 59221, Nov. 3, 1998]

§814.114 Timeframes for reviewing an HDE.

Within 75 days after receipt of an HDE that is accepted for filing and to which the applicant does not submit a major amendment, FDA shall send the applicant an approval order, an approvable letter, a not approvable letter (under §814.116), or an order denying approval (under §814.118).

[63 FR 59221, Nov. 3, 1998]

§814.116 Procedures for review of an HDE.

(a) *Substantive review*. FDA will begin substantive review of an HDE after the HDE is accepted for filing under §814.112. FDA may refer an original HDE application to a panel on its own initiative, and shall do so upon the request of an applicant, unless FDA determines that the application substantially duplicates information previously reviewed by a panel. If the HDE is referred to a panel, the agency shall follow the procedures set forth under §814.44, with the exception that FDA will complete its review of the HDE and the advisory committee report and recommendations within 75 days from receipt of an HDE that is accepted for filing under §814.112 or the date of filing as determined under §814.106, whichever is later. Within the later of these two timeframes, FDA will issue an approval order under paragraph (b) of this section, an approvable letter under paragraph (c) of this section, a not approvable letter under paragraph (d) of this section, or an order denying approval of the application under §814.118(a).

(b) *Approval order*. FDA will issue to the applicant an order approving an HDE if none of the reasons in §814.118 for denying approval of the application applies. FDA will approve an application on the basis of draft final labeling if the only deficiencies in the application concern editorial or similar minor deficiencies in the draft final labeling. Such approval will be conditioned upon the applicant incorporating the specified labeling changes exactly as directed and upon the applicant submitting to FDA a copy of the final printed labeling before marketing. The notice of approval of an HDE will be published in the FEDERAL REGISTER in accordance with the rules and policies applicable to PMA's submitted under §814.20. Following the issuance of an approval order, data and information in the HDE file will be available for public disclosure in accordance with §814.9(b) through (h), as applicable.

(c) *Approvable letter*. FDA will send the applicant an approvable letter if the application substantially meets the requirements of this subpart and the agency believes it can approve the application if specific additional information is submitted or specific conditions are agreed to by the applicant. The approvable letter will describe the information FDA requires to be provided by the applicant or the conditions the applicant is required to meet to obtain approval. For example, FDA may require as a condition to approval:

(1) The submission of certain information identified in the approvable letter, e.g., final labeling;

(2) Restrictions imposed on the device under section 520(e) of the act;

(3) Postapproval requirements as described in subpart E of this part; and

(4) An FDA inspection that finds the manufacturing facilities, methods, and controls in compliance with part 820 of this chapter and, if applicable, that verifies records pertinent to the HDE.

(d) *Not approvable letter*. FDA will send the applicant a not approvable letter if the agency believes that the application may not be approved for one or more of the reasons given in §814.118. The not approvable letter will describe the deficiencies in the application and, where practical, will identify measures required to place the HDE in approvable form. The applicant may respond to the not approvable letter in the same manner as permitted for not approvable letters for PMA's under §814.44(f), with the exception that if a major HDE amendment is submitted, the review period may be extended up to 75 days.

(e) FDA will consider an HDE to have been withdrawn voluntarily if:

(1) The applicant fails to respond in writing to a written request for an amendment within 75 days after the date FDA issues such request;

(2) The applicant fails to respond in writing to an approvable or not approvable letter within 75 days after the date FDA issues such letter; or

(3) The applicant submits a written notice to FDA that the HDE has been withdrawn.

[61 FR 33244, June 26, 1996, as amended at 63 FR 59221, Nov. 3, 1998]

§ 814.118 Denial of approval or withdrawal of approval of an HDE.

(a) FDA may deny approval or withdraw approval of an application if the applicant fails to meet the requirements of section 520(m) of the act or of this part, or of any condition of approval imposed by an IRB or by FDA, or any postapproval requirements imposed under § 814.126. In addition, FDA may deny approval or withdraw approval of an application if, upon the basis of the information submitted in the HDE or any other information before the agency, FDA determines that:

(1) There is a lack of a showing of reasonable assurance that the device is safe under the conditions of use prescribed, recommended, or suggested in the labeling thereof;

(2) The device is ineffective under the conditions of use prescribed, recommended, or suggested in the labeling thereof;

(3) The applicant has not demonstrated that there is a reasonable basis from which to conclude that the probable benefit to health from the use of the device outweighs the risk of injury or illness, taking into account the probable risks and benefits of currently available devices or alternative forms of treatment;

(4) The application or a report submitted by or on behalf of the applicant contains an untrue statement of material fact, or omits material information;

(5) The device's labeling does not comply with the requirements in part 801 or part 809 of this chapter;

(6) A nonclinical laboratory study that is described in the HDE and that is essential to show that the device is safe for use under the conditions prescribed, recommended, or suggested in its proposed labeling, was not conducted in compliance with the good laboratory practice regulations in part 58 of this chapter and no reason for the noncompliance is provided or, if it is, the differences between the practices used in conducting the study and the good laboratory practice regulations do not support the validity of the study;

(7) Any clinical investigation involving human subjects described in the HDE, subject to the institutional review board regulations in part 56 of this chapter or the informed consent regulations in part 50 of this chapter, was not conducted in compliance with those regulations such that the rights or safety of human subjects were not adequately protected;

(8) The applicant does not permit an authorized FDA employee an opportunity to inspect at a reasonable time and in a reasonable manner the facilities and controls, and to have access to and to copy and verify all records pertinent to the application; or

(9) The device's HUD designation should be revoked in accordance with § 814.102(c).

(b) If FDA issues an order denying approval of an application, the agency will comply with the same notice and disclosure provisions required for PMA's under § 814.45(b) and (d), as applicable.

(c) FDA will issue an order denying approval of an HDE after an approvable or not approvable letter has been sent and the applicant:

(1) Submits a requested amendment but any ground for denying approval of the application under § 814.118(a) still applies;

(2) Notifies FDA in writing that the requested amendment will not be submitted; or

(3) Petitions for review under section 515(d)(3) of the act by filing a petition in the form of a petition for reconsideration under § 10.33 of this chapter.

(d) Before issuing an order withdrawing approval of an HDE, FDA will provide the applicant with notice and an opportunity for a hearing as required for PMA's under § 814.46(c) and (d), and will provide the public with notice in accordance with §814.46(e), as applicable.

[61 FR 33244, June 26, 1996, as amended at 63 FR 59221, Nov. 3, 1998]

§814.120 Temporary suspension of approval of an HDE.

An HDE or HDE supplement may be temporarily suspended for the same reasons and in the same manner as prescribed for PMA's in §814.47.

[63 FR 59221, Nov. 3, 1998]

§814.122 Confidentiality of data and information.

(a) *Requirement for disclosure.* The "HDE file" includes all data and information submitted with or referenced in the HDE, any IDE incorporated into the HDE, any HDE amendment or supplement, any report submitted under §814.126, any master file, or any other related submission. Any record in the HDE file will be available for public disclosure in accordance with the provisions of this section and part 20 of this chapter.

(b) *Extent of disclosure.* Disclosure by FDA of the existence and contents of an HDE file shall be subject to the same rules that pertain to PMA's under §814.9(b) through (h), as applicable.

§814.124 Institutional Review Board requirements.

(a) *IRB approval.* The HDE holder is responsible for ensuring that a HUD approved under this subpart is administered only in facilities having an Institutional Review Board (IRB) constituted and acting pursuant to part 56 of this chapter, including continuing review of use of the device. In addition, a HUD may be administered only if such use has been approved by the IRB located at the facility or by a similarly constituted IRB that has agreed to oversee such use and to which the local IRB has deferred in a letter to the HDE holder, signed by the IRB chair or an authorized designee. If, however, a physician in an emergency situation determines that approval from an IRB cannot be obtained in time to prevent serious harm or death to a patient, a HUD may be administered without prior approval by the IRB located at the facility or by a similarly constituted IRB that has agreed to oversee such use. In such an emergency situation, the physician shall, within 5 days after the use of the device, provide written notification to the chairman of the IRB of such use. Such written notification shall include the identification of the patient involved, the date on which the device was used, and the reason for the use.

(b) *Withdrawal of IRB approval.* A holder of an approved HDE shall notify FDA of any withdrawal of approval for the use of a HUD by a reviewing IRB within 5 working days after being notified of the withdrawal of approval.

[61 FR 33244, June 26, 1996, as amended at 63 FR 59221, Nov. 3, 1998]

§814.126 Postapproval requirements and reports.

(a) An HDE approved under this subpart H shall be subject to the postapproval requirements and reports set forth under subpart E of this part, as applicable, with the exception of §814.82(a)(7). In addition, medical device reports submitted to FDA in compliance with the requirements of part 803 of this chapter shall also be submitted to the IRB of record.

(b) In addition to the reports identified in paragraph (a) of this section, the holder of an approved HDE shall prepare and submit the following complete, accurate, and timely reports:

(1) *Periodic reports.* An HDE applicant is required to submit reports in accordance with the approval order. Unless FDA specifies otherwise, any periodic report shall include:

(i) An update of the information required under §814.102(a) in a separately bound volume;

(ii) An update of the information required under §814.104(b)(2), (b)(3), and (b)(5);

(iii) The number of devices that have been shipped or sold since initial marketing approval under this subpart H and, if the number shipped or sold exceeds 4,000, an explanation and estimate of the number of devices used per patient. If a single device is used on multiple patients, the applicant shall submit an estimate of the number of patients treated or diagnosed using the device together with an explanation of the basis for the estimate;

(iv) Information describing the applicant's clinical experience with the device since the HDE was initially approved. This information shall include safety information that is known or reasonably should be known to the applicant, medical device reports made under part 803 of this chapter, any data generated from the postmarketing studies, and information (whether published or unpublished) that is known or reasonably expected to be known by the applicant that may affect an evaluation of the safety of the device or that may affect the statement of contraindications, warnings, precautions, and adverse reactions in the device's labeling; and

(v) A summary of any changes made to the device in accordance with supplements submitted under § 814.108. If information provided in the periodic reports, or any other information in the possession of FDA, gives the agency reason to believe that a device raises public health concerns or that the criteria for exemption are no longer met, the agency may require the HDE holder to submit additional information to demonstrate continued compliance with the HDE requirements.

(2) *Other.* An HDE holder shall maintain records of the names and addresses of the facilities to which the HUD has been shipped, correspondence with reviewing IRB's, as well as any other information requested by a reviewing IRB or FDA. Such records shall be maintained in accordance with the HDE approval order.

[61 FR 33244, June 26, 1996, as amended at 63 FR 59221, Nov. 3, 1998, 71 FR 16228, Mar. 31, 2006]

PART 820—QUALITY SYSTEM REGULATION

Subpart O—Statistical Techniques

AUTHORITY: 21 U.S.C. 351, 352, 360, 360c, 360d, 360e, 360h, 360i, 360j, 360l, 371, 374, 381, 383; 42 U.S.C. 216, 262, 263a, 264.

SOURCE: 61 FR 52654, Oct. 7, 1996, unless otherwise noted.

Subpart A—General Provisions

§ 820.1 Scope.

(a) *Applicability.* (1) Current good manufacturing practice (CGMP) requirements are set forth in this quality system regulation. The requirements in this part govern the methods used in, and the facilities and controls used for, the design, manufacture, packaging, labeling, storage, installation, and servicing of all finished devices intended for human use. The requirements in this part are intended to ensure that finished devices will be safe and effective and otherwise in compliance with the Federal Food, Drug, and Cosmetic Act (the act). This part establishes basic requirements applicable to manufacturers of finished medical devices. If a manufacturer engages in only some operations subject to the requirements in this part, and not in others, that manufacturer need only comply with those requirements applicable to the operations in which it is engaged. With respect to class I devices, design controls apply only to those devices listed in § 820.30(a)(2). This regulation does not apply to manufacturers of components or parts of finished devices, but such manufacturers are encouraged to use appropriate provisions of this regulation as guidance. Manufacturers of human blood and blood components are not subject to this part, but are subject to part 606 of this chapter. Manufacturers of human cells, tissues, and cellular and tissue-based products (HCT/Ps), as defined in § 1271.3(d) of this chapter, that are medical devices (subject to premarket review or notification, or exempt from notification, under an application submitted under the device provisions of the act or under a biological product license application under section 351 of the Public Health Service Act) are subject to this part and are also subject to the donor-eligibility procedures set forth in part 1271 subpart C of this chapter and applicable current good tissue practice procedures in part 1271 subpart D of this chapter. In the event of a conflict between applicable regulations in part 1271 and in other parts of this chapter, the regulation specifically applicable to the device in question shall supersede the more general.

(2) The provisions of this part shall be applicable to any finished device as defined in this part, intended for human use, that is manufactured, imported, or offered for import in any State or Territory of the United States, the District of Columbia, or the Commonwealth of Puerto Rico.

(3) In this regulation the term "where appropriate" is used several times. When a requirement is qualified by "where appropriate," it is deemed to be "appropriate" unless the manufacturer can document justification otherwise. A requirement is "appropriate" if nonimplementation could reasonably be expected to result in the product not meeting its specified requirements or the manufacturer not being able to carry out any necessary corrective action.

(b) The quality system regulation in this part supplements regulations in other parts of this chapter except where explicitly stated otherwise. In the event of a conflict between applicable regulations in this part and in other parts of this chapter, the regulations specifically applicable to the device in question shall supersede any other generally applicable requirements.

(c) *Authority.* Part 820 is established and issued under authority of sections 501, 502, 510, 513, 514, 515, 518, 519, 520, 522, 701, 704, 801, 803 of the act (21 U.S.C. 351, 352, 360, 360c, 360d, 360e, 360h, 360i, 360j, 360l, 371, 374, 381, 383). The failure to comply with any applicable provision in this part renders a device adulterated under section 501(h) of the act. Such a device, as well as any person responsible for the failure to comply, is subject to regulatory action.

(d) *Foreign manufacturers.* If a manufacturer who offers devices for import into the United States refuses to permit or allow the completion of a Food and Drug Administration (FDA) inspection of the foreign facility for the purpose of determining compliance with this part, it shall appear for purposes of section 801(a) of the act, that the methods used in, and the facilities and controls used for, the design, manufacture, packaging, labeling, storage, installation, or servicing of any devices produced at such facility that are offered for import into the United States do not conform to the requirements of section 520(f) of the act and this part and that the devices manufactured at that facility are adulterated under section 501(h) of the act.

(e) *Exemptions or variances.* (1) Any person who wishes to petition for an exemption or variance from any device quality system requirement is subject to the requirements of section 520(f)(2) of the act. Petitions for an exemption or variance shall be submitted according to the procedures set forth in § 10.30 of this chapter, the FDA's administrative procedures. Guidance is available from the Center for Devices and Radiological Health, Division of Small Manufacturers Assistance (HFZ–220), 1350 Piccard Dr., Rockville, MD 20850, U.S.A., telephone 1–800–638–2041 or 1–301–443–6597, FAX 301–443–8818.

(2) FDA may initiate and grant a variance from any device quality system requirement when the agency determines that such variance is in the best interest of the public health. Such variance will remain in effect only so long as there remains a public health need for the device and the device would not likely be made sufficiently available without the variance.

[61 FR 52654, Oct. 7, 1996, as amended at 65 FR 17136, Mar. 31, 2000; 65 FR 66636, Nov. 7, 2000; 69 FR 29829, May 25, 2005]

§ 820.3 Definitions.

(a) *Act* means the Federal Food, Drug, and Cosmetic Act, as amended (secs. 201–903, 52 Stat. 1040 *et seq.*, as amended (21 U.S.C. 321–394)). All definitions in section 201 of the act shall apply to the regulations in this part.

(b) *Complaint* means any written, electronic, or oral communication that alleges deficiencies related to the identity, quality, durability, reliability, safety, effectiveness, or performance of a device after it is released for distribution.

(c) *Component* means any raw material, substance, piece, part, software, firmware, labeling, or assembly which is intended to be included as part of the finished, packaged, and labeled device.

(d) *Control number* means any distinctive symbols, such as a distinctive combination of letters or numbers, or both, from which the history of the manufacturing, packaging, labeling, and distribution of a unit, lot, or batch of finished devices can be determined.

(e) *Design history file* (*DHF*) means a compilation of records which describes the design history of a finished device.

(f) *Design input* means the physical and performance requirements of a device that are used as a basis for device design.

(g) *Design output* means the results of a design effort at each design phase and at the end of the total design effort. The finished design output is the basis for the device master record. The total finished design output consists of the device, its packaging and labeling, and the device master record.

(h) *Design review* means a documented, comprehensive, systematic examination of a design to evaluate the adequacy of the design requirements, to evaluate the capability of the design to meet these requirements, and to identify problems.

(i) *Device history record* (*DHR*) means a compilation of records containing the production history of a finished device.

(j) *Device master record* (*DMR*) means a compilation of records containing the procedures and specifications for a finished device.

(k) *Establish* means define, document (in writing or electronically), and implement.

(l) *Finished device* means any device or accessory to any device that is suitable for use or capable of functioning, whether or not it is packaged, labeled, or sterilized.

(m) *Lot or batch* means one or more components or finished devices that consist of a single type, model, class, size, composition, or software version that are manufactured under essentially the same conditions and that are intended to have uniform characteristics and quality within specified limits.

(n) *Management with executive responsibility* means those senior employees of a manufacturer who have the authority to establish or make changes to the manufacturer's quality policy and quality system.

(o) *Manufacturer* means any person who designs, manufactures, fabricates, assembles, or processes a finished device. Manufacturer includes but is not limited to those who perform the functions of contract sterilization, installation, relabeling, remanufacturing, repacking, or specification development, and initial distributors of foreign entities performing these functions.

(p) *Manufacturing material* means any material or substance used in or used to facilitate the manufacturing process, a concomitant constituent, or a byproduct constituent produced during the manufacturing process, which is present in or on the finished device as a residue or impurity not by design or intent of the manufacturer.

(q) *Nonconformity* means the nonfulfillment of a specified requirement.

(r) *Product* means components, manufacturing materials, in- process devices, finished devices, and returned devices.

(s) *Quality* means the totality of features and characteristics that bear on the ability of a device to satisfy fitness-for-use, including safety and performance.

(t) *Quality audit* means a systematic, independent examination of a manufacturer's quality system that is performed at defined intervals and at sufficient frequency to determine whether both quality system activities and the results of such activities comply with quality system procedures, that these procedures are implemented effectively, and that these procedures are suitable to achieve quality system objectives.

(u) *Quality policy* means the overall intentions and direction of an organization with respect to quality, as established by management with executive responsibility.

(v) *Quality system* means the organizational structure, responsibilities, procedures, processes, and resources for implementing quality management.

(w) *Remanufacturer* means any person who processes, conditions, renovates, repackages, restores, or does any other act to a finished device that significantly changes the finished device's performance or safety specifications, or intended use.

(x) *Rework* means action taken on a nonconforming product so that it will fulfill the specified DMR requirements before it is released for distribution.

(y) *Specification* means any requirement with which a product, process, service, or other activity must conform.

(z) *Validation* means confirmation by examination and provision of objective evidence that the particular requirements for a specific intended use can be consistently fulfilled.

(1) *Process validation* means establishing by objective evidence that a process consistently produces a result or product meeting its predetermined specifications.

(2) *Design validation* means establishing by objective evidence that device specifications conform with user needs and intended use(s).

(aa) *Verification* means confirmation by examination and provision of objective evidence that specified requirements have been fulfilled.

§ 820.5 Quality system.

Each manufacturer shall establish and maintain a quality system that is appropriate for the specific medical device(s) designed or manufactured, and that meets the requirements of this part.

Subpart B—Quality System Requirements

§ 820.20 Management responsibility.

(a) *Quality policy*. Management with executive responsibility shall establish its policy and objectives for, and commitment to, quality. Management with executive responsibility shall ensure that the quality policy is understood, implemented, and maintained at all levels of the organization.

(b) *Organization*. Each manufacturer shall establish and maintain an adequate organizational structure to ensure that devices are designed and produced in accordance with the requirements of this part.

(1) *Responsibility and authority.* Each manufacturer shall establish the appropriate responsibility, authority, and interrelation of all personnel who manage, perform, and assess work affecting quality, and provide the independence and authority necessary to perform these tasks.

(2) *Resources.* Each manufacturer shall provide adequate resources, including the assignment of trained personnel, for management, performance of work, and assessment activities, including internal quality audits, to meet the requirements of this part.

(3) *Management representative.* Management with executive responsibility shall appoint, and document such appointment of, a member of management who, irrespective of other responsibilities, shall have established authority over and responsibility for:

(i) Ensuring that quality system requirements are effectively established and effectively maintained in accordance with this part; and

(ii) Reporting on the performance of the quality system to management with executive responsibility for review.

(c) *Management review.* Management with executive responsibility shall review the suitability and effectiveness of the quality system at defined intervals and with sufficient frequency according to established procedures to ensure that the quality system satisfies the requirements of this part and the manufacturer's established quality policy and objectives. The dates and results of quality system reviews shall be documented.

(d) *Quality planning.* Each manufacturer shall establish a quality plan which defines the quality practices, resources, and activities relevant to devices that are designed and manufactured. The manufacturer shall establish how the requirements for quality will be met.

(e) *Quality system procedures.* Each manufacturer shall establish quality system procedures and instructions. An outline of the structure of the documentation used in the quality system shall be established where appropriate.

§ 820.22 Quality audit.

Each manufacturer shall establish procedures for quality audits and conduct such audits to assure that the quality system is in compliance with the established quality system requirements and to determine the effectiveness of the quality system. Quality audits shall be conducted by individuals who do not have direct responsibility for the matters being audited. Corrective action(s), including a reaudit of deficient matters, shall be taken when necessary. A report of the results of each quality audit, and reaudit(s) where taken, shall be made and such reports shall be reviewed by management having responsibility for the matters audited. The dates and results of quality audits and reaudits shall be documented.

§ 820.25 Personnel.

(a) *General.* Each manufacturer shall have sufficient personnel with the necessary education, background, training, and experience to assure that all activities required by this part are correctly performed.

(b) *Training.* Each manufacturer shall establish procedures for identifying training needs and ensure that all personnel are trained to adequately perform their assigned responsibilities. Training shall be documented.

(1) As part of their training, personnel shall be made aware of device defects which may occur from the improper performance of their specific jobs.

(2) Personnel who perform verification and validation activities shall be made aware of defects and errors that may be encountered as part of their job functions.

Subpart C—Design Controls

§ 820.30 Design controls.

(a) *General.* (1) Each manufacturer of any class III or class II device, and the class I devices listed in paragraph (a)(2) of this section, shall establish and maintain procedures to control the design of the device in order to ensure that specified design requirements are met.

(2) The following class I devices are subject to design controls:

(i) Devices automated with computer software; and

(ii) The devices listed in the following chart.

Section	Device
868.6810	Catheter, Tracheobronchial Suction.
878.4460	Glove, Surgeon's.
880.6760	Restraint, Protective.
892.5650	System, Applicator, Radionuclide, Manual.
892.5740	Source, Radionuclide Teletherapy.

(b) *Design and development planning.* Each manufacturer shall establish and maintain plans that describe or reference the design and development activities and define responsibility for implementation. The plans shall identify and describe the interfaces with different groups or activities that provide, or result in, input to the design and development process. The plans shall be reviewed, updated, and approved as design and development evolves.

(c) *Design input.* Each manufacturer shall establish and maintain procedures to ensure that the design requirements relating to a device are appropriate and address the intended use of the device, including the needs of the user and patient. The procedures shall include a mechanism for addressing incomplete, ambiguous, or conflicting requirements. The design input requirements shall be documented and shall be reviewed and approved by a designated individual(s). The approval, including the date and signature of the individual(s) approving the requirements, shall be documented.

(d) *Design output.* Each manufacturer shall establish and maintain procedures for defining and documenting design output in terms that allow an adequate evaluation of conformance to design input requirements. Design output procedures shall contain or make reference to acceptance criteria and shall ensure that those design outputs that are essential for the proper functioning of the device are identified. Design output shall be documented, reviewed, and approved before release. The approval, including the date and signature of the individual(s) approving the output, shall be documented.

(e) *Design review.* Each manufacturer shall establish and maintain procedures to ensure that formal documented reviews of the design results are planned and conducted at appropriate stages of the device's design development. The procedures shall ensure that participants at each design review include representatives of all functions concerned with the design stage being reviewed and an individual(s) who does not have direct responsibility for the design stage being reviewed, as well as any specialists needed. The results of a design review, including identification of the design, the date, and the individual(s) performing the review, shall be documented in the design history file (the DHF).

(f) *Design verification.* Each manufacturer shall establish and maintain procedures for verifying the device design. Design verification shall confirm that the design output meets the design input requirements. The results of the design verification, including identification of the design, method(s), the date, and the individual(s) performing the verification, shall be documented in the DHF.

(g) *Design validation.* Each manufacturer shall establish and maintain procedures for validating the device design. Design validation shall be performed under defined operating conditions on initial production units, lots, or batches, or their equivalents. Design validation shall ensure that devices conform to defined user needs and intended uses and shall include testing of production units under actual or simulated use conditions. Design validation shall include software validation and risk analysis, where appropriate. The results of the design validation, including identification of the design, method(s), the date, and the individual(s) performing the validation, shall be documented in the DHF.

(h) *Design transfer.* Each manufacturer shall establish and maintain procedures to ensure that the device design is correctly translated into production specifications.

(i) *Design changes.* Each manufacturer shall establish and maintain procedures for the identification, documentation, validation or where appropriate verification, review, and approval of design changes before their implementation.

(j) *Design history file.* Each manufacturer shall establish and maintain a DHF for each type of device. The DHF shall contain or reference the records necessary to demonstrate that the design was developed in accordance with the approved design plan and the requirements of this part.

Subpart D—Document Controls

§ 820.40 Document controls.

Each manufacturer shall establish and maintain procedures to control all documents that are required by this part. The procedures shall provide for the following:

(a) *Document approval and distribution.* Each manufacturer shall designate an individual(s) to review for adequacy and approve prior to issuance all documents established to meet the requirements of this part. The approval, including the date and signature of the individual(s) approving the document, shall be documented. Documents established to meet the requirements of this part shall be available at all locations for which they are designated, used, or otherwise necessary, and all obsolete documents shall be promptly removed from all points of use or otherwise prevented from unintended use.

(b) *Document changes.* Changes to documents shall be reviewed and approved by an individual(s) in the same function or organization that performed the original review and approval, unless specifically designated otherwise. Approved changes shall be communicated to the appropriate personnel in a timely manner. Each manufacturer shall maintain records of changes to documents. Change records shall include a description of the change, identification of the affected documents, the signature of the approving individual(s), the approval date, and when the change becomes effective.

Subpart E—Purchasing Controls

§ 820.50 Purchasing controls.

Each manufacturer shall establish and maintain procedures to ensure that all purchased or otherwise received product and services conform to specified requirements.

(a) *Evaluation of suppliers, contractors, and consultants.* Each manufacturer shall establish and maintain the requirements, including quality requirements, that must be met by suppliers, contractors, and consultants. Each manufacturer shall:

(1) Evaluate and select potential suppliers, contractors, and consultants on the basis of their ability to meet specified requirements, including quality requirements. The evaluation shall be documented.

(2) Define the type and extent of control to be exercised over the product, services, suppliers, contractors, and consultants, based on the evaluation results.

(3) Establish and maintain records of acceptable suppliers, contractors, and consultants.

(b) *Purchasing data.* Each manufacturer shall establish and maintain data that clearly describe or reference the specified requirements, including quality requirements, for purchased or otherwise received product and services. Purchasing documents shall include, where possible, an agreement that the suppliers, contractors, and consultants agree to notify the manufacturer of changes in the product or service so that manufacturers may determine whether the changes may affect the quality of a finished device. Purchasing data shall be approved in accordance with § 820.40.

Subpart F—Identification and Traceability

§ 820.60 Identification.

Each manufacturer shall establish and maintain procedures for identifying product during all stages of receipt, production, distribution, and installation to prevent mixups.

§ 820.65 Traceability.

Each manufacturer of a device that is intended for surgical implant into the body or to support or sustain life and whose failure to perform when properly used in accordance with instructions for use provided in the labeling can be reasonably expected to result in a significant injury to the user shall establish and maintain procedures for identifying with a control number each unit, lot, or batch of finished devices and where appropriate components. The procedures shall facilitate corrective action. Such identification shall be documented in the DHR.

Subpart G—Production and Process Controls

§ 820.70 Production and process controls.

(a) *General.* Each manufacturer shall develop, conduct, control, and monitor production processes to ensure that a device conforms to its specifications. Where deviations from device specifications could occur as a result of the manufacturing process, the manufacturer shall establish and maintain process control procedures that describe any process controls necessary to ensure conformance to specifications. Where process controls are needed they shall include:

(1) Documented instructions, standard operating procedures (SOP's), and methods that define and control the manner of production;

(2) Monitoring and control of process parameters and component and device characteristics during production;

(3) Compliance with specified reference standards or codes;

(4) The approval of processes and process equipment; and

(5) Criteria for workmanship which shall be expressed in documented standards or by means of identified and approved representative samples.

(b) *Production and process changes.* Each manufacturer shall establish and maintain procedures for changes to a specification, method, process, or procedure. Such changes shall be verified or where appropriate validated according to § 820.75, before implementation and these activities shall be documented. Changes shall be approved in accordance with § 820.40.

(c) *Environmental control.* Where environmental conditions could reasonably be expected to have an adverse effect on product quality, the manufacturer shall establish and maintain procedures to adequately control these environmental conditions. Environmental control system(s) shall be periodically inspected to verify that the system, including necessary equipment, is adequate and functioning properly. These activities shall be documented and reviewed.

(d) *Personnel.* Each manufacturer shall establish and maintain requirements for the health, cleanliness, personal practices, and clothing of personnel if contact between such personnel and product or environment could reasonably be expected to have an adverse effect on product quality. The manufacturer shall ensure that maintenance and other personnel who are required to work temporarily under special environmental conditions are appropriately trained or supervised by a trained individual.

(e) *Contamination control.* Each manufacturer shall establish and maintain procedures to prevent contamination of equipment or product by substances that could reasonably be expected to have an adverse effect on product quality.

(f) *Buildings.* Buildings shall be of suitable design and contain sufficient space to perform necessary operations, prevent mixups, and assure orderly handling.

(g) *Equipment.* Each manufacturer shall ensure that all equipment used in the manufacturing process meets specified requirements and is appropriately designed, constructed, placed, and installed to facilitate maintenance, adjustment, cleaning, and use.

(1) *Maintenance schedule.* Each manufacturer shall establish and maintain schedules for the adjustment, cleaning, and other maintenance of equipment to ensure that manufacturing specifications are met. Maintenance activities, including the date and individual(s) performing the maintenance activities, shall be documented.

(2) *Inspection.* Each manufacturer shall conduct periodic inspections in

accordance with established procedures to ensure adherence to applicable equipment maintenance schedules. The inspections, including the date and individual(s) conducting the inspections, shall be documented.

(3) *Adjustment.* Each manufacturer shall ensure that any inherent limitations or allowable tolerances are visibly posted on or near equipment requiring periodic adjustments or are readily available to personnel performing these adjustments.

(h) *Manufacturing material.* Where a manufacturing material could reasonably be expected to have an adverse effect on product quality, the manufacturer shall establish and maintain procedures for the use and removal of such manufacturing material to ensure that it is removed or limited to an amount that does not adversely affect the device's quality. The removal or reduction of such manufacturing material shall be documented.

(i) *Automated processes.* When computers or automated data processing systems are used as part of production or the quality system, the manufacturer shall validate computer software for its intended use according to an established protocol. All software changes shall be validated before approval and issuance. These validation activities and results shall be documented.

§ 820.72 Inspection, measuring, and test equipment.

(a) *Control of inspection, measuring, and test equipment.* Each manufacturer shall ensure that all inspection, measuring, and test equipment, including mechanical, automated, or electronic inspection and test equipment, is suitable for its intended purposes and is capable of producing valid results. Each manufacturer shall establish and maintain procedures to ensure that equipment is routinely calibrated, inspected, checked, and maintained. The procedures shall include provisions for handling, preservation, and storage of equipment, so that its accuracy and fitness for use are maintained. These activities shall be documented.

(b) *Calibration.* Calibration procedures shall include specific directions and limits for accuracy and precision. When accuracy and precision limits are not met, there shall be provisions for remedial action to reestablish the limits and to evaluate whether there was any adverse effect on the device's quality. These activities shall be documented.

(1) *Calibration standards.* Calibration standards used for inspection, measuring, and test equipment shall be traceable to national or international standards. If national or international standards are not practical or available, the manufacturer shall use an independent reproducible standard. If no applicable standard exists, the manufacturer shall establish and maintain an in-house standard.

(2) *Calibration records.* The equipment identification, calibration dates, the individual performing each calibration, and the next calibration date shall be documented. These records shall be displayed on or near each piece of equipment or shall be readily available to the personnel using such equipment and to the individuals responsible for calibrating the equipment.

§ 820.75 Process validation.

(a) Where the results of a process cannot be fully verified by subsequent inspection and test, the process shall be validated with a high degree of assurance and approved according to established procedures. The validation activities and results, including the date and signature of the individual(s) approving the validation and where appropriate the major equipment validated, shall be documented.

(b) Each manufacturer shall establish and maintain procedures for monitoring and control of process parameters for validated processes to ensure that the specified requirements continue to be met.

(1) Each manufacturer shall ensure that validated processes are performed by qualified individual(s).

(2) For validated processes, the monitoring and control methods and data, the date performed, and, where appropriate, the individual(s) performing the process or the major equipment used shall be documented.

(c) When changes or process deviations occur, the manufacturer shall review and evaluate the process and perform revalidation where appropriate. These activities shall be documented.

Subpart H—Acceptance Activities

§ 820.80 Receiving, in-process, and finished device acceptance.

(a) *General.* Each manufacturer shall establish and maintain procedures for acceptance activities. Acceptance activities include inspections, tests, or other verification activities.

(b) *Receiving acceptance activities.* Each manufacturer shall establish and maintain procedures for acceptance of incoming product. Incoming product shall be inspected, tested, or otherwise verified as conforming to specified requirements. Acceptance or rejection shall be documented.

(c) *In-process acceptance activities.* Each manufacturer shall establish and maintain acceptance procedures, where appropriate, to ensure that specified requirements for in-process product are met. Such procedures shall ensure that in-process product is controlled until the required inspection and tests or other verification activities have been completed, or necessary approvals are received, and are documented.

(d) *Final acceptance activities.* Each manufacturer shall establish and maintain procedures for finished device acceptance to ensure that each production run, lot, or batch of finished devices meets acceptance criteria. Finished devices shall be held in quarantine or otherwise adequately controlled until released. Finished devices shall not be released for distribution until:

(1) The activities required in the DMR are completed;

(2) the associated data and documentation is reviewed;

(3) the release is authorized by the signature of a designated individual(s); and

(4) the authorization is dated.

(e) *Acceptance records.* Each manufacturer shall document acceptance activities required by this part. These records shall include:

(1) The acceptance activities performed;

(2) the dates acceptance activities are performed;

(3) the results;

(4) the signature of the individual(s) conducting the acceptance activities; and

(5) where appropriate the equipment used. These records shall be part of the DHR.

§ 820.86 Acceptance status.

Each manufacturer shall identify by suitable means the acceptance status of product, to indicate the conformance or nonconformance of product with acceptance criteria. The identification of acceptance status shall be maintained throughout manufacturing, packaging, labeling, installation, and servicing of the product to ensure that only product which has passed the required acceptance activities is distributed, used, or installed.

Subpart I—Nonconforming Product

§ 820.90 Nonconforming product.

(a) *Control of nonconforming product.* Each manufacturer shall establish and maintain procedures to control product that does not conform to specified requirements. The procedures shall address the identification, documentation, evaluation, segregation, and disposition of nonconforming product. The evaluation of nonconformance shall include a determination of the need for an investigation and notification of the persons or organizations responsible for the nonconformance. The evaluation and any investigation shall be documented.

(b) *Nonconformity review and disposition.* (1) Each manufacturer shall establish and maintain procedures that define the responsibility for review and the authority for the disposition of nonconforming product. The procedures shall set forth the review and disposition process. Disposition of nonconforming product shall be documented. Documentation shall include the justification for use of nonconforming product and the signature of the individual(s) authorizing the use.

(2) Each manufacturer shall establish and maintain procedures for rework, to include retesting and reevaluation of the nonconforming product after rework, to ensure that the product meets

its current approved specifications. Rework and reevaluation activities, including a determination of any adverse effect from the rework upon the product, shall be documented in the DHR.

Subpart J—Corrective and Preventive Action

§ 820.100 Corrective and preventive action.

(a) Each manufacturer shall establish and maintain procedures for implementing corrective and preventive action. The procedures shall include requirements for:

(1) Analyzing processes, work operations, concessions, quality audit reports, quality records, service records, complaints, returned product, and other sources of quality data to identify existing and potential causes of nonconforming product, or other quality problems. Appropriate statistical methodology shall be employed where necessary to detect recurring quality problems;

(2) Investigating the cause of nonconformities relating to product, processes, and the quality system;

(3) Identifying the action(s) needed to correct and prevent recurrence of nonconforming product and other quality problems;

(4) Verifying or validating the corrective and preventive action to ensure that such action is effective and does not adversely affect the finished device;

(5) Implementing and recording changes in methods and procedures needed to correct and prevent identified quality problems;

(6) Ensuring that information related to quality problems or nonconforming product is disseminated to those directly responsible for assuring the quality of such product or the prevention of such problems; and

(7) Submitting relevant information on identified quality problems, as well as corrective and preventive actions, for management review.

(b) All activities required under this section, and their results, shall be documented.

Subpart K—Labeling and Packaging Control

§ 820.120 Device labeling.

Each manufacturer shall establish and maintain procedures to control labeling activities.

(a) *Label integrity.* Labels shall be printed and applied so as to remain legible and affixed during the customary conditions of processing, storage, handling, distribution, and where appropriate use.

(b) *Labeling inspection.* Labeling shall not be released for storage or use until a designated individual(s) has examined the labeling for accuracy including, where applicable, the correct expiration date, control number, storage instructions, handling instructions, and any additional processing instructions. The release, including the date and signature of the individual(s) performing the examination, shall be documented in the DHR.

(c) *Labeling storage.* Each manufacturer shall store labeling in a manner that provides proper identification and is designed to prevent mixups.

(d) *Labeling operations.* Each manufacturer shall control labeling and packaging operations to prevent labeling mixups. The label and labeling used for each production unit, lot, or batch shall be documented in the DHR.

(e) *Control number.* Where a control number is required by § 820.65, that control number shall be on or shall accompany the device through distribution.

§ 820.130 Device packaging.

Each manufacturer shall ensure that device packaging and shipping containers are designed and constructed to protect the device from alteration or damage during the customary conditions of processing, storage, handling, and distribution.

Subpart L—Handling, Storage, Distribution, and Installation

§ 820.140 Handling.

Each manufacturer shall establish and maintain procedures to ensure that mixups, damage, deterioration, contamination, or other adverse effects to product do not occur during handling.

§ 820.150 Storage.

(a) Each manufacturer shall establish and maintain procedures for the control of storage areas and stock rooms for product to prevent mixups, damage, deterioration, contamination, or other adverse effects pending use or distribution and to ensure that no obsolete, rejected, or deteriorated product is used or distributed. When the quality of product deteriorates over time, it shall be stored in a manner to facilitate proper stock rotation, and its condition shall be assessed as appropriate.

(b) Each manufacturer shall establish and maintain procedures that describe the methods for authorizing receipt from and dispatch to storage areas and stock rooms.

§ 820.160 Distribution.

(a) Each manufacturer shall establish and maintain procedures for control and distribution of finished devices to ensure that only those devices approved for release are distributed and that purchase orders are reviewed to ensure that ambiguities and errors are resolved before devices are released for distribution. Where a device's fitness for use or quality deteriorates over time, the procedures shall ensure that expired devices or devices deteriorated beyond acceptable fitness for use are not distributed.

(b) Each manufacturer shall maintain distribution records which include or refer to the location of:

(1) The name and address of the initial consignee;

(2) The identification and quantity of devices shipped;

(3) The date shipped; and

(4) Any control number(s) used.

§ 820.170 Installation.

(a) Each manufacturer of a device requiring installation shall establish and maintain adequate installation and inspection instructions, and where appropriate test procedures. Instructions and procedures shall include directions for ensuring proper installation so that the device will perform as intended after installation. The manufacturer shall distribute the instructions and procedures with the device or otherwise make them available to the person(s) installing the device.

(b) The person installing the device shall ensure that the installation, inspection, and any required testing are performed in accordance with the manufacturer's instructions and procedures and shall document the inspection and any test results to demonstrate proper installation.

Subpart M—Records

§ 820.180 General requirements.

All records required by this part shall be maintained at the manufacturing establishment or other location that is reasonably accessible to responsible officials of the manufacturer and to employees of FDA designated to perform inspections. Such records, including those not stored at the inspected establishment, shall be made readily available for review and copying by FDA employee(s). Such records shall be legible and shall be stored to minimize deterioration and to prevent loss. Those records stored in automated data processing systems shall be backed up.

(a) *Confidentiality.* Records deemed confidential by the manufacturer may be marked to aid FDA in determining whether information may be disclosed under the public information regulation in part 20 of this chapter.

(b) *Record retention period.* All records required by this part shall be retained for a period of time equivalent to the design and expected life of the device, but in no case less than 2 years from the date of release for commercial distribution by the manufacturer.

(c) *Exceptions.* This section does not apply to the reports required by § 820.20(c) Management review, § 820.22 Quality audits, and supplier audit reports used to meet the requirements of § 820.50(a) Evaluation of suppliers, contractors, and consultants, but does apply to procedures established under these provisions. Upon request of a designated employee of FDA, an employee in management with executive responsibility shall certify in writing that the management reviews and quality audits required under this part, and supplier audits where applicable, have been performed and documented, the dates on which they were performed, and that any required corrective action has been undertaken.

§ 820.181 Device master record.

Each manufacturer shall maintain device master records (DMR's). Each manufacturer shall ensure that each DMR is prepared and approved in accordance with § 820.40. The DMR for each type of device shall include, or refer to the location of, the following information:

(a) Device specifications including appropriate drawings, composition, formulation, component specifications, and software specifications;

(b) Production process specifications including the appropriate equipment specifications, production methods, production procedures, and production environment specifications;

(c) Quality assurance procedures and specifications including acceptance criteria and the quality assurance equipment to be used;

(d) Packaging and labeling specifications, including methods and processes used; and

(e) Installation, maintenance, and servicing procedures and methods.

§ 820.184 Device history record.

Each manufacturer shall maintain device history records (DHR's). Each manufacturer shall establish and maintain procedures to ensure that DHR's for each batch, lot, or unit are maintained to demonstrate that the device is manufactured in accordance with the DMR and the requirements of this part. The DHR shall include, or refer to the location of, the following information:

(a) The dates of manufacture;

(b) The quantity manufactured;

(c) The quantity released for distribution;

(d) The acceptance records which demonstrate the device is manufactured in accordance with the DMR;

(e) The primary identification label and labeling used for each production unit; and

(f) Any device identification(s) and control number(s) used.

§ 820.186 Quality system record.

Each manufacturer shall maintain a quality system record (QSR). The QSR shall include, or refer to the location of, procedures and the documentation of activities required by this part that are not specific to a particular type of device(s), including, but not limited to, the records required by § 820.20. Each manufacturer shall ensure that the QSR is prepared and approved in accordance with § 820.40.

§ 820.198 Complaint files.

(a) Each manufacturer shall maintain complaint files. Each manufacturer shall establish and maintain procedures for receiving, reviewing, and evaluating complaints by a formally designated unit. Such procedures shall ensure that:

(1) All complaints are processed in a uniform and timely manner;

(2) Oral complaints are documented upon receipt; and

(3) Complaints are evaluated to determine whether the complaint represents an event which is required to be reported to FDA under part 803 of this chapter, Medical Device Reporting.

(b) Each manufacturer shall review and evaluate all complaints to determine whether an investigation is necessary. When no investigation is made, the manufacturer shall maintain a record that includes the reason no investigation was made and the name of the individual responsible for the decision not to investigate.

(c) Any complaint involving the possible failure of a device, labeling, or packaging to meet any of its specifications shall be reviewed, evaluated, and investigated, unless such investigation has already been performed for a similar complaint and another investigation is not necessary.

(d) Any complaint that represents an event which must be reported to FDA under part 803 of this chapter shall be promptly reviewed, evaluated, and investigated by a designated individual(s) and shall be maintained in a separate portion of the complaint files or otherwise clearly identified. In addition to the information required by § 820.198(e), records of investigation under this paragraph shall include a determination of:

(1) Whether the device failed to meet specifications;

(2) Whether the device was being used for treatment or diagnosis; and

(3) The relationship, if any, of the device to the reported incident or adverse event.

(e) When an investigation is made under this section, a record of the investigation shall be maintained by the formally designated unit identified in paragraph (a) of this section. The record of investigation shall include:

(1) The name of the device;

(2) The date the complaint was received;

(3) Any device identification(s) and control number(s) used;

(4) The name, address, and phone number of the complainant;

(5) The nature and details of the complaint;

(6) The dates and results of the investigation;

(7) Any corrective action taken; and

(8) Any reply to the complainant.

(f) When the manufacturer's formally designated complaint unit is located at a site separate from the manufacturing establishment, the investigated complaint(s) and the record(s) of investigation shall be reasonably accessible to the manufacturing establishment.

(g) If a manufacturer's formally designated complaint unit is located outside of the United States, records required by this section shall be reasonably accessible in the United States at either:

(1) A location in the United States where the manufacturer's records are regularly kept; or

(2) The location of the initial distributor.

[61 FR 52654, Oct. 7, 1996, as amended at 69 FR 11313, Mar. 10, 2004; 71 FR 16228, Mar. 31, 2006]

Subpart N—Servicing

§ 820.200 Servicing.

(a) Where servicing is a specified requirement, each manufacturer shall establish and maintain instructions and procedures for performing and verifying that the servicing meets the specified requirements.

(b) Each manufacturer shall analyze service reports with appropriate statistical methodology in accordance with § 820.100.

(c) Each manufacturer who receives a service report that represents an event which must be reported to FDA under part 803 of this chapter shall automatically consider the report a complaint and shall process it in accordance with the requirements of § 820.198.

(d) Service reports shall be documented and shall include:

(1) The name of the device serviced;

(2) Any device identification(s) and control number(s) used;

(3) The date of service;

(4) The individual(s) servicing the device;

(5) The service performed; and

(6) The test and inspection data.

[61 FR 52654, Oct. 7, 1996, as amended at 69 FR 11313, Mar. 10, 2004]

Subpart O—Statistical Techniques

§ 820.250 Statistical techniques.

(a) Where appropriate, each manufacturer shall establish and maintain procedures for identifying valid statistical techniques required for establishing, controlling, and verifying the acceptability of process capability and product characteristics.

(b) Sampling plans, when used, shall be written and based on a valid statistical rationale. Each manufacturer shall establish and maintain procedures to ensure that sampling methods are adequate for their intended use and to ensure that when changes occur the sampling plans are reviewed. These activities shall be documented.

PART 821—MEDICAL DEVICE TRACKING REQUIREMENTS

Subpart A—General Provisions

Sec.

Subpart B—Tracking Requirements

208-072 D-6

Subpart C—Additional Requirements and Responsibilities

821.30 Tracking obligations of persons other than device manufacturers: distributor requirements.

Subpart D—Records and Inspections

821.50 Availability.
821.55 Confidentiality.
821.60 Retention of records.

AUTHORITY: 21 U.S.C. 331, 351, 352, 360, 360e, 360h, 360i, 371, 374.

SOURCE: 58 FR 43447, Aug. 16, 1993, unless otherwise noted.

Subpart A—General Provisions

§ 821.1 Scope.

(a) The regulations in this part implement section 519(e) of the Federal Food, Drug, and Cosmetic Act (the act), which provides that the Food and Drug Administration may require a manufacturer to adopt a method of tracking a class II or class III device, if the device meets one of the following three criteria and FDA issues an order to the manufacturer: the failure of the device would be reasonably likely to have serious adverse health consequences; or the device is intended to be implanted in the human body for more than 1 year; or the device is a life-sustaining or life-supporting device used outside a device user facility. A device that meets one of these criteria and is the subject of an FDA order must comply with this part and is referred to, in this part, as a "tracked device."

(b) These regulations are intended to ensure that tracked devices can be traced from the device manufacturing facility to the person for whom the device is indicated, that is, the patient. Effective tracking of devices from the manufacturing facility, through the distributor network (including distributors, retailers, rental firms and other commercial enterprises, device user facilities, and licensed practitioners) and, ultimately, to the patient is necessary for the effectiveness of remedies prescribed by the act, such as patient notification (section 518(a) of the act) or device recall (section 518(e) of the act). Although these regulations do not preclude a manufacturer from involving outside organizations in that manufacturer's device tracking effort, the legal responsibility for complying with this part rests with manufacturers who are subject to tracking orders, and that responsibility cannot be altered, modified, or in any way abrogated by contracts or other agreements.

(c) The primary burden for ensuring that the tracking system works rests upon the manufacturer. A manufacturer or any other person, including a distributor, final distributor, or multiple distributor, who distributes a device subject to tracking, who fails to comply with any applicable requirement of section 519(e) of the act or of this part, or any person who causes such failure, misbrands the device within the meaning of section 501(t)(2) of the act and commits a prohibited act within the meaning of sections 301(e) and 301(q)(1)(B) of the act.

(d) Any person subject to this part who permanently discontinues doing business is required to notify FDA at the time the person notifies any government agency, court, or supplier, and provide FDA with a complete set of its tracking records and information. However, if a person ceases distribution of a tracked device but continues to do other business, that person continues to be responsible for compliance with this part unless another person, affirmatively and in writing, assumes responsibility for continuing the tracking of devices previously distributed under this part. Further, if a person subject to this part goes out of business completely, but other persons acquire the right to manufacture or distribute tracked devices, those other persons are deemed to be responsible for continuing the tracking responsibility of the previous person under this part.

[58 FR 43447, Aug. 16, 1993, as amended at 67 FR 5951, Feb. 8, 2002]

§ 821.2 Exemptions and variances.

(a) A manufacturer, importer, or distributor may seek an exemption or variance from one or more requirements of this part.

(b) A request for an exemption or variance shall be submitted in the form of a petition under § 10.30 of this chapter and shall comply with the requirements set out therein, except that a response shall be issued in 90 days. The Director or Deputy Directors, CDRH, or the Director, Office of Compliance, CDRH, shall issue responses to requests under this section. The petition shall also contain the following:

(1) The name of the device and device class and representative labeling showing the intended use(s) of the device;

(2) The reasons that compliance with the tracking requirements of this part is unnecessary;

(3) A complete description of alternative steps that are available, or that the petitioner has already taken, to ensure that an effective tracking system is in place; and

(4) Other information justifying the exemption or variance.

(c) An exemption or variance is not effective until the Director, Office of Compliance and Surveillance, CDRH, approves the request under § 10.30(e)(2)(i) of this chapter.

[58 FR 43447, Aug. 16, 1993, as amended at 59 FR 31138, June 17, 1994; 67 FR 5951, Feb. 8, 2002]

§ 821.3 Definitions.

The following definitions and terms apply to this part:

(a) *Act* means the Federal Food, Drug, and Cosmetic Act, 21 U.S.C. 321 *et seq.*, as amended.

(b) *Importer* means the initial distributor of an imported device who is subject to a tracking order. "Importer" does not include anyone who only furthers the marketing, e.g., brokers, jobbers, or warehousers.

(c) *Manufacturer* means any person, including any importer, repacker and/or relabeler, who manufactures, prepares, propagates, compounds, assembles, or processes a device or engages in any of the activities described in § 807.3(d) of this chapter.

(d) *Device failure* means the failure of a device to perform or function as intended, including any deviations from the device's performance specifications or intended use.

(e) *Serious adverse health consequences* means any significant adverse experience related to a device, including device-related events which are life-threatening or which involve permanent or long-term injuries or illnesses.

(f) *Device intended to be implanted in the human body for more than 1 year* means a device that is intended to be placed into a surgically or naturally formed cavity of the human body for more than 1 year to continuously assist, restore, or replace the function of an organ system or structure of the human body throughout the useful life of the device. The term does not include a device that is intended and used only for temporary purposes or that is intended for explantation in 1 year or less.

(g) *Life-supporting or life-sustaining device used outside a device user facility* means a device which is essential, or yields information that is essential, to the restoration or continuation of a bodily function important to the continuation of human life that is intended for use outside a hospital, nursing home, ambulatory surgical facility, or diagnostic or outpatient treatment facility. Physicians' offices are not device user facilities and, therefore, devices used therein are subject to tracking if they otherwise satisfy the statutory and regulatory criteria.

(h) *Distributor* means any person who furthers the distribution of a device from the original place of manufacture to the person who makes delivery or sale to the ultimate user, i.e., the final or multiple distributor, but who does not repackage or otherwise change the container, wrapper, or labeling of the device or device package.

(i) *Final distributor* means any person who distributes a tracked device intended for use by a single patient over the useful life of the device to the patient. This term includes, but is not limited to, licensed practitioners, retail pharmacies, hospitals, and other types of device user facilities.

(j) *Distributes* means any distribution of a tracked device, including the charitable distribution of a tracked device. This term does not include the distribution of a device under an effective investigational device exemption in accordance with section 520(g) of the act and part 812 of this chapter or the distribution of a device for teaching, law enforcement, research, or analysis as specified in § 801.125 of this chapter.

(k) *Multiple distributor* means any device user facility, rental company, or any other entity that distributes a life-sustaining or life-supporting device intended for use by more than one patient over the useful life of the device.

(l) *Licensed practitioner* means a physician, dentist, or other health care practitioner licensed by the law of the State in which he or she practices to use or order the use of the tracked device.

(m) Any term defined in section 201 of the act shall have the same definition in this part.

[58 FR 43447, Aug. 16, 1993, as amended at 67 FR 5951, Feb. 8, 2002]

§ 821.4 Imported devices.

For purposes of this part, the importer of a tracked device shall be considered the manufacturer and shall be required to comply with all requirements of this part applicable to manufacturers. Importers must keep all information required under this part in the United States.

Subpart B—Tracking Requirements

§ 821.20 Devices subject to tracking.

(a) A manufacturer of any class II or class III device that fits within one of the three criteria within § 821.1(a) must track that device in accordance with this part, if FDA issues a tracking order to that manufacturer.

(b) When responding to premarket notification submissions and remarket approval applications, FDA will notify the sponsor by issuing an order that states that FDA believes the device meets the criteria of section 519(e)(1) of the act and, by virtue of the order, the sponsor must track the device.

[67 FR 5951, Feb. 8, 2002]

§ 821.25 Device tracking system and content requirements: manufacturer requirements.

(a) A manufacturer of a tracked device shall adopt a method of tracking for each such type of device that it distributes that enables a manufacturer to provide FDA with the following information in writing for each tracked device distributed:

(1) Except as required by order under section 518(e) of the act, within 3 working days of a request from FDA, prior to the distribution of a tracked device to a patient, the name, address, and telephone number of the distributor, multiple distributor, or final distributor holding the device for distribution and the location of the device;

(2) Within 10 working days of a request from FDA for tracked devices that are intended for use by a single patient over the life of the device, after distribution to or implantation in a patient:

(i) The lot number, batch number, model number, or serial number of the device or other identifier necessary to provide for effective tracking of the devices;

(ii) The date the device was shipped by the manufacturer;

(iii) The name, address, telephone number, and social security number (if available) of the patient receiving the device, unless not released by the patient under § 821.55(a);

(iv) The date the device was provided to the patient;

(v) The name, mailing address, and telephone number of the prescribing physician;

(vi) The name, mailing address, and telephone number of the physician regularly following the patient if different than the prescribing physician; and

(vii) If applicable, the date the device was explanted and the name, mailing address, and telephone number of the explanting physician; the date of the patient's death; or the date the device was returned to the manufacturer, permanently retired from use, or otherwise permanently disposed of.

(3) Except as required by order under section 518(e) of the act, within 10 working days of a request from FDA for tracked devices that are intended for use by more than one patient, after the distribution of the device to the multiple distributor:

(i) The lot model number, batch number, serial number of the device or other identifier necessary to provide for effective tracking of the device;

(ii) The date the device was shipped by the manufacturer;

(iii) The name, address, and telephone number of the multiple distributor;

(iv) The name, address, telephone number, and social security number (if available) of the patient using the device, unless not released by the patient under § 821.55(a);

(v) The location of the device;

(vi) The date the device was provided for use by the patient;

(vii) The name, address, and telephone number of the prescribing physician; and

(viii) If and when applicable, the date the device was returned to the manufacturer, permanently retired from use, or otherwise permanently disposed of.

(b) A manufacturer of a tracked device shall keep current records in accordance with its standard operating procedure of the information identified in paragraphs (a)(1), (a)(2) and (a)(3)(i) through (a)(3)(iii) of this section on each tracked device released for distribution for as long as such device is in use or in distribution for use.

(c) A manufacturer of a tracked device shall establish a written standard operating procedure for the collection, maintenance, and auditing of the data specified in paragraphs (a) and (b) of this section. A manufacturer shall make this standard operating procedure available to FDA upon request. A manufacturer shall incorporate the following into the standard operating procedure:

(1) Data collection and recording procedures, which shall include a procedure for recording when data which is required under this part is missing and could not be collected and the reason why such required data is missing and could not be collected;

(2) A method for recording all modifications or changes to the tracking system or to the data collected and maintained under the tracking system, reasons for any modification or change, and dates of any modification or change. Modification and changes included under this requirement include modifications to the data (including termination of tracking), the data format, the recording system, and the file maintenance procedures system; and

(3) A quality assurance program that includes an audit procedure to be run for each device product subject to tracking, at not less than 6-month intervals for the first 3 years of distribution and at least once a year thereafter. This audit procedure shall provide for statistically relevant sampling of the data collected to ensure the accuracy of data and performance testing of the functioning of the tracking system.

(d) When a manufacturer becomes aware that a distributor, final distributor, or multiple distributor has not collected, maintained, or furnished any record or information required by this part, the manufacturer shall notify the FDA district office responsible for the area in which the distributor, final distributor, or multiple distributor is located of the failure of such persons to comply with the requirements of this part. Manufacturers shall have taken reasonable steps to obtain compliance by the distributor, multiple distributor, or final distributor in question before notifying FDA.

(e) A manufacturer may petition for an exemption or variance from one or more requirements of this part according to the procedures in § 821.2 of this chapter.

[58 FR 43447, Aug. 16, 1993, as amended at 67 FR 5951, Feb. 8, 2002]

Subpart C—Additional Requirements and Responsibilities

§ 821.30 Tracking obligations of persons other than device manufacturers: distributor requirements.

(a) A distributor, final distributor, or multiple distributor of any tracked device shall, upon purchasing or otherwise acquiring any interest in such a device, promptly provide the manufacturer tracking the device with the following information:

(1) The name and address of the distributor, final distributor or multiple distributor;

(2) The lot number, batch number, model number, or serial number of the device or other identifier used by the manufacturer to track the device;

(3) The date the device was received;

(4) The person from whom the device was received;

(5) If and when applicable, the date the device was explanted, the date of the patient's death, or the date the device was returned to the distributor, permanently retired from use, or otherwise permanently disposed of.

(b) A final distributor, upon sale or other distribution of a tracked device for use in or by the patient, shall promptly provide the manufacturer tracking the device with the following information:

(1) The name and address of the final distributor,

(2) The lot number, batch number, model number, or serial number of the device or other identifier used by the manufacturer to track the device;

(3) The name, address, telephone number, and social security number (if available) of the patient receiving the device, unless not released by the patient under § 821.55(a);

(4) The date the device was provided to the patient or for use in the patient;

(5) The name, mailing address, and telephone number of the prescribing physician;

(6) The name, mailing address, and telephone number of the physician regularly following the patient if different than the prescribing physician; and

(7) When applicable, the date the device was explanted and the name, mailing address, and telephone number of the explanting physician, the date of the patient's death, or the date the device was returned to the manufacturer, permanently retired from use, or otherwise permanently disposed of.

(c)(1) A multiple distributor shall keep written records of the following each time such device is distributed for use by a patient:

(i) The lot number, batch number, or model number, or serial number of the device or other identifier used by the manufacturer to track the device;

(ii) The name, address, telephone number, and social security number (if available) of the patient using the device;

(iii) The location of the device, unless not released by the patient under § 821.55(a);

(iv) The date the device was provided for use by the patient;

(v) The name, address, and telephone number of the prescribing physician;

(vi) The name, address, and telephone number of the physician regularly following the patient if different than the prescribing physician; and

(vii) When applicable, the date the device was permanently retired from use or otherwise permanently disposed of.

(2) Except as required by order under section 518(e) of the act, any person who is a multiple distributor subject to the recordkeeping requirement of paragraph (c)(1) of this section shall, within 5 working days of a request from the manufacturer or within 10 working days of a request from FDA for the information identified in paragraph (c)(1) of this section, provide such information to the manufacturer or FDA.

(d) A distributor, final distributor, or multiple distributor shall make any records required to be kept under this part available to the manufacturer of the tracked device for audit upon written request by an authorized representative of the manufacturer.

(e) A distributor, final distributor, or multiple distributor may petition for an exemption or variance from one or more requirements of this part according to the procedures in § 821.2.

[58 FR 43447, Aug. 16, 1993, as amended at 67 FR 5951, Feb. 8, 2002]

Subpart D—Records and Inspections

§ 821.50 Availability.

(a) Manufacturers, distributors, multiple distributors, and final distributors shall, upon the presentation by an FDA representative of official credentials and the issuance of Form FDA 482 at the initiation of an inspection of an establishment or person under section 704 of the act, make each record and all information required to be collected and maintained under this part and all records and information related to the events and persons identified in such records available to FDA personnel.

(b) Records and information referenced in paragraph (a) of this section shall be available to FDA personnel for purposes of reviewing, copying, or any other use related to the enforcement of the act and this part. Records required to be kept by this part shall be kept in a centralized point for each manufacturer or distributor within the United States.

[58 FR 43447, Aug. 16, 1993, as amended at 65 FR 43690, July 14, 2000]

§ 821.55 Confidentiality.

(a) Any patient receiving a device subject to tracking requirements under this part may refuse to release, or refuse permission to release, the patient's name, address, telephone number, and social security number, or other identifying information for the purpose of tracking.

(b) Records and other information submitted to FDA under this part shall be protected from public disclosure to the extent permitted under part 20 of this chapter, and in accordance with § 20.63 of this chapter, information contained in such records that would identify patient or research subjects shall not be available for public disclosure except as provided in those parts.

(c) Patient names or other identifiers may be disclosed to a manufacturer or other person subject to this part or to a physician when the health or safety of the patient requires that such persons have access to the information. Such notification will be pursuant to agreement that the record or information will not be further disclosed except as the health aspects of the patient requires. Such notification does not constitute public disclosure and will not trigger the availability of the same information to the public generally.

[58 FR 43447, Aug. 16, 1993, as amended at 67 FR 5951, Feb. 8, 2002]

§ 821.60 Retention of records.

Persons required to maintain records under this part shall maintain such records for the useful life of each tracked device they manufacture or distribute. The useful life of a device is the time a device is in use or in distribution for use. For example, a record may be retired if the person maintaining the record becomes aware of the fact that the device is no longer in use, has been explanted, returned to the manufacturer, or the patient has died.

PART 822—POSTMARKET SURVEILLANCE

Subpart A—General Provisions

Sec.

Subpart B—Notification

Subpart C—Postmarket Surveillance Plan

Subpart D—FDA Review and Action

Subpart E—Responsibilities of Manufacturers

822.24 What are my responsibilities once I am notified that I am required to conduct postmarket surveillance?
822.25 What are my responsibilities after my postmarket surveillance plan has been approved?
822.26 If my company changes ownership, what must I do?
822.27 If I go out of business, what must I do?
822.28 If I stop marketing the device subject to postmarket surveillance, what must I do?

Subpart F—Waivers and Exemptions

822.29 May I request a waiver of a specific requirement of this part?
822.30 May I request exemption from the requirement to conduct postmarket surveillance?

Subpart G—Records and Reports

822.31 What records am I required to keep?
822.32 What records are the investigators in my surveillance plan required to keep?
822.33 How long must we keep the records?
822.34 What must I do with the records if the sponsor of the plan or an investigator in the plan changes?
822.35 Can you inspect my manufacturing site or other sites involved in my postmarket surveillance plan?
822.36 Can you inspect and copy the records related to my postmarket surveillance plan?
822.37 Under what circumstances would you inspect records identifying subjects?
822.38 What reports must I submit to you?

AUTHORITY: 21 U.S.C. 331, 352, 360i, 360l, 371, 374.

SOURCE: 67 FR 38887, June 6, 2002, unless otherwise noted.

Subpart A—General Provisions

§ 822.1 What does this part cover?

This part implements section 522 of the Federal Food, Drug, and Cosmetic Act (the act) by providing procedures and requirements for postmarket surveillance of class II and class III devices that meet any of the following criteria:

(a) Failure of the device would be reasonably likely to have serious adverse health consequences;

(b) The device is intended to be implanted in the human body for more than 1 year; or

(c) The device is intended to be used outside a user facility to support or sustain life. If you fail to comply with requirements that we order under section 522 of the act and this part, your device is considered misbranded under section 502(t)(3) of the act and you are in violation of section 301(q)(1)(C) of the act.

§ 822.2 What is the purpose of this part?

The purpose of this part is to implement our postmarket surveillance authority to maximize the likelihood that postmarket surveillance plans will result in the collection of useful data. These data can reveal unforeseen adverse events, the actual rate of anticipated adverse events, or other information necessary to protect the public health.

§ 822.3 How do you define the terms used in this part?

Some of the terms we use in this part are specific to postmarket surveillance and reflect the language used in the statute (law). Other terms are more general and reflect our interpretation of the law. This section of the part defines the following terms:

(a) *Act* means the Federal Food, Drug, and Cosmetic Act, 21 U.S.C. 301 *et seq.*, as amended.

(b) *Designated person* means the individual who conducts or supervises the conduct of your postmarket surveillance. If your postmarket surveillance plan includes a team of investigators, as defined below, the designated person is the responsible leader of that team.

(c) *Device failure* means a device does not perform or function as intended, and includes any deviation from the device's performance specifications or intended use.

(d) *General plan guidance* means agency guidance that provides information about the requirement to conduct postmarket surveillance, the submission of a plan to us for approval, the content of the submission, and the conduct and reporting requirements of the surveillance.

(e) *Investigator* means an individual who collects data or information in support of a postmarket surveillance plan.

(f) *Life-supporting or life-sustaining device used outside a device user facility* means that a device is essential to, or yields information essential to, the restoration or continuation of a bodily function important to the continuation of human life and is used outside a hospital, nursing home, ambulatory surgical facility, or diagnostic or outpatient treatment facility. A physician's office is not a device user facility.

(g) *Manufacturer* means any person, including any importer, repacker, and/or relabeler, who manufactures, prepares, propagates, compounds, assembles, processes a device, or engages in any of the activities described in § 807.3(d) of this chapter.

(h) *Postmarket surveillance* means the active, systematic, scientifically valid collection, analysis, and interpretation of data or other information about a marketed device.

(i) *Prospective surveillance* means that the subjects are identified at the beginning of the surveillance and data or other information will be collected from that time forward (as opposed to retrospective surveillance).

(j) *Serious adverse health consequences* means any significant adverse experience related to a device, including device-related events that are life-threatening or that involve permanent or long-term injuries or illnesses.

(k) *Specific guidance* means guidance that provides information regarding postmarket surveillance for specific types or categories of devices or specific postmarket surveillance issues. This type of guidance may be used to supplement general guidance and may address such topics as the type of surveillance approach that is appropriate for the device and the postmarket surveillance question, sample size, or specific reporting requirements.

(l) *Surveillance question* means the issue or issues to be addressed by the postmarket surveillance.

(m) *Unforeseen adverse event* means any serious adverse health consequence that either is not addressed in the labeling of the device or occurs at a rate higher than anticipated.

§ 822.4 Does this part apply to me?

If we have ordered you to conduct postmarket surveillance of a medical device under section 522 of the act, this part applies to you. We have the authority to order postmarket surveillance of any class II or class III medical device, including a device reviewed under the licensing provisions of section 351 of the Public Health Service Act, that meets any of the following criteria:

(a) Failure of the device would be reasonably likely to have serious adverse health consequences;

(b) The device is intended to be implanted in the human body for more than 1 year; or

(c) The device is intended to be used to support or sustain life and to be used outside a user facility.

Subpart B—Notification

§ 822.5 How will I know if I must conduct postmarket surveillance?

We will send you a letter (the postmarket surveillance order) notifying you of the requirement to conduct postmarket surveillance. Before we send the order, or as part of the order, we may require that you submit information about your device that will allow us better to define the scope of a surveillance order. We will specify the device(s) subject to the surveillance order and the reason that we are requiring postmarket surveillance of the device under section 522 of the act. We will also provide you with any general or specific guidance that is available to help you develop your plan for conducting postmarket surveillance.

§ 822.6 When will you notify me that I am required to conduct postmarket surveillance?

We will notify you as soon as we have determined that postmarket surveillance of your device is necessary, based on the identification of a surveillance question. This may occur during the review of a marketing application for your device, as your device goes to market, or after your device has been marketed for a period of time.

§ 822.7 What should I do if I do not agree that postmarket surveillance is appropriate?

(a) If you do not agree with our decision to order postmarket surveillance for a particular device, you may request review of our decision by:

(1) Requesting a meeting with the Director, Office of Surveillance and Biometrics, who generally issues the order for postmarket surveillance;

(2) Seeking internal review of the order under § 10.75 of this chapter;

(3) Requesting an informal hearing under part 16 of this chapter; or

(4) Requesting review by the Medical Devices Dispute Resolution Panel of the Medical Devices Advisory Committee.

(b) You may obtain guidance documents that discuss these mechanisms from the Center for Devices and Radiological Health's (CDRH's) Web site (*www.fda.gov/cdrh/resolvingdisputes*), and from the CDRH Facts-on-Demand system (800–899–0381 or 301–827–0111).

Subpart C—Postmarket Surveillance Plan

§ 822.8 When, where, and how must I submit my postmarket surveillance plan?

You must submit your plan to conduct postmarket surveillance within 30 days of the date you receive the postmarket surveillance order. For devices regulated by the Center for Biologics Evaluation and Research, send three copies of your submission to the Document Control Center (HFM–99), Center for Biologics Evaluation and Research, Food and Drug Administration, 1401 Rockville Pike, suite 200N, Rockville, MD 20852–1448. For devices regulated by the Center for Drug Evaluation and Research, send three copies of your submission to the Central Document Room, Center for Drug Evaluation and Research, Food and Drug Administration, 5901–B Ammendale Rd., Beltsville, MD 20705–1266. When we receive your original submission, we will send you an acknowledgment letter identifying the unique document number assigned to your submission. You must use this number in any correspondence related to this submission.

[67 FR 38887, June 6, 2002, as amended at 70 FR 14986, Mar. 24, 2005]

§ 822.9 What must I include in my submission?

Your submission must include the following:

(a) Organizational/administrative information:

(1) Your name and address;

(2) Generic and trade names of your device;

(3) Name and address of the contact person for the submission;

(4) Premarket application/submission numbers for your device;

(5) Table of contents identifying the page numbers for each section of the submission;

(6) Description of the device (this may be incorporated by reference to the appropriate premarket application/submission);

(7) Product codes and a list of all relevant model numbers; and

(8) Indications for use and claims for the device;

(b) Postmarket surveillance plan;

(c) Designated person information;

(1) Name, address, and telephone number; and

(2) Experience and qualifications.

§ 822.10 What must I include in my surveillance plan?

Your surveillance plan must include a discussion of:

(a) The plan objective(s) addressing the surveillance question(s) identified in our order;

(b) The subject of the study, e.g., patients, the device, animals;

(c) The variables and endpoints that will be used to answer the surveillance question, e.g., clinical parameters or outcomes;

(d) The surveillance approach or methodology to be used;

(e) Sample size and units of observation;

(f) The investigator agreement, if applicable;

(g) Sources of data, e.g., hospital records;

(h) The data collection plan and forms;

(i) The consent document, if applicable;

(j) Institutional Review Board information, if applicable;

(k) The patient followup plan, if applicable;

(l) The procedures for monitoring conduct and progress of the surveillance;

(m) An estimate of the duration of surveillance;

(n) All data analyses and statistical tests planned;

(o) The content and timing of reports.

§ 822.11 What should I consider when designing my plan to conduct postmarket surveillance?

You must design your surveillance to address the postmarket surveillance question identified in the order you received. You should consider what, if any, patient protection measures should be incorporated into your plan. You should also consider the function, operating characteristics, and intended use of your device when designing a surveillance approach.

§ 822.12 Do you have any information that will help me prepare my submission or design my postmarket surveillance plan?

Guidance documents that discuss our current thinking on preparing a postmarket surveillance submission and designing a postmarket surveillance plan are available on the Center for Devices and Radiological Health's Web site and from the Center for Devices and Radiological Health, Office of Surveillance and Biometrics (HFZ-510), 1350 Piccard Dr., Rockville, MD 20850. Guidance documents represent our current interpretation of, or policy on, a regulatory issue. They do not establish legally enforceable rights or responsibilities and do not legally bind you or FDA. You may choose to use an approach other than the one set forth in a guidance document, as long as your alternative approach complies with the relevant statutes (laws) and regulations. If you wish, we will meet with you to discuss whether an alternative approach you are considering will satisfy the requirements of the act and regulations.

§ 822.13 [Reserved]

§ 822.14 May I reference information previously submitted instead of submitting it again?

Yes, you may reference information that you have submitted in premarket submissions as well as other postmarket surveillance submissions. You must specify the information to be incorporated and the document number and pages where the information is located.

§ 822.15 How long must I conduct postmarket surveillance of my device?

The length of postmarket surveillance will depend on the postmarket surveillance question identified in our order. We may order prospective surveillance for a period up to 36 months; longer periods require your agreement. If we believe that a prospective period of greater than 36 months is necessary to address the surveillance question, and you do not agree, we will use the Medical Devices Dispute Resolution Panel to resolve the matter. You may obtain guidance regarding dispute resolution procedures from the Center for Devices and Radiological Health's (CDRH) Web site (*www.fda.gov/cdrh/resolvingdisputes/ombudsman.html*) and from the CDRH Facts-on-Demand system (800–899–0381 or 301–827–0111, document number 1121). The 36-month period refers to the surveillance period, not the length of time from the issuance of the order.

Subpart D—FDA Review and Action

§ 822.16 What will you consider in the review of my submission?

First, we will determine that the submission is administratively complete. Then, in accordance with the law, we must determine whether the designated person has appropriate qualifications and experience to conduct the surveillance and whether the surveillance plan will result in the collection of useful data that will answer the surveillance question.

§ 822.17 How long will your review of my submission take?

We will review your submission within 60 days of receipt.

§ 822.18 How will I be notified of your decision?

We will send you a letter notifying you of our decision and identifying any action you must take.

§ 822.19 What kinds of decisions may you make?

If your plan:	Then we will send you:	And you must:
(a) Should result in the collection of useful data that will address the postmarket surveillance question	An approval order, identifying any specific requirements related to your postmarket surveillance	Conduct postmarket surveillance of your device in accordance with the approved plan
(b) Should result in the collection of useful data that will address the postmarket surveillance question after specific revisions are made or specific information is provided	An approvable letter identifying the specific revisions or information that must be submitted before your plan can be approved	Revise your postmarket surveillance submission to address the concerns in the approvable letter and submit it to us within the specified timeframe. We will determine the timeframe case-by-case, based on the types of revisions or information that you must submit
(c) Does not meet the requirements specified in this part	A letter disapproving your plan and identifying the reasons for disapproval	Revise your postmarket surveillance submission and submit it to us within the specified timeframe. We will determine the timeframe case-by-case, based on the types of revisions or information that you must submit
(d) Is not likely to result in the collection of useful data that will address the postmarket surveillance question	A letter disapproving your plan and identifying the reasons for disapproval	Revise your postmarket surveillance submission and submit it to us within the specified timeframe. We will determine the timeframe case-by-case, based on the types of revisions or information that you must submit

§ 822.20 What are the consequences if I fail to submit a postmarket surveillance plan, my plan is disapproved and I fail to submit a new plan, or I fail to conduct surveillance in accordance with my approved plan?

The failure to have an approved postmarket surveillance plan or failure to conduct postmarket surveillance in accordance with the approved plan constitutes failure to comply with section 522 of the act. Your failure would be a prohibited act under section 301(q)(1)(C) of the act, and your device would be misbranded under section 502(t)(3) of the act. We have the authority to initiate actions against products that are adulterated or misbranded, and against persons who commit prohibited acts. Adulterated or misbranded devices can be seized. Persons who commit prohibited acts can be enjoined from committing such acts, required to pay civil money penalties, or prosecuted.

§ 822.21 What must I do if I want to make changes to my postmarket surveillance plan after you have approved it?

You must receive our approval in writing before making changes in your plan that will affect the nature or validity of the data collected in accordance with the plan. To obtain our approval, you must submit three copies of the request to make the proposed change and revised postmarket surveillance plan to the applicable address listed in § 822.8. You may reference information already submitted in accordance with § 822.14. In your cover letter, you must identify your submission as a supplement and cite the unique document number that we assigned in our acknowledgment letter for your original submission, specifically identify the changes to the plan, and identify the reasons and justification for making the changes. You must report changes in your plan that will not affect the nature or validity of the data collected in accordance with the plan in the next interim report required by your approval order.

§ 822.22 What recourse do I have if I do not agree with your decision?

(a) If you disagree with us about the content of your plan or if we disapprove your plan, or if you believe there is a less burdensome approach that will answer the surveillance question, you may request review of our decision by:

(1) Requesting a meeting with the Director, Office of Surveillance and Biometrics, Center for Devices and Radiological Health (CDRH), who generally issues the order for postmarket surveillance;

(2) Seeking internal review of the order under § 10.75 of this chapter;

(3) Requesting an informal hearing under part 16 of this chapter; or

(4) Requesting review by the Medical Devices Dispute Resolution Panel of the Medical Devices Advisory Committee.

(b) You may obtain guidance documents that discuss these mechanisms from the CDRH Web site and from the CDRH Facts-on-Demand System (800–899–0381 or 301–827–0111).

§ 822.23 Is the information in my submission considered confidential?

We consider the content of your submission confidential until we have approved your postmarket surveillance plan. After we have approved your plan, the contents of the original submission and any amendments, supplements, or reports may be disclosed in accordance with the Freedom of Information Act. We will continue to protect trade secret and confidential commercial information after your plan is approved. We will not disclose information identifying individual patients. You may wish to indicate in your submission which information you consider trade secret or confidential commercial.

Subpart E—Responsibilities of Manufacturers

§ 822.24 What are my responsibilities once I am notified that I am required to conduct postmarket surveillance?

You must submit your plan to conduct postmarket surveillance to us within 30 days from receipt of the order (letter) notifying you that you are required to conduct postmarket surveillance of a device.

§ 822.25 What are my responsibilities after my postmarket surveillance plan has been approved?

After we have approved your plan, you must conduct the postmarket surveillance of your device in accordance with your approved plan. This means that you must ensure that:

(a) Postmarket surveillance is initiated in a timely manner;

(b) The surveillance is conducted with due diligence;

(c) The data identified in the plan is collected;

(d) Any reports required as part of your approved plan are submitted to us in a timely manner; and

(e) Any information that we request prior to your submission of a report or in response to our review of a report is provided in a timely manner.

§ 822.26 If my company changes ownership, what must I do?

You must notify us within 30 days of any change in ownership of your company. Your notification should identify any changes to the name or address of the company, the contact person, or the designated person (as defined in § 822.3(b)). Your obligation to conduct postmarket surveillance will generally transfer to the new owner, unless you and the new owner have both agreed that you will continue to conduct the surveillance. If you will continue to conduct the postmarket surveillance, you still must notify us of the change in ownership.

§ 822.27 If I go out of business, what must I do?

You must notify us within 30 days of the date of your decision to close your business. You should provide the expected date of closure and discuss your plans to complete or terminate postmarket surveillance of your device. You must also identify who will retain the records related to the surveillance (described in subpart G of this part) and where the records will be kept.

§ 822.28 If I stop marketing the device subject to postmarket surveillance, what must I do?

You must continue to conduct postmarket surveillance in accordance with your approved plan even if you no longer market the device. You may request that we allow you to terminate postmarket surveillance or modify your postmarket surveillance because you no longer market the device. We will make these decisions on a case-by-case basis, and you must continue to conduct the postmarket surveillance unless we notify you that you may stop your surveillance study.

Subpart F—Waivers and Exemptions

§ 822.29 May I request a waiver of a specific requirement of this part?

You may request that we waive any specific requirement of this part. You may submit your request, with supporting documentation, separately or as a part of your postmarket surveillance submission to the address in § 822.8.

§ 822.30 May I request exemption from the requirement to conduct postmarket surveillance?

You may request exemption from the requirement to conduct postmarket surveillance for your device or any specific model of that device at any time. You must comply with the requirements of this part unless and until we grant an exemption for your device. Your request for exemption must explain why you believe we should exempt the device or model from postmarket surveillance. You should demonstrate why the surveillance question does not apply to your device or does not need to be answered for the device for which you are requesting exemption. Alternatively, you may provide information that answers the surveillance question for your device, with supporting documentation, to the address in § 822.8.

Subpart G—Records and Reports

§ 822.31 What records am I required to keep?

You must keep copies of:

(a) All correspondence with your investigators or FDA, including required reports;

(b) Signed agreements from each of your investigators, if your surveillance plan uses investigators, stating the commitment to conduct the surveillance in accordance with the approved plan, any applicable FDA regulations, and any conditions of approval for your plan, such as reporting requirements;

(c) Your approved postmarket surveillance plan, with documentation of the date and reason for any deviation from the plan;

(d) All data collected and analyses conducted in support of your postmarket surveillance plan; and

(e) Any other records that we require to be maintained by regulation or by order, such as copies of signed consent documents, evidence of Institutional Review Board review and approval, etc.

§ 822.32 What records are the investigators in my surveillance plan required to keep?

Your investigator must keep copies of:

(a) All correspondence between investigators, FDA, the manufacturer, and the designated person, including required reports.

(b) The approved postmarket surveillance plan, with documentation of the date and reason for any deviation from the plan.

(c) All data collected and analyses conducted at that site for postmarket surveillance.

(d) Any other records that we require to be maintained by regulation or by order.

§ 822.33 How long must we keep the records?

You, the designated person, and your investigators must keep all records for a period of 2 years after we have accepted your final report, unless we specify otherwise.

§ 822.34 What must I do with the records if the sponsor of the plan or an investigator in the plan changes?

If the sponsor of the plan or an investigator in the plan changes, you must ensure that all records related to the postmarket surveillance have been transferred to the new sponsor or investigator and notify us within 10 working days of the effective date of the change. You must provide the name, address, and telephone number of the new sponsor or investigator, certify that all records have been transferred, and provide the date of transfer.

§ 822.35 Can you inspect my manufacturing site or other sites involved in my postmarket surveillance plan?

We can review your postmarket surveillance programs during regularly scheduled inspections, inspections initiated to investigate recalls or other similar actions, and inspections initiated specifically to review your postmarket surveillance plan. We may also inspect any other person or site involved in your postmarket surveillance, such as investigators or contractors. Any person authorized to grant access to a facility must permit authorized FDA employees to enter and inspect any facility where the device is held or where records regarding postmarket surveillance are held.

§ 822.36 Can you inspect and copy the records related to my postmarket surveillance plan?

We may, at a reasonable time and in a reasonable manner, inspect and copy any records pertaining to the conduct of postmarket surveillance that are required to be kept by this regulation. You must be able to produce records and information required by this regulation that are in the possession of others under contract with you to conduct the postmarket surveillance. Those who have signed agreements or are under contract with you must also produce the records and information upon our request. This information must be produced within 72 hours of the initiation of the inspection. We generally will redact information pertaining to individual subjects prior to copying those records, unless there are extenuating circumstances.

§ 822.37 Under what circumstances would you inspect records identifying subjects?

We can inspect and copy records identifying subjects under the same circumstances that we can inspect any records relating to postmarket surveillance. We are likely to be interested in such records if we have reason to believe that required reports have not been submitted, or are incomplete, inaccurate, false, or misleading.

§ 822.38 What reports must I submit to you?

You must submit interim and final reports as specified in your approved postmarket surveillance plan. In addition, we may ask you to submit additional information when we believe that the information is necessary for the protection of the public health and implementation of the act. We will also state the reason or purpose for the request and how we will use the information.

PART 860—MEDICAL DEVICE CLASSIFICATION PROCEDURES

Subpart A—General

Subpart B—Classification

Subpart C—Reclassification

860.132 Procedures when the Commissioner initiates a performance standard or premarket approval proceeding under section 514(b) or 515(b) of the act.
860.134 Procedures for "new devices" under section 513(f) of the act and reclassification of certain devices.
860.136 Procedures for transitional products under section 520(l) of the act.

AUTHORITY: 21 U.S.C. 360c, 360d, 360e, 360i, 360j, 371, 374.

SOURCE: 43 FR 32993, July 28, 1978, unless otherwise noted.

Subpart A—General

§ 860.1 Scope.

(a) This part implements sections 513, 514(b), 515(b), and 520(l) of the act with respect to the classification and reclassification of devices intended for human use.

(b) This part prescribes the criteria and procedures to be used by classification panels in making their recommendations and by the Commissioner in making the Commissioner's determinations regarding the class of regulatory control (class I, class II, or class III) appropriate for particular devices. Supplementing the general Food and Drug Administration procedures governing advisory committees (part 14 of this chapter), this part also provides procedures for manufacturers, importers, and other interested persons to participate in proceedings to classify and reclassify devices. This part also describes the kind of data required for determination of the safety and effectiveness of a device, and the circumstances under which information submitted to classification panels or to the Commissioner in connection with classification and reclassification proceedings will be available to the public.

§ 860.3 Definitions.

For the purposes of this part:

(a) *Act* means the Federal Food, Drug, and Cosmetic Act.

(b) *Commissioner* means the Commissioner of Food and Drugs, Food and Drug Administration, United States Department of Health and Human Services, or the Commissioner's designee.

(c) *Class* means one of the three categories of regulatory control for medical devices, defined below:

(1) *Class I* means the class of devices that are subject to only the general controls authorized by or under sections 501 (adulteration), 502 (misbranding), 510 (registration), 516 (banned devices), 518 (notification and other remedies), 519 (records and reports), and 520 (general provisions) of the act. A device is in class I if (i) general controls are sufficient to provide reasonable assurance of the safety and effectiveness of the device, or (ii) there is insufficient information from which to determine that general controls are sufficient to provide reasonable assurance of the safety and effectiveness of the device or to establish special controls to provide such assurance, but the device is not life-supporting or life-sustaining or for a use which is of substanial importance in preventing impairment of human health, and which does not present a potential unreasonable risk of illness of injury.

(2) *Class II* means the class of devices that is or eventually will be subject to special controls. A device is in class II if general controls alone are insufficient to provide reasonable assurance of its safety and effectiveness and there is sufficient information to establish special controls, including the promulgation of performance standards, postmarket surveillance, patient registries, development and dissemination of guidance documents (including guidance on the submission of clinical data in premarket notification submissions in accordance with section 510(k) of the act), recommendations, and other appropriate actions as the Commissioner deems necessary to provide such assurance. For a device that is purported or represented to be for use in supporting or sustaining human life, the Commissioner shall examine and identify the special controls, if any, that are necessary to provide adequate assurance of safety and effectiveness and describe how such controls provide such assurance.

(3) *Class III* means the class of devices for which premarket approval is or will be required in accordance with section 515 of the act. A device is in class III if

insufficient information exists to determine that general controls are sufficient to provide reasonable assurance of its safety and effectiveness or that application of special controls described in paragraph (c)(2) of this section would provide such assurance and if, in addition, the device is life-supporting or life-sustaining, or for a use which is of substantial importance in preventing impairment of human health, or if the device presents a potential unreasonable risk of illness or injury.

(d) *Implant* means a device that is placed into a surgically or naturally formed cavity of the human body. A device is regarded as an implant for the purpose of this part only if it is intended to remain implanted continuously for a period of 30 days or more, unless the Commissioner determines otherwise in order to protect human health.

(e) *Life-supporting or life-sustaining device* means a device that is essential to, or that yields information that is essential to, the restoration or continuation of a bodily function important to the continuation of human life.

(f) *Classification questionnaire* means a specific series of questions prepared by the Commissioner for use as guidelines by classification panels preparing recommendations to the Commissioner regarding classification and by petitioners submitting petitions for reclassification. The questions relate to the safety and effectiveness characteristics of a device and the answers are designed to help the Commissioner determine the proper classification of the device.

(g) *Supplemental data sheet* means information compiled by a classification panel or submitted in a petition for reclassification, including:

(1) A summary of the reasons for the recommendation (or petition);

(2) A summary of the data upon which the recommendation (or petition) is based;

(3) An identification of the risks to health (if any) presented by the device;

(4) To the extent practicable in the case of a class II or class III device, a recommendation for the assignment of a priority for the application of the requirements of performance standards or premarket approval;

(5) In the case of a class I device, a recommendation whether the device should be exempted from any of the requirements of registration, recordkeeping and reporting, or good manufacturing practice regulations;

(6) In the case of an implant or a life-supporting or life-sustaining device for which classification in class III is not recommended, a statement of the reasons for not recommending that the device be classified in class III;

(7) Identification of any needed restrictions on the use of the device, e.g., whether the device requires special labeling, should be banned, or should be used only upon authorization of a practitioner licensed by law to administer or use such device; and

(8) Any known existing standards applicable to the device, device components, or device materials.

(h) *Classification panel* means one of the several advisory committees established by the Commissioner under section 513 of the act and part 14 of this chapter for the purpose of making recommendations to the Commissioner on the classification and reclassification of devices and for other purposes prescribed by the act or by the Commissioner.

(i) *Generic type of device* means a grouping of devices that do not differ significantly in purpose, design, materials, energy source, function, or any other feature related to safety and effectiveness, and for which similar regulatory controls are sufficient to provide reasonable assurance of safety and effectiveness.

(j) *Petition* means a submission seeking reclassification of a device in accordance with §860.123.

[43 FR 32993, July 28, 1978, as amended at 57 FR 58403, Dec. 10, 1992; 65 FR 56480, Sept. 19, 2000]

§860.5 Confidentiality and use of data and information submitted in connection with classification and reclassification.

(a) This section governs the availability for public disclosure and the use by the Commissioner of data and information submitted to classification panels or to the Commissioner in connection with the classification or reclassification of devices under this part.

(b) In general, data and information submitted to classification panels in connection with the classification of devices under § 860.84 will be available immediately for public disclosure upon request. However, except as provided by the special rules in paragraph (c) of this section, this provision does not apply to data and information exempt from public disclosure in accordance with part 20 of this chapter: Such data and information will be available only in accordance with part 20.

(c)(1) Safety and effectiveness data submitted to classification panels or to the Commissioner in connection with the classification of a device under § 860.84, which have not been disclosed previously to the public, as described in § 20.81 of this chapter, shall be regarded as confidential if the device is classified in to class III. Because the classification of a device under § 860.84 may be ascertained only upon publication of a final regulation, all safety and effectiveness data that have not been disclosed previously are not available for public disclosure unless and until the device is classified into class I or II, in which case the procedure in paragraph (c)(2) of this section applies.

(2) Thirty days after publication of a final regulation under § 860.84 classifying a device into class I or class II, safety and effectiveness data submitted for that device that had been regarded as confidential under paragraph (c)(1) of this section will be available for public disclosure and placed on public display in the office of the Division of Dockets Management, Food and Drug Administration unless, within that 30-day period, the person who submitted the data demonstrates that the data still fall within the exemption for trade secrets and confidential commercial information described in § 20.61 of this chapter. Safety and effectiveness data submitted for a device that is classified into class III by regulation in accordance with § 860.84 will remain confidential and unavailable for public disclosure so long as such data have not been disclosed to the public as described in § 20.81 of this chapter.

(3) Because device classification affects generic types of devices, in making determinations under § 860.84 concerning the initial classification of a device, the classification panels and the Commissioner may consider safety and effectiveness data developed for another device in the same generic type, regardless of whether such data are regarded currently as confidential under paragraph (c)(1) of this section.

(d)(1) The fact of its existence and the contents of a petition for reclassification filed in accordance with § 860.130 or § 860.132 are available for public disclosure at the time the petition is received by the Food and Drug Administration.

(2) The fact of the existence of a petition for reclassification filed in accordance with § 860.134 or § 860.136 is available for public disclosure at the time the petition is received by the Food and Drug Administration. The contents of such a petition are not available for public disclosure for the period of time following its receipt (not longer than 30 days) during which the petition is reviewed for any deficiencies preventing the Commissioner from making a decision on it. Once it is determined that the petition contains no deficiencies preventing the Commissioner from making a decision on it, the petition will be filed with the Division of Dockets Management and its entire contents will be available for public disclosure and subject to consideration by classification panels and by the Commissioner in making a decision on the petition. If, during this 30-day period of time, the petition is found to contain deficiencies that prevent the Commissioner from making a decision on it, the petitioner will be so notified and afforded an opportunity to correct the deficiencies.

Thirty days after notice to the petitioner of deficiencies in the petition, the contents of the petition will be available for public disclosure unless, within that 30 days, the petitioner submits supplemental material intended to correct the deficiencies in the petition. The Commissioner, in the Commissioner's discretion, may allow withdrawal of a deficient petition during the 30-day period provided for correcting deficiencies. Any supplemental material submitted by the petitioner, together with the material in the original petition, is considered as a new petition. The new petition is reviewed for deficiencies in the same manner as the original petition, and the same procedures for notification and correction of deficiencies are followed. Once the petitioner has corrected the deficiencies, the entire contents of the petition will be available for public disclosure and subject to consideration by classification panels and by the Commissioner in making a decision on the petition. Deficient petitions which have not been corrected within 180 days after notification of deficiency will be returned to the petitioner and will not be considered further unless resubmitted.

(e) The Commissioner may not disclose, or use as the basis for reclassification of a device from class III to class II, any information reported to or otherwise obtained by the Commissioner under section 513, 514, 515, 516, 518, 519, 520(f), 520(g), or 704 of the act that falls within the exemption described in §20.61 of this chapter for trade secrets and confidential commercial information. The exemption described in §20.61 does not apply to data or information contained in a petition for reclassification submitted in accordance with §860.130 or §860.132, or in a petition submitted in accordance with §860.134 or §860.136 that has been determined to contain no deficiencies that prevent the Commissioner from making a decision on it. Accordingly, all data and information contained in such petitions may be disclosed by the Commissioner and used as the basis for reclassification of a device from class III to class II.

(f) For purposes of this section, safety and effectiveness data include data and results derived from all studies and tests of a device on animals and humans and from all studies and tests of the device itself intended to establish or determine its safety and effectiveness.

§860.7 Determination of safety and effectiveness.

(a) The classification panels, in reviewing evidence concerning the safety and effectiveness of a device and in preparing advice to the Commissioner, and the Commissioner, in making determinations concerning the safety and effectiveness of a device, will apply the rules in this section.

(b) In determining the safety and effectiveness of a device for purposes of classification, establishment of performance standards for class II devices, and premarket approval of class III devices, the Commissioner and the classification panels will consider the following, among other relevant factors:

(1) The persons for whose use the device is represented or intended;

(2) The conditions of use for the device, including conditions of use prescribed, recommended, or suggested in the labeling or advertising of the device, and other intended conditions of use;

(3) The probable benefit to health from the use of the device weighed against any probable injury or illness from such use; and

(4) The reliability of the device.

(c)(1) Although the manufacturer may submit any form of evidence to the Food and Drug Administration in an attempt to substantiate the safety and effectiveness of a device, the agency relies upon only valid scientific evidence to determine whether there is reasonable assurance that the device is safe and effective. After considering the nature of the device and the rules in this section, the Commissioner will determine whether the evidence submitted or otherwise available to the Commissioner is valid scientific evidence for the purpose of determining the safety or effectiveness of a particular device and whether the available evidence, when taken as a whole, is adequate to support a determination that there is reasonable assurance that the device is safe and effective for its conditions of use.

(2) Valid scientific evidence is evidence from well-controlled investigations, partially controlled studies, studies and objective trials without matched controls, well-documented case histories conducted by qualified experts, and reports of significant

human experience with a marketed device, from which it can fairly and responsibly be concluded by qualified experts that there is reasonable assurance of the safety and effectiveness of a device under its conditions of use. The evidence required may vary according to the characteristics of the device, its conditions of use, the existence and adequacy of warnings and other restrictions, and the extent of experience with its use. Isolated case reports, random experience, reports lacking sufficient details to permit scientific evaluation, and unsubstantiated opinions are not regarded as valid scientific evidence to show safety or effectiveness. Such information may be considered, however, in identifying a device the safety and effectiveness of which is questionable.

(d)(1) There is reasonable assurance that a device is safe when it can be determined, based upon valid scientific evidence, that the probable benefits to health from use of the device for its intended uses and conditions of use, when accompanied by adequate directions and warnings against unsafe use, outweigh any probable risks. The valid scientific evidence used to determine the safety of a device shall adequately demonstrate the absence of unreasonable risk of illness or injury associated with the use of the device for its intended uses and conditions of use.

(2) Among the types of evidence that may be required, when appropriate, to determine that there is reasonable assurance that a device is safe are investigations using laboratory animals, investigations involving human subjects, and nonclinical investigations including in vitro studies.

(e)(1) There is reasonable assurance that a device is effective when it can be determined, based upon valid scientific evidence, that in a significant portion of the target population, the use of the device for its intended uses and conditions of use, when accompanied by adequate directions for use and warnings against unsafe use, will provide clinically significant results.

(2) The valid scientific evidence used to determine the effectiveness of a device shall consist principally of well-controlled investigations, as defined in paragraph (f) of this section, unless the Commissioner authorizes reliance upon other valid scientific evidence which the Commissioner has determined is sufficient evidence from which to determine the effectiveness of a device, even in the absence of well-controlled investigations. The Commissioner may make such a determination where the requirement of well-controlled investigations in paragraph (f) of this section is not reasonably applicable to the device.

(f) The following principles have been developed over a period of years and are recognized by the scientific community as the essentials of a well-controlled clinical investigation. They provide the basis for the Commissioner's determination whether there is reasonable assurance that a device is effective based upon well-controlled investigations and are also useful in assessing the weight to be given to other valid scientific evidence permitted under this section.

(1) The plan or protocol for the study and the report of the results of a well-controlled investigation shall include the following:

(i) A clear statement of the objectives of the study;

(ii) A method of selection of the subjects that:

(*a*) Provides adequate assurance that the subjects are suitable for the purposes of the study, provides diagnostic criteria of the condition to be treated or diagnosed, provides confirmatory laboratory tests where appropriate and, in the case of a device to prevent a disease or condition, provides evidence of susceptibility and exposure to the condition against which prophylaxis is desired;

(*b*) Assigns the subjects to test groups, if used, in such a way as to minimize any possible bias;

(*c*) Assures comparability between test groups and any control groups of pertinent variables such as sex, severity or duration of the disease, and use of therapy other than the test device;

(iii) An explanation of the methods of observation and recording of results utilized, including the variables measured, quantitation, assessment of any subject's response, and steps taken to minimize any possible bias of subjects and observers;

(iv) A comparison of the results of treatment or diagnosis with a control in such a fashion as to permit quantitative evaluation. The precise nature of the control must be specified and an explanation provided of the methods employed to minimize any possible bias of the observers and analysts of the data. Level and methods of "blinding," if appropriate and used, are to be documented. Generally, four types of comparisons are recognized:

(*a*) *No treatments.* Where objective measurements of effectiveness are available and placebo effect is negligible, comparison of the objective results in comparable groups of treated and untreated patients;

(*b*) *Placebo control.* Where there may be a placebo effect with the use of a device, comparison of the results of use of the device with an ineffective device used under conditions designed to resemble the conditions of use under investigation as far as possible;

(*c*) *Active treatment control.* Where an effective regimen of therapy may be used for comparison, e.g., the condition being treated is such that the use of a placebo or the withholding of treatment would be inappropriate or contrary to the interest of the patient;

(*d*) *Historical control.* In certain circumstances, such as those involving diseases with high and predictable mortality or signs and symptoms of predictable duration or severity, or in the case of prophylaxis where morbidity is predictable, the results of use of the device may be compared quantitatively with prior experience historically derived from the adequately documented natural history of the disease or condition in comparable patients or populations who received no treatment or who followed an established effective regimen (therapeutic, diagnostic, prophylactic).

(v) A summary of the methods of analysis and an evaluation of the data derived from the study, including any appropriate statistical methods utilized.

(2) To insure the reliability of the results of an investigation, a well-controlled investigation shall involve the use of a test device that is standardized in its composition or design and performance.

(g)(1) It is the responsibility of each manufacturer and importer of a device to assure that adequate, valid scientific evidence exists, and to furnish such evidence to the Food and Drug Administration to provide reasonable assurance that the device is safe and effective for its intended uses and conditions of use. The failure of a manufacturer or importer of a device to present to the Food and Drug Administration adequate, valid scientific evidence showing that there is reasonable assurance of the safety and effectiveness of the device, if regulated by general controls alone, or by general controls and performance standards, may support a determination that the device be classified into class III.

(2) The Commissioner may require that a manufacturer, importer, or distributor make reports or provide other information bearing on the classification of a device and indicating whether there is reasonable assurance of the safety and effectiveness of the device or whether it is adulterated or misbranded under the act.

(3) A requirement for a report or other information under this paragraph will comply with section 519 of the act. Accordingly, the requirement will state the reason or purpose for such request; will describe the required report or information as clearly as possible; will not be imposed on a manufacturer, importer, or distributor of a classified device that has been exempted from such a requirement in accordance with § 860.95; will prescribe the time for compliance with the requirement; and will prescribe the form and manner in which the report or information is to be provided.

(4) Required information that has been submitted previously to the Center for Devices and Radiological Health need not be resubmitted, but may be incorporated by reference.

[43 FR 32993, July 28, 1978, as amended at 53 FR 11253, Apr. 6, 1988]

Subpart B—Classification

§ 860.84 Classification procedures for "old devices."

(a) This subpart sets forth the procedures for the original classification of a device that either was in commercial distribution before May 28, 1976, or is substantially equivalent to a device that was in commercial distribution before that date. Such a device will be classified by regulation into either class I (general controls), class II (special controls) or class III (premarket approval), depending upon the level of regulatory control required to provide reasonable assurance of the safety and effectiveness of the device (§ 860.3(c)). This subpart does not apply to a device that is classified into class III by statute under section 513(f) of the act because the Food and Drug Administration has determined that the device is not "substantially equivalent" to any device subject to this subpart or under section 520(l) (1) through (3) of the act because the device was regarded previously as a new drug. In classifying a device under this section, the Food and Drug Administration will follow the procedures described in paragraphs (b) through (g) of this section.

(b) The Commissioner refers the device to the appropriate classification panel organized and operated in accordance with section 513 (b) and (c) of the act and part 14 of this chapter.

(c) In order to make recommendations to the Commissioner on the class of regulatory control (class I, class II, or class III) appropriate for the device, the panel reviews the device for safety and effectiveness. In so doing, the panel:

(1) Considers the factors set forth in § 860.7 relating to the determination of safety and effectiveness;

(2) Determines the safety and effectiveness of the device on the basis of the types of scientific evidence set forth in § 860.7;

(3) Answers the questions in the classification questionnaire applicable to the device being classified;

(4) Completes a supplemental data sheet for the device;

(5) Provides, to the maximum extent practicable, an opportunity for interested persons to submit data and views on the classification of the device in accordance with part 14 of this chapter.

(d) Based upon its review of evidence of the safety and effectiveness of the device, and applying the definition of each class in § 860.3(c), the panel submits to the Commissioner a recommendation regarding the classification of the device. The recommendation will include:

(1) A summary of the reasons for the recommendation;

(2) A summary of the data upon which the recommendation is based, accompanied by references to the sources containing such data;

(3) An identification of the risks to health (if any) presented by the device;

(4) In the case of a recommendation for classification into class I, a recommendation as to whether the device should be exempted from the requirements of one or more of the following sections of the act: section 510 (registration, product listing, and premarket notification) section 519 (records and reports) and section 520(f) (good manufacturing practice regulations) in accordance with § 860.95;

(5) In the case of a recommendation for classification into class II or class III, to the extent practicable, a recommendation for the assignment to the device of a priority for the application of a performance standard or a premarket approval requirement;

(6) In the case of a recommendation for classification of an implant or a life-supporting or life-sustaining device into class I or class II, a statement of why premarket approval is not necessary to provide reasonable assurance of the safety and effectiveness of the device, accompanied by references to supporting documentation and data satisfying the requirements of § 860.7, and an identification of the risks to health, if any, presented by the device.

(e) A panel recommendation is regarded as preliminary until the Commissioner has reviewed it, discussed it with the panel if appropriate, and published a proposed regulation classifying the device. Preliminary panel recommendations are filed in the Division of Dockets Management's office upon receipt and are available to the public upon request.

(f) The Commissioner publishes the panel's recommendation in the FEDERAL REGISTER, together with a proposed regulation classifying the device, and other devices of that generic type, and provides interested persons an opportunity to submit comments on the recommendation and proposed regulation.

(g) The Commissioner reviews the comments and issues a final regulation classifying the device and other devices of that generic type. The regulation will:

(1) If classifying the device into class I, prescribe which, if any, of the requirements of sections 510, 519, and 520(f) of the act will not apply to the device and state the reasons for making the requirements inapplicable, in accordance with § 860.95;

(2) If classifying the device into class II or class III, at the discretion of the Commissioner, establish priorities for the application to the device of a performance standard or a premarket approval requirement;

(3) If classifying an implant, or life-supporting or life-sustaining device, comply with § 860.93(b).

[43 FR 32993, July 28, 1978, as amended at 57 FR 58404, Dec. 10, 1992; 64 FR 404, Jan. 5, 1999]

§ 860.93 Classification of implants, life-supporting or life-sustaining devices.

(a) The classification panel will recommend classification into class III of any implant or life-supporting or life-sustaining device unless the panel determines that such classification is not necessary to provide reasonable assurance of the safety and effectiveness of the device. If the panel recommends classification or reclassification of such a device into a class other than class III, it shall set forth in its recommendation the reasons for so doing together with references to supporting documentation and data satisfying the requirements of § 860.7, and an identification of the risks to health, if any, presented by the device.

(b) The Commissioner will classify an implant or life-supporting or life-sustaining device into class III unless the Commissioner determines that such classification is not necessary to provide reasonable assurance of the safety and effectiveness of the device. If the Commissioner proposes to classify or reclassify such a device into a class other than class III, the regulation or order effecting such classification or reclassification will be accompanied by a full statement of the reasons for so doing. A statement of the reasons for not classifying or retaining the device in class III may be in the form of concurrence with the reasons for the recommendation of the classification panel, together with supporting documentation and data satisfying the requirements of § 860.7 and an identification of the risks to health, if any, presented by the device.

§ 860.95 Exemptions from sections 510, 519, and 520(f) of the act.

(a) A panel recommendation to the Commissioner that a device be classified or reclassified into class I will include a recommendation as to whether the device should be exempted from some or all of the requirements of one or more of the following sections of the act: Section 510 (registration, product listing and premarket notification), section 519 (records and reports), and section 520(f) (good manufacturing practice regulations).

(b) A regulation or an order classifying or reclassifying a device into class I will specify which requirements, if any, of sections 510, 519, and 520(f) of the act the device is to be exempted from, together with the reasons for such exemption.

(c) The Commissioner will grant exemptions under this section only if the Commissioner determines that the requirements from which the device is exempted are not necessary to provide reasonable assurance of the safety and effectiveness of the device.

Subpart C—Reclassification

§ 860.120 General.

(a) Sections 513(e) and (f), 514(b), 515(b), and 520(l) of the act provide for reclassification of a device and prescribe the procedures to be followed to effect reclassification. The purposes of subpart C are to:

(1) Set forth the requirements as to form and content of petitions for reclassification;

(2) Describe the circumstances in which each of the five statutory reclassification provisions applies; and

(3) Explain the procedure for reclassification prescribed in the five statutory reclassification provisions.

(b) The criteria for determining the proper class for a device are set forth in § 860.3(c). The reclassification of any device within a generic type of device causes the reclassification of all substantially equivalent devices within that generic type. Accordingly, a petition for the reclassification of a specific device will be considered a petition for reclassification of all substantially equivalent devices within the same generic type.

(c) Any interested person may submit a petition for reclassification under section 513(e), 514(b), or 515(b). A manufacturer or importer may submit a petition for reclassification under section 513(f) or 520(l). The Commissioner may initiate the reclassification of a device classified into class III under sections 513(f) and 520(l) of the act.

[43 FR 32993, July 28, 1978, as amended at 57 FR 58404, Dec. 10, 1992]

§ 860.123 Reclassification petition: Content and form.

(a) Unless otherwise provided in writing by the Commissioner, any petition for reclassification of a device, regardless of the section of the act under which it is filed, shall include the following:

(1) A specification of the type of device for which reclassification is requested;

(2) A statement of the action requested by the petitioner, e.g., "It is requested that __ device(s) be reclassified from class III to a class II";

(3) A completed supplemental data sheet applicable to the device for which reclassification is requested;

(4) A completed classification questionnaire applicable to the device for which reclassification is requested;

(5) A statement of the basis for disagreement with the present classification status of the device;

(6) A full statement of the reasons, together with supporting data satisfying the requirements of § 860.7, why the device should not be classified into its present classification and how the proposed classification will provide reasonable assurance of the safety and effectiveness of the device;

(7) Representative data and information known by the petitioner that are unfavorable to the petitioner's position;

(8) If the petition is based upon new information under section 513(e), 514(b), or 515(b) of the act, a summary of the new information;

(9) Copies of source documents from which new information used to support the petition has been obtained (attached as appendices to the petition).

(10) A financial certification or disclosure statement or both as required by part 54 of this chapter.

(b) Each petition submitted pursuant to this section shall be:

(1) Addressed to the Food and Drug Administration, Center for Devices and Radiological Health, Regulations Staff (HFZ–215), 1350 Piccard Dr., Rockville, MD 20857;

(2) Marked clearly with the section of the act under which the petition is being submitted, i.e., "513(e)," "513(f)," "514(b)," "515(b)," or "520(l) Petition";

(3) Bound in a volume or volumes, where necessary; and

(4) Submitted in an original and two copies.

[43 FR 32993, July 28, 1978, as amended at 49 FR 14505, Apr. 12, 1984; 53 FR 11253, Apr. 6, 1988; 55 FR 11169, Mar. 27, 1990; 63 FR 5254, Feb. 2, 1998; 65 FR 17137, Mar. 31, 2000]

§ 860.125 Consultation with panels.

(a) When the Commissioner is required to refer a reclassification petition to a classification panel for its recommendation under § 860.134, or is required, or chooses, to consult with a panel concerning a reclassification petition, such as under § 860.130, § 860.132, or § 860.136, the Commissioner will distribute a copy of the petition, or its relevant portions, to each panel member and will consult with the panel in one of the following ways:

(1) Consultation by telephone with at least a majority of current voting panel members and, when possible, nonvoting panel members;

(2) Consultation by mail with at least a majority of current voting panel members and, when possible, nonvoting panel members; and

(3) Discussion at a panel meeting.

(b) The method of consultation chosen by the Commissioner will depend upon the importance and complexity of the subject matter involved and the time available for action. When time and circumstances permit, the Commissioner will consult with a panel through discussion at a panel meeting.

(c) When a petition is submitted under § 860.134 for a post-enactment, not substantially equivalent device ("new device"), in consulting with the panel the Commissioner will obtain a recommendation that includes the information described in § 860.84(d). In consulting with a panel about a petition submitted under § 860.130, § 860.132, or § 860.136, the Commissioner may or may not obtain a formal recommendation.

§ 860.130 General procedures under section 513(e) of the act.

(a) Section 513(e) of the act applies to reclassification proceedings under the act based upon new information.

(b) A proceeding to reclassify a device under section 513(e) may be initiated:

(1) On the initiative of the Commissioner alone;

(2) On the initiative of the Commissioner in response to a request for change in classification based upon new information, under section 514(b) or 515(b) of the act (see § 860.132); or

(3) In response to the petition of an interested person, based upon new information, filed in accordance with § 860.123.

(c) By regulation promulgated under this section, the Commissioner may change the classification from class III into:

(1) Class II if the Commissioner determines that special controls in addition to general controls would provide reasonable assurance of the safety and effectiveness of the device and there is sufficient information to establish special controls to provide assurance; or

(2) Class I if the Commissioner determines that general controls would provide reasonable assurance of the safety and effectiveness of the device.

(d) The rulemaking procedures in § 10.40 of this chapter apply to proceedings to reclassify a device under section 513(e), except that the Commissioner may secure a recommendation with respect to a proposed reclassification from the classification panel to which the device was last referred. The panel will consider a proposed reclassification submitted to it by the Commissioner in accordance with the consultation procedures of § 860.125. Any recommendation submitted to the Commissioner by the panel will be published in the FEDERAL REGISTER when the Commissioner promulgates a regulation under this section.

(e) Within 180 days after the filing of a petition for reclassification under this section, the Commissioner, by order published in the FEDERAL REGISTER, will either deny the petition or give notice of his intent to initiate a change in the classification of the device.

(f) If a device is reclassified under this section, the regulation effecting the reclassification may revoke any special control or premarket approval requirement that previously applied to the device but that is no longer applicable because of the change in classification.

(g) A regulation under this section changing the classification of a device from class III to class II may provide that such classification will not take effect until the effective date of a special control for the device established under section 514 of the act.

[43 FR 32993, July 28, 1978, as amended at 57 FR 58404, Dec. 10, 1992]

§ 860.132 Procedures when the Commissioner initiates a performance standard or premarket approval proceeding under section 514(b) or 515(b) of the act.

(a) Sections 514(b) and 515(b) of the act require the Commissioner to provide, by notice in the FEDERAL REGISTER, an opportunity for interested parties to request a change in the classification of a device based upon new information relevant to its classification when the Commissioner initiates a proceeding either to develop a performance standard for the device if in class II, or to promulgate a regulation requiring premarket approval for the device if in class III. In either case, if the Commissioner agrees that the new information warrants a change in classification, the Commissioner will publish in the FEDERAL REGISTER notice of the Commissioner's intent to initiate a proceeding under section 513(e) of the act and § 860.130 to effect such a change.

(b) The procedures for effecting a change in classification under sections 514(b) and 515(b) of the act are as follows:

(1) Within 15 days after publication of the Commissioner's notice referred to in paragraph (a) of this section, an interested person files a petition for reclassification in accordance with § 860.123.

(2) The Commissioner consults with the appropriate classification panel with regard to the petition in accordance with § 860.125.

(3) Within 60 days after publication of the notice referred to in paragraph (a) of this section, the Commissioner, by order published in the FEDERAL REGISTER, either denies the petition or gives notice of his intent to initiate a change in classification in accordance with § 860.130.

§ 860.134 Procedures for "new devices" under section 513(f) of the act and reclassification of certain devices.

(a) Section 513(f)(2) of the act applies to proceedings for reclassification of a device currently in class III by operation of section 513(f)(1) of the act. This category includes any device that is to be first introduced or delivered for introduction into interstate commerce for commercial distribution after May 28, 1976, unless:

(1) It is substantially equivalent to another device that was in commercial distribution before that date and had not been regulated before that date as a new drug; or

(2) It is substantially equivalent to another device that was not in commercial distribution before such date but which has been classified into class I or class II; or

(3) The Commissioner has classified the device into class I or class II in response to a petition for reclassification under this section.

The Commissioner determines whether a device is "substantially equivalent" for purposes of the application of this section. If a manufacturer or importer believes that a device is not "substantially equivalent" but that it should not be in class III under the criteria in § 860.3(c), the manufacturer or importer may petition for reclassification under this section. A manufacturer or importer who believes that a device is "substantially equivalent" and wishes to proceed to market the device shall submit a premarket notification in accordance with part 807 of this chapter. After considering a premarket notification, the Commissioner will determine whether the device is "substantially equivalent" and will notify the manufacturer or importer of such determination in accordance with part 807 of this chapter.

(b) The procedures for effecting reclassification under section 513(f) of the act are as follows:

(1) The manufacturer or importer of the device petitions for reclassification of the device in accordance with § 860.123.

(2) Within 30 days after the petition is filed, the Commissioner notifies the petitioner of any deficiencies in the petition that prevent the Commissioner from making a decision on it and allows the petitioner to supplement a deficient petition. Within 30 days after any supplemental material is received, the Commissioner notifies the petitioner whether the petition, as supplemented, is adequate for review.

(3) After determining that the petition contains no deficiencies precluding a decision on it, the Commissioner may for good cause shown refer the petition to the appropriate classification panel for its review and recommendation whether to approve or deny the petition.

(4) Within 90 days after the date the petition is referred to the panel, following the review procedures set forth in § 860.84(c) for the original classification of an "old" device, the panel submits to the Commissioner its recommendation containing the information set forth in § 860.84(d). A panel recommendation is regarded as preliminary until the Commissioner has reviewed it, discussed it with the panel, if appropriate, and developed a proposed reclassification order. Preliminary panel recommendations are filed in the Division of Dockets Management upon receipt and are available to the public upon request.

(5) The panel recommendation is published in the FEDERAL REGISTER as

soon as practicable and interested persons are provided an opportunity to comment on the recommendation.

(6) Within 90 days after the panel's recommendation is received (and no more than 210 days after the date the petition was filed), the Commissioner denies or approves the petition by order in the form of a letter to the petitioner. If the Commissioner approves the petition, the order will classify the device into class I or class II in accordance with the criteria set forth in § 860.3(c) and subject to the applicable requirements of § 860.93, relating to the classification of implants, life-supporting or life-sustaining devices, and § 860.95, relating to exemptions from certain requirements of the act.

(7) Within a reasonable time after issuance of an order under this section, the Commissioner announces the order by notice published in the FEDERAL REGISTER.

[43 FR 32993, July 28, 1978, as amended at 57 FR 58404, Dec. 10, 1992]

§ 860.136 Procedures for transitional products under section 520(l) of the act.

(a) Section 520(l)(2) of the act applies to reclassification proceedings initiated by a manufacturer or importer for reclassification of a device currently in class III by operation of section 520(l)(1) of the act. This section applies only to devices that the Food and Drug Administration regarded as "new drugs" before May 28, 1976.

(b) The procedures for effecting reclassification under section 520(l) are as follows:

(1) The manufacturer or importer of the device files a petition for reclassification of the device in accordance with § 860.123.

(2) Within 30 days after the petition is filed, the Commissioner notifies the petitioner of any deficiencies in the petition that prevent the Commissioner from making a decision on it, allowing the petitioner to supplement a deficient petition. Within 30 days after any supplemental material is received, the Commissioner notifies the petitioner whether the petition, as supplemented, is adequate for review.

(3) The Commissioner provides the petitioner an opportunity for a regulatory hearing conducted in accordance with part 16 of this chapter.

(4) The Commissioner consults with the appropriate classification panel with regard to the petition in accordance with § 860.125.

(5) Within 180 days after the petition is filed (where the Commissioner has determined it to be adequate for review), the Commissioner, by order in the form of a letter to the petitioner, either denies the petition or classifies the device into class I or class II in accordance with the criteria set forth in § 860.3(c).

(6) Within a reasonable time after issuance of an order under this section, the Commissioner announces the order by notice published in the FEDERAL REGISTER.

PART 861—PROCEDURES FOR PERFORMANCE STANDARDS DEVELOPMENT

Subpart A—General

Subpart B—Procedures for Performance Standards Development and Publication

AUTHORITY: 21 U.S.C. 351, 352, 360c, 360d, 360gg–360ss, 371, 374; 42 U.S.C. 262, 264.

SOURCE: 45 FR 7484, Feb. 1, 1980, unless otherwise noted.

Subpart A—General

§ 861.1 Purpose and scope.

(a) This part implements section 514 of the Federal Food, Drug, and Cosmetic Act (the act) with respect to the establishment, amendment, and revocation of performance standards applicable to devices intended for human use.

(b) The Food and Drug Administration may determine that a performance standard, as described under special controls for class II devices in §860.7(b) of this chapter, is necessary to provide reasonable assurance of the safety and effectiveness of the device. Performance standards may be established for:

(1) A class II device;

(2) A class III device which, upon the effective date of the standard, is reclassified into class II; and

(3) A class III device, as a condition to premarket approval under section 515 of the act, to reduce or eliminate a risk or risks associated with such device.

(c) References in this part to regulatory sections of the Code of Federal Regulations are to chapter I of title 21 unless otherwise noted.

[45 FR 7484, Feb. 1, 1980, as amended at 45 FR 23686, Apr. 8, 1980; 57 FR 58404, Dec. 10, 1992]

§861.5 Statement of policy.

In carrying out its duties under this section, the Food and Drug Administration will, to the maximum extent practical:

(a) Use personnel, facilities, and other technical support available in other Federal agencies;

(b) Consult with other Federal agencies concerned with standard setting and other nationally or internationally recognized standard-setting entities; and

(c) Invite participation, through conferences, workshops, or other means, by representatives of scientific, professional, industry, or consumer organizations who can make a significant contribution.

§861.7 Contents of standards.

Any performance standard established under this part will include such provisions as the Food and Drug Administration determines are necessary to provide reasonable assurance of the safety and effectiveness of the device or devices for which it is established. Where necessary to provide such assurance, a standard will address (but need not be limited to):

(a) Performance characteristics of the device;

(b) The design, construction, components, ingredients, and properties of the device, and its compatibility with power systems and connections to such systems;

(c) The manufacturing processes and quality control procedures applicable to the device;

(d) Testing of the device on either a sample or a 100-percent basis by the manufacturer, or, if it is determined that no other more practical means are available to the Food and Drug Administration to assure the conformity of the device to the standard, providing for testing by the Food and Drug Administration or a third person to ensure that the device conforms to the standard;

(e) The publication of the results of each test or of certain tests of the device to show that the device conforms to the portions of the standard for which the test or tests were required;

(f) Manufacturers' certification to purchasers or to the Food and Drug Administration that the device conforms to the applicable performance standard;

(g) Restrictions on the sale and distribution of the device, but only to the extent authorized under section 520(e) of the act;

(h) The use, and the form and content, of labeling for the proper installation, maintenance, operation, and use of the device. Among the provisions that may be required in the labeling are warnings; storage and transportation information; expiration dates; the date and place of manufacture; the results that may be expected if the device is used properly; the ranges of accuracy of diagnostic information; instructions regarding the proper care of, and the proper components, accessories, or other equipment to be used with the device; and statements concerning the appropriate patient population, for example, a statement that the device is considered safe and effective only when used by, or in the treatment of, a patient who has been tested by particular designated procedures and found to have an illness or condition for which use of the device is indicated by a person skilled in the use of the device.

Subpart B—Procedures for Performance Standards Development and Publication

§861.20 Summary of standards development process.

The procedure by which a performance standard for a device may be established, amended, or revoked is as follows:

(a) The Food and Drug Administration (FDA) will publish in the FEDERAL REGISTER a notice of proposed rulemaking for the establishment, amendment, or revocation of any performance standard for a device.

(1) A notice of proposed rulemaking for the establishment or amendment of a performance standard for a device will:

(i) Set forth a finding, with supporting justification, that the performance standard is appropriate and necessary to provide reasonable assurance of the safety and effectiveness of the device;

(ii) Set forth proposed findings with respect to the risk of illness or injury that the performance standard is intended to reduce or eliminate;

(iii) Invite interested persons to submit to the Food and Drug Administration, within 30 days of the publication of the notice, requests for changes in the classification of the device pursuant to §860.132 of this chapter, based on new information relevant to the classification; and

(iv) Invite interested persons to submit an existing performance standard for the device, including a draft or proposed performance standard, for consideration by the Commissioner of Food and Drugs.

(2) A notice of proposed rulemaking for the revocation of a performance standard will set forth a finding, with supporting justification, that the performance standard is no longer necessary to provide reasonable assurance of the safety and effectiveness of a device.

(b) A notice under this section will provide for a comment period of not less than 60 days.

(c) If, after publication of a notice under paragraph (a) of this section, FDA receives a request to change the classification of the device, FDA will, within 60 days of the publication of the notice and after consultation with the appropriate panel under §860.125 of this chapter, either deny the request or give notice of its intent to initiate a change in the classification under §860.130.

(d) If FDA initiates a rulemaking proceeding under paragraph (a) of this section, FDA will:

(1) Complete the proceeding and establish the performance standard for the device in accordance with this part and §10.40 of this chapter; or

(2) Terminate the proceeding by publishing in the FEDERAL REGISTER a notice announcing such termination and the reasons therefor and, unless the proceeding is terminated because the device is a banned device, initiate a proceeding in accordance with section 513(e) of the act to reclassify the device; or

(3) Take other appropriate action.

[57 FR 58404, Dec. 10, 1992]

§861.24 Existing standard as a proposed standard.

(a) The Food and Drug Administration may accept an existing standard or a proposed or draft standard if it includes:

(1) A description of the procedures used to develop the standard and a list of the persons and organizations that participated in its development, to the extent that such information is available or reasonably obtainable;

(2) An identification of the specific portions of the existing standard that the person submitting the standard believes are appropriate for adoption as, or inclusion in, the proposed standard; and

(3) A summary of the test data, or, if requested by the Food and Drug Administration, all such data or other information supporting the specific portions of the standard identified by the person submitting the standard.

(b) The Food and Drug Administration will publish a notice in the FEDERAL REGISTER stating either that it has accepted, or accepted with modification, as a proposed standard, an existing standard or one that has been developed, or that an existing standard is not acceptable, together with the reasons therefor.

[45 FR 7484, Feb. 1, 1980, as amended at 57 FR 58405, Dec. 10, 1992]

§ 861.30 Development of standards.

The Food and Drug Administration (FDA), while engaged in the development of a proposed standard under this section will:

(a) Support its proposed performance standard by such test data or other documents or materials as may reasonably be required;

(b) Provide interested persons an opportunity to participate in the development of the standard by accepting comments and, where appropriate, holding public meetings on issues relating to development of the standard. Notice of the opportunity to participate in the development of the standard will be furnished in a manner reasonably calculated to reach the majority of persons interested in the development of the standard. This requirement shall be satisfied by publishing such a notice in the FEDERAL REGISTER. Whenever it is appropriate, FDA will use the FEDERAL REGISTER to make announcements about the standard development process of standard developers other than Federal agencies.

(c) Maintain records disclosing the course of development of the proposed standard, the comments and other information submitted by a person in connection with such development (including comments and information regarding the need for a standard), and such other information as may be required to evaluate the standard.

[45 FR 7484, Feb. 1, 1980, as amended at 57 FR 58405, Dec. 10, 1992]

§ 861.34 Amendment or revocation of a standard.

(a) The Food and Drug Administration will provide for periodic evaluation of performance standards to determine whether such standards should be changed to reflect new medical, scientific, or other technological data.

(b) The Food and Drug Administration may, on its own initiative or upon petition of an interested party, amend or revoke by regulation a standard established under this part.

(c) Any petition to amend or revoke a standard shall:

(1) Identify the specific device and standard for which the amendment or revocation is sought; and

(2) Be submitted in accordance with the requirements of § 10.30.

(d) Proceedings to amend or revoke a performance standard shall be conducted in accordance with the rulemaking procedures of § 10.40. In addition, a notice of proposed rulemaking to amend or revoke a standard shall set forth proposed findings with respect to the degree of risk or illness to be eliminated or reduced and the benefit the public will derive from the proposed amendment or revocation.

§ 861.36 Effective dates.

(a) A regulation establishing, amending, or revoking a performance standard will set forth the date upon which it will take effect. To the extent practical, consistent with the public health and safety, such effective date will be established so as to minimize economic loss to, and disruption or dislocation of, domestic and international trade.

(b) Except as provided in paragraph (c) of this section, no regulation establishing, amending, or revoking a standard may take effect before 1 year after the date of its publication unless:

(1) The Food and Drug Administration determines that an earlier effective date is necessary to protect the public health and safety; or

(2) The standard has been established for a device that, by the effective date of the standard, has been reclassified from class III to class II.

(c) The Food and Drug Administration may declare a proposed regulation amending a standard effective on publication in the FEDERAL REGISTER if it determines that making the regulation so effective is in the public interest. A proposed amendment of a performance standard made effective upon publication may not prohibit the introduction or delivery for introduction into interstate commerce of a device that conforms to the standard without the change or changes provided in the proposed amendment until the effective date of any final action on the proposal.

[45 FR 7484, Feb. 1, 1980, as amended at 57 FR 58405, Dec. 10, 1992]

§ 861.38 Standards advisory committees.

(a) The Food and Drug Administration will establish advisory committees to which proposed regulations may be referred, and these committees shall consider such referrals in accordance with this section and part 14 of this chapter. Such advisory committees, which may not be classification panels, shall be considered ad hoc advisory committees. Their members shall be selected in accordance with §§ 14.82 and 14.84, except that no member may be a regular full-time FDA employee. Each advisory committee established under this section shall include as nonvoting members a representative of consumer interests and a representative of interests of the device manufacturing industry.

(b) A proposed regulation to establish, amend, or revoke a performance standard shall be referred to an advisory committee for a report and recommendation with respect to any matter involved in the proposed regulation which requires the exercise of scientific judgment if:

(1) The Food and Drug Administration determines that such referral is necessary or appropriate under the circumstances; or

(2) Requested by an interested person, in the form of a citizen petition in accordance with § 10.30 of this chapter, which is made within the period provided for comment on the proposed regulation and which demonstrates good cause for referral.

(c) When a proposed regulation is referred to an advisory committee, the Food and Drug Administration will furnish the committee with the data and information upon which the proposed regulation is based. After independently reviewing the materials furnished by the Food and Drug Administration and any other available data and information, the advisory committee shall, within 60 days of the referral, submit a report and recommendation on the proposed regulation, together with all underlying data and information and a statement of the reason or basis for the recommendation. A copy of the report and recommendation will be publicly displayed in the office of the Division of Dockets Management, Food and Drug Administration.

(d) Where appropriate, each proposed regulation establishing a standard published in the FEDERAL REGISTER will include a call for nominations to the advisory committee for that particular standard.

[45 FR 7484, Feb. 1, 1980, as amended at 57 FR 58405, Dec. 10, 1992]

PART 862—CLINICAL CHEMISTRY AND CLINICAL TOXICOLOGY DEVICES

Subpart A—General Provisions

Sec.

Subpart B—Clinical Chemistry Test Systems

862.1113 Bilirubin (total and unbound) in the neonate test system.
862.1115 Urinary bilirubin and its conjugates (nonquantitative) test system.
862.1117 B-type natriuretic peptide test system.
862.1118 Biotinidase test system.
862.1120 Blood gases ($P_{CO}2P_{O}2$) and blood pH test system.
862.1130 Blood volume test system.
862.1135 C-peptides of proinsulin test system.
862.1140 Calcitonin test system.
862.1145 Calcium test system.
862.1150 Calibrator.
862.1155 Human chorionic gonadotropin (HCG) test system.
862.1160 Bicarbonate/carbon dioxide test system.
862.1165 Catecholamines (total) test system.
862.1170 Chloride test system.
862.1175 Cholesterol (total) test system.
862.1177 Cholylglycine test system.
862.1180 Chymotrypsin test system.
862.1185 Compound S (11-deoxycortisol) test system.
862.1187 Conjugated sulfolithocholic acid (SLCG) test system.
862.1190 Copper test system.
862.1195 Corticoids test system.
862.1200 Corticosterone test system.
862.1205 Cortisol (hydrocortisone and hydroxycorticosterone) test system.
862.1210 Creatine test system.
862.1215 Creatine phosphokinase/creatine kinase or isoenzymes test system.
862.1225 Creatinine test system.
862.1230 Cyclic AMP test system.
862.1235 Cyclosporine test system.
862.1240 Cystine test system.
862.1245 Dehydroepiandrosterone (free and sulfate) test system.
862.1250 Desoxycorticosterone test system.
862.1255 2,3-Diphosphoglyceric acid test system.
862.1260 Estradiol test system.
862.1265 Estriol test system.
862.1270 Estrogens (total, in pregnancy) test system.
862.1275 Estrogens (total, nonpregnancy) test system.
862.1280 Estrone test system.
862.1285 Etiocholanolone test system.
862.1290 Fatty acids test system.
862.1295 Folic acid test system.
862.1300 Follicle-stimulating hormone test system.
862.1305 Formiminoglutamic acid (FIGLU) test system.
862.1310 Galactose test system.
862.1315 Galactose-1-phosphate uridyl transferase test system.
862.1320 Gastric acidity test system.
862.1325 Gastrin test system.
862.1330 Globulin test system.
862.1335 Glucagon test system.
862.1340 Urinary glucose (nonquantitative) test system.
862.1345 Glucose test system.
862.1360 Gamma-glutamyl transpeptidase and isoenzymes test system.
862.1365 Glutathione test system.
862.1370 Human growth hormone test system.
862.1375 Histidine test system.
862.1377 Urinary homocystine (nonquantitative) test system.
862.1380 Hydroxybutyric dehydrogenase test system.
862.1385 17-Hydroxycorticosteroids (17-ketogenic steroids) test system.
862.1390 5-Hydroxyindole acetic acid/serotonin test system.
862.1395 17-Hydroxyprogesterone test system.
862.1400 Hydroxyproline test system.
862.1405 Immunoreactive insulin test system.
862.1410 Iron (non-heme) test system.
862.1415 Iron-binding capacity test system.
862.1420 Isocitric dehydrogenase test system.
862.1430 17-Ketosteroids test system.
862.1435 Ketones (nonquantitative) test system.
862.1440 Lactate dehydrogenase test system.
862.1445 Lactate dehydrogenase isoenzymes test system.
862.1450 Lactic acid test system.
862.1455 Lecithin/sphingomyelin ratio in amniotic fluid test system.
862.1460 Leucine aminopeptidase test system.
862.1465 Lipase test system.
862.1470 Lipid (total) test system.
862.1475 Lipoprotein test system.
862.1485 Luteinizing hormone test system.
862.1490 Lysozyme (muramidase) test system.
862.1495 Magnesium test system.
862.1500 Malic dehydrogenase test system.
862.1505 Mucopolysaccharides (nonquantitative) test system.
862.1509 Methylmalonic acid (nonquantitative) test system.
862.1510 Nitrite (nonquantitative) test system.
862.1515 Nitrogen (amino-nitrogen) test system.
862.1520 5′-Nucleotidase test system.
862.1530 Plasma oncometry test system.
862.1535 Ornithine carbamyl transferase test system.
862.1540 Osmolality test system.
862.1542 Oxalate test system.
862.1545 Parathyroid hormone test system.
862.1550 Urinary pH (nonquantitative) test system.
862.1555 Phenylalanine test system.
862.1560 Urinary phenylketones (nonquantitative) test system.
862.1565 6-Phosphogluconate dehydrogenase test system.

862.1570 Phosphohexose isomerase test system.
862.1575 Phospholipid test system.
862.1580 Phosphorus (inorganic) test system.
862.1585 Human placental lactogen test system.
862.1590 Porphobilinogen test system.
862.1595 Porphyrins test system.
862.1600 Potassium test system.
862.1605 Pregnanediol test system.
862.1610 Pregnanetriol test system.
862.1615 Pregnenolone test system.
862.1620 Progesterone test system.
862.1625 Prolactin (lactogen) test system.
862.1630 Protein (fractionation) test system.
862.1635 Total protein test system.
862.1640 Protein-bound iodine test system.
862.1645 Urinary protein or albumin (nonquantitative) test system.
862.1650 Pyruvate kinase test system.
862.1655 Pyruvic acid test system.
862.1660 Quality control material (assayed and unassayed).
862.1665 Sodium test system.
862.1670 Sorbitol dehydrogenase test system.
862.1675 Blood specimen collection device.
862.1678 Tacrolimus test system.
862.1680 Testosterone test system.
862.1685 Thyroxine-binding globulin test system.
862.1690 Thyroid-stimulating hormone test system.
862.1695 Free thyroxine test system.
862.1700 Total thyroxine test system.
862.1705 Triglyceride test system.
862.1710 Total triiodothyronine test system.
862.1715 Triiodothyronine uptake test system.
862.1720 Triose phosphate isomerase test system.
862.1725 Trypsin test system.
862.1730 Free tyrosine test system.
862.1770 Urea nitrogen test system.
862.1775 Uric acid test system.
862.1780 Urinary calculi (stones) test system.
862.1785 Urinary urobilinogen (nonquantitative) test system.
862.1790 Uroporphyrin test system.
862.1795 Vanilmandelic acid test system.
862.1805 Vitamin A test system.
862.1810 Vitamin B_{12} test system.
862.1815 Vitamin E test system.
862.1820 Xylose test system.
862.1825 Vitamin D test system.

Subpart C—Clinical Laboratory Instruments

862.2050 General purpose laboratory equipment labeled or promoted for a specific medical use.
862.2100 Calculator/data processing module for clinical use.
862.2140 Centrifugal chemistry analyzer for clinical use.
862.2150 Continuous flow sequential multiple chemistry analyzer for clinical use.
862.2160 Discrete photometric chemistry analyzer for clinical use.
862.2170 Micro chemistry analyzer for clinical use.
862.2230 Chromatographic separation material for clinical use.
862.2250 Gas liquid chromatography system for clinical use.
862.2260 High pressure liquid chromatography system for clinical use.
862.2270 Thin-layer chromatography system for clinical use.
862.2300 Colorimeter, photometer, or spectrophotometer for clinical use.
862.2310 Clinical sample concentrator.
862.2320 Beta or gamma counter for clinical use.
862.2400 Densitometer/scanner (integrating, reflectance, TLC, or radiochromatogram) for clinical use.
862.2485 Electrophoresis apparatus for clinical use.
862.2500 Enzyme analyzer for clinical use.
862.2540 Flame emission photometer for clinical use.
862.2560 Fluorometer for clinical use.
862.2570 Instrumentation for clinical multiplex test systems.
862.2680 Microtitrator for clinical use.
862.2700 Nephelometer for clinical use.
862.2720 Plasma oncometer for clinical use.
862.2730 Osmometer for clinical use.
862.2750 Pipetting and diluting system for clinical use.
862.2800 Refractometer for clinical use.
862.2850 Atomic absorption spectrophotometer for clinical use.
862.2860 Mass spectrometer for clinical use.
862.2900 Automated urinalysis system.
862.2920 Plasma viscometer for clinical use.

Subpart D—Clinical Toxicology Test Systems

862.3030 Acetaminophen test system.
862.3035 Amikacin test system.
862.3040 Alcohol test system.
862.3050 Breath-alcohol test system.
862.3080 Breath nitric oxide test system.
862.3100 Amphetamine test system.
862.3110 Antimony test system.
862.3120 Arsenic test system.
862.3150 Barbiturate test system.
862.3170 Benzodiazepine test system.
862.3200 Clinical toxicology calibrator.
862.3220 Carbon monoxide test system.
862.3240 Cholinesterase test system.
862.3250 Cocaine and cocaine metabolite test system.
862.3270 Codeine test system.
862.3280 Clinical toxicology control material.
862.3300 Digitoxin test system.
862.3320 Digoxin test system.
862.3350 Diphenylhydantoin test system.
862.3360 Drug metabolizing enzyme genotyping system.

862.3380 Ethosuximide test system.
862.3450 Gentamicin test system.
862.3520 Kanamycin test system.
862.3550 Lead test system.
862.3555 Lidocaine test system.
862.3560 Lithium test system.
862.3580 Lysergic acid diethylamide (LSD) test system.
862.3600 Mercury test system.
862.3610 Methamphetamine test system.
862.3620 Methadone test system.
862.3630 Methaqualone test system.
862.3640 Morphine test system.
862.3645 Neuroleptic drugs radioreceptor assay test system.
862.3650 Opiate test system.
862.3660 Phenobarbital test system.
862.3670 Phenothiazine test system.
862.3680 Primidone test system.
862.3700 Propoxyphene test system.
862.3750 Quinine test system.
862.3830 Salicylate test system.
862.3840 Sirolimus test system.
862.3850 Sulfonamide test system.
862.3870 Cannabinoid test system.
862.3880 Theophylline test system.
862.3900 Tobramycin test system.
862.3910 Tricyclic antidepressant drugs test system.
862.3950 Vancomycin test system.

AUTHORITY: 21 U.S.C. 351, 360, 360c, 360e, 360j, 371.

SOURCE: 52 FR 16122, May 1, 1987, unless otherwise noted.

Subpart A—General Provisions

§ 862.1 Scope.

(a) This part sets forth the classification of clinical chemistry and clinical toxicology devices intended for human use that are in commercial distribution.

(b) The identification of a device in a regulation in this part is not a precise description of every device that is, or will be, subject to the regulation. A manufacturer who submits a premarket notification submission for a device under part 807 cannot show merely that the device is accurately described by the section title and identification provisions of a regulation in this part, but shall state why the device is substantially equivalent to other devices, as required in § 807.87.

(c) References in this part to regulatory sections of the Code of Federal Regulations are to chapter I of title 21 unless otherwise noted.

(d) Guidance documents referenced in this part are available on the Internet at *http://www.fda.gov/cdrh/guidance.html.*

[52 FR 16122, May 1, 1987, as amended at 67 FR 58329, Sept. 16, 2002]

§ 862.2 Regulation of calibrators.

Many devices classified in this part are intended to be used with a calibrator. A calibrator has a reference value assigned to it which serves as the basis by which test results of patients are derived or calculated. The calibrator for a device may be (a) manufactured and distributed separately from the device with which it is intended to be used, (b) manufactured and distributed as one of several device components, such as in a kit of reagents, or (c) built-in as an integral part of the device. Because of the central role that a calibrator plays in the measurement process and the critical effect calibrators have on accuracy of test results, elsewhere in this part, all three of these types of calibrators (§§ 862.1150 and 862.3200 of this part) are classified into class II, notwithstanding the classification of the device with which it is intended to be used. Thus, a device and its calibrator may have different classifications, even if the calibrator is built into the device.

§ 862.3 Effective dates of requirement for premarket approval.

A device included in this part that is classified into class III (premarket approval) shall not be commercially distributed after the date shown in the regulation classifying the device unless the manufacturer has an approval under section 515 of the act (unless an exemption has been granted under section 520(g)(2) of the act). An approval under section 515 of the act consists of FDA's issuance of an order approving an application for premarket approval (PMA) for the device or declaring completed a product development protocol (PDP) for the device.

(a) Before FDA requires that a device commercially distributed before the enactment date of the amendments, or a device that has been found substantially equivalent to such a device, has an approval under section 515 of the act FDA must promulgate a regulation under section 515(b) of the act requiring such approval, except as provided in paragraph (b) of this section. Such a regulation under section 515(b) of the act shall not be effective during the grace period ending on the 90th day after its promulgation or on the last day of the 30th full calendar month after the regulation that classifies the device into class III is effective, whichever is later. See section 501(f)(2)(B) of the act. Accordingly, unless an effective date of the requirement for premarket approval is shown in the regulation for a device classified into class III in this part, the device may be commercially distributed without FDA's issuance of an order approving a PMA or declaring completed a PDP for the device. If FDA promulgates a regulation under section 515(b) of the act requiring premarket approval for a device, section 501(f)(1)(A) of the act applies to the device.

(b) Any new, not substantially equivalent, device introduced into commercial distribution on or after May 28, 1976, including a device formerly marketed that has been substantially altered, is classified by statute (section 513(f) of the act) into class III without any grace period and FDA must have issued an order approving a PMA or declaring completed a PDP for the device before the device is commercially distributed unless it is reclassified. If FDA knows that a device being commercially distributed may be a "new" device as defined in this section because of any new intended use or other reasons, FDA may codify the statutory classification of the device into class III for such new use. Accordingly, the regulation for such a class III device states that as of the enactment date of the amendments, May 28, 1976, the device must have an approval under section 515 of the act before commercial distribution.

§862.9 Limitations of exemptions from section 510(k) of the Federal Food, Drug, and Cosmetic Act (the act).

The exemption from the requirement of premarket notification (section 510(k) of the act) for a generic type of class I or II device is only to the extent that the device has existing or reasonably foreseeable characteristics of commercially distributed devices within that generic type or, in the case of in vitro diagnostic devices, only to the extent that misdiagnosis as a result of using the device would not be associated with high morbidity or mortality. Accordingly, manufacturers of any commercially distributed class I or II device for which FDA has granted an exemption from the requirement of premarket notification must still submit a premarket notification to FDA before introducing or delivering for introduction into interstate commerce for commercial distribution the device when:

(a) The device is intended for a use different from the intended use of a legally marketed device in that generic type of device; e.g., the device is intended for a different medical purpose, or the device is intended for lay use where the former intended use was by health care professionals only;

(b) The modified device operates using a different fundamental scientific technology than a legally marketed device in that generic type of device; e.g., a surgical instrument cuts tissue with a laser beam rather than with a sharpened metal blade, or an in vitro diagnostic device detects or identifies infectious agents by using deoxyribonucleic acid (DNA) probe or nucleic acid hybridization technology rather than culture or immunoassay technology; or

(c) The device is an in vitro device that is intended:

(1) For use in the diagnosis, monitoring, or screening of neoplastic diseases with the exception of immunohistochemical devices;

(2) For use in screening or diagnosis of familial or acquired genetic disorders, including inborn errors of metabolism;

(3) For measuring an analyte that serves as a surrogate marker for screening, diagnosis, or monitoring life-threatening diseases such as acquired immune deficiency syndrome (AIDS), chronic or active hepatitis, tuberculosis, or myocardial infarction or to monitor therapy;

(4) For assessing the risk of cardiovascular diseases;

(5) For use in diabetes management;

(6) For identifying or inferring the identity of a microorganism directly from clinical material;

(7) For detection of antibodies to microorganisms other than immunoglobulin G (IgG) or IgG assays when the results are not qualitative, or are used to determine immunity, or the assay is intended for use in matrices other than serum or plasma;

(8) For noninvasive testing as defined in § 812.3(k) of this chapter; and

(9) For near patient testing (point of care).

[65 FR 2304, Jan. 14, 2000]

Subpart B—Clinical Chemistry Test Systems

§ 862.1020 Acid phosphatase (total or prostatic) test system.

(a) *Identification.* An acid phosphatase (total or prostatic) test system is a device intended to measure the activity of the acid phosphatase enzyme in plasma and serum.

(b) *Classification.* Class II.

§ 862.1025 Adrenocorticotropic hormone (ACTH) test system.

(a) *Identification.* An adrenocorticotropic hormone (ACTH) test system is a device intended to measure adrenocorticotropic hormone in plasma and serum. ACTH measurements are used in the differential diagnosis and treatment of certain disorders of the adrenal glands such as Cushing's syndrome, adrenocortical insufficiency, and the ectopic ACTH syndrome.

(b) *Classification.* Class II.

§ 862.1030 Alanine amino transferase (ALT/SGPT) test system.

(a) *Identification.* An alanine amino transferase (ALT/SGPT) test system is a device intended to measure the activity of the enzyme alanine amino transferase (ALT) (also known as a serum glutamic pyruvic transaminase or SGPT) in serum and plasma. Alanine amino transferase measurements are used in the diagnosis and treatment of certain liver diseases (e.g., viral hepatitis and cirrhosis) and heart diseases.

(b) *Classification.* Class I (general controls). The device is exempt from the premarket notification procedures in subpart E of part 807 of this chapter subject to § 862.9.

[52 FR 16122, May 1, 1987, as amended at 65 FR 2305, Jan. 14, 2000]

§ 862.1035 Albumin test system.

(a) *Identification.* An albumin test system is a device intended to measure the albumin concentration in serum and plasma. Albumin measurements are used in the diagnosis and treatment of numerous diseases involving primarily the liver or kidneys.

(b) *Classification.* Class II.

§ 862.1040 Aldolase test system.

(a) *Identification.* An aldolase test system is a device intended to measure the activity of the enzyme aldolase in serum or plasma. Aldolase measurements are used in the diagnosis and treatment of the early stages of acute hepatitis and for certain muscle diseases such as progressive Duchenne-type muscular dystrophy.

(b) *Classification.* Class I (general controls). The device is exempt from the premarket notification procedures in subpart E of part 807 of this chapter subject to § 862.9.

[52 FR 16122, May 1, 1987, as amended at 65 FR 2305, Jan. 14, 2000]

§ 862.1045 Aldosterone test system.

(a) *Identification.* An aldosterone test system is a device intended to measure the hormone aldosterone in serum and urine. Aldosterone measurements are used in the diagnosis and treatment of primary aldosteronism (a disorder caused by the excessive secretion of aldosterone by the adrenal gland), hypertension caused by primary aldosteronism, selective hypoaldosteronism, edematous states, and other conditions of electrolyte imbalance.

(b) *Classification.* Class II.

§ 862.1050 Alkaline phosphatase or isoenzymes test system.

(a) *Identification.* An alkaline phosphatase or isoenzymes test system is a device intended to measure alkaline phosphatase or its isoenzymes (a group of enzymes with similar biological activity) in serum or plasma. Measurements of alkaline phosphatase or its isoenzymes are used in the diagnosis and treatment of liver, bone, parathyroid, and intestinal diseases.

(b) *Classification.* Class II.

§ 862.1055 Newborn screening test system for amino acids, free carnitine, and acylcarnitines using tandem mass spectrometry.

(a) *Identification.* A newborn screening test system for amino acids, free carnitine, and acylcarnitines using tandem mass spectrometry is a device that consists of stable isotope internal standards, control materials, extraction solutions, flow solvents, instrumentation, software packages, and other reagents and materials. The device is intended for the measurement and evaluation of amino acids, free carnitine, and acylcarnitine concentrations from newborn whole blood filter paper samples. The quantitative analysis of amino acids, free carnitine, and acylcarnitines and their relationship with each other provides analyte concentration profiles that may aid in screening newborns for one or more inborn errors of amino acid, free carnitine, and acyl-carnitine metabolism.

(b) *Classification.* Class II (special controls). The special control is FDA's guidance document entitled "Class II Special Controls Guidance Document: Newborn Screening Test Systems for Amino Acids, Free Carnitine, and Acylcarnitines Using Tandem Mass Spectrometry." See § 862.1(d) for the availability of this guidance document.

[69 FR 68255, Nov. 24, 2004]

§ 862.1060 Delta-aminolevulinic acid test system.

(a) *Identification.* A *delta*-aminolevulinic acid test system is a device intended to measure the level of *delta*-aminolevulinic acid (a precursor of porphyrin) in urine. *Delta*-aminolevulinic acid measurements are used in the diagnosis and treatment of lead poisoning and certain porphyrias (diseases affecting the liver, gastrointestinal, and nervous systems that are accompanied by increased urinary excretion of various heme compounds including *delta*-aminolevulinic acid).

(b) *Classification.* Class I (general controls). The device is exempt from premarket notification procedures in subpart E of part 807 of this chapter subject to § 862.9.

[52 FR 16122, May 1, 1987, as amended at 65 FR 2305, Jan. 14, 2000]

§ 862.1065 Ammonia test system.

(a) *Identification.* An ammonia test system is a device intended to measure ammonia levels in blood, serum, and plasma, Ammonia measurements are used in the diagnosis and treatment of severe liver disorders, such as cirrhosis, hepatitis, and Reye's syndrome.

(b) *Classification.* Class I.

§ 862.1070 Amylase test system.

(a) *Identification.* An amylase test system is a device intended to measure the activity of the enzyme amylase in serum and urine. Amylase measurements are used primarily for the diagnosis and treatment of pancreatitis (inflammation of the pancreas).

(b) *Classification.* Class II.

§ 862.1075 Androstenedione test system.

(a) *Identification.* An androstenedione test system is a device intended to measure androstenedione (a substance secreted by the testes, ovary, and adrenal glands) in serum. Adrostenedione measurements are used in the diagnosis and treatment of females with excessive levels of androgen (male sex hormone) production.

(b) *Classification.* Class I (general controls). The device is exempt from the premarket notification procedures in subpart E of part 807 of this chapter subject to § 862.9.

[52 FR 16122, May 1, 1987, as amended at 65 FR 2305, Jan. 14, 2000]

§ 862.1080 Androsterone test system.

(a) *Identification.* An androsterone test system is a device intended to measure the hormone adrosterone in serum, plasma, and urine. Androsterone measurements are used in the diagnosis and treatment of gonadal and adrenal diseases.

(b) *Classification.* Class I (general controls). The device is exempt from the premarket notification procedures in

subpart E of part 807 of this chapter subject to § 862.9.

[52 FR 16122, May 1, 1987, as amended at 65 FR 2305, Jan. 14, 2000]

§ 862.1085 Angiotensin I and renin test system.

(a) *Identification.* An angiotensin I and renin test system is a device intended to measure the level of angiotensin I generated by renin in plasma. Angiotensin I measurements are used in the diagnosis and treatment of certain types of hypertension.

(b) *Classification.* Class II.

§ 862.1090 Angiotensin converting enzyme (A.C.E.) test system.

(a) *Identification.* An angiotensin converting enzyme (A.C.E.) test system is a device intended to measure the activity of angiotensin converting enzyme in serum and plasma. Measurements obtained by this device are used in the diagnosis and treatment of diseases such as sarcoidosis, a disease characterized by the formation of nodules in the lungs, bones, and skin, and Gaucher's disease, a hereditary disorder affecting the spleen.

(b) *Classification.* Class II.

§ 862.1095 Ascorbic acid test system.

(a) *Identification.* An ascorbic acid test system is a device intended to measure the level of ascorbic acid (vitamin C) in plasma, serum, and urine. Ascorbic acid measurements are used in the diagnosis and treatment of ascorbic acid dietary deficiencies.

(b) *Classification.* Class I (general controls). The device is exempt from the premarket notification procedures in subpart E of part 807 of this chapter subject to § 862.9.

[52 FR 16122, May 1, 1987, as amended at 65 FR 2305, Jan. 14, 2000]

§ 862.1100 Aspartate amino transferase (AST/SGOT) test system.

(a) *Identification.* An aspartate amino transferase (AST/SGOT) test system is a device intended to measure the activity of the enzyme aspartate amino transferase (AST) (also known as a serum glutamic oxaloacetic transferase or SGOT) in serum and plasma. Aspartate amino transferase measurements are used in the diagnosis and treatment of certain types of liver and heart disease.

(b) *Classification.* Class II.

§ 862.1110 Bilirubin (total or direct) test system.

(a) *Identification.* A bilirubin (total or direct) test system is a device intended to measure the levels of bilirubin (total or direct) in plasma or serum. Measurements of the levels of bilirubin, an organic compound formed during the normal and abnormal distruction of red blood cells, if used in the diagnosis and treatment of liver, hemolytic hematological, and metabolic disorders, including hepatitis and gall bladder block.

(b) *Classification.* Class II.

§ 862.1113 Bilirubin (total and unbound) in the neonate test system.

(a) *Identification.* A bilirubin (total and unbound) in the neonate test system is a device intended to measure the levels of bilirubin (total and unbound) in the blood (serum) of newborn infants to aid in indicating the risk of bilirubin encephalopathy (kernicterus).

(b) *Classification.* Class I.

[54 FR 30206, July 19, 1989]

§ 862.1115 Urinary bilirubin and its conjugates (nonquantitative) test system.

(a) *Identification.* A urinary bilirubin and its conjugates (nonquantitative) test system is a device intended to measure the levels of bilirubin conjugates in urine. Measurements of urinary bilirubin and its conjugates (nonquantitative) are used in the diagnosis and treatment of certain liver diseases.

(b) *Classification.* Class I (general controls). The device is exempt from the premarket notification procedures in subpart E of part 807 of this chapter subject to § 862.9.

[52 FR 16122, May 1, 1987, as amended at 65 FR 2305, Jan. 14, 2000]

§ 862.1117 B-type natriuretic peptide test system.

(a) *Identification.* The B-type natriuretic peptide (BNP) test system is an in vitro diagnostic device intended to measure BNP in whole blood and plasma. Measurements of BNP are used as an aid in the diagnosis of patients with congestive heart failure.

(b) *Classification.* Class II (special controls). The special control is "Class II Special Control Guidance Document for B-Type Natriuretic Peptide Premarket Notifications; Final Guidance for Industry and FDA Reviewers."

[66 FR 12734, Feb. 28, 2001]

§ 862.1118 Biotinidase test system.

(a) *Identification.* The biotinidase test system is an in vitro diagnostic device intended to measure the activity of the enzyme biotinidase in blood. Measurements of biotinidase are used in the treatment and diagnosis of biotinidase deficiency, an inborn error of metabolism in infants, characterized by the inability to utilize dietary protein bound vitamin or to recycle endogenous biotin. The deficiency may result in irreversible neurological impairment.

(b) *Classification.* Class II (special controls). The special control is sale, distribution, and use in accordance with the prescription device requirements in § 801.109 of this chapter.

[65 FR 16521, Mar. 29, 2000]

§ 862.1120 Blood gases ($P_{CO}2$, $P_{O}2$) and blood pH test system.

(a) *Identification.* A blood gases ($P_{CO}2$, $P_{O}2$) and blood pH test system is a device intended to measure certain gases in blood, serum, plasma or pH of blood, serum, and plasma. Measurements of blood gases ($P_{CO}2$, $P_{O}2$) and blood pH are used in the diagnosis and treatment of life-threatening acid-base disturbances.

(b) *Classification.* Class II.

§ 862.1130 Blood volume test system.

(a) *Identification.* A blood volume test system is a device intended to measure the circulating blood volume. Blood volume measurements are used in the diagnosis and treatment of shock, hemorrhage, and polycythemia vera (a disease characterized by an absolute increase in erythrocyte mass and total blood volume).

(b) *Classification.* Class I (general controls). The device is exempt from the premarket notification procedures in subpart E of part 807 of this chapter subject to § 862.9.

[52 FR 16122, May 1, 1987, as amended at 65 FR 2305, Jan. 14, 2000]

§ 862.1135 C-peptides of proinsulin test system.

(a) *Identification.* A C-peptides of proinsulin test system is a device intended to measure C-peptides of proinsulin levels in serum, plasma, and urine. Measurements of C-peptides of proinsulin are used in the diagnosis and treatment of patients with abnormal insulin secretion, including diabetes mellitus.

(b) *Classification.* Class I (general controls). The device is exempt from the premarket notification procedures in subpart E of part 807 of this chapter subject to § 862.9.

[52 FR 16122, May 1, 1987, as amended at 65 FR 2305, Jan. 14, 2000]

§ 862.1140 Calcitonin test system.

(a) *Identification.* A calcitonin test system is a device intended to measure the thyroid hormone calcitonin (thyrocalcitonin) levels in plasma and serum. Calcitonin measurements are used in the diagnosis and treatment of diseases involving the thyroid and parathyroid glands, including carcinoma and hyperparathyroidism (excessive activity of the parathyroid gland).

(b) *Classification.* Class II.

§ 862.1145 Calcium test system.

(a) *Identification.* A calcium test system is a device intended to measure the total calcium level in serum. Calcium measurements are used in the diagnosis and treatment of parathyroid disease, a variety of bone diseases, chronic renal disease and tetany (intermittent muscular contractions or spasms).

(b) *Classification.* Class II.

§ 862.1150 Calibrator.

(a) *Identification.* A calibrator is a device intended for medical purposes for use in a test system to establish points of reference that are used in the determination of values in the measurement of substances in human specimens. (See also § 862.2 in this part.)

(b) *Classification.* Class II.

§ 862.1155 Human chorionic gonadotropin (HCG) test system.

(a) *Human chorionic gonadotropin (HCG) test system intended for the early detection of pregnancy*—(1) *Identification.* A human chorionic gonadotropin (HCG) test system is a device intended for the early detection of pregnancy is intended to measure HCG, a placental hormone, in plasma or urine.

(2) *Classification.* Class II.

(b) *Human chorionic gonadotropin (HCG) test system intended for any uses other than early detection of pregnancy*—(1) *Identification.* A human chorionic goadotropin (HCG) test system is a device intended for any uses other than early detection of pregnancy (such as an aid in the diagnosis, prognosis, and management of treatment of persons with certain tumors or carcinomas) is intended to measure HCG, a placental hormone, in plasma or urine.

(2) *Classification.* Class III.

(3) *Date PMA or notice of completion of a PDP is required.* As of the enactment date of the amendments, May 28, 1976, an approval under section 515 of the act is required before the device described in paragraph (b)(1) may be commercially distributed. See § 862.3.

§ 862.1160 Bicarbonate/carbon dioxide test system.

(a) *Identification.* A bicarbonate/carbon dioxide test system is a device intended to measure bicarbonate/carbon dioxide in plasma, serum, and whole blood. Bicarbonate/carbon dioxide measurements are used in the diagnosis and treatment of numerous potentially serious disorders associated with changes in body acid-base balance.

(b) *Classification.* Class II.

§ 862.1165 Catecholamines (total) test system.

(a) *Identification.* A catecholamines (total) test system is a device intended to determine whether a group of similar compounds (epinephrine, norepinephrine, and dopamine) are present in urine and plasma. Catecholamine determinations are used in the diagnosis and treatment of adrenal medulla and hypertensive disorders, and for catecholamine-secreting tumors (pheochromo-cytoma, neuroblastoma, ganglioneuroma, and retinoblastoma).

(b) *Classification.* Class I (general controls). The device is exempt from the premarket notification procedures in subpart E of part 807 of this chapter subject to § 862.9.

[52 FR 16122, May 1, 1987, as amended at 65 FR 2305, Jan. 14, 2000]

§ 862.1170 Chloride test system.

(a) *Identification.* A chloride test system is a device intended to measure the level of chloride in plasma, serum, sweat, and urine. Chloride measurements are used in the diagnosis and treatment of electrolyte and metabolic disorders such as cystic fibrosis and diabetic acidosis.

(b) *Classification.* Class II.

§ 862.1175 Cholesterol (total) test system.

(a) *Identification.* A cholesterol (total) test system is a device intended to measure cholesterol in plasma and serum. Cholesterol measurements are used in the diagnosis and treatment of disorders involving excess cholesterol in the blood and lipid and lipoprotein metabolism disorders.

(b) *Classification.* Class I (general controls). The device is exempt from the premarket notification procedures in subpart E of part 807 of this chapter subject to § 862.9.

[52 FR 16122, May 1, 1987, as amended at 65 FR 2305, Jan. 14, 2000]

§ 862.1177 Cholylglycine test system.

(a) *Identification.* A cholylglycine test system is a device intended to measure the bile acid cholylglycine in serum. Measurements obtained by this device are used in the diagnosis and treatment of liver disorders, such as cirrhosis or obstructive liver disease.

(b) *Classification.* Class II.

§ 862.1180 Chymotrypsin test system.

(a) *Identification.* A chymotrypsin test system is a device intended to measure the activity of the enzyme chymotrypsin in blood and other body fluids and in feces. Chymotrypsin measurements are used in the diagnosis and treatment of pancreatic exocrine insufficiency.

(b) *Classification.* Class I (general controls). The device is exempt from the premarket notification procedures in subpart E of part 807 of this chapter subject to §862.9.

[52 FR 16122, May 1, 1987, as amended at 65 FR 2305, Jan. 14, 2000]

§862.1185 Compound S (11-deoxycortisol) test system.

(a) *Identification.* A compound S (11-dioxycortisol) test system is a device intended to measure the level of compound S (11-dioxycortisol) in plasma. Compound S is a steroid intermediate in the biosynthesis of the adrenal hormone cortisol. Measurements of compound S are used in the diagnosis and treatment of certain adrenal and pituitary gland disorders resulting in clinical symptoms of masculinization and hypertension.

(b) *Classification.* Class I (general controls). The device is exempt from the premarket notification procedures in subpart E of part 807 of this chapter subject to §862.9.

[52 FR 16122, May 1, 1987, as amended at 65 FR 2305, Jan. 14, 2000]

§862.1187 Conjugated sulfolithocholic acid (SLCG) test system.

(a) *Identification.* A conjugated sulfolithocholic acid (SLCG) test system is a device intended to measure the bile acid SLCG in serum. Measurements obtained by this device are used in the diagnosis and treatment of liver disorders, such as cirrhosis or obstructive liver disease.

(b) *Classification.* Class II.

§862.1190 Copper test system.

(a) *Identification.* A copper test system is a device intended to measure copper levels in plasma, serum, and urine. Measurements of copper are used in the diagnosis and treatment of anemia, infections, inflammations, and Wilson's disease (a hereditary disease primarily of the liver and nervous system). Test results are also used in monitoring patients with Hodgkin's disease (a disease primarily of the lymph system).

(b) *Classification.* Class I (general controls). The device is exempt from the premarket notification procedures in subpart E of part 807 of this chapter subject to the limitations in §862.9.

[52 FR 16122, May 1, 1987, as amended at 53 FR 21449, June 8, 1988; 66 FR 38787, July 25, 2001]

§862.1195 Corticoids test system.

(a) *Identification.* A corticoids test system is a device intended to measure the levels of corticoids (hormones of the adrenal cortex) in serum and p lasma. Measurements of corticoids are used in the diagnosis and treatment of disorders of the cortex of the adrenal glands, especially those associated with hypertension and electrolyte disturbances.

(b) *Classification.* Class I (general controls). The device is exempt from the premarket notification procedures in subpart E of part 807 of this chapter subject to §862.9.

[52 FR 16122, May 1, 1987, as amended at 65 FR 2305, Jan. 14, 2000]

§862.1200 Corticosterone test system.

(a) *Identification.* A corticosterone test system is a device intended to measure corticosterone (a steroid secreted by the adrenal gland) levels in plasma. Measurements of corticosterone are used in the diagnosis and treatment of adrenal disorders such as adrenal cortex disorders and blocks in cortisol synthesis.

(b) *Classification.* Class I (general controls). The device is exempt from the premarket notification procedures in subpart E of part 807 of this chapter subject to §862.9.

[52 FR 16122, May 1, 1987, as amended at 65 FR 2305, Jan. 14, 2000]

§862.1205 Cortisol (hydrocortisone and hydroxycorticosterone) test system.

(a) *Identification.* A cortisol (hydrocortisone and hydroxycorticosterone) test system is a device intended to measure the cortisol hormones secreted by the adrenal gland in plasma and urine. Measurements of cortisol are used in the diagnosis and treatment of disorders of the adrenal gland.

(b) *Classification.* Class II.

§ 862.1210 Creatine test system.

(a) *Identification.* A creatine test system is a device intended to measure creatine (a substance synthesized in the liver and pancreas and found in biological fluids) in plasma, serum, and urine. Measurements of creatine are used in the diagnosis and treatment of muscle diseases and endocrine disorders including hyperthyroidism.

(b) *Classification.* Class I (general controls). The device is exempt from the premarket notification procedures in subpart E of part 807 of this chapter subject to the limitations in § 862.9.

[52 FR 16122, May 1, 1987, as amended at 53 FR 21449, June 8, 1988; 66 FR 38787, July 25, 2001]

§ 862.1215 Creatine phosphokinase/creatine kinase or isoenzymes test system.

(a) *Identification.* A creatine phosphokinase/creatine kinase or isoenzymes test system is a device intended to measure the activity of the enzyme creatine phosphokinase or its isoenzymes (a group of enzymes with similar biological activity) in plasma and serum. Measurements of creatine phosphokinase and its isoenzymes are used in the diagnosis and treatment of myocardial infarction and muscle diseases such as progressive, Duchenne-type muscular dystrophy.

(b) *Classification.* Class II.

§ 862.1225 Creatinine test system.

(a) *Identification.* A creatinine test system is a device intended to measure creatinine levels in plasma and urine. Creatinine measurements are used in the diagnosis and treatment of renal diseases, in monitoring renal dialysis, and as a calculation basis for measuring other urine analytes.

(b) *Classification.* Class II.

§ 862.1230 Cyclic AMP test system.

(a) *Identification.* A cyclic AMP test system is a device intended to measure the level of adenosine 3′, 5′-monophosphate (cyclic AMP) in plasma, urine, and other body fluids. Cyclic AMP measurements are used in the diagnosis and treatment of endocrine disorders, including hyperparathyroidism (overactivity of the parathyroid gland). Cyclic AMP measurements may also be used in the diagnosis and treatment of Graves' disease (a disorder of the thyroid) and in the differentiation of causes of hypercalcemia (elevated levels of serum calcium.)

(b) *Classification.* Class II.

§ 862.1235 Cyclosporine test system.

(a) *Identification.* A cyclosporine test system is a device intended to quantitatively determine cyclosporine concentrations as an aid in the management of transplant patients receiving therapy with this drug. This generic type of device includes immunoassays and chromatographic assays for cyclosporine.

(b) *Classification.* Class II (special controls). The special control is "Class II Special Controls Guidance Document: Cyclosporine and Tacrolimus Assays; Guidance for Industry and FDA." See § 862.1(d) for the availability of this guidance document.

[67 FR 58329, Sept. 16, 2002]

§ 862.1240 Cystine test system.

(a) *Identification.* A cystine test system is a device intended to measure the amino acid cystine in urine. Cystine measurements are used in the diagnosis of cystinuria (occurrence of cystine in urine). Patients with cystinuria frequently develop kidney calculi (stones).

(b) *Classification.* Class I (general controls). The device is exempt from the premarket notification procedures in subpart E of part 807 of this chapter subject to § 862.9.

[52 FR 16122, May 1, 1987, as amended at 65 FR 2305, Jan. 14, 2000]

§ 862.1245 Dehydroepiandrosterone (free and sulfate) test system.

(a) *Identification.* A dehydroepiandrosterone (free and sulfate) test system is a device intended to measure dehydroepiandrosterone (DHEA) and its sulfate in urine, serum, plasma, and amniotic fluid. Dehydroepiandrosterone measurements are used in the diagnosis and treatment of DHEA-secreting adrenal carcinomas.

(b) *Classification.* Class I (general controls). The device is exempt from the premarket notification procedures in subpart E of part 807 of this chapter subject to § 862.9.

[52 FR 16122, May 1, 1987, as amended at 65 FR 2306, Jan. 14, 2000]

§ 862.1250 Desoxycorticosterone test system.

(a) *Identification.* A desoxycorticosterone test system is a device intended to measure desoxycorticosterone (DOC) in plasma and urine. DOC measurements are used in the diagnosis and treatment of patients with hypermineralocorticoidism (excess retention of sodium and loss of potassium) and other disorders of the adrenal gland.

(b) *Classification.* Class I (general controls). The device is exempt from the premarket notification procedures in subpart E of part 807 of this chapter subject to § 862.9.

[52 FR 16122, May 1, 1987, as amended at 65 FR 2306, Jan. 14, 2000]

§ 862.1255 2,3-Diphosphoglyceric acid test system.

(a) *Identification.* A 2,3-diphosphoglyceric acid test system is a device intended to measure 2,3-diphosphoglyceric acid (2,3-DPG) in erythrocytes (red blood cells). Measurements of 2,3-diphosphoglyceric acid are used in the diagnosis and treatment of blood disorders that affect the delivery of oxygen by erythrocytes to tissues and in monitoring the quality of stored blood.

(b) *Classification.* Class I (general controls). The device is exempt from the premarket notification procedures in subpart E of part 807 of this chapter subject to the limitations in § 862.9.

[52 FR 16122, May 1, 1987, as amended at 53 FR 21449, June 8, 1988; 66 FR 38787, July 25, 2001]

§ 862.1260 Estradiol test system.

(a) *Identification.* An estradiol test system is a device intended to measure estradiol, an estrogenic steroid, in plasma. Estradiol measurements are used in the diagnosis and treatment of various hormonal sexual disorders and in assessing placental function in complicated pregnancy.

(b) *Classification.* Class I (general controls). The device is exempt from the premarket notification procedures in subpart E of part 807 of this chapter subject to § 862.9.

[52 FR 16122, May 1, 1987, as amended at 65 FR 2306, Jan. 14, 2000]

§ 862.1265 Estriol test system.

(a) *Identification.* An estriol test system is a device intended to measure estriol, an estrogenic steroid, in plasma, serum, and urine of pregnant females. Estriol measurements are used in the diagnosis and treatment of fetoplacental distress in certain cases of high-risk pregnancy.

(b) *Classification.* Class I (general controls). The device is exempt from the premarket notification procedures in subpart E of part 807 of this chapter subject to § 862.9.

[52 FR 16122, May 1, 1987, as amended at 65 FR 2306, Jan. 14, 2000]

§ 862.1270 Estrogens (total, in pregnancy) test system.

(a) *Identification.* As estrogens (total, in pregnancy) test system is a device intended to measure total estrogens in plasma, serum, and urine during pregnancy. The device primarily measures estrone plus estradiol. Measurements of total estrogens are used to aid in the diagnosis and treatment of fetoplacental distress in certain cases of high-risk pregnancy.

(b) *Classification.* Class I (general controls). The device is exempt from the premarket notification procedures in subpart E of part 807 of this chapter subject to § 862.9.

[52 FR 16122, May 1, 1987, as amended at 65 FR 2306, Jan. 14, 2000]

§ 862.1275 Estrogens (total, nonpregnancy) test system.

(a) *Identification.* As estrogens (total, nonpregnancy) test system is a device intended to measure the level of estrogens (total estrone, estradiol, and estriol) in plasma, serum, and urine of males and nonpregnant females. Measurement of estrogens (total, nonpregnancy) is used in the diagnosis and treatment of numerous disorders, including infertility, amenorrhea (absence of menses) differentiation of primary and secondary ovarian malfunction, estrogen secreting testicular and ovarian tumors, and precocious puberty in females.

(b) *Classification.* Class I (general controls). The device is exempt from the premarket notification procedures in subpart E of part 807 of this chapter subject to § 862.9.

[52 FR 16122, May 1, 1987, as amended at 65 FR 2306, Jan. 14, 2000]

§ 862.1280 Estrone test system.

(a) *Identification.* An estrone test system is a device intended to measure estrone, an estrogenic steroid, in plasma. Estrone measurements are used in the diagnosis and treatment of numerous disorders, including infertility, amenorrhea, differentiation of primary and secondary ovarian malfunction, estrogen secreting testicular and ovarian tumors, and precocious puberty in females.

(b) *Classification.* Class I (general controls). The device is exempt from the premarket notification procedures in subpart E of part 807 of this chapter subject to § 862.9.

[52 FR 16122, May 1, 1987, as amended at 65 FR 2306, Jan. 14, 2000]

§ 862.1285 Etiocholanolone test system.

(a) *Identification.* An etiocholanolone test system is a device intended to measure etiocholanolone in serum and urine. Etiocholanolone is a metabolic product of the hormone testosterone and is excreted in the urine. Etiocholanolone measurements are used in the diagnosis and treatment of disorders of the testes and ovaries.

(b) *Classification.* Class I (general controls). The device is exempt from the premarket notification procedures in subpart E of part 807 of this chapter subject to § 862.9.

[52 FR 16122, May 1, 1987, as amended at 65 FR 2306, Jan. 14, 2000]

§ 862.1290 Fatty acids test system.

(a) *Identification.* A fatty acids test system is a device intended to measure fatty acids in plasma and serum. Measurements of fatty acids are used in the diagnosis and treatment of various disorders of lipid metabolism.

(b) *Classification.* Class I (general controls). The device is exempt from the premarket notification procedures in subpart E of part 807 of this chapter subject to the limitations in § 862.9.

[52 FR 16122, May 1, 1987, as amended at 53 FR 21449, June 8, 1988; 66 FR 38787, July 25, 2001]

§ 862.1295 Folic acid test system.

(a) *Identification.* A folic acid test system is a device intended to measure the vitamin folic acid in plasma and serum. Folic acid measurements are used in the diagnosis and treatment of megaloblastic anemia, which is characterized by the presence of megaloblasts (an abnormal red blood cell series) in the bone marrow.

(b) *Classification.* Class II.

[52 FR 16122, May 1, 1987; 53 FR 11645, Apr. 8, 1988]

§ 862.1300 Follicle-stimulating hormone test system.

(a) *Identification.* A follicle-stimulating hormone test system is a device intended to measure follicle-stimulating hormone (FSH) in plasma, serum, and urine. FSH measurements are used in the diagnosis and treatment of pituitary gland and gonadal disorders.

(b) *Classification.* Class I (general controls). The device is exempt from the premarket notification procedures in subpart E of part 807 of this chapter subject to § 862.9.

[52 FR 16122, May 1, 1987, as amended at 65 FR 2306, Jan. 14, 2000]

§ 862.1305 Formiminoglutamic acid (FIGLU) test system.

(a) *Identification.* A formiminoglutamic acid (FIGLU) test system is a device intended to measure formiminolutamic acid in urine. FIGLU measurements obtained by this device are used in the diagnosis of anemias, such as pernicious anemia and congenital hemolytic anemia.

(b) *Classification.* Class I (general controls). The device is exempt from the premarket notification procedures in

subpart E of part 807 of this chapter subject to the limitations in § 862.9.

[52 FR 16122, May 1, 1987, as amended at 53 FR 21449, June 8, 1988; 66 FR 38787, July 25, 2001]

§ 862.1310 Galactose test system.

(a) *Identification.* A galactose test system is a device intended to measure galactose in blood and urine. Galactose measurements are used in the diagnosis and treatment of the hereditary disease galactosemia (a disorder of galactose metabolism) in infants.

(b) *Classification.* Class I.

§ 862.1315 Galactose-1-phosphate uridyl transferase test system.

(a) *Identification.* A galactose-1-phosphate uridyl transferase test system is a device intended to measure the activity of the enzyme galactose-1-phosphate uridyl transferase in erythrocytes (red blood cells). Measurements of galactose-1-phosphate uridyl transferase are used in the diagnosis and treatment of the hereditary disease galactosemia (disorder of galactose metabolism) in infants.

(b) *Classification.* Class II.

§ 862.1320 Gastric acidity test system.

(a) *Identification.* A gastric acidity test system is a device intended to measure the acidity of gastric fluid. Measurements of gastric acidity are used in the diagnosis and treatment of patients with peptic ulcer, Zollinger-Ellison syndrome (peptic ulcer due to gastrin-secreting tumor of the pancreas), and related gastric disorders.

(b) *Classification.* Class I (general controls). The device is exempt from the premarket notification procedures in subpart E of part 807 of this chapter subject to the limitations in § 862.9.

[52 FR 16122, May 1, 1987, as amended at 53 FR 21449, June 8, 1988; 66 FR 38787, July 25, 2001]

§ 862.1325 Gastrin test system.

(a) *Identification.* A gastrin test system is a device intended to measure the hormone gastrin in plasma and serum. Measurements of gastrin are used in the diagnosis and treatment of patients with ulcers, pernicious anemia, and the Zollinger-Ellison syndrome (peptic ulcer due to a gastrin-secreting tumor of the pancreas).

(b) *Classification.* Class I (general controls). The device is exempt from the premarket notification procedures in subpart E of part 807 of this chapter subject to § 862.9.

[52 FR 16122, May 1, 1987, as amended at 65 FR 2306, Jan. 14, 2000]

§ 862.1330 Globulin test system.

(a) *Identification.* A globulin test system is a device intended to measure globulins (proteins) in plasma and serum. Measurements of globulin are used in the diagnosis and treatment of patients with numerous illnesses including severe liver and renal disease, multiple myeloma, and other disorders of blood globulins.

(b) *Classification.* Class I (general controls). The device is exempt from the premarket notification procedures in subpart E of part 807 of this chapter subject to § 862.9.

[52 FR 16122, May 1, 1987, as amended at 65 FR 2306, Jan. 14, 2000]

§ 862.1335 Glucagon test system.

(a) *Identification.* A glucagon test system is a device intended to measure the pancreatic hormone glucagon in plasma and serum. Glucagon measurements are used in the diagnosis and treatment of patients with various disorders of carbohydrate metabolism, including diabetes mellitus, hypoglycemia, and hyperglycemia.

(b) *Classification.* Class I (general controls). The device is exempt from the premarket notification procedures in subpart E of part 807 of this chapter subject to § 862.9.

[52 FR 16122, May 1, 1987, as amended at 65 FR 2306, Jan. 14, 2000]

§ 862.1340 Urinary glucose (nonquantitative) test system.

(a) *Identification.* A urinary glucose (nonquantitative) test system is a device intended to measure glucosuria (glucose in urine). Urinary glucose (nonquantitative) measurements are used in the diagnosis and treatment of carbohydrate metabolism disorders including diabetes mellitus, hypoglycemia, and hyperglycemia.

(b) *Classification.* Class II.

§ 862.1345 Glucose test system.

(a) *Identification.* A glucose test system is a device intended to measure glucose quantitatively in blood and other body fluids. Glucose measurements are used in the diagnosis and treatment of carbohydrate metabolism disorders including diabetes mellitus, neonatal hypoglycemia, and idiopathic hypoglycemia, and of pancreatic islet cell carcinoma.

(b) *Classification.* Class II.

§ 862.1360 Gamma-glutamyl transpeptidase and isoenzymes test system.

(a) *Identification.* A gamma-glutamyl transpeptidase and isoenzymes test system is a device intended to measure the activity of the enzyme gamma-glutamyl transpeptidase (GGTP) in plasma and serum. Gamma-glutamyl transpeptidase and isoenzymes measurements are used in the diagnosis and treatment of liver diseases such as alcoholic cirrhosis and primary and secondary liver tumors.

(b) *Classification.* Class I (general controls). The device is exempt from the premarket notification procedures in subpart E of part 807 of this chapter subject to § 862.9.

[52 FR 16122, May 1, 1987, as amended at 65 FR 2306, Jan. 14, 2000]

§ 862.1365 Glutathione test system.

(a) *Identification.* A glutathione test system is a device intended to measure glutathione (the tripeptide of glycine, cysteine, and glutamic acid) in erythrocytes (red blood cells). Glutathione measurements are used in the diagnosis and treatment of certain drug-induced hemolytic (erythrocyte destroying) anemias due to an inherited enzyme deficiency.

(b) *Classification.* Class I (general controls). The device is exempt from the premarket notification procedures in subpart E of part 807 of this chapter subject to the limitations in § 862.9

[52 FR 16122, May 1, 1987, as amended at 53 FR 21449, June 8, 1988; 66 FR 38787, July 25, 2001]

§ 862.1370 Human growth hormone test system.

(a) *Identification.* A human growth hormone test system is a device intended to measure the levels of human growth hormone in plasma. Human growth hormone measurements are used in the diagnosis and treatment of disorders involving the anterior lobe of the pituitary gland.

(b) *Classification.* Class I (general controls). The device is exempt from the premarket notification procedures in subpart E of part 807 of this chapter subject to § 862.9.

[52 FR 16122, May 1, 1987, as amended at 65 FR 2306, Jan. 14, 2000]

§ 862.1375 Histidine test system.

(a) *Identification.* A histidine test system is a device intended to measure free histidine (an amino acid) in plasma and urine. Histidine measurements are used in the diagnosis and treatment of hereditary histidinemia characterized by excess histidine in the blood and urine often resulting in mental retardation and disordered speech development.

(b) *Classification.* Class I (general controls). The device is exempt from the premarket notification procedures in subpart E of part 807 of this chapter subject to § 862.9.

[52 FR 16122, May 1, 1987, as amended at 65 FR 2306, Jan. 14, 2000]

§ 862.1377 Urinary homocystine (nonquantitative) test system.

(a) *Identification.* A urinary homocystine (nonquantitative) test system is a device intended to identify homocystine (an analogue of the amino acid cystine) in urine. The identification of urinary homocystine is used in the diagnosis and treatment of homocystinuria (homosystine in urine), a heritable metabolic disorder which may cause mental retardation.

(b) *Classification.* Class II.

§ 862.1380 Hydroxybutyric dehydrogenase test system.

(a) *Identification.* A hydroxybutyric dehydrogenase test system is a device intended to measure the activity of the enzyme alpha-hydroxybutric dehydrogenase (HBD) in plasma or serum. HBD measurements are used in the diagnosis and treatment of myocardial infarction, renal damage (such as rejection of transplants), certain hematological diseases (such as acute leukemias and megaloblastic anemias) and, to a lesser degree, liver disease.

(b) *Classification.* Class I (general controls). The device is exempt from the premarket notification procedures in subpart E of part 807 of this chapter subject to the limitations in § 862.9.

[52 FR 16122, May 1, 1987, as amended at 53 FR 21449, June 8, 1988; 66 FR 38787, July 25, 2001]

§ 862.1385 17-Hydroxycorticosteroids (17-ketogenic steroids) test system.

(a) *Identification.* A 17-hydroxycorticosteroids (17-ketogenic steroids) test system is a device intended to measure corticosteroids that possess a dihydroxyacetone

$$(HOCH_2-\underset{\overset{\|}{O}}{C}-CH_2OH)$$

moiety on the steroid nucleus in urine. Corticosteroids with this chemical configuration include cortisol, cortisone 11-desoxycortisol, desoxycorticosterone, and their tetrahydroderivatives. This group of hormones is synthesized by the adrenal gland. Measurements of 17-hydroxycorticosteroids (17-ketogenic steroids) are used in the diagnosis and treatment of various diseases of the adrenal or pituitary glands and gonadal disorders.

(b) *Classification.* Class I (general controls). The device is exempt from the premarket notification procedures in subpart E of part 807 of this chapter subject to § 862.9.

[52 FR 16122, May. 1, 1987; 52 FR 29468, Aug. 7, 1987, as amended at 65 FR 2306, Jan. 14, 2000]

§ 862.1390 5-Hydroxyindole acetic acid/serotonin test system.

(a) *Identification.* A 5-hydroxyindole acetic acid/serotonin test system is a device intended to measure 5-hydroxyindole acetic acid/serotonin in urine. Measurements of 5-hydroxyindole acetic acid/serotonin are used in the diagnosis and treatment of carcinoid tumors of endocrine tissue.

(b) *Classification.* Class I (general controls). The device is exempt from the premarket notification procedures in subpart E of part 807 of this chapter subject to § 862.9.

[52 FR 16122, May 1, 1987, as amended at 65 FR 2306, Jan. 14, 2000]

§ 862.1395 17-Hydroxyprogesterone test system.

(a) *Identification.* A 17-hydroxyprogesterone test system is a device intended to measure 17-hydroxyprogesterone (a steroid) in plasma and serum. Measurements of 17-hydroxyprogesterone are used in the diagnosis and treatment of various disorders of the adrenal glands or the ovaries.

(b) *Classification.* Class I (general controls). The device is exempt from the premarket notification procedures in subpart E of part 807 of this chapter subject to § 862.9.

[52 FR 16122, May 1, 1987, as amended at 65 FR 2306, Jan. 14, 2000]

§ 862.1400 Hydroxyproline test system.

(a) *Identification.* A hydroxyproline test system is a device intended to measure the amino acid hydroxyproline in urine. Hydroxyproline measurements are used in the diagnosis and treatment of various collagen (connective tissue) diseases, bone disease such as Paget's disease, and endocrine disorders such as hyperparathyroidism and hyperthyroidism.

(b) *Classification.* Class I (general controls). The device is exempt from the premarket notification procedures in subpart E of part 807 of this chapter subject to § 862.9.

[52 FR 16122, May 1, 1987, as amended at 65 FR 2306, Jan. 14, 2000]

§ 862.1405 Immunoreactive insulin test system.

(a) *Identification.* An immunoreactive insulin test system is a device intended to measure immunoreactive insulin in serum and plasma. Immunoreactive insulin measurements are used in the diagnosis and treatment of various carbohydrate metabolism disorders, including diabetes mellitus, and hypoglycemia.

(b) *Classification.* Class I (general controls). The device is exempt from the premarket notification procedures in subpart E of part 807 of this chapter subject to § 862.9.

[52 FR 16122, May 1, 1987, as amended at 65 FR 2306, Jan. 14, 2000]

§ 862.1410 Iron (non-heme) test system.

(a) *Identification.* An iron (non-heme) test system is a device intended to measure iron (non-heme) in serum and plasma. Iron (non-heme) measurements are used in the diagnosis and treatment of diseases such as iron deficiency anemia, hemochromatosis (a disease associated with widespread deposit in the tissues of two iron-containing pigments, hemosiderin and hemofuscin, and characterized by pigmentation of the skin), and chronic renal disease.

(b) *Classification.* Class I.

§ 862.1415 Iron-binding capacity test system.

(a) *Identification.* An iron-binding capacity test system is a device intended to measure iron-binding capacity in serum. Iron-binding capacity measurements are used in the diagnosis and treatment of anemia.

(b) *Classification.* Class I.

§ 862.1420 Isocitric dehydrogenase test system.

(a) *Identification.* An isocitric dehydrogenase test system is a device intended to measure the activity of the enzyme isocitric dehydrogenase in serum and plasma. Isocitric dehydrogenase measurements are used in the diagnosis and treatment of liver disease such as viral hepatitis, cirrhosis, or acute inflammation of the biliary tract; pulmonary disease such as pulmonary infarction (local arrest or sudden insufficiency of the blood supply to the lungs), and diseases associated with pregnancy.

(b) *Classification.* Class I (general controls). The device is exempt from the premarket notification procedures in subpart E of part 807 of this chapter subject to the limitations in § 862.9.

[52 FR 16122, May 1, 1987, as amended at 53 FR 21449, June 8, 1988; 66 FR 38788, July 25, 2001]

§ 862.1430 17-Ketosteroids test system.

(a) *Identification.* A 17-ketosteroids test system is a device intended to measure 17-ketosteroids in urine. Measurements of 17-ketosteroids are used in the diagnosis and treatment of disorders of the adrenal cortex and gonads and of other endocrine disorders, including hypertension, diabetes, and hypothyroidism.

(b) *Classification.* Class I (general controls). The device is exempt from the premarket notification procedures in subpart E of part 807 of this chapter subject to § 862.9.

[52 FR 16122, May 1, 1987, as amended at 65 FR 2307, Jan. 14, 2000]

§ 862.1435 Ketones (nonquantitative) test system.

(a) *Identification.* A ketones (nonquantitative) test system is a device intended to identify ketones in urine and other body fluids. Identification of ketones is used in the diagnosis and treatment of acidosis (a condition characterized by abnormally high acidity of body fluids) or ketosis (a condition characterized by increased production of ketone bodies such as acetone) and for monitoring patients on ketogenic diets and patients with diabetes.

(b) *Classification.* Class I (general controls). The device is exempt from the premarket notification procedures in subpart E of part 807 of this chapter subject to § 862.9.

[52 FR 16122, May 1, 1987, as amended at 65 FR 2307, Jan. 14, 2000]

§ 862.1440 Lactate dehydrogenase test system.

(a) *Identification.* A lactate dehydrogenase test system is a device intended to measure the activity of the enzyme lactate dehydrogenase in serum. Lactate dehydrogenase measurements are used in the diagnosis and treatment of liver diseases such as acute viral hepatitis, cirrhosis, and metastatic carcinoma of the liver, cardiac diseases such as myocardial infarction, and tumors of the lung or kidneys.

(b) *Classification.* Class II (special controls). The device is exempt from the premarket notification procedures

in subpart E of part 807 of this chapter subject to § 862.9.

[52 FR 16122, May 1, 1987, as amended at 63 FR 59225, Nov. 3, 1998]

§ 862.1445 Lactate dehydrogenase isoenzymes test system.

(a) *Identification.* A lactate dehydrogenase isoenzymes test system is a device intended to measure the activity of lactate dehydrogenase isoenzymes (a group of enzymes with similar biological activity) in serum. Measurements of lactate dehydrogenase isoenzymes are used in the diagnosis and treatment of liver diseases, such as viral hepatitis, and myocardial infarction.

(b) *Classification.* Class II.

§ 862.1450 Lactic acid test system.

(a) *Identification.* A lactic acid test system is a device intended to measure lactic acid in whole blood and plasma. Lactic acid measurements that evaluate the acid-base status are used in the diagnosis and treatment of lactic acidosis (abnormally high acidity of the blood).

(b) *Classification.* Class I (general controls). The device is exempt from the premarket notification procedures in subpart E of part 807 of this chapter subject to § 862.9.

[52 FR 16122, May 1, 1987, as amended at 65 FR 2307, Jan. 14, 2000]

§ 862.1455 Lecithin/sphingomyelin ratio in amniotic fluid test system.

(a) *Identification.* A lecithin/sphingomyelin ratio in amniotic fluid test system is a device intended to measure the lecithin/sphingomyelin ratio in amniotic fluid. Lecithin and sphingomyelin are phospholipids (fats or fat-like substances containing phosphorus). Measurements of the lecithin/sphingomyelin ratio in amniotic fluid are used in evaluating fetal maturity.

(b) *Classification.* Class II.

§ 862.1460 Leucine aminopeptidase test system.

(a) *Identification.* A leucine aminopeptidase test system is a device intended to measure the activity of the enzyme leucine amino-peptidase in serum, plasma, and urine. Leucine aminopeptidase measurements are used in the diagnosis and treatment of liver diseases such as viral hepatitis and obstructive jaundice.

(b) *Classification.* Class I (general controls). The device is exempt from the premarket notification procedures in subpart E of part 807 of this chapter subject to § 862.9.

[52 FR 16122, May 1, 1987, as amended at 65 FR 2307, Jan. 14, 2000]

§ 862.1465 Lipase test system.

(a) *Identification.* A lipase test system is a device intended to measure the activity of the enzymes lipase in serum. Lipase measurements are used in diagnosis and treatment of diseases of the pancreas such as acute pancreatitis and obstruction of the pancreatic duct.

(b) *Classification.* Class I (general controls). The device is exempt from the premarket notification procedures in subpart E of part 807 of this chapter subject to § 862.9.

[52 FR 16122, May 1, 1987, as amended at 65 FR 2307, Jan. 14, 2000]

§ 862.1470 Lipid (total) test system.

(a) *Identification.* A lipid (total) test system is a device intended to measure total lipids (fats or fat-like substances) in serum and plasma. Lipid (total) measurements are used in the diagnosis and treatment of various diseases involving lipid metabolism and atherosclerosis.

(b) *Classification.* Class I (general controls). The device is exempt from the premarket notification procedures in subpart E of part 807 of this chapter subject to the limitations in § 862.9.

[52 FR 16122, May 1, 1987, as amended at 53 FR 21449, June 8, 1988; 66 FR 38788, July 25, 2001]

§ 862.1475 Lipoprotein test system.

(a) *Identification.* A lipoprotein test system is a device intended to measure lipoprotein in serum and plasma. Lipoprotein measurements are used in the diagnosis and treatment of lipid disorders (such as diabetes mellitus), atherosclerosis, and various liver and renal diseases.

(b) *Classification.* Class I (general controls). The device is exempt from the premarket notification procedures in

subpart E of part 807 of this chapter subject to § 862.9.

[52 FR 16122, May 1, 1987, as amended at 65 FR 2307, Jan. 14, 2000]

§ 862.1485 Luteinizing hormone test system.

(a) *Identification.* A luteinizing hormone test system is a device intended to measure luteinizing hormone in serum and urine. Luteinizing hormone measurements are used in the diagnosis and treatment of gonadal dysfunction.

(b) *Classification.* Class I (general controls). The device is exempt from the premarket notification procedures in subpart E of part 807 of this chapter subject to § 862.9.

[52 FR 16122, May 1, 1987, as amended at 65 FR 2307, Jan. 14, 2000]

§ 862.1490 Lysozyme (muramidase) test system.

(a) *Identification.* A lysozyme (muramidase) test system is a device intended to measure the activity of the bacteriolytic enzyme lysozyme (muramidase) in serum, plasma, leukocytes, and urine. Lysozyme measurements are used in the diagnosis and treatment of monocytic leukemia and kidney disease.

(b) *Classification.* Class I (general controls). The device is exempt from the premarket notification procedures in subpart E of part 807 of this chapter subject to the limitations in § 862.9.

[52 FR 16122, May 1, 1987, as amended at 53 FR 21449, June 8, 1988; 66 FR 38788, July 25, 2001]

§ 862.1495 Magnesium test system.

(a) *Identification.* A magnesium test system is a device intended to measure magnesium levels in serum and plasma. Magnesium measurements are used in the diagnosis and treatment of hypomagnesemia (abnormally low plasma levels of magnesium) and hypermagnesemia (abnormally high plasma levels of magnesium).

(b) *Classification.* Class I.

§ 862.1500 Malic dehydrogenase test system.

(a) *Identification.* A malic dehydrogenase test system is a device that is intended to measure the activity of the enzyme malic dehydrogenase in serum and plasma. Malic dehydrogenase measurements are used in the diagnosis and treatment of muscle and liver diseases, myocardial infarctions, cancer, and blood disorders such as myelogenous (produced in the bone marrow) leukemia.

(b) *Classification.* Class I (general controls). The device is exempt from the premarket notification procedures in subpart E of part 807 of this chapter subject to § 862.9.

[52 FR 16122, May 1, 1987, as amended at 65 FR 2307, Jan. 14, 2000]

§ 862.1505 Mucopolysaccharides (nonquantitative) test system.

(a) *Identification.* A mucopolysaccharides (nonquantitative) test system is a device intended to measure the levels of mucopolysaccharides in urine. Mucopolysaccharide measurements in urine are used in the diagnosis and treatment of various inheritable disorders that affect bone and connective tissues, such as Hurler's, Hunter's, Sanfilippo's, Scheie's Morquio's and Maroteaux-Lamy syndromes.

(b) *Classification.* Class I (general controls). The device is exempt from the premarket notification procedures in subpart E of part 807 of this chapter subject to § 862.9.

[52 FR 16122, May 1, 1987, as amended at 65 FR 2307, Jan. 14, 2000]

§ 862.1509 Methylmalonic acid (nonquantitative) test system.

(a) *Identification.* A methylmalonic acid (nonquantitative) test system is a device intended to identify methylmalonic acid in urine. The identification of methylmalonic acid in urine is used in the diagnosis and treatment of methylmalonic aciduria, a heritable metabolic disorder which, if untreated, may cause mental retardation.

(b) *Classification.* Class II.

§ 862.1510 Nitrite (nonquantitative) test system.

(a) *Identification.* A nitrite (nonquantitative) test system is a device intended to identify nitrite in urine.

Nitrite identification is used in the diagnosis and treatment of uninary tract infection of bacterial origin.

(b) *Classification.* Class I (general controls). The device is exempt from the premarket notification procedures in subpart E of part 807 of this chapter subject to §862.9.

[52 FR 16122, May 1, 1987, as amended at 65 FR 2307, Jan. 14, 2000]

§862.1515 Nitrogen (amino-nitrogen) test system.

(a) *Identification.* A nitrogen (amino-nitrogen) test system is a device intended to measure amino acid nitrogen levels in serum, plasma, and urine. Nitrogen (amino-nitrogen) measurements are used in the diagnosis and treatment of certain forms of severe liver disease and renal disorders.

(b) *Classification.* Class I (general controls). The device is exempt from the premarket notification procedures in subpart E of part 807 of this chapter subject to the limitations in §862.9.

[52 FR 16122, May 1, 1987, as amended at 53 FR 21449, June 8, 1988; 66 FR 38788, July 25, 2001]

§862.1520 5′-Nucleotidase test system.

(a) *Identification.* A 5′-nucleotidase test system is a device intended to measure the activity of the enzyme 5′-nucleotidase in serum and plasma. Measurements of 5′-nucleotidase are used in the diagnosis and treatment of liver diseases and in the differentiations between liver and bone diseases in the presence of elevated serum alkaline phosphatase activity.

(b) *Classification.* Class I (general controls). The device is exempt from the premarket notification procedures in subpart E of part 807 of this chapter subject to §862.9.

[52 FR 16122, May 1, 1987, as amended at 65 FR 2307, Jan. 14, 2000]

§862.1530 Plasma oncometry test system.

(a) *Identification.* A plasma oncometry test system is a device intended to measure plasma oncotic pressure. Plasma oncotic pressure is that portion of the total fluid pressure contributed by proteins and other molecules too large to pass through a specified membrane. Measurements of plasma oncotic pressure are used in the diagnosis and treatment of dehydration and circulatory disorders related to low serum protein levels and increased capillary permeability, such as edema and shock.

(b) *Classification.* Class I (general controls). The device is exempt from the premarket notification procedures in subpart E of part 807 of this chapter subject to §862.9.

[52 FR 16122, May 1, 1987, as amended at 65 FR 2307, Jan. 14, 2000]

§862.1535 Ornithine carbamyl transferase test system.

(a) *Identification.* An ornithine carbamyl transferase test system is a device intended to measure the activity of the enzyme ornithine carbamyl transferase (OCT) in serum. Ornithine carbamyl transferase measurements are used in the diagnosis and treatment of liver diseases, such as infectious hepatitis, acute cholecystitis (inflammation of the gall bladder), cirrhosis, and liver metastases.

(b) *Classification.* Class I (general controls). The device is exempt from the premarket notification procedures in subpart E of part 807 of this chapter subject to §862.9.

[52 FR 16122, May 1, 1987, as amended at 65 FR 2307, Jan. 14, 2000]

§862.1540 Osmolality test system.

(a) *Identification.* An osmolality test system is a device intended to measure ionic and nonionic solute concentration in body fluids, such as serum and urine. Osmolality measurement is used as an adjunct to other tests in the evaluation of a variety of diseases, including kidney diseases (e.g., chronic progressive renal failure), diabetes insipidus, other endocrine and metabolic disorders, and fluid imbalances.

(b) *Classification.* Class I (general controls). The device is exempt from the premarket notification procedures in subpart E of part 807 of this chapter subject to §862.9.

[52 FR 16122, May 1, 1987, as amended at 65 FR 2307, Jan. 14, 2000]

§ 862.1542 Oxalate test system.

(a) *Identification.* An oxalate test system is a device intended to measure the concentration of oxalate in urine. Measurements of oxalate are used to aid in the diagnosis or treatment of urinary stones or certain other metabolic disorders.

(b) *Classification.* Class I (general controls). The device is exempt from the premarket notification procedures in subpart E of part 807 of this chapter subject to § 862.9.

[52 FR 16122, May 1, 1987, as amended at 65 FR 2307, Jan. 14, 2000]

§ 862.1545 Parathyroid hormone test system.

(a) *Identification.* A parathyroid hormone test system is a device intended to measure the levels of parathyroid hormone in serum and plasma. Measurements of parathyroid hormone levels are used in the differential diagnosis of hypercalcemia (abnormally high levels of calcium in the blood) and hypocalcemia (abnormally low levels of calcium in the blood) resulting from disorders of calcium metabolism.

(b) *Classification.* Class II.

§ 862.1550 Urinary pH (nonquantitative) test system.

(a) *Identification.* A urinary pH (nonquantitative) test system is a device intended to estimate the pH of urine. Estimations of pH are used to evaluate the acidity or alkalinity of urine as it relates to numerous renal and metabolic disorders and in the monitoring of patients with certain diets.

(b) *Classification.* Class I (general controls). The device is exempt from the premarket notification procedures in subpart E of part 807 of this chapter subject to § 862.9.

[52 FR 16122, May 1, 1987, as amended at 65 FR 2307, Jan. 14, 2000]

§ 862.1555 Phenylalanine test system.

(a) *Identification.* A phenylalanine test system is a device intended to measure free phenylalanine (an amino acid) in serum, plasma, and urine. Measurements of phenylalanine are used in the diagnosis and treatment of congenital phenylketonuria which, if untreated, may cause mental retardation.

(b) *Classification.* Class II.

§ 862.1560 Urinary phenylketones (nonquantitative) test system.

(a) *Identification.* A urinary phenylketones (nonquantitative) test system is a device intended to identify phenylketones (such as phenylpyruvic acid) in urine. The identification of urinary phenylketones is used in the diagnosis and treatment of congenital phenylketonuria which, if untreated, may cause mental retardation.

(b) *Classification.* Class I (general controls). The device is exempt from the premarket notification procedures in subpart E of part 807 of this chapter subject to § 862.9.

[52 FR 16122, May 1, 1987, as amended at 65 FR 2307, Jan. 14, 2000]

§ 862.1565 6-Phosphogluconate dehydrogenase test system.

(a) *Identification.* A 6-phosphogluconate dehydrogenase test system is a device intended to measure the activity of the enzyme 6-phosphogluconate dehydrogenase (6 PGD) in serum and erythrocytes. Measurements of 6-phosphogluconate dehydrogenase are used in the diagnosis and treatment of certain liver diseases (such as hepatitis) and anemias.

(b) *Classification.* Class I (general controls). The device is exempt from the premarket notification procedures in subpart E of part 807 of this chapter subject to the limitations in § 862.9.

[52 FR 16122, May 1, 1987, as amended at 53 FR 21449, June 8, 1988; 66 FR 38788, July 25, 2001]

§ 862.1570 Phosphohexose isomerase test system.

(a) *Identification.* A phosphohexose isomerase test system is a device intended to measure the activity of the enzyme phosphohexose isomerase in serum. Measurements of phosphohexose isomerase are used in the diagnosis and treatment of muscle diseases such as muscular dystrophy, liver diseases such as hepatitis or cirrhosis, and metastatic carcinoma.

(b) *Classification.* Class I (general controls). The device is exempt from the premarket notification procedures in

subpart E of part 807 of this chapter subject to § 862.9.

[52 FR 16122, May 1, 1987, as amended at 65 FR 2307, Jan. 14, 2000]

§ 862.1575 Phospholipid test system.

(a) *Identification.* A phospholipid test system is a device intended to measure phospholipids in serum and plasma. Measurements of phospholipids are used in the diagnosis and treatment of disorders involving lipid (fat) metabolism.

(b) *Classification.* Class I (general controls). The device is exempt from the premarket notification procedures in subpart E of part 807 of this chapter subject to the limitations in § 862.9.

[52 FR 16122, May 1, 1987, as amended at 53 FR 21449, June 8, 1988; 66 FR 38788, July 25, 2001]

§ 862.1580 Phosphorus (inorganic) test system.

(a) *Identification.* A phosphorus (inorganic) test system is a device intended to measure inorganic phosphorus in serum, plasma, and urine. Measurements of phosphorus (inorganic) are used in the diagnosis and treatment of various disorders, including parathyroid gland and kidney diseases, and vitamin D imbalance.

(b) *Classification.* Class I.

§ 862.1585 Human placental lactogen test system.

(a) *Identification.* A human placental lactogen test system is a device intended to measure the hormone human placental lactogen (HPL), (also known as human chorionic somatomammotrophin (HCS)), in maternal serum and maternal plasma. Measurements of human placental lactogen are used in the diagnosis and clinical management of high-risk pregnancies involving fetal distress associated with placental insufficiency. Measurements of HPL are also used in pregnancies complicated by hypertension, proteinuria, edema, post-maturity, placental insufficiency, or possible miscarriage.

(b) *Classification.* Class II.

§ 862.1590 Porphobilinogen test system.

(a) *Identification.* A porphobilinogen test system is a device intended to measure porphobilinogen (one of the derivatives of hemoglobin which can make the urine a red color) in urine. Measurements obtained by this device are used in the diagnosis and treatment of porphyrias (primarily inherited diseases associated with disturbed porphyrine metabolism), lead poisoning, and other diseases characterized by alterations in the heme pathway.

(b) *Classification.* Class I (general controls). The device is exempt from the premarket notification procedures in subpart E of part 807 of this chapter subject to § 862.9.

[52 FR 16122, May 1, 1987, as amended at 65 FR 2307, Jan. 14, 2000]

§ 862.1595 Porphyrins test system.

(a) *Identification.* A porphyrins test system is a device intended to measure porphyrins (compounds formed during the biosynthesis of heme, a constituent of hemoglobin, and related compounds) in urine and feces. Measurements obtained by this device are used in the diagnosis and treatment of lead poisoning, porphyrias (primarily inherited diseases associated with disturbed porphyrin metabolism), and other diseases characterized by alterations in the heme pathway.

(b) *Classification.* Class I (general controls). The device is exempt from the premarket notification procedures in subpart E of part 807 of this chapter subject to § 862.9.

[52 FR 16122, May 1, 1987, as amended at 65 FR 2308, Jan. 14, 2000]

§ 862.1600 Potassium test system.

(a) *Identification.* A potassium test system is a device intended to measure potassium in serum, plasma, and urine. Measurements obtained by this device are used to monitor electrolyte balance in the diagnosis and treatment of diseases conditions characterized by low or high blood potassium levels.

(b) *Classification.* Class II.

§ 862.1605 Pregnanediol test system.

(a) *Identification.* A pregnanediol test system is a device intended to measure pregnanediol (a major urinary metabolic product of progesterone) in urine. Measurements obtained by this device are used in the diagnosis and treatment of disorders of the ovaries or placenta.

(b) *Classification.* Class I (general controls). The device is exempt from the premarket notification procedures in subpart E of part 807 of this chapter subject to § 862.9.

[52 FR 16122, May 1, 1987, as amended at 65 FR 2308, Jan. 14, 2000]

§ 862.1610 Pregnanetriol test system.

(a) *Identification.* A pregnanetriol test system is a device intended to measure pregnanetriol (a precursor in the biosynthesis of the adrenal hormone cortisol) in urine. Measurements obtained by this device are used in the diagnosis and treatment of congenital adrenal hyperplasia (congenital enlargement of the adrenal gland).

(b) *Classification.* Class I (general controls). The device is exempt from the premarket notification procedures in subpart E of part 807 of this chapter subject to § 862.9.

[52 FR 16122, May 1, 1987, as amended at 65 FR 2308, Jan. 14, 2000]

§ 862.1615 Pregnenolone test system.

(a) *Identification.* A pregnenolone test system is a device intended to measure pregnenolone (a precursor in the biosynthesis of the adrenal hormone cortisol and adrenal androgen) in serum and plasma. Measurements obtained by this device are used in the diagnosis and treatment of diseases of the adrenal cortex or the gonads.

(b) *Classification.* Class I (general controls). The device is exempt from the premarket notification procedures in subpart E of part 807 of this chapter subject to § 862.9.

[52 FR 16122, May 1, 1987, as amended at 65 FR 2308, Jan. 14, 2000]

§ 862.1620 Progesterone test system.

(a) *Identification.* A progesterone test system is a device intended to measure progesterone (a female hormone) in serum and plasma. Measurements obtained by this device are used in the diagnosis and treatment of disorders of the ovaries or placenta.

(b) *Classification.* Class I (general controls). The device is exempt from the premarket notification procedures in subpart E of part 807 of this chapter subject to § 862.9.

[52 FR 16122, May 1, 1987, as amended at 65 FR 2308, Jan. 14, 2000]

§ 862.1625 Prolactin (lactogen) test system.

(a) *Identification.* A prolactin (lactogen) test system is a device intended to measure the anterior pituitary polypeptide hormone prolactin in serum and plasma. Measurements obtained by this device are used in the diagnosis and treatment of disorders of the anterior pituitary gland or of the hypothalamus portion of the brain.

(b) *Classification.* Class I (general controls). The device is exempt from the premarket notification procedures in subpart E of part 807 of this chapter subject to § 862.9.

[52 FR 16122, May 1, 1987, as amended at 65 FR 2308, Jan. 14, 2000]

§ 862.1630 Protein (fractionation) test system.

(a) *Identification.* A protein (fractionation) test system is a device intended to measure protein fractions in blood, urine, cerebrospinal fluid, and other body fluids. Protein fractionations are used as an aid in recognizing abnormal proteins in body fluids and genetic variants of proteins produced in diseases with tissue destruction.

(b) *Classification.* Class I (general controls). The device is exempt from the premarket notification procedures in subpart E of part 807 of this chapter subject to § 862.9.

[52 FR 16122, May 1, 1987, as amended at 65 FR 2308, Jan. 14, 2000]

§ 862.1635 Total protein test system.

(a) *Identification.* A total protein test system is a device intended to measure total protein(s) in serum or plasma. Measurements obtained by this device are used in the diagnosis and treatment of a variety of diseases involving the liver, kidney, or bone marrow as well as other metabolic or nutritional disorders.

(b) *Classification.* Class II (special controls). The device is exempt from the premarket notification procedures in subpart E of part 807 of this chapter subject to § 862.9.

[52 FR 16122, May 1, 1987, as amended at 63 FR 59225, Nov. 3, 1998]

§ 862.1640 Protein-bound iodine test system.

(a) *Identification.* A protein-bound iodine test system is a device intended to measure protein-bound iodine in serum. Measurements of protein-bound iodine obtained by this device are used in the diagnosis and treatment of thyroid disorders.

(b) *Classification.* Class I (general controls). The device is exempt from the premarket notification procedures in subpart E of part 807 of this chapter subject to the limitations in § 862.9.

[52 FR 16122, May 1, 1987, as amended at 53 FR 21449, June 8, 1988; 66 FR 38788, July 25, 2001]

§ 862.1645 Urinary protein or albumin (nonquantitative) test system.

(a) *Identification.* A urinary protein or albumin (nonquantitative) test system is a device intended to identify proteins or albumin in urine. Identification of urinary protein or albumin (nonquantitative) is used in the diagnosis and treatment of disease conditions such as renal or heart diseases or thyroid disorders, which are characterized by proteinuria or albuminuria.

(b) *Classification.* Class I (general controls). The device is exempt from the premarket notification procedures in subpart E of part 807 of this chapter subject to § 862.9.

[52 FR 16122, May 1, 1987, as amended at 65 FR 2308, Jan. 14, 2000]

§ 862.1650 Pyruvate kinase test system.

(a) *Identification.* A pyruvate kinase test system is a device intended to measure the activity of the enzyme pyruvate kinase in erythrocytes (red blood cells). Measurements obtained by this device are used in the diagnosis and treatment of various inherited anemias due to pyruvate kinase deficiency or of acute leukemias.

(b) *Classification.* Class I (general controls). The device is exempt from the premarket notification procedures in subpart E of part 807 of this chapter subject to § 862.9.

[52 FR 16122, May 1, 1987, as amended at 65 FR 2308, Jan. 14, 2000]

§ 862.1655 Pyruvic acid test system.

(a) *Identification.* A pyruvic acid test system is a device intended to measure pyruvic acid (an intermediate compound in the metabolism of carbohydrate) in plasma. Measurements obtained by this device are used in the evaluation of electrolyte metabolism and in the diagnosis and treatment of acid-base and electrolyte disturbances or anoxia (the reduction of oxygen in body tissues).

(b) *Classification.* Class I (general controls). The device is exempt from the premarket notification procedures in subpart E of part 807 of this chapter subject to § 862.9.

[52 FR 16122, May 1, 1987, as amended at 65 FR 2308, Jan. 14, 2000]

§ 862.1660 Quality control material (assayed and unassayed).

(a) *Identification.* A quality control material (assayed and unassayed) for clinical chemistry is a device intended for medical purposes for use in a test system to estimate test precision and to detect systematic analytical deviations that may arise from reagent or analytical instrument variation. A quality control material (assayed and unassayed) may be used for proficiency testing in interlaboratory surveys. This generic type of device includes controls (assayed and unassayed) for blood gases, electrolytes, enzymes, multianalytes (all kinds), single (specified) analytes, or urinalysis controls.

(b) *Classification.* Class I (general controls). Except when used in donor screening tests, unassayed material is exempt from the premarket notification procedures in subpart E of part 807 of this chapter subject to § 862.9.

[52 FR 16122, May 1, 1987, as amended at 65 FR 2308, Jan. 14, 2000]

§ 862.1665 Sodium test system.

(a) *Identification.* A sodium test system is a device intended to measure sodium in serum, plasma, and urine. Measurements obtained by this device are used in the diagnosis and treatment of aldosteronism (excessive secretion of the hormone aldosterone), diabetes insipidus (chronic excretion of large amounts of dilute urine, accompanied by extreme thirst), adrenal hypertension, Addison's disease (caused by destruction of the adrenal glands), dehydration, inappropriate antidiuretic hormone secretion, or other diseases involving electrolyte imbalance.

(b) *Classification.* Class II.

§ 862.1670 Sorbitol dehydrogenase test system.

(a) *Identification.* A sorbitol dehydrogenase test system is a device intended to measure the activity of the enzyme sorbitol dehydrogenase in serum. Measurements obtained by this device are used in the diagnosis and treatment of liver disorders such as cirrhosis or acute hepatitis.

(b) *Classification.* Class I (general controls). The device is exempt from the premarket notification procedures in subpart E of part 807 of this chapter subject to the limitations in § 862.9.

[52 FR 16122, May 1, 1987, as amended at 53 FR 21449, June 8, 1988; 66 FR 38788, July 25, 2001]

§ 862.1675 Blood specimen collection device.

(a) *Identification.* A blood specimen collection device is a device intended for medical purposes to collect and to handle blood specimens and to separate serum from nonserum (cellular) components prior to further testing. This generic type device may include blood collection tubes, vials, systems, serum separators, blood collection trays, or vacuum sample tubes.

(b) *Classification.* Class II.

§ 862.1678 Tacrolimus test system.

(a) *Identification.* A tacrolimus test system is a device intended to quantitatively determine tacrolimus concentrations as an aid in the management of transplant patients receiving therapy with this drug. This generic type of device includes immunoassays and chromatographic assays for tacrolimus.

(b) *Classification.* Class II (special controls). The special control is "Class II Special Controls Guidance Document: Cyclosporine and Tacrolimus Assays; Guidance for Industry and FDA." See § 862.1(d) for the availability of this guidance document.

[67 FR 58329, Sept. 16, 2002]

§ 862.1680 Testosterone test system.

(a) *Identification.* A testosterone test system is a device intended to measure testosterone (a male sex hormone) in serum, plasma, and urine. Measurement of testosterone are used in the diagnosis and treatment of disorders involving the male sex hormones (androgens), including primary and secondary hypogonadism, delayed or precocious puberty, impotence in males and, in females hirsutism (excessive hair) and virilization (masculinization) due to tumors, polycystic ovaries, and adrenogenital syndromes.

(b) *Classification.* Class I.

[52 FR 16122, May 1, 1987; 53 FR 11645, Apr. 8, 1988]

§ 862.1685 Thyroxine-binding globulin test system.

(a) *Identification.* A thyroxine-binding globulin test system is a device intended to measure thyroxine (thyroid)-binding globulin (TBG), a plasma protein which binds thyroxine, in serum and plasma. Measurements obtained by this device are used in the diagnosis and treatment of thyroid diseases.

(b) *Classification.* Class II.

§ 862.1690 Thyroid stimulating hormone test system.

(a) *Identification.* A thyroid stimulating hormone test system is a device intended to measure thyroid stimulating hormone, also known as thyrotrophin and thyrotrophic hormone, in serum and plasma. Measurements of thyroid stimulating hormone produced by the anterior pituitary are used in the diagnosis of thyroid or pituitary disorders.

(b) *Classification.* Class II.

§ 862.1695 Free thyroxine test system.

(a) *Identification.* A free thyroxine test system is a device intended to measure free (not protein bound) thyroxine (thyroid hormone) in serum or plasma. Levels of free thyroxine in plasma are thought to reflect the amount of thyroxine hormone available to the cells and may therefore determine the clinical metabolic status of thyroxine. Measurements obtained by this device are used in the diagnosis and treatment of thyroid diseases.

(b) *Classification.* Class II.

§ 862.1700 Total thyroxine test system.

(a) *Identification.* A total thyroxine test system is a device intended to measure total (free and protein bound) thyroxine (thyroid hormone) in serum and plasma. Measurements obtained by this device are used in the diagnosis and treatment of thyroid diseases.

(b) *Classification.* Class II.

§ 862.1705 Triglyceride test system.

(a) *Identification.* A triglyceride test system is a device intended to measure triglyceride (neutral fat) in serum and plasma. Measurements obtained by this device are used in the diagnosis and treatment of patients with diabetes mellitus, nephrosis, liver obstruction, other diseases involving lipid metabolism, or various endocrine disorders.

(b) *Classification.* Class I (general controls). The device is exempt from the premarket notification procedures in subpart E of part 807 of this chapter subject to § 862.9.

[52 FR 16122, May 1, 1987, as amended at 65 FR 2308, Jan. 14, 2000]

§ 862.1710 Total triiodothyronine test system.

(a) *Identification.* A total triiodothyronine test system is a device intended to measure the hormone triiodothyronine in serum and plasma. Measurements obtained by this device are used in the diagnosis and treatment of thyroid diseases such as hyperthyroidism.

(b) *Classification.* Class II. This device is exempt from the premarket notification procedures in subpart E of part 807 of this chapter subject to the limitations in § 862.9.

[52 FR 16122, May 1, 1987, as amended at 65 FR 62286, Oct. 18, 2000]

§ 862.1715 Triiodothyronine uptake test system.

(a) *Identification.* A triiodothyronine uptake test system is a device intended to measure the total amount of binding sites available for binding thyroid hormone on the thyroxine-binding proteins, thyroid-binding globulin, thyroxine-binding prealbumin, and albumin of serum and plasma. The device provides an indirect measurement of thyrkoxine levels in serum and plasma. Measurements of triiodothyronine uptake are used in the diagnosis and treatment of thyroid disorders.

(b) *Classification.* Class II. The device is exempt from the premarket notification procedures in subpart E of part 807 of this chapter subject to the limitations in § 862.9.

[52 FR 16122, May 1, 1987, as amended at 64 FR 1124, Jan. 8, 1999]

§ 862.1720 Triose phosphate isomerase test system.

(a) *Identification.* A triose phosphate isomerase test system is a device intended to measure the activity of the enzyme triose phosphate isomerase in erythrocytes (red blood cells). Triose phosphate isomerase is an enzyme important in glycolysis (the energy-yielding conversion of glucose to lactic acid in various tissues). Measurements obtained by this device are used in the diagnosis and treatment of congenital triose phosphate isomerase enzyme deficiency, which causes a type of hemolytic anemia.

(b) *Classification.* Class I (general controls). The device is exempt from the premarket notification procedures in subpart E of part 807 subject to the limitations in § 862.9.

[52 FR 16122, May 1, 1987, as amended at 53 FR 21449, June 8, 1988; 66 FR 38788, July 25, 2001]

§ 862.1725 Trypsin test system.

(a) *Identification.* A trypsin test system is a device intended to measure the activity of trypsin (a pancreatic enzyme important in digestion for the breakdown of proteins) in blood and other body fluids and in feces. Measurements obtained by this device are used in the diagnosis and treatment of pancreatic disease.

(b) *Classification.* Class I (general controls). The device is exempt from the premarket notification procedures in subpart E of part 807 of this chapter subject to § 862.9.

[52 FR 16122, May 1, 1987, as amended at 65 FR 2308, Jan. 14, 2000]

§ 862.1730 Free tyrosine test system.

(a) *Identification.* A free tyrosine test system is a device intended to measure free tyrosine (an amono acid) in serum and urine. Measurements obtained by this device are used in the diagnosis and treatment of diseases such as congenital tyrosinemia (a disease that can cause liver/kidney disorders) and as an adjunct to the measurement of phenylalanine in detecting congenital phenylketonuria (a disease that can cause brain damage).

(b) *Classification.* Class I.

§ 862.1770 Urea nitrogen test system.

(a) *Identification.* A urea nitrogen test system is a device intended to measure urea nitrogen (an end-product of nitrogen metabolism) in whole blood, serum, plasma, and urine. Measurements obtained by this device are used in the diagnosis and treatment of certain renal and metabolic diseases.

(b) *Classification.* Class II.

§ 862.1775 Uric acid test system.

(a) *Identification.* A uric acid test system is a device intended to measure uric acid in serum, plasma, and urine. Measurements obtained by this device are used in the diagnosis and treatment of numerous renal and metabolic disorders, including renal failure, gout, leukemia, psoriasis, starvation or other wasting conditions, and of patients receiving cytotoxic drugs.

(b) *Classification.* Class I.

§ 862.1780 Urinary calculi (stones) test system.

(a) *Identification.* A urinary calculi (stones) test system is a device intended for the analysis of urinary calculi. Analysis of urinary calculi is used in the diagnosis and treatment of calculi of the urinary tract.

(b) *Classification.* Class I (general controls). The device is exempt from the premarket notification procedures in subpart E of part 807 of this chapter subject to § 862.9.

[52 FR 16122, May 1, 1987, as amended at 65 FR 2308, Jan. 14, 2000]

§ 862.1785 Urinary urobilinogen (nonquantitative) test system.

(a) *Identification.* A urinary urobilinogen (nonquantitative) test system is a device intended to detect and estimate urobilinogen (a bile pigment degradation product of red cell hemoglobin) in urine. Estimations obtained by this device are used in the diagnosis and treatment of liver diseases and hemolytic (red cells) disorders.

(b) *Classification.* Class I (general controls). The device is exempt from the premarket notification procedures in subpart E of part 807 of this chapter subject to § 862.9.

[52 FR 16122, May 1, 1987, as amended at 65 FR 2308, Jan. 14, 2000]

§ 862.1790 Uroporphyrin test system.

(a) *Identification.* A uroporphyrin test system is a device intended to measure uroporphyrin in urine. Measurements obtained by this device are used in the diagnosis and treatment of porphyrias (primarily inherited diseases associated with disturbed porphyrin metabolism), lead poisoning, and other diseases characterized by alterations in the heme pathway.

(b) *Classification.* Class I (general controls). The device is exempt from the premarket notification procedures in subpart E of part 807 of this chapter subject to § 862.9.

[52 FR 16122, May 1, 1987, as amended at 65 FR 2308, Jan. 14, 2000]

§ 862.1795 Vanilmandelic acid test system.

(a) *Identification.* A vanilmandelic acid test system is a device intended to measure vanilmandelic acid in urine. Measurements of vanilmandelic acid obtained by this device are used in the diagnosis and treatment of neuroblastoma, pheochromocytoma, and certain hypertensive conditions.

(b) *Classification.* Class I (general controls). The device is exempt from the premarket notification procedures in subpart E of part 807 of this chapter subject to § 862.9.

[52 FR 16122, May 1, 1987, as amended at 65 FR 2308, Jan. 14, 2000]

§ 862.1805 Vitamin A test system.

(a) *Identification.* A vitamin A test system is a device intended to measure vitamin A in serum or plasma. Measurements obtained by this device are used in the diagnosis and treatment of vitamin A deficiency conditions, including night blindness, or skin, eye, or intestinal disorders.

(b) *Classification.* Class I (general controls). The device is exempt from the premarket notification procedures in subpart E of part 807 of this chapter subject to § 862.9.

[52 FR 16122, May 1, 1987, as amended at 65 FR 2308, Jan. 14, 2000]

§ 862.1810 Vitamin B_{12} test system.

(a) *Identification.* A vitamin B_{12} test system is a device intended to measure vitamin B_{12} in serum, plasma, and urine. Measurements obtained by this device are used in the diagnosis and treatment of anemias of gastrointestinal malabsorption.

(b) *Classification.* Class II.

§ 862.1815 Vitamin E test system.

(a) *Identification.* A vitamin E test system is a device intended to measure vitamin E (tocopherol) in serum. Measurements obtained by this device are used in the diagnosis and treatment of infants with vitamin E deficiency syndrome.

(b) *Classification.* Class I (general controls). The device is exempt from the premarket notification procedures in subpart E of part 807 subject to the limitations in § 862.9.

[52 FR 16122, May 1, 1987, as amended at 53 FR 21449, June 8, 1988; 66 FR 38788, July 25, 2001]

§ 862.1820 Xylose test system.

(a) *Identification.* A xylose test system is a device intended to measure xylose (a sugar) in serum, plasma, and urine. Measurements obtained by this device are used in the diagnosis and treatment of gastrointestinal malabsorption syndrome (a group of disorders in which there is subnormal absorption of dietary constituents and thus excessive loss from the body of the nonabsorbed substances).

(b) *Classification.* Class I (general controls). The device is exempt from the premarket notification procedures in subpart E of part 807 of this chapter subject to § 862.9.

[52 FR 16122, May 1, 1987, as amended at 65 FR 2308, Jan. 14, 2000]

§ 862.1825 Vitamin D test system.

(a) *Identification.* A vitamin D test system is a device intended for use in clinical laboratories for the quantitative determination of 25-hydroxyvitamin D (25-OH-D) and other hydroxylated metabolites of vitamin D in serum or plasma to be used in the assessment of vitamin D sufficiency.

(b) *Classification.* Class II (special controls). Vitamin D test systems must comply with the following special controls:

(1) Labeling in conformance with 21 CFR 809.10 and

(2) Compliance with existing standards of the National Committee on Clinical Laboratory Standards.

[63 FR 40366, July 29, 1998]

Subpart C—Clinical Laboratory Instruments

§ 862.2050 General purpose laboratory equipment labeled or promoted for a specific medical use.

(a) *Identification.* General purpose laboratory equipment labeled or promoted for a specific medical use is a device that is intended to prepare or examine specimens from the human body and that is labeled or promoted for a specific medical use.

(b) *Classification.* Class I (general controls). The device is identified in paragraph (a) of this section and is exempt from the premarket notification procedures in subpart E of part 807 of this chapter subject to the limitations in § 862.9. The device is also exempt from the current good manufacturing practice regulations in part 820 of this chapter, with the exception of § 820.180, with respect to general requirements

concerning records, and § 820.198, with respect to complaint files.

[52 FR 16122, May 1, 1987, as amended at 66 FR 38788, July 25, 2001]

§ 862.2100 Calculator/data processing module for clinical use.

(a) *Identification.* A calculator/data processing module for clinical use is an electronic device intended to store, retrieve, and process laboratory data.

(b) *Classification.* Class I (general controls). The device is exempt from the premarket notification procedures in subpart E of part 807 of this chapter subject to the limitations in § 862.9.

[52 FR 16122, May 1, 1987, as amended at 53 FR 21449, June 8, 1988; 66 FR 38788, July 25, 2001]

§ 862.2140 Centrifugal chemistry analyzer for clinical use.

(a) *Identification.* A centrifugal chemistry analyzer for clinical use is an automatic device intended to centrifugally mix a sample and a reagent and spectrophotometrically measure concentrations of the sample constituents. This device is intended for use in conjunction with certain materials to measure a variety of analytes.

(b) *Classification.* Class I (general controls). The device is exempt from the premarket notification procedures in subpart E of part 807 of this chapter subject to § 862.9.

[52 FR 16122, May 1, 1987, as amended at 65 FR 2308, Jan. 14, 2000]

§ 862.2150 Continuous flow sequential multiple chemistry analyzer for clinical use.

(a) *Identification.* A continuous flow sequential multiple chemistry analyzer for clinical use is a modular analytical instrument intended to simultaneously perform multiple chemical procedures using the principles of automated continuous flow systems. This device is intended for use in conjunction with certain materials to measure a variety of analytes.

(b) *Classification.* Class I (general controls). The device is exempt from the premarket notification procedures in subpart E of part 807 of this chapter subject to § 862.9.

[52 FR 16122, May 1, 1987, as amended at 65 FR 2308, Jan. 14, 2000]

§ 862.2160 Discrete photometric chemistry analyzer for clinical use.

(a) *Identification.* A discrete photometric chemistry analyzer for clinical use is a device intended to duplicate manual analytical procedures by performing automatically various steps such as pipetting, preparing filtrates, heating, and measuring color intensity. This device is intended for use in conjunction with certain materials to measure a variety of analytes. Different models of the device incorporate various instrumentation such as micro analysis apparatus, double beam, single, or dual channel photometers, and bichromatic 2-wavelength photometers. Some models of the device may include reagent-containing components that may also serve as reaction units.

(b) *Classification.* Class I (general controls). The device is exempt from the premarket notification procedures in subpart E of part 807 of this chapter subject to § 862.9.

[52 FR 16122, May 1, 1987, as amended at 65 FR 2309, Jan. 14, 2000]

§ 862.2170 Micro chemistry analyzer for clinical use.

(a) *Identification.* A micro chemistry analyzer for clinical use is a device intended to duplicate manual analytical procedures by performing automatically various steps such as pipetting, preparing filtrates, heating, and measuring color intensity. The distinguishing characteristic of the device is that it requires only micro volume samples obtainable from pediatric patients. This device is intended for use in conjunction with certain materials to measure a variety of analytes.

(b) *Classification.* Class I (general controls). The device is exempt from the premarket notification procedures in subpart E of part 807 of this chapter subject to § 862.9.

[52 FR 16122, May 1, 1987, as amended at 65 FR 2309, Jan. 14, 2000]

§ 862.2230 Chromatographic separation material for clinical use.

(a) *Identification.* A chromatographic separation material for clinical use is a device accessory (e.g., ion exchange absorbents, ion exchagne resins, and ion papers) intended for use in ion exchange chromatography, a procedure in which a compound is separated from a solution.

(b) *Classification.* Class I (general controls). The device is exempt from the premarket notification procedures in subpart E of part 807 of this chapter subject to the limitations in § 862.9.

[52 FR 16122, May 1, 1987, as amended at 61 FR 1119, Jan. 16, 1996; 66 FR 38788, July 25, 2001]

§ 862.2250 Gas liquid chromatography system for clinical use.

(a) *Identification.* A gas liquid chromatography system for clinical use is a device intended to separate one or more drugs or compounds from a mixture. Each of the constituents in a vaporized mixture of compounds is separated according to its vapor pressure. The device may include accessories such as columns, gases, column supports, and liquid coating.

(b) *Classification.* Class I (general controls). The device is exempt from the premarket notification procedures in subpart E of part 807 of this chapter subject to § 862.9.

[52 FR 16122, May 1, 1987, as amended at 65 FR 2309, Jan. 14, 2000]

§ 862.2260 High pressure liquid chromatography system for clinical use.

(a) *Identification.* A high pressure liquid chromatography system for clinical use is a device intended to separate one or more drugs or compounds from a solution by processing the mixture of compounds (solutes) through a column packed with materials of uniform size (stationary phase) under the influence of a high pressure liquid (mobile phase). Separation of the solutes occurs either by absorption, sieving, partition, or selective affinity.

(b) *Classification.* Class I (general controls). The device is exempt from the premarket notification procedures in subpart E of part 807 of this chapter subject to § 862.9.

[52 FR 16122, May 1, 1987, as amended at 65 FR 2309, Jan. 14, 2000]

§ 862.2270 Thin-layer chromatography system for clinical use.

(a) *Identification.* A thin-layer chromatography (TLC) system for clinical use is a device intended to separate one or more drugs or compounds from a mixture. The mixture of compounds is absorbed onto a stationary phase or thin layer of inert material (e.g., cellulose, alumina, etc.) and eluted off by a moving solvent (moving phase) until equilibrium occurs between the two phases.

(b) *Classification.* Class I (general controls). The device is exempt from the premarket notification procedures in subpart E of part 807 of this chapter subject to § 862.9. Particular components of TLC systems, i.e., the thin-layer chromatography apparatus, TLC atomizer, TLC developing tanks, and TLC ultraviolet light, are exempt from the current good manufacturing practice regulations in part 820 of this chapter, with the exception of § 820.180 of this chapter, with respect to general requirements concerning records, and § 820.198 of this chapter, with respect to complaint files.

[52 FR 16122, May 1, 1987, as amended at 65 FR 2309, Jan. 14, 2000]

§ 862.2300 Colorimeter, photometer, or spectrophotometer for clinical use.

(a) *Identification.* A colorimeter, a photometer, or a spectrophotometer for clinical use is an instrument intended to measure radiant energy emitted, transmitted, absorbed, or reflected under controlled conditions. The device may include a monochromator to produce light of a specific wavelength.

(b) *Classification.* Class I (general controls). The device is exempt from the premarket notification procedures in subpart E of part 807 of this chapter subject to § 862.9.

[52 FR 16122, May 1, 1987, as amended at 65 FR 2309, Jan. 14, 2000]

§ 862.2310 Clinical sample concentrator.

(a) *Identification.* A clinical sample concentrator is a device intended to concentrate (by dialysis, evaporation, etc.) serum, urine, cerebrospinal fluid, and other body fluids before the fluids are analyzed.

(b) *Classification.* Class I (general controls). The device is exempt from the premarket notification procedures in subpart E of part 807 of this chapter subject to the limitations in § 862.9.

[52 FR 16122, May 1, 1987, as amended at 60 FR 38899, July 28, 1995; 66 FR 38788, July 25, 2001]

§ 862.2320 Beta or gamma counter for clinical use.

(a) *Identification.* A beta or gamma counter for clinical use is a device intended to detect and count beta or gamma radiation emitted by clinical samples. Clinical samples are prepared by addition of a radioactive reagent to the sample. These measurements are useful in the diagnosis and treatment of various disorders.

(b) *Classification.* Class I (general controls). The device is exempt from the premarket notification procedures in subpart E of part 807 of this chapter subject to the limitations in § 862.9.

[52 FR 16122, May 1, 1987, as amended at 60 FR 38900, July 28, 1995; 66 FR 38788, July 25, 2001]

§ 862.2400 Densitometer/scanner (integrating, reflectance, TLC, or radiochromatogram) for clinical use.

(a) *Identification.* A densitometer/scanner (integrating, reflectance, thin-layer chromatography, or radiochromatogram) for clinical use is device intended to measure the concentration of a substance on the surface of a film or other support media by either a photocell measurement of the light transmission through a given area of the medium or, in the case of the radiochromatogram scanner, by measurement of the distribution of a specific radio-active element on a radiochromatogram.

(b) *Classification.* Class I (general controls). The device is exempt from the premarket notification procedures in subpart E of part 807 of this chapter subject to § 862.9.

[52 FR 16122, May 1, 1987, as amended at 65 FR 2309, Jan. 14, 2000]

§ 862.2485 Electrophoresis apparatus for clinical use.

(a) *Identification.* An electrophoresis apparatus for clinical use is a device intended to separate molecules or particles, including plasma proteins, lipoproteins, enzymes, and hemoglobulins on the basis of their net charge in specified buffered media. This device is used in conjunction with certain materials to measure a variety of analytes as an aid in the diagnosis and treatment of certain disorders.

(b) *Classification.* Class I (general controls). The device is exempt from the premarket notification procedures in subpart E of part 807 of this chapter subject to the limitations in § 862.9.

[52 FR 16122, May 1, 1987, as amended at 60 FR 38900, July 28, 1995; 66 FR 38788, July 25, 2001]

§ 862.2500 Enzyme analyzer for clinical use.

(a) *Identification.* An enzyme analyzer for clinical use is a device intended to measure enzymes in plasma or serum by nonkinetic or kinetic measurement of enzyme-catalyzed reactions. This device is used in conjunction with certain materials to measure a variety of enzymes as an aid in the diagnosis and treatment of certain enzyme-related disorders.

(b) *Classification.* Class I (general controls). The device is exempt from the premarket notification procedures in subpart E of part 807 of this chapter subject to § 862.9.

[52 FR 16122, May 1, 1987, as amended at 65 FR 2309, Jan. 14, 2000]

§ 862.2540 Flame emission photometer for clinical use.

(a) *Identification.* A flame emission photometer for clinical use is a device intended to measure the concentration of sodium, potassium, lithium, and other metal ions in body fluids. Abnormal variations in the concentration of these substances in the body are indicative of certain disorders (e.g., electrolyte imbalance and heavy metal intoxication) and are, therefore, useful in diagnosis and treatment of those disorders.

(b) *Classification.* Class I (general controls). The device is exempt from the premarket notification procedures in subpart E of part 807 of this chapter subject to § 862.9.

[52 FR 16122, May 1, 1987, as amended at 65 FR 2309, Jan. 14, 2000]

§ 862.2560 Fluorometer for clinical use.

(a) *Identification.* A fluorometer for clinical use is a device intended to measure by fluorescence certain analytes. Fluorescence is the property of certain substances of radiating, when illuminated, a light of a different wavelength. This device is used in conjunction with certain materials to measure a variety of analytes.

(b) *Classification.* Class I (general controls). The device is exempt from the premarket notification procedures in subpart E of part 807 of this chapter subject to § 862.9.

[52 FR 16122, May 1, 1987, as amended at 65 FR 2309, Jan. 14, 2000]

§ 862.2570 Instrumentation for clinical multiplex test systems.

(a) *Identification.* Instrumentation for clinical multiplex test systems is a device intended to measure and sort multiple signals generated by an assay from a clinical sample. This instrumentation is used with a specific assay to measure multiple similar analytes that establish a single indicator to aid in diagnosis. Such instrumentation may be compatible with more than one specific assay. The device includes a signal reader unit, and may also integrate reagent handling, hybridization, washing, dedicated instrument control, and other hardware components, as well as raw data storage mechanisms, data acquisition software, and software to process detected signals.

(b) *Classification.* Class II (special controls). The special control is FDA's guidance document entitled "Class II Special Controls Guidance Document: Instrumentation for Clinical Multiplex Test Systems." See § 862.1(d) for the availability of this guidance document.

[70 FR 11868, Mar. 10, 2005]

§ 862.2680 Microtitrator for clinical use.

(a) *Identification.* A microtitrator for clinical use is a device intended for use in micronanalysis to measure the concentration of a substance by reacting it with a measure "micro" volume of a known standardized solution.

(b) *Classification.* Class I (general controls). The device is exempt from the premarket notification procedures in subpart E of part 807 of this chapter subject to § 862.9.

[52 FR 16122, May 1, 1987, as amended at 65 FR 2309, Jan. 14, 2000]

§ 862.2700 Nephelometer for clinical use.

(a) *Identification.* A nephelometer for clinical use is a device intended to estimate the concentration of particles in a suspension by measuring their light scattering properties (the deflection of light rays by opaque particles in their path). The device is used in conjunction with certain materials to measure the concentration of a variety of analytes.

(b) *Classification.* Class I (general controls). The device is exempt from the premarket notification procedures in subpart E of part 807 of this chapter subject to § 862.9.

[52 FR 16122, May 1, 1987, as amended at 65 FR 2309, Jan. 14, 2000]

§ 862.2720 Plasma oncometer for clinical use.

(a) *Identification.* A plasma oncometer for clinical use is a device intended to measure plasma oncotic pressure, which is that portion of the total plasma osmotic pressure contributed by protein and other molecules too large to pass through a specified semipermeable membrane. Because variations in plasma oncotic pressure are indications of certain disorders, measurements of the variations are useful in the diagnosis and treatment of these disorders.

(b) *Classification.* Class I (general controls). The device is exempt from the premarket notification procedures in

subpart E of part 807 of this chapter subject to the limitations in § 862.9.

[52 FR 16122, May 1, 1987, as amended at 60 FR 38900, July 28, 1995; 66 FR 38788, July 25, 2001]

§ 862.2730 Osmometer for clinical use.

(a) *Identification.* An osmometer for clinical use is a device intended to measure the osmotic pressure of body fluids. Osmotic pressure is the pressure required to prevent the passage of a solution with a lesser solute concentration into a solution with greater solute concentration when the two solutions are separated by a semipermeable membrane. The concentration of a solution affects its osmotic pressure, freezing point, and other physiochemical properties. Osmometers determine osmotic pressure by methods such as the measurement of the freezing point. Measurements obtained by this device are used in the diagnosis and treatment of body fluid disorders.

(b) *Classification.* Class I (general controls). The device is exempt from the premarket notification procedures in subpart E of part 807 of this chapter subject to § 862.9.

[52 FR 16122, May 1, 1987, as amended at 65 FR 2309, Jan. 14, 2000]

§ 862.2750 Pipetting and diluting system for clinical use.

(a) *Identification.* A pipetting and diluting system for clinical use is a device intended to provide an accurately measured volume of liquid at a specified temperature for use in certain test procedures. This generic type of device system includes serial, manual, automated, and semi-automated dilutors, pipettors, dispensers, and pipetting stations.

(b) *Classification.* Class I (general controls). The device is exempt from the premarket notification procedures in subpart E of part 807 of this chapter subject to § 862.9.

[52 FR 16122, May 1, 1987, as amended at 65 FR 2309, Jan. 14, 2000]

§ 862.2800 Refractometer for clinical use.

(a) *Identification.* A refractometer for clinical use is a device intended to determine the amount of solute in a solution by measuring the index of refraction (the ratio of the velocity of light in a vacuum to the velocity of light in the solution). The index of refraction is used to measure the concentration of certain analytes (solutes), such a plasma total proteins and urinary total solids. Measurements obtained by this device are used in the diagnosis and treatment of certain conditions.

(b) *Classification.* Class I (general controls). The device is exempt from the premarket notification procedures in subpart E of part 807 of this chapter subject to the limitations in § 862.9.

[52 FR 16122, May 1, 1987, as amended at 60 FR 38900, July 28, 1995; 66 FR 38788, July 25, 2001]

§ 862.2850 Atomic absorption spectrophotometer for clinical use.

(a) *Identification.* An atomic absorption spectrophotometer for clinical use is a device intended to identify and measure elements and metals (e.g., lead and mercury) in human specimens. The metal elements are identified according to the wavelength and intensity of the light that is absorbed when the specimen is converted to the atomic vapor phase. Measurements obtained by this device are used in the diagnosis and treatment of certain conditions.

(b) *Classification.* Class I (general controls). The device is exempt from the premarket notification procedures in subpart E of part 807 of this chapter subject to § 862.9.

[52 FR 16122, May 1, 1987, as amended at 65 FR 2309, Jan. 14, 2000]

§ 862.2860 Mass spectrometer for clinical use.

(a) *Identification.* A mass spectrometer for clinical use is a device intended to identify inorganic or organic compounds (e.g., lead, mercury, and drugs) in human specimens by ionizing the compound under investigation and separating the resulting ions by means of an electrical and magnetic field according to their mass.

(b) *Classification.* Class I (general controls). The device is exempt from the premarket notification procedures in

subpart E of part 807 of this chapter subject to § 862.9.

[52 FR 16122, May 1, 1987, as amended at 65 FR 2309, Jan. 14, 2000]

§ 862.2900 Automated urinalysis system.

(a) *Identification.* An automated urinalysis system is a device intended to measure certain of the physical properties and chemical constituents of urine by procedures that duplicate manual urinalysis systems. This device is used in conjunction with certain materials to measure a variety of urinary analytes.

(b) *Classification.* Class I (general controls). The device is exempt from the premarket notification procedures in subpart E of part 807 of this chapter subject to § 862.9.

[52 FR 16122, May 1, 1987, as amended at 65 FR 2309, Jan. 14, 2000]

§ 862.2920 Plasma viscometer for clinical use.

(a) *Identification.* A plasma viscometer for clinical use is a device intended to measure the viscosity of plasma by determining the time period required for the plasma to flow a measured distance through a calibrated glass tube. Measurements obtained by this device are used to monitor changes in the amount of solids present in plasma in various disorders.

(b) *Classification.* Class I (general controls). The device is exempt from the premarket notification procedures in subpart E of part 807 of this chapter subject to the limitations in § 862.9.

[52 FR 16122, May 1, 1987, as amended at 60 FR 38900, July 28, 1995; 66 FR 38788, July 25, 2001]

Subpart D—Clinical Toxicology Test Systems

§ 862.3030 Acetaminophen test system.

(a) *Identification.* An acetaminophen test system is a device intended to measure acetaminophen, an analgestic and fever reducing drug, in serum. Measurements obtained by this device are used in the diagnosis and treatment of acetaminophen overdose.

(b) *Classification.* Class II.

§ 862.3035 Amikacin test system.

(a) *Identification.* An amikacin test system is a device intended to measure amikacin, an aminoglycoside antibiotic drug, in serum and plasma. Measurements obtained by this device are used in the diagnosis and treatment of amikacin overdose and in monitoring levels of amikacin to ensure appropriate therapy.

(b) *Classification.* Class II.

§ 862.3040 Alcohol test system.

(a) *Identification.* An alcohol test system is a device intented to measure alcohol (e.g., ethanol, methanol, isopropanol, etc.) in human body fluids (e.g., serum, whole blood, and urine). Measurements obtained by this device are used in the diagnosis and treatment of alcohol intoxication and poisoning.

(b) *Classification.* Class II.

§ 862.3050 Breath-alcohol test system.

(a) *Identification.* A breath-alcohol test system is a device intened to measure alcohol in the human breath. Measurements obtained by this device are used in the diagnosis of alcohol intoxication.

(b) *Classification.* Class I.

§ 862.3080 Breath nitric oxide test system.

(a) *Identification.* A breath nitric oxide test system is a device intended to measure fractional nitric oxide in human breath. Measurement of changes in fractional nitric oxide concentration in expired breath aids in evaluating an asthma patient's response to anti-inflammatory therapy, as an adjunct to established clinical and laboratory assessments of asthma. A breath nitric oxide test system combines chemiluminescence detection of nitric oxide with a pneumotachograph, display, and dedicated software.

(b) *Classification.* Class II (special controls). The special control is FDA's guidance entitled "Class II Special Controls Guidance Document: Breath Nitric Oxide Test System." See § 862.1(d) for the availability of this guidance document.

[68 FR 40127, July 7, 2003]

§ 862.3100 Amphetamine test system.

(a) *Identification.* An amphetamine test system is a device intended to measure amphetamine, a central nervous system stimulating drug, in plasma and urine. Measurements obtained by this device are used in the diagnosis and treatment of amphetamine use or overdose and in monitoring levels of amphetamine to ensure appropriate therapy.

(b) *Classification.* Class II.

§ 862.3110 Antimony test system.

(a) *Identification.* An antimony test system is a device intended to measure antimony, a heavy metal, in urine, blood, vomitus, and stomach contents. Measurements obtained by this device are used in the diagnosis and treatment of antimony poisoning.

(b) *Classification.* Class I.

§ 862.3120 Arsenic test system.

(a) *Identification.* An arsenic test system is a device intended to measure arsenic, a poisonous heavy metal, in urine, vomitus, stomach contents, nails, hair, and blood. Measurements obtained by this device are used in the diagnosis and treatment of arsenic poisoning.

(b) *Classification.* Class I.

§ 862.3150 Barbiturate test system.

(a) *Identification.* A barbiturate test system is a device intended to measure barbiturates, a class of hypnotic and sedative drugs, in serum, urine, and gastric contents. Measurements obtained by this device are used in the diagnosis and treatment of barbiturate use or overdose and in monitoring levels of barbiturate to ensure appropriate therapy.

(b) *Classification.* Class II.

§ 862.3170 Benzodiazepine test system.

(a) *Identification.* A benzodiazepine test system is a device intended to measure any of the benzodiazepine compounds, sedative and hypnotic drugs, in blood, plasma, and urine. The benzodiazepine compounds include chlordiazepoxide, diazepam, oxazepam, chlorzepate, flurazepam, and nitrazepam. Measurements obtained by this device are used in the diagnosis and treatment of benzodiazepine use or overdose and in monitoring levels of benzodiazepines to ensure appropriate therapy.

(b) *Classification.* Class II.

§ 862.3200 Clinical toxicology calibrator.

(a) *Identification.* A clinical toxicology calibrator is a device intended for medical purposes for use in a test system to establish points of reference that are used in the determination of values in the measurement of substances in human specimens. A clinical toxicology calibrator can be a mixture of drugs or a specific material for a particular drug (e.g., ethanol, lidocaine, etc.). (See also § 862.2 in this part.)

(b) *Classification.* Class II.

§ 862.3220 Carbon monoxide test system.

(a) *Identification.* A carbon monoxide test system is a device intended to measure carbon monoxide or carboxyhemoglobin (carbon monoxide bound to the hemoglobin in the blood) in blood. Measurements obtained by this device are used in the diagnosis and treatment of or confirmation of carbon monoxide poisoning.

(b) *Classification.* Class I.

§ 862.3240 Cholinesterase test system.

(a) *Identification.* A cholinesterase test system is a device intended to measure cholinesterase (an enzyme that catalyzes the hydrolysis of acetylcholine to choline) in human specimens. There are two principal types of cholinesterase in human tissues. True cholinesterase is present at nerve endings and in erythrocytes (red blood cells) but is not present in plasma. Pseudo cholinesterase is present in plasma and liver but is not present in erythrocytes. Measurements obtained by this device are used in the diagnosis and treatment of cholinesterase inhibition disorders (e.g., insecticide poisoning and succinylcholine poisoning).

(b) *Classification.* Class I.

§ 862.3250 Cocaine and cocaine metabolite test system.

(a) *Identification.* A cocaine and cocaine metabolite test system is a device intended to measure cocaine and a cocaine metabolite (benzoylecgonine) in serum, plasma, and urine. Measurements obtained by this device are used in the diagnosis and treatment of cocaine use or overdose.

(b) *Classification.* Class II.

§ 862.3270 Codeine test system.

(a) *Identification.* A codeine test system is a device intended to measure codeine (a narcotic pain-relieving drug) in serum and urine. Measurements obtained by this device are used in the diagnosis and treatment of codeine use or overdose and in monitoring levels of codeine to ensure appropriate therapy.

(b) *Classification.* Class II.

§ 862.3280 Clinical toxicology control material.

(a) *Identification.* A clinical toxicology control material is a device intended to provide an estimation of the precision of a device test system and to detect and monitor systematic deviations from accuracy resulting from reagent or instrument defects. This generic type of device includes various single, and multi-analyte control materials.

(b) *Classification.* Class I (general controls). Except when used in donor screening, unassayed material is exempt from the premarket notification procedures in subpart E of part 807 of this chapter subject to § 862.9.

[52 FR 16122, May 1, 1987, as amended at 65 FR 2309, Jan. 14, 2000]

§ 862.3300 Digitoxin test system.

(a) *Identification.* A digitoxin test system is a device intended to measure digitoxin, a cardiovascular drug, in serum and plasma. Measurements obtained by this device are used in the diagnosis and treatment of digitoxin overdose and in monitoring levels of digitoxin to ensure appropriate therapy.

(b) *Classification.* Class II.

§ 862.3320 Digoxin test system.

(a) *Identification.* A digoxin test system is a device intended to measure digoxin, a cardiovascular drug, in serum and plasma. Measurements obtained by this device are used in the diagnosis and treatment of digoxin overdose and in monitoring levels of digoxin to ensure appropriate therapy.

(b) *Classification.* Class II.

§ 862.3350 Diphenylhydantoin test system.

(a) *Identification.* A diphenylhydantoin test system is a device intended to measure diphenylhydantoin, an antiepileptic drug, in human specimens. Measurements obtained by this device are used in the diagnosis and treatment of diphenylhydantoin overdose and in monitoring levels of diphenylhydantoin to ensure appropriate therapy.

(b) *Classification.* Class II.

§ 862.3360 Drug metabolizing enzyme genotyping system.

(a) *Identification.* A drug metabolizing enzyme genotyping system is a device intended for use in testing deoxyribonucleic acid (DNA) extracted from clinical samples to identify the presence or absence of human genotypic markers encoding a drug metabolizing enzyme. This device is used as an aid in determining treatment choice and individualizing treatment dose for therapeutics that are metabolized primarily by the specific enzyme about which the system provides genotypic information.

(b) *Classification.* Class II (special controls). The special control is FDA's guidance document entitled "Class II Special Controls Guidance Document: Drug Metabolizing Enzyme Genotyping Test System." See § 862.1(d) for the availability of this guidance document.

[70 FR 11867, Mar. 10, 2005]

§ 862.3380 Ethosuximide test system.

(a) *Identification.* An ethosuximide test system is a device intended to measure ethosuximide, an antiepileptic drug, in human specimens. Measurements obtained by this device are used in the diagnosis and treatment of ethosuximide overdose and in monitoring levels of ethosuximide to ensure appropriate therapy.

(b) *Classification*. Class II.

§ 862.3450 Gentamicin test system.

(a) *Identification*. A gentamicin test system is a device intended to measure gentamicin, an antibiotic drug, in human specimens. Measurements obtained by this device are used in the diagnosis and treatment of gentamicin overdose and in monitoring levels of gentamicin to ensure appropriate therapy.

(b) *Classification*. Class II.

§ 862.3520 Kanamycin test system.

(a) *Identification*. A kanamycin test system is a device intended to measure kanamycin, an antibiotic drug, in plasma and serum. Measurements obtained by this device are used in the diagnosis and treatment of kanamycin overdose and in monitoring levels of kanamycin to ensure appropriate therapy.

(b) *Classification*. Class II.

§ 862.3550 Lead test system.

(a) *Identification*. A lead test system is a device intended to measure lead, a heavy metal, in blood and urine. Measurements obtained by this device are used in the diagnosis and treatment of lead poisoning.

(b) *Classification*. Class II.

§ 862.3555 Lidocaine test system.

(a) *Identification*. A lidocaine test system is a device intended to measure lidocaine, an antiarrythmic and anticonvulsant drug, in serum and plasma. Measurements obtained by this device are used in the diagnosis and treatment of lidocaine overdose or in monitoring levels of lidocaine to ensure appropriate therapy.

(b) *Classification*. Class II.

§ 862.3560 Lithium test system.

(a) *Identification*. A lithium test system is a device intended to measure lithium (from the drug lithium carbonate) in serum or plasma. Measurements of lithium are used to assure that the proper drug dosage is administered in the treatment of patients with mental disturbances, such as manic-depressive illness (bipolar disorder).

(b) *Classification*. Class II.

§ 862.3580 Lysergic acid diethylamide (LSD) test system.

(a) *Identification*. A lysergic acid diethylamide (LSD) test system is a device intended to measure lysergic acid diethylamide, a hallucinogenic drug, in serum, urine, and gastric contents. Measurements obtained by this device are used in the diagnosis and treatment of LSD use or overdose.

(b) *Classification*. Class II.

§ 862.3600 Mercury test system.

(a) *Identification*. A mercury test system is a device intended to measure mercury, a heavy metal, in human specimens. Measurements obtained by this device are used in the diagnosis and treatment of mercury poisoning.

(b) *Classification*. Class I.

§ 862.3610 Methamphetamine test system.

(a) *Identification*. A methamphetamine test system is a device intended to measure methamphetamine, a central nervous system stimulating drug, in serum, plasma, and urine. Measurements obtained by this device are used in the diagnosis and treatment of methamphetamine use or overdose.

(b) *Classification*. Class II.

§ 862.3620 Methadone test system.

(a) *Identification*. A methadone test system is a device intended to measure methadone, an addictive narcotic pain-relieving drug, in serum and urine. Measurements obtained by this device are used in the diagnosis and treatment of methadone use or overdose and to determine compliance with regulations in methadone maintenance treatment.

(b) *Classification*. Class II.

§ 862.3630 Methaqualone test system.

(a) *Identification*. A methaqualone test system is a device intended to measure methaqualone, a hypnotic and sedative drug, in urine. Measurements obtained by this device are used in the diagnosis and treatment of methaqualone use or overdose.

(b) *Classification*. Class II.

§ 862.3640 Morphine test system.

(a) *Identification.* A morphine test system is a device intended to measure morphine, an addictive narcotic pain-relieving drug, and its analogs in serum, urine, and gastric contents. Measurements obtained by this device are used in the diagnosis and treatment of morphine use or overdose and in monitoring levels of morphine and its analogs to ensure appropriate therapy.

(b) *Classification.* Class II.

§ 862.3645 Neuroleptic drugs radioreceptor assay test system.

(a) *Identification.* A neuroleptic drugs radioceptor assay test system is a device intended to measure in serum or plasma the dopamine receptor blocking activity of neuroleptic drugs and their active metabolites. A neuroleptic drug has anti-psychotic action affecting principally psychomotor activity, is generally without hypnotic effects, and is a tranquilizer. Measurements obtained by this device are used to aid in determining whether a patient is taking the prescribed dosage level of such drugs.

(b) *Classification.* Class II.

§ 862.3650 Opiate test system.

(a) *Identification.* An opiate test system is a device intended to measure any of the addictive narcotic pain-relieving opiate drugs in blood, serum, urine, gastric contents, and saliva. An opiate is any natural or synthetic drug that has morphine-like pharmocological actions. The opiates include drugs such as morphine, morphine glucoronide, heroin, codeine, nalorphine, and meperedine. Measurements obtained by this device are used in the diagnosis and treatment of opiate use or overdose and in monitoring the levels of opiate administration to ensure appropriate therapy.

(b) *Classification.* Class II.

§ 862.3660 Phenobarbital test system.

(a) *Identification.* A phenobarbitol test system is a device intended to measure phenobarbital, an antiepileptic and sedative-hypnotic drug, in human specimens. Measurements obtained by this device are used in the diagnosis and treatment of phenobarbital use or overdose and in monitoring levels of phenobarbital to ensure appropriate therapy.

(b) *Classification.* Class II.

§ 862.3670 Phenothiazine test system.

(a) *Identification.* A phenothiazine test system is a device intended to measure any of the drugs of the phenothiazine class in human specimens. Measurements obtained by this device are used in the diagnosis and treatment of phenothiazine use or overdose.

(b) *Classification.* Class II.

§ 862.3680 Primidone test system.

(a) *Identification.* A primidone test system is a device intended to measure primidone, an antiepileptic drug, in human specimens. Measurements obtained by this device are used in the diagnosis and treatment of primidone overdose and in monitoring levels of primidone to ensure appropriate therapy.

(b) *Classification.* Class II.

§ 862.3700 Propoxyphene test system.

(a) *Identification.* A propoxyphene test system is a device intended to measure propoxyphene, a pain-relieving drug, in serum, plasma, and urine. Measurements obtained by this device are used in the diagnosis and treatment of propoxyphene use or overdose or in monitoring levels of propoxyphene to ensure appropriate therapy.

(b) *Classification.* Class II.

§ 862.3750 Quinine test system.

(a) *Identification.* A quinine test system is a device intended to measure quinine, a fever-reducing and pain-relieving drug intended in the treatment of malaria, in serum and urine. Measurements obtained by this device are used in the diagnosis and treatment of quinine overdose and malaria.

(b) *Classification.* Class I.

[52 FR 16122, May 1, 1987, as amended at 53 FR 21450, June 8, 1988; 65 FR 2310, Jan. 14, 2000]

§ 862.3830 Salicylate test system.

(a) *Identification.* A salicylate test system is a device intended to measure salicylates, a class of analgesic, antipyretic and anti-inflammatory drugs that includes aspirin, in human specimens. Measurements obtained by this device are used in diagnosis and treatment of salicylate overdose and in monitoring salicylate levels to ensure appropriate therapy.

(b) *Classification.* Class II.

§ 862.3840 Sirolimus test system.

(a) *Identification.* A sirolimus test system is a device intended to quantitatively determine sirolimus concentrations in whole blood. Measurements are used as an aid in management of transplant patients receiving therapy with sirolimus.

(b) *Classification.* Class II (special controls). The special control is FDA's guidance document entitled "Class II Special Controls Guidance Document: Sirolimus Test Systems." See § 862.1(d) for the availability of this guidance document.

[69 FR 58259, Sept. 30, 2004]

§ 862.3850 Sulfonamide test system.

(a) *Identification.* A sulfonamide test system is a device intended to measure sulfonamides, any of the antibacterial drugs derived from sulfanilamide, in human specimens. Measurements obtained by this device are used in the diagnosis and treatment of sulfonamide overdose and in monitoring sulfonamide levels to ensure appropriate therapy.

(b) *Classification.* Class I.

[52 FR 16122, May 1, 1987, as amended at 53 FR 21450, June 8, 1988; 65 FR 2310, Jan. 14, 2000]

§ 862.3870 Cannabinoid test system.

(a) *Identification.* A cannabinoid test system is a device intended to measure any of the cannabinoids, hallucinogenic compounds endogenous to marihuana, in serum, plasma, saliva, and urine. Cannabinoid compounds include *delta*-9-tetrahydrocannabinol, cannabidiol, cannabinol, and cannabichromene. Measurements obtained by this device are used in the diagnosis and treatment of cannabinoid use or abuse and in monitoring levels of cannabinoids during clinical investigational use.

(b) *Classification.* Class II.

§ 862.3880 Theophylline test system.

(a) *Identification.* A theophylline test system is a device intended to measure theophylline (a drug used for stimulation of the muscles in the cardiovascular, respiratory, and central nervous systems) in serum and plasma. Measurements obtained by this device are used in the diagnosis and treatment of theophylline overdose or in monitoring levels of theophylline to ensure appropriate therapy.

(b) *Classification.* Class II.

§ 862.3900 Tobramycin test system.

(a) *Identification.* A tobramycin test system is a device intended to measure tobramycin, an aminoglycoside antibiotic drug, in plasma and serum. Measurements obtained by this device are used in the diagnosis and treatment of tobramycin overdose and in monitoring levels of tobramycin to ensure appropriate therapy.

(b) *Classification.* Class II.

§ 862.3910 Tricyclic antidepressant drugs test system.

(a) *Identification.* A tricyclic antidepressant drugs test system is a device intended to measure any of the tricyclic antidepressant drugs in serum. The tricyclic antidepressant drugs include imipramine, desipramine, amitriptyline, nortriptyline, protriptyline, and doxepin. Measurements obtained by this device are used in the diagnosis and treatment of chronic depression to ensure appropriate therapy.

(b) *Classification.* Class II.

§ 862.3950 Vancomycin test system.

(a) *Identification.* A vancomycin test system is a device intended to measure vancomycin, an antibiotic drug, in serum. Measurements obtained by this device are used in the diagnosis and treatment of vancomycin overdose and in monitoring the level of vancomycin to ensure appropriate therapy.

(b) *Classification.* Class II.

PART 864—HEMATOLOGY AND PATHOLOGY DEVICES

Subpart A—General Provisions

Sec.
864.1 Scope.
864.3 Effective dates of requirement for premarket approval.
864.9 Limitations of exemptions from section 510(k) of the Federal Food, Drug, and Cosmetic Act (the act).

Subpart B—Biological Stains

864.1850 Dye and chemical solution stains.
864.1860 Immunohistochemistry reagents and kits.

Subpart C—Cell and Tissue Culture Products

864.2220 Synthetic cell and tissue culture media and components.
864.2240 Cell and tissue culture supplies and equipment.
864.2260 Chromosome culture kit.
864.2280 Cultured animal and human cells.
864.2360 Mycoplasma detection media and components.
864.2800 Animal and human sera.
864.2875 Balanced salt solutions or formulations.

Subpart D—Pathology Instrumentation and Accessories

864.3010 Tissue processing equipment.
864.3250 Specimen transport and storage container.
864.3260 OTC test sample collection systems for drugs of abuse testing.
864.3300 Cytocentrifuge.
864.3400 Device for sealing microsections.
864.3600 Microscopes and accessories.
864.3800 Automated slide stainer.
864.3875 Automated tissue processor.

Subpart E—Specimen Preparation Reagents

864.4010 General purpose reagent.
864.4020 Analyte specific reagents.
864.4400 Enzyme preparations.

Subpart F—Automated and Semi-Automated Hematology Devices

864.5200 Automated cell counter.
864.5220 Automated differential cell counter.
864.5240 Automated blood cell diluting apparatus.
864.5260 Automated cell-locating device.
864.5300 Red cell indices device.
864.5350 Microsedimentation centrifuge.
864.5400 Coagulation instrument.
864.5425 Multipurpose system for in vitro coagulation studies.
864.5600 Automated hematocrit instrument.
864.5620 Automated hemoglobin system.
864.5680 Automated heparin analyzer.
864.5700 Automated platelet aggregation system.
864.5800 Automated sedimentation rate device.
864.5850 Automated slide spinner.
864.5950 Blood volume measuring device.

Subpart G—Manual Hematology Devices

864.6100 Bleeding time device.
864.6150 Capillary blood collection tube.
864.6160 Manual blood cell counting device.
864.6400 Hematocrit measuring device.
864.6550 Occult blood test.
864.6600 Osmotic fragility test.
864.6650 Platelet adhesion test.
864.6675 Platelet aggregometer.
864.6700 Erythrocyte sedimentation rate test.

Subpart H—Hematology Kits and Packages

864.7040 Adenosine triphosphate release assay.
864.7060 Antithrombin III assay.
864.7100 Red blood cell enzyme assay.
864.7140 Activated whole blood clotting time tests.
864.7250 Erythropoietin assay.
864.7275 Euglobulin lysis time tests.
864.7280 Factor V Leiden DNA mutation detection systems.
864.7290 Factor deficiency test.
864.7300 Fibrin monomer paracoagulation test.
864.7320 Fibrinogen/fibrin degradation products assay.
864.7340 Fibrinogen determination system.
864.7360 Erythrocytic glucose-6-phosphate dehydrogenase assay.
864.7375 Glutathione reductase assay.
864.7400 Hemoglobin A_2 assay.
864.7415 Abnormal hemoglobin assay.
864.7425 Carboxyhemoglobin assay.
864.7440 Electrophoretic hemoglobin analysis system.
864.7455 Fetal hemoglobin assay.
864.7470 Glycosylated hemoglobin assay.
864.7490 Sulfhemoglobin assay.
864.7500 Whole blood hemoglobin assays.
864.7525 Heparin assay.
864.7660 Leukocyte alkaline phosphatase test.
864.7675 Leukocyte peroxidase test.
864.7695 Platelet factor 4 radioimmunoassay.
864.7720 Prothrombin consumption test.
864.7735 Prothrombin-proconvertin test and thrombotest.
864.7750 Prothrombin time test.
864.7825 Sickle cell test.

AUTHORITY: 21 U.S.C. 351, 360, 360c, 360e, 360j, 371.

Subpart A—General Provisions

§ 864.1 Scope.

(a) This part sets forth the classification of hematology and pathology devices intended for human use that are in commercial distribution.

(b) The identification of a device in a regulation in this part is not a precise description of every device that is, or will be, subject to the regulation. A manufacturer who submits a premarket notification submission for a device under part 807 may not show merely that the device is accurately described by the section title and identification provisions of a regulation in this part, but shall state why the device is substantially equivalent to other devices, as required by § 807.87.

(c) References in this part to regulatory sections of the Code of Federal Regulations are to chapter I of title 21, unless otherwise noted.

(d) Guidance documents referenced in this part are available on the Internet at *http://www.fda.gov/cdrh/guidance.html.*

[52 FR 17732, May 11, 1987, as amended at 69 FR 12273, Mar. 16, 2004]

§ 864.3 Effective dates of requirement for premarket approval.

A device included in this part that is classified into class III (premarket approval) shall not be commercially distributed after the date shown in the regulation classifying the device unless the manufacturer has an approval under section 515 of the act (unless an exemption has been granted under section 520(g)(2) of the act). An approval under section 515 of the act consists of FDA's issuance of an order approving an application for premarket approval (PMA) for the device or declaring completed a product development protocol (PDP) for the device.

(a) Before FDA requires that a device commercially distributed before the enactment date of the amendments, or a device that has been found substantially equivalent to such a device, has an approval under section 515 of the act FDA must promulgate a regulation under section 515(b) of the act requiring such approval, except as provided in paragraph (b) of this section. Such a regulation under section 515(b) of the act shall not be effective during the grace period ending on the 90th day after its promulgation or on the last day of the 30th full calendar month after the regulation that classifies the device into class III is effective, whichever is later. See section 501(f)(2)(B) of the act. Accordingly, unless an effective date of the requirement for premarket approval is shown in the regulation for a device classified into class III in this part, the device may be commercially distributed without FDA's issuance of an order approving a PMA or declaring completed a PDP for the device. If FDA promulgates a regulation under section 515(b) of the act requiring premarket approval for a device, section 501(f)(1)(A) of the act applies to the device.

(b) Any new, not substantially equivalent, device introduced into commercial distribution on or after May 28, 1976, including a device formerly marketed that has been substantially altered, is classified by statute (section 513(f) of the act) into class III without any grace period and FDA must have issued an order approving a PMA or declaring completed a PDP for the device before the device is commercially distributed unless it is reclassified. If FDA knows that a device being commercially distributed may be a "new" device as defined in this section because of any new intended use or other reasons, FDA may codify the statutory classification of the device into class III for such new use. Accordingly, the regulation for such a class III device states that as of the enactment date of the amendments, May 28, 1976, the device must have an approval under section 515 of the act before commercial distribution.

[52 FR 17732, May 11, 1987]

§ 864.9 Limitations of exemptions from section 510(k) of the Federal Food, Drug, and Cosmetic Act (the act).

The exemption from the requirement of premarket notification (section 510(k) of the act) for a generic type of class I or II device is only to the extent that the device has existing or reasonably foreseeable characteristics of commercially distributed devices within that generic type or, in the case of in vitro diagnostic devices, only to the extent that misdiagnosis as a result of using the device would not be associated with high morbidity or mortality. Accordingly, manufacturers of any commercially distributed class I or II device for which FDA has granted an exemption from the requirement of premarket notification must still submit a premarket notification to FDA before introducing or delivering for introduction into interstate commerce for commercial distribution the device when:

(a) The device is intended for a use different from the intended use of a legally marketed device in that generic type of device; e.g., the device is intended for a different medical purpose, or the device is intended for lay use where the former intended use was by health care professionals only;

(b) The modified device operates using a different fundamental scientific technology than a legally marketed device in that generic type of device; e.g., a surgical instrument cuts tissue with a laser beam rather than with a sharpened metal blade, or an in vitro diagnostic device detects or identifies infectious agents by using deoxyribonucleic acid (DNA) probe or nucleic acid hybridization technology rather than culture or immunoassay technology; or

(c) The device is an in vitro device that is intended:

(1) For use in the diagnosis, monitoring, or screening of neoplastic diseases with the exception of immunohistochemical devices;

(2) For use in screening or diagnosis of familial or acquired genetic disorders, including inborn errors of metabolism;

(3) For measuring an analyte that serves as a surrogate marker for screening, diagnosis, or monitoring life-threatening diseases such as acquired immune deficiency syndrome (AIDS), chronic or active hepatitis, tuberculosis, or myocardial infarction or to monitor therapy;

(4) For assessing the risk of cardiovascular diseases;

(5) For use in diabetes management;

(6) For identifying or inferring the identity of a microorganism directly from clinical material;

(7) For detection of antibodies to microorganisms other than immunoglobulin G (IgG) or IgG assays when the results are not qualitative, or are used to determine immunity, or the assay is intended for use in matrices other than serum or plasma;

(8) For noninvasive testing as defined in § 812.3(k) of this chapter; and

(9) For near patient testing (point of care).

[65 FR 2310, Jan. 14, 2000]

Subpart B—Biological Stains

§ 864.1850 Dye and chemical solution stains.

(a) *Identification.* Dye and chemical solution stains for medical purposes are mixtures of synthetic or natural dyes or nondye chemicals in solutions used in staining cells and tissues for diagnostic histopathology, cytopathology, or hematology.

(b) *Classification.* Class I (general controls). These devices are exempt from the premarket notification procedures in subpart E of part 807 of this chapter subject to the limitations in § 864.9. These devices are also exempt from the current good manufacturing practice regulations in part 820 of this chapter, with the exception of § 820.180, with respect to general requirements concerning records, and § 820.198, with respect to complaint files.

[45 FR 60583, Sept. 12, 1980, as amended at 54 FR 25044, June 12, 1989; 66 FR 38789, July 25, 2001]

§ 864.1860 Immunohistochemistry reagents and kits.

(a) *Identification.* Immunohistochemistry test systems (IHC's) are in vitro diagnostic devices consisting of polyclonal or monoclonal antibodies labeled with directions for use and performance claims, which may be packaged with ancillary reagents in kits. Their intended use is to identify, by immunological techniques, antigens in tissues or cytologic specimens. Similar devices intended for use with flow cytometry devices are not considered IHC's.

(b) *Classification of immunohistochemistry devices.* (1) Class I (general controls). Except as described in paragraphs (b)(2) and (b)(3) of this section, these devices are exempt from the premarket notification requirements in part 807, subpart E of this chapter. This exemption applies to IHC's that provide the pathologist with adjunctive diagnostic information that may be incorporated into the pathologist's report, but that is not ordinarily reported to the clinician as an independent finding. These IHC's are used after the primary diagnosis of tumor (neoplasm) has been made by conventional histopathology using nonimmunologic histochemical stains, such as hematoxylin and eosin. Examples of class I IHC's are differentiation markers that are used as adjunctive tests to subclassify tumors, such as keratin.

(2) Class II (special control, guidance document: "FDA Guidance for Submission of Immunohistochemistry Applications to the FDA," Center for Devices and Radiologic Health, 1998). These IHC's are intended for the detection and/or measurement of certain target analytes in order to provide prognostic or predictive data that are not directly confirmed by routine histopathologic internal and external control specimens. These IHC's provide the pathologist with information that is ordinarily reported as independent diagnostic information to the ordering clinician, and the claims associated with these data are widely accepted and supported by valid scientific evidence. Examples of class II IHC's are those intended for semiquantitative measurement of an analyte, such as hormone receptors in breast cancer.

(3) Class III (premarket approval). IHC's intended for any use not described in paragraphs (b)(1) or (b)(2) of this section.

(c) *Date of PMA or notice of completion of a PDP is required.* As of May 28, 1976, an approval under section 515 of the Federal Food, Drug, and Cosmetic Act is required for any device described in paragraph (b)(3) of this section before this device may be commercially distributed. See § 864.3.

[63 FR 30142, June 3, 1998]

Subpart C—Cell And Tissue Culture Products

§ 864.2220 Synthetic cell and tissue culture media and components.

(a) *Identification.* Synthetic cell and tissue culture media and components are substances that are composed entirely of defined components (e.g., amino acids, vitamins, inorganic salts) that are essential for the survival and development of cell lines of humans and other animals. This does not include tissue culture media for human ex vivo tissue and cell culture processing applications as described in § 876.5885 of this chapter.

(b) *Classification.* Class I (general controls). The device is exempt from the premarket notification procedures in subpart E of part 807 of this chapter subject to the limitations in § 864.9.

[45 FR 60583, Sept. 12, 1980, as amended at 54 FR 25044, June 12, 1989; 66 FR 27024, May 16, 2001; 66 FR 38789, July 25, 2001]

§ 864.2240 Cell and tissue culture supplies and equipment.

(a) *Identification.* Cell and tissue culture supplies and equipment are devices that are used to examine, propagate, nourish, or grow cells and tissue cultures. These include such articles as slide culture chambers, perfusion and roller apparatus, cell culture suspension systems, and tissue culture flasks, disks, tubes, and roller bottles.

(b) *Classification.* Class I (general controls). These devices are exempt from the premarket notification procedures in subpart E of part 807 of this chapter subject to the limitations in § 864.9. If the devices are not labeled or otherwise represented as sterile, they are exempt from the current good manufacturing practice regulations in part 820 of this chapter, with the exception of § 820.180, with respect to general requirements concerning records, and § 820.198, with respect to complaint files.

[45 FR 60584, Sept. 12, 1980, as amended at 54 FR 25044, June 12, 1989; 66 FR 38789, July 25, 2001]

§ 864.2260 Chromosome culture kit.

(a) *Identification.* A chromosome culture kit is a device containing the necessary ingredients (e.g., Minimum Essential Media (MEM) of McCoy's 5A culture media, phytohemagglutinin, fetal calf serum, antibiotics, and heparin) used to culture tissues for diagnosis of congenital chromosome abnormalities.

(b) *Classification.* Class I (general controls). The device is exempt from the premarket notification procedures in subpart E of part 807 of this chapter subject to the limitations in § 864.9.

[45 FR 60585, Sept. 12, 1980, as amended at 54 FR 25044, June 12, 1989; 66 FR 38789, July 25, 2001]

§ 864.2280 Cultured animal and human cells.

(a) *Identification.* Cultured animal and human cells are in vitro cultivated cell lines from the tissue of humans or other animals which are used in various diagnostic procedures, particularly diagnostic virology and cytogenetic studies.

(b) *Classification.* Class I (general controls). The device is exempt from the premarket notification procedures in subpart E of part 807 of this chapter subject to § 864.9.

[45 FR 60585, Sept. 12, 1980, as amended at 65 FR 2310, Jan. 14, 2000]

§ 864.2360 Mycoplasma detection media and components.

(a) *Identification.* Mycoplasma detection media and components are used to detect and isolate mycoplasma pleuropneumonia-like organisms (PPLO), a common microbial contaminant in cell cultures.

(b) *Classification.* Class I (general controls). These devices are exempt from the premarket notification procedures in subpart E of part 807 of this chapter subject to the limitations in § 864.9.

[45 FR 60586, Sept. 12, 1980, as amended at 54 FR 25044, June 12, 1989; 66 FR 38789, July 25, 2001]

§ 864.2800 Animal and human sera.

(a) *Identification.* Animal and human sera are biological products, obtained from the blood of humans or other animals, that provide the necessary growth-promoting nutrients in a cell culture system.

(b) *Classification.* Class I (general controls). These devices are exempt from the premarket notification procedures in subpart E of part 807 of this chapter subject to the limitations in § 864.9.

[45 FR 60586, Sept. 12, 1980, as amended at 54 FR 25044, June 12, 1989; 66 FR 38789, July 25, 2001]

§ 864.2875 Balanced salt solutions or formulations.

(a) *Identification.* A balanced salt solution or formulation is a defined mixture of salts and glucose in a simple medium. This device is included as a necessary component of most cell culture systems. This media component controls for pH, osmotic pressure, energy source, and inorganic ions.

(b) *Classification.* Class I (general controls). These devices are exempt from the premarket notification procedures in subpart E of part 807 of this chapter subject to the limitations in § 864.9.

[45 FR 60586, Sept. 12, 1980, as amended at 54 FR 25044, June 12, 1989; 66 FR 38789, July 25, 2001]

Subpart D—Pathology Instrumentation and Accessories

§ 864.3010 Tissue processing equipment.

(a) *Identification.* Tissue processing equipment consists of devices used to prepare human tissue specimens for diagnostic histological examination by processing specimens through the various stages of decalcifying, infiltrating, sectioning, and mounting on microscope slides.

(b) *Classification.* Class I (general controls). These devices are exempt from the premarket notification procedures in subpart E of part 807 of this chapter subject to the limitations in § 864.9. The devices are also exempt from the current good manufacturing practice regulations in part 820 of this chapter, with the exception of § 820.180, with respect to general requirements concerning records, and § 820.198, with respect to complaint files.

[45 FR 60587, Sept. 12, 1980, as amended at 54 FR 25044, June 12, 1989; 66 FR 38789, July 25, 2001]

§ 864.3250 Specimen transport and storage container.

(a) *Identification.* A specimen transport and storage container, which may be empty or prefilled, is a device intended to contain biological specimens, body waste, or body exudate during storage and transport in order that the matter contained therein can be destroyed or used effectively for diagnostic examination. If prefilled, the device contains a fixative solution or other general purpose reagent to preserve the condition of a biological specimen added to the container. This section does not apply to specimen transport and storage containers that are intended for use as part of an over-the-counter test sample collection system for drugs of abuse testing.

(b) *Classification.* Class I (general controls). The device is exempt from the premarket notification procedures in subpart E of part 807 of this chapter subject to § 864.9. If the device is not labeled or otherwise represented as sterile, it is exempt from the current good manufacturing practice regulations in part 820 of this chapter, with the exception of § 820.180 of this chapter, with respect to general requirements concerning records, and § 820.198 of this chapter, with respect to complaint files.

[54 FR 47206, Nov. 13, 1989, as amended at 65 FR 2310, Jan. 14, 2000; 65 FR 18234, Apr. 7, 2000]

§ 864.3260 OTC test sample collection systems for drugs of abuse testing.

(a) *Identification.* An over-the-counter (OTC) test sample collection system for drugs of abuse testing is a device intended to: Collect biological specimens (such as hair, urine, sweat, or saliva), outside of a medical setting and not on order of a health care professional (e.g., in the home, insurance, sports, or workplace setting); maintain the integrity of such specimens during storage and transport in order that the matter contained therein can be tested in a laboratory for the presence of drugs of abuse or their metabolites; and provide access to test results and counseling. This section does not apply to collection, transport, or laboratory testing of biological specimens for the presence of drugs of abuse or their metabolites that is performed to develop evidence for law enforcement purposes.

(b) *Classification*. Class I (general controls). The device is exempt from the premarket notification requirements in part 807, subpart E of this chapter subject to the limitations in § 864.9 if it is sold, distributed, and used in accordance with the restrictions set forth in § 809.40 of this chapter. If the device is not labeled or otherwise represented as sterile, it is exempt from the current good manufacturing practice regulations in part 820 of this chapter, with the exception of § 820.198 of this chapter with respect to complaint files.

[65 FR 18234, Apr. 7, 2000]

§ 864.3300 Cytocentrifuge.

(a) *Identification*. A cytocentrifuge is a centrifuge used to concentrate cells from biological cell suspensions (e.g., cerebrospinal fluid) and to deposit these cells on a glass microscope slide for cytological examination.

(b) *Classification*. Class I (general controls). This device is exempt from the premarket notification procedures in subpart E of part 807 of this chapter subject to the limitations in § 864.9.

[45 FR 60588, Sept. 12, 1980, as amended at 54 FR 25044, June 12, 1989; 66 FR 38789, July 25, 2001]

§ 864.3400 Device for sealing microsections.

(a) *Identification*. A device for sealing microsections is an automated instrument used to seal stained cells and microsections for histological and cytological examination.

(b) *Classification*. Class I (general controls). This device is exempt from the premarket notification procedures in subpart E of part 807 of this chapter subject to the limitations in § 864.9.

[45 FR 60589, Sept. 12, 1980, as amended at 54 FR 25044, June 12, 1989; 66 FR 38789, July 25, 2001]

§ 864.3600 Microscopes and accessories.

(a) *Identification*. Microscopes and accessories are optical instruments used to enlarge images of specimens, preparations, and cultures for medical purposes. Variations of microscopes and accessories (through a change in the light source) used for medical purposes include the following:

(1) Phase contrast microscopes, which permit visualization of unstained preparations by altering the phase relationship of light that passes around the object and through the object.

(2) Fluorescense microscopes, which permit examination of specimens stained with fluorochromes that fluoresce under ultraviolet light.

(3) Inverted stage microscopes, which permit examination of tissue cultures or other biological specimens contained in bottles or tubes with the light source mounted above the specimen.

(b) *Classification*. Class I (general controls). These devices are exempt from the premarket notification procedures in subpart E of part 807 of this chapter subject to the limitations in § 864.9. If the device is not labeled or otherwise represented as sterile, it is exempt from the current good manufacturing practice regulations in part 820 of this chapter, with the exception of § 820.180, with respect to general requirements concerning records, and § 820.198, with respect to complaint files.

[45 FR 60590, Sept. 12, 1980, as amended at 54 FR 25044, June 12, 1989; 66 FR 38789, July 25, 2001]

§ 864.3800 Automated slide stainer.

(a) *Identification*. An automated slide stainer is a device used to stain histology, cytology, and hematology slides for diagnosis.

(b) *Classification*. Class I (general controls). This device is exempt from the premarket notification procedures in subpart E of part 807 of this chapter subject to the limitations in § 864.9.

[45 FR 60591, Sept. 12, 1980, as amended at 54 FR 25044, June 12, 1989; 66 FR 38789, July 25, 2001]

§ 864.3875 Automated tissue processor.

(a) *Identification*. An automated tissue processor is an automated system used to process tissue specimens for examination through fixation, dehydration, and infiltration.

(b) *Classification*. Class I (general controls). This device is exempt from the premarket notification procedures in

subpart E of part 807 of this chapter subject to the limitations in § 864.9.

[45 FR 60591, Sept. 12, 1980, as amended at 54 FR 25045, June 12, 1989; 66 FR 38789, July 25, 2001]

Subpart E—Specimen Preparation Reagents

§ 864.4010 General purpose reagent.

(a) A general purpose reagent is a chemical reagent that has general laboratory application, that is used to collect, prepare, and examine specimens from the human body for diagnostic purposes, and that is not labeled or otherwise intended for a specific diagnostic application. It may be either an individual substance, or multiple substances reformulated, which, when combined with or used in conjunction with an appropriate analyte specific reagent (ASR) and other general purpose reagents, is part of a diagnostic test procedure or system constituting a finished in vitro diagnostic (IVD) test. General purpose reagents are appropriate for combining with one or more than one ASR in producing such systems and include labware or disposable constituents of tests; but they do not include laboratory machinery, automated or powered systems. General purpose reagents include cytological preservatives, decalcifying reagents, fixative and adhesives, tissue processing reagents, isotonic solutions and pH buffers. Reagents used in tests for more than one individual chemical substance or ligand are general purpose reagents (e.g., *Thermus aquaticus* (TAQ) polymerase, substrates for enzyme immunoassay (EIA)).

(b) *Classification.* Class I (general controls). The device is exempt from the premarket notification procedures in subpart E of part 807 of this chapter subject to the limitations in § 864.9. If the device is not labeled or otherwise represented as sterile, it is exempt from the current good manufacturing practice regulations in part 820 of this chapter, with the exception of § 820.180, with respect to general requirements concerning records, and § 820.198, with respect to complaint files.

[45 FR 60592, Sept. 12, 1980, as amended at 54 FR 25045, June 12, 1989; 62 FR 62260, Nov. 21, 1997; 66 FR 38789, July 25, 2001]

§ 864.4020 Analyte specific reagents.

(a) *Identification.* Analyte specific reagents (ASR's) are antibodies, both polyclonal and monoclonal, specific receptor proteins, ligands, nucleic acid sequences, and similar reagents which, through specific binding or chemical reaction with substances in a specimen, are intended for use in a diagnostic application for identification and quantification of an individual chemical substance or ligand in biological specimens. ASR's that otherwise fall within this definition are not within the scope of subpart E of this part when they are sold to:

(1) In vitro diagnostic manufacturers; or

(2) Organizations that use the reagents to make tests for purposes other than providing diagnostic information to patients and practitioners, e.g., forensic, academic, research, and other nonclinical laboratories.

(b) *Classification.* (1) Class I (general controls). Except as described in paragraphs (b)(2) and (b)(3) of this section, these devices are exempt from the premarket notification requirements in part 807, subpart E of this chapter.

(2) Class II (special controls/guidance documents), when the analyte is used in blood banking tests that have been classified as class II devices (e.g., certain cytomegalovirus serological and treponema pallidum nontreponemal test reagents). Guidance Documents:

1. "Specifications for Immunological Testing for Infectious Disease; Approved Guideline," NCCLS Document I/LA18–A, December 1994.

2. "Assessment of the Clinical Accuracy of Laboratory Tests Using Receiver Operating Characteristic (ROC) Plots; Tentative Guideline," NCCLS Document KGP10–T, December 1993.

3. "Review Criteria for Assessment of In Vitro Diagnostic Devices for Direct Detection of Mycobacterium spp," FDA, July 6, 1993, and its "Attachment 1," February 28, 1994.

4. "Draft Review Criteria for Nucleic Acid Amplification-Based In Vitro Diagnostic Devices for Direct Detection of Infectious Microorganisms," FDA, July 6, 1993.

5. The Center for Biologics Evaluation and Research, FDA, "Points to Consider in the Manufacture and Clinical Evaluation of In Vitro Tests to Detect Antibodies to the Human Immunodeficiency Virus, Type I" (54 FR 48943, November 28, 1989).

(3) Class III (premarket approval), when:

(i) The analyte is intended as a component in a test intended for use in the diagnosis of a contagious condition that is highly likely to result in a fatal outcome and prompt, accurate diagnosis offers the opportunity to mitigate the public health impact of the condition (e.g., human immunodeficiency virus (HIV/AIDS)or tuberculosis (TB)); or

(ii) The analyte is intended as a component in a test intended for use in donor screening for conditions for which FDA has recommended or required testing in order to safeguard the blood supply or establish the safe use of blood and blood products (e.g., tests for hepatitis or tests for identifying blood groups).

(c) *Date of 510(k), or date of PMA or notice of completion of a product development protocol is required.* (1) Preamendments ASR's; No effective date has been established for the requirement for premarket approval for the device described in paragraph (b)(3) of this section. See § 864.3.

(2) For postamendments ASR's; November 23, 1998.

(d) *Restrictions.* Restrictions on the sale, distribution and use of ASR's are set forth in § 809.30 of this chapter.

[62 FR 62260, Nov. 21, 1997]

§ 864.4400 Enzyme preparations.

(a) *Identification.* Enzyme preparations are products that are used in the histopathology laboratory for the following purposes:

(1) To disaggregate tissues and cells already in established cultures for preparation into subsequent cultures (e.g., trypsin);

(2) To disaggregate fluid specimens for cytological examination (e.g., papain for gastric lavage or trypsin for sputum liquefaction);

(3) To aid in the selective staining of tissue specimens (e.g., diastase for glycogen determination).

(b) *Classification.* Class I (general controls). This device is exempt from the premarket notification procedures in subpart E of part 807 of this chapter subject to the limitations in § 864.9.

[45 FR 60592, Sept. 12, 1980, as amended at 54 FR 25045, June 12, 1989; 66 FR 38789, July 25, 2001]

Subpart F—Automated and Semi-Automated Hematology Devices

§ 864.5200 Automated cell counter.

(a) *Identification.* An automated cell counter is a fully-automated or semi-automated device used to count red blood cells, white blood cells, or blood platelets using a sample of the patient's peripheral blood (blood circulating in one of the body's extremities, such as the arm). These devices may also measure hemoglobin or hematocrit and may also calculate or measure one or more of the red cell indices (the erythrocyte mean corpuscular volume, the mean corpuscular hemoglobin, or the mean corpuscular hemoglobin concentration). These devices may use either an electronic particle counting method or an optical counting method.

(b) *Classification.* Class II (performance standards).

[45 FR 60593, Sept. 12, 1980]

§ 864.5220 Automated differential cell counter.

(a) *Identification.* An automated differential cell counter is a device used to identify one or more of the formed elements of the blood. The device may also have the capability to flag, count, or classify immature or abnormal hematopoietic cells of the blood, bone marrow, or other body fluids. These devices may combine an electronic particle counting method, optical method, or a flow cytometric method utilizing monoclonal CD (cluster designation) markers. The device includes accessory CD markers.

(b) *Classification.* Class II (special controls). The special control for this device is the FDA document entitled "Class II Special Controls Guidance Document: Premarket Notifications for

Automated Differential Cell Counters for Immature or Abnormal Blood Cells; Final Guidance for Industry and FDA.''

[67 FR 1607, Jan. 14, 2002]

§ 864.5240 Automated blood cell diluting apparatus.

(a) *Identification.* An automated blood cell diluting apparatus is a fully automated or semi-automated device used to make appropriate dilutions of a blood sample for further testing.

(b) *Classification.* Class I (general controls). The device is exempt from the premarket notification procedures in subpart E of part 807 of this chapter subject to § 864.9.

[45 FR 60596, Sept. 12, 1980, as amended at 65 FR 2310, Jan. 14, 2000]

§ 864.5260 Automated cell-locating device.

(a) *Identification.* An automated cell-locating device is a device used to locate blood cells on a peripheral blood smear, allowing the operator to identify and classify each cell according to type. (Peripheral blood is blood circulating in one of the body's extremities, such as the arm.)

(b) *Classification.* Class II (performance standards).

[45 FR 60597, Sept. 12, 1980]

§ 864.5300 Red cell indices device.

(a) *Identification.* A red cell indices device, usually part of a larger system, calculates or directly measures the erythrocyte mean corpuscular volume (MCV), the mean corpuscular hemoglobin (MCH), and the mean corpuscular hemoglobin concentration (MCHC). The red cell indices are used for the differential diagnosis of anemias.

(b) *Classification.* Class II (performance standards).

[45 FR 60597, Sept. 12, 1980]

§ 864.5350 Microsedimentation centrifuge.

(a) *Identification.* A microsedimentation centrifuge is a device used to sediment red cells for the microsedimentation rate test.

(b) *Classification.* Class I (general controls). This device is exempt from the premarket notification procedures in subpart E of part 807 of this chapter subject to the limitations in § 864.9.

[45 FR 60598, Sept. 12, 1980, as amended at 59 FR 63007, Dec. 7, 1994; 66 FR 38789, July 25, 2001]

§ 864.5400 Coagulation instrument.

(a) *Identification.* A coagulation instrument is an automated or semiautomated device used to determine the onset of clot formation for in vitro coagulation studies.

(b) *Classification.* Class II (performance standards).

[45 FR 60598, Sept. 12, 1980]

§ 864.5425 Multipurpose system for in vitro coagulation studies.

(a) *Identification.* A multipurpose system for in vitro coagulation studies is a device consisting of one automated or semiautomated instrument and its associated reagents and controls. The system is used to perform a series of coagulation studies and coagulation factor assays.

(b) *Classification.* Class II (performance standards).

[45 FR 60599, Sept. 12, 1980]

§ 864.5600 Automated hematocrit instrument.

(a) *Identification.* An automated hematocrit instrument is a fully automated or semi-automated device which may or may not be part of a larger system. This device measures the packed red cell volume of a blood sample to distinguish normal from abnormal states, such as anemia and erythrocytosis (an increase in the number of red cells).

(b) *Classification.* Class II (performance standards).

[45 FR 60600, Sept. 12, 1980]

§ 864.5620 Automated hemoglobin system.

(a) *Identification.* An automated hemoglobin system is a fully automated or semi-automated device which may or may not be part of a larger system. The generic type of device consists of the reagents, calibrators, controls, and instrumentation used to determine the hemoglobin content of human blood.

(b) *Classification.* Class II (performance standards).

[45 FR 60601, Sept. 12, 1980]

§ 864.5680 Automated heparin analyzer.

(a) *Identification.* An automated heparin analyzer is a device used to determine the heparin level in a blood sample by mixing the sample with protamine (a heparin-neutralizing substance) and determining photometrically the onset of air-activated clotting. The analyzer also determines the amount of protamine necessary to neutralize the heparin in the patient's circulation.

(b) *Classification.* Class II (special controls).

[45 FR 60601, Sept. 12, 1980, as amended at 52 FR 17733, May 11, 1987; 58 FR 51571, Oct. 4, 1993]

§ 864.5700 Automated platelet aggregation system.

(a) *Identification.* An automated platelet aggregation system is a device used to determine changes in platelet shape and platelet aggregation following the addition of an aggregating reagent to a platelet-rich plasma.

(b) *Classification.* Class II (performance standards).

[45 FR 60602, Sept. 12, 1980]

§ 864.5800 Automated sedimentation rate device.

(a) *Identification.* An automated sedimentation rate device is an instrument that measures automatically the erythrocyte sedimentation rate in whole blood. Because an increased sedimentation rate indicates tissue damage or inflammation, the erythrocyte sedimentation rate device is useful in monitoring treatment of a disease.

(b) *Classification.* Class I (general controls). This device is exempt from the premarket notification procedures in subpart E of part 807 of this chapter subject to the limitations in § 864.9.

[45 FR 60602, Sept. 12, 1980, as amended at 54 FR 25045, June 12, 1989; 66 FR 38790, July 25, 2001]

§ 864.5850 Automated slide spinner.

(a) *Identification.* An automated slide spinner is a device that prepares automatically a blood film on a microscope slide using a small amount of peripheral blood (blood circulating in one of the body's extremities, such as the arm).

(b) *Classification.* Class I (general controls). This device is exempt from the premarket notification procedures in subpart E of part 807 of this chapter subject to the limitations in § 864.9.

[45 FR 60603, Sept. 12, 1980, as amended at 54 FR 25045, June 12, 1989; 66 FR 38790, July 25, 2001]

§ 864.5950 Blood volume measuring device.

(a) *Identification.* A blood volume measuring device is a manual, semiautomated, or automated system that is used to calculate the red cell mass, plasma volume, and total blood volume.

(b) *Classification.* Class II (performance standards).

[45 FR 60603, Sept. 12, 1980]

Subpart G—Manual Hematology Devices

§ 864.6100 Bleeding time device.

(a) *Identification.* A bleeding time device is a device, usually employing two spring-loaded blades, that produces two small incisions in the patient's skin. The length of time required for the bleeding to stop is a measure of the effectiveness of the coagulation system, primarily the platelets.

(b) *Classification.* Class II (special controls). The device is exempt from the premarket notification procedures in subpart E of part 807 of this chapter subject to § 864.9.

[45 FR 60604, Sept. 12, 1980, as amended at 63 FR 59225, Nov. 3, 1998]

§ 864.6150 Capillary blood collection tube.

(a) *Identification.* A capillary blood collection tube is a plain or heparinized glass tube of very small diameter used to collect blood by capillary action.

(b) *Classification.* Class I (general controls). The device is exempt from the premarket notification procedures in

subpart E of part 807 of this chapter subject to § 864.9.

[45 FR 60604, Sept. 12, 1980, as amended at 54 FR 25045, June 12, 1989; 65 FR 2310, Jan. 14, 2000]

§ 864.6160 Manual blood cell counting device.

(a) *Identification.* A manual blood cell counting device is a device used to count red blood cells, white blood cells, or blood platelets.

(b) *Classification.* Class I (general controls). This device is exempt from the premarket notification procedures in subpart E of part 807 of this chapter subject to the limitations in § 864.9.

[45 FR 60605, Sept. 12, 1980, as amended at 54 FR 25045, June 12, 1989; 66 FR 38790, July 25, 2001]

§ 864.6400 Hematocrit measuring device.

(a) *Identification.* A hematocrit measuring device is a system consisting of instruments, tubes, racks, and a sealer and a holder. The device is used to measure the packed red cell volume in blood to determine whether the patient's total red cell volume is normal or abnormal. Abnormal states include anemia (an abnormally low total red cell volume) and erythrocytosis (an abnormally high total red cell mass). The packed red cell volume is produced by centrifuging a given volume of blood.

(b) *Classification.* Class II (special controls). The device is exempt from the premarket notification procedures in subpart E of part 807 of this chapter subject to § 864.9.

[45 FR 60606, Sept. 12, 1980, as amended at 63 FR 59225, Nov. 3, 1998]

§ 864.6550 Occult blood test.

(a) *Identification.* An occult blood test is a device used to detect occult blood in urine or feces. (Occult blood is blood present in such small quantities that it can be detected only by chemical tests of suspected material, or by microscopic or spectroscopic examination.)

(b) *Classification.* Class II (performance standards).

[45 FR 60606, Sept. 12, 1980]

§ 864.6600 Osmotic fragility test.

(a) *Identification.* An osmotic fragility test is a device used to determine the resistance of red blood cells to hemolysis (destruction) in varying concentrations of hypotonic saline solutions.

(b) *Classification.* Class I (general controls). This device is exempt from the premarket notification procedures in subpart E of part 807 of this chapter subject to the limitations in § 864.9.

[45 FR 60607, Sept. 12, 1980, as amended at 54 FR 25045, June 12, 1989; 66 FR 38790, July 25, 2001]

§ 864.6650 Platelet adhesion test.

(a) *Identification.* A platelet adhesion test is a device used to determine in vitro platelet function.

(b) *Classification.* Class II (performance standards).

[45 FR 60608, Sept. 12, 1980]

§ 864.6675 Platelet aggregometer.

(a) *Identification.* A platelet aggregometer is a device, used to determine changes in platelet shape and platelet aggregation following the addition of an aggregating reagent to a platelet rich plasma.

(b) *Classification.* Class II (performance standards).

[45 FR 60608, Sept. 12, 1980]

§ 864.6700 Erythrocyte sedimentation rate test.

(a) *Identification.* An erythrocyte sedimentation rate test is a device that measures the length of time required for the red cells in a blood sample to fall a specified distance or a device that measures the degree of sedimentation taking place in a given length of time. An increased rate indicates tissue damage or inflammation.

(b) *Classification.* Class I (general controls). This device is exempt from the premarket notification procedures in subpart E of part 807 of this chapter subject to the limitations in § 864.9.

[45 FR 60608, Sept. 12, 1980, as amended at 54 FR 25045, June 12, 1989; 66 FR 38790, July 25, 2001]

Subpart H—Hematology Kits and Packages

§ 864.7040 Adenosine triphosphate release assay.

(a) *Identification.* An adenosine triphosphate release assay is a device that measures the release of adenosine triphosphate (ATP) from platelets following aggregation. This measurement is made on platelet-rich plasma using a photometer and a luminescent firefly extract. Simultaneous measurements of platelet aggregation and ATP release are used to evaluate platelet function disorders.

(b) *Classification.* Class I (general controls).

[45 FR 60609, Sept. 12, 1980]

§ 864.7060 Antithrombin III assay.

(a) *Identification.* An antithrombin III assay is a device that is used to determine the plasma level of antithrombin III (a substance which acts with the anticoagulant heparin to prevent coagulation). This determination is used to monitor the administration of heparin in the treatment of thrombosis. The determination may also be used in the diagnosis of thrombophilia (a congenital deficiency of antithrombin III).

(b) *Classification.* Class II (performance standards).

[45 FR 60609, Sept. 12, 1980]

§ 864.7100 Red blood cell enzyme assay.

(a) *Identification.* Red blood cell enzyme assay is a device used to measure the activity in red blood cells of clinically important enzymatic reactions and their products, such as pyruvate kinase or 2,3-diphosphoglycerate. A red blood cell enzyme assay is used to determine the enzyme defects responsible for a patient's hereditary hemolytic anemia.

(b) *Classification.* Class II (performance standards).

[45 FR 60610, Sept. 12, 1980]

§ 864.7140 Activated whole blood clotting time tests.

(a) *Identification.* An activated whole blood clotting time tests is a device, used to monitor heparin therapy for the treatment of venous thrombosis or pulmonary embolism by measuring the coagulation time of whole blood.

(b) *Classification.* Class II (performance standards).

[45 FR 60611, Sept. 12, 1980]

§ 864.7250 Erythropoietin assay.

(a) *Identification.* A erythropoietin assay is a device that measures the concentration of erythropoietin (an enzyme that regulates the production of red blood cells) in serum or urine. This assay provides diagnostic information for the evaluation of erythrocytosis (increased total red cell mass) and anemia.

(b) *Classification.* Class II. The special control for this device is FDA's "Document for Special Controls for Erythropoietin Assay Premarket Notification (510(k)s)."

[45 FR 60612, Sept. 12, 1980, as amended at 52 FR 17733, May 11, 1987; 65 FR 17144, Mar. 31, 2000]

§ 864.7275 Euglobulin lysis time tests.

(a) *Identification.* A euglobulin lysis time test is a device that measures the length of time required for the lysis (dissolution) of a clot formed from fibrinogen in the euglobulin fraction (that fraction of the plasma responsible for the formation of plasmin, a clot lysing enzyme). This test evaluates natural fibrinolysis (destruction of a blood clot after bleeding has been arrested). The test also will detect accelerated fibrinolysis.

(b) *Classification.* Class II (performance standards).

[45 FR 60612, Sept. 12, 1980]

§ 864.7280 Factor V Leiden DNA mutation detection systems.

(a) *Identification.* Factor V Leiden deoxyribonucleic acid (DNA) mutation detection systems are devices that consist of different reagents and instruments which include polymerase chain reaction (PCR) primers, hybridization matrices, thermal cyclers, imagers, and software packages. The detection of the Factor V Leiden mutation aids in the diagnosis of patients with suspected thrombophilia.

(b) *Classification.* Class II (special controls). The special control is FDA's guidance entitled "Class II Special Controls Guidance Document: Factor V Leiden DNA Mutation Detection Systems." (See § 864.1(d) for the availability of this guidance document.)

[69 FR 12273, Mar. 16, 2004]

§ 864.7290 Factor deficiency test.

(a) *Identification.* A factor deficiency test is a device used to diagnose specific coagulation defects, to monitor certain types of therapy, to detect coagulation inhibitors, and to detect a carrier state (a person carrying both a recessive gene for a coagulation factor deficiency such as hemophilia and the corresponding normal gene).

(b) *Classification.* Class II (performance standards).

[45 FR 60613, Sept. 12, 1980]

§ 864.7300 Fibrin monomer paracoagulation test.

(a) *Identification.* A fibrin monomer paracoagulation test is a device used to detect fibrin monomer in the diagnosis of disseminated intravascular coagulation (nonlocalized clotting within a blood vessel) or in the differential diagnosis between disseminated intravascular coagulation and primary fibrinolysis (dissolution of the fibrin in a blood clot).

(b) *Classification.* Class II. The special control for this device is FDA's "In Vitro Diagnostic Fibrin Monomer Paracoagulation Test."

[45 FR 60614, Sept. 12, 1980, as amended at 52 FR 17733, May 11, 1987; 65 FR 17144, Mar. 31, 2000]

§ 864.7320 Fibrinogen/fibrin degradation products assay.

(a) *Identification.* A fibrinogen/fibrin degradation products assay is a device used to detect and measure fibrinogen degradation products and fibrin degradation products (protein fragments produced by the enzymatic action of plasmin on fibrinogen and fibrin) as an aid in detecting the presence and degree of intravascular coagulation and fibrinolysis (the dissolution of the fibrin in a blood clot) and in monitoring therapy for disseminated intravascular coagulation (nonlocalized clotting in the blood vessels).

(b) *Classification.* Class II (performance standards).

[45 FR 60615, Sept. 12, 1980]

§ 864.7340 Fibrinogen determination system.

(a) *Identification.* A fibrinogen determination system is a device that consists of the instruments, reagents, standards, and controls used to determine the fibrinogen levels in disseminated intravascular coagulation (nonlocalized clotting within the blood vessels) and primary fibrinolysis (the dissolution of fibrin in a blood clot).

(b) *Classification.* Class II (performance standards).

[45 FR 60615, Sept. 12, 1980]

§ 864.7360 Erythrocytic glucose-6-phosphate dehydrogenase assay.

(a) *Identification.* An erythrocytic glucose-6-phosphate dehydrogenase assay is a device used to measure the activity of the enzyme glucose-6-phosphate dehydrogenase or of glucose-6-phosphate dehydrogenase isoenzymes. The results of this assay are used in the diagnosis and treatment of nonspherocytic congenital hemolytic anemia or drug-induced hemolytic anemia associated with a glucose-6-phosphate dehydrogenase deficiency. This generic device includes assays based on fluorescence, electrophoresis, methemoglobin reduction, catalase inhibition, and ultraviolet kinetics.

(b) *Classification.* Class II (performance standards).

[45 FR 60616, Sept. 12, 1980]

§ 864.7375 Glutathione reductase assay.

(a) *Identification.* A glutathione reductase assay is a device used to determine the activity of the enzyme glutathione reductase in serum, plasma, or erythrocytes by such techniques as fluorescence and photometry. The results of this assay are used in the diagnosis of liver disease, glutathione reductase deficiency, or riboflavin deficiency.

(b) *Classification.* Class II (performance standards).

[45 FR 60616, Sept. 12, 1980]

§ 864.7400 Hemoglobin A_2 assay.

(a) *Identification.* A hemoglobin A_2 assay is a device used to determine the hemoglobin A_2 content of human blood. The measurement of hemoglobin A_2 is used in the diagnosis of the thalassemias (hereditary hemolytic anemias characterized by decreased synthesis of one or more types of hemoglobin polypeptide chains).

(b) *Classification.* Class II (performance standards).

[45 FR 60617, Sept. 12, 1980]

§ 864.7415 Abnormal hemoglobin assay.

(a) *Identification.* An abnormal hemoglobin assay is a device consisting of the reagents, apparatus, instrumentation, and controls necessary to isolate and identify abnormal genetically determined hemoglobin types.

(b) *Classification.* Class II (performance standards).

[45 FR 60618, Sept. 12, 1980]

§ 864.7425 Carboxyhemoglobin assay.

(a) *Identification.* A carboxyhemoglobin assay is a device used to determine the carboxyhemoglobin (the compound formed when hemoglobin is exposed to carbon monoxide) content of human blood as an aid in the diagnosis of carbon monoxide poisoning. This measurement may be made using methods such as spectroscopy, colorimetry, spectrophotometry, and gasometry.

(b) *Classification.* Class II (performance standards).

[45 FR 60619, Sept. 12, 1980]

§ 864.7440 Electrophoretic hemoglobin analysis system.

(a) *Identification.* An electrophoretic hemoglobin analysis system is a device that electrophoretically separates and identifies normal and abnormal hemoglobin types as an aid in the diagnosis of anemia or erythrocytosis (increased total red cell mass) due to a hemoglobin abnormality.

(b) *Classification.* Class II (performance standards).

[45 FR 60620, Sept. 12, 1980]

§ 864.7455 Fetal hemoglobin assay.

(a) *Identification.* A fetal hemoglobin assay is a device that is used to determine the presence and distribution of fetal hemoglobin (hemoglobin F) in red cells or to measure the amount of fetal hemoglobin present. The assay may be used to detect fetal red cells in the maternal circulation or to detect the elevated levels of fetal hemoglobin exhibited in cases of hemoglobin abnormalities such as thalassemia (a hereditary hemolytic anemia characterized by a decreased synthesis of one or more types of hemoglobin polypeptide chains). The hemoglobin determination may be made by methods such as electrophoresis, alkali denaturation, column chromatography, or radial immunodiffusion.

(b) *Classification.* Class II (performance standards).

[45 FR 60620, Sept. 12, 1980]

§ 864.7470 Glycosylated hemoglobin assay.

(a) *Identification.* A glycosylated hemoglobin assay is a device used to measure the glycosylated hemoglobins (A_{1a}, A_{1b}, and A_{1c}) in a patient's blood by a column chromatographic procedure. Measurement of glycosylated hemoglobin is used to assess the level of control of a patient's diabetes and to determine the proper insulin dosage for a patient. Elevated levels of glycosylated hemoglobin indicate uncontrolled diabetes in a patient.

(b) *Classification.* Class II (performance standards).

[45 FR 60621, Sept. 12, 1980]

§ 864.7490 Sulfhemoglobin assay.

(a) *Identification.* A sulfhemoglobin assay is a device consisting of the reagents, calibrators, controls, and instrumentation used to determine the sulfhemoglobin (a compound of sulfur and hemoglobin) content of human blood as an aid in the diagnosis of sulfhemoglobinemia (presence of sulfhemoglobin in the blood due to drug administration or exposure to a poison). This measurement may be made using methods such as spectroscopy, colorimetry, spectrophotometry, or gasometry.

(b) *Classification.* Class II (performance standards).

[45 FR 60621, Sept. 12, 1980]

§ 864.7500 Whole blood hemoglobin assays.

(a) *Identification.* A whole blood hemoglobin assay is a device consisting or reagents, calibrators, controls, or photometric or spectrophotometric instrumentation used to measure the hemoglobin content of whole blood for the detection of anemia. This generic device category does not include automated hemoglobin systems.

(b) *Classification.* Class II (performance standards).

[45 FR 60622, Sept. 12, 1980]

§ 864.7525 Heparin assay.

(a) *Identification.* A heparin assay is a device used to determine the level of the anticoagulant heparin in the patient's circulation. These assays are quantitative clotting time procedures using the effect of heparin on activated coagulation factor X (Stuart factor) or procedures based on the neutralization of heparin by protamine sulfate (a protein that neutralizes heparin).

(b) *Classification.* Class II (performance standards).

[45 FR 60623, Sept. 12, 1980]

§ 864.7660 Leukocyte alkaline phosphatase test.

(a) *Identification.* A leukocyte alkaline phosphatase test is a device used to identify the enzyme leukocyte alkaline phosphatase in neutrophilic granulocytes (granular leukocytes stainable by neutral dyes). The cytochemical identification of alkaline phosphatase depends on the formation of blue granules in cells containing alkaline phosphatase. The results of this test are used to differentiate chronic granulocytic leukemia (a malignant disease characterized by excessive overgrowth of granulocytes in the bone marrow) and reactions that resemble true leukemia, such as those occuring in severe infections and polycythemia (increased total red cell mass).

(b) *Classification.* Class I (general controls). This device is exempt from the premarket notification procedures in subpart E of part 807 of this chapter subject to the limitations in § 864.9.

[45 FR 60623, Sept. 12, 1980, as amended at 59 FR 63007, Dec. 7, 1994; 66 FR 38790, July 25, 2001]

§ 864.7675 Leukocyte peroxidase test.

(a) *Identification.* A leukocyte peroxidase test is a device used to distinguish certain myeloid cells derived from the bone marrow, i.e., neutrophils, eosinophils, and monocytes, from lymphoid cells of the lymphatic system and erythroid cells of the red blood cell series on the basis of their peroxidase activity as evidenced by staining. The results of this test are used in the differential diagnosis of the leukemias.

(b) *Classification.* Class I (general controls). This device is exempt from the premarket notification procedures in subpart E of part 807 of this chapter subject to the limitations in § 864.9.

[45 FR 60624, Sept. 12, 1980, as amended at 59 FR 63007, Dec. 7, 1994; 66 FR 38790, July 25, 2001]

§ 864.7695 Platelet factor 4 radioimmunoassay.

(a) *Identification.* A platelet factor 4 radioimmunoassay is a device used to measure the level of platelet factor 4, a protein released during platelet activation by radioimmunoassay. This device measures platelet activiation, which may indicate a coagulation disorder, such as myocardial infarction or coronary artery disease.

(b) *Classification.* Class II (performance standards).

[45 FR 60625, Sept. 12, 1980; 46 FR 14890, Mar. 3, 1981]

§ 864.7720 Prothrombin consumption test.

(a) *Identification.* A prothrombin consumption tests is a device that measures the patient's capacity to generate thromboplastin in the coagulation process. The test also is an indirect indicator of qualitative or quantitative platelet abnormalities. It is a screening test for thrombocytopenia (decreased number of blood platelets) and hemophilia A and B.

(b) *Classification.* Class II (performance standards).

[45 FR 60625, Sept. 12, 1980]

§864.7735 Prothrombin-proconvertin test and thrombotest.

(a) *Identification.* The prothrombin-proconvertin test and thrombotest are devices used in the regulation of coumarin therapy (administration of a coumarin anticoagulant such as sodium warfarin in the treatment of venous thrombosis and pulmonary embolism) and as a diagnostic test in conjunction with, or in place of, the Quick prothrombin time test to detect coagulation disorders.

(b) *Classification.* Class II (performance standards).

[45 FR 60626, Sept. 12, 1980]

§864.7750 Prothrombin time test.

(a) *Identification.* A prothrombin time test is a device used as a general screening procedure for the detection of possible clotting factor deficiencies in the extrinsic coagulation pathway, which involves the reaction between coagulation factors III and VII, and to monitor patients receiving coumarin therapy (the administration of one of the coumarin anticoagulants in the treatment of venous thrombosis or pulmonary embolism).

(b) *Classification.* Class II (performance standards).

[45 FR 60626, Sept. 12, 1980]

§864.7825 Sickle cell test.

(a) *Identification.* A sickle cell test is a device used to determine the sickle cell hemoglobin content of human blood to detect sickle cell trait or sickle cell diseases.

(b) *Classification.* Class II (performance standards).

[45 FR 60627, Sept. 12, 1980]

§864.7875 Thrombin time test.

(a) *Identification.* A thrombin time test is a device used to measure fibrinogen concentration and detect fibrin or fibrinogen split products for the evaluation of bleeding disorders.

(b) *Classification.* Class II (performance standards).

[45 FR 60628, Sept. 12, 1980]

§864.7900 Thromboplastin generation test.

(a) *Identification.* A thromboplastin generation test is a device used to detect and identify coagulation factor deficiencies and coagulation inhibitors.

(b) *Classification.* Class I (general controls). This device is exempt from the premarket notification procedures in subpart E of part 807 of this chapter subject to the limitations in §864.9.

[45 FR 60628, Sept. 12, 1980, as amended at 59 FR 63007, Dec. 7, 1994; 66 FR 38790, July 25, 2001]

§864.7925 Partial thromboplastin time tests.

(a) *Identification.* A partial thromboplastin time test is a device used for primary screening for coagulation abnormalities, for evaluation of the effect of therapy on procoagulant disorders, and as an assay for coagulation factor deficiencies of the intrinsic coagulation pathway.

(b) *Classification.* Class II (performance standards).

[45 FR 60629, Sept. 12, 1980]

Subpart I—Hematology Reagents

§864.8100 Bothrops atrox reagent.

(a) *Identification.* A Bothrops atrox reagent is a device made from snake venom and used to determine blood fibrinogen levels to aid in the evaluation of disseminated intravascular coagulation (nonlocalized clotting in the blood vessels) in patients receiving heparin therapy (the administration of the anticoagulant heparin in the treatment of thrombosis) or as an aid in the classification of dysfibrinogenemia (presence in the plasma of functionally defective fibrinogen).

(b) *Classification.* Class II (performance standards).

[45 FR 60629, Sept. 12, 1980]

§864.8150 Calibrator for cell indices.

(a) *Identification.* A calibrator for cell indices is a device that approximates whole blood or certain blood cells and that is used to set an instrument intended to measure mean cell volume (MCV), mean corpuscular hemoglobin

(MCH), and mean corpuscular hemoglobin concentration (MCHC), or other cell indices. It is a suspension of particles or cells whose size, shape, concentration, and other characteristics have been precisely and accurately determined.

(b) *Classification.* Class II (performance standards).

[45 FR 60631, Sept. 12, 1980]

§ 864.8165 Calibrator for hemoglobin or hematocrit measurement.

(a) *Identification.* A calibrator for hemoglobin or hematocrit measurement is a device that approximates whole blood, red blood cells, or a hemoglobin derivative and that is used to set instruments intended to measure hemoglobin, the hematocrit, or both. It is a material whose characteristics have been precisely and accurately determined.

(b) *Classification.* Class II (performance standards).

[45 FR 60632, Sept. 12, 1980]

§ 864.8175 Calibrator for platelet counting.

(a) *Identification.* A calibrator for platelet counting is a device that resembles platelets in plasma or whole blood and that is used to set a platelet counting instrument. It is a suspension of particles or cells whose size, shape concentration, and other characteristics have been precisely and accurately determined.

(b) *Classification.* Class II (performance standards).

[45 FR 60633, Sept. 12, 1980]

§ 864.8185 Calibrator for red cell and white cell counting.

(a) *Identification.* A calibrator for red cell and white cell counting is a device that resembles red or white blood cells and that is used to set instruments intended to count red cells, white cells, or both. It is a suspension of particles or cells whose size, shape, concentration, and other characteristics have been precisely and accurately determined.

(b) *Classification.* Class II (performance standards).

[45 FR 60634, Sept. 12, 1980]

§ 864.8200 Blood cell diluent.

(a) *Identification.* A blood cell diluent is a device used to dilute blood for further testing, such as blood cell counting.

(b) *Classification.* Class I (general controls). This device is exempt from the premarket notification procedures in subpart E of part 807 of this chapter subject to the limitations in § 864.9.

[45 FR 60635, Sept. 12, 1980, as amended at 54 FR 25045, June 12, 1989; 66 FR 38790, July 25, 2001]

§ 864.8500 Lymphocyte separation medium.

(a) *Identification.* A lymphocyte separation medium is a device used to isolate lymphocytes from whole blood.

(b) *Classification.* Class I (general controls). This device is exempt from the premarket notification procedures in subpart E of part 807 of this chapter subject to the limitations in § 864.9.

[45 FR 60636, Sept. 12, 1980, as amended at 59 FR 63007, Dec. 7, 1994; 66 FR 38790, July 25, 2001]

§ 864.8540 Red cell lysing reagent.

(a) *Identification.* A red cell lysing reagent is a device used to lyse (destroy) red blood cells for hemoglobin determinations or aid in the counting of white blood cells.

(b) *Classification.* Class I (general controls). This device is exempt from the premarket notification procedures in subpart E of part 807 of this chapter subject to the limitations in § 864.9.

[45 FR 60636, Sept. 12, 1980, as amended at 54 FR 25045, June 12, 1989; 66 FR 38790, July 25, 2001]

§ 864.8625 Hematology quality control mixture.

(a) *Identification.* A hematology quality control mixture is a device used to ascertain the accuracy and precision of manual, semiautomated, and automated determinations of cell parameters such as white cell count (WBC), red cell count (RBC), platelet count (PLT), hemoglobin, hematocrit (HCT), mean corpuscular volume (MCV), mean corpuscular hemoglobin (MCH), and mean corpuscular hemoglobin concentration (MCHC).

(b) *Classification.* Class II (performance standards).

[45 FR 60637, Sept. 12, 1980]

§ 864.8950 Russell viper venom reagent.

(a) *Identification.* Russell viper venom reagent is a device used to determine the cause of an increase in the prothrombin time.

(b) *Classification.* Class I (general controls).

[45 FR 60637, Sept. 12, 1980]

Subpart J—Products Used In Establishments That Manufacture Blood and Blood Products

§ 864.9050 Blood bank supplies.

(a) *Identification.* Blood bank supplies are general purpose devices intended for in vitro use in blood banking. This generic type of device includes products such as blood bank pipettes, blood grouping slides, blood typing tubes, blood typing racks, and cold packs for antisera reagents. The device does not include articles that are licensed by the Center for Biologics Evaluation and Research of the Food and Drug Administration.

(b) *Classification.* Class I (general controls).

[45 FR 60638, Sept. 12, 1980, as amended at 53 FR 11253, Apr. 6, 1988]

§ 864.9100 Empty container for the collection and processing of blood and blood components.

(a) *Identification.* An empty container for the collection and processing of blood and blood components is a device intended for medical purposes that is an empty plastic bag or plastic or glass bottle used to collect, store, or transfer blood and blood components for further processing.

(b) *Classification.* Class II (performance standards).

[45 FR 60638, Sept. 12, 1980]

§ 864.9125 Vacuum-assisted blood collection system.

(a) *Identification.* A vacuum-assisted blood collection system is a device intended for medical purposes that uses a vacuum to draw blood for subsequent reinfusion.

(b) *Classification.* Class I (general controls). The manual device is exempt from the premarket notification procedures in subpart E of part 807 of this chapter subject to § 864.9.

[45 FR 60639, Sept. 12, 1980, as amended at 65 FR 2310, Jan. 14, 2000]

§ 864.9145 Processing system for frozen blood.

(a) *Identification.* A processing system for frozen blood is a device used to glycerolize red blood cells prior to freezing to minimize hemolysis (disruption of the red cell membrane accompanied by the release of hemoglobin) due to freezing and thawing of red blood cells and to deglycerolize and wash thawed cells for subsequent reinfusion.

(b) *Classification.* Class II (performance standards).

[45 FR 60639, Sept. 12, 1980]

§ 864.9160 Blood group substances of nonhuman origin for in vitro diagnostic use.

(a) *Identification.* Blood group substances of nonhuman origin for in vitro diagnostic use are materials, such as blood group specific substances prepared from nonhuman sources (e.g., pigs, cows, and horses) used to detect, identify, or neutralize antibodies to various human blood group antigens. This generic type of device does not include materials that are licensed by the Center for Biologics Evaluation and Research of the Food and Drug Administration.

(b) *Classification.* Class II (special controls). The device is exempt from the premarket notification procedures in subpart E of part 807 of this chapter subject to § 864.9.

[45 FR 60640, Sept. 12, 1980, as amended at 53 FR 11253, Apr. 6, 1988; 63 FR 59225, Nov. 3, 1998]

§ 864.9175 Automated blood grouping and antibody test system.

(a) *Identification.* An automated blood grouping and antibody test system is a device used to group erythrocytes (red blood cells) and to detect antibodies to blood group antigens.

(b) *Classification.* Class II (performance standards).

[45 FR 60641, Sept. 12, 1980]

§ 864.9185 Blood grouping view box.

(a) *Identification.* A blood grouping view box is a device with a glass or plastic viewing surface, which may be illuminated and heated, that is used to view cell reactions in antigen-antibody testing.

(b) *Classification.* Class I (general controls). The device is exempt from the premarket notification procedures in subpart E of part 807 of this chapter subject to § 864.9.

[45 FR 60641, Sept. 12, 1980, as amended at 65 FR 2310, Jan. 14, 2000]

§ 864.9195 Blood mixing devices and blood weighing devices.

(a) *Identification.* A blood mixing device is a device intended for medical purposes that is used to mix blood or blood components by agitation. A blood weighing device is a device intended for medical purposes that is used to weigh blood or blood components as they are collected.

(b) *Classification.* Class I (general controls). The manual device is exempt from the premarket notification procedures in subpart E of part 807 of this chapter subject to § 864.9.

[45 FR 60642, Sept. 12, 1980, as amended at 65 FR 2310, Jan. 14, 2000]

§ 864.9205 Blood and plasma warming device.

(a) *Nonelectromagnetic blood or plasma warming device*—(1) *Identification.* A nonelectromagnetic blood and plasma warming device is a device that warms blood or plasma, by means other than electromagnetic radiation, prior to administration.

(2) *Classification.* Class II (performance standards).

(b) *Electromagnetic blood and plasma warming device*—(1) *Identification.* An electromagnetic blood and plasma warming device is a device that employs electromagnetic radiation (radiowaves or microwaves) to warm a bag or bottle of blood or plasma prior to administration.

(2) *Classfication.* Class III (premarket approval).

(c) *Date PMA or notice of completion of a PDP is required.* No effective date has been established of the requirement for premarket approval for the device described in paragraph (b)(1). See § 864.3.

[45 FR 60642, Sept. 12, 1980, as amended at 52 FR 17733, May 11, 1987]

§ 864.9225 Cell-freezing apparatus and reagents for in vitro diagnostic use.

(a) *Identification.* Cell-freezing apparatus and reagents for in vitro diagnostic use are devices used to freeze human red blood cells for in vitro diagnostic use.

(b) *Classification.* Class I (general controls). The device is exempt from the premarket notification procedures in subpart E of part 807 of this chapter subject to § 864.9.

[45 FR 60643, Sept. 12, 1980, as amended at 65 FR 2310, Jan. 14, 2000]

§ 864.9245 Automated blood cell separator.

(a) *Identification.* An automated blood cell separator is a device that automatically removes whole blood from a donor, separates the blood into components (red blood cells, white blood cells, plasma, and platelets), retains one or more of the components, and returns the remainder of the blood to the donor. The components obtained are transfused or used to prepare blood products for administration. These devices operate on either a centrifugal separation principle or a filtration principle. The separation bowls of centrifugal blood cell separators may be reusable or disposable.

(b) *Classification of device operating by filtration separation principle.* Class II (special controls). The special controls for the device are that the manufacturer must file an annual report with FDA for 3 consecutive years. Each annual report must include the following:

(1) A summary of adverse donor reactions reported by the users to the manufacturer that do not meet the threshold for medical device reporting under part 803 of this chapter;

(2) Any change to the device, including but not limited to:

(i) New indications for use of the device;

(ii) Labeling changes, including operation manual changes;

(iii) Computer software changes, hardware changes, and disposable item changes, e.g., collection bags, tubing, filters;

(3) Equipment failures, including software, hardware, and disposable item failures, e.g., collection bags, tubing, filters.

(c) *Classification of device operating by centrifugal separation principle.* Class III (premarket approval).

(d) *Date PMA or notice of completion of a PDP is required.* No effective date has been established of the requirement for premarket approval for the device described in paragraph (c) of this section. See § 864.3.

[45 FR 60645, Sept. 12, 1980, as amended at 52 FR 17733, May 11, 1987; 68 FR 9532, Feb. 28, 2003]

§ 864.9275 Blood bank centrifuge for in vitro diagnostic use.

(a) *Identification.* A blood bank centrifuge for in vitro diagnostic use is a device used only to separate blood cells for further diagnostic testing.

(b) *Classification.* Class I (general controls). The device is exempt from the premarket notification procedures in subpart E of part 807 of this chapter subject to § 864.9.

[45 FR 60645, Sept. 12, 1980, as amended at 65 FR 2310, Jan. 14, 2000]

§ 864.9285 Automated cell-washing centrifuge for immuno-hematology.

(a) *Identification.* An automated cell-washing centrifuge for immuno-hematology is a device used to separate and prepare cells and sera for further in vitro diagnostic testing.

(b) *Classification.* Class II (performance standards).

[45 FR 60646, Sept. 12, 1980]

§ 864.9300 Automated Coombs test systems.

(a) *Identification.* An automated Coombs test system is a device used to detect and identify antibodies in patient sera or antibodies bound to red cells. The Coombs test is used for the diagnosis of hemolytic disease of the newborn, and autoimmune hemolytic anemia. The test is also used in crossmatching and in investigating transfusion reactions and drug-induced red cell sensitization.

(b) *Classification.* Class II (performance standards).

[45 FR 60646, Sept. 12, 1980]

§ 864.9320 Copper sulfate solution for specific gravity determinations.

(a) *Identification.* A copper sulfate solution for specific gravity determinations is a device used to determine whether the hemoglobin content of a potential donor's blood meets the required level (12.5 grams per 100 milliliters of blood for women and 13.5 grams per 100 milliliters of blood for men).

(b) *Classification.* Class I (general controls). The device is exempt from the premarket notification procedures in subpart E of part 807 of this chapter subject to § 864.9.

[45 FR 60647, Sept. 12, 1980, as amended at 65 FR 2310, Jan. 14, 2000]

§ 864.9400 Stabilized enzyme solution.

(a) *Identification.* A stabilized enzyme solution is a reagent intended for medical purposes that is used to enhance the reactivity of red blood cells with certain antibodies, including antibodies that are not detectable by other techniques. These enzyme solutions include papain, bromelin, ficin, and trypsin.

(b) *Classification.* Class II (performance standards).

[45 FR 60647, Sept. 12, 1980]

§ 864.9550 Lectins and protectins.

(a) *Identification.* Lectins and protectins are proteins derived from plants and lower animals that cause cell agglutination in the presence of certain antigens. These substances are used to detect blood group antigens for in vitro diagnostic purposes.

(b) *Classification.* Class II (special controls). The device is exempt from the premarket notification procedures in subpart E of part 807 of this chapter subject to § 864.9.

[45 FR 60648, Sept. 12, 1980, as amended at 63 FR 59226, Nov. 3, 1998]

§ 864.9575 Environmental chamber for storage of platelet concentrate.

(a) *Identification.* An environmental chamber for storage of platelet concentrate is a device used to hold platelet-rich plasma within a preselected temperature range.

(b) *Classification.* Class II (special controls). The device is exempt from the premarket notification procedures in subpart E of part 807 of this chapter subject to § 864.9.

[45 FR 60648, Sept. 12, 1980, as amended at 63 FR 59226, Nov. 3, 1998]

§ 864.9600 Potentiating media for in vitro diagnostic use.

(a) *Identification.* Potentiating media for in vitro diagnostic use are media, such as bovine albumin, that are used to suspend red cells and to enhance cell reactions for antigen-antibody testing.

(b) *Classification.* Class II (special controls). The device is exempt from the premarket notification procedures in subpart E of part 807 of this chapter subject to § 864.9.

[45 FR 60649, Sept. 12, 1980, as amended at 63 FR 59226, Nov. 3, 1998]

§ 864.9650 Quality control kit for blood banking reagents.

(a) *Identification.* A quality control kit for blood banking reagents is a device that consists of sera, cells, buffers, and antibodies used to determine the specificity, potency, and reactivity of the cells and reagents used for blood banking.

(b) *Classification.* Class II (performance standards).

[45 FR 60649, Sept. 12, 1980]

§ 864.9700 Blood storage refrigerator and blood storage freezer.

(a) *Identification.* A blood storage refrigerator and a blood storage freezer are devices intended for medical purposes that are used to preserve blood and blood products by storing them at cold or freezing temperatures.

(b) *Classification.* Class II (special controls). The device is exempt from the premarket notification procedures in subpart E of part 807 of this chapter subject to § 864.9.

[45 FR 60650, Sept. 12, 1980, as amended at 63 FR 59226, Nov. 3, 1998]

§ 864.9750 Heat-sealing device.

(a) *Identification.* A heat-sealing device is a device intended for medical purposes that uses heat to seal plastic bags containing blood or blood components.

(b) *Classification.* Class I (general controls). The device is exempt from the premarket notification procedures in subpart E of part 807 of this chapter subject to § 864.9.

[45 FR 60650, Sept. 12, 1980, as amended at 65 FR 2311, Jan. 14, 2000]

§ 864.9875 Transfer set.

(a) *Identification.* A transfer set is a device intended for medical purposes that consists of a piece of tubing with suitable adaptors used to transfer blood or plasma from one container to another.

(b) *Classification.* Class II (performance standards).

[45 FR 60651, Sept. 12, 1980]

PART 866—IMMUNOLOGY AND MICROBIOLOGY DEVICES

Subpart A—General Provisions

Sec.

Subpart B—Diagnostic Devices

Subpart C—Microbiology Devices

866.2170 Automated colony counter.
866.2180 Manual colony counter.
866.2300 Multipurpose culture medium.
866.2320 Differential culture medium.
866.2330 Enriched culture medium.
866.2350 Microbiological assay culture medium.
866.2360 Selective culture medium.
866.2390 Transport culture medium.
866.2410 Culture medium for pathogenic *Neisseria* spp.
866.2420 Oxidase screening test for gonorrhea.
866.2440 Automated medium dispensing and stacking device.
866.2450 Supplement for culture media.
866.2480 Quality control kit for culture media.
866.2500 Microtiter diluting and dispensing device.
866.2540 Microbiological incubator.
866.2560 Microbial growth monitor.
866.2580 Gas-generating device.
866.2600 Wood's fluorescent lamp.
866.2660 Microorganism differentiation and identification device.
866.2850 Automated zone reader.
866.2900 Microbiological specimen collection and transport device.

Subpart D—Serological Reagents

866.3010 *Acinetobacter calcoaceticus* serological reagents.
866.3020 Adenovirus serological reagents.
866.3035 *Arizona* spp. serological reagents.
866.3040 *Aspergillus* spp. serological reagents.
866.3050 Beta-glucan serological assays.
866.3060 *Blastomyces dermatitidis* serological reagents.
866.3065 *Bordetella* spp. serological reagents.
866.3085 *Brucella* spp. serological reagents.
866.3110 *Campylobacter fetus* serological reagents.
866.3120 Chlamydia serological reagents.
866.3125 *Citrobacter* spp. serological reagents.
866.3135 *Coccidioides immitis* serological reagents.
866.3140 *Corynebacterium* spp. serological reagents.
866.3145 Coxsackievirus serological reagents.
866.3165 *Cryptococcus neoformans* serological reagents.
866.3175 Cytomegalovirus serological reagents.
866.3200 *Echinococcus* spp. serological reagents.
866.3205 Echovirus serological reagents.
866.3210 Endotoxin assay.
866.3220 *Entamoeba histolytica* serological reagents.
866.3235 Epstein-Barr virus serological reagents.
866.3240 Equine encephalomyelitis virus serological reagents.
866.3250 *Erysipelothrix rhusiopathiae* serological reagents.
866.3255 *Escherichia coli* serological reagents.
866.3270 *Flavobacterium* spp. serological reagents.
866.3280 *Francisella tularensis* serological reagents.
866.3290 Gonococcal antibody test (GAT).
866.3300 *Haemophilus* spp. serological reagents.
866.3305 Herpes simplex virus serological reagents.
866.3310 Hepatitis A virus (HAV) serological assays.
866.3320 *Histoplasma capsulatum* serological reagents.
866.3330 Influenza virus serological reagents.
866.3332 Reagents for detection of specific novel influenza A viruses.
866.3340 *Klebsiella* spp. serological reagents.
866.3350 *Leptospira* spp. serological reagents.
866.3355 *Listeria* spp. serological reagents.
866.3360 Lymphocytic choriomeningitis virus serological reagents.
866.3370 *Mycobacterium tuberculosis* immunofluorescent reagents.
866.3375 *Mycoplasma* spp. serological reagents.
866.3380 Mumps virus serological reagents.
866.3390 *Neisseria* spp. direct serological test reagents.
866.3400 Parainfluenza virus serological reagents.
866.3405 Poliovirus serological reagents.
866.3410 *Proteus* spp. (Weil-Felix) serological reagents.
866.3415 *Pseudomonas* spp. serological reagents.
866.3460 Rabiesvirus immunofluorescent reagents.
866.3470 Reovirus serological reagents.
866.3480 Respiratory syncytial virus serological reagents.
866.3490 Rhinovirus serological reagents.
866.3500 Rickettsia serological reagents.
866.3510 Rubella virus serological reagents.
866.3520 Rubeola (measles) virus serological reagents.
866.3550 *Salmonella* spp. serological reagents.
866.3600 *Schistosoma* spp. serological reagents.
866.3630 *Serratia* spp. serological reagents.
866.3660 *Shigella* spp. serological reagents.
866.3680 *Sporothrix schenckii* serological reagents.
866.3700 *Staphylococcus aureus* serological reagents.
866.3720 *Streptococcus* spp. exoenzyme reagents.
866.3740 *Streptococcus* spp. serological reagents.
866.3780 *Toxoplasma gondii* serological reagents.
866.3820 *Treponema pallidum* nontreponemal test reagents.

866.3830 *Treponema pallidum* treponemal test reagents.
866.3850 *Trichinella spiralis* serological reagents.
866.3870 *Trypanosoma* spp. serological reagents.
866.3900 Varicella-zoster virus serological reagents.
866.3930 *Vibrio cholerae* serological reagents.
866.3940 West Nile virus serological reagents.

Subpart E—Immunology Laboratory Equipment and Reagents

866.4070 RNA Preanalytical Systems.
866.4100 Complement reagent.
866.4500 Immunoelectrophoresis equipment.
866.4520 Immunofluorometer equipment.
866.4540 Immunonephelometer equipment.
866.4600 Ouchterlony agar plate.
866.4700 Automated fluorescence *in situ* hybridization (FISH) enumeration systems.
866.4800 Radial immunodiffusion plate.
866.4830 Rocket immunoelectrophoresis equipment.
866.4900 Support gel.

Subpart F—Immunological Test Systems

866.5040 Albumin immunological test system.
866.5060 Prealbumin immunological test system.
866.5065 Human allotypic marker immunological test system.
866.5080 *Alpha*-1-antichymotrypsin immunological test system.
866.5090 Antimitochondrial antibody immunological test system.
866.5100 Antinuclear antibody immunological test system.
866.5110 Antiparietal antibody immunological test system.
866.5120 Antismooth muscle antibody immunological test system.
866.5130 *Alpha*-1-antitrypsin immunological test system.
866.5150 Bence-Jones proteins immunological test system.
866.5160 *Beta-globulin* immunological test system.
866.5170 Breast milk immunological test system.
866.5200 Carbonic anhydrase B and C immunological test system.
866.5210 Ceruloplasmin immunological test system.
866.5220 Cohn fraction II immunological test system.
866.5230 Colostrum immunological test system.
866.5240 Complement components immunological test system.
866.5250 Complement C_1 inhibitor (inactivator) immunological test system.
866.5260 Complement C_{3b} inactivator immunological test system.
866.5270 C-reactive protein immunological test system.
866.5320 Properidin factor B immunological test system.
866.5330 Factor XIII, A, S, immunological test system.
866.5340 Ferritin immunological test system.
866.5350 Fibrinopeptide A immunological test system.
866.5360 Cohn fraction IV immunological test system.
866.5370 Cohn fraction V immunological test system.
866.5380 Free secretory component immunological test system.
866.5400 *Alpha*-globulin immunological test system.
866.5420 *Alpha*-1-glycoproteins immunological test system.
866.5425 *Alpha*-2-glycoproteins immunological test system.
866.5430 *Beta*-2-glycoprotein I immunological test system.
866.5440 *Beta*-2-glycoprotein III immunological test system.
866.5460 Haptoglobin immunological test system.
866.5470 Hemoglobin immunological test system.
866.5490 Hemopexin immunological test system.
866.5500 Hypersensitivity pneumonitis immunological test system.
866.5510 Immunoglobulins A, G, M, D, and E immunological test system.
866.5520 Immunoglobulin G (Fab fragment specific) immunological test system.
866.5530 Immunoglobulin G (Fc fragment specific) immunological test system.
866.5540 Immunoglobulin G (Fd fragment specific) immunological test system.
866.5550 Immunoglobulin (light chain specific) immunological test system.
866.5560 Lactic dehydrogenase immunological test system.
866.5570 Lactoferrin immunological test system.
866.5580 *Alpha*-1-lipoprotein immunological test system.
866.5590 Lipoprotein X immunological test system.
866.5600 Low-density lipoprotein immunological test system.
866.5620 *Alpha*-2-macroglobulin immunological test system.
866.5630 *Beta*-2-microglobulin immunological test system.
866.5640 Infectious mononucleosis immunological test system.
866.5660 Multiple autoantibodies immunological test system.
866.5680 Myoglobin immunological test system.

Subpart G—Tumor Associated Antigen Immunological Test Systems

AUTHORITY: 21 U.S.C. 351, 360, 360c, 360e, 360j, 371.

SOURCE: 47 FR 50823, Nov. 9, 1982, unless otherwise noted.

Subpart A—General Provisions

§ 866.1 Scope.

(a) This part sets forth the classification of immunology and microbiology devices intended for human use that are in commercial distribution.

(b) The indentification of a device in a regulation in this part is not a precise description of every device that is, or will be, subject to the regulation. A manufacturer who submits a premarket notification submission for a device under part 807 may not show merely that the device is accurately described by the section title and identification provisions of a regulation in this part, but shall state why the device is substantially equivalent to other devices, as required by § 807.87.

(c) To avoid duplicative listings, an immunology and microbiology device that has two or more types of uses (e.g., used both as a diagnostic device and as a microbiology device) is listed only in one subpart.

(d) References in this part to regulatory sections of the Code of Federal Regulations are to chapter I of title 21, unless otherwise noted.

(e) Guidance documents referenced in this part are available on the Internet at *http://www.fda.gov/cdrh.guidance.html*.

[52 FR 17733, May 11, 1987, as amended at 68 FR 5827, Feb. 5, 2003]

§ 866.3 Effective dates of requirement for premarket approval.

A device included in this part that is classified into class III (Premarket approval) shall not be commercially distributed after the date shown in the regulation classifying the device unless the manufacturer has an approval under section 515 of the act (unless an exemption has been granted under section 520(g)(2) of the act). An approval under section 515 of the act consists of FDA's issuance of an order approving an application for premarket approval (PMA) for the device or declaring completed a product development protocol (PDP) for the device.

(a) Before FDA requires that a device commercially distributed before the enactment date of the amendments, or a device that has been found substantially equivalent to such a device, has an approval under section 515 of the act FDA must promulgate a regulation under section 515(b) of the act requiring such approval, except as provided in paragraphs (b) and (c) of this section. Such a regulation under section 515(b) of the act shall not be effective during the grace period ending on the 90th day after its promulgation or on the last day of the 30th full calendar month after the regulation that classifies the device into class III is effective, whichever is later. See section 501(f)(2)(B) of the act. Accordingly, unless an effective date of the requirement for premarket approval is shown in the regulation for a device classified into class III in this part, the device may be commercially distributed without FDA's issuance of an order approving a PMA or declaring completed a PDP for the device. If FDA promulgates a regulation under section 515(b) of the act requiring premarket approval for a device, section 501(f)(1)(A) of the act applies to the device.

(b) Any new, not substantially equivalent, device introduced into commercial distribution on or after May 28, 1976, including a device formerly marketed that has been substantially altered, is classified by statute (section 513(f) of the act) into class III without any grace period and FDA must have issued an order approving a PMA or declaring completed a PDP for the device before the device is commercially distributed unless it is reclassified. If FDA knows that a device being commercially distributed may be a "new" device as defined in this section because of any new intended use or other reasons, FDA may codify the statutory classification of the device into class III for such new use. Accordingly, the regulation for such a class III device states that as of the enactment date of the amendments, May 28, 1976, the device must have an approval under section 515 of the act before commercial distribution.

(c) A device identified in a regulation in this part that is classified into class III and that is subject to the transitional provisions of section 520(l) of the act is automatically classified by statute into class III and must have an approval under section 515 of the act before being commercially distributed. Accordingly, the regulation for such a class III transitional device states that as of the enactment date of the amendments, May 28, 1976, the device must have an approval under section 515 of the act before commercial distribution.

[52 FR 17733, May 11, 1987; 52 FR 22577, June 12, 1987]

§ 866.9 Limitations of exemptions from section 510(k) of the Federal Food, Drug, and Cosmetic Act (the act).

The exemption from the requirement of premarket notification (section 510(k) of the act) for a generic type of class I or II device is only to the extent that the device has existing or reasonably foreseeable characteristics of commercially distributed devices within that generic type or, in the case of in vitro diagnostic devices, only to the extent that misdiagnosis as a result of using the device would not be associated with high morbidity or mortality. Accordingly, manufacturers of any commercially distributed class I or II device for which FDA has granted an exemption from the requirement of premarket notification must still submit a premarket notification to FDA before introducing or delivering for introduction into interstate commerce for commercial distribution the device when:

(a) The device is intended for a use different from the intended use of a legally marketed device in that generic type of device; e.g., the device is intended for a different medical purpose, or the device is intended for lay use where the former intended use was by health care professionals only;

(b) The modified device operates using a different fundamental scientific technology than a legally marketed device in that generic type of device; e.g., a surgical instrument cuts tissue with a laser beam rather than with a sharpened metal blade, or an in vitro diagnostic device detects or identifies infectious agents by using deoxyribonucleic acid (DNA) probe or nucleic acid hybridization technology rather than culture or immunoassay technology; or

(c) The device is an in vitro device that is intended:

(1) For use in the diagnosis, monitoring, or screening of neoplastic diseases with the exception of immunohistochemical devices;

(2) For use in screening or diagnosis of familial or acquired genetic disorders, including inborn errors of metabolism;

(3) For measuring an analyte that serves as a surrogate marker for screening, diagnosis, or monitoring life-threatening diseases such as acquired immune deficiency syndrome (AIDS), chronic or active hepatitis, tuberculosis, or myocardial infarction or to monitor therapy;

(4) For assessing the risk of cardiovascular diseases;

(5) For use in diabetes management;

(6) For identifying or inferring the identity of a microorganism directly from clinical material;

(7) For detection of antibodies to microorganisms other than immunoglobulin G (IgG) or IgG assays when the results are not qualitative, or are used to determine immunity, or the assay is intended for use in matrices other than serum or plasma;

(8) For noninvasive testing as defined in § 812.3(k) of this chapter; and

(9) For near patient testing (point of care).

[65 FR 2311, Jan. 14, 2000]

Subpart B—Diagnostic Devices

§ 866.1620 Antimicrobial susceptibility test disc.

(a) *Identification.* An antimicrobial susceptibility test disc is a device that consists of antimicrobic-impregnated paper discs used to measure by a disc-agar diffusion technique or a disc-broth elution technique the in vitro susceptibility of most clinically important bacterial pathogens to antimicrobial agents. In the disc-agar diffusion technique, bacterial susceptibility is ascertained by directly measuring the magnitude of a zone of bacterial inhibition around the disc on an agar surface. The disc-broth elution technique is associated with an automated rapid susceptibility test system and employs a fluid medium in which susceptibility is ascertained by photometrically measuring changes in bacterial growth resulting when antimicrobial material is eluted from the disc into the fluid medium. Test results are used to determine the antimicrobial agent of choice in the treatment of bacterial diseases.

(b) *Classification.* Class II (performance standards).

§ 866.1640 Antimicrobial susceptibility test powder.

(a) *Identification.* An antimicrobial susceptibility test powder is a device that consists of an antimicrobial drug powder packaged in vials in specified amounts and intended for use in clinical laboratories for determining in vitro susceptibility of bacterial pathogens to these therapeutic agents. Test results are used to determine the antimicrobial agent of choice in the treatment of bacterial diseases.

(b) *Classification.* Class II (performance standards).

§ 866.1645 Fully automated short-term incubation cycle antimicrobial susceptibility system.

(a) *Identification.* A fully automated short-term incubation cycle antimicrobial susceptibility system is a device that incorporates concentrations of antimicrobial agents into a system for the purpose of determining in vitro susceptibility of bacterial pathogens isolated from clinical specimens. Test results obtained from short-term (less than 16 hours) incubation are used to determine the antimicrobial agent of choice to treat bacterial diseases.

(b) *Classification.* Class II (special controls). The special control for this device is FDA's guidance document entitled "Class II Special Controls Guidance Document: Antimicrobial Susceptibility Test (AST) Systems; Guidance for Industry and FDA."

[68 FR 5827, Feb. 5, 2003]

§ 866.1700 Culture medium for antimicrobial susceptibility tests.

(a) *Identification.* A culture medium for antimicrobial susceptibility tests is a device intended for medical purposes that consists of any medium capable of supporting the growth of many of the bacterial pathogens that are subject to antimicrobial susceptibility tests. The medium should be free of components known to be antagonistic to the common agents for which susceptibility tests are performed in the treatment of disease.

(b) *Classification.* Class II (performance standards).

Subpart C—Microbiology Devices

§ 866.2050 Staphylococcal typing bacteriophage.

(a) *Identification.* A staphylococcal typing bacteriophage is a device consisting of a bacterial virus intended for medical purposes to identify pathogenic staphylococcal bacteria through use of the bacteria's susceptibility to destruction by the virus. Test results are used principally for the collection of epidemiological information.

208-072 D-9

(b) *Classification.* Class I (general controls). The device is exempt from the premarket notification procedures in subpart E of part 807 of this chapter subject to the limitations in § 866.9.

[47 FR 50823, Nov. 9, 1982, as amended at 54 FR 25045, June 12, 1989; 66 FR 38790, July 25, 2001]

§ 866.2120 Anaerobic chamber.

(a) *Identification.* An anaerobic chamber is a device intended for medical purposes to maintain an anaerobic (oxygen free) environment. It is used to isolate and cultivate anaerobic microorganisms.

(b) *Classification.* Class I (general controls). The device is exempt from the premarket notification procedures in subpart E of part 807 of this chapter subject to the limitations in § 866.9. The device is also exempt from the good manufacturing practice regulations in part 820 of this chapter, with the exception of § 820.180, with respect to general requirements concerning records, and § 820.198, with respect to complaint files.

[47 FR 50823, Nov. 9, 1982, as amended at 66 FR 38790, July 25, 2001]

§ 866.2160 Coagulase plasma.

(a) *Identification.* Coagulase plasma is a device that consists of freeze-dried animal or human plasma that is intended for medical purposes to perform coagulase tests primarily on staphylococcal bacteria. When reconstituted, the fluid plasma is clotted by the action of the enzyme coagulase which is produced by pathogenic staphylococci. Test results are used primarily as an aid in the diagnosis of disease caused by pathogenic bacteria belonging to the genus *Staphylococcus* and provide epidemiological information on disease caused by these microorganisms.

(b) *Classification.* Class I (general controls). The device is exempt from the premarket notification procedures in subpart E of part 807 of this chapter subject to the limitations in § 866.9.

[47 FR 50823, Nov. 9, 1982, as amended at 61 FR 1119, Jan. 16, 1996; 66 FR 38790, July 25, 2001]

§ 866.2170 Automated colony counter.

(a) *Identification.* An automated colony counter is a mechanical device intended for medical purposes to determine the number of bacterial colonies present on a bacteriological culture medium contained in a petri plate. The number of colonies counted is used in the diagnosis of disease as a measure of the degree of bacterial infection.

(b) *Classification.* Class I (general controls). The device is exempt from the premarket notification procedures in subpart E of part 807 of this chapter subject to the limitations in § 866.9.

[47 FR 50823, Nov. 9, 1982, as amended at 54 FR 25045, June 12, 1989; 66 FR 38790, July 25, 2001]

§ 866.2180 Manual colony counter.

(a) *Identification.* A manual colony counter is a device intended for medical purposes that consists of a printed grid system superimposed on an illuminated screen. Petri plates containing bacterial colonies to be counted are placed on the screen for better viewing and ease of counting. The number of colonies counted is used in the diagnosis of disease as a measure of the degree of bacterial infection.

(b) *Classification.* Class I (general controls). The device is exempt from the premarket notification procedures in subpart E of part 807 of this chapter subject to the limitations in § 866.9. The device is also exempt from the good manufacturing practice regulations in part 820 of this chapter, with the exception of § 820.180, with respect to general requirements concerning records, and § 820.198, with respect to complaint files.

[47 FR 50823, Nov. 9, 1982, as amended at 66 FR 38790, July 25, 2001]

§ 866.2300 Multipurpose culture medium.

(a) *Identification.* A multipurpose culture medium is a device that consists primarily of liquid or solid biological materials intended for medical purposes for the cultivation and identification of several types of pathogenic microorganisms without the need of additional nutritional supplements. Test results aid in the diagnosis of disease and also provide epidemiological information on diseases caused by these microorganisms.

(b) *Classification.* Class I (general controls). The device is exempt from the premarket notification procedures in subpart E of part 807 of this chapter subject to the limitations in § 866.9.

[47 FR 50823, Nov. 9, 1982, as amended at 54 FR 25046, June 12, 1989; 66 FR 38790, July 25, 2001]

§ 866.2320 Differential culture medium.

(a) *Identification.* A differential culture medium is a device that consists primarily of liquid biological materials intended for medical purposes to cultivate and identify different types of pathogenic microorganisms. The identification of these microorganisms is accomplished by the addition of a specific biochemical component(s) to the medium. Microorganisms are identified by a visible change (e.g., a color change) in a specific biochemical component(s) which indicates that specific metabolic reactions have occurred. Test results aid in the diagnosis of disease and also provide epidemiological information on diseases caused by these microorganisms.

(b) *Classification.* Class I (general controls). The device is exempt from the premarket notification procedures in subpart E of part 807 of this chapter subject to the limitations in § 866.9.

[47 FR 50823, Nov. 9, 1982, as amended at 54 FR 25046, June 12, 1989; 66 FR 38790, July 25, 2001]

§ 866.2330 Enriched culture medium.

(a) *Identification.* An enriched culture medium is a device that consists primarily of liquid or solid biological materials intended for medical purposes to cultivate and identify fastidious microorganisms (those having complex nutritional requirements). The device consists of a relatively simple basal medium enriched by the addition of such nutritional components as blood, blood serum, vitamins, and extracts of plant or animal tissues. The device is used in the diagnosis of disease caused by pathogenic microorganisms and also provides epidemiological information on these diseases.

(b) *Classification.* Class I (general controls). The device is exempt from the premarket notification procedures in subpart E of part 807 of this chapter subject to the limitations in § 866.9.

[47 FR 50823, Nov. 9, 1982, as amended at 54 FR 25046, June 12, 1989; 66 FR 38791, July 25, 2001]

§ 866.2350 Microbiological assay culture medium.

(a) *Identification.* A microbiological assay culture medium is a device that consists primarily of liquid or solid biological materials intended for medical purposes to cultivate selected test microorganisms in order to measure by microbiological procedures the concentration in a patient's serum of certain substances, such as amino acids, antimicrobial agents, and vitamins. The concentration of these substances is measured by their ability to promote or inhibit the growth of the test organism in the innoculated medium. Test results aid in the diagnosis of disease resulting from either deficient or excessive amounts of these substances in a patient's serum. Tests results may also be used to monitor the effects of the administration of certain antimicrobial drugs.

(b) *Classification.* Class I (general controls). The device is exempt from the premarket notification procedures in subpart E of part 807 of this chapter subject to the limitations in § 866.9.

[47 FR 50823, Nov. 9, 1982, as amended at 54 FR 25046, June 12, 1989; 66 FR 38791, July 25, 2001]

§ 866.2360 Selective culture medium.

(a) *Identification.* A selective culture medium is a device that consists primarily of liquid or solid biological materials intended for medical purposes to cultivate and identify certain pathogenic microorganisms. The device contains one or more components that suppress the growth of certain microorganisms while either promoting or not affecting the growth of other microorganisms. The device aids in the diagnosis of disease caused by pathogenic microorganisms and also provides epidemiological information on these diseases.

(b) *Classification.* Class I (general controls). The device is exempt from the premarket notification procedures in

subpart E of part 807 of this chapter subject to the limitations in § 866.9.

[47 FR 50823, Nov. 9, 1982, as amended at 54 FR 25046, June 12, 1989; 66 FR 38791, July 25, 2001]

§ 866.2390 Transport culture medium.

(a) *Identification.* A transport culture medium is a device that consists of a semisolid, usually non-nutrient, medium that maintains the viability of suspected pathogens contained in patient specimens while in transit from the specimen collection area to the laboratory. The device aids in the diagnosis of disease caused by pathogenic microorganisms and also provides epidemiological information on these diseases.

(b) *Classification.* Class I (general controls).

§ 866.2410 Culture medium for pathogenic *Neisseria* spp.

(a) *Identification.* A culture medium for pathogenic *Neisseria* spp. is a device that consists primarily of liquid or solid biological materials used to cultivate and identify pathogenic *Neisseria* spp. The identification aids in the diagnosis of disease caused by bacteria belonging to the genus *Neisseria*, such as epidemic cerebrospinal meningitis, other meningococcal disease, and gonorrhea, and also provides epidemiological information on these microorganisms.

(b) *Classification.* Class II (performance standards).

§ 866.2420 Oxidase screening test for gonorrhea.

(a) *Identification.* An oxidase screening test for gonorrhea is an in vitro device that consists of the articles intended to identify by chemical reaction, cytochrome oxidase, an oxidizing enzyme that is associated with certain bacteria including *Neisseria gonorrhoeae*. A sample of a male's urethral discharge is obtained on a swab which is placed into a wetting agent containing an ingredient that will react with cytochrome oxidase. When cytochrome oxidase is present, the swab turns a dark purple color within 3 minutes. Because it is unlikely that cytochrome oxidase-positive organisms other than *Neisseria gonorrhoeae* are present in the urethral discharge of males, the identification of cytochrome oxidase with this device indicates presumptive infection of the patient with the causative agent of gonorrhea.

(b) *Classification.* Class III (premarket approval) (transitional device).

(c) *Date PMA or notice of completion of a PDP is required.* As of May 28, 1976, an approval under section 515 of the act is required before this device may be commercially distributed. See § 866.3.

[47 FR 50823, Nov. 9, 1982, as amended at 52 FR 17734, May 11, 1987]

§ 866.2440 Automated medium dispensing and stacking device.

(a) *Identification.* An automated medium dispensing and stacking device is a device intended for medical purposes to dispense a microbiological culture medium into petri dishes and then mechanically stack the petri dishes.

(b) *Classification.* Class I (general controls). The device is exempt from the premarket notification procedures in subpart E of part 807 of this chapter subject to the limitations in § 866.9. The device is also exempt from the good manufacturing practice regulations in part 820 of this chapter, with the exception of § 820.180, with respect to general requirements concerning records, and § 820.198, with respect to complaint files.

[47 FR 50823, Nov. 9, 1982, as amended at 66 FR 38791, July 25, 2001]

§ 866.2450 Supplement for culture media.

(a) *Identification.* A supplement for culture media is a device, such as a vitamin or sugar mixture, that is added to a solid or liquid basal culture medium to produce a desired formulation and that is intended for medical purposes to enhance the growth of fastidious microorganisms (those having complex nutritional requirements). This device aids in the diagnosis of diseases caused by pathogenic microorganisms.

(b) *Classification.* Class I (general controls). The device is exempt from the premarket notification procedures in

subpart E of part 807 of this chapter subject to the limitations in § 866.9.

[47 FR 50823, Nov. 9, 1982, as amended at 54 FR 25046, June 12, 1989; 66 FR 38791, July 25, 2001]

§ 866.2480 Quality control kit for culture media.

(a) *Identification.* A quality control kit for culture media is a device that consists of paper discs (or other suitable materials), each impregnated with a specified, freeze-dried, viable microorganism, intended for medical purposes to determine if a given culture medium is able to support the growth of that microorganism. The device aids in the diagnosis of disease caused by pathogenic microorganisms and also provides epidemiological information on these diseases.

(b) *Classification.* Class I (general controls). The device is exempt from the premarket notification procedures in subpart E of part 807 of this chapter subject to the limitations in § 866.9.

[47 FR 50823, Nov. 9, 1982, as amended at 54 FR 25046, June 12, 1989; 66 FR 38791, July 25, 2001]

§ 866.2500 Microtiter diluting and dispensing device.

(a) *Identification.* A microtiter diluting and dispensing device is a mechanical device intended for medical purposes to dispense or serially dilute very small quantities of biological or chemical reagents for use in various diagnostic procedures.

(b) *Classification.* Class I (general controls). The device is exempt from the premarket notification procedures in subpart E of part 807 of this chapter subject to the limitations in § 866.9.

[47 FR 50823, Nov. 9, 1982, as amended at 54 FR 25046, June 12, 1989; 66 FR 38791, July 25, 2001]

§ 866.2540 Microbiological incubator.

(a) *Identification.* A microbiological incubator is a device with various chambers or water-filled compartments in which controlled environmental conditions, particularly temperature, are maintained. It is intended for medical purposes to cultivate microorganisms and aid in the diagnosis of disease.

(b) *Classification.* Class I (general controls). The device is exempt from the premarket notification procedures in subpart E of part 807 of this chapter subject to the limitations in § 866.9. The device is also exempt from the good manufacturing practice regulations in part 820 of this chapter, with the exception of § 820.180, with respect to general requirements concerning records, and § 820.198, with respect to complaint files.

[47 FR 50823, Nov. 9, 1982, as amended at 66 FR 38791, July 25, 2001]

§ 866.2560 Microbial growth monitor.

(a) *Identification.* A microbial growth monitor is a device intended for medical purposes that measures the concentration of bacteria suspended in a liquid medium by measuring changes in light scattering properties, optical density, electrical impedance, or by making direct bacterial counts. The device aids in the diagnosis of disease caused by pathogenic microorganisms.

(b) *Classification.* Class I. With the exception of automated blood culturing system devices that are used in testing for bacteria, fungi, and other microorganisms in blood and other normally sterile body fluids, this device is exempt from the premarket notification procedures in subpart E of part 807 of this chapter.

[47 FR 50823, Nov. 9, 1982, as amended at 60 FR 38482, July 27, 1995]

§ 866.2580 Gas-generating device.

(a) *Identification.* A gas-generating device is a device intended for medical purposes that produces predetermined amounts of selected gases to be used in a closed chamber in order to establish suitable atmospheric conditions for cultivation of microorganisms with special atmospheric requirements. The device aids in the diagnosis of disease.

(b) *Classification.* Class I (general controls). The device is exempt from the premarket notification procedures in subpart E of part 807 of this chapter subject to the limitations in § 866.9.

[47 FR 50823, Nov. 9, 1982, as amended at 54 FR 25046, June 12, 1989; 66 FR 38791, July 25, 2001]

§ 866.2600 Wood's fluorescent lamp.

(a) *Identification.* A Wood's fluorescent lamp is a device intended for medical purposes to detect fluorescent materials (e.g., fluorescein pigment produced by certain microorganisms) as an aid in the identification of these microorganisms. The device aids in the diagnosis of disease.

(b) *Classification.* Class I (general controls). The device is exempt from the premarket notification procedures in subpart E of part 807 of this chapter subject to the limitations in § 866.9. The device is also exempt from the good manufacturing practice regulations in part 820 of this chapter, with the exception of § 820.180, with respect to general requirements concerning records, and § 820.198, with respect to complaint files.

[47 FR 50823, Nov. 9, 1982, as amended at 66 FR 38791, July 25, 2001]

§ 866.2660 Microorganism differentiation and identification device.

(a) *Identification.* A microorganism differentiation and identification device is a device intended for medical purposes that consists of one or more components, such as differential culture media, biochemical reagents, and paper discs or paper strips impregnated with test reagents, that are usually contained in individual compartments and used to differentiate and identify selected microorganisms. The device aids in the diagnosis of disease.

(b) *Classification.* Class I (general controls). The device is exempt from the premarket notification procedures in subpart E of part 807 of this chapter subject to § 866.9.

[47 FR 50823, Nov. 9, 1982, as amended at 65 FR 2311, Jan. 14, 2000]

§ 866.2850 Automated zone reader.

(a) *Identification.* An automated zone reader is a mechanical device intended for medical purposes to measure zone diameters of microbial growth inhibition (or exhibition), such as those observed on the surface of certain culture media used in disc-agar diffusion antimicrobial susceptibility tests. The device aids in decisionmaking respecting the treatment of disease.

(b) *Classification.* Class I (general controls).

§ 866.2900 Microbiological specimen collection and transport device.

(a) *Identification.* A microbiological specimen collection and transport device is a specimen collecting chamber intended for medical purposes to preserve the viability or integrity of microorganisms in specimens during storage of specimens after their collection and during their transport from the collecting area to the laboratory. The device may be labeled or otherwise represented as sterile. The device aids in the diagnosis of disease caused by pathogenic microorganisms.

(b) *Classification.* Class I (general controls).

Subpart D—Serological Reagents

§ 866.3010 *Acinetobacter calcoaceticus* serological reagents.

(a) *Identification. Acinetobacter calcoaceticus* serological reagents are devices that consist of *Acinetobacter calcoaceticus* antigens and antisera used to identify this bacterium from cultured isolates derived from clinical specimens. The identification aids in the diagnosis of disease caused by the bacterium *Acinetobacter calcoaceticus* and provides epidemiological information on disease caused by this microorganism. This organism becomes pathogenic in patients with burns or with immunologic deficiency, and infection can result in sepsis (blood poisoning).

(b) *Classification.* Class I (general controls). The device is exempt from the premarket notification procedures in subpart E of part 807 of this chapter subject to the limitations in § 866.9.

[47 FR 50823, Nov. 9, 1982, as amended at 54 FR 25046, June 12, 1989; 66 FR 38791, July 25, 2001]

§ 866.3020 Adenovirus serological reagents.

(a) *Identification.* Adenovirus serological reagents are devices that consist of antigens and antisera used in serological tests to identify antibodies to adenovirus in serum. Additionally, some of these reagents consist of adenovirus antisera conjugated with a fluorescent dye and are used to identify adenoviruses directly from clinical specimens. The identification aids in the diagnosis of disease caused by adenoviruses and provides epidemiological information on these diseases. Adenovirus infections may cause pharyngitis (inflammation of the throat), acute respiratory diseases, and certain external diseases of the eye (e.g., conjunctivitis).

(b) *Classification.* Class I (general controls). The device is exempt from the premarket notification procedures in subpart E of part 807 of this chapter subject to the limitations in § 866.9.

[47 FR 50823, Nov. 9, 1982, as amended at 54 FR 25046, June 12, 1989; 66 FR 38791, July 25, 2001]

§ 866.3035 *Arizona* spp. serological reagents.

(a) *Identification. Arizona* spp. serological reagents are devices that consist of antisera and antigens used to identify *Arizona* spp. in cultured isolates derived from clinical specimens. The identification aids in the diagnosis of disease caused by bacteria belonging to the genus *Arizona* and provides epidemiological information on diseases caused by these microorganisms. *Arizona* spp. can cause gastroenteritis (food poisoning) and sepsis (blood poisoning).

(b) *Classification.* Class I (general controls). The device is exempt from the premarket notification procedures in subpart E of part 807 of this chapter subject to the limitations in § 866.9.

[47 FR 50823, Nov. 9, 1982, as amended at 54 FR 25046, June 12, 1989; 66 FR 38791, July 25, 2001]

§ 866.3040 *Aspergillus* spp. serological reagents.

(a) *Identification. Aspergillus* spp. serological reagents are devices that consist of antigens and antisera used in various serological tests to identify antibodies to *Aspergillus* spp. in serum. The identification aids in the diagnosis of aspergillosis caused by fungi belonging to the genus *Aspergillus.* Aspergillosis is a disease marked by inflammatory granulomatous (tumorlike) lessions in the skin, ear, eyeball cavity, nasal sinuses, lungs, and occasionally the bones.

(b) *Classification.* Class I (general controls). The device is exempt from the premarket notification procedures in subpart E of part 807 of this chapter subject to § 866.9.

[47 FR 50823, Nov. 9, 1982, as amended at 65 FR 2311, Jan. 14, 2000]

§ 866.3050 Beta-glucan serological assays.

(a) *Identification.* Beta-glucan serological assays are devices that consist of antigens or proteases used in serological assays. The device is intended for use for the presumptive diagnosis of fungal infection. The assay is indicated for use in patients with symptoms of, or medical conditions predisposing the patient to invasive fungal infection. The device can be used as an aid in the diagnosis of deep seated mycoses and fungemias.

(b) *Classification.* Class II (special controls). The special control is FDA's guidance document entitled "Class II Special Controls Guidance Document: Serological Assays for the Detection of Beta-Glucan." See § 866.1(e) for the availability of this guidance document.

[69 FR 56936, Sept. 23, 2004]

§ 866.3060 *Blastomyces dermatitidis* serological reagents.

(a) *Identification. Blastomyces dermatitidis* serological reagents are devices that consist of antigens and antisera used in serological tests to identify antibodies to *Blastomyces determatitidis* in serum. The identification aids in the diagnosis of blastomycosis caused by the fungus *Blastomyces dermatitidis.* Blastomycosis is a chronic granulomatous (tumorlike) disease, which may be limited to the skin or lung or may be widely disseminated in the body resulting in lesions of the bones, liver, spleen, and kidneys.

(b) *Classification.* Class II (special controls). The device is exempt from the premarket notification procedures in subpart E of part 807 of this chapter subject to § 866.9.

[47 FR 50823, Nov. 9, 1982, as amended at 63 FR 59226, Nov. 3, 1998]

§ 866.3065 *Bordetella* spp. serological reagents.

(a) *Identification. Bordetella* spp. serological reagents are devices that consist of antigens and antisera, including antisera conjugated with a fluorescent dye, used in serological tests to identify *Bordetella* spp. from cultured isolates or directly from clinical specimens. The identification aids in the diagnosis of diseases caused by bacteria belonging to the genus *Bordetella* and provides epidemiological information on these diseases. *Bordetella* spp. cause whooping cough (*Bordetella pertussis*) and other similiarly contagious and acute respiratory infections characterized by pneumonitis (inflammation of the lungs).

(b) *Classification.* Class I (general controls). The device is exempt from the premarket notification procedures in subpart E of part 807 of this chapter subject to the limitations in § 866.9.

[47 FR 50823, Nov. 9, 1982, as amended at 54 FR 25046, June 12, 1989; 66 FR 38791, July 25, 2001]

§ 866.3085 *Brucella* spp. serological reagents.

(a) *Identification. Brucella* spp. serological reagents are devices that consist of antigens and antisera used for serological identification of *Brucella* spp. from cultured isolates derived from clinical specimens or to identify antibodies to *Brucella* spp. in serum. Additionally, some of these reagents consist of antisera conjugated with a fluorescent dye (immunofluorescent reagents) used to identify *Brucella* spp. directly from clinical specimens or cultured isolates derived from clinical specimens. The identification aids in the diagnosis of brucellosis (e.g., undulant fever, Malta fever) caused by bacteria belonging to the genus *Brucella* and provides epidemiological information on diseases caused by these microorganisms.

(b) *Classification.* Class II (special controls). The device is exempt from the premarket notification procedures in subpart E of part 807 of this chapter subject to § 866.9.

[47 FR 50823, Nov. 9, 1982, as amended at 63 FR 59226, Nov. 3, 1998]

§ 866.3110 *Campylobacter fetus* serological reagents.

(a) *Identification. Campylobacter fetus* serological reagents are devices that consist of antisera conjugated with a fluorescent dye used to identify *Campylobacter fetus* from clinical specimens or cultured isolates derived from clinical specimens. The identification aids in the diagnosis of diseases caused by this bacterium and provides epidemiological information on these diseases. *Campylobacter fetus* is a frequent cause of abortion in sheep and cattle and is sometimes responsible for endocarditis (inflammation of certain membranes of the heart) and enteritis (inflammation of the intestines) in humans.

(b) *Classification.* Class I (general controls).

§ 866.3120 Chlamydia serological reagents.

(a) *Identification.* Chlamydia serological reagents are devices that consist of antigens and antisera used in serological tests to identify antibodies to chlamydia in serum. Additionally, some of these reagents consist of chlamydia antisera conjugated with a fluorescent dye used to identify chlamydia directly from clinical specimens or cultured isolates derived from clinical specimens. The identification aids in the diagnosis of disease caused by bacteria belonging to the genus *Chlamydia* and provides epidemiological information on these diseases. Chlamydia are the causative agents of psittacosis (a form of pneumonia), lymphogranuloma venereum (a venereal disease), and trachoma (a chronic disease of the eye and eyelid).

(b) *Classification.* Class I (general controls).

§ 866.3125 *Citrobacter* spp. serological reagents.

(a) *Identification. Citrobacter* spp. serological reagents are devices that consist of antigens and antisera used in serological tests to identify *Citrobacter* spp. from cultured isolates derived from clinical specimens. The identification aids in the diagnosis of disease caused by bacteria belonging to the genus *Citrobacter* and provides epidemiological information on diseases caused by these microorganisms. *Citrobacter* spp. have occasionally been associated with urinary tract infections.

(b) *Classification.* Class I (general controls). The device is exempt from the premarket notification procedures in subpart E of part 807 of this chapter subject to the limitations in § 866.9.

[47 FR 50823, Nov. 9, 1982, as amended at 54 FR 25046, June 12, 1989; 66 FR 38791, July 25, 2001]

§ 866.3135 *Coccidioides immitis* serological reagents.

(a) *Identification. Coccidioides immitis* serological reagents are devices that consist of antigens and antisera used in serological tests to identify antibodies to *Coccidioides immitis* in serum. The identification aids in the diagnosis of coccidioidomycosis caused by a fungus belonging to the genus *Coccidioides* and provides epidemiological information on diseases caused by this microorganism. An infection with *Coccidioides immitis* produces symptoms varying in severity from those accompanying the common cold to those of influenza.

(b) *Classification.* Class II (special controls). The device is exempt from the premarket notification procedures in subpart E of part 807 of this chapter subject to § 866.9.

[47 FR 50823, Nov. 9, 1982, as amended at 63 FR 59226, Nov. 3, 1998]

§ 866.3140 *Corynebacterium* spp. serological reagents.

(a) *Identification. Corynebacterium* spp. serological reagents are devices that consist of antisera conjugated with a fluorescent dye used to identify *Corynebacterium* spp. from clinical specimens. The identification aids in the diagnosis of disease caused by bacteria belonging to the genus *Corynebacterium* and provides epidemiological information on diseases caused by these microorganisms. The principal human pathogen of this genus, *Corynebacterium diphtheriae,* causes diphtheria. However, many other types of corynebacteria form part of the normal flora of the human respiratory tract, other mucus membranes, and skin, and are either nonpathogenic or have an uncertain role.

(b) *Classification.* Class I (general controls). The device is exempt from the premarket notification procedures in subpart E of part 807 of this chapter subject to § 866.9.

[47 FR 50823, Nov. 9, 1982, as amended at 65 FR 2311, Jan. 14, 2000]

§ 866.3145 Coxsackievirus serological reagents.

(a) *Identification.* Coxsackievirus serological reagents are devices that consist of antigens and antisera used in serological tests to identify antibodies to coxsackievirus in serum. Additionally, some of these reagents consist of coxsackievirus antisera conjugated with a fluorescent dye that are used to identify coxsackievirus from clinical specimens or from tissue culture isolates derived from clinical specimens. The identification aids in the diagnosis of coxsackievirus infections and provides epidemiological information on diseases caused by these viruses. Coxsackieviruses produce a variety of infections, including common colds, meningitis (inflammation of brain and spinal cord membranes), herpangina (brief fever accompanied by ulcerated lesions of the throat), and myopericarditis (inflammation of heart tissue).

(b) *Classification.* Class I (general controls). The device is exempt from the premarket notification procedures in subpart E of part 807 of this chapter subject to § 866.9.

[47 FR 50823, Nov. 9, 1982, as amended at 65 FR 2311, Jan. 14, 2000]

§ 866.3165 *Cryptococcus neoformans* serological reagents.

(a) *Identification. Cryptococcus neoformans* serological reagents are devices that consist of antigens used in serological tests to identify antibodies to *Cryptococcus neoformans* in serum. Additionally, some of these reagents consist of antisera conjugated with a fluorescent dye (immunofluorescent reagents) and are used to identify *Cryptococcus neoformans* directly from clinical specimens or from cultured isolates derived from clinical specimens. The identification aids in the diagnosis of cryptococcosis and provides epidemiological information on this type of disease. Cryptococcosis infections are found most often as chronic meningitis (inflammation of brain membranes) and, if not treated, are usually fatal.

(b) *Classification.* Class II (special controls). The device is exempt from the premarket notification procedures in subpart E of part 807 of this chapter subject to § 866.9.

[47 FR 50823, Nov. 9, 1982, as amended at 63 FR 59226, Nov. 3, 1998]

§ 866.3175 Cytomegalovirus serological reagents.

(a) *Identification.* Cytomegalovirus serological reagents are devices that consist of antigens and antisera used in serological tests to identify antibodies to cytomegalovirus in serum. The identification aids in the diagnosis of diseases caused by cytomegaloviruses (principally cytomegalic inclusion disease) and provides epidemiological information on these diseases. Cytomegalic inclusion disease is a generalized infection of infants and is caused by intrauterine or early postnatal infection with the virus. The disease may cause severe congenital abnormalities, such as microcephaly (abnormal smallness of the head), motor disability, and mental retardation. Cytomegalovirus infection has also been associated with acquired hemolytic anemia, acute and chronic hepatitis, and an infectious mononucleosis-like syndrome.

(b) *Classification.* Class II (performance standards).

§ 866.3200 *Echinococcus* spp. serological reagents.

(a) *Identification. Echinococcus* spp. serological reagents are devices that consist of *Echinococcus* spp. antigens and antisera used in serological tests to identify antibodies to *Echinococcus* spp. in serum. The identification aids in the diagnosis of echinococcosis, caused by parasitic tapeworms belonging to the genus *Echinococcus* and provides epidemiological information on this disease. Echinococcosis is characterized by the development of cysts in the liver, lung, kidneys, and other organs formed by the larva of the infecting organisms.

(b) *Classification.* Class I (general controls). The device is exempt from the premarket notification procedures in subpart E of part 807 of this chapter subject to § 866.9.

[47 FR 50823, Nov. 9, 1982, as amended at 65 FR 2311, Jan. 14, 2000]

§ 866.3205 Echovirus serological reagents.

(a) *Identification.* Echovirus serological reagents are devices that consist of antigens and antisera used in serological tests to identify antibodies to echovirus in serum. Additionally, some of these reagents consist of echovirus antisera conjugated with a fluorescent dye used to identify echoviruses from clinical specimens or from tissue culture isolates derived from clinical specimens. The identification aids in the diagnosis of echovirus infections and provides epidemiological information on diseases caused by these viruses. Echoviruses cause illnesses such as meningitis (inflammation of the brain and spinal cord membranes), febrile illnesses (accompanied by fever) with or without rash, and the common cold.

(b) *Classification.* Class I (general controls). The device is exempt from the premarket notification procedures in subpart E of part 807 of this chapter subject to the limitations in § 866.9.

[47 FR 50823, Nov. 9, 1982, as amended at 54 FR 25046, June 12, 1989; 66 FR 38791, July 25, 2001]

§ 866.3210 Endotoxin assay.

(a) *Identification.* An endotoxin assay is a device that uses serological techniques in whole blood. The device is intended for use in conjunction with other laboratory findings and clinical assessment of the patient to aid in the risk assessment of critically ill patients for progression to severe sepsis.

(b) *Classification.* Class II (special controls). The special control for this device is the FDA guidance entitled "Class II Special Controls Guidance Document: Endotoxin Assay." See § 866.1(e) for the availability of this guidance document.

[68 FR 62008, Oct. 31, 2003. Redesignated at 70 FR 53069, Sept. 7, 2005]

§ 866.3220 *Entamoeba histolytica* serological reagents.

(a) *Identification. Entamoeba histolytica* serological reagents are devices that consist of antigens and antisera used in serological tests to identify antibodies to *Entamoeba histolytica* in serum. Additionally, some of these reagents consist of antisera conjugated with a fluorescent dye (immunofluorescent reagents) used to identify *Entamoeba histolytica* directly from clinical specimens. The identification aids in the diagnosis of amebiasis caused by the microscopic protozoan parasite *Entamoeba histolytica* and provides epidemiological information on diseases caused by this parasite. The parasite may invade the skin, liver, intestines, lungs, and diaphragm, causing disease conditions such as indolent ulcers, an amebic hepatitis, amebic dysentery, and pulmonary lesions.

(b) *Classification.* Class II (special controls). The device is exempt from the premarket notification procedures in subpart E of part 807 of this chapter subject to § 866.9.

[47 FR 50823, Nov. 9, 1982; 47 FR 56846, Dec. 21, 1982, as amended at 63 FR 59226, Nov. 3, 1998]

§ 866.3235 Epstein-Barr virus serological reagents.

(a) *Identification.* Epstein-Barr virus serological reagents are devices that consist of antigens and antisera used in serological tests to identify antibodies to Epstein-Barr virus in serum. The identification aids in the diagnosis of Epstein-Barr virus infections and provides epidemiological information on diseases caused by these viruses. Epstein-Barr viruses are thought to cause infectious mononucleosis and have been associated with Burkitt's lymphoma (a tumor of the jaw in African children and young adults) and postnasal carcinoma (cancer).

(b) *Classification.* Class I (general controls).

§ 866.3240 Equine encephalomyelitis virus serological reagents.

(a) *Identification.* Equine encephalomyelitis virus serological reagents are devices that consist of antigens and antisera used in serological tests to identify antobodies to equine encephalomyelitis virus in serum. The identification aids in the diagnosis of diseases caused by equine encephalomyelitis viruses and provides epidemiological information on these viruses. Equine encephalomyelitis viruses are transmitted to humans by the bite of insects, such as mosquitos and ticks, and may cause encephalitis (inflammation of the brain), rash, acute arthritis, or hepatitis.

(b) *Classification.* Class I (general controls). The device is exempt from the premarket notification procedures in subpart E of part 807 of this chapter subject to § 866.9.

[47 FR 50823, Nov. 9, 1982, as amended at 65 FR 2311, Jan. 14, 2000]

§ 866.3250 *Erysipelothrix rhusiopathiae* serological reagents.

(a) *Identification. Erysipelothrix rhusiopathiae* serological reagents are devices that consist of antigens and antisera used in serological tests to identify *Erysipelothrix rhusiopathiae* from cultured isolates derived from clinical specimens. The identification aids in the diagnosis of disease caused by this bacterium belonging to the genus *Erysipelothrix.* This organism is responsible for a variety of inflammations of the skin following skin abrasions from contact with fish, shellfish, or poultry.

(b) *Classification.* Class I (general controls). The device is exempt from the premarket notification procedures in subpart E of part 807 of this chapter subject to the limitations in § 866.9.

[47 FR 50823, Nov. 9, 1982, as amended at 54 FR 25046, June 12, 1989; 66 FR 38791, July 25, 2001]

§ 866.3255 *Escherichia coli* serological reagents.

(a) *Identification. Escherichia coli* serological reagents are devices that consist of antigens and antisera used in serological tests to identify *Escherichia coli* from cultured isolates derived from clinical specimens. Additionally, some of these reagents consist of *Escherichia coli* antisera conjugated with a fluorescent dye used to identify *Escherichia coli* directly from clinical specimens or cultured isolates derived from clinical specimens. The identification aids in the diagnosis of diseases caused by this bacterium belonging to the genus *Escherichia*, and provides epidemiological information on diseases caused by this microorganism. Although *Escherichia coli* constitutes the greater part of the microorganisms found in the intestinal tract in humans and is usually nonpathogenic, those strains which are pathogenic may cause urinary tract infections or epidemic diarrheal disease, especially in children.

(b) *Classification*. Class I (general controls). The device is exempt from the premarket notification procedures in subpart E of part 807 of this chapter subject to the limitations in § 866.9.

[47 FR 50823, Nov. 9, 1982, as amended at 54 FR 25046, June 12, 1989; 66 FR 38791, July 25, 2001]

§ 866.3270 *Flavobacterium* spp. serological reagents.

(a) *Identification*. *Flavobacterium* spp. serological reagents are devices that consist of antigens and antisera used in serological tests to identify *Flavobacteriuim* spp. from cultured isolates derived from clinical specimens. The identification aids in the diagnosis of disease caused by bacteria belonging to the genus *Flavobacterium* and provides epidemiological information on diseases caused by these microorganisms. Most members of this genus are found in soil and water and, under certain conditions, may become pathogenic to humans. *Flavobacterium meningosepticum* is highly virulent for the newborn, in whom it may cause epidemics of septicemia (blood poisoning) and meningitis (inflammation of the membranes of the brain) and is usually attributable to contaminated hospital equipment.

(b) *Classification*. Class I (general controls). The device is exempt from the premarket notification procedures in subpart E of part 807 of this chapter subject to the limitations in § 866.9.

[47 FR 50823, Nov. 9, 1982, as amended at 54 FR 25046, June 12, 1989; 66 FR 38792, July 25, 2001]

§ 866.3280 *Francisella tularensis* serological reagents.

(a) *Identification*. *Francisella tularensis* serological reagents are devices that consist of antigens and antisera used in serological tests to identify antibodies to *Francisella tularensis* in serum or to identify *Francisella tularensis* in cultured isolates derived from clinical specimens. Additionally, some of these reagents consist of antisera conjugated with a fluorescent dye (immunofluorescent reagents) used to identify *Francisella tularensis* directly from clinical specimens. The identification aids in the diagnosis of tularemia caused by *Francisella tularensis* and provides epidemiological information on this disease. Tularemia is a desease principally of rodents, but may be transmitted to humans through handling of infected animals, animal products, or by the bites of fleas and ticks. The disease takes on several forms depending upon the site of infection, such as skin lesions, lymph node enlargements, or pulmonary infection.

(b) *Classification*. Class II (special controls). The device is exempt from the premarket notification procedures in subpart E of part 807 of this chapter subject to § 866.9.

[47 FR 50823, Nov. 9, 1982, as amended at 63 FR 59226, Nov. 3, 1998]

§ 866.3290 Gonococcal antibody test (GAT).

(a) *Identification*. A gonococcal antibody test (GAT) is an in vitro device that consists of the reagents intended to identify by immunochemical techniques, such as latex agglutination, indirect fluorescent antibody, or radioimmunoassay, antibodies to *Neisseria gonorrhoeae* in sera of asymptomatic females at low risk of infection. Identification of antibodies with this device may indicate past or present infection of the patient with *Neisseria gonorrhoeae*.

(b) *Classification*. Class III (premarket approval) (transitional device).

(c) *Date PMA or notice of completion of a PDP is required*. As of May 28, 1976, an approval under section 515 of the act is required before this device may be commercially distributed. See § 866.3.

[47 FR 50823, Nov. 9, 1982, as amended at 52 FR 17734, May 11, 1987]

§ 866.3300 *Haemophilus* spp. serological reagents.

(a) *Identification*. *Haemophilus* spp. serological reagents are devices that consist of antigens and antisera, including antisera conjugated with a fluorescent dye, that are used in serological tests to identify *Haemophilus* spp. directly from clinical specimens or tissue culture isolates derived from clinical specimens. The identification aids in the diagnosis of diseases caused by bacteria belonging to the genus *Haemophilus* and provides epidemiological information on diseases cause by these microorganisms. Diseases most often caused by *Haemophilus* spp. include pneumonia, pharyngitis, sinusitis, vaginitis, chancroid venereal disease, and a contagious form of conjunctivitis (inflammation of eyelid membranes).

(b) *Classification.* Class II (special controls). The device is exempt from the premarket notification procedures in subpart E of part 807 of this chapter subject to § 866.9.

[47 FR 50823, Nov. 9, 1982, as amended at 63 FR 59226, Nov. 3, 1998]

§ 866.3305 Herpes simplex virus serological reagents.

(a) *Identification.* Herpes simplex virus serological reagents are devices that consist of antigens and antisera used in various serological tests to identify antibodies to herpes simplex virus in serum. Additionally, some of the reagents consist of herpes simplex virus antisera conjugated with a fluorescent dye (immunofluorescent reagents) used to identify herpes simplex virus directly from clinical specimens or tissue culture isolates derived from clinical specimens. The identification aids in the diagnosis of diseases caused by herpes simplex viruses and provides epidemiological information on these diseases. Herpes simplex viral infections range from common and mild lesions of the skin and mucous membranes to a severe form of encephalitis (inflammation of the brain). Neonatal herpes virus infections range from a mild infection to a severe generalized disease with a fatal outcome.

(b) *Classification.* Class III (premarket approval).

(c) *Date PMA or notice of completion of a PDP is required.* No effective date has been established of the requirement for premarket approval. See § 866.3.

[47 FR 50823, Nov. 9, 1982, as amended at 52 FR 17734, May 11, 1987]

§ 866.3310 Hepatitis A virus (HAV) serological assays.

(a) *Identification.* HAV serological assays are devices that consist of antigens and antisera for the detection of hepatitis A virus-specific IgM, IgG, or total antibodies (IgM and IgG), in human serum or plasma. These devices are used for testing specimens from individuals who have signs and symptoms consistent with acute hepatitis to determine if an individual has been previously infected with HAV, or as an aid to identify HAV-susceptible individuals. The detection of these antibodies aids in the clinical laboratory diagnosis of an acute or past infection by HAV in conjunction with other clinical laboratory findings. These devices are not intended for screening blood or solid or soft tissue donors.

(b) *Classification.* Class II (special controls). The special control is "Guidance for Industry and FDA Staff: Class II Special Controls Guidance Document: Hepatitis A Virus Serological Assays." See § 866.1(e) for the availability of this guidance document.

[FR 6679, Feb. 9, 2006]

§ 866.3320 *Histoplasma capsulatum* serological reagents.

(a) *Identification. Histoplasma capsulatum* serological reagents are devices that consist of antigens and antisera used in serological tests to identify antibodies to *Histoplasma capsulatum* in serum. Additionally, some of these reagents consist of *Histoplasma capsulatum* antisera conjugated with a fluorescent dye (immunofluorescent reagents) used to identify *Histoplasma capsulatum* from clinical specimens or cultured isolates derived from clinical specimens. The identification aids in the diagnosis of histoplasmosis caused by this fungus belonging to the genus *Histoplasma* and provides epidemiological information on the diseases caused by this fungus. Histoplasmosis usually is a mild and often asymptomatic respiratory infection, but in a small number of infected individuals the lesions may spread to practically all tissues and organs.

(b) *Classification.* Class II (special controls). The device is exempt from the premarket notification procedures

in subpart E of part 807 of this chapter subject to § 866.9.

[47 FR 50823, Nov. 9, 1982, as amended at 63 FR 59227, Nov. 3, 1998]

§ 866.3330 Influenza virus serological reagents.

(a) *Identification.* Influenza virus serological reagents are devices that consist of antigens and antisera used in serological tests to identify antibodies to influenza in serum. The identification aids in the diagnosis of influenza (flu) and provides epidemiological information on influenza. Influenza is an acute respiratory tract disease, which is often epidemic.

(b) *Classification.* Class I (general controls). The device is exempt from the premarket notification procedures in subpart E of part 807 of this chapter subject to the limitations in § 866.9.

[47 FR 50823, Nov. 9, 1982, as amended at 54 FR 25047, June 12, 1989; 66 FR 38792, July 25, 2001]

§ 866.3332 Reagents for detection of specific novel influenza A viruses.

(a) *Identification.* Reagents for detection of specific novel influenza A viruses are devices that are intended for use in a nucleic acid amplification test to directly detect specific virus RNA in human respiratory specimens or viral cultures. Detection of specific virus RNA aids in the diagnosis of influenza caused by specific novel influenza A viruses in patients with clinical risk of infection with these viruses, and also aids in the presumptive laboratory identification of specific novel influenza A viruses to provide epidemiological information on influenza. These reagents include primers, probes, and specific influenza A virus controls.

(b) *Classification.* Class II (special controls). The special controls are:

(1) FDA's guidance document entitled "Class II Special Controls Guidance Document: Reagents for Detection of Specific Novel Influenza A Viruses." See § 866.1(e) for information on obtaining this document.

(2) The distribution of these devices is limited to laboratories with experienced personnel who have training in standardized molecular testing procedures and expertise in viral diagnosis, and appropriate biosafety equipment and containment.

EFFECTIVE DATE NOTE: At 71 FR 14379, Mar. 22, 2006, § 866.3332 was added, effective April 21, 2006.

§ 866.3340 *Klebsiella* spp. serological reagents.

(a) *Identification. Klebsiella* spp. serological reagents are devices that consist of antigens and antisera, including antisera conjugated with a fluorescent dye (immunofluorescent reagents), that are used in serological tests to identify *Klebsiella* spp. from cultured isolates derived from clinical specimens. The identification aids in the diagnosis of diseases caused by bacteria belonging to the genus *Klebsiella* and provides epidemiological information on these diseases. These organisms can cause serious urinary tract and pulmonary infections, particularly in hospitalized patients.

(b) *Classification.* Class I (general controls). The device is exempt from the premarket notification procedures in subpart E of part 807 of this chapter subject to the limitations in § 866.9.

[47 FR 50823, Nov. 9, 1982, as amended at 54 FR 25047, June 12, 1989; 66 FR 38792, July 25, 2001]

§ 866.3350 *Leptospira* spp. serological reagents.

(a) *Identification. Leptospira* spp. serological reagents are devices that consist of antigens and antisera used in serological tests to identify antibodies to *Leptospira* spp. in serum or identify *Leptospira* spp. from cultured isolates derived from clinical specimens. Additionally, some of these antisera are conjugated with a fluorescent dye (immunofluorescent reagents) and used to identify *Leptospira* spp. directly from clinical specimens. The identification aids in the diagnosis of leptospirosis caused by bacteria belonging to the genus *Leptospira* and provides epidemiological information on this disease. *Leptospira* infections range from mild fever-producing illnesses to severe liver and kidney involvement producing hemorrhage and dysfunction of these organs.

(b) *Classification.* Class II (special controls). The device is exempt from the premarket notification procedures

in subpart E of part 807 of this chapter subject to § 866.9.

[47 FR 50823, Nov. 9, 1982, as amended at 63 FR 59227, Nov. 3, 1998]

§ 866.3355 *Listeria* spp. serological reagents.

(a) *Identification. Listeria* spp. serological reagents are devices that consist of antigens and antisera used in serological tests to identify *Listeria* spp. from cultured isolates derived from clinical specimens. Additionally, some of these reagents consist of *Listeria* spp. antisera conjugated with a fluorescent dye (immunofluorescent reagents) used to identify *Listeria* spp. directly from clinical specimens. The identification aids in the diagnosis of listeriosis, a disease caused by bacteria belonging to the genus *Listeria,* and provides epidemiological information on diseases caused by these microorganisms. *Listeria monocytogenes,* the most common human pathogen of this genus, causes meningitis (inflammation of the brain membranes) and meningoencephalitis (inflammation of the brain and brain membranes) and is often fatal if untreated. A second form of human listeriosis is an intrauterine infection in pregnant women that results in a high mortality rate for infants before or after birth.

(b) *Classification.* Class I (general controls). The device is exempt from the premarket notification procedures in subpart E of part 807 of this chapter subject to § 866.9.

[47 FR 50823, Nov. 9, 1982, as amended at 65 FR 2311, Jan. 14, 2000]

§ 866.3360 Lymphocytic choriomeningitis virus serological reagents.

(a) *Identification.* Lymphocytic choriomeningitis virus serological reagents are devices that consist of antigens and antisera used in serological tests to identify antibodies to lymphocytic choriomeningitis virus in serum. The identification aids in the diagnosis of lymphocytic choriomeningitis virus infections and provides epidemiological information on diseases caused by these viruses. Lymphocytic choriomeningitis viruses usually cause a mild cerebral meningitis (inflammation of membranes that envelop the brain) and occasionally a mild pneumonia, but in rare instances may produce severe and even fatal illnesses due to complications from cerebral meningitis and pneumonia.

(b) *Classification.* Class I (general controls). The device is exempt from the premarket notification procedures in subpart E of part 807 of this chapter subject to § 866.9.

[47 FR 50823, Nov. 9, 1982, as amended at 65 FR 2311, Jan. 14, 2000]

§ 866.3370 *Mycobacterium tuberculosis* immunofluorescent reagents.

(a) *Identification. Mycobacterium tuberculosis* immunofluorescent reagents are devices that consist of antisera conjugated with a fluorescent dye used to identify *Mycobacterium tuberculosis* directly from clinical specimens. The identification aids in the diagnosis of tuberculosis and provides epidemiological information on this disease. *Mycobacterium tuberculosis* is the common causative organism in human tuberculosis, a chronic infectious disease characterized by formation of tubercles (small rounded nodules) and tissue necrosis (destruction), usually occurring in the lung.

(b) *Classification.* Class I (general controls).

§ 866.3375 *Mycoplasma* spp. serological reagents.

(a) *Identification. Mycoplasma* spp. serological reagents are devices that consist of antigens and antisera used in serological tests to identify antibodies to *Mycoplasma* spp. in serum. Additionally, some of these reagents consist of *Mycoplasma* spp. antisera conjugated with a fluorescent dye (immunofluorescent reagents) used to identify *Mycoplasma* spp. directly from clinical specimens. The identification aids in the diagnosis of disease caused by bacteria belonging to the genus *Mycoplasma* and provides epidemiological information on diseases caused by these microorganisms. *Mycoplasma* spp. are associated with inflammatory conditions of the urinary and respiratory tracts, the genitals, and the mouth. The effects in humans of infection with *Mycoplasma pneumoniae* range from inapparent infection to mild or severe upper respiratory disease, ear infection, and bronchial pneumonia.

(b) *Classification.* Class I (general controls). The device is exempt from the premarket notification procedures in subpart E of part 807 of this chapter subject to § 866.9.

[47 FR 50823, Nov. 9, 1982, as amended at 65 FR 2311, Jan. 14, 2000]

§ 866.3380 Mumps virus serological reagents.

(a) *Identification.* Mumps virus serological reagents consist of antigens and antisera used in serological tests to identify antibodies to mumps virus in serum. Additionally, some of these reagents consist of antisera conjugated with a fluorescent dye (immunofluorescent reagents) used in serological tests to identify mumps viruses from tissue culture isolates derived from clinical specimens. The identification aids in the diagnosis of mumps and provides epidemiological information on mumps. Mumps is an acute contagious disease, particularly in children, characterized by an enlargement of one or both of the parotid glands (glands situated near the ear), although other organs may also be involved.

(b) *Classification.* Class I (general controls). The device is exempt from the premarket notification procedures in subpart E of part 807 of this chapter subject to § 866.9.

[47 FR 50823, Nov. 9, 1982, as amended at 65 FR 2311, Jan. 14, 2000]

§ 866.3390 *Neisseria* spp. direct serological test reagents.

(a) *Identification. Neisseria* spp. direct serological test reagents are devices that consist of antigens and antisera used in serological tests to identify *Neisseria* spp. from cultured isolates. Additionally, some of these reagents consist of *Neisseria* spp. antisera conjugated with a fluorescent dye (immunofluorescent reagents) which may be used to detect the presence of *Neisseria* spp. directly from clinical specimens. The identification aids in the diagnosis of disease caused by bacteria belonging to the genus *Neisseria,* such as epidemic cerebrospinal meningitis, meningococcal disease, and gonorrhea, and also provides epidemiological information on diseases caused by these microorganisms. The device does not include products for the detection of gonorrhea in humans by indirect methods, such as detection of antibodies or of oxidase produced by gonococcal organisms.

(b) *Classification.* Class II (performance standards).

§ 866.3400 Parainfluenza virus serological reagents.

(a) *Identification.* Parainfluenza virus serological reagents are devices that consist of antigens and antisera used in serological tests to identify antibodies to parainfluenza virus in serum. The identification aids in the diagnosis of parainfluenza virus infections and provides epidemiological information on diseases caused by these viruses. Parainfluenza viruses cause a variety of respiratory illnesses ranging from the common cold to pneumonia.

(b) *Classification.* Class I (general controls). The device is exempt from the premarket notification procedures in subpart E of part 807 of this chapter subject to the limitations in § 866.9.

[47 FR 50823, Nov. 9, 1982, as amended at 54 FR 25047, June 12, 1989; 66 FR 38792, July 25, 2001]

§ 866.3405 Poliovirus serological reagents.

(a) *Identification.* Poliovirus serological reagents are devices that consist of antigens and antisera used in serological tests to identify antibodies to poliovirus in serum. Additionally, some of these reagents consist of poliovirus antisera conjugated with a fluorescent dye (immunofluorescent reagents) used to identify polioviruses from clinical specimens or from tissue culture isolates derived from clinical specimens. The identification aids in the diagnosis of poliomyelitis (polio) and provides epidemiological information on this disease. Poliomyelitis is an acute infectious disease which in its serious form affects the central nervous system resulting in atrophy (wasting away) of groups of muscles, ending in contraction and permanent deformity.

(b) *Classification.* Class I (general controls). The device is exempt from the premarket notification procedures in

subpart E of part 807 of this chapter subject to § 866.9.

[47 FR 50823, Nov. 9, 1982, as amended at 65 FR 2312, Jan. 14, 2000]

§ 866.3410 *Proteus* spp. (Weil-Felix) serological reagents.

(a) *Identification. Proteus* spp. (Weil-Felix) serological reagents are devices that consist of antigens and antisera, including antisera conjugated with a fluorescent dye (immunofluorescent reagents), derived from the bacterium *Proteus vulgaris* used in agglutination tests (a specific type of antigen-antibody reaction) for the detection of antibodies to rickettsia (virus-like bacteria) in serum. Test results aid in the diagnosis of diseases caused by bacteria belonging to the genus *Rickettsiae* and provide epidemiological information on these diseases. Rickettsia are generally transmitted by arthropods (e.g., ticks and mosquitoes) and produce infections in humans characterized by rash and fever (e.g., typhus fever, spotted fever, Q fever, and trench fever).

(b) *Classification.* Class I (general controls). The device is exempt from the premarket notification procedures in subpart E of part 807 of this chapter subject to the limitations in § 866.9.

[47 FR 50823, Nov. 9, 1982, as amended at 54 FR 25047, June 12, 1989; 66 FR 38792, July 25, 2001]

§ 866.3415 *Pseudomonas* spp. serological reagents.

(a) *Identification. Pseudomonas* spp. serological reagents are devices that consist of antigens and antisera, including antisera conjugated with a fluorescent dye (immunofluorescent reagents), used to identify *Pseudomonas* spp. from clinical specimens or from cultured isolates derived from clinical specimens. The identification aids in the diagnosis of disease caused by bacteria belonging to the genus *Pseudomonas. Pseudomonas aeruginosa* is a major cause of hospital-acquired infections, and has been associated with urinary tract infections, eye infections, burn and wound infections, blood poisoning, abscesses, and meningitis (inflammation of brain membranes). *Pseudomonas pseudomallei* causes melioidosis, a chronic pneumonia.

(b) *Classification.* Class II (special controls). The device is exempt from the premarket notification procedures in subpart E of part 807 of this chapter subject to § 866.9.

[47 FR 50823, Nov. 9, 1982, as amended at 63 FR 59227, Nov. 3, 1998]

§ 866.3460 Rabiesvirus immunofluorescent reagents.

(a) *Identification.* Rabiesvirus immunofluorescent reagents are devices that consist of rabiesvirus antisera conjugated with a fluorescent dye used to identify rabiesvirus in specimens taken from suspected rabid animals. The identification aids in the diagnosis of rabies in patients exposed by animal bites and provides epidemiological information on rabies. Rabies is an acute infectious disease of the central nervous system which, if undiagnosed, may be fatal. The disease is commonly transmitted to humans by a bite from a rabid animal.

(b) *Classification.* Class II (performance standards).

§ 866.3470 Reovirus serological reagents.

(a) *Identification.* Reovirus serological reagents are devices that consist of antigens and antisera used in serological tests to identify antibodies to reovirus in serum. The identification aids in the diagnosis of reovirus infections and provides epidemiological information on diseases caused by these viruses. Reoviruses are thought to cause only mild respiratory and gastrointestinal illnesses.

(b) *Classification.* Class I (general controls). The device is exempt from the premarket notification procedures in subpart E of part 807 of this chapter subject to the limitations in § 866.9.

[47 FR 50823, Nov. 9, 1982, as amended at 54 FR 25047, June 12, 1989; 66 FR 38792, July 25, 2001]

§ 866.3480 Respiratory syncytial virus serological reagents.

(a) *Identification.* Respiratory syncytial virus serological reagents are devices that consist of antigens and antisera used in serological tests to identify antibodies to respiratory syncytial virus in serum. Additionally, some of these reagents consist of respiratory syncytial virus antisera conjugated with a fluorescent dye (immunofluorescent reagents) and used to identify respiratory syncytial viruses from clinical specimens or from tissue culture isolates derived from clinical specimens. The identification aids in the diagnosis of respiratory syncytial virus infections and provides epidemiological information on diseases caused by these viruses. Respiratory syncytial viruses cause a number of respiratory tract infections, including the common cold, pharyngitis, and infantile bronchopneumonia.

(b) *Classification.* Class I (general controls). The device is exempt from the premarket notification procedures in subpart E of part 807 of this chapter subject to § 866.9.

[47 FR 50823, Nov. 9, 1982, as amended at 65 FR 2312, Jan. 14, 2000]

§ 866.3490 Rhinovirus serological reagents.

(a) *Identification.* Rhinovirus serological reagents are devices that consist of antigens and antisera used in serological tests to identify antibodies to rhinovirus in serum. The identification aids in the diagnosis of rhinovirus infections and provides epidemiological information on diseases caused by these viruses. Rhinoviruses cause common colds.

(b) *Classification.* Class I (general controls). The device is exempt from the premarket notification procedures in subpart E of part 807 of this chapter subject to the limitations in § 866.9.

[47 FR 50823, Nov. 9, 1982, as amended at 54 FR 25047, June 12, 1989; 66 FR 38792, July 25, 2001]

§ 866.3500 Rickettsia serological reagents.

(a) *Identification.* Rickettsia serological reagents are devices that consist of antigens and antisera used in serological tests to identify antibodies to rickettsia in serum. Additionally, some of these reagents consist of rickettsial antisera conjugated with a fluorescent dye (immunofluorescent reagents) used to identify rickettsia directly from clinical specimens. The identification aids in the diagnosis of diseases caused by virus-like bacteria belonging to the genus *Rickettsiae* and provides epidemiological information on these diseases. Rickettsia are generally transmitted by arthropods (e.g., ticks and mosquitoes) and produce infections in humans characterized by rash and fever (e.g., typhus fever, spotted fever, Q fever, and trench fever).

(b) *Classification.* Class I (general controls). The device is exempt from the premarket notification procedures in subpart E of part 807 of this chapter subject to § 866.9.

[47 FR 50823, Nov. 9, 1982, as amended at 65 FR 2312, Jan. 14, 2000]

§ 866.3510 Rubella virus serological reagents.

(a) *Identification.* Rubella virus serological reagents are devices that consist of antigens and antisera used in serological tests to identify antibodies to rubella virus in serum. The identification aids in the diagnosis of rubella (German measles) or confirmation of a person's immune status from past infections or immunizations and provides epidemiological information on German measles. Newborns infected in the uterus with rubella virus may be born with multiple congenital defects (rubella syndrome).

(b) *Classification.* Class II. The special controls for this device are:

(1) National Committee for Clinical Laboratory Standards':

(i) 1/LA6 "Detection and Quantitation of Rubella IgG Antibody: Evaluation and Performance Criteria for Multiple Component Test Products, Speciment Handling, and Use of the Test Products in the Clinical Laboratory, October 1997,"

(ii) 1/LA18 "Specifications for Immunological Testing for Infectious Diseases, December 1994,"

(iii) D13 "Agglutination Characteristics, Methodology, Limitations, and Clinical Validation, October 1993,"

(iv) EP5 "Evaluation of Precision Performance of Clinical Chemistry Devices, February 1999," and

(v) EP10 "Preliminary Evaluation of the Linearity of Quantitive Clinical Laboratory Methods, May 1998,"

(2) Centers for Disease Control's:

(i) Low Titer Rubella Standard,

(ii) Reference Panel of Well Characterized Rubella Sera, and

(3) World Health Organization's International Rubella Standard.

[47 FR 50823, Nov. 9, 1982, as amended at 52 FR 17734, May 11, 1987; 65 FR 17144, Mar. 31, 2000]

§ 866.3520 Rubeola (measles) virus serological reagents.

(a) *Identification.* Rubeola (measles) virus serological reagents are devices that consist of antigens and antisera used in serological tests to identify antibodies to rubeola virus in serum. The identification aids in the diagnosis of measles and provides epidemiological information on the disease. Measles is an acute, highly infectious disease of the respiratory and reticuloendothelial tissues, particularly in children, characterized by a confluent and blotchy rash.

(b) *Classification.* Class I (general controls). The device is exempt from the premarket notification procedures in subpart E of part 807 of this chapter subject to the limitations in § 866.9.

[47 FR 50823, Nov. 9, 1982, as amended at 54 FR 25047, June 12, 1989; 66 FR 38792, July 25, 2001]

§ 866.3550 *Salmonella* spp. serological reagents.

(a) *Identification. Salmonella* spp. serological reagents are devices that consist of antigens and antisera used in serological tests to identify *Salmonella* spp. from cultured isolates derived from clinical specimens. Additionally, some of these reagents consist of antisera conjugated with a fluorescent dye (immunofluorescent reagents) used to identify *Salmonella* spp. directly from clinical specimens or cultured isolates derived from clinical specimens. The identification aids in the diagnosis of salmonellosis caused by bacteria belonging to the genus *Salmonella* and provides epidemiological information on this disease. Salmonellosis is characterized by high grade fever (''enteric fever''), severe diarrhea, and cramps.

(b) *Classification.* Class II (special controls). The device is exempt from the premarket notification procedures in subpart E of part 807 of this chapter subject to § 866.9.

[47 FR 50823, Nov. 9, 1982, as amended at 63 FR 59227, Nov. 3, 1998]

§ 866.3600 *Schistosoma* spp. serological reagents.

(a) *Identification. Schistosoma* spp. serological reagents are devices that consist of antigens and antisera used in serological tests to identify antibodies to *Schistosoma* spp. in serum. The identification aids in the diagnosis of schistosomiasis caused by parasitic flatworms of the genus *Schistosoma.* Schistosomiasis is characterized by a variety of acute and chronic infections. Acute infection is marked by fever, allergic symptoms, and diarrhea. Chronic effects are usually severe and are caused by fibrous degeneration of tissue around deposited eggs of the parasite in the liver, lungs, and central nervous system. Schistosomes can also cause schistosome dermatitis (e.g., swimmer's itch), a skin disease marked by intense itching.

(b) *Classification.* Class I (general controls). The device is exempt from the premarket notification procedures in subpart E of part 807 of this chapter subject to § 866.9.

[47 FR 50823, Nov. 9, 1982, as amended at 65 FR 2312, Jan. 14, 2000]

§ 866.3630 *Serratia* spp. serological reagents.

(a) *Identification. Serratia* spp. serological reagents are devices that consist of antigens and antisera used in serological tests to identify *Serratia* spp. from cultured isolates. The identification aids in the diagnosis of disease caused by bacteria belonging to the genus *Serratia* and provides epidemiological information on these diseases. *Serratia* spp. are occasionally associated with gastroenteritis (food poisoning) and wound infections.

(b) *Classification.* Class I (general controls). The device is exempt from the premarket notification procedures in subpart E of part 807 of this chapter subject to the limitations in § 866.9.

[47 FR 50823, Nov. 9, 1982 as amended at 54 FR 25047, June 12, 1989; 66 FR 38792, July 25, 2001]

§ 866.3660 *Shigella* spp. serological reagents.

(a) *Identification. Shigella* spp. serological reagents are devices that consist of antigens and antisera, including antisera conjugated with a fluorescent dye (immunofluorescent reagents), used in serological tests to identify *Shigella* spp. from cultured isolates. The identification aids in the diagnosis of shigellosis caused by bacteria belonging to the genus *Shigella* and provides epidemiological information on this disease. Shigellosis is characterized by abdominal pain, cramps, diarrhea, and fever.

(b) *Classification.* Class II (special controls). The device is exempt from the premarket notification procedures in subpart E of part 807 of this chapter subject to § 866.9.

[47 FR 50823, Nov. 9, 1982, as amended at 63 FR 59227, Nov. 3, 1998]

§ 866.3680 *Sporothrix schenckii* serological reagents.

(a) *Identification. Sporothrix schenckii* serological reagents are devices that consist of antigens and antisera used in serological tests to identify antibodies to *Sporothrix schenckii* in serum. The identification aids in the diagnosis of sporothrichosis caused by a fungus belonging to the genus *Sporothrix* and provides epidemiological information on this disease. Sporothrichosis is a chronic tumorlike infection primarily of the skin.

(b) *Classification.* Class I (general controls). The device is exempt from the premarket notification procedures in subpart E of part 807 of this chapter subject to § 866.9.

[47 FR 50823, Nov. 9, 1982, as amended at 65 FR 2312, Jan. 14, 2000]

§ 866.3700 *Staphylococcus aureus* serological reagents.

(a) *Identification. Staphylococcus aureus* serological reagents are devices that consist of antigens and antisera used in serological tests to identify enterotoxin (toxin affecting the intestine) producing staphylococci from cultured isolates. The identification aids in the diagnosis of disease caused by this bacterium belonging to the genus *Staphylococcus* and provides epidemiological information on these diseases. Certain strains of *Staphylococcus aureus* produce an enterotoxin while growing in meat, dairy, or bakery products. After ingestion, this enterotoxin is absorbed in the gut and causes destruction of the intestinal lining (gastroenteritis).

(b) *Classification.* Class I (general controls). The device is exempt from the premarket notification procedures in subpart E of part 807 of this chapter subject to the limitations in § 866.9.

[47 FR 50823, Nov. 9, 1982, as amended at 54 FR 25047, June 12, 1989; 66 FR 38792, July 25, 2001]

§ 866.3720 *Streptococcus* spp. exoenzyme reagents.

(a) *Identification. Streptococcus* spp. exoenzyme reagents are devices used to identify antibodies to *Streptococcus* spp. exoenzyme in serum. The identification aids in the diagnosis of disease caused by bacteria belonging to the genus *Streptococcus* and provides epidemiological information on these diseases. Pathogenic streptococci are associated with infections, such as sore throat, impetigo (an infection characterized by small pustules on the skin), urinary tract infections, rheumatic fever, and kidney disease.

(b) *Classification.* Class I (general controls). The device is exempt from the premarket notification procedures in subpart E of part 807 of this chapter subject to the limitations in § 866.9.

[47 FR 50823, Nov. 9, 1982, as amended at 61 FR 1119, Jan. 16, 1996; 66 FR 38792, July 25, 2001]

§ 866.3740 *Streptococcus* spp. serological reagents.

(a) *Identification. Streptococcus* spp. serological reagents are devices that consist of antigens and antisera (excluding streptococcal exoenzyme reagents made from enzymes secreted by streptococci) used in serological tests to identify *Streptococcus* spp. from cultured isolates derived from clinical specimens. The identification aids in the diagnosis of diseases caused by bacteria belonging to the genus *Streptococcus* and provides epidemiological information on these diseases. Pathogenic streptococci are associated with infections, such as sore throat, impetigo (an infection characterized by small pustules on the skin), urinary tract infections, rheumatic fever, and kidney disease.

(b) *Classification*. Class I (general controls). The device is exempt from the premarket notification procedures in subpart E of part 807 of this chapter subject to § 866.9.

[47 FR 50823, Nov. 9, 1982, as amended at 65 FR 2312, Jan. 14, 2000]

§ 866.3780 *Toxoplasma gondii* serological reagents.

(a) *Identification*. *Toxoplasma gondii* serological reagents are devices that consist of antigens and antisera used in serological tests to identify antibodies to *Toxoplasma gondii* in serum. Additionally, some of these reagents consist of antisera conjugated with a fluorescent dye (immunofluorescent reagents) used to identify *Toxoplasma gondii* from clinical specimens. The identification aids in the diagnosis of toxoplasmosis caused by the parasitic protozoan *Toxoplasma gondii* and provides epidemiological information on this disease. Congenital toxoplasmosis is characterized by lesions of the central nervous system, which if undetected and untreated may lead to brain defects, blindness, and death of an unborn fetus. The disease is characterized in children by inflammation of the brain and spinal cord.

(b) *Classification*. Class II (performance standards).

§ 866.3820 *Treponema pallidum* nontreponemal test reagents.

(a) *Identification*. *Treponema pallidum* nontreponemal test reagents are devices that consist of antigens derived from nontreponemal sources (sources not directly associated with treponemal organisms) and control sera (standardized sera with which test results are compared) used in serological tests to identify reagin, an antibody-like agent, which is produced from the reaction of treponema microorganisms with body tissues. The identification aids in the diagnosis of syphilis caused by microorganisms belonging to the genus *Treponema* and provides epidemiological information on syphilis.

(b) *Classification*. Class II (performance standards).

§ 866.3830 *Treponema pallidum* treponemal test reagents.

(a) *Identification*. *Treponema pallidum* treponemal test reagents are devices that consist of the antigens, antisera and all control reagents (standardized reagents with which test results are compared) which are derived from treponemal sources and that are used in the fluorescent treponemal antibody absorption test (FTA-ABS), the *Treponema pallidum* immobilization test (T.P.I.), and other treponemal tests used to identify antibodies to *Treponema pallidum* directly from infecting treponemal organisms in serum. The identification aids in the diagnosis of syphilis caused by bacteria belonging to the genus *Treponema* and provides epidemiological information on syphilis.

(b) *Classification*. Class II (performance standards).

§ 866.3850 *Trichinella spiralis* serological reagents.

(a) *Identification*. *Trichinella spiralis* serological reagents are devices that consist of antigens and antisera used in serological tests to identify antibodies to *Trichinella spiralis* in serum. The identification aids in the diagnosis of trichinosis caused by parasitic roundworms belonging to the genus *Trichinella* and provides epidemiological information on trichinosis. Trichinosis is caused by ingestion of undercooked, infested meat, especially pork, and characterized by fever, muscle weakness, and diarrhea.

(b) *Classification*. Class I (general controls). The device is exempt from the premarket notification procedures in subpart E of part 807 of this chapter subject to § 866.9.

[47 FR 50823, Nov. 9, 1982, as amended at 65 FR 2312, Jan. 14, 2000]

§ 866.3870 *Trypanosoma* spp. serological reagents.

(a) *Identification*. *Trypanosoma* spp. serological reagents are devices that consist of antigens and antisera used in serological tests to identify antibodies to *Trypanosoma* spp. in serum. The identification aids in the diagnosis of trypanosomiasis, a disease caused by parasitic protozoans belonging to the genus *Trypanosoma.* Trypanosomiasis in adults is a chronic disease characterized by fever, chills, headache, and vomiting. Central nervous system involvement produces typical sleeping sickness syndrome: physical exhaustion, inability to eat, tissue wasting, and eventual death. Chagas disease, an acute form of trypanosomiasis in children, most seriously affects the central nervous system and heart muscle.

(b) *Classification.* Class I (general controls).

§ 866.3900 Varicella-zoster virus serological reagents.

(a) *Identification.* Varicella-zoster virus serological reagents are devices that consist of antigens and antisera used in serological tests to identify antibodies to varicella-zoster in serum. The identification aids in the diagnosis of diseases caused by varicella-zoster viruses and provides epidemiological information on these diseases. Varicella (chicken pox) is a mild, highly infectious disease, chiefly of children. Zoster (shingles) is the recurrent form of the disease, occurring in adults who were previously infected with varicella-zoster viruses. Zoster is the response (characterized by a rash) of the partially immune host to a reactivation of varicella viruses present in latent form in the patient's body.

(b) *Classification.* Class II (performance standards).

§ 866.3930 *Vibrio cholerae* serological reagents.

(a) *Identification. Vibrio cholerae* serological reagents are devices that are used in the agglutination (an antigen-antibody clumping reaction) test to identify *Vibrio cholerae* from cultured isolates derived from clinical specimens. The identification aids in the diagnosis of cholera caused by the bacterium *Vibrio cholerae* and provides epidemiological information on cholera. Cholera is an acute infectious disease characterized by severe diarrhea with extreme fluid and electrolyte (salts) depletion, and by vomiting, muscle cramps, and prostration. If untreated, the severe dehydration may lead to shock, renal failure, cardiovascular collapse, and death.

(b) *Classification.* Class II (special controls). The device is exempt from the premarket notification procedures in subpart E of part 807 of this chapter subject to § 866.9.

[47 FR 50823, Nov. 9, 1982, as amended at 63 FR 59227, Nov. 3, 1998]

§ 866.3940 West Nile virus serological reagents.

(a) *Identification.* West Nile virus serological reagents are devices that consist of antigens and antisera for the detection of anti-West Nile virus IgM antibodies, in human serum, from individuals who have signs and symptoms consistent with viral meningitis/encephalitis. The detection aids in the clinical laboratory diagnosis of viral meningitis/encephalitis caused by West Nile virus.

(b) *Classification.* Class II (special controls). The special control is FDA's guidance entitled "Class II Special Controls Guidance Document: Serological Reagents for the Laboratory Diagnosis of West Nile Virus." See § 866.1(e) for the availability of this guidance document.

[68 FR 61745, Oct. 30, 2003]

Subpart E—Immunology Laboratory Equipment and Reagents

§866.4070 RNA Preanalytical Systems.

(a) *Identification.* RNA Preanalytical Systems are devices intended to collect, store, and transport patient specimens, and stabilize intracellular RNA from the specimens, for subsequent isolation and purification of the intracellular RNA for RT–PCR used in in vitro molecular diagnostic testing.

(b) *Classification.* Class II (special controls). The special control is FDA's guidance document entitled "Class II Special Controls Guidance Document: RNA Preanalytical Systems (RNA Collection, Stabilization and Purification System for RT–PCR Used in Molecular Diagnostic Testing)." See § 866.1(e) for the availability of this guidance document.

[70 FR 49863, Aug. 25, 2005]

§ 866.4100 Complement reagent.

(a) *Identification.* A complement reagent is a device that consists of complement, a naturally occurring serum protein from any warm-blooded animal such as guinea pigs, that may be included as a component part of serological test kits used in the diagnosis of disease.

(b) *Classification.* Class I (general controls). The device is exempt from the premarket notification procedures in subpart E of part 807 of this chapter subject to the limitations in § 866.9.

[47 FR 50823, Nov. 9, 2001, as amended at 66 FR 38792, July 25, 2001]

§ 866.4500 Immunoelectrophoresis equipment.

(a) *Identification.* Immunoelectrophoresis equipment for clinical use with its electrical power supply is a device used for separating protein molecules. Immunoelectrophoresis is a procedure in which a complex protein mixture is placed in an agar gel and the various proteins are separated on the basis of their relative mobilities under the influence of an electric current. The separated proteins are then permitted to diffuse through the agar toward a multispecific antiserum, allowing precipitation and visualization of the separate complexes.

(b) *Classification.* Class I (general controls). The device is exempt from the premarket notification procedures in subpart E of part 807 of this chapter subject to the limitations in § 866.9.

[47 FR 50823, Nov. 9, 1982, as amended at 54 FR 25047, June 12, 1989; 66 FR 38792, July 25, 2001]

§ 866.4520 Immunofluorometer equipment.

(a) *Identification.* Immunofluorometer equipment for clinical use with its electrical power supply is a device used to measure the fluorescence of fluorochrome-labeled antigen-antibody complexes. The concentration of these complexes may be measured by means of reflected light. A beam of light is passed through a solution in which a fluorochrome has been selectively attached to serum protein antibody molecules in suspension. The amount of light emitted by the fluorochrome label is detected by a photodetector, which converts light energy into electrical energy. The amount of electrical energy registers on a readout system such as a digital voltmeter or a recording chart. This electrical readout is called the fluorescence value and is used to measure the concentration of antigen-antibody complexes.

(b) *Classification.* Class I (general controls). The device is exempt from the premarket notification procedures in subpart E of part 807 of this chapter subject to the limitations in § 866.9.

[47 FR 50823, Nov. 9, 1982, as amended at 54 FR 25047, June 12, 1989; 66 FR 38792, July 25, 2001]

§ 866.4540 Immunonephelometer equipment.

(a) *Identification.* Immunonephelometer equipment for clinical use with its electrical power supply is a device that measures light scattering from antigen-antibody complexes. The concentration of these complexes may be measured by means of reflected light. A beam of light passed through a solution is scattered by the particles in suspension. The amount of light is detected by a photodetector, which converts light energy into electrical energy. The amount of electrical energy registers on a readout system such as a digital voltmeter or a recording chart. This electrical readout is called the light-scattering value and is used to measure the concentration of antigen-antibody complexes. This generic type of device includes devices with various kinds of light sources, such as laser equipment.

(b) *Classification.* Class I (general controls). The device is exempt from the premarket notification procedures in subpart E of part 807 of this chapter subject to the limitations in § 866.9.

[47 FR 50823, Nov. 9, 1982, as amended at 54 FR 25047, June 12, 1989; 66 FR 38792, July 25, 2001]

§ 866.4600 Ouchterlony agar plate.

(a) *Identification.* An ouchterlony agar plate for clinical use is a device containing an agar gel used to examine antigen-antibody reactions. In immunodiffusion, antibodies and antigens migrate toward each other through gel which originally contained neither of these reagents. As the reagents come in contact with each other, they combine to form a precipitate that is trapped in the gel matrix and is immobilized.

(b) *Classification.* Class I (general controls). The device is exempt from the premarket notification procedures in subpart E of part 807 of this chapter subject to the limitations in § 866.9.

[47 FR 50823, Nov. 9, 1982, as amended at 54 FR 25047, June 12, 1989; 66 FR 38792, July 25, 2001]

§ 866.4700 Automated fluorescence *in situ* hybridization (FISH) enumeration systems.

(a) *Identification.* An automated FISH enumeration system is a device that consists of an automated scanning microscope, image analysis system, and customized software applications for FISH assays. This device is intended for in vitro diagnostic use with FISH assays as an aid in the detection, counting and classification of cells based on recognition of cellular color, size, and shape, and in the detection and enumeration of FISH signals in interphase nuclei of formalin-fixed, paraffin-embedded human tissue specimens.

(b) *Classification.* Class II (special controls). The special control is FDA's guidance document entitled "Class II Special Controls Guidance Document: Automated Fluorescence *in situ* Hybridization (FISH) Enumeration Systems." See § 866.1(e) for the availability of this guidance document.

[70 FR 14534, Mar. 23, 2005]

§ 866.4800 Radial immunodiffusion plate.

(a) *Identification.* A radial immunodiffusion plate for clinical use is a device that consists of a plastic plate to which agar gel containing antiserum is added. In radial immunodiffusion, antigens migrate through gel which originally contains specific antibodies. As the reagents come in contact with each other, they combine to form a precipitate that is trapped in the gel matrix and immobilized.

(b) *Classification.* Class I (general controls). The device is exempt from the premarket notification procedures in subpart E of part 807 of this chapter subject to the limitations in § 866.9.

[47 FR 50823, Nov. 9, 1982, as amended at 66 FR 38792, July 25, 2001]

§ 866.4830 Rocket immunoelectrophoresis equipment.

(a) *Identification.* Rocket immunoelectrophoresis equipment for clinical use is a device used to perform a specific test on proteins by using a procedure called rocket immunoelectrophoresis. In this procedure, an electric current causes the protein in solution to migrate through agar gel containing specific antisera. The protein precipitates with the antisera in a rocket-shaped pattern, giving the name to the device. The height of the peak (or the area under the peak) is proportional to the concentration of the protein.

(b) *Classification.* Class I (general controls). The device is exempt from the premarket notification procedures in subpart E of part 807 of this chapter subject to the limitations in § 866.9.

[47 FR 50823, Nov. 9, 1982, as amended at 54 FR 25047, June 12, 1989; 66 FR 38792, July 25, 2001]

§ 866.4900 Support gel.

(a) *Identification.* A support gel for clinical use is a device that consists of an agar or agarose preparation that is used while measuring various kinds of, or parts of, protein molecules by various immunochemical techniques, such as immunoelectrophoresis, immunodiffusion, or chromatography.

(b) *Classification.* Class I (general controls). The device is exempt from the premarket notification procedures in subpart E of part 807 of this chapter subject to the limitations in § 866.9.

[47 FR 50823, Nov. 9, 1982, as amended at 54 FR 25047, June 12, 1989; 66 FR 38792, July 25, 2001]

Subpart F—Immunological Test Systems

§ 866.5040 Albumin immunological test system.

(a) *Identification.* An albumin immunological test system is a device that consists of the reagents used to measure by immunochemical techniques the albumin (a plasma protein) in serum and other body fluids. Measurement of albumin aids in the diagnosis of kidney and intestinal diseases.

(b) *Classification.* Class II (special controls). The device is exempt from the premarket notification procedures in subpart E of part 807 of this chapter subject to § 866.9.

[47 FR 50823, Nov. 9, 1982, as amended at 63 FR 59227, Nov. 3, 1998]

§ 866.5060 Prealbumin immunological test system.

(a) *Identification.* A prealbumin immunological test system is a device that consists of the reagents used to measure by immunochemical techniques the prealbumin (a plasma protein) in serum and other body fluids. Measurement of prealbumin levels in serum may aid in the assessment of the patient's nutritional status.

(b) *Classification.* Class I (general controls). The device is exempt from the premarket notification procedures in subpart E of part 807 of this chapter subject to § 866.9.

[47 FR 50823, Nov. 9, 1982, as amended at 65 FR 2312, Jan. 14, 2000]

§ 866.5065 Human allotypic marker immunological test system.

(a) *Identification.* A human allotypic marker immunological test system is a device that consists of the reagents used to identify by immunochemical techniques the inherited human protein allotypic markers (such as nGm, nA_2 m, and Km allotypes) in serum and other body fluids. The identification may be used while studying population genetics.

(b) *Classification.* Class I (general controls). The device is exempt from the premarket notification procedures in subpart E of part 807 of this chapter subject to § 866.9.

[47 FR 50823, Nov. 9, 1982, as amended at 65 FR 2312, Jan. 14, 2000]

§ 866.5080 *Alpha*-1-antichymotrypsin immunological test system.

(a) *Identification.* An *alpha*-1-antichymotrypsin immunological test system is a device that consists of the reagents used to measure by immunochemical techniques *alpha*-1-antichymotrypsin (a protein) in serum, other body fluids, and tissues. *Alpha*-1-antichymotrypsin helps protect tissues against proteolytic (protein-splitting) enzymes released during infection.

(b) *Classification.* Class II (performance standards).

§ 866.5090 Antimitochondrial antibody immunological test system.

(a) *Identification.* An antimitochondrial antibody immunological test system is a device that consists of the reagents used to measure by immunochemical techniques the antimitochondrial antibodies in human serum. The measurements aid in the diagnosis of diseases that produce a spectrum of autoantibodies (antibodies produced against the body's own tissue), such as primary biliary cirrhosis (degeneration of liver tissue) and chronic active hepatitis (inflammation of the liver).

(b) *Classification.* Class II (performance standards).

§ 866.5100 Antinuclear antibody immunological test system.

(a) *Identification.* An antinuclear antibody immunological test system is a device that consists of the reagents used to measure by immunochemical techniques the autoimmune antibodies in serum, other body fluids, and tissues that react with cellular nuclear constituents (molecules present in the nucleus of a cell, such as ribonucleic acid, deoxyribonucleic acid, or nuclear proteins). The measurements aid in the diagnosis of systemic lupus erythematosus (a multisystem autoimmune disease in which antibodies attack the victim's own tissues), hepatitis (a liver disease), rheumatoid arthritis, Sjogren's syndrome (arthritis with inflammation of the eye, eyelid, and salivary glands), and systemic sclerosis (chronic hardening and shrinking of many body tissues).

(b) *Classification.* Class II (performance standards).

§ 866.5110 Antiparietal antibody immunological test system.

(a) *Identification.* An antiparietal antibody immunological test system is a device that consists of the reagents used to measure by immunochemical techniques the specific antibody for gastric parietal cells in serum and other body fluids. Gastric parietal cells are those cells located in the stomach that produce a protein that enables vitamin B_{12} to be absorbed by the body. The measurements aid in the diagnosis of vitamin B_{12} deficiency (or pernicious anemia), atrophic gastritis (inflammation of the stomach), and autoimmune connective tissue diseases (diseases resulting when the body produces antibodies against its own tissues).

(b) *Classification.* Class II (performance standards).

§ 866.5120 Antismooth muscle antibody immunological test system.

(a) *Identification.* An antismooth muscle antibody immunological test system is a device that consists of the reagents used to measure by immunochemical techniques the antismooth muscle antibodies (antibodies to nonstriated, involuntary muscle) in serum. The measurements aid in the diagnosis of chronic hepatitis (inflammation of the liver) and autoimmune connective tissue diseases (diseases resulting from antibodies produced against the body's own tissues).

(b) *Classification* Class II (performance standards).

§ 866.5130 *Alpha*-1-antitrypsin immunological test system.

(a) *Identification.* An *alpha*-1-antitrypsin immunological test system is a device that consists of the reagents used to measure by immunochemical techniques the *alpha*-1-antitrypsin (a plasma protein) in serum, other body fluids, and tissues. The measurements aid in the diagnosis of several conditions including juvenile and adult cirrhosis of the liver. In addition, *alpha*-1-antitrypsin deficiency has been associated with pulmonary emphysema.

(b) *Classification.* Class II (performance standards).

§ 866.5150 Bence-Jones proteins immunological test system.

(a) *Identification.* A Bence-Jones proteins immunological test system is a device that consists of the reagents used to measure by immunochemical techniques the Bence-Jones proteins in urine and plasma. Immunoglobulin molecules normally consist of pairs of polypeptide chains (subunits) of unequal size (light chains and heavy chains) bound together by several disulfide bridges. In some cancerous conditions, there is a proliferation of one plasma cell (antibody-producing cell) with excess production of light chains of one specific kind (monoclonal light chains). These free homogeneous light chains not associated with an immunoglobulin molecule can be found in urine and plasma, and have been called Bence-Jones proteins. Measurement of Bence-Jones proteins and determination that they are monoclonal aid in the diagnosis of multiple myeloma (malignant proliferation of plasma cells), Waldenstrom's macroglobulinemia (increased production of large immunoglobulins by spleen and bone marrow cells), leukemia (cancer of the blood-forming organs), and lymphoma (cancer of the lymphoid tissue).

(b) *Classification.* Class II (performance standards).

§ 866.5160 *Beta*-globulin immunological test system.

(a) *Identification.* A *beta*-globulin immunological test system is a device that consists of reagents used to measure by immunochemical techniques beta globulins (serum protein) in serum and other body fluids. *Beta*-globulin proteins include *beta*-lipoprotein, transferrin, glycoproteins, and complement, and are rarely associated with specific pathologic disorders.

(b) *Classification.* Class I (general controls). The device is exempt from the premarket notification procedures in subpart E of part 807 of this chapter subject to § 866.9.

[47 FR 50823, Nov. 9, 1982, as amended at 65 FR 2312, Jan. 14, 2000]

§ 866.5170 Breast milk immunological test system.

(a) *Identification.* A breast milk immunological test system is a device that consists of the reagents used to measure by immunochemical techniques the breast milk proteins.

(b) *Classification.* Class I (general controls). The device is exempt from the premarket notification procedures in

subpart E of part 807 of this chapter subject to the limitations in § 866.9.

[47 FR 50823, Nov. 9, 1982, as amended at 59 FR 63007, Dec. 7, 1994; 66 FR 38793, July 25, 2001]

§ 866.5200 Carbonic anhydrase B and C immunological test system.

(a) *Identification.* A carbonic anhydrase B and C immunological test system is a device that consists of the reagents used to measure by immunochemical techniques specific carbonic anhydrase protein molecules in serum and other body fluids. Measurements of carbonic anhydrase B and C aid in the diagnosis of abnormal hemoglobin metabolism.

(b) *Classification.* Class I (general controls). The device is exempt from the premarket notification procedures in subpart E of part 807 of this chapter subject to § 866.9.

[47 FR 50823, Nov. 9, 1982, as amended at 65 FR 2312, Jan. 14, 2000]

§ 866.5210 Ceruloplasmin immunological test system.

(a) *Identification.* A ceruloplasmin immunological test system is a device that consists of the reagents used to measure by immunochemical techniques the ceruloplasmin (copper-transporting serum protein) in serum, other body fluids, or tissues. Measurements of ceruloplasmin aid in the diagnosis of copper metabolism disorders.

(b) *Classification.* Class II (performance standards).

§ 866.5220 Cohn fraction II immunological test system.

(a) *Identification.* A Cohn fraction II immunological test system is a device that consists of the reagents that contain or are used to measure that fraction of plasma containing protein gamma globulins, predominantly of the IgG class. The device may be used as a coprecipitant in radioimmunoassay methods, as raw material for the purification of IgG subclasses, and to reduce nonspecific adsorption of plasma proteins in immunoassay techniques. Measurement of these proteins aids in the diagnosis of any disease concerned with abnormal levels of IgG gamma globulins such as agammaglobulinemia or multiple myeloma.

(b) *Classification.* Class I (general controls). The device is exempt from the premarket notification procedures in subpart E of part 807 of this chapter subject to the limitations in § 866.9.

[47 FR 50823, Nov. 9, 1982, as amended at 59 FR 63007, Dec. 7, 1994; 66 FR 38793, July 25, 2001]

§ 866.5230 Colostrum immunological test system.

(a) *Identification.* A colostrum immunological test system is a device that consists of the reagents used to measure by immunochemical techniques the specific proteins in colostrum. Colostrum is a substance excreted by the mammary glands during pregnancy and until production of breast milk begins 1 to 5 days after childbirth.

(b) *Classification.* Class I (general controls). The device is exempt from the premarket notification procedures in subpart E of part 807 of this chapter subject to the limitations in § 866.9.

[47 FR 50823, Nov. 9, 1982, as amended at 59 FR 63007, Dec. 7, 1994; 66 FR 38793, July 25, 2001]

§ 866.5240 Complement components immunological test system.

(a) *Identification.* A complement components immunological test system is a device that consists of the reagents used to measure by immunochemical techniques complement components C_{1q}, C_{1r}, C_{1s}, C_2, C_3, C_4, C_5, C_6, C_7, C_8, and C_9, in serum, other body fluids, and tissues. Complement is a group of serum proteins which destroy infectious agents. Measurements of these proteins aids in the diagnosis of immunologic disorders, especially those associated with deficiencies of complement components.

(b) *Classification.* Class II (performance standards).

[47 FR 50823, Nov. 9, 1982, as amended at 53 FR 11253, Apr. 6, 1988]

§ 866.5250 Complement C_2 inhibitor (inactivator) immunological test system.

(a) *Identification.* A complement C_1 inhibitor (inactivator) immunological test system is a device that consists of the reagents used to measure by immunochemical techniques the complement C_1 inhibitor (a plasma protein) in serum. Complement C_1 inhibitor occurs normally in plasma and blocks the action of the C_1 component of complement (a group of serum proteins which destroy infectious agents). Measurement of complement C_1 inhibitor aids in the diagnosis of hereditary angioneurotic edema (increased blood vessel permeability causing swelling of tissues) and a rare form of angioedema associated with lymphoma (lymph node cancer).

(b) *Classification.* Class II (performance standards).

§ 866.5260 Complement C_{3b} inactivator immunological test system.

(a) *Identification.* A complement C_{3b} inactivator immunological test system is a device that consists of the reagents used to measure by immunochemical techniques the complement C_{3b} inactivator (a plasma protein) in serum. Complement is a group of serum proteins that destroy infectious agents. Measurement of complement C_{3b} inactivator aids in the diagnosis of inherited antibody dysfunction.

(b) *Classification.* Class II (performance standards).

§ 866.5270 C-reactive protein immunological test system.

(a) *Identification.* A C-reactive protein immunological test system is a device that consists of the reagents used to measure by immunochemical techniques the C-reactive protein in serum and other body fluids. Measurement of C-reactive protein aids in evaluation of the amount of injury to body tissues.

(b) *Classification.* Class II (performance standards).

§ 866.5320 Properdin factor B immunological test system.

(a) *Identification.* A properdin factor B immunological test system is a device that consists of the reagents used to measure by immunochemical techniques properdin factor B in serum and other body fluids. The deposition of properdin factor B in body tissues or a corresponding depression in the amount of properdin factor B in serum and other body fluids is evidence of the involvement of the alternative to the classical pathway of activation of complement (a group of plasma proteins which cause the destruction of cells which are foreign to the body). Measurement of properdin factor B aids in the diagnosis of several kidney diseases, e.g., chronic glomerulonephritis (inflammation of the glomeruli of the kidney), lupus nephritis (kidney disease associated with a multisystem autoimmune disease, systemic lupus erythematosus), as well as several skin diseases, e.g., dermititis herpetiformis (presence of vesicles on the skin that burn and itch), and pemphigus vulgaris (large vesicles on the skin). Other diseases in which the alternate pathway of complement activation has been implicated include rheumatoid arthritis, sickle cell anemia, and gram-negative bacteremia.

(b) *Classification.* Class II (special controls). The device is exempt from the premarket notification procedures in subpart E of part 807 of this chapter subject to § 866.9.

[47 FR 50823, Nov. 9, 1982, as amended at 63 FR 59227, Nov. 3, 1998]

§ 866.5330 Factor XIII, A, S, immunological test system.

(a) *Identification.* A factor XIII, A, S, immunological test system is a device that consists of the reagents used to measure by immunochemical techniques the factor XIII (a bloodclotting factor), in platelets (A) or serum (S). Measurements of factor XIII, A, S, aid in the diagnosis and treatment of certain bleeding disorders resulting from a deficiency of this factor.

(b) *Classification.* Class I (general controls). The device is exempt from the premarket notification procedures in subpart E of part 807 of this chapter subject to § 866.9. This exemption does not apply to factor deficiency tests classified under § 864.7290 of this chapter.

[47 FR 50823, Nov. 9, 1982, as amended at 65 FR 2312, Jan. 14, 2000]

§ 866.5340 Ferritin immunological test system.

(a) *Identification.* A ferritin immunological test system is a device that consists of the reagents used to measure by immunochemical techniques the ferritin (an iron-storing protein) in serum and other body fluids. Measurements of ferritin aid in the diagnosis of diseases affecting iron metabolism, such as hemochromatosis (iron overload) and iron deficiency amemia.

(b) *Classification.* Class II (performance standards).

§ 866.5350 Fibrinopeptide A immunological test system.

(a) *Identification.* A fibrinopeptide A immunological test system is a device that consists of the reagents used to measure by immunochemical techniques the fibrinopeptide A (a blood-clotting factor) in plasma and other body fluids. Measurement of fibrinopeptide A may aid in the diagnosis and treatment of certain blood-clotting disorders.

(b) *Classification.* Class II (performance standards).

§ 866.5360 Cohn fraction IV immunological test system.

(a) *Identification.* A Cohn fraction IV immunological test system is a device that consists of or measures that fraction of plasma proteins, predominantly *alpha-* and *beta-* globulins, used as a raw material for the production of pure *alpha-* or *beta-* globulins. Measurement of specific *alpha-* or *beta-* globulins aids in the diagnosis of many diseases, such as Wilson's disease (an inherited disease affecting the liver and brain), Tangier's disease (absence of *alpha*-1-lipoprotein), malnutrition, iron deficiency anemia, red blood cell disorders, and kidney disease.

(b) *Classification.* Class I (general controls). The device is exempt from the premarket notification procedures in subpart E of part 807 of this chapter subject to the limitations in § 866.9.

[47 FR 50823, Nov. 9, 1982; 47 FR 56846, Dec. 21, 1982, as amended at 59 FR 63007, Dec. 7, 1994; 66 FR 38793, July 25, 2001]

§ 866.5370 Cohn fraction V immunological test system.

(a) *Identification.* A Cohn fraction V immunological test system is a device that consists of or measures that fraction of plasma containing predominantly albumin (a plasma protein). This test aids in the diagnosis of diseases where albumin levels may be depressed, e.g., nephrosis (disease of the kidney), proteinuria (protein in the urine), gastroenteropathy (disease of the stomach and small intestine), rheumatoid arthritis, and viral hepatitis.

(b) *Classification.* Class I (general controls). The device is exempt from the premarket notification procedures in subpart E of part 807 of this chapter subject to the limitations in § 866.9.

[47 FR 50823, Nov. 9, 1982, as amended at 59 FR 63007, Dec. 7, 1994; 66 FR 38793, July 25, 2001]

§ 866.5380 Free secretory component immunological test system.

(a) *Identification.* A free secretory component immunological test system is a device that consists of the reagents used to measure by immunochemical techniques free secretory component (normally a portion of the secretory IgA antibody molecule) in body fluids. Measurement of free secretory component (protein molecules) aids in the diagnosis or repetitive lung infections and other hypogammaglobulinemic conditions (low antibody levels).

(b) *Classification.* Class II (special controls). The device is exempt from the premarket notification procedures in subpart E of part 807 of this chapter subject to § 866.9.

[47 FR 50823, Nov. 9, 1982, as amended at 63 FR 59227, Nov. 3, 1998]

§ 866.5400 *Alpha*-globulin immunological test system.

(a) *Identification.* An *alpha*-globulin immunological test system is a device that consists of the reagents used to measure by immunochemical techniques the *alpha*-globulin (a serum protein) in serum and other body fluids. Measurement of *alpha*-globulin may aid in the diagnosis of inflammatory lesions, infections, severe burns, and a variety of other conditions.

(b) *Classification.* Class I (general controls). The device is exempt from the premarket notification procedures in subpart E of part 807 of this chapter subject to § 866.9.

[47 FR 50823, Nov. 9, 1982, as amended at 65 FR 2312, Jan. 14, 2000]

§ 866.5420 *Alpha*-1-glycoproteins immunological test system.

(a) *Identification.* An *alpha*-1-glycoproteins immunological test system is a device that consists of the reagents used to measure by immunochemical techniques *alpha*-1-glycoproteins (a group of plasma proteins found in the *alpha*-1 group when subjected to electrophoresis) in serum and other body fluids. Measurement of specific *alpha*-1-glycoproteins may aid in the diagnosis of collagen (connective tissue) disorders, tuberculosis, infections, extensive malignancy, and diabetes.

(b) *Classification.* Class I (general controls). The device is exempt from the premarket notification procedures in subpart E of part 807 of this chapter subject to § 866.9.

[47 FR 50823, Nov. 9, 1982, as amended at 65 FR 2312, Jan. 14, 2000]

§ 866.5425 *Alpha*-2-glycoproteins immunological test system.

(a) *Identification.* An *alpha*-2-glycoproteins immunolgical test system is a device that consists of the reagents used to measure by immunochemical techniques the *alpha*-2-glycoproteins (a group of plasma proteins found in the *alpha*-2 group when subjected to electrophoresis) in serum and other body fluids. Measurement of *alpha*-2-glycoproteins aids in the diagnosis of some cancers and genetically inherited deficiencies of these plasma proteins.

(b) *Classification.* Class I (general controls). The device is exempt from the premarket notification procedures in subpart E of part 807 of this chapter subject to § 866.9.

[47 FR 50823, Nov. 9, 1982, as amended at 65 FR 2312, Jan. 14, 2000]

§ 866.5430 *Beta*-2-glycoprotein I immunological test system.

(a) *Identification.* A *beta*-2-glycoprotein I immunological test system is a device that consists of the reagents used to measure by immunochemical techniques the *beta*-2-glycoprotein I (a serum protein) in serum and other body fluids. Measurement of *beta*-2-glycoprotein I aids in the diagnosis of an inherited deficiency of this serum protein.

(b) *Classification.* Class I (general controls). The device is exempt from the premarket notification procedures in subpart E of part 807 of this chapter subject to § 866.9.

[47 FR 50823, Nov. 9, 1982, as amended at 65 FR 2312, Jan. 14, 2000]

§ 866.5440 *Beta*-2-glycoprotein III immunological test system.

(a) *Identification.* A *beta*-2-glycoprotein III immunological test system is a device that consists of the reagents used to measure by immunochemical techniques the *beta*-2-glycoprotein III (a serum protein) in serum and other body fluids. Measurement of *beta*-2-glycoprotein III aids in the diagnosis of an inherited deficiency of this serum protein and a variety of other conditions.

(b) *Classification.* Class I (general controls). The device is exempt from the premarket notification procedures in subpart E of part 807 of this chapter subject to § 866.9.

[47 FR 50823, Nov. 9, 1982, as amended at 65 FR 2312, Jan. 14, 2000]

§ 866.5460 Haptoglobin immunological test system.

(a) *Identification.* A haptoglobin immunological test system is a device that consists of the reagents used to measure by immunochemical techniques the haptoglobin (a protein that binds hemoglobin, the oxygen-carrying pigment in red blood cells) in serum. Measurement of haptoglobin may aid in the diagnosis of hemolytic diseases (diseases in which the red blood cells rupture and release hemoglobin) related to the formation of hemoglobin-haptoglobin complexes and certain kidney diseases.

(b) *Classification.* Class II (special controls). The device is exempt from the premarket notification procedures in subpart E of part 807 of this chapter subject to § 866.9.

[47 FR 50823, Nov. 9, 1982, as amended at 63 FR 59227, Nov. 3, 1998]

§ 866.5470 Hemoglobin immunological test system.

(a) *Indentification.* A hemoglobin immunological test system is a device that consists of the reagents used to measure by immunochemical techniques the different types of free hemoglobin (the oxygen-carrying pigment in red blood cells) in blood, urine, plasma, or other body fluids. Measurements of free hemoglobin aid in the diagnosis of various hematologic disorders, such as sickle cell anemia, Fanconi's anemia (a rare inherited disease), aplastic anemia (bone marrow does not produce enough blood cells), and leukemia (cancer of the blood-forming organs).

(b) *Classification.* Class II (performance standards).

§ 866.5490 Hemopexin immunological test system.

(a) *Indentification.* A hemopexin immunological test system is a device that consists of the reagents used to measure by immunochemical techniques the hemopexin (a serum protein that binds heme, a component of hemoglobin) in serum. Measurement of hemopexin aids in the diagnosis of various hematologic disorders, such as hemolytic anemia (anemia due to shortened in vivo survival of mature red blood cells and inability of the bone marrow to compensate for their decreased life span) and sickle cell anemia.

(b) *Classification.* Class II (special controls). The device is exempt from the premarket notification procedures in subpart E of part 807 of this chapter subject to § 866.9.

[47 FR 50823, Nov. 9, 1982, as amended at 63 FR 59227, Nov. 3, 1998]

§ 866.5500 Hypersensitivity pneumonitis immunological test system.

(a) *Identification.* A hypersensitivity pneumonitis immunological test system is a device that consists of the reagents used to measure by immunochemical techniques the immunoglobulin antibodies in serum which react specifically with organic dust derived from fungal or animal protein sources. When these antibodies react with such dusts in the lung, immune complexes precipitate and trigger an inflammatory reaction (hypersensitivity pneumonitis). Measurement of these immunoglobulin G antibodies aids in the diagnosis of hypersensitivity pneumonitis and other allergic respiratory disorders.

(b) *Classification.* Class II (performance standards).

§ 866.5510 Immunoglobulins A, G, M, D, and E immunological test system.

(a) *Identification.* An immunoglobulins A, G, M, D, and E immunological test system is a device that consists of the reagents used to measure by immunochemical techniques the immunoglobulins A, G, M, D, an E (serum antibodies) in serum. Measurement of these immunoglobulins aids in the diagnosis of abnormal protein metabolism and the body's lack of ability to resist infectious agents.

(b) *Classification.* Class II (performance standards).

§ 866.5520 Immunoglobulin G (Fab fragment specific) immunological test system.

(a) *Identification.* An immunoglobulin G (Fab fragment specific) immunological test system is a device that consists of the reagents used to measure by immunochemical techniques the Fab antigen-binding fragment resulting from breakdown of immunoglobulin G antibodies in urine, serum, and other body fluids. Measurement of Fab fragments of immunoglobulin G aids in the diagnosis of lymphoproliferative disorders, such as multiple myeloma (tumor of bone marrow cells), Waldenstrom's macroglobulinemia (increased immunoglobulin production by the spleen and bone marrow cells), and lymphoma (tumor of the lymphoid tissues).

(b) *Classification.* Class I (general controls). The device is exempt from the premarket notification procedures in subpart E of part 807 of this chapter subject to the limitations in § 866.9.

[47 FR 50823, Nov. 9, 1982, as amended at 61 FR 1119, Jan. 16, 1996; 66 FR 38793, July 25, 2001]

§ 866.5530 Immunoglobulin G (Fc fragment specific) immunological test system.

(a) *Identification.* An immunoglobulin G (Fc fragment specific) immunological test system is a device that consists of the reagents used to measure by immunochemical techniques the Fc (carbohydrate containing) fragment of immunoglobulin G (resulting from breakdown of immunoglobulin G antibodies) in urine, serum, and other body fluids. Measurement of immunoglobulin G Fc fragments aids in the diagnosis of plasma cell antibody-forming abnormalities, e.g., gamma heavy chain disease.

(b) *Classification.* Class I (general controls). The device is exempt from the premarket notification procedures in subpart E of part 807 of this chapter subject to the limitations in § 866.9.

[47 FR 50823, Nov. 9, 1982, as amended at 61 FR 1119, Jan. 16, 1996; 66 FR 38793, July 25, 2001]

§ 866.5540 Immunoglobulin G (Fd fragment specific) immunological test system.

(a) *Identification.* An immunoglobulin G (Fd fragment specific) immunological test system is a device that consists of the reagents used to measure by immunochemical techniques the amino terminal (antigen-binding) end (Fd fragment) of the heavy chain (a subunit) of the immunoglobulin antibody molecule in serum. Measurement of immunoglobulin G Fd fragments aids in the diagnosis of plasma antibody-forming cell abnormalities.

(b) *Classification.* Class I (general controls). The device is exempt from the premarket notification procedures in subpart E of part 807 of this chapter subject to the limitations in § 866.9.

[47 FR 50823, Nov. 9, 1982, as amended at 59 FR 63007, Dec. 7, 1994; 66 FR 38793, July 25, 2001]

§ 866.5550 Immunoglobulin (light chain specific) immunological test system.

(a) *Identification.* An immunoglobulin (light chain specific) immunological test system is a device that consists of the reagents used to measure by immunochemical techniques both kappa and lambda types of light chain portions of immunoglobulin molecules in serum, other body fluids, and tissues. In some disease states, an excess of light chains are produced by the antibody-forming cells. These free light chains, unassociated with gamma globulin molecules, can be found in a patient's body fluids and tissues. Measurement of the various amounts of the different types of light chains aids in the diagnosis of multiple myeloma (cancer of antibody-forming cells), lymphocytic neoplasms (cancer of lymphoid tissue), Waldenstrom's macroglobulinemia (increased production of large immunoglobulins), and connective tissue diseases such as rheumatoid arthritis or systemic lupus erythematosus.

(b) *Classification.* Class II (performance standards).

§ 866.5560 Lactic dehydrogenase immunological test system.

(a) *Identification.* A lactic dehydrogenase immunological test system is a device that consists of the reagents used to measure by immunochemical techniques the activity of the lactic dehydrogenase enzyme in serum. Increased levels of lactic dehydrogenase are found in a variety of conditions, including megaloblastic anemia (decrease in the number of mature red blood cells), myocardial infarction (heart disease), and some forms of leukemia (cancer of the blood-forming organs). However, the diagnostic usefulness of this device is limited because of the many conditions known to cause increased lactic dehydrogenase levels.

(b) *Classification.* Class I (general controls). The device is exempt from the premarket notification procedures in subpart E of part 807 of this chapter subject to § 866.9.

[47 FR 50823, Nov. 9, 1982, as amended at 65 FR 2312, Jan. 14, 2000]

§ 866.5570 Lactoferrin immunological test system.

(a) *Identification.* A lactoferrin immunological test system is a device that consists of the reagents used to measure by immunochemical techniques the lactoferrin (an iron-binding protein with the ability to inhibit the growth of bacteria) in serum, breast milk, other body fluids, and tissues. Measurement of lactoferrin may aid in the diagnosis of an inherited deficiency of this protein.

(b) *Classification.* Class I (general controls). The device is exempt from the premarket notification procedures in subpart E of part 807 of this chapter subject to § 866.9.

[47 FR 50823, Nov. 9, 1982, as amended at 65 FR 2312, Jan. 14, 2000]

§ 866.5580 *Alpha*-1-lipoprotein immunological test system.

(a) *Identification.* An *alpha*-1-lipoprotein immunological test system is a device that consists of the reagents used to measure by immunochemical techniques the *alpha*-1-lipoprotein (high-density lipoprotein) in serum and plasma. Measurement of *alpha*-1-lipoprotein may aid in the diagnosis of Tangier disease (a hereditary disorder of fat metabolism).

(b) *Classification.* Class II (performance standards).

§ 866.5590 Lipoprotein X immunological test system.

(a) *Identification.* A lipoprotein X immunological test system is a device that consists of the reagents used to measure by immunochemical techniques lipoprotein X (a high-density lipoprotein) in serum and other body fluids. Measurement of lipoprotein X aids in the diagnosis of obstructive liver disease.

(b) *Classification.* Class I (general controls). The device is exempt from the premarket notification procedures in subpart E of part 807 of this chapter subject to § 866.9.

[47 FR 50823, Nov. 9, 1982, as amended at 65 FR 2313, Jan. 14, 2000]

§ 866.5600 Low-density lipoprotein immunological test system.

(a) *Identification.* A low-density lipoprotein immunological test system is a device that consists of the reagents used to measure by immunochemical techniques the low-density lipoprotein in serum and other body fluids. Measurement of low-density lipoprotein in serum may aid in the diagnosis of disorders of lipid (fat) metabolism and help to identify young persons at risk from cardiovascular diseases.

(b) *Classification.* Class II (performance standards).

§ 866.5620 *Alpha*-2-macroglobulin immunological test system.

(a) *Identification.* An *alpha*-2-macroglobulin immunological test system is a device that consists of the reagents used to measure by immunochemical techniques the *alpha*-2-macroglobulin (a serum protein) in plasma. Measurement of *alpha*-2-macroglobulin may aid in the diagnosis of blood-clotting or clot lysis disorders.

(b) *Classification.* Class II (performance standards).

§ 866.5630 *Beta*-2-microglobulin immunological test system.

(a) *Identification.* A *beta*-2-microglobulin immunological test system is a device that consists of the reagents used to measure by immunochemical techniques *beta*-2-microglobulin (a protein molecule) in serum, urine, and other body fluids. Measurement of *beta*-2-microglobulin aids in the diagnosis of active rheumatoid arthritis and kidney disease.

(b) *Classification.* Class II (performance standards).

§ 866.5640 Infectious mononucleosis immunological test system.

(a) *Identification.* An infectious mononucleosis immunological test system is a device that consists of the reagents used to measure by immunochemical techniques heterophile antibodies frequently associated with infectious mononucleosis in serum, plasma, and other body fluids. Measurements of these antibodies aid in the diagnosis of infectious mononucleosis.

(b) *Classification.* Class II (performance standards).

[47 FR 50823, Nov. 9, 1982; 47 FR 56846, Dec. 21, 1982]

§ 866.5660 Multiple autoantibodies immunological test system.

(a) *Identification.* A multiple autoantibodies immunological test system is a device that consists of the reagents used to measure by immunochemical techniques the autoantibodies (antibodies produced against the body's own tissues) in

208-072 D-10

serum and other body fluids. Measurement of multiple autoantibodies aids in the diagnosis of autoimmune disorders (disease produced when the body's own tissues are injured by autoantibodies).

(b) *Classification.* Class II (performance standards).

§ 866.5680 Myoglobin immunological test system.

(a) *Identification.* A myoglobin immunological test system is a device that consists of the reagents used to measure by immunochemical techniques the myoglobin (an oxygen storage protein found in muscle) in serum and other body fluids. Measurement of myoglobin aids in the rapid diagnosis of heart or renal disease.

(b) *Classification.* Class II (performance standards).

§ 866.5700 Whole human plasma or serum immunological test system.

(a) *Identification.* A whole human plasma or serum immunological test system is a device that consists of reagents used to measure by immunochemical techniques the proteins in plasma or serum. Measurements of proteins in plasma or serum aid in the diagnosis of any disease concerned with abnormal levels of plasma or serum proteins, e.g., agammaglobulinemia, allergies, multiple myeloma, rheumatoid vasculitis, or hereditary angioneurotic edema.

(b) *Classification.* Class I (general controls). The device is exempt from the premarket notification procedures in subpart E of part 807 of this chapter subject to the limitations in § 866.9.

[47 FR 50823, Nov. 9, 1982, as amended at 59 FR 63007, Dec. 7, 1994; 66 FR 38793, July 25, 2001]

§ 866.5715 Plasminogen immunological test system.

(a) *Identification.* A plasminogen immunological test system is a device that consists of the reagents used to measure by immunochemical techniques the plasminogen (an inactive substance from which plasmin, a blood-clotting factor, is formed) in serum, other body fluids, and tissues. Measurement of plasminogen levels may aid in the diagnosis of fibrinolytic (blood-clotting) disorders.

(b) *Classification.* Class I (general controls). The device is exempt from the premarket notification procedures in subpart E of part 807 of this chapter subject to § 866.9.

[47 FR 50823, Nov. 9, 1982, as amended at 65 FR 2313, Jan. 14, 2000]

§ 866.5735 Prothrombin immunological test system.

(a) *Identification.* A prothrombin immunological test system is a device that consists of the reagents used to measure by immunochemical techniques the prothrombin (clotting factor II) in serum. Measurements of the amount of antigenically competent (ability to react with protein antibodies) prothrombin aid in the diagnosis of blood-clotting disorders.

(b) *Classification.* Class I (general controls). The device is exempt from the premarket notification procedures in subpart E of part 807 of this chapter subject to § 866.9. This exemption does not apply to multipurpose systems for in vitro coagulation studies classified under § 864.5425 of this chapter or prothrombin time tests classified under § 864.7750 of this chapter.

[47 FR 50823, Nov. 9, 1982, as amended at 65 FR 2313, Jan. 14, 2000]

§ 866.5750 Radioallergosorbent (RAST) immunological test system.

(a) *Identification.* A radioallergosorbent immunological test system is a device that consists of the reagents used to measure by immunochemical techniques the allergen antibodies (antibodies which cause an allergic reaction) specific for a given allergen. Measurement of specific allergen antibodies may aid in the diagnosis of asthma, allergies, and other pulmonary disorders.

(b) *Classification.* Class II (performance standards).

§ 866.5765 Retinol-binding protein immunological test system.

(a) *Identification.* A retinol-binding protein immunological test system is a device that consists of the reagents used to measure by immunochemical techniques the retinol-binding protein that binds and transports vitamin A in serum and urine. Measurement of this protein may aid in the diagnosis of kidney disease and in monitoring patients with kidney transplants.

(b) *Classification.* Class I (general controls). The device is exempt from the premarket notification procedures in subpart E of part 807 of this chapter subject to § 866.9.

[47 FR 50823, Nov. 9, 1982, as amended at 65 FR 2313, Jan. 14, 2000]

§ 866.5775 Rheumatoid factor immunological test system.

(a) *Identification.* A rheumatoid factor immunological test system is a device that consists of the reagents used to measure by immunochemical techniques the rheumatoid factor (antibodies to immunoglobulins) in serum, other body fluids, and tissues. Measurement of rheumatoid factor may aid in the diagnosis of rheumatoid arthritis.

(b) *Classification.* Class II (performance standards).

§ 866.5785 Anti-Saccharomyces cerevisiae (S. cerevisiae) antibody (ASCA) test systems.

(a) *Identification.* The Anti-*Saccharomyces cerevisiae* (*S. cerevisiae*) antibody (ASCA) test system is an in vitro diagnostic device that consists of the reagents used to measure, by immunochemical techniques, antibodies to *S. cerevisiae* (baker's or brewer's yeast) in human serum or plasma. Detection of *S. cerevisiae* antibodies may aid in the diagnosis of Crohn's disease.

(b) *Classification.* Class II (special controls). The special control is FDA's "Guidance for Industry and FDA Reviewers: Class II Special Control Guidance Document for Anti-*Saccharomyces cerevisiae* (*S. cerevisiae*) Antibody (ASCA) Premarket Notifications."

[65 FR 70307, Nov. 22, 2000]

§ 866.5800 Seminal fluid (sperm) immunological test system.

(a) *Identification.* A seminal fluid (sperm) immunological test system is a device that consists of the reagents used for legal purposes to identify and differentiate animal and human semen. The test results may be used as court evidence in alleged instances of rape and other sex-related crimes.

(b) *Classification.* Class I (general controls). The device is exempt from the premarket notification procedures in subpart E of part 807 of this chapter subject to the limitations in § 866.9.

[54 FR 25047, June 12, 1989, as amended at 66 FR 38793, July 25, 2001]

§ 866.5820 Systemic lupus erythematosus immunological test system.

(a) *Identification.* A systemic lupus erythematosus (SLE) immunological test system is a device that consists of the reagents used to measure by immunochemical techniques the autoimmune antibodies in serum and other body fluids that react with cellular nuclear double-stranded deoxyribonucleic acid (DNA) or other nuclear constituents that are specifically diagnostic of SLE. Measurement of nuclear double-stranded DNA antibodies aids in the diagnosis of SLE (a multisystem autoimmune disease in which tissues are attacked by the person's own antibodies).

(b) *Classification.* Class II (performance standards).

§ 866.5860 Total spinal fluid immunological test system.

(a) *Identification.* A total spinal fluid immunological test system is a device that consists of the reagents used to measure by immunochemical techniques the total protein in cerebrospinal fluid. Measurement of spinal fluid proteins may aid in the diagnosis of multiple sclerosis and other diseases of the nervous system.

(b) *Classification.* Class I (general controls). The device is exempt from the premarket notification procedures in subpart E of part 807 of this chapter subject to the limitations in § 866.9.

[47 FR 50823, Nov. 9, 1982, as amended at 61 FR 1119, Jan. 16, 1996; 66 FR 38793, July 25, 2001]

§ 866.5870 Thyroid autoantibody immunological test system.

(a) *Identification.* A thyroid autoantibody immunological test system is a device that consists of the reagents used to measure by immunochemical techniques the thyroid autoantibodies (antibodies produced against the body's own tissues). Measurement of thyroid autoantibodies may aid in the diagnosis of certain thyroid disorders, such as Hashimoto's disease (chronic lymphocytic thyroiditis), nontoxic goiter (enlargement of thyroid gland), Grave's disease (enlargement of the thyroid gland with protrusion of the eyeballs), and cancer of the thyroid.

(b) *Classification.* Class II (performance standards).

§ 866.5880 Transferrin immunological test system.

(a) *Identification.* A transferrin immunological test system is a device that consists of the reagents used to measure by immunochemical techniques the transferrin (an iron-binding and transporting serum protein) in serum, plasma, and other body fluids. Measurement of transferrin levels aids in the diagnosis of malnutrition, acute inflammation, infection, and red blood cell disorders, such as iron deficiency anemia.

(b) *Classification.* Class II (performance standards).

§ 866.5890 Inter-*alpha* trypsin inhibitor immunological test system.

(a) *Identification.* An inter-*alpha* trypsin inhibitor immunological test system is a device that consists of the reagents used to measure by immunochemical techniques the inter-*alpha* trypsin inhibitor (a protein) in serum and other body fluids. Measurement of inter-*alpha* trypsin inhibitor may aid in the diagnosis of acute bacterial infection and inflammation.

(b) *Classification.* Class I (general controls). The device is exempt from the premarket notification procedures in subpart E of part 807 of this chapter subject to § 866.9.

[47 FR 50823, Nov. 9, 1982, as amended at 53 FR 11253, Apr. 6, 1988; 65 FR 2313, Jan. 14, 2000]

§ 866.5900 Cystic fibrosis transmembrane conductance regulator (CFTR) gene mutation detection system.

(a) *Identification.* The CFTR gene mutation detection system is a device used to simultaneously detect and identify a panel of mutations and variants in the CFTR gene. It is intended as an aid in confirmatory diagnostic testing of individuals with suspected cystic fibrosis (CF), carrier identification, and newborn screening. This device is not intended for stand-alone diagnostic purposes, prenatal diagnostic, pre-implantation, or population screening.

(b) *Classification.* Class II (special controls). The special control is FDA's guidance document entitled "Class II Special Controls Guidance Document: CFTR Gene Mutation Detection System." See § 866.1(e) for the availability of this guidance document.

[70 FR 61738, Oct. 26, 2005]

Subpart G—Tumor Associated Antigen immunological Test Systems

§ 866.6010 Tumor-associated antigen immunological test system.

(a) *Identification.* A tumor-associated antigen immunological test system is a device that consists of reagents used to qualitatively or quantitatively measure, by immunochemical techniques, tumor-associated antigens in serum, plasma, urine, or other body fluids. This device is intended as an aid in monitoring patients for disease progress or response to therapy or for the detection of recurrent or residual disease.

(b) *Classification.* Class II (special controls). Tumor markers must comply with the following special controls: (1) A guidance document entitled "Guidance Document for the Submission of Tumor Associated Antigen Premarket Notifications (510(k)s) to FDA," and (2) voluntary assay performance standards issued by the National Committee on Clinical Laboratory Standards.

[62 FR 66005, Dec. 17, 1997]

§ 866.6020 Immunomagnetic circulating cancer cell selection and enumeration system.

(a) *Identification.* An immunomagnetic circulating cancer cell selection and enumeration system is a device that consists of biological probes, fluorochromes, and other reagents; preservation and preparation devices; and a semiautomated analytical instrument to select and count circulating cancer cells in a prepared sample of whole blood. This device is intended for adjunctive use in monitoring or predicting cancer disease progression, response to therapy, and for the detection of recurrent disease.

(b) *Classification.* Class II (special controls). The special control for this device is FDA's guidance document entitled "Class II Special Controls Guidance Document: Immunomagnetic Circulating Cancer Cell Selection and Enumeration System." See § 866.1(e) for availability of this guidance document.

[69 FR 26038, May 11, 2004]

§ 866.6030 AFP-L3% immunological test system.

(a) *Identification.* An AFP-L3% immunological test system is an in vitro device that consists of reagents and an automated instrument used to quantitatively measure, by immunochemical techniques, AFP and AFP-L3 subfraction in human serum. The device is intended for in vitro diagnostic use as an aid in the risk assessment of patients with chronic liver disease for development of hepatocellular carcinoma, in conjunction with other laboratory findings, imaging studies, and clinical assessment.

(b) *Classification.* Class II (special controls). The special control is FDA's guidance document entitled "Class II Special Controls Guidance Document: AFP-L3% Immunological Test Systems." See § 866.1(e) for the availability of this guidance document.

[70 FR 57749, Oct. 4, 2005]

PART 868—ANESTHESIOLOGY DEVICES

Subpart A—General Provisions

Sec.

Subpart B—Diagnostic Devices

Subpart C—Monitoring Devices

Subparts D–E [Reserved]

Subpart F—Therapeutic Devices

868.5090 Emergency airway needle.
868.5100 Nasopharyngeal airway.
868.5110 Oropharyngeal airway.
868.5115 Device to relieve acute upper airway obstruction.
868.5120 Anesthesia conduction catheter.
868.5130 Anesthesia conduction filter.
868.5140 Anesthesia conduction kit.
868.5150 Anesthesia conduction needle.
868.5160 Gas machine for anesthesia or analgesia.
868.5165 Nitric oxide administration apparatus.
868.5170 Laryngotracheal topical anesthesia applicator.
868.5180 Rocking bed.
868.5220 Blow bottle.
868.5240 Anesthesia breathing circuit.
868.5250 Breathing circuit circulator.
868.5260 Breathing circuit bacterial filter.
868.5270 Breathing system heater.
868.5280 Breathing tube support.
868.5300 Carbon dioxide absorbent.
868.5310 Carbon dioxide absorber.
868.5320 Reservoir bag.
868.5330 Breathing gas mixer.
868.5340 Nasal oxygen cannula.
868.5350 Nasal oxygen catheter.
868.5365 Posture chair for cardiac or pulmonary treatment.
868.5375 Heat and moisture condenser (artificial nose).
868.5400 Electroanesthesia apparatus.
868.5420 Ether hook.
868.5430 Gas-scavenging apparatus.
868.5440 Portable oxygen generator.
868.5450 Respiratory gas humidifier.
868.5460 Therapeutic humidifier for home use.
868.5470 Hyperbaric chamber.
868.5530 Flexible laryngoscope.
868.5540 Rigid laryngoscope.
868.5550 Anesthetic gas mask.
868.5560 Gas mask head strap.
868.5570 Nonrebreathing mask.
868.5580 Oxygen mask.
868.5590 Scavenging mask.
868.5600 Venturi mask.
868.5610 Membrane lung for long-term pulmonary support.
868.5620 Breathing mouthpiece.
868.5630 Nebulizer.
868.5640 Medicinal nonventilatory nebulizer (atomizer).
868.5650 Esophageal obturator.
868.5655 Portable liquid oxygen unit.
868.5665 Powered percussor.
868.5675 Rebreathing device.
868.5690 Incentive spirometer.
868.5700 Nonpowered oxygen tent.
868.5710 Electrically powered oxygen tent.
868.5720 Bronchial tube.
868.5730 Tracheal tube.
868.5740 Tracheal/bronchial differential ventilation tube.
868.5750 Inflatable tracheal tube cuff.
868.5760 Cuff spreader.
868.5770 Tracheal tube fixation device.
868.5780 Tube introduction forceps.
868.5790 Tracheal tube stylet.
868.5795 Tracheal tube cleaning brush.
868.5800 Tracheostomy tube and tube cuff.
868.5810 Airway connector.
868.5820 Dental protector.
868.5830 Autotransfusion apparatus.
868.5860 Pressure tubing and accessories.
868.5870 Nonrebreathing valve.
868.5880 Anesthetic vaporizer.
868.5895 Continuous ventilator.
868.5905 Noncontinuous ventilator (IPPB).
868.5915 Manual emergency ventilator.
868.5925 Powered emergency ventilator.
868.5935 External negative pressure ventilator.
868.5955 Intermittent mandatory ventilation attachment.
868.5965 Positive end expiratory pressure breathing attachment.
868.5975 Ventilator tubing.
868.5995 Tee drain (water trap).

Subpart G—Miscellaneous

868.6100 Anesthetic cabinet, table, or tray.
868.6175 Cardiopulmonary emergency cart.
868.6225 Nose clip.
868.6250 Portable air compressor.
868.6400 Calibration gas.
868.6700 Anesthesia stool.
868.6810 Tracheobronchial suction catheter.
868.6820 Patient position support.
868.6885 Medical gas yoke assembly.

AUTHORITY: 21 U.S.C. 351, 360, 360c, 360e, 360j, 371.

SOURCE: 47 FR 31142, July 16, 1982, unless otherwise noted.

Subpart A—General Provisions

§ 868.1 Scope.

(a) This part sets forth the classification of anesthesiology devices intended for human use that are in commercial distribution.

(b) The identification of a device in a regulation in this part is not a precise description of every device that is, or will be, subject to the regulation. A manufacturer who submits a premarket notification submission for a device under part 807 may not show merely that the device is accurately described by the section title and identification provisions of a regulation in this part, but shall state why the device is substantially equivalent to other devices, as required by § 807.87.

(c) To avoid duplicative listings, an anesthesiology device that has two or more types of uses (e.g., used both as a diagnostic device and as a therapeutic device) is listed only in one subpart.

(d) References in this part to regulatory sections of the Code of Federal Regulations are to chapter I of title 21, unless otherwise noted.

(e) Guidance documents referenced in this part are available on the Internet at *http://www.fda.gov/cdrh/guidance.html.*

[52 FR 17734, May 11, 1987, as amended at 67 FR 76681, Dec. 13, 2002]

§ 868.3 Effective dates of requirement for premarket approval.

A device included in this part that is classified into class III (premarket approval) shall not be commercially distributed after the date shown in the regulation classifying the device unless the manufacturer has an approval under section 515 of the act (unless an exemption has been granted under section 520(g)(2) of the act). An approval under section 515 of the act consists of FDA's issuance of an order approving an application for premarket approval (PMA) for the device or declaring completed a product development protocol (PDP) for the device.

(a) Before FDA requires that a device commercially distributed before the enactment date of the amendments, or a device that has been found substantially equivalent to such a device, has an approval under section 515 of the act FDA must promulgate a regulation under section 515(b) of the act requiring such approval, except as provided in paragraph (b) of this section. Such a regulation under section 515(b) of the act shall not be effective during the grace period ending on the 90th day after its promulgation or on the last day of the 30th full calendar month after the regulation that classifies the device into class III is effective, whichever is later. See section 501(f)(2)(B) of the act. Accordingly, unless an effective date of the requirement for premarket approval is shown in the regulation for a device classified into class III in this part, the device may be commercially distributed without FDA's issuance of an order approving a PMA or declaring completed a PDP for the device. If FDA promulgates a regulation under section 515(b) of the act requiring premarket approval for a device, section 501(f)(1)(A) of the act applies to the device.

(b) Any new, not substantially equivalent, device introduced into commercial distribution on or after May 28, 1976, including a device formerly marketed that has been substantially altered, is classified by statute (section 513(f) of the act) into class III without any grace period and FDA must have issued an order approving a PMA or declaring completed a PDP for the device before the device is commercially distributed unless it is reclassified. If FDA knows that a device being commercially distributed may be a "new" device as defined in this section because of any new intended use or other reasons, FDA may codify the statutory classification of the device into class III for such new use. Accordingly, the regulation for such a class III device states that as of the enactment date of the amendments, May 28, 1976, the device must have an approval under section 515 of the act before commercial distribution.

[52 FR 17734, May 11, 1987]

§ 868.9 Limitations of exemptions from section 510(k) of the Federal Food, Drug, and Cosmetic Act (the act).

The exemption from the requirement of premarket notification (section 510(k) of the act) for a generic type of class I or II device is only to the extent that the device has existing or reasonably foreseeable characteristics of commercially distributed devices within that generic type or, in the case of in vitro diagnostic devices, only to the extent that misdiagnosis as a result of using the device would not be associated with high morbidity or mortality. Accordingly, manufacturers of any commercially distributed class I or II device for which FDA has granted an exemption from the requirement of premarket notification must still submit a premarket notification to FDA before introducing or delivering for introduction into interstate commerce for commercial distribution the device when:

(a) The device is intended for a use different from the intended use of a legally marketed device in that generic type of device; e.g., the device is intended for a different medical purpose, or the device is intended for lay use where the former intended use was by health care professionals only;

(b) The modified device operates using a different fundamental scientific technology than a legally marketed device in that generic type of device; e.g., a surgical instrument cuts tissue with a laser beam rather than with a sharpened metal blade, or an in vitro diagnostic device detects or identifies infectious agents by using deoxyribonucleic acid (DNA) probe or nucleic acid hybridization technology rather than culture or immunoassay technology; or

(c) The device is an in vitro device that is intended:

(1) For use in the diagnosis, monitoring, or screening of neoplastic diseases with the exception of immunohistochemical devices;

(2) For use in screening or diagnosis of familial or acquired genetic disorders, including inborn errors of metabolism;

(3) For measuring an analyte that serves as a surrogate marker for screening, diagnosis, or monitoring life-threatening diseases such as acquired immune deficiency syndrome (AIDS), chronic or active hepatitis, tuberculosis, or myocardial infarction or to monitor therapy;

(4) For assessing the risk of cardiovascular diseases;

(5) For use in diabetes management;

(6) For identifying or inferring the identity of a microorganism directly from clinical material;

(7) For detection of antibodies to microorganisms other than immunoglobulin G (IgG) or IgG assays when the results are not qualitative, or are used to determine immunity, or the assay is intended for use in matrices other than serum or plasma;

(8) For noninvasive testing as defined in § 812.3(k) of this chapter; and

(9) For near patient testing (point of care).

[65 FR 2313, Jan. 14, 2000]

Subpart B—Diagnostic Devices

§ 868.1030 Manual algesimeter.

(a) *Identification.* A manual algesimeter is a mechanical device intended to determine a patient's sensitivity to pain after administration of an anesthetic agent, e.g., by pricking with a sharp point.

(b) *Classification.* Class I (general controls). The device is exempt from the premarket notification procedures in subpart E of part 807 of this chapter subject to the limitations in § 868.9. The device is also exempt from the current good manufacturing practice regulations in part 820 of this chapter, with the exception of § 820.180, with respect to general requirements concerning records, and § 820.198, with respect to complaint files.

[54 FR 25048, June 12, 1989, as amended at 66 FR 38793, July 25, 2001]

§ 868.1040 Powered algesimeter.

(a) *Identification.* A powered algesimeter is a device using electrical stimulation intended to determine a patient's sensitivity to pain after administration of an anesthetic agent.

(b) *Classification.* Class II (performance standards).

§ 868.1075 Argon gas analyzer.

(a) *Identification.* An argon gas analyzer is a device intended to measure the concentration of argon in a gas mixture to aid in determining the patient's ventilatory status. The device may use techniques such as mass spectrometry or thermal conductivity.

(b) *Classification.* Class II (performance standards).

§ 868.1100 Arterial blood sampling kit.

(a) *Identification.* An arterial blood sampling kit is a device, in kit form, used to obtain arterial blood samples from a patient for blood gas determinations. The kit may include a syringe, needle, cork, and heparin.

(b) *Classification.* Class I (general controls). The device is exempt from the premarket notification procedures in

subpart E of part 807 of this chapter subject to the limitations in § 868.9.

[47 FR 31142, July 16, 1982, as amended at 61 FR 1119, Jan. 16, 1996; 66 FR 38793, July 25, 2001]

§ 868.1120 Indwelling blood oxyhemoglobin concentration analyzer.

(a) *Identification.* An indwelling blood oxyhemoglobin concentration analyzer is a photoelectric device used to measure, in vivo, the oxygen-carrying capacity of hemoglobin in blood to aid in determining the patient's physiological status.

(b) *Classification.* Class III (premarket approval).

(c) Date PMA or notice of completion of PDP is required. A PMA or notice of completion of a PDP is required to be filed with the Food and Drug Administration on or before September 21, 2004, for any indwelling blood oxyhemoglobin concentration analyzer that was in commercial distribution before May 28, 1976, or that has, on or before September 21, 2004, been found to be substantially equivalent to an indwelling blood oxyhemoglobin concentration analyzer that was in commercial distribution before May 28, 1976. Any other indwelling blood oxyhemoglobin concentration analyzer shall have an approved PMA or declared completed PDP in effect before being placed in commercial distribution.

[47 FR 31142, July 16, 1982, as amended at 52 FR 17735, May 11, 1987; 52 FR 22577, June 12, 1987; 69 FR 34920, June 23, 2004]

§ 868.1150 Indwelling blood carbon dioxide partial pressure (P_{CO2}) analyzer.

(a) *Identification.* An indwelling blood carbon dioxide partial pressure P_{CO2} analyzer is a device that consists of a catheter-tip P_{CO2} transducer (e.g., P_{CO2} electrode) and that is used to measure, in vivo, the partial pressure of carbon dioxide in blood to aid in determining the patient's circulatory, ventilatory, and metabolic status.

(b) *Classification.* Class II (special controls). The special control for this device is FDA's "Class II Special Controls Guidance Document: Indwelling Blood Gas Analyzers; Final Guidance for Industry and FDA."

[47 FR 31142, July 16, 1982; 47 FR 40410, Sept. 14, 1982, as amended at 52 FR 17735, May 11, 1987; 66 FR 57368, Nov. 15, 2001]

§ 868.1170 Indwelling blood hydrogen ion concentration (pH) analyzer.

(a) *Identification.* An indwelling blood hydrogen ion concentration (pH) analyzer is a device that consists of a catheter-tip pH electrode and that is used to measure, in vivo, the hydrogen ion concentration (pH) in blood to aid in determining the patient's acid-base balance.

(b) *Classification.* Class II (special controls). The special control for this device is FDA's "Class II Special Controls Guidance Document: Indwelling Blood Gas Analyzers; Final Guidance for Industry and FDA."

[47 FR 31142, July 16, 1982, as amended at 52 FR 17735, May 11, 1987; 66 FR 57368, Nov. 15, 2001]

§ 868.1200 Indwelling blood oxygen partial pressure (P_{O2}) analyzer.

(a) *Identification.* An indwelling blood oxygen partial pressure (P_{O2}) analyzer is a device that consists of a catheter-tip P_{O2} transducer (e.g., P_{O2} electrode) and that is used to measure, in vivo, the partial pressure of oxygen in blood to aid in determining the patient's circulatory, ventilatory, and metabolic status.

(b) *Classification.* Class II (special controls). The special control for this device is FDA's "Class II Special Controls Guidance Document: Indwelling Blood Gas Analyzers; Final Guidance for Industry and FDA."

[47 FR 31142, July 16, 1982; 47 FR 40410, Sept. 14, 1982, as amended at 52 FR 17735, May 11, 1987; 66 FR 57368, Nov. 15, 2001]

§ 868.1400 Carbon dioxide gas analyzer.

(a) *Identification.* A carbon dioxide gas analyzer is a device intended to measure the concentration of carbon dioxide in a gas mixture to aid in determining the patient's ventilatory, circulatory, and metabolic status. The device may use techniques such as chemical titration, absorption of infrared radiation, gas chromatography, or mass spectrometry.

(b) *Classification.* Class II (performance standards).

§ 868.1430 Carbon monoxide gas analyzer.

(a) *Identification.* A carbon monoxide gas analyzer is a device intended to measure the concentration of carbon monoxide in a gas mixture to aid in determining the patient's ventilatory status. The device may use techniques such as infrared absorption or gas chromatography.

(b) *Classification.* Class II (performance standards).

§ 868.1500 Enflurane gas analyzer.

(a) *Identification.* An enflurane gas analyzer is a device intended to measure the concentration of enflurane anesthetic in a gas mixture.

(b) *Classification.* Class II (performance standards).

§ 868.1575 Gas collection vessel.

(a) *Identification.* A gas collection vessel is a container-like device intended to collect a patient's exhaled gases for subsequent analysis. It does not include a sampling pump.

(b) *Classification.* Class I (general controls). The device is exempt from the premarket notification procedures in subpart E of part 807 of this chapter subject to the limitations in § 868.9.

[47 FR 31142, July 16, 1982, as amended at 61 FR 1119, Jan. 16, 1996; 66 FR 38793, July 25, 2001]

§ 868.1620 Halothane gas analyzer.

(a) *Identification.* A halothane gas analyzer is a device intended to measure the concentration of halothane anesthetic in a gas mixture. The device may use techniques such as mass spectrometry or absorption of infrared or ultraviolet radiation.

(b) *Classification.* Class II (performance standards).

§ 868.1640 Helium gas analyzer.

(a) *Identification.* A helium gas analyzer is a device intended to measure the concentration of helium in a gas mixture during pulmonary function testing. The device may use techniques such as thermal conductivity, gas chromatography, or mass spectrometry.

(b) *Classification.* Class II (performance standards).

§ 868.1670 Neon gas analyzer.

(a) *Identification.* A neon gas analyzer is a device intended to measure the concentration of neon in a gas mixture exhaled by a patient. The device may use techniques such as mass spectrometry or thermal conductivity.

(b) *Classification.* Class II (performance standards).

§ 868.1690 Nitrogen gas analyzer.

(a) *Identification.* A nitrogen gas analyzer is a device intended to measure the concentration of nitrogen in respiratory gases to aid in determining a patient's ventilatory status. The device may use techniques such as gas chromatography or mass spectrometry.

(b) *Classification.* Class II (performance standards).

§ 868.1700 Nitrous oxide gas analyzer.

(a) *Identification.* A nitrous oxide gas analyzer is a device intended to measure the concentration of nitrous oxide anesthetic in a gas mixture. The device may use techniques such as infrared absorption or mass spectrometry.

(b) *Classification.* Class II (performance standards).

§ 868.1720 Oxygen gas analyzer.

(a) *Identification.* An oxygen gas analyzer is a device intended to measure the concentration of oxygen in respiratory gases by techniques such as mass spectrometry, polarography, thermal conductivity, or gas chromatography. This generic type of device also includes paramagnetic analyzers.

(b) *Classification.* Class II (performance standards).

§ 868.1730 Oxygen uptake computer.

(a) *Identification.* An oxygen uptake computer is a device intended to compute the amount of oxygen consumed by a patient and may include components for determining expired gas volume and composition.

(b) *Classification.* Class II (performance standards).

§ 868.1750 Pressure plethysmograph.

(a) *Identification.* A pressure plethysmograph is a device used to determine a patient's airway resistance and lung volumes by measuring pressure changes while the patient is in an airtight box.

(b) *Classification.* Class II (performance standards).

§ 868.1760 Volume plethysmograph.

(a) *Identification.* A volume plethysmograph is an airtight box, in which a patient sits, that is used to determine the patient's lung volume changes.

(b) *Classification.* Class II (performance standards).

§ 868.1780 Inspiratory airway pressure meter.

(a) *Identification.* An inspiratory airway pressure meter is a device used to measure the amount of pressure produced in a patient's airway during maximal inspiration.

(b) *Classification.* Class II (performance standards).

§ 868.1800 Rhinoanemometer.

(a) *Identification.* A rhinoanemometer is a device used to quantify the amount of nasal congestion by measuring the airflow through, and differential pressure across, a patient's nasal passages.

(b) *Classification.* Class II (performance standards).

§ 868.1840 Diagnostic spirometer.

(a) *Identification.* A diagnostic spirometer is a device used in pulmonary function testing to measure the volume of gas moving in or out of a patient's lungs.

(b) *Classification.* Class II (performance standards).

§ 868.1850 Monitoring spirometer.

(a) *Identification.* A monitoring spirometer is a device used to measure continuously a patient's tidal volume (volume of gas inhaled by the patient during each respiration cycle) or minute volume (the tidal volume multiplied by the rate of respiration for 1 minute) for the evaluation of the patient's ventilatory status.

(b) *Classification.* Class II (performance standards).

§ 868.1860 Peak-flow meter for spirometry.

(a) *Identification.* A peak-flow meter for spirometry is a device used to measure a patient's maximum ventilatory flow rate.

(b) *Classification.* Class II (performance standards).

§ 868.1870 Gas volume calibrator.

(a) *Identification.* A gas volume calibrator is a device that is intended for medical purposes and that is used to calibrate the output of gas volume measurement instruments by delivering a known gas volume.

(b) *Classification.* Class I (general controls). The device is exempt from the premarket notification procedures in subpart E of part 807 of this chapter subject to the limitations in § 868.9.

[47 FR 31142, July 16, 1982, as amended at 61 FR 1119, Jan. 16, 1996; 66 FR 38793, July 25, 2001]

§ 868.1880 Pulmonary-function data calculator.

(a) *Identification.* A pulmonary-function data calculator is a device used to calculate pulmonary-function values based on actual physical data obtained during pulmonary-function testing.

(b) *Classification.* Class II (performance standards).

§ 868.1890 Predictive pulmonary-function value calculator.

(a) *Identification.* A predictive pulmonary-function value calculator is a device used to calculate normal pulmonary-function values based on empirical equations.

(b) *Classification.* Class II (performance standards).

§ 868.1900 Diagnostic pulmonary-function interpretation calculator.

(a) *Identification.* A diagnostic pulmonary-function interpretation calculator is a device that interprets pulmonary study data to determine clinical significance of pulmonary-function values.

(b) *Classification.* Class II (performance standards).

§ 868.1910 Esophageal stethoscope.

(a) *Identification.* An esophageal stethoscope is a nonpowered device that is inserted into a patient's esophagus to enable the user to listen to heart and breath sounds.

(b) *Classification.* Class I (general controls). The device is exempt from the premarket notification procedures in subpart E of part 807 of this chapter subject to § 868.9.

[47 FR 31142, July 16, 1982, as amended at 65 FR 2313, Jan. 14, 2000]

§ 868.1920 Esophageal stethoscope with electrical conductors.

(a) *Identification.* An esophageal stethoscope with electrical conductors is a device that is inserted into the esophagus to listen to a patient's heart and breath sounds and to monitor electrophysiological signals. The device may also incorporate a thermistor for temperature measurement.

(b) *Classification.* Class II (performance standards).

§ 868.1930 Stethoscope head.

(a) *Identification.* A stethoscope head is a weighted chest piece used during anesthesia to listen to a patient's heart, breath, and other physiological sounds.

(b) *Classification.* Class I (general controls). The device is exempt from the premarket notification procedures in subpart E of part 807 of this chapter subject to the limitations in § 868.9.

[47 FR 31142, July 16, 1982, as amended at 54 FR 25048, June 12, 1989; 66 FR 38793, July 25, 2001]

§ 868.1965 Switching valve (ploss).

(a) *Identification.* A switching valve (ploss) is a three-way valve located between a stethoscope placed over the heart, a blood pressure cuff, and an earpiece. The valve allows the user to eliminate one sound channel and listen only to a patient's heart or korotkoff (blood pressure) sounds through the other channel.

(b) *Classification.* Class I (general controls). The device is exempt from the premarket notification procedures in subpart E of part 807 of this chapter subject to the limitations in § 868.9. The device is also exempt from the current good manufacturing practice regulations in part 820 of this chapter, with the exception of § 820.180, with respect to general requirements concerning records, and § 820.198, with respect to complaint files.

[47 FR 31142, July 16, 1982, as amended at 54 FR 25048, June 12, 1989; 66 FR 38793, July 25, 2001]

§ 868.1975 Water vapor analyzer.

(a) *Identification.* A water vapor analyzer is a device intended to measure the concentration of water vapor in a patient's expired gases by using techniques such as mass spectrometry.

(b) *Classification.* Class I (general controls). The device is exempt from the premarket notification procedures in subpart E of part 807 of this chapter subject to the limitations in § 868.9.

[47 FR 31142, July 16, 1982, as amended at 61 FR 1119, Jan. 16, 1996; 66 FR 38793, July 25, 2001]

Subpart C—Monitoring Devices

§ 868.2025 Ultrasonic air embolism monitor.

(a) *Identification.* An ultrasonic air embolism monitor is a device used to detect air bubbles in a patient's blood stream. It may use Doppler or other ultrasonic principles.

(b) *Classification.* Class II (performance standards).

§ 868.2300 Bourdon gauge flowmeter.

(a) *Identification.* A bourdon gauge flowmeter is a device intended for medical purposes that is used in conjunction with respiratory equipment to sense gas pressure. The device is calibrated to indicate gas flow rate when the outflow is open to the atmosphere.

(b) *Classification.* Class I (general controls). The device is exempt from the premarket notification procedures in subpart E of part 807 of this chapter subject to the limitations in § 868.9.

[47 FR 31142, July 16, 1982, as amended at 61 FR 1119, Jan. 16, 1996; 66 FR 38794, July 25, 2001]

§ 868.2320 Uncompensated thorpe tube flowmeter.

(a) *Identification.* An uncompensated thorpe tube flowmeter is a device intended for medical purposes that is used to indicate and control gas flow rate accurately. The device includes a vertically mounted tube and is calibrated when the outlet of the flowmeter is open to the atmosphere.

(b) *Classification.* Class I (general controls). The device is exempt from the premarket notification procedures in subpart E of part 807 of this chapter subject to the limitations in § 868.9.

[47 FR 31142, July 16, 1982, as amended at 61 FR 1119, Jan. 16, 1996; 66 FR 38794, July 25, 2001]

§ 868.2340 Compensated thorpe tube flowmeter.

(a) *Identification.* A compensated thorpe tube flowmeter is a device intended for medical purposes that is used to control and measure gas flow rate accurately. The device includes a vertically mounted tube, with the outlet of the flowmeter calibrated to a reference pressure.

(b) *Classification.* Class I (general controls). The device is exempt from the premarket notification procedures in subpart E of part 807 of this chapter subject to the limitations in § 868.9.

[47 FR 31142, July 16, 1982, as amended at 61 FR 1119, Jan. 16, 1996; 66 FR 38794, July 25, 2001]

§ 868.2350 Gas calibration flowmeter.

(a) *Identification.* A gas calibration flowmeter is a device intended for medical purposes that is used to calibrate flowmeters and accurately measure gas flow.

(b) *Classification.* Class I (general controls). The device is exempt from the premarket notification procedures in subpart E of part 807 of this chapter subject to the limitations in § 868.9.

[47 FR 31142, July 16, 1982, as amended at 61 FR 1119, Jan. 16, 1996; 66 FR 38794, July 25, 2001]

§ 868.2375 Breathing frequency monitor.

(a) *Identification.* A breathing (ventilatory) frequency monitor is a device intended to measure or monitor a patient's respiratory rate. The device may provide an audible or visible alarm when the respiratory rate, averaged over time, is outside operator settable alarm limits. This device does not include the apnea monitor classified in § 868.2377.

(b) *Classification.* Class II (performance standards).

[47 FR 31142, July 16, 1982, as amended at 67 FR 46852, July 17, 2002]

§ 868.2377 Apnea monitor.

(a) *Identification.* An apnea monitor is a complete system intended to alarm primarily upon the cessation of breathing timed from the last detected breath. The apnea monitor also includes indirect methods of apnea detection such as monitoring of heart rate and other physiological parameters linked to the presence or absence of adequate respiration.

(b) *Classification.* Class II (special controls). The special control for this device is the FDA guidance document entitled "Class II Special Controls Guidance Document: Apnea Monitors; Guidance for Industry and FDA."

[67 FR 46852, July 17, 2002]

§ 868.2380 Nitric oxide analyzer.

(a) *Identification.* The nitric oxide analyzer is a device intended to measure the concentration of nitric oxide in respiratory gas mixtures during administration of nitric oxide.

(b) *Classification.* Class II. The special control for this device is FDA's "Guidance Document for Premarket Notification Submissions for Nitric Oxide Administration Apparatus, Nitric Oxide Analyzer, and Nitrogen Dioxide Analyzer."

[65 FR 14465, Mar. 3, 2000]

§ 868.2385 Nitrogen dioxide analyzer.

(a) *Identification.* The nitrogen dioxide analyzer is a device intended to measure the concentration of nitrogen dioxide in respiratory gas mixtures during administration of nitric oxide.

(b) *Classification.* Class II. The special control for this device is FDA's "Guidance Document for Premarket Notification Submissions for Nitric Oxide Administration Apparatus, Nitric

Oxide Analyzer, and Nitrogen Dioxide Analyzer."

[65 FR 11465, Mar. 3, 2000]

§ 868.2450 Lung water monitor.

(a) *Identification.* A lung water monitor is a device used to monitor the trend of fluid volume changes in a patient's lung by measuring changes in thoracic electrical impedance (resistance to alternating current) by means of electrodes placed on the patient's chest.

(b) *Classification.* Class III (premarket approval).

(c) *Date PMA or notice of completion of a PDP is required.* A PMA or a notice of completion of a PDP for a device is required to be filed with the Food and Drug Administration on or before July 12, 2000, for any lung water monitor that was in commercial distribution before May 28, 1976, or that has, on or before July 12, 2000, been found to be substantially equivalent to a lung water monitor that was in commercial distribution before May 28, 1976. Any other lung water monitor device shall have an approved PMA or declared completed PDP in effect before being placed in commercial distribution.

[47 FR 31142, July 16, 1982, as amended at 52 FR 17735, May 11, 1987; 65 FR 19834, Apr. 13, 2000]

§ 868.2480 Cutaneous carbon dioxide ($PcCO_2$) monitor.

(a) *Identification.* A cutaneous carbon dioxide ($PcCO_2$) monitor is a noninvasive heated sensor and a pH-sensitive glass electrode placed on a patient's skin, which is intended to monitor relative changes in a hemodynamically stable patient's cutaneous carbon dioxide tension as an adjunct to arterial carbon dioxide tension measurement.

(b) *Classification.* Class II (special controls). The special control for this device is FDA's "Class II Special Controls Guidance Document: Cutaneous Carbon Dioxide ($PcCO_2$) and Oxygen (PcO_2) Monitors; Guidance for Industry and FDA." See § 868.1(e) for the availability of this guidance document.

[54 FR 27160, June 28, 1989, as amended at 67 FR 76681, Dec. 13, 2002]

§ 868.2500 Cutaneous oxygen (PcO_2) monitor.

(a) *Identification.* A cutaneous oxygen (PcO_2) monitor is a noninvasive, heated sensor (e.g., a Clark-type polargraphic electrode) placed on the patient's skin that is intended to monitor relative changes in the cutaneous oxygen tension.

(b) *Classification.* Class II (special controls). The special control for this device is FDA's "Class II Special Controls Guidance Document: Cutaneous Carbon Dioxide ($PcCO_2$) and Oxygen (PcO_2) Monitors; Guidance for Industry and FDA." See § 868.1(e) for the availability of this guidance document.

[67 FR 76681, Dec. 13, 2002]

§ 868.2550 Pneumotachometer.

(a) *Identification.* A pneumotachometer is a device intended for medical purposes that is used to determine gas flow by measuring the pressure differential across a known resistance. The device may use a set of capillaries or a metal screen for the resistive element.

(b) *Classification.* Class II (performance standards).

§ 868.2600 Airway pressure monitor.

(a) *Identification.* An airway pressure monitor is a device used to measure the pressure in a patient's upper airway. The device may include a pressure gauge and an alarm.

(b) *Classification.* Class II (performance standards).

§ 868.2610 Gas pressure gauge.

(a) *Identification.* A gas pressure gauge (e.g., bourdon tube pressure gauge) is a device intended for medical purposes that is used to measure gas pressure in a medical gas delivery system.

(b) *Classification.* Class I (general controls). The device is exempt from the premarket notification procedures in subpart E of part 807 of this chapter subject to the limitations in § 868.9.

[47 FR 31142, July 16, 1982, as amended at 61 FR 1119, Jan. 16, 1996; 66 FR 38794, July 25, 2001]

§ 868.2620 Gas pressure calibrator.

(a) *Identification.* A gas pressure calibrator is a device intended for medical purposes that is used to calibrate pressure-measuring instruments by generating a known gas pressure.

(b) *Classification.* Class I (general controls). The device is exempt from the premarket notification procedures in subpart E of part 807 of this chapter subject to the limitations in § 868.9.

[47 FR 31142, July 16, 1982, as amended at 61 FR 1119, Jan. 16, 1996; 66 FR 38794, July 25, 2001]

§ 868.2700 Pressure regulator.

(a) *Identification.* A pressure regulator is a device, often called a pressure-reducing valve, that is intended for medical purposes and that is used to convert a medical gas pressure from a high variable pressure to a lower, more constant working pressure. This device includes mechanical oxygen regulators.

(b) *Classification.* Class I (general controls). The device is exempt from the premarket notification procedures in subpart E of part 807 of this chapter subject to the limitations in § 868.9.

[47 FR 31142, July 16, 1982, as amended at 61 FR 1119, Jan. 16, 1996; 66 FR 38794, July 25, 2001]

§ 868.2775 Electrical peripheral nerve stimulator.

(a) *Identification.* An electrical peripheral nerve stimulator (neuromuscular blockade monitor) is a device used to apply an electrical current to a patient to test the level of pharmacological effect of anesthetic drugs and gases.

(b) *Classification.* Class II (performance standards).

§ 868.2875 Differential pressure transducer.

(a) *Identification.* A differential pressure transducer is a two-chambered device intended for medical purposes that is often used during pulmonary function testing. It generates an electrical signal for subsequent display or processing that is proportional to the difference in gas pressures in the two chambers.

(b) *Classification.* Class I (general controls). The device is exempt from the premarket notification procedures in subpart E of part 807 of this chapter subject to the limitations in § 868.9.

[47 FR 31142, July 16, 1982, as amended at 61 FR 1119, Jan. 16, 1996; 66 FR 38794, July 25, 2001]

§ 868.2885 Gas flow transducer.

(a) *Identification.* A gas flow transducer is a device intended for medical purposes that is used to convert gas flow rate into an electrical signal for subsequent display or processing.

(b) *Classification.* Class I (general controls). The device is exempt from the premarket notification procedures in subpart E of part 807 of this chapter subject to the limitations in § 868.9.

[47 FR 31142, July 16, 1982, as amended at 61 FR 1119, Jan. 16, 1996; 66 FR 38794, July 25, 2001]

§ 868.2900 Gas pressure transducer.

(a) *Identification.* A gas pressure transducer is a device intended for medical purposes that is used to convert gas pressure into an electrical signal for subsequent display or processing.

(b) *Classification.* Class I. The device is exempt from the premarket notification procedures in subpart E of part 807 of this chapter.

[47 FR 31142, July 16, 1982, as amended at 61 FR 1120, Jan. 16, 1996]

Subparts D–E [Reserved]

Subpart F—Therapeutic Devices

§ 868.5090 Emergency airway needle.

(a) *Identification.* An emergency airway needle is a device intended to puncture a patient's cricothyroid membrane to provide an emergency airway during upper airway obstruction.

(b) *Classification.* Class II (performance standards).

§ 868.5100 Nasopharyngeal airway.

(a) *Identification.* A nasopharyngeal airway is a device used to aid breathing by means of a tube inserted into a patient's pharynx through the nose to provide a patent airway.

(b) *Classification.* Class I (general controls). The device is exempt from the premarket notification procedures in subpart E of part 807 of this chapter subject to the limitations in § 868.9.

[47 FR 31142, July 16, 1982, as amended at 61 FR 1120, Jan. 16, 1996; 66 FR 38794, July 25, 2001]

§ 868.5110 Oropharyngeal airway.

(a) *Identification.* An oropharyngeal airway is a device inserted into a patient's pharynx through the mouth to provide a patent airway.

(b) *Classification.* Class I (general controls). The device is exempt from the premarket notification procedures in subpart E of part 807 of this chapter subject to the limitations in § 868.9.

[47 FR 31142, July 16, 1982, as amended at 61 FR 1120, Jan. 16, 1996; 66 FR 38794, July 25, 2001]

§ 868.5115 Device to relieve acute upper airway obstruction.

(a) *Identification.* The device is a raised, rounded pad that, in the event of choking on a foreign body, can be applied to the abdomen and pushed upward to generate expulsion pressure to remove the obstruction to relieve acute upper airway obstruction.

(b) *Classification.* Class II (special controls) ("Class II Special Control Guidance Document for Acute Upper Airway Obstruction Devices"). The device is exempt from the premarket notification procedures in subpart E of part 807 of this chapter, subject to § 868.9.

[65 FR 39099, June 23, 2000; 65 FR 47669, Aug. 3, 2000]

§ 868.5120 Anesthesia conduction catheter.

(a) *Identification.* An anesthesia conduction catheter is a flexible tubular device used to inject local anesthetics into a patient and to provide continuous regional anesthesia.

(b) *Classification.* Class II (performance standards).

§ 868.5130 Anesthesia conduction filter.

(a) *Identification.* An anesthesia conduction filter is a microporous filter used while administering to a patient injections of local anesthetics to minimize particulate (foreign material) contamination of the injected fluid.

(b) *Classification.* Class II (performance standards).

§ 868.5140 Anesthesia conduction kit.

(a) *Identification.* An anesthesia conduction kit is a device used to administer to a patient conduction, regional, or local anesthesia. The device may contain syringes, needles, and drugs.

(b) *Classification.* Class II (performance standards).

§ 868.5150 Anesthesia conduction needle.

(a) *Identification.* An anesthesia conduction needle is a device used to inject local anesthetics into a patient to provide regional anesthesia.

(b) *Classification.* Class II (performance standards).

§ 868.5160 Gas machine for anesthesia or analgesia.

(a) *Gas machine for anesthesia*—(1) *Identification.* A gas machine for anesthesia is a device used to administer to a patient, continuously or intermittently, a general inhalation anesthetic and to maintain a patient's ventilation. The device may include a gas flowmeter, vaporizer, ventilator, breathing circuit with bag, and emergency air supply.

(2) *Classification.* Class II (performance standards).

(b) *Gas machine for analgesia*—(1) *Identification.* A gas machine for analgesia is a device used to administer to a patient an analgesic agent, such as a nitrous oxide-oxygen mixture (maximum concentration of 70 percent nitrous oxide).

(2) *Classification.* Class II (performance standards).

§ 868.5165 Nitric oxide administration apparatus.

(a) *Identification.* The nitric oxide administration apparatus is a device used to add nitric oxide to gases that are to be breathed by a patient. The nitric oxide administration apparatus is to be used in conjunction with a ventilator or other breathing gas administration system.

(b) *Classification.* Class II. The special control for this device is FDA's "Guidance Document for Premarket Notification Submissions for Nitric Oxide Administration Apparatus, Nitric Oxide Analyzer, and Nitrogen Dioxide Analyzer."

[65 FR 11465, Mar. 3, 2000]

§ 868.5170 Laryngotracheal topical anesthesia applicator.

(a) *Identification.* A laryngotracheal topical anesthesia applicator is a device used to apply topical anesthetics to a patient's laryngotracheal area.

(b) *Classification.* Class II (performance standards).

§ 868.5180 Rocking bed.

(a) *Identification.* A rocking bed is a device intended for temporary use to help patient ventilation (breathing) by repeatedly tilting the patient, thereby using the weight of the abdominal contents to move the diaphragm.

(b) *Classification.* Class II (performance standards).

§ 868.5220 Blow bottle.

(a) *Identification.* A blow bottle is a device that is intended for medical purposes to induce a forced expiration from a patient. The patient blows into the device to move a column of water from one bottle to another.

(b) *Classification.* Class I (general controls). The device is exempt from the premarket notification procedures in subpart E of part 807 of this chapter subject to the limitations in § 868.9. If the device is not labeled or otherwise represented as sterile, it is exempt from the current good manufacturing practice regulations in part 820 of this chapter, with the exception of § 820.180, with respect to general requirements concerning records, and § 820.198, with respect to complaint files.

[47 FR 31142, July 16, 1982, as amended at 54 FR 25048, June 12, 1989; 66 FR 38794, July 25, 2001]

§ 868.5240 Anesthesia breathing circuit.

(a) *Identification.* An anesthesia breathing circuit is a device that is intended to administer medical gases to a patient during anesthesia. It provides both an inhalation and exhalation route and may include a connector, adaptor, and Y-piece.

(b) *Classification.* Class I (general controls). The device is exempt from the premarket notification procedures in subpart E of part 807 of this chapter subject to the limitations in § 868.9.

[47 FR 31142, July 16, 1982, as amended at 61 FR 1120, Jan. 16, 1996; 66 FR 38794, July 25, 2001]

§ 868.5250 Breathing circuit circulator.

(a) *Identification.* A breathing circuit circulator is a turbine device that is attached to a closed breathing circuit and that is intended to circulate anesthetic gases continuously by maintaining the unidirectional valves in an open position and reducing mechanical dead space and resistance in the breathing circuit.

(b) *Classification.* Class II (performance standards).

§ 868.5260 Breathing circuit bacterial filter.

(a) *Identification.* A breathing circuit bacterial filter is a device that is intended to remove microbiological and particulate matter from the gases in the breathing circuit.

(b) *Classification.* Class II (performance standards).

§ 868.5270 Breathing system heater.

(a) *Identification.* A breathing system heater is a device that is intended to warm breathing gases before they enter a patient's airway. The device may include a temperature controller.

(b) *Classification.* Class II (performance standards).

§ 868.5280 Breathing tube support.

(a) *Identification.* A breathing tube support is a device that is intended to support and anchor a patient's breathing tube(s).

(b) *Classification.* Class I (general controls). The device is exempt from the premarket notification procedures in subpart E of part 807 of this chapter subject to the limitations in § 868.9.

[47 FR 31142, July 16, 1982, as amended at 54 FR 25048, June 12, 1989; 66 FR 38794, July 25, 2001]

§ 868.5300 Carbon dioxide absorbent.

(a) *Identification.* A carbon dioxide absorbent is a device intended for medical purposes that consists of an absorbent material (e.g., soda lime) that is intended to remove carbon dioxide from the gases in the breathing circuit.

(b) *Classification.* Class I (general controls). The device is exempt from the premarket notification procedures in subpart E of part 807 of this chapter subject to the limitations in § 868.9.

[47 FR 31142, July 16, 1982, as amended at 61 FR 1120, Jan. 16, 1996; 66 FR 38794, July 25, 2001]

§ 868.5310 Carbon dioxide absorber.

(a) *Identification.* A carbon dioxide absorber is a device that is intended for medical purposes and that is used in a breathing circuit as a container for carbon dioxide absorbent. It may include a canister and water drain.

(b) *Classification.* Class I (general controls). The device is exempt from the premarket notification procedures in subpart E of part 807 of this chapter subject to the limitations in § 868.9.

[47 FR 31142, July 16, 1982, as amended at 61 FR 1120, Jan. 16, 1996; 66 FR 38794, July 25, 2001]

§ 868.5320 Reservoir bag.

(a) *Identification.* A reservoir bag is a device, usually made of conductive rubber, intended for use in a breathing circuit as a reservoir for breathing gas and to assist, control, or monitor a patient's ventilation.

(b) *Classification.* Class I (general controls). The device is exempt from the premarket notification procedures in subpart E of part 807 of this chapter subject to the limitations in § 868.9.

[47 FR 31142, July 16, 1982, as amended at 61 FR 1120, Jan. 16, 1996; 66 FR 38794, July 25, 2001]

§ 868.5330 Breathing gas mixer.

(a) *Identification.* A breathing gas mixer is a device intended for use in conjunction with a respiratory support apparatus to control the mixing of gases that are to be breathed by a patient.

(b) *Classification.* Class II (performance standards).

§ 868.5340 Nasal oxygen cannula.

(a) *Identification.* A nasal oxygen cannula is a two-pronged device used to administer oxygen to a patient through both nostrils.

(b) *Classification.* Class I (general controls). The device is exempt from the premarket notification procedures in subpart E of part 807 of this chapter subject to the limitations in § 868.9.

[47 FR 31142, July 16, 1982, as amended at 59 FR 63007, Dec. 7, 1994; 66 FR 38794, July 25, 2001]

§ 868.5350 Nasal oxygen catheter.

(a) *Identification.* A nasal oxygen catheter is a device intended to be inserted through a patient's nostril to administer oxygen.

(b) *Classification.* Class I (general controls). The device is exempt from the premarket notification procedures in subpart E of part 807 of this chapter subject to the limitations in § 868.9.

[47 FR 31142, July 16, 1982, as amended at 59 FR 63007, Dec. 7, 1994; 66 FR 38794, July 25, 2001]

§ 868.5365 Posture chair for cardiac or pulmonary treatment.

(a) *Identification.* A posture chair for cardiac or pulmonary treatment is a device intended to assist in the rehabilitation and mobilization of patients with chronic heart or lung disease.

(b) *Classification.* Class I (general controls). The device is exempt from the premarket notification procedures in subpart E of part 807 of this chapter subject to the limitations in § 868.9.

[47 FR 31142, July 16, 1982, as amended at 54 FR 25048, June 12, 1989; 66 FR 38794, July 25, 2001]

§ 868.5375 Heat and moisture condenser (artificial nose).

(a) *Identification.* A heat and moisture condenser (artificial nose) is a device intended to be positioned over a tracheotomy (a surgically created opening in the throat) or tracheal tube (a tube inserted into the trachea) to warm and humidify gases breathed in by a patient.

(b) *Classification.* Class I (general controls). The device is exempt from the premarket notification procedures in

subpart E of part 807 of this chapter subject to the limitations in § 868.9.

[47 FR 31142, July 16, 1982, as amended at 61 FR 1120, Jan. 16, 1996; 66 FR 38795, July 25, 2001]

§ 868.5400 Electroanesthesia apparatus.

(a) *Identification.* An electroanesthesia apparatus is a device used for the induction and maintenance of anesthesia during surgical procedures by means of an alternating or pulsed electric current that is passed through electrodes fixed to a patient's head.

(b) *Classification.* Class III (premarket approval).

(c) *Date PMA or notice of completion of a PDP is required.* A PMA or notice of completion of a PDP is required to be filed with the Food and Drug Administration on or before December 26, 1996 for any electroanesthesia apparatus that was in commercial distribution before May 28, 1976, or that has, on or before December 26, 1996 been found to be substantially equivalent to an electroanesthesia apparatus that was in commercial distribution before May 28, 1976. Any other electroanesthesia apparatus shall have an approved PMA or a declared completed PDP in effect before being placed in commercial distribution.

[47 FR 31142, July 16, 1982, as amended at 52 FR 17735, May 11, 1987; 61 FR 50706, Sept. 27, 1996]

§ 868.5420 Ether hook.

(a) *Identification.* An ether hook is a device that fits inside a patient's mouth and that is intended to deliver vaporized ether.

(b) *Classification.* Class I (general controls). The device is exempt from the premarket notification procedures in subpart E of part 807 of this chapter subject to the limitations in § 868.9. If the device is not labeled or otherwise represented as sterile, it is exempt from the current good manufacturing practice regulations in part 820 of this chapter, with the exception of § 820.180, with respect to general requirements concerning records, and § 820.198, with respect to complaint files.

[47 FR 31142, July 16, 1982, as amended at 54 FR 25048, June 12, 1989; 66 FR 38795, July 25, 2001]

§ 868.5430 Gas-scavenging apparatus.

(a) *Identification.* A gas-scavenging apparatus is a device intended to collect excess anesthetic, analgesic, or trace gases or vapors from a patient's breathing system, ventilator, or extracorporeal pump-oxygenator, and to conduct these gases out of the area by means of an exhaust system.

(b) *Classification.* Class II (performance standards).

§ 868.5440 Portable oxygen generator.

(a) *Identification.* A portable oxygen generator is a device that is intended to release oxygen for respiratory therapy by means of either a chemical reaction or physical means (e.g., a molecular sieve).

(b) *Classification.* Class II (performance standards).

§ 868.5450 Respiratory gas humidifier.

(a) *Identification.* A respiratory gas humidifier is a device that is intended to add moisture to, and sometimes to warm, the breathing gases for administration to a patient. Cascade, gas, heated, and prefilled humidifiers are included in this generic type of device.

(b) *Classification.* Class II (performance standards).

§ 868.5460 Therapeutic humidifier for home use.

(a) *Identification.* A therapeutic humidifier for home use is a device that adds water vapor to breathing gases and that is intended for respiratory therapy or other medical purposes. The vapor produced by the device pervades the area surrounding the patient, who breathes the vapor during normal respiration.

(b) *Classification.* Class I (general controls). The device is exempt from the premarket notification procedures in subpart E of part 807 of this chapter subject to the limitations in § 868.9.

[47 FR 31142, July 16, 1982; 47 FR 40410, Sept. 14, 1982, as amended at 61 FR 1120, Jan. 16, 1996; 66 FR 38795, July 25, 2001]

§ 868.5470 Hyperbaric chamber.

(a) *Identification.* A hyperbaric chamber is a device that is intended to increase the environmental oxygen pressure to promote the movement of oxygen from the environment to a patient's tissue by means of pressurization that is greater than atmospheric pressure. This device does not include topical oxygen chambers for extremities (§ 878.5650).

(b) *Classification.* Class II (performance standards).

§ 868.5530 Flexible laryngoscope.

(a) *Identification.* A flexible laryngoscope is a fiberoptic device used to examine and visualize a patient's upper airway and aid placement of a tracheal tube.

(b) *Classification.* Class I (general controls). The device is exempt from the premarket notification procedures in subpart E of part 807 of this chapter subject to the limitations in § 868.9.

[47 FR 41107, Sept. 17, 1982, as amended at 61 FR 1120, Jan. 16, 1996; 66 FR 38795, July 25, 2001]

§ 868.5540 Rigid laryngoscope.

(a) *Identification.* A rigid laryngoscope is a device used to examine and visualize a patient's upper airway and aid placement of a tracheal tube.

(b) *Classification.* Class I (general controls). The device is exempt from the premarket notification procedures in subpart E of part 807 of this chapter subject to the limitations in § 868.9

[47 FR 41107, Sept. 17, 1982, as amended at 61 FR 1120, Jan. 16, 1996; 66 FR 38795, July 25, 2001]

§ 868.5550 Anesthetic gas mask.

(a) *Identification.* An anesthetic gas mask is a device, usually made of conductive rubber, that is positioned over a patient's nose or mouth to direct anesthetic gases to the upper airway.

(b) *Classification.* Class I (general controls). The device is exempt from the premarket notification procedures in subpart E of part 807 of this chapter subject to the limitations in § 868.9.

[47 FR 41107, Sept. 17, 1982, as amended at 61 FR 1120, Jan. 16, 1996; 66 FR 38795, July 25, 2001]

§ 868.5560 Gas mask head strap.

(a) *Identification.* A gas mask head strap is a device used to hold an anesthetic gas mask in position on a patient's face.

(b) *Classification.* Class I (general controls). The device is exempt from the premarket notification procedures in subpart E of part 807 of this chapter subject to the limitations in § 868.9.

[47 FR 41107, Sept. 17, 1982, as amended at 54 FR 25048, June 12, 1989; 66 FR 38795, July 25, 2001]

§ 868.5570 Nonrebreathing mask.

(a) *Identification.* A nonrebreathing mask is a device fitting over a patient's face to administer oxygen. It utilizes one-way valves to prevent the patient from rebreathing previously exhaled gases.

(b) *Classification.* Class I (general controls). The device is exempt from the premarket notification procedures in subpart E of part 807 of this chapter subject to the limitations in § 868.9.

[47 FR 31142, July 16, 1982, as amended at 61 FR 1120, Jan. 16, 1996; 66 FR 38795, July 25, 2001]

§ 868.5580 Oxygen mask.

(a) *Identification.* An oxygen mask is a device placed over a patient's nose, mouth, or tracheostomy to administer oxygen or aerosols.

(b) *Classification.* Class I (general controls). The device is exempt from the premarket notification procedures in subpart E of part 807 of this chapter subject to the limitations in § 868.9.

[47 FR 31142, July 16, 1982, as amended at 61 FR 1120, Jan. 16, 1996; 66 FR 38795, July 25, 2001]

§ 868.5590 Scavenging mask.

(a) *Identification.* A scavenging mask is a device positioned over a patient's nose to deliver anesthetic or analgesic gases to the upper airway and to remove excess and exhaled gas. It is usually used during dentistry.

(b) *Classification.* Class I (general controls). The device is exempt from the premarket notification procedures in

subpart E of part 807 of this chapter subject to the limitations in § 868.9.

[47 FR 31142, July 16, 1982, as amended at 61 FR 1120, Jan. 16, 1996; 66 FR 38795, July 25, 2001]

§ 868.5600 Venturi mask.

(a) *Identification.* A venturi mask is a device containing an air-oxygen mixing mechanism that dilutes 100 percent oxygen to a predetermined concentration and delivers the mixed gases to a patient.

(b) *Classification.* Class I (general controls). The device is exempt from the premarket notification procedures in subpart E of part 807 of this chapter subject to the limitations in § 868.9.

[47 FR 31142, July 16, 1982, as amended at 61 FR 1120, Jan. 16, 1996; 66 FR 38795, July 25, 2001]

§ 868.5610 Membrane lung for long-term pulmonary support.

(a) *Identification.* A membrane lung for long-term pulmonary support is a device used to provide to a patient extracorporeal blood oxygenation for longer than 24 hours.

(b) *Classification.* Class III (premarket approval).

(c) *Date PMA or notice of completion of a PDP is required.* No effective date has been established of the requirement for premarket approval. See § 868.3.

[47 FR 31142, July 16, 1982, as amended at 52 FR 17735, May 11, 1987]

§ 868.5620 Breathing mouthpiece.

(a) *Identification.* A breathing mouthpiece is a rigid device that is inserted into a patient's mouth and that connects with diagnostic or therapeutic respiratory devices.

(b) *Classification.* Class I (general controls). The device is exempt from the premarket notification procedures in subpart E of part 807 of this chapter subject to § 868.9.

[47 FR 31142, July 16, 1982, as amended at 65 FR 2313, Jan. 14, 2000]

§ 868.5630 Nebulizer.

(a) *Identification.* A nebulizer is a device intended to spray liquids in aerosol form into gases that are delivered directly to the patient for breathing. Heated, ultrasonic, gas, venturi, and refillable nebulizers are included in this generic type of device.

(b) *Classification.* Class II (performance standards).

§ 868.5640 Medicinal nonventilatory nebulizer (atomizer).

(a) *Identification.* A medicinal nonventilatory nebulizer (atomizer) is a device that is intended to spray liquid medication in aerosol form into the air that a patient will breathe.

(b) *Classification.* Class I (general controls). The device is exempt from the premarket notification procedures in subpart E of part 807 of this chapter subject to § 868.9.

[47 FR 31142, July 16, 1982, as amended at 65 FR 2313, Jan. 14, 2000]

§ 868.5650 Esophageal obturator.

(a) *Identification.* An esophageal obturator is a device inserted through a patient's mouth to aid ventilation of the patient during emergency resuscitation by occluding (blocking) the esophagus, thereby permitting positive pressure ventilation through the trachea. The device consists of a closed-end semirigid esophageal tube that is attached to a face mask.

(b) *Classification.* Class II (performance standards).

§ 868.5655 Portable liquid oxygen unit.

(a) *Identification.* A portable liquid oxygen unit is a portable, thermally insulated container of liquid oxygen that is intended to supplement gases to be inhaled by a patient, is sometimes accompanied by tubing and an oxygen mask. An empty portable liquid oxygen unit is a device, while the oxygen contained therein is a drug.

(b) *Classification.* Class II (performance standards).

§ 868.5665 Powered percussor.

(a) *Identification.* A powered percussor is a device that is intended to transmit vibration through a patient's chest wall to aid in freeing mucus deposits in the lung in order to improve bronchial drainage and that may be powered by electricity or compressed gas.

(b) *Classification.* Class II (performance standards).

§ 868.5675 Rebreathing device.

(a) *Identification.* A rebreathing device is a device that enables a patient to rebreathe exhaled gases. It may be used in conjunction with pulmonary function testing or for increasing minute ventilation.

(b) *Classification.* Class I (general controls). The device is exempt from the premarket notification procedures in subpart E of part 807 of this chapter subject to § 868.9.

[47 FR 31142, July 16, 1982, as amended at 65 FR 2313, Jan. 14, 2000]

§ 868.5690 Incentive spirometer.

(a) *Identification.* An incentive spirometer is a device that indicates a patient's breathing volume or flow and that provides an incentive to the patient to improve his or her ventilation.

(b) *Classification.* Class II (performance standards).

§ 868.5700 Nonpowered oxygen tent.

(a) *Identification.* A nonpowered oxygen tent is a device that encloses a patient's head and upper body to contain oxygen delivered to the patient for breathing. This generic type of device includes infant oxygen hoods.

(b) *Classification.* Class I (general controls). The device is exempt from the premarket notification procedures in subpart E of part 807 of this chapter subject to § 868.9.

[47 FR 31142, July 16, 1982, as amended at 65 FR 2313, Jan. 14, 2000]

§ 868.5710 Electrically powered oxygen tent.

(a) *Identification.* An electrically powered oxygen tent is a device that encloses a patient's head and, by means of an electrically powered unit, administers breathing oxygen and controls the temperature and humidity of the breathing gases. This generic type device includes the pediatric aerosol tent.

(b) *Classification.* Class II (performance standards).

§ 868.5720 Bronchial tube.

(a) *Identification.* A bronchial tube is a device used to differentially intubate a patient's bronchus (one of the two main branches of the trachea leading directly to the lung) in order to isolate a portion of lung distal to the tube.

(b) *Classification.* Class II (performance standards).

§ 868.5730 Tracheal tube.

(a) *Identification.* A tracheal tube is a device inserted into a patient's trachea via the nose or mouth and used to maintain an open airway.

(b) *Classification.* Class II (performance standards).

§ 868.5740 Tracheal/bronchial differential ventilation tube.

(a) *Identification.* A tracheal/bronchial differential ventilation tube is a device used to isolate the left or the right lung of a patient for anesthesia or pulmonary function testing.

(b) *Classification.* Class II (performance standards).

§ 868.5750 Inflatable tracheal tube cuff.

(a) *Identification.* An inflatable tracheal tube cuff is a device used to provide an airtight seal between a tracheal tube and a patient's trachea.

(b) *Classification.* Class II (performance standards).

§ 868.5760 Cuff spreader.

(a) *Identification.* A cuff spreader is a device used to install tracheal tube cuffs on tracheal or tracheostomy tubes.

(b) *Classification.* Class I (general controls). The device is exempt from the premarket notification procedures in subpart E of part 807 of this chapter subject to the limitations in § 868.9. If the device is not labeled or otherwise represented as sterile, it is exempt from the current good manufacturing practice regulations in part 820 of this chapter, with the exception of § 820.180, with respect to general requirements concerning records, and § 820.198, with respect to complaint files.

[47 FR 31142, July 16, 1982, as amended at 54 FR 25048, June 12, 1989; 66 FR 38795, July 25, 2001]

§ 868.5770 Tracheal tube fixation device.

(a) *Identification.* A tracheal tube fixation device is a device used to hold a tracheal tube in place, usually by means of straps or pinch rings.

(b) *Classification.* Class I (general controls). The device is exempt from the premarket notification procedures in subpart E of part 807 of this chapter subject to the limitations in § 868.9.

[47 FR 31142, July 16, 1982, as amended at 61 FR 1120, Jan. 16, 1996; 66 FR 38795, July 25, 2001]

§ 868.5780 Tube introduction forceps.

(a) *Identification.* Tube introduction forceps (e.g., Magill forceps) are a right-angled device used to grasp a tracheal tube and place it in a patient's trachea.

(b) *Classification.* Class I (general controls). The device is exempt from the premarket notification procedures in subpart E of part 807 of this chapter subject to the limitations in § 868.9.

[47 FR 31142, July 16, 1982, as amended at 61 FR 1120, Jan. 16, 1996; 66 FR 38795, July 25, 2001]

§ 868.5790 Tracheal tube stylet.

(a) *Identification.* A tracheal tube stylet is a device used temporarily to make rigid a flexible tracheal tube to aid its insertion into a patient.

(b) *Classification.* Class I (general controls). The device is exempt from the premarket notification procedures in subpart E of part 807 of this chapter subject to the limitations in § 868.9.

[47 FR 31142, July 16, 1982, as amended at 61 FR 1120, Jan. 16, 1996; 66 FR 38795, July 25, 2001]

§ 868.5795 Tracheal tube cleaning brush.

(a) *Identification.* A tracheal tube cleaning brush is a device consisting of a brush with plastic bristles intended to clean tracheal cannula devices after their removal from patients.

(b) *Classification.* Class I (general controls). The device is exempt from the premarket notification procedures in subpart E of part 807 of this chapter subject to the limitations in § 868.9. If the device is not labeled or otherwise represented as sterile, it is exempt from the current good manufacturing practice regulations in part 820 of this chapter, with the exception of § 820.180, with respect to general requirements concerning records, and § 820.198, with respect to complaint files.

[51 FR 40388, Nov. 6, 1986, as amended at 66 FR 38795, July 25, 2001]

§ 868.5800 Tracheostomy tube and tube cuff.

(a) *Identification.* A tracheostomy tube and tube cuff is a device intended to be placed into a surgical opening of the trachea to facilitate ventilation to the lungs. The cuff may be a separate or integral part of the tracheostomy tube and is, when inflated, intended to establish a seal between the tracheal wall and the tracheostomy tube. The cuff is used to prevent the patient's aspiration of substances, such as blood or vomit, or to provide a means for positive-pressure ventilation of the patient. This device is made of either stainless steel or plastic.

(b) *Classification.* Class II.

[51 FR 40389, Nov. 6, 1986]

§ 868.5810 Airway connector.

(a) *Identification.* An airway connector is a device intended to connect a breathing gas source to a tracheal tube, tracheostomy tube, or mask.

(b) *Classification.* Class I (general controls). The device is exempt from the premarket notification procedures in subpart E of part 807 of this chapter subject to the limitations in § 868.9.

[47 FR 31142, July 16, 1982, as amended at 61 FR 1120, Jan. 16, 1996; 66 FR 38795, July 25, 2001]

§ 868.5820 Dental protector.

(a) *Identification.* A dental protector is a device intended to protect a patient's teeth during manipulative procedures within a patient's oral cavity.

(b) *Classification.* Class I (general controls). The device is exempt from the premarket notification procedures in subpart E of part 807 of this chapter subject to the limitations in § 868.9.

[47 FR 31142, July 16, 1982, as amended at 61 FR 1120, Jan. 16, 1996; 66 FR 38795, July 25, 2001]

§ 868.5830 Autotransfusion apparatus.

(a) *Identification.* An autotransfusion apparatus is a device used to collect and reinfuse the blood lost by a patient due to surgery or trauma.

(b) *Classification.* Class II (performance standards).

§ 868.5860 Pressure tubing and accessories.

(a) *Identification.* Pressure tubing and accessories are flexible or rigid devices intended to deliver pressurized medical gases.

(b) *Classification.* Class I (general controls). The device is exempt from the premarket notification procedures in subpart E of part 807 of this chapter subject to the limitations in § 868.9.

[47 FR 31142, July 16, 1982, as amended at 61 FR 1120, Jan. 16, 1996; 66 FR 38796, July 25, 2001]

§ 868.5870 Nonrebreathing valve.

(a) *Identification.* A nonrebreathing valve is a one-way valve that directs breathing gas flow to the patient and vents exhaled gases into the atmosphere.

(b) *Classification.* Class II (performance standards).

§ 868.5880 Anesthetic vaporizer.

(a) *Identification.* An anesthetic vaporizer is a device used to vaporize liquid anesthetic and deliver a controlled amount of the vapor to the patient.

(b) *Classification.* Class II (performance standards).

§ 868.5895 Continuous ventilator.

(a) *Identification.* A continuous ventilator (respirator) is a device intended to mechanically control or assist patient breathing by delivering a predetermined percentage of oxygen in the breathing gas. Adult, pediatric, and neonatal ventilators are included in this generic type of device.

(b) *Classification.* Class II (performance standards).

§ 868.5905 Noncontinuous ventilator (IPPB).

(a) *Identification.* A noncontinuous ventilator (intermittent positive pressure breathing-IPPB) is a device intended to deliver intermittently an aerosol to a patient's lungs or to assist a patient's breathing.

(b) *Classification.* Class II (performance standards).

§ 868.5915 Manual emergency ventilator.

(a) *Identification.* A manual emergency ventilator is a device, usually incorporating a bag and valve, intended to provide emergency respiratory support by means of a face mask or a tube inserted into a patient's airway.

(b) *Classification.* Class II (performance standards).

§ 868.5925 Powered emergency ventilator.

(a) *Identification.* A powered emergency ventilator is a demand valve or inhalator intended to provide emergency respiratory support by means of a face mask or a tube inserted into a patient's airway.

(b) *Classification.* Class II (performance standards).

§ 868.5935 External negative pressure ventilator.

(a) *Identification.* An external negative pressure ventilator (e.g., iron lung, cuirass) is a device chamber that is intended to support a patient's ventilation by alternately applying and releasing external negative pressure over the diaphragm and upper trunk of the patient.

(b) *Classification.* Class II (performance standards).

§ 868.5955 Intermittent mandatory ventilation attachment.

(a) *Identification.* An intermittent mandatory ventilation (IMV) attachment is a device attached to a mechanical ventilator that allows spontaneous breathing by a patient while providing mechanical ventilation at a preset rate.

(b) *Classification.* Class II (performance standards).

§ 868.5965 Positive end expiratory pressure breathing attachment.

(a) *Identification.* A positive end expiratory pressure (PEEP) breathing attachment is a device attached to a ventilator that is used to elevate pressure in a patient's lungs above atmospheric pressure at the end of exhalation.

(b) *Classification.* Class II (performance standards).

§ 868.5975 Ventilator tubing.

(a) *Identification.* Ventilator tubing is a device intended for use as a conduit for gases between a ventilator and a patient during ventilation of the patient.

(b) *Classification.* Class I (general controls). The device is exempt from the premarket notification procedures in subpart E of part 807 of this chapter subject to the limitations in § 868.9.

[47 FR 31142, July 16, 1982, as amended at 61 FR 1120, Jan. 16, 1996; 66 FR 38796, July 25, 2001]

§ 868.5995 Tee drain (water trap).

(a) *Identification.* A tee drain (water trap) is a device intended to trap and drain water that collects in ventilator tubing during respiratory therapy, thereby preventing an increase in breathing resistance.

(b) *Classification.* Class I (general controls). The device is exempt from the premarket notification procedures in subpart E of part 807 of this chapter subject to the limitations in § 868.9.

[47 FR 31142, July 16, 1982, as amended at 61 FR 1120, Jan. 16, 1996; 66 FR 38796, July 25, 2001]

Subpart G—Miscellaneous

§ 868.6100 Anesthetic cabinet, table, or tray.

(a) *Identification.* An anesthetic cabinet, table, or tray is a device intended to store anesthetic equipment and drugs. The device is usually constructed to eliminate build-up of static electrical charges.

(b) *Classification.* Class I (general controls). The device is exempt from the premarket notification procedures in subpart E of part 807 of this chapter subject to the limitations in § 868.9.

[47 FR 31142, July 16, 1982, as amended at 54 FR 25048, June 12, 1989; 66 FR 38796, July 25, 2001]

§ 868.6175 Cardiopulmonary emergency cart.

(a) *Identification.* A cardiopulmonary emergency cart is a device intended to store and transport resuscitation supplies for emergency treatment. The device does not include any equipment used in cardiopulmonary resuscitation.

(b) *Classification.* Class I (general controls). The device is exempt from the premarket notification procedures in subpart E of part 807 of this chapter subject to the limitations in § 868.9. The device is also exempt from the current good manufacturing practice regulations in part 820 of this chapter, with the exception of § 820.180, with respect to general requirements concerning records, and § 820.198, with respect to complaint files.

[47 FR 31142, July 16, 1982, as amended at 54 FR 25048, June 12, 1989; 66 FR 38796, July 25, 2001]

§ 868.6225 Nose clip.

(a) *Identification.* A nose clip is a device intended to close a patient's external nares (nostrils) during diagnostic or therapeutic procedures.

(b) *Classification.* Class I (general controls). The device is exempt from the premarket notification procedures in subpart E of part 807 of this chapter subject to the limitations in § 868.9. The device is also exempt from the current good manufacturing practice regulations in part 820 of this chapter, with the exception of § 820.180, with respect to general requirements concerning records, and § 820.198, with respect to complaint files.

[47 FR 31142, July 16, 1982, as amended at 54 FR 25048, June 12, 1989; 66 FR 38796, July 25, 2001]

§ 868.6250 Portable air compressor.

(a) *Identification.* A portable air compressor is a device intended to provide compressed air for medical purposes, e.g., to drive ventilators and other respiratory devices.

(b) *Classification.* Class II (performance standards).

§ 868.6400 Calibration gas.

(a) *Identification.* A calibration gas is a device consisting of a container of gas of known concentration intended to calibrate medical gas concentration measurement devices.

(b) *Classification.* Class I (general controls). The device is exempt from the premarket notification procedures in

subpart E of part 807 of this chapter subject to the limitations in § 868.9.

[47 FR 31142, July 16, 1982, as amended at 61 FR 1121, Jan. 16, 1996; 66 FR 38796, July 25, 2001]

§ 868.6700 Anesthesia stool.

(a) *Identification.* An anesthesia stool is a device intended for use as a stool for the anesthesiologist in the operating room.

(b) *Classification.* Class I (general controls). The device is exempt from the premarket notification procedures in subpart E of part 807 of this chapter subject to the limitations in § 868.9.

[47 FR 31142, July 16, 1982, as amended at 54 FR 25049, June 12, 1989; 66 FR 38796, July 25, 2001]

§ 868.6810 Tracheobronchial suction catheter.

(a) *Identification.* A tracheobronchial suction catheter is a device used to aspirate liquids or semisolids from a patient's upper airway.

(b) *Classification.* Class 1 (general controls). The device is exempt from the premarket notification procedures in subpart E of part 807 of this chapter subject to § 868.9.

[47 FR 31142, July 16, 1982, as amended at 65 FR 2314, Jan. 14, 2000]

§ 868.6820 Patient position support.

(a) *Identification.* A patient position support is a device intended to maintain the position of an anesthetized patient during surgery.

(b) *Classification.* Class I (general controls). The device is exempt from the premarket notification procedures in subpart E of part 807 of this chapter subject to the limitations in § 868.9.

[47 FR 31142, July 16, 1982, as amended at 61 FR 1121, Jan. 16, 1996; 66 FR 38796, July 25, 2001]

§ 868.6885 Medical gas yoke assembly.

(a) *Identification.* A medical gas yoke assembly is a device intended to connect medical gas cylinders to regulators or needle valves to supply gases for anesthesia or respiratory therapy. The device may include a particulate filter.

(b) *Classification.* Class I (general controls). The device is exempt from the premarket notification procedures in subpart E of part 807 of this chapter subject to the limitations in § 868.9.

[47 FR 31142, July 16, 1982, as amended at 61 FR 1121, Jan. 16, 1996; 66 FR 38796, July 25, 2001]

PART 870—CARDIOVASCULAR DEVICES

Subpart A—General Provisions

Sec.
870.1 Scope.
870.3 Effective dates of requirement for premarket approval.
870.9 Limitations of exemptions from section 510(k) of the Federal Food, Drug, and Cosmetic Act (the act).

Subpart B—Cardiovascular Diagnostic Devices

870.1025 Arrhythmia detector and alarm (including ST-segment measurement and alarm).
870.1100 Blood pressure alarm.
870.1110 Blood pressure computer.
870.1120 Blood pressure cuff.
870.1130 Noninvasive blood pressure measurement system.
870.1140 Venous blood pressure manometer.
870.1200 Diagnostic intravascular catheter.
870.1210 Continuous flush catheter.
870.1220 Electrode recording catheter or electrode recording probe.
870.1230 Fiberoptic oximeter catheter.
870.1240 Flow-directed catheter.
870.1250 Percutaneous catheter.
870.1270 Intracavitary phonocatheter system.
870.1280 Steerable catheter.
870.1290 Steerable catheter control system.
870.1300 Catheter cannula.
870.1310 Vessel dilator for percutaneous catheterization.
870.1330 Catheter guide wire.
870.1340 Catheter introducer.
870.1350 Catheter balloon repair kit.
870.1360 Trace microsphere.
870.1370 Catheter tip occluder.
870.1380 Catheter stylet.
870.1390 Trocar.
870.1425 Programmable diagnostic computer.
870.1435 Single-function, preprogrammed diagnostic computer.
870.1450 Densitometer.
870.1650 Angiographic injector and syringe.
870.1660 Indicator injector.
870.1670 Syringe actuator for an injector.
870.1750 External programmable pacemaker pulse generator.
870.1800 Withdrawal-infusion pump.
870.1875 Stethoscope.

870.1915 Thermodilution probe.

Subpart C—Cardiovascular Monitoring Devices

870.2050 Biopotential amplifier and signal conditioner.
870.2060 Transducer signal amplifier and signal conditioner.
870.2100 Cardiovascular blood flowmeter.
870.2120 Extravascular blood flow probe.
870.2300 Cardiac monitor (including cardiotachometer and rate alarm).
870.2310 Apex cardiograph (vibrocardiograph).
870.2320 Ballistocardiograph.
870.2330 Echocardiograph.
870.2340 Electrocardiograph.
870.2350 Electrocardiograph lead switching adaptor.
870.2360 Electrocardiograph electrode.
870.2370 Electrocardiograph surface electrode tester.
870.2390 Phonocardiograph.
870.2400 Vectorcardiograph.
870.2450 Medical cathode-ray tube display.
870.2600 Signal isolation system.
870.2620 Line isolation monitor.
870.2640 Portable leakage current alarm.
870.2675 Oscillometer.
870.2700 Oximeter.
870.2710 Ear oximeter.
870.2750 Impedance phlebograph.
870.2770 Impedance plethysmograph.
870.2780 Hydraulic, pneumatic, or photoelectric plethysmographs.
870.2800 Medical magnetic tape recorder.
870.2810 Paper chart recorder.
870.2840 Apex cardiographic transducer.
870.2850 Extravascular blood pressure transducer.
870.2855 Implantable Intra-aneurysm Pressure Measurement System.
870.2860 Heart sound transducer.
870.2870 Catheter tip pressure transducer.
870.2880 Ultrasonic transducer.
870.2890 Vessel occlusion transducer.
870.2900 Patient transducer and electrode cable (including connector).
870.2910 Radiofrequency physiological signal transmitter and receiver.
870.2920 Telephone electrocardiograph transmitter and receiver.

Subpart D—Cardiovascular Prosthetic Devices

870.3250 Vascular clip.
870.3260 Vena cava clip.
870.3300 Vascular embolization device.
870.3375 Cardiovascular intravascular filter.
870.3450 Vascular graft prosthesis.
870.3470 Intracardiac patch or pledget made of polypropylene, polyethylene terephthalate, or polytetrafluoroethylene.
870.3535 Intra-aortic balloon and control system.
870.3545 Ventricular bypass (assist) device.
870.3600 External pacemaker pulse generator.
870.3610 Implantable pacemaker pulse generator.
870.3620 Pacemaker lead adaptor.
870.3630 Pacemaker generator function analyzer.
870.3640 Indirect pacemaker generator function analyzer.
870.3650 Pacemaker polymeric mesh bag.
870.3670 Pacemaker charger.
870.3680 Cardiovascular permanent or temporary pacemaker electrode.
870.3690 Pacemaker test magnet.
870.3700 Pacemaker programmers.
870.3710 Pacemaker repair or replacement material.
870.3720 Pacemaker electrode function tester.
870.3730 Pacemaker service tools.
870.3800 Annuloplasty ring.
870.3850 Carotid sinus nerve stimulator.
870.3925 Replacement heart valve.
870.3935 Prosthetic heart valve holder.
870.3945 Prosthetic heart valve sizer.

Subpart E—Cardiovascular Surgical Devices

870.4075 Endomyocardial biopsy device.
870.4200 Cardiopulmonary bypass accessory equipment.
870.4205 Cardiopulmonary bypass bubble detector.
870.4210 Cardiopulmonary bypass vascular catheter, cannula, or tubing.
870.4220 Cardiopulmonary bypass heart-lung machine console.
870.4230 Cardiopulmonary bypass defoamer.
870.4240 Cardiopulmonary bypass heat exchanger.
870.4250 Cardiopulmonary bypass temperature controller.
870.4260 Cardiopulmonary bypass arterial line blood filter.
870.4270 Cardiopulmonary bypass cardiotomy suction line blood filter.
870.4280 Cardiopulmonary prebypass filter.
870.4290 Cardiopulmonary bypass adaptor, stopcock, manifold, or fitting.
870.4300 Cardiopulmonary bypass gas control unit.
870.4310 Cardiopulmonary bypass coronary pressure gauge.
870.4320 Cardiopulmonary bypass pulsatile flow generator.
870.4330 Cardiopulmonary bypass on-line blood gas monitor.
870.4340 Cardiopulmonary bypass level sensing monitor and/or control.
870.4350 Cardiopulmonary bypass oxygenator.
870.4360 Nonroller-type cardiopulmonary bypass blood pump.

Subpart F—Cardiovascular Therapeutic Devices

AUTHORITY: 21 U.S.C. 351, 360, 360c, 360e, 360j, 371.

SOURCE: 45 FR 7907-7971, Feb. 5, 1980, unless otherwise noted.

Subpart A—General Provisions

§ 870.1 Scope.

(a) This part sets forth the classification of cardiovascular devices intended for human use that are in commercial distribution.

(b) The identification of a device in a regulation in this part is not a precise description of every device that is, or will be, subject to the regulation. A manufacturer who submits a premarket notification submission for a device under part 807 may not show merely that the device is accurately described by the section title and identification provisions of a regulation in this part, but shall state why the device is substantially equivalent to other devices, as required by § 807.87.

(c) To avoid duplicative listings, a cardiovascular device that has two or more types of uses (e.g., used both as a diagnostic device and as a therapeutic device) is listed only in one subpart.

(d) References in this part to regulatory sections of the Code of Federal Regulations are to chapter I of title 21, unless otherwise noted.

(e) Guidance documents referenced in this part are available on the Internet at *http://www.fda.gov/cdrh/guidance.html.*

[52 FR 17735, May 11, 1987, as amended at 68 FR 61344, Oct. 28, 2003]

§ 870.3 Effective dates of requirement for premarket approval.

A device included in this part that is classified into class III (premarket approval) shall not be commercially distributed after the date shown in the regulation classifying the device unless the manufacturer has an approval under section 515 of the act (unless an exemption has been granted under section 520(g)(2) of the act). An approval under section 515 of the act consists of FDA's issuance of an order approving an application for premarket approval (PMA) for the device or declaring completed a product development protocol (PDP) for the device.

(a) Before FDA requires that a device commercially distributed before the enactment date of the amendments, or a device that has been found substantially equivalent to such a device, has an approval under section 515 of the act FDA must promulgate a regulation under section 515(b) of the act requiring such approval, except as provided in paragraph (b) of this section. Such a regulation under section 515(b) of the act shall not be effective during the grace period ending on the 90th day after its promulgation or on the last day of the 30th full calendar month after the regulation that classifies the device into class III is effective, whichever is later. See section 501(f)(2)(B) of the act. Accordingly, unless an effective date of the requirement for premarket approval is shown in the regulation for a device classified into class III in this part, the device may be commercially distributed without FDA's issuance of an order approving a PMA or declaring completed a PDP for the device. If FDA promulgates a regulation under section 515(b) of the act requiring premarket approval for a device, section 501(f)(1)(A) of the act applies to the device.

(b) Any new, not substantially equivalent, device introduced into commercial distribution on or after May 28, 1976, including a device formerly marketed that has been substantially altered, is classified by statute (section 513(f) of the act) into class III without any grace period and FDA must have issued an order approving a PMA or declaring completed a PDP for the device before the device is commercially distributed unless it is reclassified. If FDA knows that a device being commercially distributed may be a "new" device as defined in this section because of any new intended use or other reasons, FDA may codify the statutory classification of the device into class III for such new use. Accordingly, the regulation for such a class III device states that as of the enactment date of the amendments, May 28, 1976, the device must have an approval under section 515 of the act before commercial distribution.

[52 FR 17735, May 11, 1987]

§ 870.9 Limitations of exemptions from section 510(k) of the Federal Food, Drug, and Cosmetic Act (the act).

The exemption from the requirement of premarket notification (section 510(k) of the act) for a generic type of class I or II device is only to the extent that the device has existing or reasonably foreseeable characteristics of commercially distributed devices within that generic type or, in the case of in vitro diagnostic devices, only to the extent that misdiagnosis as a result of using the device would not be associated with high morbidity or mortality. Accordingly, manufacturers of any commercially distributed class I or II device for which FDA has granted an exemption from the requirement of premarket notification must still submit a premarket notification to FDA before introducing or delivering for introduction into interstate commerce for commercial distribution the device when:

(a) The device is intended for a use different from the intended use of a legally marketed device in that generic type of device; e.g., the device is intended for a different medical purpose, or the device is intended for lay use where the former intended use was by health care professionals only;

(b) The modified device operates using a different fundamental scientific technology than a legally marketed device in that generic type of device; e.g., a surgical instrument cuts tissue with a laser beam rather than with a sharpened metal blade, or an in vitro diagnostic device detects or identifies infectious agents by using deoxyribonucleic acid (DNA) probe or nucleic acid hybridization technology rather than culture or immunoassay technology; or

(c) The device is an in vitro device that is intended:

(1) For use in the diagnosis, monitoring, or screening of neoplastic diseases with the exception of immunohistochemical devices;

(2) For use in screening or diagnosis of familial or acquired genetic disorders, including inborn errors of metabolism;

(3) For measuring an analyte that serves as a surrogate marker for screening, diagnosis, or monitoring life-threatening diseases such as acquired immune deficiency syndrome (AIDS), chronic or active hepatitis, tuberculosis, or myocardial infarction or to monitor therapy;

(4) For assessing the risk of cardiovascular diseases;

(5) For use in diabetes management;

(6) For identifying or inferring the identity of a microorganism directly from clinical material;

(7) For detection of antibodies to microorganisms other than immunoglobulin G (IgG) or IgG assays when the results are not qualitative, or are used to determine immunity, or the assay is intended for use in matrices other than serum or plasma;

(8) For noninvasive testing as defined in § 812.3(k) of this chapter; and

(9) For near patient testing (point of care).

[65 FR 2314, Jan. 14, 2000]

Subpart B—Cardiovascular Diagnostic Devices

§ 870.1025 Arrhythmia detector and alarm (including ST-segment measurement and alarm).

(a) *Identification.* The arrhythmia detector and alarm device monitors an electrocardiogram and is designed to produce a visible or audible signal or alarm when atrial or ventricular arrhythmia, such as premature contraction or ventricular fibrillation, occurs.

(b) *Classification.* Class II (special controls). The guidance document entitled "Class II Special Controls Guidance Document: Arrhythmia Detector and Alarm" will serve as the special control. See § 870.1 for the availability of this guidance document.

[68 FR 61344, Oct. 28, 2003]

§ 870.1100 Blood pressure alarm.

(a) *Identification.* A blood pressure alarm is a device that accepts the signal from a blood pressure transducer amplifier, processes the signal, and emits an alarm when the blood pressure falls outside a pre-set upper or lower limit.

(b) *Classification.* Class II (performance standards).

§ 870.1110 Blood pressure computer.

(a) *Identification.* A blood pressure computer is a device that accepts the electrical signal from a blood pressure transducer amplifier and indicates the systolic, diastolic, or mean pressure based on the input signal.

(b) *Classification.* Class II (performance standards).

§ 870.1120 Blood pressure cuff.

(a) *Identification.* A blood pressure cuff is a device that has an inflatable bladder in an inelastic sleeve (cuff) with a mechanism for inflating and deflating the bladder. The cuff is used in conjunction with another device to determine a subject's blood pressure.

(b) *Classification.* Class II (performance standards).

§ 870.1130 Noninvasive blood pressure measurement system.

(a) *Identification.* A noninvasive blood pressure measurement system is a device that provides a signal from which systolic, diastolic, mean, or any combination of the three pressures can be derived through the use of tranducers placed on the surface of the body.

(b) *Classification.* Class II (performance standards).

§ 870.1140 Venous blood pressure manometer.

(a) *Identification.* A venous blood pressure manometer is a device attached to a venous catheter to indicate manometrically the central or peripheral venous pressure.

(b) *Classification.* Class II (performance standards).

§ 870.1200 Diagnostic intravascular catheter.

(a) *Identification.* An intravascular diagnostic catheter is a device used to record intracardiac pressures, to sample blood, and to introduce substances into the heart and vessels. Included in this generic device are right-heart catheters, left-heart catheters, and angiographic catheters, among others.

(b) *Classification.* Class II (performance standards).

§ 870.1210 Continuous flush catheter.

(a) *Identification.* A continuous flush catheter is an attachment to a catheter-transducer system that permits continuous intravascular flushing at a slow infusion rate for the purpose of eliminating clotting, back-leakage, and waveform damping.

(b) *Classification.* Class II (performance standards).

§ 870.1220 Electrode recording catheter or electrode recording probe.

(a) *Identification.* An electrode recording catheter or an electrode recording probe is a device used to detect an intracardiac electrocardiogram, or to detect cardiac output or left-to-right heart shunts. The device may be unipolar or multipolar for electrocardiogram detection, or may be a platinum-tipped catheter which senses the presence of a special indicator for cardiac output or left-to-right heart shunt determinations.

(b) *Classification.* Class II (performance standards).

§870.1230 Fiberoptic oximeter catheter.

(a) *Identification.* A fiberoptic oximeter catheter is a device used to estimate the oxygen saturation of the blood. It consists of two fiberoptic bundles that conduct light at a desired wavelength through blood and detect the reflected and scattered light at the distal end of the catheter.

(b) *Classification.* Class II (performance standards).

§870.1240 Flow-directed catheter.

(a) *Identification.* A flow-directed catheter is a device that incorporates a gas-filled balloon to help direct the catheter to the desired position.

(b) *Classification.* Class II (performance standards).

§870.1250 Percutaneous catheter.

(a) *Identification.* A percutaneous catheter is a device that is introduced into a vein or artery through the skin using a dilator and a sheath (introducer) or guide wire.

(b) *Classification.* Class II (performance standards).

§870.1270 Intracavitary phonocatheter system.

(a) *Identification.* An intracavitary phonocatheter system is a system that includes a catheter with an acoustic transducer and the associated device that processes the signal from the transducer; this device records bioacoustic phenomena from a transducer placed within the heart, blood vessels, or body cavities.

(b) *Classification.* Class II (performance standards).

§870.1280 Steerable catheter.

(a) *Identification.* A steerable catheter is a catheter used for diagnostic and monitoring purposes whose movements are directed by a steering control unit.

(b) *Classification.* Class II (performance standards).

§870.1290 Steerable catheter control system.

(a) *Identification.* A steerable catheter control system is a device that is connected to the proximal end of a steerable guide wire that controls the motion of the steerable catheter.

(b) *Classification.* Class II (performance standards).

§870.1300 Catheter cannula.

(a) *Identification.* A catheter cannula is a hollow tube which is inserted into a vessel or cavity; this device provides a rigid or semirigid structure which can be connected to a tube or connector.

(b) *Classification.* Class II (performance standards).

§870.1310 Vessel dilator for percutaneous catheterization.

(a) *Identification.* A vessel dilator for percutaneous catheterization is a device which is placed over the guide wire to enlarge the opening in the vessel, and which is then removed before sliding the catheter over the guide wire.

(b) *Classification.* Class II (performance standards).

§870.1330 Catheter guide wire.

(a) *Identification.* A catheter guide wire is a coiled wire that is designed to fit inside a percutaneous catheter for the purpose of directing the catheter through a blood vessel.

(b) *Classification.* Class II (performance standards).

§870.1340 Catheter introducer.

(a) *Identification.* A catheter introducer is a sheath used to facilitate placing a catheter through the skin into a vein or artery.

(b) *Classification.* Class II (performance standards).

§870.1350 Catheter balloon repair kit.

(a) *Identification.* A catheter balloon repair kit is a device used to repair or replace the balloon of a balloon catheter. The kit contains the materials, such as glue and balloons, necessary to effect the repair or replacement.

(b) *Classification.* Class III (premarket approval).

(c) *Date PMA or notice of completion of a PDP is required.* A PMA or notice of completion of a PDP is required to be filed with the Food and Drug Administration on or before December 26, 1996 for any catheter balloon repair kit that was in commercial distribution before May 28, 1976, or that has, on or before

December 26, 1996 been found to be substantially equivalent to a catheter balloon repair kit that was in commercial distribution before May 28, 1976. Any other catheter balloon repair kit shall have an approved PMA or a declared completed PDP in effect before being placed in commercial distribution.

[45 FR 7907–7971, Feb. 5, 1980, as amended at 52 FR 17736, May 11, 1987; 61 FR 50706, Sept. 27, 1996]

§ 870.1360 Trace microsphere.

(a) *Identification.* A trace microsphere is a radioactively tagged nonbiodegradable particle that is intended to be injected into an artery or vein and trapped in the capillary bed for the purpose of studying blood flow within or to an organ.

(b) *Classification.* Class III (premarket approval).

(c) *Date PMA or notice of completion of a PDP is required.* A PMA or notice of completion of a PDP is required to be filed with the Food and Drug Administration on or before December 26, 1996 for any trace microsphere that was in commercial distribution before May 28, 1976, or that has, on or before December 26, 1996 been found to be substantially equivalent to a trace microsphere that was in commercial distribution before May 28, 1976. Any other trace microsphere shall have an approved PMA or a declared completed PDP in effect before being placed in commercial distribution.

[45 FR 7907–7971, Feb. 5, 1980, as amended at 52 FR 17736, May 11, 1987; 61 FR 50706, Sept. 27, 1996]

§ 870.1370 Catheter tip occluder.

(a) *Identification.* A catheter tip occluder is a device that is inserted into certain catheters to prevent flow through one or more orifices.

(b) *Classification.* Class II (performance standards).

§ 870.1380 Catheter stylet.

(a) *Identification.* A catheter stylet is a wire that is run through a catheter or cannula to render it stiff.

(b) *Classification.* Class II (performance standards).

§ 870.1390 Trocar.

(a) *Identification.* A trocar is a sharp-pointed instrument used with a cannula for piercing a vessel or chamber to facilitate insertion of the cannula.

(b) *Classification.* Class II (performance standards).

§ 870.1425 Programmable diagnostic computer.

(a) *Identification.* A programmable diagnostic computer is a device that can be programmed to compute various physiologic or blood flow parameters based on the output from one or more electrodes, transducers, or measuring devices; this device includes any associated commercially supplied programs.

(b) *Classification.* Class II (performance standards).

§ 870.1435 Single-function, preprogrammed diagnostic computer.

(a) *Identification.* A single-function, preprogrammed diagnostic computer is a hard-wired computer that calculates a specific physiological or blood-flow parameter based on information obtained from one or more electrodes, transducers, or measuring devices.

(b) *Classification.* Class II (performance standards).

§ 870.1450 Densitometer.

(a) *Identification.* A densitometer is a device used to measure the transmission of light through an indicator in a sample of blood.

(b) *Classification.* Class II (performance standards).

§ 870.1650 Angiographic injector and syringe.

(a) *Identification.* An angiographic injector and syringe is a device that consists of a syringe and a high-pressure injector which are used to inject contrast material into the heart, great vessels, and coronary arteries to study the heart and vessels by x-ray photography.

(b) *Classification.* Class II (performance standards).

§ 870.1660 Indicator injector.

(a) *Identification.* An indicator injector is an electrically or gas-powered device designed to inject accurately an indicator solution into the blood stream. This device may be used in conjuction with a densitometer or thermodilution device to determine cardiac output.

(b) *Classification.* Class II (performance standards).

§ 870.1670 Syringe actuator for an injector.

(a) *Identification.* A syringe actuator for an injector is an electrical device that controls the timing of an injection by an angiographic or indicator injector and synchronizes the injection with the electrocardiograph signal.

(b) *Classification.* Class II (performance standards).

§ 870.1750 External programmable pacemaker pulse generator.

(a) *Identification.* An external programmable pacemaker pulse generators is a device that can be programmed to produce one or more pulses at preselected intervals; this device is used in electrophysiological studies.

(b) *Classification.* Class II (performance standards).

§ 870.1800 Withdrawal-infusion pump.

(a) *Identification.* A withdrawal-infusion pump is a device designed to inject accurately drugs into the bloodstream and to withdraw blood samples for use in determining cardiac output.

(b) *Classification.* Class II (performance standards).

§ 870.1875 Stethoscope.

(a) *Manual stethoscope*—(1) *Identification.* A manual stethoscope is a mechanical device used to project the sounds associated with the heart, arteries, and veins and other internal organs.

(2) *Classification.* Class I (general controls). The device is exempt from the premarket notification procedures in subpart E of part 807 of this chapter subject to the limitations in § 870.9.

(b) *Electronic stethoscope*—(1) *Identification.* An electronic stethoscope is an electrically amplified device used to project the sounds associated with the heart, arteries, and veins and other internal organs.

(2) *Classification.* Class II (performance standards).

[45 FR 7907-7971, Feb. 5, 1980, as amended at 59 FR 63007, Dec. 7, 1994; 66 FR 38796, July 25, 2001]

§ 870.1915 Thermodilution probe.

(a) *Identification.* A thermodilution probe is a device that monitors cardiac output by use of thermodilution techniques; this device is commonly attached to a catheter that may have one or more probes.

(b) *Classification.* Class II (performance standards).

Subpart C—Cardiovascular Monitoring Devices

§ 870.2050 Biopotential amplifier and signal conditioner.

(a) *Identification.* A biopotential amplifier and signal conditioner is a device used to amplify or condition an electrical signal of biologic origin.

(b) *Classification.* Class II (performance standards).

§ 870.2060 Transducer signal amplifier and conditioner.

(a) *Identification.* A transducer signal amplifier and conditioner is a device used to provide the excitation energy for the transducer and to amplify or condition the signal emitted by the transducer.

(b) *Classification.* Class II (performance standards).

§ 870.2100 Cardiovascular blood flowmeter.

(a) *Identification.* A cardiovascular blood flowmeter is a device that is connected to a flow transducer that energizes the transducer and processes and displays the blood flow signal.

(b) *Classification.* Class II (performance standards).

§ 870.2120 Extravascular blood flow probe.

(a) *Identification.* An extravascular blood flow probe is an extravascular ultrasonic or electromagnetic probe used in conjunction with a blood flowmeter to measure blood flow in a chamber or vessel.

(b) *Classification.* Class II (performance standards).

§ 870.2300 Cardiac monitor (including cardiotachometer and rate alarm).

(a) *Identification.* A cardiac monitor (including cardiotachometer and rate alarm) is a device used to measure the heart rate from an analog signal produced by an electrocardiograph, vectorcardiograph, or blood pressure monitor. This device may sound an alarm when the heart rate falls outside preset upper and lower limits.

(b) *Classification.* Class II (performance standards).

§ 870.2310 Apex cardiograph (vibrocardiograph).

(a) *Identification.* An apex cardiograph (vibrocardiograph) is a device used to amplify or condition the signal from an apex cardiographic transducer and to produce a visual display of the motion of the heart; this device also provides any excitation energy required by the transducer.

(b) *Classification.* Class II (performance standards).

§ 870.2320 Ballistocardiograph.

(a) *Identification.* A ballistocardiograph is a device, including a supporting structure on which the patient is placed, that moves in response to blood ejection from the heart. The device often provides a visual display.

(b) *Classification.* Class II (performance standards).

§ 870.2330 Echocardiograph.

(a) *Identification.* An echocardiograph is a device that uses ultrasonic energy to create images of cardiovascular structures. It includes phased arrays and two-dimensional scanners.

(b) *Classification.* Class II (performance standards).

§ 870.2340 Electrocardiograph.

(a) *Identification.* An electrocardiograph is a device used to process the electrical signal transmitted through two or more electrocardiograph electrodes and to produce a visual display of the electrical signal produced by the heart.

(b) *Classification.* Class II (performance standards).

§ 870.2350 Electrocardiograph lead switching adaptor.

(a) *Identification.* An electrocardiograph lead switching adaptor is a passive switching device to which electrocardiograph limb and chest leads may be attached. This device is used to connect various combinations of limb and chest leads to the output terminals in order to create standard lead combinations such as leads I, II, and III.

(b) *Classification.* Class II (performance standards).

§ 870.2360 Electrocardiograph electrode.

(a) *Identification.* An electrocardiograph electrode is the electrical conductor which is applied to the surface of the body to transmit the electrical signal at the body surface to a processor that produces an electrocardiogram or vectorcardiogram.

(b) *Classification.* Class II (performance standards).

§ 870.2370 Electrocardiograph surface electrode tester.

(a) *Identification.* An electrocardiograph surface electrode tester is a device used to test the function and application of electrocardiograph electrodes.

(b) *Classification.* Class II (performance standards).

§ 870.2390 Phonocardiograph.

(a) *Identification.* A phonocardiograph is a device used to amplify or condition the signal from a heart sound transducer. This device furnishes the excitation energy for the transducer and provides a visual or audible display of the heart sounds.

(b) *Classification.* Class I (general controls). The device is exempt from the premarket notification procedures in subpart E of part 807 of this chapter subject to the limitations in § 870.9.

[45 FR 7907-7971, Feb. 5, 1980, as amended at 61 FR 1121, Jan. 16, 1996; 66 FR 38796, July 25, 2001]

§870.2400 Vectorcardiograph.

(a) *Identification.* A vectorcardiograph is a device used to process the electrical signal transmitted through electrocardiograph electrodes and to produce a visual display of the magnitude and direction of the electrical signal produced by the heart.

(b) *Classification.* Class II (performance standards).

§870.2450 Medical cathode-ray tube display.

(a) *Identification.* A medical cathode-ray tube display is a device designed primarily to display selected biological signals. This device often incorporates special display features unique to a specific biological signal.

(b) *Classification.* Class II (performance standards).

§870.2600 Signal isolation system.

(a) *Identification.* A signal isolation system is a device that electrically isolates the patient from equipment connected to the commercial power supply received from a utility company. This isolation may be accomplished, for example, by transformer coupling, acoustic coupling, or optical coupling.

(b) *Classification.* Class I (general controls). The device is exempt from the premarket notification procedures in subpart E of part 807 of this chapter subject to the limitations in §870.9.

[45 FR 7907-7971, Feb. 5, 1980, as amended at 61 FR 1121, Jan. 16, 1996; 66 FR 38796, July 25, 2001]

§870.2620 Line isolation monitor.

(a) *Identification.* A line isolation monitor is a device used to monitor the electrical leakage current from a power supply electrically isolated from the commercial power supply received from a utility company.

(b) *Classification.* Class I (general controls). The device is exempt from the premarket notification procedures in subpart E of part 807 of this chapter subject to the limitations in §870.9.

[45 FR 7907-7971, Feb. 5, 1980, as amended at 61 FR 1121, Jan. 16, 1996; 66 FR 38796, July 25, 2001]

§870.2640 Portable leakage current alarm.

(a) *Identification.* A portable leakage current alarm is a device used to measure the electrical leakage current between any two points of an electrical system and to sound an alarm if the current exceeds a certain threshold.

(b) *Classification.* Class I (general controls). The device is exempt from the premarket notification procedures in subpart E of part 807 of this chapter subject to the limitations in §870.9.

[45 FR 7907-7971, Feb. 5, 1980, as amended at 61 FR 1121, Jan. 16, 1996; 66 FR 38796, July 25, 2001]

§870.2675 Oscillometer.

(a) *Identification.* An oscillometer is a device used to measure physiological oscillations of any kind, e.g., changes in the volume of arteries.

(b) *Classification.* Class II (performance standards).

§870.2700 Oximeter.

(a) *Identification.* An oximeter is a device used to transmit radiation at a known wavelength(s) through blood and to measure the blood oxygen saturation based on the amount of reflected or scattered radiation. It may be used alone or in conjunction with a fiberoptic oximeter catheter.

(b) *Classification.* Class II (performance standards).

§870.2710 Ear oximeter.

(a) *Identification.* An ear oximeter is an extravascular device used to transmit light at a known wavelength(s) through blood in the ear. The amount of reflected or scattered light as indicated by this device is used to measure the blood oxygen saturation.

(b) *Classification.* Class II (performance standards).

§870.2750 Impedance phlebograph.

(a) *Identification.* An impedance phlebograph is a device used to provide a visual display of the venous pulse or drainage by measuring electrical impedance changes in a region of the body.

(b) *Classification.* Class II (performance standards).

§ 870.2770 Impedance plethysmograph.

(a) *Identification.* An impedance plethysmograph is a device used to estimate peripheral blood flow by measuring electrical impedance changes in a region of the body such as the arms and legs.

(b) *Classification.* Class II (performance standards).

§ 870.2780 Hydraulic, pneumatic, or photoelectric plethysmographs.

(a) *Identification.* A hydraulic, pneumatic, or photoelectric plethysmograph is a device used to estimate blood flow in a region of the body using hydraulic, pneumatic, or photoelectric measurement techniques.

(b) *Classification.* Class II (performance standards).

§ 870.2800 Medical magnetic tape recorder.

(a) *Identification.* A medical magnetic tape recorder is a device used to record and play back signals from, for example, physiological amplifiers, signal conditioners, or computers.

(b) *Classification.* Class II (performance standards).

§ 870.2810 Paper chart recorder.

(a) *Identification.* A paper chart recorder is a device used to print on paper, and create a permanent record of the signal from, for example, a physiological amplifier, signal conditioner, or computer.

(b) *Classification.* Class I (general controls). The device is exempt from the premarket notification procedures in subpart E of part 807 of this chapter subject to the limitations in § 870.9.

[45 FR 7907–7971, Feb. 5, 1980, as amended at 61 FR 1121, Jan. 16, 1996; 66 FR 38796, July 25, 2001]

§ 870.2840 Apex cardiographic transducer.

(a) *Identification.* An apex cardiographic transducer is a device used to detect motion of the heart (acceleration, velocity, or displacement) by changes in the mechanical or electrical properties of the device.

(b) *Classification.* Class II (performance standards).

§ 870.2850 Extravascular blood pressure transducer.

(a) *Identification.* An extravascular blood pressure transducer is a device used to measure blood pressure by changes in the mechanical or electrical properties of the device. The proximal end of the transducer is connected to a pressure monitor that produces an analog or digital electrical signal related to the electrical or mechanical changes produced in the transducer.

(b) *Classification.* Class II (performance standards).

§ 870.2855 Implantable Intra-aneurysm Pressure Measurement System.

(a) *Identification.* Implantable intra-aneurysm pressure measurement system is a device used to measure the intra-sac pressure in a vascular aneurysm. The device consists of a pressure transducer that is implanted into the aneurysm and a monitor that reads the pressure from the transducer.

(b) *Classification.* Class II (special controls). The special control is FDA's guidance document entitled "Class II Special Controls Guidance Document: Implantable Intra-Aneurysm Pressure Measurement System." See § 870.1 (e) for the availability of this guidance document.

[71 FR 7871, Feb. 15, 2006]

§ 870.2860 Heart sound transducer.

(a) *Identification.* A heart sound transducer is an external transducer that exhibits a change in mechanical or electrical properties in relation to sounds produced by the heart. This device may be used in conjunction with a phonocardiograph to record heart sounds.

(b) *Classification.* Class II (performance standards).

§ 870.2870 Catheter tip pressure transducer.

(a) *Identification.* A catheter tip pressure transducer is a device incorporated into the distal end of a catheter. When placed in the bloodstream, its mechanical or electrical properties change in relation to changes in blood pressure. These changes are transmitted to accessory equipment for processing.

(b) *Classification.* Class II (performance standards).

§ 870.2880 Ultrasonic transducer.

(a) *Identification.* An ultrasonic transducer is a device applied to the skin to transmit and receive ultrasonic energy that is used in conjunction with an echocardiograph to provide imaging of cardiovascular structures. This device includes phased arrays and two-dimensional scanning transducers.

(b) *Classification.* Class II (performance standards).

§ 870.2890 Vessel occlusion transducer.

(a) *Identification.* A vessel occlusion transducer is a device used to provide an electrical signal corresponding to sounds produced in a partially occluded vessel. This device includes motion, sound, and ultrasonic transducers.

(b) *Classification.* Class II (performance standards).

§ 870.2900 Patient transducer and electrode cable (including connector).

(a) *Identification.* A patient transducer and electrode cable (including connector) is an electrical conductor used to transmit signals from, or power or excitation signals to, patient-connected electrodes or transducers.

(b) *Classification.* Class II (performance standards).

§ 870.2910 Radiofrequency physiological signal transmitter and receiver.

(a) *Identification.* A radiofrequency physiological signal transmitter and receiver is a device used to condition a physiological signal so that it can be transmitted via radiofrequency from one location to another, e.g., a central monitoring station. The received signal is reconditioned by the device into its original format so that it can be displayed.

(b) *Classification.* Class II (performance standards).

§ 870.2920 Telephone electrocardiograph transmitter and receiver.

(a) *Identification.* A telephone electrocardiograph transmitter and receiver is a device used to condition an electrocardiograph signal so that it can be transmitted via a telephone line to another location. This device also includes a receiver that reconditions the received signal into its original format so that it can be displayed. The device includes devices used to transmit and receive pacemaker signals.

(b) *Classification.* Class II (performance standards).

Subpart D—Cardiovascular Prosthetic Devices

§ 870.3250 Vascular clip.

(a) *Identification.* A vascular clip is an implanted extravascular device designed to occlude, by compression, blood flow in small blood vessels other than intracranial vessels.

(b) *Classification.* Class II (performance standards).

§ 870.3260 Vena cava clip.

(a) *Identification.* A vena cava clip is an implanted extravascular device designed to occlude partially the vena cava for the purpose of inhibiting the flow of thromboemboli through that vessel.

(b) *Classification.* Class II (performance standards).

§ 870.3300 Vascular embolization device.

(a) *Identification.* A vascular embolization device is an intravascular implant intended to control hemorrhaging due to aneurysms, certain types of tumors (e.g., nephroma, hepatoma, uterine fibroids), and arteriovenous malformations. This does not include cyanoacrylates and other embolic agents, which act by polymerization or precipitation. Embolization devices used in neurovascular applications are also not included in this classification, see § 882.5950 of this chapter.

(b) *Classification.* Class II (special controls.) The special control for this device is the FDA guidance document entitled "Class II Special Controls Guidance Document: Vascular and Neurovascular Embolization Devices." For availability of this guidance document, see § 870.1(e).

[69 FR 77899, Dec. 29, 2004]

§ 870.3375 Cardiovascular intravascular filter.

(a) *Identification.* A cardiovascular intravascular filter is an implant that is placed in the inferior vena cava for the purpose of preventing pulmonary thromboemboli (blood clots generated in the lower limbs and broken loose into the blood stream) from flowing into the right side of the heart and the pulmonary circulation.

(b) *Classification.* Class II. The special controls for this device are:

(1) "Use of International Standards Organization's ISO 10993 'Biological Evaluation of Medical Devices Part I: Evaluation and Testing,' " and

(2) FDA's:

(i) "510(k) Sterility Review Guidance and Revision of 2/12/90 (K90–1)" and

(ii) "Guidance for Cardiovascular Intravascular Filter 510(k) Submissions."

[45 FR 7907–7971, Feb. 5, 1980, as amended at 52 FR 17736, May 11, 1987; 65 FR 17144, Mar. 31, 2000]

§ 870.3450 Vascular graft prosthesis.

(a) *Identification.* A vascular graft prosthesis is an implanted device intended to repair, replace, or bypass sections of native or artificial vessels, excluding coronary or cerebral vasculature, and to provide vascular access. It is commonly constructed of materials such as polyethylene terephthalate and polytetrafluoroethylene, and it may be coated with a biological coating, such as albumin or collagen, or a synthetic coating, such as silicone. The graft structure itself is not made of materials of animal origin, including human umbilical cords.

(b) *Classification.* Class II (special controls). The special control for this device is the FDA guidance document entitled "Guidance Document for Vascular Prostheses 510(k) Submissions."

[66 FR 18542, Apr. 10, 2001]

§ 870.3470 Intracardiac patch or pledget made of polypropylene, polyethylene terephthalate, or polytetrafluoroethylene.

(a) *Identification.* An intracardiac patch or pledget made of polypropylene, polyethylene terephthalate, or polytetrafluoroethylene is a fabric device placed in the heart that is used to repair septal defects, for patch grafting, to repair tissue, and to buttress sutures.

(b) *Classification.* Class II (performance standards).

§ 870.3535 Intra-aortic balloon and control system

(a) *Identification.* A intra-aortic balloon and control system is a device that consists of an inflatable balloon, which is placed in the aorta to improve cardiovascular functioning during certain life-threatening emergencies, and a control system for regulating the inflation and deflation of the balloon. The control system, which monitors and is synchronized with the electrocardiogram, provides a means for setting the inflation and deflation of the balloon with the cardiac cycle.

(b) *Classification.* Class III (premarket approval).

(c) *Date PMA or notice of completion of a PDP is required.* No effective date has been established of the requirement for premarket approval. See § 870.3.

[45 FR 7907–7971, Feb. 5, 1980, as amended at 52 FR 17736, May 11, 1987]

§ 870.3545 Ventricular bypass (assist) device.

(a) *Identification.* A ventricular bypass (assist) device is a device that assists the left or right ventricle in maintaining circulatory blood flow. The device is either totally or partially implanted in the body.

(b) *Classification.* Class III (premarket approval).

(c) *Date PMA or notice of completion of a PDP is required.* No effective date has been established of the requirement for premarket approval. See § 870.3.

[45 FR 7907–7971, Feb. 5, 1980, as amended at 52 FR 17736, May 11, 1987]

§ 870.3600 External pacemaker pulse generator.

(a) *Identification.* An external pacemaker pulse generator is a device that has a power supply and electronic circuits that produce a periodic electrical pulse to stimulate the heart. This device, which is used outside the body, is used as a temporary substitute for the heart's intrinsic pacing sytem until a permanent pacemaker can be implanted, or to control irregular heartbeats in patients following cardiac surgery or a myocardial infarction. The device may have adjustments for impulse strength, duration, R-wave sensitivity, and other pacing variables.

(b) *Classification.* Class III (premarket approval).

(c) *Date PMA or notice of completion of a PDP is required.* No effective date has been established of the requirement for premarket approval. See § 870.3.

[45 FR 7907–7971, Feb. 5, 1980, as amended at 52 FR 17736, May 11, 1987]

§ 870.3610 Implantable pacemaker pulse generator.

(a) *Identification.* An implantable pacemaker pulse generator is a device that has a power supply and electronic circuits that produce a periodic electrical pulse to stimulate the heart. This device is used as a substitute for the heart's intrinsic pacing system to correct both intermittent and continuous cardiac rhythm disorders. This device includes triggered, inhibited, and asynchronous devices implanted in the human body.

(b) *Classification.* Class III (premarket approval).

(c) *Date PMA or notice of completion of a PDP is required.* No effective date has been established of the requirement for premarket approval. See § 870.3.

[45 FR 7907–7971, Feb. 5, 1980, as amended at 52 FR 17736, May 11, 1987]

§ 870.3620 Pacemaker lead adaptor.

(a) *Identification.* A pacemaker lead adaptor is a device used to adapt a pacemaker lead so that it can be connected to a pacemaker pulse generator produced by a different manufacturer.

(b) *Classification.* Class II (special controls). The special control for this device is the FDA guidance document entitled "Guidance for the Submission of Research and Marketing Applications for Permanent Pacemaker Leads and for Pacemaker Lead Adaptor 510(k) Submissions."

[45 FR 7907–7971, Feb. 5, 1980, as amended at 52 FR 17736, May 11, 1987; 66 FR 18542, Apr. 10, 2001]

§ 870.3630 Pacemaker generator function analyzer.

(a) *Identification.* A pacemaker generator function analyzer is a device that is connected to a pacemaker pulse generator to test any or all of the generator's parameters, including pulse duration, pulse amplitude, pulse rate, and sensing threshold.

(b) *Classification.* Class II (performance standards).

§ 870.3640 Indirect pacemaker generator function analyzer.

(a) *Identification.* An indirect pacemaker generator function analyzer is an electrically powered device that is used to determine pacemaker function or pacemaker battery function by periodically monitoring an implanted pacemaker's pulse rate and pulse width. The device is noninvasive, and it detects pacemaker pulse rate and width via external electrodes in contact with the patient's skin.

(b) *Classification.* Class II (performance standards).

§ 870.3650 Pacemaker polymeric mesh bag.

(a) *Identification.* A pacemaker polymeric mesh bag is an implanted device used to hold a pacemaker pulse generator. The bag is designed to create a stable implant environment for the pulse generator.

(b) *Classification.* Class I (general controls). The device is exempt from the premarket notification procedures in subpart E of part 807 of this chapter subject to the limitations in § 870.9.

[45 FR 7907-7971, Feb. 5, 1980, as amended at 61 FR 1121, Jan. 16, 1996; 66 FR 38796, July 25, 2001]

§ 870.3670 Pacemaker charger.

(a) *Identification.* A pacemaker charger is a device used transcutaneously to recharge the batteries of a rechargeable pacemaker.

(b) *Classification.* Class I (general controls). The device is exempt from the premarket notification procedures in subpart E of part 807 of this chapter subject to the limitations in § 870.9.

[45 FR 7907-7971, Feb. 5, 1980, as amended at 61 FR 1121, Jan. 16, 1996; 66 FR 38796, July 25, 2001]

§ 870.3680 Cardiovascular permanent or temporary pacemaker electrode.

(a) *Temporary pacemaker electrode*—(1) *Identification.* A temporary pacemaker electrode is a device consisting of flexible insulated electrical conductors with one end connected to an *external* pacemaker pulse generator and the other end applied to the heart. The device is used to transmit a pacing electrical stimulus from the pulse generator to the heart and/or to transmit the electrical signal of the heart to the pulse generator.

(2) *Classification.* Class II (performance standards).

(b) *Permanent pacemaker electrode*—(1) *Identification.* A permanent pacemaker electrode is a device consisting of flexible insulated electrical conductors with one end connected to an implantable pacemaker pulse generator and the other end applied to the heart. The device is used to transmit a pacing electrical stimulus from the pulse generator to the heart and/or to transmit the electrical signal of the heart to the pulse generator.

(2) *Classification.* Class III (premarket approval).

(c) *Date PMA or notice of completion of a PDP is required.* No effective date has been established of the requirement for premarket approval for the device described in paragraph (b)(1). See § 870.3.

[45 FR 7907–7971, Feb. 5, 1980, as amended at 52 FR 17736, May 11, 1987]

§ 870.3690 Pacemaker test magnet.

(a) *Identification.* A pacemaker test magnet is a device used to test an inhibited or triggered type of pacemaker pulse generator and cause an inhibited or triggered generator to revert to asynchronous operation.

(b) *Classification.* Class I (general controls). The device is exempt from the premarket notification procedures in subpart E of part 807 of this chapter subject to the limitations in § 870.9.

[45 FR 7907-7971, Feb. 5, 1980, as amended at 61 FR 1121, Jan. 16, 1996; 66 FR 38796, July 25, 2001]

§ 870.3700 Pacemaker programmers.

(a) *Identification.* A pacemaker programmer is a device used to change noninvasively one or more of the electrical operating characteristics of a pacemaker.

(b) *Classification.* Class III (premarket approval).

(c) *Date PMA or notice of completion of a PDP is required.* No effective date has been established of the requirement for premarket approval. See § 870.3.

[45 FR 7907–7971, Feb. 5, 1980, as amended at 52 FR 17736, May 11, 1987]

§ 870.3710 Pacemaker repair or replacement material.

(a) *Identification.* A pacemaker repair or replacement material is an adhesive, a sealant, a screw, a crimp, or any other material used to repair a pacemaker lead or to reconnect a pacemaker lead to a pacemaker pulse generator.

(b) *Classification.* Class III (premarket approval).

(c) *Date PMA or notice of completion of a PDP is required.* No effective date has been established of the requirement for premarket approval. See § 870.3.

[45 FR 7907–7971, Feb. 5, 1980, as amended at 52 FR 17736, May 11, 1987]

§ 870.3720 Pacemaker electrode function tester.

(a) *Identification.* A pacemaker electrode function tester is a device which is connected to an implanted pacemaker lead that supplies an accurately calibrated, variable pacing pulse for measuring the patient's pacing threshold and intracardiac R-wave potential.

(b) *Classification.* Class II (performance standards).

§ 870.3730 Pacemaker service tools.

(a) *Identification.* Pacemaker service tools are devices such as screwdrivers and Allen wrenches, used to repair a pacemaker lead or to reconnect a pacemaker lead to a pacemaker generator.

(b) *Classification.* Class I (general controls). The device is exempt from the premarket notification procedures in subpart E of part 807 of this chapter subject to the limitations in § 870.9.

[45 FR 7907–7971, Feb. 5, 1980, as amended at 54 FR 25049, June 12, 1989; 66 FR 38797, July 25, 2001]

§ 870.3800 Annuloplasty ring.

(a) *Identification.* An annuloplasty ring is a rigid or flexible ring implanted around the mitral or tricuspid heart valve for reconstructive treatment of valvular insufficiency.

(b) *Classification.* Class II (special controls). The special control for this device is the FDA guidance document entitled "Guidance for Annuloplasty Rings 510(k) Submissions."

[45 FR 7907–7971, Feb. 5, 1980, as amended at 52 FR 17736, May 11, 1987; 66 FR 18542, Apr. 10, 2001]

§ 870.3850 Carotid sinus nerve stimulator.

(a) *Identification.* A carotid sinus nerve stimulator is an implantable device used to decrease arterial pressure by stimulating Hering's nerve at the carotid sinus.

(b) *Classification.* Class III (premarket approval).

(c) *Date PMA or notice of completion of a PDP is required.* A PMA or a notice of completion of a PDP is required to be filed with the Food and Drug Administration on or before December 26, 1996 for any carotid sinus nerve stimulator that was in commercial distribution before May 28, 1976, or that has, on or before December 26, 1996 been found to be substantially equivalent to a carotid sinus nerve stimulator that was in commercial distribution before May 28, 1976. Any other carotid sinus nerve stimulator shall have an approved PMA or a declared completed PDP in effect before being placed in commercial distribution.

[45 FR 7907–7971, Feb. 5, 1980, as amended at 52 FR 17736, May 11, 1987; 61 FR 50706, Sept. 27, 1996]

§ 870.3925 Replacement heart valve.

(a) *Identification.* A replacement heart valve is a device intended to perform the function of any of the heart's natural valves. This device includes valves constructed of prosthetic materials, biologic valves (e.g., porcine valves), or valves constructed of a combination of prosthetic and biologic materials.

(b) *Classification.* Class III (premarket approval).

(c) *Date premarket approval application (PMA) or notice of completion of a product development protocol (PDP) is required.* A PMA or a notice of completion of a PDP is required to be filed with the Food and Drug Administration on or before December 9, 1987 for any replacement heart valve that was in commercial distribution before May 28, 1976, or that has on or before December 9, 1987 been found to be substantially equivalent to a replacement heart valve that was in commercial distribution before May 28, 1976. Any other replacement heart valve shall have an approved PMA or a declared completed PDP in effect before being placed in commercial distribution.

[45 FR 7907–7971, Feb. 5, 1980, as amended at 52 FR 18163, May 13, 1987; 52 FR 23137, June 17, 1987]

§ 870.3935 Prosthetic heart valve holder.

(a) *Identification.* A prosthetic heart valve holder is a device used to hold a replacement heart valve while it is being sutured into place.

(b) *Classification.* Class I. The device is exempt from the premarket notification procedures in subpart E of part 807 of this chapter.

[45 FR 7907-7971, Feb. 5, 1980, as amended at 61 FR 1121, Jan. 16, 1996]

§ 870.3945 Prosthetic heart valve sizer.

(a) *Identification.* A prosthetic heart valve sizer is a device used to measure the size of the natural valve opening to determine the size of the appropriate replacement heart valve.

(b) *Classification.* Class I (general controls). The device is exempt from the premarket notification procedures in subpart E of part 807 of this chapter subject to the limitations in § 870.9.

[45 FR 7907-7971, Feb. 5, 1980, as amended at 61 FR 1121, Jan. 16, 1996; 66 FR 38797, July 25, 2001]

Subpart E—Cardiovascular Surgical Devices

§ 870.4075 Endomyocardial biopsy device.

(a) *Identification.* An endomyocardial biopsy device is a device used in a catheterization procedure to remove samples of tissue from the inner wall of the heart.

(b) *Classification.* Class II (performance standards).

§ 870.4200 Cardiopulmonary bypass accessory equipment.

(a) *Identification.* Cardiopulmonary bypass accessory equipment is a device that has no contact with blood and that is used in the cardiopulmonary bypass circuit to support, adjoin, or connect components, or to aid in the setup of the extracorporeal line, e.g., an oxygenator mounting bracket or system-priming equipment.

(b) *Classification.* (1) Class I. The device is classified as class I if it does not involve an electrical connection to the patient. The device is exempt from the premarket notification procedures in subpart E of part 807 of this chapter subject to § 870.9.

(2) Class II (special controls). The device is classified as class II if it involves an electrical connection to the patient. The special controls are as follows:

(i) The performance standard under part 898 of this chapter, and

(ii) The guidance document entitled "Guidance on the Performance Standard for Electrode Lead Wires and Patient Cables." The device is exempt from the premarket notification procedures in subpart E of part 807 of this chapter subject to § 870.9.

[65 FR 19319, Apr. 11, 2000]

§ 870.4205 Cardiopulmonary bypass bubble detector.

(a) *Identification.* A cardiopulmonary bypass bubble detector is a device used to detect bubbles in the arterial return line of the cardiopulmonary bypass circuit.

(b) *Classification.* Class II (performance standards).

§ 870.4210 Cardiopulmonary bypass vascular catheter, cannula, or tubing.

(a) *Identification.* A cardiopulmonary bypass vascular catheter, cannula, or tubing is a device used in cardiopulmonary surgery to cannulate the vessels, perfuse the coronary arteries, and to interconnect the catheters and cannulas with an oxygenator. The device includes accessory bypass equipment.

(b) *Classification.* Class II (performance standards).

§ 870.4220 Cardiopulmonary bypass heart-lung machine console.

(a) *Identification.* A cardiopulmonary bypass heart-lung machine console is a device that consists of a control panel and the electrical power and control circuitry for a heart-lung machine. The console is designed to interface with the basic units used in a gas exchange system, including the pumps, oxygenator, and heat exchanger.

(b) *Classification.* Class II (performance standards).

§ 870.4230 Cardiopulmonary bypass defoamer.

(a) *Identification.* A cardiopulmonary bypass defoamer is a device used in conjunction with an oxygenator during cardiopulmonary bypass surgery to remove gas bubbles from the blood.

(b) *Classification.* Class II (special controls). The special control for this device is the FDA guidance document entitled "Guidance for Extracorporeal Blood Circuit Defoamer 510(k) Submissions."

[45 FR 7907–7971, Feb. 5, 1980, as amended at 52 FR 17737, May 11, 1987; 66 FR 18542, Apr. 10, 2001]

§ 870.4240 Cardiopulmonary bypass heat exchanger.

(a) *Identification.* A cardiopulmonary bypass heat exchanger is a device, consisting of a heat exchange system used in extracorporeal circulation to warm or cool the blood or perfusion fluid flowing through the device.

(b) *Classification.* Class II (performance standards).

§ 870.4250 Cardiopulmonary bypass temperature controller.

(a) *Identification.* A cardiopulmonary bypass temperature controller is a device used to control the temperature of the fluid entering and leaving a heat exchanger.

(b) *Classification.* Class II (performance standards).

§ 870.4260 Cardiopulmonary bypass arterial line blood filter.

(a) *Identification.* A cardiopulmonary bypass arterial line blood filter is a device used as part of a gas exchange (oxygenator) system to filter nonbiologic particles and emboli (blood clots or pieces of foreign material flowing in the bloodstream which will obstruct circulation by blocking a vessel) out of the blood. It is used in the arterial return line.

(b) *Classification.* Class II (special controls). The special control for this device is the FDA guidance document entitled "Guidance for Cardiopulmonary Bypass Arterial Line Blood Filter 510(k) Submissions."

[45 FR 7907–7971, Feb. 5, 1980, as amended at 52 FR 17737, May 11, 1987; 66 FR 18542, Apr. 10, 2001]

§ 870.4270 Cardiopulmonary bypass cardiotomy suction line blood filter.

(a) *Identification.* A cardiopulmonary bypass cardiotomy suction line blood filter is a device used as part of a gas exchange (oxygenator) system to filter nonbiologic particles and emboli (a blood clot or a piece of foreign material flowing in the bloodstream which will obstruct circulation by blocking a vessel) out of the blood. This device is intended for use in the cardiotomy suction line.

(b) *Classification.* Class II (performance standards).

§ 870.4280 Cardiopulmonary prebypass filter.

(a) *Identification.* A cardiopulmonary prebypass filter is a device used during priming of the oxygenator circuit to remove particulates or other debris from the circuit prior to initiating bypass. The device is not used to filter blood.

(b) *Classification.* Class II (performance standards).

§ 870.4290 Cardiopulmonary bypass adaptor, stopcock, manifold, or fitting.

(a) *Identification.* A cardiopulmonary bypass adaptor, stopcock, manifold, or fitting is a device used in cardiovascular diagnostic, surgical, and therapeutic applications to interconnect tubing, catheters, or other devices.

(b) *Classification.* Class II (performance standards).

§ 870.4300 Cardiopulmonary bypass gas control unit.

(a) *Identification.* A cardiopulmonary bypass gas control unit is a device used to control and measure the flow of gas into the oxygenator. The device is calibrated for a specific gas.

(b) *Classification.* Class II (performance standards).

§ 870.4310 Cardiopulmonary bypass coronary pressure gauge.

(a) *Identification.* A cardiopulmonary bypass coronary pressure gauge is a device used in cardiopulmonary bypass surgery to measure the pressure of the blood perfusing the coronary arteries.

(b) *Classification.* Class II (performance standards).

§ 870.4320 Cardiopulmonary bypass pulsatile flow generator.

(a) *Identification.* A cardiopulmonary bypass pulsatile flow generator is an electrically and pneumatically operated device used to create pulsatile blood flow. The device is placed in a cardiopulmonary bypass circuit downstream from the oxygenator.

(b) *Classification.* Class III (premarket approval).

(c) Date PMA or notice of completion of PDP is required. A PMA or notice of completion of a PDP is required to be filed with the Food and Drug Administration on or before September 21, 2004, for any cardiopulmonary bypass pulsatile flow generator that was in commercial distribution before May 28, 1976, or that has, on or before September 21, 2004, been found to be substantially equivalent to any cardiopulmonary bypass pulsatile flow generator that was in commercial distribution before May 28, 1976. Any other cardiopulmonary bypass pulsatile flow generator shall have an approved PMA or declared completed PDP in effect before being placed in commercial distribution.

[45 FR 7907–7971, Feb. 5, 1980, as amended at 52 FR 17737, May 11, 1987; 69 FR 34920, June 23, 2004]

§ 870.4330 Cardiopulmonary bypass on-line blood gas monitor.

(a) *Identification.* A cardiopulmonary bypass on-line blood gas monitor is a device used in conjunction with a blood gas sensor to measure the level of gases in the blood.

(b) *Classification.* Class II (performance standards).

§ 870.4340 Cardiopulmonary bypass level sensing monitor and/or control.

(a) *Identification.* A cardiopulmonary bypass level sensing monitor and/or control is a device used to monitor and/or control the level of blood in the blood reservoir and to sound an alarm when the level falls below a predetermined value.

(b) *Classification.* Class II (performance standards).

§ 870.4350 Cardiopulmonary bypass oxygenator.

(a) *Identification.* A cardiopulmonary bypass oxygenator is a device used to exchange gases between blood and a gaseous environment to satisfy the gas exchange needs of a patient during open-heart surgery.

(b) *Classification.* Class II (special controls). The special control for this device is the FDA guidance document entitled "Guidance for Cardiopulmonary Bypass Oxygenators 510(k) Submissions."

[45 FR 7907–7971, Feb. 5, 1980, as amended at 52 FR 17737, May 11, 1987; 66 FR 18542, Apr. 10, 2001]

§ 870.4360 Nonroller-type cardiopulmonary bypass blood pump.

(a) *Identification.* A nonroller-type cardiopulmonary bypass blood pump is a device that uses a method other than revolving rollers to pump the blood through the cardiopulmonary bypass circuit during bypass surgery.

(b) *Classification.* Class III (premarket approval).

(c) *Date PMA or notice of completion of a PDP is required.* No effective date has been established of the requirement for premarket approval. See § 870.3.

[45 FR 7907–7971, Feb. 5, 1980, as amended at 52 FR 17737, May 11, 1987]

§ 870.4370 Roller-type cardiopulmonary bypass blood pump.

(a) *Identification.* A roller-type cardiopulmonary bypass blood pump is a device that uses a revolving roller mechanism to pump the blood through the cardiopulmonary bypass circuit during bypass surgery.

(b) *Classification.* Class II (performance standards).

§ 870.4380 Cardiopulmonary bypass pump speed control.

(a) *Identification.* A cardiopulmonary bypass pump speed control is a device used that incorporates an electrical system or a mechanical system, or both, and is used to control the speed of blood pumps used in cardiopulmonary bypass surgery.

(b) *Classification.* Class II (performance standards).

§ 870.4390 Cardiopulmonary bypass pump tubing.

(a) *Identification.* A cardiopulmonary bypass pump tubing is polymeric tubing which is used in the blood pump head and which is cyclically compressed by the pump to cause the blood to flow through the cardiopulmonary bypass circuit.

(b) *Classification.* Class II (performance standards).

§ 870.4400 Cardiopulmonary bypass blood reservoir.

(a) *Identification.* A cardiopulmonary bypass blood reservoir is a device used in conjunction with short-term extracorporeal circulation devices to hold a reserve supply of blood in the bypass circulation.

(b) *Classification.* Class II (performance standards), except that a reservoir that contains a defoamer or filter is classified into the same class as the defoamer or filter.

§ 870.4410 Cardiopulmonary bypass in-line blood gas sensor.

(a) *Identification.* A cardiopulmonary bypass in-line blood gas sensor is a transducer that measures the level of gases in the blood.

(b) *Classification.* Class II (performance standards).

§ 870.4420 Cardiopulmonary bypass cardiotomy return sucker.

(a) *Identification.* A cardiopulmonary bypass cardiotomy return sucker is a device that consists of tubing, a connector, and a probe or tip that is used to remove blood from the chest or heart during cardiopulmonary bypass surgery.

(b) *Classification.* Class II (performance standards).

§ 870.4430 Cardiopulmonary bypass intracardiac suction control.

(a) *Identification.* A cardiopulmonary bypass intracardiac suction control is a device which provides the vacuum and control for a cardiotomy return sucker.

(b) *Classification.* Class II (performance standards).

§ 870.4450 Vascular clamp.

(a) *Identification.* A vascular clamp is a surgical instrument used to occlude a blood vessel temporarily.

(b) *Classification.* Class II (performance standards).

§ 870.4475 Surgical vessel dilator.

(a) *Identification.* A surgical vessel dilator is a device used to enlarge or calibrate a vessel.

(b) *Classification.* Class II (performance standards).

§ 870.4500 Cardiovascular surgical instruments.

(a) *Identification.* Cardiovascular surgical instruments are surgical instruments that have special features for use in cardiovascular surgery. These devices include, e.g., forceps, retractors, and scissors.

(b) *Classification.* Class I (general controls). The device is exempt from the premarket notification procedures in subpart E of part 807 of this chapter subject to the limitations in § 870.9.

[45 FR 7907–7971, Feb. 5, 1980, as amended at 54 FR 25049, June 12, 1989; 66 FR 38797, July 25, 2001]

§ 870.4875 Intraluminal artery stripper.

(a) *Identification.* An intraluminal artery stripper is a device used to perform an endarterectomy (removal of plaque deposits from arterisclerotic arteries.)

(b) *Classification.* Class II (performance standards).

§ 870.4885 External vein stripper.

(a) *Identification.* An external vein stripper is an extravascular device used to remove a section of a vein.

(b) *Classification.* Class II (performance standards).

Subpart F—Cardiovascular Therapeutic Devices

§ 870.5050 Patient care suction apparatus.

(a) *Identification.* A patient care suction apparatus is a device used with an intrathoracic catheter to withdraw fluid from the chest during the recovery period following surgery.

(b) *Classification.* Class II (performance standards).

§ 870.5150 Embolectomy catheter.

(a) *Identification.* An embolectomy catheter is a balloon-tipped catheter that is used to remove thromboemboli, i.e., blood clots which have migrated in blood vessels from one site in the vascular tree to another.

(b) *Classification.* Class II (performance standards).

§ 870.5175 Septostomy catheter.

(a) *Identification.* A septostomy catheter is a special balloon catheter that is used to create or enlarge the atrial septal defect found in the heart of certain infants.

(b) *Classification.* Class II (performance standards).

§ 870.5200 External cardiac compressor.

(a) *Identification.* An external cardiac compressor is an external device that is electrically, pneumatically, or manually powered and is used to compress the chest periodically in the region of the heart to provide blood flow during cardiac arrest.

(b) *Classification.* Class III (premarket approval).

(c) *Date PMA or notice of completion of a PDP is required.* No effective date has been established of the requirement for premarket approval. See § 870.3.

[45 FR 7907–7971, Feb. 5, 1980, as amended at 52 FR 17737, May 11, 1987]

§ 870.5225 External counter-pulsating device.

(a) *Identification.* An external counter-pulsating device is a noninvasive device used to assist the heart by applying positive or negative pressure to one or more of the body's limbs in synchrony with the heart cycle.

(b) *Classification.* Class III (premarket approval).

(c) *Date PMA or notice of completion of a PDP is required.* No effective date has been established of the requirement for premarket approval. See § 870.3.

[45 FR 7907–7971, Feb. 5, 1980, as amended at 52 FR 17737, May 11, 1987]

§ 870.5300 DC-defribrillator (including paddles).

(a) *Low-energy DC-defibrillator*—(1) *Identification.* A low-energy DC-defibrillator is a device that delivers into a 50 ohm test load an electrical shock of a maximum of 360 joules of energy used for defibrillating (restoring normal heart rhythm) the atria or ventricles of the heart or to terminate other cardiac arrhythmias. This generic type of device includes low energy defibrillators with a maximum electrical output of less than 360 joules of energy that are used in pediatric defibrillation or in cardiac surgery. The device may either synchronize the shock with the proper phase of the electrocardiogram or may operate asynchronously. The device delivers the electrical shock through paddles placed either directly across the heart or on the surface of the body.

(2) *Classification.* Class II (performance standards).

(b) *High-energy DC-defibrillator*—(1) *Identification.* A high-energy DC-defibrillator is a device that delivers into a 50 ohm test load an electrical shock of greater than 360 joules of energy used for defibrillating the atria or ventricles of the heart or to terminate other cardiac arrhythmias. The device may either synchronize the shock with the proper phase of the electrocardiogram or may operate asynchronously. The device delivers the electrical shock through paddles placed either directly across the heart or on the surface of the body.

(2) *Classification.* Class III (premarket approval).

(c) *Date PMA or notice of completion of a PDP is required.* A PMA or a notice of completion of a PDP is required to be filed with the Food and Drug Administration on or before December 26, 1996 for any DC-defibrillator (including paddles) described in paragraph (b)(1) of this section that was in commercial distribution before May 28, 1976, or that has, on or before December 26, 1996 been found to be substantially equivalent to a DC-defibrillator (including paddles) described in paragraph (b)(1) of this section that was in commercial distribution before May 28, 1976. Any other DC-defibrillator (including paddles) described in paragraph (b)(1) of this section shall have an approved PMA or declared completed PDP in effect before being placed in commercial distribution.

[45 FR 7907–7971, Feb. 5, 1980, as amended at 52 FR 17737, May 11, 1987; 61 FR 50706, Sept. 27, 1996]

§ 870.5310 Automated external defibrillator.

(a) *Identification.* An automated external defibrillator (AED) is a low-energy device with a rhythm recognition detection system that delivers into a 50 ohm test load an electrical shock of a maximum of 360 joules of energy used for defibrillating (restoring normal heart rhythm) the atria or ventricles of the heart. An AED analyzes the patient's electrocardiogram, interprets the cardiac rhythm, and automatically delivers an electrical shock (fully automated AED), or advises the user to deliver the shock (semi-automated or shock advisory AED) to treat ventricular fibrillation or pulseless ventricular tachycardia.

(b) *Classification.* Class III (premarket approval)

(c) *Date PMA or notice of PDP is required.* No effective date has been established of the requirement for premarket approval. See § 870.3.

[68 FR 61344, Oct. 28, 2003; 69 FR 10615, Mar. 8, 2004]

§ 870.5325 Defibrillator tester.

(a) *Identification.* A defibrillator tester is a device that is connected to the output of a defibrillator and is used to measure the energy delivered by the defibrillator into a standard resistive load. Some testers also provide waveform information.

(b) *Classification.* Class II (performance standards).

§ 870.5550 External transcutaneous cardiac pacemaker (noninvasive).

(a) *Identification.* An external transcutaneous cardiac pacemaker (noninvasive) is a device used to supply a periodic electrical pulse intended to pace the heart. The pulse from the device is usually applied to the surface of the chest through electrodes such as defibrillator paddles.

(b) *Classification.* Class II. The special controls for this device are:

(1) "American National Standards Institute/American Association for Medical Instrumentation's DF–21 'Cardiac Defibrillator Devices' " 2d ed., 1996, and

(2) "The maximum pulse amplitude should not exceed 200 milliamperes. The maximum pulse duration should not exceed 50 milliseconds."

[45 FR 7907–7971, Feb. 5, 1980, as amended at 52 FR 17737, May 11, 1987; 65 FR 17144, Mar. 31, 2000]

§ 870.5800 Compressible limb sleeve.

(a) *Identification.* A compressible limb sleeve is a device that is used to prevent pooling of blood in a limb by inflating periodically a sleeve around the limb.

(b) *Classification.* Class II (performance standards).

§ 870.5900 Thermal regulating system.

(a) *Identification.* A thermal regulating system is an external system consisting of a device that is placed in contact with the patient and a temperature controller for the device. The system is used to regulate patient temperature.

(b) *Classification.* Class II (performance standards).

§ 870.5925 Automatic rotating tourniquet.

(a) *Identification.* An automatic rotating tourniquet is a device that prevents blood flow in one limb at a time, which temporarily reduces the total blood volume, thereby reducing the normal workload of the heart.

(b) *Classification.* Class II (performance standards).

PART 872—DENTAL DEVICES

Subpart A—General Provisions

Subpart B—Diagnostic Devices

Subpart C [Reserved]

Subpart D—Prosthetic Devices

872.3130 Preformed anchor.
872.3140 Resin applicator.
872.3150 Articulator.
872.3165 Precision attachment.
872.3200 Resin tooth bonding agent.
872.3220 Facebow.
872.3240 Dental bur.
872.3250 Calcium hydroxide cavity liner.
872.3260 Cavity varnish.
872.3275 Dental cement.
872.3285 Preformed clasp.
872.3300 Hydrophilic resin coating for dentures.
872.3310 Coating material for resin fillings.
872.3330 Preformed crown.
872.3350 Gold or stainless steel cusp.
872.3360 Preformed cusp.
872.3400 Karaya and sodium borate with or without acacia denture adhesive.
872.3410 Ethylene oxide homopolymer and/or carboxymethylcellulose sodium denture adhesive.
872.3420 Carboxymethylcellulose sodium and cationic polyacrylamide polymer denture adhesive.
872.3450 Ethylene oxide homopolymer and/or karaya denture adhesive.
872.3480 Polyacrylamide polymer (modified cationic) denture adhesive.
872.3490 Carboxymethylcellulose sodium and/or polyvinylmethylether maleic acid calcium-sodium double salt denture adhesive.
872.3500 Polyvinylmethylether maleic anhydride (PVM-MA), acid copolymer, and carboxymethylcellulose sodium (NACMC) denture adhesive.
872.3520 OTC denture cleanser.
872.3530 Mechanical denture cleaner.
872.3540 OTC denture cushion or pad.
872.3560 OTC denture reliner.
872.3570 OTC denture repair kit.
872.3580 Preformed gold denture tooth.
872.3590 Preformed plastic denture tooth.
872.3600 Partially fabricated denture kit.
872.3630 Endosseous dental implant abutment.
872.3640 Endosseous dental implant.
872.3645 Subperiosteal implant material.
872.3660 Impression material.
872.3661 Optical Impression Systems for CAD/CAM.
872.3670 Resin impression tray material.
872.3680 Polytetrafluoroethylene (PTFE) vitreous carbon materials.
872.3690 Tooth shade resin material.
872.3700 Dental mercury.
872.3710 Base metal alloy.
872.3730 Pantograph.
872.3740 Retentive and splinting pin.
872.3750 Bracket adhesive resin and tooth conditioner.
872.3760 Denture relining, repairing, or rebasing resin.
872.3765 Pit and fissure sealant and conditioner.
872.3770 Temporary crown and bridge resin.
872.3810 Root canal post.
872.3820 Root canal filling resin.
872.3830 Endodontic paper point.
872.3840 Endodontic silver point.
872.3850 Gutta percha.
872.3890 Endodontic stabilizing splint.
872.3900 Posterior artificial tooth with a metal insert.
872.3910 Backing and facing for an artificial tooth.
872.3920 Porcelain tooth.
872.3930 Bone grafting material.
872.3940 Total temporomandibular joint prosthesis.
872.3950 Glenoid fossa prosthesis.
872.3960 Mandibular condyle prosthesis.
872.3970 Interarticular disc prosthesis (interpositional implant).
872.3980 Endosseous dental implant accessories.

Subpart E—Surgical Devices

872.4120 Bone cutting instrument and accessories.
872.4130 Intraoral dental drill.
872.4200 Dental handpiece and accessories.
872.4465 Gas-powered jet injector.
872.4475 Spring-powered jet injector.
872.4535 Dental diamond instrument.
872.4565 Dental hand instrument.
872.4600 Intraoral ligature and wire lock.
872.4620 Fiber optic dental light.
872.4630 Dental operating light.
872.4730 Dental injecting needle.
872.4760 Bone plate.
872.4840 Rotary scaler.
872.4850 Ultrasonic scaler.
872.4880 Intraosseous fixation screw or wire.
872.4920 Dental electrosurgical unit and accessories.

Subpart F—Therapeutic Devices

872.5410 Orthodontic appliance and accessories.
872.5470 Orthodontic plastic bracket.
872.5500 Extraoral orthodontic headgear.
872.5525 Preformed tooth positioner.
872.5550 Teething ring.
872.5570 Intraoral devices for snoring and intraoral devices for snoring and obstructive sleep apnea.
872.5580 Oral rinse to reduce the adhesion of dental plaque.

Subpart G—Miscellaneous Devices

872.6010 Abrasive device and accessories.
872.6030 Oral cavity abrasive polishing agent.
872.6050 Saliva absorber.
872.6070 Ultraviolet activator for polymerization.
872.6080 Airbrush.
872.6100 Anesthetic warmer.
872.6140 Articulation paper.
872.6200 Base plate shellac.

872.6250	Dental chair and accessories.
872.6290	Prophylaxis cup.
872.6300	Rubber dam and accessories.
872.6350	Ultraviolet detector.
872.6390	Dental floss.
872.6475	Heat source for bleaching teeth.
872.6510	Oral irrigation unit.
872.6570	Impression tube.
872.6640	Dental operative unit and accessories.
872.6650	Massaging pick or tip for oral hygiene.
872.6660	Procelain powder for clinical use.
872.6670	Silicate protector.
872.6710	Boiling water sterilizer.
872.6730	Endodontic dry heat sterilizer.
872.6770	Cartridge syringe.
872.6855	Manual toothbrush.
872.6865	Powered toothbrush.
872.6870	Disposable fluoride tray.
872.6880	Preformed impression tray.
872.6890	Intraoral dental wax.

AUTHORITY: 21 U.S.C. 351, 360, 360c, 360e, 360j, 371.

SOURCE: 52 FR 30097, Aug. 12, 1987, unless otherwise noted.

Subpart A—General Provisions

§ 872.1 Scope.

(a) This part sets forth the classification of dental devices intended for human use that are in commercial distribution.

(b) The identification of a device in a regulation in this part is not a precise description of every device that is, or will be, subject to the regulation. A manufacturer who submits a premarket notification submission for a device under part 807 cannot show merely that the device is accurately described by the section title and identification provisions of a regulation in this part, but shall state why the device is substantially equivalent to other devices, as required by § 807.87.

(c) To avoid duplicative listings, a dental device that has two or more types of uses (e.g., used both as a diagnostic device and as a therapeutic device) is listed in one subpart only.

(d) References in this part to regulatory sections of the Code of Federal Regulations are to chapter I of title 21 unless otherwise noted.

(e) Guidance documents referenced in this part are available on the Internet at *http://www.fda.gov/cdrh.guidance.html.*

[52 FR 30097, Aug. 12, 1987, as amended at 68 FR 19737, Apr. 22, 2003]

§ 872.3 Effective dates of requirement for premarket approval.

A device included in this part that is classified into class III (premarket approval) shall not be commercially distributed after the date shown in the regulation classifying the device unless the manufacturer has an approval under section 515 of the act (unless an exemption has been granted under section 520(g)(2) of the act). An approval under section 515 of the act consists of FDA's issuance of an order approving an application for premarket approval (PMA) for the device or declaring completed a product development protocol (PDP) for the device.

(a) Before FDA requires that a device commercially distributed before the enactment date of the amendments, or a device that has been found substantially equivalent to such a device, has an approval under section 515 of the act, FDA must promulgate a regulation under section 515(b) of the act requiring such approval, except as provided in paragraphs (b) and (c) of this section. Such a regulation under section 515(b) of the act shall not be effective during the grace period ending on the 90th day after its promulgation or on the last day of the 30th full calendar month after the regulation that classifies the device into class III is effective, whichever is later. See section 501(f)(2)(B) of the act. Accordingly, unless an effective date of the requirement for premarket approval is shown in the regulation for a device classified into class III in this part, the device may be commercially distributed without FDA's issuance of an order approving a PMA or declaring completed a PDP for the device. If FDA promulgates a regulation under section 515(b) of the act requiring premarket approval for a device, section 501(f)(1)(A) of the act applies to the device.

(b) Any new, not substantially equivalent, device introduced into commercial distribution on or after May 28, 1976, including a device formerly marketed that has been substantially altered, is classified by statute (section

513(f) of the act) into class III without any grace period and FDA must have issued an order approving a PMA or declaring completed a PDP for the device before the device is commercially distributed unless it is reclassified. If FDA knows that a device being commercially distributed may be a "new" device as defined in this section because of any new intended use or other reasons, FDA may codify the statutory classification of the device into class III for such new use. Accordingly, the regulation for such a class III device states that as of the enactment date of the amendments, May 28, 1976, the device must have an approval under section 515 of the act before commercial distribution.

(c) A device identified in a regulation in this part that is classified into class III and that is subject to the transitional provisions of section 520(l) of the act is automatically classified by statute into class III and must have an approval under section 515 of the act before being commercially distributed. Accordingly, the regulation for such a class III transitional device states that as of the enactment date of the amendments, May 28, 1976, the device must have an approval under section 515 of the act before commercial distribution.

§ 872.9 Limitations of exemptions from section 510(k) of the Federal Food, Drug, and Cosmetic Act (the act).

The exemption from the requirement of premarket notification (section 510(k) of the act) for a generic type of class I or II device is only to the extent that the device has existing or reasonably foreseeable characteristics of commercially distributed devices within that generic type or, in the case of in vitro diagnostic devices, only to the extent that misdiagnosis as a result of using the device would not be associated with high morbidity or mortality. Accordingly, manufacturers of any commercially distributed class I or II device for which FDA has granted an exemption from the requirement of premarket notification must still submit a premarket notification to FDA before introducing or delivering for introduction into interstate commerce for commercial distribution the device when:

(a) The device is intended for a use different from the intended use of a legally marketed device in that generic type of device; e.g., the device is intended for a different medical purpose, or the device is intended for lay use where the former intended use was by health care professionals only;

(b) The modified device operates using a different fundamental scientific technology than a legally marketed device in that generic type of device; e.g., a surgical instrument cuts tissue with a laser beam rather than with a sharpened metal blade, or an in vitro diagnostic device detects or identifies infectious agents by using deoxyribonucleic acid (DNA) probe or nucleic acid hybridization technology rather than culture or immunoassay technology; or

(c) The device is an in vitro device that is intended:

(1) For use in the diagnosis, monitoring, or screening of neoplastic diseases with the exception of immunohistochemical devices;

(2) For use in screening or diagnosis of familial or acquired genetic disorders, including inborn errors of metabolism;

(3) For measuring an analyte that serves as a surrogate marker for screening, diagnosis, or monitoring life-threatening diseases such as acquired immune deficiency syndrome (AIDS), chronic or active hepatitis, tuberculosis, or myocardial infarction or to monitor therapy;

(4) For assessing the risk of cardiovascular diseases;

(5) For use in diabetes management;

(6) For identifying or inferring the identity of a microorganism directly from clinical material;

(7) For detection of antibodies to microorganisms other than immunoglobulin G (IgG) or IgG assays when the results are not qualitative, or are used to determine immunity, or the assay is intended for use in matrices other than serum or plasma;

(8) For noninvasive testing as defined in § 812.3(k) of this chapter; and

(9) For near patient testing (point of care).

[65 FR 2314, Jan. 14, 2000]

Subpart B—Diagnostic Devices

§ 872.1500 Gingival fluid measurer.

(a) *Identification.* A gingival fluid measurer is a gauge device intended to measure the amount of fluid in the gingival sulcus (depression between the tooth and gums) to determine if there is a gingivitis condition.

(b) *Classification.* Class I (general controls). The device is exempt from the premarket notification procedures in subpart E of part 807 of this chapter subject to the limitations in § 872.9.

[52 FR 30097, Aug. 12, 1987, as amended at 59 FR 63007, Dec. 7, 1994; 66 FR 38797, July 25, 2001]

§ 872.1720 Pulp tester.

(a) *Identification.* A pulp tester is an AC or battery powered device intended to evaluate the pulpal vitality of teeth by employing high frequency current transmitted by an electrode to stimulate the nerve tissue in the dental pulp.

(b) *Classification.* Class II.

§ 872.1730 Electrode gel for pulp testers.

(a) *Identification.* An electrode gel for pulp testers is a device intended to be applied to the surface of a tooth before use of a pulp tester to aid conduction of electrical current.

(b) *Classification.* Class I (general controls). The device is exempt from the premarket notification procedures in subpart E of part 807 of this chapter subject to the limitations in § 872.9.

[52 FR 30097, Aug. 12, 1987, as amended at 54 FR 13830, Apr. 5, 1989; 66 FR 38797, July 25, 2001]

§ 872.1740 Caries detection device.

(a) *Identification.* The caries detection device is a device intended to show the existence of decay in a patient's tooth by use of electrical current.

(b) *Classification.* Class II.

§ 872.1745 Laser fluorescence caries detection device.

(a) *Identification.* A laser fluorescence caries detection device is a laser, a fluorescence detector housed in a dental handpiece, and a control console that performs device calibration, as well as variable tone emitting and fluorescence measurement functions. The intended use of the device is to aid in the detection of tooth decay by measuring increased laser induced fluorescence.

(b) *Classification.* Class II, subject to the following special controls:

(1) Sale, distribution, and use of this device are restricted to prescription use in accordance with § 801.109 of this chapter;

(2) Premarket notifications must include clinical studies, or other relevant information, that demonstrates that the device aids in the detection of tooth decay by measuring increased laser induced fluorescence; and

(3) The labeling must include detailed use instructions with precautions that urge users to:

(i) Read and understand all directions before using the device,

(ii) Store probe tips under proper conditions,

(iii) Properly sterilize the emitter-detector handpick before each use, and

(iv) Properly maintain and handle the instrument in the specified manner and condition.

[65 FR 18235, Apr. 7, 2000]

§ 872.1800 Extraoral source x-ray system.

(a) *Identification.* An extraoral source x-ray system is an AC-powered device that produces x-rays and is intended for dental radiographic examination and diagnosis of diseases of the teeth, jaw, and oral structures. The x-ray source (a tube) is located outside the mouth. This generic type of device may include patient and equipment supports and component parts.

(b) *Classification.* Class II.

§ 872.1810 Intraoral source x-ray system.

(a) *Identification.* An intraoral source x-ray system is an electrically powered device that produces x-rays and is intended for dental radiographic examination and diagnosis of diseases of the teeth, jaw, and oral structures. The x-ray source (a tube) is located inside the mouth. This generic type of device may include patient and equipment supports and component parts.

(b) *Classification.* Class II.

§ 872.1820 Dental x-ray exposure alignment device.

(a) *Identification.* A dental x-ray exposure alignment device is a device intended to position x-ray film and to align the examination site with the x-ray beam.

(b) *Classification.* Class I (general controls). The device is exempt from the premarket notification procedures in subpart E of part 807 of this chapter subject to the limitations in § 872.9.

[52 FR 30097, Aug. 12, 1987, as amended at 59 FR 63008, Dec. 7, 1994; 66 FR 38797, July 25, 2001]

§ 872.1830 Cephalometer.

(a) *Identification.* A cephalometer is a device used in dentistry during x-ray procedures. The device is intended to place and to hold a patient's head in a standard position during dental x-rays.

(b) *Classification.* Class II.

§ 872.1840 Dental x-ray position indicating device.

(a) *Identification.* A dental x-ray position indicating device is a device, such as a collimator, cone, or aperture, that is used in dental radiographic examination. The device is intended to align the examination site with the x-ray beam and to restrict the dimensions of the dental x-ray field by limiting the size of the primary x-ray beam.

(b) *Classification.* Class I (general controls). The device is exempt from the premarket notification procedures in subpart E of part 807 of this chapter subject to the limitations in § 872.9.

[52 FR 30097, Aug. 12, 1987, as amended at 61 FR 1121, Jan. 16, 1996; 66 FR 38797, July 25, 2001]

§ 872.1850 Lead-lined position indicator.

(a) *Identification.* A lead-lined position indicator is a cone-shaped device lined with lead that is attached to a dental x-ray tube and intended to aid in positioning the tube, to prevent the misfocusing of the x-rays by absorbing divergent radiation, and to prevent leakage of radiation.

(b) *Classification.* Class I (general controls). The device is exempt from the premarket notification procedures in subpart E of part 807 of this chapter subject to the limitations in § 872.9.

[52 FR 30097, Aug. 12, 1987, as amended at 61 FR 1121, Jan. 16, 1996; 66 FR 38797, July 25, 2001]

§ 872.1870 Sulfide detection device.

(a) *Identification.* A sulfide detection device is a device consisting of an AC-powered control unit, probe handle, probe tips, cables, and accessories. This device is intended to be used in vivo, to manually measure periodontal pocket probing depths, detect the presence or absence of bleeding on probing, and detect the presence of sulfides in periodontal pockets, as an adjunct in the diagnosis of periodontal diseases in adult patients.

(b) *Classification.* Class II (special controls) prescription use in accordance with § 801.109 of this chapter; conformance with recognized standards of biocompatibility, electrical safety, and sterility; clinical and analytical performance testing, and proper labeling.

[63 FR 59717, Nov. 5, 1998]

§ 872.1905 Dental x-ray film holder.

(a) *Identification.* A dental x-ray film holder is a device intended to position and to hold x-ray film inside the mouth.

(b) *Classification.* Class I (general controls). The device is exempt from the premarket notification procedures in subpart E of part 807 of this chapter subject to the limitations in § 872.9. If the device is not labeled or otherwise represented as sterile, it is also exempt from the current good manufacturing practice regulations in part 820 of this chapter, with the exceptions of § 820.180, with respect to general requirements concerning records, and § 820.198, with respect to complaint files.

[52 FR 30097, Aug. 12, 1987, as amended at 54 FR 13830, Apr. 5, 1989; 66 FR 38797, July 25, 2001]

§ 872.2050 Dental sonography device.

(a) *Dental sonography device for monitoring*—(1) *Identification.* A dental sonography device for monitoring is an electrically powered device, intended to be used to monitor temporomandibular joint sounds. The device detects and records sounds made by the temporomandibular joint.

(2) *Classification.* Class I. The device is exempt from the premarket notification provisions of subpart E of part 807 of this chapter subject to § 872.9.

(b) *Dental sonography device for interpretation and diagnosis*—(1) *Identification.* A dental sonography device for interpretation and diagnosis is an electrically powered device, intended to interpret temporomandibular joint sounds for the diagnosis of temporomandibular joint disorders and associated orofacial pain. The device detects, records, displays, and stores sounds made by the temporomandibular joint during jaw movement. The device interprets these sounds to generate meaningful output, either directly or by connection to a personal computer. The device may be part of a system of devices, contributing joint sound information to be considered with data from other diagnostic components.

(2) *Classification.* Class II (special controls). The special control for this device is FDA's guidance document entitled "Class II Special Controls Guidance Document: Dental Sonography and Jaw Tracking Devices."

[68 FR 67367, Dec. 2, 2003]

§ 872.2060 Jaw tracking device.

(a) *Jaw tracking device for monitoring mandibular jaw positions relative to the maxilla*—(1) *Identification.* A jaw tracking device for monitoring mandibular jaw positions relative to the maxilla is a nonpowered or electrically powered device that measures and records anatomical distances and angles in three dimensional space, to determine the relative position of the mandible with respect to the location and position of the maxilla, while at rest and during jaw movement.

(2) *Classification.* Class I (general controls). The device is exempt from the premarket notification provisions of subpart E of part 807 of this chapter subject to § 872.9.

(b) *Jaw tracking device for interpretation of mandibular jaw positions for the diagnosis*—(1) *Identification.* A jaw tracking device for interpretation of mandibular jaw positions relative to the maxilla for the diagnosis of temporomandibular joint disorders and associated orofacial pain is a nonpowered or electrically powered device that measures and records anatomical distances and angles to determine the relative position of the mandible in three dimensional space, with respect to the location and position of the maxilla, while at rest and during jaw movement. The device records, displays, and stores information about jaw position. The device interprets jaw position to generate meaningful output, either directly or by connection to a personal computer. The device may be a part of a system of devices, contributing jaw position information to be considered with data from other diagnostic components.

(2) *Classification.* Class II (special controls). The special control for this device is FDA's guidance document entitled "Class II Special Controls Guidance Document: Dental Sonography and Jaw Tracking Devices."

[68 FR 67367, Dec. 2, 2003]

Subpart C [Reserved]

Subpart D—Prosthetic Devices

§ 872.3050 Amalgam alloy.

(a) *Identification.* An amalgam alloy is a device that consists of a metallic substance intended to be mixed with mercury to form filling material for treatment of dental caries.

(b) *Classification.* Class II.

§ 872.3060 Noble metal alloy.

(a) *Identification.* A noble metal alloy is a device composed primarily of noble metals, such as gold, palladium, platinum, or silver, that is intended for use in the fabrication of cast or porcelain-fused-to-metal crown and bridge restorations.

(b) *Classification.* Class II (special controls). The special control for these devices is FDA's "Class II Special Controls Guidance Document: Dental Noble Metal Alloys." The devices are exempt from the premarket notification procedures in subpart E of part 807 of this chapter subject to the limitations in § 872.9. See § 872.1(e) for availability of guidance information.

[69 FR 51766, Aug. 23, 2004]

§ 872.3080 Mercury and alloy dispenser.

(a) *Identification.* A mercury and alloy dispenser is a device with a spring-activated valve intended to measure and dispense into a mixing capsule a predetermined amount of dental mercury in droplet form and a premeasured amount of alloy pellets.

(b) *Classification.* Class I (general controls). The device is exempt from the premarket notification procedures in subpart E of part 807 of this chapter subject to the limitations in § 872.9.

[52 FR 30097, Aug. 12, 1987, as amended at 54 FR 13830, Apr. 5, 1989; 66 FR 38797, July 25, 2001]

§ 872.3100 Dental amalgamator.

(a) *Identification.* A dental amalgamator is a device, usually AC-powered, intended to mix, by shaking, amalgam capsules containing mercury and dental alloy particles, such as silver, tin, zinc, and copper. The mixed dental amalgam material is intended for filling dental caries.

(b) *Classification.* Class I (general controls). The device is exempt from the premarket notification procedures in subpart E of part 807 of this chapter subject to the limitations in § 872.9.

[55 FR 48439, Nov. 20, 1990, as amended at 59 FR 63008, Dec. 7, 1994; 66 FR 38797, July 25, 2001]

§ 872.3110 Dental amalgam capsule.

(a) *Identification.* A dental amalgam capsule is a container device in which silver alloy is intended to be mixed with mercury to form dental amalgam.

(b) *Classification.* Class I (general controls). The device is exempt from the premarket notification procedures in subpart E of part 807 of this chapter subject to the limitations in § 872.9.

[52 FR 30097, Aug. 12, 1987, as amended at 54 FR 13830, Apr. 5, 1989; 66 FR 38797, July 25, 2001]

§ 872.3130 Preformed anchor.

(a) *Identification.* A preformed anchor is a device made of austenitic alloys or alloys containing 75 percent or greater gold or metals of the platinum group intended to be incorporated into a dental appliance, such as a denture, to help stabilize the appliance in the patient's mouth.

(b) *Classification.* Class I (general controls). The device is exempt from the premarket notification procedures in subpart E of part 807 of this chapter subject to the limitations in § 872.9.

[52 FR 30097, Aug. 12, 1987, as amended at 59 FR 63008, Dec. 7, 1994; 66 FR 38797, July 25, 2001]

§ 872.3140 Resin applicator.

(a) *Identification.* A resin applicator is a brushlike device intended for use in spreading dental resin on a tooth during application of tooth shade material.

(b) *Classification.* Class I (general controls). The device is exempt from the premarket notification procedures in subpart E of part 807 of this chapter subject to the limitations in § 872.9. If the device is not labeled or otherwise represented as sterile, the device is exempt from the current good manufacturing practice regulations in part 820 of this chapter, with the exceptions of § 820.180, with respect to general requirements concerning records, and § 820.198, with respect to complaint files.

[52 FR 30097, Aug. 12, 1987, as amended at 54 FR 13830, Apr. 5, 1989; 66 FR 38797, July 25, 2001]

§ 872.3150 Articulator.

(a) *Identification.* An articulator is a mechanical device intended to simulate movements of a patient's upper and lower jaws. Plaster casts of the patient's teeth and gums are placed in the device to reproduce the occlusion (bite) and articulation of the patient's jaws. An articulator is intended to fit dentures or provide orthodontic treatment.

(b) *Classification.* Class I (general controls). The device is exempt from the premarket notification procedures in subpart E of part 807 of this chapter subject to the limitations in § 872.9. If the device is not labeled or otherwise represented as sterile, the device is exempt from the current good manufacturing practice regulations in part 820 of this chapter, with the exceptions of § 820.180, with respect to general requirements concerning records, and

§ 820.198, with respect to complaint files.

[52 FR 30097, Aug. 12, 1987, as amended at 54 FR 13830, Apr. 5, 1989; 66 FR 38797, July 25, 2001]

§ 872.3165 Precision attachment.

(a) *Identification.* A precision attachment or preformed bar is a device made of austenitic alloys or alloys containing 75 percent or greater gold and metals of the platinum group intended for use in prosthetic dentistry in conjunction with removable partial dentures. Various forms of the device are intended to connect a lower partial denture with another lower partial denture, to connect an upper partial denture with another upper partial denture, to connect either an upper or lower partial denture to a tooth or a crown, or to connect a fixed bridge to a partial denture.

(b) *Classification.* Class I (general controls). The device is exempt from the premarket notification procedures in subpart E of part 807 of this chapter subject to the limitations in § 872.9.

[52 FR 30097, Aug. 12, 1987, as amended at 59 FR 63008, Dec. 7, 1994; 66 FR 38797, July 25, 2001]

§ 872.3200 Resin tooth bonding agent.

(a) *Identification.* A resin tooth bonding agent is a device material, such as methylmethacrylate, intended to be painted on the interior of a prepared cavity of a tooth to improve retention of a restoration, such as a filling.

(b) *Classification.* Class II.

§ 872.3220 Facebow.

(a) *Identification.* A facebow is a device intended for use in denture fabrication to determine the spatial relationship between the upper and lower jaws. This determination is intended for use in placing denture casts accurately into an articulator (§ 872.3150) and thereby aiding correct placement of artificial teeth into a denture base.

(b) *Classification.* Class I (general controls). The device is exempt from the premarket notification procedures in subpart E of part 807 of this chapter subject to the limitations in § 872.9. If the device is not labeled or otherwise represented as sterile, the device is exempt from the current good manufacturing practice regulations in part 820 of this chapter, with the exceptions of § 820.180, with respect to general requirements concerning records, and § 820.198, with respect to complaint files.

[52 FR 30097, Aug. 12, 1987, as amended at 54 FR 13830, Apr. 5, 1989; 66 FR 38797, July 25, 2001]

§ 872.3240 Dental bur.

(a) *Identification.* A dental bur is a rotary cutting device made from carbon steel or tungsten carbide intended to cut hard structures in the mouth, such as teeth or bone. It is also intended to cut hard metals, plastics, porcelains, and similar materials intended for use in the fabrication of dental devices.

(b) *Classification.* Class I (general controls). The device is exempt from the premarket notification procedures in subpart E of part 807 of this chapter subject to the limitations in § 872.9.

[52 FR 30097, Aug. 12, 1987, as amended at 59 FR 63008, Dec. 7, 1994; 66 FR 38798, July 25, 2001]

§ 872.3250 Calcium hydroxide cavity liner.

(a) *Identification.* A calcium hydroxide cavity liner is a device material intended to be applied to the interior of a prepared cavity before insertion of restorative material, such as amalgam, to protect the pulp of a tooth.

(b) *Classification.* Class II.

§ 872.3260 Cavity varnish.

(a) *Identification.* Cavity varnish is a device that consists of a compound intended to coat a prepared cavity of a tooth before insertion of restorative materials. The device is intended to prevent penetration of restorative materials, such as amalgam, into the dentinal tissue.

(b) *Classification.* Class II.

§ 872.3275 Dental cement.

(a) *Zinc oxide-eugenol*—(1) *Identification.* Zinc oxide-eugenol is a device composed of zinc oxide-eugenol intended to serve as a temporary tooth filling or as a base cement to affix a temporary tooth filling, to affix dental devices such as crowns or bridges, or to be applied to a tooth to protect the tooth pulp.

(2) *Classification.* Class I (general controls). The device is exempt from the premarket notification procedures in subpart E of part 807 of this chapter subject to § 872.9.

(b) *Dental cement other than zinc oxide-eugenol*—(1) *Identification.* Dental cement other than zinc oxide-eugenol is a device composed of various materials other than zinc oxide-eugenol intended to serve as a temporary tooth filling or as a base cement to affix a temporary tooth filling, to affix dental devices such as crowns or bridges, or to be applied to a tooth to protect the tooth pulp.

(2) *Classification.* Class II.

[52 FR 30097, Aug. 12, 1987, as amended at 65 FR 2314, Jan. 14, 2000]

§ 872.3285 Preformed clasp.

(a) *Identification.* A preformed clasp or a preformed wire clasp is a prefabricated device made of austenitic alloys or alloys containing 75 percent or greater gold and metals of the platinum group intended to be incorporated into a dental appliance, such as a partial denture, to help stabilize the appliance in the patient's mouth by fastening the appliance to an adjacent tooth.

(b) *Classification.* Class I (general controls). The device is exempt from the premarket notification procedures in subpart E of part 807 of this chapter subject to the limitations in § 872.9.

[52 FR 30097, Aug. 12, 1987, as amended at 59 FR 63008, Dec. 7, 1994; 66 FR 38798, July 25, 2001]

§ 872.3300 Hydrophilic resin coating for dentures.

(a) *Identification.* A hydrophilic resin coating for dentures is a device that consists of a water-retaining polymer that is intended to be applied to the base of a denture before the denture is inserted into the patient's mouth to improve denture retention and comfort.

(b) *Classification.* Class II.

§ 872.3310 Coating material for resin fillings.

(a) *Identification.* A coating material for resin fillings is a device intended to be applied to the surface of a restorative resin dental filling to attain a smooth, glaze-like finish on the surface of the filling.

(b) *Classification.* Class II.

§ 872.3330 Preformed crown.

(a) *Identification.* A preformed crown is a prefabricated device made of plastic or austenitic alloys or alloys containing 75 percent or greater gold and metals of the platinum group intended to be affixed temporarily to a tooth after removal of, or breakage of, the natural crown (that portion of the tooth that normally protrudes above the gums). It is intended for use as a functional restoration until a permanent crown is constructed. The device also may be intended for use as a functional restoration for a badly decayed deciduous (baby) tooth until the adult tooth erupts.

(b) *Classification.* Class I (general controls). The device is exempt from the premarket notification procedures in subpart E of part 807 of this chapter subject to the limitations in § 872.9.

[52 FR 30097, Aug. 12, 1987, as amended at 59 FR 63008, Dec. 7, 1994; 66 FR 38798, July 25, 2001]

§ 872.3350 Gold or stainless steel cusp.

(a) *Identification.* A gold or stainless steel cusp is a prefabricated device made of austenitic alloys or alloys containing 75 percent or greater gold and metals of the platinum group or stainless steel intended to provide a permanent cusp (a projection on the chewing surface of a tooth) to achieve occlusal harmony (a proper bite) between the teeth and a removable denture.

(b) *Classification.* Class I (general controls). The device is exempt from the premarket notification procedures in subpart E of part 807 of this chapter subject to the limitations in § 872.9.

[52 FR 30097, Aug. 12, 1987, as amended at 59 FR 63008, Dec. 7, 1994; 66 FR 38798, July 25, 2001]

§ 872.3360 Preformed cusp.

(a) *Identification.* A performed cusp is a prefabricated device made of plastic or austenitic alloys or alloys containing 75 percent or greater gold and metals of the platinum group intended to be used as a temporary cusp (a projection on the chewing surface of a tooth) to achieve occlusal harmony (a proper bite) before permanent restoration of a tooth.

(b) *Classification.* Class I (general controls). The device is exempt from the premarket notification procedures in subpart E of part 807 of this chapter subject to the limitations in § 872.9.

[52 FR 30097, Aug. 12, 1987, as amended at 59 FR 63008, Dec. 7, 1994; 66 FR 38798, July 25, 2001]

§ 872.3400 Karaya and sodium borate with or without acacia denture adhesive.

(a) *Identification.* A karaya and sodium borate with or without acacia denture adhesive is a device composed of karaya and sodium borate with or without acacia intended to be applied to the base of a denture before the denture is inserted into patient's mouth to improve denture retention and comfort.

(b) *Classification.* (1) Class I (general controls) if the device contains less than 12 percent by weight of sodium borate. The class I device is exempt from the premarket notification procedures in subpart E of part 807 of this chapter subject to § 872.9.

(2) Class III if the device contains 12 percent or more by weight of sodium borate.

(c) *Date PMA or notice of completion of a PDP is required.* A PMA or a notice of completion of a PDP is required to be filed with the Food and Drug Administration on or before December 26, 1996 for any karaya and sodium borate with or without acacia denture adhesive that was in commercial distribution before May 28, 1976, or that has, on or before December 26, 1996 been found to be substantially equivalent to a karaya and sodium borate with or without acacia denture adhesive that was in commercial distribution before May 28, 1976. Any other karaya and sodium borate with or without acacia denture adhesive shall have an approved PMA or a declared completed PDP in effect before being placed in commercial distribution.

[52 FR 30097, Aug. 12, 1987, as amended at 61 FR 50706, Sept. 27, 1996; 65 FR 2315, Jan. 14, 2000]

§ 872.3410 Ethylene oxide homopolymer and/or carboxymethylcellulose sodium denture adhesive.

(a) *Identification.* An ethylene oxide homopolymer and/or carboxymethylcellulose sodium denture adhesive is a device containing ethylene oxide homopolymer and/or carboxymethylcellulose sodium intended to be applied to the base of a denture before the denture is inserted in a patient's mouth to improve denture retention and comfort.

(b) *Classification.* Class I (general controls). The device is exempt from the premarket notification procedures in subpart E of part 807 of this chapter subject to the limitations in § 872.9.

[52 FR 30097, Aug. 12, 1987, as amended at 59 FR 63008, Dec. 7, 1994; 66 FR 38798, July 25, 2001]

§ 872.3420 Carboxymethylcellulose sodium and cationic polyacrylamide polymer denture adhesive.

(a) *Identification.* A carboxymethylcellulose sodium and cationic polyacrylamide polymer denture adhesive is a device composed of carboxymethylcellulose sodium and cationic polyacrylamide polymer intended to be applied to the base of a denture before the denture is inserted in a patient's mouth to improve denture retention and comfort.

(b) *Classification.* Class III.

(c) *Date PMA or notice of completion of a PDP is required.* A PMA or a notice of completion of a PDP is required to be filed with the Food and Drug Administration on or before December 26, 1996 for any carboxymethylcellulose sodium and cationic polyacrylamide polymer denture adhesive that was in commercial distribution before May 28, 1976, or that has, on or before December 26, 1996 been found to be substantially equivalent to a carboxymethylcellulose sodium and cationic polyacrylamide polymer denture adhesive that was in commercial distribution before May 28, 1976. Any other carboxymethylcellulose sodium and cationic polyacrylamide polymer denture adhesive shall have an approved PMA or a declared completed

PDP in effect before being placed in commercial distribution.

[52 FR 30097, Aug. 12, 1987, as amended at 61 FR 50707, Sept. 27, 1996]

§ 872.3450 Ethylene oxide homopolymer and/or karaya denture adhesive.

(a) *Identification.* Ethylene oxide homopolymer and/or karaya denture adhesive is a device composed of ethylene oxide homopolymer and/or karaya intended to be applied to the base of a denture before the denture is inserted in a patient's mouth to improve denture retention and comfort.

(b) *Classification.* (1) Class I if the device is made of wax-impregnated cotton cloth that the patient applies to the base or inner surface of a denture before inserting the denture into the mouth. The device is intended to be discarded following 1 day's use. The class I device is exempt from the premarket notification procedures in subpart E of part 807 of this chapter subject to § 872.9.

[52 FR 30097, Aug. 12, 1987, as amended at 59 FR 63008, Dec. 7, 1994; 65 FR 2315, Jan. 14, 2000]

§ 872.3480 Polyacrylamide polymer (modified cationic) denture adhesive.

(a) *Identification.* A polyacrylamide polymer (modified cationic) denture adhesive is a device composed of polyacrylamide polymer (modified cationic) intended to be applied to the base of a denture before the denture is inserted in a patient's mouth to improve denture retention and comfort.

(b) *Classification.* Class III.

(c) *Date PMA or notice of completion of a PDP is required.* A PMA or a notice of completion of a PDP is required to be filed with the Food and Drug Administration on or before December 26, 1996 for any polyacrylamide polymer (modified cationic) denture adhesive that was in commercial distribution before May 28, 1976, or that has, on or before December 26, 1996 been found to be substantially equivalent to a polyacrylamide polymer (modified cationic) denture adhesive that was in commercial distribution before May 28, 1976. Any other polyacrylamide polymer (modified cationic) denture adhesive shall have an approved PMA or a declared completed PDP in effect before being place in commercial distribution.

[52 FR 30097, Aug. 12, 1987, as amended at 61 FR 50707, Sept. 27, 1996]

§ 872.3490 Carboxymethylcellulose sodium and/or polyvinylmethylether maleic acid calcium-sodium double salt denture adhesive.

(a) *Identification.* A carboxymethylcellulose sodium and/or polyvinylmethylether maleic acid calcium-sodium double salt denture adhesive is a device composed of carboxymethylcellulose sodium and/or polyvinylmethylether maleic acid calcium-sodium double salt intended to be applied to the base of a denture before the denture is inserted in a patient's mouth to improve denture retention and comfort.

(b) *Classification.* Class I (general controls). The device is exempt from the premarket notification procedures in subpart E of part 807 of this chapter subject to the limitations in § 872.9.

[52 FR 30097, Aug. 12, 1987, as amended at 59 FR 63008, Dec. 7, 1994; 66 FR 38798, July 25, 2001]

§ 872.3500 Polyvinylmethylether maleic anhydride (PVM-MA), acid copolymer, and carboxymethylcellulose sodium (NACMC) denture adhesive.

(a) *Identification.* Polyvinylmethylether maleic anhydride (PVM-MA), acid copolymer, and carboxymethylcellulose sodium (NACMC) denture adhesive is a device composed of polyvinylmethylether maleic anhydride, acid copolymer, and carboxymethylcellulose sodium intended to be applied to the base of a denture before the denture is inserted in a patient's mouth to improve denture retention and comfort.

(b) *Classification.* Class III.

(c) *Date PMA or notice of completion of a PDP is required.* A PMA or a notice of completion of a PDP is required to be filed with the Food and Drug Administration on or before December 26, 1996 for any polyvinylmethylether maleic anhydride (PVM-MA), acid copolymer, and carboxymethylcellulose sodium (NACMC) denture adhesive that was in commercial distribution before May 28, 1976, or that has, on or before December 26, 1996 been found to be substantially equivalent to a polyvinylmethylether maleic anhydride (PVM-MA), acid copolymer, and carboxymethylcellulose sodium (NACMC) denture adhesive that was in commercial distribution before May 28, 1976. Any other polyvinylmethylether maleic anhydride (PVM-MA), acid copolymer, and carboxymethylcellulose sodium (NACMC) denture adhesive shall have an approved PMA or a declared completed PDP in effect before being placed in commercial distribution.

[52 FR 30097, Aug. 12, 1987, as amended at 61 FR 50707, Sept. 27, 1996]

§ 872.3520 OTC denture cleanser.

(a) *Identification.* An OTC denture cleanser is a device that consists of material in the form of a powder, tablet, or paste that is intended to remove debris from removable prosthetic dental appliances, such as bridges or dentures. The dental appliance is removed from the patient's mouth when the appliance is cleaned.

(b) *Classification.* Class I (general controls). The device is exempt from the premarket notification procedures in subpart E of part 807 of this chapter subject to the limitations in § 872.9.

[52 FR 30097, Aug. 12, 1987, as amended at 59 FR 63008, Dec. 7, 1994; 66 FR 38798, July 25, 2001]

§ 872.3530 Mechanical denture cleaner.

(a) *Identification.* A mechanical denture cleaner is a device, usually AC-powered, that consists of a container for mechanically agitating a denture cleansing solution. The device is intended to clean a denture by submersion in the agitating cleansing solution in the container.

(b) *Classification.* Class I (general controls). The device is exempt from the premarket notification procedures in subpart E of part 807 of this chapter subject to the limitations in § 872.9.

[55 FR 48439, Nov. 20, 1990, as amended at 59 FR 63008, Dec. 7, 1994; 66 FR 38798, July 25, 2001]

§ 872.3540 OTC denture cushion or pad.

(a) *Identification.* An OTC denture cushion or pad is a prefabricated or noncustom made disposable device that is intended to improve the fit of a loose or uncomfortable denture, and may be available for purchase over-the-counter.

(b) *Classification.* (1) Class I if the device is made of wax-impregnated cotton cloth that the patient applies to the base or inner surface of a denture before inserting the denture into the mouth. The device is intended to be discarded following 1 day's use. The class I device is exempt from the premarket notification procedures in subpart E of part 807 of this chapter subject to § 872.9.

(2) Class II if the OTC denture cushion or pad is made of a material other than wax-impregnated cotton cloth or if the intended use of the device differs from that described in paragraph (b)(1) of this section. The special controls for this device are FDA's:

(i) "Use of International Standard ISO 10993 'Biological Evaluation of Medical—Devices Part I: Evaluation and Testing,' " and

(ii) "OTC Denture Reliners, Repair Kits, and Partially Fabricated Denture Kits."

[52 FR 30097, Aug. 12, 1987, as amended at 65 FR 2315, 2000; 65 FR 17144, Mar. 31, 2000]

§ 872.3560 OTC denture reliner.

(a) *Identification.* An OTC denture reliner is a device consisting of a material such as plastic resin that is intended to be applied as a permanent coating or lining on the base or tissue-contacting surface of a denture. The device is intended to replace a worn denture lining and may be available for purchase over the counter.

(b) *Classification.* Class II. The special controls for this device are FDA's:

(1) "Use of International Standard ISO 10993 'Biological Evaluation of Medical Devices—Part I: Evaluation and Testing,' " and

(2) "OTC Denture Reliners, Repair Kits, and Partially Fabricated Denture Kits."

[52 FR 30097, Aug. 12, 1987, as amended at 61 FR 50707, Sept. 27, 1996; 65 FR 17144, Mar. 31, 2000]

§ 872.3570 OTC denture repair kit.

(a) *Identification.* An OTC denture repair kit is a device consisting of a material, such as a resin monomer system of powder and liquid glues, that is intended to be applied permanently to a denture to mend cracks or breaks. The device may be available for purchase over-the counter.

(b) *Classification.* Class II. The special controls for this device are FDA's:

(1) "Use of International Standard ISO 10993 'Biological Evaluation of Medical Devices—Part I: Evaluation and Testing,' " and

(2) "OTC Denture Reliners, Repair Kits, and Partially Fabricated Denture Kits."

[52 FR 30097, Aug. 12, 1987, as amended at 65 FR 17144, Mar. 31, 2000]

§ 872.3580 Preformed gold denture tooth.

(a) *Identification.* A preformed gold denture tooth is a device composed of austenitic alloys or alloys containing 75 percent or greater gold and metals of the platinum group intended for use as a tooth or a portion of a tooth in a fixed or removable partial denture.

(b) *Classification.* Class I (general controls). The device is exempt from the premarket notification procedures in subpart E of part 807 of this chapter subject to the limitations in § 872.9.

[52 FR 30097, Aug. 12, 1987, as amended at 59 FR 63008, Dec. 7, 1994; 66 FR 38798, July 25, 2001]

§ 872.3590 Preformed plastic denture tooth.

(a) *Identification.* A preformed plastic denture tooth is a prefabricated device, composed of materials such as methyl methacrylate, that is intended for use as a tooth in a denture.

(b) *Classification.* Class II.

§ 872.3600 Partially fabricated denture kit.

(a) *Identification.* A partially fabricated denture kit is a device composed of connected preformed teeth that is intended for use in construction of a denture. A denture base is constructed using the patient's mouth as a mold, by partially polymerizing the resin denture base materials while the materials are in contact with the oral tissues. After the denture base is constructed, the connected preformed teeth are chemically bonded to the base.

(b) *Classification.* Class II. The special controls for this device are FDA's:

(1) "Use of International Standard ISO 10993 'Biological Evaluation of Medical Devices—Part I: Evaluation and Testing,' " and

(2) "OTC Denture Reliners, Repair Kits, and Partially Fabricated Denture Kits."

[52 FR 30097, Aug. 12, 1987, as amended at 65 FR 17144, Mar. 31, 2000]

§ 872.3630 Endosseous dental implant abutment.

(a) *Identification.* An endosseous dental implant abutment is a premanufactured prosthetic component directly connected to the endosseous dental implant and is intended for use as an aid in prosthetic rehabilitation.

(b) *Classification.* Class II (special controls). The guidance document entitled "Class II Special Controls Guidance Document: Root-Form Endosseous Dental Implants and Endosseous Dental Implant Abutments" will serve as the special control. (See § 872.1(e) for the availability of this guidance document.)

[69 FR 26304, May 12, 2004]

§ 872.3640 Endosseous dental implant.

(a) *Identification.* An endosseous dental implant is a device made of a material such as titanium or titanium alloy, that is intended to be surgically placed in the bone of the upper or lower jaw arches to provide support for prosthetic devices, such as artificial teeth, in order to restore a patient's chewing function.

(b) *Classification.* (1) Class II (special controls). The device is classified as class II if it is a root-form endosseous dental implant. The root-form endosseous dental implant is characterized by four geometrically distinct types: Basket, screw, solid cylinder, and hollow cylinder. The guidance document entitled "Class II Special Controls Guidance Document: Root-Form Endosseous Dental Implants and Endosseous Dental Implant Abutments" will serve as the special control. (See § 872.1(e) for the availability of this guidance document.)

(2) Class III (premarket approval). The device is classified as class III if it is a blade-form endosseous dental implant.

[69 FR 26304, May 12, 2004]

§ 872.3645 Subperiosteal implant material.

(a) *Identification.* Subperiosteal implant material is a device composed of titanium or cobalt chrome molybdenum intended to construct custom prosthetic devices which are surgically implanted into the lower or upper jaw between the periosteum (connective tissue covering the bone) and supporting bony structures. The device is intended to provide support for prostheses, such as dentures.

(b) *Classification.* Class II.

§ 872.3660 Impression material.

(a) *Identification.* Impression material is a device composed of materials such as alginate or polysulfide intended to be placed on a preformed impression tray and used to reproduce the structure of a patient's teeth and gums. The device is intended to provide models for study and for production of restorative prosthetic devices, such as gold inlays and dentures.

(b) *Classification.* Class II (Special Controls).

[52 FR 30097, Aug. 12, 1987, as amended at 68 FR 19738, Apr. 22, 2003]

§ 872.3661 Optical Impression Systems for CAD/CAM.

(a) *Identification.* An optical impression system for computer assisted design and manufacturing (CAD/CAM) is a device used to record the topographical characteristics of teeth, dental impressions, or stone models by analog or digital methods for use in the computer-assisted design and manufacturing of dental restorative prosthetic devices. Such systems may consist of a camera, scanner, or equivalent type of sensor and a computer with software.

(b) *Classification.* Class II (Special Controls). The device is exempt from the premarket notification procedures in subpart E of part 807 of the chapter subject to the limitations in § 872.9. The special control for these devices is the FDA guidance document entitled "Class II Special Controls Guidance Document: Optical Impression Systems for Computer Assisted Design and Manufacturing (CAD/CAM) of Dental Restorations; Guidance for Industry and FDA." For the availability of this guidance document, see § 872.1(e).

[68 FR 19738, Apr. 22, 2003]

§ 872.3670 Resin impression tray material.

(a) *Identification.* Resin impression tray material is a device intended for use in a two-step dental mold fabricating process. The device consists of a resin material, such as methyl methacrylate, and is used to form a custom impression tray for use in cases in which a preformed impression tray is not suitable, such as the fabrication of crowns, bridges, or full dentures. A preliminary plaster or stone model of the patient's teeth and gums is made. The resin impression tray material is applied to this preliminary study model to form a custom tray. This tray is then filled with impression material and inserted into the patient's mouth to make an impression, from which a final, more precise, model of the patient's mouth is cast.

(b) *Classification.* Class I (general controls). The device is exempt from the premarket notification procedures in subpart E of part 807 of this chapter subject to the limitations in § 872.9. If the device is not labeled or otherwise represented as sterile, it is exempt from the current good manufacturing practice regulations in part 820 of this chapter, with the exception of § 820.180, with respect to general requirements

concerning records, and § 820.198, with respect to complaint files.

[52 FR 30097, Aug. 12, 1987, as amended at 59 FR 63008, Dec. 7, 1994; 66 FR 38798, July 25, 2001]

§ 872.3680 Polytetrafluoroethylene (PTFE) vitreous carbon materials.

(a) *Identification.* Polytetrafluoroethylene (PTFE) vitreous carbon material is a device composed of polytetrafluoroethylene (PTFE) vitreous carbon intended for use in maxillofacial alveolar ridge augmentation (building up the upper or lower jaw area that contains the sockets in which teeth are rooted) or intended to coat metal surgical implants to be placed in the alveoli (sockets in which the teeth are rooted) or the temporomandibular joints (the joint between the upper and lower jaws).

(b) *Classification.* Class II.

[52 FR 30097, Aug. 12, 1987; 52 FR 34456, Sept. 11, 1987]

§ 872.3690 Tooth shade resin material.

(a) *Identification.* Tooth shade resin material is a device composed of materials such as bisphenol-A glycidyl methacrylate (Bis-GMA) intended to restore carious lesions or structural defects in teeth.

(b) *Classification.* Class II.

§ 872.3700 Dental mercury.

(a) *Identification.* Dental mercury is a device composed of mercury intended for use as a component of amalgam alloy in the restoration of a dental cavity or a broken tooth.

(b) *Classification.* Class I.

§ 872.3710 Base metal alloy.

(a) *Identification.* A base metal alloy is a device composed primarily of base metals, such as nickel, chromium, or cobalt, that is intended for use in fabrication of cast or porcelain-fused-to-metal crown and bridge restorations.

(b) *Classification.* Class II (special controls). The special control for this device is FDA's "Class II Special Controls Guidance Document: Dental Base Metal Alloys." The device is exempt from the premarket notification procedures in subpart E of part 807 of this chapter subject to the limitations in § 872.9. See § 872.1(e) for availability of guidance information.

[69 FR 51766, Aug. 23, 2004]

§ 872.3730 Pantograph.

(a) *Identification.* A pantograph is a device intended to be attached to a patient's head to duplicate lower jaw movements to aid in construction of restorative and prosthetic dental devices. A marking pen is attached to the lower jaw component of the device and, as the patient's mouth opens, the pen records on graph paper the angle between the upper and the lower jaw.

(b) *Classification.* Class I (general controls). The device is exempt from the premarket notification procedures in subpart E of part 807 of this chapter subject to the limitations in § 872.9. If the device is not labeled or otherwise represented as sterile, it is exempt from the current good manufacturing practice regulations in part 820 of this chapter, with the exception of § 820.180, with respect to general requirements concerning records, and § 820.198, with respect to complaint files.

[52 FR 30097, Aug. 12, 1987, as amended at 66 FR 38798, July 25, 2001]

§ 872.3740 Retentive and splinting pin.

(a) *Identification.* A retentive and splinting pin is a device made of austenitic alloys or alloys containing 75 percent or greater gold and metals of the platinum group intended to be placed permanently in a tooth to provide retention and stabilization for a restoration, such as a crown, or to join two or more teeth together.

(b) *Classification.* Class I (general controls). The device is exempt from the premarket notification procedures in subpart E of part 807 of this chapter subject to the limitations in § 872.9

[52 FR 30097, Aug. 12, 1987, as amended at 60 FR 38900, July 28, 1995; 66 FR 38798, July 25, 2001]

§ 872.3750 Bracket adhesive resin and tooth conditioner.

(a) *Identification.* A bracket adhesive resin and tooth conditioner is a device composed of an adhesive compound, such as polymethylmethacrylate, intended to cement an orthodontic bracket to a tooth surface.

(b) *Classification.* Class II.

§ 872.3760 Denture relining, repairing, or rebasing resin.

(a) *Identification.* A denture relining, repairing, or rebasing resin is a device composed of materials such as methylmethacrylate, intended to reline a denture surface that contacts tissue, to repair a fractured denture, or to form a new denture base. This device is not available for over-the-counter (OTC) use.

(b) *Classification.* Class II.

§ 872.3765 Pit and fissure sealant and conditioner.

(a) *Identification.* A pit and fissure sealant and conditioner is a device composed of resin, such as polymethylmethacrylate, intended for use primarily in young children to seal pit and fissure depressions (faults in the enamel) in the biting surfaces of teeth to prevent cavities.

(b) *Classification.* Class II.

§ 872.3770 Temporary crown and bridge resin.

(a) *Identification.* A temporary crown and bridge resin is a device composed of a material, such as polymethylmethacrylate, intended to make a temporary prosthesis, such as a crown or bridge, for use until a permanent restoration is fabricated.

(b) *Classification.* Class II.

§ 872.3810 Root canal post.

(a) *Identification.* A root canal post is a device made of austenitic alloys or alloys containing 75 percent or greater gold and metals of the platinum group intended to be cemented into the root canal of a tooth to stabilize and support a restoration.

(b) *Classification.* Class I (general controls). The device is exempt from the premarket notification procedures in subpart E of part 807 of this chapter subject to the limitations in § 872.9.

[52 FR 30097, Aug. 12, 1987, as amended at 60 FR 38900, July 28, 1995; 66 FR 38798, July 25, 2001]

§ 872.3820 Root canal filling resin.

(a) *Identification.* A root canal filling resin is a device composed of material, such as methylmethacrylate, intended for use during endodontic therapy to fill the root canal of a tooth.

(b) *Classification.* (1) Class II if chloroform is not used as an ingredient in the device.

(2) Class III if chloroform is used as an ingredient in the device.

(c) *Date PMA or notice of completion of a PDP is required.* A PMA or a notice of completion of a PDP is required to be filed with the Food and Drug Administration on or before December 26, 1996 for any root canal filling resin described in paragraph (b)(2) of this section that was in commercial distribution before May 28, 1976, or that has, on or before December 26, 1996 been found to be substantially equivalent to a root canal filling resin described in paragraph (b)(2) of this section that was in commercial distribution before May 28, 1976. Any other root canal filling resin shall have an approved PMA or a declared completed PDP in effect before being placed in commercial distribution.

[52 FR 30097, Aug. 12, 1987, as amended at 61 FR 50707, Sept. 27, 1996]

§ 872.3830 Endodontic paper point.

(a) *Identification.* An endodontic paper point is a device made of paper intended for use during endodontic therapy to dry, or apply medication to, the root canal of a tooth.

(b) *Classification.* Class I (general controls). The device is exempt from the premarket notification procedures in subpart E of part 807 of this chapter subject to the limitations in § 872.9.

[52 FR 30097, Aug. 12, 1987, as amended at 54 FR 13830, Apr. 5, 1989; 66 FR 38798, July 25, 2001]

§ 872.3840 Endodontic silver point.

(a) *Identification.* An endodontic silver point is a device made of silver intended for use during endodontic therapy to fill permanently the root canal of a tooth.

(b) *Classification.* Class I (general controls). The device is exempt from the premarket notification procedures in subpart E of part 807 of this chapter subject to the limitations in § 872.9.

[52 FR 30097, Aug. 12, 1987, as amended at 54 FR 13830, Apr. 5, 1989; 66 FR 38798, July 25, 2001]

§ 872.3850 Gutta percha.

(a) *Identification.* Gutta percha is a device made from coagulated sap of certain tropical trees intended to fill the root canal of a tooth. The gutta percha is softened by heat and inserted into the root canal, where it hardens as it cools.

(b) *Classification.* Class I (general controls). The device is exempt from the premarket notification procedures in subpart E of part 807 of this chapter subject to the limitations in § 872.9.

[52 FR 30097, Aug. 12, 1987, as amended at 54 FR 13830, Apr. 5, 1989; 66 FR 38798, July 25, 2001]

§ 872.3890 Endodontic stabilizing splint.

(a) *Identification.* An endodontic stabilizing splint is a device made of a material, such as titanium, intended to be inserted through the root canal into the upper or lower jaw bone to stabilize a tooth.

(b) *Classification.* Class II.

§ 872.3900 Posterior artificial tooth with a metal insert.

(a) *Identification.* A posterior artificial tooth with a metal insert is a porcelain device with an insert made of austenitic alloys or alloys containing 75 percent or greater gold and metals of the platinum group intended to replace a natural tooth. The device is attached to surrounding teeth by a bridge and is intended to provide both an improvement in appearance and functional occlusion (bite).

(b) *Classification.* Class I (general controls). The device is exempt from the premarket notification procedures in subpart E of part 807 of this chapter subject to the limitations in § 872.9.

[52 FR 30097, Aug. 12, 1987, as amended at 59 FR 63008, Dec. 7, 1994; 66 FR 38798, July 25, 2001]

§ 872.3910 Backing and facing for an artificial tooth.

(a) *Identification.* A backing and facing for an artificial tooth is a device intended for use in fabrication of a fixed or removable dental appliance, such as a crown or bridge. The backing, which is made of gold, is attached to the dental appliance and supports the tooth-colored facing, which is made of porcelain or plastic.

(b) *Classification.* Class I (general controls). The device is exempt from the premarket notification procedures in subpart E of part 807 of this chapter subject to the limitations in § 872.9.

[52 FR 30097, Aug. 12, 1987, as amended at 59 FR 63008, Dec. 7, 1994; 66 FR 38799, July 25, 2001]

§ 872.3920 Porcelain tooth.

(a) *Identification.* A porcelain tooth is a prefabricated device made of porcelain powder for clinical use (§ 872.6660) intended for use in construction of fixed or removable prostheses, such as crowns and partial dentures.

(b) *Classification.* Class II.

§ 872.3930 Bone grafting material.

(a) *Identification.* Bone grafting material is a material such as hydroxyapatite, tricalcium phosphate, polylactic and polyglycolic acids, or collagen, that is intended to fill, augment, or reconstruct periodontal or bony defects of the oral and maxillofacial region.

(b) *Classification.* (1) Class II (special controls) for bone grafting materials that do not contain a drug that is a therapeutic biologic. The special control is FDA's "Class II Special Controls Guidance Document: Dental Bone Grafting Material Devices." (See § 872.1(e) for the availability of this guidance document.)

(2) Class III (premarket approval) for bone grafting materials that contain a drug that is a therapeutic biologic. Bone grafting materials that contain a drug that is a therapeutic biologic, such as biological response modifiers, require premarket approval.

(c) *Date premarket approval application (PMA) or notice of product development protocol (PDP) is required.* Devices described in paragraph (b)(2) of this section shall have an approved PMA or a declared completed PDP in effect before being placed in commercial distribution.

[70 FR 21949, Apr. 28, 2005]

§ 872.3940 Total temporomandibular joint prosthesis.

(a) *Identification.* A total temporomandibular joint prosthesis is a device that is intended to be implanted in the human jaw to replace the mandibular condyle and augment the glenoid fossa to functionally reconstruct the temporomandibular joint.

(b) *Classification.* Class III.

(c) *Date PMA or notice of completion of a PDP is required.* A PMA or a notice of completion of a PDP is required to be filed with the Food and Drug Administration on or before March 30, 1999, for any total temporomandibular joint prosthesis that was in commercial distribution before May 28, 1976, or that has, on or before March 30, 1999, been found to be substantially equivalent to a total temporomandibular joint prosthesis that was in commercial distribution before May 28, 1976. Any other total temporomandibular joint prosthesis shall have an approved PMA or a declared completed PDP in effect before being placed in commercial distribution.

[59 FR 65478, Dec. 20, 1994, as amended at 63 FR 71746, Dec. 30, 1998]

§ 872.3950 Glenoid fossa prosthesis.

(a) *Identification.* A glenoid fossa prosthesis is a device that is intended to be implanted in the temporomandibular joint to augment a glenoid fossa or to provide an articulation surface for the head of a mandibular condyle.

(b) *Classification.* Class III.

(c) *Date PMA or notice of completion of a PDP is required.* A PMA or a notice of completion of a PDP is required to be filed with the Food and Drug Administration on or before March 30, 1999, for any glenoid fossa prosthesis that was in commercial distribution before May 28, 1976, or that has on or before March 30, 1999, been found to be substantially equivalent to a glenoid fossa prosthesis that was in commercial distribution before May 28, 1976. Any other glenoid fossa prosthesis shall have an approved PMA or a declared completed PDP in effect before being placed in commercial distribution.

[59 FR 65478, Dec. 20, 1994, as amended at 63 FR 71746, Dec. 30, 1998]

§ 872.3960 Mandibular condyle prosthesis.

(a) *Identification.* A mandibular condyle prosthesis is a device that is intended to be implanted in the human jaw to replace the mandibular condyle and to articulate within a glenoid fossa.

(b) *Classification.* Class III.

(c) *Date PMA or notice of completion of a PDP is required.* (1) Except as described in paragraph (c)(2) of this section, a PMA or a notice of completion of a PDP is required to be filed with the Food and Drug Administration on or before March 30, 1999, for any mandibular condyle prosthesis that was in commercial distribution before May 28, 1976, or that has, on or before March 30, 1999, been found to be substantially equivalent to a mandibular condyle prosthesis that was in commercial distribution before May 28, 1976. Any other mandibular condyle prosthesis shall have an approved PMA or a declared completed PDP in effect before being placed in commercial distribution.

(2) No effective date has been established of the requirement for premarket approval for any mandibular condyle prosthesis intended to be implanted in the human jaw for temporary reconstruction of the mandibular condyle in patients who have undergone resective procedures to remove malignant or benign tumors, requiring the removal of the mandibular condyle. See § 870.3 of this chapter.

[59 FR 65478, Dec. 20, 1994, as amended at 63 FR 71746, Dec. 30, 1998]

§ 872.3970 Interarticular disc prosthesis (interpositional implant).

(a) *Identification.* An interarticular disc prosthesis (interpositional implant) is a device that is intended to be an interface between the natural articulating surface of the mandibular condyle and glenoid fossa.

(b) *Classification.* Class III.

(c) *Date PMA or notice of completion of a PDP is required.* A PMA or a notice of completion of a PDP is required to be filed with the Food and Drug Administration on or before March 30, 1999, for any interarticular disc prosthesis (interpositional implant) that was in commercial distribution before May 28,

208-072 D-12

1976, or that has on or before March 30, 1999, been found to be substantially equivalent to an interarticular disc prosthesis (interpositional implant) that was in commercial distribution before May 28, 1976. Any other interarticular disc prosthesis (interpositional implant) shall have an approved PMA or a declared completed PDP in effect before being placed in commercial distribution.

[59 FR 65478, Dec. 20, 1994, as amended at 63 FR 71746, Dec. 30, 1998]

§ 872.3980 Endosseous dental implant accessories.

(a) *Identification.* Endosseous dental implant accessories are manually powered devices intended to aid in the placement or removal of endosseous dental implants and abutments, prepare the site for placement of endosseous dental implants or abutments, aid in the fitting of endosseous dental implants or abutments, aid in the fabrication of dental prosthetics, and be used as an accessory with endosseous dental implants when tissue contact will last less than 1 hour. These devices include drill bits, screwdrivers, countertorque devices, placement and removal tools, laboratory pieces used for fabrication of dental prosthetics, and trial abutments.

(b) *Classification.* Class I (general controls). The device is exempt from the premarket notification procedures in subpart E of part 807 of this chapter subject to the limitations in § 872.9.

[65 FR 60099, Oct. 10, 2000]

Subpart E—Surgical Devices

§ 872.4120 Bone cutting instrument and accessories.

(a) *Identification.* A bone cutting instrument and accessories is a metal device intended for use in reconstructive oral surgery to drill or cut into the upper or lower jaw and may be used to prepare bone to insert a wire, pin, or screw. The device includes the manual bone drill and wire driver, powered bone drill, rotary bone cutting handpiece, and AC-powered bone saw.

(b) *Classification.* Class II.

§ 872.4130 Intraoral dental drill.

(a) *Identification.* An intraoral dental drill is a rotary device intended to be attached to a dental handpiece to drill holes in teeth to secure cast or preformed pins to retain operative dental appliances.

(b) *Classification.* Class I (general controls). The device is exempt from the premarket notification procedures in subpart E of part 807 of this chapter subject to the limitations in § 872.9.

[52 FR 30097, Aug. 12, 1987, as amended at 59 FR 63008, Dec. 7, 1994; 66 FR 38799, July 25, 2001]

§ 872.4200 Dental handpiece and accessories.

(a) *Identification.* A dental handpiece and accessories is an AC-powered, water-powered, air-powered, or belt-driven, hand-held device that may include a foot controller for regulation of speed and direction of rotation or a contra-angle attachment for difficult to reach areas intended to prepare dental cavities for restorations, such as fillings, and for cleaning teeth.

(b) *Classification.* Class I.

[55 FR 48439, Nov. 20, 1990]

§ 872.4465 Gas-powered jet injector.

(a) *Identification.* A gas-powered jet injector is a syringe device intended to administer a local anesthetic. The syringe is powered by a cartridge containing pressurized carbon dioxide which provides the pressure to force the anesthetic out of the syringe.

(b) *Classification.* Class II.

§ 872.4475 Spring-powered jet injector.

(a) *Identification.* A spring-powered jet injector is a syringe device intended to administer a local anesthetic. The syringe is powered by a spring mechanism which provides the pressure to force the anesthetic out of the syringe.

(b) *Classification.* Class II.

§ 872.4535 Dental diamond instrument.

(a) *Identification.* A dental diamond instrument is an abrasive device intended to smooth tooth surfaces during the fitting of crowns or bridges. The device consists of a shaft which is inserted into a handpiece and a head which has diamond chips imbedded into it. Rotation of the diamond instrument provides an abrasive action when it contacts a tooth.

(b) *Classification.* Class I. The device is exempt from the premarket notification procedures in subpart E of part 807 of this chapter.

[52 FR 30097, Aug. 12, 1987, as amended at 59 FR 63008, Dec. 7, 1994]

§ 872.4565 Dental hand instrument.

(a) *Identification.* A dental hand instrument is a hand-held device intended to perform various tasks in general dentistry and oral surgery procedures. The device includes the operative burnisher, operative amalgam carrier, operative dental amalgam carver, surgical bone chisel, operative amalgam and foil condenser, endodontic curette, operative curette, periodontic curette, surgical curette, dental surgical elevator, operative dental excavator, operative explorer surgical bone file, operative margin finishing file, periodontic file, periodontic probe, surgical rongeur forceps, surgical tooth extractor forceps, surgical hemostat, periodontic hoe, operative matrix contouring instrument, operative cutting instrument, operative margin finishing periodontic knife, periodontic marker, operative pliers, endodontic root canal plugger, endodontic root canal preparer, surgical biopsy punch, endodontic pulp canal reamer, crown remover, periodontic scaler, collar and crown scissors, endodontic pulp canal filling material spreader, surgical osteotome chisel, endodontic broach, dental wax carver, endodontic pulp canal file, hand instrument for calculus removal, dental depth gauge instrument, plastic dental filling instrument, dental instrument handle, surgical tissue scissors, mouth mirror, orthodontic band driver, orthodontic band pusher, orthodontic band setter, orthodontic bracket aligner, orthodontic pliers, orthodontic ligature tucking instrument, forceps, for articulation paper, forceps for dental dressing, dental matrix band, matrix retainer, dental retractor, dental retractor accessories, periodontic or endodontic irrigating syringe, and restorative or impression material syringe.

(b) *Classification.* Class I (general controls). If the device is made of the same materials that were used in the device before May 28, 1976, it is exempt from the premarket notification procedures in subpart E of part 807 of this chapter subject to the limitations in § 872.9.

[52 FR 30097, Aug. 12, 1987, as amended at 54 FR 13830, Apr. 5, 1989; 66 FR 38799, July 25, 2001]

§ 872.4600 Intraoral ligature and wire lock.

(a) *Identification.* An intraoral ligature and wire lock is a metal device intended to constrict fractured bone segments in the oral cavity. The bone segments are stabilized by wrapping the ligature (wire) around the fractured bone segments and locking the ends together.

(b) *Classification.* Class II.

§ 872.4620 Fiber optic dental light.

(a) *Identification.* A fiber optic dental light is a device that is a light, usually AC-powered, that consists of glass or plastic fibers which have special optical properties. The device is usually attached to a dental handpiece and is intended to illuminate a patient's oral structures.

(b) *Classification.* Class I (general controls). The device is exempt from the premarket notification procedures in subpart E of part 807 of this chapter subject to the limitations in § 872.9.

[55 FR 48439, Nov. 20, 1990, as amended at 59 FR 63008, Dec. 7, 1994; 66 FR 38799, July 25, 2001]

§ 872.4630 Dental operating light.

(a) *Identification.* A dental operating light, including the surgical headlight, is an AC-powered device intended to illuminate oral structures and operating areas.

(b) *Classification.* Class I (general controls). The device is exempt from the premarket notification procedures in subpart E of part 807 of this chapter subject to the limitations in § 872.9.

[52 FR 30097, Aug. 12, 1987, as amended at 61 FR 1121, Jan. 16, 1996; 66 FR 38799, July 25, 2001]

§ 872.4730 Dental injecting needle.

(a) *Identification.* A dental injecting needle is a slender, hollow metal device with a sharp point intended to be attached to a syringe to inject local anesthetics and other drugs.

(b) *Classification.* Class I (general controls). The device is exempt from the premarket notification procedures in subpart E of part 807 of this chapter subject to the limitations in § 872.9.

[52 FR 30097, Aug. 12, 1987, as amended at 59 FR 63008, Dec. 7, 1994; 66 FR 38799, July 25, 2001]

§ 872.4760 Bone plate.

(a) *Identification.* A bone plate is a metal device intended to stabilize fractured bone structures in the oral cavity. The bone segments are attached to the plate with screws to prevent movement of the segments.

(b) *Classification.* Class II.

§ 872.4840 Rotary scaler.

(a) *Identification.* A rotary scaler is an abrasive device intended to be attached to a powered handpiece to remove calculus deposits from teeth during dental cleaning and periodontal (gum) therapy.

(b) *Classification.* Class II.

§ 872.4850 Ultrasonic scaler.

(a) *Identification.* An ultrasonic scaler is a device intended for use during dental cleaning and periodontal (gum) therapy to remove calculus deposits from teeth by application of an ultrasonic vibrating scaler tip to the teeth.

(b) *Classification.* Class II.

§ 872.4880 Intraosseous fixation screw or wire.

(a) *Identification.* An intraosseous fixation screw or wire is a metal device intended to be inserted into fractured jaw bone segments to prevent their movement.

(b) *Classification.* Class II.

§ 872.4920 Dental electrosurgical unit and accessories.

(a) *Identification.* A dental electrosurgical unit and accessories is an AC-powered device consisting of a controlled power source and a set of cutting and coagulating electrodes. This device is intended to cut or remove soft tissue or to control bleeding during surgical procedures in the oral cavity. An electrical current passes through the tip of the electrode into the tissue and, depending upon the operating mode selected, cuts through soft tissue or coagulates the tissue.

(b) *Classification.* Class II.

Subpart F—Therapeutic Devices

§ 872.5410 Orthodontic appliance and accessories.

(a) *Identification.* An orthodontic appliance and accessories is a device intended for use in orthodontic treatment. The device is affixed to a tooth so that pressure can be exerted on the teeth. This device includes the preformed orthodontic band, orthodontic band material, orthodontic elastic band, orthodontic metal bracket, orthodontic wire clamp, preformed orthodontic space maintainer, orthodontic expansion screw retainer, orthodontic spring, orthodontic tube, and orthodontic wire.

(b) *Classification.* Class I (general controls). The device is exempt from the premarket notification procedures in subpart E of part 807 of this chapter subject to the limitations in § 872.9.

[52 FR 30097, Aug. 12, 1987, as amended at 59 FR 63009, Dec. 7, 1994; 66 FR 38799, July 25, 2001]

§ 872.5470 Orthodontic plastic bracket.

(a) *Identification.* An orthodontic plastic bracket is a plastic device intended to be bonded to a tooth to apply pressure to a tooth from a flexible orthodontic wire to alter its position.

(b) *Classification.* Class II.

§ 872.5500 Extraoral orthodontic headgear.

(a) *Identification.* An extraoral orthodontic headgear is a device intended for use with an orthodontic appliance to exert pressure on the teeth from outside the mouth. The headgear has a strap intended to wrap around the patient's neck or head and an inner bow portion intended to be fastened to the orthodontic appliance in the patient's mouth.

(b) *Classification.* Class II.

§ 872.5525 Preformed tooth positioner.

(a) *Identification.* A preformed tooth positioner is a plastic device that is an impression of a perfected bite intended to prevent a patient's teeth from shifting position or to move teeth to a final position after orthodontic appliances (braces) have been removed. The patient bites down on the device for several hours a day to force the teeth into a final position or to maintain the teeth in their corrected position.

(b) *Classification.* Class I (general controls). The device is exempt from the premarket notification procedures in subpart E of part 807 of this chapter subject to the limitations in §872.9.

[52 FR 30097, Aug. 12, 1987, as amended at 59 FR 63009, Dec. 7, 1994; 66 FR 38799, July 25, 2001]

§872.5550 Teething ring.

(a) *Identification.* A teething ring is a divice intended for use by infants for medical purposes to soothe gums during the teething process.

(b)(1) *Classification.* Class I if the teething ring does not contain a fluid, such as water. The device is exempt from the premarket notification procedures in subpart E of part 807 of this chapter.

(2) Class II if the teething ring contains a fluid, such as water.

[52 FR 30097, Aug. 12, 1987, as amended at 59 FR 63009, Dec. 7, 1994]

§872.5570 Intraoral devices for snoring and intraoral devices for snoring and obstructive sleep apnea.

(a) *Identification.* Intraoral devices for snoring and intraoral devices for snoring and obstructive sleep apnea are devices that are worn during sleep to reduce the incidence of snoring and to treat obstructive sleep apnea. The devices are designed to increase the patency of the airway and to decrease air turbulence and airway obstruction. The classification includes palatal lifting devices, tongue retaining devices, and mandibular repositioning devices.

(b) *Classification.* Class II (special controls). The special control for these devices is the FDA guidance document entitled "Class II Special Controls Guidance Document: Intraoral Devices for Snoring and/or Obstructive Sleep Apnea; Guidance for Industry and FDA."

[67 FR 68512, Nov. 12, 2002]

§872.5580 Oral rinse to reduce the adhesion of dental plaque.

(a) *Identification.* The device is assigned the generic name oral rinse to reduce the adhesion of dental plaque and is identified as a device intended to reduce the presence of bacterial plaque on teeth and oral mucosal surfaces by physical means. The device type includes those devices that act by reducing the attachment and inhibiting the growth of bacterial plaque.

(b) *Classification.* Class II (special controls). The special control is FDA's guidance document entitled "Class II Special Controls Guidance Document: Oral Rinse to Reduce the Adhesion of Dental Plaque." See §872.1(e) for the availability of this guidance document.

[70 FR 55028, Sept. 20, 2005]

Subpart G—Miscellaneous Devices

§872.6010 Abrasive device and accessories.

(a) *Identification.* An abrasive device and accessories is a device constructed of various abrasives, such as diamond chips, that are glued to shellac-based paper. The device is intended to remove excessive restorative materials, such as gold, and to smooth rough surfaces from oral restorations, such as crowns. The device is attached to a shank that is held by a handpiece. The device includes the abrasive disk, guard for an abrasive disk, abrasive point, polishing agent strip, and polishing wheel.

(b) *Classification.* Class I (general controls). The device is exempt from the premarket notification procedures in subpart E of part 807 of this chapter subject to the limitations in §872.9. If the device is not labeled or otherwise represented as sterile, it is exempt from the current good manufacturing practice regulations in part 820 of this chapter, with the exception of §820.180, with respect to general requirements concerning records, and §820.198, with respect to complaint files.

[52 FR 30097, Aug. 12, 1987, as amended at 54 FR 13830, Apr. 5, 1989; 66 FR 38799, July 25, 2001]

§ 872.6030 Oral cavity abrasive polishing agent.

(a) *Identification.* An oral cavity abrasive polishing agent is a device in paste or powder form that contains an abrasive material, such as silica pumice, intended to remove debris from the teeth. The abrasive polish is applied to the teeth by a handpiece attachment (prophylaxis cup).

(b) *Classification.* Class I (general controls). The device is exempt from the premarket notification procedures in subpart E of part 807 of this chapter subject to the limitations in § 872.9.

[52 FR 30097, Aug. 12, 1987, as amended at 59 FR 63009, Dec. 7, 1994; 66 FR 38799, July 25, 2001]

§ 872.6050 Saliva absorber.

(a) *Identification.* A saliva absorber is a device made of paper or cotton intended to absorb moisture from the oral cavity during dental procedures.

(b) *Classification.* Class I (general controls). The device is exempt from the premarket notification procedures in subpart E of part 807 of this chapter subject to the limitations in § 872.9. If the device is not labeled or otherwise represented as sterile, it is exempt from the current good manufacturing practice regulations in part 820 of this chapter, with the exception of § 820.180, with respect to general requirements concerning records, and § 820.198, with respect to complaint files.

[52 FR 30097, Aug. 12, 1987, as amended at 54 FR 13830, Apr. 5, 1989; 66 FR 38799, July 25, 2001]

§ 872.6070 Ultraviolet activator for polymerization.

(a) *Identification.* An ultraviolet activator for polymerization is a device that produces ultraviolet radiation intended to polymerize (set) resinous dental pit and fissure sealants or restorative materials by transmission of light through a rod.

(b) *Classification.* Class II.

§ 872.6080 Airbrush.

(a) *Identification.* An airbrush is an AC-powered device intended for use in conjunction with articulation paper. The device uses air-driven particles to roughen the surfaces of dental restorations. Uneven areas of the restorations are then identified by use of articulation paper.

(b) *Classification.* Class II. The special control for this device is International Electrotechnical Commission's IEC 60601–1–AM2 (1995–03), Amendment 2, "Medical Electrical Equipment—Part 1: General Requirements for Safety."

[52 FR 30097, Aug. 12, 1987; 52 FR 49250, Dec. 30, 1987]

§ 872.6100 Anesthetic warmer.

(a) *Identification.* An anesthetic warmer is an AC-powered device into which tubes containing anesthetic solution are intended to be placed to warm them prior to administration of the anesthetic.

(b) *Classification.* Class I (general controls). The device is exempt from the premarket notification procedures in subpart E of part 807 of this chapter subject to the limitations in § 872.9.

[52 FR 30097, Aug. 12, 1987, as amended at 60 FR 38900, July 28, 1995; 66 FR 38799, July 25, 2001]

§ 872.6140 Articulation paper.

(a) *Identification.* Articulation paper is a device composed of paper coated with an ink dye intended to be placed between the patient's upper and lower teeth when the teeth are in the bite position to locate uneven or high areas.

(b) *Classification.* Class I (general controls). The device is exempt from the premarket notification procedures in subpart E of part 807 of this chapter subject to the limitations in § 872.9. If the device is not labeled or otherwise represented as sterile, it is exempt from the current good manufacturing practice regulations in part 820 of this chapter, with the exception of § 820.180, with respect to general requirements concerning records, and § 820.198, with respect to complaint files.

[52 FR 30097, Aug. 12, 1987, as amended at 59 FR 63009, Dec. 7, 1994; 66 FR 38799, July 25, 2001]

§ 872.6200 Base plate shellac.

(a) *Identification.* Base plant shellac is a device composed of shellac intended to rebuild the occlusal rim of full or partial dentures.

(b) *Classification.* Class I (general controls). The device is exempt from the premarket notification procedures in subpart E of part 807 of this chapter subject to the limitations in § 872.9. If the device is not labeled or otherwise represented as sterile, it is exempt from the current good manufacturing practice regulations in part 820 of this chapter, with the exception of § 820.180, with respect to general requirements concerning records, and § 820.198, with respect to complaint files.

[52 FR 30097, Aug. 12, 1987, as amended at 54 FR 13830, Apr. 5, 1989; 66 FR 38799, July 25, 2001]

§ 872.6250 Dental chair and accessories.

(a) *Identification.* A dental chair and accessories is a device, usually AC-powered, in which a patient sits. The device is intended to properly position a patient to perform dental procedures. A dental operative unit may be attached.

(b) *Classification.* Class I. The dental chair without the operative unit device is exempt from the premarket notification procedures in subpart E of part 807 of this chapter.

[55 FR 48439, Nov. 20, 1990, as amended at 59 FR 63009, Dec. 7, 1994]

§ 872.6290 Prophylaxis cup.

(a) *Identification.* A prophylaxis cup is a device made of rubber intended to be held by a dental handpiece and used to apply polishing agents during prophylaxis (cleaning). The dental handpiece spins the rubber cup holding the polishing agent and the user applies it to the teeth to remove debris.

(b) *Classification.* Class I (general controls). The device is exempt from the premarket notification procedures in subpart E of part 807 of this chapter subject to the limitations in § 872.9. If the device is not labeled or otherwise represented as sterile, it is exempt from the current good manufacturing practice regulations in part 820 of this chapter, with the exception of § 820.180, with respect to general requirements concerning records, and § 820.198, with respect to complaint files.

[52 FR 30097, Aug. 12, 1987, as amended at 54 FR 13831, Apr. 5, 1989; 66 FR 38799, July 25, 2001]

§ 872.6300 Rubber dam and accessories.

(a) *Identification.* A rubber dam and accessories is a device composed of a thin sheet of latex with a hole in the center intended to isolate a tooth from fluids in the mouth during dental procedures, such as filling a cavity preparation. The device is stretched around a tooth by inserting a tooth through a hole in the center. The device includes the rubber dam, rubber dam clamp, rubber dam frame, and forceps for a rubber dam clamp. This classification does not include devices intended for use in preventing transmission of sexually transmitted diseases through oral sex; those devices are classified as condoms in § 884.5300 of this chapter.

(b) *Classification.* Class I (general controls). The device is exempt from the premarket notification procedures in subpart E of part 807 of this chapter subject to § 872.9. If the device is not labeled or otherwise represented as sterile, it is exempt from the current good manufacturing practice regulations in part 820 of this chapter, with the exception of § 820.180 of this chapter, with respect to general requirements concerning records, and § 820.198 of this chapter, with respect to complaint files.

[65 FR 2315, Jan. 14, 2000]

§ 872.6350 Ultraviolet detector.

(a) *Identification.* An ultraviolet detector is a device intended to provide a source of ultraviolet light which is used to identify otherwise invisible material, such as dental plaque, present in or on teeth.

(b) *Classification.* Class II.

§ 872.6390 Dental floss.

(a) *Identification.* Dental floss is a string-like device made of cotton or other fibers intended to remove plaque and food particles from between the teeth to reduce tooth decay. The fibers of the device may be coated with wax for easier use.

(b) *Classification.* Class I (general controls). The device is exempt from the premarket notification procedures in subpart E of part 807 of this chapter subject to § 872.9.

[52 FR 30097, Aug. 12, 1987, as amended at 61 FR 1121, Jan. 16, 1996; 65 FR 2315, Jan. 14, 2000]

§ 872.6475 Heat source for bleaching teeth.

(a) *Identification.* A heat source for bleaching teeth is an AC-powered device that consists of a light or an electric heater intended to apply heat to a tooth after it is treated with a bleaching agent.

(b) *Classification.* Class I (general controls). The device is exempt from the premarket notification procedures in subpart E of part 807 of this chapter subject to the limitations in § 872.9.

[55 FR 48439, Nov. 20, 1990, as amended at 59 FR 63009, Dec. 7, 1994; 66 FR 38799, July 25, 2001]

§ 872.6510 Oral irrigation unit.

(a) *Identification.* An oral irrigation unit is an AC-powered device intended to provide a pressurized stream of water to remove food particles from between the teeth and promote good periodontal (gum) condition.

(b) *Classification.* Class I (general controls). The device is exempt from the premarket notification procedures in subpart E of part 807 of this chapter subject to the limitations in § 872.9.

[55 FR 48439, Nov. 20, 1990, as amended at 59 FR 63009, Dec. 7, 1994; 66 FR 38800, July 25, 2001]

§ 872.6570 Impression tube.

(a) *Identification.* An impression tube is a device consisting of a hollow copper tube intended to take an impression of a single tooth. The hollow tube is filled with impression material. One end of the tube is sealed with a softened material, such as wax, the remaining end is slipped over the tooth to make the impression.

(b) *Classification.* Class I (general controls). The device is exempt from the premarket notification procedures in subpart E of part 807 of this chapter subject to the limitations in § 872.9. If the device is not labeled or otherwise represented as sterile, it is exempt from the current good manufacturing practice regulations in part 820 of this chapter, with the exception of § 820.180, with respect to general requirements concerning records, and § 820.198, with respect to complaint files.

[52 FR 30097, Aug. 12, 1987, as amended at 54 FR 13831, Apr. 5, 1989; 66 FR 38800, July 25, 2001]

§ 872.6640 Dental operative unit and accessories.

(a) *Identification.* A dental operative unit and accessories is an AC-powered device that is intended to supply power to and serve as a base for other dental devices, such as a dental handpiece, a dental operating light, an air or water syringe unit, and oral cavity evacuator, a suction operative unit, and other dental devices and accessories. The device may be attached to a dental chair.

(b) *Classification.* Class I (general controls). Except for dental operative unit, accessories are exempt from premarket notification procedures in subpart E of part 807 of this chapter subject to § 872.9.

[55 FR 48439, Nov. 20, 1990, as amended at 59 FR 63009, Dec. 7, 1994; 65 FR 2315, Jan. 14, 2000]

§ 872.6650 Massaging pick or tip for oral hygiene.

(a) *Identification.* A massaging pick or tip for oral hygiene is a rigid, pointed device intended to be used manually to stimulate and massage the gums to promote good periodontal (gum) condition.

(b) *Classification.* Class I (general controls). The device is exempt from the premarket notification procedures in subpart E of part 807 of this chapter subject to the limitations in § 872.9. If the device is not labeled or otherwise represented as sterile, it is exempt from the current good manufacturing practice regulations in part 820 of this chapter, with the exception of § 820.180, with respect to general requirements concerning records, and § 820.198, with respect to complaint files.

[52 FR 30097, Aug. 12, 1987, as amended at 54 FR 13831, Apr. 5, 1989; 66 FR 38800, July 25, 2001]

§ 872.6660 Porcelain powder for clinical use.

(a) *Identification.* Porcelain powder for clinical use is a device consisting of a mixture of kaolin, felspar, quartz, or other substances intended for use in the production of artificial teeth in fixed or removable dentures, of jacket crowns, facings, and veneers. The device is used in prosthetic dentistry by heating the powder mixture to a high temperature in an oven to produce a hard prosthesis with a glass-like finish.

(b) *Classification.* Class II.

§ 872.6670 Silicate protector.

(a) *Identification.* A silicate protector is a device made of silicone intended to be applied with an absorbent tipped applicator to the surface of a new restoration to exclude temporarily fluids from its surface.

(b) *Classification.* Class I (general controls). The device is exempt from the premarket notification procedures in subpart E of part 807 of this chapter subject to the limitations in § 872.9. If the device is not labeled or otherwise represented as sterile, it is exempt from the current good manufacturing practice regulations in part 820 of this chapter, with the exception of § 820.180, with respect to general requirements concerning records, and § 820.198, with respect to complaint files.

[52 FR 30097, Aug. 12, 1987, as amended at 54 FR 13831, Apr. 5, 1989; 66 FR 38800, July 25, 2001]

§ 872.6710 Boiling water sterilizer.

(a) *Identification.* A boiling water sterilizer is an AC-powered device that consists of a container for boiling water. The device is intended to sterilize dental and surgical instruments by submersion in the boiling water in the container.

(b) *Classification.* Class I (general controls).

[55 FR 48439, Nov. 20, 1990, as amended at 66 FR 46952, Sept. 10, 2001]

§ 872.6730 Endodontic dry heat sterilizer.

(a) *Identification.* An endodontic dry heat sterilizer is a device intended to sterilize endodontic and other dental instruments by the application of dry heat. The heat is supplied through glass beads which have been heated by electricity.

(b) *Classification.* Class III.

(c) *Date premarket approval application (PMA) or notice of completion of product development protocol (PDP) is required.* A PMA or notice of completion of a PDP is required to be filed with the Food and Drug Administration on or before April 21, 1997, for any endodontic dry heat sterilizer that was in commercial distribution before May 28, 1976, or that has on or before April 21, 1997, been found to be substantially equivalent to the endodontic dry heat sterilizer that was in commercial distribution before May 28, 1976. Any other endodontic dry heat sterilizer shall have an approved PMA or declared completed PDP in effect before being placed in commercial distribution.

[52 FR 30097, Aug. 12, 1987, as amended at 62 FR 2902, Jan. 21, 1997; 62 FR 31512, June 10, 1997]

§ 872.6770 Cartridge syringe.

(a) *Identification.* A cartridge syringe is a device intended to inject anesthetic agents subcutaneously or intramuscularly. The device consists of a metal syringe body into which a disposable, previously filled, glass carpule (a cylindrical cartridge) containing anesthetic is placed. After attaching a needle to the syringe body and activating the carpule by partially inserting the plunger on the syringe, the device is used to administer an injection to the patient.

(b) *Classification.* Class II.

§ 872.6855 Manual toothbrush.

(a) *Identification.* A manual toothbrush is a device composed of a shaft with either natural or synthetic bristles at one end intended to remove adherent plaque and food debris from the teeth to reduce tooth decay.

(b) *Classification.* Class I (general controls). The device is exempt from the premarket notification procedures in subpart E of part 807 of this chapter subject to the limitations in § 872.9. If the device is not labeled or otherwise represented as sterile, it is exempt from the current good manufacturing practice regulations in part 820 of this chapter, with the exception of § 820.180, with respect to general requirements concerning records, and § 820.198, with respect to complaint files.

[52 FR 30097, Aug. 12, 1987, as amended at 54 FR 13831, Apr. 5, 1989; 66 FR 38800, July 25, 2001]

§ 872.6865 Powered toothbrush.

(a) *Identification.* A powered toothbrush is an AC-powered or battery-powered device that consists of a handle containing a motor that provides mechanical movement to a brush intended to be applied to the teeth. The device is intended to remove adherent plaque and food debris from the teeth to reduce tooth decay.

(b) *Classification.* Class I (general controls). The device is exempt from the premarket notification procedures in subpart E of part 807 of this chapter subject to the limitations in § 872.9.

[55 FR 48440, Nov. 20, 1990, as amended at 59 FR 63009, Dec. 7, 1994; 66 FR 38800, July 25, 2001]

§ 872.6870 Disposable flouride tray.

(a) *Identification.* A disposable fluoride tray is a device made of styrofoam intended to apply fluoride topically to the teeth. To use the tray, the patient bites down on the tray which has been filled with a fluoride solution.

(b) *Classification.* Class I (general controls). The device is exempt from the premarket notification procedures in subpart E of part 807 of this chapter subject to the limitations in § 872.9. If the device is not labeled or otherwise represented as sterile, it is exempt from the current good manufacturing practice regulations in part 820 of this chapter, with the exception of § 820.180, with respect to general requirements concerning records, and § 820.198, with respect to complaint files.

[52 FR 30097, Aug. 12, 1987, as amended at 54 FR 13831, Apr. 5, 1989; 66 FR 38800, July 25, 2001]

§ 872.6880 Preformed impression tray.

(a) *Identification.* A preformed impression tray is a metal or plastic device intended to hold impression material, such as alginate, to make an impression of a patient's teeth or alveolar process (bony tooth sockets) to reproduce the structure of a patient's teeth and gums.

(b) *Classification.* Class I (general controls). The device is exempt from the premarket notification procedures in subpart E of part 807 of this chapter subject to the limitations in § 872.9. If the device is not labeled or otherwise represented as sterile, it is exempt from the current good manufacturing practice regulations in part 820 of this chapter, with the exception of § 820.180, with respect to general requirements concerning records, and § 820.198, with respect to complaint files.

[52 FR 30097, Aug. 12, 1987, as amended at 54 FR 13832, Apr. 5, 1989; 66 FR 38800, July 25, 2001]

§ 872.6890 Intraoral dental wax.

(a) *Identification.* Intraoral dental wax is a device made of wax intended to construct patterns from which custom made metal dental prostheses, such as crowns and bridges, are cast. In orthodontic dentistry, the device is intended to make a pattern of a patient's bite to make study models of the teeth.

(b) *Classification.* Class I (general controls). The device is exempt from the premarket notification procedures in subpart E of part 807 of this chapter subject to the limitations in § 872.9. If the device is not labeled or otherwise represented as sterile, it is exempt from the current good manufacturing practice regulations in part 820 of this chapter, with the exception of § 820.180, with respect to general requirements concerning records, and § 820.198, with respect to complaint files.

[52 FR 30097, Aug. 12, 1987, as amended at 59 FR 63009, Dec. 7, 1994; 66 FR 38800, July 25, 2001]

PART 874—EAR, NOSE, AND THROAT DEVICES

Subpart A—General Provisions

Sec.

Subpart B—Diagnostic Devices

874.1050 Audiometer.
874.1060 Acoustic chamber for audiometric testing.
874.1070 Short increment sensitivity index (SISI) adapter.
874.1080 Audiometer calibration set.
874.1090 Auditory impedance tester.
874.1100 Earphone cushion for audiometric testing.
874.1120 Electronic noise generator for audiometric testing.
874.1325 Electroglottograph.
874.1500 Gustometer.
874.1800 Air or water caloric stimulator.
874.1820 Surgical nerve stimulator/locator.
874.1925 Toynbee diagnostic tube.

Subpart C [Reserved]

Subpart D—Prosthetic Devices

874.3300 Hearing aid.
874.3310 Hearing aid calibrator and analysis system.
874.3320 Group hearing aid or group auditory trainer.
874.3330 Master hearing aid.
874.3375 Battery-powered artificial larynx.
874.3400 Tinnitus masker.
874.3430 Middle ear mold.
874.3450 Partial ossicular replacement prosthesis.
874.3495 Total ossicular replacement prosthesis.
874.3540 Prosthesis modification instrument for ossicular replacement surgery.
874.3620 Ear, nose, and throat synthetic polymer material.
874.3695 Mandibular implant facial prosthesis.
874.3730 Laryngeal prosthesis (Taub design).
874.3760 Sacculotomy tack (Cody tack).
874.3820 Endolymphatic shunt.
874.3850 Endolymphatic shunt tube with valve.
874.3880 Tympanostomy tube.
874.3900 Nasal dilator.
874.3930 Tympanostomy tube with semipermeable membrane.
874.3950 Transcutaneous air conduction hearing aid system.

Subpart E—Surgical Devices

874.4100 Epistaxis balloon.
874.4140 Ear, nose, and throat bur.
874.4175 Nasopharyngeal catheter.
874.4250 Ear, nose, and throat electric or pneumatic surgical drill.
874.4350 Ear, nose, and throat fiberoptic light source and carrier.
874.4420 Ear, nose, and throat manual surgical instrument.
874.4490 Argon laser for otology, rhinology, and laryngology.
874.4500 Ear, nose, and throat microsurgical carbon dioxide laser.
874.4680 Bronchoscope (flexible or rigid) and accessories.
874.4710 Esophagoscope (flexible or rigid) and accessories.
874.4720 Mediastinoscope and accessories.
874.4750 Laryngostroboscope.
874.4760 Nasopharyngoscope (flexible or rigid) and accessories.
874.4770 Otoscope.
874.4780 Intranasal splint.
874.4800 Bone particle collector.

Subpart F—Therapeutic Devices

874.5220 Ear, nose, and throat drug administration device.
874.5300 Ear, nose, and throat examination and treatment unit.
874.5350 Suction antichoke device.
874.5370 Tongs antichoke device.
874.5550 Powered nasal irrigator.
874.5800 External nasal splint.
874.5840 Antistammering device.

AUTHORITY: 21 U.S.C. 351, 360, 360c, 360e, 360j, 371.

SOURCE: 51 FR 40389, Nov. 6, 1986, unless otherwise noted.

Subpart A—General Provisions

§ 874.1 Scope.

(a) This part sets forth the classification of ear, nose, and throat devices intended for human use that are in commercial distribution.

(b) The identification of a device in a regulation in this part is not a precise description of every device that is, or will be, subject to the regulation. A manufacturer who submits a premarket notification submission for a device under part 807 cannot show merely that the device is accurately described by the section title and identification provision of a regulation in this part, but shall state why the device is substantially equivalent to other devices, as required by § 807.87.

(c) To avoid duplicative listings, an ear, nose, and throat device that has two or more types of uses (e.g., used both as a diagnostic device and as a therapeutic device) is listed in one subpart only.

(d) References in this part to regulatory sections of the Code of Federal Regulations are to chapter I of title 21 unless otherwise noted.

(e) Guidance documents referenced in this part are available on the Internet at *http://www.fda.gov/cdrh/guidance.html*.

[51 FR 40389, Nov. 6, 1986, as amended at 67 FR 67790, Nov. 7, 2002]

§ 874.3 Effective dates of requirement for premarket approval.

A device included in this part that is classified into class III (premarket approval) shall not be commercially distributed after the date shown in the regulation classifying the device unless the manufacturer has an approval under section 515 of the act (unless an exemption has been granted under section 520(g)(2) of the act). An approval under section 515 of the act consists of FDA's issuance of an order approving an application for premarket approval (PMA) for the device or declaring completed a product development protocol (PDP) for the device.

(a) Before FDA requires that a device commercially distributed before the enactment date of the amendments, or a device that has been found substantially equivalent to such a device, has an approval under section 515 of the act FDA must promulgate a regulation under section 515(b) of the act requiring such approval, except as provided in paragraph (b) of this section. Such a regulation under section 515(b) of the act shall not be effective during the grace period ending on the 90th day after its promulgation or on the last day of the 30th full calendar month after the regulation that classifies the device into class III is effective, whichever is later. See section 501(f)(2)(B) of the act. Accordingly, unless an effective date of the requirement for premarket approval is shown in the regulation for a device classified into class III in this part, the device may be commercially distributed without FDA's issuance of an order approving a PMA declaring completed a PDP for the device. If FDA promulgates a regulation under section 515(b) of the act requiring premarket approval for a device, section 501(f)(1)(A) of the act applies to the device.

(b) Any new, not substantially equivalent, device introduced into commercial distribution on or after May 28, 1976, including a device formerly marketed that has been substantially altered, is classified by statute (section 513(f) of the act) into class III without any grace period and FDA must have issued an order approving a PMA or declaring completed a PDP for the device before the device is commercially distributed unless it is reclassified. If FDA knows that a device being commercially distributed may be a "new" device as defined in this section because of any new intended use or other reasons, FDA may codify the statutory classification of the device into class III for such new use. Accordingly, the regulation for such a class III device states that as of the enactment date of the amendments, May 28, 1976, the device must have an approval under section 515 of the act before commercial distribution.

§ 874.9 Limitations of exemptions from section 510(k) of the Federal Food, Drug, and Cosmetic Act (the act).

The exemption from the requirement of premarket notification (section 510(k) of the act) for a generic type of class I or II device is only to the extent that the device has existing or reasonably foreseeable characteristics of commercially distributed devices within that generic type or, in the case of in vitro diagnostic devices, only to the extent that misdiagnosis as a result of using the device would not be associated with high morbidity or mortality. Accordingly, manufacturers of any commercially distributed class I or II device for which FDA has granted an exemption from the requirement of premarket notification must still submit a premarket notification to FDA before introducing or delivering for introduction into interstate commerce for commercial distribution the device when:

(a) The device is intended for a use different from the intended use of a legally marketed device in that generic type of device; e.g., the device is intended for a different medical purpose, or the device is intended for lay use where the former intended use was by health care professionals only;

(b) The modified device operates using a different fundamental scientific technology than a legally marketed device in that generic type of device; e.g., a surgical instrument cuts tissue with a laser beam rather than with a sharpened metal blade, or an in vitro diagnostic device detects or identifies infectious agents by using deoxyribonucleic acid (DNA) probe or nucleic acid hybridization technology rather than culture or immunoassay technology; or

(c) The device is an in vitro device that is intended:

(1) For use in the diagnosis, monitoring, or screening of neoplastic diseases with the exception of immunohistochemical devices;

(2) For use in screening or diagnosis of familial or acquired genetic disorders, including inborn errors of metabolism;

(3) For measuring an analyte that serves as a surrogate marker for screening, diagnosis, or monitoring life-threatening diseases such as acquired immune deficiency syndrome (AIDS), chronic or active hepatitis, tuberculosis, or myocardial infarction or to monitor therapy;

(4) For assessing the risk of cardiovascular diseases;

(5) For use in diabetes management;

(6) For identifying or inferring the identity of a microorganism directly from clinical material;

(7) For detection of antibodies to microorganisms other than immunoglobulin G (IgG) or IgG assays when the results are not qualitative, or are used to determine immunity, or the assay is intended for use in matrices other than serum or plasma;

(8) For noninvasive testing as defined in §812.3(k) of this chapter; and

(9) For near patient testing (point of care).

[65 FR 2315, Jan. 14, 2000]

Subpart B—Diagnostic Devices

§874.1050 Audiometer.

(a) *Identification.* An audiometer or automated audiometer is an electroacoustic device that produces controlled levels of test tones and signals intended for use in conducting diagnostic hearing evaluations and assisting in the diagnosis of possible otologic disorders.

(b) *Classification.* Class II. Except for the otoacoustic emission device, the device is exempt from the premarket notification procedures in subpart E of part 807 of this chapter, if it is in compliance with American National Standard Institute S3.6–1996, "Specification for Audiometers," and subject to the limitations in §874.9.

[51 FR 40389, Nov. 6, 1986, as amended at 64 FR 14831, Mar. 29, 1999]

§874.1060 Acoustic chamber for audiometric testing.

(a) *Identification.* An acoustic chamber for audiometric testing is a room that is intended for use in conducting diagnostic hearing evaluations and that eliminates sound reflections and provides isolation from outside sounds.

(b) *Classification.* Class I (general controls). The device is exempt from the premarket notification procedures in subpart E of part 807 of this chapter subject to the limitations in §874.9.

[51 FR 40389, Nov. 6, 1986, as amended at 61 FR 1121, Jan. 16, 1996; 66 FR 38800, July 25, 2001]

§874.1070 Short increment sensitivity index (SISI) adapter.

(a) *Identification.* A short increment sensitivity index (SISI) adapter is a device used with an audiometer in diagnostic hearing evaluations. A SISI adapter provides short periodic sound pulses in specific small decibel increments that are intended to be superimposed on the audiometer's output tone frequency.

(b) *Classification.* Class I (general controls). The device is exempt from the premarket notification procedures in subpart E of part 807 of this chapter subject to §874.9.

[55 FR 48440, Nov. 20, 1990, as amended at 65 FR 2315, Jan. 14, 2000]

§874.1080 Audiometer calibration set.

(a) *Identification.* An audiometer calibration set is an electronic reference device that is intended to calibrate an audiometer. It measures the sound frequency and intensity characteristics that emanate from an audiometer earphone. The device consists of an acoustic cavity of known volume, a sound level meter, a microphone with calibration traceable to the National Bureau of Standards, oscillators, frequency counters, microphone amplifiers, and a recorder. The device can measure selected audiometer test frequencies at a given intensity level, and selectable audiometer attenuation settings at a given test frequency.

(b) *Classification.* Class I (general controls). The device is exempt from the premarket notification procedures in subpart E of part 807 of this chapter subject to the limitations in § 874.9.

[51 FR 40389, Nov. 6, 1986, as amended at 61 FR 1121, Jan. 16, 1996; 66 FR 38800, July 25, 2001]

§ 874.1090 Auditory impedance tester.

(a) *Identification.* An auditory impedance tester is a device that is intended to change the air pressure in the external auditory canal and measure and graph the mobility characteristics of the tympanic membrane to evaluate the functional condition of the middle ear. The device is used to determine abnormalities in the mobility of the tympanic membrane due to stiffness, flaccidity, or the presence of fluid in the middle ear cavity. The device is also used to measure the acoustic reflex threshold from contractions of the stapedial muscle, to monitor healing of tympanic membrane grafts or stapedectomies, or to monitor followup treatment for inflammation of the middle ear.

(b) *Classification.* Class II.

§ 874.1100 Earphone cushion for audiometric testing.

(a) *Identification.* An earphone cushion for audiometric testing is a device that is used to cover an audiometer earphone during audiometric testing to provide an acoustic coupling (sound connection path) between the audiometer earphone and the patient's ear.

(b) *Classification.* Class I (general controls). The device is exempt from the premarket notification procedures in subpart E of part 807 of this chapter subject to § 874.9.

[51 FR 40389, Nov. 9, 1986; 52 FR 18495, May 15, 1987, as amended at 52 FR 32111, Aug. 25, 1987; 65 FR 2315, Jan. 14, 2000]

§ 874.1120 Electronic noise generator for audiometric testing.

(a) *Identification.* An electronic noise generator for audiometric testing is a device that consists of a swept frequency generator, an amplifier, and an earphone. It is intended to introduce a masking noise into the non-test ear during an audiometric evaluation. The device minimizes the non-test ear's sensing of test tones and signals being generated for the ear being tested.

(b) *Classification.* Class II.

§ 874.1325 Electroglottograph.

(a) *Identification.* An electroglottograph is an AC-powered device that employs a pair of electrodes that are placed in contact with the skin on both sides of the larynx and held in place by a collar. It is intended to measure the electrical impedance of the larynx to aid in assessing the degree of closure of the vocal cords, confirm larygeal diagnosis, aid behavioral treatment of voice disorders, and aid research concerning the laryngeal mechanism.

(b) *Classification.* Class II.

§ 874.1500 Gustometer.

(a) *Identification.* A gustometer is a battery-powered device that consists of two electrodes that are intended to be placed on both sides of the tongue at different taste centers and that provides a galvanic stimulus resulting in taste sensation. It is used for assessing the sense of taste.

(b) *Classification.* Class I (general controls). The device is exempt from the premarket notification procedures in subpart E of part 807 of this chapter subject to § 874.9. If the device is not labeled or otherwise represented as sterile, it is exempt from the current good manufacturing practice regulations in part 820 of this chapter, with the exception of § 820.180 of this chapter, with respect to general requirements concerning records, and § 820.198 of this chapter, with respect to complaint files.

[51 FR 40389, Nov. 6, 1986, as amended at 65 FR 2316, Jan. 14, 2000]

§ 874.1800 Air or water caloric stimulator.

(a) *Identification.* An air or water caloric stimulator is a device that delivers a stream of air or water to the ear canal at controlled rates of flow and temperature and that is intended for vestibular function testing of a patient's body balance system. The vestibular stimulation of the semicircular canals produce involuntary eye movements that are measured and recorded by a nystagmograph.

(b) *Classification.* Class I (general controls). The device is exempt from the premarket notification procedures in subpart E of part 807 of this chapter subject to § 874.9.

[55 FR 48440, Nov. 20, 1990, as amended at 65 FR 2316, Jan. 14, 2000]

§ 874.1820 Surgical nerve stimulator/locator.

(a) *Identification.* A surgical nerve stimulator/locator is a device that is intended to provide electrical stimulation to the body to locate and identify nerves and to test their excitability.

(b) *Classification.* Class II.

§ 874.1925 Toynbee diagnostic tube.

(a) *Identification.* The toynbee diagnostic tube is a listening device intended to determine the degree of openness of the eustachian tube.

(b) *Classification.* Class I (general controls). The device is exempt from the premarket notification procedures in subpart E of part 807 of this chapter subject to § 874.9.

[51 FR 40389, Nov. 6, 1986, as amended at 65 FR 2316, Jan. 14, 2000]

Subpart C [Reserved]

Subpart D—Prosthetic Devices

§ 874.3300 Hearing Aid.

(a) *Identification.* A hearing aid is wearable sound-amplifying device that is intended to compensate for impaired hearing. This generic type of device includes the air-conduction hearing aid and the bone-conduction hearing aid, but excludes the group hearing aid or group auditory trainer (§ 874.3320), master hearing aid (§ 874.3330), and tinnitus masker (§ 874.3400).

(b) *Classification.* (1) Class I (general controls) for the air-conduction hearing aid. The air-conduction hearing aid is exempt from the premarket notification procedures in subpart E of part 807 of this chapter subject to § 874.9.

(2) Class II for the bone-conduction hearing aid.

[51 FR 40389, Nov. 6, 1986, as amended at 65 FR 2316, Jan. 14, 2000]

§ 874.3310 Hearing aid calibrator and analysis system.

(a) *Identification.* A hearing aid calibrator and analysis system is an electronic reference device intended to calibrate and assess the electroacoustic frequency and sound intensity characteristics emanating from a hearing aid, master hearing aid, group hearing aid or group auditory trainer. The device consists of an acoustic complex of known cavity volume, a sound level meter, a microphone, oscillators, frequency counters, microphone amplifiers, a distoration analyzer, a chart recorder, and a hearing aid test box.

(b) *Classification.* Class II.

§ 874.3320 Group hearing aid or group auditory trainer.

(a) *Identification.* A group hearing aid or group auditory trainer is a hearing aid that is intended for use in communicating simultaneously with one or more listeners having hearing impairment. The device is used with an associated transmitter microphone. It may be either monaural or binaural, and it provides coupling to the ear through either earphones or earmolds. The generic type of device includes three types of applications: hardwire systems, inductance loop systems, and wireless systems.

(b) *Classification.* Class II.

§ 874.3330 Master hearing aid.

(a) *Identification.* A master hearing aid is an electronic device intended to simulate a hearing aid during audiometric testing. It has adjustable acoustic output levels, such as those for gain, output, and frequency response. The device is used to select and adjust a person's wearable hearing aid.

(b) *Classification.* Class II.

§ 874.3375 Battery-powered artificial larynx.

(a) *Identification.* A battery-powered artificial larynx is an externally applied device intended for use in the absence of the larynx to produce sound.

When held against the skin in the area of the voicebox, the device generates mechanical vibrations which resonate in the oral and nasal cavities and can be modulated by the tongue and lips in a normal manner, thereby allowing the production of speech.

(b) *Classification.* Class I (general controls). The device is exempt from the premarket notification procedures in subpart E of part 807 of this chapter subject to the limitations in § 874.9.

[51 FR 40389, Nov. 6, 1986, as amended at 59 FR 63009, Dec. 7, 1994; 66 FR 38800, July 25, 2001]

§ 874.3400 Tinnitus masker.

(a) *Identification.* A tinnitus masker is an electronic device intended to generate noise of sufficient intensity and bandwidth to mask ringing in the ears or internal head noises. Because the device is able to mask internal noises, it is also used as an aid in hearing external noises and speech.

(b) *Classification.* Class II. The special control for this device is patient labeling regarding:

(1) Hearing health care professional diagnosis, fitting of the device, and followup care,

(2) Risks,

(3) Benefits,

(4) Warnings for safe use, and

(5) Specifications.

[51 FR 40389, Nov. 6, 1986, as amended at 65 FR 17145, Mar. 31, 2000]

§ 874.3430 Middle ear mold.

(a) *Identification.* A middle ear mold is a preformed device that is intended to be implanted to reconstruct the middle ear cavity during repair of the tympanic membrane. The device permits an ample air-filled cavity to be maintained in the middle ear and promotes regeneration of the mucous membrane lining of the middle ear cavity. A middle ear mold is made of materials such as polyamide, polytetrafluoroethylene, silicone elastomer, or polyethylene, but does not contain porous polyethylene.

(b) *Classification.* Class II.

§ 874.3450 Partial ossicular replacement prosthesis.

(a) *Identification.* A partial ossicular replacement prosthesis is a device intended to be implanted for the functional reconstruction of segments of the ossicular chain and facilitates the conduction of sound wave from the tympanic membrane to the inner ear. The device is made of materials such as stainless steel, tantalum, polytetrafluoroethylene, polyethylene, polytetrafluoroethylene with carbon fibers composite, absorbable gelatin material, porous polyethylene, or from a combination of these materials.

(b) *Classification.* Class II.

§ 874.3495 Total ossicular replacement prosthesis.

(a) *Identification.* A total ossicular replacement prosthesis is a device intended to be implanted for the total functional reconstruction of the ossicular chain and facilitates the conduction of sound waves from the tympanic membrance to the inner ear. The device is made of materials such as polytetrafluoroethylene, polytetrafluoroethylene with vitreous carbon fibers composite, porous polyethylene, or from a combination of these materials.

(b) *Classification.* Class II.

§ 874.3540 Prosthesis modification instrument for ossicular replacement surgery.

(a) *Identification.* A prosthesis modification instrument for ossicular replacement surgery is a device intended for use by a surgeon to construct ossicular replacements. This generic type of device includes the ear, nose, and throat cutting block; wire crimper, wire bending die; wire closure forceps; piston cutting jib; gelfoam™ punch; wire cutting scissors; and ossicular finger vise.

(b) *Classification.* Class I (general controls). The device is exempt from the premarket notification procedures in subpart E of part 807 of this chapter subject to § 874.9. If the device is not labeled or otherwise represented as sterile, it is exempt from the current good manufacturing practice regulations in part 820 of this chapter, with the exception of § 820.180 of this chapter, with respect to general requirements concerning records, and § 820.198 of this chapter, with respect to complaint files.

[51 FR 40389, Nov. 9, 1986, as amended at 52 FR 32111, Aug. 25, 1987; 65 FR 2316, Jan. 14, 2000]

§874.3620 Ear, nose, and throat synthetic polymer material.

(a) *Identification.* Ear, nose, and throat synthetic polymer material is a device material that is intended to be implanted for use as a space-occupying substance in the reconstructive surgery of the head and neck. The device is used, for example, in augmentation rhinoplasty and in tissue defect closures in the esophagus. The device is shaped and formed by the suregon to conform to the patient's needs. This generic type of device is made of material such as polyamide mesh or foil and porous polyethylene.

(b) *Classification.* Class II.

§874.3695 Mandibular implant facial prosthesis.

(a) *Identification.* A mandibular implant facial prosthesis is a device that is intended to be implanted for use in the functional reconstruction of mandibular deficits. The device is made of materials such as stainless steel, tantalum, titanium, cobalt-chromium based alloy, polytetrafluoroethylene, silicone elastomer, polyethylene, polyurethane, or polytetrafluoroethylene with carbon fibers composite.

(b) *Classification.* Class II.

§874.3730 Laryngeal prosthesis (Taub design).

(a) *Identification.* A laryngeal prosthesis (Taub design) is a device intended to direct pulmonary air flow to the pharynx in the absence of the larynx, thereby permitting esophageal speech. The device is interposed between openings in the trachea and the esophagus and may be removed and replaced each day by the patient. During phonation, air from the lungs is directed to flow through the device and over the esophageal mucosa to provide a sound source that is articulated as speech.

(b) *Classification.* Class II.

§874.3760 Sacculotomy tack (Cody tack)

(a) *Identification.* A sacculotomy tack (Cody tack) is a device that consists of a pointed stainless steel tack intended to be implanted to relieve the symptoms of vertigo. The device repetitively ruptures the utricular membrane as the membrane expands under increased endolymphatic pressure.

(b) *Classification.* Class II.

§874.3820 Endolymphatic shunt.

(a) *Identification.* An endolymphatic shunt is a device that consists of a tube or sheet intended to be implanted to relieve the symptons of vertigo. The device permits the unrestricted flow of excess endolymph from the distended end of the endolymphatic system into the mastoid cavity where resorption occurs. This device is made of polytetrafluoroethylene or silicone elastomer.

(b) *Classification.* Class II.

§874.3850 Endolymphatic shunt tube with valve.

(a) *Identification.* An endolymphatic shunt tube with valve is a device that consists of a pressure-limiting valve associated with a tube intended to be implanted in the inner ear to relieve symptoms of vertigo and hearing loss due to endolymphatic hydrops (increase in endolymphatic fluid) of Meniere's disease.

(b) *Classification.* Class II (special controls). The special control for this device is the FDA guidance document "Class II Special Controls Guidance Document: Endolymphatic Shunt Tube With Valve; Guidance for Industry and FDA."

[67 FR 20894, Apr. 29, 2002]

§874.3880 Tympanostomy tube.

(a) *Identification.* A tympanostomy tube is a device that is intended to be implanted for ventilation or drainage of the middle ear. The device is inserted through the tympanic membrane to permit a free exchange of air between the outer ear and middle ear. A type of tympanostomy tube known as the malleous clip tube attaches to the malleous to provide middle ear ventilation. The device is made of materials such as polytetrafluoroethylene, polyethylene, silicon elastomer, or porous polyethylene.

(b) *Classification.* Class II.

§ 874.3900 Nasal dilator.

(a) *Identification.* A nasal dilator is a device intended to provide temporary relief from transient causes of breathing difficulties resulting from structural abnormalities and/or transient causes of nasal congestion associated with reduced nasal airflow. The device decreases airway resistance and increases nasal airflow. The external nasal dilator is constructed from one or more layers of material upon which a spring material is attached, with a skin adhesive applied to adhere to the skin of the nose; it acts with a pulling action to open the nares. The internal nasal dilator is constructed from metal or plastic and is placed inside the nostrils; it acts by pushing the nostrils open or by gently pressing on the columella.

(b) *Classification.* Class I (general controls). The device is exempt from the premarket notification procedures in subpart E of part 807 of this chapter subject to the limitations in § 874.9.

[64 FR 10949, Mar. 8, 1999]

§ 874.3930 Tympanostomy tube with semipermeable membrane.

(a) *Identification.* A tympanostomy tube with a semipermeable membrane is a device intended to be implanted for ventilation or drainage of the middle ear and for preventing fluids from entering the middle ear cavity. The device is inserted through the tympanic membrane to permit a free exchange of air between the outer ear and middle ear. The tube portion of the device is made of silicone elastomer or porous polyethylene, and the membrane portion is made of polytetrafluoroethylene.

(b) *Classification.* Class II. The special control for this device is FDA's "Tympanostomy Tubes, Submission Guidance for a 510(k)."

[51 FR 40389, Nov. 6, 1986, as amended at 65 FR 17145, Mar. 31, 2000]

§ 874.3950 Transcutaneous air conduction hearing aid system.

(a) *Identification.* A transcutaneous air conduction hearing aid system is a wearable sound-amplifying device intended to compensate for impaired hearing without occluding the ear canal. The device consists of an air conduction hearing aid attached to a surgically fitted tube system, which is placed through soft tissue between the post auricular region and the outer ear canal.

(b) *Classification.* Class II (special controls). The special control for this device is FDA's guidance document entitled "Class II Special Controls Guidance Document: Transcutaneous Air Conduction Hearing Aid System (TACHAS); Guidance for Industry and FDA." See § 874.1 for the availability of this guidance document.

[67 FR 67790, Nov. 7, 2002]

Subpart E—Surgical Devices

§ 874.4100 Epistaxis balloon.

(a) *Identification.* An epistaxis balloon is a device consisting of an inflatable balloon intended to control internal nasal bleeding by exerting pressure against the sphenopalatine artery.

(b) *Classification* Class I (general controls). The device is exempt from the premarket notification procedures in subpart E of part 807 of this chapter subject to § 874.9.

[51 FR 40389, Nov. 6, 1986, as amended at 65 FR 2316, Jan. 14, 2000]

§ 874.4140 Ear, nose, and throat bur.

(a) *Identification.* An ear, nose, and throat bur is a device consisting of an interchangeable drill bit that is intended for use in an ear, nose, and throat electric or pneumatic surgical drill (§ 874.4250) for incising or removing bone in the ear, nose, or throat area. The bur consists of a carbide cutting tip on a metal shank or a coating of diamond on a metal shank. The device is used in mastoid surgery, frontal sinus surgery, and surgery of the facial nerves.

(b) *Classification.* Class I (general controls). The device is exempt from the premarket notification procedures in

subpart E of part 807 of this chapter subject to the limitations in § 874.9.

[51 FR 40389, Nov. 6, 1986, as amended at 61 FR 1122, Jan. 16, 1996; 66 FR 38800, July 25, 2001]

§ 874.4175 Nasopharyngeal catheter.

(a) *Identification.* A nasopharyngeal catheter is a device consisting of a bougie or filiform catheter that is intended for use in probing or dilating the eustachian tube. This generic type of device includes eustachian catheters.

(b) *Classification.* Class I (general controls). The device is exempt from the premarket notification procedures in subpart E of part 807 of this chapter subject to the limitations in § 874.9.

[51 FR 40389, Nov. 6, 1986, as amended at 61 FR 1122, Jan. 16, 1996; 66 FR 38801, July 25, 2001]

§ 874.4250 Ear, nose, and throat electric or pneumatic surgical drill.

(a) *Identification.* An ear, nose, and throat electric or pneumatic surgical drill is a rotating drilling device, including the handpiece, that is intended to drive various accessories, such as an ear, nose, and throat bur (§ 874.4140), for the controlled incision or removal of bone in the ear, nose, and throat area.

(b) *Classification.* Class II.

§ 874.4350 Ear, nose, and throat fiberoptic light source and carrier.

(a) *Identification.* An ear, nose, and throat fiberoptic light source and carrier is an AC-powered device that generates and transmits light through glass of plastic fibers and that is intended to provide illumination at the tip of an ear, nose, or throat endoscope. Endoscopic devices which utilize fiberoptic light sources and carriers include the bronchoscope, esophagoscope, laryngoscope, mediastinoscope, laryngeal-bronchial telescope, and nasopharyngoscope.

(b) *Classification.* Class I (general controls). The device is exempt from the premarket notification procedures in subpart E of part 807 of this chapter subject to the limitations in § 874.9.

[51 FR 40389, Nov. 6, 1986, as amended at 61 FR 1122, Jan. 16, 1996; 66 FR 38801, July 25, 2001]

§ 874.4420 Ear, nose, and throat manual surgical instrument.

(a) *Identification.* An ear, nose, and throat manual surgical instrument is one of a variety of devices intended for use in surgical procedures to examine or treat the bronchus, esophagus, trachea, larynx, pharynx, nasal and paranasal sinus, or ear. This generic type of device includes the esophageal dilator; tracheal bistour (a long, narrow surgical knife); tracheal dilator; tracheal hook; laryngeal injection set; laryngeal knife; laryngeal saw; laryngeal trocar; laryngectomy tube; adenoid curette; adenotome; metal tongue depressor; mouth gag; oral screw; salpingeal curette; tonsillectome; tonsil guillotine; tonsil screw; tonsil snare; tonsil suction tub; tonsil suturing hook; antom reforator; ethmoid curette; frontal sinus-rasp; nasal curette; nasal rasp; nasal rongeur; nasal saw; nasal scissors; nasal snare; sinus irrigator; sinus trephine; ear curette; ear excavator; ear rasp; ear scissor, ear snare; ear spoon; ear suction tub; malleous ripper; mastoid gauge; microsurgical ear chisel; myringotomy tube inserter; ossici holding clamp; sacculotomy tack inserter; vein press; wire ear loop; microrule; mirror; mobilizer; ear, nose, and throat punch; ear, nose and throat knife; and ear, nose, and throat trocar.

(b) *Classification* Class I (general controls). The device is exempt from the premarket notification procedures in subpart E of part 807 of this chapter subject to § 874.9.

[51 FR 40389, Nov. 9, 1986, as amended at 52 FR 32111, Aug. 25, 1987; 65 FR 2316, Jan. 14, 2000]

§ 874.4490 Argon laser for otology, rhinology, and laryngology.

(a) *Identification.* The argon laser device for use in otology, rhinology, and laryngology is an electro-optical device which produces coherent, electromagnetic radiation with principal wavelength peaks of 488 and 514 nanometers. In otology, the device is used for the purpose of coagulating and vaporizing soft and fibrous tissues, including osseous tissue. In rhinology and laryngology, the device is used to coagulate and vaporize soft and fibrous tissues, but not including osseous tissues.

(b) *Classification.* Class II.

[58 FR 29534, May 21, 1993]

§ 874.4500 Ear, nose, and throat microsurgical carbon dioxide laser.

(a) *Identification.* An ear, nose, and throat microsurgical carbon dioxide laser is a device intended for the surgical excision of tissue from the ear, nose, and throat area. The device is used, for example, in microsurgical procedures to excise lesions and tumors of the vocal cords and adjacent areas.

(b) *Classification.* Class II.

§ 874.4680 Bronchoscope (flexible or rigid) and accessories.

(a) *Identification.* A bronchoscope (flexible or rigid) and accessories is a tubular endoscopic device with any of a group of accessory devices which attach to the bronchoscope and is intended to examine or treat the larynx and tracheobronchial tree. It is typically used with a fiberoptic light source and carrier to provide illumination. The device is made of materials such as stainless steel or flexible plastic. This generic type of device includes the rigid ventilating bronchoscope, rigid nonventilating bronchoscope, nonrigid bronchoscope, laryngeal-bronchial telescope, flexible foreign body claw, bronchoscope tubing, flexible biopsy forceps, rigid biopsy curette, flexible biopsy brush, rigid biopsy forceps, flexible biopsy curette, and rigid bronchoscope aspirating tube, but excludes the fiberoptic light source and carrier.

(b) *Classification.* Class II.

§ 874.4710 Esophagoscope (flexible or rigid) and accessories.

(a) *Identification.* An esophagoscope (flexible or rigid) and accessories is a tubular endoscopic device with any of a group of accessory devices which attach to the esophagoscope and is intended to examine or treat esophageal malfunction symptoms, esophageal or mediastinal disease, or to remove foreign bodies from the esophagus. When inserted, the device extends from the area of the hypopharynx to the stomach. It is typically used with a fiberoptic light source and carrier to provide illumination. The device is made of materials such as stainless steel or flexible plastic. This generic type of device includes the flexible foreign body claw, flexible biopsy forceps, rigid biopsy curette, flexible biopsy brush, rigid biopsy forceps and flexible biopsy curette, but excludes the fiberoptic light source and carrier.

(b) *Classification.* Class II.

§ 874.4720 Mediastinoscope and accessories.

(a) *Identification.* A mediastinoscope and accessories is a tubular tapered electrical endoscopic device with any of a group of accessory devices which attach to the mediastinoscope and is intended to examine or treat tissue in the area separating the lungs. The device is inserted transthoracicly and is used in diagnosis of tumors and lesions and to determine whether excision of certain organs or tissues is indicated. It is typically used with a fiberoptic light source and carrier to provide illumination. The device is made of materials such as stainless steel. This generic type of device includes the flexible foreign body claw, flexible biopsy forceps, rigid biopsy curette, flexible biopsy brush, rigid biopsy forceps, and flexible biopsy curette, but excludes the fiberoptic light source and carrier.

(b) *Classification.* Class II.

§ 874.4750 Laryngostroboscope.

(a) *Identification.* A laryngostroboscope is a device that is intended to allow observation of glottic action during phonation. The device operates by focusing a stroboscopic light through a lens for direct or mirror reflected viewing of glottic action. The light and microphone that amplifies acoustic signals from the glottic area may or may not contact the patient.

(b) *Classification.* Class I (general controls). The device is exempt from the premarket notification procedures in subpart E of part 807 of this chapter subject to the limitations in § 874.9.

[55 FR 48440, Nov. 20, 1990, as amended at 59 FR 63009, Dec. 7, 1994; 66 FR 38801, July 25, 2001]

§ 874.4760 Nasopharyngoscope (flexible or rigid) and accessories.

(a) *Identification.* A nasopharyngoscope (flexible or rigid) and accessories is a tubular endoscopic device with any of a group of accessory devices which attach to the nasopharyngoscope and is intended to examine or treat the nasal cavity and nasal pharynx. It is typically used with a fiberoptic light source and carrier to provide illumination. The device is made of materials such as stainless steel and flexible plastic. This generic type of device includes the antroscope, nasopharyngolaryngoscope, nasosinuscope, nasoscope, postrhinoscope, rhinoscope, salpingoscope, flexible foreign body claw, flexible biopsy forceps, rigid biopsy curette, flexible biospy brush, rigid biopsy forceps and flexible biopsy curette, but excludes the fiberoptic light source and carrier.

(b) *Classification.* Class II.

§ 874.4770 Otoscope.

(a) *Identification.* An otoscope is a device intended to allow inspection of the external ear canal and tympanic membrane under magnification. The device provides illumination of the ear canal for observation by using an AC- or battery-powered light source and an optical magnifying system.

(b) *Classification.* Class I (general controls). The device is exempt from the premarket notification procedures in subpart E of part 807 of this chapter subject to the limitations in § 874.9 only when used in the external ear canal.

[55 FR 48440, Nov. 20, 1990, as amended at 61 FR 1122, Jan. 16, 1996; 66 FR 38801, July 25, 2001]

§ 874.4780 Intranasal splint.

(a) *Identification.* An intranasal splint is intended to minimize bleeding and edema and to prevent adhesions between the septum and the nasal cavity. It is placed in the nasal cavity after surgery or trauma. The intranasal splint is constructed from plastic, silicone, or absorbent material.

(b) *Classification.* Class I (general controls). The device is exempt from the premarket notification procedures in subpart E of part 807 of this chapter subject to the limitations in § 874.9.

[64 FR 10949, Mar. 8, 1999]

§ 874.4800 Bone particle collector.

(a) *Identification.* A bone particle collector is a filtering device intended to be inserted into a suction tube during the early stages of otologic surgery to collect bone particles for future use.

(b) *Classification.* Class I (general controls). The device is exempt from premarket notification procedures in subpart E of part 807 of this chapter subject to the limitations in § 874.9.

[64 FR 10949, Mar. 8, 1999]

Subpart F—Therapeutic Devices

§ 874.5220 Ear, nose, and throat drug administration device.

(a) *Identification.* An ear, nose, and throat drug administration device is one of a group of ear, nose, and throat devices intended specifically to administer medicinal substances to treat ear, nose, and throat disorders. These instruments include the powder blower, dropper, ear wick, manual nebulizer pump, and nasal inhaler.

(b) *Classification.* Class I (general controls). The device is exempt from the premarket notification procedures in subpart E of part 807 of this chapter subject to the limitations in § 874.9. If the device is not labeled or otherwise represented as sterile, it is exempt from the current good manufacturing practice regulations in part 820 of this chapter, with the exception of § 820.180, with respect to general requirements concerning records, and § 820.198, with respect to complaint files.

[51 FR 40389, Nov. 6, 1986, as amended at 59 FR 63009, Dec. 7, 1994; 66 FR 38801, July 25, 2001]

§ 874.5300 Ear, nose, and throat examination and treatment unit.

(a) *Identification.* An ear, nose, and throat examination and treatment unit is an AC-powered device intended to support a patient during an otologic examination while providing specialized features for examination and treatment. The unit consists of a patient chair and table, drawers for equipment, suction and blowing apparatus, and receptacles for connection of specialized lights and examining instruments.

(b) *Classification.* Class I (general controls). The device is exempt from the premarket notification procedures in subpart E of part 807 of this chapter subject to § 874.9.

[55 FR 48440, Nov. 20, 1990, as amended at 65 FR 2316, Jan. 14, 2000]

§ 874.5350 Suction antichoke device.

(a) *Identification.* A suction antichoke device is a device intended to be used in an emergency situation to remove, by the application of suction, foreign objects that obstruct a patient's airway to prevent asphyxiation to the patient.

(b) *Classification.* Class III.

(c) *Date PMA or notice of completion of PDP is required.* A PMA or a notice of completion of a PDP for a device is required to be filed with the Food and Drug Administration on or before July 13, 1999 for any suction antichoke device that was in commercial distribution before May 28, 1976, or that has, on or before July 13, 1999, been found to be substantially equivalent to a suction antichoke device that was in commercial distribution before May 28, 1976. Any other suction antichoke device shall have an approved PMA or declared completed PDP in effect before being placed in commercial distribution.

[51 FR 40389, Nov. 6, 1986, as amended at 64 FR 18329, Apr. 14, 1999; 65 FR 2316, Jan. 14, 2000]

§ 874.5370 Tongs antichoke device.

(a) *Identification.* A tongs antichoke device is a device that is intended to be used in an emergency situation to grasp and remove foreign objects that obstruct a patient's airway to prevent asphyxiation of the patient. This generic type of device includes a plastic instrument with serrated ends that is inserted into the airway in a blind manner to grasp and extract foreign objects, and a stainless steel forceps with spoon ends that is inserted under tactile guidance to grasp and extract foreign objects from the airway.

(b) *Classification.* Class III.

(c) *Date PMA or notice of completion of PDP is required.* A PMA or a notice of completion of a PDP for a device is required to be filed with the Food and Drug Administration on or before July 13, 1999 for any tongs antichoke device that was in commercial distribution before May 28, 1976, or that has, on or before July 13, 1999, been found to be substantially equivalent to a tongs antichoke device that was in commercial distribution before May 28, 1976. Any other tongs antichoke device shall have an approved PMA or declared completed PDP in effect before being placed in commercial distribution.

[51 FR 40389, Nov. 6, 1986, as amended at 64 FR 18329, Apr. 14, 1999]

§ 874.5550 Powered nasal irrigator.

(a) *Identification.* A powered nasal irrigator is an AC-powered device intended to wash the nasal cavity by means of a pressure-controlled pulsating stream of water. The device consists of a control unit and pump connected to a spray tube and nozzle.

(b) *Classification.* Class I (general controls). The device is exempt from the premarket notification procedures in subpart E of part 807 of this chapter subject to § 874.9.

[55 FR 48440, Nov. 20, 1990, as amended at 65 FR 2316, Jan. 14, 2000]

§ 874.5800 External nasal splint.

(a) *Identification.* An external nasal splint is a rigid or partially rigid device intended for use externally for immobilization of parts of the nose.

(b) *Classification.* Class I (general controls). The device is exempt from the premarket notification procedures in subpart E of part 807 of this chapter subject to the limitations in § 874.9.

[51 FR 40389, Nov. 9, 1986, as amended at 52 FR 32111, Aug. 25, 1987; 59 FR 63009, Dec. 7, 1994; 66 FR 38801, July 25, 2001]

§ 874.5840 Antistammering device.

(a) *Identification.* An antistammering device is a device that electronically generates a noise when activated or when it senses the user's speech and that is intended to prevent the user from hearing the sounds of his or her own voice. The device is used to minimize a user's involuntary hesitative or repetitive speech.

(b) *Classification.* Class I (general controls). The device is exempt from the premarket notification procedures in

subpart E of part 807 of this chapter subject to § 874.9.

[51 FR 40389, Nov. 6, 1986, as amended at 65 FR 2316, Jan. 14, 2000]

PART 876—GASTROENTEROLOGY-UROLOGY DEVICES

Subpart A—General Provisions

Subpart B—Diagnostic Devices

Subpart C—Monitoring Devices

Subpart D—Prosthetic Devices

Subpart E—Surgical Devices

Subpart F—Therapeutic Devices

AUTHORITY: 21 U.S.C. 351, 360, 360c, 360e, 360j, 3601, 371.

SOURCE: 48 FR 53023, Nov. 23, 1983, unless otherwise noted.

Subpart A—General Provisions

§ 876.1 Scope.

(a) This part sets forth the classification of gastroenterology-urology devices intended for human use that are in commercial distribution.

(b) The identification of a device in a regulation in this part is not a precise description of every device that is, or will be, subject to the regulation. A manufacturer who submits a premarket notification submission for a device under part 807 may not show merely that the device is accurately described by the section title and identification provisions of a regulation in this part, but shall state why the device is substantially equivalent to other devices, as required by § 807.87.

(c) To avoid duplicative listings, a gastroenterology-urology device that has two or more types of uses (e.g., used both as a diagnostic device and as a therapeutic device) is listed only in one subpart.

(d) References in this part to regulatory sections of the Code of Federal Regulations are to chapter I of title 21, unless otherwise noted.

(e) Guidance documents referenced in this part are available on the Internet at *http://www.fda.gov/cdrh/guidance.html.*

[52 FR 17737, May 11, 1987; 52 FR 22577, June 12, 1987, as amended at 69 FR 77623, Dec. 28, 2004]

§ 876.3 Effective dates of requirement for premarket approval.

A device included in this part that is classified into class III (premarket approval) shall not be commercially distributed after the date shown in the regulation classifying the device unless the manufacturer has an approval under section 515 of the act (unless an exemption has been granted under section 520(g)(2) of the act). An approval under section 515 of the act consists of FDA's issuance of an order approving an application for premarket approval (PMA) for the device or declaring completed a product development protocol (PDP) for the device.

(a) Before FDA requires that a device commercially distributed before the enactment date of the amendments, or a device that has been found substantially equivalent to such a device, has an approval under section 515 of the act FDA must promulgate a regulation under section 515(b) of the act requiring such approval, except as provided in paragraph (b) of this section. Such a regulation under section 515(b) of the act shall not be effective during the grace period ending on the 90th day after its promulgation or on the last day of the 30th full calendar month after the regulation that classifies the device into class III is effective, whichever is later. See section 501(f)2)(B) of the act. Accordingly, unless an effective date of the requirement for premarket approval is shown in the regulation for a device classified into class III in this part, the device may be commerically distributed without FDA's issuance of an order approving a PMA or declaring completed a PDP for the device. If FDA promulgates a regulation under section 515(b) of the act requiring premarket approval for a device, section 501(f)(1)(A) of the act applies to the device.

(b) Any new, not substantially equivalent, device introduced into commercial distribution on or after May 28, 1976, including a device formerly marketed that has been substantially altered, is classified by statute (section 513(f) of the act) into class III without any grace period and FDA must have issued an order approving a PMA or declaring completed a PDP for the device before the device is commercially distributed unless it is reclassified. If FDA knows that a device being commercially distributed may be a ''new'' device as defined in this section because of any new intended use or other reasons, FDA may codify the statutory classification of the device into class III for such new use. Accordingly, the regulation for such a class III device states that as of the enactment date of the amendments, May 28, 1976, the device must have an approval under section 515 of the act before commercial distribution.

[52 FR 17737, May 11, 1987]

§ 876.9 Limitations of exemptions from section 510(k) of the Federal Food, Drug, and Cosmetic Act (the act).

The exemption from the requirement of premarket notification (section 510(k) of the act) for a generic type of class I or II device is only to the extent that the device has existing or reasonably foreseeable characteristics of commercially distributed devices within that generic type or, in the case of in vitro diagnostic devices, only to the extent that misdiagnosis as a result of using the device would not be associated with high morbidity or mortality. Accordingly, manufacturers of any commercially distributed class I or II device for which FDA has granted an exemption from the requirement of premarket notification must still submit a premarket notification to FDA before introducing or delivering for introduction into interstate commerce for commercial distribution the device when:

(a) The device is intended for a use different from the intended use of a legally marketed device in that generic type of device; e.g., the device is intended for a different medical purpose, or the device is intended for lay use where the former intended use was by health care professionals only;

(b) The modified device operates using a different fundamental scientific technology than a legally marketed device in that generic type of device; e.g., a surgical instrument cuts tissue with a laser beam rather than with a sharpened metal blade, or an in vitro diagnostic device detects or identifies infectious agents by using deoxyribonucleic acid (DNA) probe or nucleic acid hybridization technology rather than culture or immunoassay technology; or

(c) The device is an in vitro device that is intended:

(1) For use in the diagnosis, monitoring, or screening of neoplastic diseases with the exception of immunohistochemical devices;

(2) For use in screening or diagnosis of familial or acquired genetic disorders, including inborn errors of metabolism;

(3) For measuring an analyte that serves as a surrogate marker for screening, diagnosis, or monitoring life-threatening diseases such as acquired immune deficiency syndrome (AIDS), chronic or active hepatitis, tuberculosis, or myocardial infarction or to monitor therapy;

(4) For assessing the risk of cardiovascular diseases;

(5) For use in diabetes management;

(6) For identifying or inferring the identity of a microorganism directly from clinical material;

(7) For detection of antibodies to microorganisms other than immunoglobulin G (IgG) or IgG assays when the results are not qualitative, or are used to determine immunity, or the assay is intended for use in matrices other than serum or plasma;

(8) For noninvasive testing as defined in § 812.3(k) of this chapter; and

(9) For near patient testing (point of care).

[65 FR 2316, Jan. 14, 2000]

Subpart B—Diagnostic Devices

§ 876.1075 Gastroenterology-urology biopsy instrument.

(a) *Identification.* A gastroenterology-urology biopsy instrument is a device used to remove, by cutting or aspiration, a specimen of tissue for microscopic examination. This generic type of device includes the biopsy punch, gastrointestinal mechanical biopsy instrument, suction biopsy instrument, gastro-urology biopsy needle and needle set, and nonelectric biopsy forceps. This section does not apply to biopsy instruments that have specialized uses in other medical specialty areas and that are covered by classification regulations in other parts of the device classification regulations.

(b) *Classification.* (1) Class II (performance standards).

(2) Class I for the biopsy forceps cover and the non-electric biopsy forceps. The devices subject to this paragraph (b)(2) are exempt from the premarket notification procedures in subpart E of part 807 of this chapter subject to the limitations in § 876.9.

[48 FR 53023, Nov. 23, 1983, as amended at 61 FR 1122, Jan. 16, 1996; 66 FR 38801, July 25, 2001]

§ 876.1300 Ingestible telemetric gastrointestinal capsule imaging system.

(a) *Identification.* An ingestible telemetric gastrointestinal capsule imaging system is used for visualization of the small bowel mucosa as an adjunctive tool in the detection of abnormalities of the small bowel. The device captures images of the small bowel with a wireless camera contained in a capsule. This device includes an ingestible capsule (containing a light source, camera, transmitter, and battery), an antenna array, a receiving/recording unit, a data storage device, computer software to process the images, and accessories.

(b) *Classification.* Class II (special controls). The special control is FDA's guidance, "Class II Special Controls Guidance Document: Ingestible Telemetric Gastrointestinal Capsule Imaging Systems; Final Guidance for Industry and FDA."

[67 FR 3433, Jan. 24, 2002]

§ 876.1400 Stomach pH electrode.

(a) *Identification.* A stomach pH electrode is a device used to measure intragastric and intraesophageal pH (hydrogen ion concentration). The pH electrode is at the end of a flexible lead which may be inserted into the esophagus or stomach through the patient's mouth. The device may include an integral gastrointestinal tube.

(b) *Classification.* Class I. The device is exempt from the premarket notification procedures in subpart E of part 807 of this chapter.

[48 FR 53023, Nov. 23, 1983, as amended at 61 FR 1122, Jan. 16, 1996]

§ 876.1500 Endoscope and accessories.

(a) *Identification.* An endoscope and accessories is a device used to provide access, illumination, and allow observation or manipulation of body cavities, hollow organs, and canals. The device consists of various rigid or flexible instruments that are inserted into body spaces and may include an optical system for conveying an image to the user's eye and their accessories may assist in gaining access or increase the versatility and augment the capabilities of the devices. Examples of devices that are within this generic type of device include cleaning accessories for endoscopes, photographic accessories for endoscopes, nonpowered anoscopes, binolcular attachments for endoscopes, pocket battery boxes, flexible or rigid choledochoscopes, colonoscopes, diagnostic cystoscopes, cystourethroscopes, enteroscopes, esophagogastroduodenoscopes, rigid esophagoscopes, fiberoptic illuminators for endoscopes, incandescent endoscope lamps, biliary pancreatoscopes, proctoscopes, resectoscopes, nephroscopes, sigmoidoscopes, ureteroscopes, urethroscopes, endomagnetic retrievers, cytology brushes for endoscopes, and lubricating jelly for transurethral surgical instruments. This section does not apply to endoscopes that have specialized uses in other medical specialty areas and that are covered by classification regulations in other parts of the device classification regulations.

(b) *Classification.* (1) Class II (performance standards).

(2) Class I for the photographic accessories for endoscope, miscellaneous bulb adapter for endoscope, binocular attachment for endoscope, eyepiece attachment for prescription lens, teaching attachment, inflation bulb, measuring device for panendoscope, photographic equipment for physiologic function monitor, special lens instrument for endoscope, smoke removal tube, rechargeable battery box, pocket battery box, bite block for endoscope, and cleaning brush for endoscope. The devices subject to this paragraph (b)(2) are exempt from the premarket notification procedures in subpart E of part 807of this chapter, subject to the limitations in § 876.9.

[48 FR 53023, Nov. 23, 1983, as amended at 61 FR 1122, Jan. 16, 1996; 66 FR 38801, July 25, 2001]

§ 876.1620 Urodynamics measurement system.

(a) *Identification.* A urodynamics measurement system is a device used to measure volume and pressure in the urinary bladder when it is filled through a catheter with carbon dioxide or water. The device controls the supply of carbon dioxide or water and may also record the electrical activity of the muscles associated with urination. The device system may include transducers, electronic signal conditioning and display equipment, a catheter withdrawal device to enable a urethral pressure profile to be obtained, and special catheters for urethral profilometry and electrodes for electromyography. This generic type of device includes the cystometric gas (carbon dioxide) device, the cystometric hydrualic device, and the electrical recording cystometer, but excludes any device that uses air to fill the bladder.

(b) *Classification.* Class II (special controls). The device is exempt from the premarket notification procedures in subpart E of part 807 of this chapter subject to § 876.9.

[48 FR 53023, Nov. 23, 1983, as amended at 63 FR 59228, Nov. 3, 1998]

§ 876.1725 Gastrointestinal motility monitoring system.

(a) *Identification.* A gastrointestinal motility monitoring system is a device used to measure peristalic activity or pressure in the stomach or esophagus by means of a probe with transducers that is introduced through the mouth into the gastrointestinal tract. The device may include signal conditioning, amplifying, and recording equipment. This generic type of device includes the esophageal motility monitor and tube, the gastrointestinal motility (electrical) system, and certain accessories, such as a pressure transducer, amplifier, and external recorder.

(b) *Classification.* Class II (performance standards).

§ 876.1735 Electrogastrography system.

(a) *Identification.* An electrogastrography system (EGG) is a device used to measure gastric myoelectrical activity as an aid in the diagnosis of gastric motility disorders. The device system includes the external recorder, amplifier, skin electrodes, strip chart, cables, analytical software, and other accessories.

(b) *Classification.* Class II (Special Controls). The special controls are as follows:

(1) The sale, distribution and use of this device are restricted to prescription use in accordance with § 801.109 of this chapter.

(2) The labeling must include specific instructions:

(i) To describe proper patient set-up prior to the start of the test, including the proper placement of electrodes;

(ii) To describe how background data should be gathered and used to eliminate artifact in the data signal;

(iii) To describe the test protocol (including the measurement of baseline data) that may be followed to obtain the EGG signal; and

(iv) To explain how data results may be interpreted.

(3) The device design should ensure that the EGG signal is distinguishable from background noise that may interfere with the true gastric myoelectric signal.

(4) Data should be collected to demonstrate that the device has adequate precision and the EGG signal is reproducible and is interpretable.

[64 FR 51444, Sept. 23, 1999]

§ 876.1800 Urine flow or volume measuring system.

(a) *Identification.* A urine flow or volume measuring system is a device that measures directly or indirectly the volume or flow of urine from a patient, either during the course of normal urination or while the patient is catheterized. The device may include a drip chamber to reduce the risk of retrograde bacterial contamination of the bladder and a transducer and electrical signal conditioning and display equipment. This generic type of device includes the electrical urinometer, mechanical urinometer, nonelectric urinometer, disposable nonelectric urine flow rate measuring device, and uroflowmeter.

(b) *Classification.* (1) Class II (special controls). The device is exempt from the premarket notification procedures in subpart E of part 807 of this chapter subject to § 876.9.

[48 FR 53023, Nov. 23, 1983, as amended at 61 FR 1122, Jan. 16, 1996; 63 FR 59228, Nov. 3, 1998]

Subpart C—Monitoring Devices

§ 876.2040 Enuresis alarm.

(a) *Identification.* An enuresis alarm is a device intended for use in treatment of bedwetting. Through an electrical trigger mechanism, the device sounds an alarm when a small quantity of urine is detected on a sensing pad. This generic type of device includes conditioned response enuresis alarms.

(b) *Classification.* Class II (special controls). The device is exempt from the premarket notification procedures

in subpart E of part 807 of this chapter subject to § 876.9.

[48 FR 53023, Nov. 23, 1983, as amended at 63 FR 59228, Nov. 3, 1998]

Subpart D—Prosthetic Devices

§ 876.3350 Penile inflatable implant.

(a) *Identification.* A penile inflatable implant is a device that consists of two inflatable cylinders implanted in the penis, connected to a reservoir filled with radiopaque fluid implanted in the abdomen, and a subcutaneous manual pump implanted in the scrotum. When the cylinders are inflated, they provide rigidity to the penis. This device is used in the treatment of erectile impotence.

(b) *Classification.* Class III (premarket approval).

(c) *Date premarket approval application (PMA) or notice of completion of a product development protocol (PDP) is required.* A PMA or a notice of completion of a PDP is required to be filed with the Food and Drug Administration on or before July 11, 2000, for any penile inflatable implant that was in commercial distribution before May 28, 1976, or that has, on or before July 11, 2000, been found to be substantially equivalent to a penile inflatable implant that was in commercial distribution before May 28, 1976. Any other penile inflatable implant shall have an approved PMA or a declared completed PDP in effect before being placed in commercial distribution.

[48 FR 53023, Nov. 23, 1983, as amended at 52 FR 17738, May 11, 1987; 65 FR 19658, Apr. 12, 2000]

§ 876.3630 Penile rigidity implant.

(a) *Identification.* A penile rigidity implant is a device that consists of a pair of semi-rigid rods implanted in the corpora cavernosa of the penis to provide rigidity. It is intended to be used in men diagnosed as having erectile dysfunction.

(b) *Classification.* Class II. The special control for this device is the FDA guidance entitled "Guidance for the Content of Premarket Notifications for Penile Rigidity Implants."

[65 FR 4882, Feb. 2, 2000]

§ 876.3750 Testicular prosthesis.

(a) *Identification.* A testicular prosthesis is an implanted device that consists of a solid or gel-filled silicone rubber prosthesis that is implanted surgically to resemble a testicle.

(b) *Classification.* Class III (premarket approval).

(c) *Date premarket approval application (PMA) or notice of product development protocol (PDP) is required.* A PMA or notice of completion of a PDP is required to be filed with the Food and Drug Administration on or before July 5, 1995, for any testicular prosthesis that was in commercial distribution before May 28, 1976, or that has on or before July 5, 1995, been found to be substantially equivalent to a testicular prosthesis that was in commercial distribution before May 28, 1976. Any other testicular prosthesis shall have an approved PMA or a declared completed PDP in effect before being placed in commercial distribution.

[48 FR 53023, Nov. 23, 1983, as amended at 52 FR 17738, May 11, 1987; 60 FR 17216, Apr. 5, 1995]

Subpart E—Surgical Devices

§ 876.4020 Fiberoptic light ureteral catheter.

(a) *Identification.* A fiberoptic light ureteral catheter is a device that consists of a fiberoptic bundle that emits light throughout its length and is shaped so that it can be inserted into the ureter to enable the path of the ureter to be seen during lower abdominal or pelvic surgery.

(b) *Classification.* Class II (performance standards).

§ 876.4270 Colostomy rod.

(a) *Identification.* A colostomy rod is a device used during the loop colostomy procedure. A loop of colon is surgically brought out through the abdominal wall and the stiff colostomy rod is placed through the loop temporarily to keep the colon from slipping back through the surgical opening.

(b) *Classification.* Class II (performance standards).

§ 876.4300 Endoscopic electrosurgical unit and accessories.

(a) *Identification.* An endoscopic electrosurgical unit and accessories is a device used to perform electrosurgical procedures through an endoscope. This generic type of device includes the electrosurgical generator, patient plate, electric biopsy forceps, electrode, flexible snare, electrosurgical alarm system, electrosurgical power supply unit, electrical clamp, self-opening rigid snare, flexible suction coagulator electrode, patient return wristlet, contact jelly, adaptor to the cord for transurethral surgical instruments, the electric cord for transurethral surgical instruments, and the transurethral desiccator.

(b) *Classification.* Class II (performance standards).

§ 876.4370 Gastroenterology-urology evacuator.

(a) *Identification.* A gastroenterology-urology evacuator is a device used to remove debris and fluids during gastroenterological and urological procedures by drainage, aspiration, or irrigation. This generic type of device includes the fluid evacuator system, manually powered bladder evacuator, and the AC-powered vacuum pump.

(b) *Classification.* (1) Class II (special controls) for the gastroenterology-urology evacuator when other than manually powered. The device is exempt from the premarket notification procedures in subpart E of part 807 of this chapter subject to § 876.9.

(2) Class I for the gastroenterology-urology evacuator when manually powered. The device subject to this paragraph (b)(2) is exempt from the premarket notification procedures in subpart E of part 807 of this chapter.

[48 FR 53023, Nov. 23, 1983, as amended at 54 FR 25049, June 12, 1989; 63 FR 59228, Nov. 3, 1998]

§ 876.4400 Hemorrhoidal ligator.

(a) *Identification.* A hemorrhoidal ligator is a device used to cut off the blood flow to hemorrhoidal tissue by means of a ligature or band placed around the hemorrhoid.

(b) *Classification.* Class II (performance standards).

§ 876.4480 Electrohydraulic lithotriptor.

(a) *Identification.* An electrohydraulic lithotriptor is an AC-powered device used to fragment urinary bladder stones. It consists of a high voltage source connected by a cable to a bipolar electrode that is introduced into the urinary bladder through a cystoscope. The electrode is held against the stone in a water-filled bladder and repeated electrical discharges between the two poles of the electrode cause electrohydraulic shock waves which disintegrate the stone.

(b) *Classification.* Class II. The special control for this device is FDA's "Guidance for the Content of Premarket Notifications for Intracorporeal Lithotripters."

[48 FR 53023, Nov. 23, 1983, as amended at 52 FR 17738, May 11, 1987; 65 FR 17145, Mar. 31, 2000]

§ 876.4500 Mechanical lithotriptor.

(a) *Identification.* A mechanical lithotriptor is a device with steel jaws that is inserted into the urinary bladder through the urethra to grasp and crush bladder stones.

(b) *Classification.* Class II (performance standards).

§ 876.4530 Gastroenterology-urology fiberoptic retractor.

(a) *Identification.* A gastroenterology-urology fiberoptic retractor is a device that consists of a mechanical retractor with a fiberoptic light system that is used to illuminate deep surgical sites.

(b) *Classification.* Class I (general controls). The device is exempt from the premarket notification procedures in subpart E of part 807 of this chapter subject to the limitations in § 876.9.

[48 FR 53023, Nov. 23, 1983, as amended at 54 FR 25049, June 12, 1989; 66 FR 38801, July 25, 2001]

§ 876.4560 Ribdam.

(a) *Identification.* A ribdam is a device that consists of a broad strip of latex with supporting ribs used to drain surgical wounds where copious urine drainage is expected.

(b) *Classification.* Class I (general controls). The device is exempt from the premarket notification procedures in

subpart E of part 807 of this chapter subject to the limitations in § 876.9.

[48 FR 53023, Nov. 23, 1983, as amended at 54 FR 25049, June 12, 1989; 66 FR 38801, July 25, 2001]

§ 876.4590 Interlocking urethral sound.

(a) *Identification.* An interlocking urethral sound is a device that consists of two metal sounds (elongated instruments for exploring or sounding body cavities) with interlocking ends, such as with male and female threads or a rounded point and mating socket, used in the repair of a ruptured urethra. The device may include a protective cap to fit over the metal threads.

(b) *Classification.* Class I (general controls). The device is exempt from the premarket notification procedures in subpart E of part 807 of this chapter subject to the limitations in § 876.9.

[48 FR 53023, Nov. 23, 1983, as amended at 61 FR 1122, Jan. 16, 1996; 66 FR 38801, July 25, 2001]

§ 876.4620 Ureteral stent.

(a) *Identification.* A ureteral stent is a tube-like implanted device that is inserted into the ureter to provide ureteral rigidity and allow the passage of urine. The device may have finger-like protrusions or hooked ends to keep the tube in place. It is used in the treatment of ureteral injuries and ureteral obstruction.

(b) *Classification.* Class II (performance standards).

§ 876.4650 Water jet renal stone dislodger system.

(a) *Identification.* A water jet renal stone dislodger system is a device used to dislodge stones from renal calyces (recesses of the pelvis of the kidney) by means of a pressurized stream of water through a conduit. The device is used in the surgical removal of kidney stones.

(b) *Classification.* Class II (special controls). The device is exempt from the premarket notification procedures in subpart E of part 807 of this chapter subject to § 876.9.

[48 FR 53023, Nov. 23, 1983, as amended at 63 FR 59228, Nov. 3, 1998]

§ 876.4680 Ureteral stone dislodger.

(a) *Identification.* A ureteral stone dislodger is a device that consists of a bougie or a catheter with an expandable wire basket near the tip, a special flexible tip, or other special construction. It is inserted through a cystoscope and used to entrap and remove stones from the ureter. This generic type of device includes the metal basket and the flexible ureteral stone dislodger.

(b) *Classification.* Class II (special controls). The device is exempt from the premarket notification procedures in subpart E of part 807 of this chapter subject to § 876.9.

[48 FR 53023, Nov. 23, 1983, as amended at 63 FR 59228, Nov. 3, 1998]

§ 876.4730 Manual gastroenterology-urology surgical instrument and accessories.

(a) *Identification.* A manual gastroenterology-urology surgical instrument and accessories is a device designed to be used for gastroenterological and urological surgical procedures. The device may be nonpowered, hand-held, or hand-manipulated. Manual gastroenterology-urology surgical instruments include the biopsy forceps cover, biopsy tray without biopsy instruments, line clamp, nonpowered rectal probe, nonelectrical clamp, colostomy spur-crushers, locking device for intestinal clamp, needle holder, gastro-urology hook, gastro-urology probe and director, nonself-retaining retractor, laparotomy rings, nonelectrical snare, rectal specula, bladder neck spreader, self-retaining retractor, and scoop.

(b) *Classification.* Class I (general controls). The device is exempt from the premarket notification procedures in subpart E of part 807 of this chapter subject to the limitations in § 876.9.

[48 FR 53023, Nov. 23, 1983, as amended at 54 FR 25049, June 12, 1989; 66 FR 38801, July 25, 2001]

§ 876.4770 Urethrotome.

(a) *Identification.* A urethrotome is a device that is inserted into the urethra and used to cut urethral strictures and enlarge the urethra. It is a metal instrument equipped with a dorsal-fin cutting blade which can be elevated from its sheath. Some urethrotomes incorporate an optical channel for visual control.

(b) *Classification.* Class II (performance standards).

§ 876.4890 Urological table and accessories.

(a) *Identification.* A urological table and accessories is a device that consists of a table, stirrups, and belts used to support a patient in a suitable position for endoscopic procedures of the lower urinary tract. The table can be adjusted into position manually or electrically.

(b) *Classification.* (1) Class II (special controls) for the electrically powered urological table and accessories. The device is exempt from the premarket notification procedures in subpart E of part 807 of this chapter subject to § 876.9.

(2) Class I for the manually powered table and accessories, and for stirrups for electrically powered table. The device subject to this paragraph (b)(2) is exempt from the premarket notification procedures in subpart E of part 807 of this chapter subject to the limitations in § 876.9.

[48 FR 53023, Nov. 23, 1983, as amended at 61 FR 1122, Jan. 16, 1996; 63 FR 59228, Nov. 3, 1998; 66 FR 38801, July 25, 2001]

Subpart F—Therapeutic Devices

§ 876.5010 Biliary catheter and accessories.

(a) *Identification.* A biliary catheter and accessories is a tubular flexible device used for temporary or prolonged drainage of the biliary tract, for splinting of the bile duct during healing, or for preventing stricture of the bile duct. This generic type of device may include a bile collecting bag that is attached to the biliary catheter by a connector and fastened to the patient with a strap.

(b) *Classification.* Class II (performance standards).

§ 876.5020 External penile rigidity devices.

(a) *Identification.* External penile rigidity devices are devices intended to create or maintain sufficient penile rigidity for sexual intercourse. External penile rigidity devices include vacuum pumps, constriction rings, and penile splints which are mechanical, powered, or pneumatic devices.

(b) *Classification.* Class II (special controls). The devices are exempt from the premarket notification procedures in subpart E of part 807 of this chapter subject to the limitations in § 876.9. The special control for these devices is the FDA guidance document entitled "Class II Special Controls Guidance Document: External Penile Rigidity Devices." See § 876.1(e) for the availability of this guidance document.

[69 FR 77623, Dec. 28, 2004]

§ 876.5030 Continent ileostomy catheter.

(a) *Identification.* A continent ileostomy catheter is a flexible tubular device used as a form during surgery for continent ileostomy and it provides drainage after surgery. Additionally, the device may be inserted periodically by the patient for routine care to empty the ileal pouch. This generic type of device includes the rectal catheter for continent ileostomy.

(b) *Classification.* Class I (general controls). The device is exempt from the premarket notification procedures in subpart E of part 807 of this chapter subject to the limitations in § 876.9.

[48 FR 53023, Nov. 23, 1983, as amended at 54 FR 25050, June 12, 1989; 66 FR 38801, July 25, 2001]

§ 876.5090 Suprapubic urological catheter and accessories.

(a) *Identification.* A suprapubic urological catheter and accessories is a flexible tubular device that is inserted through the abdominal wall into the urinary bladder with the aid of a trocar and cannula. The device is used to pass fluids to and from the urinary tract. This generic type of device includes the suprapubic catheter and tube, Malecot catheter, catheter punch instrument, suprapubic drainage tube, and the suprapubic cannula and trocar.

(b) *Classification.* (1) Class II (performance standards).

(2) Class I for the catheter punch instrument, nondisposable cannula and trocar, and gastro-urological trocar. The devices subject to this paragraph

(b)(2) are exempt from the premarket notification procedures in subpart E of part 807 of this chapter subject to the limitations in § 876.9.

[48 FR 53023, Nov. 23, 1983, as amended at 61 FR 1122, Jan. 16, 1996; 66 FR 38801, July 25, 2001]

§ 876.5130 Urological catheter and accessories.

(a) *Identification.* A urological catheter and accessories is a flexible tubular device that is inserted through the urethra and used to pass fluids to or from the urinary tract. This generic type of device includes radiopaque urological catheters, ureteral catheters, urethral catheters, coudé catheters, balloon retention type catheters, straight catheters, upper urinary tract catheters, double lumen female urethrographic catheters, disposable ureteral catheters, male urethrographic catheters, and urological catheter accessories including ureteral catheter stylets, ureteral catheter adapters, ureteral catheter holders, ureteral catheter stylets, ureteral catheterization trays, and the gastro-urological irrigation tray (for urological use).

(b) *Classification.* (1) Class II (performance standards).

(2) Class I for the ureteral stylet (guidewire), stylet for gastrourological catheter, ureteral catheter adapter, ureteral catheter connector, and ureteral catheter holder. The devices subject to this paragraph (b)(2) are exempt from the premarket notification procedures in subpart E of part 807 of this chapter subject to the limitations in § 876.9.

[48 FR 53023, Nov. 23, 1983, as amended at 61 FR 1122, Jan. 16, 1996; 66 FR 38801, July 25, 2001]

§ 876.5160 Urological clamp for males.

(a) *Identification.* A urological clamp for males is a device used to close the urethra of a male to control urinary incontinence or to hold anesthetic or radiography contrast media in the urethra temporarily. It is an external clamp.

(b) *Classification.* Class I (general controls). Except when intended for internal use or use on females, the device is exempt from the premarket notification procedures in subpart E of part 807 of this chapter subject to § 876.9.

[48 FR 53023, Nov. 23, 1963, as amended at 65 FR 2317, Jan. 14, 2000]

§ 876.5210 Enema kit.

(a) *Identification.* An enema kit is a device intended to instill water or other fluids into the colon through a nozzle inserted into the rectum to promote evacuation of the contents of the lower colon. The device consists of a container for fluid connected to the nozzle either directly or via tubing. This device does not include the colonic irrigation system (§ 876.5220).

(b) *Classification.* Class I (general controls). The device is exempt from the premarket notification procedures in subpart E of part 807 of this chapter subject to § 876.9. The device is exempt from the current good manufacturing practice regulations in part 820 of this chapter, with the exception of § 820.180 of this chapter, with respect to general requirements concerning records, and § 820.198 of this chapter, with respect to complaint files.

[48 FR 53023, Nov. 23, 1963, as amended at 65 FR 2317, Jan. 14, 2000]

§ 876.5220 Colonic irrigation system.

(a) *Identification.* A colonic irrigation system is a device intended to instill water into the colon through a nozzle inserted into the rectum to cleanse (evacuate) the contents of the lower colon. The system is designed to allow evacuation of the contents of the colon during the administration of the colonic irrigation. The device consists of a container for fluid connected to the nozzle via tubing and includes a system which enables the pressure, temperature, or flow of water through the nozzle to be controlled. The device may include a console-type toilet and necessary fittings to allow the device to be connected to water and sewer pipes. The device may use electrical power to heat the water. The device does not include the enema kit (§ 876.5210).

(b) *Classification.* (1) Class II (performance standards) when the device is intended for colon cleansing when medically indicated, such as before radiological or endoscopic examinations.

(2) Class III (premarket approval) when the device is intended for other uses, including colon cleansing routinely for general well being.

(c) *Date PMA or notice of completion of a PDP is required.* A PMA or a notice of completion of a PDP is required to be filed with the Food and Drug Administration on or before December 26, 1996 for any colonic irrigation system described in paragraph (b)(2) of this section that was in commercial distribution before May 28, 1976, or that has, on or before December 26, 1996 been found to be substantially equivalent to a colonic irrigation system described in paragraph (b)(2) of this section that was in commercial distribution before May 28, 1976. Any other colonic irrigation system shall have an approved PMA in effect before being placed in commercial distribution.

[48 FR 53023, Nov. 23, 1983, as amended at 52 FR 17738, May 11, 1987; 61 FR 50707, Sept. 27, 1996]

§876.5250 Urine collector and accessories.

(a) *Identification.* A urine collector and accessories is a device intended to collect urine. The device and accessories consist of tubing, a suitable receptacle, connectors, mechanical supports, and may include a means to prevent the backflow of urine or ascent of infection. The two kinds of urine collectors are:

(1) A urine collector and accessories intended to be connected to an indwelling catheter, which includes the urinary drainage collection kit and the closed urine drainage system and drainage bag; and

(2) A urine collector and accessories not intended to be connected to an indwelling catheter, which includes the corrugated rubber sheath, pediatric urine collector, leg bag for external use, urosheath type incontinence device, and the paste-on device for incontinence.

(b) *Classification.* (1) Class II (special controls) for a urine collector and accessories intended to be connected to an indwelling catheter. The device is exempt from the premarket notification procedures in subpart E of part 807 of this chapter subject to §876.9.

(2) Class I (general controls). For a urine collector and accessories not intended to be connected to an indwelling catheter, subject to the limitations in §876.9. If the device is not labeled or otherwise represented as sterile, it is exempt from the current good manufacturing practice regulations in part 820 of this chapter, with the exception of §820.180, with respect to general requirements concerning records, and §820.198, with respect to complaint files.

[48 FR 53023, Nov. 23, 1983, as amended at 63 FR 59228, Nov. 3, 1998; 65 FR 2317, Jan. 14, 2000; 66 FR 38802, July 25, 2001]

§876.5270 Implanted electrical urinary continence device.

(a) *Identification.* An implanted electrical urinary device is a device intended for treatment of urinary incontinence that consists of a receiver implanted in the abdomen with electrodes for pulsed-stimulation that are implanted either in the bladder wall or in the pelvic floor, and a battery-powered transmitter outside the body.

(b) *Classification.* Class III (premarket approval).

(c) *Date PMA or notice of completion of a PDP is required.* A PMA or a notice of completion of a PDP is required to be filed with the Food and Drug Administration on or before December 26, 1996 for any implanted electrical urinary continence device that was in commercial distribution before May 28, 1976, or that has, on or before December 26, 1996 been found to be substantially equivalent to an implanted electrical urinary continence device that was in commercial distribution before May 28, 1976. Any other implanted electrical urinary continence device shall have an approved PMA or a declared completed PDP in effect before being placed in commercial distribution.

[48 FR 53023, Nov. 23, 1983, as amended at 52 FR 17738, May 11, 1987; 61 FR 50707, Sept. 27, 1996]

§876.5280 Implanted mechanical/hydraulic urinary continence device.

(a) *Identification.* An implanted mechanical/hydraulic urinary continence device is a device used to treat urinary incontinence by the application of continuous or intermittent pressure to occlude the urethra. The totally implanted device may consist of a static pressure pad, or a system with a container of radiopaque fluid in the abdomen and a manual pump and valve under the skin surface that is connected by tubing to an adjustable pressure pad or to a cuff around the urethra. The fluid is pumped as needed from the container to inflate the pad or cuff to pass on the urethra.

(b) *Classification.* Class III (premarket approval).

(c) *Date PMA or notice of completion of a PDP is required.* A PMA or a notice of completion of a PDP is required to be filed with the Food and Drug Administration on or before December 26, 2000, for any implanted mechanical/hydraulic urinary continence device that was in commercial distribution before May 28, 1976, or that has, on or before December 26, 2000, been found to be substantially equivalent to an implanted mechanical/hydraulic urinary continence device that was in commercial distribution before May 28, 1976. Any other implanted mechanical/hydraulic urinary continence device shall have an approved PMA or a declared completed PDP in effect before being placed in commercial distribution.

[48 FR 53023, Nov. 23, 1983, as amended at 52 FR 17738, May 11, 1987; 65 FR 57731, Sept. 26, 2000]

§ 876.5310 Nonimplanted, peripheral electrical continence device.

(a) *Identification.* A nonimplanted, peripheral electrical continence device is a device that consists of an electrode that is connected by an electrical cable to a battery-powered pulse source. The electrode is placed onto or inserted into the body at a peripheral location and used to stimulate the nerves associated with pelvic floor function to maintain urinary continence. When necessary, the electrode may be removed by the user.

(b) *Classification.* Class II, subject to the following special controls:

(1) That sale, distribution, and use of this device are restricted to prescription use in accordance with § 801.109 of this chapter.

(2) That the labeling must bear all information required for the safe and effective use of the device as outlined in § 801.109(c) of this chapter, including a detailed summary of the clinical information upon which the instructions are based.

[65 FR 18237, Apr. 7, 2000]

§ 876.5320 Nonimplanted electrical continence device.

(a) *Identification.* A nonimplanted electrical continence device is a device that consists of a pair of electrodes on a plug or a pessary that are connected by an electrical cable to a battery-powered pulse source. The plug or pessary is inserted into the rectum or into the vagina and used to stimulate the muscles of the pelvic floor to maintain urinary or fecal continence. When necessary, the plug or pessary may be removed by the user. This device excludes an AC-powered nonimplanted electrical continence device and the powered vaginal muscle stimulator for therapeutic use (§ 884.5940).

(b) *Classification.* Class II (performance standards).

§ 876.5365 Esophageal dilator.

(a) *Identification.* An esophageal dilator is a device that consists of a cylindrical instrument that may be hollow and weighted with mercury or a metal olive-shaped weight that slides on a guide, such as a string or wire and is used to dilate a stricture of the esophagus. This generic type of device includes esophageal or gastrointestinal bougies and the esophageal dilator (metal olive).

(b) *Classification.* Class II (performance standards).

§ 876.5450 Rectal dilator.

(a) *Identification.* A rectal dilator is a device designed to dilate the anal sphincter and canal when the size of the anal opening may interfere with its function or the passage of an examining instrument.

(b) *Classification.* Class I (general controls). The device is exempt from the premarket notification procedures in

subpart E of part 807 of this chapter subject to the limitations in § 876.9.

[48 FR 53023, Nov. 23, 1983, as amended at 61 FR 1122, Jan. 16, 1996; 66 FR 38802, July 25, 2001]

§ 876.5470 Ureteral dilator.

(a) *Identification.* A ureteral dilator is a device that consists of a specially shaped catheter or bougie and is used to dilate the ureter at the place where a stone has become lodged or to dilate a ureteral stricture.

(b) *Classification.* Class II (performance standards).

§ 876.5520 Urethral dilator.

(a) *Identification.* A urethral dilator is a device that consists of a slender hollow or solid instrument made of metal, plastic, or other suitable material in a cylindrical form and in a range of sizes and flexibilities. The device may include a mechanism to expand the portion of the device in the urethra and indicate the degree of expansion on a dial. It is used to dilate the urethra. This generic type of device includes the mechanical urethral dilator, urological bougies, metal or plastic urethral sound, urethrometer, filiform, and filiform follower.

(b) *Classification.* (1) Class II (performance standards).

(2) Class I for the urethrometer, urological bougie, filiform and filiform follower, and metal or plastic urethral sound. The devices subject to this paragraph (b)(2) are exempt from the premarket notification procedures in subpart E of part 807 of this chapter subject to the limitations in § 876.9.

[48 FR 53023, Nov. 23, 1983, as amended at 61 FR 1122, Jan. 16, 1996; 66 FR 38802, July 25, 2001]

§ 876.5540 Blood access device and accessories.

(a) *Identification.* A blood access device and accessories is a device intended to provide access to a patient's blood for hemodialysis or other chronic uses. When used in hemodialysis, it is part of an artificial kidney system for the treatment of patients with renal failure or toxemic conditions and provides access to a patient's blood for hemodialysis. The device includes implanted blood access devices, nonimplanted blood access devices, and accessories for both the implanted and nonimplanted blood access devices.

(1) The implanted blood access device consists of various flexible or rigid tubes, which are surgically implanted in appropriate blood vessels, may come through the skin, and are intended to remain in the body for 30 days or more. This generic type of device includes various shunts and connectors specifically designed to provide access to blood, such as the arteriovenous (A-V) shunt cannula and vessel tip.

(2) The nonimplanted blood access device consists of various flexible or rigid tubes, such as catheters, cannulae or hollow needles, which are inserted into appropriate blood vessels or a vascular graft prosthesis (§§ 870.3450 and 870.3460), and are intended to remain in the body for less than 30 days. This generic type of device includes fistula needles, the single needle dialysis set (coaxial flow needle), and the single needle dialysis set (alternating flow needle).

(3) Accessories common to either type include the shunt adaptor, cannula clamp, shunt connector, shunt stabilizer, vessel dilator, disconnect forceps, shunt guard, crimp plier, tube plier, crimp ring, joint ring, fistula adaptor, and declotting tray (including contents).

(b) *Classification.* (1) Class III (premarket approval) for the implanted blood access device.

(2) Class II (performance standards) for the nonimplanted blood access device.

(3) Class II (performance standards) for accessories for both the implanted and the nonimplanted blood access devices not listed in paragraph (b)(4) of this section.

(4) Class I for the cannula clamp, disconnect forceps, crimp plier, tube plier, crimp ring, and joint ring, accessories for both the implanted and nonimplanted blood access device. The devices subject to this paragraph (b)(4) are exempt from the premarket notification procedures in subpart E of part 807 of this chapter subject to the limitations in § 876.9.

(c) *Date PMA or notice of completion of a PDP is required.* No effective date has been established of the requirement for premarket approval for the device described in paragraph (b)(1). See § 876.3.

[48 FR 53023, Nov. 23, 1983, as amended at 52 FR 17738, May 11, 1987; 61 FR 1122, Jan. 16, 1996; 66 FR 38802, July 25, 2001]

§ 876.5600 Sorbent regenerated dialysate delivery system for hemodialysis.

(a) *Identification.* A sorbent regenerated dialysate delivery system for hemodialysis is a device that is part of an artificial kidney system for the treatment of patients with renal failure or toxemic conditions, and that consists of a sorbent cartridge and the means to circulate dialysate through this cartridge and the dialysate compartment of the dialyzer. The device is used with the extracorporeal blood system and the dialyzer of the hemodialysis system and accessories (§ 876.5820). The device includes the means to maintain the temperature, conductivity, electrolyte balance, flow rate and pressure of the dialysate, and alarms to indicate abnormal dialysate conditions. The sorbent cartridge may include absorbent, ion exchange and catalytic materials.

(b) *Classification.* Class II (performance standards).

§ 876.5630 Peritoneal dialysis system and accessories.

(a) *Identification.* (1) A peritoneal dialysis system and accessories is a device that is used as an artificial kidney system for the treatment of patients with renal failure or toxemic conditions, and that consists of a peritoneal access device, an administration set for peritoneal dialysis, a source of dialysate, and, in some cases, a water purification mechanism. After the dialysate is instilled into the patient's peritoneal cavity, it is allowed to dwell there so that undesirable substances from the patient's blood pass through the lining membrane of the peritoneal cavity into this dialysate. These substances are then removed when the dialysate is drained from the patient. The peritoneal dialysis system may regulate and monitor the dialysate temperature, volume, and delivery rate together with the time course of each cycle of filling, dwell time, and draining of the peritoneal cavity or manual controls may be used. This generic device includes the semiautomatic and the automatic peritoneal delivery system.

(2) The peritoneal access device is a flexible tube that is implanted through the abdominal wall into the peritoneal cavity and that may have attached cuffs to provide anchoring and a skin seal. The device is either a single use peritoneal catheter, intended to remain in the peritoneal cavity for less than 30 days, or a long term peritoneal catheter. Accessories include stylets and trocars to aid in the insertion of the catheter and an obturator to maintain the patency of the surgical fistula in the abdominal wall between treatments.

(3) The disposable administration set for peritoneal dialysis consists of tubing, an optional reservoir bag, and appropriate connectors. It may include a peritoneal dialysate filter to trap and remove contaminating particles.

(4) The source of dialysate may be sterile prepackaged dialysate (for semiautomatic peritoneal dialysate delivery systems or "cycler systems") or dialysate prepared from dialysate concentrate and sterile purified water (for automatic peritoneal dialysate delivery systems or "reverse osmosis" systems). Prepackaged dialysate intended for use with either of the peritoneal dialysate delivery systems is regulated by FDA as a drug.

(b) *Classification.* Class II (performance standards).

§ 876.5665 Water purification system for hemodialysis.

(a) *Identification.* A water purification system for hemodialysis is a device that is intended for use with a hemodialysis system and that is intended to remove organic and inorganic substances and microbial contaminants from water used to dilute dialysate concentrate to form dialysate. This generic type of device may include a water softener, sediment filter, carbon filter, and water distillation system.

(b) *Classification.* Class II (performance standards).

§ 876.5820 Hemodialysis system and accessories.

(a) *Identification.* A hemodialysis system and accessories is a device that is used as an artificial kidney system for the treatment of patients with renal failure or toxemic conditions and that consists of an extracorporeal blood system, a conventional dialyzer, a dialysate delivery system, and accessories. Blood from a patient flows through the tubing of the extracorporeal blood system and accessories to the blood compartment of the dialyzer, then returns through further tubing of the extracorporeal blood system to the patient. The dialyzer has two compartments that are separated by a semipermeable membrane. While the blood is in the blood compartment, undesirable substances in the blood pass through the semipermeable membrane into the dialysate in the dialysate compartment. The dialysate delivery system controls and monitors the dialysate circulating through the dialysate compartment of the dialyzer.

(1) The extracorporeal blood system and accessories consists of tubing, pumps, pressure monitors, air foam or bubble detectors, and alarms to keep blood moving safely from the blood access device and accessories for hemodialysis (§ 876.5540) to the blood compartment of the dialyzer and back to the patient.

(2) The conventional dialyzer allows a transfer of water and solutes between the blood and the dialysate through the semipermeable membrane. The semipermeable membrane of the conventional dialyzer has a sufficiently low permeability to water that an ultrafiltration controller is not required to prevent excessive loss of water from the patient's blood. This conventional dialyzer does not include hemodialyzers with the disposable inserts (Kiil type) (§ 876.5830) or dialyzers of high permeability (§ 876.5860).

(3) The dialysate delivery system consists of mechanisms that monitor and control the temperature, conductivity, flow rate, and pressure of the dialysate and circulates dialysate through the dialysate compartment of the dialyzer. The dialysate delivery system includes the dialysate concentrate for hemodialysis (liquid or powder) and alarms to indicate abnormal dialysate conditions. This dialysate delivery system does not include the sorbent regenerated dialysate delivery system for hemodialysis (§ 876.5600), the dialysate delivery system of the peritoneal dialysis system and accessories (§ 876.5630), or the controlled dialysate delivery system of the high permeability hemodialysis system § 876.5860).

(4) Remote accessories to the hemodialysis system include the unpowered dialysis chair without a scale, the powered dialysis chair without a scale, the dialyzer holder set, dialysis tie gun and ties, and hemodialysis start/stop tray.

(b) *Classification.* (1) Class II (performance standards) for hemodialysis systems and all accessories directly associated with the extracorporeal blood system and the dialysate delivery system.

(2) Class I for other accessories of the hemodialysis system remote from the extracorporeal blood system and the dialysate delivery system, such as the unpowered dialysis chair, hemodialysis start/stop tray, dialyzer holder set, and dialysis tie gun and ties. The devices subject to this paragraph (b)(2) are exempt from the premarket notification procedures in subpart E of part 807 of this chapter subject to the limitations in § 876.9.

[48 FR 53023, Nov. 23, 1983, as amended at 54 FR 25050, June 12, 1989; 66 FR 38802, July 25, 2001]

§ 876.5830 Hemodialyzer with disposable insert (Kiil type).

(a) *Identification.* A hemodialyzer with disposable inserts (Kiil type) is a device that is used as a part of an artificial kidney system for the treatment of patients with renal failure or toxemic conditions and that includes disposable inserts consisting of layers of semipermeable membranes which are sandwiched between support plates. The device is used with the extracorporeal blood system and the dialysate delivery system of the hemodialysis system and accessories (§ 876.5820).

(b) *Classification.* Class II (performance standards).

[48 FR 53023, Nov. 23, 1983, as amended at 53 FR 11253, Apr. 6, 1988]

§ 876.5860 High permeability hemodialysis system.

(a) *Identification.* A high permeability hemodialysis system is a device intended for use as an artificial kidney system for the treatment of patients with renal failure, fluid overload, or toxemic conditions by performing such therapies as hemodialysis, hemofiltration, hemoconcentration, and hemodiafiltration. Using a hemodialyzer with a semipermeable membrane that is more permeable to water than the semipermeable membrane of the conventional hemodialysis system (§ 876.5820), the high permeability hemodialysis system removes toxins or excess fluid from the patient's blood using the principles of convection (via a high ultrafiltration rate) and/or diffusion (via a concentration gradient in dialysate). During treatment, blood is circulated from the patient through the hemodialyzer's blood compartment, while the dialysate solution flows countercurrent through the dialysate compartment. In this process, toxins and/or fluid are transferred across the membrane from the blood to the dialysate compartment. The hemodialysis delivery machine controls and monitors the parameters related to this processing, including the rate at which blood and dialysate are pumped through the system, and the rate at which fluid is removed from the patient. The high permeability hemodialysis system consists of the following devices:

(1) The hemodialyzer consists of a semipermeable membrane with an in vitro ultrafiltration coefficient (K_{uf}) greater than 8 milliliters per hour per conventional millimeter of mercury, as measured with bovine or expired human blood, and is used with either an automated ultrafiltration controller or anther method of ultrafiltration control to prevent fluid imbalance.

(2) The hemodialysis delivery machine is similar to the extracorporeal blood system and dialysate delivery system of the hemodialysis system and accessories (§ 876.5820), with the addition of an ultrafiltration controller and mechanisms that monitor and/or control such parameters as fluid balance, dialysate composition, and patient treatment parameters (e.g., blood pressure, hematocrit, urea, etc.).

(3) The high permeability hemodialysis system accessories include, but are not limited to, tubing lines and various treatment related monitors (e.g., dialysate pH, blood pressure, hematocrit, and blood recirculation monitors).

(b) *Classification.* Class II. The special controls for this device are FDA's:

(1) "Use of International Standard ISO 10993 'Biological Evaluation of Medical Device—Part I: Evaluation and Testing,' "

(2) "Guidance for the Content of 510(k)s for Conventional and High Permeability Hemodialyzers,"

(3) "Guidance for Industry and CDRH Reviewers on the Content of Premarket Notifications for Hemodialysis Delivery Systems,"

(4) "Guidance for the Content of Premarket Notifications for Water Purification Components and Systems for Hemodialysis," and

(5) "Guidance for Hemodialyzer Reuse Labeling."

[65 FR 17145, Mar. 31, 2000]

§ 876.5870 Sorbent hemoperfusion system.

(a) *Identification.* A sorbent hemoperfusion system is a device that consists of an extracorporeal blood system similar to that identified in the hemodialysis system and accessories (§ 876.5820) and a container filled with adsorbent material that removes a wide range of substances, both toxic and normal, from blood flowing through it. The adsorbent materials are usually activated-carbon or resins which may be coated or immobilized to prevent fine particles entering the patient's blood. The generic type of device may include lines and filters specifically designed to connect the device to the extracorporeal blood system. The device is used in the treatment of poisoning, drug overdose, hepatic coma, or metabolic disturbances.

(b) *Classification.* Class III (premarket approval).

(c) *Date PMA or notice of completion of a PDP is required.* No effective date has been established of the requirement for premarket approval. See §876.3.

[48 FR 53023, Nov. 23, 1983, as amended at 52 FR 17738, May 11, 1987]

§876.5880 Isolated kidney perfusion and transport system and accessories.

(a) *Identification.* An isolated kidney perfusion and transport system and accesssories is a device that is used to support a donated or a cadaver kidney and to maintain the organ in a near-normal physiologic state until it is transplanted into a recipient patient. This generic type of device may include tubing, catheters, connectors, an ice storage or freezing container with or without bag or preservatives, pulsatile or nonpulsatile hypothermic isolated organ perfusion apparatus with or without oxygenator, and disposable perfusion set.

(b) *Classification.* Class II (performance standards).

§876.5885 Tissue culture media for human ex vivo tissue and cell culture processing applications.

(a) *Identification.* Tissue culture media for human ex vivo tissue and cell culture processing applications consist of cell and tissue culture media and components that are composed of chemically defined components (e.g., amino acids, vitamins, inorganic salts) that are essential for the ex vivo development, survival, and maintenance of tissues and cells of human origin. The solutions are indicated for use in human ex vivo tissue and cell culture processing applications.

(b) *Classification.* Class II (special controls): FDA guidance document, "Class II Special Controls Guidance Document: Tissue Culture Media for Human Ex Vivo Processing Applications; Final Guidance for Industry and FDA Reviewers."

[66 FR 27025, May 16, 2001]

§876.5895 Ostomy irrigator.

(a) *Identification.* An ostomy irrigator is a device that consists of a container for fluid, tubing with a cone-shaped tip or a soft and flexible catheter with a retention shield and that is used to wash out the colon through a colostomy, a surgically created opening of the colon on the surface of the body.

(b) *Classification.* Class II (performance standards).

§876.5900 Ostomy pouch and accessories.

(a) *Identification.* An ostomy pouch and accessories is a device that consists of a bag that is attached to the patient's skin by an adhesive material and that is intended for use as a receptacle for collection of fecal material or urine following an ileostomy, colostomy, or ureterostomy (a surgically created opening of the small intestine, large intestine, or the ureter on the surface of the body). This generic type of device and its accessories includes the ostomy pouch, ostomy adhesive, the disposable colostomy appliance, ostomy collector, colostomy pouch, urinary ileostomy bag, urine collecting ureterostomy bag, ostomy drainage bag with adhesive, stomal bag, ostomy protector, and the ostomy size selector, but excludes ostomy pouches which incorporate arsenic-containing compounds.

(b) *Classification.* Class I (general controls). The device is exempt from the premarket notification procedures in subpart E of part 807 of this chapter subject to the limitations in §876.9.

[48 FR 53023, Nov. 23, 1983, as amended at 54 FR 25050, June 12, 1989; 66 FR 38802, July 25, 2001]

§876.5920 Protective garment for incontinence.

(a) *Identification.* A protective garment for incontinence is a device that consists of absorbent padding and a fluid barrier and that is intended to protect an incontinent patient's garment from the patient's excreta. This generic type of device does not include diapers for infants.

(b) *Classification.* Class I (general controls). The device is exempt from the premarket notification procedures in subpart E of part 807 of this chapter subject to the limitations in §876.9. The device is also exempt from the current good manufacturing practice regulations in part 820 of this chapter, with the exception of §820.180, regarding general requirements concerning records, and § 820.198, regarding complaint files.

[48 FR 53023, Nov. 23, 1983, as amended at 54 FR 25050, June 12, 1989; 66 FR 38802, July 25, 2001]

§ 876.5955 Peritoneo-venous shunt.

(a) *Identification.* A peritoneo-venous shunt is an implanted device that consists of a catheter and a pressure activated one-way valve. The catheter is implanted with one end in the peritoneal cavity and the other in a large vein. This device enables ascitic fluid in the peritoneal cavity to flow into the venous system for the treatment of intractable ascites.

(b) *Classification.* Class II. The special controls for this device are FDA's:

(1) "Use of International Standard ISO 10993 'Biological Evaluation of Medical Devices—Part I: Evaluation and Testing,' "

(2) "510(k) Sterility Review Guidance of 2/12/90 (K90–1)," and

(3) Backflow specification and testing to prevent reflux of blood into the shunt.

[48 FR 53023, Nov. 23, 1983, as amended at 52 FR 17738, May 11, 1987; 65 FR 17145, Mar. 31, 2000]

§ 876.5970 Hernia support.

(a) *Identification.* A hernia support is a device, usually made of elastic, canvas, leather, or metal, that is intended to be placed over a hernial opening (a weakness in the abdominal wall) to prevent protrusion of the abdominal contents. This generic type of device includes the umbilical truss.

(b) *Classification.* Class I (general controls). The device is exempt from the premarket notification procedures in subpart E of part 807 of this chapter subject to the limitations in § 876.9. The device is also exempt from the current good manufacturing practice regulations in part 820 of this chapter, with the exception of § 820.180, regarding general requirements concerning records, and § 820.198, regarding complaint files.

[48 FR 53023, Nov. 23, 1983, as amended at 59 FR 63010, Dec. 7, 1994; 66 FR 38802, July 25, 2001]

§ 876.5980 Gastrointestinal tube and accessories.

(a) *Identification.* A gastrointestinal tube and accessories is a device that consists of flexible or semi-rigid tubing used for instilling fluids into, withdrawing fluids from, splinting, or suppressing bleeding of the alimentary tract. This device may incorporate an integral inflatable balloon for retention or hemostasis. This generic type of device includes the hemostatic bag, irrigation and aspiration catheter (gastric, colonic, etc.), rectal catheter, sterile infant gavage set, gastrointestinal string and tubes to locate internal bleeding, double lumen tube for intestinal decompression or intubation, feeding tube, gastroenterostomy tube, Levine tube, nasogastric tube, single lumen tube with mercury weight balloon for intestinal intubation or decompression, and gastro-urological irrigation tray (for gastrological use).

(b) *Classification.* (1) Class II (special controls). The barium enema retention catheter and tip with or without a bag that is a gastrointestinal tube and accessory is exempt from the premarket notification procedures in subpart E of this part subject to the limitations in § 876.9.

(2) Class I (general controls) for the dissolvable nasogastric feed tube guide for the nasogastric tube. The class I device is exempt from the premarket notification procedures in subpart E of part 807 of this chapter subject to § 876.9.

[49 FR 573, Jan. 5, 1984, as amended at 65 FR 2317, Jan. 14, 2000; 65 FR 76932, Dec. 8, 2000]

§ 876.5990 Extracorporeal shock wave lithotripter.

(a) *Identification.* An extracorporeal shock wave lithotripter is a device that focuses ultrasonic shock waves into the body to noninvasively fragment urinary calculi within the kidney or ureter. The primary components of the device are a shock wave generator, high voltage generator, control console, imaging/localization system, and patient table. Prior to treatment, the urinary stone is targeted using either an integral or stand-alone localization/imaging system. Shock waves are typically generated using electrostatic spark discharge (spark gap), electromagnetically repelled membranes, or piezoelectric crystal arrays, and focused onto the stone with either a specially designed reflector, dish, or acoustic lens. The shock waves are created under water within the shock wave generator, and are transferred to the patient's body using an appropriate acoustic interface. After the stone has been fragmented by the focused shock waves, the fragments pass out of the body with the patient's urine.

(b) *Classification.* Class II (special controls) (FDA guidance document: "Guidance for the Content of Premarket Notifications (510(k)'s) for Extracorporeal Shock Wave Lithotripters Indicated for the Fragmentation of Kidney and Ureteral Calculi.")

[65 FR 48612, Aug. 9, 2000]

PART 878—GENERAL AND PLASTIC SURGERY DEVICES

Subpart A—General Provisions

Sec.
878.1 Scope.
878.3 Effective dates of requirement for premarket approval.
878.9 Limitations of exemptions from section 510(k) of the Federal Food, Drug, and Cosmetic Act (the act).

Subpart B—Diagnostic Devices

878.1800 Speculum and accessories.

Subpart C [Reserved]

Subpart D—Prosthetic Devices

878.3250 External facial fracture fixation appliance.
878.3300 Surgical mesh.
878.3500 Polytetrafluoroethylene with carbon fibers composite implant material.
878.3530 Silicone inflatable breast prosthesis.
878.3540 Silicone gel-filled breast prosthesis.
878.3550 Chin prosthesis.
878.3590 Ear prosthesis.
878.3610 Esophageal prosthesis.
878.3680 Nose prosthesis.
878.3720 Tracheal prosthesis.
878.3750 External prosthesis adhesive.
878.3800 External aesthetic restoration prosthesis.
878.3900 Inflatable extremity splint.
878.3910 Noninflatable extremity splint.
878.3925 Plastic surgery kit and accessories.

Subpart E—Surgical Devices

878.4014 Nonresorbable gauze/sponge for external use.
878.4018 Hydrophilic wound dressing.
878.4020 Occlusive wound dressing.
878.4022 Hydrogel wound dressing and burn dressing.
878.4025 Silicone sheeting.
878.4040 Surgical apparel.
878.4100 Organ bag.
878.4160 Surgical camera and accessories.
878.4200 Introduction/drainage catheter and accessories.
878.4300 Implantable clip.
878.4320 Removable skin clip.
878.4350 Cryosurgical unit and accessories.
878.4370 Surgical drape and drape accessories.
878.4380 Drape adhesive.
878.4400 Electrosurgical cutting and coagulation device and accessories.
878.4410 Low energy ultrasound wound cleaner.
878.4440 Eye pad.
878.4450 Nonabsorbable gauze for internal use.
878.4460 Surgeon's glove.
878.4470 Surgeon's gloving cream.
878.4480 Absorbable powder for lubricating a surgeon's glove.
878.4490 Absorbable hemostatic agent and dressing.
878.4493 Absorbable poly(glycolide/L-lactide) surgical suture.
878.4495 Stainless steel suture.
878.4520 Polytetrafluoroethylene injectable.
878.4580 Surgical lamp.
878.4630 Ultraviolet lamp for dermatologic disorders.
878.4635 Ultraviolet lamp for tanning.
878.4660 Skin marker.
878.4680 Nonpowered, single patient, portable suction apparatus.
878.4700 Surgical microscope and accessories.
878.4730 Surgical skin degreaser or adhesive tape solvent.
878.4750 Implantable staple.
878.4760 Removable skin staple.
878.4780 Powered suction pump.
878.4800 Manual surgical instrument for general use.
878.4810 Laser surgical instrument for use in general and plastic surgery and in dermatology.
878.4820 Surgical instrument motors and accessories/attachments.
878.4830 Absorbable surgical gut suture.
878.4840 Absorbable polydioxanone surgical suture.
878.4930 Suture retention device.
878.4950 Manual operating table and accessories and manual operating chair and accessories.

AUTHORITY: 21 U.S.C. 351, 360, 360c, 360e, 360j, 360l, 371.

SOURCE: 53 FR 23872, June 24, 1988, unless otherwise noted.

Subpart A—General Provisions

§ 878.1 Scope.

(a) This part sets forth the classification of general and plastic surgery devices intended for human use that are in commercial distribution.

(b) The identification of a device in a regulation in this part is not a precise description of every device that is, or will be, subject to the regulation. A manufacturer who submits a premarket notification submission for a device under part 807 cannot show merely that the device is accurately described by the section title and identification provision of a regulation in this part, but shall state why the device is substantially equivalent to other devices, as required by § 807.87 of this chapter.

(c) To avoid duplicative listings, a general and plastic surgery device that has two or more types of uses (e.g., used both as a diagnostic device and as a therapeutic device) is listed in one subpart only.

(d) References in this part to regulatory sections of the Code of Federal Regulations are to chapter I of title 21 unless otherwise noted.

(e) Guidance documents referenced in this part are available on the Internet at *http://www.fda.gov/cdrh/guidance.html.*

[53 FR 23872, June 24, 1988, as amended at 67 FR 77676, Dec. 19, 2002]

§ 878.3 Effective dates of requirement for premarket approval.

A device included in this part that is classified into class III (premarket approval) shall not be commercially distributed after the date shown in the regulation classifying the device unless the manufacturer has an approval under section 515 of the act (unless an exemption has been granted under section 520(g)(2) of the act). An approval under section 515 of the act consists of FDA's issuance of an order approving an application for premarket approval (PMA) for the device or declaring completed a product development protocol (PDP) for the device.

(a) Before FDA requires that a device commercially distributed before the enactment date of the amendments, or a device that has been found substantially equivalent to such a device, has an approval under section 515 of the act, FDA must promulgate a regulation under section 515(b) of the act requiring such approval, except as provided in paragraphs (b) and (c) of this section. Such a regulation under section 515(b) of the act shall not be effective during the grace period ending on the 90th day after its promulgation or on the last day of the 30th full calendar month after the regulation that classifies the device into class III is effective, whichever is later. See section 501(f)(2)(B) of the act. Accordingly, unless an effective date of the requirement for premarket approval is shown in the regulation for a device classified into class III in this part, the device may be commercially distributed without FDA's issuance of an order approving a PMA or declaring completed a PDP for the device. If FDA promulgates a regulation under section 515(b) of the act requiring premarket approval for a device, section 501(f)(1)(A) of the act applies to the device.

(b) Any new, not substantially equivalent, device introduced into commercial distribution on or after May 28,

1976, including a device formerly marketed that has been substantially altered, is classified by statute (section 513(f) of the act) into class III without any grace period and FDA must have issued an order approving a PMA or declaring completed a PDP for the device before the device is commercially distributed unless it is reclassified. If FDA knows that a device being commercially distributed may be a "new" device as defined in this section because of any new intended use or other reasons, FDA may codify the statutory classification of the device into class III for such new use. Accordingly, the regulation for such a class III device states that as of the enactment date of the amendments, May 28, 1976, the device must have an approval under section 515 of the act before commercial distribution.

(c) A device identified in a regulation in this part that is classified into class III and that is subject to the transitional provisions of section 520(l) of the act is automatically classified by statute into class III and must have an approval under section 515 of the act before being commercially distributed. Accordingly, the regulation for such a class III transitional device states that as of the enactment date of the amendments, May 28, 1976, the device must have an approval under section 515 of the act before commercial distribution.

§878.9 Limitations of exemptions from section 510(k) of the Federal Food, Drug, and Cosmetic Act (the act).

The exemption from the requirement of premarket notification (section 510(k) of the act) for a generic type of class I or II device is only to the extent that the device has existing or reasonably foreseeable characteristics of commercially distributed devices within that generic type or, in the case of in vitro diagnostic devices, only to the extent that misdiagnosis as a result of using the device would not be associated with high morbidity or mortality. Accordingly, manufacturers of any commercially distributed class I or II device for which FDA has granted an exemption from the requirement of premarket notification must still submit a premarket notification to FDA before introducing or delivering for introduction into interstate commerce for commercial distribution the device when:

(a) The device is intended for a use different from the intended use of a legally marketed device in that generic type of device; e.g., the device is intended for a different medical purpose, or the device is intended for lay use where the former intended use was by health care professionals only;

(b) The modified device operates using a different fundamental scientific technology than a legally marketed device in that generic type of device; e.g., a surgical instrument cuts tissue with a laser beam rather than with a sharpened metal blade, or an in vitro diagnostic device detects or identifies infectious agents by using deoxyribonucleic acid (DNA) probe or nucleic acid hybridization technology rather than culture or immunoassay technology; or

(c) The device is an in vitro device that is intended:

(1) For use in the diagnosis, monitoring, or screening of neoplastic diseases with the exception of immunohistochemical devices;

(2) For use in screening or diagnosis of familial or acquired genetic disorders, including inborn errors of metabolism;

(3) For measuring an analyte that serves as a surrogate marker for screening, diagnosis, or monitoring life-threatening diseases such as acquired immune deficiency syndrome (AIDS), chronic or active hepatitis, tuberculosis, or myocardial infarction or to monitor therapy;

(4) For assessing the risk of cardiovascular diseases;

(5) For use in diabetes management;

(6) For identifying or inferring the identity of a microorganism directly from clinical material;

(7) For detection of antibodies to microorganisms other than immunoglobulin G (IgG) or IgG assays when the results are not qualitative, or are used to determine immunity, or the assay is intended for use in matrices other than serum or plasma;

(8) For noninvasive testing as defined in §812.3(k) of this chapter; and

(9) For near patient testing (point of care).

[65 FR 2317, Jan. 14, 2000]

Subpart B—Diagnostic Devices

§ 878.1800 Speculum and accessories.

(a) *Identification.* A speculum is a device intended to be inserted into a body cavity to aid observation. It is either nonilluminated or illuminated and may have various accessories.

(b) *Classification.* Class I (general controls). The device is exempt from the premarket notification procedures in subpart E of part 807 of this chapter, subject to the limitations in § 878.9.

[53 FR 23872, June 24, 1988, as amended at 54 FR 13827, Apr. 5, 1989; 59 FR 63010, Dec. 7, 1994; 66 FR 38802, July 25, 2001]

Subpart C [Reserved]

Subpart D—Prosthetic Devices

§ 878.3250 External facial fracture fixation appliance.

(a) *Identification.* An external facial fracture fixation appliance is a metal apparatus intended to be used during surgical reconstruction and repair to immobilize maxillofacial bone fragments in their proper facial relationship.

(b) *Classification.* Class I (general controls). The device is exempt from the premarket notification procedures in subpart E of part 807 of this chapter subject to § 878.9.

[53 FR 23872, June 24, 1988, as amended at 54 FR 13827, Apr. 5, 1989; 65 FR 2317, Jan. 14, 2000]

§ 878.3300 Surgical mesh.

(a) *Identification.* Surgical mesh is a metallic or polymeric screen intended to be implanted to reinforce soft tissue or bone where weakness exists. Examples of surgical mesh are metallic and polymeric mesh for hernia repair, and acetabular and cement restrictor mesh used during orthopedic surgery.

(b) *Classification.* Class II.

§ 878.3500 Polytetrafluoroethylene with carbon fibers composite implant material.

(a) *Identification.* A polytetrafluoroethylene with carbon fibers composite implant material is a porous device material intended to be implanted during surgery of the chin, jaw, nose, or bones or tissue near the eye or ear. The device material serves as a space-occupying substance and is shaped and formed by the surgeon to conform to the patient's need.

(b) *Classification.* Class II.

§ 878.3530 Silicone inflatable breast prosthesis.

(a) *Identification.* A silicone inflatable breast prosthesis is a silicone rubber shell made of polysiloxane(s), such as polydimethylsiloxane and polydiphenylsiloxane, that is inflated to the desired size with sterile isotonic saline before or after implantation. The device is intended to be implanted to augment or reconstruct the female breast.

(b) *Classification.* Class III.

(c) *Date PMA or notice of completion of a PDP is required.* A PMA or a notice of completion of a PDP is required to be filed with the Food and Drug Administration on or before November 17, 1999, for any silicone inflatable breast prosthesis that was in commercial distribution before May 28, 1976, or that has, on or before November 17, 1999, been found to be substantially equivalent to a silicone inflatable breast prosthesis that was in commercial distribution before May 28, 1976. Any other silicone inflatable breast prosthesis shall have an approved PMA or a declared completed PDP in effect before being placed in commercial distribution.

[53 FR 23872, June 24, 1988, as amended at 64 FR 45161, Aug. 19, 1999]

§ 878.3540 Silicone gel-filled breast prosthesis.

(a) *Identification*—(1) *Single-lumen silicone gel-filled breast prosthesis.* A single-lumen silicone gel-filled breast prosthesis is a silicone rubber shell made of polysiloxane(s), such as polydimethylsiloxane and polydiphenylsiloxane. The shell either contains a fixed amount cross-linked polymerized silicone gel, filler, and stabilizers or is filled to the desired size with injectable silicone gel at time of implantation. The device is intended to be implanted to augment or reconstruct the female breast.

(2) *Double-lumen silicone gel-filled breast prosthesis.* A double lumen silicone gel-filled breast prosthesis is a silicone rubber inner shell and a silicone rubber outer shell, both shells made of polysiloxane(s), such as polydimethylsiloxane and polydiphenylsiloxane. The inner shell contains fixed amounts of cross-linked polymerized silicone gel, fillers, and stabilizers. The outer shell is inflated to the desired size with sterile isotonic saline before or after implantation. The device is intended to be implanted to augment or reconstruct the female breast.

(3) *Polyurethane covered silicone gel-filled breast prosthesis.* A polyurethane covered silicone gel-filled breast prosthesis is an inner silicone rubber shell made of polysiloxane(s), such as polydimethylsiloxane and polydiphenylsiloxane, with an outer silicone adhesive layer and an outer covering of polyurethane; contained within the inner shell is a fixed amount of cross-linked polymerized silicone gel, fillers, and stabilizers and an inert support structure compartmentalizing the silicone gel. The device is intended to be implanted to augment or reconstruct the female breast.

(b) *Classification.* Class III.

(c) *Date premarket approval application (PMA) is required.* A PMA is required to be filed with the Food and Drug Administration on or before July 9, 1991 for any silicone gel-filled breast prosthesis that was in commercial distribution before May 28, 1976, or that has on or before July 9, 1991 been found to be substantially equivalent to a silicone gel-filled breast prosthesis that was in commercial distribution before May 28, 1976. Any other silicone gel-filled breast prosthesis shall have an approved PMA in effect before being placed in commercial distribution.

[53 FR 23872, June 24, 1988, as amended at 56 FR 14627, Apr. 10, 1991]

§ 878.3550 Chin prosthesis.

(a) *Identification.* A chin prosthesis is a silicone rubber solid device intended to be implanted to augment or reconstruct the chin.

(b) *Classification.* Class II.

§ 878.3590 Ear prosthesis.

(a) *Identification.* An ear prosthesis is a silicone rubber solid device intended to be implanted to reconstruct the external ear.

(b) *Classification.* Class II.

§ 878.3610 Esophageal prosthesis.

(a) *Identification.* An esophageal prosthesis is a rigid, flexible, or expandable tubular device made of a plastic, metal, or polymeric material that is intended to be implanted to restore the structure and/or function of the esophagus. The metal esophageal prosthesis may be uncovered or covered with a polymeric material. This device may also include a device delivery system.

(b) *Classification.* Class II. The special control for this device is FDA's "Guidance for the Content of Premarket Notification Submissions for Esophageal and Tracheal Prostheses."

[65 FR 17145, Mar. 31, 2000]

§ 878.3680 Nose prosthesis.

(a) *Identification.* A nose prosthesis is a silicone rubber solid device intended to be implanted to augment or reconstruct the nasal dorsum.

(b) *Classification.* Class II.

§ 878.3720 Tracheal prosthesis.

(a) *Identification.* The tracheal prosthesis is a rigid, flexible, or expandable tubular device made of a silicone, metal, or polymeric material that is intended to be implanted to restore the structure and/or function of the trachea or trachealbronchial tree. It may be unbranched or contain one or two branches. The metal tracheal prosthesis may be uncovered or covered with a polymeric material. This device may also include a device delivery system.

(b) *Classification.* Class II. The special control for this device is FDA's "Guidance for the Content of Premarket Notification Submissions for Esophageal and Tracheal Prostheses."

[65 FR 17146, Mar. 31, 2000]

§ 878.3750 External prosthesis adhesive.

(a) *Identification.* An external prosthesis adhesive is a silicone-type adhesive intended to be used to fasten to the body an external aesthetic restoration prosthesis, such as an artificial nose.

(b) *Classification.* Class I (general controls). The device is exempt from the premarket notification procedures in subpart E of part 807 of this chapter, subject to the limitations in § 878.9.

[53 FR 23872, June 24, 1988, as amended at 59 FR 63010, Dec. 7, 1994; 66 FR 38802, July 25, 2001]

§ 878.3800 External aesthetic restoration prosthesis.

(a) *Identification.* An external aesthetic restoration prosthesis is a device intended to be used to construct an external artificial body structure, such as an ear, breast, or nose. Usually the device is made of silicone rubber and it may be fastened to the body with an external prosthesis adhesive. The device is not intended to be implanted.

(b) *Classification.* Class I (general controls). The device is exempt from the premarket notification procedures in subpart E of part 807 of this chapter, subject to the limitations in § 878.9. If the device is intended for use without an external prosthesis adhesive to fasten it to the body, the device is exempt from the current good manufacturing practice regulations in part 820 of this chapter, with the exception of § 820.180, with respect to general requirements concerning records, and § 820.198, with respect to complaint files.

[53 FR 23872, June 24, 1988, as amended at 59 FR 63010, Dec. 7, 1994; 66 FR 38802, July 25, 2001]

§ 878.3900 Inflatable extremity splint.

(a) *Identification.* An inflatable extremity splint is a device intended to be inflated to immobilize a limb or an extremity.

(b) *Classification.* Class I (general controls). The device is exempt from the premarket notification procedures in subpart E of part 807 of this chapter, subject to the limitations in § 878.9.

[53 FR 23872, June 24, 1988, as amended at 59 FR 63010, Dec. 7, 1994; 66 FR 38802, July 25, 2001]

§ 878.3910 Noninflatable extremity splint.

(a) *Identification.* A noninflatable extremity splint is a device intended to immobilize a limb or an extremity. It is not inflatable.

(b) *Classification.* Class I (general controls). The device is exempt from the premarket notification procedures in subpart E of part 807 of this chapter subject to § 878.9. If the device is not labeled or otherwise represented as sterile, it is exempt from the current good manufacturing practice regulations in part 820 of this chapter, with the exception of § 820.180 of this chapter, with respect to general requirements concerning records, and § 820.198 of this chapter, with respect to complaint files.

[53 FR 23872, June 24, 1988, as amended at 54 FR 13827, Apr. 5, 1989; 65 FR 2317, Jan. 14, 2000]

§ 878.3925 Plastic surgery kit and accessories.

(a) *Identification.* A plastic surgery kit and accessories is a device intended to be used to reconstruct maxillofacial deficiencies. The kit contains surgical instruments and materials used to make maxillofacial impressions before molding an external prosthesis.

(b) *Classification.* Class I (general controls). The device is exempt from the premarket notification procedures in subpart E of part 807 of this chapter subject to § 878.9.

[53 FR 23872, June 24, 1988, as amended at 54 FR 13827, Apr. 5, 1989; 65 FR 2317, Jan. 14, 2000]

Subpart E—Surgical Devices

§ 878.4014 Nonresorbable gauze/sponge for external use.

(a) *Identification.* A nonresorbable gauze/sponge for external use is a sterile or nonsterile device intended for medical purposes, such as to be placed directly on a patient's wound to absorb exudate. It consists of a strip, piece, or pad made from open woven or nonwoven mesh cotton cellulose or a simple chemical derivative of cellulose. This classification does not include a nonresorbable gauze/sponge for external use that contains added drugs such as antimicrobial agents, added biologics such as growth factors, or is composed of materials derived from animal sources.

(b) *Classification.* Class I (general controls). The device is exempt from the premarket notification procedures in part 807, subpart E of this chapter subject to the limitations in § 878.9.

[64 FR 53929, Oct. 5, 1999]

§ 878.4018 Hydrophilic wound dressing.

(a) *Identification.* A hydrophilic wound dressing is a sterile or non-sterile device intended to cover a wound and to absorb exudate. It consists of nonresorbable materials with hydrophilic properties that are capable of absorbing exudate (e.g., cotton, cotton derivatives, alginates, dextran, and rayon). This classification does not include a hydrophilic wound dressing that contains added drugs such as antimicrobial agents, added biologics such as growth factors, or is composed of materials derived from animal sources.

(b) *Classification.* Class I (general controls). The device is exempt from the premarket notification procedures in part 807, subpart E of this chapter subject to the limitations in § 878.9.

[64 FR 53929, Oct. 5, 1999]

§ 878.4020 Occlusive wound dressing.

(a) *Identification.* An occlusive wound dressing is a nonresorbable, sterile or non-sterile device intended to cover a wound, to provide or support a moist wound environment, and to allow the exchange of gases such as oxygen and water vapor through the device. It consists of a piece of synthetic polymeric material, such as polyurethane, with or without an adhesive backing. This classification does not include an occlusive wound dressing that contains added drugs such as antimicrobial agents, added biologics such as growth factors, or is composed of materials derived from animal sources.

(b) *Classification.* Class I (general controls). The device is exempt from the premarket notification procedures in part 807, subpart E of this chapter subject to the limitations in § 878.9.

[64 FR 53929, Oct. 5, 1999]

§ 878.4022 Hydrogel wound dressing and burn dressing.

(a) *Identification.* A hydrogel wound dressing is a sterile or non-sterile device intended to cover a wound, to absorb wound exudate, to control bleeding or fluid loss, and to protect against abrasion, friction, desiccation, and contamination. It consists of a nonresorbable matrix made of hydrophilic polymers or other material in combination with water (at least 50 percent) and capable of absorbing exudate. This classification does not include a hydrogel wound dressing that contains added drugs such as antimicrobial agents, added biologics such as growth factors, or is composed of materials derived from animal sources.

(b) *Classification.* Class I (general controls). The device is exempt from the premarket notification procedures in part 807, subpart E of this chapter subject to the limitations in § 878.9.

[64 FR 53929, Oct. 5, 1999]

§ 878.4025 Silicone sheeting.

(a) *Identification.* Silicone sheeting is intended for use in the management of closed hyperproliferative (hypertrophic and keloid) scars.

(b) *Classification.* Class I (general controls). The device is exempt from the premarket notification procedures in subpart E of part 807 of this chapter subject to the limitations in § 878.9.

[69 FR 48148, Aug. 9, 2004]

§ 878.4040 Surgical apparel.

(a) *Identification.* Surgical apparel are devices that are intended to be worn by operating room personnel during surgical procedures to protect both the surgical patient and the operating room personnel from transfer of microorganisms, body fluids, and particulate material. Examples include surgical caps, hoods, masks, gowns, operating room shoes and shoe covers, and isolation masks and gowns. Surgical suits and dresses, commonly known as scrub suits, are excluded.

(b) *Classification.* (1) Class II (special controls) for surgical gowns and surgical masks.

(2) Class I (general controls) for surgical apparel other than surgical gowns and surgical masks. The class I device is exempt from the premarket notification procedures in subpart E of part 807 of this chapter subject to § 878.9.

[53 FR 23872, June 24, 1988, as amended at 65 FR 2317, Jan. 14, 2000]

§ 878.4100 Organ bag.

(a) *Identification.* An organ bag is a device that is a flexible plastic bag intended to be used as a temporary receptacle for an organ during surgical procedures to prevent moisture loss.

(b) *Classification.* Class I (general controls). The device is exempt from the premarket notification procedures in subpart E of part 807 of this chapter subject to § 878.9.

[53 FR 23872, June 24, 1988, as amended at 59 FR 63010, Dec. 7, 1994; 65 FR 2318, Jan. 14, 2000]

§ 878.4160 Surgical camera and accessories.

(a) *Identification.* A surgical camera and accessories is a device intended to be used to record operative procedures.

(b) *Classification.* Class I (general controls). The device is exempt from the premarket notification procedures in subpart E of part 807 of this chapter, subject to the limitations in § 878.9.

[53 FR 23872, June 24, 1988, as amended at 54 FR 13827, Apr. 5, 1989; 66 FR 38802, July 25, 2001]

§ 878.4200 Introduction/drainage catheter and accessories.

(a) *Identification.* An introduction/drainage catheter is a device that is a flexible single or multilumen tube intended to be used to introduce nondrug fluids into body cavities other than blood vessels, drain fluids from body cavities, or evaluate certain physiologic conditions. Examples include irrigation and drainage catheters, pediatric catheters, peritoneal catheters (including dialysis), and other general surgical catheters. An introduction/drainage catheter accessory is intended to aid in the manipulation of or insertion of the device into the body. Examples of accessories include adaptors, connectors, and catheter needles.

(b) *Classification.* Class I (general controls). The device is exempt from the premarket notification procedures in subpart E of part 807 of this chapter subject to § 878.9.

[53 FR 23872, June 24, 1988, as amended at 65 FR 2318, Jan. 14, 2000]

§ 878.4300 Implantable clip.

(a) *Identification.* An implantable clip is a clip-like device intended to connect internal tissues to aid healing. It is not absorbable.

(b) *Classification.* Class II.

§ 878.4320 Removable skin clip.

(a) *Identification.* A removable skin clip is a clip-like device intended to connect skin tissues temporarily to aid healing. It is not absorbable.

(b) *Classification.* Class I (general controls). The device is exempt from the premarket notification procedures in subpart E of part 807 of this chapter subject to § 878.9.

[53 FR 23872, June 24, 1988, as amended at 65 FR 2318, Jan. 14, 2000]

§ 878.4350 Cryosurgical unit and accessories.

(a) *Identification*—(1) *Cryosurgical unit with a liquid nitrogen cooled cryoprobe and accessories.* A cryosurgical unit with a liquid nitrogen cooled cryoprobe and accessories is a device intended to destroy tissue during surgical procedures by applying extreme cold.

(2) *Cryosurgical unit with a nitrous oxide cooled cryoprobe and accessories.* A cryosurgical unit with a nitrous oxide cooled cryoprobe and accessories is a device intended to destroy tissue during surgical procedures, including urological applications, by applying extreme cold.

(3) *Cryosurgical unit with a carbon dioxide cooled cryoprobe or a carbon dioxide dry ice applicator and accessories.* A cryosurgical unit with a carbon dioxide cooled cryoprobe or a carbon dioxide dry ice applicator and accessories is a

device intended to destroy tissue during surgical procedures by applying extreme cold. The device is intended to treat disease conditions such as tumors, skin cancers, acne scars, or hemangiomas (benign tumors consisting of newly formed blood vessels) and various benign or malignant gynecological conditions affecting vulvar, vaginal, or cervical tissue. The device is not intended for urological applications.

(b) *Classification.* Class II.

§ 878.4370 Surgical drape and drape accessories.

(a) *Identification.* A surgical drape and drape accessories is a device made of natural or synthetic materials intended to be used as a protective patient covering, such as to isolate a site of surgical incision from microbial and other contamination. The device includes a plastic wound protector that may adhere to the skin around a surgical incision or be placed in a wound to cover its exposed edges, and a latex drape with a self-retaining finger cot that is intended to allow repeated insertion of the surgeon's finger into the rectum during performance of a transurethral prostatectomy.

(b) *Classification.* Class II.

§ 878.4380 Drape adhesive.

(a) *Identification.* A drape adhesive is a device intended to be placed on the skin to attach a surgical drape.

(b) *Classification.* Class I (general controls). The device is exempt from the premarket notification procedures in subpart E of part 807 of this chapter, subject to the limitations in § 878.9.

[53 FR 23872, June 24, 1988, as amended at 59 FR 63010, Dec. 7, 1994; 66 FR 38802, July 25, 2001]

§ 878.4400 Electrosurgical cutting and coagulation device and accessories.

(a) *Identification.* An electrosurgical cutting and coagulation device and accessories is a device intended to remove tissue and control bleeding by use of high-frequency electrical current.

(b) *Classification.* Class II.

§ 878.4410 Low energy ultrasound wound cleaner.

(a) *Identification.* A low energy ultrasound wound cleaner is a device that uses ultrasound energy to vaporize a solution and generate a mist that is used for the cleaning and maintenance debridement of wounds. Low levels of ultrasound energy may be carried to the wound by the saline mist.

(b) *Classification.* Class II (special controls). The special control is FDA's guidance document entitled "Class II Special Controls Guidance Document: Low Energy Ultrasound Wound Cleaner." See § 878.1(e) for the availability of this guidance document.

[70 FR 67355, Nov. 7, 2005]

§ 878.4440 Eye pad.

(a) *Identification.* An eye pad is a device that consists of a pad made of various materials, such as gauze and cotton, intended for use as a bandage over the eye for protection or absorption of secretions.

(b) *Classification.* Class I (general controls). The device is exempt from the premarket notification procedures in subpart E of part 807 of this chapter, subject to the limitations in § 878.9.

[53 FR 23872, June 24, 1988, as amended at 59 FR 63010, Dec. 7, 1994; 66 FR 38803, July 25, 2001]

§ 878.4450 Nonabsorbable gauze for internal use.

(a) *Identification.* Nonabsorbable gauze for internal use is a device made of an open mesh fabric intended to be used inside the body or a surgical incision or applied to internal organs or structures, to control bleeding, absorb fluid, or protect organs or structures from abrasion, drying, or contamination. The device is woven from material made of not less than 50 percent by mass cotton, cellulose, or a simple chemical derivative of cellulose, and contains x-ray detectable elements.

(b) *Classification.* Class I (general controls). The device is exempt from the premarket notification procedures in subpart E of part 807 of this chapter, subject to the limitations in § 878.9.

[53 FR 23872, June 24, 1988, as amended at 61 FR 1123, Jan. 16, 1996; 66 FR 38803, July 25, 2001]

§ 878.4460 Surgeon's glove.

(a) *Identification.* A surgeon's glove is a device made of natural or synthetic rubber intended to be worn by operating room personnel to protect a surgical wound from contamination. The lubricating or dusting powder used in the glove is excluded.

(b) *Classification.* Class I (general controls).

[53 FR 23872, June 24, 1988, as amended at 66 FR 46952, Sept. 10, 2001]

§ 878.4470 Surgeon's gloving cream.

(a) *Identification.* Surgeon's gloving cream is an ointment intended to be used to lubricate the user's hand before putting on a surgeon's glove.

(b) *Classification.* Class I (general controls). The device is exempt from the premarket notification procedures in subpart E of part 807 of this chapter, subject to the limitations in § 878.9.

[53 FR 23872, June 24, 1988, as amended at 59 FR 63010, Dec. 7, 1994; 66 FR 38803, July 25, 2001]

§ 878.4480 Absorbable powder for lubricating a surgeon's glove.

(a) *Identification.* Absorbable powder for lubricating a surgeon's glove is a powder made from corn starch that meets the specifications for absorbable powder in the United States Pharmacopeia (U.S.P.) and that is intended to be used to lubricate the surgeon's hand before putting on a surgeon's glove. The device is absorbable through biological degradation.

(b) *Classification.* Class III.

(c) *Date PMA or notice of completion of a PDP is required.* As of May 28, 1976, an approval under section 515 of the act is required before this device may be commercially distributed. See § 878.3.

§ 878.4490 Absorbable hemostatic agent and dressing.

(a) *Identification.* An absorbable hemostatic agent or dressing is a device intended to produce hemostasis by accelerating the clotting process of blood. It is absorbable.

(b) *Classification.* Class III.

(c) *Date PMA or notice of completion of a PDP is required.* As of May 28, 1976, an approval under section 515 of the act is required before this device may be commercially distributed. See § 878.3.

§ 878.4493 Absorbable poly(glycolide/L-lactide) surgical suture.

(a) *Identification.* An absorbable poly(glycolide/L-lactide) surgical suture (PGL suture) is an absorbable sterile, flexible strand as prepared and synthesized from homopolymers of glycolide and copolymers made from 90 percent glycolide and 10 percent L-lactide, and is indicated for use in soft tissue approximation. A PGL suture meets United States Pharmacopeia (U.S.P.) requirements as described in the U.S.P. "Monograph for Absorbable Surgical Sutures;" it may be monofilament or multifilament (braided) in form; it may be uncoated or coated; and it may be undyed or dyed with an FDA-approved color additive. Also, the suture may be provided with or without a standard needle attached.

(b) *Classification.* Class II (special controls). The special control for this device is FDA's "Class II Special Controls Guidance Document: Surgical Sutures; Guidance for Industry and FDA." See § 878.1(e) for the availability of this guidance document.

[56 FR 47151, Sept. 18, 1991, as amended at 68 FR 32984, June 3, 2003]

§ 878.4495 Stainless steel suture.

(a) *Identification.* A stainless steel suture is a needled or unneedled nonabsorbable surgical suture composed of 316L stainless steel, in USP sizes 12–0 through 10, or a substantially equivalent stainless steel suture, intended for use in abdominal wound closure, intestinal anastomosis, hernia repair, and sternal closure.

(b) *Classification.* Class II (special controls). The special control for this device is FDA's "Class II Special Controls Guidance Document: Surgical Sutures; Guidance for Industry and FDA." See § 878.1(e) for the availability of this guidance document.

[65 FR 19836, Apr. 13, 2000, as amended at 68 FR 32984, June 3, 2003]

§ 878.4520 Polytetrafluoroethylene injectable.

(a) *Identification.* Polytetrafluoroethylene injectable is an injectable paste prosthetic device composed of polytetrafluoroethylene intended to be used to augment or reconstruct a vocal cord.

(b) *Classification.* Class III.

(c) *Date PMA or notice of completion of a PDP is required.* As of May 28, 1976, an approval under section 515 of the act is required before this device may be commercially distributed. See § 878.3.

§ 878.4580 Surgical lamp.

(a) *Identification.* A surgical lamp (including a fixture) is a device intended to be used to provide visible illumination of the surgical field or the patient.

(b) *Classification.* Class II.

§ 878.4630 Ultraviolet lamp for dermatologic disorders.

(a) *Identification.* An ultraviolet lamp for dermatologic disorders is a device (including a fixture) intended to provide ultraviolet radiation of the body to photoactivate a drug in the treatment of a dermatologic disorder if the labeling of the drug intended for use with the device bears adequate directions for the device's use with that drug.

(b) *Classification.* Class II.

§ 878.4635 Ultraviolet lamp for tanning.

(a) *Identification.* An ultraviolet lamp for tanning is a device that is a lamp (including a fixture) intended to provide ultraviolet radiation to tan the skin. See § 1040.20 of this chapter.

(b) *Classification.* Class I (general controls). The device is exempt from the premarket notification procedures in subpart E of part 807 of this chapter, subject to the limitations in § 878.9.

[55 FR 48440, Nov. 20, 1990, as amended at 59 FR 63010, Dec. 7, 1994; 66 FR 38803, July 25, 2001]

§ 878.4660 Skin marker.

(a) *Identification.* A skin marker is a pen-like device intended to be used to write on the patient's skin, e.g., to outline surgical incision sites or mark anatomical sites for accurate blood pressure measurement.

(b) *Classification.* Class I (general controls). The device is exempt from the premarket notification procedures in subpart E of part 807 of this chapter, subject to the limitations in § 878.9.

[53 FR 23872, June 24, 1988, as amended at 59 FR 63010, Dec. 7, 1994; 66 FR 38803, July 25, 2001]

§ 878.4680 Nonpowered, single patient, portable suction apparatus.

(a) *Identification.* A nonpowered, single patient, portable suction apparatus is a device that consists of a manually operated plastic, disposable evacuation system intended to provide a vacuum for suction drainage of surgical wounds.

(b) *Classification.* Class I (general controls). The device is exempt from the premarket notification procedures in subpart E of part 807 of this chapter subject to § 878.9.

[53 FR 23872, June 24, 1988, as amended at 65 FR 2318, Jan. 14, 2000]

§ 878.4700 Surgical microscope and accessories.

(a) *Identification.* A surgical microscope and accessories is an AC-powered device intended for use during surgery to provide a magnified view of the surgical field.

(b) *Classification.* Class I (general controls). The device is exempt from the premarket notification procedures in subpart E of part 807 of this chapter, subject to the limitations in § 878.9.

[55 FR 48440, Nov. 20, 1990, as amended at 59 FR 63010, Dec. 7, 1994; 66 FR 38803, July 25, 2001]

§ 878.4730 Surgical skin degreaser or adhesive tape solvent.

(a) *Identification.* A surgical skin degreaser or an adhesive tape solvent is a device that consists of a liquid such as 1,1,2-trichloro-1,2,2-trifluoroethane; 1,1,1-trichloroethane; and 1,1,1-trichloroethane with mineral spirits intended to be used to dissolve surface skin oil or adhesive tape.

(b) *Classification.* Class I (general controls). The device is exempt from the premarket notification procedures in subpart E of part 807 of this chapter, subject to the limitations in § 878.9.

[53 FR 23872, June 24, 1988, as amended at 59 FR 63010, Dec. 7, 1994; 66 FR 38803, July 25, 2001]

§ 878.4750 Implantable staple.

(a) *Identification.* An implantable staple is a staple-like device intended to connect internal tissues to aid healing. It is not absorbable.

(b) *Classification.* Class II.

§ 878.4760 Removable skin staple.

(a) *Identification.* A removable skin staple is a staple-like device intended to connect external tissues temporarily to aid healing. It is not absorbable.

(b) *Classification.* Class I (general controls). The device is exempt from the premarket notification procedures in subpart E of part 807 of this chapter subject to § 878.9.

[53 FR 23872, June 24, 1988, as amended at 65 FR 2318, Jan. 14, 2000]

§ 878.4780 Powered suction pump.

(a) *Identification.* A powered suction pump is a portable, AC-powered or compressed air-powered device intended to be used to remove infectious materials from wounds or fluids from a patient's airway or respiratory support system. The device may be used during surgery in the operating room or at the patient's bedside. The device may include a microbial filter.

(b) *Classification.* Class II.

§ 878.4800 Manual surgical instrument for general use.

(a) *Identification.* A manual surgical instrument for general use is a nonpowered, hand-held, or hand-manipulated device, either reusable or disposable, intended to be used in various general surgical procedures. The device includes the applicator, clip applier, biopsy brush, manual dermabrasion brush, scrub brush, cannula, ligature carrier, chisel, clamp, contractor, curette, cutter, dissector, elevator, skin graft expander, file, forceps, gouge, instrument guide, needle guide, hammer, hemostat, amputation hook, ligature passing and knot-tying instrument, knife, blood lancet, mallet, disposable or reusable aspiration and injection needle, disposable or reusable suturing needle, osteotome, pliers, rasp, retainer, retractor, saw, scalpel blade, scalpel handle, one-piece scalpel, snare, spatula, stapler, disposable or reusable stripper, stylet, suturing apparatus for the stomach and intestine, measuring tape, and calipers. A surgical instrument that has specialized uses in a specific medical specialty is classified in separate regulations in parts 868 through 892.

(b) *Classification.* Class I (general controls). The device is exempt from the premarket notification procedures in subpart E of part 807 of this chapter, subject to the limitations in § 878.9.

[53 FR 23872, June 24, 1988, as amended at 54 FR 13828, Apr. 5, 1989; 59 FR 63010, Dec. 7, 1994; 66 FR 38803, July 25, 2001]

§ 878.4810 Laser surgical instrument for use in general and plastic surgery and in dermatology.

(a) *Identification.* (1) A carbon dioxide laser for use in general surgery and in dermatology is a laser device intended to cut, destroy, or remove tissue by light energy emitted by carbon dioxide.

(2) An argon laser for use in dermatology is a laser device intended to destroy or coagulate tissue by light energy emitted by argon.

(b) *Classification.* (1) Class II.

(2) Class I for special laser gas mixtures used as a lasing medium for this class of lasers. The devices subject to this paragraph (b)(2) are exempt from the premarket notification procedures in subpart E of part 807 of this chapter, subject to the limitations in § 878.9.

[53 FR 23872, June 24, 1988, as amended at 61 FR 1123, Jan. 16, 1996; 66 FR 38803, July 25, 2001]

§ 878.4820 Surgical instrument motors and accessories/attachments.

(a) *Identification.* Surgical instrument motors and accessories are AC-powered, battery-powered, or air-powered devices intended for use during surgical procedures to provide power to operate various accessories or attachments to cut hard tissue or bone and soft tissue. Accessories or attachments may include a bur, chisel (osteotome), dermabrasion brush, dermatome, drill bit, hammerhead, pin driver, and saw blade.

(b) *Classification.* Class I (general controls). The device is exempt from the premarket notification procedures in

subpart E of part 807 of this chapter subject to § 878.9.

[55 FR 48440, Nov. 20, 1990, as amended at 65 FR 2318, 2000]

§ 878.4830 Absorbable surgical gut suture.

(a) *Identification.* An absorbable surgical gut suture, both plain and chromic, is an absorbable, sterile, flexible thread prepared from either the serosal connective tissue layer of beef (bovine) or the submucosal fibrous tissue of sheep (ovine) intestine, and is intended for use in soft tissue approximation.

(b) *Classification.* Class II (special controls). The special control for this device is FDA's "Class II Special Controls Guidance Document: Surgical Sutures; Guidance for Industry and FDA." See § 878.1(e) for the availability of this guidance document.

[54 FR 50738, Dec. 11, 1989, as amended at 68 FR 32984, June 3, 2003]

§ 878.4840 Absorbable polydioxanone surgical suture.

(a) *Identification.* An absorbable polydioxanone surgical suture is an absorbable, flexible, sterile, monofilament thread prepared from polyester polymer poly (p-dioxanone) and is intended for use in soft tissue approximation, including pediatric cardiovascular tissue where growth is expected to occur, and ophthalmic surgery. It may be coated or uncoated, undyed or dyed, and with or without a standard needle attached.

(b) *Classification.* Class II (special controls). The special control for the device is FDA's "Class II Special Controls Guidance Document: Surgical Sutures; Guidance for Industry and FDA." See § 878.1(e) for the availability of this guidance document.

[67 FR 77676, Dec. 19, 2002]

§ 878.4930 Suture retention device.

(a) *Identification.* A suture retention device is a device, such as a retention bridge, a surgical button, or a suture bolster, intended to aid wound healing by distributing suture tension over a larger area in the patient.

(b) *Classification.* Class I (general controls). The device is exempt from the premarket notification procedures in subpart E of part 807 of this chapter, subject to the limitations in § 878.9.

[53 FR 23872, June 24, 1988, as amended at 59 FR 63010, Dec. 7, 1994; 66 FR 38803, July 25, 2001]

§ 878.4950 Manual operating table and accessories and manual operating chair and accessories.

(a) *Identification.* A manual operating table and accessories and a manual operating chair and accessories are nonpowered devices, usually with movable components, intended to be used to support a patient during diagnostic examinations or surgical procedures.

(b) *Classification.* Class I (general controls). The device is exempt from the premarket notification procedures in subpart E of part 807 of this chapter, subject to the limitations in § 878.9.

[53 FR 23872, June 24, 1988, as amended at 54 FR 13828, Apr. 5, 1989; 59 FR 63010, Dec. 7, 1994; 66 FR 38803, July 25, 2001]

§ 878.4960 Operating tables and accessories and operating chairs and accessories.

(a) *Identification.* Operating tables and accessories and operating chairs and accessories are AC-powered or air-powered devices, usually with movable components, intended for use during diagnostic examinations or surgical procedures to support and position a patient.

(b) *Classification.* Class I (general controls). The device is exempt from the premarket notification procedures in subpart E of part 807 of this chapter subject to § 878.9.

[55 FR 48440, Nov. 20, 1990, as amended at 65 FR 2318, Jan. 14, 2000]

§ 878.5000 Nonabsorbable poly(ethylene terephthalate) surgical suture.

(a) *Identification.* Nonabsorbable poly(ethylene terephthalate) surgical suture is a multifilament, nonabsorbable, sterile, flexible thread prepared from fibers of high molecular weight, long-chain, linear polyesters having recurrent aromatic rings as an integral component and is indicated for use in soft tissue approximation. The poly(ethylene terephthalate) surgical suture meets U.S.P. requirements as described in the U.S.P. Monograph for

Nonabsorbable Surgical Sutures; it may be provided uncoated or coated; and it may be undyed or dyed with an appropriate FDA listed color additive. Also, the suture may be provided with or without a standard needle attached.

(b) *Classification.* Class II (special controls). The special control for this device is FDA's "Class II Special Controls Guidance Document: Surgical Sutures; Guidance for Industry and FDA." See § 878.1(e) for the availability of this guidance document.

[56 FR 24685, May 31, 1991, as amended at 68 FR 32984, June 3, 2003]

§ 878.5010 Nonabsorbable polypropylene surgical suture.

(a) *Identification.* Nonabsorbable polypropylene surgical suture is a monofilament, nonabsorbable, sterile, flexible thread prepared from long-chain polyolefin polymer known as polypropylene and is indicated for use in soft tissue approximation. The polypropylene surgical suture meets United States Pharmacopeia (U.S.P.) requirements as described in the U.S.P. Monograph for Nonabsorbable Surgical Sutures; it may be undyed or dyed with an FDA approved color additive; and the suture may be provided with or without a standard needle attached.

(b) *Classification.* Class II (special controls). The special control for this device is FDA's "Class II Special Controls Guidance Document: Surgical Sutures; Guidance for Industry and FDA." See § 878.1(e) for the availability of this guidance document.

[56 FR 24685, May 31, 1991, as amended at 68 FR 32984, June 3, 2003]

§ 878.5020 Nonabsorbable polyamide surgical suture.

(a) *Identification.* Nonabsorbable polyamide surgical suture is a nonabsorbable, sterile, flexible thread prepared from long-chain aliphatic polymers Nylon 6 and Nylon 6,6 and is indicated for use in soft tissue approximation. The polyamide surgical suture meets United States Pharmacopeia (U.S.P.) requirements as described in the U.S.P. monograph for nonabsorbable surgical sutures; it may be monofilament or multifilament in form; it may be provided uncoated or coated; and it may be undyed or dyed with an appropriate FDA listed color additive. Also, the suture may be provided with or without a standard needle attached.

(b) *Classification.* Class II (special controls). The special control for this device is FDA's "Class II Special Controls Guidance Document: Surgical Sutures; Guidance for Industry and FDA." See § 878.1(e) for the availability of this guidance document.

[56 FR 24685, May 31, 1991, as amended at 68 FR 32985, June 3, 2003]

§ 878.5030 Natural nonabsorbable silk surgical suture.

(a) *Identification.* Natural nonabsorbable silk surgical suture is a nonabsorbable, sterile, flexible multifilament thread composed of an organic protein called fibroin. This protein is derived from the domesticated species *Bombyx mori* (*B. mori*) of the family *Bombycidae.* Natural nonabsorbable silk surgical suture is indicated for use in soft tissue approximation. Natural nonabsorbable silk surgical suture meets the United States Pharmacopeia (U.S.P.) monograph requirements for Nonabsorbable Surgical Suture (class I). Natural nonabsorbable silk surgical suture may be braided or twisted; it may be provided uncoated or coated; and it may be undyed or dyed with an FDA listed color additive.

(b) *Classification.* Class II (special controls). The special control for this device is FDA's "Class II Special Controls Guidance Document: Surgical Sutures; Guidance for Industry and FDA." See § 878.1(e) for the availability of this guidance document.

[58 FR 57558, Oct. 26, 1993, as amended at 68 FR 32985, June 3, 2003]

§ 878.5035 Nonabsorbable expanded polytetrafluoroethylene surgical suture.

(a) *Identification.* Nonabsorbable expanded polytetrafluoroethylene (ePTFE) surgical suture is a monofilament, nonabsorbable, sterile, flexible thread prepared from ePTFE and is intended for use in soft tissue approximation and ligation, including cardiovascular surgery. It may be undyed or dyed with an approved color additive and may be provided with or without an attached needle(s).

(b) *Classification.* Class II (special controls). The special control for this device is FDA's "Class II Special Controls Guidance Document: Surgical Sutures; Guidance for Industry and FDA." See § 878.1(e) for the availability of this guidance document.

[65 FR 20735, Apr. 18, 2000, as amended at 68 FR 32985, June 3, 2003]

§ 878.5040 Suction lipoplasty system.

(a) *Identification.* A suction lipoplasty system is a device intended for aesthetic body contouring. The device consists of a powered suction pump (containing a microbial filter on the exhaust and a microbial in-line filter in the connecting tubing between the collection bottle and the safety trap), collection bottle, cannula, and connecting tube. The microbial filters, tubing, collection bottle, and cannula must be capable of being changed between patients. The powered suction pump has a motor with a minimum of 1/3 horsepower, a variable vacuum range from 0 to 29.9 inches of mercury, vacuum control valves to regulate the vacuum with accompanying vacuum gauges, a single or double rotary vane (with or without oil), a single or double diaphragm, a single or double piston, and a safety trap.

(b) *Classification.* Class II (special controls). Consensus standards and labeling restrictions.

[63 FR 7705, Feb. 17, 1998]

Subpart F—Therapeutic Devices

§ 878.5070 Air-handling apparatus for a surgical operating room.

(a) *Identification.* Air-handling apparatus for a surgical operating room is a device intended to produce a directed, nonturbulent flow of air that has been filtered to remove particulate matter and microorganisms to provide an area free of contaminants to reduce the possibility of infection in the patient.

(b) *Classification.* Class II.

§ 878.5350 Needle-type epilator.

(a) *Identification.* A needle-type epilator is a device intended to destroy the dermal papilla of a hair by applying electric current at the tip of a fine needle that has been inserted close to the hair shaft, under the skin, and into the dermal papilla. The electric current may be high-frequency AC current, high-frequency AC combined with DC current, or DC current only.

(b) *Classification.* Class I (general controls). The device is exempt from the premarket notification procedures in subpart E of part 807 of this chapter, subject to the limitations in § 878.9.

[53 FR 23872, June 24, 1988, as amended at 61 FR 1123, Jan. 16, 1996; 66 FR 38803, July 25, 2001]

§ 878.5360 Tweezer-type epilator.

(a) *Identification.* The tweezer-type epilator is an electrical device intended to remove hair. The energy provided at the tip of the tweezer used to remove hair may be radio frequency, galvanic (direct current), or a combination of radio frequency and galvanic energy.

(b) *Classification.* Class I (general controls). The device is exempt from premarket notification procedures in subpart E of part 807 of this chapter subject to § 878.9.

[63 FR 57060, Oct. 26, 1998]

§ 878.5650 Topical oxygen chamber for extremities.

(a) *Identification.* A topical oxygen chamber for extremities is a device intended to surround hermetically a patient's limb and apply humidified oxygen topically at a pressure slightly greater than atmospheric pressure to aid healing of chronic skin ulcers or bed sores.

(b) *Classification.* Class III.

(c) *Date PMA or notice of completion of a PDP is required.* No effective date has been established of the requirement for premarket approval. See § 878.3.

§ 878.5900 Nonpneumatic tourniquet.

(a) *Identification.* A nonpneumatic tourniquet is a device consisting of a strap or tubing intended to be wrapped around a patient's limb and tightened to reduce circulation.

(b) *Classification.* Class I (general controls). The device is exempt from the premarket notification procedures in

subpart E of part 807 of this chapter, subject to the limitations in § 878.9.

[53 FR 23872, June 24, 1988, as amended at 54 FR 13828, Apr. 5, 1989; 59 FR 63010, Dec. 7, 1994; 66 FR 38803, July 25, 2001]

§ 878.5910 Pneumatic tourniquet.

(a) *Identification.* A pneumatic tourniquet is an air-powered device consisting of a pressure-regulating unit, connecting tubing, and an inflatable cuff. The cuff is intended to be wrapped around a patient's limb and inflated to reduce or totally occlude circulation during surgery.

(b) *Classification.* Class I (general controls). The device is exempt from the premarket notification procedures in subpart E of part 807 of this chapter, subject to the limitations in § 878.9.

[53 FR 23872, June 24, 1988, as amended at 61 FR 1123, Jan. 16, 1996; 66 FR 38803, July 25, 2001]

PART 880—GENERAL HOSPITAL AND PERSONAL USE DEVICES

Subpart A—General Provisions

Sec.

880.1 Scope.
880.3 Effective dates of requirement for premarket approval.
880.9 Limitations of exemptions from section 510(k) of the Federal Food, Drug, and Cosmetic Act (the act).

Subpart B [Reserved]

Subpart C—General Hospital and Personal Use Monitoring Devices

880.2200 Liquid crystal forehead temperature strip.
880.2400 Bed-patient monitor.
880.2420 Electronic monitor for gravity flow infusion systems.
880.2460 Electrically powered spinal fluid pressure monitor.
880.2500 Spinal fluid manometer.
880.2700 Stand-on patient scale.
880.2720 Patient scale.
880.2740 Surgical sponge scale.
880.2800 Sterilization process indicator.
880.2900 Clinical color change thermometer.
880.2910 Clinical electronic thermometer.
880.2920 Clinical mercury thermometer.
880.2930 Apgar timer.

Subparts D–E [Reserved]

Subpart F—General Hospital and Personal Use Therapeutic Devices

880.5025 I.V. container.
880.5045 Medical recirculating air cleaner.
880.5075 Elastic bandage.
880.5090 Liquid bandage.
880.5100 AC-powered adjustable hospital bed.
880.5110 Hydraulic adjustable hospital bed.
880.5120 Manual adjustable hospital bed.
880.5130 Infant radiant warmer.
880.5140 Pediatric hospital bed.
880.5150 Nonpowered flotation therapy mattress.
880.5160 Therapeutic medical binder.
880.5180 Burn sheet.
880.5200 Intravascular catheter.
880.5210 Intravascular catheter securement device.
880.5240 Medical adhesive tape and adhesive bandage.
880.5270 Neonatal eye pad.
880.5300 Medical absorbent fiber.
880.5400 Neonatal incubator.
880.5410 Neonatal transport incubator.
880.5420 Pressure infusor for an I.V. bag.
880.5430 Nonelectrically powered fluid injector.
880.5440 Intravascular administration set.
880.5450 Patient care reverse isolation chamber.
880.5475 Jet lavage.
880.5500 AC-powered patient lift.
880.5510 Non-AC-powered patient lift.
880.5550 Alternating pressure air flotation mattress.
880.5560 Temperature regulated water mattress.
880.5570 Hypodermic single lumen needle.
880.5580 Acupuncture needle.
880.5630 Nipple shield.
880.5640 Lamb feeding nipple.
880.5680 Pediatric position holder.
880.5700 Neonatal phototherapy unit.
880.5725 Infusion pump.
880.5740 Suction snakebite kit.
880.5760 Chemical cold pack snakebite kit.
880.5780 Medical support stocking.
880.5820 Therapeutic scrotal support.
880.5860 Piston syringe.
880.5950 Umbilical occlusion device.
880.5960 Lice removal kit.
880.5965 Subcutaneous, implanted, intravascular infusion port and catheter.
880.5970 Percutaneous, implanted, long-term intravascular catheter.

Subpart G—General Hospital and Personal Use Miscellaneous Devices

880.6025 Absorbent tipped applicator.
880.6050 Ice bag.
880.6060 Medical disposable bedding.
880.6070 Bed board.
880.6080 Cardiopulmonary resuscitation board.
880.6085 Hot/cold water bottle.

880.6100 Ethylene oxide gas aerator cabinet.
880.6140 Medical chair and table.
880.6150 Ultrasonic cleaner for medical instruments.
880.6175 [Reserved]
880.6185 Cast cover.
880.6190 Mattress cover for medical purposes.
880.6200 Ring cutter.
880.6230 Tongue depressor.
880.6250 Patient examination glove.
880.6265 Examination gown.
880.6280 Medical insole.
880.6300 Implantable radiofrequency transponder system for patient identification and health information.
880.6320 AC-powered medical examination light.
880.6350 Battery-powered medical examination light.
880.6375 Patient lubricant.
880.6430 Liquid medication dispenser.
880.6450 Skin pressure protectors.
880.6500 Medical ultraviolet air purifier.
880.6710 Medical ultraviolet water purifier.
880.6730 Body waste receptacle.
880.6740 Vacuum-powered body fluid suction apparatus.
880.6760 Protective restraint.
880.6775 Powered patient transfer device.
880.6785 Manual patient transfer device.
880.6800 Washers for body waste receptacles.
880.6820 Medical disposable scissors.
880.6850 Sterilization wrap.
880.6860 Ethylene oxide gas sterilizer.
880.6870 Dry-heat sterilizer.
880.6880 Steam sterilizer.
880.6885 Liquid chemical sterilants/high level disinfectants.
880.6890 General purpose disinfectants.
880.6900 Hand-carried stretcher.
880.6910 Wheeled stretcher.
880.6920 Syringe needle introducer.
880.6960 Irrigating syringe.
880.6970 Liquid crystal vein locator.
880.6980 Vein stabilizer.
880.6990 Infusion stand.
880.6991 Medical washer.
880.6992 Medical washer-disinfector.

AUTHORITY: 21 U.S.C. 351, 360, 360c, 360e, 360j, 371.

SOURCE: 45 FR 69682, Oct. 21, 1980, unless otherwise noted.

Subpart A—General Provisions

§ 880.1 Scope.

(a) This part sets forth the classification of general hospital and personal use devices intended for human use that are in commercial distribution.

(b) The identification of a device in a regulation in this part is not a precise description of every device that is, or will be, subject to the regulation. A manufacturer who submits a premarket notification submission for a device under part 807 may not show merely that the device is accurately described by the section title and identification provisions of a regulation in this part, but shall state why the device is substantially equivalent to other devices, as required by § 807.87.

(c) To avoid duplicative listings, a general hospital and personal use device that has two or more types of uses (e.g., used both as a diagnostic device and as a therapeutic device) is listed only in one subpart.

(d) References in this part to regulatory sections of the Code of Federal Regulations are to chapter I of title 21, unless otherwise noted.

(e) Guidance documents referenced in this part are available on the Internet at *http://www.fda.gov/cdrh/guidance.html.*

[52 FR 17738, May 11, 1987, as amended at 69 FR 71704, Dec. 8, 2004]

§ 880.3 Effective dates of requirement for premarket approval.

A device included in this part that is classified into class III (premarket approval) shall not be commercially distributed after the date shown in the regulation classifying the device unless the manufacturer has an approval under section 515 of the act (unless an exemption has been granted under section 520(g)(2) of the act). An approval under section 515 of the act consists of FDA's issuance of an order approving an application for premarket approval (PMA) for the device or declaring completed a product development protocol (PDP) for the device.

(a) Before FDA requires that a device commercially distributed before the enactment date of the amendments, or a device that has been found substantially equivalent to such a device, has an approval under section 515 of the act FDA must promulgate a regulation under section 515(b) of the act requiring such approval, except as provided in paragraph (b) of this section. Such a regulation under section 515(b) of the act shall not be effective during the grace period ending on the 90th day after its promulgation or on the last day of the 30th full calendar month after the regulation that classifies the device into class III is effective, whichever is later. See section 501(f)(2)(B) of the act. Accordingly, unless an effective date of the requirement for premarket approval is shown in the regulation for a device classified into class III in this part, the device may be commercially distributed without FDA's issuance of an order approving a PMA or declaring completed a PDP for the device. If FDA promulgates a regulation under section 515(b) of the act requiring premarket approval for a device, section 501(f)(1)(A) of the act applies to the device.

(b) Any new, not substantially equivalent, device introduced into commercial distribution on or after May 28, 1976, including a device formerly marketed that has been substantially altered, is classified by statute (section 513(f) of the act) into class III without any grace period and FDA must have issued an order approving a PMA or declaring completed a PDP for the device before the device is commercially distributed unless it is reclassified. If FDA knows that a device being commercially distributed may be a "new" devices defined in this section because of any new intended use or other reasons, FDA may codify the statutory classification of the device into class III for such new use. Accordingly, the regulation for such a class III device states that as of the enactment date of the amendments, May 28, 1976, the device must have an approval under section 515 of the act before commercial distribution.

[52 FR 17738, May 11, 1987]

§ 880.9 Limitations of exemptions from section 510(k) of the Federal Food, Drug, and Cosmetic Act (the act).

The exemption from the requirement of premarket notification (section 510(k) of the act) for a generic type of class I or II device is only to the extent that the device has existing or reasonably foreseeable characteristics of commercially distributed devices within that generic type or, in the case of in vitro diagnostic devices, only to the extent that misdiagnosis as a result of using the device would not be associated with high morbidity or mortality. Accordingly, manufacturers of any commercially distributed class I or II device for which FDA has granted an exemption from the requirement of premarket notification must still submit a premarket notification to FDA before introducing or delivering for introduction into interstate commerce for commercial distribution the device when:

(a) The device is intended for a use different from the intended use of a legally marketed device in that generic type of device; e.g., the device is intended for a different medical purpose, or the device is intended for lay use where the former intended use was by health care professionals only;

(b) The modified device operates using a different fundamental scientific technology than a legally marketed device in that generic type of device; e.g., a surgical instrument cuts tissue with a laser beam rather than with a sharpened metal blade, or an in vitro diagnostic device detects or identifies infectious agents by using deoxyribonucleic acid (DNA) probe or nucleic acid hybridization technology rather than culture or immunoassay technology; or

(c) The device is an in vitro device that is intended:

(1) For use in the diagnosis, monitoring, or screening of neoplastic diseases with the exception of immunohistochemical devices;

(2) For use in screening or diagnosis of familial or acquired genetic disorders, including inborn errors of metabolism;

(3) For measuring an analyte that serves as a surrogate marker for screening, diagnosis, or monitoring life-threatening diseases such as acquired immune deficiency syndrome (AIDS), chronic or active hepatitis, tuberculosis, or myocardial infarction or to monitor therapy;

(4) For assessing the risk of cardiovascular diseases;

(5) For use in diabetes management;

(6) For identifying or inferring the identity of a microorganism directly from clinical material;

(7) For detection of antibodies to microorganisms other than immunoglobulin G (IgG) or IgG assays when the results are not qualitative, or are used to determine immunity, or the

assay is intended for use in matrices other than serum or plasma;

(8) For noninvasive testing as defined in § 812.3(k) of this chapter; and

(9) For near patient testing (point of care).

[65 FR 2318, Jan. 14, 2000]

Subpart B [Reserved]

Subpart C—General Hospital and Personal Use Monitoring Devices

§ 880.2200 Liquid crystal forehead temperature strip.

(a) *Identification.* A liquid crystal forehead temperature strip is a device applied to the forehead that is used to indicate the presence or absence of fever, or to monitor body temperature changes. The device displays the color changes of heat sensitive liquid crystals corresponding to the variation in the surface temperature of the skin. The liquid crystals, which are cholesteric esters, are sealed in plastic.

(b) *Classification.* Class II (special controls). The device is exempt from the premarket notification procedures in subpart E of part 807 of this chapter subject to § 880.9.

[45 FR 69682–69737, Oct. 21, 1980, as amended at 63 FR 59228, Nov. 3, 1998]

§ 880.2400 Bed-patient monitor.

(a) *Identification.* A bed-patient monitor is a battery-powered device placed under a mattress and used to indicate by an alarm or other signal when a patient attempts to leave the bed.

(b) *Classification.* Class I (general controls). The device is exempt from the premarket notification procedures in subpart E of part 807 of this chapter subject to the limitations in § 880.9.

[45 FR 69682–69737, Oct. 21, 1980, as amended at 59 FR 63010, Dec. 7, 1994; 66 FR 38803, July 25, 2001]

§ 880.2420 Electronic monitor for gravity flow infusion systems.

(a) *Identification.* An electronic monitor for gravity flow infusion systems is a device used to monitor the amount of fluid being infused into a patient. The device consists of an electronic transducer and equipment for signal amplification, conditioning, and display.

(b) *Classification.* Class II (performance standards).

§ 880.2460 Electrically powered spinal fluid pressure monitor.

(a) *Identification.* An electrically powered spinal fluid pressure monitor is an electrically powered device used to measure spinal fluid pressure by the use of a transducer which converts spinal fluid pressure into an electrical signal. The device includes signal amplification, conditioning, and display equipment.

(b) *Classification.* Class II (performance standards).

§ 880.2500 Spinal fluid manometer.

(a) *Identification.* A spinal fluid manometer is a device used to measure spinal fluid pressure. The device uses a hollow needle, which is inserted into the spinal column fluid space, to connect the spinal fluid to a graduated column so that the pressure can be measured by reading the height of the fluid.

(b) *Classification.* Class II (performance standards).

§ 880.2700 Stand-on patient scale.

(a) *Identification.* A stand-on patient scale is a device intended for medical purposes that is used to weigh a patient who is able to stand on the scale platform.

(b) *Classification.* Class I (general controls). The device is exempt from the premarket notification procedures in subpart E of part 807 of this chapter, subject to the limitations in § 880.9. The device also is exempt from the current good manufacturing practice regulations in part 820 of this chapter, with the exception of § 820.180, with respect to general requirements concerning records, and § 820.198, with respect to complaint files.

[45 FR 69682–69737, Oct. 21, 1980, as amended at 66 FR 38803, July 25, 2001

§ 880.2720 Patient scale.

(a) *Identification.* A patient scale is a device intended for medical purposes that is used to measure the weight of a patient who cannot stand on a scale. This generic device includes devices placed under a bed or chair to weigh both the support and the patient, devices where the patient is lifted by a sling from a bed to be weighed, and devices where the patient is placed on the scale platform to be weighed. The device may be mechanical, battery powered, or AC-powered and may include transducers, electronic signal amplification, conditioning and display equipment.

(b) *Classification.* Class I (general controls). The device is exempt from the premarket notification procedures in subpart E of part 807 of this chapter subject to the limitations in § 880.9.

[45 FR 69682–69737, Oct. 21, 1980, as amended at 61 FR 1123, Jan. 16, 1996; 66 FR 38803, July 25, 2001]

§ 880.2740 Surgical sponge scale.

(a) *Identification.* A surgical sponge scale is a nonelectrically powered device used to weigh surgical sponges that have been used to absorb blood during surgery so that, by comparison with the known dry weight of the sponges, an estimate may be made of the blood lost by the patient during surgery.

(b) *Classification.* Class I (general controls). The device is exempt from the premarket notification procedures in subpart E of part 807 of this chapter, subject to the limitations in § 880.9. The device also is exempt from the current good manufacturing practice regulations in part 820 of this chapter, with the exception of § 820.180, with respect to general requirements concerning records, and § 820.198, with respect to complaint files.

[45 FR 69682–69737, Oct. 21, 1980, as amended at 66 FR 38804, July 25, 2001

§ 880.2800 Sterilization process indicator.

(a) *Biological sterilization process indicator*—(1) *Identification.* A biological sterilization process indicator is a device intended for use by a health care provider to accompany products being sterilized through a sterilization procedure and to monitor adequacy of sterilization. The device consists of a known number of microorganisms, of known resistance to the mode of sterilization, in or on a carrier and enclosed in a protective package. Subsequent growth or failure of the microorganisms to grow under suitable conditions indicates the adequacy of sterilization.

(2) *Classification.* Class II (performance standards).

(b) *Physical/chemical sterilization process indicator*—(1) *Identification.* A physical/chemical sterilization process indicator is a device intended for use by a health care provider to accompany products being sterilized through a sterilization procedure and to monitor one or more parameters of the sterilization process. The adequacy of the sterilization conditions as measured by these parameters is indicated by a visible change in the device.

(2) *Classification.* Class II (performance standards).

§ 880.2900 Clinical color change thermometer.

(a) *Identification.* A clinical color change thermometer is a disposable device used to measure a patient's oral, rectal, or axillary (armpit) body temperature. The device records body temperature by use of heat sensitive chemicals which are sealed at the end of a plastic or metal strip. Body heat causes a stable color change in the heat sensitive chemicals.

(b) *Classification.* Class I (general controls). The device is exempt from the premarket notification procedures in subpart E of part 807 of this chapter, subject to the limitations in § 880.9.

[45 FR 69682–69737, Oct. 21, 1980, as amended at 61 FR 1123, Jan. 16, 1996; 66 FR 38804, July 25, 2001]

§ 880.2910 Clinical electronic thermometer.

(a) *Identification.* A clinical electronic thermometer is a device used to measure the body temperature of a patient by means of a transducer coupled with an electronic signal amplification, conditioning, and display unit. The transducer may be in a detachable probe with or without a disposable cover.

(b) *Classification.* Class II (performance standards).

§ 880.2920 Clinical mercury thermometer.

(a) *Identification.* A clinical mercury thermometer is a device used to measure oral, rectal, or axillary (armpit) body temperature using the thermal expansion of mercury.

(b) *Classification.* Class II (special controls). The device is exempt from the premarket notification procedures in subpart E of part 807 of this chapter subject to § 880.9.

[45 FR 69682–69737, Oct. 21, 1980, as amended at 63 FR 59228, Nov. 3, 1998]

§ 880.2930 Apgar timer.

(a) *Identification.* The Apgar timer is a device intended to alert a health care provider to take the Apgar score of a newborn infant.

(b) *Classification.* Class I (general controls). The device is exempt from the premarket notification procedures in subpart E of part 807 of this chapter subject to the limitations in § 880.9. The device is also exempt from the current good manufacturing practice requirements in part 820 of this chapter, with the exception of § 820.180 of this chapter, with respect to general requirements concerning records, and § 820.198 of this chapter, with respect to complaint files.

[63 FR 59718, Nov. 5, 1998]

Subparts D–E [Reserved]

Subpart F—General Hospital and Personal Use Therapeutic Devices

§ 880.5025 I.V. container.

(a) *Identification.* An I.V. container is a container made of plastic or glass used to hold a fluid mixture to be administered to a patient through an intravascular administration set.

(b) *Classification.* Class II (performance standards).

§ 880.5045 Medical recirculating air cleaner.

(a) *Identification.* A medical recirculating air cleaner is a device used to remove particles from the air for medical purposes. The device may function by electrostatic precipitation or filtration.

(b) *Classification.* Class II (performance standards).

§ 880.5075 Elastic bandage.

(a) *Identification.* An elastic bandage is a device consisting of either a long flat strip or a tube of elasticized material that is used to support and compress a part of a patient's body.

(b) *Classification.* Class I (general controls). The device is exempt from the premarket notification procedures in subpart E of part 807 of this chapter, subject to the limitations in § 880.9. The device also is exempt from the current good manufacturing practice regulations in part 820 of this chapter, with the exception of § 820.180, with respect to general requirements concerning records, and § 820.198, with respect to complaint files.

[45 FR 69682–69737, Oct. 21, 1980, as amended at 66 FR 38804, July 25, 2001]

§ 880.5090 Liquid bandage.

(a) *Identification.* A liquid bandage is a sterile device that is a liquid, semiliquid, or powder and liquid combination used to cover an opening in the skin or as a dressing for burns. The device is also used as a topical skin protectant.

(b) *Classification.* Class I (general controls). When used only as a skin protectant, the device is exempt from the premarket notification procedures in subpart E of part 807 of this chapter subject to § 880.9.

[45 FR 69682–69737, Oct. 21, 1980, as amended at 65 FR 2318, Jan. 14, 2000]

§ 880.5100 AC-powered adjustable hospital bed.

(a) *Identification.* An AC-powered adjustable hospital bed is a device intended for medical purposes that consists of a bed with a built-in electric motor and remote controls that can be operated by the patient to adjust the height and surface contour of the bed. The device includes movable and latchable side rails.

(b) *Classification.* Class II (special controls). The device is exempt from the premarket notification procedures in subpart E of part 807 of this chapter subject to § 880.9.

[45 FR 69682–69737, Oct. 21, 1980, as amended at 63 FR 59229, Nov. 3, 1998]

§ 880.5110 Hydraulic adjustable hospital bed.

(a) *Identification.* A hydraulic adjustable hospital bed is a device intended for medical purposes that consists of a bed with a hydraulic mechanism operated by an attendant to adjust the height and surface contour of the bed. The device includes movable and latchable side rails.

(b) *Classification.* Class I (general controls). The device is exempt from the premarket notification procedures in subpart E of part 807 of this chapter, subject to the limitations in § 880.9.

[45 FR 69682–69737, Oct. 21, 1980, as amended at 66 FR 38804, July 25, 2001]

§ 880.5120 Manual adjustable hospital bed.

(a) *Identification.* A manual adjustable hospital bed is a device intended for medical purposes that consists of a bed with a manual mechanism operated by an attendant to adjust the height and surface contour of the bed. The device includes movable and latchable side rails.

(b) *Classification.* Class I (general controls). The device is exempt from the premarket notification procedures in subpart E of part 807 of this chapter, subject to the limitations in § 880.9. The device is also exempt from the current good manufacturing practice regulations in part 820 of this chapter, with the exception of § 820.180, with respect to general requirements concerning records, and § 820.198, with respect to complaint files.

[45 FR 69682–69737, Oct. 21, 1980, as amended at 54 FR 25050, June 12, 1989; 66 FR 38804, July 25, 2001]

§ 880.5130 Infant radiant warmer.

(a) *Identification.* The infant radiant warmer is a device consisting of an infrared heating element intended to be placed over an infant to maintain the infant's body temperature by means of radiant heat. The device may also contain a temperature monitoring sensor, a heat output control mechanism, and an alarm system (infant temperature, manual mode if present, and failure alarms) to alert operators of a temperature condition over or under the set temperature, manual mode time limits, and device component failure, respectively. The device may be placed over a pediatric hospital bed or it may be built into the bed as a complete unit.

(b) *Classification.* Class II (Special Controls):

(1) The Association for the Advancement of Medical Instrumentation (AAMI) Voluntary Standard for the Infant Radiant Warmer;

(2) A prescription statement in accordance with § 801.109 of this chapter (restricted to use by or upon the order of qualified practitioners as determined by the States); and

(3) Labeling for use only in health care facilities and only by persons with specific training and experience in the use of the device.

[62 FR 33350, June 19, 1997]

§ 880.5140 Pediatric hospital bed.

(a) *Identification.* A pediatric hospital bed is a device intended for medical purposes that consists of a bed or crib designed for the use of a pediatric patient, with fixed end rails and movable and latchable side rails. The contour of the bed surface may be adjustable.

(b) *Classification.* Class II (special controls). The device is exempt from the premarket notification procedures in subpart E of part 807 of this chapter subject to § 880.9.

[45 FR 69682–69737, Oct. 21, 1980, as amended at 63 FR 59229, Nov. 3, 1998]

§ 880.5150 Nonpowered flotation therapy mattress.

(a) *Identification.* A nonpowered flotation therapy mattress is a mattress intended for medical purposes which contains air, fluid, or other materials that have the functionally equivalent effect of supporting a patient and avoiding excess pressure on local body areas. The device is intended to treat or prevent decubitus ulcers (bed sores).

(b) *Classification.* Class I (general controls). The device is exempt from the premarket notification procedures in subpart E of part 807 of this chapter, subject to the limitations in § 880.9. The device also is exempt from the current good manufacturing practice regulations in part 820 of this chapter, with the exception of § 820.180, with respect to general requirements concerning records, and § 820.198, with respect to complaint files.

[45 FR 69682–69737, Oct. 21, 1980, as amended at 66 FR 38804, July 25, 2001]

§880.5160 Therapeutic medical binder.

(a) *Identification.* A therapeutic medical binder is a device, usually made of cloth, that is intended for medical purposes and that can be secured by ties so that it supports the underlying part of the body or holds a dressing in place. This generic type of device includes the abdominal binder, breast binder, and perineal binder.

(b) *Classification.* Class I (general controls). The device is exempt from the premarket notification procedures in subpart E of part 807 of this chapter, subject to the limitations in §880.9. If the device is not labeled or otherwise represented as sterile, it is also exempt from the current good manufacturing practice regulations in part 820 of this chapter, with the exception of §820.180, with respect to general requirements concerning records, and §820.198, with respect to complaint files.

[45 FR 69682–69737, Oct. 21, 1980, as amended at 66 FR 38804, July 25, 2001]

§880.5180 Burn sheet.

(a) *Identification.* A burn sheet is a device made of a porous material that is wrapped aroung a burn victim to retain body heat, to absorb wound exudate, and to serve as a barrier against contaminants.

(b) *Classification.* Class I (general controls). The device is exempt from the premarket notification procedures in subpart E of part 807 of this chapter, subject to the limitations in §880.9.

[45 FR 69682–69737, Oct. 21, 1980, as amended at 59 FR 63011, Dec. 7, 1994; 66 FR 38804, July 25, 2001]

§880.5200 Intravascular catheter.

(a) *Identification.* An intravascular catheter is a device that consists of a slender tube and any necessary connecting fittings and that is inserted into the patient's vascular system for short term use (less than 30 days) to sample blood, monitor blood pressure, or administer fluids intravenously. The device may be constructed of metal, rubber, plastic, or a combination of these materials.

(b) *Classification.* Class II (performance standards).

§880.5210 Intravascular catheter securement device.

(a) *Identification.* An intravascular catheter securement device is a device with an adhesive backing that is placed over a needle or catheter and is used to keep the hub of the needle or the catheter flat and securely anchored to the skin.

(b) *Classification.* Class I (general controls). The device is exempt from the premarket notification procedures in subpart E of part 807 of this chapter, subject to the limitations in §880.9.

[45 FR 69682–69737, Oct. 21, 1980, as amended at 59 FR 63011, Dec. 7, 1994; 66 FR 38804, July 25, 2001]

§880.5240 Medical adhesive tape and adhesive bandage.

(a) *Identification.* A medical adhesive tape or adhesive bandage is a device intended for medical purposes that consists of a strip of fabric material or plastic, coated on one side with an adhesive, and may include a pad of surgical dressing without a disinfectant. The device is used to cover and protect wounds, to hold together the skin edges of a wound, to support an injured part of the body, or to secure objects to the skin.

(b) *Classification.* Class I (general controls). The device is exempt from the premarket notification procedures in subpart E of part 807 of this chapter, subject to the limitations in §880.9.

[45 FR 69682–69737, Oct. 21, 1980, as amended at 59 FR 63011, Dec. 7, 1994; 66 FR 38804, July 25, 2001]

§880.5270 Neonatal eye pad.

(a) *Identification.* A neonatal eye pad is an opaque device used to cover and protect the eye of an infant during therapeutic procedures, such as phototherapy.

(b) *Classification.* Class I (general controls). The device is exempt from the premarket notification procedures in subpart E of part 807 of this chapter subject to §880.9. If the device is not labeled or otherwise represented as sterile, it is exempt from the current good manufacturing practice regulations in part 820 of this chapter, with the exception of § 820.180 of this chapter, with respect to general requirements concerning records, and § 820.198 of this chapter, with respect to complaint files.

[45 FR 69682–69737, Oct. 21, 1980, as amended at 65 FR 2318, Jan. 14, 2000]

§ 880.5300 Medical absorbent fiber.

(a) *Identification.* A medical absorbent fiber is a device intended for medical purposes that is made from cotton or synthetic fiber in the shape of a ball or a pad and that is used for applying medication to, or absorbing small amounts of body fluids from, a patient's body surface. Absorbent fibers intended solely for cosmetic purposes are not included in this generic device category.

(b) *Classification.* Class I (general controls). The device is exempt from the premarket notification procedures in subpart E of part 807 of this chapter, subject to the limitations in § 880.9. If the device is not labeled or otherwise represented as sterile, it is also exempt from the current good manufacturing practice regulations in part 820 of this chapter, with the exception of § 820.180, with respect to general requirements concerning records, and § 820.198, with respect to complaint files.

[45 FR 69682–69737, Oct. 21, 1980, as amended at 66 FR 38804, July 25, 2001]

§ 880.5400 Neonatal incubator.

(a) *Identification.* A neonatal incubator is a device consisting of a rigid boxlike enclosure in which an infant may be kept in a controlled environment for medical care. The device may include an AC-powered heater, a fan to circulate the warmed air, a container for water to add humidity, a control valve through which oxygen may be added, and access ports for nursing care.

(b) *Classification.* Class II (performance standards).

§ 880.5410 Neonatal transport incubator.

(a) *Identification.* A neonatal transport incubator is a device consisting of a portable rigid boxlike enclosure with insulated walls in which an infant may be kept in a controlled environment while being transported for medical care. The device may include straps to secure the infant, a battery-operated heater, an AC-powered battery charger, a fan to circulate the warmed air, a container for water to add humidity, and provision for a portable oxygen bottle.

(b) *Classification.* Class II (performance standards).

§ 880.5420 Pressure infusor for an I.V. bag.

(a) *Identification.* A pressure infusor for an I.V. bag is a device consisting of an inflatable cuff which is placed around an I.V. bag. When the device is inflated, it increases the pressure on the I.V. bag to assist the infusion of the fluid.

(b) *Classification.* Class I (general controls). The device is exempt from the premarket notification procedures in subpart E of part 807 of this chapter subject to § 880.9.

[45 FR 69682–69737, Oct. 21, 1980, as amended at 65 FR 2318, Jan. 14, 2000]

§ 880.5430 Nonelectrically powered fluid injector.

(a) *Identification.* A nonelectrically powered fluid injector is a nonelectrically powered device used by a health care provider to give a hypodermic injection by means of a narrow, high velocity jet of fluid which can penetrate the surface of the skin and deliver the fluid to the body. It may be used for mass inoculations.

(b) *Classification.* Class II (performance standards).

§ 880.5440 Intravascular administration set.

(a) *Identification.* An intravascular administration set is a device used to administer fluids from a container to a patient's vascular system through a needle or catheter inserted into a vein. The device may include the needle or catheter, tubing, a flow regulator, a drip chamber, an infusion line filter, an I.V. set stopcock, fluid delivery tubing, connectors between parts of the set, a side tube with a cap to serve as an injection site, and a hollow spike to penetrate and connect the tubing to an

I.V. bag or other infusion fluid container.

(b) *Classification.* Class II (special controls). The special control for pharmacy compounding systems within this classification is the FDA guidance document entitled "Class II Special Controls Guidance Document: Pharmacy Compounding Systems; Final Guidance for Industry and FDA Reviewers." Pharmacy compounding systems classified within the intravascular administration set are exempt from the premarket notification procedures in subpart E of this part and subject to the limitations in § 880.9.

[45 FR 69682–69737, Oct. 21, 1980, as amended at 66 FR 15798, Mar. 21, 2001]

§ 880.5450 Patient care reverse isolation chamber.

(a) *Identification.* A patient care reverse isolation chamber is a device consisting of a roomlike enclosure designed to prevent the entry of harmful airborne material. This device protects a patient who is undergoing treatment for burns or is lacking a normal immunosuppressive defense due to therapy or congenital abnormality. The device includes fans and air filters which maintain an atmosphere of clean air at a pressure greater than the air pressure outside the enclosure.

(b) *Classification.* Class II (performance standards).

§ 880.5475 Jet lavage.

(a) *Identification.* A jet lavage is a device used to clean a wound by a pulsatile jet of sterile fluid. The device consists of the pulsing head, tubing to connect to a container of sterile fluid, and a means of propelling the fluid through the tubing, such as an electric roller pump.

(b) *Classification.* Class II (special controls). The device is exempt from the premarket notification procedures in subpart E of part 807 of this chapter subject to § 880.9.

[45 FR 69682–69737, Oct. 21, 1980, as amended at 63 FR 59229, Nov. 3, 1998]

§ 880.5500 AC-powered patient lift.

(a) *Identification.* An AC-powered lift is an electrically powered device either fixed or mobile, used to lift and transport patients in the horizontal or other required position from one place to another, as from a bed to a bath. The device includes straps and slings to support the patient.

(b) *Classification.* Class II (special controls). The device is exempt from the premarket notification procedures in subpart E of part 807 of this chapter subject to § 880.9.

[45 FR 69682–69737, Oct. 21, 1980, as amended at 63 FR 59229, Nov. 3, 1998]

§ 880.5510 Non-AC-powered patient lift.

(a) *Identification.* A non-AC-powered patient lift is a hydraulic, battery, or mechanically powered device, either fixed or mobile, used to lift and transport a patient in the horizontal or other required position from one place to another, as from a bed to a bath. The device includes straps and a sling to support the patient.

(b) *Classification.* Class I (general controls). The device is exempt from the premarket notification procedures in subpart E of part 807 of this chapter, subject to the limitations in § 880.9.

[45 FR 69682–69737, Oct. 21, 1980, as amended at 54 FR 25050, June 12, 1989; 66 FR 38804, July 25, 2001]

§ 880.5550 Alternating pressure air flotation mattress.

(a) *Identification.* An alternating pressure air flotation mattress is a device intended for medical purposes that consists of a mattress with multiple air cells that can be filled and emptied in an alternating pattern by an associated control unit to provide regular, frequent, and automatic changes in the distribution of body pressure. The device is used to prevent and treat decubitus ulcers (bed sores).

(b) *Classification.* Class II (special controls). The device is exempt from the premarket notification procedures in subpart E of part 807 of this chapter subject to § 880.9.

[45 FR 69682–69737, Oct. 21, 1980, as amended at 63 FR 59229, Nov. 3, 1998]

208-072 D-14

§ 880.5560 Temperature regulated water mattress.

(a) *Identification.* A temperature regulated water mattress is a device intended for medical purposes that consists of a mattress of suitable size, filled with water which can be heated or in some cases cooled. The device includes electrical heating and water circulating components, and an optional cooling component. The temperature control may be manual or automatic.

(b) *Classification.* Class I (general controls). The device is exempt from the premarket notification procedures in subpart E of part 807 of this chapter, subject to the limitations in § 880.9.

[45 FR 69682–69737, Oct. 21, 1980, as amended at 61 FR 1123, Jan. 16, 1996; 66 FR 38804, July 25, 2001]

§ 880.5570 Hypodermic single lumen needle.

(a) *Identification.* A hypodermic single lumen needle is a device intended to inject fluids into, or withdraw fluids from, parts of the body below the surface of the skin. The device consists of a metal tube that is sharpened at one end and at the other end joined to a female connector (hub) designed to mate with a male connector (nozzle) of a piston syringe or an intravascular administration set.

(b) *Classification.* Class II (performance standards).

§ 880.5580 Acupuncture needle.

(a) *Identification.* An acupuncture needle is a device intended to pierce the skin in the practice of acupuncture. The device consists of a solid, stainless steel needle. The device may have a handle attached to the needle to facilitate the delivery of acupuncture treatment.

(b) *Classification.* Class II (special controls). Acupuncture needles must comply with the following special controls:

(1) Labeling for single use only and conformance to the requirements for prescription devices set out in 21 CFR 801.109,

(2) Device material biocompatibility, and

(3) Device sterility.

[61 FR 64617, Dec. 6, 1996]

§ 880.5630 Nipple shield.

(a) *Identification.* A nipple shield is a device consisting of a cover used to protect the nipple of a nursing woman. This generic device does not include nursing pads intended solely to protect the clothing of a nursing woman from milk.

(b) *Classification.* Class I (general controls). The device is exempt from the premarket notification procedures in subpart E of part 807 of this chapter, subject to the limitations in § 880.9.

[45 FR 69682–69737, Oct. 21, 1980, as amended at 59 FR 63011, Dec. 7, 1994; 66 FR 33804, July 25, 2001]

§ 880.5640 Lamb feeding nipple.

(a) *Identification.* A lamb feeding nipple is a device intended for use as a feeding nipple for infants with oral or facial abnormalities.

(b) *Classification.* Class I (general controls). The device is exempt from the premarket notification procedures in subpart E of part 807 of this chapter, subject to the limitations in § 880.9. If the device is not labeled or otherwise represented as sterile, it is also exempt from the current good manufacturing practice regulations in part 820 of this chapter, with the exception of § 820.180, with respect to general requirements concerning records, and § 820.198, with respect to complaint files.

[45 FR 69682–69737, Oct. 21, 1980, as amended at 66 FR 38804, July 25, 2001]

§ 880.5680 Pediatric position holder.

(a) *Identification.* A pediatric position holder is a device used to hold an infant or a child in a desired position for therapeutic or diagnostic purposes, e.g., in a crib under a radiant warmer, or to restrain a child while an intravascular injection is administered.

(b) *Classification.* Class I (general controls). The device is exempt from the good manufacturing practice regulation in part 820 of this chapter, with the exception of § 820.180, with respect to general requirements concerning records, and § 820.198, with respect to complaint files.

[45 FR 69682–69737, Oct. 21, 1980, as amended at 66 FR 46952, Sept. 10, 2001]

§ 880.5700 Neonatal phototherapy unit.

(a) *Identification.* A neonatal phototherapy unit is a device used to treat or prevent hyperbilirubinemia (elevated serum bilirubin level). The device consists of one or more lamps that emit a specific spectral band of light, under which an infant is placed for therapy. This generic type of device may include supports for the patient and equipment and component parts.

(b) *Classification.* Class II (performance standards).

§ 880.5725 Infusion pump.

(a) *Identification.* An infusion pump is a device used in a health care facility to pump fluids into a patient in a controlled manner. The device may use a piston pump, a roller pump, or a peristaltic pump and may be powered electrically or mechanically. The device may also operate using a constant force to propel the fluid through a narrow tube which determines the flow rate. The device may include means to detect a fault condition, such as air in, or blockage of, the infusion line and to activate an alarm.

(b) *Classification.* Class II (performance standards).

§ 880.5740 Suction snakebite kit.

(a) *Identification.* A suction snakebite kit is a device consisting of a knife, suction device, and tourniquet used for first-aid treatment of snakebites by removing venom from the wound.

(b) *Classification.* Class I (general controls). The device is exempt from the premarket notification procedures in subpart E of part 807 of this chapter, subject to the limitations in § 880.9.

[45 FR 69682–69737, Oct. 21, 1980, as amended at 59 FR 63011, Dec. 7, 1994; 66 FR 38805, July 25, 2001]

§ 880.5760 Chemical cold pack snakebite kit.

(a) *Identification.* A chemical cold pack snakebit kit is a device consisting of a chemical cold pack and tourniquet used for first-aid treatment of snakebites.

(b) *Classification.* Class III (premarket approval).

(c) *Date PMA or notice of completion of a PDP is required.* A PMA or a notice of completion of a PDP is required to be filed with the Food and Drug Administration on or before December 26, 1996 for any chemical cold pack snakebite kit that was in commercial distribution before May 28, 1976, or that has, on or before December 26, 1996 been found to be substantially equivalent to a chemical cold pack snakebite kit that was in commercial distribution before May 28, 1976. Any other chemical cold pack snakebite kit shall have an approved PMA or a declared completed PDP in effect before being placed in commercial distribution.

[45 FR 69682–69737, Oct. 21, 1980, as amended at 52 FR 17739, May 11, 1987; 61 FR 50708, Sept. 27, 1996]

§ 880.5780 Medical support stocking.

(a) *Medical support stocking to prevent the pooling of blood in the legs*—(1) *Identification.* A medical support stocking to prevent the pooling of blood in the legs is a device that is constructed of elastic material and designed to apply controlled pressure to the leg and that is intended for use in the prevention of pooling of blood in the leg.

(2) *Classification.* Class II (performance standards).

(b) *Medical support stocking for general medical purposes*—(1) *Identification.* A medical support stocking for general medical purposes is a device that is constructed of elastic material and designed to apply controlled pressure to the leg and that is intended for medical purposes other than the prevention of pooling of blood in the leg.

(2) *Classification.* Class I. The device is exempt from the premarket notification procedures in subpart E of part 807 of this chapter, subject to the limitations in § 880.9. The device is also exempt from the current good manufacturing practice regulations in part 820 of this chapter, with the exception of § 820.180, with respect to general requirements concerning records, and § 820.198, with respect to complaint files.

[45 FR 69682–69737, Oct. 21, 1980, as amended at 59 FR 63011, Dec. 7, 1994; 66 FR 38805, July 25, 2001]

§ 880.5820 Therapeutic scrotal support.

(a) *Identification.* A therapeutic scrotal support is a device intended for medical purposes that consist of a pouch attached to an elastic waistband and that is used to support the scrotum (the sac that contains the testicles).

(b) *Classification.* Class I (general controls). The device is exempt from the premarket notification procedures in subpart E of part 807 of this chapter, subject to the limitations in § 880.9. The device also is exempt from the current good manufacturing practice regulations in part 820 of this chapter, with the exception of § 820.180, with respect to general requirements concerning records, and § 820.198, with respect to complaint files.

[45 FR 69682–69737, Oct. 21, 1980, as amended at 66 FR 38805, July 25, 2001]

§ 880.5860 Piston syringe.

(a) *Identification.* A piston syringe is a device intended for medical purposes that consists of a calibrated hollow barrel and a movable plunger. At one end of the barrel there is a male connector (nozzle) for fitting the female connector (hub) of a hypodermic single lumen needle. The device is used to inject fluids into, or withdraw fluids from, the body.

(b) *Classification.* Class II (performance standards).

§ 880.5950 Umbilical occlusion device.

(a) *Identification.* An umbilical occlusion device is a clip, tie, tape, or other article used to close the blood vessels in the umbilical cord of a newborn infant.

(b) *Classification.* Class I. The device is exempt from the premarket notification procedures in subpart E of part 807 of this chapter.

[45 FR 69682–69737, Oct. 21, 1980, as amended at 59 FR 63011, Dec. 7, 1994; 66 FR 38805, July 25, 2001]

§ 880.5960 Lice removal kit.

(a) *Identification.* The lice removal kit is a comb or comb-like device intended to remove and/or kill lice and nits from head and body hair. It may or may not be battery operated.

(b) *Classification.* Class I (general controls). The device is exempt from the premarket notification procedures in subpart E of part 807 of this chapter subject to the limitations in § 880.9.

[63 FR 59718, Nov. 5, 1998]

§ 880.5965 Subcutaneous, implanted, intravascular infusion port and catheter.

(a) *Identification.* A subcutaneous, implanted, intravascular infusion port and catheter is a device that consists of a subcutaneous, implanted reservoir that connects to a long-term intravascular catheter. The device allows for repeated access to the vascular system for the infusion of fluids and medications and the sampling of blood. The device consists of a portal body with a resealable septum and outlet made of metal, plastic, or combination of these materials and a long-term intravascular catheter is either preattached to the port or attached to the port at the time of device placement. The device is available in various profiles and sizes and can be of a single or multiple lumen design.

(b) *Classification.* Class II (special controls) Guidance Document: "Guidance on 510(k) Submissions for Implanted Infusion Ports," FDA October 1990.

[65 FR 37043, June 13, 2000]

§ 880.5970 Percutaneous, implanted, long-term intravascular catheter.

(a) *Identification.* A percutaneous, implanted, long-term intravascular catheter is a device that consists of a slender tube and any necessary connecting fittings, such as luer hubs, and accessories that facilitate the placement of the device. The device allows for repeated access to the vascular system for long-term use of 30 days or more, and it is intended for administration of fluids, medications, and nutrients; the sampling of blood; and monitoring blood pressure and temperature. The device may be constructed of metal, rubber, plastic, composite materials, or any combination of these materials and may be of single or multiple lumen design.

(b) *Classification.* Class II (special controls) Guidance Document: "Guidance on Premarket Notification

[510(k)] Submission for Short-Term and Long-Term Intravascular Catheters.''

[65 FR 37043, June 13, 2000]

Subpart G—General Hospital and Personal Use Miscellaneous Devices

§ 880.6025 Absorbent tipped applicator.

(a) *Identification.* An absorbent tipped applicator is a device intended for medical purposes that consists of an absorbent swab on a wooden, paper, or plastic stick. The device is used to apply medications to, or to take specimens from, a patient.

(b) *Classification.* Class I (general controls). The device is exempt from the premarket notification procedures in subpart E of part 807 of this chapter, subject to the limitations in § 880.9. If the device is not labeled or otherwise represented as sterile, it is also exempt from the current good manufacturing practice regulations in part 820 of this chapter, with the exception of § 820.180, with respect to general requirements concerning records, and § 820.198, with respect to complaint files.

[45 FR 69682–69737, Oct. 21, 1980, as amended at 66 FR 38805, July 25, 2001]

§ 880.6050 Ice bag.

(a) *Identification.* An ice bag is a device intended for medical purposes that is in the form of a container intended to be filled with ice that is used to apply dry cold therapy to an area of the body. The device may include a holder that keeps the bag in place against an external area of the patient.

(b) *Classification.* Class I (general controls). The device is exempt from the premarket notification procedures in subpart E of part 807 of this chapter, subject to the limitations in § 880.9. If the device is not labeled or otherwise represented as sterile, it is also exempt from the current good manufacturing practice regulations in part 820 of this chapter, with the exception of § 820.180, with respect to general requirements concerning records, and § 820.198, with respect to complaint files.

[45 FR 69682–69737, Oct. 21, 1980, as amended at 66 FR 38805, July 25, 2001]

§ 880.6060 Medical disposable bedding.

(a) *Identification.* Medical disposable bedding is a device intended for medical purposes to be used by one patient for a period of time and then discarded. This generic type of device may include disposable bedsheets, bedpads, pillows and pillowcases, blankets, emergency rescue blankets, or waterproof sheets.

(b) *Classification.* Class I (general controls). The device is exempt from the premarket notification procedures in subpart E of part 807 of this chapter, subject to the limitations in § 880.9. If the device is not labeled or otherwise represented as sterile, it is also exempt from the current good manufacturing practice regulations in part 820 of this chapter, with the exception of § 820.180, with respect to general requirements concerning records, and § 820.198, with respect to complaint files.

[45 FR 69682–69737, Oct. 21, 1980, as amended at 59 FR 63011, Dec. 7, 1994; 66 FR 38805, July 25, 2001]

§ 880.6070 Bed board.

(a) *Identification.* A bed board is a device intended for medical purposes that consists of a stiff board used to increase the firmness of a bed.

(b) *Classification.* Class I (general controls). The device is exempt from the premarket notification procedures in subpart E of part 807 of this chapter, subject to the limitations in § 880.9. The device is also exempt from the current good manufacturing practice regulations in part 820 of this chapter, with the exception of § 820.180, with respect to general requirements concerning records, and § 820.198, with respect to complaint files.

[45 FR 69682–69737, Oct. 21, 1980, as amended at 66 FR 38805, July 25, 2001]

§ 880.6080 Cardiopulmonary resuscitation board.

(a) *Identification.* A cardiopulmonary resuscitation board is a device consisting of a rigid board which is placed under a patient to act as a support during cardiopulmonary resuscitation.

(b) *Classification.* Class I (general controls). The device is exempt from the premarket notification procedures in subpart E of part 807 of this chapter, subject to the limitations in § 880.9. The device is also exempt from the current good manufacturing practice regulations in part 820 of this chapter, with the exception of § 820.180, with respect to general requirements concerning records, and § 820.198, with respect to complaint files.

[45 FR 69682–69737, Oct. 21, 1980, as amended at 66 FR 38805, July 25, 2001]

§ 880.6085 Hot/cold water bottle.

(a) *Identification.* A hot/cold water bottle is a device intended for medical purposes that is in the form of a container intended to be filled with hot or cold water to apply heat or cold to an area of the body.

(b) *Classification.* Class I (general controls). The device is exempt from the premarket notification procedures in subpart E of part 807 of this chapter, subject to the limitations in § 880.9. The device is also exempt from the current good manufacturing practice regulations in part 820 of this chapter, with the exception of § 820.180, with respect to general requirements concerning records, and § 820.198, with respect to complaint files.

[45 FR 69682–69737, Oct. 21, 1980, as amended at 66 FR 38805, July 25, 2001]

§ 880.6100 Ethylene oxide gas aerator cabinet.

(a) *Identification.* An ethyene oxide gas aerator cabinet is a device that is intended for use by a health care provider and consists of a cabinet with a ventilation system designed to circulate and exchange the air in the cabinet to shorten the time required to remove residual ethylene oxide (ETO) from wrapped medical devices that have undergone ETO sterilization. The device may include a heater to warm the circulating air.

(b) *Classification.* Class II (performance standards).

§ 880.6140 Medical chair and table.

(a) *Identification.* A medical chair or table is a device intended for medical purposes that consists of a chair or table without wheels and not electrically powered which, by reason of special shape or attachments, such as food trays or headrests, or special features such as a built-in raising and lowering mechanism or removable arms, is intended for use of blood donors, geriatric patients, or patients undergoing treatment or examination.

(b) *Classification.* Class I (general controls). The device is exempt from the premarket notification procedures in subpart E of part 807 of this chapter, subject to the limitations in § 880.9. The device is also exempt from the current good manufacturing practice regulations in part 820 of this chapter, with the exception of § 820.180, with respect to general requirements concerning records, and § 820.198, with respect to complaint files.

[45 FR 69682–69737, Oct. 21, 1980, as amended at 66 FR 38805, July 25, 2001]

§ 880.6150 Ultrasonic cleaner for medical instruments.

(a) *Identification.* An ultrasonic cleaner for medical instruments is a device intended for cleaning medical instruments by the emission of high frequency soundwaves.

(b) *Classification.* Class I. The device, including any solutions intended for use with the device for cleaning and sanitizing the instruments, is exempt from the premarket notification procedures in subpart E of part 807 of this chapter, subject to the limitations in § 880.9.

[45 FR 69682–69737, Oct. 21, 1980, as amended at 54 FR 25050, June 12, 1989; 59 FR 63011, Dec. 7, 1994; 66 FR 38805, July 25, 2001]

§ 880.6175 [Reserved]

§ 880.6185 Cast cover.

(a) *Identification.* A cast cover is a device intended for medical purposes that is made of waterproof material and placed over a cast to protect it from getting wet during a shower or a bath.

(b) *Classification.* Class I (general controls). The device is exempt from the premarket notification procedures in subpart E of part 807 of this chapter, subject to the limitations in § 880.9. If the device is not labeled or otherwise represented as sterile, it is also exempt from the current good manufacturing practice regulations in part 820 of this chapter, with the exception of § 820.180, with respect to general requirements

concerning records, and § 820.198, with respect to complaint files.

[45 FR 69682–69737, Oct. 21, 1980, as amended at 66 FR 38806, July 25, 2001]

§ 880.6190 Mattress cover for medical purposes.

(a) *Identification.* A mattress cover for medical purposes is a device intended for medical purposes that is used to protect a mattress. It may be electrically conductive or contain a germicide.

(b) *Classification.* Class I (general controls). The device is exempt from the premarket notification procedures in subpart E of part 807 of this chapter, subject to the limitations in § 880.9. If the device is not labeled or otherwise represented as sterile, it is also exempt from the current good manufacturing practice regulations in part 820 of this chapter, with the exception of § 820.180, with respect to general requirements concerning records, and § 820.198, with respect to complaint files.

[45 FR 69682–69737, Oct. 21, 1980, as amended at 59 FR 63011, Dec. 7, 1994; 66 FR 38806, July 25, 2001]

§ 880.6200 Ring cutter.

(a) *Identification.* A ring cutter is a device intended for medical purposes that is used to cut a ring on a patient's finger so that the ring can be removed. The device incorporates a guard to prevent injury to the patient's finger.

(b) *Classification.* Class I (general controls). The device is exempt from the premarket notification procedures in subpart E of part 807 of this chapter, subject to the limitations in § 880.9. The device also is exempt from the current good manufacturing practice regulations in part 820 of this chapter, with the exception of § 820.180, with respect to general requirements concerning records, and § 820.198, with respect to complaint files.

[45 FR 69682–69737, Oct. 21, 1980, as amended at 66 FR 38806, July 25, 2001]

§ 880.6230 Tongue depressor.

(a) *Identification.* A tongue depressor is a device intended to displace the tongue to facilitate examination of the surrounding organs and tissues.

(b) *Classification.* Class I (general controls). The device is exempt from the premarket notification procedures in subpart E of part 807 of this chapter, subject to the limitations in § 880.9. If the device is not labeled or otherwise represented as sterile, it is also exempt from the current good manufacturing practice regulations in part 820 of this chapter, with the exception of § 820.180, with respect to general requirements concerning records, and § 820.198, with respect to complaint files.

[45 FR 69682–69737, Oct. 21, 1980, as amended at 66 FR 38806, July 25, 2001]

§ 880.6250 Patient examination glove.

(a) *Identification.* A patient examination glove is a disposable device intended for medical purposes that is worn on the examiner's hand or finger to prevent contamination between patient and examiner.

(b) *Classification.* Class I (general controls).

[45 FR 69682–69737, Oct. 21, 1980, as amended at 53 FR 1604, Jan. 13, 1989; 66 FR 46952, Sept. 10, 2001]

§ 880.6265 Examination gown.

(a) *Identification.* An examination gown is a device intended for medical purposes that is made of cloth, paper, or other material that is draped over or worn by a patient as a body covering during a medical examination.

(b) *Classification.* Class I (general controls). The device is exempt from the premarket notification procedures in subpart E of part 807 of this chapter, subject to the limitations in § 880.9. If the device is not labeled or otherwise represented as sterile, it is also exempt from the current good manufacturing practice regulations in part 820 of this chapter, with the exception of § 820.180, with respect to general requirements concerning records, and § 820.198, with respect to complaint files.

[45 FR 69682–69737, Oct. 21, 1980, as amended at 66 FR 38806, July 25, 2001]

§ 880.6280 Medical insole.

(a) *Identification.* A medical insole is a device intended for medical purposes that is placed inside a shoe to relieve the symptoms of athlete's foot infection by absorbing moisture.

(b) *Classification.* Class I (general controls). The device is exempt from the premarket notification procedures in subpart E of part 807 of this chapter, subject to the limitations in § 880.9.

[45 FR 69682–69737, Oct. 21, 1980, as amended at 54 FR 25050, June 12, 1989; 66 FR 38806, July 25, 2001]

§ 880.6300 Implantable radiofrequency transponder system for patient identification and health information.

(a) *Identification.* An implantable radiofrequency transponder system for patient identification and health information is a device intended to enable access to secure patient identification and corresponding health information. This system may include a passive implanted transponder, inserter, and scanner. The implanted transponder is used only to store a unique electronic identification code that is read by the scanner. The identification code is used to access patient identity and corresponding health information stored in a database.

(b) *Classification.* Class II (special controls). The special control is FDA's guidance document entitled "Class II Special Controls Guidance Document: Implantable Radiofrequency Transponder System for Patient Identification and Health Information." See § 880.1(e) for the availability of this guidance document. This device is exempt from the premarket notification procedures in subpart E of part 807 of this chapter subject to the limitations in § 880.9.

[69 FR 71704, Dec. 10, 2004]

§ 880.6320 AC-powered medical examination light.

(a) *Identification.* An AC-powered medical examination light is an AC-powered device intended for medical purposes that is used to illuminate body surfaces and cavities during a medical examination.

(b) *Classification.* Class I (general controls). The device is exempt from the premarket notification procedures in subpart E of part 807 of this chapter, subject to the limitations in § 880.9.

[45 FR 69682–69737, Oct. 21, 1980, as amended at 61 FR 1123, Jan. 16, 1996; 66 FR 38806, July 25, 2001]

§ 880.6350 Battery-powered medical examination light.

(a) *Identification.* A battery-powered medical examination light is a battery-powered device intended for medical purposes that is used to illuminate body surfaces and cavities during a medical examination.

(b) *Classification.* Class I (general controls). The device is exempt from the premarket notification procedures in subpart E of part 807 of this chapter, subject to the limitations in § 880.9. The device also is exempt from the current good manufacturing practice regulations in part 820 of this chapter, with the exception of § 820.180, with respect to general requirements concerning records, and § 820.198, with respect to complaint files.

[45 FR 69682–69737, Oct. 21, 1980, as amended at 66 FR 38806, July 25, 2001]

§ 880.6375 Patient lubricant.

(a) *Identification.* A patient lubricant is a device intended for medical purposes that is used to lubricate a body orifice to facilitate entry of a diagnostic or therapeutic device.

(b) *Classification.* Class I (general controls).

[45 FR 69682–69737, Oct. 21, 1980, as amended at 66 FR 46952, Sept. 10, 2001]

§ 880.6430 Liquid medication dispenser.

(a) *Identification.* A Liquid medication dispenser is a device intended for medical purposes that is used to issue a measured amount of liquid medication.

(b) *Classification.* Class I (general controls). The device is exempt from the premarket notification procedures in subpart E of part 807 of this chapter, subject to the limitations in § 880.9. The device is also exempt from the current good manufacturing practice regulations in part 820 of this chapter, with the exception of § 820.180, with respect to general requirements concerning records, and § 820.198, with respect to complaint files.

[45 FR 69682–69737, Oct. 21, 1980, as amended at 66 FR 38806, July 25, 2001]

§ 880.6450 Skin pressure protectors.

(a) *Identification.* A skin pressure protector is a device intended for medical purposes that is used to reduce pressure on the skin over a bony prominence to reduce the likelihood of the patient's developing decubitus ulcers (bedsores).

(b) *Classification.* Class I (general controls). The device is exempt from the premarket notification procedures in subpart E of part 807 of this chapter, subject to the limitations in § 880.9. The device is also exempt from the current good manufacturing practice regulations in part 820 of this chapter, with the exception of § 820.180, with respect to general requirements concerning records, and § 820.198, with respect to complaint files.

[45 FR 69682–69737, Oct. 21, 1980, as amended at 66 FR 38806, July 25, 2001]

§ 880.6500 Medical ultraviolet air purifier.

(a) *Identification.* A medical ultraviolet air purifier is a device intended for medical purposes that is used to destroy bacteria in the air by exposure to ultraviolet radiation.

(b) *Classification.* Class II (performance standards).

§ 880.6710 Medical ultraviolet water purifier.

(a) *Identification.* A medical ultraviolet water purifier is a device intended for medical purposes that is used to destroy bacteria in water by exposure to ultraviolet radiation.

(b) *Classification.* Class II (performance standards).

§ 880.6730 Body waste receptacle.

(a) *Identification.* A body waste receptacle is a device intended for medical purposes that is not attached to the body and that is used to collect the body wastes of a bed patient.

(b) *Classification.* Class I (general controls). The device is exempt from the premarket notification procedures in subpart E of part 807 of this chapter, subject to the limitations in § 880.9. The device also is exempt from the current good manufacturing practice regulations in part 820 of this chapter, with the exception of § 820.180, with respect to general requirements concerning records, and § 820.198, with respect to complaint files.

[66 FR 38806, July 25, 2001]

§ 880.6740 Vacuum-powered body fluid suction apparatus.

(a) *Identification.* A vacuum-powered body fluid suction apparatus is a device used to aspirate, remove, or sample body fluids. The device is powered by an external source of vacuum. This generic type of device includes vacuum regulators, vacuum collection bottles, suction catheters and tips, connecting flexible aspirating tubes, rigid suction tips, specimen traps, noninvasive tubing, and suction regulators (with gauge).

(b) *Classification.* Class II (special controls). The device is exempt from the premarket notification procedures in subpart E of part 807 of this chapter subject to § 880.9.

[45 FR 69682–69737, Oct. 21, 1980, as amended at 63 FR 59229, Nov. 3, 1998]

§ 880.6760 Protective restraint.

(a) *Identification.* A protective restraint is a device, including but not limited to a wristlet, anklet, vest, mitt, straight jacket, body/limb holder, or other type of strap, that is intended for medical purposes and that limits the patient's movements to the extent necessary for treatment, examination, or protection of the patient or others.

(b) *Classification.* Class I (general controls).

[61 FR 8439, Mar. 4, 1996, as amended at 66 FR 46952, Sept. 10, 2001]

§ 880.6775 Powered patient transfer device.

(a) *Identification.* A powered patient transfer device is a device consisting of a wheeled stretcher and a powered mechanism that has a broad, flexible band stretched over long rollers that can advance itself under a patient and transfer the patient with minimal disturbance in a horizontal position to the stretcher.

(b) *Classification.* Class II (special controls). The device is exempt from the premarket notification procedures

in subpart E of part 807 of this chapter subject to § 880.9.

[45 FR 69682–69737, Oct. 21, 1980, as amended at 63 FR 59229, Nov. 3, 1998]

§ 880.6785 Manual patient transfer device.

(a) *Identification.* A manual patient transfer device is a device consisting of a wheeled stretcher and a mechanism on which a patient can be placed so that the patient can be transferred with minimal disturbance in a horizontal position to the stretcher.

(b) *Classification.* Class I (general controls). The device is exempt from the premarket notification procedures in subpart E of part 807 of this chapter, subject to the limitations in § 880.9. The device is also exempt from the current good manufacturing practice regulations in part 820 of this chapter, with the exception of § 820.180, with respect to general requirements concerning records, and § 820.198, with respect to complaint files.

[45 FR 69682–69737, Oct. 21, 1980, as amended at 66 FR 38807, July 25, 2001]

§ 880.6800 Washers for body waste receptacles.

(a) *Identification.* A washer for body waste receptacles is a device intended for medical purposes that is used to clean and sanitize a body waste receptacle, such as a bedpan. The device consists of a wall-mounted plumbing fixture with a door through which a body waste receptacle is inserted. When the door is closed the body waste receptacle is cleaned by hot water, steam, or germicide.

(b) *Classification.* Class I (general controls). The device is exempt from the premarket notification procedures in subpart E of part 807 of this chapter, subject to the limitations in § 880.9. The device also is exempt from the current good manufacturing practice regulations in part 820 of this chapter, with the exception of § 820.180, with respect to general requirements concerning records, and § 820.198, with respect to complaint files.

[45 FR 69682–69737, Oct. 21, 1980, as amended at 66 FR 38807, July 25, 2001]

§ 880.6820 Medical disposable scissors.

(a) *Identification.* Medical disposable scissors are disposable type general cutting devices intended for medical purposes. This generic type of device does not include surgical scissors.

(b) *Classification.* Class I (general controls). The device is exempt from the premarket notification procedures in subpart E of part 807 of this chapter, subject to the limitations in § 880.9.

[45 FR 69682–69737, Oct. 21, 1980, as amended at 66 FR 38807, July 25, 2001]

§ 880.6850 Sterilization wrap.

(a) *Identification.* A sterilization wrap (pack, sterilization wrapper, bag, or accessories, is a device intended to be used to enclose another medical device that is to be sterilized by a health care provider. It is intended to allow sterilization of the enclosed medical device and also to maintain sterility of the enclosed device until used.

(b) *Classification.* Class II (performance standards).

§ 880.6860 Ethylene oxide gas sterilizer.

(a) *Identification.* An ethylene gas sterilizer is a nonportable device intended for use by a health care provider that uses ethylene oxide (ETO) to sterilize medical products.

(b) *Classification.* Class II (performance standards).

§ 880.6870 Dry-heat sterilizer.

(a) *Identification.* A dry-heat sterilizer is a device that is intended for use by a health care provider to sterilize medical products by means of dry heat.

(b) *Classification.* Class II (performance standards).

§ 880.6880 Steam sterilizer.

(a) *Identification.* A steam sterilizer (autoclave) is a device that is intended for use by a health care provider to sterilize medical products by means of pressurized steam.

(b) *Classification.* Class II (performance standards).

§ 880.6885 Liquid chemical sterilants/high level disinfectants.

(a) *Identification.* A liquid chemical sterilant/high level disinfectant is a germicide that is intended for use as the terminal step in processing critical and semicritical medical devices prior to patient use. Critical devices make contact with normally sterile tissue or body spaces during use. Semicritical devices make contact during use with mucous membranes or nonintact skin.

(b) *Classification.* Class II (special controls). Guidance on the Content and Format of Premarket Notification (510(k)) Submissions for Liquid Chemical Sterilants/High Level Disinfectants, and user information and training.

[65 FR 36325, June 8, 2000]

§ 880.6890 General purpose disinfectants.

(a) *Identification.* A general purpose disinfectant is a germicide intended to process noncritical medical devices and equipment surfaces. A general purpose disinfectant can be used to preclean or decontaminate critical or semicritical medical devices prior to terminal sterilization or high level disinfection. Noncritical medical devices make only topical contact with intact skin.

(b) *Classification.* Class I (general controls). The device is exempt from the premarket notification procedures in subpart E of part 807 of this chapter subject to the limitations in § 880.9.

[65 FR 36326, June 8, 2000]

§ 880.6900 Hand-carried stretcher.

(a) *Identification.* A hand-carried stretcher is a device consisting of a lightweight frame, or of two poles with a cloth or metal platform, on which a patient can be carried.

(b) *Classification.* Class I (general controls). The device is exempt from the premarket notification procedures in subpart E of part 807 of this chapter, subject to the limitations in § 880.9. The device is also exempt from the current good manufacturing practice regulations in part 820 of this chapter, with the exception of § 820.180, with respect to general requirements concerning records, and § 820.198, with respect to complaint files.

[45 FR 69682–69737, Oct. 21, 1980, as amended at 59 FR 63011, Dec. 7, 1994; 66 FR 38807, July 25, 2001]

§ 880.6910 Wheeled stretcher.

(a) *Identification.* A wheeled stretcher is a device consisting of a platform mounted on a wheeled frame that is designed to transport patients in a horizontal position. The device may have side rails, supports for fluid infusion equipment, and patient securement straps. The frame may be fixed or collapsible for use in an ambulance.

(b) *Classification.* Class II (special controls). The device is exempt from the premarket notification procedures in subpart E of part 807 of this chapter subject to § 880.9.

[45 FR 69682–69737, Oct. 21, 1980, as amended at 63 FR 59229, Nov. 3, 1998]

§ 880.6920 Syringe needle introducer.

(a) *Identification.* A syringe needle introducer is a device that uses a spring-loaded mechanism to drive a hypodermic needle into a patient to a predetermined depth below the skin surface.

(b) *Classification.* Class II (performance standards).

§ 880.6960 Irrigating syringe.

(a) *Identification.* An irrigating syringe is a device intended for medical purposes that consists of a bulb or a piston syringe with an integral or a detachable tube. The device is used to irrigate, withdraw fluid from, or instill fluid into, a body cavity or wound.

(b) *Classification.* Class I (general controls). The device is exempt from the premarket notification procedures in subpart E of part 807 of this chapter, subject to the limitations in § 880.9. If the device is not labeled or otherwise represented as sterile, it is also exempt from the current good manufacturing practice regulations in part 820 of this chapter, with the exception of § 820.180, with respect to general requirements concerning records, and § 820.198, with respect to complaint files.

[45 FR 69682–69737, Oct. 21, 1980, as amended at 66 FR 38807, July 25, 2001]

§ 880.6970 Liquid crystal vein locator.

(a) *Identification.* A liquid crystal vein locator is a device used to indicate the location of a vein by revealing variations in the surface temperature of the skin by displaying the color changes of heat sensitive liquid crystals (cholesteric esters).

(b) *Classification.* Class I (general controls). The device is exempt from the premarket notification procedures in subpart E of part 807 of this chapter, subject to the limitations in § 880.9.

[45 FR 69682–69737, Oct. 21, 1980, as amended at 54 FR 25050, June 12, 1989; 66 FR 38807, July 25, 2001]

§ 880.6980 Vein stabilizer.

(a) *Identification.* A vein stabilizer is a device consisting of a flat piece of plastic with two noninvasive prongs. The device is placed on the skin so that the prongs are on either side of a vein and hold it stable while a hypodermic needle is inserted into the vein.

(b) *Classification.* Class I (general controls). The device is exempt from the premarket notification procedures in subpart E of part 807 of this chapter, subject to the limitations in § 880.9. If the device is not labeled or otherwise represented as sterile, it is also exempt from the current good manufacturing practice regulations in part 820 of this chapter, with the exception of § 820.180, with respect to general requirements concerning records, and § 820.198, with respect to complaint files.

[45 FR 69682–69737, Oct. 21, 1980, as amended at 66 FR 38807, July 25, 2001]

§ 880.6990 Infusion stand.

(a) *Identification.* The infusion stand is a stationary or movable stand intended to hold infusion liquids, infusion accessories, and other medical devices.

(b) *Classification.* Class I (general controls). The device is exempt from the premarket notification procedures in subpart E of part 807 of this chapter subject to the limitations in § 880.9.

[63 FR 59718, Nov. 5, 1998]

§ 880.6991 Medical washer.

(a) *Identification.* A medical washer is a device that is intended for general medical purposes to clean and dry surgical instruments, anesthesia equipment, hollowware, and other medical devices.

(b) *Classification.* Class II (special controls). The special control for this device is the FDA guidance document entitled "Class II Special Controls Guidance Document: Medical Washers and Medical Washer-Disinfectors." The device is exempt from the premarket notification procedures in subpart E of part 807 of this chapter subject to § 880.9.

[67 FR 69121, Nov. 15, 2002]

§ 880.6992 Medical washer-disinfector.

(a) *Identification.* A medical washer-disinfector is a device that is intended for general medical purposes to clean, decontaminate, disinfect, and dry surgical instruments, anesthesia equipment, hollowware, and other medical devices.

(b) *Classification.* Class II (special controls). The special control for this device is the FDA guidance document entitled "Class II Special Controls Guidance Document: Medical Washers and Medical Washer-Disinfectors."

(1) Medical washer-disinfectors that are intended to clean, high level disinfect, and dry surgical instruments, anesthesia equipment, hollowware, and other medical devices.

(2) Medical washer-disinfectors that are intended to clean, low or intermediate level disinfect, and dry surgical instruments, anesthesia equipment, hollowware, and other medical devices are exempt from the premarket notification procedures in subpart E of part 807 of this chapter subject to § 880.9.

[67 FR 69121, Nov. 15, 2002]

PART 882—NEUROLOGICAL DEVICES

Subpart A—General Provisions

Subpart B—Neurological Diagnostic Devices

882.1310 Cortical electrode.
882.1320 Cutaneous electrode.
882.1330 Depth electrode.
882.1340 Nasopharyngeal electrode.
882.1350 Needle electrode.
882.1400 Electroencephalograph.
882.1410 Electroencephalograph electrode/lead tester.
882.1420 Electroencephalogram (EEG) signal spectrum analyzer.
882.1430 Electroencephalograph test signal generator.
882.1460 Nystagmograph.
882.1480 Neurological endoscope.
882.1500 Esthesiometer.
882.1525 Tuning fork.
882.1540 Galvanic skin response measurement device.
882.1550 Nerve conduction velocity measurement device.
882.1560 Skin potential measurement device.
882.1570 Powered direct-contact temperature measurement device.
882.1610 Alpha monitor.
882.1620 Intracranial pressure monitoring device.
882.1700 Percussor.
882.1750 Pinwheel.
882.1790 Ocular plethysmograph.
882.1825 Rheoencephalograph.
882.1835 Physiological signal amplifier.
882.1845 Physiological signal conditioner.
882.1855 Electroencephalogram (EEG) telemetry system.
882.1870 Evoked response electrical stimulator.
882.1880 Evoked response mechanical stimulator.
882.1890 Evoked response photic stimulator.
882.1900 Evoked response auditory stimulator.
882.1925 Ultrasonic scanner calibration test block.
882.1950 Tremor transducer.

Subparts C–D [Reserved]

Subpart E—Neurological Surgical Devices

882.4030 Skull plate anvil.
882.4060 Ventricular cannula.
882.4100 Ventricular catheter.
882.4125 Neurosurgical chair.
882.4150 Scalp clip.
882.4175 Aneurysm clip applier.
882.4190 Clip forming/cutting instrument.
882.4200 Clip removal instrument.
882.4215 Clip rack.
882.4250 Cryogenic surgical device.
882.4275 Dowel cutting instrument.
882.4300 Manual cranial drills, burrs, trephines, and their accessories.
882.4305 Powered compound cranial drills, burrs, trephines, and their accessories.
882.4310 Powered simple cranial drills, burrs, trephines, and their accessories.
882.4325 Cranial drill handpiece (brace).
882.4360 Electric cranial drill motor.
882.4370 Pneumatic cranial drill motor.
882.4400 Radiofrequency lesion generator.
882.4440 Neurosurgical headrests.
882.4460 Neurosurgical head holder (skull clamp).
882.4500 Cranioplasty material forming instrument.
882.4525 Microsurgical instrument.
882.4535 Nonpowered neurosurgical instrument.
882.4545 Shunt system implantation instrument.
882.4560 Stereotaxic instrument.
882.4600 Leukotome.
882.4650 Neurosurgical suture needle.
882.4700 Neurosurgical paddle.
882.4725 Radiofrequency lesion probe.
882.4750 Skull punch.
882.4800 Self-retaining retractor for neurosurgery.
882.4840 Manual rongeur.
882.4845 Powered rongeur.
882.4900 Skullplate screwdriver.

Subpart F—Neurological Therapeutic Devices

882.5030 Methyl methacrylate for aneurysmorrhaphy.
882.5050 Biofeedback device.
882.5070 Bite block.
882.5150 Intravascular occluding catheter.
882.5175 Carotid artery clamp.
882.5200 Aneurysm clip.
882.5225 Implanted malleable clip.
882.5235 Aversive conditioning device.
882.5250 Burr hole cover.
882.5275 Nerve cuff.
882.5300 Methyl methacrylate for cranioplasty.
882.5320 Preformed alterable cranioplasty plate.
882.5330 Preformed nonalterable cranioplasty plate.
882.5360 Cranioplasty plate fastener.
882.5500 Lesion temperature monitor.
882.5550 Central nervous system fluid shunt and components.
882.5800 Cranial electrotheraphy stimulator.
882.5810 External functional neuromuscular stimulator.
882.5820 Implanted cerebellar stimulator.
882.5830 Implanted diaphragmatic/phrenic nerve stimulator.
882.5840 Implanted intracerebral/subcortical stimulator for pain relief.
882.5850 Implanted spinal cord stimulator for bladder evacuation.
882.5860 Implanted neuromuscular stimulator.
882.5870 Implanted peripheral nerve stimulator for pain relief.
882.5880 Implanted spinal cord stimulator for pain relief.

882.5890 Transcutaneous electrical nerve stimulator for pain relief.
882.5900 Preformed craniosynostosis strip.
882.5910 Dura substitute.
882.5940 Electroconvulsive therapy device.
882.5950 Neurovascular embolization device.
882.5960 Skull tongs for traction.
882.5970 Cranial orthosis.
882.5975 Human dura mater.

AUTHORITY: 21 U.S.C. 351, 360, 360c, 360e, 360j, 371.

SOURCE: 44 FR 51730, Sept. 4, 1979, unless otherwise noted.

Subpart A—General Provisions

§ 882.1 Scope.

(a) This part sets forth the classification of neurological devices intended for human use that are in commercial distribution.

(b) The identification of a device in a regulation in this part is not a precise description of every device that is, or will be, subject to the regulation. A manufacturer who submits a premarket notification submission for a device under part 807 may not show merely that the device is accurately described by the section title and identification provisions of a regulation in this part, but shall state why the device is substantially equivalent to other devices, as required by § 807.87.

(c) To avoid duplicative listings, a neurological device that has two or more types of uses (e.g., used both as a diagnostic device and as a therapeutic device) is listed only in one subpart.

(d) References in this part to regulatory sections of the Code of Federal Regulations are to chapter I of title 21, unless otherwise noted.

(e) Guidance documents referenced in this part are available on the Internet at *http://www.fda.gov/cdrh/guidance.html.*

[52 FR 17739, May 11, 1987, as amended at 68 FR 70436, Dec. 18, 2003]

§ 882.3 Effective dates of requirement for premarket approval.

A device included in this part that is classified into class III (premarket approval) shall not be commercially distributed after the date shown in the regulation classifying the device unless the manufacturer has an approval under section 515 of the act (unless an exemption has been granted under section 520(g)(2) of the act). An approval under section 515 of the act consists of FDA's issuance of an order approving an application for premarket approval (PMA) for the device or declaring completed a product development protocol (PDP) for the device.

(a) Before FDA requires that a device commercially distributed before the enactment date of the amendments, or a device that has been found substantially equivalent to such a device, has an approval under section 515 of the act FDA must promulgate a regulation under section 515(b) of the act requiring such approval, except as provided in paragraph (b) of this section. Such a regulation under section 515(b) of the act shall not be effective during the grace period ending on the 90th day after its promulgation or on the last day of the 30th full calendar month after the regulation that classifies the device into class III is effective, whichever is later. See section 501(f)(2)(B) of the act. Accordingly, unless an effective date of the requirement for premarket approval is shown in the regulation for a device classified into class III in this part, the device may be commercially distributed without FDA's issuance of an order approving a PMA or declaring completed a PDP for the device. If FDA promulgates a regulation under section 515(b) of the act requiring premarket approval for a device, section, 501(f)(1)(A) of the act applies to the device.

(b) Any new, not substantially equivalent, device introduced into commercial distribution on or after May 28, 1976, including a device formerly marketed that has been substantially altered, is classified by statute (section 513(f) of the act) into class III without any grace period and FDA must have issued an order approving a PMA or declaring completed a PDP for the device before the device is commercially distributed unless it is reclassified. If FDA knows that a device being commercially distributed may be a "new" device as defined in this section because of any new intended use or other reasons, FDA may codify the statutory classification of the device into class III for such new use. Accordingly, the regulation for such a class III device states that as of the enactment date of the amendments, May 28, 1976, the device must have an approval under section 515 of the act before commercial distribution.

[52 FR 17739, May 11, 1987]

§ 882.9 Limitations of exemptions from section 510(k) of the Federal Food, Drug, and Cosmetic Act (the act).

The exemption from the requirement of premarket notification (section 510(k) of the act) for a generic type of class I or II device is only to the extent that the device has existing or reasonably foreseeable characteristics of commercially distributed devices within that generic type or, in the case of in vitro diagnostic devices, only to the extent that misdiagnosis as a result of using the device would not be associated with high morbidity or mortality. Accordingly, manufacturers of any commercially distributed class I or II device for which FDA has granted an exemption from the requirement of premarket notification must still submit a premarket notification to FDA before introducing or delivering for introduction into interstate commerce for commercial distribution the device when:

(a) The device is intended for a use different from the intended use of a legally marketed device in that generic type of device; e.g., the device is intended for a different medical purpose, or the device is intended for lay use where the former intended use was by health care professionals only;

(b) The modified device operates using a different fundamental scientific technology than a legally marketed device in that generic type of device; e.g., a surgical instrument cuts tissue with a laser beam rather than with a sharpened metal blade, or an in vitro diagnostic device detects or identifies infectious agents by using deoxyribonucleic acid (DNA) probe or nucleic acid hybridization technology rather than culture or immunoassay technology; or

(c) The device is an in vitro device that is intended:

(1) For use in the diagnosis, monitoring, or screening of neoplastic diseases with the exception of immunohistochemical devices;

(2) For use in screening or diagnosis of familial or acquired genetic disorders, including inborn errors of metabolism;

(3) For measuring an analyte that serves as a surrogate marker for screening, diagnosis, or monitoring life-threatening diseases such as acquired immune deficiency syndrome (AIDS), chronic or active hepatitis, tuberculosis, or myocardial infarction or to monitor therapy;

(4) For assessing the risk of cardiovascular diseases;

(5) For use in diabetes management;

(6) For identifying or inferring the identity of a microorganism directly from clinical material;

(7) For detection of antibodies to microorganisms other than immunoglobulin G (IgG) or IgG assays when the results are not qualitative, or are used to determine immunity, or the assay is intended for use in matrices other than serum or plasma;

(8) For noninvasive testing as defined in § 812.3(k) of this chapter; and

(9) For near patient testing (point of care).

[65 FR 2319, Jan. 14, 2000]

Subpart B—Neurological Diagnostic Devices

§ 882.1020 Rigidity analyzer.

(a) *Identification.* A rigidity analyzer is a device for quantifying the extent of the rigidity of a patient's limb to determine the effectiveness of drugs or other treatments.

(b) *Classification.* Class II (performance standards).

§ 882.1030 Ataxiagraph.

(a) *Identification.* An ataxiagraph is a device used to determine the extent of ataxia (failure of muscular coordination) by measuring the amount of swaying of the body when the patient is standing erect and with eyes closed.

(b) *Classification.* Class I (general controls).

[44 FR 51730–51778, Sept. 4, 1979, as amended at 66 FR 46952, Sept. 10, 2001]

§ 882.1200 Two-point discriminator.

(a) *Identification.* A two-point discriminator is a device with points used for testing a patient's touch discrimination.

(b) *Classification.* Class I (general controls). The device is exempt from the premarket notification procedures in subpart E of part 807 of this chapter subject to § 882.9. The device is also exempt from the current good manufacturing practice regulations in part 820 of this chapter, with the exception of § 820.180 of this chapter, with respect to general requirements concerning records, and § 820.198 of this chapter, with respect to complaint files.

[44 FR 51730–51778, Sept. 4, 1979, as amended at 54 FR 25051, June 12, 1989; 65 FR 2319, Jan. 14, 2000]

§ 882.1240 Echoencephalograph.

(a) *Identification.* An echoencephalograph is an ultrasonic scanning device (including A-scan, B-scan, and doppler systems) that uses noninvasive transducers for measuring intracranial interfaces and blood flow velocity to and in the head.

(b) *Classification.* Class II (performance standards).

§ 882.1275 Electroconductive media.

(a) *Identification.* Electroconductive media are the conductive creams or gels used with external electrodes to reduce the impedance (resistance to alternating current) of the contact between the electrode surface and the skin.

(b) *Classification.* Class II (performance standards).

§ 882.1310 Cortical electrode.

(a) *Identification.* A cortical electrode is an electrode which is temporarily placed on the surface of the brain for stimulating the brain or recording the brain's electrical activity.

(b) *Classification.* Class II (performance standards).

§ 882.1320 Cutaneous electrode.

(a) *Identification.* A cutaneous electrode is an electrode that is applied directly to a patient's skin either to record physiological signals (e.g., the electroencephalogram) or to apply electrical stimulation.

(b) *Classification.* Class II (performance standards).

§ 882.1330 Depth electrode.

(a) *Identification.* A depth electrode is an electrode used for temporary stimulation of, or recording electrical signals at, subsurface levels of the brain.

(b) *Classification.* Class II (performance standards).

§ 882.1340 Nasopharyngeal electrode.

(a) *Identification.* A nasopharyngeal electrode is an electrode which is temporarily placed in the nasopharyngeal region for the purpose of recording electrical activity.

(b) *Classification.* Class II (performance standards).

§ 882.1350 Needle electrode.

(a) *Identification.* A needle electrode is a device which is placed subcutaneously to stimulate or to record electrical signals.

(b) *Classification.* Class II (performance standards).

§ 882.1400 Electroencephalograph.

(a) *Identification.* An electroencephalograph is a device used to measure and record the electrical activity of the patient's brain obtained by placing two or more electrodes on the head.

(b) *Classification.* Class II (performance standards).

§ 882.1410 Electroencephalograph electrode/lead tester.

(a) *Identification.* An electroencephalograph electrode/lead tester is a device used for testing the impedance (resistance to alternating current) of the electrode and lead system of an electroencephalograph to assure that an adequate contact is made between the electrode and the skin.

(b) *Classification.* Class I (general controls). The device is exempt from the premarket notification procedures in subpart E of part 807 of this chapter subject to the limitations in § 882.9.

[44 FR 51730–51778, Sept. 4, 1979, as amended at 61 FR 1123, Jan. 16, 1996; 66 FR 38807, July 25, 2001]

§ 882.1420 Electroencephalogram (EEG) signal spectrum analyzer.

(a) *Identification.* An electroencephalogram (EEG) signal spectrum analyzer is a device used to display the frequency content or power spectral density of the electroencephalogram (EEG) signal.

(b) *Classification.* Class I (general controls).

[44 FR 51730–51778, Sept. 4, 1979, as amended at 66 FR 46953, Sept. 10, 2001]

§ 882.1430 Electroencephalograph test signal generator.

(a) *Identification.* An electroencephalograph test signal generator is a device used to test or calibrate an electroencephalograph.

(b) *Classification.* Class I (general controls). The device is exempt from the premarket notification procedures in subpart E of part 807 of this chapter subject to the limitations in § 882.9.

[44 FR 51730–51778, Sept. 4, 1979, as amended at 59 FR 63011, Dec. 7, 1994; 66 FR 38807, July 25, 2001]

§ 882.1460 Nystagmograph.

(a) *Identification.* A nystagmograph is a device used to measure, record, or visually display the involuntary movements (nystagmus) of the eyeball.

(b) *Classification.* Class II (performance standards).

§ 882.1480 Neurological endoscope.

(a) *Identification.* A neurological endoscope is an instrument with a light source used to view the inside of the ventricles of the brain.

(b) *Classification.* Class II (performance standards).

§ 882.1500 Esthesiometer.

(a) *Identification.* An esthesiometer is a mechanical device which usually consists of a single rod or fiber which is held in the fingers of the physician or other examiner and which is used to determine whether a patient has tactile sensitivity.

(b) *Classification.* Class I (general controls). The device is exempt from the premarket notification procedures in subpart E of part 807 of this chapter subject to § 882.9. The device is also exempt from the current good manufacturing practice regulations in part 820 of this chapter, with the exception of § 820.180 of this chapter, with respect to general requirements concerning records, and § 820.198 of this chapter, with respect to complaint files.

[44 FR 51730–51778, Sept. 4, 1979, as amended at 54 FR 25051, June 12, 1989; 65 FR 2319, Jan. 14, 2000]

§ 882.1525 Tuning fork.

(a) *Identification.* A tuning fork is a mechanical device which resonates at a given frequency and is used to diagnose hearing disorders and to test for vibratory sense.

(b) *Classification.* Class I (general controls). The device is exempt from the premarket notification procedures in subpart E of part 807 of this chapter subject to the limitations in § 882.9. The device is also exempt from the current good manufacturing practice regulations in part 820 of this chapter, of this chapter, with the exception of § 820.180, with respect to general requirements concerning records, and § 820.198, with respect to complaint files.

[44 FR 51730–51778, Sept. 4, 1979, as amended at 54 FR 25051, June 12, 1989; 66 FR 38807, July 25, 2001]

§ 882.1540 Galvanic skin response measurement device.

(a) *Identification.* A galvanic skin response measurement device is a device used to determine autonomic responses as psychological indicators by measuring the electrical resistance of the skin and the tissue path between two electrodes applied to the skin.

(b) *Classification.* Class II (performance standards).

§ 882.1550 Nerve conduction velocity measurement device.

(a) *Identification.* A nerve conduction velocity measurement device is a device which measures nerve conduction time by applying a stimulus, usually to a patient's peripheral nerve. This device includes the stimulator and the electronic processing equipment for measuring and displaying the nerve conduction time.

(b) *Classification.* Class II (performance standards).

§ 882.1560 Skin potential measurement device.

(a) *Identification.* A skin potential measurement device is a general diagnostic device used to measure skin voltage by means of surface skin electrodes.

(b) *Classification.* Class II (performance standards).

§ 882.1570 Powered direct-contact temperature measurement device.

(a) *Identification.* A powered direct-contact temperature measurement device is a device which contains a power source and is used to measure differences in temperature between two points on the body.

(b) *Classification.* Class II (performance standards).

§ 882.1610 Alpha monitor.

(a) *Identification.* An alpha monitor is a device with electrodes that are placed on a patient's scalp to monitor that portion of the electroencephalogram which is referred to as the alpha wave.

(b) *Classification.* Class II (performance standards).

§ 882.1620 Intracranial pressure monitoring device.

(a) *Identification.* An intracranial pressure monitoring device is a device used for short-term monitoring and recording of intracranial pressures and pressure trends. The device includes the transducer, monitor, and interconnecting hardware.

(b) *Classification.* Class II (performance standards).

§ 882.1700 Percussor.

(a) *Identification.* A percussor is a small hammerlike device used by a physician to provide light blows to a body part. A percussor is used as a diagnostic aid during physical examinations.

(b) *Classification.* Class I (general controls). The device is exempt from the premarket notification procedures in subpart E of part 807 of this chapter subject to the limitations in § 882.9. The device is also exempt from the current good manufacturing practice regulations in part 820 of this chapter, with the exception of § 820.180, with respect to general requirements concerning records, and § 820.198, with respect to complaint files.

[44 FR 51730–51778, Sept. 4, 1979, as amended at 54 FR 25051, June 12, 1989; 59 FR 63011, Dec. 7, 1994; 66 FR 38807, July 25, 2001]

§ 882.1750 Pinwheel.

(a) *Identification.* A pinwheel is a device with sharp points on a rotating wheel used for testing pain sensation.

(b) *Classification.* Class I (general controls). The device is exempt from the premarket notification procedures in subpart E of part 807 of this chapter subject to § 882.9.

[44 FR 51730–51778, Sept. 4, 1979, as amended at 54 FR 25051, June 12, 1989; 65 FR 2319, Jan. 14, 2000]

§ 882.1790 Ocular plethysmograph.

(a) *Identification.* An ocular plethysmograph is a device used to measure or detect volume changes in the eye produced by pulsations of the artery, to diagnose carotid artery occlusive disease (restrictions on blood flow in the carotid artery).

(b) *Classification.* Class III (premarket approval).

(c) Date PMA or notice of completion of PDP is required. A PMA or notice of completion of a PDP is required to be filed with the Food and Drug Administration on or before September 21, 2004, for any ocular plethysmograph that was in commercial distribution before May 28, 1976. Any other ocular plethysmograph shall have an approved PMA or declared completed PDP in effect before being placed in commercial distribution.

[44 FR 51730–51778, Sept. 4, 1979, as amended at 52 FR 17739, May 11, 1987; 69 FR 34920, June 23, 2004]

§ 882.1825 Rheoencephalograph.

(a) *Identification.* A rheoencephalograph is a device used to estimate a patient's cerebral circulation (blood flow in the brain) by electrical impedance methods with direct electrical connections to the scalp or neck area.

(b) *Classification.* Class III (premarket approval).

(c) *Date PMA or notice of completion of a PDP is required.* A PMA or a notice of completion of a PDP is required to be filed with the Food and Drug Administration on or before December 26, 1996 for any rheoencephalograph that was in commercial distribution before May 28, 1976, or that has, on or before December 26, 1996 been found to be substantially equivalent to a rheoencephalograph that was in commercial distribution before May 28, 1976. Any other rheoencephalograph shall have an approved PMA or a declared completed PDP in effect before being placed in commercial distribution.

[44 FR 51730–51778, Sept. 4, 1979, as amended at 52 FR 17740, May 11, 1987; 61 FR 50708, Sept. 27, 1996]

§882.1835 Physiological signal amplifier.

(a) *Identification.* A physiological signal amplifier is a general purpose device used to electrically amplify signals derived from various physiological sources (e.g., the electroencephalogram).

(b) *Classification.* Class II (performance standards).

§882.1845 Physiological signal conditioner.

(a) *Identification.* A physiological signal conditioner is a device such as an integrator or differentiator used to modify physiological signals for recording and processing.

(b) *Classification.* Class II (performance standards).

§882.1855 Electroencephalogram (EEG) telemetry system.

(a) *Identification.* An electroencephalogram (EEG) telemetry system consists of transmitters, receivers, and other components used for remotely monitoring or measuring EEG signals by means of radio or telephone transmission systems.

(b) *Classification.* Class II (performance standards).

§882.1870 Evoked response electrical stimulator.

(a) *Identification.* An evoked response electrical stimulator is a device used to apply an electrical stimulus to a patient by means of skin electrodes for the purpose of measuring the evoked response.

(b) *Classification.* Class II (performance standards).

§882.1880 Evoked response mechanical stimulator.

(a) *Identification.* An evoked response mechanical stimulator is a device used to produce a mechanical stimulus or a series of mechanical stimuli for the purpose of measuring a patient's evoked response.

(b) *Classification.* Class II (performance standards).

§882.1890 Evoked response photic stimulator.

(a) *Identification.* An evoked response photic stimulator is a device used to generate and display a shifting pattern or to apply a brief light stimulus to a patient's eye for use in evoked response measurements or for electroencephalogram (EEG) activation.

(b) *Classification.* Class II (performance standards).

§882.1900 Evoked response auditory stimulator.

(a) *Identification.* An evoked response auditory stimulator is a device that produces a sound stimulus for use in evoked response measurements or electroencephalogram activation.

(b) *Classification.* Class II (performance standards).

§882.1925 Ultrasonic scanner calibration test block.

(a) *Identification.* An ultrasonic scanner calibration test block is a block of material with known properties used to calibrate ultrasonic scanning devices (e.g., the echoencephalograph).

(b) *Classification.* Class I (general controls). The device is exempt from the premarket notification procedures in subpart E of part 807 of this chapter subject to the limitations in §882.9.

[44 FR 51730–51778, Sept. 4, 1979, as amended at 59 FR 63011, Dec. 7, 1994; 66 FR 38807, July 25, 2001]

§882.1950 Tremor transducer.

(a) *Identification.* A tremor transducer is a device used to measure the degree of tremor caused by certain diseases.

(b) *Classification.* Class II (performance standards).

Subparts C–D [Reserved]

Subpart E—Neurological Surgical Devices

§ 882.4030 Skull plate anvil.

(a) *Identification.* A skull plate anvil is a device used to form alterable skull plates in the proper shape to fit the curvature of a patient's skull.

(b) *Classification.* Class I (general controls). The device is exempt from the premarket notification procedures in subpart E of part 807 of this chapter subject to the limitations in § 882.9.

[44 FR 51730–51778, Sept. 4, 1979, as amended at 59 FR 63011, Dec. 7, 1994; 66 FR 38808, July 25, 2001]

§ 882.4060 Ventricular cannula.

(a) *Identification.* A ventricular cannula is a device used to puncture the ventricles of the brain for aspiration or for injection. This device is frequently referred to as a ventricular needle.

(b) *Classification.* Class I (general controls). When made only of surgical grade stainless steel, the device is exempt from the premarket notification procedures in subpart E of part 807 of this chapter subject to § 882.9.

[44 FR 51730–51778, Sept. 4, 1979, as amended at 65 FR 2319, Jan. 14, 2000]

§ 882.4100 Ventricular catheter.

(a) *Identification.* A ventricular catheter is a device used to gain access to the cavities of the brain for injection of material into, or removal of material from, the brain.

(b) *Classification.* Class II (performance standards).

§ 882.4125 Neurosurgical chair.

(a) *Identification.* A neurosurgical chair is an operating room chair used to position and support a patient during neurosurgery.

(b) *Classification.* Class I (general controls). The device is exempt from the premarket notification procedures in subpart E of part 807 of this chapter subject to the limitations in § 882.9.

[44 FR 51730–51778, Sept. 4, 1979, as amended at 59 FR 63012, Dec. 7, 1994; 66 FR 38808, July 25, 2001]

§ 882.4150 Scalp clip.

(a) *Identification.* A scalp clip is a plastic or metal clip used to stop bleeding during surgery on the scalp.

(b) *Classification.* Class II (performance standards).

§ 882.4175 Aneurysm clip applier.

(a) *Identification.* An aneurysm clip applier is a device used by the surgeon for holding and applying intracranial aneurysm clips.

(b) *Classification.* Class II (performance standards).

§ 882.4190 Clip forming/cutting instrument.

(a) *Identification.* A clip forming/cutting instrument is a device used by the physician to make tissue clips from wire stock.

(b) *Classification.* Class I. The device is exempt from the premarket notification procedures in subpart E of part 807 of this chapter.

[44 FR 51730–51778, Sept. 4, 1979, as amended at 59 FR 63012, Dec. 7, 1994]

§ 882.4200 Clip removal instrument.

(a) *Identification.* A clip removal instrument is a device used to remove surgical clips from the patient.

(b) *Classification.* Class I (general controls). The device is exempt from the premarket notification procedures in subpart E of part 807 of this chapter subject to the limitations in § 882.9.

[44 FR 51730–51778, Sept. 4, 1979, as amended at 59 FR 63012, Dec. 7, 1994; 66 FR 38808, July 25, 2001]

§ 882.4215 Clip rack.

(a) *Identification.* A clip rack is a device used to hold or store surgical clips during surgery.

(b) *Classification.* Class I (general controls). The device is exempt from the premarket notification procedures in subpart E of part 807 of this chapter subject to the limitations in § 882.9.

[44 FR 51730–51778, Sept. 4, 1979, as amended at 54 FR 25051, June 12, 1989; 59 FR 63012, Dec. 7, 1994; 66 FR 38808, July 25, 2001]

§ 882.4250 Cryogenic surgical device.

(a) *Identification.* A cryogenic surgical device is a device used to destroy nervous tissue or produce lesions in nervous tissue by the application of extreme cold to the selected site.

(b) *Classification.* Class II (performance standards).

§ 882.4275 Dowel cutting instrument.

(a) *Identification.* A dowel cutting instrument is a device used to cut dowels of bone for bone grafting.

(b) *Classification.* Class II (performance standards).

§ 882.4300 Manual cranial drills, burrs, trephines, and their accessories

(a) *Identification.* Manual cranial drills, burrs, trephines, and their accessories are bone cutting and drilling instruments that are used without a power source on a patient's skull.

(b) *Classification.* Class II (performance standards).

§ 882.4305 Powered compound cranial drills, burrs, trephines, and their accessories.

(a) *Identification.* Powered compound cranial drills, burrs, trephines, and their accessories are bone cutting and drilling instruments used on a patient's skull. The instruments employ a clutch mechanism to disengage the tip of the instrument after penetrating the skull to prevent plunging of the tip into the brain.

(b) *Classification.* Class II (performance standards).

§ 882.4310 Powered simple cranial drills, burrs, trephines, and their accessories.

(a) *Identification.* Powered simple cranial drills, burrs, trephines, and their accessories are bone cutting and drilling instruments used on a patient's skull. The instruments are used with a power source but do not have a clutch mechanism to disengage the tip after penetrating the skull.

(b) *Classification.* Class II (performance standards).

§ 882.4325 Cranial drill handpiece (brace).

(a) *Identification.* A cranial drill handpiece (brace) is a hand holder, which is used without a power source, for drills, burrs, trephines, or other cutting tools that are used on a patient's skull.

(b) *Classification.* Class I (general controls). The device is exempt from the premarket notification procedures in subpart E of part 807 of this chapter subject to the limitations in § 882.9.

[44 FR 51730–51778, Sept. 4, 1979, as amended at 61 FR 1123, Jan. 16, 1996; 66 FR 38808, July 25, 2001]

§ 882.4360 Electric cranial drill motor.

(a) *Identification.* An electric cranial drill motor is an electrically operated power source used with removable rotating surgical cutting tools or drill bits on a patient's skull.

(b) *Classification.* Class II (performance standards).

§ 882.4370 Pneumatic cranial drill motor.

(a) *Identification.* A pneumatic cranial drill motor is a pneumatically operated power source used with removable rotating surgical cutting tools or drill bits on a patient's skull.

(b) *Classification.* Class II (performance standards).

§ 882.4400 Radiofrequency lesion generator.

(a) *Identification.* A radiofrequency lesion generator is a device used to produce lesions in the nervous system or other tissue by the direct application of radiofrequency currents to selected sites.

(b) *Classification.* Class II (performance standards).

§ 882.4440 Neurosurgical headrests.

(a) *Identification.* A neurosurgical headrest is a device used to support the patient's head during a surgical procedure.

(b) *Classification.* Class I (general controls). The device is exempt from the premarket notification procedures in subpart E of part 807 of this chapter subject to the limitations in § 882.9.

[44 FR 51730–51778, Sept. 4, 1979, as amended at 59 FR 63012, Dec. 7, 1994; 66 FR 38808, July 25, 2001]

§ 882.4460 Neurosurgical head holder (skull clamp).

(a) *Identification.* A neurosurgical head holder (skull clamp) is a device used to clamp the patient's skull to hold head and neck in a particular position during surgical procedures.

(b) *Classification*. Class II (performance standards).

§ 882.4500 Cranioplasty material forming instrument.

(a) *Identification*. A cranioplasty material forming instrument is a roller used in the preparation and forming of cranioplasty (skull repair) materials.

(b) *Classification*. Class I (general controls). The device is exempt from the premarket notification procedures in subpart E of part 807 of this chapter subject to the limitations in § 882.9.

[44 FR 51730–51778, Sept. 4, 1979, as amended at 59 FR 63012, Dec. 7, 1994; 66 FR 38808, July 25, 2001]

§ 882.4525 Microsurgical instrument.

(a) *Identification*. A microsurgical instrument is a nonpowered surgical instrument used in neurological microsurgery procedures.

(b) *Classification*. Class I (general controls). The device is exempt from the premarket notification procedures in subpart E of part 807 of this chapter subject to the limitations in § 882.9.

[44 FR 51730–51778, Sept. 4, 1979, as amended at 59 FR 63012, Dec. 7, 1994; 66 FR 38808, July 25, 2001]

§ 882.4535 Nonpowered neurosurgical instrument.

(a) *Identification*. A nonpowered neurosurgical instrument is a hand instrument or an accessory to a hand instrument used during neurosurgical procedures to cut, hold, or manipulate tissue. It includes specialized chisels, osteotomes, curettes, dissectors, elevators, forceps, gouges, hooks, surgical knives, rasps, scissors, separators, spatulas, spoons, blades, blade holders, blade breakers, probes, etc.

(b) *Classification*. Class I (general controls). The device is exempt from the premarket notification procedures in subpart E of part 807 of this chapter subject to the limitations in § 882.9.

[44 FR 51730–51778, Sept. 4, 1979, as amended at 59 FR 63012, Dec. 7, 1994; 66 FR 38808, July 25, 2001]

§ 882.4545 Shunt system implantation instrument.

(a) *Identification*. A shunt system implantation instrument is an instrument used in the implantation of cerebrospinal fluid shunts, and includes tunneling instruments for passing shunt components under the skin.

(b) *Classification*. Class I (general controls). When made only of surgical grade stainless steel, the device is exempt from the premarket notification procedures in subpart E of part 807 of this chapter subject to § 882.9.

[44 FR 51730–51778, Sept. 4, 1979, as amended at 65 FR 2319, Jan. 14, 2000]

§ 882.4560 Stereotaxic instrument.

(a) *Identification*. A stereotaxic instrument is a device consisting of a rigid frame with a calibrated guide mechanism for precisely positioning probes or other devices within a patient's brain, spinal cord, or other part of the nervous system.

(b) *Classification*. Class II (performance standards).

§ 882.4600 Leukotome.

(a) *Identification*. A leukotome is a device used to cut sections out of the brain.

(b) *Classification*. Class I (general controls). The device is exempt from the premarket notification procedures in subpart E of part 807 of this chapter subject to the limitations in § 882.9.

[44 FR 51730–51778, Sept. 4, 1979, as amended at 59 FR 63012, Dec. 7, 1994; 66 FR 38808, July 25, 2001]

§ 882.4650 Neurosurgical suture needle.

(a) *Identification*. A neurosurgical suture needle is a needle used in suturing during neurosurgical procedures or in the repair of nervous tissue.

(b) *Classification*. Class I (general controls). The device is exempt from the premarket notification procedures in subpart E of part 807 of this chapter subject to § 882.9.

[44 FR 51730–51778, Sept. 4, 1979, as amended at 54 FR 25051, June 12, 198965 FR 2319, Jan. 14, 2000]

§ 882.4700 Neurosurgical paddie.

(a) A neurosurgical paddie is a pad used during surgery to protect nervous tissue, absorb fluids, or stop bleeding.

(b) *Classification.* Class II (performance standards).

[44 FR 51730–51778, Sept. 4, 1979, as amended at 69 FR 10332, Mar. 5, 2004]

§ 882.4725 Radiofrequency lesion probe.

(a) *Identification.* A radiofrequency lesion probe is a device connected to a radiofrequency (RF) lesion generator to deliver the RF energy to the site within the nervous system where a lesion is desired.

(b) *Classification.* Class II (performance standards).

§ 882.4750 Skull punch.

(a) *Identification.* A skull punch is a device used to punch holes through a patient's skull to allow fixation of cranioplasty plates or bone flaps by wire or other means.

(b) *Classification.* Class I (general controls). The device is exempt from the premarket notification procedures in subpart E of part 807 of this chapter subject to § 882.9. This exemption does not apply to powered compound cranial drills, burrs, trephines, and their accessories classified under § 882.4305.

[44 FR 51730–51778, Sept. 4, 1979, as amended at 65 FR 2319, Jan. 14, 2000]

§ 882.4800 Self-retaining retractor for neurosurgery.

(a) *Identification.* A self-retaining retractor for neurosurgery is a self-locking device used to hold the edges of a wound open during neurosurgery.

(b) *Classification.* Class II (performance standards).

§ 882.4840 Manual rongeur.

(a) *Identification.* A manual rongeur is a manually operated instrument used for cutting or biting bone during surgery involving the skull or spinal column.

(b) *Classification.* Class II (performance standards).

§ 882.4845 Powered rongeur.

(a) *Identification.* A powered rongeur is a powered instrument used for cutting or biting bone during surgery involving the skull or spinal column.

(b) *Classification.* Class II (performance standards).

§ 882.4900 Skullplate screwdriver.

(a) *Identification.* A skullplate screwdriver is a tool used by the surgeon to fasten cranioplasty plates or skullplates to a patient's skull by screws.

(b) *Classification.* Class I (general controls). The device is exempt from the premarket notification procedures in subpart E of part 807 of this chapter subject to the limitations in § 882.9.

[44 FR 51730–51778, Sept. 4, 1979, as amended at 59 FR 63012, Dec. 7, 1994; 66 FR 38808, July 25, 2001]

Subpart F—Neurological Therapeutic Devices

§ 882.5030 Methyl methacrylate for aneurysmorrhaphy.

(a) *Identification.* Methyl methacrylate for aneurysmorrhaphy (repair of aneurysms, which are balloonlike sacs formed on blood vessels) is a self-curing acrylic used to encase and reinforce intracranial aneurysms that are not amenable to conservative management, removal, or obliteration by aneurysm clip.

(b) *Classification.* Class II (performance standards).

§ 882.5050 Biofeedback device.

(a) *Identification.* A biofeedback device is an instrument that provides a visual or auditory signal corresponding to the status of one or more of a patient's physiological parameters (e.g., brain alpha wave activity, muscle activity, skin temperature, etc.) so that the patient can control voluntarily these physiological parameters.

(b) *Classification.* Class II (special controls). The device is exempt from the premarket notification procedures in subpart E of part 807 of this chapter when it is a prescription battery powered device that is indicated for relaxation training and muscle reeducation and prescription use, subject to § 882.9.

[44 FR 51730–51778, Sept. 4, 1979, as amended at 63 FR 59229, Nov. 3, 1998]

§ 882.5070 Bite block.

(a) *Identification.* A bite block is a device inserted into a patient's mouth to protect the tongue and teeth while the patient is having convulsions.

(b) *Classification.* Class II (performance standards).

§ 882.5150 Intravascular occluding catheter.

(a) *Identification.* An intravascular occluding catheter is a catheter with an inflatable or detachable balloon tip that is used to block a blood vessel to treat malformations, e.g., aneurysms (balloonlike sacs formed on blood vessels) of intracranial blood vessels.

(b) *Classification.* Class III (premarket approval).

(c) *Date PMA or notice of completion of a PDP is required.* A PMA or a notice of completion of a PDP is required to be filed with the Food and Drug Administration on or before December 26, 1996 for any intravascular occluding catheter that was in commercial distribution before May 28, 1976, or that has, on or before December 26, 1996 been found to be substantially equivalent to an intravascular occluding catheter that was in commercial distribution before May 28, 1976. Any other intravascular occluding catheter shall have an approved PMA or a declared completed PDP in effect before being placed in commercial distribution.

[44 FR 51730–51778, Sept. 4, 1979, as amended at 52 FR 17740, May 11, 1987; 61 FR 50708, Sept. 27, 1996]

§ 882.5175 Carotid artery clamp.

(a) *Identification.* A carotid artery clamp is a device that is surgically placed around a patient's carotid artery (the principal artery in the neck that supplies blood to the brain) and has a removable adjusting mechanism that protrudes through the skin of the patient's neck. The clamp is used to occlude the patient's carotid artery to treat intracranial aneurysms (balloonlike sacs formed on blood vessels) or other intracranial vascular malformations that are difficult to attach directly by reducing the blood pressure and blood flow to the aneurysm or malformation.

(b) *Classification.* Class II (performance standards).

§ 882.5200 Aneurysm clip.

(a) *Identification.* An aneurysm clip is a device used to occlude an intracranial aneurysm (a balloonlike sac formed on a blood vessel) to prevent it from bleeding or bursting.

(b) *Classification.* Class II (performance standards).

§ 882.5225 Implanted malleable clip.

(a) *Identification.* An implanted malleable clip is a bent wire or staple that is forcibly closed with a special instrument to occlude an intracranial blood vessel or aneurysm (a balloonlike sac formed on a blood vessel), stop bleeding, or hold tissue or a mechanical device in place in a patient.

(b) *Classification.* Class II (performance standards).

§ 882.5235 Aversive conditioning device.

(a) *Identification.* An aversive conditioning device is an instrument used to administer an electrical shock or other noxious stimulus to a patient to modify undesirable behavioral characteristics.

(b) *Classification.* Class II (performance standards).

§ 882.5250 Burr hole cover.

(a) *Identification.* A burr hole cover is a plastic or metal device used to cover or plug holes drilled into the skull during surgery and to reattach cranial bone removed during surgery.

(b) *Classification.* Class II (performance standards).

§ 882.5275 Nerve cuff.

(a) *Identification.* A nerve cuff is a tubular silicone rubber sheath used to encase a nerve for aid in repairing the nerve (e.g., to prevent ingrowth of scar tissue) and for capping the end of the nerve to prevent the formation of neuroma (tumors).

(b) *Classification.* Class II (performance standards).

§ 882.5300 Methyl methacrylate for cranioplasty.

(a) *Identification.* Methyl methacrylate for cranioplasty (skull repair) is a self-curing acrylic that a surgeon uses to repair a skull defect in a patient. At the time of surgery, the surgeon initiates polymerization of the material and forms it into a plate or other appropriate shape to repair the defect.

(b) *Classification.* Class II (performance standards).

§ 882.5320 Preformed alterable cranioplasty plate.

(a) *Identification.* A preformed alterable cranioplasty plate is a device that is implanted into a patient to repair a skull defect. It is constructed of a material, e.g., tantalum, that can be altered or reshaped at the time of surgery without changing the chemical behavior of the material.

(b) *Classification.* Class II (performance standards).

§ 882.5330 Preformed nonalterable cranioplasty plate.

(a) *Identification.* A preformed nonalterable cranioplasty plate is a device that is implanted in a patient to repair a skull defect and is constructed of a material, e.g., stainless steel or vitallium, that cannot be altered or reshaped at the time of surgery without changing the chemical behavior of the material.

(b) *Classification.* Class II (performance standards).

§ 882.5360 Cranioplasty plate fastener.

(a) *Identification.* A cranioplasty plate fastener is a screw, wire, or other article made of tantalum, vitallium, or stainless steel used to secure a plate to the patient's skull to repair a skull defect.

(b) *Classification.* Class II (performance standards).

§ 882.5500 Lesion temperature monitor.

(a) *Identification.* A lesion temperature monitor is a device used to monitor the tissue temperature at the site where a lesion (tissue destruction) is to be made when a surgeon uses a radiofrequency (RF) lesion generator and probe.

(b) *Classification.* Class II (performance standards).

§ 882.5550 Central nervous system fluid shunt and components.

(a) *Identification.* A central nervous system fluid shunt is a device or combination of devices used to divert fluid from the brain or other part of the central nervous system to an internal delivery site or an external receptacle for the purpose of relieving elevated intracranial pressure or fluid volume (e.g., due to hydrocephalus). Components of a central nervous system shunt include catheters, valved catheters, valves, connectors, and other accessory components intended to facilitate use of the shunt or evaluation of a patient with a shunt.

(b) *Classification.* Class II (performance standards).

§ 882.5800 Cranial electrotherapy stimulator.

(a) *Identification.* A cranial electrotherapy stimulator is a device that applies electrical current to a patient's head to treat insomnia, depression, or anxiety.

(b) *Classification.* Class III (premarket approval).

(c) *Date a PMA or notice of completion of a PDP is required.* No effective date has been established of the requirement for premarket approval. See § 882.3.

[44 FR 51730–51778, Sept. 4, 1979, as amended at 52 FR 17740, May 11, 1987; 60 FR 43969, Aug. 24, 1995; 62 FR 30457, June 4, 1997]

§ 882.5810 External functional neuromuscular stimulator.

(a) *Identification.* An external functional neuromuscular stimulator is an electrical stimulator that uses external electrodes for stimulating muscles in the leg and ankle of partially paralyzed patients (e.g., after stroke) to provide flexion of the foot and thus improve the patient's gait.

(b) *Classification.* Class II (performance standards).

§ 882.5820 Implanted cerebellar stimulator.

(a) *Identification.* An implanted cerebellar stimulator is a device used to stimulate electrically a patient's cerebellar cortex for the treatment of intractable epilepsy, spasticity, and some movement disorders. The stimulator consists of an implanted receiver with electrodes that are placed on the patient's cerebellum and an external transmitter for transmitting the stimulating pulses across the patient's skin to the implanted receiver.

(b) *Classification.* Class III (premarket approval).

(c) *Date premarket approval application (PMA) or notice of completion of a product development protocol (PDP) is required.* A PMA or notice of completion of a PDP is required to be filed with the Food and Drug Administration on or before September 26, 1984. Any implanted cerebellar stimulator that was not in commercial distribution before May 28, 1976, or that has not on or before September 26, 1984 been found by FDA to be substantially equivalent to an implanted cerebellar stimulator that was in commercial distribution before May 28, 1976 shall have an approved PMA or declared completed PDP in effect before beginning commercial distribution.

[44 FR 51730–51778, Sept. 4, 1979 and 49 FR 26574, June 28, 1984]

§ 882.5830 Implanted diaphragmatic/phrenic nerve stimulator.

(a) *Identification.* An implanted diaphragmatic/phrenic nerve stimulator is a device that provides electrical stimulation of a patient's phrenic nerve to contract the diaphragm rhythmically and produce breathing in patients who have hypoventilation (a state in which an abnormally low amount of air enters the lungs) caused by brain stem disease, high cervical spinal cord injury, or chronic lung disease. The stimulator consists of an implanted receiver with electrodes that are placed around the patient's phrenic nerve and an external transmitter for transmitting the stimulating pulses across the patient's skin to the implanted receiver.

(b) *Classification.* Class III (premarket approval).

(c) *Date premarket approval application (PMA) or notice of completion of a product development protocol (PDP) is required.* A PMA or a notice of completion of a PDP is required to be filed with the Food and Drug Administration on or before July 7, 1986 for any implanted diaphragmatic/phrenic nerve stimulator that was in commercial distribution before May 28, 1976, or that has on or before July 7, 1986 been found to be substantially equivalent to an implanted diaphragmatic/phrenic nerve stimulator that was in commercial distribution before May 28, 1976. Any other implanted diaphragmatic/phrenic nerve stimulator shall have an approved PMA or a declared completed PDP in effect before being placed in commercial distribution.

[44 FR 51730–51778, Sept. 4, 1979, as amended at 51 FR 12101, Apr. 8, 1986]

§ 882.5840 Implanted intracerebral/subcortical stimulator for pain relief.

(a) *Identification.* An implanted intracerebral/subcortical stimulator for pain relief is a device that applies electrical current to subsurface areas of a patient's brain to treat severe intractable pain. The stimulator consists of an implanted receiver with electrodes that are placed within a patient's brain and an external transmitter for transmitting the stimulating pulses across the patient's skin to the implanted receiver.

(b) *Classification.* Class III (premarket approval).

(c) *Date premarket approval application (PMA) or notice of completion of a product development protocol (PDP) is required.* A PMA or a notice of completion of a PDP is required to be filed with the Food and Drug Administration on or before March 1, 1989, for any implanted intracerebral/subcortical stimulator for pain relief that was in commercial distribution before May 28, 1976, or that has on or before March 1, 1989, been found to be substantially equivalent to an implanted intracerebral/subcortical stimulator for pain relief that was in commercial distribution before May 28, 1976. Any other implanted intracerebral/subcortical stimulator for pain relief shall have an approved PMA or a declared completed PDP in effect before being placed in commercial distribution.

[44 FR 51730–51778, Sept. 4, 1979, as amended at 53 FR 48621, Dec. 1, 1988]

§ 882.5850 Implanted spinal cord stimulator for bladder evacuation.

(a) *Identification.* An implanted spinal cord stimulator for bladder evacuation is an electrical stimulator used to empty the bladder of a paraplegic patient who has a complete transection of the spinal cord and who is unable to empty his or her bladder by reflex means or by the intermittent use of catheters. The stimulator consists of an implanted receiver with electrodes that are placed on the conus medullaris portion of the patient's spinal cord and an external transmitter for transmitting the stimulating pulses across the patient's skin to the implanted receiver.

(b) *Classification.* Class III (premarket approval).

(c) *Date PMA or notice of completion of a PDP is required.* A PMA or a notice of completion of a PDP is required to be filed with the Food and Drug Administration on or before December 26, 1996 for any implanted spinal cord stimulator for bladder evacuation that was in commercial distribution before May 28, 1976, or that has, on or before December 26, 1996 been found to be substantially equivalent to an implanted spinal cord stimulator for bladder evacuation that was in commercial distribution before May 28, 1976. Any other implanted spinal cord stimulator for bladder evacuation shall have an approved PMA or a declared completed PDP in effect before being placed in commercial distribution.

[44 FR 51730–51778, Sept. 4, 1979, as amended at 52 FR 17740, May 11, 1987; 61 FR 50708, Sept. 27, 1996]

§ 882.5860 Implanted neuromuscular stimulator.

(a) *Identification.* An implanted neuromuscular stimulator is a device that provides electrical stimulation to a patient's peroneal or femoral nerve to cause muscles in the leg to contract, thus improving the gait in a patient with a paralyzed leg. The stimulator consists of an implanted receiver with electrodes that are placed around a patient's nerve and an external transmitter for transmitting the stimulating pulses across the patient's skin to the implanted receiver. The external transmitter is activated by a switch in the heel in the patient's shoe.

(b) *Classification.* Class III (premarket approval).

(c) *Date PMA or notice of completion of PDP is required.* A PMA or notice of completion of a PDP for a device described in paragraph (b) of this section is required to be filed with the Food and Drug Administration on or before July 13, 1999 for any implanted neuromuscular stimulator that was in commercial distribution before May 28, 1976, or that has, on or before July 13, 1999, been found to be substantially equivalent to an implanted neuromuscular stimulator that was in commercial distribution before May 28, 1976. Any other implanted neuromuscular stimulator shall have an approved PMA or declared completed PDP in effect before being placed in commercial distribution.

[44 FR 51730–51778, Sept. 4, 1979, as amended at 52 FR 17740, May 11, 1987; 64 FR 18329, Apr. 14, 1999]

§ 882.5870 Implanted peripheral nerve stimulator for pain relief.

(a) *Identification.* An implanted peripheral nerve stimulator for pain relief is a device that is used to stimulate electrically a peripheral nerve in a patient to relieve severe intractable pain. The stimulator consists of an inplanted receiver with electrodes that are placed around a peripheral nerve and an external transmitter for transmitting the stimulating pulses across the patient's skin to the implanted receiver.

(b) *Classification.* Class II (performance standards).

§ 882.5880 Implanted spinal cord stimulator for pain relief.

(a) *Identification.* An implanted spinal cord stimulator for pain relief is a device that is used to stimulate electrically a patient's spinal cord to relieve severe intractable pain. The stimulator consists of an implanted receiver with electrodes that are placed on the patient's spinal cord and an external transmitter for transmitting the stimulating pulses across the patient's skin to the implanted receiver.

(b) *Classification.* Class II (performance standards).

§ 882.5890 Transcutaneous electrical nerve stimulator for pain relief.

(a) *Identification.* A transcutaneous electrical nerve stimulator for pain relief is a device used to apply an electrical current to electrodes on a patient's skin to treat pain.

(b) *Classification.* Class II (performance standards).

§ 882.5900 Preformed craniosynostosis strip.

(a) *Identification.* A preformed craniosynostosis strip is a plastic strip used to cover bone edges of craniectomy sites (sites where the skull has been cut) to prevent the bone from regrowing in patients whose skull sutures are abnormally fused together.

(b) *Classification.* Class II (performance standards).

§ 882.5910 Dura substitute.

(a) *Identification.* A dura substitute is a sheet or material that is used to repair the dura mater (the membrane surrounding the brain).

(b) *Classification.* Class II (performance standards).

§ 882.5940 Electroconvulsive therapy device.

(a) *Identification.* An electroconvulsive therapy device is a device used for treating severe psychiatric disturbances (e.g., severe depression) by inducing in the patient a major motor seizure by applying a brief intense electrical current to the patient's head.

(b) *Classification.* Class III (premarket approval).

(c) *Date PMA or notice of completion of a PDP is required.* No effective date has been established of the requirement for premarket approval. See § 882.3.

[44 FR 51730–51778, Sept. 4, 1979, as amended at 52 FR 17740, May 11, 1987]

§ 882.5950 Neurovascular embolization device.

(a) *Identification.* A neurovascular embolization device is an intravascular implant intended to permanently occlude blood flow to cerebral aneurysms and cerebral ateriovenous malformations. This does not include cyanoacrylates and other embolic agents, which act by polymerization or precipitation. Embolization devices used in other vascular applications are also not included in this classification, see § 870.3300.

(b) *Classification.* Class II (special controls.) The special control for this device is the FDA guidance document entitled "Class II Special Controls Guidance Document: Vascular and Neurovascular Embolization Devices." For availability of this guidance document, see § 882.1(e).

[69 FR 77900, Dec. 29, 2004]

§ 882.5960 Skull tongs for traction.

(a) *Identification.* Skull tongs for traction is an instrument used to immobilize a patient with a cervical spine injury (e.g., fracture or dislocation). The device is caliper shaped with tips that penetrate the skin. It is anchored to the skull and has a heavy weight attached to it that maintains, by traction, the patient's position.

(b) *Classification.* Class II (performance standards).

§ 882.5970 Cranial orthosis.

(a) *Identification.* A cranial orthosis is a device that is intended for medical purposes to apply pressure to prominent regions of an infant's cranium in order to improve cranial symmetry and/or shape in infants from 3 to 18 months of age, with moderate to severe nonsynostotic positional plagiocephaly, including infants with plagiocephalic-, brachycephalic-, and scaphocephalic-shaped heads.

(b) *Classification.* Class II (special controls) (prescription use in accordance with § 801.109 of this chapter, biocompatibility testing, and labeling (contraindications, warnings, precautions, adverse events, instructions for physicians and parents)).

[63 FR 40651, July 30, 1998]

§ 882.5975 Human dura mater.

(a) *Identification.* Human dura mater is human pachymeninx tissue intended to repair defects in human dura mater.

(b) *Classification.* Class II (special controls). The special control for this device is the FDA guidance document entitled "Class II Special Controls Guidance Document: Human Dura

Mater." See §882.1(e) for the availability of this guidance.

[68 FR 70436, Dec. 18, 2003]

PART 884—OBSTETRICAL AND GYNECOLOGICAL DEVICES

Subpart A—General Provisions

Subpart B—Obstetrical and Gynecological Diagnostic Devices

Subpart C—Obstetrical and Gynecological Monitoring Devices

Subpart D—Obstetrical and Gynecological Prosthetic Devices

Subpart E—Obstetrical and Gynecological Surgical Devices

Subpart F—Obstetrical and Gynecological Therapeutic Devices

884.5390 Perineal heater.
884.5400 Menstrual cup.
884.5425 Scented or scented deodorized menstrual pad.
884.5435 Unscented menstrual pad.
884.5460 Scented or scented deodorized menstrual tampon.
884.5470 Unscented menstrual tampon.
884.5900 Therapeutic vaginal douche apparatus.
884.5920 Vaginal insufflator.
884.5940 Powered vaginal muscle stimulator for therapeutic use.
884.5960 Genital vibrator for therapeutic use.
884.5970 Clitoral engorgement device.

Subpart G—Assisted Reproduction Devices

884.6100 Assisted reproduction needles.
884.6110 Assisted reproduction catheters.
884.6120 Assisted reproduction accessories.
884.6130 Assisted reproduction microtools.
884.6140 Assisted reproduction micropipette fabrication instruments.
884.6150 Assisted reproduction micromanipulators and microinjectors.
884.6160 Assisted reproduction labware.
884.6170 Assisted reproduction water and water purification systems.
884.6180 Reproductive media and supplements.
884.6190 Assisted reproductive microscopes and microscope accessories.
884.6200 Assisted reproduction laser system.

AUTHORITY: 21 U.S.C. 351, 360, 360c, 360e, 360j, 371.

SOURCE: 45 FR 12684, Feb. 26, 1980, unless otherwise noted.

Subpart A—General Provisions

§ 884.1 Scope.

(a) This part sets forth the classification of obstetrical and gynecological devices intended for human use that are in commercial distribution.

(b) The identification of a device in a regulation in this part is not a precise description of every device that is, or will be, subject to the regulation. A manufacturer who submits a premarket notification submission for a device under part 807 may not show merely that the device is accurately described by the section title and identification provisions of a regulation in this part, but shall state why the device is substantially equivalent to other devices, as required by § 807.87.

(c) To avoid duplicative listings, an obstetrical and gynecological device that has two or more types of uses (e.g., used both as a diagnostic device and as a therapeutic device) is listed only in one subpart.

(d) References in this part to regulatory sections of the Code of Federal Regulations are to chapter I of title 21, unless otherwise noted.

(e) Guidance documents referenced in this part are available on the Internet at *http://www.fda.gov/cdrh/guidance.html.*

[52 FR 17740, May 11, 1987, as amended at 68 FR 44415, Aug. 27, 2003]

§ 884.3 Effective dates of requirement for premarket approval.

A device included in this part that is classified into class III (premarket approval) shall not be commercially distributed after the date shown in the regulation classifying the device unless the manufacturer has an approval under section 515 of the act (unless an exemption has been granted under section 520(g)(2) of the act). An approval under section 515 of the act consists of FDA's issuance of an order approving an application for premarket approval (PMA) for the device or declaring completed a product development protocol (PDP) for the device.

(a) Before FDA requires that a device commercially distributed before the enactment date of the amendments, or a device that has been found substantially equivalent to such a device, has an approval under section 515 of the act FDA must promulgate a regulation under section 515(b) of the act requiring such approval, except as provided in paragraph (b) of this section. Such a regulation under section 515(b) of the act shall not be effective during the grace period ending on the 90th day after its promulgation or on the last day of the 30th full calendar month after the regulation that classifies the device into class III is effective, whichever is later. See section 501(f)(2)(B) of the act. Accordingly, unless an effective date of the requirement for premarket approval is shown in the regulation for a device classified into class III in this part, the device may be commercially distributed without FDA's issuance of an order approving a PMA or declaring completed a PDP for the device. If FDA promulgates a regulation under section 515(b) of the act requiring premarket approval for a device, section 501(f)(1)(A) of the act applies to the device.

(b) Any new, not substantially equivalent, device introduced into commercial distribution on or after May 28, 1976, including a device formerly marketed that has been substantially altered, is classified by statute (section 513(f) of the act) into class III without any grace period and FDA must have issued an order approving a PMA or declaring completed a PDP for the device before the device is commercially distributed unless it is reclassified. If FDA knows that a device being commercially distributed may be a "new" device as defined in this section because of any new intended use or other reasons, FDA may codify the statutory classification of the device into class III for such new use. Accordingly, the regulation for such a class III device states that as of the enactment date of the amendments, May 28, 1976, the device must have an approval under section 515 of the act before commercial distribution.

[52 FR 17740, May 11, 1987]

§ 884.9 Limitations of exemptions from section 510(k) of the Federal Food, Drug, and Cosmetic Act (the act).

The exemption from the requirement of premarket notification (section 510(k) of the act) for a generic type of class I or II device is only to the extent that the device has existing or reasonably foreseeable characteristics of commercially distributed devices within that generic type or, in the case of in vitro diagnostic devices, only to the extent that misdiagnosis as a result of using the device would not be associated with high morbidity or mortality. Accordingly, manufacturers of any commercially distributed class I or II device for which FDA has granted an exemption from the requirement of premarket notification must still submit a premarket notification to FDA before introducing or delivering for introduction into interstate commerce for commercial distribution the device when:

(a) The device is intended for a use different from the intended use of a legally marketed device in that generic type of device; e.g., the device is intended for a different medical purpose, or the device is intended for lay use where the former intended use was by health care professionals only;

(b) The modified device operates using a different fundamental scientific technology than a legally marketed device in that generic type of device; e.g., a surgical instrument cuts tissue with a laser beam rather than with a sharpened metal blade, or an in vitro diagnostic device detects or identifies infectious agents by using deoxyribonucleic acid (DNA) probe or nucleic acid hybridization technology rather than culture or immunoassay technology; or

(c) The device is an in vitro device that is intended:

(1) For use in the diagnosis, monitoring, or screening of neoplastic diseases with the exception of immunohistochemical devices;

(2) For use in screening or diagnosis of familial or acquired genetic disorders, including inborn errors of metabolism;

(3) For measuring an analyte that serves as a surrogate marker for screening, diagnosis, or monitoring life-threatening diseases such as acquired immune deficiency syndrome (AIDS), chronic or active hepatitis, tuberculosis, or myocardial infarction or to monitor therapy;

(4) For assessing the risk of cardiovascular diseases;

(5) For use in diabetes management;

(6) For identifying or inferring the identity of a microorganism directly from clinical material;

(7) For detection of antibodies to microorganisms other than immunoglobulin G (IgG) or IgG assays when the results are not qualitative, or are used to determine immunity, or the assay is intended for use in matrices other than serum or plasma;

(8) For noninvasive testing as defined in § 812.3(k) of this chapter; and

(9) For near patient testing (point of care).

[65 FR 2319, Jan. 14, 2000]

Subpart B—Obstetrical and Gynecological Diagnostic Devices

§ 884.1040 Viscometer for cervical mucus.

(a) *Identification.* A viscometer for cervical mucus is a device that is intended to measure the relative viscoelasticity of cervical mucus collected from a female patient. Measurements of relative viscoelasticity are intended for use as an adjunct in the clinical evaluation of a female with chronic infertility, to determine the time of ovulation and the penetrability of cervical mucus to motile sperm.

(b) *Classification.* Class I (general controls). The device is exempt from the premarket notification procedures in subpart E of part 807 of this chapter subject to § 884.9.

[47 FR 14706, Apr. 6, 1982, as amended at 65 FR 2320, Jan. 14, 2000]

§ 884.1050 Endocervical aspirator.

(a) *Identification.* An endocervical aspirator is a device designed to remove tissue from the endocervix (mucous membrane lining the canal of the cervix of the uterus) by suction with a syringe, bulb and pipette, or catheter. This device is used to evaluate endocervical tissue to detect malignant and premalignant lesions.

(b) *Classification.* Class II (performance standards).

§ 884.1060 Endometrial aspirator.

(a) *Identification.* An endometrial aspirator is a device designed to remove materials from the endometrium (the mucosal lining of the uterus) by suction with a syringe, bulb and pipette, or catheter. This device is used to study endometrial cytology (cells).

(b) *Classification.* Class II. The special controls for this device are:

(1) FDA's:

(i) ''Use of International Standard ISO 10993 'Biological Evaluation of Medical Devices—Part I: Evaluation and Testing,' '' and

(ii) ''510(k) Sterility Review Guidance of 2/12/90 (K90–1),''

(2) Labeling:

(i) Indication: Only to evaluate the endometrium, and

(ii) Contraindications: Pregnancy, history of uterine perforation, or a recent cesarean section, and

(3) The sampling component is covered within vagina.

[45 FR 12684–12720, Feb. 26, 1980, as amended at 52 FR 17741, May 11, 1987; 65 FR 17146, Mar. 31, 2000]

§ 884.1100 Endometrial brush.

(a) *Identification.* An endometrial brush is a device designed to remove samples of the endometrium (the mucosal lining of the uterus) by brushing its surface. This device is used to study endometrial cytology (cells).

(b) *Classification.* Class II. The special controls for this device are:

(1) FDA's:

(i) ''Use of International Standard ISO 10993 'Biological Evaluation of Medical Devices—Part I: Evaluation and Testing,' '' and

(ii) ''510(k) Sterility Review Guidance of 2/12/90 (K90–1),''

(2) Labeling:

(i) Indication: Only to evaluate the endometrium, and

(ii) Contraindications: Pregnancy, history of uterine perforation, or a recent cesarean section, and

(3) Design and testing:

(i) The sampling component is covered within the vagina, and

(ii) For adherence of the bristles and brush head.

[45 FR 12684–12720, Feb. 26, 1980, as amended at 52 FR 17741, May 11, 1987; 65 FR 17146, Mar. 31, 2000]

§ 884.1175 Endometrial suction curette and accessories.

(a) *Identification.* An endometrial suction curette is a device used to remove material from the uterus and from the mucosal lining of the uterus by scraping and vacuum suction. This device is used to obtain tissue for biopsy or for menstrual extraction. This generic type of device may include catheters, syringes, and tissue filters or traps.

(b) *Classification.* Class II (performance standards).

§ 884.1185 Endometrial washer.

(a) *Identification.* An endometrial washer is a device used to remove materials from the endometrium (the mucosal lining of the uterus) by washing with water or saline solution and then aspirating with negative pressure. This device is used to study endometrial cytology (cells).

(b) *Classification.* Class II. The special controls for this device are:

(1) FDA's:

(i) "Use of International Organization for Standardization's ISO 10993 'Biological Evaluation of Medical Devices—Part I: Evaluation and Testing,' " and

(ii) "510(k) Sterility Review Guidance of 2/12/90 (K90–1),"

(2) Labeling:

(i) Indication: Only to evaluate the endometrium,

(ii) Contraindications: Pregnancy, history of uterine perforation, or a recent cesarean section, and

(iii) Warning: Do not attach to a wall or any external suction, and

(3) Design and Testing:

(i) The sampling component is covered within the vagina, and

(ii) Intrauterine pressure should not exceed 50 millimeters of mercury.

[45 FR 12684–12720, Feb. 26, 1980, as amended at 52 FR 17741, May 11, 1987; 65 FR 17146, Mar. 31, 2000]

§ 884.1300 Uterotubal carbon dioxide insufflator and accessories.

(a) *Identification.* A uterotubal carbon dioxide insufflator and accessories is a device used to test the patency (lack of obstruction) of the fallopian tubes by pressurizing the uterus and fallopian tubes and filling them with carbon dioxide gas.

(b) *Classification.* Class II (performance standards).

§ 884.1425 Perineometer.

(a) *Identification.* A perineometer is a device consisting of a fluid-filled sack for intravaginal use that is attached to an external manometer. The devices measure the strength of the perineal muscles by offering resistence to a patient's voluntary contractions of these muscles and is used to diagnose and to correct, through exercise, uninary incontinence or sexual dysfunction.

(b) *Classification.* Class II (performance standards).

§ 884.1550 Amniotic fluid sampler (amniocentesis tray).

(a) *Identification.* The amniotic fluid sampler (amniocentesis tray) is a collection of devices used to aspirate amniotic fluid from the amniotic sac via a transabdominal approach. Components of the amniocentesis tray include a disposable 3 inch 20 gauge needle with stylet and a 30 cc. syringe, as well as the various sample collection accessories, such as vials, specimen containers, medium, drapes, etc. The device is used at 16–18 weeks gestation for antepartum diagnosis of certain congenital abnormalities or anytime after 24 weeks gestation when used to assess fetal maturity.

(b) *Classification.* Class I (general controls). The device is exempt from the premarket notification procedures in subpart E of part 807 of this chapter subject to the limitations in § 884.9.

[61 FR 1123, Jan. 16, 1996, as amended at 66 FR 33808, July 25, 2001]

§ 884.1560 Fetal blood sampler.

(a) *Identification.* A fetal blood sampler is a device used to obtain fetal blood transcervically through an endoscope by puncturing the fetal skin with a short blade and drawing blood into a heparinized tube. The fetal blood pH is determined and used in the diagnosis of fetal distress and fetal hypoxia.

(b) *Classification.* Class II (performance standards).

§ 884.1600 Transabdominal amnioscope (fetoscope) and accessories.

(a) *Identification.* A transabdominal amnioscope is a device designed to permit direct visual examination of the fetus by a telescopic system via abdominal entry. The device is used to ascertain fetal abnormalities, to obtain fetal blood samples, or to obtain fetal tissue. This generic type of device may include the following accessories: trocar and cannula, instruments used through an operating channel or through a separate cannula associated with the amnioscope, light source and cables, and component parts.

(b) *Classification.* Class III (premarket approval).

(c) *Date premarket approval application (PMA) or notice of completion of a product development protocol (PDP) is required.* A PMA or a notice of completion of a PDP is required to be filed with the Food and Drug Administration on or before January 29, 1987 for any transabdominal amnioscope (fetoscope) and accessories that was in commercial distribution before May 28, 1976, or that has on or before January 29, 1987 been found to be substantially equivalent to a transabdominal amnioscope (fetoscope) and accessories that was in commercial distribution before May 28, 1976. Any other transabdominal amnioscope (fetoscope) and accessories shall have an approved PMA or a declared completed PDP in effect before being placed in commercial distribution.

[45 FR 12684—12720, Feb. 26, 1980, as amended at 51 FR 39845, Oct. 31, 1986]

§ 884.1630 Colposcope.

(a) *Identification.* A colposcope is a device designed to permit direct viewing of the tissues of the vagina and cervix by a telescopic system located outside the vagina. It is used to diagnose abnormalities and select areas for biopsy. This generic type of device may include a light source, cables, and component parts.

(b) *Classification.* Class II (performance standards).

§ 884.1640 Culdoscope and accessories.

(a) *Identification.* A culdoscope is a device designed to permit direct viewing of the organs within the peritoneum by a telescopic system introduced into the pelvic cavity through the posterior vaginal fornix. It is used to perform diagnostic and surgical procedures on the female genital organs. This generic type of device may include trocar and cannula, instruments used through an operating channel, scope preheaters, light source and cables, and component parts.

(b) *Classification.* (1) Class II (performance standards).

(2) Class I for culdoscope accessories that are not part of a specialized instrument or device delivery system; do not have adapters, connectors, channels, or do not have portals for electrosurgical, laser, or other power sources. Such culdoscope accessory instruments include: lens cleaning brush, biopsy brush, clip applier (without clips), applicator, cannula (without trocar or valves), ligature carrier/needle holder, clamp/hemostat/grasper, curette, instrument guide, ligature passing and knotting instrument, suture needle (without suture), retractor, mechanical (noninflatable), snare, stylet, forceps, dissector, mechanical (noninflatable) scissors, and suction/irrigation probe. The devices subject to this paragraph (b)(2) are exempt from the premarket notification procedures in subpart E of part 807 of this chapter, subject to the limitations in § 884.9.

[45 FR 12684–12720, Feb. 26, 1980, as amended at 61 FR 1123, Jan. 16, 1996; 66 FR 38808, July 25, 2001]

§ 884.1660 Transcervical endoscope (amnioscope) and accessories.

(a) *Identification.* A transcervical endoscope is a device designed to permit direct viewing of the fetus and amniotic sac by means of an open tube introduced into the uterus through the cervix. The device may be used to visualize the fetus or amniotic fluid and to sample fetal blood or amniotic fluid. This generic type of device may include obturators, instruments used through an operating channel, light sources and cables, and component parts.

(b) *Classification.* Class II (performance standards).

§ 884.1690 Hysteroscope and accessories.

(a) *Identification.* A hysteroscope is a device used to permit direct viewing of the cervical canal and the uterine cavity by a telescopic system introduced into the uterus through the cervix. It is used to perform diagnostic and surgical procedures other than sterilization. This generic type of device may include obturators and sheaths, instruments used through an operating channel, scope preheaters, light sources and cables, and component parts.

(b) *Classification.* (1) Class II (performance standards).

(2) Class I for hysteroscope accessories that are not part of a specialized instrument or device delivery system;

do not have adapters, connectors, channels, or do not have portals for electrosurgical, laser, or other power sources. Such hysteroscope accessory instruments include: lens cleaning brush, cannula (without trocar or valves), clamp/hemostat/grasper, curette, instrument guide, forceps, dissector, mechanical (noninflatable), and scissors. The devices subject to this paragraph (b)(2) are exempt from the premarket notification procedures in subpart E of part 807 of this chapter, subject to the limitations in § 884.9.

[45 FR 12684–12720, Feb. 26, 1980, as amended at 61 FR 1123, Jan. 16, 1996; 66 FR 38808, July 25, 2001]

§ 884.1700 Hysteroscopic insufflator.

(a) *Identification.* A hysteroscopic insufflator is a device designed to distend the uterus by filling the uterine cavity with a liquid or gas to facilitate viewing with a hysteroscope.

(b) *Classification.* (1) Class II (performance standards).

(2) Class I for tubing and tubing/filter fits which only include accessory instruments that are not used to effect intrauterine access, e.g., hysteroscopic introducer sheaths, etc.; and single-use tubing kits used for only intrauterine insufflation. The devices subject to this paragraph (b)(2) are exempt from the premarket notification procedures in subpart E of part 807 of this chapter, subject to the limitations in § 884.9.

[45 FR 12684–12720, Feb. 26, 1980, as amended at 61 FR 1124, Jan. 16, 1996; 66 FR 38808, July 25, 2001]

§ 884.1720 Gynecologic laparoscope and accessories.

(a) *Identification.* A gynecologic laparoscope is a device used to permit direct viewing of the organs within the peritoneum by a telescopic system introduced through the abdominal wall. It is used to perform diagnostic and surgical procedures on the female genital organs. This generic type of device may include: Trocar and cannula, instruments used through an operating channel, scope preheater, light source and cables, and component parts.

(b) *Classification.* (1) Class II (performance standards).

(2) Class I for gynecologic laparoscope accessories that are not part of a specialized instrument or device delivery system, do not have adapters, connector channels, or do not have portals for electrosurgical, lasers, or other power sources. Such gynecologic laparosope accessory instruments include: the lens cleaning brush, biopsy brush, clip applier (without clips), applicator, cannula (without trocar or valves), ligature carrier/needle holder, clamp/hemostat/grasper, curette, instrument guide, ligature passing and knotting instrument, suture needle (without suture), retractor, mechanical (noninflatable), snare, stylet, forceps, dissector, mechanical (noninflatable), scissors, and suction/irrigation probe. The devices subject to this paragraph (b)(2) are exempt from the premarket notification procedures in subpart E of part 807 of this chapter, subject to the limitations in § 884.9.

[45 FR 12684–12720, Feb. 26, 1980, as amended at 61 FR 1124, Jan. 16, 1996; 66 FR 38808, July 25, 2001]

§ 884.1730 Laparoscopic insufflator.

(a) *Identification.* A laparoscopic insufflator is a device used to facilitate the use of the laparoscope by filling the peritoneal cavity with gas to distend it.

(b) *Classification.* (1) Class II (performance standards).

(2) Class I for tubing and tubing/filter kits which include accessory instruments that are not used to effect intra-abdominal insufflation (pneumoperitoneum). The devices subject to this paragraph (b)(2) are exempt from the premarket notification procedures in subpart E of part 807 of this chapter, subject to the limitations in § 884.9.

[45 FR 12684–12720, Feb. 26, 1980, as amended at 61 FR 1124, Jan. 16, 1996; 66 FR 38809, July 25, 2001]

Subpart C—Obstetrical and Gynecological Monitoring Devices

§ 884.2050 Obstetric data analyzer.

(a) *Identification.* An obstetric data analyzer (fetal status data analyzer) is a device used during labor to analyze electronic signal data obtained from fetal and maternal monitors. The obstetric data analyzer provides clinical diagnosis of fetal status and recommendations for labor management and clinical interventions. This generic type of device may include signal analysis and display equipment, electronic interfaces for other equipment, and power supplies and component parts.

(b) *Classification:* Class III (premarket approval).

(c) *Date PMA or notice of completion of PDP is required.* A PMA or a notice of completion of a PDP is required to be filed with the Food and Drug Administration on or before October 3, 2000, for any obstetric data analyzer described in paragraph (a) of this section that was in commercial distribution before May 28, 1976, or that has been found, on or before October 3, 2000, to be substantially equivalent to an obstetric data analyzer described in paragraph (a) of this section that was in commercial distribution before May 28, 1976. Any other obstetric data analyzer described in paragraph (a) of this section shall have an approved PMA or declared completed PDP in effect before being placed in commercial distribution.

[65 FR 41332, July 5, 2000]

§ 884.2225 Obstetric-gynecologic ultrasonic imager.

(a) *Identification.* An obstetric-gynecologic ultrasonic imager is a device designed to transmit and receive ultrasonic energy into and from a female patient by pulsed echoscopy. This device is used to provide a visual representation of some physiological or artificial structure, or of a fetus, for diagnostic purposes during a limited period of time. This generic type of device may include the following: signal analysis and display equipment, electronic interfaces for other equipment, patient and equipment supports, coupling gel, and component parts. This generic type of device does not include devices used to monitor the changes in some physiological condition over long periods of time.

(b) *Classification.* Class II (performance standards).

§ 884.2600 Fetal cardiac monitor.

(a) *Identification.* A fetal cardiac monitor is a device used to ascertain fetal heart activity during pregnancy and labor. The device is designed to separate fetal heart signals from maternal heart signals by analyzing electrocardiographic signals (electrical potentials generated during contraction and relaxation of heart muscle) obtained from the maternal abdomen with external electrodes. This generic type of device may include an alarm that signals when the heart rate crosses a preset threshold. This generic type of device includes the "fetal cardiotachometer (with sensors)" and the "fetal electrocardiographic monitor."

(b) *Classification.* Class II (performance standards).

§ 884.2620 Fetal electroencephalographic monitor.

(a) *Identification.* A fetal electroencephalographic monitor is a device used to detect, measure, and record in graphic form (by means of one or more electrodes placed transcervically on the fetal scalp during labor) the rhythmically varying electrical skin potentials produced by the fetal brain.

(b) *Classification.* Class III (premarket approval).

(c) *Date PMA or notice of completion of a PDP is required.* A PMA or a notice of completion of a PDP is required to be filed with the Food and Drug Administration on or before December 26, 1996 for any fetal electroencephalographic monitor that was in commercial distribution before May 28, 1976, or that has, on or before December 26, 1996 been found to be substantially equivalent to a fetal electroencephalographic monitor in commercial distribution before May 28, 1976. Any other fetal electroencephalographic monitor shall have an approved PMA or a declared completed PDP in effect before being placed in commercial distribution.

[45 FR 12684–12720, Feb. 26, 1980, as amended at 52 FR 17741, May 11, 1987; 61 FR 50708, Sept. 27, 1996]

§ 884.2640 Fetal phonocardiographic monitor and accessories.

(a) *Identification.* A fetal phonocardiographic monitor is a device designed to detect, measure, and record fetal heart sounds electronically, in graphic form, and noninvasively, to ascertain fetal condition during labor.

This generic type of device includes the following accessories: signal analysis and display equipment, patient and equipment supports, and other component parts.

(b) *Classification.* Class II (performance standards).

§884.2660 Fetal ultrasonic monitor and accessories.

(a) *Identification.* A fetal ultrasonic monitor is a device designed to transmit and receive ultrasonic energy into and from the pregnant woman, usually by means of continuous wave (doppler) echoscopy. The device is used to represent some physiological condition or characteristic in a measured value over a period of time (e.g., perinatal monitoring during labor) or in an immediately perceptible form (e.g., use of the ultrasonic stethoscope). This generic type of device may include the following accessories: signal analysis and display equipment, electronic interfaces for other equipment, patient and equipment supports, and component parts. This generic type of device does not include devices used to image some relatively unchanging physiological structure or interpret a physiological condition, but does include devices which may be set to alarm automatically at a predetermined threshold value.

(b) *Classification.* Class II (performance standards).

§884.2675 Fetal scalp circular (spiral) electrode and applicator.

(a) *Identification.* A fetal scalp circular (spiral) electrode and applicator is a device used to obtain a fetal electrocardiogram during labor and delivery. It establishes electrical contact between fetal skin and an external monitoring device by a shallow subcutaneous puncture of fetal scalp tissue with a curved needle or needles. This generic type of device includes nonreusable spiral electrodes and reusable circular electrodes.

(b) *Classification.* Class II (performance standards).

§884.2685 Fetal scalp clip electrode and applicator.

(a) *Identification.* A fetal scalp clip electrode and applicator is a device designed to establish electrical contact between fetal skin and an external monitoring device by means of pinching skin tissue with a nonreusable clip. This device is used to obtain a fetal electrocardiogram. This generic type of device may include a clip electrode applicator.

(b) *Classification.* Class III (premarket approval).

(c) *Date PMA or notice of completion of a PDP is required.* A PMA or a notice of completion of a PDP is required to be filed with the Food and Drug Administration on or before December 26, 1996 for any fetal scalp clip electrode and applicator that was in commercial distribution before May 28, 1976, or that has, on or before December 26, 1996 been found to be substantially equivalent to a fetal scalp clip electrode and applicator that was in commercial distribution before May 28, 1976. Any other fetal scalp clip electrode and applicator shall have an approved PMA or a declared completed PDP in effect before being placed in commercial distribution.

[45 FR 12684–12720, Feb. 26, 1980, as amended at 52 FR 17741, May 11, 1987; 61 FR 50708, Sept. 27, 1996]

§884.2700 Intrauterine pressure monitor and accessories.

(a) *Identification.* An intrauterine pressure monitor is a device designed to detect and measure intrauterine and amniotic fluid pressure with a catheter placed transcervically into the uterine cavity. The device is used to monitor intensity, duration, and frequency of uterine contractions during labor. This generic type of device may include the following accessories: signal analysis and display equipment, patient and equipment supports, and component parts.

(b) *Classification.* Class II (performance standards).

§884.2720 External uterine contraction monitor and accessories.

(a) *Identification.* An external uterine contraction monitor (i.e., the tokodynamometer) is a device used to monitor the progress of labor. It measures the duration, frequency, and relative pressure of uterine contractions with a transducer strapped to the maternal abdomen. This generic type of device may include an external pressure transducer, support straps, and other patient and equipment supports.

(b) *Classification.* Class II (performance standards).

§ 884.2730 Home uterine activity monitor.

(a) *Identification.* A home uterine activity monitor (HUAM) is an electronic system for at home antepartum measurement of uterine contractions, data transmission by telephone to a clinical setting, and for receipt and display of the uterine contraction data at the clinic. The HUAM system comprises a tocotransducer, an at-home recorder, a modem, and a computer and monitor that receive, process, and display data. This device is intended for use in women with a previous preterm delivery to aid in the detection of preterm labor.

(b) *Classification.* Class II (special controls); guidance document (Class II Special Controls Guidance for Home Uterine Activity Monitors).

[66 FR 14076, Mar. 9, 2001]

§ 884.2740 Perinatal monitoring system and accessories.

(a) *Identification.* A perinatal monitoring system is a device used to show graphically the relationship between maternal labor and the fetal heart rate by means of combining and coordinating uterine contraction and fetal heart monitors with appropriate displays of the well-being of the fetus during pregnancy, labor, and delivery. This generic type of device may include any of the devices subject to §§ 884.2600, 884.2640, 884.2660, 884.2675, 884.2700, and 884.2720. This generic type of device may include the following accessories: Central monitoring system and remote repeaters, signal analysis and display equipment, patient and equipment supports, and component parts.

(b) *Classification.* Class II (performance standards).

§ 884.2900 Fetal stethoscope.

(a) *Identification.* A fetal stethoscope is a device used for listening to fetal heart sounds. It is designed to transmit the fetal heart sounds not only through sound channels by air conduction, but also through the user's head by tissue conduction into the user's ears. It does not use ultrasonic energy. This device is designed to eliminate noise interference commonly caused by handling conventional stethoscopes.

(b) *Classification.* Class I (general controls). The device is exempt from the premarket notification procedures in subpart E of part 807 of this chapter, subject to the limitations in § 884.9.

[45 FR 12684–12720, Feb. 26, 1980, as amended at 66 FR 38809, July 25, 2001]

§ 884.2960 Obstetric ultrasonic transducer and accessories.

(a) *Identification.* An obstetric ultrasonic transducer is a device used to apply ultrasonic energy to, and to receive ultrasonic energy from, the body in conjunction with an obstetric monitor or imager. The device converts electrical signals into ultrasonic energy, and vice versa, by means of an assembly distinct from an ultrasonic generator. This generic type of device may include the following accessories: coupling gel, preamplifiers, amplifiers, signal conditioners with their power supply, connecting cables, and component parts. This generic type of device does not include devices used to generate the ultrasonic frequency electrical signals for application.

(b) *Classification.* Class II (performance standards).

§ 884.2980 Telethermographic system.

(a) *Telethermographic system intended for adjunctive diagnostic screening for detection of breast cancer or other uses*—(1) *Identification.* A telethermographic system for adjunctive diagnostic screening for detection of breast cancer or other uses is an electrically powered device with a detector that is intended to measure, without touching the patient's skin, the self-emanating infrared radiation that reveals the temperature variations of the surface of the body. This generic type of device may include signal analysis and display equipment, patient and equipment supports, component parts, and accessories.

(2) *Classification.* Class I (general controls).

(b) *Telethermographic system intended for use alone in diagnostic screening for detection of breast cancer or other uses*—(1) *Identification.* A telethermographic system for use as the sole diagnostic screening tool for detection of breast cancer or other uses is an electrically powered device with a detector that is intended to measure, without touching the patient's skin, the self-emanating infrared radiation that reveals the temperature variations of the surface of the body. This generic type of device may include signal analysis and display equipment, patient and equipment supports, component parts, and accessories.

(2) *Classification.* Class III.

(3) *Date PMA or notice of completion of a PDP is required.* As of the enactment date of the amendments, May 28, 1976, an approval under section 515 of the act is required before the device described in paragraph (b)(1) may be commercially distributed. See § 884.3.

[53 FR 1566, Jan. 20, 1988, as amended at 55 FR 48440, Nov. 20, 1990; 66 FR 46953, Sept. 10, 2001]

§ 884.2982 Liquid crystal thermographic system.

(a) *A nonelectrically powered or an AC-powered liquid crystal thermographic system intended for adjunctive use in diagnostic screening for detection of breast cancer or other uses*—(1) *Identification.* A nonelectrically powered or an AC-powered liquid crystal thermographic system intended for use as an adjunct to physical palpation or mammography in diagnostic screening for detection of breast cancer or other uses is a nonelectrically powered or an AC-powered device applied to the skin that displays the color patterns of heat sensitive cholesteric liquid crystals that respond to temperature variations of the surface of the body. This generic type of device may include patient and equipment supports, a means to ensure thermal contact between the patient's skin and the liquid crystals, component parts, and accessories.

(2) *Classification.* Class I (general controls).

(b) *A nonelectrically powered or an AC-powered liquid crystal thermographic system intended for use alone in diagnostic screening for detection of breast cancer or other uses*—(1) *Identification.* A nonelectrically powered or an AC-powered liquid crystal thermographic system intended for use as the sole diagnostic screening tool for detection of breast cancer or other uses is a nonelectrically powered or an AC-powered device applied to the skin that displays the color patterns of heat sensitive cholesteric liquid crystals that respond to temperature variations of the surface of the body. This generic type of device may include image display and recording equipment, patient and equipment supports, a means to ensure thermal contact between the patient's skin and the liquid crystals, component parts, and accessories.

(2) *Classification.* Class III.

(3) *Date PMA or notice of completion of a PDP is required.* As of the enactment date of the amendments, May 28, 1976, an approval under section 515 of the act is required before the device described in paragraph (b)(1) may be commercially distributed. See § 884.3.

[53 FR 1566, Jan. 20, 1988, as amended at 55 FR 48441, Nov. 20, 1990; 66 FR 46953, Sept. 10, 2001]

§ 884.2990 Breast lesion documentation system.

(a) *Identification.* A breast lesion documentation system is a device for use in producing a surface map of the breast as an aid to document palpable breast lesions identified during a clinical breast examination.

(b) *Classification.* Class II (special controls). The special control is FDA's guidance entitled "Class II Special Controls Guidance Document: Breast Lesion Documentation System." See § 884.1(e) for the availability of this guidance document.

[68 FR 44415, Aug. 27, 2003]

Subpart D—Obstetrical and Gynecological Prosthetic Devices

§ 884.3200 Cervical drain.

(a) *Identification.* A cervical drain is a device designed to provide an exit channel for draining discharge from the cervix after pelvic surgery.

(b) *Classification.* Class II (performance standards).

§ 884.3575 Vaginal pessary.

(a) *Identification.* A vaginal pessary is a removable structure placed in the vagina to support the pelvic organs and is used to treat conditions such as uterine prolapse (falling down of uterus), uterine retroposition (backward displacement), or gynecologic hernia.

(b) *Classification.* Class II (performance standards).

§ 884.3650 Fallopian tube prosthesis.

(a) *Identification.* A fallopian tube prosthesis is a device designed to maintain the patency (openness) of the fallopian tube and is used after reconstructive surgery.

(b) *Classification.* Class II (performance standards).

§ 884.3900 Vaginal stent.

(a) *Identification.* A vaginal stent is a device used to enlarge the vagina by stretching, or to support the vagina and to hold a skin graft after reconstructive surgery.

(b) *Classification.* Class II (performance standards).

Subpart E—Obstetrical and Gynecological Surgical Devices

§ 884.4100 Endoscopic electrocautery and accessories.

(a) *Identification.* An endoscopic electrocautery is a device used to perform female sterilization under endoscopic observation. It is designed to coagulate fallopian tube tissue with a probe heated by low-voltage energy. This generic type of device may include the following accessories: electrical generators, probes, and electrical cables.

(b) *Classification.* Class II. The special controls for this device are:

(1) FDA's:

(i) ''Use of International Standard ISO 10993 'Biological Evaluation of Medical Devices—Part I: Evaluation and Testing,' ''

(ii) ''510(k) Sterility Review Guidance 2/12/90 (K–90),'' and

(iii) ''Guidance ('Guidelines') for Evaluation of Laproscopic Bipolar and Thermal Coagulators (and Accessories),''

(2) International Electrotechnical Commission's IEC 60601–1–AM2 (1995–03), Amendment 2, ''Medical Electrical Equipment—Part 1: General Requirements for Safety,''

(3) American National Standards Institute/American Association for Medical Instrumentation's HF–18, 1993, ''Electrosurgical Devices,''

(4) Labeling:

(i) Indication: For female tubal sterilization, and

(ii) Instructions for use:

(A) Destroy at least 2 centimeters of the fallopian tubes,

(B) Use a cut or undampened sinusoidal waveform,

(C) Use a minimum power of 25 watts, and

(D) For devices with ammeters: continue electrode activation for 5 seconds after the visual endpoint (tissue blanching) is reached or current flow ceases indicating adequate tissue destruction.

[45 FR 12684–12720, Feb. 26, 1980, as amended at 52 FR 17741, May 11, 1987; 65 FR 17146, Mar. 31, 2000]

§ 884.4120 Gynecologic electrocautery and accessories.

(a) *Identification.* A gynecologic electrocautery is a device designed to destroy tissue with high temperatures by tissue contact with an electrically heated probe. It is used to excise cervical lesions, perform biopsies, or treat chronic cervicitis under direct visual observation. This generic type of device may include the following accessories: an electrical generator, a probe, and electrical cables.

(b) *Classification.* Class II (performance standards).

§ 884.4150 Bipolar endoscopic coagulator-cutter and accessories.

(a) *Identification.* A bipolar endoscopic coagulator-cutter is a device used to perform female sterilization and other operative procedures under endoscopic observation. It destroys tissue with high temperatures by directing a high frequency electrical current through tissue between two electrical contacts of a probe. This generic type of device may include the following accessories: an electrical generator, probes, and electrical cables.

(b) *Classification*. Class II. The special controls for this device are:

(1) FDA's:

(i) "Use of International Standard ISO 10993 'Biological Evaluation of Medical Devices—Part I: Evaluation and Testing,' "

(ii) "510(k) Sterility Review Guidance 2/12/90 (K–90)," and

(iii) "Guidance ('Guidelines') for Evaluation of Laproscopic Bipolar and Thermal Coagulators (and Accessories),"

(2) International Electrotechnical Commission's IEC 60601–1–AM2 (1995–03), Amendment 2, "Medical Electrical Equipment—Part 1: General Requirements for Safety,"

(3) American National Standards Institute/American Association for Medical Instrumentation's HF–18, 1993, "Electrosurgical Devices,"

(4) Labeling:

(i) Indication: For female tubal sterilization, and

(ii) Instructions for use:

(A) Destroy at least 2 centimeters of the fallopian tubes,

(B) Use a cut or undampened sinusoidal waveform,

(C) Use a minimum power of 25 watts, and

(D) For devices with ammeters: continue electrode activation for 5 seconds after the visual endpoint (tissue blanching) is reached or current flow ceases indicating adequate tissue destruction.

[45 FR 12684–12720, Feb. 26, 1980, as amended at 52 FR 17741, May 11, 1987; 65 FR 17146, Mar. 31, 2000]

§ 884.4160 Unipolar endoscopic coagulator-cutter and accessories.

(a) *Identification*. A unipolar endoscopic coagulator-cutter is a device designed to destroy tissue with high temperatures by directing a high frequency electrical current through the tissue between an energized probe and a grounding plate. It is used in female sterilization and in other operative procedures under endoscopic observation. This generic type of device may include the following accessories: an electrical generator, probes and electrical cables, and a patient grounding plate. This generic type of device does not include devices used to perform female sterilization under hysteroscopic observation.

(b) *Classification*. Class II (performance standards).

§ 884.4250 Expandable cervical dilator.

(a) *Identification*. An expandable cervical dilator is an instrument with two handles and two opposing blades used manually to dilate (stretch open) the cervical os.

(b) *Classification*. Class III (premarket approval).

(c) *Date PMA or notice of completion of a PDP is required*. A PMA or a notice of completion of a PDP is required to be filed with the Food and Drug Administration on or before December 26, 1996 for any expandable cervical dilator that was in commercial distribution before May 28, 1976, or that has, on or before December 26, 1996 been found to be substantially equivalent to an expandable cervical dilator that was in commercial distribution before May 28, 1976. Any other expandable cervical dilator shall have an approved PMA or a declared completed PDP in effect before being placed in commercial distribution.

[45 FR 12684–12720, Feb. 26, 1980, as amended at 52 FR 17741, May 11, 1987; 61 FR 50708, Sept. 27, 1996]

§ 884.4260 Hygroscopic Laminaria cervical dilator.

(a) *Identification*. A hygroscopic *Laminaria* cervical dilator is a device designed to dilate (stretch open) the cervical os by cervical insertion of a conical and expansible material made from the root of a seaweed (*Laminaria* digitata or *Laminaria* japonica). The device is used to induce abortion.

(b) *Classification*. Class II (performance standards).

§ 884.4270 Vibratory cervical dilators.

(a) *Identification*. A vibratory cervical dilator is a device designed to dilate the cervical os by stretching it with a power-driven vibrating probe head. The device is used to gain access to the uterus or to induce abortion, but is not to be used during labor when a viable fetus is desired or anticipated.

(b) *Classification.* Class III (premarket approval).

(c) *Date PMA or notice of completion of a PDP is required.* A PMA or a notice of completion of a PDP is required to be filed with the Food and Drug Administration on or before December 26, 1996 for any vibratory cervical dilator that was in commercial distribution before May 28, 1976, or that has, on or before December 26, 1996 been found to be substantially equivalent to a vibratory cervical dilator that was in commercial distribution before May 28, 1976. Any other vibratory cervical dilator shall have an approved PMA or a declared completed PDP in effect before being placed in commercial distribution.

[45 FR 12684–12720, Feb. 26, 1980, as amended at 52 FR 17741, May 11, 1987; 61 FR 50708, Sept. 27, 1996]

§ 884.4340 Fetal vacuum extractor.

(a) *Identification.* A fetal vacuum extractor is a device used to facilitate delivery. The device enables traction to be applied to the fetal head (in the birth canal) by means of a suction cup attached to the scalp and is powered by an external vacuum source. This generic type of device may include the cup, hosing, vacuum source, and vacuum control.

(b) *Classification.* Class II (performance standards).

§ 884.4400 Obstetric forceps.

(a) *Identification.* An obstetric forceps is a device consisting of two blades, with handles, designed to grasp and apply traction to the fetal head in the birth passage and facilitate delivery.

(b) *Classification.* Class II (performance standards).

§ 884.4500 Obstetric fetal destructive instrument.

(a) *Identification.* An obstetric fetal destructive instrument is a device designed to crush or pull the fetal body to facilitate the delivery of a dead or anomalous (abnormal) fetus. This generic type of device includes the cleidoclast, cranioclast, craniotribe, and destructive hook.

(b) *Classification.* Class II (performance standards).

§ 884.4520 Obstetric-gynecologic general manual instrument.

(a) *Identification.* An obstetric-gynecologic general manual instrument is one of a group of devices used to perform simple obstetric and gynecologic manipulative functions. This generic type of device consists of the following:

(1) An episiotomy scissors is a cutting instrument, with two opposed shearing blades, used for surgical incision of the vulvar orifice for obstetrical purposes.

(2) A fiberoptic metal vaginal speculum is a metal instrument, with fiberoptic light, used to expose and illuminate the interior of the vagina.

(3) A metal vaginal speculum is a metal instrument used to expose the interior of the vagina.

(4) An umbilical scissors is a cutting instrument, with two opposed shearing blades, used to cut the umbilical cord.

(5) A uterine clamp is an instrument used to hold the uterus by compression.

(6) A uterine packer is an instrument used to introduce dressing into the uterus or vagina.

(7) A vaginal applicator is an instrument used to insert medication into the vagina.

(8) A vaginal retractor is an instrument used to maintain vaginal exposure by separating the edges of the vagina and holding back the tissue.

(9) A gynecological fibroid hook is an instrument used to exert traction upon a fibroid.

(10) A pelvimeter (external) is an instrument used to measure the external diameters of the pelvis.

(b) *Classification.* Class I (general controls). The devices are exempt from the premarket notification procedures in subpart E of part 807 of this chapter, subject to the limitations in § 884.9.

[45 FR 12684–12720, Feb. 26, 1980, as amended at 54 FR 25052, June 12, 1989; 66 FR 38809, July 25, 2001]

§ 884.4530 Obstetric-gynecologic specialized manual instrument.

(a) *Identification.* An obstetric-gynecologic specialized manual instrument is one of a group of devices used during obstetric-gynecologic procedures to perform manipulative diagnostic and surgical functions (e.g., dilating, grasping, measuring, and scraping), where structural integrity is the chief criterion of device performance. This type of device consists of the following:

(1) An amniotome is an instrument used to rupture the fetal membranes.

(2) A circumcision clamp is an instrument used to compress the foreskin of the penis during circumcision of a male infant.

(3) An umbilical clamp is an instrument used to compress the umbilical cord.

(4) A uterine curette is an instrument used to scrape and remove material from the uterus.

(5) A fixed-size cervical dilator is any of a series of bougies of various sizes used to dilate the cervical os by stretching the cervix.

(6) A uterine elevator is an instrument inserted into the uterus used to lift and manipulate the uterus.

(7) A gynecological surgical forceps is an instrument with two blades and handles used to pull, grasp, or compress during gynecological examination.

(8) A cervical cone knife is a cutting instrument used to excise and remove tissue from the cervix.

(9) A gynecological cerclage needle is a looplike instrument used to suture the cervix.

(10) A hook-type contraceptive intrauterine device (IUD) remover is an instrument used to remove an IUD from the uterus.

(11) A gynecological fibroid screw is an instrument used to hold onto a fibroid.

(12) A uterine sound is an instrument used to determine the depth of the uterus by inserting it into the uterine cavity.

(13) A cytological cervical spatula is a blunt instrument used to scrape and remove cytological material from the surface of the cervix or vagina.

(14) A gynecological biopsy forceps is an instrument with two blades and handles used for gynecological biopsy procedures.

(15) A uterine tenaculum is a hooklike instrument used to seize and hold the cervix or fundus.

(16) An internal pelvimeter is an instrument used within the vagina to measure the diameter and capacity of the pelvis.

(17) A nonmetal vaginal speculum is a nonmetal instrument used to expose the interior of the vagina.

(18) A fiberoptic nonmetal vaginal speculum is a nonmetal instrument, with fiberoptic light, used to expose and illuminate the interior of the vagina.

(b) *Classification.* (1) Class II (performance standards).

(2) Class I for the amniotome, uterine curette, cervical dilator (fixed-size bougies), cerclage needle, IUD remover, uterine sound, and gynecological biopsy forceps. The devices subject to this paragraph (b)(2) are exempt from the premarket notification procedures in subpart E of part 807 of this chapter, subject to the limitations in § 884.9.

[45 FR 12684–12720, Feb. 26, 1980, as amended at 61 FR 1124, Jan. 16, 1996; 66 FR 38809, July 25, 2001]

§ 884.4550 Gynecologic surgical laser.

(a) *Identification.* A gynecologic surgical laser is a continuous wave carbon dioxide laser designed to destroy tissue thermally or to remove tissue by radiant light energy. The device is used only in conjunction with a colposcope as part of a gynecological surgical system. A colposcope is a magnifying lens system used to examine the vagina and cervix.

(b) *Classification.* Class II (performance standards).

§ 884.4900 Obstetric table and accessories.

(a) *Identification.* An obstetric table is a device with adjustable sections designed to support a patient in the various positions required during obstetric and gynecologic procedures. This generic type of device may include the following accessories: patient equipment, support attachments, and cabinets for warming instruments and disposing of wastes.

(b) *Classification.* Class II (performance standards).

Subpart F—Obstetrical and Gynecological Therapeutic Devices

§ 884.5050 Metreurynter-balloon abortion system.

(a) *Identification.* A metreurynter-balloon abortion system is a device used to induce abortion. The device is inserted into the uterine cavity, inflated, and slowly extracted. The extraction of the balloon from the uterus causes dilation of the cervical os. This generic type of device may include pressure sources and pressure controls.

(b) *Classification.* Class III (premarket approval).

(c) *Date PMA or notice of completion of a PDP is required.* A PMA or a notice of completion of a PDP is required to be filed with the Food and Drug Administration on or before December 26, 1996 for any metreurynter-balloon abortion system that was in commercial distribution before May 28, 1976, or that has, on or before December 26, 1996 been found to be substantially equivalent to a metreurynter-balloon abortion system that was in commercial distribution before May 28, 1976. Any other metreurynter-balloon abortion system shall have an approved PMA or a declared completed PDP in effect before being placed in commercial distribution.

[45 FR 12684–12720, Feb. 26, 1980, as amended at 52 FR 17741, May 11, 1987; 61 FR 50709, Sept. 27, 1996]

§ 884.5070 Vacuum abortion system.

(a) *Identification.* A vacuum abortion system is a device designed to aspirate transcervically the products of conception or menstruation from the uterus by using a cannula connected to a suction source. This device is used for pregnancy termination or menstrual regulation. This type of device may include aspiration cannula, vacuum source, and vacuum controller.

(b) *Classification.* Class II (performance standards).

§ 884.5100 Obstetric anesthesia set.

(a) *Identification.* An obstetric anesthesia set is an assembly of antiseptic solution, needles, needle guides, syringes, and other accessories, intended for use with an anesthetic drug. This device is used to administer regional blocks (e.g., paracervical, uterosacral, and pudendal) that may be used during labor, delivery, or both.

(b) *Classification.* Class II (performance standards).

§ 884.5150 Nonpowered breast pump.

(a) *Identification.* A nonpowered breast pump is a manual suction device used to express milk from the breast.

(b) *Classification.* Class I. The device is exempt from the premarket notification procedures in subpart E of part 807 of this chapter, subject to the limitations in § 884.9, if the device is using either a bulb or telescoping mechanism which does not develop more than 250 mm Hg suction, and the device materials that contact breast or breast milk do not produce cytotoxicity, irritation, or sensitization effects.

[45 FR 12684–12720, Feb. 26, 1980, as amended at 61 FR 1124, Jan. 16, 1996; 66 FR 38809, July 25, 2001]

§ 884.5160 Powered breast pump.

(a) *Identification.* A powered breast pump in an electrically powered suction device used to express milk from the breast.

(b) *Classification.* Class II (performance standards).

§ 884.5225 Abdominal decompression chamber.

(a) *Identification.* An abdominal decompression chamber is a hoodlike device used to reduce pressure on the pregnant patient's abdomen for the relief of abdominal pain during pregnancy or labor.

(b) *Classification.* Class III (premarket approval).

(c) *Date PMA or notice of completion of a PDP is required.* A PMA or a notice of completion of a PDP is required to be filed with the Food and Drug Administration on or before December 26, 1996 for any abdominal decompression chamber that was in commercial distribution before May 28, 1976, or that has, on or before December 26, 1996 been found to be substantially equivalent to an abdominal decompression chamber that was in commercial distribution before May 28, 1976. Any other abdominal decompression chamber shall have an approved PMA or a declared completed PDP in effect before being placed in commercial distribution.

[45 FR 12684–12720, Feb. 26, 1980, as amended at 52 FR 17741, May 11, 1987; 61 FR 50709, Sept. 27, 1996]

§ 884.5250 Cervical cap.

(a) *Identification.* A cervical cap is a flexible cuplike receptacle that fits over the cervix to collect menstrual flow or to aid artificial insemination. This generic type of device is not for contraceptive use.

(b) *Classification.* Class II (performance standards).

§ 884.5300 Condom.

(a) *Identification.* A condom is a sheath which completely covers the penis with a closely fitting membrane. The condom is used for contraceptive and for prophylactic purposes (preventing transmission of venereal disease). The device may also be used to collect semen to aid in the diagnosis of infertility.

(b) *Classification.* Class II (performance standards).

§ 884.5310 Condom with spermicidal lubricant.

(a) *Identification.* A condom with spermicidal lubricant is a sheath which completely covers the penis with a closely fitting membrane with a lubricant that contains a spermicidal agent, nonoxynol–9. This condom is used for contraceptive and prophylactic purposes (preventing transmission of venereal disease).

(b) *Classification.* Class II (performance standards).

[47 FR 49022, Oct. 29, 1982]

§ 884.5320 Glans sheath.

(a) *Identification.* A glans sheath device is a sheath which covers only the glans penis or part thereof and may also cover the area in the immediate proximity thereof, the corona and frenulum, but not the entire shaft of the penis. It is indicated only for the prevention of pregnancy and not for the prevention of sexually-transmitted diseases.

(b) *Classification.* Class III (premarket approval).

(c) *Date premarket approval application (PMA) or notice of completion of a product development protocol (PDP) is required.* A PMA or a notice of completion of a PDP is required to be filed with the Food and Drug Administration on or before September 12, 2002, for any glans sheath that was in commercial distribution before May 28, 1976, or that has, on or before September 12, 2002, been found to be substantially equivalent to a glans sheath that was in commercial distribution before May 28, 1976. Any other glans sheath shall have an approved PMA or declared completed PDP in effect before being placed in commercial distribution.

[59 FR 67187, Dec. 29, 1994, as amended at 67 FR 40849, June 14, 2002]

§ 884.5330 Female condom.

(a) *Identification.* A female condom is a sheath-like device that lines the vaginal wall and is inserted into the vagina prior to the initiation of coitus. It is indicated for contraceptive and prophylactic (preventing the transmission of sexually transmitted diseases) purposes.

(b) *Classification.* Class III (premarket approval).

(c) *Date premarket approval application (PMA) or notice of completion of a product development protocol (PDP) is required.* No effective date has been established of the requirement for premarket approval for the devices described in paragraph (b) of this section. See § 884.3 for effective dates of requirement for premarket approval.

[65 FR 31455, May 18, 2000]

§ 884.5350 Contraceptive diaphragm and accessories.

(a) *Identification.* A contraceptive diaphragm is a closely fitting membrane placed between the posterior aspect of the pubic bone and the posterior vaginal fornix. The device covers the cervix completely and is used with a spermicide to prevent pregnancy. This generic type of device may include an introducer.

(b) *Classification.* Class II (performance standards).

§ 884.5360 Contraceptive intrauterine device (IUD) and introducer.

(a) *Identification.* A contraceptive intrauterine device (IUD) is a device used to prevent pregnancy. The device is placed high in the uterine fundus with a string extending from the device through the cervical os into the vagina. This generic type of device includes the introducer, but does not include contraceptive IUD's that function by drug activity, which are subject to the new drug provisions of the Federal Food, Drug, and Cosmetic Act (see § 310.502).

(b) *Classification.* Class III (premarket approval).

(c) *Labeling.* Labeling requirements for contraceptive IUD's are set forth in § 801.427.

(d) *Date premarket approval application (PMA) or notice of completion of a product development protocol (PDP) is required.* A PMA or a notice of completion of a PDP is required to be filed with the Food and Drug Administration on or before August 4, 1986, for any IUD and introducer that was in commercial distribution before May 28, 1976, or that has on or before August 4, 1986, been found to be substantially equivalent to an IUD and introducer that was in commercial distribution before May 28, 1976. Any other IUD and introducer shall have an approved PMA or a declared completed PDP in effect before being placed in commercial distribution.

[45 FR 12684–12720, Feb. 26, 1980, as amended at 51 FR 16649, May 5, 1986]

§ 884.5380 Contraceptive tubal occlusion device (TOD) and introducer.

(a) *Identification.* A contraceptive tubal occlusion device (TOD) and introducer is a device designed to close a fallopian tube with a mechanical structure, e.g., a band or clip on the outside of the fallopian tube or a plug or valve on the inside. The devices are used to prevent pregnancy.

(b) *Classification.* Class III (premarket approval).

(c) *Date premarket approval application (PMA) or notice of completion of a product development protocol (PDP) is required.* A PMA or a notice of completion of a PDP is required to be filed with the Food and Drug Administration on or before December 30, 1987, for any TOD and introducer that was in commercial distribution before May 28, 1976, or that has on or before December 30, 1987, been found to be substantially equivalent to a TOD and introducer that was in commercial distribution before May 28, 1976. Any other TOD and introducer shall have an approved PMA or a declared completed PDP in effect before being placed in commercial distribution.

[45 FR 12684–12720, Feb. 26, 1980, as amended at 52 FR 36883, Oct. 1, 1987]

§ 884.5390 Perineal heater.

(a) *Identification.* A perineal heater is a device designed to apply heat directly by contact, or indirectly from a radiant source, to the surface of the perineum (the area between the vulva and the anus) and is used to soothe or to help heal the perineum after an episiotomy (incision of the vulvar orifice for obstetrical purposes).

(b) *Classification.* Class II (performance standards).

§ 884.5400 Menstrual cup.

(a) *Identification.* A menstrual cup is a receptacle placed in the vagina to collect menstrual flow.

(b) *Classification.* Class II (performance standards).

§ 884.5425 Scented or scented deodorized menstrual pad.

(a) *Identification.* A scented or scented deodorized menstrual pad is a device that is a pad made of cellulosic or synthetic material which is used to absorb menstrual or other vaginal discharge. It has scent (*i.e.*, fragrance materials) added for aesthetic purposes (scented menstrual pad) or for deodorizing purposes (scented deodorized menstrual pad). This generic type of device includes sterile scented menstrual pads used for medically indicated conditions, but does not include menstrual pads treated with added antimicrobial agents or other drugs.

(b) *Classification.* (1) Class I (general controls) for menstrual pads made of common cellulosic and synthetic material with an established safety profile. The devices subject to this paragraph (b)(1) are exempt from the premarket notification procedures in subpart E of part 807 of this chapter, subject to the limitations in § 884.9. This exemption does not include the intralabial pads and reusable menstrual pads.

(2) Class II (special controls) for scented or scented deodorized menstrual pads made of materials not described in paragraph (b)(1).

[45 FR 12684–12720, Feb. 26, 1980, as amended at 45 FR 51185, Aug. 1, 1980; 61 FR 67714, Dec. 24, 1996; 66 FR 38809, July 25, 2001]

§ 884.5435 Unscented menstrual pad.

(a) *Identification.* An unscented menstrual pad is a device that is a pad made of cellulosic or synthetic material which is used to absorb menstrual or other vaginal discharge. This generic type of device includes sterile unscented menstrual pads used for medically indicated conditions, but does not include menstrual pads treated with scent (i.e., fragrance materials) or those with added antimicrobial agents or other drugs.

(b) *Classification.* Class I (general controls). The device is exempt from the premarket notification procedures in subpart E of part 807 of this chapter only when the device is made of common cellulosic and synthetic material with an established safety profile. This exemption does not include the interlabial pads and reusable menstrual pads.

[45 FR 12684–12720, Feb. 26, 1980, as amended at 61 FR 67714, Dec. 24, 1996; 65 FR 2320, Jan. 14, 2000]

§ 884.5460 Scented or scented deodorized menstrual tampon.

(a) *Identification.* A scented or scented deodorized menstrual tampon is a device that is a plug made of cellulosic or synthetic material that is inserted into the vagina and used to absorb menstrual or other vaginal discharge. It has scent (*i.e.*, fragrance materials) added for aesthetic purposes (scented menstrual tampon) or for deodorizing purposes (scented deodorized menstrual tampon). This generic type of device does not include menstrual tampons treated with added antimicrobial agents or other drugs.

(b) *Classification.* Class II (performance standards).

[45 FR 12684–12720, Feb. 26, 1980, as amended at 45 FR 51186, Aug. 1, 1980]

§ 884.5470 Unscented menstrual tampon.

(a) *Identification.* An unscented menstrual tampon is a device that is a plug made of cellulosic or synthetic material that is inserted into the vagina and used to absorb menstrual or other vaginal discharge. This generic type of device does not include menstrual tampons treated with scent (i.e., fragrance materials) or those with added antimicrobial agents or other drugs.

(b) *Classification.* Class II (performance standards).

§ 884.5900 Therapeutic vaginal douche apparatus.

(a) *Identification.* A therapeutic vaginal douche apparatus is a device that is a bag or bottle with tubing and a nozzle. The apparatus does not include douche solutions. The apparatus is intended and labeled for use in the treatment of medical conditions except it is not for contraceptive use. After filling the therapeutic vaginal douche apparatus with a solution, the patient uses the device to direct a stream of solution into the vaginal cavity.

(b) *Classification.* (1) Class II (performance standards).

(2) Class I if the device is operated by gravity feed. Devices subject to this paragraph (b)(2) are exempt from the premarket notification procedures in subpart E of part 807 of this chapter, subject to the limitations in § 884.9.

[45 FR 12684–12720, Feb. 26, 1980, as amended at 61 FR 1124, Jan. 16, 1996; 66 FR 38809, July 25, 2001]

§ 884.5920 Vaginal insufflator.

(a) *Identification.* A vaginal insufflator is a device used to treat vaginitis by introducing medicated powder from a hand-held bulb into the vagina through an open speculum.

(b) *Classification.* Class I. The device is exempt from the premarket notification procedures in subpart E of part 807 of this chapter, subject to the limitations in § 884.9.

[45 FR 12684–12720, Feb. 26, 1980, as amended at 54 FR 25052, June 12, 1989; 66 FR 38809, July 25, 2001]

§ 884.5940 Powered vaginal muscle stimulator for therapeutic use.

(a) *Identification.* A powered vaginal muscle stimulator is an electrically powered device designed to stimulate directly the muscles of the vagina with pulsating electrical current. This device is intended and labeled for therapeutic use in increasing muscular tone and strength in the treatment of sexual dysfunction. This generic type of device does not include devices used to treat urinary incontinence.

(b) *Classification.* Class III (premarket approval).

(c) *Date PMA or notice of completion of a PDP is required.* A PMA or a notice of completion of a PDP for a device is required to be filed with the Food and Drug Administration on or before July 12, 2000, for any powered vaginal muscle stimulator for therapeutic use that was in commercial distribution before May 28, 1976, or that has, on or before July 12, 2000, been found to be substantially equivalent to a powered vaginal muscle stimulator that was in commercial distribution before May 28, 1976. Any other powered vaginal muscle stimulator for therapeutic use shall have an approved PMA or declared completed PDP in effect before being placed in commercial distribution.

[45 FR 12684–12720, Feb. 26, 1980, as amended at 52 FR 17741, May 11, 1987; 65 FR 19834, Apr. 13, 2000]

§ 884.5960 Genital vibrator for therapeutic use.

(a) *Identification.* A genital vibrator for therapeutic use is an electrically operated device intended and labeled for therapeutic use in the treatment of sexual dysfunction or as an adjunct to Kegel's exercise (tightening of the muscles of the pelvic floor to increase muscle tone).

(b) *Classification.* Class II (performance standards).

§ 884.5970 Clitoral engorgement device.

(a) *Identification.* A clitoral engorgement device is designed to apply a vacuum to the clitoris. It is intended for use in the treatment of female sexual arousal disorder.

(b) *Classification.* Class II (special controls). The special control is a guidance document entitled: "Guidance for Industry and FDA Reviewers: Class II Special Controls Guidance Document for Clitoral Engorgement Devices."

[65 FR 47306, Aug. 2, 2000]

Subpart G—Assisted Reproduction Devices

SOURCE: 63 FR 48436, Sept. 10, 1998, unless otherwise noted.

§ 884.6100 Assisted reproduction needles.

(a) *Identification.* Assisted reproduction needles are devices used in in vitro fertilization (IVF), gamete intrafallopian transfer (GIFT), or other assisted reproduction procedures to obtain gametes from the body or introduce gametes, zygote(s), preembryo(s) and/or embryo(s) into the body. This generic type of device may include a single or double lumen needle and component parts, including needle guides, such as those used with ultrasound.

(b) *Classification.* Class II (special controls) (mouse embryo assay information, endotoxin testing, sterilization validation, design specifications, labeling requirements, biocompatibility testing, and clinical testing).

§ 884.6110 Assisted reproduction catheters.

(a) *Identification.* Assisted reproduction catheters are devices used in in vitro fertilization (IVF), gamete intrafallopian transfer (GIFT), or other assisted reproduction procedures to introduce or remove gametes, zygote(s), preembryo(s), and/or embryo(s) into or from the body. This generic type of device may include catheters, cannulae, introducers, dilators, sheaths, stylets, and component parts.

(b) *Classification.* Class II (special controls) (mouse embryo assay information, endotoxin testing, sterilization validation, design specifications, labeling requirements, biocompatibility testing, and clinical testing).

§884.6120 Assisted reproduction accessories.

(a) *Identification.* Assisted reproduction accessories are a group of devices used during assisted reproduction procedures, in conjunction with assisted reproduction needles and/or assisted reproduction catheters, to aspirate, incubate, infuse, and/or maintain temperature. This generic type of device may include:

(1) Powered aspiration pumps used to provide low flow, intermittent vacuum for the aspiration of eggs (ova).

(2) Syringe pumps (powered or manual) used to activate a syringe to infuse or aspirate small volumes of fluid during assisted reproduction procedures.

(3) Collection tube warmers, used to maintain the temperature of egg (oocyte) collection tubes at or near body temperature. A dish/plate/microscope stage warmer is a device used to maintain the temperature of the egg (oocyte) during manipulation.

(4) Embryo incubators, used to store and preserve gametes and/or embryos at or near body temperature.

(5) Cryopreservation instrumentation and devices, used to contain, freeze, and maintain gametes and/or embryos at an appropriate freezing temperature.

(b) *Classification.* Class II (special controls) (design specifications, labeling requirements, and clinical testing).

§884.6130 Assisted reproduction microtools.

(a) *Identification.* Assisted reproduction microtools are pipettes or other devices used in the laboratory to denude, micromanipulate, hold, or transfer human gametes or embryos for assisted hatching, intracytoplasmic sperm injection (ICSI), or other assisted reproduction methods.

(b) *Classification.* Class II (special controls) (mouse embryo assay information, endotoxin testing, sterilization validation, design specifications, labeling requirements, and clinical testing).

§884.6140 Assisted reproduction micropipette fabrication instruments.

(a) *Identification.* Assisted reproduction micropipette fabrication devices are instruments intended to pull, bevel, or forge a micropipette or needle for intracytoplasmic sperm injection (ICSI), in vitro fertilization (IVF) or other similar assisted reproduction procedures.

(b) *Classification.* Class II (special controls) (design specifications, labeling requirements, and clinical testing).

§884.6150 Assisted reproduction micromanipulators and microinjectors.

(a) *Identification.* Assisted reproduction micromanipulators are devices intended to control the position of an assisted reproduction microtool. Assisted reproduction microinjectors are any device intended to control aspiration or expulsion of the contents of an assisted reproduction microtool.

(b) *Classification.* Class II (special controls) (design specifications, labeling requirements, and clinical testing).

§884.6160 Assisted reproduction labware.

(a) *Identification.* Assisted reproduction labware consists of laboratory equipment or supplies intended to prepare, store, manipulate, or transfer human gametes or embryos for in vitro fertilization (IVF), gamete intrafallopian transfer (GIFT), or other assisted reproduction procedures. These include syringes, IVF tissue culture dishes, IVF tissue culture plates, pipette tips, dishes, plates, and other vessels that come into physical contact with gametes, embryos or tissue culture media.

(b)*Classification.* Class II (special controls) (mouse embryo assay information, endotoxin testing, sterilization validation, design specifications, labeling requirements, and clinical testing).

§884.6170 Assisted reproduction water and water purification systems.

(a) *Identification.* Assisted reproduction water purification systems are devices specifically intended to generate high quality, sterile, pyrogen-free water for reconstitution of media used for aspiration, incubation, transfer or storage of gametes or embryos for in vitro fertilization (IVF) or other assisted reproduction procedures. These devices may also be intended as the final rinse for labware or other assisted reproduction devices that will contact the gametes or embryos. These devices also include bottled water ready for reconstitution available from a vendor that is specifically intended for reconstitution of media used for aspiration, incubation, transfer, or storage of gametes or embryos for IVF or other assisted reproduction procedures.

(b) *Classification.* Class II (special controls) (mouse embryo assay information, endotoxin testing, sterilization validation, water quality testing, design specifications, labeling requirements, biocompatibility testing, and clinical testing).

§ 884.6180 Reproductive media and supplements.

(a) *Identification.* Reproductive media and supplement are products that are used for assisted reproduction procedures. Media include liquid and powder versions of various substances that come in direct physical contact with human gametes or embryos (including water, acid solutions used to treat gametes or embryos, rinsing solutions, sperm separation media, supplements, or oil used to cover the media) for the purposes of preparation, maintenance, transfer or storage. Supplements are specific reagents added to media to enhance specific properties of the media (e.g., proteins, sera, antibiotics, etc.).

(b) *Classification.* Class II (special controls) (mouse embryo assay information, endotoxin testing, sterilization validation, design specifications, labeling requirements, biocompatibility testing, and clinical testing).

§ 884.6190 Assisted reproductive microscopes and microscope accessories.

(a) *Identification.* Assisted reproduction microscopes and microscope accessories (excluding microscope stage warmers, which are classified under assisted reproduction accessories) are optical instruments used to enlarge images of gametes or embryos. Variations of microscopes and accessories used for these purposes would include phase contrast microscopes, dissecting microscopes and inverted stage microscopes.

(b) *Classification.* Class I. The device is exempt from the premarket notification procedures in subpart E of part 807 of this chapter, subject to the limitations in § 884.9.

[63 FR 48436, Sept. 10, 1998, as amended at 64 FR 62977, Nov. 18, 1999; 66 FR 38809, July 25, 2001]

§ 884.6200 Assisted reproduction laser system.

(a) *Identification.* The assisted reproduction laser system is a device that images, targets, and controls the power and pulse duration of a laser beam used to ablate a small tangential hole in, or to thin, the zona pellucida of an embryo for assisted hatching or other assisted reproduction procedures.

(b) *Classification.* Class II (special controls). The special control is FDA's guidance document entitled "Class II Special Controls Guidance Document: Assisted Reproduction Laser Systems." See § 884.1(e) for the availability of this guidance document.

[69 FR 77624, Dec. 28, 2004]

PART 886—OPHTHALMIC DEVICES

Subpart A—General Provisions

Subpart B—Diagnostic Devices

886.1320 Fornixscope.
886.1330 Amsler grid.
886.1340 Haploscope.
886.1350 Keratoscope.
886.1360 Visual field laser instrument.
886.1375 Bagolini lens.
886.1380 Diagnostic condensing lens.
886.1385 Polymethylmethacrylate (PMMA) diagnostic contact lens.
886.1390 Flexible diagnostic Fresnel lens.
886.1395 Diagnostic Hruby fundus lens.
886.1400 Maddox lens.
886.1405 Ophthalmic trial lens set.
886.1410 Ophthalmic trial lens clip.
886.1415 Ophthalmic trial lens frame.
886.1420 Ophthalmic lens gauge.
886.1425 Lens measuring instrument.
886.1430 Ophthalmic contact lens radius measuring device.
886.1435 Maxwell spot.
886.1450 Corneal radius measuring device.
886.1460 Stereopsis measuring instrument.
886.1500 Headband mirror.
886.1510 Eye movement monitor.
886.1570 Ophthalmoscope.
886.1605 Perimeter.
886.1630 AC-powered photostimulator.
886.1640 Ophthalmic preamplifier.
886.1650 Ophthalmic bar prism.
886.1655 Ophthalmic Fresnel prism.
886.1660 Gonioscopic prism.
886.1665 Ophthalmic rotary prism.
886.1670 Ophthalmic isotope uptake probe.
886.1680 Ophthalmic projector.
886.1690 Pupillograph.
886.1700 Pupillometer.
886.1750 Skiascopic rack.
886.1760 Ophthalmic refractometer.
886.1770 Manual refractor.
886.1780 Retinoscope.
886.1790 Nearpoint ruler.
886.1800 Schirmer strip.
886.1810 Tangent screen (campimeter).
886.1840 Simulatan (including crossed cylinder).
886.1850 AC-powered slitlamp biomicroscope.
886.1860 Ophthalmic instrument stand.
886.1870 Stereoscope.
886.1880 Fusion and stereoscopic target.
886.1905 Nystagmus tape.
886.1910 Spectacle dissociation test system.
886.1930 Tonometer and accessories.
886.1940 Tonometer sterilizer.
886.1945 Transilluminator.

Subpart C [Reserved]

Subpart D—Prosthetic Devices

886.3100 Ophthalmic tantalum clip.
886.3130 Ophthalmic conformer.
886.3200 Artificial eye.
886.3300 Absorbable implant (scleral buckling method).
886.3320 Eye sphere implant.
886.3340 Extraocular orbital implant.
886.3400 Keratoprosthesis.
886.3600 Intraocular lens.
886.3800 Scleral shell.
886.3920 Aqueous shunt.

Subpart E—Surgical Devices

886.4070 Powered corneal burr.
886.4100 Radiofrequency electrosurgical cautery apparatus.
886.4115 Thermal cautery unit.
886.4150 Vitreous aspiration and cutting instrument.
886.4170 Cryophthalmic unit.
886.4230 Ophthalmic knife test drum.
886.4250 Ophthalmic electrolysis unit.
886.4270 Intraocular gas.
886.4275 Intraocular fluid.
886.4280 Intraocular pressure measuring device.
886.4300 Intraocular lens guide.
886.4335 Operating headlamp.
886.4350 Manual ophthalmic surgical instrument.
886.4360 Ocular surgery irrigation device.
886.4370 Keratome.
886.4390 Ophthalmic laser.
886.4392 Nd:YAG laser for posterior capsulotomy and peripheral iridotomy.
886.4400 Electronic metal locator.
886.4440 AC-powered magnet.
886.4445 Permanent magnet.
886.4570 Ophthalmic surgical marker.
886.4610 Ocular pressure applicator.
886.4670 Phacofragmentation system.
886.4690 Ophthalmic photocoagulator.
886.4750 Ophthalmic eye shield.
886.4770 Ophthalmic operating spectacles (loupes).
886.4790 Ophthalmic sponge.
886.4855 Ophthalmic instrument table.

Subpart F—Therapeutic Devices

886.5100 Ophthalmic beta radiation source.
886.5120 Low-power binocular loupe.
886.5420 Contact lens inserter/remover.
886.5540 Low-vision magnifier.
886.5600 Ptosis crutch.
886.5800 Ophthalmic bar reader.
886.5810 Ophthalmic prism reader.
886.5820 Closed-circuit television reading system.
886.5840 Magnifying spectacles.
886.5842 Spectacle frame.
886.5844 Prescription spectacle lens.
886.5850 Sunglasses (nonprescription).
886.5870 Low-vision telescope.
886.5900 Electronic vision aid.
886.5910 Image intensification vision aid.
886.5915 Optical vision aid.
886.5916 Rigid gas permeable contact lens.
886.5918 Rigid gas permeable contact lens care products.
886.5925 Soft (hydrophilic) contact lens.
886.5928 Soft (hydrophilic) contact lens care products.

886.5933 [Reserved]

AUTHORITY: 21 U.S.C. 351, 360, 360c, 360e, 360j, 371.

SOURCE: 52 FR 33355, Sept. 2, 1987, unless otherwise noted.

Subpart A—General Provisions

§ 886.1 Scope.

(a) This part sets forth the classification of ophthalmic devices intended for human use that are in commercial distribution.

(b) The identification of a device in a regulation in this part is not a precise description of every device that is, or will be, subject to the regulation. A manufacturer who submits a premarket notification submission for a device under part 807 cannot show merely that the device is accurately described by the section title and identification provision of a regulation in this part but shall state why the device is substantially equivalent to other devices, as required by § 807.87.

(c) To avoid duplicative listings, an ophthalmic device that has two or more types of uses (e.g., used both as a diagnostic device and as a therapeutic device) is listed in one subpart only.

(d) References in this part to regulatory sections of the Code of Federal Regulations are to chapter I of title 21 unless otherwise noted.

§ 886.3 Effective dates of requirement for premarket approval.

A device included in this part that is classified into class III (premarket approval) shall not be commercially distributed after the date shown in the regulation classifying the device unless the manufacturer has an approval under section 515 of the act (unless an exemption has been granted under section 520(g)(2) of the act). An approval under section 515 of the act consists of FDA's issuance of an order approving an application for premarket approval (PMA) for the device or declaring completed a product development protocol (PDP) for the device.

(a) Before FDA requires that a device commercially distributed before the enactment date of the amendments, or a device that has been found substantially equivalent to such a device, has an approval under section 515 of the act, FDA must promulgate a regulation under section 515(b) of the act requiring such approval, except as provided in paragraphs (b) and (c) of this section. Such a regulation under section 515(b) of the act shall not be effective during the grace period ending on the 90th day after its promulgation or on the last day of the 30th full calendar month after the regulation that classifies the device into class III is effective, whichever is later. See section 501(f)(2)(B) of the act. Accordingly, unless an effective date of the requirement for premarket approval is shown in the regulation for a device classified into class III in this part, the device may be commercially distributed without FDA's issuance of an order approving a PMA or declaring completed a PDP for the device. If FDA promulgates a regulation under section 515(b) of the act requiring premarket approval for a device, section 501(f)(1)(A) of the act applies to the device.

(b) Any new, not substantially equivalent, device introduced into commercial distribution on or after May 28, 1976, including a device formerly marketed that has been substantially altered, is classified by statute (section 513(f) of the act) into class III without any grace period and FDA must have issued an order approving a PMA or declaring completed a PDP for the device before the device is commercially distributed unless it is reclassified. If FDA knows that a device being commercially distributed may be a "new" device as defined in this section because of any new intended use or other reasons, FDA may codify the statutory classification of the device into class III for such new use. Accordingly, the regulation for such a class III device states that as of the enactment date of the amendments, May 28, 1976, the device must have an approval under section 515 of the act before commercial distribution.

(c) A device identified in a regulation in this part that is classified into class III and that is subject to the transitional provisions of section 520(1) of the act is automatically classified by statute into class III and must have an approval under section 515 of the act before being commercially distributed. Accordingly, the regulation for such a class III transitional device states that as of the enactment date of the amendments, May 28, 1976, the device must have an approval under section 515 of the act before commercial distribution.

§886.9 Limitations of exemptions from section 510(k) of the Federal Food, Drug, and Cosmetic Act (the act).

The exemption from the requirement of premarket notification (section 510(k) of the act) for a generic type of class I or II device is only to the extent that the device has existing or reasonably foreseeable characteristics of commercially distributed devices within that generic type or, in the case of in vitro diagnostic devices, only to the extent that misdiagnosis as a result of using the device would not be associated with high morbidity or mortality. Accordingly, manufacturers of any commercially distributed class I or II device for which FDA has granted an exemption from the requirement of premarket notification must still submit a premarket notification to FDA before introducing or delivering for introduction into interstate commerce for commercial distribution the device when:

(a) The device is intended for a use different from the intended use of a legally marketed device in that generic type of device; e.g., the device is intended for a different medical purpose, or the device is intended for lay use where the former intended use was by health care professionals only;

(b) The modified device operates using a different fundamental scientific technology than a legally marketed device in that generic type of device; e.g., a surgical instrument cuts tissue with a laser beam rather than with a sharpened metal blade, or an in vitro diagnostic device detects or identifies infectious agents by using deoxyribonucleic acid (DNA) probe or nucleic acid hybridization technology rather than culture or immunoassay technology; or

(c) The device is an in vitro device that is intended:

(1) For use in the diagnosis, monitoring, or screening of neoplastic diseases with the exception of immunohistochemical devices;

(2) For use in screening or diagnosis of familial or acquired genetic disorders, including inborn errors of metabolism;

(3) For measuring an analyte that serves as a surrogate marker for screening, diagnosis, or monitoring life-threatening diseases such as acquired immune deficiency syndrome (AIDS), chronic or active hepatitis, tuberculosis, or myocardial infarction or to monitor therapy;

(4) For assessing the risk of cardiovascular diseases;

(5) For use in diabetes management;

(6) For identifying or inferring the identity of a microorganism directly from clinical material;

(7) For detection of antibodies to microorganisms other than immunoglobulin G (IgG) or IgG assays when the results are not qualitative, or are used to determine immunity, or the assay is intended for use in matrices other than serum or plasma;

(8) For noninvasive testing as defined in §812.3(k) of this chapter; and

(9) For near patient testing (point of care).

[65 FR 2320, Jan. 14, 2000]

Subpart B—Diagnostic Devices

§886.1040 Ocular esthesiometer.

(a) *Identification.* An ocular esthesiometer is a device, such as a single-hair brush, intended to touch the cornea to assess corneal sensitivity.

(b) *Classification.* Class I (general controls). The device is exempt from the premarket notification procedures in subpart E of part 807 of this chapter, subject to the limitations in §886.9.

[52 FR 33355, Sept. 2, 1987, as amended at 53 FR 35603, Sept. 14, 1988; 59 FR 63012, Dec. 7, 1994; 66 FR 38809, July 25, 2001]

§886.1050 Adaptometer (biophotometer).

(a) *Identification.* An adaptometer (biophotometer) is an AC-powered device that provides a stimulating light source which has various controlled intensities intended to measure the time required for retinal adaptation (regeneration of the visual purple) and the minimum light threshold.

(b) *Classification.* Class I (general controls). The device is exempt from the premarket notification procedures in subpart E of part 807 of this chapter, subject to the limitations in § 886.9.

[55 FR 48441, Nov. 20, 1990, as amended at 59 FR 63012, Dec. 7, 1994; 66 FR 38809, July 25, 2001]

§ 886.1070 Anomaloscope.

(a) *Identification.* An anomaloscope is an AC-powered device intended to test for anomalies of color vision by displaying mixed spectral lines to be matched by the patient.

(b) *Classification.* Class I (general controls). The device is exempt from the premarket notification procedures in subpart E of part 807 of this chapter, subject to the limitations in § 886.9.

[55 FR 48441, Nov. 20, 1990, as amended at 59 FR 63012, Dec. 7, 1994; 66 FR 38810, July 25, 2001]

§ 886.1090 Haidlinger brush.

(a) *Identification.* A Haidlinger brush is an AC-powered device that provides two conical brushlike images with apexes touching which are viewed by the patient through a Nicol prism and intended to evaluate visual function. It may include a component for measuring macular integrity.

(b) *Classification.* Class I (general controls). The device is exempt from the premarket notification procedures in subpart E of part 807 of this chapter, subject to the limitations in § 886.9.

[55 FR 48441, Nov. 20, 1990, as amended at 59 FR 63012, Dec. 7, 1994; 66 FR 38810, July 25, 2001]

§ 886.1120 Opthalmic camera.

(a) *Identification.* An ophthalmic camera is an AC-powered device intended to take photographs of the eye and the surrounding area.

(b) *Classification.* Class II.

[55 FR 48441, Nov. 20, 1990]

§ 886.1140 Ophthalmic chair.

(a) *Identification.* An ophthalmic chair is an AC-powered or manual device with adjustable positioning in which a patient is to sit or recline during ophthalmological examination or treatment.

(b) *Classification.* Class I. The AC-powered device and the manual device are exempt from the premarket notification procedures in subpart E of part 807 of this chapter, subject to the limitations in § 886.9. The manual device is also exempt from the current good manufacturing practice regulations in part 820 of this chapter, with the exception of § 820.180, with respect to general requirements concerning records, and § 820.198, with respect to complaint files.

[55 FR 48441, Nov. 20, 1990, as amended at 59 FR 63012, Dec. 7, 1994; 66 FR 38810, July 25, 2001]

§ 886.1150 Visual acuity chart.

(a) *Identification.* A visual acuity chart is a device that is a chart, such as a Snellen chart with block letters or other symbols in graduated sizes, intended to test visual acuity.

(b) *Classification.* Class I (general controls). The device is exempt from the premarket notification procedures in subpart E of part 807 of this chapter, subject to the limitations in § 886.9. The device is also exempt from the current good manufacturing practice regulations in part 820 of this chapter, with the exception of § 820.180, with respect to general requirements concerning records, and § 820.198, with respect to complaint files.

[52 FR 33355, Sept. 2, 1987, as amended at 53 FR 35603, Sept. 14, 1988; 53 FR 40825, Oct. 18, 1988; 66 FR 38810, July 25, 2001]

§ 886.1160 Color vision plate illuminator.

(a) *Identification.* A color vision plate illuminator is an AC-powered device that is a lamp intended to properly illuminate color vision testing plates. It may include a filter.

(b) *Classification.* Class I (general controls). The device is exempt from the premarket notification procedures in subpart E of part 807 of this chapter, subject to the limitations in § 886.9.

[55 FR 48441, Nov. 20, 1990, as amended at 59 FR 63012, Dec. 7, 1994; 66 FR 38810, July 25, 2001]

§886.1170 Color vision tester.

(a) *Identification.* A color vision tester is a device that consists of various colored materials, such as colored yarns or color vision plates (multicolored plates which patients with color vision deficiency would perceive as being of one color), intended to evaluate color vision.

(b) *Classification.* Class I (general controls). The device is exempt from the premarket notification procedures in subpart E of part 807 of this chapter, subject to the limitations in §886.9. The device is also exempt from the current good manufacturing practice regulations in part 820 of this chapter, with the exception of §820.180, with respect to general requirements concerning records, and §820.198, with respect to complaint files.

[52 FR 33355, Sept. 2, 1987, as amended at 53 FR 35603, Sept. 14, 1988; 66 FR 38810, July 25, 2001]

§886.1190 Distometer.

(a) *Identification.* A distometer is a device intended to measure the distance between the cornea and a corrective lens during refraction to help measure the change of the visual image when a lens is in place.

(b) *Classification.* Class I (general controls). The device is exempt from the premarket notification procedures in subpart E of part 807 of this chapter, subject to the limitations in §886.9. The device is also exempt from the current good manufacturing practice regulations in part 820 of this chapter, with the exception of §820.180, with respect to general requirements concerning records, and §820.198, with respect to complaint files.

[52 FR 33355, Sept. 2, 1987, as amended at 53 FR 35603, Sept. 14, 1988; 66 FR 38810, July 25, 2001]

§886.1200 Optokinetic drum.

(a) *Identification.* An optokinetic drum is a drum-like device covered with alternating white and dark stripes or pictures that can be rotated on its handle. The device is intended to elicit and evaluate nystagmus (involuntary rapid movement of the eyeball) in patients.

(b) *Classification.* Class I (general controls). The device is exempt from the premarket notification procedures in subpart E of part 807 of this chapter, subject to the limitations in §886.9. The device is also exempt from the current good manufacturing practice regulations in part 820 of this chapter, with the exception of §820.180, with respect to general requirements concerning records, and §820.198, with respect to complaint files.

[52 FR 33355, Sept. 2, 1987, as amended at 53 FR 35604, Sept. 14, 1988; 66 FR 38810, July 25, 2001]

§886.1220 Corneal electrode.

(a) *Identification.* A corneal electrode is an AC-powered device, usually part of a special contact lens, intended to be applied directly to the cornea to provide data showing the changes in electrical potential in the retina after electroretinography (stimulation by light).

(b) *Classification.* Class II.

§886.1250 Euthyscope.

(a) *Identification.* A euthyscope is a device that is a modified AC-powered or battery-powered ophthalmoscope (a perforated mirror device intended to inspect the interior of the eye) that projects a bright light encompassing an arc of about 30 degrees onto the fundus of the eye. The center of the light bundle is blocked by a black disk covering the fovea (the central depression of the macular retinae where only cones are present and blood vessels are lacking). The device is intended for use in the treatment of amblyopia (dimness of vision without apparent disease of the eye).

(b) *Classification.* Class I for the battery powered device. The battery powered device is exempt from the premarket notification procedures in subpart E of part 807 of this chapter, subject to the limitations in §886.9. Class II for the AC-powered device.

[55 FR 48441, Nov. 20, 1990, as amended at 59 FR 63012, Dec. 7, 1994; 66 FR 38810, July 25, 2001]

§886.1270 Exophthalmometer.

(a) *Identification.* An exophthalmometer is a device, such as a ruler, gauge, or caliper, intended to measure the degree of exophthalmos (abnormal protrusion of the eyeball).

(b) *Classification.* Class I (general controls). The device is exempt from the premarket notification procedures in subpart E of part 807 of this chapter, subject to the limitations in § 886.9.

[52 FR 33355, Sept. 2, 1987, as amended at 53 FR 35604, Sept. 14, 1988; 66 FR 38810, July 25, 2001]

§ 886.1290 Fixation device.

(a) *Identification.* A fixation device is an AC-powered device intended for use as a fixation target for the patient during ophthalmological examination. The patient directs his or her gaze so that the visual image of the object falls on the fovea centralis (the center of the macular retina of the eye.)

(b) *Classification.* Class I (general controls). The device is exempt from the premarket notification procedures in subpart E of part 807 of this chapter, subject to the limitations in § 886.9.

[55 FR 48441, Nov. 20, 1990, as amended at 59 FR 63012, Dec. 7, 1994; 66 FR 38810, July 25, 2001]

§ 886.1300 Afterimage flasher.

(a) *Identification.* An afterimage flasher is an AC-powered light that automatically switches on and off to allow performance of an afterimage test in which the patient indicates the positions of afterimages after the light is off. The device is intended to determine harmonious/anomalous retinal correspondence (the condition in which corresponding points on the retina have the same directional value).

(b) *Classification.* Class II.

[55 FR 48441, Nov. 20, 1990]

§ 886.1320 Fornixscope.

(a) *Identification.* A fornixscope is a device intended to pull back and hold open the eyelid to aid examination of the conjunctiva.

(b) *Classification.* Class I (general controls). The device is exempt from the premarket notification procedures in subpart E of part 807 of this chapter, subject to the limitations in § 886.9. The device is also exempt from the current good manufacturing practice regulations in part 820 of this chapter, with the exception of § 820.180, with respect to general requirements concerning records, and § 820.198, with respect to complaint files.

[52 FR 33355, Sept. 2, 1987, as amended at 53 FR 35604, Sept. 14, 1988; 66 FR 38810, July 25, 2001]

§ 886.1330 Amsler grid.

(a) *Identification.* An Amsler grid is a device that is a series of charts with grids of different sizes that are held at 30 centimeters distance from the patient and intended to rapidly detect central and paracentral irregularities in the visual field.

(b) *Classification.* Class I (general controls). The device is exempt from the premarket notification procedures in subpart E of part 807 of this chapter, subject to the limitations in § 886.9. The device is also exempt from the current good manufacturing practice regulations in part 820 of this chapter, with the exception of § 820.180, with respect to general requirements concerning records, and § 820.198, with respect to complaint files.

[52 FR 33355, Sept. 2, 1987, as amended at 53 FR 35604, Sept. 14, 1988; 66 FR 38810, July 25, 2001]

§ 886.1340 Haploscope.

(a) *Identification.* A haploscope is an AC-powered device that consists of two movable viewing tubes, each containing a slide carrier, a low-intensity light source for the illumination of the slides, and a high-intensity light source for creating afterimages. The device is intended to measure strabismus (eye muscle imbalance), to assess binocular vision (use of both eyes to see), and to treat suppression and amblyopia (dimness of vision without any apparent disease of the eye).

(b) *Classification.* Class I (general controls). The device is exempt from the premarket notification procedures in subpart E of part 807 of this chapter, subject to the limitations in § 886.9.

[55 FR 48441, Nov. 20, 1990, as amended at 59 FR 63012, Dec. 7, 1994; 66 FR 38810, July 25, 2001]

§ 886.1350 Keratoscope.

(a) *Identification.* A keratoscope is an AC-powered or battery-powered device intended to measure and evaluate the corneal curvature of the eye. Lines and circles within the keratoscope are used to observe the corneal reflex. This generic type of device includes the photokeratoscope which records corneal curvature by taking photographs of the cornea.

(b) The device is exempt from the premarket notification procedures in subpart E of part 807 of this chapter subject to § 886.9. The battery-powered device is exempt from the current good manufacturing practice regulations in part 820 of this chapter, with the exception of § 820.180 of this chapter, with respect to general requirements concerning records, and § 820.198 of this chapter, with respect to complaint files

[55 FR 48441, Nov. 20, 1990, as amended at 59 FR 63012, Dec. 7, 1994; 65 FR 2320, Jan. 14, 2000]

§ 886.1360 Visual field laser instrument.

(a) *Identification.* A visual field laser instrument is an AC-powered device intended to provide visible laser radiation that produces an interference pattern on the retina to evaluate retinal function.

(b) *Classification.* Class II.

§ 886.1375 Bagolini lens.

(a) *Identification.* A Bagolini lens is a device that consists of a plane lens containing almost imperceptible striations that do not obscure visualization of objects. The device is placed in a trial frame and intended to determine harmonious/anomalous retinal correspondence (a condition in which corresponding points on the retina have the same directional values).

(b) *Classification.* Class I (general controls). The device is exempt from the premarket notification procedures in subpart E of part 807 of this chapter, subject to the limitations in § 886.9. The device is also exempt from the current good manufacturing practice regulations in part 820 of this chapter, with the exception of § 820.180, with respect to general requirements concerning records, and § 820.198, with respect to complaint files.

[52 FR 33355, Sept. 2, 1987, as amended at 53 FR 35604, Sept. 14, 1988; 66 FR 38810, July 25, 2001]

§ 886.1380 Diagnostic condensing lens.

(a) *Identification.* A diagnostic condensing lens is a device used in binocular indirect ophthalmoscopy (a procedure that produces an inverted or reversed direct magnified image of the eye) intended to focus reflected light from the fundus of the eye.

(b) *Classification.* Class I (general controls). The device is exempt from the premarket notification procedures in subpart E of part 807 of this chapter, subject to the limitations in § 886.9. The device is also exempt from the current good manufacturing practice regulations in part 820 of this chapter, with the exception of § 820.180, with respect to general requirements concerning records, and § 820.198, with respect to complaint files.

[52 FR 33355, Sept. 2, 1987, as amended at 53 FR 35604, Sept. 14, 1988; 66 FR 38810, July 25, 2001]

§ 886.1385 Polymethylmethacrylate (PMMA) diagnostic contact lens.

(a) *Identification.* A polymethylmethacrylate (PMMA) diagnostic contact lens is a device that is a curved shell of PMMA intended to be applied for a short period of time directly on the globe or cornea of the eye for diagnosis or therapy of intraocular abnormalities.

(b) *Classification.* Class II.

§ 886.1390 Flexible diagnostic Fresnel lens.

(a) *Identification.* A flexible diagnostic Fresnel lens is a device that is a very thin lens which has its surface a concentric series of increasingly refractive zones. The device is intended to be applied to the back of the spectacle lenses of patients with aphakia (absence of the lens of the eye).

(b) *Classification.* Class I (general controls). The device is exempt from the premarket notification procedures in subpart E of part 807 of this chapter, subject to the limitations in § 886.9. The device is also exempt from the current good manufacturing practice regulations in part 820 of this chapter, with the exception of § 820.180, with respect to general requirements concerning

records, and § 820.198, with respect to complaint files.

[52 FR 33355, Sept. 2, 1987, as amended at 53 FR 35604, Sept. 14, 1988; 66 FR 38811, July 25, 2001]

§ 886.1395 Diagnostic Hruby fundus lens.

(a) *Identification.* A diagnostic Hruby fundus lens is a device that is a 55 diopter lens intended for use in the examination of the vitreous body and the fundus of the eye under slitlamp illumination and magnification.

(b) *Classification.* Class I (general controls). The device is exempt from the premarket notification procedures in subpart E of part 807 of this chapter, subject to the limitations in § 886.9. The device is also exempt from the current good manufacturing practice regulations in part 820 of this chapter, with the exception of § 820.180, with respect to general requirements concerning records, and § 820.198, with respect to complaint files.

[52 FR 33355, Sept. 2, 1987, as amended at 53 FR 35604, Sept. 14, 1988; 66 FR 38811, July 25, 2001]

§ 886.1400 Maddox lens.

(a) *Identification.* A Maddox lens is a device that is a series of red cylinders that change the size, shape, and color of an image. The device is intended to be handheld or placed in a trial frame to evaluate eye muscle dysfunction.

(b) *Classification.* Class I (general controls). The device is exempt from the premarket notification procedures in subpart E of part 807 of this chapter, subject to the limitations in § 886.9. The device is also exempt from the current good manufacturing practice regulations in part 820 of this chapter, with the exception of § 820.180, with respect to general requirements concerning records, and § 820.198, with respect to complaint files.

[52 FR 33355, Sept. 2, 1987, as amended at 53 FR 35604, Sept. 14, 1988; 66 FR 38811, July 25, 2001]

§ 886.1405 Ophthalmic trial lens set.

(a) *Identification.* An ophthalmic trial lens set is a device that is a set of lenses of various dioptric powers intended to be handheld or inserted in a trial frame for vision testing to determine refraction.

(b) *Classification.* Class I (general controls). The device is exempt from the premarket notification procedures in subpart E of part 807 of this chapter, subject to the limitations in § 886.9.

[52 FR 33355, Sept. 2, 1987, as amended at 61 FR 1124, Jan. 16, 1996; 66 FR 38811, July 25, 2001]

§ 886.1410 Ophthalmic trial lens clip.

(a) *Identification.* An ophthalmic trial lens clip is a device intended to hold prisms, spheres, cylinders, or occluders on a trial frame or spectacles for vision testing.

(b) *Classification.* Class I (general controls). The device is exempt from the premarket notification procedures in subpart E of part 807 of this chapter, subject to the limitations in § 886.9.

[52 FR 33355, Sept. 2, 1987, as amended at 53 FR 35604, Sept. 14, 1988; 66 FR 38811, July 25, 2001]

§ 886.1415 Ophthalmic trial lens frame.

(a) *Identification.* An opthalmic trial lens frame is a mechanical device intended to hold trial lenses for vision testing.

(b) *Classification.* Class I (general controls). The device is exempt from the premarket notification procedures in subpart E of part 807 of this chapter, subject to the limitations in § 886.9. The device is also exempt from the current good manufacturing practice regulations in part 820 of this chapter, with the exception of § 820.180, with respect to general requirements concerning records, and § 820.198, with respect to complaint files.

[52 FR 33355, Sept. 2, 1987, as amended at 53 FR 35604, Sept. 14, 1988; 66 FR 38811, July 25, 2001]

§ 886.1420 Ophthalmic lens gauge.

(a) *Identification.* An ophthalmic lens gauge is a calibrated device intended to manually measure the curvature of a spectacle lens.

(b) *Classification.* Class I (general controls). The device is exempt from the premarket notification procedures in

subpart E of part 807 of this chapter, subject to the limitations in §886.9.

[52 FR 33355, Sept. 2, 1987, as amended at 53 FR 35604, Sept. 14, 1988; 66 FR 38811, July 25, 2001]

§886.1425 Lens measuring instrument.

(a) *Identification.* A lens measuring instrument is an AC-powered device intended to measure the power of lenses, prisms, and their centers (e.g., lensometer).

(b) *Classification.* Class I (general controls). The device is exempt from the premarket notification procedures in subpart E of part 807 of this chapter, subject to the limitations in §886.9.

[55 FR 48442, Nov. 20, 1990, as amended at 59 FR 63013, Dec. 7, 1994; 66 FR 38811, July 25, 2001]

§886.1430 Ophthalmic contact lens radius measuring device.

(a) *Identification.* An ophthalmic contact lens radius measuring device is an AC-powered device that is a microscope and dial gauge intended to measure the radius of a contact lens.

(b) *Classification.* Class I (general controls). The device is exempt from the premarket notification procedures in subpart E of part 807 of this chapter, subject to the limitations in §886.9.

[55 FR 48442, Nov. 20, 1990, as amended at 59 FR 63013, Dec. 7, 1994; 66 FR 38811, July 25, 2001]

§886.1435 Maxwell spot.

(a) *Identification.* A Maxwell spot is an AC-powered device that is a light source with a red and blue filter intended to test macular function.

(b) *Classification.* Class I (general controls). The device is exempt from the premarket notification procedures in subpart E of part 807 of this chapter, subject to the limitations in §886.9.

[55 FR 48442, Nov. 20, 1990, as amended at 59 FR 63013, Dec. 7, 1994; 66 FR 38811, July 25, 2001]

§886.1450 Corneal radius measuring device.

(a) *Identification.* A corneal radius measuring device is an AC-powered device intended to measure corneal size by superimposing the image of the cornea on a scale at the focal length of the lens of a small, hand held, single tube penscope or eye gauge magnifier.

(b) *Classification.* Class I (general controls). The device is exempt from the premarket notification procedures in subpart E of part 807 of this chapter, subject to the limitations in §886.9, only when the device does not include computer software in the unit or topographers.

[55 FR 48442, Nov. 20, 1990, as amended at 59 FR 63013, Dec. 7, 1994; 66 FR 38811, July 25, 2001]

§886.1460 Stereopsis measuring instrument.

(a) *Identification.* A stereopsis measuring instrument is a device intended to measure depth perception by illumination of objects placed on different planes.

(b) *Classification.* Class I (general controls). The device is exempt from the premarket notification procedures in subpart E of part 807 of this chapter, subject to the limitations in §886.9. The device is also exempt from the current good manufacturing practice regulations in part 820 of this chapter, with the exception of §820.180, with respect to general requirements concerning records, and §820.198, with respect to complaint files.

[52 FR 33355, Sept. 2, 1987, as amended at 53 FR 35605, Sept. 14, 1988; 66 FR 38811, July 25, 2001]

§886.1500 Headband mirror.

(a) *Identification.* A headband mirror is a device intended to be strapped to the head of the user to reflect light for use in examination of the eye.

(b) *Classification.* Class I (general controls). The device is exempt from the premarket notification procedures in subpart E of part 807 of this chapter, subject to the limitations in §886.9. The device is also exempt from the current good manufacturing practice regulations in part 820 of this chapter, with the exception of §820.180, with respect to general requirements concerning records, and §820.198, with respect to complaint files.

[52 FR 33355, Sept. 2, 1987, as amended at 53 FR 35605, Sept. 14, 1988; 66 FR 38811, July 25, 2001]

§ 886.1510 Eye movement monitor.

(a) *Identification.* An eye movement monitor is an AC-powered device with an electrode intended to measure and record ocular movements.

(b) *Classification.* Class II.

§ 886.1570 Ophthalmoscope.

(a) *Identification.* An ophthalmoscope is an AC-powered or battery-powered device containing illumination and viewing optics intended to examine the media (cornea, aqueous, lens, and vitreous) and the retina of the eye.

(b) *Classification.* Class II.

§ 886.1605 Perimeter.

(a) *Identification.* A perimeter is an AC-powered or manual device intended to determine the extent of the peripheral visual field of a patient. The device projects light on various points of a curved surface, and the patient indicates whether he or she sees the light.

(b) *Classification.* Class I (general controls). The device is exempt from the premarket notification procedures in subpart E of part 807 of this chapter, subject to the limitations in § 886.9. The device is also exempt from the current good manufacturing practice regulations in part 820 of this chapter, with the exception of § 820.180, with respect to general requirements concerning records, and § 820.198, with respect to complaint files.

[55 FR 48442, Nov. 20, 1990, as amended at 66 FR 38811, July 25, 2001]

§ 886.1630 AC-powered photostimulator.

(a) *Identification.* An AC-powered photostimulator is an AC-powered device intended to provide light stimulus which allows measurement of retinal or visual function by perceptual or electrical methods (e.g., stroboscope).

(b) *Classification.* Class II.

§ 886.1640 Ophthalmic preamplifier.

(a) *Identification.* An ophthalmic preamplifier is an AC-powered or battery-powered device intended to amplify electrical signals from the eye in electroretinography (recording retinal action currents from the surface of the eyeball after stimulation by light), electrooculography (testing for retinal dysfunction by comparing the standing potential in the front and the back of the eyeball), and electromyography (recording electrical currents generated in active muscle).

(b) *Classification.* Class II.

§ 886.1650 Ophthalmic bar prism.

(a) *Identification.* An ophthalmic bar prism is a device that is a bar composed of fused prisms of gradually increasing strengths intended to measure latent and manifest strabismus (eye muscle deviation) or the power of fusion of a patient's eyes.

(b) *Classification.* Class I (general controls). The device is exempt from the premarket notification procedures in subpart E of part 807 of this chapter, subject to the limitations in § 886.9. The device is also exempt from the current good manufacturing practice regulations in part 820 of this chapter, with the exception of § 820.180, with respect to general requirements concerning records, and § 820.198, with respect to complaint files.

[52 FR 33355, Sept. 2, 1987, as amended at 53 FR 35605, Sept. 14, 1988; 66 FR 38812, July 25, 2001]

§ 886.1655 Ophthalmic Fresnel prism.

(a) *Identification.* An ophthalmic Fresnel prism is a device that is a thin plastic sheet with embossed rulings which provides the optical effect of a prism. The device is intended to be applied to spectacle lenses to give a prismatic effect.

(b) *Classification.* Class I (general controls). The device is exempt from the premarket notification procedures in subpart E of part 807 of this chapter, subject to the limitations in § 886.9. The device is also exempt from the current good manufacturing practice regulations in part 820 of this chapter, with the exception of § 820.180, with respect to general requirements concerning records, and § 820.198, with respect to complaint files.

[52 FR 33355, Sept. 2, 1987, as amended at 53 FR 35605, Sept. 14, 1988; 66 FR 38812, July 25, 2001]

§ 886.1660 Gonioscopic prism.

(a) *Identification.* A gonioscopic prism is a device that is a prism intended to be placed on the eye to study the anterior chamber. The device may have angled mirrors to facilitate visualization of anatomical features.

(b) *Classification.* Class I (general controls). The device is exempt from the premarket notification procedures in subpart E of part 807 of this chapter, subject to the limitations in § 886.9.

[52 FR 33355, Sept. 2, 1987, as amended at 53 FR 35605, Sept. 14, 1988; 59 FR 63013, Dec. 7, 1994; 66 FR 38812, July 25, 2001]

§ 886.1665 Ophthalmic rotary prism.

(a) *Identification.* An ophthalmic rotary prism is a device with various prismatic powers intended to be handheld and used to measure ocular deviation in patients with latent or manifest strabismus (eye muscle deviation).

(b) *Classification.* Class I (general controls). The device is exempt from the premarket notification procedures in subpart E of part 807 of this chapter, subject to the limitations in § 886.9. The device is also exempt from the current good manufacturing practice regulations in part 820 of this chapter, with the exception of § 820.180, with respect to general requirements concerning records, and § 820.198, with respect to complaint files.

[52 FR 33355, Sept. 2, 1987, as amended at 53 FR 35605, Sept. 14, 1988; 66 FR 38812, July 25, 2001]

§ 886.1670 Ophthalmic isotope uptake probe.

(a) *Identification.* An ophthalmic isotope uptake probe is an AC-powered device intended to measure, by a probe which is placed in close proximity to the eye, the uptake of a radioisotope (phosphorus 32) by tumors to detect tumor masses on, around, or within the eye.

(b) *Classification.* Class II.

§ 886.1680 Ophthalmic projector.

(a) *Identification.* An ophthalmic projector is an AC-powered device intended to project an image on a screen for vision testing.

(b) *Classification.* Class I (general controls). The device is exempt from the premarket notification procedures in subpart E of part 807 of this chapter, subject to the limitations in § 886.9.

[55 FR 48442, Nov. 20, 1990, as amended at 59 FR 63013, Dec. 7, 1994; 66 FR 38812, July 25, 2001]

§ 886.1690 Pupillograph.

(a) *Identification.* A pupillograph is an AC-powered device intended to measure the pupil of the eye by reflected light and record the responses of the pupil.

(b) *Classification.* Class I (general controls). The device is exempt from the premarket notification procedures in subpart E of part 807 of this chapter, subject to the limitations in § 886.9.

[55 FR 48442, Nov. 20, 1990, as amended at 59 FR 63013, Dec. 7, 1994; 66 FR 38812, July 25, 2001]

§ 886.1700 Pupillometer.

(a) *Identification.* A pupillometer is an AC-powered or manual device intended to measure by reflected light the width or diameter of the pupil of the eye.

(b) *Classification.* Class I (general controls). The AC-powered device and the manual device are exempt from the premarket notification procedures in subpart E of part 807 of this chapter, subject to the limitations in § 886.9. The manual device is also exempt from the current good manufacturing practice regulations in part 820 of this chapter, with the exception of § 820.180, with respect to general requirements concerning records, and § 820.198, with respect to complaint files.

[55 FR 48442, Nov. 20, 1990, as amended at 59 FR 63013, Dec. 7, 1994; 66 FR 38812, July 25, 2001]

§ 886.1750 Skiascopic rack.

(a) *Identification.* A skiascopic rack is a device that is a rack and a set of attached ophthalmic lenses of various dioptric strengths intended as an aid in refraction.

(b) *Classification.* Class I (general controls). The device is exempt from the premarket notification procedures in subpart E of part 807 of this chapter, subject to the limitations in § 886.9.

[52 FR 33355, Sept. 2, 1987, as amended at 61 FR 1124, Jan. 16, 1996; 66 FR 38812, July 25, 2001]

§ 886.1760 Ophthalmic refractometer.

(a) *Identification.* An ophthalmic refractometer is an automatic AC-powered device that consists of a fixation system, a measurement and recording system, and an alignment system intended to measure the refractive power of the eye by measuring light reflexes from the retina.

(b) *Classification.* Class I (general controls). The device is exempt from the premarket notification procedures in subpart E of part 807 of this chapter, subject to the limitations in § 886.9.

[52 FR 33355, Sept. 2, 1987, as amended at 61 FR 1124, Jan. 16, 1996; 66 FR 38812, July 25, 2001]

§ 886.1770 Manual refractor.

(a) *Identification.* A manual refractor is a device that is a set of lenses of varous dioptric powers intended to measure the refractive error of the eye.

(b) *Classification.* Class I (general controls). The device is exempt from the premarket notification procedures in subpart E of part 807 of this chapter, subject to the limitations in § 886.9. The device is also exempt from the current good manufacturing practice regulations in part 820 of this chapter, with the exception of § 820.180, with respect to general requirements concerning records, and § 820.198, with respect to complaint files.

[52 FR 33355, Sept. 2, 1987, as amended at 53 FR 35605, Sept. 14, 1988; 66 FR 38812, July 25, 2001]

§ 886.1780 Retinoscope.

(a) *Identification.* A retinoscope is an AC-powered or battery-powered device intended to measure the refraction of the eye by illuminating the retina and noting the direction of movement of the light on the retinal surface and of the refraction by the eye of the emergent rays.

(b) *Classification.* (1) Class II (special controls) for the AC-powered device.

(2) Class I (general controls) for the battery-powered device. The class I battery-powered device is exempt from the premarket notification procedures in subpart E of part 807 of this chapter subject to § 886.9. The battery-powered device is exempt from the current good manufacturing practice regulations in part 820 of this chapter, with the exception of § 820.180 of this chapter, with respect to general requirements concerning records, and § 820.198 of this chapter, with respect to complaint files.

[55 FR 48442, Nov. 20, 1990; 55 FR 51799, Dec. 17, 1990, as amended at 65 FR 2320, Jan. 14, 2000]

§ 886.1790 Nearpoint ruler.

(a) *Identification.* A nearpoint ruler is a device calibrated in centimeters intended to measure the nearpoint of convergence (the point to which the visual lines are directed when convergence is at its maximum).

(b) *Classification.* Class I (general controls). The device is exempt from the premarket notification procedures in subpart E of part 807 of this chapter, subject to the limitations in § 886.9. The device is also exempt from the current good manufacturing practice regulations in part 820 of this chapter, with the exception of § 820.180, with respect to general requirements concerning records, and § 820.198, with respect to complaint files.

[52 FR 33355, Sept. 2, 1987, as amended at 53 FR 35605, Sept. 14, 1988; 53 FR 40825, Oct. 18, 1988; 66 FR 38812, July 25, 2001]

§ 886.1800 Schirmer strip.

(a) *Identification.* A Schirmer strip is a device made of filter paper or similar material intended to be inserted under a patient's lower eyelid to stimulate and evaluate formation of tears.

(b) *Classification.* Class I (general controls). If the device is made of the same materials that were used in the device before May 28, 1976, the device is exempt from the premarket notification procedures in subpart E of part 807 of this chapter, subject to the limitations in § 886.9.

[52 FR 33355, Sept. 2, 1987, as amended at 53 FR 35605, Sept. 14, 1988; 66 FR 38812, July 25, 2001]

§ 886.1810 Tangent screen (campimeter).

(a) *Identification.* A tangent screen (campimeter) is an AC-powered or battery-powered device that is a large square cloth chart with a central mark of fixation intended to map on a flat surface the central 30 degrees of a patient's visual field. This generic type of device includes projection tangent screens, target tangent screens and targets, felt tangent screens, and stereo campimeters.

(b) *Classification.* Class I (general controls). The AC-powered device and the battery-powered device are exempt from the premarket notification procedures in subpart E of part 807 of this chapter, subject to the limitations in §886.9. The battery-powered device is also exempt from the current good manufacturing practice regulations in part 820 of this chapter, with the exception of §820.180, with respect to general requirements concerning records, and §820.198, with respect to complaint files.

[55 FR 48442, Nov. 20, 1990, as amended at 59 FR 63013, Dec. 7, 1994; 66 FR 38812, July 25, 2001]

§886.1840 Simulatan (including crossed cylinder).

(a) *Identification.* A simulatan (including crossed cylinder) is a device that is a set of pairs of cylinder lenses that provides various equal plus and minus refractive strengths. The lenses are arranged so that the user can exchange the positions of plus and minus cylinder lenses of equal strengths. The device is intended for subjective refraction (refraction in which the patient judges whether a given object is clearly in focus, as the examiner uses different lenses).

(b) *Classification.* Class I (general controls). The device is exempt from the premarket notification procedures in subpart E of part 807 of this chapter, subject to the limitations in §886.9. The device is also exempt from the current good manufacturing practice regulations in part 820 of this chapter, with the exception of §820.180, with respect to general requirements concerning records, and §820.198, with respect to complaint files.

[52 FR 33355, Sept. 2, 1987, as amended at 53 FR 35605, Sept. 14, 1988; 66 FR 38812, July 25, 2001]

§886.1850 AC-powered slitlamp biomicroscope.

(a) *Identification.* An AC-powered slitlamp biomicroscope is an AC-powered device that is a microscope intended for use in eye examination that projects into a patient's eye through a control diaphragm a thin, intense beam of light.

(b) *Classification.* Class II.

§886.1860 Ophthalmic instrument stand.

(a) *Identification.* An ophthalmic instrument stand is an AC-powered or nonpowered device intended to store ophthalmic instruments in a readily accessible position.

(b) *Classification.* Class I (general controls). The AC-powered device and the battery-powered device are exempt from the premarket notification procedures in subpart E of part 807 of this chapter, subject to the limitations in §886.9. The battery-powered device is also exempt from the current good manufacturing practice regulations in part 820 of this chapter, with the exception of §820.180, with respect to general requirements concerning records, and §820.198, with respect to complaint files.

[55 FR 48442, Nov. 20, 1990, as amended at 59 FR 63013, Dec. 7, 1994; 66 FR 38812, July 25, 2001]

§886.1870 Stereoscope.

(a) *Identification.* A stereoscope is an AC-powered or battery-powered device that combines the images of two similar objects to produce a three-dimensional appearance of solidity and relief. It is intended to measure the angle of strabismus (eye muscle deviation), evaluate binocular vision (usage of both eyes to see), and guide a patient's corrective exercises of eye muscles.

(b) *Classification.* Class I (general controls). The AC-powered device and the battery-powered device are exempt from the premarket notification procedures in subpart E of part 807 of this chapter, subject to the limitations in §886.9. The battery-powered device is also exempt from the current good manufacturing practice regulations in part 820 of this chapter, with the exception of §820.180, with respect to general requirements concerning records, and

§ 820.198, with respect to complaint files.

[55 FR 48442, Nov. 20, 1990, as amended at 59 FR 63013, Dec. 7, 1994; 66 FR 38813, July 25, 2001]

§ 886.1880 Fusion and stereoscopic target.

(a) *Identification.* A fusion and stereoscopic target is a device intended for use as a viewing object with a stereoscope (§ 886.1870).

(b) *Classification.* Class I (general controls). The device is exempt from the premarket notification procedures in subpart E of part 807 of this chapter, subject to the limitations in § 886.9. The device is also exempt from the current good manufacturing practice regulations in part 820 of this chapter, with the exception of § 820.180, with respect to general requirements concerning records, and § 820.198, with respect to complaint files.

[52 FR 33355, Sept. 2, 1987, as amended at 53 FR 35606, Sept. 14, 1988; 66 FR 38813, July 25, 2001]

§ 886.1905 Nystagmus tape.

(a) *Identification.* Nystagmus tape is a device that is a long, narrow strip of fabric or other flexible material on which a series of objects are printed. The device is intended to be moved across a patient's field of vision to elicit optokinetic nystagmus (abnormal and irregular eye movements) and to test for blindness.

(b) *Classification.* Class I (general controls). The device is exempt from the premarket notification procedures in subpart E of part 807 of this chapter, subject to the limitations in § 886.9. The device is also exempt from the current good manufacturing practice regulations in part 820 of this chapter, with the exception of § 820.180, with respect to general requirements concerning records, and § 820.198, with respect to complaint files.

[52 FR 33355, Sept. 2, 1987, as amended at 53 FR 35606, Sept. 14, 1988; 66 FR 38813, July 25, 2001]

§ 886.1910 Spectacle dissociation test system.

(a) *Identification.* A spectacle dissociation test system is an AC-powered or battery-powered device, such as a Lancaster test system, that consists of a light source and various filters, usually red or green filters, intended to subjectively measure imbalance of ocular muscles.

(b) *Classification.* Class I (general controls). The AC-powered device and the battery-powered device are exempt from the premarket notification procedures in subpart E of part 807 of this chapter, subject to the limitations in § 886.9. The battery-powered device is also exempt from the current good manufacturing practice regulations in part 820 of this chapter, with the exception of § 820.180, with respect to general requirements concerning records, and § 820.198, with respect to complaint files.

[55 FR 48442, Nov. 20, 1990; 55 FR 51799, Dec. 17, 1990, as amended at 59 FR 63013, Dec. 7, 1994; 66 FR 38813, July 25, 2001]

§ 886.1930 Tonometer and accessories.

(a) *Identification.* A tonometer and accessories is a manual device intended to measure intraocular pressure by applying a known force on the globe of the eye and measuring the amount of indentation produced (Schiotz type) or to measure intraocular tension by applanation (applying a small flat disk to the cornea). Accessories for the device may include a tonometer calibrator or a tonograph recording system. The device is intended for use in the diagnosis of glaucoma.

(b) *Classification.* Class II.

§ 886.1940 Tonometer sterilizer.

(a) *Identification.* A tonometer sterilizer is an AC-powered device intended to heat sterilize a tonometer (a device used to measure intraocular pressure).

(b) *Classification.* Class I (general controls). The device is exempt from the premarket notification procedures in subpart E of part 807 of this chapter subject to § 886.9.

[55 FR 48443, Nov. 20, 1990, as amended at 65 FR 2321, Jan. 14, 2000]

§ 886.1945 Transilluminator.

(a) *Identification.* A transilluminator is an AC-powered or battery-powered device that is a light source intended to transmit light through tissues to aid examination of patients.

(b) *Classification.* Class I for the battery-powered device. The battery-powered device is also exempt from the premarket notification procedures in subpart E of part 807 of this chapter, subject to the limitations in § 886.9. Class II for the AC-powered device.

[55 FR 48443, Nov. 20, 1990, as amended at 59 FR 63013, Dec. 7, 1994; 66 FR 38813, July 25, 2001]

Subpart C [Reserved]

Subpart D—Prosthetic Devices

§ 886.3100 Ophthalmic tantalum clip.

(a) *Identification.* An ophthalmic tantalum clip is a malleable metallic device intended to be implanted permanently or temporarily to bring together the edges of a wound to aid healing or prevent bleeding from small blood vessels in the eye.

(b) *Classification.* Class II (special controls). The device is exempt from the premarket notification procedures in subpart E of part 807 of this chapter subject to § 886.9.

[52 FR 33355, Sept. 2, 1987, as amended at 63 FR 59230, Nov. 3, 1998]

§ 886.3130 Ophthalmic conformer.

(a) *Identification.* An ophthalmic conformer is a device usually made of molded plastic intended to be inserted temporarily between the eyeball and eyelid to maintain space in the orbital cavity and prevent closure or adhesions during the healing process following surgery.]

(b) *Classification.* Class II (special controls). The device is exempt from the premarket notification procedures in subpart E of part 807 of this chapter subject to § 886.9.

[52 FR 33355, Sept. 2, 1987, as amended at 63 FR 59230, Nov. 3, 1998]

§ 886.3200 Artificial eye.

(a) *Identification.* An artificial eye is a device resembling the anterior portion of the eye, usually made of glass or plastic, intended to be inserted in a patient's eye socket anterior to an orbital implant, or the eviscerated eyeball, for cosmetic purposes. The device is not intended to be implanted.

(b) *Classification.* Class I (general controls). The device is exempt from the premarket notification procedures in subpart E of part 807 of this chapter, subject to the limitations in § 886.9, if the device is made from the same materials, has the same chemical composition, and uses the same manufacturing processes as currently legally marketed devices.

[61 FR 1124, Jan. 16, 1996, as amended at 66 FR 38813, July 25, 2001]

§ 886.3300 Absorbable implant (scleral buckling method).

(a) *Identification.* An absorbable implant (scleral buckling method) is a device intended to be implanted on the sclera to aid retinal reattachment.

(b) *Classification.* Class II.

§ 886.3320 Eye sphere implant.

(a) *Identification.* An eye sphere implant is a device intended to be implanted in the eyeball to occupy space following the removal of the contents of the eyeball with the sclera left intact.

(b) *Classification.* Class II.

§ 886.3340 Extraocular orbital implant.

(a) *Identification.* An extraocular orbital implant is a nonabsorbable device intended to be implanted during scleral surgery for buckling or building up the floor of the eye, usually in conjunction with retinal reattachment. Injectable substances are excluded.

(b) *Classification.* Class II.

§ 886.3400 Keratoprosthesis.

(a) *Identification.* A keratoprosthesis is a device intended to provide a transparent optical pathway through an opacified cornea, either intraoperatively or permanently, in an eye that is not a reasonable candidate for a corneal transplant.

(b) *Classification.* Class II. The special controls for this device are FDA's:

(1) "Use of International Standard ISO 10993 'Biological Evaluation of Medical Devices—Part I: Evaluation and Testing,' "

(2) "510(k) Sterility Review Guidance of 2/12/90 (K90–1)," and

(3) "Guidance on 510(k) Submissions for Keratoprostheses."

[65 FR 17147, Mar. 31, 2000]

§ 886.3600 Intraocular lens.

(a) *Identification.* An intraocular lens is a device made of materials such as glass or plastic intended to be implanted to replace the natural lens of an eye.

(b) *Classification.* Class III.

(c) *Date PMA or notice of completion of a PDP is required.* As of May 28, 1976, an approval under section 515 of the act is required before this device may be commercially distributed. See § 886.3.

§ 886.3800 Scleral shell.

(a) *Identification.* A scleral shell is a device made of glass or plastic that is intended to be inserted for short time periods over the cornea and proximal-cornea sclera for cosmetic or reconstructive purposes. An artificial eye is usually painted on the device. The device is not intended to be implanted.

(b) *Classification.* Class II (special controls). The device is exempt from the premarket notification procedures in subpart E of part 807 of this chapter subject to § 886.9.

[52 FR 33355, Sept. 2, 1987, as amended at 63 FR 59230, Nov. 3, 1998]

§ 886.3920 Aqueous shunt.

(a) *Identification.* An aqueous shunt is an implantable device intended to reduce intraocular pressure in the anterior chamber of the eye in patients with neovascular glaucoma or with glaucoma when medical and conventional surgical treatments have failed.

(b) *Classification.* Class II. The special controls for this device are FDA's:

(1) "Use of International Standard ISO 10993 'Biological Evaluation of Medical Devices—Part I: Evaluation and Testing,' "

(2) "510(k) Sterility Review Guidance of 2/12/90 (K90–1)," and

(3) "Aqueous Shunts—510(k) Submissions."

[65 FR 17147, Mar. 31, 2000, as amended at 66 FR 18542, Apr. 10, 2001]

Subpart E—Surgical Devices

§ 886.4070 Powered corneal burr.

(a) *Identification.* A powered corneal burr is an AC-powered or battery-powered device that is a motor and drilling tool intended to remove rust rings from the cornea of the eye.

(b) *Classification.* Class I (general controls). When intended only for rust ring removal, the device is exempt from the premarket notification procedures in subpart E of part 807 of this chapter subject to § 886.9.

[55 FR 48443, Nov. 20, 1990; 55 FR 51799, Dec. 17, 1990, as amended at 65 FR 2321, Jan. 14, 2000]

§ 886.4100 Radiofrequency electrosurgical cautery apparatus.

(a) *Identification.* A radiofrequency electrosurgical cautery apparatus is an AC-powered or battery-powered device intended for use during ocular surgery to coagulate tissue or arrest bleeding by a high frequency electric current.

(b) *Classification.* Class II.

§ 886.4115 Thermal cautery unit.

(a) *Identification.* A thermal cautery unit is an AC-powered or battery-powered device intended for use during ocular surgery to coagulate tissue or arrest bleeding by heat conducted through a wire tip.

(b) *Classification.* Class II.

§ 886.4150 Vitreous aspiration and cutting instrument.

(a) *Identification.* A vitreous aspiration and cutting instrument is an electrically powered device, which may use ultrasound, intended to remove vitreous matter from the vitreous cavity or remove a crystalline lens.

(b) *Classification.* Class II.

§ 886.4170 Cryophthalmic unit.

(a) *Identification.* A cryophthalmic unit is a device that is a probe with a small tip that becomes extremely cold through the controlled use of a refrigerant or gas. The device may be AC-powered. The device is intended to remove cataracts by the formation of an adherent ice ball in the lens, to freeze the eye and adjunct parts for surgical removal of scars, and to freeze tumors.

(b) *Classification.* Class II.

§ 886.4230 Ophthalmic knife test drum.

(a) *Identification.* An ophthalmic knife test drum is a device intended to test the keenness of ophthalmic surgical knives to determine whether resharpening is needed.

(b) *Classification.* Class I (general controls). The device is exempt from the premarket notification procedures in subpart E of part 807 of this chapter, subject to the limitations in § 886.9. The device is also exempt from the current good manufacturing practice regulations in part 820 of this chapter, with the exception of § 820.180, with respect to general requirements concerning records, and § 820.198, with respect to complaint files.

[52 FR 33355, Sept. 2, 1987, as amended at 53 FR 35606, Sept. 14, 1988; 66 FR 38813, July 25, 2001]

§ 886.4250 Ophthalmic electrolysis unit.

(a) *Identification.* An ophthalmic electrolysis unit is an AC-powered or battery-powered device intended to destroy ocular hair follicles by applying a galvanic electrical current.

(b) *Classification.* Class I for the battery-powered device. Class II for the AC-powered device. The battery-powered device is exempt from the premarket notification procedures in subpart E of part 807 of this chapter, subject to the limitations in § 886.9.

[55 FR 48443, Nov. 20, 1990, as amended at 59 FR 63013, Dec. 7, 1994; 66 FR 38813, July 25, 2001]

§ 886.4270 Intraocular gas.

(a) *Identification.* An intraocular gas is a device consisting of a gaseous fluid intended to be introduced into the eye to place pressure on a detached retina.

(b) *Classification.* Class III.

(c) *Date PMA or notice of completion of a PDP is required.* As of May 28, 1976, an approval under section 515 of the act is required before this device may be commercially distributed. See § 886.3.

§ 886.4275 Intraocular fluid.

(a) *Identification.* An intraocular fluid is a device consisting of a nongaseous fluid intended to be introduced into the eye to aid performance of surgery, such as to maintain anterior chamber depth, preserve tissue integrity, protect tissue from surgical trauma, or function as a tamponade during retinal reattachment.

(b) *Classification.* Class III.

(c) *Date PMA or notice of completion of a PDP is required.* As of May 28, 1976, an approval under section 515 of the act is required before this device may be commercially distributed. See § 886.3.

§ 886.4280 Intraocular pressure measuring device.

(a) *Identification.* An intraocular pressure measuring device is a manual or AC-powered device intended to measure intraocular pressure. Also included are any devices found by FDA to be substantially equivalent to such devices. Accessories for the device may include calibrators or recorders. The device is intended for use in the diagnosis of glaucoma.

(b) *Classification.* Class III.

(c) *Date PMA or notice of completion of PDP is required.* As of May 28, 1976, an approval under section 515 of the act is required before this device may be commercially distributed. See § 886.3.

§ 886.4300 Intraocular lens guide.

(a) *Identification.* An intraocular lens guide is a device intended to be inserted into the eye during surgery to direct the insertion of an intraocular lens and be removed after insertion is completed.

(b) *Classification.* Class I (general controls). Except when used as folders or injectors for soft or foldable intraocular lenses, the device is exempt from the premarket notification procedures in subpart E of part 807 of this chapter subject to § 886.9.

[52 FR 33355, Sept. 2, 1987, as amended at 65 FR 2321, 2000]

§ 886.4335 Operating headlamp.

(a) *Identification.* An operating headlamp is an AC-powered or battery-powered device intended to be worn on the user's head to provide a light source to aid visualization during surgical, diagnostic, or therapeutic procedures.

(b) *Classification.* Class I for the battery-powered device. Class II for the

AC-powered device. The battery-powered device is exempt from the premarket notification procedures in subpart E of part 807 of this chapter, subject to the limitations in § 886.9.

[55 FR 48443, Nov. 20, 1990, as amended at 66 FR 38813, July 25, 2001]

§ 886.4350 Manual ophthalmic surgical instrument.

(a) *Identification.* A manual ophthalmic surgical instrument is a nonpowered, handheld device intended to aid or perform ophthalmic surgical procedures. This generic type of device includes the manual corneal burr, ophthalmic caliper, ophthalmic cannula, eyelid clamp, ophthalmic muscle clamp, iris retractor clip, orbital compressor, ophthalmic curette, cystotome, orbital depressor, lachrymal dilator, erisophake, expressor, ophthalmic forcep, ophthalmic hook, sphere introducer, ophthalmic knife, ophthalmic suturing needle, lachrymal probe, trabeculotomy probe, cornea-sclera punch, ophthalmic retractor, ophthalmic ring (Flieringa), lachrymal sac rongeur, ophthalmic scissors, enucleating snare, ophthalmic spatula, ophthalmic specula, ophthalmic spoon, ophthalmic spud, trabeculotome or ophthalmic manual trephine.

(b) *Classification.* Class I (general controls). The device is exempt from the premarket notification procedures in subpart E of part 807 of this chapter, subject to the limitations in § 886.9.

[52 FR 33355, Sept. 2, 1987, as amended at 53 FR 35606, Sept. 14, 1988; 59 FR 63013, Dec. 7, 1994; 60 FR 15872, Mar. 28, 1995; 66 FR 38813, July 25, 2001]

§ 886.4360 Ocular surgery irrigation device.

(a) *Identification.* An ocular surgery irrigation device is a device intended to be suspended over the ocular area during ophthalmic surgery to deliver continuous, controlled irrigation to the surgical field.

(b) *Classification.* Class I (general controls). The device is exempt from the premarket notification procedures in subpart E of part 807 of this chapter, subject to the limitations in § 886.9.

[52 FR 33355, Sept. 2, 1987, as amended at 53 FR 35606, Sept. 14, 1988; 59 FR 63013, Dec. 7, 1994; 66 FR 38813, July 25, 2001]

§ 886.4370 Keratome.

(a) *Identification.* A keratome is an AC-powered or battery-powered device intended to shave tissue from sections of the cornea for a lamellar (partial thickness) transplant.

(b) *Classification.* Class I.

[55 FR 48443, Nov. 20, 1990]

§ 886.4390 Ophthalmic laser.

(a) *Identification.* An ophthalmic laser is an AC-powered device intended to coagulate or cut tissue of the eye, orbit, or surrounding skin by a laser beam.

(b) *Classification.* Class II.

§ 886.4392 Nd:YAG laser for posterior capsulotomy and peripheral iridotomy.

(a) *Identification.* The Nd:YAG laser for posterior capsulotomy and peripheral iridotomy consists of a mode-locked or Q-switched solid state Nd:YAG laser intended for disruption of the posterior capsule or the iris via optical breakdown. The Nd:YAG laser generates short pulse, low energy, high power, coherent optical radiation. When the laser output is combined with focusing optics, the high irradiance at the target causes tissue disruption via optical breakdown. A visible aiming system is utilized to target the invisible Nd:YAG laser radiation on or in close proximity to the target tissue.

(b) *Classification.* Class II (special controls). Design Parameters: Device must emit a laser beam with the following parameters: wavelength = 1064 nanometers; spot size = 50 to 100 micros; pulse width = 3 to 30 nanoseconds; output energy per pulse = 0.5 to 15 millijoules (mJ); repetition rate = 1 to 10 pulses; and total energy = 20 to 120 mJ.

[65 FR 6894, Feb. 11, 2000]

§ 886.4400 Electronic metal locator.

(a) *Identification.* An electronic metal locator is an AC-powered device with probes intended to locate metallic foreign bodies in the eye or eye socket.

(b) *Classification.* Class II.

§ 886.4440 AC-powered magnet.

(a) *Identification.* An AC-powered magnet is an AC-powered device that generates a magnetic field intended to find and remove metallic foreign bodies from eye tissue.

(b) *Classification.* Class II.

§ 886.4445 Permanent magnet.

(a) *Identification.* A permanent magnet is a nonelectric device that generates a magnetic field intended to find and remove metallic foreign bodies from eye tissue.

(b) *Classification.* Class I (general controls). The device is exempt from the premarket notification procedures in subpart E of part 807 of this chapter, subject to the limitations in § 886.9. The device is also exempt from the current good manufacturing practice regulations in part 820 of this chapter, with the exception of § 820.180, with respect to general requirements concerning records, and § 820.198, with respect to complaint files.

[52 FR 33355, Sept. 2, 1987, as amended at 53 FR 35606, Sept. 14, 1988; 66 FR 38813, July 25, 2001]

§ 886.4570 Ophthalmic surgical marker.

(a) *Identification.* An ophthalmic surgical marker is a device intended to mark by use of ink, dye, or indentation the location of ocular or scleral surgical manipulation.

(b) *Classification.* Class I (general controls). The device is exempt from the premarket notification procedures in subpart E of part 807 of this chapter, subject to the limitations in § 886.9.

[52 FR 33355, Sept. 2, 1987, as amended at 53 FR 35606, Sept. 14, 1988; 59 FR 63013, Dec. 7, 1994; 66 FR 38813, July 25, 2001]

§ 886.4610 Ocular pressure applicator.

(a) *Identification.* An ocular pressure applicator is a manual device that consists of a sphygmomanometer-type squeeze bulb, a dial indicator, a band, and bellows, intended to apply pressure on the eye in preparation for ophthalmic surgery.

(b) *Classification.* Class II.

§ 886.4670 Phacofragmentation system.

(a) *Identification.* A phacofragmentation system is an AC-powered device with a fragmenting needle intended for use in cataract surgery to disrupt a cataract with ultrasound and extract the cataract.

(b) *Classification.* Class II.

§ 886.4690 Ophthalmic photocoagulator.

(a) *Identification.* An ophthalmic photocoagulator is an AC-powered device intended to use the energy from an extended noncoherent light source to occlude blood vessels of the retina, choroid, or iris.

(b) *Classification.* Class II.

§ 886.4750 Ophthalmic eye shield.

(a) *Identification.* An ophthalmic eye shield is a device that consists of a plastic or aluminum eye covering intended to protect the eye or retain dressing materials in place.

(b) *Classification.* Class I (general controls). When made only of plastic or aluminum, the device is exempt from the premarket notification procedures in subpart E of part 807 of this chapter subject to § 886.9. When made only of plastic or aluminum, the devices are exempt from the current good manufacturing practice regulations in part 820 of this chapter, with the exception of § 820.180 of this chapter, with respect to general requirements concerning records, and § 820.198 of this chapter, with respect to complaint files.

[52 FR 33355, Sept. 2, 1987, as amended at 59 FR 63014, Dec. 7, 1994; 65 FR 2321, Jan. 14, 2000]

§ 886.4770 Ophthalmic operating spectacles (loupes).

(a) *Identification.* Ophthalmic operating spectacles (loupes) are devices that consist of convex lenses or lens systems intended to be worn by a surgeon to magnify the surgical site during ophthalmic surgery.

(b) *Classification.* Class I (general controls). The device is exempt from the premarket notification procedures in subpart E of part 807 of this chapter, subject to the limitations in § 886.9. The device is also exempt from the current good manufacturing practice regulations in part 820 of this chapter, with the exception of § 820.180, with respect to general requirements concerning

records, and § 820.198, with respect to complaint files.

[52 FR 33355, Sept. 2, 1987, as amended at 53 FR 35606, Sept. 14, 1988; 66 FR 38813, July 25, 2001]

§ 886.4790 Ophthalmic sponge.

(a) *Identification.* An ophthalmic sponge is a device that is an absorbant sponge, pad, or spear made of folded gauze, cotton, cellulose, or other material intended to absorb fluids from the operative field in ophthalmic surgery.

(b) *Classification.* Class II.

§ 886.4855 Ophthalmic instrument table.

(a) *Identification.* An ophthalmic instrument table is an AC-powered or manual device on which ophthalmic instruments are intended to be placed.

(b) *Classification.* Class I (general controls). The AC-powered device and the manual device are exempt from the premarket notification procedures in subpart E of part 807 of this chapter, subject to the limitations in § 886.9. The manual device is also exempt from the current good manufacturing practice regulations in part 820 of this chapter, with the exception of § 820.180, with respect to general requirements concerning records, and § 820.198, with respect to complaint files.

[55 FR 48443, Nov. 20, 1990, as amended at 59 FR 63014, Dec. 7, 1994; 66 FR 38814, July 25, 2001]

Subpart F—Therapeutic Devices

§ 886.5100 Ophthalmic beta radiation source.

(a) *Identification.* An ophthalmic beta radiation source is a device intended to apply superficial radiation to benign and malignant ocular growths.

(b) *Classification.* Class II.

§ 886.5120 Low-power binocular loupe.

(a) *Identification.* A low-power binocular loupe is a device that consists of two eyepieces, each with a lens or lens system, intended for medical purposes to magnify the appearance of objects.

(b) *Classification.* Class I (general controls). The device is exempt from the premarket notification procedures in subpart E of part 807 of this chapter, subject to the limitations in § 886.9. The device is also exempt from the current good manufacturing practice regulations in part 820 of this chapter, with the exception of § 820.180, with respect to general requirements concerning records, and § 820.198, with respect to complaint files.

[52 FR 33355, Sept. 2, 1987, as amended at 53 FR 35607, Sept. 14, 1988; 66 FR 38814, July 25, 2001]

§ 886.5420 Contact lens inserter/remover.

(a) *Identification.* A contact lens inserter/remover is a handheld device intended to insert or remove contact lenses by surface adhesion or suction.

(b) *Classification.* Class I (general controls). The device is exempt from the premarket notification procedures in subpart E of part 807 of this chapter, subject to the limitations in § 886.9.

[52 FR 33355, Sept. 2, 1987, as amended at 53 FR 35607, Sept. 14, 1988; 66 FR 38814, July 25, 2001]

§ 886.5540 Low-vision magnifier.

(a) *Identification.* A low-vision magnifier is a device that consists of a magnifying lens intended for use by a patient who has impaired vision. The device may be held in the hand or attached to spectacles.

(b) *Classification.* Class I (general controls). The device is exempt from the premarket notification procedures in subpart E of part 807 of this chapter, subject to the limitations in § 886.9. The device is also exempt from the current good manufacturing practice regulations in part 820 of this chapter, with the exception of § 820.180, with respect to general requirements concerning records, and § 820.198, with respect to complaint files.

[52 FR 33355, Sept. 2, 1987, as amended at 53 FR 35607, Sept. 14, 1988; 66 FR 38814, July 25, 2001]

§ 886.5600 Ptosis crutch.

(a) *Identification.* A ptosis crutch is a device intended to be mounted on the spectacles of a patient who has ptosis (drooping of the upper eyelid as a result of faulty development or paralysis) to hold the upper eyelid open.

(b) *Classification.* Class I (general controls). The device is exempt from the premarket notification procedures in subpart E of part 807 of this chapter, subject to the limitations in § 886.9. The device is also exempt from the current good manufacturing practice regulations in part 820 of this chapter, with the exception of § 820.180, with respect to general requirements concerning records, and § 820.198, with respect to complaint files.

[52 FR 33355, Sept. 2, 1987, as amended at 53 FR 35607, Sept. 14, 1988; 66 FR 38814, July 25, 2001]

§ 886.5800 Ophthalmic bar reader.

(a) *Identification.* An ophthalmic bar reader is a device that consists of a magnifying lens intended for use by a patient who has impaired vision. The device is placed directly onto reading material to magnify print.

(b) *Classification.* Class I (general controls). The device is exempt from the premarket notification procedures in subpart E of part 807 of this chapter, subject to the limitations in § 886.9. The device is also exempt from the current good manufacturing practice regulations in part 820 of this chapter, with the exception of § 820.180, with respect to general requirements concerning records, and § 820.198, with respect to complaint files.

[52 FR 33355, Sept. 2, 1987, as amended at 53 FR 35607, Sept. 14, 1988; 66 FR 38814, July 25, 2001]

§ 886.5810 Ophthalmic prism reader.

(a) *Identification.* An ophthalmic prism reader is a device intended for use by a patient who is in a supine position to change the angle of print to aid reading.

(b) *Classification.* Class I (general controls). The device is exempt from the premarket notification procedures in subpart E of part 807 of this chapter, subject to the limitations in § 886.9. The device is also exempt from the current good manufacturing practice regulations in part 820 of this chapter, with the exception of § 820.180, with respect to general requirements concerning records, and § 820.198, with respect to complaint files.

[52 FR 33355, Sept. 2, 1987, as amended at 53 FR 35607, Sept. 14, 1988; 66 FR 38814, July 25, 2001]

§ 886.5820 Closed-circuit television reading system.

(a) *Identification.* A closed-circuit television reading system is a device that consists of a lens, video camera, and video monitor that is intended for use by a patient who has subnormal vision to magnify reading material.

(b) *Classification.* Class I (general controls). The device is exempt from the premarket notification procedures in subpart E of part 807 of this chapter, subject to the limitations in § 886.9.

[55 FR 48443, Nov. 20, 1990, as amended at 59 FR 63014, Dec. 7, 1994; 66 FR 38814, July 25, 2001]

§ 886.5840 Magnifying spectacles.

(a) *Identification.* Magnifying spectacles are devices that consist of spectacle frames with convex lenses intended to be worn by a patient who has impaired vision to enlarge images.

(b) *Classification.* Class I (general controls). The device is exempt from the premarket notification procedures in subpart E of part 807 of this chapter, subject to the limitations in § 866.9.

[52 FR 33355, Sept. 2, 1987, as amended at 53 FR 35607, Sept. 14, 1988; 59 FR 63014, Dec. 7, 1994; 66 FR 38814, July 25, 2001]

§ 886.5842 Spectacle frame.

(a) *Identification.* A spectacle frame is a device made of metal or plastic intended to hold prescription spectacle lenses worn by a patient to correct refractive errors.

(b) *Classification.* Class I (general controls). The device is exempt from the premarket notification procedures in subpart E of part 807 of this chapter, subject to the limitations in § 886.9.

[52 FR 33355, Sept. 2, 1987, as amended at 59 FR 63014, Dec. 7, 1994; 66 FR 38814, July 25, 2001]

§ 886.5844 Prescription spectacle lens.

(a) *Identification.* A prescription spectacle lens is a glass or plastic device that is a lens intended to be worn by a patient in a spectacle frame to provide refractive corrections in accordance with a prescription for the patient. The device may be modified to protect the eyes from bright sunlight (i.e., prescription sunglasses). Prescription sunglass lenses may be reflective, tinted, polarizing, or photosensitized.

(b) *Classification.* Class I (general controls). The device is exempt from the premarket notification procedures in subpart E of part 807 of this chapter, subject to the limitations in § 886.9.

[52 FR 33355, Sept. 2, 1987, as amended at 53 FR 35607, Sept. 14, 1988; 59 FR 63014, Dec. 7, 1994; 66 FR 38814, July 25, 2001]

§ 886.5850 Sunglasses (nonprescription).

(a) *Identification.* Sunglasses (nonprescription) are devices that consist of spectacle frames or clips with absorbing, reflective, tinted, polarizing, or photosensitized lenses intended to be worn by a person to protect the eyes from bright sunlight but not to provide refractive corrections. This device is usually available over-the-counter.

(b) *Classification.* Class I (general controls). The device is exempt from the premarket notification procedures in subpart E of part 807 of this chapter subject to § 886.9.

[52 FR 33355, Sept. 2, 1987, as amended at 65 FR 2321, 2000]

§ 886.5870 Low-vision telescope.

(a) *Identification.* A low-vision telescope is a device that consists of an arrangement of lenses or mirrors intended for use by a patient who has impaired vision to increase the apparent size of objects. This generic type of device includes handheld or spectacle telescopes.

(b) *Classification.* Class I (general controls). The device is exempt from the premarket notification procedures in subpart E of part 807 of this chapter, subject to the limitations in § 886.9. The device is also exempt from the current good manufacturing practice regulations in part 820 of this chapter, with the exception of § 820.180, with respect to general requirements concerning records, and § 820.198, with respect to complaint files.

[52 FR 33355, Sept. 2, 1987, as amended at 53 FR 35607, Sept. 14, 1988; 66 FR 38814, July 25, 2001]

§ 886.5900 Electronic vision aid.

(a) *Identification.* An electronic vision aid is an AC-powered or battery-powered device that consists of an electronic sensor/transducer intended for use by a patient who has impaired vision or blindness to translate visual images of objects into tactile or auditory signals.

(b) *Classification.* Class I (general controls). The device is exempt from the premarket notification procedures in subpart E of part 807 of this chapter, subject to the limitations in § 886.9.

[55 FR 48443, Nov. 20, 1990, as amended at 59 FR 63014, Dec. 7, 1994; 66 FR 38814, July 25, 2001]

§ 886.5910 Image intensification vision aid.

(a) *Identification.* An image intensification vision aid is a battery-powered device intended for use by a patient who has limited dark adaptation or impaired vision to amplify ambient light.

(b) *Classification.* Class I (general controls). The device is exempt from the premarket notification procedures in subpart E of part 807 of this chapter, subject to the limitations in § 886.9. The device is also exempt from the current good manufacturing practice regulations in part 820 of this chapter, with the exception of § 820.180, with respect to general requirements concerning records, and § 820.198, with respect to complaint files.

[52 FR 33355, Sept. 2, 1987, as amended at 53 FR 35607, Sept. 14, 1988; 66 FR 38814, July 25, 2001]

§ 886.5915 Optical vision aid.

(a) *Identification.* An optical vision aid is a device that consists of a magnifying lens with an accompanying AC-powered or battery-powered light source intended for use by a patient who has impaired vision to increase the apparent size of object detail.

(b) *Classification.* Class I (general controls). The AC-powered device and the battery-powered device are exempt from the premarket notification procedures in subpart E of part 807 of this chapter, subject to the limitations in §886.9. The battery-powered device is also exempt from the current good manufacturing practice regulations in part 820 of this chapter, with the exception of §820.180, with respect to general requirements concerning records, and §820.198, with respect to complaint files.

[55 FR 48443, Nov. 20, 1990, as amended at 59 FR 63014, Dec. 7, 1994; 66 FR 38815, July 25, 2001]

§886.5916 Rigid gas permeable contact lens.

(a) *Identification.* A rigid gas permeable contact lens is a device intended to be worn directly against the cornea of the eye to correct vision conditions. The device is made of various materials, such as cellulose acetate butyrate, polyacrylate-silicone, or silicone elastomers, whose main polymer molecules generally do not absorb or attract water.

(b) *Classification.* (1) Class II if the device is intended for daily wear only.

(2) Class III if the device is intended for extended wear.

(c) *Date PMA or notice of completion of a PDP is required.* As of May 28, 1976, an approval under section 515 of the act is required before a device described in paragraph (b)(2) of this section may be commercially distributed. See §886.3.

[52 FR 33355, Sept. 2, 1987, as amended at 59 FR 10284, Mar. 4, 1994]

§886.5918 Rigid gas permeable contact lens care products.

(a) *Identification.* A rigid gas permeable contact lens care product is a device intended for use in the cleaning, conditioning, rinsing, lubricating/rewetting, or storing of a rigid gas permeable contact lens. This includes all solutions and tablets used together with rigid gas permeable contact lenses.

(b) *Classification.* Class II (Special Controls) Guidance Document: "Guidance for Industry Premarket Notification (510(k)) Guidance Document for Contact Lens Care Products."

[62 FR 30987, June 6, 1997]

§886.5925 Soft (hydrophilic) contact lens.

(a) *Identification.* A soft (hydrophilic) contact lens is a device intended to be worn directly against the cornea and adjacent limbal and scleral areas of the eye to correct vision conditions or act as a therapeutic bandage. The device is made of various polymer materials the main polymer molecules of which absorb or attract a certain volume (percentage) of water.

(b) *Classification.* (1) Class II if the device is intended for daily wear only.

(2) Class III if the device is intended for extended wear.

(c) *Date PMA or notice of completion of a PDP is required.* As of May 28, 1976, an approval under section 515 of the act is required before a device described in paragraph (b)(2) of this section may be commercially distributed. See §886.3.

[52 FR 33355, Sept. 2, 1987, as amended at 59 FR 10284, Mar. 4, 1994]

§886.5928 Soft (hydrophilic) contact lens care products.

(a) *Identification.* A soft (hydrophilic) contact lens care product is a device intended for use in the cleaning, rinsing, disinfecting, lubricating/rewetting, or storing of a soft (hydrophilic) contact lens. This includes all solutions and tablets used together with soft (hydrophilic) contact lenses and heat disinfecting units intended to disinfect a soft (hydrophilic) contact lens by means of heat.

(b) *Classification.* Class II (Special Controls) Guidance Document: "Guidance for Industry Premarket Notification (510(k)) Guidance Document for Contact Lens Care Products."

[62 FR 30988, June 6, 1997]

§886.5933 [Reserved]

PART 888—ORTHOPEDIC DEVICES

Subpart A—General Provisions

Subpart B—Diagnostic Devices

888.1100 Arthroscope.
888.1240 AC-powered dynamometer.
888.1250 Nonpowered dynamometer.
888.1500 Goniometer.
888.1520 Nonpowered goniometer.

Subpart C [Reserved]

Subpart D—Prosthetic Devices

888.3000 Bone cap.
888.3010 Bone fixation cerclage.
888.3015 Bone heterograft.
888.3020 Intramedullary fixation rod.
888.3025 Passive tendon prosthesis.
888.3027 Polymethylmethacrylate (PMMA) bone cement.
888.3030 Single/multiple component metallic bone fixation appliances and accessories.
888.3040 Smooth or threaded metallic bone fixation fastener.
888.3045 Resorbable calcium salt bone void filler device.
888.3050 Spinal interlaminal fixation orthosis.
888.3060 Spinal intervertebral body fixation orthosis.
888.3070 Pedicle screw spinal system.
888.3100 Ankle joint metal/composite semi-constrained cemented prosthesis.
888.3110 Ankle joint metal/polymer semi-constrained cemented prosthesis.
888.3120 Ankle joint metal/polymer non-constrained cemented prosthesis.
888.3150 Elbow joint metal/polymer constrained cemented prosthesis.
888.3160 Elbow joint metal/polymer semi-constrained cemented prosthesis.
888.3170 Elbow joint radial (hemi-elbow) polymer prosthesis.
888.3180 Elbow joint humeral (hemi-elbow) metallic uncemented prosthesis.
888.3200 Finger joint metal/metal constrained uncemented prosthesis.
888.3210 Finger joint metal/metal constrained cemented prosthesis.
888.3220 Finger joint metal/polymer constrained cemented prosthesis.
888.3230 Finger joint polymer constrained prosthesis.
888.3300 Hip joint metal constrained cemented or uncemented prosthesis.
888.3310 Hip joint metal/polymer constrained cemented or uncemented prosthesis.
888.3320 Hip joint metal/metal semi-constrained, with a cemented acetabular component, prosthesis.
888.3330 Hip joint metal/metal semi-constrained, with an uncemented acetabular component, prosthesis.
888.3340 Hip joint metal/composite semi-constrained cemented prosthesis.
888.3350 Hip joint metal/polymer semi-constrained cemented prosthesis.
888.3353 Hip joint metal/ceramic/polymer semi-constrained cemented or nonporous uncemented prosthesis.
888.3358 Hip joint metal/polymer/metal semi-constrained porous-coated uncemented prosthesis.
888.3360 Hip joint femoral (hemi-hip) metallic cemented or uncemented prosthesis.
888.3370 Hip joint (hemi-hip) acetabular metal cemented prosthesis.
888.3380 Hip joint femoral (hemi-hip) trunnion-bearing metal/polyacetal cemented prosthesis.
888.3390 Hip joint femoral (hemi-hip) metal/polymer cemented or uncemented prosthesis.
888.3400 Hip joint femoral (hemi-hip) metallic resurfacing prosthesis.
888.3410 Hip joint metal/polymer or ceramic/polymer semiconstrained resurfacing
888.3480 Knee joint femorotibial metallic constrained cemented prosthesis.
888.3490 Knee joint femorotibial metal/composite non-constrained cemented prosthesis.
888.3500 Knee joint femorotibial metal/composite semi-constrained cemented prosthesis.
888.3510 Knee joint femorotibial metal/polymer constrained cemented prosthesis.
888.3520 Knee joint femorotibial metal/polymer non-constrained cemented prosthesis.
888.3530 Knee joint femorotibial metal/polymer semi-constrained cemented prosthesis.
888.3535 Knee joint femorotibial (uni-compartmental) metal/polymer porous-coated uncemented prosthesis.
888.3540 Knee joint patellofemoral polymer/metal semi-constrained cemented prosthesis.
888.3550 Knee joint patellofemorotibial polymer/metal/metal constrained cemented prosthesis.
888.3560 Knee joint patellofemorotibial polymer/metal/polymer semi-constrained cemented prosthesis.
888.3565 Knee joint patellofemorotibial metal/polymer porous-coated uncemented prosthesis.
888.3570 Knee joint femoral (hemi-knee) metallic uncemented prosthesis.
888.3580 Knee joint patellar (hemi-knee) metallic resurfacing uncemented prosthesis.
888.3590 Knee joint tibial (hemi-knee) metallic resurfacing uncemented prosthesis.
888.3640 Shoulder joint metal/metal or metal/polymer constrained cemented prosthesis.
888.3650 Shoulder joint metal/polymer non-constrained cemented prosthesis.
888.3660 Shoulder joint metal/polymer semi-constrained cemented prosthesis.
888.3670 Shoulder joint metal/polymer/metal nonconstrained or semi-constrained porous-coated uncemented prosthesis.

Subpart E—Surgical Devices

AUTHORITY: 21 U.S.C. 351, 360, 360c, 360e, 360j, 371.

SOURCE: 52 FR 33702, Sept. 4, 1987, unless otherwise noted.

Subpart A—General Provisions

§ 888.1 Scope.

(a) This part sets forth the classification of orthopedic devices intended for human use that are in commercial distribution.

(b) The identification of a device in a regulation in this part is not a precise description of every device that is, or will be, subject to the regulation. A manufacturer who submits a premarket notification submission for a device under part 807 cannot show merely that the device is accurately described by the section title and identification provision of a regulation in this part, but shall state why the device is substantially equivalent to other devices, as required by § 807.87.

(c) To avoid duplicative listings, an orthopedic device that has two or more types of uses (e.g., used both as a diagnostic device and as a surgical device) is listed in one subpart only.

(d) References in this part to regulatory sections of the Code of Federal Regulations are to chapter I of title 21 unless otherwise noted.

(e) Guidance documents referenced in this part are available on the Internet at *http://www.fda.gov/cdrh/guidance.html.*

[52 FR 33702, Sept. 4, 1987, as amended at 68 FR 14137, Mar. 24, 2003]

§ 888.3 Effective dates of requirement for premarket approval.

A device included in this part that is classified into class III (premarket approval) shall not be commercially distributed after the date shown in the regulation classifying the device unless the manufacturer has an approval under section 515 of the act (unless an exemption has been granted under section 520(g)(2) of the act). An approval under section 515 of the act consists of FDA's issuance of an order approving an application for premarket approval (PMA) for the device or declaring completed a product development protocol (PDP) for the device.

(a) Before FDA requires that a device commercially distributed before the enactment date of the amendments, or a device that has been found substantially equivalent to such a device, has an approval under section 515 of the act, FDA must promulgate a regulation under section 515(b) of the act requiring such approval, except as provided in paragraphs (b) and (c) of this section. Such a regulation under section 515(b) of the act shall not be effective during the grace period ending on the 90th day after its promulgation or on the last day of the 30th full calendar month after the regulation that classifies the device into class III is effective, whichever is later. See section

501(f)(2)(B) of the act. Accordingly, unless an effective date of the requirement for premarket approval is shown in the regulation for a device classified into class III in this part, the device may be commercially distributed without FDA's issuance of an order approving a PMA or declaring completed a PDP for the device. If FDA promulgates a regulation under section 515(b) of the act requiring premarket approval for a device, section 501(f)(1)(A) of the act applies to the device.

(b) Any new, not substantially equivalent, device introduced into commercial distribution on or after May 28, 1976, including a device formerly marketed that has been substantially altered, is classified by statute (section 513(f) of the act) into class III without any grace period and FDA must have issued an order approving a PMA or declaring completed a PDP for the device before the device is commercially distributed unless it is reclassified. If FDA knows that a device being commercially distributed may be a "new" device as defined in this section because of any new intended use or other reasons, FDA may codify the statutory classification of the device into class III for such new use. Accordingly, the regulation for such a class III device states that as of the enactment date of the amendments, May 28, 1976, the device must have an approval under section 515 of the act before commercial distribution.

(c) A device identified in a regulation in this part that is classified into class III and that is subject to the transitional provisions of section 520(1) of the act is automatically classified by statute into class III and must have an approval under section 515 of the act before being commercially distributed. Accordingly, the regulation for such a class III transitional device states that as of the enactment date of the amendments, May 28, 1976, the device must have an approval under section 515 of the act before commercial distribution.

§ 888.5 Resurfacing technique.

Because of resurfacing techniques, certain joint prostheses require far less bone resection than other devices intended to repair or replace the same joint. The amount of bone resection may or may not affect the safety and effectiveness of the implantation of the prosthesis. When a resurfacing technique is used, the name of the prosthesis includes this information.

§ 888.6 Degree of constraint.

Certain joint prostheses provide more constraint of joint movement than others. FDA believes that the degree of constraint is an important factor affecting the safety and effectiveness of orthopedic prostheses. FDA is defining the following standard terms for categorizing the degree of constraint.

(a) A "constrained" joint prosthesis is used for joint replacement and prevents dislocation of the prosthesis in more than one anatomic plane and consists of either a single, flexible, across-the-joint component or more than one component linked together or affined.

(b) A "semi-constrained" joint prosthesis is used for partial or total joint replacement and limits translation and rotation of the prosthesis in one or more planes via the geometry of its articulating surfaces. It has no across-the-joint linkage.

(c) A "non-constrained" joint prosthesis is used for partial or total joint replacement and restricts minimally prosthesis movement in one or more planes. Its components have no across-the-joint linkage.

§ 888.9 Limitations of exemptions from section 510(k) of the Federal Food, Drug, and Cosmetic Act (the act).

The exemption from the requirement of premarket notification (section 510(k) of the act) for a generic type of class I or II device is only to the extent that the device has existing or reasonably foreseeable characteristics of commercially distributed devices within that generic type or, in the case of in vitro diagnostic devices, only to the extent that misdiagnosis as a result of using the device would not be associated with high morbidity or mortality. Accordingly, manufacturers of any commercially distributed class I or II device for which FDA has granted an exemption from the requirement of premarket notification must still submit a premarket notification to FDA before introducing or delivering for introduction into interstate commerce for commercial distribution the device when:

(a) The device is intended for a use different from the intended use of a legally marketed device in that generic type of device; e.g., the device is intended for a different medical purpose, or the device is intended for lay use where the former intended use was by health care professionals only;

(b) The modified device operates using a different fundamental scientific technology than a legally marketed device in that generic type of device; e.g., a surgical instrument cuts tissue with a laser beam rather than with a sharpened metal blade, or an in vitro diagnostic device detects or identifies infectious agents by using deoxyribonucleic acid (DNA) probe or nucleic acid hybridization technology rather than culture or immunoassay technology; or

(c) The device is an in vitro device that is intended:

(1) For use in the diagnosis, monitoring, or screening of neoplastic diseases with the exception of immunohistochemical devices;

(2) For use in screening or diagnosis of familial or acquired genetic disorders, including inborn errors of metabolism;

(3) For measuring an analyte that serves as a surrogate marker for screening, diagnosis, or monitoring life-threatening diseases such as acquired immune deficiency syndrome (AIDS), chronic or active hepatitis, tuberculosis, or myocardial infarction or to monitor therapy;

(4) For assessing the risk of cardiovascular diseases;

(5) For use in diabetes management;

(6) For identifying or inferring the identity of a microorganism directly from clinical material;

(7) For detection of antibodies to microorganisms other than immunoglobulin G (IgG) or IgG assays when the results are not qualitative, or are used to determine immunity, or the assay is intended for use in matrices other than serum or plasma;

(8) For noninvasive testing as defined in §812.3(k) of this chapter; and

(9) For near patient testing (point of care).

[65 FR 2321, Jan. 14, 2000]

Subpart B—Diagnostic Devices

§888.1100 Arthroscope.

(a) *Identification.* An arthroscope is an electrically powered endoscope intended to make visible the interior of a joint. The arthroscope and accessories also is intended to perform surgery within a joint.

(b) *Classification.* (1) Class II (performance standards).

(2) Class I for the following manual arthroscopic instruments: cannulas, currettes, drill guides, forceps, gouges, graspers, knives, obturators, osteotomes, probes, punches, rasps, retractors, rongeurs, suture passers, suture knotpushers, suture punches, switching rods, and trocars. The devices subject to this paragraph (b)(2) are exempt from the premarket notification procedures in subpart E of part 807 of this chapter, subject to the limitations in §888.9.

[52 FR 33702, Sept. 4, 1987, as amended at 61 FR 1124, Jan. 16, 1996; 66 FR 38815, July 25, 2001]

§888.1240 AC-powered dynamometer.

(a) *Identification.* An AC-powered dynamometer is an AC-powered device intended for medical purposes to assess neuromuscular function or degree of neuromuscular blockage by measuring, with a force transducer (a device that translates force into electrical impulses), the grip-strength of a patient's hand.

(b) *Classification.* Class II.

§888.1250 Nonpowered dynamometer.

(a) *Identification.* A nonpowered dynamometer is a mechanical device intended for medical purposes to measure the pinch and grip muscle strength of a patient's hand.

(b) *Classification.* Class I. The device is exempt from the premarket notification procedures in subpart E of part 807.

§888.1500 Goniometer.

(a) *Identification.* A goniometer is an AC-powered or battery powered device intended to evaluate joint function by measuring and recording ranges of motion, acceleration, or forces exerted by a joint.

(b) *Classification.* (1) Class I (general controls) for a goniometer that does not use electrode lead wires and patient cables. This device is exempt from the premarket notification procedures of subpart E of part 807 of this chapter subject to § 888.9.

(2) Class II (special controls) for a goniometer that uses electrode lead wires and patient cables. The special controls consist of:

(i) The performance standard under part 898 of this chapter, and

(ii) The guidance entitled "Guidance on the Performance Standard for Electrode Lead Wires and Patient Cables." This device is exempt from the premarket notification procedures of subpart E of part 807 of this chapter subject to § 888.9.

[65 FR 19319, Apr. 11, 2000]

§ 888.1520 Nonpowered goniometer.

(a) *Identification.* A nonpowered goniometer is a mechanical device intended for medical purposes to measure the range of motion of joints.

(b) *Classification.* Class I (general controls). The device is exempt from the premarket notification procedures in subpart E of part 807 of this chapter, subject to the limitations in § 888.9.

[52 FR 33702, Sept. 4, 1987, as amended at 66 FR 38815, July 25, 2001]

Subpart C [Reserved]

Subpart D—Prosthetic Devices

§ 888.3000 Bone cap.

(a) *Identification.* A bone cap is a mushroom-shaped device intended to be implanted made of either silicone elastomer or ultra-high molecular weight polyethylene. It is used to cover the severed end of a long bone, such as the humerus or tibia, to control bone overgrowth in juvenile amputees.

(b) *Classification.* Class I (general controls). The device is exempt from the premarket notification procedures in subpart E of part 807 of this chapter, subject to the limitations in § 888.9.

[52 FR 33702, Sept. 4, 1987, as amended at 61 FR 1124, Jan. 16, 1996; 66 FR 38815, July 25, 2001]

§ 888.3010 Bone fixation cerclage.

(a) *Identification.* A bone fixation cerclage is a device intended to be implanted that is made of alloys, such as cobalt-chromium-molybdenum, and that consists of a metallic ribbon or flat sheet or a wire. The device is wrapped around the shaft of a long bone, anchored to the bone with wire or screws, and used in the fixation of fractures.

(b) *Classification.* Class II.

§ 888.3015 Bone heterograft.

(a) *Identification.* Bone heterograft is a device intended to be implanted that is made from mature (adult) bovine bones and used to replace human bone following surgery in the cervical region of the spinal column.

(b) *Classification.* Class III.

(c) *Date PMA or notice of completion of a PDP is required.* As of May 28, 1976, an approval under section 515 of the act is required before this device may be commercially distributed. See § 888.3.

§ 888.3020 Intramedullary fixation rod.

(a) *Identification.* An intramedullary fixation rod is a device intended to be implanted that consists of a rod made of alloys such as cobalt-chromium-molybdenum and stainless steel. It is inserted into the medullary (bone marrow) canal of long bones for the fixation of fractures.

(b) *Classification.* Class II.

§ 888.3025 Passive tendon prosthesis.

(a) *Identification.* A passive tendon prosthesis is a device intended to be implanted made of silicon elastomer or a polyester reinforced medical grade silicone elastomer intended for use in the surgical reconstruction of a flexor tendon of the hand. The device is implanted for a period of 2 to 6 months to aid growth of a new tendon sheath. The device is not intended as a permanent implant nor to function as a replacement for the ligament or tendon nor to function as a scaffold for soft tissue ingrowth.

(b) *Classification.* Class II.

§ 888.3027 Polymethylmethacrylate (PMMA) bone cement.

(a) *Identification.* Polymethylmethacrylate (PMMA) bone cement is a device intended to be implanted that is made from methylmethacrylate, polymethylmethacrylate, esters of methacrylic acid, or copolymers containing polymethylmethacrylate and polystyrene. The device is intended for use in arthroplastic procedures of the hip, knee, and other joints for the fixation of polymer or metallic prosthetic implants to living bone.

(b) *Classification.* Class II (special controls). The special control for this device is the FDA guidance document entitled "Class II Special Controls Guidance Document: Polymethylmethacrylate (PMMA) Bone Cement."

[67 FR 46855, July 17, 2002]

§ 888.3030 Single/multiple component metallic bone fixation appliances and accessories.

(a) *Identification.* Single/multiple component metallic bone fixation appliances and accessories are devices intended to be implanted consisting of one or more metallic components and their metallic fasteners. The devices contain a plate, a nail/plate combination, or a blade/plate combination that are made of alloys, such as cobalt-chromium-molybdenum, stainless steel, and titanium, that are intended to be held in position with fasteners, such as screws and nails, or bolts, nuts, and washers. These devices are used for fixation of fractures of the proximal or distal end of long bones, such as intracapsular, intertrochanteric, intercervical, supracondylar, or condylar fractures of the femur; for fusion of a joint; or for surgical procedures that involve cutting a bone. The devices may be implanted or attached through the skin so that a pulling force (traction) may be applied to the skeletal system.

(b) *Classification.* Class II.

§ 888.3040 Smooth or threaded metallic bone fixation fastener.

(a) *Identification.* A smooth or threaded metallic bone fixation fastener is a device intended to be implanted that consists of a stiff wire segment or rod made of alloys, such as cobalt-chromium-molybdenum and stainless steel, and that may be smooth on the outside, fully or partially threaded, straight or U-shaped; and may be either blunt pointed, sharp pointed, or have a formed, slotted head on the end. It may be used for fixation of bone fractures, for bone reconstructions, as a guide pin for insertion of other implants, or it may be implanted through the skin so that a pulling force (traction) may be applied to the skeletal system.

(b) *Classification.* Class II.

§ 888.3045 Resorbable calcium salt bone void filler device.

(a) *Identification.* A resorbable calcium salt bone void filler device is a resorbable implant intended to fill bony voids or gaps of the extremities, spine, and pelvis that are caused by trauma or surgery and are not intrinsic to the stability of the bony structure.

(b) *Classification.* Class II (special controls). The special control for this device is the FDA guidance document entitled "Class II Special Controls Guidance: Resorbable Calcium Salt Bone Void Filler Device; Guidance for Industry and FDA." See § 888.1(e) of this chapter for the availability of this guidance.

[68 FR 32636, June 2, 2003]

§ 888.3050 Spinal interlaminal fixation orthosis.

(a) *Identification.* A spinal interlaminal fixation orthosis is a device intended to be implanted made of an alloy, such as stainless steel, that consists of various hooks and a posteriorly placed compression or distraction rod. The device is implanted, usually across three adjacent vertebrae, to straighten and immobilize the spine to allow bone grafts to unite and fuse the vertebrae together. The device is used primarily in the treatment of scoliosis (a lateral curvature of the spine), but it also may be used in the treatment of fracture or dislocation of the spine, grades 3 and 4 of spondylolisthesis (a dislocation of the spinal column), and lower back syndrome.

(b) *Classification.* Class II.

§ 888.3060 Spinal intervertebral body fixation orthosis.

(a) *Identification.* A spinal intervertebral body fixation orthosis is a device intended to be implanted made of titanium. It consists of various vertebral plates that are punched into each of a series of vertebral bodies. An eye-type screw is inserted in a hole in the center of each of the plates. A braided cable is threaded through each eye-type screw. The cable is tightened with a tension device and it is fastened or crimped at each eye-type screw. The device is used to apply force to a series of vertebrae to correct "sway back," scoliosis (lateral curvature of the spine), or other conditions.

(b) *Classification.* Class II.

§ 888.3070 Pedicle screw spinal system.

(a) *Identification.* Pedicle screw spinal systems are multiple component devices, made from a variety of materials, including alloys such as 316L stainless steel, 316LVM stainless steel, 22Cr-13Ni-5Mn stainless steel, Ti-6Al-4V, and unalloyed titanium, that allow the surgeon to build an implant system to fit the patient's anatomical and physiological requirements. Such a spinal implant assembly consists of a combination of anchors (e.g., bolts, hooks, and/or screws); interconnection mechanisms incorporating nuts, screws, sleeves, or bolts; longitudinal members (e.g., plates, rods, and/or plate/rod combinations); and/or transverse connectors.

(b) *Classification.* (1) Class II (special controls), when intended to provide immobilization and stabilization of spinal segments in skeletally mature patients as an adjunct to fusion in the treatment of the following acute and chronic instabilities or deformities of the thoracic, lumbar, and sacral spine: severe spondylolisthesis (grades 3 and 4) of the L5–S1 vertebra; degenerative spondylolisthesis with objective evidence of neurologic impairment; fracture; dislocation; scoliosis; kyphosis; spinal tumor; and failed previous fusion (pseudarthrosis). These pedicle screw spinal systems must comply with the following special controls:

(i) Compliance with material standards;

(ii) Compliance with mechanical testing standards;

(iii) Compliance with biocompatibility standards; and

(iv) Labeling that contains these two statements in addition to other appropriate labeling information:

"Warning: The safety and effectiveness of pedicle screw spinal systems have been established only for spinal conditions with significant mechanical instability or deformity requiring fusion with instrumentation. These conditions are significant mechanical instability or deformity of the thoracic, lumbar, and sacral spine secondary to severe spondylolisthesis (grades 3 and 4) of the L5–S1 vertebra, degenerative spondylolisthesis with objective evidence of neurologic impairment, fracture, dislocation, scoliosis, kyphosis, spinal tumor, and failed previous fusion (pseudarthrosis). The safety and effectiveness of these devices for any other conditions are unknown."

"Precaution: The implantation of pedicle screw spinal systems should be performed only by experienced spinal surgeons with specific training in the use of this pedicle screw spinal system because this is a technically demanding procedure presenting a risk of serious injury to the patient."

(2) Class III (premarket approval), when intended to provide immobilization and stabilization of spinal segments in the thoracic, lumbar, and sacral spine as an adjunct to fusion in the treatment of degenerative disc disease and spondylolisthesis other than either severe spondylolisthesis (grades 3 and 4) at L5–S1 or degenerative spondylolisthesis with objective evidence of neurologic impairment.

(c) *Date PMA or notice of completion of a PDP is required.* No effective date has been established of the requirement for premarket approval for the devices described in paragraph (b)(2) of this section. See § 888.3.

[66 FR 28053, May 22, 2001]

§ 888.3100 Ankle joint metal/composite semi-constrained cemented prosthesis.

(a) *Identification.* An ankle joint metal/composite semi-constrained cemented prosthesis is a device intended to be implanted to replace an ankle joint. The device limits translation and rotation: in one or more planes via the geometry of its articulating surfaces. It has no linkage across-the-joint. This generic type of device includes prostheses that consist of a talar resurfacing component made of alloys, such as cobalt-chromium-molybdenum, and a tibial resurfacing component fabricated from ultra-high molecular weight polyethylene with carbon fibers composite, and is limited to those prostheses intended for use with bone cement (§ 888.3027).

(b) *Classification.* Class II.

§ 888.3110 Ankle joint metal/polymer semi-constrained cemented prosthesis.

(a) *Identification.* An ankle joint metal/polymer semi-constrained cemented prosthesis is a device intended to be implanted to replace an ankle joint. The device limits translation and rotation in one or more planes via the geometry of its articulating surfaces and has no linkage across-the-joint. This generic type of device includes prostheses that have a talar resurfacing component made of alloys, such as cobalt-chromium-molybdenum, and a tibial resurfacing component made of ultra-high molecular weight polyethylene and is limited to those prostheses intended for use with bone cement (§ 888.3027).

(b) *Classification.* Class II.

§ 888.3120 Ankle joint metal/polymer non-constrained cemented prosthesis.

(a) *Identification.* An ankle joint metal/polymer non-constrained cemented prosthesis is a device intended to be implanted to replace an ankle joint. The device limits minimally (less than normal anatomic constraints) translation in one or more planes. It has no linkage across-the-joint. This generic type of device includes prostheses that have a tibial component made of alloys, such as cobalt-chromium-molybdenum, and a talar component made of ultra-high molecular weight polyethylene, and is limited to those prostheses intended for use with bone cement (§ 888.3027).

(b) *Classification.* Class III.

(c) *Date PMA or notice of completion of a PDP is required.* A PMA or a notice of completion of a PDP is required to be filed with the Food and Drug Administration on or before December 26, 1996 for any ankle joint metal/polymer non-constrained cemented prosthesis that was in commercial distribution before May 28, 1976, or that has, on or before December 26, 1996, been found to be substantially equivalent to an ankle joint metal/polymer non-constrained cemented prosthesis that was in commercial distribution before May 28, 1976. Any other ankle joint metal/polymer non-constrained cemented prosthesis shall have an approved PMA or a declared completed PDP in effect before being placed in commercial distribution.

[52 FR 33702, Sept. 4, 1987, as amended at 61 FR 50709, Sept. 27, 1996]

§ 888.3150 Elbow joint metal/polymer constrained cemented prosthesis.

(a) *Identification.* An elbow joint metal/polymer constrained cemented prosthesis is a device intended to be implanted to replace an elbow joint. It is made of alloys, such as cobalt-chromium-molybdenum, or of these alloys and of an ultra-high molecular weight polyethylene bushing. The device prevents dislocation in more than one anatomic plane and consists of two components that are linked together. This generic type of device is limited to those prostheses intended for use with bone cement (§ 888.3027).

(b) *Classification.* Class II. The special controls for this device are:

(1) FDA's:

(i) "Use of International Standard ISO 10993 'Biological Evaluation of Medical Devices—Part I: Evaluation and Testing,' "

(ii) "510(k) Sterility Review Guidance of 2/12/90 (K90–1),"

(iii) "Guidance Document for Testing Orthopedic Implants with Modified Metallic Surfaces Apposing Bone or Bone Cement,"

(iv) "Guidance Document for the Preparation of Premarket Notification (510(k)) Application for Orthopedic Devices,"

(v) "Guidance Document for Testing Non-articulating, 'Mechanically Locked' Modular Implant Components,"

(2) International Organization for Standardization's (ISO):

(i) ISO 5832–3:1996 "Implants for Surgery—Metallic Materials—Part 3:

Wrought Titanium 6-Aluminum 4-Vandium Alloy,''

(ii) ISO 5832–4:1996 ''Implants for Surgery—Metallic Materials—Part 4: Cobalt-Chromium-Molybdenum Casting Alloy,''

(iii) ISO 5832–12:1996 ''Implants for Surgery—Metallic Materials—Part 12: Wrought Cobalt-Chromium-Molybdenum Alloy,''

(iv) ISO 5833:1992 ''Implants for Surgery—Acrylic Resin Cements,''

(v) ISO 5834–2:1998 ''Implants for Surgery—Ultra High Molecular Weight Polyethylene—Part 2: Moulded Forms,''

(vi) ISO 6018:1987 ''Orthopaedic Implants—General Requirements for Marking, Packaging, and Labeling,''

(vii) ISO 9001:1994 ''Quality Systems—Model for Quality Assurance in Design/Development, Production, Installation, and Servicing,'' and

(viii) ISO 14630:1997 ''Non-active Surgical Implants—General Requirements,''

(3) American Society for Testing and Materials':

(i) F 75–92 ''Specification for Cast Cobalt-28 Chromium-6 Molybdenum Alloy for Surgical Implant Material,''

(ii) F 648–98 ''Specification for Ultra-High-Molecular-Weight Polyethylene Powder and Fabricated Form for Surgical Implants,''

(iii) F 799–96 ''Specification for Cobalt-28 Chromium-6 Molybdenum Alloy Forgings for Surgical Implants,''

(iv) F 981–93 ''Practice for Assessment of Compatibility of Biomaterials (Nonporous) for Surgical Implant with Respect to Effect of Material on Muscle and Bone,''

(v) F 1044–95 ''Test Method for Shear Testing of Porous Metal Coatings,''

(vi) F 1108–97 ''Specification for Titanium-6 Aluminum-4 Vanadium Alloy Castings for Surgical Implants,''

(vii) F 1147–95 ''Test Method for Tension Testing of Porous Metal Coatings, '' and

(viii) F 1537–94 ''Specification for Wrought Cobalt-28 Chromium-6 Molybdenum Alloy for Surgical Implants.''

[65 FR 17147, Mar. 31, 2000]

§ 888.3160 Elbow joint metal/polymer semi-constrained cemented prosthesis.

(a) *Identification.* An elbow joint metal/polymer semi-constrained cemented prosthesis is a device intended to be implanted to replace an elbow joint. The device limits translation and rotation in one or more planes via the geometry of its articulating surfaces. It has no linkage across-the-joint. This generic type of device includes prostheses that consist of a humeral resurfacing component made of alloys, such as cobalt-chromium-molybdenum, and a radial resurfacing component made of ultra-high molecular weight polyethylene. This generic type of device is limited to those prostheses intended for use with bone cement (§ 888.3027).

(b) *Classification.* Class II.

§ 888.3170 Elbow joint radial (hemi-elbow) polymer prosthesis.

(a) *Identification.* An elbow joint radial (hemi-elbow) polymer prosthesis is a device intended to be implanted made of medical grade silicone elastomer used to replace the proximal end of the radius.

(b) *Classification.* Class II.

§ 888.3180 Elbow joint humeral (hemi-elbow) metallic uncemented prosthesis.

(a) *Identification.* An elbow joint humeral (hemi-elbow) metallic uncemented prosthesis is a device intended to be implanted made of alloys, such as cobalt-chromium-molybdenum, that is used to replace the distal end of the humerus formed by the trochlea humeri and the capitulum humeri. The generic type of device is limited to prostheses intended for use without bone cement (§ 888.3027).

(b) *Classification.* Class III.

(c) *Date PMA or notice of completion of a PDP is required.* A PMA or a notice of completion of a PDP is required to be filed with the Food and Drug Administration on or before December 26, 1996 for any elbow joint humeral (hemi-elbow) metallic uncemented prosthesis that was in commercial distribution before May 28, 1976, or that has, on or before December 26, 1996 been found to be substantially equivalent to an elbow joint humeral (hemi-elbow) metallic uncemented prosthesis that was in commercial distribution before May 28, 1976. Any other elbow joint humeral (hemi-elbow) metallic uncemented prosthesis shall have an approved PMA or a declared completed PDP in effect before being placed in commercial distribution.

[52 FR 33702, Sept. 4, 1987, as amended at 61 FR 50709, Sept. 27, 1996]

§ 888.3200 Finger joint metal/metal constrained uncemented prosthesis.

(a) *Identification.* A finger joint metal/metal constrained uncemented prosthesis is a device intended to be implanted to replace a metacarpophalangeal or proximal interphalangeal (finger) joint. The device prevents dislocation in more than one anatomic plane and consists of two components which are linked together. This generic type of device includes prostheses made of alloys, such as cobalt-chromium-molybdenum, or protheses made from alloys and ultra-high molecular weight polyethylene. This generic type of device is limited to prostheses intended for use without bone cement (§ 888.3027).

(b) *Classification.* Class III.

(c) *Date PMA or notice of completion of a PDP is required.* A PMA or a notice of completion of a PDP is required to be filed with the Food and Drug Administration on or before December 26, 1996 for any finger joint metal/metal constrained uncemented prosthesis that was in commercial distribution before May 28, 1976, or that has, on or before December 26, 1996 been found to be substantially equivalent to a finger joint metal/metal constrained uncemented prosthesis that was in commercial distribution before May 28, 1976. Any other finger joint metal/metal constrained uncemented prosthesis shall have an approved PMA or a declared completed PDP in effect before being placed in commercial distribution.

[52 FR 33702, Sept. 4, 1987, as amended at 61 FR 50709, Sept. 27, 1996]

§ 888.3210 Finger joint metal/metal constrained cemented prosthesis.

(a) *Identification.* A finger joint metal/metal constrained cemented prosthesis is a device intended to be implanted to replace a metacarpophalangeal (finger) joint. This device prevents dislocation in more than one anatomic plane and has components which are linked together. This generic type of device includes prostheses that are made of alloys, such as cobalt-chromium-molybdenum, and is limited to those prostheses intended for use with bone cement (§ 888.3027).

(b) *Classification.* Class III.

(c) *Date PMA or notice of completion of a PDP is required.* A PMA or a notice of completion of a PDP is required to be filed with the Food and Drug Administration on or before December 26, 1996 for any finger joint metal/metal constrained cemented prosthesis that was in commercial distribution before May 28, 1976, or that has, on or before December 26, 1996 been found to be substantially equivalent to a finger joint metal/metal constrained cemented prosthesis that was in commercial distribution before May 28, 1976. Any other finger joint metal/metal constrained cemented prosthesis shall have an approved PMA or a declared completed PDP in effect before being placed in commercial distribution.

[52 FR 33702, Sept. 4, 1987, as amended at 61 FR 50709, Sept. 27, 1996]

§ 888.3220 Finger joint metal/polymer constrained cemented prosthesis.

(a) *Identification.* A finger joint metal/polymer constrained cemented prosthesis is a device intended to be implanted to replace a metacarpophalangeal or proximal interphalangeal (finger) joint. The device prevents dislocation in more than one anatomic plane, and consists of two components which are linked together. This generic type of device includes prostheses that are made of alloys, such as cobalt-chromium-molybdenum, and ultra-high molecular weight polyethylene, and is limited to those prostheses intended for use with bone cement (§ 888.3027).

(b) *Classification.* Class III.

(c) *Date PMA or notice of completion of a PDP is required.* A PMA or a notice of completion of a PDP is required to be filed with the Food and Drug Administration on or before December 26, 1996 for any finger joint metal/polymer constrained cemented prosthesis that was in commercial distribution before May 28, 1976, or that has, on or before December 26, 1996 been found to be substantially equivalent to a finger joint metal/polymer constrained cemented prosthesis that was in commercial distribution before May 28, 1976. Any other finger joint metal/polymer constrained cemented prosthesis shall have an approved PMA or a declared completed PDP in effect before being placed in commercial distribution.

[52 FR 33702, Sept. 4, 1987, as amended at 61 FR 50709, Sept. 27, 1996]

§ 888.3230 Finger joint polymer constrained prosthesis.

(a) *Identification.* A finger joint polymer constrained prosthesis is a device intended to be implanted to replace a metacarpophalangeal or proximal interphalangeal (finger) joint. This generic type of device includes prostheses that consist of a single flexible across-the-joint component made from either a silicone elastomer or a combination pf polypropylene and polyester material. The flexible across-the-joint component may be covered with a silicone rubber sleeve.

(b) *Classification.* Class II.

§ 888.3300 Hip joint metal constrained cemented or uncemented prosthesis.

(a) *Identification.* A hip joint metal constrained cemented or uncemented prosthesis is a device intended to be implanted to replace a hip joint. The device prevents dislocation in more than one anatomic plane and has components that are linked together. This generic type of device includes prostheses that have components made of alloys, such as cobalt-chromium-molybdenum, and is intended for use with or without bone cement (§ 888.3027). This device is not intended for biological fixation.

(b) *Classification.* Class III.

(c) *Date PMA or notice of completion of a PDP is required.* A PMA or a notice of completion of a PDP is required to be filed with the Food and Drug Administration on or before December 26, 1996 for any hip joint metal constrained cemented or uncemented prosthesis that was in commercial distribution before May 28, 1976, or that has, on or before December 26, 1996 been found to be substantially equivalent to a hip joint metal constrained cemented or uncemented prosthesis that was in commercial distribution before May 28, 1976. Any other hip joint metal constrained cemented or uncemented prosthesis shall have an approved PMA or a declared completed PDP in effect before being placed in commercial distribution.

[52 FR 33702, Sept. 4, 1987, as amended at 61 FR 50709, Sept. 27, 1996]

§ 888.3310 Hip joint metal/polymer constrained cemented or uncemented prosthesis.

(a) *Identification.* A hip joint metal/polymer constrained cemented or uncemented prosthesis is a device intended to be implanted to replace a hip joint. The device prevents dislocation in more than one anatomic plane and has components that are linked together. This generic type of device includes prostheses that have a femoral component made of alloys, such as cobalt-chromium-molybdenum, and an acetabular component made of ultra-high-molecular-weight polyethylene with or without a metal shell, made of alloys, such as cobalt-chromium-molybdenum and titanium alloys. This generic type of device is intended for use with or without bone cement (§ 888.3027).

(b) *Classification.* Class II (special controls). The special control for this device is the FDA guidance document entitled "Class II Special Controls Guidance: Hip Joint Metal/Polymer Constrained Cemented or Uncemented Prosthesis."

[67 FR 21173, Apr. 30, 2002]

§ 888.3320 Hip joint metal/metal semi-constrained, with a cemented acetabular component, prosthesis.

(a) *Identification.* A hip joint metal/metal semi-constrained, with a cemented acetabular component, prosthesis is a two-part device intended to be implanted to replace a hip joint. The device limits translation and rotation in one or more planes via the geometry of its articulating surfaces. It has no linkage across-the-joint. This generic type of device includes prostheses that consist of a femoral and an acetabular component, both made of alloys, such as cobalt-chromium-molybdenum. This generic type of device is limited to those prostheses intended for use with bone cement (§ 888.3027).

(b) *Classification.* Class III.

(c) *Date PMA or notice of completion of a PDP is required.* No effective date has been established of the requirement for premarket approval. See § 888.3.

§ 888.3330 Hip joint metal/metal semi-constrained, with an uncemented acetabular component, prosthesis.

(a) *Identification.* A hip joint metal/metal semi-constrained, with an uncemented acetabular component, prosthesis is a two-part device intended to be implanted to replace a hip joint. The device limits translation and rotation in one or more planes via the geometry of its articulating surfaces. It has no linkage across-the-joint. This generic type of device includes prostheses that consist of a femoral and an acetabular component, both made of alloys, such as cobalt-chromium-molybdenum. The femoral component is intended to be fixed with bone cement. The acetabular component is intended for use without bone cement (§ 888.3027).

(b) *Classification.* Class III.

(c) *Date PMA or notice of completion of a PDP is required.* No effective date has been established of the requirement for premarket approval. See § 888.3.

§ 888.3340 Hip joint metal/composite semi-constrained cemented prosthesis.

(a) *Identification.* A hip joint metal/composite semi-constrained cemented prosthesis is a two-part device intended to be implanted to replace a hip joint. The device limits translation and rotation in one or more planes via the geometry of its articulating surfaces. It has no linkage across-the-joint. This generic type of device includes prostheses that consist of a femoral component made of alloys, such as cobalt-chromium-molybdenum, and an acetabular component made of ultra-high molecular weight polyethylene with carbon fibers composite. Both components are intended for use with bone cement (§ 888.3027).

(b) *Classification.* Class II.

§ 888.3350 Hip joint metal/polymer semi-constrained cemented prosthesis.

(a) *Identification.* A hip joint metal/polymer semi-constrained cemented prosthesis is a device intended to be implanted to replace a hip joint. The device limits translation and rotation in one or more planes via the geometry of its articulating surfaces. It has no linkage across-the-joint. This generic type of device includes prostheses that have a femoral component made of alloys, such as cobalt-chromium-molybdenum, and an acetabular resurfacing component made of ultra-high molecular weight polyethylene and is limited to those prostheses intended for use with bone cement (§ 888.3027).

(b) *Classification.* Class II.

§ 888.3353 Hip joint metal/ceramic/polymer semi-constrained cemented or nonporous uncemented prosthesis.

(a) *Identification.* A hip joint metal/ceramic/polymer semi-constrained cemented or nonporous uncemented prosthesis is a device intended to be implanted to replace a hip joint. This device limits translation and rotation in one or more planes via the geometry of its articulating surfaces. It has no linkage across-the-joint. The two-part femoral component consists of a femoral stem made of alloys to be fixed in the intramedullary canal of the femur by impaction with or without use of bone cement. The proximal end of the femoral stem is tapered with a surface that ensures positive locking with the spherical ceramic (aluminium oxide, $A1_20_3$) head of the femoral component. The acetabular component is made of ultra-high molecular weight polyethylene or ultra-high molecular weight polyethylene reinforced with nonporous metal alloys, and used with or without bone cement.

(b) *Classification.* Class II.

[54 FR 48239, Nov. 22, 1989; 54 FR 51342, Dec. 14, 1989]

§ 888.3358 Hip joint metal/polymer/metal semi-constrained porous-coated uncemented prosthesis.

(a) *Identification.* A hip joint metal/polymer/metal semi-constrained porous-coated uncemented prosthesis is a device intended to be implanted to replace a hip joint. The device limits translation and rotation in one or more planes via the geometry of its articulating surfaces. It has no linkage across the joint. This generic type of device has a femoral component made of a cobalt-chromium-molybdenum (Co-Cr-Mo) alloy or a titanium-aluminum-vanadium (Ti-6Al-4V) alloy and an acetabular component composed of an ultra-high molecular weight polyethylene articulating bearing surface fixed in a metal shell made of Co-Cr-Mo or Ti-6Al-4V. The femoral stem and acetabular shell have a porous coating made of, in the case of Co-Cr-Mo substrates, beads of the same alloy, and in the case of Ti-6Al-4V substrates, fibers of commercially pure titanium or Ti-6Al-4V alloy. The porous coating has a volume porosity between 30 and 70 percent, an average pore size between 100 and 1,000 microns, interconnecting porosity, and a porous coating thickness between 500 and 1,500 microns. The generic type of device has a design to achieve biological fixation to bone without the use of bone cement.

(b) *Classification.* Class II.

[58 FR 3228, Jan. 8, 1993]

§ 888.3360 Hip joint femoral (hemi-hip) metallic cemented or uncemented prosthesis.

(a) *Identification.* A hip joint femoral (hemi-hip) metallic cemented or uncemented prosthesis is a device intended to be implanted to replace a portion of the hip joint. This generic type of device includes prostheses that have a femoral component made of alloys, such as cobalt-chromium-molybdenum. This generic type of device includes designs which are intended to be fixed to the bone with bone cement (§ 888.3027) as well as designs which have large window-like holes in the stem of the device and which are intended for use without bone cement. However, in these latter designs, fixation of the device is not achieved by means of bone ingrowth.

(b) *Classification.* Class II.

§ 888.3370 Hip joint (hemi-hip) acetabular metal cemented prosthesis.

(a) *Identification.* A hip joint (hemi-hip) acetabular metal cemented prosthesis is a device intended to be implanted to replace a portion of the hip joint. This generic type of device includes prostheses that have an acetabular component made of alloys, such as cobalt-chromium-molybdenum. This generic type of device is limited to those prostheses intended for use with bone cement (§ 888.3027).

(b) *Classification.* Class III.

(c) *Date PMA or notice of completion of a PDP is required.* A PMA or a notice of completion of a PDP is required to be filed with the Food and Drug Administration on or before December 26, 1996 for any hip joint (hemi-hip) acetabular metal cemented prosthesis that was in commercial distribution before May 28, 1976, or that has, on or before December 26, 1996 been found to be substantially equivalent to a hip joint (hemi-hip) acetabular metal cemented prosthesis that was in commercial distribution before May 28, 1976. Any other hip joint metal (hemi-hip) acetabular metal cemented prosthesis shall have an approved PMA or a declared completed PDP in effect before being placed in commercial distribution.

[52 FR 33702, Sept. 4, 1987, as amended at 61 FR 50710, Sept. 27, 1996]

§ 888.3380 Hip joint femoral (hemi-hip) trunnion-bearing metal/polyacetal cemented prosthesis.

(a) *Identification.* A hip joint femoral (hemi-hip) trunnion-bearing metal/polyacetal cemented prosthesis is a two-part device intended to be implanted to replace the head and neck of the femur. This generic type of device includes prostheses that consist of a metallic stem made of alloys, such as cobalt-chromium-molybdenum, with an integrated cylindrical trunnion bearing at the upper end of the stem that fits into a recess in the head of the device. The head of the device is made of polyacetal (polyoxymethylene) and it is covered by a metallic alloy, such as cobalt-chromium-molybdenum. The trunnion bearing allows the head of the device to rotate on its stem. The prosthesis is intended for use with bone cement (§ 888.3027).

(b) *Classification.* Class III.

(c) *Date PMA or notice of completion of a PDP is required.* A PMA or a notice of completion of a PDP is required to be filed with the Food and Drug Administration on or before December 26, 1996 for any hip joint femoral (hemi-hip) trunnion-bearing metal/polyacetal cemented prosthesis that was in commercial distribution before May 28, 1976, or that has, on or before December 26, 1996 been found to be substantially equivalent to a hip joint femoral (hemi-hip) trunnion-bearing metal/polyacetal cemented prosthesis that was in commercial distribution before May 28, 1976. Any other hip joint femoral (hemi-hip) trunnion-bearing metal/polyacetal cemented prosthesis shall have an approved PMA or a declared completed PDP in effect before being placed in commercial distribution.

[52 FR 33702, Sept. 4, 1987, as amended at 61 FR 50710, Sept. 27, 1996]

§ 888.3390 Hip joint femoral (hemi-hip) metal/polymer cemented or uncemented prosthesis.

(a) *Identification.* A hip joint femoral (hemi-hip) metal/polymer cemented or uncemented prosthesis is a two-part device intended to be implanted to replace the head and neck of the femur. This generic type of device includes prostheses that have a femoral component made of alloys, such as cobalt-chromium-molybdenum, and a snap-fit acetabular component made of an alloy, such as cobalt-chromium-molybdenum, and ultra-high molecular weight polyethylene. This generic type of device may be fixed to the bone with bone cement (§ 888.3027) or implanted by impaction.

(b) *Classification.* Class II.

§ 888.3400 Hip joint femoral (hemi-hip) metallic resurfacing prosthesis.

(a) *Identification.* A hip joint femoral (hemi-hip) metallic resurfacing prosthesis is a device intended to be implanted to replace a portion of the hip joint. This generic type of device includes prostheses that have a femoral resurfacing component made of alloys, such as cobalt-chromium-molybdenum.

(b) *Classification.* Class II.

§ 888.3410 Hip joint metal/polymer or ceramic/polymer semiconstrained resurfacing cemented prosthesis.

(a) *Identification.* A hip joint metal/polymer or ceramic/polymer semi-constrained resurfacing cemented prosthesis is a two-part device intended to be implanted to replace the articulating surfaces of the hip while preserving the femoral head and neck. The device limits translation and rotation in one or more planes via the geometry of its articulating surfaces. It has no linkage across the joint. This generic type of device includes prostheses that consist of a femoral cap component made of a metal alloy, such as cobalt-chromium-molybdenum, or a ceramic material, that is placed over a surgically prepared femoral head, and an acetabular resurfacing polymer component. Both components are intended for use with bone cement (§ 888.3027).

(b) *Classification.* Class III.

(c) *Date PMA or notice of completion of a PDP is required.* A PMA or a notice of completion of a PDP is required to be filed with the Food and Drug Administration on or before January 3, 2005, for any hip joint metal/polymer or ceramic/polymer semiconstrained resurfacing cemented prosthesis that was in commercial distribution before May 28, 1976, or that has, on or before January 3, 2005, been found to be substantially equivalent to a hip joint metal/polymer or ceramic/polymer semiconstrained resurfacing cemented prosthesis that was in commercial distribution before May 28, 1976. Any other hip joint metal/polymer or ceramic/polymer semiconstrained resurfacing cemented prosthesis must have an approved PMA or a declared completed PDP in effect before being placed in commercial distribution.

[69 FR 59134, Oct. 4, 2004]

§ 888.3480 Knee joint femorotibial metallic constrained cemented prosthesis.

(a) *Identification.* A knee joint femorotibial metallic constrained cemented prosthesis is a device intended to be implanted to replace part of a knee joint. The device prevents dislocation in more than one anatomic plane and has components that are linked together. The only knee joint movement allowed by the device is in the sagittal plane. This generic type of device includes prostheses that have an intramedullary stem at both the proximal and distal locations. The upper and lower components may be joined either by a solid bolt or pin, an internally threaded bolt with locking screw, or a bolt retained by circlip. The components of the device are made of alloys, such as cobalt-chromium-molybdenum. The stems of the device may be perforated, but are intended for use with bone cement (§ 888.3027).

(b) *Classification.* Class III.

(c) *Date PMA or notice of completion of a PDP is required.* A PMA or a notice of completion of a PDP is required to be filed with the Food and Drug Administration on or before December 26, 1996 for any knee joint femorotibial metallic constrained cemented prosthesis that was in commercial distribution before May 28, 1976, or that has, on or before December 26, 1996 been found to be substantially equivalent to a knee joint femorotibial metallic constrained cemented prosthesis that was in commercial distribution before May 28, 1976. Any other knee joint femorotibial metallic constrained cemented prosthesis shall have an approved PMA or a declared completed PDP in effect before being placed in commercial distribution.

[52 FR 33702, Sept. 4, 1987, as amended at 61 FR 50710, Sept. 27, 1996]

§ 888.3490 Knee joint femorotibial metal/composite non-constrained cemented prosthesis.

(a) *Identification.* A knee joint femorotibial metal/composite non-constrained cemented prosthesis is a device intended to be implanted to replace part of a knee joint. The device limits minimally (less than normal anatomic constraints) translation in one or more planes. It has no linkage across-the-joint. This generic type of device includes prostheses that have a femoral condylar resurfacing component or components made of alloys, such as cobalt-chromium-molybdenum, and a tibial condylar component or components made of ultra-high molecular weight polyethylene with carbon fibers composite and are intended for use with bone cement (§ 888.3027).

(b) *Classification.* Class II.

§ 888.3500 Knee joint femorotibial metal/composite semi-constrained cemented prosthesis.

(a) *Identification.* A knee joint femorotibial metal/composite semi-constrained cemented prosthesis is a two-part device intended to be implanted to replace part of a knee joint. The device limits translation and rotation in one or more planes via the geometry of its articulating surfaces. It has no linkage across-the-joint. This generic type of device includes prostheses that have a femoral component made of alloys, such as cobalt-chromium-molybdenum, and a tibial component with the articulating surfaces made of ultra-high molecular weight polyethylene with carbon-fibers composite and is limited to those prostheses intended for use with bone cement (§ 888.3027).

(b) *Classification.* Class II.

§ 888.3510 Knee joint femorotibial metal/polymer constrained cemented prosthesis.

(a) *Identification.* A knee joint femorotibial metal/polymer constrained cemented prosthesis is a device intended to be implanted to replace part of a knee joint. The device limits translation or rotation in one or more planes and has components that are linked together or affined. This generic type of device includes prostheses composed of a ball-and-socket joint located between a stemmed femoral and a stemmed tibial component and a runner and track joint between each pair of femoral and tibial condyles. The ball-and-socket joint is composed of a ball at the head of a column rising from the stemmed tibial component. The ball, the column, the tibial plateau, and the stem for fixation of the tibial component are made of an alloy, such as cobalt-chromium-molybdenum. The ball of the tibial component is held within the socket of the femoral component by the femoral component's flat outer surface. The flat outer surface of the tibial component abuts both a reciprocal flat surface within the cavity of the femoral component and flanges on the femoral component designed to prevent distal displacement. The stem of the femoral component is made of an alloy, such as cobalt-chromium-molybdenum, but the socket of the component is made of ultra-high molecular weight polyethylene. The femoral component has metallic runners which align with the ultra-high molecular weight polyethylene tracks that press-fit into the metallic tibial component. The generic class also includes devices whose upper and lower components are linked with a solid bolt passing through a journal bearing of greater radius, permitting some rotation in the transverse plane, a minimal arc of abduction/adduction. This generic type of device is limited to those prostheses intended for use with bone cement (§ 888.3027).

(b) *Classification.* Class II.

§ 888.3520 Knee joint femorotibial metal/polymer non-constrained cemented prosthesis.

(a) *Identification.* A knee joint femorotibial metal/polymer non-constrained cemented prosthesis is a device intended to be implanted to replace part of a knee joint. The device limits minimally (less than normal anatomic constraints) translation in one or more planes. It has no linkage across-the-joint. This generic type of device includes prostheses that have a femoral condylar resurfacing component or components made of alloys, such as cobalt-chromium-molybdenum, and a tibial component or components made of ultra-high molecular weight polyethylene and are intended for use with bone cement (§ 888.3027).

(b) *Classification.* Class II.

§ 888.3530 Knee joint femorotibial metal/polymer semi-constrained cemented prosthesis.

(a) *Identification.* A knee joint femorotibial metal/polymer semi-constrained cemented prosthesis is a device intended to be implanted to replace part of a knee joint. The device limits translation and rotation in one or more planes via the geometry of its articulating surfaces. It has no linkage across-the-joint. This generic type of device includes prostheses that consist of a femoral component made of alloys, such as cobalt-chromium-molybdenum, and a tibial component made of ultra-high molecular weight polyethylene and is limited to those prostheses intended for use with bone cement (§ 888.3027).

(b) *Classification.* Class II.

§ 888.3535 Knee joint femorotibial (uni-compartmental) metal/polymer porous-coated uncemented prosthesis.

(a) *Identification.* A knee joint femorotibial (uni-compartmental) metal/polymer porous-coated uncemented prosthesis is a device intended to be implanted to replace part of a knee joint. The device limits translation and rotation in one or more planes via the geometry of its articulating surface. It has no linkage across-the-joint. This generic type of device is designed to achieve biological fixation to bone without the use of bone cement. This identification includes fixed-bearing knee prostheses where the ultra-high molecular weight polyethylene tibial bearing is rigidly secured to the metal tibial baseplate.

(b) *Classification.* Class II (special controls). The special control is FDA's guidance: "Class II Special Controls Guidance Document: Knee Joint Patellofemorotibial and Femorotibial Metal/Polymer Porous-Coated Uncemented Prostheses; Guidance for Industry and FDA." See § 888.1 for the availability of this guidance.

[68 FR 14137, Mar. 24, 2003]

§ 888.3540 Knee joint patellofemoral polymer/metal semi-constrained cemented prosthesis.

(a) *Identification.* A knee joint patellofemoral polymer/metal semi-constrained cemented prosthesis is a two-part device intended to be implanted to replace part of a knee joint in the treatment of primary patellofemoral arthritis or chondromalacia. The device limits translation and rotation in one or more planes via the geometry of its articulating surfaces. It has no linkage across-the-joint. This generic type of device includes a component made of alloys, such as cobalt-chromium-molybdenum or austenitic steel, for resurfacing the intercondylar groove (femoral sulcus) on the anterior aspect of the distal femur, and a patellar component made of ultra-high molecular weight polyethylene. This generic type of device is limited to those devices intended for use with bone cement (§ 888.3027). The patellar component is designed to be implanted only with its femoral component.

(b) *Classification.* Class II. The special controls for this device are:

(1) FDA's:

(i) "Use of International Standard ISO 10993 'Biological Evaluation of Medical Devices—Part I: Evaluation and Testing,"

(ii) "510(k) Sterility Review Guidance of 2/12/90 (K90–1),"

(iii) "Guidance Document for Testing Orthopedic Implants with Modified Metallic Surfaces Apposing Bone or Bone Cement,"

(iv) "Guidance Document for the Preparation of Premarket Notification (510(k)) Applications for Orthopedic Devices," and

(v) "Guidance Document for Testing Non-articulating, 'Mechanically Locked' Modular Implant Components," and

(2) International Organization for Standardization's (ISO):

(i) ISO 5832–3:1996 "Implants for Surgery—Metallic Materials—Part 3: Wrought Titanium 6-Aluminum 4-Vandium Alloy,"

(ii) ISO 5832–4:1996 "Implants for Surgery—Metallic Materials—Part 4: Cobalt-Chromium-Molybdenum Casting Alloy,"

(iii) ISO 5832–12:1996 "Implants for Surgery—Metallic Materials—Part 12: Wrought Cobalt-Chromium-Molybdenum Alloy,"

(iv) ISO 5833:1992 "Implants for Surgery—Acrylic Resin Cements,"

(v) ISO 5834–2:1998 "Implants for Surgery—Ultra-high Molecular Weight Polyethylene—Part 2: Moulded Forms,"

(vi) ISO 6018:1987 "Orthopaedic Implants—General Requirements for Marking, Packaging, and Labeling,"

(vii) ISO 7207–2:1998 "Implants for Surgery—Components for Partial and Total Knee Joint Prostheses—Part 2: Articulating Surfaces Made of Metal, Ceramic and Plastic Materials," and

(viii) ISO 9001:1994 "Quality Systems—Model for Quality Assurance in Design/Development, Production, Installation, and Servicing," and

(3) American Society for Testing and Materials':

(i) F 75–92 "Specification for Cast Cobalt-28 Chromium-6 Molybdenum Alloy for Surgical Implant Material,"

(ii) F 648–98 "Specification for Ultra-High-Molecular-Weight Polyethylene Powder and Fabricated Form for Surgical Implants,"

(iii) F 799–96 "Specification for Cobalt-28 Chromium-6 Molybdenum Alloy Forgings for Surgical Implants,"

(iv) F 1044–95 "Test Method for Shear Testing of Porous Metal Coatings,"

(v) F 1108–97 "Titanium-6 Aluminum-4 Vanadium Alloy Castings for Surgical Implants,"

(vi) F 1147–95 "Test Method for Tension Testing of Porous Metal Coatings,"

(vii) F 1537–94 "Specification for Wrought Cobalt-28 Chromium-6 Molybdenum Alloy for Surgical Implants," and

(viii) F 1672–95 "Specification for Resurfacing Patellar Prosthesis."

[52 FR 33702, Sept. 4, 1987, as amended at 61 FR 50710, Sept. 27, 1996; 65 FR 17147, Mar. 31, 2000]

§ 888.3550 Knee joint patellofemorotibial polymer/metal/metal constrained cemented prosthesis.

(a) *Identification.* A knee joint patellofemorotibial polymer/metal/metal constrained cemented prosthesis is a device intended to be implanted to replace a knee joint. The device prevents dislocation in more than one anatomic plane and has components that are linked together. This generic type of device includes prostheses that have a femoral component, a tibial component, a cylindrical bolt and accompanying locking hardware that are all made of alloys, such as cobalt-chromium-molybdenum, and a retropatellar resurfacing component made of ultra-high molecular weight polyethylene. The retropatellar surfacing component may be attached to the resected patella either with a metallic screw or bone cement. All stemmed metallic components within this generic type are intended for use with bone cement (§ 888.3027).

(b) *Classification.* Class III.

(c) *Date PMA or notice of completion of a PDP is required.* A PMA or a notice of completion of a PDP is required to be filed with the Food and Drug Administration on or before December 26, 1996 for any knee joint patellofemorotibial polymer/metal/metal constrained cemented prosthesis that was in commercial distribution before May 28, 1976, or that has, on or before December 26, 1996 been found to be substantially equivalent to a knee joint patellofemorotibial polymer/metal/metal constrained cemented prosthesis that was in commercial distribution before May 28, 1976. Any other knee joint patellofemorotibial polymer/metal/metal constrained cemented prosthesis shall have an approved PMA or a declared completed PDP in effect before being placed in commercial distribution.

[52 FR 33702, Sept. 4, 1987, as amended at 61 FR 50710, Sept. 27, 1996]

§ 888.3560 Knee joint patellofemorotibial polymer/metal/polymer semi-constrained cemented prosthesis.

(a) *Identification.* A knee joint patellofemorotibial polymer/metal/polymer semi-constrained cemented prosthesis is a device intended to be implanted to replace a knee joint. The device limits translation and rotation in one or more planes via the geometry of its articulating surfaces. It has no linkage across-the-joint. This generic type of device includes prostheses that have a femoral component made of alloys, such as cobalt-chromium-molybdenum, and a tibial component or components and a retropatellar resurfacing component made of ultra-high molecular weight polyethylene. This generic type of device is limited to those prostheses intended for use with bone cement (§ 888.3027).

(b) *Classification.* Class II.

§ 888.3565 Knee joint patellofemorotibial metal/polymer porous-coated uncemented prosthesis.

(a) *Identification.* A knee joint patellofemorotibial metal/polymer porous-coated uncemented prosthesis is a device intended to be implanted to replace a knee joint. The device limits translation and rotation in one or more planes via the geometry of its articulating surfaces. It has no linkage across-the-joint. This generic type of device is designed to achieve biological fixation to bone without the use of bone cement. This identification includes fixed-bearing knee prostheses where the ultra high molecular weight polyethylene tibial bearing is rigidly secured to the metal tibial base plate.

(b) *Classification.* Class II (special controls). The special control is FDA's guidance: "Class II Special Controls Guidance Document: Knee Joint Patellofemorotibial and Femorotibial Metal/Polymer Porous-Coated Uncemented Prostheses; Guidance for Industry and FDA." See § 888.1 for the availability of this guidance.

[68 FR 14137, Mar. 24, 2003]

§ 888.3570 Knee joint femoral (hemi-knee) metallic uncemented prosthesis.

(a) *Identification.* A knee joint femoral (hemi-knee) metallic uncemented prosthesis is a device made of alloys, such as cobalt-chromium-molybdenum, intended to be implanted to replace part of a knee joint. The device limits translation and rotation in one or more planes via the geometry of its articulating surfaces. It has no linkage across-the-joint. This generic type of device includes prostheses that consist of a femoral component with or without protuberance(s) for the enhancement of fixation and is limited to those prostheses intended for use without bone cement (§ 888.3027).

(b) *Classification.* Class III.

(c) *Date PMA or notice of completion of a PDP is required.* A PMA or a notice of completion of a PDP is required to be filed with the Food and Drug Administration on or before December 26, 1996 for any knee joint femoral (hemi-knee) metallic uncemented prosthesis that was in commercial distribution before May 28, 1976, or that has, on or before December 26, 1996 been found to be substantially equivalent to a knee joint femoral (hemi-knee) metallic uncemented prosthesis that was in commercial distribution before May 28, 1976. Any other knee joint femoral (hemi-knee) metallic uncemented prosthesis shall have an approved PMA or a declared completed PDP in effect before being placed in commercial distribution.

[52 FR 33702, Sept. 4, 1987, as amended at 61 FR 50710, Sept. 27, 1996]

§ 888.3580 Knee joint patellar (hemi-knee) metallic resurfacing uncemented prosthesis.

(a) *Identification.* A knee joint patellar (hemi-knee) metallic resurfacing uncemented prosthesis is a device made of alloys, such as cobalt-chromium-molybdenum, intended to be implanted to replace the retropatellar articular surface of the patellofemoral joint. The device limits minimally (less than normal anatomic constraints) translation in one or more planes. It has no linkage across-the-joint. This generic type of device includes prostheses that have a retropatellar resurfacing component and an orthopedic screw to transfix the patellar remnant. This generic type of device is limited to those prostheses intended for use without bone cement (§ 888.3027).

(b) *Classification.* (1) Class II when intended for treatment of degenerative and posttraumatic patellar arthritis.

(2) Class III when intended for uses other than treatment of degenerative and posttraumatic patellar arthritis.

(c) *Date PMA or notice of completion of a PDP is required.* A PMA or a notice of completion of a PDP is required to be filed with the Food and Drug Administration on or before December 26, 1996 for any knee joint patellar (hemi-knee) metallic resurfacing uncemented prosthesis described in paragraph (b)(2) of this section that was in commercial distribution before May 28, 1976, or that has, on or before December 26, 1996 been found to be substantially equivalent to a knee joint patellar (hemi-knee) metallic resurfacing uncemented prosthesis that was in commercial distribution before May 28, 1976. Any other knee joint patellar (hemi-knee) metallic resurfacing uncemented prosthesis shall have an approved PMA or a declared completed PDP in effect before being placed in commercial distribution.

[52 FR 33702, Sept. 4, 1987, as amended at 61 FR 50711, Sept. 27, 1996]

§ 888.3590 Knee joint tibial (hemi-knee) metallic resurfacing uncemented prosthesis.

(a) *Identification.* A knee joint tibial (hemi-knee) metallic resurfacing uncemented prosthesis is a device intended to be implanted to replace part of a knee joint. The device limits minimally (less than normal anatomic constraints) translation in one or more planes. It has no linkage across-the-joint. This prosthesis is made of alloys, such as cobalt-chromium-molybdenum, and is intended to resurface one tibial condyle. The generic type of device is limited to those prostheses intended for use without bone cement (§ 888.3027).

(b) *Classification.* Class II.

§ 888.3640 Shoulder joint metal/metal or metal/polymer constrained cemented prosthesis.

(a) *Identification.* A shoulder joint metal/metal or metal/polymer constrained cemented prosthesis is a device intended to be implanted to replace a shoulder joint. The device prevents dislocation in more than one anatomic plane and has components that are linked together. This generic type of device includes prostheses that have a humeral component made of alloys, such as cobalt-chromium-molybdenum, and a glenoid component made of this alloy or a combination of this alloy and ultra-high molecular weight polyethylene. This generic type of device is limited to those prostheses intended for use with bone cement (§ 888.3027).

(b) *Classification.* Class III.

(c) *Date PMA or notice of completion of a PDP is required.* A PMA or a notice of completion of a PDP is required to be filed with the Food and Drug Administration on or before December 26, 1996 for any shoulder joint metal/metal or metal/polymer constrained cemented prosthesis that was in commercial distribution before May 28, 1976, or that has, on or before December 26, 1996 been found to be substantially equivalent to a shoulder joint metal/metal or metal/polymer constrained cemented prosthesis that was in commercial distribution before May 28, 1976. Any other shoulder joint metal/metal or metal/polymer constrained cemented prosthesis shall have an approved PMA or a declared completed PDP in effect before being placed in commercial distribution.

[52 FR 33702, Sept. 4, 1987, as amended at 61 FR 50711, Sept. 27, 1996]

§ 888.3650 Shoulder joint metal/polymer non-constrained cemented prosthesis.

(a) *Identification.* A shoulder joint metal/polymer non-constrained cemented prosthesis is a device intended to be implanted to replace a shoulder joint. The device limits minimally (less than normal anatomic constraints) translation in one or more planes. It has no linkage across-the-joint. This generic type of device includes prostheses that have a humeral component made of alloys, such as cobalt-chromium-molybdenum, and a glenoid resurfacing component made of ultra-high molecular weight polyethylene, and is limited to those prostheses intended for use with bone cement (§ 888.3027).

(b) *Classification.* Class II. The special controls for this device are:

(1) FDA's:

(i) "Use of International Standard ISO 10993 'Biological Evaluation of Medical Devices—Part I: Evaluation and Testing,' "

(ii) "510(k) Sterility Review Guidance of 2/12/90 (K90–1),"

(iii) "Guidance Document for Testing Orthopedic Implants with Modified Metallic Surfaces Apposing Bone or Bone Cement,"

(iv) "Guidance Document for the Preparation of Premarket Notification (510(k)) Application for Orthopedic Devices," and

(v) "Guidance Document for Testing Non-articulating, 'Mechanically Locked' Modular Implant Components,"

(2) International Organization for Standardization's (ISO):

(i) ISO 5832–3:1996 "Implants for Surgery—Metallic Materials—Part 3: Wrought Titanium 6-Aluminum 4-Vandium Alloy,"

(ii) ISO 5832–4:1996 "Implants for Surgery—Metallic Materials—Part 4: Cobalt-Chromium-Molybdenum Casting Alloy,"

(iii) ISO 5832–12:1996 "Implants for Surgery—Metallic Materials—Part 12: Wrought Cobalt-Chromium-Molybdenum Alloy,"

(iv) ISO 5833:1992 "Implants for Surgery—Acrylic Resin Cements,"

(v) ISO 5834–2:1998 "Implants for Surgery—Ultra-high Molecular Weight Polyethylene—Part 2: Moulded Forms,"

(vi) ISO 6018:1987 "Orthopaedic Implants—General Requirements for Marking, Packaging, and Labeling," and

(vii) ISO 9001:1994 "Quality Systems—Model for Quality Assurance in Design/Development, Production, Installation, and Servicing," and

(3) American Society for Testing and Materials':

(i) F 75–92 "Specification for Cast Cobalt-28 Chromium-6 Molybdenum Alloy for Surgical Implant Material,"

(ii) F 648–98 "Specification for Ultra-High-Molecular-Weight Polyethylene Powder and Fabricated Form for Surgical Implants,"

(iii) F 799–96 "Specification for Cobalt-28 Chromium-6 Molybdenum Alloy Forgings for Surgical Implants,"

(iv) F 1044–95 "Test Method for Shear Testing of Porous Metal Coatings,"

(v) F 1108–97 "Titanium-6 Aluminum-4 Vanadium Alloy Castings for Surgical Implants,"

(vi) F 1147–95 "Test Method for Tension Testing of Porous Metal Coatings,"

(vii) F 1378–97 "Specification for Shoulder Prosthesis," and

(viii) F 1537–94 "Specification for Wrought Cobalt-28 Chromium-6 Molybdenum Alloy for Surgical Implants."

[52 FR 33702, Sept. 4, 1987, as amended at 65 FR 17148, Mar. 31, 2000]

§ 888.3660 Shoulder joint metal/polymer semi-constrained cemented prosthesis.

(a) *Identification.* A shoulder joint metal/polymer semi-constrained cemented prosthesis is a device intended to be implanted to replace a shoulder joint. The device limits translation and rotation in one or more planes via the geometry of its articulating surfaces. It has no linkage across-the-joint. This generic type of device includes prostheses that have a humeral resurfacing component made of alloys, such as cobalt-chromium-molybdenum, and a glenoid resurfacing component made of ultra-high molecular weight polyethylene, and is limited to those prostheses intended for use with bone cement (§ 888.3027).

(b) *Classification.* Class II. The special controls for this device are:

(1) FDA's:

(i) "Use of International Standard ISO 10993 'Biological Evaluation of Medical Devices—Part I: Evaluation and Testing,' "

(ii) "510(k) Sterility Review Guidance of 2/12/90 (K90–1),"

(iii) "Guidance Document for Testing Orthopedic Implants with Modified Metallic Surfaces Apposing Bone or Bone Cement,"

(iv) "Guidance Document for the Preparation of Premarket Notification (510(k)) Application for Orthopedic Devices," and

(v) "Guidance Document for Testing Non-articulating, 'Mechanically Locked' Modular Implant Components,"

(2) International Organization for Standardization's (ISO):

(i) ISO 5832–3:1996 "Implants for Surgery—Metallic Materials—Part 3: Wrought Titanium 6-aluminum 4-vandium Alloy,"

(ii) ISO 5832–4:1996 "Implants for Surgery—Metallic Materials—Part 4: Cobalt-chromium-molybdenum casting alloy,"

(iii) ISO 5832–12:1996 "Implants for Surgery—Metallic Materials—Part 12: Wrought Cobalt-chromium-molybdenum alloy,"

(iv) ISO 5833:1992 "Implants for Surgery—Acrylic Resin Cements,"

(v) ISO 5834–2:1998 "Implants for Surgery—Ultra-high Molecular Weight Polyethylene—Part 2: Moulded Forms,"

(vi) ISO 6018:1987 "Orthopaedic Implants—General Requirements for Marking, Packaging, and Labeling," and

(vii) ISO 9001:1994 "Quality Systems—Model for Quality Assurance in Design/Development, Production, Installation, and Servicing," and

(3) American Society for Testing and Materials':

(i) F 75–92 "Specification for Cast Cobalt-28 Chromium-6 Molybdenum Alloy for Surgical Implant Material,"

(ii) F 648–98 "Specification for Ultra-High-Molecular-Weight Polyethylene Powder and Fabricated Form for Surgical Implants,"

(iii) F 799–96 "Specification for Cobalt-28 Chromium-6 Molybdenum Alloy Forgings for Surgical Implants,"

(iv) F 1044–95 "Test Method for Shear Testing of Porous Metal Coatings,"

(v) F 1108–97 "Specification for Titanium-6 Aluminum-4 Vanadium Alloy Castings for Surgical Implants,"

(vi) F 1147–95 "Test Method for Tension Testing of Porous Metal,"

(vii) F 1378–97 "Standard Specification for Shoulder Prosthesis," and

(viii) F 1537–94 "Specification for Wrought Cobalt-28 Chromium-6 Molybdenum Alloy for Surgical Implants."

[52 FR 33702, Sept. 4, 1987, as amended at 65 FR 17148, Mar. 31, 2000]

§ 888.3670 Shoulder joint metal/polymer/metal nonconstrained or semi-constrained porous-coated uncemented prosthesis.

(a) *Identification.* A shoulder joint metal/polymer/metal nonconstrained or semi-constrained porous-coated uncemented prosthesis is a device intended to be implanted to replace a shoulder joint. The device limits movement in one or more planes. It has no linkage across-the-joint. This generic type of device includes prostheses that have a humeral component made of alloys such as cobalt-chromium-molybdenum (Co-Cr-Mo) and titanium-aluminum-vanadium (Ti-6Al-4V) alloys, and a glenoid resurfacing component made of ultra-high molecular weight polyethylene, or a combination of an articulating ultra-high molecular weight bearing surface fixed in a metal shell made of alloys such as Co-Cr-Mo and Ti-6Al-4V. The humeral component and glenoid backing have a porous coating made of, in the case of Co-Cr-Mo components, beads of the same alloy or commercially pure titanium powder, and in the case of Ti-6Al-4V components, beads or fibers of commercially pure titanium or Ti-6Al-4V alloy, or commercially pure titanium powder. The porous coating has a volume porosity between 30 and 70 percent, an average pore size between 100 and 1,000 microns, interconnecting porosity, and a porous coating thickness between 500 and 1,500 microns. This generic type of device is designed to achieve biological fixation to bone without the use of bone cement.

(b) *Classification.* Class II (special controls). The special control for this device is FDA's "Class II Special Controls Guidance: Shoulder Joint Metal/Polymer/Metal Nonconstrained or Semi-Constrained Porous-Coated Uncemented Prosthesis."

[66 FR 12737, Feb. 28, 2001]

§ 888.3680 Shoulder joint glenoid (hemi-shoulder) metallic cemented prosthesis.

(a) *Identification.* A shoulder joint glenoid (hemi-shoulder) metallic cemented prosthesis is a device that has a glenoid (socket) component made of alloys, such as cobalt-chromium-molybdenum, or alloys with ultra-high molecular weight polyethylene and intended to be implanted to replace part of a shoulder joint. This generic type of device is limited to those prostheses intended for use with bone cement (§ 888.3027).

(b) *Classification.* Class III.

(c) *Date PMA or notice of completion of a PDP is required.* A PMA or a notice of completion of a PDP is required to be filed with the Food and Drug Administration on or before December 26, 1996 for any shoulder joint glenoid (hemi-shoulder) metallic cemented prosthesis that was in commercial distribution before May 28, 1976, or that has, on or before December 26, 1996 been found to be substantially equivalent to a shoulder joint glenoid (hemi-shoulder) metallic cemented prosthesis that was in commercial distribution before May 28, 1976. Any other shoulder joint glenoid (hemi-shoulder) metallic cemented prosthesis shall have an approved PMA or a declared completed PDP in effect before being placed in commercial distribution.

[52 FR 33702, Sept. 4, 1987, as amended at 61 FR 50711, Sept. 27, 1996]

§ 888.3690 Shoulder joint humeral (hemi-shoulder) metallic uncemented prosthesis.

(a) *Identification.* A shoulder joint humeral (hemi-shoulder) metallic uncemented prosthesis is a device made of alloys, such as cobalt-chromium-molybdenum. It has an intramedullary stem and is intended to be implanted to replace the articular surface of the proximal end of the humerus and to be fixed without bone cement (§ 888.3027). This device is not intended for biological fixation.

(b) *Classification.* Class II.

§ 888.3720 Toe joint polymer constrained prosthesis.

(a) *Identification.* A toe joint polymer constrained prosthesis is a device made of silicone elastomer or polyester reinforced silicone elastomer intended to be implanted to replace the first metatarsophalangeal (big toe) joint. This generic type of device consists of a single flexible across-the-joint component that prevents dislocation in more than one anatomic plane.

(b) *Classification.* Class II.

§ 888.3730 Toe joint phalangeal (hemi-toe) polymer prosthesis.

(a) *Identification.* A toe joint phalangeal (hemi-toe) polymer prosthesis is a device made of silicone elastomer intended to be implanted to replace the base of the proximal phalanx of the toe.

(b) *Classification.* Class II.

§ 888.3750 Wrist joint carpal lunate polymer prosthesis.

(a) *Identification.* A wrist joint carpal lunate prosthesis is a one-piece device made of silicone elastomer intended to be implanted to replace the carpal lunate bone of the wrist.

(b) *Classification.* Class II.

§ 888.3760 Wrist joint carpal scaphoid polymer prosthesis.

(a) *Identification.* A wrist joint carpal scaphoid polymer prosthesis is a one-piece device made of silicone elastomer intended to be implanted to replace the carpal scaphoid bone of the wrist.

(b) *Classification.* Class II.

§ 888.3770 Wrist joint carpal trapezium polymer prosthesis.

(a) *Identification.* A wrist joint carpal trapezium polymer prosthesis is a one-piece device made of silicone elastomer or silicone elastomer/polyester material intended to be implanted to replace the carpal trapezium bone of the wrist.

(b) *Classification.* Class II.

§ 888.3780 Wrist joint polymer constrained prosthesis.

(a) *Identification.* A wrist joint polymer constrained prosthesis is a device made of polyester-reinforced silicone elastomer intended to be implanted to replace a wrist joint. This generic type of device consists of a single flexible across-the-joint component that prevents dislocation in more than one anatomic plane.

(b) *Classification.* Class II.

§ 888.3790 Wrist joint metal constrained cemented prosthesis.

(a) *Identification.* A wrist joint metal constrained cemented prosthesis is a device intended to be implanted to replace a wrist joint. The device prevents dislocation in more than one anatomic plane and consists of either a single flexible across-the-joint component or two components linked together. This generic type of device is limited to a device which is made of alloys, such as cobalt-chromium-molybdenum, and is limited to those prostheses intended for use with bone cement (§ 888.3027).

(b) *Classification.* Class III.

(c) *Date PMA or notice of completion of a PDP is required.* A PMA or a notice of completion of a PDP is required to be filed with the Food and Drug Administration on or before December 26, 1996 for any wrist joint metal constrained cemented prosthesis that was in commercial distribution before May 28, 1976, or that has, on or before December 26, 1996 been found to be substantially equivalent to a wrist joint metal constrained cemented prosthesis that was in commercial distribution before May 28, 1976. Any other wrist joint metal constrained cemented prosthesis shall have an approved PMA or a declared completed PDP in effect before being placed in commercial distribution.

[52 FR 33702, Sept. 4, 1987, as amended at 61 FR 50711, Sept. 27, 1996]

§ 888.3800 Wrist joint metal/polymer semi-constrained cemented prosthesis.

(a) *Identification.* A wrist joint metal/polymer semi-constrained cemented prosthesis is a device intended to be implanted to replace a wrist joint. The device limits translation and rotation in one or more planes via the geometry of its articulating surfaces. It has no linkage across-the-joint. This generic type of device includes prostheses that have either a one-part radial component made of alloys, such as cobalt-chromium-molybdenum, with an ultra-high molecular weight polyethylene bearing surface, or a two-part radial component made of alloys and an ultra-high molecular weight polyethylene ball that is mounted on the radial component with a trunnion bearing. The metallic portion of the two-part radial component is inserted into the radius. These devices have a metacarpal component(s) made of alloys, such as cobalt-chromium-molybdenum. This generic type of device is limited to those prostheses intended for use with bone cement (§ 888.3027).

(b) *Classification.* Class II.

§ 888.3810 Wrist joint ulnar (hemi-wrist) polymer prosthesis.

(a) *Identification.* A wrist joint ulnar (hemi-wrist) polymer prosthesis is a mushroom-shaped device made of a medical grade silicone elastomer or ultra-high molecular weight polyethylene intended to be implanted into the intramedullary canal of the bone and held in place by a suture. Its purpose is to cover the resected end of the distal ulna to control bone overgrowth and to provide an articular surface for the radius and carpus.

(b) *Classification.* Class II.

Subpart E—Surgical Devices

§ 888.4150 Calipers for clinical use.

(a) *Identification.* A caliper for clinical use is a compass-like device intended for use in measuring the thickness or diameter of a part of the body or the distance between two body surfaces, such as for measuring an excised skeletal specimen to determine the proper replacement size of a prosthesis.

(b) *Classification.* Class I (general controls). The device is exempt from the premarket notification procedures in subpart E of part 807 of this chapter, subject to the limitations in § 888.9.

[52 FR 33702, Sept. 4, 1987, as amended at 66 FR 38815, July 25, 2001]

§ 888.4200 Cement dispenser.

(a) *Identification.* A cement dispenser is a nonpowered syringe-like device intended for use in placing bone cement (§ 888.3027) into surgical sites.

(b) *Classification.* Class I (general controls). The device is exempt from the premarket notification procedures in subpart E of part 807 of this chapter, subject to the limitations in § 888.9.

[52 FR 33702, Sept. 4, 1987, as amended at 53 FR 52953, Dec. 29, 1988; 59 FR 63014, Dec. 7, 1994; 66 FR 38815, July 25, 2001]

§ 888.4210 Cement mixer for clinical use.

(a) *Identification.* A cement mixer for clinical use is a device consisting of a container intended for use in mixing bone cement (§ 888.3027).

(b) *Classification.* Class I (general controls). The device is exempt from the premarket notification procedures in subpart E of part 807 of this chapter, subject to the limitations in § 888.9.

[52 FR 33702, Sept. 4, 1987, as amended at 53 FR 52953, Dec. 29, 1988; 59 FR 63014, Dec. 7, 1994; 66 FR 38815, July 25, 2001]

§ 888.4220 Cement monomer vapor evacuator.

(a) *Identification.* A cement monomer vapor evacuator is a device intended for use during surgery to contain or remove undesirable fumes, such as monomer vapor from bone cement (§ 888.3027).

(b) *Classification.* Class I (general controls). The device is exempt from the premarket notification procedures in subpart E of part 807 of this chapter, subject to the limitations in § 888.9.

[52 FR 33702, Sept. 4, 1987, as amended at 53 FR 52954, Dec. 29, 1988; 66 FR 38815, July 25, 2001]

§ 888.4230 Cement ventilation tube.

(a) *Identification.* A cement ventilation tube is a tube-like device usually made of plastic intended to be inserted into a surgical cavity to allow the release of air or fluid from the cavity as it is being filled with bone cement (§ 888.3027).

(b) *Classification.* Class I (general controls). The device is exempt from the premarket notification procedures in subpart E of part 807 of this chapter, subject to the limitations in § 888.9.

[52 FR 33702, Sept. 4, 1987, as amended at 53 FR 52954, Dec. 29, 1988; 59 FR 63014, Dec. 7, 1994; 66 FR 38815, July 25, 2001]

§ 888.4300 Depth gauge for clinical use.

(a) *Identification.* A depth gauge for clinical use is a measuring device intended for various medical purposes, such as to determine the proper length of screws for fastening the ends of a fractured bone.

(b) *Classification.* Class I (general controls). The device is exempt from the premarket notification procedures in subpart E of part 807 of this chapter, subject to the limitations in § 888.9.

[52 FR 33702, Sept. 4, 1987, as amended at 66 FR 38815, July 25, 2001]

§ 888.4540 Orthopedic manual surgical instrument.

(a) *Identification.* An orthopedic manual surgical instrument is a nonpowered hand-held device intended for medical purposes to manipulate tissue, or for use with other devices in orthopedic surgery. This generic type of device includes the cerclage applier, awl, bender, drill brace, broach, burr, corkscrew, countersink, pin crimper, wire cutter, prosthesis driver, extractor, file, fork, needle holder, impactor, bending or contouring instrument, compression instrument, passer, socket positioner, probe, femoral neck punch, socket pusher, reamer, rongeur, scissors, screwdriver, bone skid, staple driver, bone screw starter, surgical stripper, tamp, bone tap, trephine, wire twister, and wrench.

(b) *Classification.* Class I (general controls). The device is exempt from the premarket notification procedures in subpart E of part 807 of this chapter, subject to the limitations in § 888.9.

[52 FR 33702, Sept. 4, 1987, as amended at 59 FR 63014, Dec. 7, 1994; 66 FR 38815, July 25, 2001]

208-072 D-17

§ 888.4580 Sonic surgical instrument and accessories/attachments.

(a) *Identification.* A sonic surgical instrument is a hand-held device with various accessories or attachments, such as a cutting tip that vibrates at high frequencies, and is intended for medical purposes to cut bone or other materials, such as acrylic.

(b) *Classification.* Class II.

§ 888.4600 Protractor for clinical use.

(a) *Identification.* A protractor for clinical use is a device intended for use in measuring the angles of bones, such as on x-rays or in surgery.

(b) *Classification.* Class I (general controls). The device is exempt from the premarket notification procedures in subpart E of part 807 of this chapter, subject to the limitations in § 888.9.

[52 FR 33702, Sept. 4, 1987, as amended at 66 FR 38815, July 25, 2001]

§ 888.4800 Template for clinical use.

(a) *Identification.* A template for clinical use is a device that consists of a pattern or guide intended for medical purposes, such as selecting or positioning orthopedic implants or guiding the marking of tissue before cutting.

(b) *Classification.* Class I (general controls). The device is exempt from the premarket notification procedures in subpart E of part 807 of this chapter, subject to the limitations in § 888.9.

[52 FR 33702, Sept. 4, 1987, as amended at 66 FR 38815, July 25, 2001]

§ 888.5850 Nonpowered orthopedic traction apparatus and accessories.

(a) *Identification.* A nonpowered orthopedic traction apparatus is a device that consists of a rigid frame with nonpowered traction accessories, such as cords, pulleys, or weights, and that is intended to apply a therapeutic pulling force to the skeletal system.

(b) *Classification.* Class I (general controls). The device is exempt from the premarket notification procedures in subpart E of part 807 of this chapter, subject to the limitations in § 888.9. The device is also exempt from the current good manufacturing practice regulations in part 820 of this chapter, with the exception of § 820.180, regarding general requirements concerning records, and § 820.198, regarding complaint files.

[52 FR 33702, Sept. 4, 1987, as amended at 66 FR 38815, July 25, 2001]

§ 888.5890 Noninvasive traction component.

(a) *Identification.* A noninvasive traction component is a device, such as a head halter, pelvic belt, or a traction splint, that does not penetrate the skin and is intended to assist in connecting a patient to a traction apparatus so that a therapeutic pulling force may be applied to the patient's body.

(b) *Classification.* Class I (general controls). The device is exempt from the premarket notification procedures in subpart E of part 807 of this chapter, subject to the limitations in § 888.9. The device is also exempt from the current good manufacturing practice regulations in part 820 of this chapter, with the exception of § 820.180, regarding general requirements concerning records, and § 820.198, regarding complaint files.

[52 FR 33702, Sept. 4, 1987, as amended at 53 FR 52954, Dec. 29, 1988; 66 FR 38815, July 25, 2001]

§ 888.5940 Cast component.

(a) *Identification.* A cast component is a device intended for medical purposes to protect or support a cast. This generic type of device includes the cast heel, toe cap, cast support, and walking iron.

(b) *Classification.* Class I (general controls). The device is exempt from the premarket notification procedures in subpart E of part 807 of this chapter, subject to the limitations in § 888.9. The device is also exempt from the current good manufacturing practice regulations in part 820 of this chapter, with the exception of § 820.180, regarding general requirements concerning records, and § 820.198, regarding complaint files.

[52 FR 33702, Sept. 4, 1987, as amended at 53 FR 52954, Dec. 29, 1988; 59 FR 63014, Dec. 7, 1994; 66 FR 38815, July 25, 2001]

§ 888.5960 Cast removal instrument.

(a) *Identification.* A cast removal instrument is an AC-powered, hand-held device intended to remove a cast from a patient. This generic type of device includes the electric cast cutter and cast vacuum.

(b) *Classification.* Class I (general controls). The device is exempt from the premarket notification procedures in subpart E of part 807 of this chapter, subject to the limitations in §888.9.

[55 FR 48443, Nov. 20, 1990, as amended at 61 FR 1125, Jan. 16, 1996; 66 FR 38816, July 25, 2001]

§888.5980 Manual cast application and removal instrument.

(a) *Identification.* A manual cast application and removal instrument is a nonpowered hand-held device intended to be used in applying or removing a cast. This generic type of device includes the cast knife, cast spreader, plaster saw, plaster dispenser, and casting stand.

(b) *Classification.* Class I (general controls). The device is exempt from the premarket notification procedures in subpart E of part 807 of this chapter, subject to the limitations in §888.9. The device is also exempt from the current good manufacturing practice regulations in part 820 of this chapter, with the exception of §820.180, regarding general requirements concerning records, and §820.198, regarding complaint files.

[52 FR 33702, Sept. 4, 1987, as amended at 53 FR 52954, Dec. 29, 1988; 66 FR 38816, July 25, 2001]

PART 890—PHYSICAL MEDICINE DEVICES

Subpart A—General Provisions

Sec.

Subpart B—Physical Medicine Diagnostic Devices

Subpart C [Reserved]

Subpart D—Physical Medicine Prosthetic Devices

Subpart E [Reserved]

Subpart F—Physical Medicine Therapeutic Devices

890.5250 Moist steam cabinet.
890.5275 Microwave diathermy.
890.5290 Shortwave diathermy.
890.5300 Ultrasonic diathermy.
890.5350 Exercise component.
890.5360 Measuring exercise equipment.
890.5370 Nonmeasuring exercise equipment.
890.5380 Powered exercise equipment.
890.5410 Powered finger exerciser.
890.5500 Infrared lamp.
890.5525 Iontophoresis device.
890.5575 Powered external limb overload warning device.
890.5650 Powered inflatable tube massager.
890.5660 Therapeutic massager.
890.5700 Cold pack.
890.5710 Hot or cold disposable pack.
890.5720 Water circulating hot or cold pack.
890.5730 Moist heat pack.
890.5740 Powered heating pad.
890.5765 Pressure-applying device.
890.5850 Powered muscle stimulator.
890.5860 Ultrasound and muscle stimulator.
890.5880 Multi-function physical therapy table.
890.5900 Powered traction equipment.
890.5925 Traction accessory.
890.5940 Chilling unit.
890.5950 Powered heating unit.
890.5975 Therapeutic vibrator.

AUTHORITY: 21 U.S.C. 351, 360, 360c, 360e, 360j, 371.

SOURCE: 48 FR 53047, Nov. 23, 1983, unless otherwise noted.

Subpart A—General Provisions

§ 890.1 Scope.

(a) This part sets forth the classification of physical medicine devices intended for human use that are in commercial distribution.

(b) The identification of a device in a regulation in this part is not a precise description of every device that is, or will be, subject to the regulation. A manufacturer who submits a premarket notification submission for a device under part 807 may not show merely that the device is accurately described by the section title and identification provisions of a regulation in this part, but shall state why the device is substantially equivalent to other devices, as required by § 807.87.

(c) To avoid duplicative listings, a physical medicine device that has two or more types of uses (e.g., used both as a diagnostic device and as a therapeutic device) is listed only in one subpart.

(d) References in this part to regulatory sections of the Code of Federal Regulations are to chapter I of title 21, unless otherwise noted.

[52 FR 17741, May 11, 1987]

§ 890.3 Effective dates of requirement for premarket approval.

A device included in this part that is classified into class III (premarket approval) shall not be commercially distributed after the date shown in the regulation classifying the device unless the manufacturer has an approval under section 515 of the act (unless an exemption has been granted under section 520(g)(2) of the act). An approval under section 515 of the act consists of FDA's issuance of an order approving an application of premarket approval (PMA) for the device or declaring completed a product development protocol (PDP) for the device.

(a) Before FDA requires that a device commercially distributed before the enactment date of the amendments, or a device that has been found substantially equivalent to such a device, has an approval under section 515 of the act FDA must promulgate a regulation under section 515(b) of the act requiring such approval, except as provided in paragraph (b) of this section. Such a regulation under section 515(b) of the act shall not be effective during the grace period ending on the 90th day after its promulgation or on the last day of the 30th full calendar month after the regulation that classifies the device into class III is effective, whichever is later. See section 501(f)(2)(B) of the act. Accordingly, unless an effective date of the requirement for premarket approval is shown in the regulation for a device classified into class III in this part, the device may be commercially distributed without FDA's issuance of an order approving a PMA or declaring completed a PDP for the device. If FDA promulgates a regulation under section 515(b) of the act requiring premarket approval for a device, section 501(f)(1)(A) of the act applies to the device.

(b) Any new, not substantially equivalent, device introduced into commercial distribution on or after May 28,

1976, includiing a device formerly marketed that has been substantially altered, is classified by statute (section 513(f) of the act) into class III without any grace period and FDA must have issued an order approving a PMA or declaring completed a PDP for the device before the device is commercially distributed unless it is reclassified. If FDA knows that a device being commercially distributed may be a "new" device as defined in this section because of any new intended use or other reasons, FDA may codify the statutory classification of the device into class III for such new use. Accordingly, the regulation for such a class III device states that as of the enactment date of the amendments, May 28, 1976, the device must have an approval under section 515 of the act before commercial distribution.

[52 FR 17741, May 11, 1987]

§ 890.9 Limitations of exemptions from section 510(k) of the Federal Food, Drug, and Cosmetic Act (the act).

The exemption from the requirement of premarket notification (section 510(k) of the act) for a generic type of class I or II device is only to the extent that the device has existing or reasonably foreseeable characteristics of commercially distributed devices within that generic type or, in the case of in vitro diagnostic devices, only to the extent that misdiagnosis as a result of using the device would not be associated with high morbidity or mortality. Accordingly, manufacturers of any commercially distributed class I or II device for which FDA has granted an exemption from the requirement of premarket notification must still submit a premarket notification to FDA before introducing or delivering for introduction into interstate commerce for commercial distribution the device when:

(a) The device is intended for a use different from the intended use of a legally marketed device in that generic type of device; e.g., the device is intended for a different medical purpose, or the device is intended for lay use where the former intended use was by health care professionals only;

(b) The modified device operates using a different fundamental scientific technology than a legally marketed device in that generic type of device; e.g., a surgical instrument cuts tissue with a laser beam rather than with a sharpened metal blade, or an in vitro diagnostic device detects or identifies infectious agents by using deoxyribonucleic acid (DNA) probe or nucleic acid hybridization technology rather than culture or immunoassay technology; or

(c) The device is an in vitro device that is intended:

(1) For use in the diagnosis, monitoring, or screening of neoplastic diseases with the exception of immunohistochemical devices;

(2) For use in screening or diagnosis of familial or acquired genetic disorders, including inborn errors of metabolism;

(3) For measuring an analyte that serves as a surrogate marker for screening, diagnosis, or monitoring life-threatening diseases such as acquired immune deficiency syndrome (AIDS), chronic or active hepatitis, tuberculosis, or myocardial infarction or to monitor therapy;

(4) For assessing the risk of cardiovascular diseases;

(5) For use in diabetes management;

(6) For identifying or inferring the identity of a microorganism directly from clinical material;

(7) For detection of antibodies to microorganisms other than immunoglobulin G (IgG) or IgG assays when the results are not qualitative, or are used to determine immunity, or the assay is intended for use in matrices other than serum or plasma;

(8) For noninvasive testing as defined in § 812.3(k) of this chapter; and

(9) For near patient testing (point of care).

[65 FR 2321, Jan. 14, 2000]

Subpart B—Physical Medicine Diagnostic Devices

§ 890.1175 Electrode cable.

(a) *Identification*. An electrode cable is a device composed of strands of insulated electrical conductors laid together around a central core and intended for medical purposes to connect an electrode from a patient to a diagnostic machine.

(b) *Classification.* Class II (special controls). The special controls consist of:

(1) The performance standard under part 898 of this chapter, and

(2) The guidance document entitled "Guidance on the Performance Standard for Electrode Lead Wires and Patient Cables." This device is exempt from the premarket notification procedures of subpart E of part 807 of this chapter subject to § 890.9.

[48 FR 53047, Nov. 23, 1983, as amended at 59 FR 63014, Dec. 7, 1994; 65 FR 19319, Apr. 11, 2000]

§ 890.1225 Chronaximeter.

(a) *Identification.* A chronaximeter is a device intended for medical purposes to measure neuromuscular excitability by means of a strength-duration curve that provides a basis for diagnosis and prognosis of neurological dysfunction.

(b) *Classification.* Class II (performance standards).

§ 890.1375 Diagnostic electromyograph.

(a) *Identification.* A diagnostic electromyograph is a device intended for medical purposes, such as to monitor and display the bioelectric signals produced by muscles, to stimulate peripheral nerves, and to monitor and display the electrical activity produced by nerves, for the diagnosis and prognosis of neuromuscular disease.

(b) *Classification.* Class II (performance standards).

§ 890.1385 Diagnostic electromyograph needle electrode.

(a) *Identification.* A diagnostic electromyograph needle electrode is a monopolar or bipolar needle intended to be inserted into muscle or nerve tissue to sense bioelectrical signals. The device is intended for medical purposes for use in connection with electromyography (recording the intrinsic electrical properties of skeletal muscle).

(b) *Classification.* Class II (performance standards).

§ 890.1450 Powered reflex hammer.

(a) *Identification.* A powered reflex hammer is a motorized device intended for medical purposes to elicit and determine controlled deep tendon reflexes.

(b) *Classification.* Class II (performance standards).

§ 890.1575 Force-measuring platform.

(a) *Identification.* A force-measuring platform is a device intended for medical purposes that converts pressure applied upon a planar surface into analog mechanical or electrical signals. This device is used to determine ground reaction force, centers of percussion, centers of torque, and their variations in both magnitude and direction with time.

(b) *Classification.* Class I (general controls). The device is exempt from the premarket notification procedures in subpart E of part 807 of this chapter, subject to the limitations in § 890.9.

[48 FR 53047, Nov. 23, 1983, as amended at 61 FR 1125, Jan. 16, 1996; 66 FR 38816, July 25, 2001]

§ 890.1600 Intermittent pressure measurement system.

(a) *Identification.* An intermittent pressure measurement system is an evaluative device intended for medical purposes, such as to measure the actual pressure between the body surface and the supporting media.

(b) *Classification.* Class I (general controls). The device is exempt from the premarket notification procedures in subpart E of part 807 of this chapter, subject to the limitations in § 890.9.

[48 FR 53047, Nov. 23, 1983, as amended at 61 FR 1125, Jan. 16, 1996; 66 FR 38816, July 25, 2001]

§ 890.1615 Miniature pressure transducer.

(a) *Identification.* A miniature pressure transducer is a device intended for medical purposes to measure the pressure between a device and soft tissue by converting mechanical inputs to analog electrical signals.

(b) *Classification.* Class I (general controls). The device is exempt from the premarket notification procedures in

subpart E of part 807 of this chapter, subject to the limitations in § 890.9.

[48 FR 53047, Nov. 23, 1983, as amended at 61 FR 1125, Jan. 16, 1996; 66 FR 38816, July 25, 2001]

§ 890.1850 Diagnostic muscle stimulator.

(a) *Identification.* A diagnostic muscle stimulator is a device used mainly with an electromyograph machine to initiate muscle activity. It is intended for medical purposes, such as to diagnose motor nerve or sensory neuromuscular disorders and neuromuscular function.

(b) *Classification.* Class II (performance standards).

§ 890.1925 Isokinetic testing and evaluation system.

(a) *Identification.* An isokinetic testing and evaluation system is a rehabilitative exercise device intended for medical purposes, such as to measure, evaluate, and increase the strength of muscles and the range of motion of joints.

(b) *Classification.* Class II (special controls). The device is exempt from the premarket notification procedures in subpart E of part 807 of this chapter subject to § 890.9.

[48 FR 53047, Nov. 23, 1983, as amended at 63 FR 59230, Nov. 3, 1998]

Subpart C [Reserved]

Subpart D—Physical Medicine Prosthetic Devices

§ 890.3025 Prosthetic and orthotic accessory.

(a) *Identification.* A prosthetic and orthotic accessory is a device intended for medical purposes to support, protect, or aid in the use of a cast, orthosis (brace), or prosthesis. Examples of prosthetic and orthotic accessories include the following: A pelvic support band and belt, a cast shoe, a cast bandage, a limb cover, a prosthesis alignment device, a postsurgical pylon, a transverse rotator, and a temporary training splint.

(b) *Classification.* Class I (general controls). The device is exempt from the premarket notification procedures in subpart E of part 807 of this chapter, subject to the limitations in § 890.9. The device is also exempt from the current good manufacturing practice regulations in part 820 of this chapter, with the exception of § 820.180, regarding general requirements concerning records and § 820.198, regarding complaint files.

[48 FR 53047, Nov. 23, 1983, as amended at 66 FR 38816, July 25, 2001]

§ 890.3075 Cane.

(a) *Identification.* A cane is a device intended for medical purposes that is used to provide minimal weight support while walking. Examples of canes include the following: A standard cane, a forearm cane, and a cane with a tripod, quad, or retractable stud on the ground end.

(b) *Classification.* Class I (general controls). The device is exempt from the premarket notification procedures in subpart E of part 807 of this chapter, subject to the limitations in § 890.9. The device is also exempt from the current good manufacturing practice regulations in part 820 of this chapter, with the exception of § 820.180, regarding general requirements concerning records and § 820.198, regarding complaint files.

[48 FR 53047, Nov. 23, 1983, as amended at 66 FR 38816, July 25, 2001]

§ 890.3100 Mechanical chair.

(a) *Identification.* A mechanical chair is a manually operated device intended for medical purposes that is used to assist a disabled person in performing an activity that the person would otherwise find difficult to do or be unable to do. Examples of mechanical chairs include the following: A chair with an elevating seat used to raise a person from a sitting position to a standing position and a chair with casters used by a person to move from one place to another while sitting.

(b) *Classification.* Class I (general controls). The device is exempt from the premarket notification procedures in subpart E of part 807 of this chapter, subject to the limitations in § 890.9.

[48 FR 53047, Nov. 23, 1983, as amended at 59 FR 63014, Dec. 7, 1994; 66 FR 38816, July 25, 2001]

§ 890.3110 Electric positioning chair.

(a) *Identification.* An electric positioning chair is a device with a motorized positioning control that is intended for medical purposes and that can be adjusted to various positions. The device is used to provide stability for patients with athetosis (involuntary spasms) and to alter postural positions.

(b) *Classification.* Class II (performance standards).

§ 890.3150 Crutch.

(a) *Identification.* A crutch is a device intended for medical purposes for use by disabled persons to provide minimal to moderate weight support while walking.

(b) *Classification.* Class I (general controls). The device is exempt from the premarket notification procedures in subpart E of part 807 of this chapter, subject to the limitations in § 890.9. The device is also exempt from the current good manufacturing practice regulations in part 820 of this chapter, with the exception of § 820.180, regarding general requirements concerning records and § 820.198, regarding complaint files.

[48 FR 53047, Nov. 23, 1983, as amended at 66 FR 38816, July 25, 2001]

§ 890.3175 Flotation cushion.

(a) *Identification.* A flotation cushion is a device intended for medical purposes that is made of plastic, rubber, or other type of covering, that is filled with water, air, gel, mud, or any other substance allowing a flotation media, used on a seat to lessen the likelihood of skin ulcers.

(b) *Classification.* Class I (general controls). The device is exempt from the premarket notification procedures in subpart E of part 807 of this chapter, subject to the limitations in § 890.9.

[48 FR 53047, Nov. 23, 1983, as amended at 61 FR 1125, Jan. 16, 1996; 66 FR 38816, July 25, 2001]

§ 890.3410 External limb orthotic component.

(a) *Identification.* An external limb orthotic component is a device intended for medical purposes for use in conjunction with an orthosis (brace) to increase the function of the orthosis for a patient's particular needs. Examples of external limb orthotic components include the following: A brace-setting twister and an external brace stirrup.

(b) *Classification.* Class I (general controls). The device is exempt from the premarket notification procedures in subpart E of part 807 of this chapter, subject to the limitations in § 890.9. The device is also exempt from the current good manufacturing practice regulations in part 820 of this chapter, with the exception of § 820.180, regarding general requirements concerning records and § 820.198, regarding complaint files.

[48 FR 53047, Nov. 23, 1983, as amended at 66 FR 38816, July 25, 2001]

§ 890.3420 External limb prosthetic component.

(a) *Identification.* An external limb prosthetic component is a device intended for medical purposes that, when put together with other appropriate components, constitutes a total prosthesis. Examples of external limb prosthetic components include the following: Ankle, foot, hip, knee, and socket components; mechanical or powered hand, hook, wrist unit, elbow joint, and shoulder joint components; and cable and prosthesis suction valves.

(b) *Classification.* Class I (general controls). The device is exempt from the premarket notification procedures in subpart E of part 807 of this chapter, subject to the limitations in § 890.9. The device is also exempt from the current good manufacturing practice regulations in part 820 of this chapter, with the exception of § 820.180, regarding general requirements concerning records and § 820.198, regarding complaint files.

[48 FR 53047, Nov. 23, 1983, as amended at 66 FR 38816, July 25, 2001]

§ 890.3475 Limb orthosis.

(a) *Identification.* A limb orthosis (brace) is a device intended for medical purposes that is worn on the upper or lower extremities to support, to correct, or to prevent deformities or to align body structures for functional improvement. Examples of limb orthoses include the following: A whole limb and joint brace, a hand splint, an elastic stocking, a knee cage, and a corrective shoe.

(b) *Classification.* Class I (general controls). The device is exempt from the premarket notification procedures in subpart E of part 807 of this chapter, subject to the limitations in §890.9. The device is also exempt from the current good manufacturing practice regulations in part 820 of this chapter, with the exception of §820.180, regarding general requirements concerning records and §820.198, regarding complaint files.

[48 FR 53047, Nov. 23, 1983, as amended at 66 FR 38816, July 25, 2001]

§890.3490 Truncal orthosis.

(a) *Identification.* A truncal orthosis is a device intended for medical purposes to support or to immobilize fractures, strains, or sprains of the neck or trunk of the body. Examples of truncal orthoses are the following: Abdominal, cervical, cervical-thoracic, lumbar, lumbo-sacral, rib fracture, sacroiliac, and thoracic orthoses and clavicle splints.

(b) *Classification.* Class I (general controls). The device is exempt from the premarket notification procedures in subpart E of part 807 of this chapter, subject to the limitations in §890.9. The device is also exempt from the current good manufacturing practice regulations in part 820 of this chapter, with the exception of §820.180, regarding general requirements concerning records and §820.198, regarding complaint files.

[48 FR 53047, Nov. 23, 1983, as amended at 66 FR 38816, July 25, 2001]

§890.3500 External assembled lower limb prosthesis.

(a) *Identification.* An external assembled lower limb prosthesis is a device that is intended for medical purposes and is a preassembled external artificial limb for the lower extremity. Examples of external assembled lower limb prostheses are the following: Knee/shank/ankle/foot assembly and thigh/knee/shank/ankle/foot assembly.

(b) *Classification.* Class II (special controls). The device is exempt from the premarket notification procedures in subpart E of part 807 of this chapter subject to §890.9.

[48 FR 53047, Nov. 23, 1983, as amended at 63 FR 59231, Nov. 3, 1998]

§890.3520 Plinth.

(a) *Identification.* A plinth is a flat, padded board with legs that is intended for medical purposes. A patient is placed on the device for treatment or examination.

(b) *Classification.* Class I (general controls). The device is exempt from the premarket notification procedures in subpart E of part 807 of this chapter, subject to the limitations in §890.9. The device is also exempt from the current good manufacturing practice regulations in part 820 of this chapter, with the exception of §820.180, regarding general requirements concerning records and §820.198, regarding complaint files.

[48 FR 53047, Nov. 23, 1983, as amended at 66 FR 38817, July 25, 2001]

§890.3610 Rigid pneumatic structure orthosis.

(a) *Identification.* A rigid pneumatic structure orthosis is a device intended for medical purposes to provide whole body support by means of a pressurized suit to help thoracic paraplegics walk.

(b) *Classification.* Class III (premarket approval).

(c) *Date PMA or notice of completion of a PDP is required.* A PMA or a notice of completion of a PDP is required to be filed with the Food and Drug Administration on or before December 26, 1996 for any rigid pneumatic structure orthosis that was in commercial distribution before May 28, 1976, or that has, on or before December 26, 1996 been found to be substantially equivalent to a rigid pneumatic structure orthosis that was in commercial distribution before May 28, 1976. Any other rigid pneumatic structure orthosis shall have an approved PMA or a declared completed PDP in effect before being placed in commercial distribution.

[48 FR 53047, Nov. 23, 1983, as amended at 52 FR 17742, May 11, 1987; 61 FR 50711, Sept. 27, 1996]

§ 890.3640 Arm sling.

(a) *Identification.* An arm sling is a device intended for medical purposes to immobilize the arm, by means of a fabric band suspended from around the neck.

(b) *Classification.* Class I (general controls). The device is exempt from the premarket notification procedures in subpart E of part 807 of this chapter, subject to the limitations in § 890.9. The device is also exempt from the current good manufacturing practice regulations in part 820 of this chapter, with the exception of § 820.180, regarding general requirements concerning records and § 820.198, regarding complaint files.

[48 FR 53047, Nov. 23, 1983, as amended at 66 FR 38817, July 25, 2001]

§ 890.3665 Congenital hip dislocation abduction splint.

(a) *Identification.* A congenital hip dislocation abduction splint is a device intended for medical purposes to stabilize the hips of a young child with dislocated hips in an abducted position (away from the midline).

(b) *Classification.* Class I (general controls). The device is exempt from the premarket notification procedures in subpart E of part 807 of this chapter, subject to the limitations in § 890.9. The device is also exempt from the current good manufacturing practice regulations in part 820 of this chapter, with the exception of § 820.180, regarding general requirements concerning records and § 820.198, regarding complaint files.

[48 FR 53047, Nov. 23, 1983, as amended at 66 FR 38817, July 25, 2001]

§ 890.3675 Denis Brown splint.

(a) *Identification.* A Denis Brown splint is a device intended for medical purposes to immobilize the foot. It is used on young children with tibial torsion (excessive rotation of the lower leg) or club foot.

(b) *Classification.* Class I (general controls). The device is exempt from the premarket notification procedures in subpart E of part 807 of this chapter, subject to the limitations in § 890.9. The device is also exempt from the current good manufacturing practice regulations in part 820 of this chapter, with the exception of § 820.180, regarding general requirements concerning records and § 820.198, regarding complaint files.

[48 FR 53047, Nov. 23, 1983, as amended at 66 FR 38817, July 25, 2001]

§ 890.3690 Powered wheeled stretcher.

(a) *Identification.* A powered wheeled stretcher is a battery-powered table with wheels that is intended for medical purposes for use by patients who are unable to propel themselves independently and who must maintain a prone or supine position for prolonged periods because of skin ulcers or contractures (muscle contractions).

(b) *Classification.* Class II (performance standards).

§ 890.3700 Nonpowered communication system.

(a) *Identification.* A nonpowered communication system is a mechanical device intended for medical purposes that is used to assist a patient in communicating when physical impairment prevents writing, telephone use, reading, or talking. Examples of nonpowered communications systems include an alphabet board and a page turner.

(b) *Classification.* Class I (general controls). The device is exempt from the premarket notification procedures in subpart E of part 807 of this chapter, subject to the limitations in § 890.9. The device is also exempt from the current good manufacturing practice regulations in part 820 of this chapter, with the exception of § 820.180, regarding general requirements concerning records and § 820.198, regarding complaint files.

[48 FR 53047, Nov. 23, 1983, as amended at 54 FR 25052, June 12, 1989; 66 FR 38817, July 25, 2001]

§ 890.3710 Powered communication system.

(a) *Identification.* A powered communication system is an AC- or battery-powered device intended for medical purposes that is used to transmit or receive information. It is used by persons unable to use normal communication methods because of physical impairment. Examples of powered communication systems include the following:

a specialized typewriter, a reading machine, and a video picture and word screen.

(b) *Classification.* Class II (special controls). The device is exempt from the premarket notification procedures in subpart E of part 807 of this chapter subject to § 890.9.

[48 FR 53047, Nov. 23, 1983, as amended at 63 FR 59231, Nov. 3, 1998]

§ 890.3725 Powered environmental control system.

(a) *Identification.* A powered environmental control system is an AC- or battery-powered device intended for medical purposes that is used by a patient to operate an environmental control function. Examples of environmental control functions include the following: to control room temperature, to answer a doorbell or telephone, or to sound an alarm for assistance.

(b) *Classification.* Class II (special controls). The device is exempt from the premarket notification procedures in subpart E of part 807 of this chapter subject to § 890.9.

[48 FR 53047, Nov. 23, 1983, as amended at 63 FR 59231, Nov. 3, 1998]

§ 890.3750 Mechanical table.

(a) *Identification.* A mechanical table is a device intended for medical purposes that has a flat surface that can be inclined or adjusted to various positions. It is used by patients with circulatory, neurological, or musculoskeletal conditions to increase tolerance to an upright or standing position.

(b) *Classification.* Class I (general controls). The device is exempt from the premarket notification procedures in subpart E of part 807 of this chapter, subject to the limitations in § 890.9.

[48 FR 53047, Nov. 23, 1983, as amended at 59 FR 63014, Dec. 7, 1994; 66 FR 38817, July 25, 2001]

§ 890.3760 Powered table.

(a) *Identification.* A powered table is a device intended for medical purposes that is an electrically operated flat surface table that can be adjusted to various positions. It is used by patients with circulatory, neurological, or musculoskeletal conditions to increase tolerance to an upright or standing position.

(b) *Classification.* Class I (general controls). The device is exempt from the premarket notification procedures in subpart E of part 807 of this chapter, subject to the limitations in § 890.9.

[48 FR 53047, Nov. 23, 1983, as amended at 61 FR 1125, Jan. 16, 1996; 66 FR 38817, July 25, 2001]

§ 890.3790 Cane, crutch, and walker tips and pads.

(a) *Identification.* Cane, crutch, and walker tips and pads are rubber (or rubber substitute) device accessories intended for medical purposes that are applied to the ground end of mobility aids to prevent skidding or that are applied to the body contact area of the device for comfort or as an aid in using an ambulatory assist device.

(b) *Classification.* Class I (general controls). The device is exempt from the premarket notification procedures in subpart E of part 807 of this chapter, subject to the limitations in § 890.9. The device is also exempt from the current good manufacturing practice regulations in part 820 of this chapter, with the exception of § 820.180, regarding general requirements concerning records and § 820.198, regarding complaint files.

[48 FR 53047, Nov. 23, 1983, as amended at 66 FR 38817, July 25, 2001]

§ 890.3800 Motorized three-wheeled vehicle.

(a) *Identification.* A motorized three-wheeled vehicle is a gasoline-fueled or battery-powered device intended for medical purposes that is used for outside transportation by disabled persons.

(b) *Classification.* Class II (performance standards).

§ 890.3825 Mechanical walker.

(a) *Identification.* A mechanical walker is a four-legged device with a metal frame intended for medical purposes to provide moderate weight support while walking. It is used by disabled persons who lack strength, good balance, or endurance.

(b) *Classification.* Class I (general controls). The device is exempt from the premarket notification procedures in

subpart E of part 807 of this chapter, subject to the limitations in § 890.9. The device is also exempt from the current good manufacturing practice regulations in part 820 of this chapter, with the exception of § 820.180, regarding general requirements concerning records and § 820.198, regarding complaint files.

[48 FR 53047, Nov. 23, 1983, as amended at 66 FR 38817, July 25, 2001]

§ 890.3850 Mechanical wheelchair.

(a) *Identification.* A mechanical wheelchair is a manually operated device with wheels that is intended for medical purposes to provide mobility to persons restricted to a sitting position.

(b) *Classification.* Class I (general controls).

§ 890.3860 Powered wheelchair.

(a) *Identification.* A powered wheelchair is a battery-operated device with wheels that is intended for medical purposes to provide mobility to persons restricted to a sitting position.

(b) *Classification.* Class II (performance standards).

§ 890.3880 Special grade wheelchair.

(a) *Identification.* A special grade wheelchair is a device with wheels that is intended for medical purposes to provide mobility to persons restricted to a sitting position. It is intended to be used in all environments for long-term use, e.g., for paraplegics, quadraplegics, and amputees.

(b) *Classification.* Class II (performance standards).

§ 890.3890 Stair-climbing wheelchair.

(a) *Identification.* A stair-climbing wheelchair is a device with wheels that is intended for medical purposes to provide mobility to persons restricted to a sitting position. The device is intended to climb stairs by means of two endless belt tracks that are lowered from under the chair and adjusted to the angle of the stairs.

(b) *Classification.* Class III (premarket approval).

(c) *Date PMA or notice of completion of a PDP is required.* A PMA or notice of completion of a PDP for a device described in paragraph (b) of this section is required to be filed with the Food and Drug Administration on or before July 12, 2000, for any stair-climbing wheelchair that was in commercial distribution before May 28, 1976, or that has, on or before July 12, 2000, been found to be substantially equivalent to a stair-climbing wheelchair that was in commercial distribution before May 28, 1976. Any other stair-climbing wheelchair shall have an approved PMA or declared completed PDP in effect before being placed in commercial distribution.

[48 FR 53047, Nov. 23, 1983, as amended at 52 FR 17742, May 11, 1987; 52 FR 22577, June 12, 1987; 65 FR 19834, Apr. 13, 2000]

§ 890.3900 Standup wheelchair.

(a) *Identification.* A standup wheelchair is a device with wheels that is intended for medical purposes to provide mobility to persons restricted to a sitting position. The device incorporates an external manually controlled mechanical system that is intended to raise a paraplegic to an upright position by means of an elevating seat.

(b) *Classification.* Class II (performance standards).

§ 890.3910 Wheelchair accessory.

(a) *Identification.* A wheelchair accessory is a device intended for medical purposes that is sold separately from a wheelchair and is intended to meet the specific needs of a patient who uses a wheelchair. Examples of wheelchair accessories include but are not limited to the following: armboard, lapboard, pusher cuff, crutch and cane holder, overhead suspension sling, head and trunk support, and blanket and leg rest strap.

(b) *Classification.* Class I (general controls). If the device is not intended for use as a protective restraint as defined in § 880.6760 of this chapter, it is exempt from the premarket notification procedures in subpart E of part 807 of this chapter, subject to the limitations in § 890.9. The device is also exempt from the current good manufacturing practice regulations in part 820 of this chapter, with the exception of § 820.180, regarding general requirements concerning records, and § 820.198, regarding complaint files.

[61 FR 8439, Mar. 4, 1996, as amended at 66 FR 38817, July 25, 2001]

§ 890.3920 Wheelchair component.

(a) *Identification.* A wheelchair component is a device intended for medical purposes that is generally sold as an integral part of a wheelchair, but may also be sold separately as a replacement part. Examples of wheelchair components are the following: Armrest, narrowing attachment, belt, extension brake, curb climber, cushion, antitip device, footrest, handrim, hill holder, leg rest, heel loops, and toe loops.

(b) *Classification.* Class I (general controls). The device is exempt from the premarket notification procedures in subpart E of part 807 of this chapter, subject to the limitations in § 890.9.

[48 FR 53047, Nov. 23, 1983, as amended at 59 FR 63014, Dec. 7, 1994; 66 FR 38817, July 25, 2001]

§ 890.3930 Wheelchair elevator.

(a) *Identification.* A wheelchair elevator is a motorized lift device intended for medical purposes to provide a means for a disabled person to move a wheelchair from one level to another.

(b) *Classification.* Class II (performance standards).

§ 890.3940 Wheelchair platform scale.

(a) *Identification.* A wheelchair platform scale is a device with a base designed to accommodate a wheelchair. It is intended for medical purposes to weigh a person who is confined to a wheelchair.

(b) *Classification.* Class I (general controls). The device is exempt from the premarket notification procedures in subpart E of part 807 of this chapter, subject to the limitations in § 890.9. The device is also exempt from the current good manufacturing practice regulations in part 820 of this chapter, with the exception of § 820.180, regarding general requirements concerning records and § 820.198, regarding complaint files.

[48 FR 53047, Nov. 23, 1983, as amended at 59 FR 63015, Dec. 7, 1994; 66 FR 38817, July 25, 2001]

Subpart E [Reserved]

Subpart F—Physical Medicine Therapeutic Devices

§ 890.5050 Daily activity assist device.

(a) *Identification.* A daily activity assist device is a modified adaptor or utensil (e.g., a dressing, grooming, recreational activity, transfer, eating, or homemaking aid) that is intended for medical purposes to assist a patient to perform a specific function.

(b) *Classification.* Class I (general controls). The device is exempt from the premarket notification procedures in subpart E of part 807 of this chapter, subject to the limitations in § 890.9. If the device is not labeled or otherwise represented as sterile, the device is also exempt from the current good manufacturing practice regulations in part 820 of this chapter, with the exception of § 820.180, regarding general requirements concerning records and § 820.198, regarding complaint files.

[48 FR 53047, Nov. 23, 1983, as amended at 66 FR 38817, July 25, 2001]

§ 890.5100 Immersion hydrobath.

(a) *Identification.* An immersion hydrobath is a device intended for medical purposes that consists of water agitators and that may include a tub to be filled with water. The water temperature may be measured by a gauge. It is used in hydrotherapy to relieve pain and itching and as an aid in the healing process of inflamed and traumatized tissue, and it serves as a setting for removal of contaminated tissue.

(b) *Classification.* Class II (performance standards).

§ 890.5110 Paraffin bath.

(a) *Identification.* A paraffin bath is a device intended for medical purposes that consists of a tub to be filled with liquid paraffin (wax) and maintained at an elevated temperature in which the patient's appendages (e.g., hands or fingers) are placed to relieve pain and stiffness.

(b) *Classification.* Class II (performance standards).

§ 890.5125 Nonpowered sitz bath.

(a) *Identification.* A nonpowered sitz bath is a device intended for medical purposes that consists of a tub to be filled with water for use in external hydrotherapy to relieve pain or pruritis and to accelerate the healing of inflamed or traumatized tissues of the perianal and perineal areas.

(b) *Classification.* Class I (general controls). The device is exempt from the premarket notification procedures in subpart E of part 807 of this chapter, subject to the limitations in § 890.9. The device is also exempt from the current good manufacturing practice regulations in part 820 of this chapter, with the exception of § 820.180, regarding general requirements concerning records and § 820.198, regarding complaint files.

[48 FR 53047, Nov. 23, 1983, as amended at 54 FR 25052, June 12, 1989; 66 FR 38818, July 25, 2001]

§ 890.5150 Powered patient transport.

(a) *Identification.* A powered patient transport is a motorized device intended for medical purposes to assist transfers of patients to and from the bath, beds, chairs, treatment modalities, transport vehicles, and up and down flights of stairs. This generic type of device does not include motorized threewheeled vehicles or wheelchairs.

(b) *Classification.* Class II (performance standards).

§ 890.5160 Air-fluidized bed.

(a) *Identification.* An air-fluidized bed is a device employing the circulation of filtered air through ceramic spherules (small, round ceramic objects) that is intended for medical purposes to treat or prevent bedsores, to treat severe or extensive burns, or to aid circulation.

(b) *Classification.* Class II (special controls). The device is exempt from the premarket notification procedures in subpart E of part 807 of this chapter subject to § 890.9.

[48 FR 53047, Nov. 23, 1983, as amended at 63 FR 59231, Nov. 3, 1998]

§ 890.5170 Powered flotation therapy bed.

(a) *Identification.* A powered flotation therapy bed is a device that is equipped with a mattress that contains a large volume of constantly moving water, air, mud, or sand. It is intended for medical purposes to treat or prevent a patient's bedsores, to treat severe or extensive burns, or to aid circulation. The mattress may be electrically heated.

(b) *Classification.* Class II (special controls). The device is exempt from the premarket notification procedures in subpart E of part 807 of this chapter subject to § 890.9.

[48 FR 53047, Nov. 23, 1983, as amended at 63 FR 59231, Nov. 3, 1998]

§ 890.5180 Manual patient rotation bed.

(a) *Identification.* A manual patient rotation bed is a device that turns a patient who is restricted to a reclining position. It is intended for medical purposes to treat or prevent bedsores, to treat severe and extensive burns, or to aid circulation.

(b) *Classification.* Class I (general controls). The device is exempt from the premarket notification procedures in subpart E of part 807 of this chapter subject to § 890.9.

[48 FR 53047, Nov. 23, 1963, as amended at 65 FR 2322, Jan. 14, 2000]

§ 890.5225 Powered patient rotation bed.

(a) *Identification.* A powered patient rotation bed is a device that turns a patient who is restricted to a reclining position. It is intended for medical purposes to treat or prevent bedsores, to treat severe and extensive burns, urinary tract blockage, and to aid circulation.

(b) *Classification.* Class II (special controls). The device is exempt from the premarket notification procedures

in subpart E of part 807 of this chapter subject to § 890.9.

[48 FR 53047, Nov. 23, 1983, as amended at 63 FR 59231, Nov. 3, 1998]

§ 890.5250 Moist steam cabinet.

(a) *Identification.* A moist steam cabinet is a device intended for medical purposes that delivers a flow of heated, moisturized air to a patient in an enclosed unit. It is used to treat arthritis and fibrosis (a formation of fibrosis tissue) and to increase local blood flow.

(b) *Classification.* Class II (performance standards).

§ 890.5275 Microwave diathermy.

(a) *Microwave diathermy for use in applying therapeutic deep heat for selected medical conditions*—(1) *Identification.* A microwave diathermy for use in applying therapeutic deep heat for selected medical conditions is a device that applies to specific areas of the body electromagnetic energy in the microwave frequency bands of 915 megahertz to 2,450 megahertz and that is intended to generate deep heat within body tissues for the treatment of selected medical conditions such as relief of pain, muscle spasms, and joint contractures, but not for the treatment of malignancies.

(2) *Classification.* Class II (performance standards).

(b) *Microwave diathermy for all other uses*—(1) *Identification.* A microwave diathermy for all other uses except for the treatment of malignancies is a device that applies to the body electromagnetic energy in the microwave frequency bands of 915 megahertz to 2,450 megahertz and that is intended for the treatment of medical conditions by means other than the generation of deep heat within body tissues as described in paragraph (a) of this section.

(2) *Classification.* Class III (premarket approval).

(c) *Date PMA or notice of completion of PDP is required.* A PMA or a notice of completion of a PDP for a device described in paragraph (b) of this section is required to be filed with the Food and Drug Administration on or before July 13, 1999, for any microwave diathermy described in paragraph (b) of this section that was in commercial distribution before May 28, 1976, or that has, on or before July 13, 1999, been found to be substantially equivalent to a microwave diathermy described in paragraph (b) of this section that was in commercial distribution before May 28, 1976. Any other microwave diathermy described in paragraph (b) of this section shall have an approved PMA or declared completed PDP in effect before being placed in commercial distribution.

[48 FR 53047, Nov. 23, 1983, as amended at 52 FR 17742, May 11, 1987; 64 FR 18331, Apr. 14, 1999]

§ 890.5290 Shortwave diathermy.

(a) *Shortwave diathermy for use in applying therapeutic deep heat for selected medical conditions*—(1) *Identification.* A shortwave diathermy for use in applying therapeutic deep heat for selected medical conditions is a device that applies to specific areas of the body electromagnetic energy in the radio frequency bands of 13 megahertz to 27.12 megahertz and that is intended to generate deep heat within body tissues for the treatment of selected medical conditions such as relief of pain, muscle spasms, and joint contractures, but not for the treatment of malignancies.

(2) *Classification.* Class II (performance standards).

(b) *Shortwave diathermy for all other uses*—(1) *Identification.* A shortwave diathermy for all other uses except for the treatment of malignancies is a device that applies to the body electromagnetic energy in the radio frequency bands of 13 megahertz to 27.12 megahertz and that is intended for the treatment of medical conditions by means other than the generation of deep heat within body tissues as described in paragraph (a) of this section.

(2) *Classification.* Class III (premarket approval).

(c) *Date PMA or notice of completion of a PDP is required.* No effective date has been established of the requirement for premarket approval for the device described in paragraph (b)(1). See § 890.3.

[48 FR 53047, Nov. 23, 1983, as amended at 52 FR 17742, May 11, 1987]

§ 890.5300 Ultrasonic diathermy.

(a) *Ultrasonic diathermy for use in applying therapeutic deep heat for selected medical conditions*—(1) *Identification.* An ultrasonic diathermy for use in applying therapeutic deep heat for selected medical conditions is a device that applies to specific areas of the body ultrasonic energy at a frequency beyond 20 kilohertz and that is intended to generate deep heat within body tissues for the treatment of selected medical conditions such as relief of pain, muscle spasms, and joint contractures, but not for the treatment of malignancies.

(2) *Classification.* Class II (performance standards).

(b) *Ultrasonic diathermy for all other uses*—(1) *Identification.* An ultrasonic diathermy for all other uses except for the treatment of malignancies is a device that applies to the body ultrasonic energy at a frequency beyond 20 kilohertz and that is intended for the treatment of medical conditions by means other than the generation of deep heat within body tissues as described in paragraph (a) of this section.

(2) *Classification.* Class III (premarket approval).

(c) *Date PMA or notice of completion of PDP is required.* A PMA or notice of completion of a PDP for a device described in paragraph (b) of this section is required to be filed with the Food and Drug Administration on or before July 13, 1999, for any ultrasonic diathermy described in paragraph (b) of this section that was in commercial distribution before May 28, 1976, or that has, on or before July 13, 1999, been found to be substantially equivalent to an ultrasonic diathermy described in paragraph (b) of this section that was in commercial distribution before May 28, 1976. Any other ultrasonic diathermy described in paragraph (b) of this section shall have an approved PMA or declared completed PDP in effect before being placed in commercial distribution.

[48 FR 53047, Nov. 23, 1983, as amended at 52 FR 17742, May 11, 1987; 64 FR 18331, Apr. 14, 1999]

§ 890.5350 Exercise component.

(a) *Identification.* An exercise component is a device that is used in conjunction with other forms of exercise and that is intended for medical purposes, such as to redevelope muscles or restore motion to joints or for use as an adjunct treatment for obesity. Examples include weights, dumbbells, straps, and adaptive hand mitts.

(b) *Classification.* Class I (general controls). The device is exempt from the premarket notification procedures in subpart E of part 807 of this chapter, subject to the limitations in § 890.9. The device is also exempt from the current good manufacturing practice regulations in part 820 of this chapter, with the exception of § 820.180, regarding general requirements concerning records and § 820.198, regarding complaint files.

[48 FR 53047, Nov. 23, 1983, as amended at 66 FR 38818, July 25, 2001]

§ 890.5360 Measuring exercise equipment.

(a) *Identification.* Measuring exercise equipment consist of manual devices intended for medical purposes, such as to redevelop muscles or restore motion to joints or for use as an adjunct treatment for obesity. These devices also include instrumentation, such as the pulse rate monitor, that provide information used for physical evaluation and physical planning purposes., Examples include a therapeutic exercise bicycle with measuring instrumentation, a manually propelled treadmill with measuring instrumentation, and a rowing machine with measuring instrumentation.

(b) *Classification.* Class II (performance standards).

§ 890.5370 Nonmeasuring exercise equipment.

(a) *Identification.* Nonmeasuring exercise equipment consist of devices intended for medical purposes, such as to redevelop muscles or restore motion to joints or for use as an adjunct treatment for obesity. Examples include a prone scooter board, parallel bars, a mechanical treadmill, an exercise table, and a manually propelled exercise bicycle.

(b) *Classification.* Class I (general controls). The device is exempt from the premarket notification procedures in subpart E of part 807 of this chapter, subject to the limitations in § 890.9. The device is also exempt from the current good manufacturing practice regulations in part 820 of this chapter, with the exception of § 820.180, regarding general requirements concerning records and § 820.198, regarding complaint files.

[48 FR 53047, Nov. 23, 1983, as amended at 66 FR 38818, July 25, 2001]

§ 890.5380 Powered exercise equipment.

(a) *Identification.* Powered exercise equipment consist of powered devices intended for medical purposes, such as to redevelop muscles or restore motion to joints or for use as an adjunct treatment for obesity. Examples include a powered treadmill, a powered bicycle, and powered parallel bars.

(b) *Classification.* Class I (general controls). The device is exempt from the premarket notification procedures in subpart E of part 807 of this chapter, subject to the limitations in § 890.9.

[48 FR 53047, Nov. 23, 1983, as amended at 61 FR 1125, Jan. 16, 1996; 66 FR 38818, July 25, 2001]

§ 890.5410 Powered finger exerciser.

(a) *Identification.* A powered finger exerciser is a device intended for medical purposes to increase flexion and the extension range of motion of the joints of the second to the fifth fingers of the hand.

(b) *Classification.* Class I (general controls). The device is exempt from the premarket notification procedures in subpart E of part 807 of this chapter, subject to the limitations in § 890.9.

[48 FR 53047, Nov. 23, 1983, as amended at 61 FR 1125, Jan. 16, 1996; 66 FR 38818, July 25, 2001]

§ 890.5500 Infrared lamp.

(a) *Identification.* An infrared lamp is a device intended for medical purposes that emits energy at infrared frequencies (approximately 700 nanometers to 50,000 nanometers) to provide topical heating.

(b) *Classification.* Class II (performance standards).

§ 890.5525 Iontophoresis device.

(a) *Iontophoresis device intended for certain specified uses*—(1) *Identification.* An iontophoresis device is a device that is intended to use a direct current to introduce ions of soluble salts or other drugs into the body and induce sweating for use in the diagnosis of cystic fibrosis or for other uses if the labeling of the drug intended for use with the device bears adequate directions for the device's use with that drug. When used in the diagnosis of cystic fibrosis, the sweat is collected and its composition and weight are determined.

(2) *Classification.* Class II (performance standards).

(b) *Iontophoresis device intended for any other purposes*—(1) *Identification.* An iontophoresis device is a device that is intended to use a direct current to introduce ions of soluble salts or other drugs into the body for medical purposes other than those specified in paragraph (a) of this section.

(2) *Classification.* Class III (premarket approval).

(c) *Date PMA or notice of completion of a PDP is required.* No effective date has been established of the requirement for premarket approval for the device described in paragraph (b)(1). See § 890.3.

[48 FR 53047, Nov. 23, 1983, as amended at 52 FR 17742, May 11, 1987]

§ 890.5575 Powered external limb overload warning device.

(a) *Identification.* A powered external limb overload warning device is a device intended for medical purposes to warn a patient of an overload or an underload in the amount of pressure placed on a leg.

(b) *Classification.* Class II (performance standards).

§ 890.5650 Powered inflatable tube massager.

(a) *Identification.* A powered inflatable tube massager is a powered device intended for medical purposes, such as to relieve minor muscle aches and pains and to increase circulation. It simulates kneading and stroking of tissues with the hands by use of an inflatable pressure cuff.

(b) *Classification.* Class II (performance standards).

§ 890.5660 Therapeutic massager.

(a) *Identification.* A therapeutic massager is an electrically powered device intended for medical purposes, such as to relieve minor muscle aches and pains.

(b) *Classification.* Class I (general controls). The device is exempt from the premarket notification procedures in subpart E of part 807 of this chapter, subject to the limitations in § 890.9.

[48 FR 53047, Nov. 23, 1983, as amended at 61 FR 1125, Jan. 16, 1996; 66 FR 38818, July 25, 2001]

§ 890.5700 Cold pack.

(a) *Identification.* A cold pack is a device intended for medical purposes that consists of a compact fabric envelope containing a specially hydrated pliable silicate gel capable of forming to the contour of the body and that provides cold therapy for body surfaces.

(b) *Classification.* Class I (general controls). The device is exempt from the premarket notification procedures in subpart E of part 807. The device also is exempt from the current good manufacturing practice regulations in part 820, with the exception of § 820.180, with respect to general requirements concerning records, and § 820.198, with respect to complaint files.

§ 890.5710 Hot or cold disposable pack.

(a) *Identification.* A hot or cold disposable pack is a device intended for medical purposes that consists of a sealed plastic bag incorporating chemicals that, upon activation, provides hot or cold therapy for body surfaces.

(b) *Classification.* Class I (general controls). Except when intended for use on infants, the device is exempt from the premarket notification procedures in subpart E of part 807 of this chapter subject to § 890.9.

[48 FR 53047, Nov. 23, 1963, as amended at 65 FR 2322, Jan. 14, 2000]

§ 890.5720 Water circulating hot or cold pack.

(a) *Identification.* A water circulating hot or cold pack is a device intended for medical purposes that operates by pumping heated or chilled water through a plastic bag and that provides hot or cold therapy for body surfaces.

(b) *Classification.* Class II (special controls). The device is exempt from the premarket notification procedures in subpart E of part 807 of this chapter subject to § 890.9.

[48 FR 53047, Nov. 23, 1983, as amended at 63 FR 59231, Nov. 3, 1998]

§ 890.5730 Moist heat pack.

(a) *Identification.* A moist heat pack is a device intended for medical purposes that consists of silica gel in a fabric container used to retain an elevated temperature and that provides moist heat therapy for body surfaces.

(b) *Classification.* Class I (general controls). The device is exempt from the premarket notification procedures in subpart E of part 807 of this chapter, subject to the limitations in § 890.9. The device is also exempt from the current good manufacturing practice regulations in part 820 of this chapter, with the exception of § 820.180, regarding general requirements concerning records and § 820.198, regarding complaint files.

[48 FR 53047, Nov. 23, 1983, as amended at 66 FR 38818, July 25, 2001]

§ 890.5740 Powered heating pad.

(a) *Identification.* A powered heating pad is an electrical device intended for medical purposes that provides dry heat therapy for body surfaces. It is capable of maintaining an elevated temperature during use.

(b) *Classification.* Class II (special controls). The device is exempt from the premarket notification procedures in subpart E part 807 of this chapter subject to § 890.9.

[48 FR 53047, Nov. 23, 1983, as amended at 63 FR 59231, Nov. 3, 1998]

§ 890.5765 Presssure-applying device.

(a) *Identification.* A presssure-applying device is a device intended for medical purposes to apply continuous pressure to the paravertebral tissues for muscular relaxation and neuro-inhibition. It consists of a table with an adjustable overhead weight that, in place of the therapist's hands, presses on the back of a prone patient.

(b) *Classification.* Class I (general controls). The device is exempt from the premarket notification procedures in

subpart E of part 807 of this chapter, subject to the limitations in § 890.9.

[48 FR 53047, Nov. 23, 1983, as amended at 59 FR 63015, Dec. 7, 1994; 66 FR 38818, July 25, 2001]

§ 890.5850 Powered muscle stimulator.

(a) *Identification.* A powered muscle stimulator is an electrically powered device intended for medical purposes that repeatedly contracts muscles by passing electrical currents through electrodes contacting the affected body area.

(b) *Classification.* Class II (performance standards).

§ 890.5860 Ultrasound and muscle stimulator.

(a) *Ultrasound and muscle stimulator for use in applying therapeutic deep heat for selected medical conditions*—(1) *Identification.* An ultrasound and muscle stimulator for use in applying therapeutic deep heat for selected medical conditions is a device that applies to specific areas of the body ultrasonic energy at a frequency beyond 20 kilohertz and that is intended to generate deep heat within body tissues for the treatment of selected medical conditions such as relief of pain, muscle spasms, and joint contractures, but not for the treatment of malignancies. The device also passes electrical currents through the body area to stimulate or relax muscles.

(2) *Classification.* Class II (performance standards).

(b) *Ultrasound and muscle stimulator for all other uses*—(1) *Identification.* An ultrasound and muscle stimulator for all other uses except for the treatment of malignancies is a device that applies to the body ultrasonic energy at a frequency beyond 20 kilohertz and applies to the body electrical currents and that is intended for the treatment of medical conditions by means other than the generation of deep heat within body tissues and the stimulation or relaxation of muscles as described in paragraph (a) of this section.

(2) *Classification.* Class III (premarket approval).

(c) *Date PMA or notice of completion of PDP is required.* A PMA or notice of completion of a PDP for a device described in paragraph (b) of this section is required to be filed with the Food and Drug Administration on or before July 13, 1999 for any ultrasound and muscle stimulator described in paragraph (b) of this section that was in commercial distribution before May 28, 1976, or that has, on or before July 13, 1999, been found to be substantially equivalent to an ultrasound and muscle stimulator described in paragraph (b) of this section that was in commercial distribution before May 28, 1976. Any other ultrasound and muscle stimulator described in paragraph (b) of this section shall have an approved PMA or declared completed PDP in effect before being placed in commercial distribution.

[48 FR 53047, Nov. 23, 1983, as amended at 52 FR 17742, May 11, 1987; 64 FR 18331, Apr. 14, 1999]

§ 890.5880 Multi-function physical therapy table.

(a) *Identification.* A multi-function physical therapy table is a device intended for medical purposes that consists of a motorized table equipped to provide patients with heat, traction, and muscle relaxation therapy.

(b) *Classification.* Class II (performance standards).

§ 890.5900 Power traction equipment.

(a) *Identification.* Powered traction equipment consists of powered devices intended for medical purposes for use in conjunction with traction accessories, such as belts and harnesses, to exert therapeutic pulling forces on the patient's body.

(b) *Classification.* Class II (performance standards).

§ 890.5925 Traction accessory.

(a) *Identification.* A traction accessory is a nonpowered accessory device intended for medical purposes to be used with powered traction equipment to aid in exerting therapeutic pulling forces on the patient's body. This generic type of device includes the pulley, strap, head halter, and pelvic belt.

(b) *Classification.* Class I (general controls). The device is exempt from the premarket notification procedures in subpart E of part 807 of this chapter, subject to the limitations in § 890.9. The device is also exempt from the current good manufacturing practice regulations in part 820 of this chapter, with the exception of § 820.180, regarding general requirements concerning records and § 820.198, regarding complaint files.

[48 FR 53047, Nov. 23, 1983, as amended at 61 FR 1125, Jan. 16, 1996; 66 FR 38818, July 25, 2001]

§ 890.5940 Chilling unit.

(a) *Identification.* A chilling unit is a refrigerative device intended for medical purposes to chill and maintain cold packs at a reduced temperature.

(b) *Classification.* Class I (general controls). The device is exempt from the premarket notification procedures in subpart E of part 807 of this chapter, subject to the limitations in § 890.9

[48 FR 53047, Nov. 23, 1983, as amended at 61 FR 1125, Jan. 16, 1996; 66 FR 38818, July 25, 2001]

§ 890.5950 Powered heating unit.

(a) *Identification.* A powered heating unit is a device intended for medical purposes that consists of an encased cabinet containing hot water and that is intended to heat and maintain hot packs at an elevated temperature.

(b) *Classification.* Class I (general controls). The device is exempt from the premarket notification procedures in subpart E of part 807 of this chapter, subject to the limitations in § 890.9.

[48 FR 53047, Nov. 23, 1983, as amended at 61 FR 1125, Jan. 16, 1996; 66 FR 38818, July 25, 2001]

§ 890.5975 Therapeutic vibrator.

(a) *Identification.* A therapeutic vibrator is an electrically powered device intended for medical purposes that incorporates various kinds of pads and that is held in the hand or attached to the hand or to a table. It is intended for various uses, such as relaxing muscles and relieving minor aches and pains.

(b) *Classification.* Class I (general controls). The device is exempt from the premarket notification procedures in subpart E of part 807 of this chapter, subject to the limitations in § 890.9.

[48 FR 53047, Nov. 23, 1983, as amended at 61 FR 1125, Jan. 16, 1996; 66 FR 38818, July 25, 2001]

PART 892—RADIOLOGY DEVICES

Subpart A—General Provisions

Sec.

892.1 Scope.
892.3 Effective dates of requirement for premarket approval.
892.9 Limitations of exemptions from section 510(k) of the Federal Food, Drug, and Cosmetic Act (the act).

Subpart B—Diagnostic Devices

892.1000 Magnetic resonance diagnostic device.
892.1100 Scintillation (gamma) camera.
892.1110 Positron camera.
892.1130 Nuclear whole body counter.
892.1170 Bone densitometer.
892.1200 Emission computed tomography system.
892.1220 Fluorescent scanner.
892.1300 Nuclear rectilinear scanner.
892.1310 Nuclear tomography system.
892.1320 Nuclear uptake probe.
892.1330 Nuclear whole body scanner.
892.1350 Nuclear scanning bed.
892.1360 Radionuclide dose calibrator.
892.1370 Nuclear anthropomorphic phantom.
892.1380 Nuclear flood source phantom.
892.1390 Radionuclide rebreathing system.
892.1400 Nuclear sealed calibration source.
892.1410 Nuclear electrocardiograph synchronizer.
892.1420 Radionuclide test pattern phantom.
892.1540 Nonfetal ultrasonic monitor.
892.1550 Ultrasonic pulsed doppler imaging system.
892.1560 Ultrasonic pulsed echo imaging system.
892.1570 Diagnostic ultrasonic transducer.
892.1600 Angiographic x-ray system.
892.1610 Diagnostic x-ray beam-limiting device.
892.1620 Cine or spot fluorographic x-ray camera.
892.1630 Electrostatic x-ray imaging system.
892.1640 Radiographic film marking system.
892.1650 Image-intensified fluoroscopic x-ray system.
892.1660 Non-image-intensified fluoroscopic x-ray system.
892.1670 Spot-film device.
892.1680 Stationary x-ray system.
892.1700 Diagnostic x-ray high voltage generator.
892.1710 Mammographic x-ray system.
892.1720 Mobile x-ray system.
892.1730 Photofluorographic x-ray system.
892.1740 Tomographic x-ray system.
892.1750 Computed tomography x-ray system.
892.1760 Diagnostic x-ray tube housing assembly.
892.1770 Diagnostic x-ray tube mount.
892.1820 Pneumoencephalographic chair.

892.1830 Radiologic patient cradle.
892.1840 Radiographic film.
892.1850 Radiographic film cassette.
892.1860 Radiographic film/cassette changer.
892.1870 Radiographic film/cassette changer programmer.
892.1880 Wall-mounted radiographic cassette holder.
892.1890 Radiographic film illuminator.
892.1900 Automatic radiographic film processor.
892.1910 Radiographic grid.
892.1920 Radiographic head holder.
892.1940 Radiologic quality assurance instrument.
892.1950 Radiographic anthropomorphic phantom.
892.1960 Radiographic intensifying screen.
892.1970 Radiographic ECG/respirator synchronizer.
892.1980 Radiologic table.
892.1990 Transilluminator for breast evaluation.
892.2010 Medical image storage device.
892.2020 Medical image communications device.
892.2030 Medical image digitizer.
892.2040 Medical image hardcopy device.
892.2050 Picture archiving and communications system.

Subparts C–E [Reserved]

Subpart F—Therapeutic Devices

892.5050 Medical charged-particle radiation therapy system.
892.5300 Medical neutron radiation therapy system.
892.5650 Manual radionuclide applicator system.
892.5700 Remote controlled radionuclide applicator system.
892.5710 Radiation therapy beam-shaping block.
892.5730 Radionuclide brachytherapy source.
892.5740 Radionuclide teletherapy source.
892.5750 Radionuclide radiation therapy system.
892.5770 Powered radiation therapy patient support assembly.
892.5780 Light beam patient position indicator.
892.5840 Radiation therapy simulation system.
892.5900 X-ray radiation therapy system.
892.5930 Therapeutic x-ray tube housing assembly.

Subpart G—Miscellaneous Devices

892.6500 Personnel protective shield.

AUTHORITY: 21 U.S.C. 351, 360, 360c, 360e, 360j, 371.

SOURCE: 53 FR 1567, Jan. 20, 1988, unless otherwise noted.

Subpart A—General Provisions

§ 892.1 Scope.

(a) This part sets forth the classification of radiology devices intended for human use that are in commercial distribution.

(b) The identification of a device in a regulation in this part is not a precise description of every device that is, or will be, subject to the regulation. A manufacturer who submits a premarket notification submission for a device under part 807 cannot show merely that the device is accurately described by the section title and identification provision of a regulation in this part but shall state why the device is substantially equivalent to other devices, as required by § 807.87.

(c) To avoid duplicative listings, a radiology device that has two or more types of uses (e.g., use both as a diagnostic device and a therapeutic device) is listed in one subpart only.

(d) References in this part to regulatory sections of the Code of Federal Regulations are to chapter I of this title 21, unless otherwise noted.

§ 892.3 Effective dates of requirement for premarket approval.

A device included in this part that is classified into class III (premarket approval) shall not be commercially distributed after the date shown in the regulation classifying the device unless the manufacturer has an approval under section 515 of the act (unless an exemption has been granted under section 520(g)(2) of the act). An approval under section 515 of the act consists of FDA's issuance of an order approving an application for premarket approval (PMA) for the device or declaring completed a product development protocol (PDP) for the device.

(a) Before FDA requires that a device commercially distributed before the enactment date of the amendments, or a device that has been found substantially equivalent to such a device, has an approval under section 515 of the act, FDA must promulgate a regulation under section 515(b) of the act requiring such approval, except as provided in paragraph (b) of this section. Such a regulation under section 515(b) of the act shall not be effective during the grace period ending on the 90th day after its promulgation or on the last day of the 30th full calendar month after the regulation that classifies the device into class III is effective, whichever is later. See section 501(f)(2)(B) of the act. Accordingly, unless an effective date of the requirement for premarket approval is shown in the regulation for a device classified into class III in this part, the device may be commercially distributed without FDA's issuance of an order approving a PMA or declaring completed a PDP for the device. If FDA promulgates a regulation under section 515(b) of the act requiring premarket approval for a device, section 501(f)(1)(A) of the act applies to the device.

(b) Any new, not substantially equivalent, device introduced into commercial distribution on or after May 28, 1976, including a device formerly marketed that has been substantially altered, is classified by statute (section 513(f) of the act) into class III without any grace period and FDA must have issued an order approving a PMA or declaring completed a PDP for the device before the device is commercially distributed unless it is reclassified. If FDA knows that a device being commercially distributed may be a "new" device as defined in this section because of any new intended use or other reasons, FDA may codify the statutory classification of the device into class III for such new use. Accordingly, the regulation for such a class III device states that as of the enactment date of the amendments, May 28, 1976, the device must have an approval under section 515 of the act before commercial distribution.

§ 892.9 Limitations of exemptions from section 510(k) of the Federal Food, Drug, and Cosmetic Act (the act).

The exemption from the requirement of premarket notification (section 510(k) of the act) for a generic type of class I or II device is only to the extent that the device has existing or reasonably foreseeable characteristics of commercially distributed devices within that generic type or, in the case of in vitro diagnostic devices, only to the extent that misdiagnosis as a result of using the device would not be associated with high morbidity or mortality. Accordingly, manufacturers of any commercially distributed class I or II device for which FDA has granted an exemption from the requirement of premarket notification must still submit a premarket notification to FDA before introducing or delivering for introduction into interstate commerce for commercial distribution the device when:

(a) The device is intended for a use different from the intended use of a legally marketed device in that generic type of device; e.g., the device is intended for a different medical purpose, or the device is intended for lay use where the former intended use was by health care professionals only;

(b) The modified device operates using a different fundamental scientific technology than a legally marketed device in that generic type of device; e.g., a surgical instrument cuts tissue with a laser beam rather than with a sharpened metal blade, or an in vitro diagnostic device detects or identifies infectious agents by using deoxyribonucleic acid (DNA) probe or nucleic acid hybridization technology rather than culture or immunoassay technology; or

(c) The device is an in vitro device that is intended:

(1) For use in the diagnosis, monitoring, or screening of neoplastic diseases with the exception of immunohistochemical devices;

(2) For use in screening or diagnosis of familial or acquired genetic disorders, including inborn errors of metabolism;

(3) For measuring an analyte that serves as a surrogate marker for screening, diagnosis, or monitoring life-threatening diseases such as acquired immune deficiency syndrome (AIDS), chronic or active hepatitis, tuberculosis, or myocardial infarction or to monitor therapy;

(4) For assessing the risk of cardiovascular diseases;

(5) For use in diabetes management;

(6) For identifying or inferring the identity of a microorganism directly from clinical material;

(7) For detection of antibodies to microorganisms other than immunoglobulin G (IgG) or IgG assays

when the results are not qualitative, or are used to determine immunity, or the assay is intended for use in matrices other than serum or plasma;

(8) For noninvasive testing as defined in §812.3(k) of this chapter; and

(9) For near patient testing (point of care).

[65 FR 2322, Jan. 14, 2000]

Subpart B—Diagnostic Devices

§892.1000 Magnetic resonance diagnostic device.

(a) *Identification.* A magnetic resonance diagnostic device is intended for general diagnostic use to present images which reflect the spatial distribution and/or magnetic resonance spectra which reflect frequency and distribution of nuclei exhibiting nuclear magnetic resonance. Other physical parameters derived from the images and/or spectra may also be produced. The device includes hydrogen-1 (proton) imaging, sodium-23 imaging, hydrogen-1 spectroscopy, phosphorus-31 spectroscopy, and chemical shift imaging (preserving simultaneous frequency and spatial information).

(b) *Classification.* Class II.

[53 FR 5078, Feb. 1, 1989]

§892.1100 Scintillation (gamma) camera.

(a) *Identification.* A scintillation (gamma) camera is a device intended to image the distribution of radionuclides in the body by means of a photon radiation detector. This generic type of device may include signal analysis and display equipment, patient and equipment supports, radionuclide anatomical markers, component parts, and accessories.

(b) *Classification.* Class I (general controls).

[55 FR 48443, Nov. 20, 1990, as amended at 66 FR 46953, Sept. 10, 2001]

§892.1110 Positron camera.

(a) *Identification.* A positron camera is a device intended to image the distribution of positron-emitting radionuclides in the body. This generic type of device may include signal analysis and display equipment, patient and equipment supports, radionuclide anatomical markers, component parts, and accessories.

(b) *Classification.* Class I (general controls).

[55 FR 48444, Nov. 20, 1990, as amended at 66 FR 46953, Sept. 10, 2001]

§892.1130 Nuclear whole body counter.

(a) *Identification.* A nuclear whole body counter is a device intended to measure the amount of radionuclides in the entire body. This generic type of device may include signal analysis and display equipment, patient and equipment supports, component parts, and accessories.

(b) *Classification.* Class I (general controls). The device is exempt from the premarket notification procedures in subpart E of part 807 of this chapter, subject to the limitations in §892.9.

[53 FR 1567, Jan. 20, 1988, as amended at 59 FR 63015, Dec. 7, 1994; 66 FR 38818, July 25, 2001]

[55 FR 48444, Nov. 20, 1990]

§892.1170 Bone densitometer.

(a) *Identification.* A bone densitometer is a device intended for medical purposes to measure bone density and mineral content by x-ray or gamma ray transmission measurements through the bone and adjacent tissues. This generic type of device may include signal analysis and display equipment, patient and equipment supports, component parts, and accessories.

(b) *Classification.* Class II.

§892.1200 Emission computed tomography system.

(a) *Identification.* An emission computed tomography system is a device intended to detect the location and distribution of gamma ray- and positron-emitting radionuclides in the body and produce cross-sectional images through computer reconstruction of the data. This generic type of device may include signal analysis and display equipment, patient and equipment supports, radionuclide anatomical markers, component parts, and accessories.

(b) *Classification.* Class II.

§ 892.1220 Fluorescent scanner.

(a) *Identification.* A fluorescent scanner is a device intended to measure the induced fluorescent radiation in the body by exposing the body to certain x-rays or low-energy gamma rays. This generic type of device may include signal analysis and display equipment, patient and equipment supports, component parts and accessories.

(b) *Classification.* Class II.

§ 892.1300 Nuclear rectilinear scanner.

(a) *Identification.* A nuclear rectilinear scanner is a device intended to image the distribution of radionuclides in the body by means of a detector (or detectors) whose position moves in two directions with respect to the patient. This generic type of device may include signal analysis and display equipment, patient and equipment supports, radionuclide anatomical markers, component parts, and accessories.

(b) *Classification.* Class I (general controls). The device is exempt from the premarket notification procedures in subpart E of part 807 of this chapter, subject to the limitations in § 892.9.

[55 FR 48444, Nov. 20, 1990, as amended at 65 FR 2322, Jan. 14, 2000; 66 FR 38818, July 25, 2001]

§ 892.1310 Nuclear tomography system.

(a) *Identification.* A nuclear tomography system is a device intended to detect nuclear radiation in the body and produce images of a specific cross-sectional plane of the body by blurring or eliminating detail from other planes. This generic type of devices may include signal analysis and display equipment, patient and equipment supports, radionuclide anatomical markers, component parts, and accessories.

(b) *Classification.* Class II.

§ 892.1320 Nuclear uptake probe.

(a) *Identification.* A nuclear uptake probe is a device intended to measure the amount of radionuclide taken up by a particular organ or body region. This generic type of device may include a single or multiple detector probe, signal analysis and display equipment, patient and equipment supports, component parts, and accessories.

(b) *Classification.* Class I (general controls). The device is exempt from the premarket notification procedures in subpart E of part 807 of this chapter subject to § 892.9.

[55 FR 48444, Nov. 20, 1990, as amended at 65 FR 2322, Jan. 14, 2000]

§ 892.1330 Nuclear whole body scanner.

(a) *Identification.* A nuclear whole body scanner is a device intended to measure and image the distribution of radionuclides in the body by means of a wide-aperture detector whose position moves in one direction with respect to the patient. This generic type of device may include signal analysis and display equipment, patient and equipment supports, radionuclide anatomical markers, component parts, and accessories.

(b) *Classification.* Class I (general controls). The device is exempt from the premarket notification procedures in subpart E of part 807 of this chapter subject to § 892.9.

[55 FR 48444, Nov. 20, 1990, as amended at 65 FR 2322, Jan. 14, 2000]

§ 892.1350 Nuclear scanning bed.

(a) *Identification.* A nuclear scanning bed is an adjustable bed intended to support a patient during a nuclear medicine procedure.

(b) *Classification.* Class I (general controls). The device is exempt from the premarket notification procedures in subpart E of part 807 of this chapter subject to § 892.9.

[55 FR 48444, Nov. 20, 1990, as amended at 59 FR 63015, Dec. 7, 1994; 65 FR 2322, Jan. 14, 2000]

§ 892.1360 Radionuclide dose calibrator.

(a) *Identification.* A radionuclide dose calibrator is a radiation detection device intended to assay radionuclides before their administration to patients.

(b) *Classification.* Class II.

§ 892.1370 Nuclear anthropomorphic phantom.

(a) *Identification.* A nuclear anthropomorphic phantom is a human tissue facsimile that contains a radioactive source or a cavity in which a radioactive sample can be inserted. It is intended to calibrate nuclear uptake probes or other medical instruments.

(b) *Classification.* Class I (general controls). The device is exempt from the premarket notification procedures in subpart E of part 807 of this chapter, subject to the limitations in § 892.9.

[53 FR 1567, Jan. 20, 1988, as amended at 54 FR 13832, Apr. 5, 1989; 66 FR 38818, July 25, 2001]

§ 892.1380 Nuclear flood source phantom.

(a) *Identification.* A nuclear flood source phantom is a device that consists of a radiolucent container filled with a uniformly distributed solution of a desired radionuclide. It is intended to calibrate a medical gamma camera-collimator system for uniformity of response.

(b) *Classification.* Class I (general controls). The device is exempt from the premarket notification procedures in subpart E of part 807 of this chapter, subject to the limitations in § 892.9.

[53 FR 1567, Jan. 20, 1988, as amended at 54 FR 13832, Apr. 5, 1989; 66 FR 38819, July 25, 2001]

§ 892.1390 Radionuclide rebreathing system.

(a) *Identification.* A radionuclide rebreathing system is a device intended to be used to contain a gaseous or volatile radionuclide or a radionuclide-labeled aerosol and permit it to be respired by the patient during nuclear medicine ventilatory tests (testing process of exchange between the lungs and the atmosphere). This generic type of device may include signal analysis and display equipment, patient and equipment supports, component parts, and accessories.

(b) *Classification.* Class II.

§ 892.1400 Nuclear sealed calibration source.

(a) *Identification.* A nuclear sealed calibration source is a device that consists of an encapsulated reference radionuclide intended for calibration of medical nuclear radiation detectors.

(b) *Classification.* Class I (general controls). The device is exempt from the premarket notification procedures in subpart E of part 807 of this chapter, subject to the limitations in § 892.9.

[53 FR 1567, Jan. 20, 1988, as amended at 54 FR 13832, Apr. 5, 1989; 66 FR 38819, July 25, 2001]

§ 892.1410 Nuclear electrocardiograph synchronizer.

(a) *Identification.* A nuclear electrocardiograph synchronizer is a device intended for use in nuclear radiology to relate the time of image formation to the cardiac cycle during the production of dynamic cardiac images.

(b) *Classification.* Class I (general controls). The device is exempt from the premarket notification procedures in subpart E of part 807 of this chapter subject to § 892.9.

[55 FR 48444, Nov. 20, 1990, as amended at 65 FR 2322, Jan. 14, 2000]

§ 892.1420 Radionuclide test pattern phantom.

(a) *Identification.* A radionuclide test pattern phantom is a device that consists of an arrangement of radiopaque or radioactive material sealed in a solid pattern intended to serve as a test for a performance characteristic of a nuclear medicine imaging device.

(b) *Classification.* Class I (general controls). The device is exempt from the premarket notification procedures in subpart E of part 807 of this chapter, subject to the limitations in § 892.9.

[53 FR 1567, Jan. 20, 1988, as amended at 54 FR 13832, Apr. 5, 1989; 66 FR 38819, July 25, 2001]

§ 892.1540 Nonfetal ultrasonic monitor.

(a) *Identification.* A nonfetal ultrasonic monitor is a device that projects a continuous high-frequency sound wave into body tissue other than a fetus to determine frequency changes (doppler shift) in the reflected wave and is intended for use in the investigation of nonfetal blood flow and other nonfetal body tissues in motion. This generic type of device may include signal analysis and display equipment, patient and equipment supports, component parts, and accessories.

(b) *Classification.* Class II.

§ 892.1550 Ultrasonic pulsed doppler imaging system.

(a) *Identification.* An ultrasonic pulsed doppler imaging system is a device that combines the features of continuous wave doppler-effect technology with pulsed-echo effect technology and is intended to determine stationary body tissue characteristics, such as depth or location of tissue interfaces or dynamic tissue characteristics such as velocity of blood or tissue motion. This generic type of device may include signal analysis and display equipment, patient and equipment supports, component parts, and accessories.

(b) *Classification.* Class II.

§ 892.1560 Ultrasonic pulsed echo imaging system.

(a) *Identification.* An ultrasonic pulsed echo imaging system is a device intended to project a pulsed sound beam into body tissue to determine the depth or location of the tissue interfaces and to measure the duration of an acoustic pulse from the transmitter to the tissue interface and back to the receiver. This generic type of device may include signal analysis and display equipment, patient and equipment supports, component parts, and accessories.

(b) *Classification.* Class II.

§ 892.1570 Diagnostic ultrasonic transducer.

(a) *Identification.* A diagnostic ultrasonic transducer is a device made of a piezoelectric material that converts electrical signals into acoustic signals and acoustic signals into electrical signals and intended for use in diagnostic ultrasonic medical devices. Accessories of this generic type of device may include transmission media for acoustically coupling the transducer to the body surface, such as acoustic gel, paste, or a flexible fluid container.

(b) *Classification.* Class II.

§ 892.1600 Angiographic x-ray system.

(a) *Identification.* An angiographic x-ray system is a device intended for radiologic visualization of the heart, blood vessels, or lymphatic system during or after injection of a contrast medium. This generic type of device may include signal analysis and display equipment, patient and equipment supports, component parts, and accessories.

(b) *Classification.* Class II.

§ 892.1610 Diagnostic x-ray beam-limiting device.

(a) *Identification.* A diagnostic x-ray beam-limiting device is a device such as a collimator, a cone, or an aperture intended to restrict the dimensions of a diagnostic x-ray field by limiting the size of the primary x-ray beam.

(b) *Classification.* Class II.

§ 892.1620 Cine or spot fluorographic x-ray camera.

(a) *Identification.* A cine or spot fluorographic x-ray camera is a device intended to photograph diagnostic images produced by x-rays with an image intensifier.

(b) *Classification.* Class II.

§ 892.1630 Electrostatic x-ray imaging system.

(a) *Identification.* An electrostatic x-ray imaging system is a device intended for medical purposes that uses an electrostatic field across a semiconductive plate, a gas-filled chamber, or other similar device to convert a pattern of x-radiation into an electrostatic image and, subsequently, into a visible image. This generic type of device may include signal analysis and display equipment, patient and equipment supports, component parts, and accessories.

(b) *Classification.* Class II.

§ 892.1640 Radiographic film marking system.

(a) *Identification.* A radiographic film marking system is a device intended for medical purposes to add identification and other information onto radiographic film by means of exposure to visible light.

(b) *Classification.* Class I (general controls). The device is exempt from the premarket notification procedures in subpart E of part 807 of this chapter, subject to the limitations in § 892.9.

[55 FR 48444, Nov. 20, 1990, as amended at 59 FR 63015, Dec. 7, 1994; 66 FR 38819, July 25, 2001]

§ 892.1650 Image-intensified fluoroscopic x-ray system.

(a) *Identification.* An image-intensified fluoroscopic x-ray system is a device intended to visualize anatomical structures by converting a pattern of x-radiation into a visible image through electronic amplification. This generic type of device may include signal analysis and display equipment, patient and equipment supports, component parts, and accessories.

(b) *Classification.* Class II. When intended as an accessory to the device described in paragraph (a) of this section, the fluoroscopic compression device is exempt from the premarket notification procedures in subpart E of part 807 of this chapter subject to § 892.9.

[53 FR 1567, Jan. 20, 1988, as amended at 66 FR 57369, Nov. 15, 2001]

§ 892.1660 Non-image-intensified fluoroscopic x-ray system.

(a) *Identification.* A non-image-intensified fluoroscopic x-ray system is a device intended to be used to visualize anatomical structures by using a fluorescent screen to convert a pattern of x-radiation into a visible image. This generic type of device may include signal analysis and display equipment, patient and equipment supports, component parts, and accessories.

(b) *Classification.* Class II.

§ 892.1670 Spot-film device.

(a) *Identification.* A spot-film device is an electromechanical component of a fluoroscopic x-ray system that is intended to be used for medical purposes to position a radiographic film cassette to obtain radiographs during fluoroscopy.

(b) *Classification.* Class II.

§ 892.1680 Stationary x-ray system.

(a) *Identification.* A stationary x-ray system is a permanently installed diagnostic system intended to generate and control x-rays for examination of various anatomical regions. This generic type of device may include signal analysis and display equipment, patient and equipment supports, component parts, and accessories.

(b) *Classification.* Class II.

§ 892.1700 Diagnostic x-ray high voltage generator.

(a) *Identification.* A diagnostic x-ray high voltage generator is a device that is intended to supply and control the electrical energy applied to a diagnostic x-ray tube for medical purposes. This generic type of device may include a converter that changes alternating current to direct current, filament transformers for the x-ray tube, high voltage switches, electrical protective devices, or other appropriate elements.

(b) *Classification.* Class I (general controls). The device is exempt from the premarket notification procedures in subpart E of part 807 of this chapter, subject to the limitations in § 892.9.

[53 FR 1567, Jan. 20, 1988, as amended at 61 FR 1125, Jan. 16, 1996; 66 FR 38819, July 25, 2001]

§ 892.1710 Mammographic x-ray system.

(a) *Identification.* A mammographic x-ray system is a device intended to be used to produce radiographs of the breast. This generic type of device may include signal analysis and display equipment, patient and equipment supports, component parts, and accessories.

(b) *Classification.* Class II.

§ 892.1720 Mobile x-ray system.

(a) *Identification.* A mobile x-ray system is a transportable device system intended to be used to generate and control x-ray for diagnostic procedures. This generic type of device may include signal analysis and display equipment, patient and equipment supports, component parts, and accessories.

(b) *Classification.* Class II.

§ 892.1730 Photofluorographic x-ray system.

(a) *Identification.* A photofluorographic x-ray system is a device that includes a fluoroscopic x-ray unit and a camera intended to be used to produce, then photograph, a fluoroscopic image of the body. This generic type of device may include signal analysis and display equipment, patient and equipment supports, component parts, and accessories.

(b) *Classification.* Class II.

§ 892.1740 Tomographic x-ray system.

(a) *Identification.* A tomographic x-ray system is an x-ray device intended to be used to produce radiologic images of a specific cross-sectional plane of the body by blurring or eliminating detail from other planes. This generic type of device may include signal analysis and display equipment, patient and equipment supports, component parts, and accessories.

(b) *Classification.* Class II.

§ 892.1750 Computed tomography x-ray system.

(a) *Identification.* A computed tomography x-ray system is a diagnostic x-ray system intended to produce cross-sectional images of the body by computer reconstruction of x-ray transmission data from the same axial plane taken at different angles. This generic type of device may include signal analysis and display equipment, patient and equipment supports, component parts, and accessories.

(b) *Classification.* Class II.

§ 892.1760 Diagnostic x-ray tube housing assembly.

(a) *Identification.* A diagnostic x-ray tube housing assembly is an x-ray generating tube encased in a radiation-shielded housing that is intended for diagnostic purposes. This generic type of device may include high voltage and filament transformers or other appropriate components.

(b) *Classification.* Class I (general controls). The device is exempt from the premarket notification procedures in subpart E of part 807 of this chapter, subject to the limitations in § 892.9.

[53 FR 1567, Jan. 20, 1988, as amended at 61 FR 1125, Jan. 16, 1996; 66 FR 38819, July 25, 2001]

§ 892.1770 Diagnostic x-ray tube mount.

(a) *Identification.* A diagnostic x-ray tube mount is a device intended to support and to position the diagnostic x-ray tube housing assembly for a medical radiographic procedure.

(b) *Classification.* Class I (general controls). The device is exempt from the premarket notification procedures in subpart E of part 807 of this chapter, subject to the limitations in § 892.9.

[53 FR 1567, Jan. 20, 1988, as amended at 61 FR 1125, Jan. 16, 1996; 66 FR 38819, July 25, 2001]

§ 892.1820 Pneumoencephalographic chair.

(a) *Identification.* A pneumoencephalographic chair is a chair intended to support and position a patient during pneumoencephalography (x-ray imaging of the brain).

(b) *Classification.* Class II.

§ 892.1830 Radiologic patient cradle.

(a) *Identification.* A radiologic patient cradle is a support device intended to be used for rotational positioning about the longitudinal axis of a patient during radiologic procedures.

(b) *Classification.* Class I (general controls). The device is exempt from the premarket notification procedures in subpart E of part 807 of this chapter, subject to the limitations in § 892.9.

[53 FR 1567, Jan. 20, 1988, as amended at 61 FR 1125, Jan. 16, 1996; 66 FR 38819, July 25, 2001]

§ 892.1840 Radiographic film.

(a) *Identification.* Radiographic film is a device that consists of a thin sheet of radiotransparent material coated on one or both sides with a photographic emulsion intended to record images during diagnostic radiologic procedures.

(b) *Classification.* Class I (general controls). The device is exempt from the premarket notification procedures in subpart E of part 807 of this chapter, subject to the limitations in § 892.9.

[53 FR 1567, Jan. 20, 1988, as amended at 66 FR 38819, July 25, 2001]

§ 892.1850 Radiographic film cassette.

(a) *Identification.* A radiographic film cassette is a device intended for use during diagnostic x-ray procedures to hold a radiographic film in close contact with an x-ray intensifying screen and to provide a light-proof enclosure for direct exposure of radiographic film.

(b) *Classification.* Class II.

§ 892.1860 Radiographic film/cassette changer.

(a) *Identification.* A radiographic film/cassette changer is a device intended to be used during a radiologic procedure to move a radiographic film or cassette between x-ray exposures and to position it during the exposure.

(b) *Classification.* Class II.

§ 892.1870 Radiographic film/cassette changer programmer.

(a) *Identification.* A radiographic film/cassette changer programmer is a device intended to be used to control the operations of a film or cassette changer during serial medical radiography.

(b) *Classification.* Class II.

§ 892.1880 Wall-mounted radiographic cassette holder.

(a) *Identification.* A wall-mounted radiographic cassette holder is a device that is a support intended to hold and position radiographic cassettes for a radiographic exposure for medical use.

(b) *Classification.* Class I (general controls). The device is exempt from the premarket notification procedures in subpart E of part 807 of this chapter, subject to the limitations in § 892.9.

[53 FR 1567, Jan. 20, 1988, as amended at 61 FR 1125, Jan. 16, 1996; 66 FR 38819, July 25, 2001]

§ 892.1890 Radiographic film illuminator.

(a) *Identification.* A radiographic film illuminator is a device containing a visible light source covered with a translucent front that is intended to be used to view medical radiographs.

(b) *Classification.* Class I (general controls). The device is exempt from the premarket notification procedures in subpart E of part 807 of this chapter subject to § 892.9.

[55 FR 48444, Nov. 20, 1990, as amended at 65 FR 2323, Jan. 14, 2000]

§ 892.1900 Automatic radiographic film processor.

(a) *Identification.* An automatic radiographic film processor is a device intended to be used to develop, fix, wash, and dry automatically and continuously film exposed for medical purposes.

(b) *Classification.* Class II.

[55 FR 48444, Nov. 20, 1990]

§ 892.1910 Radiographic grid.

(a) *Identification.* A radiographic grid is a device that consists of alternating radiolucent and radiopaque strips intended to be placed between the patient and the image receptor to reduce the amount of scattered radiation reaching the image receptor.

(b) *Classification.* Class I (general controls). The device is exempt from the premarket notification procedures in subpart E of part 807 of this chapter subject to § 892.9.

[53 FR 1567, Jan. 20, 1988, as amended at 65 FR 2323, Jan. 14, 2000]

§ 892.1920 Radiographic head holder.

(a) *Identification.* A radiographic head holder is a device intended to position the patient's head during a radiographic procedure.

(b) *Classification.* Class I (general controls). The device is exempt from the premarket notification procedures in subpart E of part 807 of this chapter, subject to the limitations in § 892.9. The device is also exempt from the current good manufacturing practice regulations in part 820 of this chapter, with the exception of § 820.180, with respect to general requirements concerning records, and § 820.198, with respect to complaint files.

[53 FR 1567, Jan. 20, 1988, as amended at 66 FR 38819, July 25, 2001]

§ 892.1940 Radiologic quality assurance instrument.

(a) *Identification.* A radiologic quality assurance instrument is a device intended for medical purposes to measure a physical characteristic associated with another radiologic device.

(b) *Classification.* Class I (general controls). The device is exempt from the premarket notification procedures in subpart E of part 807 of this chapter, subject to the limitations in § 892.9. The device is also exempt from the current good manufacturing practice regulations in part 820 of this chapter, with the exception of § 820.180, with respect to general requirements concerning

records, and § 820.198, with respect to complaint files.

[53 FR 1567, Jan. 20, 1988, as amended at 66 FR 38819, July 25, 2001]

§ 892.1950 Radiographic anthropomorphic phantom.

(a) *Identification.* A radiographic anthropomorphic phantom is a device intended for medical purposes to simulate a human body for positioning radiographic equipment.

(b) *Classification.* Class I (general controls). The device is exempt from the premarket notification procedures in subpart E of part 807 of this chapter, subject to the limitations in § 892.9. The device is also exempt from the current good manufacturing practice regulations in part 820 of this chapter, with the exception of § 820.180, with respect to general requirements concerning records, and § 820.198, with respect to complaint files.

[53 FR 1567, Jan. 20, 1988, as amended at 66 FR 38819, July 25, 2001]

§ 892.1960 Radiographic intensifying screen.

(a) *Identification.* A radiographic intensifying screen is a device that is a thin radiolucent sheet coated with a luminescent material that transforms incident x-ray photons into visible light and intended for medical purposes to expose radiographic film.

(b) *Classification.* Class I (general controls). The device is exempt from the premarket notification procedures in subpart E of part 807 of this chapter subject to § 892.9.

[53 FR 1567, Jan. 20, 1988, as amended at 65 FR 2323, Jan. 14, 2000]

§ 892.1970 Radiographic ECG/respirator synchronizer.

(a) *Identification.* A radiographic ECG/respirator synchronizer is a device intended to be used to coordinate an x-ray film exposure with the signal from an electrocardiograph (ECG) or respirator at a predetermined phase of the cardiac or respiratory cycle.

(b) *Classification.* Class I (general controls). The device is exempt from the premarket notification procedures in subpart E of part 807 of this chapter subject to § 892.9.

[55 FR 48444, Nov. 20, 1990, as amended at 65 FR 2323, Jan. 14, 2000]

§ 892.1980 Radiologic table.

(a) *Identification.* A radiologic table is a device intended for medical purposes to support a patient during radiologic procedures. The table may be fixed or tilting and may be electrically powered.

(b) *Classification.* Class II (special controls). The device is exempt from the premarket notification procedures in subpart E of part 807 of this chapter subject to § 892.9.

[53 FR 1567, Jan. 20, 1988, as amended at 63 FR 59231, Nov. 3, 1998]

§ 892.1990 Transilluminator for breast evaluation.

(a) *Identification.* A transilluminator, also known as a diaphanoscope or lightscanner, is an electrically powered device that uses low intensity emissions of visible light and near-infrared radiation (approximately 700–1050 nanometers (nm)), transmitted through the breast, to visualize translucent tissue for the diagnosis of cancer, other conditions, diseases, or abnormalities.

(b) *Classification.* Class III (premarket approval).

(c) *Date premarket approval (PMA) or notice of completion of a product development protocol (PDP) is required.* The effective date of the requirement for premarket approval has not been established. See § 892.3.

[60 FR 36639, July 18, 1995]

§ 892.2010 Medical image storage device.

(a) *Identification.* A medical image storage device is a device that provides electronic storage and retrieval functions for medical images. Examples include devices employing magnetic and optical discs, magnetic tape, and digital memory.

(b) *Classification.* Class I (general controls). The device is exempt from the premarket notification procedures in

subpart E of part 807 of this chapter subject to § 892.9.

[63 FR 23387, Apr. 29, 1998; 63 FR 44998, Aug. 24, 1998, as amended at 65 FR 2323, Jan. 14, 2000]

§ 892.2020 Medical image communications device.

(a) *Identification.* A medical image communications device provides electronic transfer of medical image data between medical devices. It may include a physical communications medium, modems, interfaces, and a communications protocol.

(b) *Classification.* Class I (general controls). The device is exempt from the premarket notification procedures in subpart E of part 807 of this chapter subject to § 892.9.

[63 FR 23387, Apr. 29, 1998; 63 FR 44998, Aug. 24, 1998, as amended at 65 FR 2323, Jan. 14, 2000]

§ 892.2030 Medical image digitizer.

(a) *Identification.* A medical image digitizer is a device intended to convert an analog medical image into a digital format. Examples include Iystems employing video frame grabbers, and scanners which use lasers or charge-coupled devices.

(b) *Classification.* Class II (special controls; voluntary standards—Digital Imaging and Communications in Medicine (DICOM) Std., Joint Photographic Experts Group (JPEG) Std.).

[63 FR 23387, Apr. 29, 1998]

§ 892.2040 Medical image hardcopy device.

(a) *Identification.* A medical image hardcopy device is a device that produces a visible printed record of a medical image and associated identification information. Examples include multiformat cameras and laser printers.

(b) *Classification.* Class II (special controls; voluntary standards—Digital Imaging and Communications in Medicine (DICOM) Std., Joint Photographic Experts Group (JPEG) Std., Society of Motion Picture and Television Engineers (SMPTE) Test Pattern).

[63 FR 23387, Apr. 29, 1998]

§ 892.2050 Picture archiving and communications system.

(a) *Identification.* A picture archiving and communications system is a device that provides one or more capabilities relating to the acceptance, transfer, display, storage, and digital processing of medical images. Its hardware components may include workstations, digitizers, communications devices, computers, video monitors, magnetic, optical disk, or other digital data storage devices, and hardcopy devices. The software components may provide functions for performing operations related to image manipulation, enhancement, compression or quantification.

(b) *Classification.* Class II (special controls; voluntary standards—Digital Imaging and Communications in Medicine (DICOM) Std., Joint Photographic Experts Group (JPEG) Std., Society of Motion Picture and Television Engineers (SMPTE) Test Pattern).

[63 FR 23387, Apr. 29, 1998]

Subparts C–E [Reserved]

Subpart F—Therapeutic Devices

§ 892.5050 Medical charged-particle radiation therapy system.

(a) *Identification.* A medical charged-particle radiation therapy system is a device that produces by acceleration high energy charged particles (e.g., electrons and protons) intended for use in radiation therapy. This generic type of device may include signal analysis and display equipment, patient and equipment supports, treatment planning computer programs, component parts, and accessories.

(b) *Classification.* Class II. When intended for use as a quality control system, the film dosimetry system (film scanning system) included as an accessory to the device described in paragraph (a) of this section, is exempt from the premarket notification procedures in subpart E of part 807 of this chapter subject to the limitations in § 892.9.

[53 FR 1567, Jan. 20, 1988, as amended at 64 FR 1125, Jan. 8, 1999]

§ 892.5300 Medical neutron radiation therapy system.

(a) *Identification.* A medical neutron radiation therapy system is a device intended to generate high-energy neutrons for radiation therapy. This generic type of device may include signal analysis and display equipment, patient and equipment support, treatment planning computer programs, component parts, and accessories.

(b) *Classification.* Class II.

§ 892.5650 Manual radionuclide applicator system.

(a) *Identification.* A manual radionuclide applicator system is a manually operated device intended to apply a radionuclide source into the body or to the surface of the body for radiation therapy. This generic type of device may include patient and equipment supports, component parts, treatment planning computer programs, and accessories.

(b) *Classification.* Class I (general controls). The device is exempt from the premarket notification procedures in subpart E of part 807 of this chapter subject to § 892.9.

[53 FR 1567, Jan. 20, 1988, as amended at 65 FR 2323, Jan. 14, 2000]

§ 892.5700 Remote controlled radionuclide applicator system.

(a) *Identification.* A remote controlled radionuclide applicator system is an electromechanical or pneumatic device intended to enable an operator to apply, by remote control, a radionuclide source into the body or to the surface of the body for radiation therapy. This generic type of device may include patient and equipment supports, component parts, treatment planning computer programs, and accessories.

(b) *Classification.* Class II.

§ 892.5710 Radiation therapy beam-shaping block.

(a) *Identification.* A radiation therapy beam-shaping block is a device made of a highly attenuating material (such as lead) intended for medical purposes to modify the shape of a beam from a radiation therapy source.

(b) *Classification.* Class II.

§ 892.5730 Radionuclide brachytherapy source.

(a) *Identification.* A radionuclide brachytherapy source is a device that consists of a radionuclide which may be enclosed in a sealed container made of gold, titanium, stainless steel, or platinum and intended for medical purposes to be placed onto a body surface or into a body cavity or tissue as a source of nuclear radiation for therapy.

(b) *Classification.* Class II.

§ 892.5740 Radionuclide teletherapy source.

(a) *Identification.* A radionuclide teletherapy source is a device consisting of a radionuclide enclosed in a sealed container. The device is intended for radiation therapy, with the radiation source located at a distance from the patient's body.

(b) *Classification.* Class I (general controls). The device is exempt from the premarket notification procedures in subpart E of part 807 of this chapter, subject to the limitations in § 892.9.

[53 FR 1567, Jan. 20, 1988, as amended at 59 FR 63015, Dec. 7, 1994; 66 FR 38819, July 25, 2001]

§ 892.5750 Radionuclide radiation therapy system.

(a) *Identification.* A radionuclide radiation therapy system is a device intended to permit an operator to administer gamma radiation therapy, with the radiation source located at a distance from the patient's body. This generic type of device may include signal analysis and display equipment, patient and equipment supports, treatment planning computer programs, component parts (including beam-limiting devices), and accessories.

(b) *Classification.* Class II.

§ 892.5770 Powered radiation therapy patient support assembly.

(a) *Identification.* A powered radiation therapy patient support assembly is an electrically powered adjustable couch intended to support a patient during radiation therapy.

(b) *Classification.* Class II.

§ 892.5780 Light beam patient position indicator.

(a) *Identification.* A light beam patient position indicator is a device that projects a beam of light (incoherent light or laser) to determine the alignment of the patient with a radiation beam. The beam of light is intended to be used during radiologic procedures to ensure proper positioning of the patient and to monitor alignment of the radiation beam with the patient's anatomy.

(b) *Classification.* Class I (general controls). The device is exempt from the premarket notification procedures in subpart E of part 807 of this chapter, subject to the limitations in § 892.9.

[53 FR 1567, Jan. 20, 1988, as amended at 61 FR 1125, Jan. 16, 1996; 66 FR 38819, July 25, 2001]

§ 892.5840 Radiation therapy simulation system.

(a) *Identification.* A radiation therapy simulation system is a fluoroscopic or radiographic x-ray system intended for use in localizing the volume to be exposed during radiation therapy and confirming the position and size of the therapeutic irradiation field produced. This generic type of device may include signal analysis and display equipment, patient and equipment supports, treatment planning computer programs, component parts, and accessories.

(b) *Classification.* Class II.

§ 892.5900 X-ray radiation therapy system.

(a) *Identification.* An x-ray radiation therapy system is a device intended to produce and control x-rays used for radiation therapy. This generic type of device may include signal analysis and display equipment, patient and equipment supports, treatment planning computer programs, component parts, and accessories.

(b) *Classification.* Class II.

§ 892.5930 Therapeutic x-ray tube housing assembly.

(a) *Identification.* A therapeutic x-ray tube housing assembly is an x-ray generating tube encased in a radiation-shielded housing intended for use in radiation therapy. This generic type of device may include high-voltage and filament transformers or other appropriate components when contained in radiation-shielded housing.

(b) *Classification.* Class II.

Subpart G—Miscellaneous Devices

§ 892.6500 Personnel protective shield.

(a) *Identification.* A personnel protective shield is a device intended for medical purposes to protect the patient, the operator, or other persons from unnecessary exposure to radiation during radiologic procedures by providing an attenuating barrier to radiation. This generic type of device may include articles of clothing, furniture, and movable or stationary structures.

(b) *Classification.* Class I (general controls). The device is exempt from the premarket notification procedures in subpart E of part 807 of this chapter subject to § 892.9.

[53 FR 1567, Jan. 20, 1988, as amended at 61 FR 1125, Jan. 16, 1996; 65 FR 2323, Jan. 14, 2000]

PART 895—BANNED DEVICES

Subpart A—General Provisions

Sec.
895.1 Scope.
895.20 General.
895.21 Procedures for banning a device.
895.22 Submission of data and information by the manufacturer, distributor, or importer.
895.25 Labeling.
895.30 Special effective date.

Subpart B—Listing of Banned Devices

895.101 Prosthetic hair fibers.

AUTHORITY: 21 U.S.C. 352, 360f, 360h, 360i, 371.

SOURCE: 44 FR 29221, May 18, 1979, unless otherwise noted.

Subpart A—General Provisions

§ 895.1 Scope.

(a) This part describes the procedures by which the Commissioner may institute proceedings to make a device intended for human use that presents

208-072 D-18

substantial deception or an unreasonable and substantial risk of illness or injury a banned device.

(b) This part applies to any "device", as defined in section 201(h) of the Federal Food, Drug, and Cosmetic Act (act) that is intended for human use.

(c) A device that is made a banned device in accordance with this part is adulterated under section 501(g) of the act. A restricted device that is banned may also be misbranded under section 502(q) of the act.

(d) Although this part does not cover devices intended for animal use, the manufacturer, distributor, importer, or any other person(s) responsible for the labeling of the device that is banned cannot avoid the ban by relabeling the device for veterinary use. A device that has been banned from human use but that also has a valid veterinary use may be marketed for use as a veterinary device only under the following conditions: The device shall comply with all requirements applicable to veterinary devices under the Federal Food, Drug, and Cosmetic Act and this chapter, and the label for the device shall bear the following statement: "For Veterinary Use Only. Caution: Federal law prohibits the distribution of this device for human use." A device so labeled, however, that is determined by the Food and Drug Administration to be intended for human use, will be considered to be a banned device. In determining whether such a device is intended for human use, the Food and Drug Administration will consider, among other things, the ultimate destination of the device.

§ 895.20 General.

The Commissioner may initiate a proceeding to make a device a banned device whenever the Commissioner finds, on the basis of all available data and information, that the device presents substantial deception or an unreasonable and substantial risk of illness or injury that the Commissioner determines cannot be, or has not been, corrected or eliminated by labeling or by a change in labeling, or by a change in advertising if the device is a restricted device.

[44 FR 29221, May 18, 1979, as amended at 57 FR 58405, Dec. 10, 1992]

§ 895.21 Procedures for banning a device.

(a) Before initiating a proceeding to make a device a banned device, the Commissioner shall find that the continued marketing of the device presents a substantial deception or an unreasonable and substantial risk of illness or injury.

(1) In determining whether the deception or risk of illness or injury is substantial, the Commissioner will consider whether the deception or risk posed by continued marketing of the device, or continued marketing of the device as presently labeled, is important, material, or significant in relation to the benefit to the public health from its continued marketing.

(2) In determining whether a device is deceptive, the Commissioner will consider whether users of the device may be deceived or otherwise harmed by the device. The Commissioner is not required to determine that there was an intent on the part of the manufacturer, distributor, importer, or any other responsible person(s) to mislead or otherwise harm users of the device or that there exists any actual proof of deception of, or injury to, an individual.

(3) In determining whether a device presents deception or risk of illness or injury, the Commissioner will consider all available data and information, including data and information that the Commissioner may obtain under other provisions of the act, data and information that may be supplied by the manufacturer, distributor, or importer of the device under § 895.22, and data and information voluntarily submitted by any other interested persons.

(b) Before initiating a proceeding to make a device a banned device, the Commissioner of Food and Drugs (the Commissioner) may consult with the panel established under section 513 of the act that has expertise with respect to the type of device under consideration. The consultation with the panel may occur at a regular or specially scheduled panel meeting or may be accomplished by correspondence or telephone conversation with panel members. The Commissioner may request that the panel submit in writing any advice on the device under consideration. The Commissioner will record in written memoranda any oral communications with a panel or its members.

(c) If the Commissioner determines that any substantial deception or unreasonable and substantial risk of illness or injury or any unreasonable, direct, and substantial danger to the health of individuals presented by a device can be corrected or eliminated by labeling or change in labeling, or change in advertising if the device is a restricted device, the Commissioner will notify the responsible person of the required labeling or change in labeling or change in advertising in accordance with § 895.25. If such required relabeling or change in advertising is not accomplished in accordance with § 895.25, the Commissioner may initiate a proceeding to ban the device in accordance with § 895.21(d) and, when appropriate, may establish a special effective date in accordance with § 895.30.

(d) If the Commissioner decides to initiate a proceeding to make a device a banned device, a notice of proposed rulemaking will be published in the FEDERAL REGISTER to this effect. The notice will briefly summarize—

(1) The Commissioner's finding under paragraph (a) of this section that the device presents substantial deception or an unreasonable and substantial risk of illness or injury, and, when appropriate, the Commissioner's determination under § 895.30 that the deception or risk of illness or injury presents an unreasonable, direct, and substantial danger to the health of individuals;

(2) The reasons why the Commissioner initiated the proceeding;

(3) The evaluation of data and information obtained under other provisions of the act, submitted by the manufacturer, distributer, or importer of the device, or voluntarily submitted by any other interested persons under paragraph (a)(3) of this section, if any;

(4) The consultation with the panel, if any, under paragraph (b) of this section;

(5) The determination as to whether the deception or risk of illness or injury or the danger to the health of individuals could be corrected by labeling or change in labeling, or change in advertising if the device is a restricted device;

(6) The determination of whether the required labeling or change of labeling, or change in advertising if the device is a restricted device, if any, has been made in accordance with paragraph (c) of this section;

(7) The determination as to whether, and the reasons why, the banning should apply to devices already in commercial distribution or those already sold to the ultimate user, or both; and

(8) Any other data and information that the Commissioner believes are pertinent to the proceeding. The notice will afford all interested persons an opportunity to submit written comments within 30 days after the date of publication of the proposed regulation. All nonconfidential information upon which the proposed finding is based, including the recommendations of the panel, will be available for public review in the Division of Dockets Management, Food and Drug Administration.

The notice will afford all interested persons an opportunity to submit written comments and request an informal hearing, as defined in section 201(x) of the act, before the Food and Drug Administration within 30 days after the date of publication of the proposed regulation. If a request for an informal hearing is granted, the hearing will be conducted as a regulatory hearing under the applicable provisions of part 16 of this chapter. All nonconfidential information upon which the proposed finding is based, including the recommendations of the panel, will be available for public review in the office of the Division of Dockets Management, Food and Drug Administration.

(e)(1) If, after reviewing the administrative record of the regulatory hearing before the Food and Drug Administration, if any, the written comments received on the proposed regulation, and any additional available data and information, the Commissioner determines to ban a device, a final regulation to this effect will be published in the FEDERAL REGISTER. The final regulation will amend subpart B by adding the name or description of the device, or both, to the list of banned devices.

(2) If the Commissioner determines not to ban the device, a notice of withdrawal and termination of rulemaking

proceedings and reasons therefor will be published in the FEDERAL REGISTER.

(f) The effective date of a final regulation to make a device a banned device, promulgated under paragraph (e) of this section, will be the date of publication of the final regulation in the FEDERAL REGISTER unless the Commissioner, for reasons stated, determines that the effective date should be later than the date of the publication and specifies that date in the notice. Each such regulation will specify whether devices already in commercial distribution or sold to the ultimate user or both are banned.

(g) A regulation promulgated under paragraph (e) of this section is final agency action, subject to judicial review under section 517 of the act.

(h) Upon petition of any interested person submitted in accordance with § 10.30 of this chapter, or as a matter of discretion, the Commissioner may institute proceedings to amend or revoke a regulation that made a device a banned device if the Commissioner finds that the conditions that constituted the basis for the regulation banning the device are no longer applicable. When appropriate, the procedures in this section will be employed in such proceedings.

[44 FR 29221, May 18, 1979, as amended at 53 FR 11254, Apr. 6, 1988; 57 FR 58405, Dec. 10, 1992; 65 FR 43690, July 14, 2000]

§ 895.22 Submission of data and information by the manufacturer, distributor, or importer.

(a) A manufacturer, distributor, or importer of a device may be required to submit to the Food and Drug Administration all relevant and available data and information to enable the Commissioner to determine whether the device presents substantial deception, unreasonable and substantial risk of illness or injury, or unreasonable, direct, and substantial danger to the health of individuals. The data and information required by the Commissioner may include scientific or test data, reports, records, or other information, including data and information on whether the device is safe and effective for its intended use or when used as directed, whether the device performs according to the claims made for the device, and information on adulteration or misbranding. Any relevant information that is voluntarily submitted will also be reviewed.

(b) A manufacturer, distributor, or importer of a device required to submit data and information as provided in paragraph (a) of this section will be notified in writing by the Food and Drug Administration that such data and information shall be submitted. The written notification will advise the manufacturer, distributor, or importer of the device that the purpose for the request is to enable the Commissioner to determine whether any of the conditions listed in paragraph (a) of this section or § 895.30(a)(1) exists with respect to the device such that a proceeding should be initiated to make the device a banned device. When the required data and information can be identified by the Food and Drug Administration at the time of the notification, the agency will provide such identification to the manufacturer, distributor, or importer of the device.

(c) The required data and information shall be submitted to the Food and Drug Administration no more than 30 days after the date of receipt of the request, unless the Commissioner determines that the data and information shall be submitted by some other date and so informs the manufacturer, distributor, or importer, in which case the data and information shall be submitted on the date specified by the Commissioner.

(d) If the data or information submitted to the Food and Drug Administration is sufficient to persuade the Commissioner that the deception or risk of illness or injury or the danger to the health of individuals presented by a device could be corrected or eliminated by labeling or change in labeling, or change in advertising if the device is a restricted device, the Commissioner will proceed in accordance with § 895.25.

(e) If the data or information submitted to the Food and Drug Administration is insufficient to show that the device does not present a substantial deception or an unreasonable and substantial risk of illness or injury, or an unreasonable, direct, and substantial danger to the health of individuals, or if the manufacturer, distributor, or importer fails to submit the required information, the Commissioner may rely upon this insufficiency or failure to submit the required information in considering whether to initiate a proceeding to make the device a banned device under § 895.21(d) and, when appropriate, to establish a special effective date in accordance with § 895.30. The Commissioner may also initiate other regulatory action as provided in the act or this chapter.

§ 895.25 Labeling.

(a) If the Commissioner determines that the substantial deception or unreasonable and substantial risk of illness or injury or the unreasonable, direct, and substantial danger to the health of individuals presented by a device can be corrected or eliminated by labeling or a change in labeling, or change in advertising if the device is a restricted device, the Commissioner will provide written notice to the manufacturer, distributor, importer, or any other person(s) responsible for the labeling or advertising of the device specifying:

(1) The deception or risk of illness or injury or the danger to the health of individuals,

(2) The labeling or change in labeling, or change in advertising if the device is a restricted device, necessary to correct the deception or eliminate or reduce such risk or danger, and

(3) The period of time within which the labeling, change in labeling, or change in advertising must be accomplished.

(b) In specifying the labeling or change in labeling or change in advertising to correct the deception or to eliminate or reduce the risk of illness or injury or the danger to the health of individuals, the Commissioner may require the manufacturer, distributor, importer, or any other person(s) responsible for the labeling or advertising of the device to include in labeling for the device, and in advertising if the device is a restricted device, a statement, notice, or warning. Such statement, notice, or warning shall be in the manner and form prescribed by the Commissioner and shall identify the deception or risk of illness or injury or the unreasonable, direct, and substantial danger to the health of individuals associated with the device as previously labeled. Such statement, notice, or warning shall be used in the labeling and advertising of the device for a time period specified by the Commissioner on the basis of the degree of deception, risk of illness or injury, or danger to health; the frequency of sale of the device; the length of time the device has been on the market; the intended uses of the device; the method of its use; and any other factors that the Commissioner considers pertinent.

(c) The Commissioner will allow a manufacturer, distributor, importer, or any other person(s) responsible for the labeling or advertising of the device a reasonable time, considering the deception or risk of illness or injury or the danger to the health of individuals presented by the device, within which to accomplish the required labeling, change in labeling, and, if the device is a restricted device, any change in advertising. The Commissioner may, however, request that no additional devices be introduced into commerce until the labeling or change in labeling, or change in advertising is accomplished by the manufacturer, distributor, importer, or other person(s) responsible for the labeling or advertising of the device.

(d) If such voluntary action is not taken, the Commissioner may take action under other sections of the act to prevent the introduction of the devices into commerce. The Commissioner may consider the failure of a manufacturer, distributor, importer, or any other person(s) responsible for the labeling or advertising of the device to accomplish the required labeling or change in labeling, or change in advertising in accordance with this section as a basis for initiating a proceeding to make a device a banned device in accordance with § 895.21(d) and when appropriate to establish a special effective date in accordance with § 895.30.

§ 895.30 Special effective date.

(a) The Commissioner may declare a proposed regulation under § 895.21(d) to be effective upon its publication in the

FEDERAL REGISTER and until the effective date of any final action taken respecting the regulation if:

(1) The Commissioner determines, on the basis of all available data and information, that the deception or risk of illness or injury associated with use of the device that is subject to the regulation presents an unreasonable, direct, and substantial danger to the health of individuals, and

(2) Before the date of the publication of such regulation, the Commissioner notifies the domestic manufacturer and importer, if any, of the device that the regulation is to be made so effective. If necessary, the Commissioner may also notify the distributor or any other responsible person(s). In addition, the Commissioner will attempt to notify any foreign manufacturer when the name and address of the foreign manufacturer are readily available.

(b) This procedure may be used when the Commissioner determines that the potential or actual injury involved is a serious one that the Commissioner believes will endanger the health of individuals who have been, or will be, exposed to the device. In assessing the degree of danger, the Commissioner need not find that the danger is immediate, and it shall be sufficient for the Commissioner to determine that the danger may involve a serious long-term risk.

(c) If the Commissioner makes a proposed regulation effective in accordance with this section, the Commissioner will, as expeditiously as possible, give interested persons prompt notice of this action in the FEDERAL REGISTER.

(d) After the hearing, if any, and after considering any written comments submitted on the proposal and any additional available information and data, the Commissioner will as expeditiously as possible either affirm, modify, or revoke the proposed regulation making the device a banned device. If the Commissioner decides to affirm or modify the proposed regulation to make a device a banned device, the Commissioner will amend subpart B by adding the name or description of the device, or both, to the list of banned devices. If the Commissioner decides to revoke a proposed regulation making a device a banned device, a notice of termination of rulemaking proceedings and reasons therefor will be published in the FEDERAL REGISTER.

(e) The Commissioner may declare the special effective date provided by this section to be in effect after the publication of a proposed regulation under § 895.21(d), if, based on new information, or upon reconsideration of previously available information, the Commissioner makes the determination and provides the appropriate notices and an opportunity for a hearing in accordance with paragraphs (a) and (c) of this section.

(f) Those devices that have been named banned devices under § 895.30 and that have already been sold to the public may be subject to relabeling by the manufacturer, distributor, importer, or any other person(s) responsible for the labeling of the device or may be subject to the provisions of section 518(a) or (b) of the act.

[44 FR 29221, May 18, 1979, as amended at 57 FR 58405, Dec. 10, 1992]

Subpart B—Listing of Banned Devices

§ 895.101 Prosthetic hair fibers.

Prosthetic hair fibers are devices intended for implantation into the human scalp to simulate natural hair or conceal baldness. Prosthetic hair fibers may consist of various materials; for example, synthetic fibers, such as modacrylic, polyacrylic, and polyester; and natural fibers, such as processed human hair. Excluded from the banned device are natural hair transplants, in which a person's hair and its surrounding tissue are surgically removed from one location on the person's scalp and then grafted onto another area of the person's scalp.

[48 FR 25136, June 3, 1983]

PART 898—PERFORMANCE STANDARD FOR ELECTRODE LEAD WIRES AND PATIENT CABLES

Sec.

AUTHORITY: 21 U.S.C. 351, 352, 360c, 360d, 360gg–360ss, 371, 374; 42 U.S.C. 262, 264.

SOURCE: 62 FR 25497, May 9, 1997, unless otherwise noted.

§898.11 Applicability.

Electrode lead wires and patient cables intended for use with a medical device shall be subject to the performance standard set forth in §898.12.

§898.12 Performance standard.

(a) Any connector in a cable or electrode lead wire having a conductive connection to a patient shall be constructed in such a manner as to comply with subclause 56.3(c) of the following standard:

International Electrotechnical Commission (IEC)

601–1: Medical Electrical Equipment

601–1 (1988) Part 1: General requirements for safety

Amendment No. 1 (1991)

Amendment No. 2 (1995).

(b) Compliance with the standard shall be determined by inspection and by applying the test requirements and test methods of subclause 56.3(c) of the standard set forth in paragraph (a) of this section.

§898.13 Compliance dates.

The dates for compliance with the standard set forth in §898.12(a) shall be as follows:

(a) For electrode lead wires and patient cables used with, or intended for use with, the following devices, the date for which compliance is required is May 11, 1998:

LISTING OF DEVICES FOR WHICH COMPLIANCE IS REQUIRED EFFECTIVE May 11, 1998

Phase	Product code	21 CFR section	Class	Device name
1	73 BZQ	868.2375	II	Monitor, Breathing Frequency.
1	73 FLS	868.2375	II	Monitor (Apnea Detector), Ventilatory Effort.
1	74 DPS	870.2340	II	Electrocardiograph.
1	74 DRG	870.2910	II	Transmitters and Receivers, Physiological Signal, Radio Frequency.
1	74 DRT	870.2300	II	Monitor, Cardiac (including Cardiotachometer and Rate Alarm).
1	74 DRX	870.2360	II	Electrode, Electrocardiograph.
1	74 DSA	870.2900	II	Cable, Transducer and Electrode, Patient (including Connector).
1	74 DSH	870.2800	II	Recorder, Magnetic Tape, Medical.
1	74 DSI	870.1025	III	Detector and Alarm, Arrhythmia.
1	74 DXH	870.2920	II	Transmitters and Receivers, Electrocardiograph, Telephone.

(b) For electrode lead wires and patient cables used with, or intended for use with, any other device, the date for which compliance is required is May 9, 2000.

§898.14 Exemptions and variances.

(a) A request for an exemption or variance shall be submitted in the form of a petition under §10.30 of this chapter and shall comply with the requirements set out therein. The petition shall also contain the following:

(1) The name of the device, the class in which the device has been classified, and representative labeling showing the intended uses(s) of the device;

(2) The reasons why compliance with the performance standard is unnecessary or unfeasible;

(3) A complete description of alternative steps that are available, or that the petitioner has already taken, to ensure that a patient will not be inadvertently connected to hazardous voltages via an unprotected patient cable or electrode lead wire for intended use with the device; and

(4) Other information justifying the exemption or variance.

(b) An exemption or variance is not effective until the agency approves the request under §10.30(e)(2)(i) of this chapter.

EFFECTIVE DATE NOTE: At 62 FR 25477, May 9, 1997, §898.14 was stayed pending Office of

Management and Budget approval of information collection and recordkeeping requirements.

SUBCHAPTER I—MAMMOGRAPHY QUALITY STANDARDS ACT

PART 900—MAMMOGRAPHY

AUTHORITY: 21 U.S.C. 360i, 360nn, 374(e); 42 U.S.C. 263b.

SOURCE: 62 FR 55976, Oct. 28, 1997, unless otherwise noted. Republished and corrected at 62 FR 60614, Nov. 10, 1997.

Subpart A—Accreditation

§900.1 Scope.

The regulations set forth in this part implement the Mammography Quality Standards Act (MQSA) (42 U.S.C. 263b). Subpart A of this part establishes procedures whereby an entity can apply to become a Food and Drug Administration (FDA)-approved accreditation body to accredit facilities to be eligible to perform screening or diagnostic mammography services. Subpart A further establishes requirements and standards for accreditation bodies to ensure that all mammography facilities under the jurisdiction of the United States are adequately and consistently evaluated for compliance with national quality standards for mammography. Subpart B of this part establishes minimum national quality standards for mammography facilities to ensure safe, reliable, and accurate mammography. The regulations set forth in this part do not apply to facilities of the Department of Veterans Affairs.

§900.2 Definitions.

The following definitions apply to subparts A, B, and C of this part:

(a) *Accreditation body* or *body* means an entity that has been approved by FDA under §900.3(d) to accredit mammography facilities.

(b) *Action limits* or *action levels* means the minimum and maximum values of a quality assurance measurement that can be interpreted as representing acceptable performance with respect to the parameter being tested. Values less than the minimum or greater than the maximum action limit or level indicate that corrective action must be taken by the facility. Action limits or levels are also sometimes called control limits or levels.

(c) *Adverse event* means an undesirable experience associated with mammography activities within the scope of 42 U.S.C. 263b. Adverse events include but are not limited to:

(1) Poor image quality;

(2) Failure to send mammography reports within 30 days to the referring physician or in a timely manner to the self-referred patient; and

(3) Use of personnel that do not meet the applicable requirements of §900.12(a).

(d) *Air kerma* means kerma in a given mass of air. The unit used to measure the quantity of air kerma is the Gray (Gy). For X-rays with energies less than 300 kiloelectron volts (keV), 1 Gy = 100 rad. In air, 1 Gy of absorbed dose is delivered by 114 roentgens (R) of exposure.

(e) *Breast implant* means a prosthetic device implanted in the breast.

(f) *Calendar quarter* means any one of the following time periods during a given year: January 1 through March 31, April 1 through June 30, July 1 through September 30, or October 1 through December 31.

(g) *Category I* means medical educational activities that have been designated as Category I by the Accreditation Council for Continuing Medical Education (ACCME), the American Osteopathic Association (AOA), a state medical society, or an equivalent organization.

(h) *Certificate* means the certificate described in § 900.11(a).

(i) *Certification* means the process of approval of a facility by FDA or a certification agency to provide mammography services.

(j) *Clinical image* means a mammogram.

(k) *Consumer* means an individual who chooses to comment or complain in reference to a mammography examination, including the patient or representative of the patient (e.g., family member or referring physician).

(l) *Continuing education unit* or *continuing education credit* means one contact hour of training.

(m) *Contact hour* means an hour of training received through direct instruction.

(n) *Direct instruction* means:

(1) Face-to-face interaction between instructor(s) and student(s), as when the instructor provides a lecture, conducts demonstrations, or reviews student performance; or

(2) The administration and correction of student examinations by an instructor(s) with subsequent feedback to the student(s).

(o) *Direct supervision* means that:

(1) During joint interpretation of mammograms, the supervising interpreting physician reviews, discusses, and confirms the diagnosis of the physician being supervised and signs the resulting report before it is entered into the patient's records; or

(2) During the performance of a mammography examination or survey of the facility's equipment and quality assurance program, the supervisor is present to observe and correct, as needed, the performance of the individual being supervised who is performing the examination or conducting the survey.

(p) *Established operating level* means the value of a particular quality assurance parameter that has been established as an acceptable normal level by the facility's quality assurance program.

(q) *Facility* means a hospital, outpatient department, clinic, radiology practice, mobile unit, office of a physician, or other facility that conducts mammography activities, including the following: Operation of equipment to produce a mammogram, processing of the mammogram, initial interpretation of the mammogram, and maintaining viewing conditions for that interpretation. This term does not include a facility of the Department of Veterans Affairs.

(r) *First allowable time* means the earliest time a resident physician is eligible to take the diagnostic radiology boards from an FDA-designated certifying body. The "first allowable time" may vary with the certifying body.

(s) *FDA* means the Food and Drug Administration.

(t) *Interim regulations* means the regulations entitled "Requirements for Accrediting Bodies of Mammography Facilities" (58 FR 67558–67565) and "Quality Standards and Certification Requirements for Mammography Facilities" (58 FR 67565–67572), published by FDA on December 21, 1993, and amended on September 30, 1994 (59 FR 49808–49813). These regulations established the standards that had to be met by mammography facilities in order to lawfully operate between October 1, 1994, and April 28, 1999.

(u) *Interpreting physician* means a licensed physician who interprets mammograms and who meets the requirements set forth in § 900.12(a)(1).

(v) *Kerma* means the sum of the initial energies of all the charged particles liberated by uncharged ionizing particles in a material of given mass.

(w) *Laterality* means the designation of either the right or left breast.

(x) *Lead interpreting physician* means the interpreting physician assigned the general responsibility for ensuring that a facility's quality assurance program meets all of the requirements of § 900.12(d) through (f). The administrative title and other supervisory responsibilities of the individual, if any, are left to the discretion of the facility.

(y) *Mammogram* means a radiographic image produced through mammography.

(z) *Mammographic modality* means a technology, within the scope of 42 U.S.C. 263b, for radiography of the breast. Examples are screen-film mammography and xeromammography.

(aa) *Mammography* means radiography of the breast, but, for the purposes of this part, does not include:

(1) Radiography of the breast performed during invasive interventions for localization or biopsy procedures; or

(2) Radiography of the breast performed with an investigational mammography device as part of a scientific study conducted in accordance with FDA's investigational device exemption regulations in part 812 of this chapter.

(bb) *Mammography equipment evaluation* means an onsite assessment of mammography unit or image processor performance by a medical physicist for the purpose of making a preliminary determination as to whether the equipment meets all of the applicable standards in § 900.12(b) and (e).

(cc) *Mammography medical outcomes audit* means a systematic collection of mammography results and the comparison of those results with outcomes data.

(dd) *Mammography unit* or *units* means an assemblage of components for the production of X-rays for use during mammography, including, at a minimum: An X-ray generator, an X-ray control, a tube housing assembly, a beam limiting device, and the supporting structures for these components.

(ee) *Mean optical density* means the average of the optical densities measured using phantom thicknesses of 2, 4, and 6 centimeters with values of kilovolt peak (kVp) clinically appropriate for those thicknesses.

(ff) *Medical physicist* means a person trained in evaluating the performance of mammography equipment and facility quality assurance programs and who meets the qualifications for a medical physicist set forth in § 900.12(a)(3).

(gg) *MQSA* means the Mammography Quality Standards Act.

(hh) *Multi-reading* means two or more physicians, at least one of whom is an interpreting physician, interpreting the same mammogram.

(ii) *Patient* means any individual who undergoes a mammography evaluation in a facility, regardless of whether the person is referred by a physician or is self-referred.

(jj) *Phantom* means a test object used to simulate radiographic characteristics of compressed breast tissue and containing components that radiographically model aspects of breast disease and cancer.

(kk) *Phantom image* means a radiographic image of a phantom.

(ll) *Physical science* means physics, chemistry, radiation science (including medical physics and health physics), and engineering.

(mm) *Positive mammogram* means a mammogram that has an overall assessment of findings that are either "suspicious" or "highly suggestive of malignancy."

(nn) *Provisional certificate* means the provisional certificate described in § 900.11(b)(2).

(oo) *Qualified instructor* means an individual whose training and experience adequately prepares him or her to carry out specified training assignments. Interpreting physicians, radiologic technologists, or medical physicists who meet the requirements of § 900.12(a) would be considered qualified instructors in their respective areas of mammography. Other examples of individuals who may be qualified instructors for the purpose of providing training to meet the regulations of this part include, but are not limited to, instructors in a post-high school training institution and manufacturer's representatives.

(pp) *Quality control technologist* means an individual meeting the requirements of § 900.12(a)(2) who is responsible for those quality assurance responsibilities not assigned to the lead interpreting physician or to the medical physicist.

(qq) *Radiographic equipment* means X-ray equipment used for the production of static X-ray images.

(rr) *Radiologic technologist* means an individual specifically trained in the use of radiographic equipment and the positioning of patients for radiographic examinations and who meets the requirements set forth in § 900.12(a)(2).

(ss) *Serious adverse event* means an adverse advent that may significantly compromise clinical outcomes, or an adverse event for which a facility fails to take appropriate corrective action in a timely manner.

(tt) *Serious complaint* means a report of a serious adverse event.

(uu) *Standard breast* means a 4.2 centimeter (cm) thick compressed breast consisting of 50 percent glandular and 50 percent adipose tissue.

(vv) *Survey* means an onsite physics consultation and evaluation of a facility quality assurance program performed by a medical physicist.

(ww) *Time cycle* means the film development time.

(xx) *Traceable to a national standard* means an instrument is calibrated at either the National Institute of Standards and Technology (NIST) or at a calibration laboratory that participates in a proficiency program with NIST at least once every 2 years and the results of the proficiency test conducted within 24 months of calibration show agreement within ±3 percent of the national standard in the mammography energy range.

(yy) *Review physician* means a physician who, by meeting the requirements set out in § 900.4(c)(5), is qualified to review clinical images on behalf of the accreditation body.

(zz) *Certification agency* means a State that has been approved by FDA under § 900.21 to certify mammography facilities.

(aaa) *Performance indicators* mean the measures used to evaluate the certification agency's ability to conduct certification, inspection, and compliance activities.

(bbb) *Authorization* means obtaining approval from FDA to utilize new or changed State regulations or procedures during the issuance, maintenance, and withdrawal of certificates by the certification agency.

[62 FR 55976, Oct. 28, 1997; 62 FR 60614, Nov. 10, 1997, as amended at 63 FR 56558, Oct. 22, 1998; 64 FR 32407, June 17, 1999; 67 FR 5467, Feb. 6, 2002]

§ 900.3 Application for approval as an accreditation body.

(a) *Eligibility*. Private nonprofit organizations or State agencies capable of meeting the requirements of this subpart A may apply for approval as accreditation bodies.

(b) *Application for initial approval*. (1) An applicant seeking initial FDA approval as an accreditation body shall inform the Division of Mammography Quality and Radiation Programs (DMQRP), Center for Devices and Radiology Health (HFZ–240), Food and Drug Administration, 1350 Piccard Dr., Rockville, MD 20850, marked Attn: Mammography Standards Branch, of its desire to be approved as an accreditation body and of its requested scope of authority.

(2) Following receipt of the request, FDA will provide the applicant with additional information to aid in submission of an application for approval as an accreditation body.

(3) The applicant shall furnish to FDA, at the address in § 900.3(b)(1), three copies of an application containing the following information, materials, and supporting documentation:

(i) Name, address, and phone number of the applicant and, if the applicant is not a State agency, evidence of nonprofit status (i.e., of fulfilling Internal Revenue Service requirements as a nonprofit organization);

(ii) Detailed description of the accreditation standards the applicant will require facilities to meet and a discussion substantiating their equivalence to FDA standards required under § 900.12;

(iii) Detailed description of the applicant's accreditation review and decisionmaking process, including:

(A) Procedures for performing accreditation and reaccreditation clinical image review in accordance with § 900.4(c), random clinical image reviews in accordance with § 900.4(f), and additional mammography review in accordance with § 900.12(j);

(B) Procedures for performing phantom image review;

(C) Procedures for assessing mammography equipment evaluations and surveys;

(D) Procedures for initiating and performing onsite visits to facilities;

(E) Procedures for assessing facility personnel qualifications;

(F) Copies of the accreditation application forms, guidelines, instructions, and other materials the applicant will send to facilities during the accreditation process, including an accreditation history form that requires each facility to provide a complete history of prior accreditation activities and a statement that all information and data submitted in the application is true and accurate, and that no material fact has been omitted;

(G) Policies and procedures for notifying facilities of deficiencies;

(H) Procedures for monitoring corrections of deficiencies by facilities;

(I) Policies and procedures for suspending or revoking a facility's accreditation;

(J) Policies and procedures that will ensure processing of accreditation applications and renewals within a timeframe approved by FDA and assurances that the body will adhere to such policies and procedures; and

(K) A description of the applicant's appeals process for facilities contesting adverse accreditation status decisions.

(iv) Education, experience, and training requirements for the applicant's professional staff, including reviewers of clinical or phantom images;

(v) Description of the applicant's electronic data management and analysis system with respect to accreditation review and decision processes and the applicant's ability to provide electronic data in a format compatible with FDA data systems;

(vi) Resource analysis that demonstrates that the applicant's staffing, funding, and other resources are adequate to perform the required accreditation activities;

(vii) Fee schedules with supporting cost data;

(viii) Statement of policies and procedures established to avoid conflicts of interest or the appearance of conflicts of interest by the applicant's board members, commissioners, professional personnel (including reviewers of clinical and phantom images), consultants, administrative personnel, and other representatives of the applicant;

(ix) Statement of policies and procedures established to protect confidential information the applicant will collect or receive in its role as an accreditation body;

(x) Disclosure of any specific brand of imaging system or component, measuring device, software package, or other commercial product used in mammography that the applicant develops, sells, or distributes;

(xi) Description of the applicant's consumer complaint mechanism;

(xii) Satisfactory assurances that the applicant shall comply with the requirements of § 900.4; and

(xiii) Any other information as may be required by FDA.

(c) *Application for renewal of approval.* An approved accreditation body that intends to continue to serve as an accreditation body beyond its current term shall apply to FDA for renewal or notify FDA of its plans not to apply for renewal in accordance with the following procedures and schedule:

(1) At least 9 months before the date of expiration of a body's approval, the body shall inform FDA, at the address given in § 900.3(b)(1), of its intent to seek renewal.

(2) FDA will notify the applicant of the relevant information, materials, and supporting documentation required under § 900.3(b)(3) that the applicant shall submit as part of the renewal procedure.

(3) At least 6 months before the date of expiration of a body's approval, the applicant shall furnish to FDA, at the address in § 900.3(b)(1), three copies of a renewal application containing the information, materials, and supporting documentation requested by FDA in accordance with § 900.3(c)(2).

(4) No later than July 28, 1998, any accreditation body approved under the interim regulations published in the FEDERAL REGISTER of December 21, 1993 (58 FR 67558), that desires to continue to serve as an accreditation body under

the final regulations shall apply for renewal of approval in accordance with the procedures set forth in paragraphs (c)(1) through (c)(3) of this section.

(5) Any accreditation body that does not plan to renew its approval shall so notify FDA at the address given in paragraph (b)(1) of this section at least 9 months before the expiration of the body's term of approval.

(d) *Rulings on applications for initial and renewed approval.* (1) FDA will conduct a review and evaluation to determine whether the applicant substantially meets the applicable requirements of this subpart and whether the accreditation standards the applicant will require facilities to meet are substantially the same as the quality standards published under subpart B of this part.

(2) FDA will notify the applicant of any deficiencies in the application and request that those deficiencies be rectified within a specified time period. If the deficiencies are not rectified to FDA's satisfaction within the specified time period, the application for approval as an accreditation body may be rejected.

(3) FDA shall notify the applicant whether the application has been approved or denied. That notification shall list any conditions associated with approval or state the bases for any denial.

(4) The review of any application may include a meeting between FDA and representatives of the applicant at a time and location mutually acceptable to FDA and the applicant.

(5) FDA will advise the applicant of the circumstances under which a denied application may be resubmitted.

(6) If FDA does not reach a final decision on a renewal application in accordance with this paragraph before the expiration of an accreditation body's current term of approval, the approval will be deemed extended until the agency reaches a final decision on the application, unless an accreditation body does not rectify deficiencies in the application within the specified time period, as required in paragraph (d)(2) of this section.

(e) *Relinquishment of authority.* An accreditation body that decides to relinquish its accreditation authority before expiration of the body's term of approval shall submit a letter of such intent to FDA, at the address in § 900.3(b)(1), at least 9 months before relinquishing such authority.

(f) *Transfer of records.* An accreditation body that does not apply for renewal of accreditation body approval, is denied such approval by FDA, or relinquishes its accreditation authority and duties before expiration of its term of approval, shall:

(1) Transfer facility records and other related information as required by FDA to a location and according to a schedule approved by FDA.

(2) Notify, in a manner and time period approved by FDA, all facilities accredited or seeking accreditation by the body that the body will no longer have accreditation authority.

(g) *Scope of authority.* An accreditation body's term of approval is for a period not to exceed 7 years. FDA may limit the scope of accreditation authority.

[62 FR 55976, Oct. 28, 1997; 62 FR 60614, Nov. 10, 1997]

§ 900.4 Standards for accreditation bodies.

(a) *Code of conduct and general responsibilities.* The accreditation body shall accept the following responsibilities in order to ensure safe and accurate mammography at the facilities it accredits and shall perform these responsibilities in a manner that ensures the integrity and impartiality of accreditation body actions.

(1)(i) When an accreditation body receives or discovers information that suggests inadequate image quality, or upon request by FDA, the accreditation body shall review a facility's clinical images or other aspects of a facility's practice to assist FDA in determining whether or not the facility's practice poses a serious risk to human health. Such reviews are in addition to the evaluation an accreditation body performs as part of the initial accreditation or renewal process for facilities.

(ii) If review by the accreditation body demonstrates that a problem does exist with respect to image quality or other aspects of a facility's compliance with quality standards, or upon request by FDA, the accreditation body shall require or monitor corrective actions, or suspend or revoke accreditation of the facility.

(2) The accreditation body shall inform FDA as soon as possible but in no case longer than 2 business days after becoming aware of equipment or practices that pose a serious risk to human health.

(3) The accreditation body shall establish and administer a quality assurance (QA) program that has been approved by FDA in accordance with § 900.3(d) or paragraph (a)(8) of this section. Such quality assurance program shall:

(i) Include requirements for clinical image review and phantom image review;

(ii) Ensure that clinical and phantom images are evaluated consistently and accurately; and

(iii) Specify the methods and frequency of training and evaluation for clinical and phantom image reviewers, and the bases and procedures for removal of such reviewers.

(4) The accreditation body shall establish measures that FDA has approved in accordance with § 900.3(d) or paragraph (a)(8) of this section to reduce the possibility of conflict of interest or facility bias on the part of individuals acting on the body's behalf. Such individuals who review clinical or phantom images under the provisions of paragraphs (c) and (d) of this section or who visit facilities under the provisions of paragraph (f) of this section shall not review clinical or phantom images from or visit a facility with which such individuals maintain a relationship, or when it would otherwise be a conflict of interest for them to do so, or when they have a bias in favor of or against the facility.

(5) The accreditation body may require specific equipment performance or design characteristics that FDA has approved. However, no accreditation body shall require, either explicitly or implicitly, the use of any specific brand of imaging system or component, measuring device, software package, or other commercial product as a condition for accreditation by the body, unless FDA determines that it is in the best interest of public health to do so.

(i) Any representation, actual or implied, either orally, in sales literature, or in any other form of representation, that the purchase or use of a particular product brand is required in order for any facility to be accredited or certified under § 900.11(b), is prohibited, unless FDA approves such representation.

(ii) Unless FDA has approved the exclusive use and promotion of a particular commercial product in accordance with this section, all products produced, distributed, or sold by an accreditation body or an organization that has a financial or other relationship with the accreditation body that may be a conflict of interest or have the appearance of a conflict of interest with the body's accreditation functions, shall bear a disclaimer stating that the purchase or use of such products is not required for accreditation or certification of any facility under § 900.11(b). Any representations about such products shall include a similar disclaimer.

(6) When an accreditation body denies accreditation to a facility, the accreditation body shall notify the facility in writing and explain the bases for its decision. The notification shall also describe the appeals process available from the accreditation body for the facility to contest the decision.

(7) No accreditation body may establish requirements that preclude facilities from being accredited under § 900.11(b) by any other accreditation body, or require accreditation by itself under MQSA if another accreditation body is available to a facility.

(8) The accreditation body shall obtain FDA authorization for any changes it proposes to make in any standards that FDA has previously accepted under § 900.3(d).

(9) An accreditation body shall establish procedures to protect confidential information it collects or receives in its role as an accreditation body.

(i) Nonpublic information collected from facilities for the purpose of carrying out accreditation body responsibilities shall not be used for any other purpose or disclosed, other than to FDA or its duly designated representatives, including State agencies, without the consent of the facility;

(ii) Nonpublic information that FDA or its duly designated representatives, including State agencies, share with the accreditation body concerning a facility that is accredited or undergoing accreditation by that body shall not be further disclosed except with the written permission of FDA.

(b) *Monitoring facility compliance with quality standards.* (1) The accreditation body shall require that each facility it accredits meet standards for the performance of quality mammography that are substantially the same as those in this subpart and in subpart B of this part.

(2) The accreditation body shall notify a facility regarding equipment, personnel, and other aspects of the facility's practice that do not meet such standards and advise the facility that such equipment, personnel, or other aspects of the practice should not be used by the facility for activities within the scope of part 900.

(3) The accreditation body shall specify the actions that facilities shall take to correct deficiencies in equipment, personnel, and other aspects of the practice to ensure facility compliance with applicable standards.

(4) If deficiencies cannot be corrected to ensure compliance with standards or if a facility is unwilling to take corrective actions, the accreditation body shall immediately so notify FDA, and shall suspend or revoke the facility's accreditation in accordance with the policies and procedures described under § 900.3(b)(3)(iii)(I).

(c) *Clinical image review for accreditation and reaccreditation*—(1) *Frequency of review.* The accreditation body shall review clinical images from each facility accredited by the body at least once every 3 years.

(2) *Requirements for clinical image attributes.* The accreditation body shall use the following attributes for all clinical image reviews, unless FDA has approved other attributes:

(i) *Positioning.* Sufficient breast tissue shall be imaged to ensure that cancers are not likely to be missed because of inadequate positioning.

(ii) *Compression.* Compression shall be applied in a manner that minimizes the potential obscuring effect of overlying breast tissue and motion artifact.

(iii) *Exposure level.* Exposure level shall be adequate to visualize breast structures. Images shall be neither underexposed nor overexposed.

(iv) *Contrast.* Image contrast shall permit differentiation of subtle tissue density differences.

(v) *Sharpness.* Margins of normal breast structures shall be distinct and not blurred.

(vi) *Noise.* Noise in the image shall not obscure breast structures or suggest the appearance of structures not actually present.

(vii) *Artifacts.* Artifacts due to lint, processing, scratches, and other factors external to the breast shall not obscure breast structures or suggest the appearance of structures not actually present.

(viii) *Examination identification.* Each image shall have the following information indicated on it in a permanent, legible, and unambiguous manner and placed so as not to obscure anatomic structures:

(A) Name of the patient and an additional patient identifier.

(B) Date of examination.

(C) *View and laterality.* This information shall be placed on the image in a position near the axilla. Standardized codes specified by the accreditation body and approved by FDA in accordance with § 900.3(d) or paragraph (a)(8) of this section shall be used to identify view and laterality.

(D) *Facility name and location.* At a minimum, the location shall include the city, State, and zip code of the facility.

(E) Technologist identification.

(F) Cassette/screen identification.

(G) Mammography unit identification, if there is more than one unit in the facility.

(3) *Scoring of clinical images.* Accreditation bodies shall establish and administer a system for scoring clinical images using all attributes specified in paragraphs (c)(2)(i) through (c)(2)(viii) of this section or an alternative system that FDA has approved in accordance with § 900.3(d) or paragraph (a)(8) of this section. The scoring system shall include an evaluation for each attribute.

(i) The accreditation body shall establish and employ criteria for acceptable and nonacceptable results for each of the 8 attributes as well as an overall pass-fail system for clinical image review that has been approved by FDA in accordance with § 900.3(d) or paragraph (a)(8) of this section.

(ii) All clinical images submitted by a facility to the accreditation body shall be reviewed independently by two or more review physicians.

(4) *Selection of clinical images for review.* Unless otherwise specified by FDA, the accreditation body shall require that for each mammography unit in the facility:

(i) The facility shall submit craniocaudal (CC) and mediolateral oblique (MLO) views from two mammographic examinations that the facility produced during a time period specified by the accreditation body;

(ii) Clinical images submitted from one such mammographic examination for each unit shall be of dense breasts (predominance of glandular tissue) and the other shall be of fat-replaced breasts (predominance of adipose tissue);

(iii) All clinical images submitted shall be images that the facility's interpreting physician(s) interpreted as negative or benign.

(iv) If the facility has no clinical images meeting the requirements in paragraphs (c)(4)(i) through (c)(4)(iii) of this section, it shall so notify the accreditation body, which shall specify alternative clinical image selection methods that do not compromise care of the patient.

(5) *Review physicians.* Accreditation bodies shall ensure that all of their review physicians:

(i) Meet the interpreting physician requirements specified in § 900.12(a)(1) and meet such additional requirements as have been established by the accreditation body and approved by FDA;

(ii) Are trained and evaluated in the clinical image review process, for the types of clinical images to be evaluated by a review physician, by the accreditation body before designation as review physicians and periodically thereafter; and

(iii) Clearly document their findings and reasons for assigning a particular score to any clinical image and provide information to the facility for use in improving the attributes for which significant deficiencies were identified.

(6) *Image management.* The accreditation body's QA program shall include a tracking system to ensure the security and return to the facility of all clinical images received and to ensure completion of all clinical image reviews by the body in a timely manner. The accreditation body shall return all clinical images to the facility within 60 days of their receipt by the body, with the following exceptions:

(i) If the clinical images are needed earlier by the facility for clinical purposes, the accreditation body shall cooperate with the facility to accommodate such needs.

(ii) If a review physician identifies a suspicious abnormality on an image submitted for clinical image review, the accreditation body shall ensure that this information is provided to the facility and that the clinical images are returned to the facility. Both shall occur no later than 10-business days after identification of the suspected abnormality.

(7) *Notification of unsatisfactory image quality.* If the accreditation body determines that the clinical images received from a facility are of unsatisfactory quality, the body shall notify the facility of the nature of the problem and its possible causes.

(d) *Phantom image review for accreditation and reaccreditation*—(1) *Frequency of review.* The accreditation body shall review phantom images from each facility accredited by the body at least once every 3 years.

(2) *Requirements for the phantom used.* The accreditation body shall require that each facility submit for review phantom images that the facility produced using a phantom and methods of use specified by the body and approved by FDA in accordance with § 900.3(d) or paragraph (a)(8) of this section.

(3) *Scoring phantom images.* The accreditation body shall use a system for scoring phantom images that has been approved by FDA in accordance with § 900.3(b) and (d) or paragraph (a)(8) of this section.

(4) *Phantom images selected for review.* For each mammography unit in the facility, the accreditation body shall require the facility to submit phantom

images that the facility produced during a time period specified by the body.

(5) *Phantom image reviewers.* Accreditation bodies shall ensure that all of their phantom image reviewers:

(i) Meet the requirements specified in § 900.12(a)(3) or alternative requirements established by the accreditation body and approved by FDA in accordance with § 900.3 or paragraph (a)(8) of this section;

(ii) Are trained and evaluated in the phantom image review process, for the types of phantom images to be evaluated by a phantom image reviewer, by the accreditation body before designation as phantom image reviewers and periodically thereafter; and

(iii) Clearly document their findings and reasons for assigning a particular score to any phantom image and provide information to the facility for use in improving its phantom image quality with regard to the significant deficiencies identified.

(6) *Image management.* The accreditation body's QA program shall include a tracking system to ensure the security of all phantom images received and to ensure completion of all phantom image reviews by the body in a timely manner. All phantom images that result in a failure of accreditation shall be returned to the facility.

(7) *Notification measures for unsatisfactory image quality.* If the accreditation body determines that the phantom images received from a facility are of unsatisfactory quality, the body shall notify the facility of the nature of the problem and its possible causes.

(e) *Reports of mammography equipment evaluation, surveys, and quality control.* The following requirements apply to all facility equipment covered by the provisions of subparts A and B:

(1) The accreditation body shall require every facility applying for accreditation to submit:

(i) With its initial accreditation application, a mammography equipment evaluation that was performed by a medical physicist no earlier than 6 months before the date of application for accreditation by the facility. Such evaluation shall demonstrate compliance of the facility's equipment with the requirements in § 900.12(e).

(ii) Prior to accreditation, a survey that was performed no earlier than 6 months before the date of application for accreditation by the facility. Such survey shall assess the facility's compliance with the facility standards referenced in paragraph (b) of this section.

(2) The accreditation body shall require that all facilities undergo an annual survey to ensure continued compliance with the standards referenced in paragraph (b) of this section and to provide continued oversight of facilities' quality control programs as they relate to such standards. The accreditation body shall require for all facilities that:

(i) Such surveys be conducted annually;

(ii) Facilities take reasonable steps to ensure that they receive reports of such surveys within 30 days of survey completion; and

(iii) Facilities submit the results of such surveys and any other information that the body may require to the body at least annually.

(3) The accreditation body shall review and analyze the information required in this section and use it to identify necessary corrective measures for facilities and to determine whether facilities should remain accredited by the body.

(f) *Accreditation body onsite visits and random clinical image reviews.* The accreditation body shall conduct onsite visits and random clinical image reviews of a sample of facilities to monitor and assess their compliance with standards established by the body for accreditation. The accreditation body shall submit annually to FDA, at the address given in § 900.3(b)(1), 3 copies of a summary report describing all facility assessments the body conducted under the provisions of this section for the year being reported.

(1) *Onsite visits*—(i) *Sample size.* Annually, each accreditation body shall visit at least 5 percent of the facilities it accredits. However, a minimum of 5 facilities shall be visited, and visits to no more than 50 facilities are required, unless problems identified in paragraph (f)(1)(i)(B) of this section indicate a need to visit more than 50 facilities.

(A) At least 50 percent of the facilities visited shall be selected randomly.

(B) Other facilities visited shall be selected based on problems identified through State or FDA inspections, serious complaints received from consumers or others, a previous history of noncompliance, or any other information in the possession of the accreditation body, inspectors, or FDA.

(C) Before, during, or after any facility visit, the accreditation body may require that the facility submit to the body for review clinical images, phantom images, or any other information relevant to applicable standards in this subpart and in subpart B of this part.

(ii) *Visit plan.* The accreditation body shall conduct facility onsite visits according to a visit plan that has been approved by FDA in accordance with § 900.3(d) or paragraph (a)(8) of this section, unless otherwise directed by FDA in particular circumstances. At a minimum, such a plan shall provide for:

(A) Assessment of overall clinical image QA activities of the facility;

(B) Review of facility documentation to determine if appropriate mammography reports are sent to patients and physicians as required;

(C) Selection of a sample of clinical images for clinical image review by the accreditation body. Clinical images shall be selected in a manner specified by the accreditation body and approved by FDA that does not compromise care of the patient as a result of the absence of the selected images from the facility;

(D) Verification that the facility has a medical audit system in place and is correlating films and pathology reports for positive cases;

(E) Verification that personnel specified by the facility are the ones actually performing designated personnel functions;

(F) Verification that equipment specified by the facility is the equipment that is actually being used to perform designated equipment functions;

(G) Verification that a consumer complaint mechanism is in place and that the facility is following its procedures; and

(H) Review of all factors related to previously identified concerns or concerns identified during that visit.

(2) *Clinical image review for random sample of facilities*—(i) *Sample size.* In addition to conducting clinical image reviews for accreditation and reaccreditation for all facilities, the accreditation body shall conduct clinical image reviews annually for a randomly selected sample as specified by FDA, but to include at least 3 percent of the facilities the body accredits. Accreditation bodies may count toward this random sample requirement all facilities selected randomly for the onsite visits described in paragraph (f)(1)(i)(A) of this section. Accreditation bodies shall not count toward the random sample requirement any facilities described in paragraph (f)(1)(i)(B) of this section that were selected for a visit because of previously identified concerns.

(ii) *Random clinical image review.* In performing clinical image reviews of the random sample of facilities, accreditation bodies shall evaluate the same attributes as those in paragraph (c) of this section for review of clinical images for accreditation and reaccreditation.

(iii) Accreditation bodies should not schedule random clinical image reviews at facilities that have received notification of the need to begin the accreditation renewal process or that have completed the accreditation renewal process within the previous 6 months.

(iv) *Selection of the random sample of clinical images for clinical image review by the accreditation body.* Clinical images shall be selected in a manner, specified by the accreditation body and approved by FDA under § 900.3(d) or paragraph (a)(8) of this section, that does not compromise care of the patient as a result of the absence of the selected images from the facility.

(g) *Consumer complaint mechanism.* The accreditation body shall develop and administer a written and documented system, including timeframes, for collecting and resolving serious consumer complaints that could not be resolved at a facility. Such system shall have been approved by FDA in accordance with§ 900.3(d) or paragraph (a)(8) of this section. Accordingly, all accreditation bodies shall:

(1) Provide a mechanism for all facilities it accredits to file serious unresolved complaints with the accreditation body;

(2) Maintain a record of every serious unresolved complaint received by the body on all facilities it accredits for a period of at least 3 years from the date of receipt of each such complaint;

(h) *Reporting and recordkeeping.* All reports to FDA specified in paragraphs (h)(1) through (h)(4) of this section shall be prepared and submitted in a format and medium prescribed by FDA and shall be submitted to a location and according to a schedule specified by FDA. The accreditation body shall:

(1) Collect and submit to FDA the information required by 42 U.S.C. 263b(d) for each facility when the facility is initially accredited and at least annually when updated, in a manner and at a time specified by FDA.

(2) Accept applications containing the information required in 42 U.S.C. 263b(c)(2) for provisional certificates and in § 900.11(b)(3) for extension of provisional certificates, on behalf of FDA, and notify FDA of the receipt of such information;

(3) Submit to FDA the name, identifying information, and other information relevant to 42 U.S.C. 263b and specified by FDA for any facility for which the accreditation body denies, suspends, or revokes accreditation, and the reason(s) for such action;

(4) Submit to FDA an annual report summarizing all serious complaints received during the previous calendar year, their resolution status, and any actions taken in response to them;

(5) Provide to FDA other information relevant to 42 U.S.C. 263b and required by FDA about any facility accredited or undergoing accreditation by the body.

(i) *Fees.* Fees charged to facilities for accreditation shall be reasonable. Costs of accreditation body activities that are not related to accreditation functions under 42 U.S.C. 263b are not recoverable through fees established for accreditation.

(1) The accreditation body shall make public its fee structure, including those factors, if any, contributing to variations in fees for different facilities.

(2) At FDA's request, accreditation bodies shall provide financial records or other material to assist FDA in assessing the reasonableness of accreditation body fees. Such material shall be provided to FDA in a manner and time period specified by the agency.

[62 FR 55976, Oct. 28, 1997; 62 FR 60614, Nov. 10, 1997, as amended at 64 FR 32407, June 17, 1999]

§ 900.5 Evaluation.

FDA shall evaluate annually the performance of each accreditation body. Such evaluation shall include an assessment of the reports of FDA or State inspections of facilities accredited by the body as well as any additional information deemed relevant by FDA that has been provided by the accreditation body or other sources or has been required by FDA as part of its oversight initiatives. The evaluation shall include a determination of whether there are major deficiencies in the accreditation body's performance that, if not corrected, would warrant withdrawal of the approval of the accreditation body under the provisions of § 900.6.

§ 900.6 Withdrawal of approval.

If FDA determines, through the evaluation activities of § 900.5, or through other means, that an accreditation body is not in substantial compliance with this subpart, FDA may initiate the following actions:

(a) *Major deficiencies.* If FDA determines that an accreditation body has failed to perform a major accreditation function satisfactorily, has demonstrated willful disregard for public health, has violated the code of conduct, has committed fraud, or has submitted material false statements to the agency, FDA may withdraw its approval of that accreditation body.

(1) FDA shall notify the accreditation body of the agency's action and the grounds on which the approval was withdrawn.

(2) An accreditation body that has lost its approval shall notify facilities accredited or seeking accreditation by it that its approval has been withdrawn. Such notification shall be made within a time period and in a manner approved by FDA.

(b) *Minor deficiencies.* If FDA determines that an accreditation body has demonstrated deficiencies in performing accreditation functions and responsibilities that are less serious or more limited than the deficiencies in paragraph (a) of this section, FDA shall notify the body that it has a specified period of time to take particular corrective measures directed by FDA or to submit to FDA for approval the body's own plan of corrective action addressing the minor deficiencies. FDA may place the body on probationary status for a period of time determined by FDA, or may withdraw approval of the body as an accreditation body if corrective action is not taken.

(1) If FDA places an accreditation body on probationary status, the body shall notify all facilities accredited or seeking accreditation by it of its probationary status within a time period and in a manner approved by FDA.

(2) Probationary status shall remain in effect until such time as the body can demonstrate to the satisfaction of FDA that it has successfully implemented or is implementing the corrective action plan within the established schedule, and that the corrective actions have substantially eliminated all identified problems.

(3) If FDA determines that an accreditation body that has been placed on probationary status is not implementing corrective actions satisfactorily or within the established schedule, FDA may withdraw approval of the accreditation body. The accreditation body shall notify all facilities accredited or seeking accreditation by it of its loss of FDA approval, within a time period and in a manner approved by FDA.

(c) *Reapplication by accreditation bodies that have had their approval withdrawn.* (1) A former accreditation body that has had its approval withdrawn may submit a new application for approval if the body can provide information to FDA to establish that the problems that were grounds for withdrawal of approval have been resolved.

(2) If FDA determines that the new application demonstrates that the body satisfactorily has addressed the causes of its previous unacceptable performance, FDA may reinstate approval of the accreditation body.

(3) FDA may request additional information or establish additional conditions that must be met by a former accreditation body before FDA approves the reapplication.

(4) FDA may refuse to accept an application from a former accreditation body whose approval was withdrawn because of fraud or willful disregard of public health.

§ 900.7 Hearings.

(a) Opportunities to challenge final adverse actions taken by FDA regarding approval or reapproval of accreditation bodies, withdrawal of approval of accreditation bodies, or rejection of a proposed fee for accreditation shall be communicated through notices of opportunity for informal hearings in accordance with part 16 of this chapter.

(b) A facility that has been denied accreditation is entitled to an appeals process from the accreditation body. The appeals process shall be specified in writing by the accreditation body and shall have been approved by FDA in accordance with § 900.3(d) or § 900.4(a)(8).

(c) A facility that cannot achieve satisfactory resolution of an adverse accreditation decision through the accreditation body's appeals process may appeal to FDA for reconsideration in accordance with § 900.15.

§§ 900.8–900.9 [Reserved]

Subpart B—Quality Standards and Certification

§ 900.10 Applicability.

The provisions of subpart B are applicable to all facilities under the regulatory jurisdiction of the United States that provide mammography services, with the exception of the Department of Veterans Affairs.

§ 900.11 Requirements for certification.

(a) *General.* After October 1, 1994, a certificate issued by FDA is required for lawful operation of all mammography facilities subject to the provisions of this subpart. To obtain a certificate from FDA, facilities are required to meet the quality standards in § 900.12 and to be accredited by an approved accreditation body or other entity as designated by FDA.

(b) *Application*—(1) *Certificates.* (i) In order to qualify for a certificate, a facility must apply to an FDA-approved accreditation body, or to another entity designated by FDA. The facility shall submit to such body or entity the information required in 42 U.S.C. 263b(d)(1).

(ii) Following the agency's receipt of the accreditation body's decision to accredit a facility, or an equivalent decision by another entity designated by FDA, the agency may issue a certificate to the facility, or renew an existing certificate, if the agency determines that the facility has satisfied the requirements for certification or recertification.

(2) *Provisional certificates.* (i) A new facility beginning operation after October 1, 1994, is eligible to apply for a provisional certificate. The provisional certificate will enable the facility to perform mammography and to obtain the clinical images needed to complete the accreditation process. To apply for and receive a provisional certificate, a facility must meet the requirements of 42 U.S.C. 263b(c)(2) and submit the necessary information to an approved accreditation body or other entity designated by FDA.

(ii) Following the agency's receipt of the accreditation body's decision that a facility has submitted the required information, FDA may issue a provisional certificate to a facility upon determination that the facility has satisfied the requirements of § 900.11(b)(2)(i). A provisional certificate shall be effective for up to 6 months from the date of issuance. A provisional certificate cannot be renewed, but a facility may apply for a 90-day extension of the provisional certificate.

(3) *Extension of provisional certificate.* (i) To apply for a 90-day extension to a provisional certificate, a facility shall submit to its accreditation body, or other entity designated by FDA, a statement of what the facility is doing to obtain certification and evidence that there would be a significant adverse impact on access to mammography in the geographic area served if such facility did not obtain an extension.

(ii) The accreditation body shall forward the request, with its recommendation, to FDA within 2 business days after receipt.

(iii) FDA may issue a 90-day extension for a provisional certificate upon determination that the extension meets the criteria set forth in 42 U.S.C. 263b(c)(2).

(iv) There can be no renewal of a provisional certificate beyond the 90-day extension.

(c) *Reinstatement policy.* A previously certified facility that has allowed its certificate to expire, that has been refused a renewal of its certificate by FDA, or that has had its certificate suspended or revoked by FDA, may apply to have the certificate reinstated so that the facility may be considered to be a new facility and thereby be eligible for a provisional certificate.

(1) Unless prohibited from reinstatement under § 900.11(c)(4), a facility applying for reinstatement shall:

(i) Contact an FDA-approved accreditation body or other entity designated by FDA to determine the requirements for reapplication for accreditation;

(ii) Fully document its history as a previously provisionally certified or certified mammography facility, including the following information:

(A) Name and address of the facility under which it was previously provisionally certified or certified;

(B) Name of previous owner/lessor;

(C) FDA facility identification number assigned to the facility under its previous certification; and

(D) Expiration date of the most recent FDA provisional certificate or certificate; and

(iii) Justify application for reinstatement of accreditation by submitting to the accreditation body or other entity designated by FDA, a corrective action plan that details how the facility has corrected deficiencies that contributed to the lapse of, denial of renewal, or revocation of its certificate.

(2) FDA may issue a provisional certificate to the facility if:

(i) The accreditation body or other entity designated by FDA notifies the agency that the facility has adequately corrected, or is in the process of correcting, pertinent deficiencies; and

(ii) FDA determines that the facility has taken sufficient corrective action since the lapse of, denial of renewal, or revocation of its previous certificate.

(3) After receiving the provisional certificate, the facility may lawfully resume performing mammography services while completing the requirements for certification.

(4) If a facility's certificate was revoked on the basis of an act described in 41 U.S.C. 263b(i)(1), no person who owned or operated that facility at the time the act occurred may own or operate a mammography facility within 2 years of the date of revocation.

§ 900.12 Quality standards.

(a) *Personnel.* The following requirements apply to all personnel involved in any aspect of mammography, including the production, processing, and interpretation of mammograms and related quality assurance activities:

(1) *Interpreting physicians.* All physicians interpreting mammograms shall meet the following qualifications:

(i) *Initial qualifications.* Unless the exemption in paragraph (a)(1)(iii)(A) of this section applies, before beginning to interpret mammograms independently, the interpreting physician shall:

(A) Be licensed to practice medicine in a State;

(B)(*1*) Be certified in an appropriate specialty area by a body determined by FDA to have procedures and requirements adequate to ensure that physicians certified by the body are competent to interpret radiological procedures, including mammography; or

(*2*) Have had at least 3 months of documented formal training in the interpretation of mammograms and in topics related to mammography. The training shall include instruction in radiation physics, including radiation physics specific to mammography, radiation effects, and radiation protection. The mammographic interpretation component shall be under the direct supervision of a physician who meets the requirements of paragraph (a)(1) of this section;

(C) Have a minimum of 60 hours of documented medical education in mammography, which shall include: Instruction in the interpretation of mammograms and education in basic breast anatomy, pathology, physiology, technical aspects of mammography, and quality assurance and quality control in mammography. All 60 of these hours shall be category I and at least 15 of the category I hours shall have been acquired within the 3 years immediately prior to the date that the physician qualifies as an interpreting physician. Hours spent in residency specifically devoted to mammography will be considered as equivalent to Category I continuing medical education credits and will be accepted if documented in writing by the appropriate representative of the training institution; and

(D) Unless the exemption in paragraph (a)(1)(iii)(B) of this section applies, have interpreted or multi-read at least 240 mammographic examinations within the 6-month period immediately prior to the date that the physician qualifies as an interpreting physician. This interpretation or multi-reading shall be under the direct supervision of an interpreting physician.

(ii) *Continuing experience and education.* All interpreting physicians shall maintain their qualifications by meeting the following requirements:

(A) Following the second anniversary date of the end of the calendar quarter in which the requirements of paragraph (a)(1)(i) of this section were completed, the interpreting physician shall have interpreted or multi-read at least 960 mammographic examinations during the 24 months immediately preceding the date of the facility's annual MQSA inspection or the last day of the calendar quarter preceding the inspection or any date in-between the two. The facility will choose one of these dates to determine the 24-month period.

(B) Following the third anniversary date of the end of the calendar quarter in which the requirements of paragraph (a)(1)(i) of this section were completed, the interpreting physician shall have taught or completed at least 15 category I continuing medical education units in mammography during the 36 months immediately preceding the date of the facility's annual MQSA inspection or the last day of the calendar quarter preceding the inspection or any date in between the two. The facility will choose one of these dates to determine the 36-month period. This training shall include at least six category I continuing medical education credits in each mammographic modality used by the interpreting physician in his or her practice; and

(C) Before an interpreting physician may begin independently interpreting mammograms produced by a new mammographic modality, that is, a mammographic modality in which the physician has not previously been trained, the interpreting physician shall have at least 8 hours of training in the new mammographic modality.

(D) Units earned through teaching a specific course can be counted only once towards the 15 required by paragraph (a)(1)(ii)(B) of this section, even if the course is taught multiple times during the previous 36 months.

(iii) *Exemptions.* (A) Those physicians who qualified as interpreting physicians under paragraph (a)(1) of this section of FDA's interim regulations prior to April 28, 1999, are considered to have met the initial requirements of paragraph (a)(1)(i) of this section. They may continue to interpret mammograms provided they continue to meet the licensure requirement of paragraph (a)(1)(i)(A) of this section and the continuing experience and education requirements of paragraph (a)(1)(ii) of this section.

(B) Physicians who have interpreted or multi-read at least 240 mammographic examinations under the direct supervision of an interpreting physician in any 6-month period during the last 2 years of a diagnostic radiology residency and who become appropriately board certified at the first allowable time, as defined by an eligible certifying body, are otherwise exempt from paragraph (a)(1)(i)(D) of this section.

(iv) *Reestablishing qualifications.* Interpreting physicians who fail to maintain the required continuing experience or continuing education requirements shall reestablish their qualifications before resuming the independent interpretation of mammograms, as follows:

(A) Interpreting physicians who fail to meet the continuing experience requirements of paragraph (a)(1)(ii)(A) of this section shall:

(*1*) Interpret or multi-read at least 240 mammographic examinations under the direct supervision of an interpreting physician, or

(*2*) Interpret or multi-read a sufficient number of mammographic examinations, under the direct supervision of an interpreting physician, to bring the physician's total up to 960 examinations for the prior 24 months, whichever is less.

(*3*) The interpretations required under paragraph (a)(1)(iv)(A)(*1*) or (a)(1)(iv)(A)(*2*) of this section shall be done within the 6 months immediately prior to resuming independent interpretation.

(B) Interpreting physicians who fail to meet the continuing education requirements of paragraph (a)(1)(ii)(B) of this section shall obtain a sufficient number of additional category I continuing medical education credits in mammography to bring their total up to the required 15 credits in the previous 36 months before resuming independent interpretation.

(2) *Radiologic technologists.* All mammographic examinations shall be performed by radiologic technologists who meet the following general requirements, mammography requirements, and continuing education and experience requirements:

(i) *General requirements.* (A) Be licensed to perform general radiographic procedures in a State; or

(B) Have general certification from one of the bodies determined by FDA to have procedures and requirements adequate to ensure that radiologic technologists certified by the body are competent to perform radiologic examinations; and

(ii) *Mammography requirements.* Have, prior to April 28, 1999, qualified as a radiologic technologist under paragraph (a)(2) of this section of FDA's interim regulations of December 21, 1993, or completed at least 40 contact hours of documented training specific to mammography under the supervision of a qualified instructor. The hours of documented training shall include, but not necessarily be limited to:

(A) Training in breast anatomy and physiology, positioning and compression, quality assurance/quality control techniques, imaging of patients with breast implants;

(B) The performance of a minimum of 25 examinations under the direct supervision of an individual qualified under paragraph (a)(2) of this section; and

(C) At least 8 hours of training in each mammography modality to be used by the technologist in performing mammography exams; and

(iii) *Continuing education requirements.* (A) Following the third anniversary date of the end of the calendar quarter in which the requirements of paragraphs (a)(2)(i) and (a)(2)(ii) of this section were completed, the radiologic technologist shall have taught or completed at least 15 continuing education units in mammography during the 36 months immediately preceding the date of the facility's annual MQSA inspection or the last day of the calendar quarter preceding the inspection or any date in between the two. The facility will choose one of these dates to determine the 36-month period.

(B) Units earned through teaching a specific course can be counted only once towards the 15 required in paragraph (a)(2)(iii)(A) of this section, even if the course is taught multiple times during the previous 36 months.

(C) At least six of the continuing education units required in paragraph (a)(2)(iii)(A) of this section shall be related to each mammographic modality used by the technologist.

(D) *Requalification.* Radiologic technologists who fail to meet the continuing education requirements of paragraph (a)(2)(iii)(A) of this section shall obtain a sufficient number of continuing education units in mammography to bring their total up to at least 15 in the previous 3 years, at least 6 of which shall be related to each modality used by the technologist in mammography. The technologist may not resume performing unsupervised mammography examinations until the continuing education requirements are completed.

(E) Before a radiologic technologist may begin independently performing mammographic examinations using a mammographic modality other than one of those for which the technologist received training under paragraph (a)(2)(ii)(C) of this section, the technologist shall have at least 8 hours of continuing education units in the new modality.

(iv) *Continuing experience requirements.* (A) Following the second anniversary date of the end of the calendar quarter in which the requirements of paragraphs (a)(2)(i) and (a)(2)(ii) of this section were completed or of April 28, 1999, whichever is later, the radiologic technologist shall have performed a minimum of 200 mammography examinations during the 24 months immediately preceding the date of the facility's annual inspection or the last day of the calendar quarter preceding the inspection or any date in between the two. The facility will choose one of these dates to determine the 24-month period.

(B) *Requalification.* Radiologic technologists who fail to meet the continuing experience requirements of paragraph (a)(2)(iv)(A) of this section shall perform a minimum of 25 mammography examinations under the direct supervision of a qualified radiologic technologist, before resuming the performance of unsupervised mammography examinations.

(3) *Medical physicists.* All medical physicists conducting surveys of mammography facilities and providing oversight of the facility quality assurance program under paragraph (e) of this section shall meet the following:

(i) *Initial qualifications.* (A) Be State licensed or approved or have certification in an appropriate specialty area by one of the bodies determined by FDA to have procedures and requirements to ensure that medical physicists certified by the body are competent to perform physics survey; and

(B)(*1*) Have a masters degree or higher in a physical science from an accredited institution, with no less than 20 semester hours or equivalent (e.g., 30 quarter hours) of college undergraduate or graduate level physics;

(*2*) Have 20 contact hours of documented specialized training in conducting surveys of mammography facilities; and

(*3*) Have the experience of conducting surveys of at least 1 mammography facility and a total of at least 10 mammography units. No more than one survey of a specific unit within a period of 60 days can be counted towards the total mammography unit survey requirement. After April 28, 1999, experience conducting surveys must be acquired under the direct supervision of a medical physicist who meets all the requirements of paragraphs (a)(3)(i) and (a)(3)(iii) of this section; or

(ii) *Alternative initial qualifications.* (A) Have qualified as a medical physicist under paragraph (a)(3) of this section of FDA's interim regulations and retained that qualification by maintenance of the active status of any licensure, approval, or certification required under the interim regulations; and

(B) Prior to the April 28, 1999, have:

(*1*) A bachelor's degree or higher in a physical science from an accredited institution with no less than 10 semester hours or equivalent of college undergraduate or graduate level physics,

(*2*) Forty contact hours of documented specialized training in conducting surveys of mammography facilities and,

(*3*) Have the experience of conducting surveys of at least 1 mammography facility and a total of at least 20 mammography units. No more than one survey of a specific unit within a period of 60 days can be counted towards the total mammography unit survey requirement. The training and experience requirements must be met after fulfilling the degree requirement.

(iii) *Continuing qualifications.* (A) Continuing education. Following the third anniversary date of the end of the calendar quarter in which the requirements of paragraph (a)(3)(i) or (a)(3)(ii) of this section were completed, the medical physicist shall have taught or completed at least 15 continuing education units in mammography during the 36 months immediately preceding the date of the facility's annual inspection or the last day of the calendar quarter preceding the inspection or any date in between the two. The facility shall choose one of these dates to determine the 36-month period. This continuing education shall include hours of training appropriate to each mammographic modality evaluated by the medical physicist during his or her surveys or oversight of quality assurance programs. Units earned through teaching a specific course can be counted only once towards the required 15 units in a 36-month period, even if the course is taught multiple times during the 36 months.

(B) *Continuing experience.* Following the second anniversary date of the end of the calendar quarter in which the requirements of paragraphs (a)(3)(i) and (a)(3)(ii) of this section were completed or of April 28, 1999, whichever is later, the medical physicist shall have surveyed at least two mammography facilities and a total of at least six mammography units during the 24 months immediately preceding the date of the facility's annual MQSA inspection or the last day of the calender quarter preceding the inspection or any date in between the two. The facility shall choose one of these dates to determine the 24-month period. No more than one survey of a specific facility within a 10-month period or a specific unit within a period of 60 days can be counted towards this requirement.

(C) Before a medical physicist may begin independently performing mammographic surveys of a new mammographic modality, that is, a mammographic modality other than one for which the physicist received training to qualify under paragraph (a)(3)(i) or (a)(3)(ii) of this section, the physicist must receive at least 8 hours of training in surveying units of the new mammographic modality.

(iv) *Reestablishing qualifications.* Medical physicists who fail to maintain the required continuing qualifications of paragraph (a)(3)(iii) of this section may not perform the MQSA surveys without the supervision of a qualified medical physicist. Before independently surveying another facility, medical physicists must reestablish their qualifications, as follows:

(A) Medical physicists who fail to meet the continuing educational requirements of paragraph (a)(3)(iii)(A) of this section shall obtain a sufficient number of continuing education units to bring their total units up to the required 15 in the previous 3 years.

(B) Medical physicists who fail to meet the continuing experience requirement of paragraph (a)(3)(iii)(B) of this section shall complete a sufficient number of surveys under the direct supervision of a medical physicist who meets the qualifications of paragraphs (a)(3)(i) and (a)(3)(iii) of this section to bring their total surveys up to the required two facilities and six units in the previous 24 months. No more than one survey of a specific unit within a period of 60 days can be counted towards the total mammography unit survey requirement.

(4) *Retention of personnel records.* Facilities shall maintain records to document the qualifications of all personnel who worked at the facility as interpreting physicians, radiologic technologists, or medical physicists. These records must be available for review by the MQSA inspectors. Records of personnel no longer employed by the facility should not be discarded until the next annual inspection has been completed and FDA has determined that the facility is in compliance with the MQSA personnel requirements.

(b) *Equipment.* Regulations published under §§ 1020.30, 1020.31, and 900.12(e) of this chapter that are relevant to equipment performance should also be consulted for a more complete understanding of the equipment performance requirements.

(1) *Prohibited equipment.* Radiographic equipment designed for general purpose or special nonmammography procedures shall not be used for mammography. This prohibition includes systems that have been modified or equipped with special attachments for mammography. This requirement supersedes the implied acceptance of such systems in § 1020.31(f)(3) of this chapter.

(2) *General.* All radiographic equipment used for mammography shall be specifically designed for mammography and shall be certified pursuant to § 1010.2 of this chapter as meeting the applicable requirements of §§ 1020.30 and 1020.31 of this chapter in effect at the date of manufacture.

(3) *Motion of tube-image receptor assembly.* (i) The assembly shall be capable of being fixed in any position where it is designed to operate. Once fixed in any such position, it shall not undergo unintended motion.

(ii) The mechanism ensuring compliance with paragraph (b)(3)(i) of this section shall not fail in the event of power interruption.

(4) *Image receptor sizes.* (i) Systems using screen-film image receptors shall provide, at a minimum, for operation with image receptors of 18×24 centimeters (cm) and 24×30 cm.

(ii) Systems using screen-film image receptors shall be equipped with moving grids matched to all image receptor sizes provided.

(iii) Systems used for magnification procedures shall be capable of operation with the grid removed from between the source and image receptor.

(5) *Light fields.* For any mammography system with a light beam that passes through the x-ray beam-limiting device, the light shall provide an average illumination of not less than 160 lux (15 foot candles) at 100 cm or the maximum source-image receptor distance (SID), whichever is less.

(6) *Magnification.* (i) Systems used to perform noninterventional problem solving procedures shall have radiographic magnification capability available for use by the operator.

(ii) Systems used for magnification procedures shall provide, at a minimum, at least one magnification value within the range of 1.4 to 2.0.

(7) *Focal spot selection.* (i) When more than one focal spot is provided, the system shall indicate, prior to exposure, which focal spot is selected.

(ii) When more than one target material is provided, the system shall indicate, prior to exposure, the preselected target material.

(iii) When the target material and/or focal spot is selected by a system algorithm that is based on the exposure or on a test exposure, the system shall display, after the exposure, the target material and/or focal spot actually used during the exposure.

(8) *Compression.* All mammography systems shall incorporate a compression device.

(i) *Application of compression.* Effective October 28, 2002, each system shall provide:

(A) An initial power-driven compression activated by hands-free controls operable from both sides of the patient; and

(B) Fine adjustment compression controls operable from both sides of the patient.

(ii) *Compression paddle.* (A) Systems shall be equipped with different sized compression paddles that match the sizes of all full-field image receptors provided for the system. Compression paddles for special purposes, including those smaller than the full size of the image receptor (for "spot compression") may be provided. Such compression paddles for special purposes are not subject to the requirements of paragraphs (b)(8)(ii)(D) and (b)(8)(ii)(E) of this section.

(B) Except as provided in paragraph (b)(8)(ii)(C) of this section, the compression paddle shall be flat and parallel to the breast support table and shall not deflect from parallel by more than 1.0 cm at any point on the surface of the compression paddle when compression is applied.

(C) Equipment intended by the manufacturer's design to not be flat and parallel to the breast support table during compression shall meet the manufacturer's design specifications and maintenance requirements.

(D) The chest wall edge of the compression paddle shall be straight and parallel to the edge of the image receptor.

(E) The chest wall edge may be bent upward to allow for patient comfort but shall not appear on the image.

(9) *Technique factor selection and display.* (i) Manual selection of milliampere seconds (mAs) or at least one of its component parts (milliapere (mA) and/or time) shall be available.

(ii) The technique factors (peak tube potential in kilovolt (kV) and either tube current in mA and exposure time in seconds or the product of tube current and exposure time in mAs) to be used during an exposure shall be indicated before the exposure begins, except when automatic exposure controls (AEC) are used, in which case the technique factors that are set prior to the exposure shall be indicated.

(iii) Following AEC mode use, the system shall indicate the actual kilovoltage peak (kVp) and mAs used during the exposure. The mAs may be displayed as mA and time.

(10) *Automatic exposure control.* (i) Each screen-film system shall provide an AEC mode that is operable in all combinations of equipment configuration provided, e.g., grid, nongrid; magnification, nonmagnification; and various target-filter combinations.

(ii) The positioning or selection of the detector shall permit flexibility in the placement of the detector under the target tissue.

(A) The size and available positions of the detector shall be clearly indicated at the X-ray input surface of the breast compression paddle.

(B) The selected position of the detector shall be clearly indicated.

(iii) The system shall provide means for the operator to vary the selected optical density from the normal (zero) setting.

(11) *X-ray film.* The facility shall use X-ray film for mammography that has been designated by the film manufacturer as appropriate for mammography.

(12) *Intensifying screens.* The facility shall use intensifying screens for mammography that have been designated by the screen manufacturer as appropriate for mammography and shall use film that is matched to the screen's spectral output as specified by the manufacturer.

(13) *Film processing solutions.* For processing mammography films, the facility shall use chemical solutions that are capable of developing the films used by the facility in a manner equivalent to the minimum requirements specified by the film manufacturer.

(14) *Lighting.* The facility shall make special lights for film illumination, i.e., hot-lights, capable of producing light levels greater than that provided by the view box, available to the interpreting physicians.

(15) *Film masking devices.* Facilities shall ensure that film masking devices that can limit the illuminated area to a region equal to or smaller than the exposed portion of the film are available to all interpreting physicians interpreting for the facility.

(c) *Medical records and mammography reports*—(1) *Contents and terminology.* Each facility shall prepare a written report of the results of each mammography examination performed under its certificate. The mammography report shall include the following information:

(i) The name of the patient and an additional patient identifier;

(ii) Date of examination;

(iii) The name of the interpreting physician who interpreted the mammogram;

(iv) Overall final assessment of findings, classified in one of the following categories:

(A) "Negative:" Nothing to comment upon (if the interpreting physician is aware of clinical findings or symptoms, despite the negative assessment, these shall be explained);

(B) "Benign:" Also a negative assessment;

(C) "Probably Benign:" Finding(s) has a high probability of being benign;

(D) "Suspicious:" Finding(s) without all the characteristic morphology of breast cancer but indicating a definite probability of being malignant;

(E) "Highly suggestive of malignancy:" Finding(s) has a high probability of being malignant;

(v) In cases where no final assessment category can be assigned due to incomplete work-up, "Incomplete: Need additional imaging evaluation" shall be assigned as an assessment and reasons why no assessment can be made shall be stated by the interpreting physician; and

(vi) Recommendations made to the health care provider about what additional actions, if any, should be taken. All clinical questions raised by the referring health care provider shall be addressed in the report to the extent possible, even if the assessment is negative or benign.

(2) *Communication of mammography results to the patients.* Each facility shall send each patient a summary of the mammography report written in lay terms within 30 days of the mammographic examination. If assessments are "Suspicious" or "Highly suggestive of malignancy," the facility shall make reasonable attempts to ensure that the results are communicated to the patient as soon as possible.

(i) Patients who do not name a health care provider to receive the mammography report shall be sent the report described in paragraph (c)(1) of this section within 30 days, in addition to the written notification of results in lay terms.

(ii) Each facility that accepts patients who do not have a health care provider shall maintain a system for referring such patients to a health care provider when clinically indicated.

(3) *Communication of mammography results to health care providers.* When the patient has a referring health care provider or the patient has named a health care provider, the facility shall:

(i) Provide a written report of the mammography examination, including the items listed in paragraph (c)(1) of this section, to that health care provider as soon as possible, but no later than 30 days from the date of the mammography examination; and

(ii) If the assessment is "Suspicious" or "Highly suggestive of malignancy," make reasonable attempts to communicate with the health care provider as soon as possible, or if the health care provider is unavailable, to a responsible designee of the health care provider.

(4) *Recordkeeping.* Each facility that performs mammograms:

(i) Shall (except as provided in paragraph (c)(4)(ii) of this section) maintain mammography films and reports in a permanent medical record of the patient for a period of not less than 5 years, or not less than 10 years if no additional mammograms of the patient are performed at the facility, or a longer period if mandated by State or local law; and

(ii) Shall upon request by, or on behalf of, the patient, permanently or temporarily transfer the original mammograms and copies of the patient's reports to a medical institution, or to a physician or health care provider of the patient, or to the patient directly;

(iii) Any fee charged to the patients for providing the services in paragraph (c)(4)(ii) of this section shall not exceed the documented costs associated with this service.

(5) *Mammographic image identification.* Each mammographic image shall have the following information indicated on

it in a permanent, legible, and unambiguous manner and placed so as not to obscure anatomic structures:

(i) Name of patient and an additional patient identifier.

(ii) Date of examination.

(iii) *View and laterality*. This information shall be placed on the image in a position near the axilla. Standardized codes specified by the accreditation body and approved by FDA in accordance with § 900.3(b) or § 900.4(a)(8) shall be used to identify view and laterality.

(iv) *Facility name and location*. At a minimum, the location shall include the city, State, and zip code of the facility.

(v) Technologist identification.

(vi) Cassette/screen identification.

(vii) Mammography unit identification, if there is more than one unit in the facility.

(d) *Quality assurance—general*. Each facility shall establish and maintain a quality assurance program to ensure the safety, reliability, clarity, and accuracy of mammography services performed at the facility.

(1) *Responsible individuals*. Responsibility for the quality assurance program and for each of its elements shall be assigned to individuals who are qualified for their assignments and who shall be allowed adequate time to perform these duties.

(i) *Lead interpreting physician*. The facility shall identify a lead interpreting physician who shall have the general responsibility of ensuring that the quality assurance program meets all requirements of paragraphs (d) through (f) of this section. No other individual shall be assigned or shall retain responsibility for quality assurance tasks unless the lead interpreting physician has determined that the individual's qualifications for, and performance of, the assignment are adequate.

(ii) *Interpreting physicians*. All interpreting physicians interpreting mammograms for the facility shall:

(A) Follow the facility procedures for corrective action when the images they are asked to interpret are of poor quality, and

(B) Participate in the facility's medical outcomes audit program.

(iii) *Medical physicist*. Each facility shall have the services of a medical physicist available to survey mammography equipment and oversee the equipment-related quality assurance practices of the facility. At a minimum, the medical physicist(s) shall be responsible for performing the surveys and mammography equipment evaluations and providing the facility with the reports described in paragraphs (e)(9) and (e)(10) of this section.

(iv) *Quality control technologist*. Responsibility for all individual tasks within the quality assurance program not assigned to the lead interpreting physician or the medical physicist shall be assigned to a quality control technologist(s). The tasks are to be performed by the quality control technologist or by other personnel qualified to perform the tasks. When other personnel are utilized for these tasks, the quality control technologist shall ensure that the tasks are completed in such a way as to meet the requirements of paragraph (e) of this section.

(2) *Quality assurance records*. The lead interpreting physician, quality control technologist, and medical physicist shall ensure that records concerning mammography technique and procedures, quality control (including monitoring data, problems detected by analysis of that data, corrective actions, and the effectiveness of the correction actions), safety, protection, and employee qualifications to meet assigned quality assurance tasks are properly maintained and updated. These quality control records shall be kept for each test specified in paragraphs (e) and (f) of this section until the next annual inspection has been completed and FDA has determined that the facility is in compliance with the quality assurance requirements or until the test has been performed two additional times at the required frequency, whichever is longer.

(e) *Quality assurance—equipment*—(1) *Daily quality control tests*. Film processors used to develop mammograms shall be adjusted and maintained to meet the technical development specifications for the mammography film in use. A processor performance test shall be performed on each day that clinical films are processed before any clinical films are processed that day. The test shall include an assessment of base plus fog density, mid-density, and density difference, using the mammography film used clinically at the facility.

(i) The base plus fog density shall be within + 0.03 of the established operating level.

(ii) The mid-density shall be within ±0.15 of the established operating level.

(iii) The density difference shall be within ±0.15 of the established operating level.

(2) *Weekly quality control tests.* Facilities with screen-film systems shall perform an image quality evaluation test, using an FDA-approved phantom, at least weekly.

(i) The optical density of the film at the center of an image of a standard FDA-accepted phantom shall be at least 1.20 when exposed under a typical clinical condition.

(ii) The optical density of the film at the center of the phantom image shall not change by more than ±0.20 from the established operating level.

(iii) The phantom image shall achieve at least the minimum score established by the accreditation body and accepted by FDA in accordance with § 900.3(d) or § 900.4(a)(8).

(iv) The density difference between the background of the phantom and an added test object, used to assess image contrast, shall be measured and shall not vary by more than ±0.05 from the established operating level.

(3) *Quarterly quality control tests.* Facilities with screen-film systems shall perform the following quality control tests at least quarterly:

(i) *Fixer retention in film.* The residual fixer shall be no more than 5 micrograms per square cm.

(ii) *Repeat analysis.* If the total repeat or reject rate changes from the previously determined rate by more than 2.0 percent of the total films included in the analysis, the reason(s) for the change shall be determined. Any corrective actions shall be recorded and the results of these corrective actions shall be assessed.

(4) *Semiannual quality control tests.* Facilities with screen-film systems shall perform the following quality control tests at least semiannually:

(i) *Darkroom fog.* The optical density attributable to darkroom fog shall not exceed 0.05 when a mammography film of the type used in the facility, which has a mid-density of no less than 1.2 OD, is exposed to typical darkroom conditions for 2 minutes while such film is placed on the counter top emulsion side up. If the darkroom has a safelight used for mammography film, it shall be on during this test.

(ii) *Screen-film contact.* Testing for screen-film contact shall be conducted using 40 mesh copper screen. All cassettes used in the facility for mammography shall be tested.

(iii) *Compression device performance.* (A) A compression force of at least 111 newtons (25 pounds) shall be provided.

(B) Effective October 28, 2002, the maximum compression force for the initial power drive shall be between 111 newtons (25 pounds) and 200 newtons (45 pounds).

(5) *Annual quality control tests.* Facilities with screen-film systems shall perform the following quality control tests at least annually:

(i) *Automatic exposure control performance.* (A) The AEC shall be capable of maintaining film optical density within ±0.30 of the mean optical density when thickness of a homogeneous material is varied over a range of 2 to 6 cm and the kVp is varied appropriately for such thicknesses over the kVp range used clinically in the facility. If this requirement cannot be met, a technique chart shall be developed showing appropriate techniques (kVp and density control settings) for different breast thicknesses and compositions that must be used so that optical densities within ±0.30 of the average under phototimed conditions can be produced.

(B) After October 28, 2002, the AEC shall be capable of maintaining film optical density (OD) within ±0.15 of the mean optical density when thickness of a homogeneous material is varied over a range of 2 to 6 cm and the kVp is varied appropriately for such thicknesses over the kVp range used clinically in the facility.

(C) The optical density of the film in the center of the phantom image shall not be less than 1.20.

(ii) *Kilovoltage peak (kVp) accuracy and reproducibility.* (A) The kVp shall be accurate within ±5 percent of the indicated or selected kVp at:

(*1*) The lowest clinical kVp that can be measured by a kVp test device;

(*2*) The most commonly used clinical kVp;

(*3*) The highest available clinical kVp, and

(B) At the most commonly used clinical settings of kVp, the coefficient of variation of reproducibility of the kVp shall be equal to or less than 0.02.

(iii) *Focal spot condition.* Until October 28, 2002, focal spot condition shall be evaluated either by determining system resolution or by measuring focal spot dimensions. After October 28, 2002, facilities shall evaluate focal spot condition only by determining the system resolution.

(A) *System resolution.* (*1*) Each X-ray system used for mammography, in combination with the mammography screen-film combination used in the facility, shall provide a minimum resolution of 11 Cycles/millimeter (mm) (line-pairs/mm) when a high contrast resolution bar test pattern is oriented with the bars perpendicular to the anode-cathode axis, and a minimum resolution of 13 line-pairs/mm when the bars are parallel to that axis.

(*2*) The bar pattern shall be placed 4.5 cm above the breast support surface, centered with respect to the chest wall edge of the image receptor, and with the edge of the pattern within 1 cm of the chest wall edge of the image receptor.

(*3*) When more than one target material is provided, the measurement in paragraph (e)(5)(iii)(A) of this section shall be made using the appropriate focal spot for each target material.

(*4*) When more than one SID is provided, the test shall be performed at SID most commonly used clinically.

(*5*) Test kVp shall be set at the value used clinically by the facility for a standard breast and shall be performed in the AEC mode, if available. If necessary, a suitable absorber may be placed in the beam to increase exposure times. The screen-film cassette combination used by the facility shall be used to test for this requirement and shall be placed in the normal location used for clinical procedures.

(B) *Focal spot dimensions.* Measured values of the focal spot length (dimension parallel to the anode cathode axis) and width (dimension perpendicular to the anode cathode axis) shall be within the tolerance limits specified in table 1.

TABLE 1

Focal Spot Tolerance Limit

Nominal Focal Spot Size (mm)	Maximum Measured Dimensions	
	Width(mm)	Length(mm)
0.10	0.15	0.15
0.15	0.23	0.23
0.20	0.30	0.30
0.30	0.45	0.65
0.40	0.60	0.85
0.60	0.90	1.30

(iv) *Beam quality and half-value layer (HVL).* The HVL shall meet the specifications of § 1020.30(m)(1) of this chapter for the minimum HVL. These values, extrapolated to the mammographic range, are shown in table 2. Values not shown in table 2 may be determined by linear interpolation or extrapolation.

TABLE 2

X-ray Tube Voltage (kilovolt peak) and Minimum HVL

Designed Operating Range (kV)	Measured Operating Voltage (kV)	Minimum HVL (millimeters of aluminum)
Below 50	20	0.20
	25	0.25
	30	0.30

(v) *Breast entrance air kerma and AEC reproducibility*. The coefficient of variation for both air kerma and mAs shall not exceed 0.05.

(vi) *Dosimetry*. The average glandular dose delivered during a single cranio-caudal view of an FDA-accepted phantom simulating a standard breast shall not exceed 3.0 milligray (mGy) (0.3 rad) per exposure. The dose shall be determined with technique factors and conditions used clinically for a standard breast.

(vii) *X-ray field/light field/image receptor/compression paddle alignment*. (A) All systems shall have beam-limiting devices that allow the entire chest wall edge of the x-ray field to extend to the chest wall edge of the image receptor and provide means to assure that the x-ray field does not extend beyond any edge of the image receptor by more than 2 percent of the SID.

(B) If a light field that passes through the X-ray beam limitation device is provided, it shall be aligned with the X-ray field so that the total of any misalignment of the edges of the light field and the X-ray field along either the length or the width of the visually defined field at the plane of the breast support surface shall not exceed 2 percent of the SID.

(C) The chest wall edge of the compression paddle shall not extend beyond the chest wall edge of the image receptor by more than one percent of the SID when tested with the compression paddle placed above the breast support surface at a distance equivalent to standard breast thickness. The shadow of the vertical edge of the compression paddle shall not be visible on the image.

(viii) *Uniformity of screen speed*. Uniformity of screen speed of all the cassettes in the facility shall be tested and the difference between the maximum and minimum optical densities shall not exceed 0.30. Screen artifacts shall also be evaluated during this test.

(ix) *System artifacts*. System artifacts shall be evaluated with a high-grade, defect-free sheet of homogeneous material large enough to cover the mammography cassette and shall be performed for all cassette sizes used in the facility using a grid appropriate for the cassette size being tested. System artifacts shall also be evaluated for all available focal spot sizes and target filter combinations used clinically.

(x) *Radiation output*. (A) The system shall be capable of producing a minimum output of 4.5 mGy air kerma per second (513 milli Roentgen (mR) per second) when operating at 28 kVp in the standard mammography (moly/moly) mode at any SID where the system is designed to operate and when measured by a detector with its center located 4.5 cm above the breast support surface with the compression paddle in place between the source and the detector. After October 28, 2002, the system, under the same measuring conditions shall be capable of producing a minimum output of 7.0 mGy air kerma per second (800 mR per second) when operating at 28 kVp in the standard (moly/moly) mammography mode at any SID where the system is designed to operate.

(B) The system shall be capable of maintaining the required minimum radiation output averaged over a 3.0 second period.

(xi) *Decompression*. If the system is equipped with a provision for automatic decompression after completion of an exposure or interruption of power to the system, the system shall be tested to confirm that it provides:

(A) An override capability to allow maintenance of compression;

(B) A continuous display of the override status; and

(C) A manual emergency compression release that can be activated in the event of power or automatic release failure.

(6) *Quality control tests—other modalities*. For systems with image receptor modalities other than screen-film, the quality assurance program shall be substantially the same as the quality assurance program recommended by the image receptor manufacturer, except that the maximum allowable dose shall not exceed the maximum allowable dose for screen-film systems in paragraph (e)(5)(vi) of this section.

(7) *Mobile units*. The facility shall verify that mammography units used to produce mammograms at more than one location meet the requirements in paragraphs (e)(1) through (e)(6) of this

section. In addition, at each examination location, before any examinations are conducted, the facility shall verify satisfactory performance of such units using a test method that establishes the adequacy of the image quality produced by the unit.

(8) *Use of test results.* (i) After completion of the tests specified in paragraphs (e)(1) through (e)(7) of this section, the facility shall compare the test results to the corresponding specified action limits; or, for nonscreen-film modalities, to the manufacturer's recommended action limits; or, for post-move, preexamination testing of mobile units, to the limits established in the test method used by the facility.

(ii) If the test results fall outside of the action limits, the source of the problem shall be identified and corrective actions shall be taken:

(A) Before any further examinations are performed or any films are processed using a component of the mammography system that failed any of the tests described in paragraphs (e)(1), (e)(2), (e)(4)(i), (e)(4)(ii), (e)(4)(iii), (e)(5)(vi), (e)(6), or (e)(7) of this section;

(B) Within 30 days of the test date for all other tests described in paragraph (e) of this section.

(9) *Surveys.* (i) At least once a year, each facility shall undergo a survey by a medical physicist or by an individual under the direct supervision of a medical physicist. At a minimum, this survey shall include the performance of tests to ensure that the facility meets the quality assurance requirements of the annual tests described in paragraphs (e)(5) and (e)(6) of this section and the weekly phantom image quality test described in paragraph (e)(2) of this section.

(ii) The results of all tests conducted by the facility in accordance with paragraphs (e)(1) through (e)(7) of this section, as well as written documentation of any corrective actions taken and their results, shall be evaluated for adequacy by the medical physicist performing the survey.

(iii) The medical physicist shall prepare a survey report that includes a summary of this review and recommendations for necessary improvements.

(iv) The survey report shall be sent to the facility within 30 days of the date of the survey.

(v) The survey report shall be dated and signed by the medical physicist performing or supervising the survey. If the survey was performed entirely or in part by another individual under the direct supervision of the medical physicist, that individual and the part of the survey that individual performed shall also be identified in the survey report.

(10) *Mammography equipment evaluations.* Additional evaluations of mammography units or image processors shall be conducted whenever a new unit or processor is installed, a unit or processor is disassembled and reassembled at the same or a new location, or major components of a mammography unit or processor equipment are changed or repaired. These evaluations shall be used to determine whether the new or changed equipment meets the requirements of applicable standards in paragraphs (b) and (e) of this section. All problems shall be corrected before the new or changed equipment is put into service for examinations or film processing. The mammography equipment evaluation shall be performed by a medical physicist or by an individual under the direct supervision of a medical physicist.

(11) *Facility cleanliness.* (i) The facility shall establish and implement adequate protocols for maintaining darkroom, screen, and view box cleanliness.

(ii) The facility shall document that all cleaning procedures are performed at the frequencies specified in the protocols.

(12) *Calibration of air kerma measuring instruments.* Instruments used by medical physicists in their annual survey to measure the air kerma or air kerma rate from a mammography unit shall be calibrated at least once every 2 years and each time the instrument is repaired. The instrument calibration must be traceable to a national standard and calibrated with an accuracy of ±6 percent (95 percent confidence level) in the mammography energy range.

(13) *Infection control.* Facilities shall establish and comply with a system specifying procedures to be followed by the facility for cleaning and disinfecting mammography equipment

after contact with blood or other potentially infectious materials. This system shall specify the methods for documenting facility compliance with the infection control procedures established and shall:

(i) Comply with all applicable Federal, State, and local regulations pertaining to infection control; and

(ii) Comply with the manufacturer's recommended procedures for the cleaning and disinfection of the mammography equipment used in the facility; or

(iii) If adequate manufacturer's recommendations are not available, comply with generally accepted guidance on infection control, until such recommendations become available.

(f) *Quality assurance-mammography medical outcomes audit.* Each facility shall establish and maintain a mammography medical outcomes audit program to followup positive mammographic assessments and to correlate pathology results with the interpreting physician's findings. This program shall be designed to ensure the reliability, clarity, and accuracy of the interpretation of mammograms.

(1) *General requirements.* Each facility shall establish a system to collect and review outcome data for all mammograms performed, including followup on the disposition of all positive mammograms and correlation of pathology results with the interpreting physician's mammography report. Analysis of these outcome data shall be made individually and collectively for all interpreting physicians at the facility. In addition, any cases of breast cancer among women imaged at the facility that subsequently become known to the facility shall prompt the facility to initiate followup on surgical and/or pathology results and review of the mammograms taken prior to the diagnosis of a malignancy.

(2) *Frequency of audit analysis.* The facility's first audit analysis shall be initiated no later than 12 months after the date the facility becomes certified, or 12 months after April 28, 1999, whichever date is the latest. This audit analysis shall be completed within an additional 12 months to permit completion of diagnostic procedures and data collection. Subsequent audit analyses will be conducted at least once every 12 months.

(3) *Audit interpreting physician.* Each facility shall designate at least one interpreting physician to review the medical outcomes audit data at least once every 12 months. This individual shall record the dates of the audit period(s) and shall be responsible for analyzing results based on this audit. This individual shall also be responsible for documenting the results and for notifying other interpreting physicians of their results and the facility aggregate results. If followup actions are taken, the audit interpreting physician shall also be responsible for documenting the nature of the followup.

(g) *Mammographic procedure and techniques for mammography of patients with breast implants.* (1) Each facility shall have a procedure to inquire whether or not the patient has breast implants prior to the actual mammographic exam.

(2) Except where contraindicated, or unless modified by a physician's directions, patients with breast implants undergoing mammography shall have mammographic views to maximize the visualization of breast tissue.

(h) *Consumer complaint mechanism.* Each facility shall:

(1) Establish a written and documented system for collecting and resolving consumer complaints;

(2) Maintain a record of each serious complaint received by the facility for at least 3 years from the date the complaint was received;

(3) Provide the consumer with adequate directions for filing serious complaints with the facility's accreditation body if the facility is unable to resolve a serious complaint to the consumer's satisfaction;

(4) Report unresolved serious complaints to the accreditation body in a manner and timeframe specified by the accreditation body.

(i) *Clinical image quality.* Clinical images produced by any certified facility must continue to comply with the standards for clinical image quality established by that facility's accreditation body.

(j) *Additional mammography review and patient notification.* (1) If FDA believes that mammography quality at a facility has been compromised and may present a serious risk to human health, the facility shall provide clinical images and other relevant information, as specified by FDA, for review by the accreditation body or other entity designated by FDA. This additional mammography review will help the agency to determine whether the facility is in compliance with this section and, if not, whether there is a need to notify affected patients, their physicians, or the public that the reliability, clarity, and accuracy of interpretation of mammograms has been compromised.

(2) If FDA determines that the quality of mammography performed by a facility, whether or not certified under § 900.11, was so inconsistent with the quality standards established in this section as to present a significant risk to individual or public health, FDA may require such facility to notify patients who received mammograms at such facility, and their referring physicians, of the deficiencies presenting such risk, the potential harm resulting, appropriate remedial measures, and such other relevant information as FDA may require. Such notification shall occur within a timeframe and in a manner specified by FDA.

[62 FR 55976, Oct. 28, 1997; 62 FR 60614, Nov. 10, 1997, as amended at 63 FR 56558, Oct. 22, 1998; 64 FR 18333, Apr. 14, 1999; 64 FR 32408, June 17, 1999; 65 FR 43690, July 14, 2000]

§ 900.13 Revocation of accreditation and revocation of accreditation body approval.

(a) *FDA action following revocation of accreditation.* If a facility's accreditation is revoked by an accreditation body, the agency may conduct an investigation into the reasons for the revocation. Following such investigation, the agency may determine that the facility's certificate shall no longer be in effect or the agency may take whatever other action or combination of actions will best protect the public health, including the establishment and implementation of a corrective plan of action that will permit the certificate to continue in effect while the facility seeks reaccreditation. A facility whose certificate is no longer in effect because it has lost its accreditation may not practice mammography.

(b) *Withdrawal of FDA approval of an accreditation body.* (1) If FDA withdraws approval of an accreditation body under § 900.6, the certificates of facilities previously accredited by such body shall remain in effect for up to 1 year from the date of the withdrawal of approval, unless FDA determines, in order to protect human health or because the accreditation body fraudulently accredited facilities, that the certificates of some or all of the facilities should be revoked or suspended or that a shorter time period should be established for the certificates to remain in effect.

(2) After 1 year from the date of withdrawal of approval of an accreditation body, or within any shorter period of time established by the agency, the affected facilities must obtain accreditation from another accreditation body, or from another entity designated by FDA.

§ 900.14 Suspension or revocation of certificates.

(a) Except as provided in paragraph (b) of this section, FDA may suspend or revoke a certificate if FDA finds, after providing the owner or operator of the facility with notice and opportunity for an informal hearing in accordance with part 16 of this chapter, that the owner, operator, or any employee of the facility:

(1) Has been guilty of misrepresentation in obtaining the certificate;

(2) Has failed to comply with the standards of § 900.12;

(3) Has failed to comply with reasonable requests of the agency or the accreditation body for records, information, reports, or materials that FDA believes are necessary to determine the continued eligibility of the facility for a certificate or continued compliance with the standards of § 900.12;

(4) Has refused a reasonable request of a duly designated FDA inspector, State inspector, or accreditation body representative for permission to inspect the facility or the operations and pertinent records of the facility;

(5) Has violated or aided and abetted in the violation of any provision of or regulation promulgated pursuant to 42 U.S.C. 263b; or

(6) Has failed to comply with prior sanctions imposed by the agency under 42 U.S.C. 263b(h).

(b) FDA may suspend the certificate of a facility before holding a hearing if FDA makes a finding described in paragraph (a) of this section and also determines that;

(1) The failure to comply with required standards presents a serious risk to human health;

(2) The refusal to permit inspection makes immediate suspension necessary; or

(3) There is reason to believe that the violation or aiding and abetting of the violation was intentional or associated with fraud.

(c) If FDA suspends a certificate in accordance with paragraph (b) of this section:

(1) The agency shall provide the facility with an opportunity for an informal hearing under part 16 of this chapter not later than 60 days from the effective date of this suspension;

(2) The suspension shall remain in effect until the agency determines that:

(i) Allegations of violations or misconduct were not substantiated;

(ii) Violations of required standards have been corrected to the agency's satisfaction; or

(iii) The facility's certificate is revoked in accordance with paragraph (d) of this section;

(d) After providing a hearing in accordance with paragraph (c)(1) of this section, the agency may revoke the facility's certificate if the agency determines that the facility:

(1) Is unwilling or unable to correct violations that were the basis for suspension; or

(2) Has engaged in fraudulent activity to obtain or continue certification.

§900.15 Appeals of adverse accreditation or reaccreditation decisions that preclude certification or recertification.

(a) The appeals procedures described in this section are available only for adverse accreditation or reaccreditation decisions that preclude certification or recertification by FDA. Agency decisions to suspend or revoke certificates that are already in effect will be handled in accordance with §900.14.

(b) Upon learning that a facility has failed to become accredited or reaccredited, FDA will notify the facility that the agency is unable to certify that facility without proof of accreditation.

(c) A facility that has been denied accreditation or reaccreditation is entitled to an appeals process from the accreditation body, in accordance with §900.7. A facility must avail itself of the accreditation body's appeal process before requesting reconsideration from FDA.

(d) A facility that cannot achieve satisfactory resolution of an adverse accreditation decision through the accreditation body's appeal process is entitled to further appeal in accordance with procedures set forth in this section and in regulations published in 42 CFR part 498.

(1) References to the Health Care Financing Administration (HCFA) in 42 CFR part 498 should be read as the Division of Mammography Quality and Radiation Programs (DMQRP), Center for Devices and Radiological Health, Food and Drug Administration.

(2) References to the Appeals Council of the Social Security Administration in 42 CFR part 498 should be read as references to the Departmental Appeals Board.

(3) In accordance with the procedures set forth in subpart B of 42 CFR part 498, a facility that has been denied accreditation following appeal to the accreditation body may request reconsideration of that adverse decision from DMQRP.

(i) A facility must request reconsideration by DMQRP within 60 days of the accreditation body's adverse appeals decision, at the following address: Division of Mammography Quality and Radiation Programs (HFZ–240), Center for Devices and Radiological Health, Food and Drug Administration, 1350 Piccard Dr., Rockville, MD 20850, Attn: Facility Accreditation Review Committee.

(ii) The request for reconsideration shall include three copies of the following records:

(A) The accreditation body's original denial of accreditation.

(B) All information the facility submitted to the accreditation body as part of the appeals process;

(C) A copy of the accreditation body's adverse appeals decision; and

(D) A statement of the basis for the facility's disagreement with the accreditation body's decision.

(iii) DMQRP will conduct its reconsideration in accordance with the procedures set forth in subpart B of 42 CFR part 498.

(4) A facility that is dissatisfied with DMQRP's decision following reconsideration is entitled to a formal hearing in accordance with procedures set forth in subpart D of 42 CFR part 498.

(5) Either the facility or FDA may request review of the hearing officer's decision. Such review will be conducted by the Departmental Appeals Board in accordance with subpart E of 42 CFR part 498.

(6) A facility cannot perform mammography services while an adverse accreditation decision is being appealed.

§ 900.16 Appeals of denials of certification.

(a) The appeals procedures described in this section are available only to facilities that are denied certification by FDA after they have been accredited by an approved accreditation body. Appeals for facilities that have failed to become accredited are governed by the procedures set forth in § 900.15.

(b) FDA may deny the application if the agency has reason to believe that:

(1) The facility will not be operated in accordance with standards established under § 900.12;

(2) The facility will not permit inspections or provide access to records or information in a timely fashion; or

(3) The facility has been guilty of misrepresentation in obtaining the accreditation.

(c)(1) If FDA denies an application for certification by a faciity that has received accreditation from an approved accreditation body, FDA shall provide the facility with a statement of the grounds on which the denial is based.

(2) A facility that has been denied accreditation may request reconsideration and appeal of FDA's determination in accordance with the applicable provisions of § 900.15(d).

§ 900.17 [Reserved]

§ 900.18 Alternative requirements for § 900.12 quality standards.

(a) *Criteria for approval of alternative standards.* Upon application by a qualified party as defined in paragraph (b) of this section, FDA may approve an alternative to a quality standard under § 900.12, when the agency determines that:

(1) The proposed alternative standard will be at least as effective in assuring quality mammography as the standard it proposes to replace, and

(2) The proposed alternative:

(i) Is too limited in its applicability to justify an amendment to the standard; or

(ii) Offers an expected benefit to human health that is so great that the time required for amending the standard would present an unjustifiable risk to the human health; and

(3) The granting of the alternative is in keeping with the purposes of 42 U.S.C. 263b.

(b) *Applicants for alternatives.* (1) Mammography facilities and accreditation bodies may apply for alternatives to the quality standards of § 900.12.

(2) Federal agencies and State governments that are not accreditation bodies may apply for alternatives to the standards of § 900.12(a).

(3) Manufacturers and assemblers of equipment used for mammography may apply for alternatives to the standards of § 900.12(b) and (e).

(c) *Applications for approval of an alternative standard.* An application for approval of an alternative standard or for an amendment or extension of the alternative standard shall be submitted in an original and two copies to the Director, Division of Mammography Quality and Radiation Programs (HFZ–240), Center for Devices and Radiological Health, Food and Drug Administration, 1350 Piccard Dr., Rockville, MD 20850. The application for approval of an alternative standard shall include the following information:

(1) Identification of the original standard for which the alternative standard is being proposed and an explanation of why the applicant is proposing the alternative;

(2) A description of the manner in which the alternative is proposed to deviate from the original standard;

(3) A description, supported by data, of the advantages to be derived from such deviation;

(4) An explanation, supported by data, of how such a deviation would ensure equal or greater quality of production, processing, or interpretation of mammograms than the original standard;

(5) The suggested period of time that the proposed alternative standard would be in effect; and

(6) Such other information required by the Director to evaluate and act on the application.

(d) *Ruling on applications.* (1) FDA may approve or deny, in whole or in part, a request for approval of an alternative standard or any amendment or extension thereof, and shall inform the applicant in writing of this action. The written notice shall state the manner in which the requested alternative standard differs from the agency standard and a summary of the reasons for approval or denial of the request. If the request is approved, the written notice shall also include the effective date and the termination date of the approval and a summary of the limitations and conditions attached to the approval and any other information that may be relevant to the approved request. Each approved alternative standard shall be assigned an identifying number.

(2) Notice of an approved request for an alternative standard or any amendment or extension thereof shall be placed in the public docket file in the Division of Dockets Management and may also be in the form of a notice published in the FEDERAL REGISTER. The notice shall state the name of the applicant, a description of the published agency standard, and a description of the approved alternative standard, including limitations and conditions attached to the approval of the alternative standard.

(3) Summaries of the approval of alternative standards, including information on their nature and number, shall be provided to the National Mammography Quality Assurance Advisory Committee.

(4) All applications for approval of alternative standards and for amendments and extensions thereof and all correspondence (including written notices of approval) on these applications shall be available for public disclosure in the Division of Dockets Management, excluding patient identifiers and confidential commercial information.

(e) *Amendment or extension of an alternative standard.* An application for amending or extending approval of an alternative standard shall include the following information:

(1) The approval number and the expiration date of the alternative standard;

(2) The amendment or extension requested and the basis for the amendment or extension; and

(3) An explanation, supported by data, of how such an amendment or extension would ensure equal or greater quality of production, processing, or interpretation of mammograms than the original standard.

(f) *Applicability of the alternative standards.* (1) Except as provided in paragraphs (f)(2) and (f)(3) of this section, any approval of an alternative standard, amendment, or extension may be implemented only by the entity to which it was granted and under the terms under which it was granted. Other entities interested in similar or identical approvals must file their own application following the procedures of paragraph (c) of this section.

(2) When an alternative standard is approved for a manufacturer of equipment, any facility using that equipment will also be covered by the alternative standard.

(3) The agency may extend the alternative standard to other entities when FDA determines that expansion of the approval of the alternative standard would be an effective means of promoting the acceptance of measures to improve the quality of mammography. All such determinations will be publicized by appropriate means.

(g) *Withdrawal of approval of alternative requirements.* FDA shall amend or withdraw approval of an alternative standard whenever the agency determines that this action is necessary to protect the human health or otherwise is justified by §900.12. Such action will become effective on the date specified in the written notice of the action sent to the applicant, except that it will become effective immediately upon notification of the applicant when FDA determines that such action is necessary to prevent an imminent health hazard.

[62 FR 55976, Oct. 28, 1997; 62 FR 60614, Nov. 10, 1997]

Subpart C—States as Certifiers

SOURCE: 67 FR 5467, Feb. 6, 2002, unless otherwise noted.

§ 900.20 Scope.

The regulations set forth in this part implement the Mammography Quality Standards Act (MQSA) (42 U.S.C. 263b). Subpart C of this part establishes procedures whereby a State can apply to become a FDA-approved certification agency to certify facilities within the State to perform mammography services. Subpart C of this part further establishes requirements and standards for State certification agencies to ensure that all mammography facilities under their jurisdiction are adequately and consistently evaluated for compliance with quality standards at least as stringent as the national quality standards established by FDA.

§ 900.21 Application for approval as a certification agency.

(a) *Eligibility.* State agencies may apply for approval as a certification agency if they have standards at least as stringent as those of § 900.12, qualified personnel, adequate resources to carry out the States as Certifiers' responsibilities, and the authority to enter into a legal agreement with FDA to accept these responsibilities.

(b) *Application for approval.* (1) An applicant seeking FDA approval as a certification agency shall inform the Division of Mammography Quality and Radiation Programs (DMQRP), Center for Devices and Radiological Health (HFZ–240), Food and Drug Administration, Rockville, MD 20850, marked Attn: SAC[1] Coordinator, in writing, of its desire to be approved as a certification agency.

(2) Following receipt of the written request, FDA will provide the applicant with additional information to aid in the submission of an application for approval as a certification agency.

(3) The applicant shall furnish to FDA, at the address in paragraph (b)(1) of this section, three copies of an application containing the following information, materials, and supporting documentation:

(i) Name, address, and phone number of the applicant;

(ii) Detailed description of the mammography quality standards the applicant will require facilities to meet and, for those standards different from FDA's quality standards, information substantiating that they are at least as stringent as FDA standards under § 900.12;

(iii) Detailed description of the applicant's review and decisionmaking process for facility certification, including:

(A) Policies and procedures for notifying facilities of certificate denials and expirations;

(B) Procedures for monitoring and enforcement of the correction of deficiencies by facilities;

(C) Policies and procedures for suspending or revoking a facility's certification;

(D) Policies and procedures that will ensure processing certificates within a timeframe approved by FDA;

(E) A description of the appeals process for facilities contesting adverse certification status decisions;

(F) Education, experience, and training requirements of the applicant's professional and supervisory staff;

(G) Description of the applicant's electronic data management and analysis system;

(H) Fee schedules;

(I) Statement of policies and procedures established to avoid conflict of interest;

(J) Description of the applicant's mechanism for handling facility inquiries and complaints;

(K) Description of a plan to ensure that certified mammography facilities will be inspected according to MQSA (42 U.S.C. 263b) and procedures and policies for notifying facilities of inspection deficiencies;

[1] SAC means States as Certifiers.

(L) Policies and procedures for monitoring and enforcing the correction of facility deficiencies discovered during inspections or by other means;

(M) Policies and procedures for additional mammography review and for requesting such reviews from accreditation bodies;

(N) Policies and procedures for patient notification;

(O) If a State has regulations that are more stringent than those of § 900.12, an explanation of how adverse actions taken against a facility under the more stringent regulations will be distinguished from those taken under the requirements of § 900.12; and

(P) Any other information that FDA identifies as necessary to make a determination on the approval of the State as a certification agency.

(c) *Rulings on applications for approval.* (1) FDA will conduct a review and evaluation to determine whether the applicant substantially meets the applicable requirements of this subpart and whether the certification standards the applicant will require facilities to meet are the quality standards published under subpart B of this part or at least as stringent as those of subpart B.

(2) FDA will notify the applicant of any deficiencies in the application and request that those deficiencies be corrected within a specified time period. If the deficiencies are not corrected to FDA's satisfaction within the specified time period, FDA may deny the application for approval as a certification agency.

(3) FDA shall notify the applicant whether the application has been approved or denied. The notification shall list any conditions associated with approval or state the bases for any denial.

(4) The review of any application may include a meeting between FDA and representatives of the applicant at a time and location mutually acceptable to FDA and the applicant.

(5) FDA will advise the applicant of the circumstances under which a denied application may be resubmitted.

(d) *Scope of authority.* FDA may limit the scope of certification authority delegated to the State in accordance with MQSA.

§ 900.22 Standards for certification agencies.

The certification agency shall accept the following responsibilities in order to ensure quality mammography at the facilities it certifies and shall perform these responsibilities in a manner that ensures the integrity and impartiality of the certification agency's actions:

(a) *Conflict of interest.* The certification agency shall establish and implement measures that FDA has approved in accordance with § 900.21(b) to reduce the possibility of conflict of interest or facility bias on the part of individuals acting on the certification agency's behalf.

(b) *Certification and inspection responsibilities.* Mammography facilities shall be certified and inspected in accordance with statutory and regulatory requirements that are at least as stringent as those of MQSA and this part.

(c) *Compliance with quality standards.* The scope, timeliness, disposition, and technical accuracy of completed inspections and related enforcement activities shall ensure compliance with facility quality standards required under § 900.12.

(d) *Enforcement actions.* (1) There shall be appropriate criteria and processes for the suspension and revocation of certificates.

(2) There shall be prompt investigation of and appropriate enforcement action for facilities performing mammography without certificates.

(e) *Appeals.* There shall be processes for facilities to appeal inspection findings, enforcement actions, and adverse certification decision or adverse accreditation decisions after exhausting appeals to the accreditation body.

(f) *Additional mammography review.* There shall be a process for the certification agency to request additional mammography review from accreditation bodies for issues related to mammography image quality and clinical practice. The certification agency should request additional mammography review only when it believes that mammography quality at a facility has been compromised and may present a serious risk to human health.

(g) *Patient notification.* There shall be processes for the certification agency to conduct, or cause to be conducted, patient notifications should the certification agency determine that mammography quality has been compromised to such an extent that it may present a serious risk to human health.

(h) *Electronic data transmission.* There shall be processes to ensure the timeliness and accuracy of electronic transmission of inspection data and facility certification status information in a format and timeframe determined by FDA.

(i) *Changes to standards.* A certification agency shall obtain FDA authorization for any changes it proposes to make in any standard that FDA has previously accepted under § 900.21 before requiring facilities to comply with the changes as a condition of obtaining or maintaining certification.

§ 900.23 Evaluation.

FDA shall evaluate annually the performance of each certification agency. The evaluation shall include the use of performance indicators that address the adequacy of program performance in certification, inspection, and enforcement activities. FDA will also consider any additional information deemed relevant by FDA that has been provided by the certification body or other sources or has been required by FDA as part of its oversight mandate. The evaluation also shall include a review of any changes in the standards or procedures in the areas listed in §§ 900.21(b) and 900.22 that have taken place since the original application or the last evaluation, whichever is most recent. The evaluation shall include a determination of whether there are major deficiencies in the certification agency's regulations or performance that, if not corrected, would warrant withdrawal of the approval of the certification agency under the provisions of § 900.24, or minor deficiencies that would require corrective action.

§ 900.24 Withdrawal of approval.

If FDA determines, through the evaluation activities of § 900.23, or through other means, that a certification agency is not in substantial compliance with this subpart, FDA may initiate the following actions:

(a) *Major deficiencies.* If, after providing notice and opportunity for corrective action, FDA determines that a certification agency has demonstrated willful disregard for public health, has committed fraud, has failed to provide adequate resources for the program, has submitted material false statements to the agency, has failed to achieve the MQSA goals of quality mammography and access, or has performed or failed to perform a delegated function in a manner that may cause serious risk to human health, FDA may withdraw its approval of that certification agency. The certification agency shall notify, within a time period and in a manner approved by FDA, all facilities certified or seeking certification by it that it has been required to correct major deficiencies.

(1) FDA shall notify the certification agency of FDA's action and the grounds on which the approval was withdrawn.

(2) A certification agency that has lost its approval shall notify facilities certified or seeking certification by it, as well as the appropriate accreditation bodies with jurisdiction in the State, that its approval has been withdrawn. Such notification shall be made within a timeframe and in a manner approved by FDA.

(b) *Minor deficiencies.* If FDA determines that a certification agency has demonstrated deficiencies in performing certification functions and responsibilities that are less serious or more limited than the deficiencies in paragraph (a) of this section, including failure to follow the certification agency's own procedures and policies as approved by FDA, FDA shall notify the certification agency that it has a specified period of time to take particular corrective measures as directed by FDA or to submit to FDA for approval the certification agency's own plan of corrective action addressing the minor deficiencies. If the approved corrective actions are not being implemented satisfactorily or within the established schedule, FDA may place the agency on probationary status for a period of time determined by FDA, or may withdraw approval of the certification agency.

(1) If FDA places a certification agency on probationary status, the certification agency shall notify all facilities

certified or seeking certification by it of its probationary status within a time period and in a manner approved by FDA.

(2) Probationary status shall remain in effect until such time as the certification agency can demonstrate to the satisfaction of FDA that it has successfully implemented or is implementing the corrective action plan within the established schedule, and that the corrective actions have substantially eliminated all identified problems, or

(3) If FDA determines that a certification agency that has been placed on probationary status is not implementing corrective actions satisfactorily or within the established schedule, FDA may withdraw approval of the certification agency. The certification agency shall notify all facilities certified or seeking certification by it, as well as the appropriate accreditation bodies with jurisdiction in the State, of its loss of FDA approval, within a timeframe and in a manner approved by FDA.

(c) *Transfer of records.* A certification agency that has its approval withdrawn shall transfer facility records and other related information as required by FDA to a location and according to a schedule approved by FDA.

§ 900.25 Hearings and appeals.

(a) Opportunities to challenge final adverse actions taken by FDA regarding approval of certification agencies or withdrawal of approval of certification agencies shall be communicated through notices of opportunity for informal hearings in accordance with part 16 of this chapter.

(b) A facility that has been denied certification is entitled to an appeals process from the certification agency. The appeals process shall be specified in writing by the certification agency and shall have been approved by FDA in accordance with §§ 900.21 and 900.22.

SUBCHAPTER J—RADIOLOGICAL HEALTH

PART 1000—GENERAL

Subpart A—General Provisions

Subpart B—Statements of Policy and Interpretation

Subpart C—Radiation Protection Recommendations

AUTHORITY: 21 U.S.C. 360hh–360ss.

SOURCE: 38 FR 28624, Oct. 15, 1973, unless otherwise noted.

Subpart A—General Provisions

§ 1000.1 General.

References in this subchapter J to regulatory sections of the Code of Federal Regulations are to chapter I of title 21 unless otherwise noted.

[50 FR 33688, Aug. 20, 1985]

§ 1000.3 Definitions.

As used in this subchapter J:

(a) *Accidental radiation occurrence* means a single event or series of events that has/have resulted in injurious or potentially injurious exposure of any person to electronic product radiation as a result of the manufacturing, testing, or use of an electronic product.

(b) *Act* means the Federal Food, Drug, and Cosmetic Act (21 U.S.C. 360hh–360ss).

(c) *Chassis family* means a group of one or more models with all of the following common characteristics:

(1) The same circuitry in the high voltage, horizontal oscillator, and power supply sections;

(2) The same worst component failures;

(3) The same type of high voltage hold-down or safety circuits; and

(4) The same design and installation.

(d) *Commerce* means:

(1) Commerce between any place in any State and any place outside thereof, and

(2) Commerce wholly within the District of Columbia.

(e) *Component,* for the purposes of this part, means an essential functional part of a subassembly or of an assembled electronic product, and which may affect the quantity, quality, direction, or radiation emission of the finished product.

(f) *Dealer* means a person engaged in the business of offering electronic products for sale to purchasers, without regard to whether such person is or has been primarily engaged in such business, and includes persons who offer such products for lease or as prizes or awards.

(g) *Director* means the Director of the Center for Devices and Radiological Health.

(h) *Distributor* means a person engaged in the business of offering electronic products for sale to dealers, without regard to whether such person is or has been primarily or customarily engaged in such business.

(i) *Electromagnetic radiation* includes the entire electromagnetic spectrum of radiation of any wavelength. The electromagnetic spectrum illustrated in figure 1 includes, but is not limited to, gamma rays, x-rays, ultra-violet, visible, infrared, microwave, radiowave, and low frequency radiation.

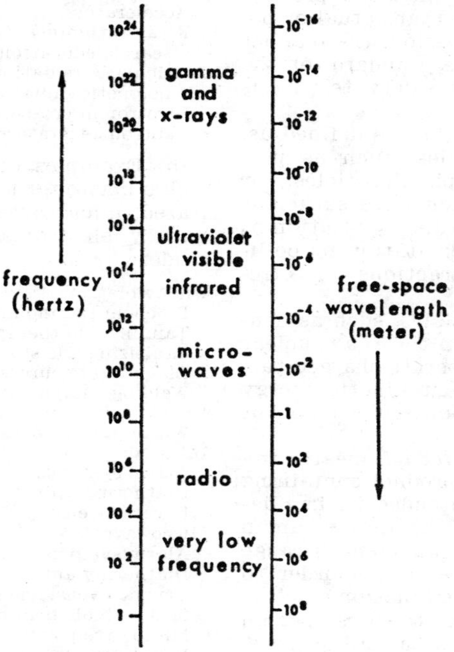

Figure 1. The Electromagnetic Spectrum

(j) *Electronic product* means:

(1) Any manufactured or assembled product which, when in operation:

(i) Contains or acts as part of an electronic circuit and

(ii) Emits (or in the absence of effective shielding or other controls would emit) electronic product radiation, or

(2) Any manufactured or assembled article that is intended for use as a component, part, or accessory of a product described in paragraph (j)(1) of this section and which, when in operation, emits (or in the absence of effective shielding or other controls would emit) such radiation.

(k) *Electronic product radiation* means:

(1) Any ionizing or nonionizing electromagnetic or particulate radiation, or

(2) Any sonic, infrasonic, or ultrasonic wave that is emitted from an electronic product as the result of the operation of an electronic circuit in such product.

(l) *Federal standard* means a performance standard issued pursuant to section 534 of the Federal Food, Drug, and Cosmetic Act.

(m) *Infrasonic, sonic (or audible) and ultrasonic waves* refer to energy transmitted as an alteration (pressure, particle displacement or density) in a property of an elastic medium (gas, liquid or solid) that can be detected by an instrument or listener.

(n) *Manufacturer* means any person engaged in the business of manufacturing, assembling, or importing electronic products.

(o) *Model* means any identifiable, unique electronic product design, and refers to products having the same structural and electrical design characteristics and to which the manufacturer has assigned a specific designation to differentiate between it and other products produced by that manufacturer.

(p) *Model family* means products having similar design and radiation characteristics but different manufacturer model numbers.

(q) *Modified model* means a product that is redesigned so that actual or potential radiation emission, the manner of compliance with a standard, or the manner of radiation safety testing is affected.

(r) *Particulate radiation* is defined as:

(1) Charged particles, such as protons, electrons, alpha particles, or heavy particles, which have sufficient kinetic energy to produce ionization or atomic or electron excitation by collision, electrical attractions or electrical repulsion; or

(2) Uncharged particles, such as neutrons, which can initiate a nuclear transformation or liberate charged particles having sufficient kinetic energy to produce ionization or atomic or electron excitation.

(s) *Phototherapy product* means any ultraviolet lamp, or product containing such lamp, that is intended for irradiation of any part of the living human body by light in the wavelength range of 200 to 400 nanometers, in order to perform a therapeutic function.

(t) *Purchaser* means the first person who, for value, or as an award or prize, acquires an electronic product for purposes other than resale, and includes a person who leases an electronic product for purposes other than subleasing.

(u) *State* means a State, the District of Columbia, the Commonwealth of Puerto Rico, the Virgin Islands, Guam, and American Samoa.

[60 FR 48380, Sept. 19, 1995; 61 FR 13422, Mar. 27, 1996]

Subpart B—Statements of Policy and Interpretation

§ 1000.15 Examples of electronic products subject to the Radiation Control for Health and Safety Act of 1968.

The following listed electronic products are intended to serve as illustrative examples of sources of electronic product radiation to which the regulations of this part apply.

(a) Examples of electronic products which may emit x-rays and other ionizing electromagnetic radiation, electrons, neutrons, and other particulate radiation include:

Ionizing electromagnetic radiation:
- Television receivers.
- Accelerators.
- X-ray machines (industrial, medical, research, educational).

Particulate radiation and ionizing electromagnetic radiation:
- Electron microscopes.
- Neutron generators.

(b) Examples of electronic products which may emit ultraviolet, visible, infrared, microwaves, radio and low frequency electromagnetic radiation include:

Ultraviolet:
- Biochemical and medical analyzers.
- Tanning and therapeutic lamps.
- Sanitizing and sterilizing devices.
- Black light sources.
- Welding equipment.

Visible:
- White light devices.

Infrared:
- Alarm systems.
- Diathermy units.
- Dryers, ovens, and heaters.

Microwave:
- Alarm systems.
- Diathermy units.
- Dryers, ovens, and heaters.
- Medico-biological heaters.
- Microwave power generating devices.
- Radar devices.
- Remote control devices.
- Signal generators.

Radio and low frequency:
- Cauterizers.
- Diathermy units.
- Power generation and transmission equipment.
- Signal generators.
- Electromedical equipment.

(c) Examples of electronic products which may emit coherent electromagnetic radiation produced by stimulated emission include:

Laser:
- Art-form, experimental and educational devices.
- Biomedical analyzers.
- Cauterizing, burning and welding devices.
- Cutting and drilling devices.
- Communications transmitters.
- Rangefinding devices.

Maser:
- Communications transmitters.

(d) Examples of electronic products which may emit infrasonic, sonic, and ultrasonic vibrations resulting from operation of an electronic circuit include:

Infrasonic:
- Vibrators.

Sonic:
- Electronic oscillators.
- Sound amplification equipment.

Ultrasonic:
- Cauterizers.
- Cell and tissue disintegrators.
- Cleaners.
- Diagnostic and nondestructive testing equipment.
- Ranging and detection equipment.

Subpart C—Radiation Protection Recommendations

§ 1000.50 Recommendation for the use of specific area gonad shielding on patients during medical diagnostic x-ray procedures.

Specific area gonad shielding covers an area slightly larger than the region of the gonads. It may therefore be used without interfering with the objectives of the examination to protect the germinal tissue of patients from radiation exposure that may cause genetic mutations during many medical x-ray procedures in which the gonads lie within or are in close proximity to the x-ray field. Such shielding should be provided when the following conditions exist:

(a) The gonads will lie within the primary x-ray field, or within close proximity (about 5 centimeters), despite proper beam limitation. Except as provided in paragraph (b) or (c) of this section:

(1) Specific area testicular shielding should always be used during those examinations in which the testes usually are in the primary x-ray field, such as examinations of the pelvis, hip, and upper femur;

(2) Specific area testicular shielding may also be warranted during other examinations of the abdominal region in which the testes may lie within or in close proximity to the primary x-ray field, depending upon the size of the patient and the examination techniques and equipment employed. Some examples of these are: Abdominal, lumbar spine and lumbosacral spine examinations, intravenous pyelograms, and abdominal scout film for barium enemas and upper GI series. Each x-ray facility should evaluate its procedures, techniques, and equipment and compile a list of such examinations for which specific area testicular shielding should be routinely considered for use. As a basis for judgment, specific area testicular shielding should be considered for all examinations of male patients in which the pubic symphysis will be visualized on the film;

(3) Specific area gonad shielding should never be used as a substitute for careful patient positioning, the use of correct technique factors and film processing, or proper beam limitation (confinement of the x-ray field to the area of diagnostic interest), because this could result in unnecessary doses to other sensitive tissues and could adversely affect the quality of the radiograph; and

(4) Specific area gonad shielding should provide attenuation of x-rays at least equivalent to that afforded by 0.25 millimeter of lead.

(b) The clinical objectives of the examination will not be compromised.

(1) Specific area testicular shielding usually does not obscure needed information except in a few cases such as oblique views of the hip, retrograde urethrograms and voiding cystourethrograms, visualization of the rectum and, occasionally, the pubic symphysis. Consequently, specific area testicular shielding should be considered for use in the majority of x-ray examinations of male patients in which the testes will lie within the primary beam or within 5 centimeters of its edge. It is not always possible to position shields on male patients so that no bone is obscured. Therefore, if all bone structure of the pelvic area must be visualized for a particular patient, the use of shielding should be carefully evaluated. The decision concerning the applicability of shielding for an individual patient is dependent upon consideration of the patient's unique anthropometric characteristics and the diagnostic information needs of the examination.

(2) The use of specific area ovarian shielding is frequently impractical at present because the exact location of the ovaries is difficult to estimate, and the shield may obscure visualization of portions of adjacent structures such as the spine, ureters, and small and large bowels. However, it may be possible for

practitioners to use specific area ovarian shielding during selected views in some examinations.

(c) The patient has a reasonable reproductive potential.

(1) Specific area shielding need not be used on patients who cannot or are not likely to have children in the future.

(2) The following table of statistical data regarding the average number of children expected by potential parents in various age categories during their remaining lifetimes is provided for x-ray facilities that wish to use it as a basis for judging reproductive potential:

EXPECTED NUMBER OF FUTURE CHILDREN VERSUS AGE OF POTENTIAL PARENT [1]

Age	Male parent	Female parent
Fetus	2.6	2.6
0 to 4	2.6	2.5
5 to 9	2.7	2.5
10 to 14	2.7	2.6
15 to 19	2.7	2.6
20 to 24	2.6	2.2
25 to 29	2.0	1.4
30 to 34	1.1	.6
35 to 39	.5	.2
40 to 44	.2	.04
45 to 49	.07	0
50 to 54	.03	0
55 to 64	.01	0
Over 65	0	0

[1] Derived from data published by the National Center for Health Statistics, "Final Natality Statistics 1970," HRA 74–1120, vol. 22, No. 12, Mar. 20, 1974.

[41 FR 30328, July 23, 1976; 41 FR 31812, July 30, 1976]

§ 1000.55 Recommendation for quality assurance programs in diagnostic radiology facilities.

(a) *Applicability.* Quality assurance programs as described in paragraph (c) of this section are recommended for all diagnostic radiology facilities.

(b) *Definitions.* As used in this section, the following definitions apply:

(1) *Diagnostic radiology facility* means any facility in which an x-ray system(s) is used in any procedure that involves irradiation of any part of the human body for the purpose of diagnosis or visualization. Offices of individual physicians, dentists, podiatrists, and chiropractors, as well as mobile laboratories, clinics, and hospitals are all examples of diagnostic radiology facilities.

(2) *Quality assurance* means the planned and systematic actions that provide adequate confidence that a diagnostic x-ray facility will produce consistently high quality images with minimum exposure of the patients and healing arts personnel. The determination of what constitutes high quality will be made by the facility producing the images. Quality assurance actions include both "quality control" techniques and "quality administration" procedures.

(3) *Quality assurance program* means an organized entity designed to provide "quality assurance" for a diagnostic radiology facility. The nature and extent of this program will vary with the size and type of the facility, the type of examinations conducted, and other factors.

(4) *Quality control techniques* are those techniques used in the monitoring (or testing) and maintenance of the components of an x-ray system. The quality control techniques thus are concerned directly with the equipment.

(5) *Quality administration procedures* are those management actions intended to guarantee that monitoring techniques are properly performed and evaluated and that necessary corrective measures are taken in response to monitoring results. These procedures provide the organizational framework for the quality assurance program.

(6) *X-ray system* means an assemblage of components for the controlled production of diagnostic images with x-rays. It includes minimally an x-ray high voltage generator, an x-ray control, a tube-housing assembly, a beam-limiting device, and the necessary supporting structures. Other components that function with the system, such as image receptors, image processors, view boxes, and darkrooms, are also parts of the system.

(c) *Elements.* A quality assurance program should contain the elements listed in paragraphs (c)(1) through (10) of this section. The extent to which each element of the quality assurance program is implemented should be determined by an analysis of the facility's objectives and resources conducted by its qualified staff or by qualified outside consultants. The extent of implementation should be determined on the basis of whether the expected benefits in radiation exposure reduction, improved image quality, and/or financial savings will compensate for the resources required for the program.

(1) *Responsibility.* (i) Responsibility and authority for the overall quality assurance program as well as for monitoring, evaluation, and corrective measures should be specified and recorded in a quality assurance manual.

(ii) The owner or practitioner in charge of the facility has primary responsibility for implementing and maintaining the quality assurance program.

(iii) Staff technologists will generally be delegated a basic quality assurance role by the practitioner in charge. Responsibility for specific quality control monitoring and maintenance techniques or quality administration procedures may be assigned, provided that the staff technologists are qualified by training or experience for these duties. The staff technologists should also be responsible for identifying problems or potential problems requiring actions beyond the level of their training. They should bring these problems to the attention of the practitioner in charge, or his or her representative, so that assistance in solving the problems may be obtained from inside or outside the facility.

(iv) In facilities where they are available, physicists, supervisory technologists, or quality control technologists should have a major role in the quality assurance program. Such specialized personnel may be assigned responsibility for day-to-day administration of the program, may carry out monitoring duties beyond the level of training of the staff technologist or, if desired by the facility, may relieve the staff technologists of some or all of their basic monitoring duties. Staff service engineers may also be assigned responsibility for certain preventive or corrective maintenance actions.

(v) Responsibility for certain quality control techniques and corrective measures may be assigned to personnel qualified by training or experience, such as consultants or industrial representatives, from outside of the facility, provided there is a written agreement clearly specifying these services.

(vi) In large facilities, responsibility for long-range planning of quality assurance goals and activities should be assigned to a quality assurance committee as described in paragraph (c)(9) of this section.

(2) *Purchase specifications.* Before purchasing new equipment, the staff of the diagnostic radiology facility should determine the desired performance specifications for the equipment. Initially, these specifications may be stated in terms of the desired performance of the equipment, or prospective vendors may be informed solely of the functions the equipment should be able to perform and asked to provide the performance specifications of items from their equipment line that can perform these functions. In either case, the responses of the prospective vendors should serve as the basis for negotiations to establish the final purchase specifications, taking into account the state of the art and balancing the need for the specified performance levels with the cost of the equipment to meet them. The final purchase specifications should be in writing and should include performance specifications. The availability of experienced service personnel should also be taken into consideration in making the final purchase decisions. Any understandings with respect to service personnel should be incorporated into the purchase specifications. After the equipment is installed, the facility should conduct a testing program, as defined in its purchase specifications, to ensure that the equipment meets the agreed upon specifications, including applicable Federal and State regulatory requirements. The equipment should not be formally accepted until any necessary corrections have been made by the vendor. The purchase specifications and the records of the acceptance testing should be retained throughout the life of the equipment for comparison with monitoring results in order to assess continued acceptability of performance.

(3) *Monitoring and maintenance.* A routine quality control monitoring and maintenance system incorporating state-of-the-art procedures should be established and conducted on a regular schedule. The purpose of monitoring is

to permit evaluation of the performance of the facility's x-ray system(s) in terms of the standards for image quality established by the facility (as described in paragraph (c)(4) of this section) and compliance with applicable Federal and State regulatory requirements. The maintenance program should include corrective maintenance to eliminate problems revealed by monitoring or other means before they have a serious deleterious impact on patient care. To the extent permitted by the training of the facility staff, the maintenance program should also include preventive maintenance, which could prevent unexpected breakdowns of equipment and disruption of departmental routine.

(i) The parameters to be monitored in a facility should be determined by that facility on the basis of an analysis of expected benefits and cost. Such factors as the size and resources of the facility, the type of examinations conducted, and the quality assurance problems that have occurred in that or similar facilities should be taken into account in establishing the monitoring system. The monitoring frequency should also be based upon need and can be different for different parameters.

(ii) Although the parameters to be monitored will vary somewhat from facility to facility, every diagnostic radiology facility should consider monitoring the following five key components of the x-ray system:

(*a*) Film processing.

(*b*) Basic performance characteristics of the x-ray unit.

(*c*) Cassettes and grids.

(*d*) View boxes.

(*e*) Darkroom.

(iii) Examples of parameters of the above-named components and of more specialized equipment that may be monitored are as follows:

(*a*) For film processing:

An index of speed.
An index of contrast.
Base plus fog.
Solution temperatures.
Film artifact identification.

(*b*) For basic performance characteristics of the x-ray unit:

(*1*) For fluoroscopic x-ray units:

Table-top exposure rates.
Centering alignment.
Collimation.
kVp accuracy and reproducibility.
mA accuracy and reproducibility.
Exposure time accuracy and reproducibility.
Reproducibility of x-ray output.
Focal spot size consistency.
Half-value layer.
Representative entrance skin exposures.

(*2*) For image-intensified systems:

Resolution.
Focusing.
Distortion.
Glare.
Low contrast performance.
Physical alignment of camera and collimating lens.

(*3*) For radiographic x-ray units:

Reproducibility of x-ray output.
Linearity and reproducibility of mA stations.
Reproducibility and accuracy of timer stations.
Reproducibility and accuracy of kVp stations.
Accuracy of source-to-film distance indicators.
Light/x-ray field congruence.
Half-value layer.
Focal spot size consistency.
Representative entrance skin exposures.

(*4*) For automatic exposure control devices:

Reproducibility.
kVp compensation.
Field sensitivity matching.
Minimum response time.
Backup timer verification.

(*c*) For cassettes and grids:

(*1*) For cassettes:

Film/screen contact.
Screen condition.
Light leaks.
Artifact identification.

(*2*) For grids:

Alignment and focal distance.
Artifact identification.

(*d*) For view boxes:

Consistency of light output with time.
Consistency of light output from one box to another.
View box surface conditions.

(*e*) For darkrooms:

Darkroom integrity.
Safe light conditions.

(*f*) For specialized equipment:

(*1*) For tomographic systems:

Accuracy of depth and cut indicator.

Thickness of cut plane.
Exposure angle.
Completeness of tomographic motion.
Flatness of tomographic field.
Resolution.
Continuity of exposure.
Flatness of cassette.
Representative entrance skin exposures.

(*2*) For computerized tomography:

Precision (noise).
Contrast scale.
High and low contrast resolution.
Alignment.
Representative entrance skin exposures.

(iv) The maintenance program should include both preventive and corrective aspects.

(*a*) *Preventive maintenance.* Preventive maintenance should be performed on a regularly scheduled basis with the goal of preventing breakdowns due to equipment failing without warning signs detectable by monitoring. Such actions have been found cost effective if responsibility is assigned to facility staff members. Possible preventive maintenance procedures are visual inspection of the mechanical and electrical characteristics of the x-ray system (covering such things as checking conditions of cables, watching the tomographic unit for smoothness of motion, assuring cleanliness with respect to spilling of contaminants in the examination room or the darkroom, and listening for unusual noises in the moving parts of the system), following the manufacturer's recommended procedures for cleaning and maintenance of the equipment, and regular inspection and replacement of switches and parts that routinely wear out or fail. The procedures included would depend upon the background of the staff members available. Obviously, a large facility with its own service engineers can do more than an individual practitioner's office.

(*b*) *Corrective maintenance.* For maximum effectiveness, the quality assurance program should make provision, as described in paragraph (c)(5) of this section, for ascertaining whether potential problems are developing. If potential or actual problems are detected, corrective maintenance should be carried out to eliminate them before they cause a major impact on patient care.

(4) *Standards for image quality.* Standards of acceptable image quality should be established. Ideally, these should be objective, e.g., acceptability limits for the variations of parameter values, but they may be subjective, e.g., the opinions of professional personnel, in cases where adequate objective standards cannot be defined. These standards should be routinely reviewed and redefined as needed, as described in paragraph (c)(10) of this section.

(5) *Evaluation.* The facility's quality assurance program should include means for two levels of evaluation.

(i) On the first level, the results of the monitoring procedures should be used to evaluate the performance of the x-ray system(s) to determine whether corrective actions are needed to adjust the equipment so that the image quality consistently meets the standards for image quality. This evaluation should include analysis of trends in the monitoring data as well as the use of the data to determine the need for corrective actions on a day-by-day basis. Comparison of monitoring data with the purchase specifications and acceptance testing results for the equipment in question is also useful.

(ii) On the second level, the facility quality assurance program should also include means for evaluating the effectiveness of the program itself. Possible means include ongoing studies of the retake rate and the causes of the repeated radiographs, examination of equipment repair and replacement costs, subjective evaluation of the radiographs being produced, occurrence and reasons for complaints by radiologists, and analysis of trends in the results of monitoring procedures such as sensitometric studies. Of these, ongoing studies of the retake rate (reject rate) and its causes are often the most useful and may also provide information of value in the first level of evaluation. Such studies can be used to evaluate potential for improvement, to make corrections, and to determine whether the corrective actions were effective. The number of rejects should be recorded daily or weekly, depending on the facility's analysis of its needs. Ideally, the reasons for the rejection should also be determined and recorded. Should determining these reasons be impossible on a regular basis with the available staff, the analysis should be done for a 2-week period after major changes have occurred in diagnostic procedures or the x-ray system and at least semi-annually.

(6) *Records*. The program should include provisions for the keeping of records on the results of the monitoring techniques, any difficulties detected, the corrective measures applied to these difficulties, and the effectiveness of these measures. The extent and form of these records should be determined by the facility on the basis of its needs. The facility should view these records as a tool for maintaining an effective quality assurance program and not view the data in them as an end in itself but rather as a beginning. For example, the records should be made available to vendors to help them provide better service. More importantly, the data should be the basis for the evaluation and the reviews suggested in paragraphs (c)(5) and (10) of this section.

(7) *Manual*. A quality assurance manual should be written in a format permitting convenient revision as needed and should be made readily available to all personnel. The content of the manual should be determined by the facility staff, but the following items are suggested as providing essential information:

(i) A list of the individuals responsible for monitoring and maintenance techniques.

(ii) A list of the parameters to be monitored and the frequency of monitoring.

(iii) A description of the standards, criteria of quality, or limits of acceptability that have been established for each of the parameters monitored.

(iv) A brief description of the procedures to be used for monitoring each parameter.

(v) A description of procedures to be followed when difficulties are detected to call these difficulties to the attention of those responsible for correcting them.

(vi) A list of the publications in which detailed instructions for monitoring and maintenance procedures can be found. Copies of these publications should also be readily available to the entire staff, but they should be separate from the manual. (Publications providing these instructions can usually be obtained from FDA or private sources, although the facility may wish to make some modifications to meet its needs more effectively.)

(vii) A list of the records, with sample forms, that the facility staff has decided should be kept. The facility staff should also determine and note in the manual the length of time each type of record should be kept before discarding.

(viii) A copy of each set of purchase specifications developed for new equipment and the results of the acceptance testing for that equipment.

(8) *Training*. The program should include provisions for appropriate training for all personnel with quality assurance responsibilities. This should include both training provided before the quality assurance responsibilities are assumed and continuing education to keep the personnel up-to-date. Practical experience with the techniques conducted under the supervision of experienced instructors, either in the facility or in a special program, is the most desirable type of training. The use of self-teaching materials can be an adequate substitute for supervised instruction, especially in continuing education programs, if supervised instruction is not available.

(9) *Committee*. A facility whose size would make it impractical for all staff members to meet for planning purposes should consider the establishment of a quality assurance committee whose primary function would be to maintain lines of communication among all groups with quality assurance and/or image production or interpretation responsibilities. For maximum communication, all departments of the facility with x-ray equipment should be represented. The committee may also be assigned policy-making duties such as some or all of the following:

Assign quality assurance responsibilities; maintain acceptable standards of quality; periodically review program effectiveness, etc. Alternatively, the duties of this committee could be assigned to an already-existing committee such as the Radiation Safety Committee. In smaller facilities, all staff members should participate in the committee's tasks. The Quality Assurance Committee should report directly to the head of the radiology department, or, in facilities where more than one department operates x-ray equipment, to the chief medical officer of the facility. The committee should meet on a regular basis.

(10) *Review.* The facility's quality assurance program should be reviewed by the Quality Assurance Committee and/or the practitioner in charge to determine whether its effectiveness could be improved. Items suggested for inclusion in the review include:

(i) The reports of the monitoring and maintenance techniques to ensure that they are being performed on schedule and effectively. These reports should be reviewed at least quarterly.

(ii) The monitoring and maintenance techniques and their schedules to ensure that they continue to be appropriate and in step with the latest developments in quality assurance. They should be made current at least annually.

(iii) The standards for image quality to ensure that they are consistent with the state-of-the-art and the needs and resources of the facility. These standards should be evaluated at least annually.

(iv) The results of the evaluations of the effectiveness of the quality assurance actions to determine whether changes need to be made. This determination should be made at least annually.

(v) The quality assurance manual should also be reviewed at least annually to determine whether revision is needed.

[44 FR 71737, Dec. 11, 1979]

§ 1000.60 Recommendation on administratively required dental x-ray examinations.

(a) The Food and Drug Administration recommends that dental x-ray examinations be performed only after careful consideration of the dental or other health needs of the patient, that is, when the patient's dentist or physician judges them to be necessary for diagnosis, treatment, or prevention of disease. Administratively required dental x-ray examinations are those required by a remote third party for reasons not related to the patient's immediate dental needs. These x-ray examinations are usually a source of unnecessary radiation exposure to the patient. Because any unnecessary radiation exposure should be avoided, third parties should not require dental x-ray examinations unless they can demonstrate that such examinations provide a direct clinical benefit to the patient, and the patient's dentist or physician agrees with that assessment.

(b) Some examples of administrative x-ray examinations that should not be required by third parties are those intended solely:

(1) To monitor insurance claims or detect fraud;

(2) To satisfy a prerequisite for reimbursement;

(3) To provide training or experience;

(4) To certify qualifications or competence.

(c) This recommendation is not intended to preclude dental x-ray examinations ordered by the attending practitioner, based on the patient's history or physical examination, or those performed on selected populations shown to have significant yields of previously undiagnosed disease. This recommendation is also not intended to preclude the administrative use by third parties of dental radiographs that are taken on the order of the patient's dentist or physician as a necessary part of the patient's clinical care.

[45 FR 40978, June 17, 1980]

PART 1002—RECORDS AND REPORTS

Subpart A—General Provisions

Sec.

Subpart B—Required Manufacturers' Reports for Listed Electronic Products

1002.10 Product reports.
1002.11 Supplemental reports.
1002.12 Abbreviated reports.
1002.13 Annual reports.

Subpart C—Manufacturers' Reports on Accidental Radiation Occurrences

1002.20 Reporting of accidental radiation occurrences.

Subpart D—Manufacturers' Records

1002.30 Records to be maintained by manufacturers.
1002.31 Preservation and inspection of records.

Subpart E—Dealer and Distributor Records

1002.40 Records to be obtained by dealers and distributors.
1002.41 Disposition of records obtained by dealers and distributors.
1002.42 Confidentiality of records furnished by dealers and distributors.

Subpart F—Exemptions From Records and Reports Requirements

1002.50 Special exemptions.
1002.51 Exemptions for manufacturers of products intended for the U.S. Government.

AUTHORITY: 21 U.S.C. 352, 360, 360i, 360j, 360hh–360ss, 371, 374.

SOURCE: 38 FR 28625, Oct. 15, 1973, unless otherwise noted.

Subpart A—General Provisions

§ 1002.1 Applicability.

The provisions of this part are applicable as follows:

(a) All manufacturers of electronic products are subject to § 1002.20.

(b) Manufacturers, dealers, and distributors of electronic products are subject to the provisions of part 1002 as set forth in table 1 of this section, unless excluded by paragraph (c) of this section, or unless an exemption has been granted under § 1002.50 or § 1002.51.

(c) The requirements of part 1002 as specified in table 1 of this section are not applicable to:

(1) Manufacturers of electronic products intended solely for export if such product is labeled or tagged to show that the product meets all the applicable requirements of the country to which such product is intended for export.

(2) Manufacturers of electronic products listed in table 1 of this section if such product is sold exclusively to other manufacturers for use as components of electronic products to be sold to purchasers, with the exception that the provisions are applicable to those manufacturers certifying components of diagnostic x-ray systems pursuant to provisions of § 1020.30(c) of this chapter.

(3) Manufacturers of electronic products that are intended for use by the U.S. Government and whose function or design cannot be divulged by the manufacturer for reasons of national security, as evidenced by government security classification.

(4) Assemblers of diagnostic x-ray equipment subject to the provisions of § 1020.30(d) of this chapter, provided the assembler has submitted the report required by § 1020.30(d)(1) or (d)(2) of this chapter and retains a copy of such report for a period of 5 years from its date.

TABLE 1—RECORD AND REPORTING REQUIREMENTS BY PRODUCT

	Manufacturer						Dealer & Distributor
Products	Product reports § 1002.10	Supplemental reports § 1002.11	Abbreviated reports § 1002.12	Annual reports § 1002.13	Test records § 1002.30(a) [1]	Distribution records § 1002.30(b) [2]	Distribution records §§ 1002.40 and 1002.41
DIAGNOSTIC X-RAY [3] (1020.30, 1020.31, 1020.32, 1020.33)							
Computed tomography	X	X		X	X	X	X
X-ray system [4]	X	X		X	X	X	X
Tube housing assembly	X	X		X	X	X	
X-ray control	X	X		X	X	X	X
X-ray high voltage generator	X	X		X	X	X	X

TABLE 1—RECORD AND REPORTING REQUIREMENTS BY PRODUCT—Continued

Products	Manufacturer: Product reports § 1002.10	Manufacturer: Supplemental reports § 1002.11	Manufacturer: Abbreviated reports § 1002.12	Manufacturer: Annual reports § 1002.13	Manufacturer: Test records § 1002.30(a) [1]	Manufacturer: Distribution records § 1002.30(b) [2]	Dealer & Distributor: Distribution records §§ 1002.40 and 1002.41
X-ray table or cradle			X		X	X	X
X-ray film changer			X		X	X	
Vertical cassette holders mounted in a fixed location and cassette holders with front panels			X		X	X	X
Beam-limiting devices	X	X		X	X	X	X
Spot-film devices and image intensifiers manufactured after April 26, 1977	X	X		X	X	X	X
Cephalometric devices manufactured after February 25, 1978			X		X	X	
Image receptor support devices for mammographic X-ray systems manufactured after September 5, 1978			X		X	X	X
CABINET X RAY (§ 1020.40)							
Baggage inspection	X	X		X	X	X	X
Other	X	X		X	X	X	
PRODUCTS INTENDED TO PRODUCE PARTICULATE RADIATION OR X-RAYS OTHER THAN DIAGNOSTIC OR CABINET DIAGNOSTIC X-RAY							
Medical			X	X	X	X	
Analytical			X	X	X	X	
Industrial			X	X	X	X	
TELEVISION PRODUCTS (§ 1020.10)							
<25 kilovolt (kV) and <0.1 milliroentgen per hour (mR/hr IRLC [5,6]			X	X [6]			
≥25kV and <0.1mR/hr IRLC [5]	X	X		X			
≥0.1mR/hr IRLC [5]	X	X		X	X	X	
MICROWAVE/RF							
MW ovens (§ 1030.10)	X	X		X	X	X	
MW diathermy			X				
MW heating, drying, security systems			X				
RF sealers, electromagnetic induction and heating equipment, dielectric heaters (2–500 megahertz)			X				
OPTICAL							
Phototherapy products	X	X					
Laser products (§§ 1040.10, 1040.11)							
Class I lasers and products containing such lasers [7]	X			X	X		
Class I laser products containing class IIa, II, IIIa, lasers [7]	X			X	X	X	
Class IIa, II, IIIa lasers and products other than class I products containing such lasers [7]	X	X		X	X	X	X
Class IIIb and IV lasers and products containing such lasers [7]	X	X		X	X	X	X
Sunlamp products (§ 1040.20)							
Lamps only	X						
Sunlamp products	X	X		X	X	X	X
Mercury vapor lamps (§ 1040.30)							
T lamps	X	X		X			
R lamps			X				
ACOUSTIC							
Ultrasonic therapy (1050.10)	X	X		X	X	X	X
Diagnostic ultrasound			X				

TABLE 1—RECORD AND REPORTING REQUIREMENTS BY PRODUCT—Continued

	Manufacturer						Dealer & Distributor
Products	Product reports § 1002.10	Supplemental reports § 1002.11	Abbreviated reports § 1002.12	Annual reports § 1002.13	Test records § 1002.30(a) [1]	Distribution records § 1002.30(b)[2]	Distribution records §§ 1002.40 and 1002.41
Medical ultrasound other than therapy or diagnostic	X	X					
Nonmedical ultrasound			X				

[1]However, authority to inspect all appropriate documents supporting the adequacy of a manufacturer's compliance testing program is retained.
[2]The requirement includes §§ 1002.31 and 1002.42, if applicable.
[3]Report of Assembly (Form FDA 2579) is required for diagnostic x-ray components; see 21 CFR 1020.30(d)(1) through (d)(3).
[4]Systems records and reports are required if a manufacturer exercises the option and certifies the system as permitted in 21 CFR 1020.30(c).
[5]Determined using the isoexposure rate limit curve (IRLC) under phase III test conditions (1020.10(c)(3)(iii)).
[6]Annual report is for production status information only.
[7]Determination of the applicable reporting category for a laser product shall be based on the worst-case hazard present within the laser product.

[60 FR 48382, Sept. 19, 1995; 61 FR 13423, Mar. 27, 1996]

§ 1002.2 [Reserved]

§ 1002.3 Notification to user of performance and technical data.

The Director and Deputy Director of the Center for Devices and Radiological Health, as authorized under delegated authority, may require a manufacturer of a radiation emitting electronic product to provide to the ultimate purchaser, at the time of original purchase, such performance data and other technical data related to safety of the product as the Director or Deputy Director finds necessary.

[69 FR 17292, Apr. 2, 2004]

§ 1002.4 Confidentiality of information.

The Secretary or his representative shall not disclose any information reported to or otherwise obtained by him, pursuant to this part, which concerns or relates to a trade secret or other matter referred to in section 1905 of title 18 of the United States Code, except that such information may be disclosed to other officers or employees of the Department and of the other agencies concerned with carrying out the requirements of the Act. Nothing in this section shall authorize the withholding of information by the Secretary, or by any officers or employees under his control, from the duly authorized committees of the Congress.

§ 1002.7 Submission of data and reports.

All submissions such as reports, test data, product descriptions, and other information required by this part, or voluntarily submitted to the Director, Center for Devices and Radiological Health, shall be filed with the number of copies as prescribed by the Director, Center for Devices and Radiological Health, and shall be signed by the person making the submission. The submissions required by this part shall be addressed to the Center for Devices and Radiological Health, Electronic Product Reports, Office of Compliance (HFZ–307), 2098 Gaither Rd., Rockville, MD 20850.

(a) In addition to the requirements of this part, all material submitted to the Director, Center for Devices and Radiological Health, shall be submitted pursuant to the provisions of part 20—Public Information, of this chapter.

(b) Where guides or instructions have been issued by the Director for the submission of material required by this part, such as test data, product reports, abbreviated reports, supplemental reports, and annual reports, the material submitted shall conform to the applicable reporting guides or instructions. Where it is not feasible or where it would not be appropriate to conform to any portion of a prescribed reporting guide or instruction, an alternate format for providing the information requested by that portion of the guide or instruction may be used provided the submitter of such information submits adequate explanation and justification for use of an alternate format. If the Director, Center for Devices and Radiological Health, determines that such justification is inadequate and that it is feasible or appropriate to conform to the prescribed reporting guide or instruction, he may require resubmission of the information in conformance with the reporting guide or instruction.

(c) Where the submission of quality control and testing information is common to more than one model, or model family of the same product category, a "common aspects report" consolidating similar information may be provided, if applicable.

[42 FR 18062, Apr. 5, 1977, as amended at 53 FR 11254, Apr. 6, 1988; 60 FR 48385, Sept. 19, 1995]

Subpart B—Required Manufacturers' Reports for Listed Electronic Products

SOURCE: 60 FR 48386, Sept. 19, 1995, unless otherwise noted.

§1002.10 Product reports.

Every manufacturer of a product or component requiring aproduct report as set forth in table 1 of §1002.1 shall submit a product report to the Center for Devices and Radiological Health, Electronic Product Reports, Office of Compliance (HFZ–307), 2098 Gaither Rd., Rockville, MD 20850, prior to the introduction of such product into commerce. The report shall be distinctly marked "Radiation Safety Product Report of (name of manufacturer)" and shall:

(a) Identify which listed product is being reported.

(b) Identify each model of the listed product together with sufficient information concerning the manufacturer's code or other system of labeling to enable the Director to determine the place of manufacture.

(c) Include information on all components and accessories provided in, on, or with the listed product that may affect the quantity, quality, or direction of the radiation emissions.

(d) Describe the function, operational characteristics affecting radiation emissions, and intended and known uses of each model of the listed product.

(e) State the standard or design specifications, if any, for each model with respect to electronic product radiation safety. Reference may be made to a Federal standard, if applicable.

(f) For each model, describe the physical or electrical characteristics, such as shielding or electronic circuitry, incorporated into the product in order to meet the standards or specifications reported pursuant to paragraph (e) of this section.

(g) Describe the methods and procedures employed, if any, in testing and measuring each model with respect to electronic product radiation safety, including the control of unnecessary, secondary, or leakage electronic product radiation, the applicable quality control procedures used for each model, and the basis for selecting such testing and quality control procedures.

(h) For those products which may produce increased radiation with aging, describe the methods and procedures used, and frequency of testing of each model for durability and stability with respect to electronic product radiation safety. Include the basis for selecting such methods and procedures, or for determining that such testing and quality control procedures are not necessary.

(i) Provide sufficient results of the testing, measuring, and quality control procedures described in accordance with paragraphs (g) and (h) of this section to enable the Director to determine the effectiveness of those test methods and procedures.

(j) Report for each model all warning signs, labels, and instructions for installation, operation, and use that relate to electronic product radiation safety.

(k) Provide, upon request, such other information as the Director may reasonably require to enable him/her to determine whether the manufacturer has acted or is acting in compliance with the Act and any standards prescribed thereunder, and to enable the

Director to carry out the purposes of the Act.

§ 1002.11 Supplemental reports.

Prior to the introduction into commerce of a new or modified model within a model or chassis family of a product listed in table 1 of § 1002.1 for which a report under § 1002.10 is required, each manufacturer shall submit a report with respect to such new or modified model describing any changes in the information previously submitted in the product report. Reports will be required for changes that:

(a) Affect actual or potential radiation emission.

(b) Affect the manner of compliance with a standard or manner of testing for radiation safety.

§ 1002.12 Abbreviated reports.

Manufacturers of products requiring abbreviated reports as specified in table 1 of § 1002.1 shall submit, prior to the introduction of such product, a report distinctly marked "Radiation Safety Abbreviated Report" which shall include:

(a) Firm and model identification.

(b) A brief description of operational characteristics that affect radiation emissions, transmission, or leakage or that control exposure.

(c) A list of applications or uses.

(d) Radiation emission, transmission, or leakage levels.

(e) If necessary, additional information as may be requested to determine compliance with the Act and this part.

§ 1002.13 Annual reports.

(a) Every manufacturer of products requiring an annual report as specified in table 1 of § 1002.1 shall submit an annual report summarizing the contents of the records required to be maintained by § 1002.30(a) and providing the volume of products produced, sold, or installed.

(b) Reports are due annually by September 1. Such reports shall cover the 12-month period ending on June 30 preceding the due date of the report.

(c) New models of a model family that do not involve changes in radiation emission or requirements of a performance standard do not require supplemental reports prior to introduction into commerce. These model numbers should be reported in quarterly updates to the annual report.

Subpart C—Manufacturers' Reports on Accidental Radiation Occurrences

§ 1002.20 Reporting of accidental radiation occurrences.

(a) Manufacturers of electronic products shall, where reasonable grounds for suspecting that such an incident has occurred, immediately report to the Director, Center for Devices and Radiological Health, all accidental radiation occurrences reported to or otherwise known to the manufacturer and arising from the manufacturing, testing, or use of any product introduced or intended to be introduced into commerce by such manufacturer. Reasonable grounds include, but are not necessarily limited to, professional, scientific, or medical facts or opinions documented or otherwise, that conclude or lead to the conclusion that such an incident has occurred.

(b) Such reports shall be addressed to the Director, Center for Devices and Radiological Health, 5600 Fishers Lane, Rockville, MD 20857, and the reports and their envelopes shall be distinctly marked "Report on § 1002.20" and shall contain all of the following information where known to the manufacturer:

(1) The nature of the accidental radiation occurrence;

(2) The location at which the accidental radiation occurrence occurred;

(3) The manufacturer, type, and model number of the electronic product or products involved;

(4) The circumstances surrounding the accidental radiation occurrence, including causes;

(5) The number of persons involved, adversely affected, or exposed during the accidental radiation occurrence, the nature and magnitude of their exposure and/or injuries and, if requested by the Director, Center for Devices and Radiological Health, the names of the persons involved;

(6) The actions, if any, which may have been taken by the manufacturer, to control, correct, or eliminate the causes and to prevent reoccurrence; and

(7) Any other pertinent information with respect to the accidental radiation occurrence.

(c) If a manufacturer is required to report to the Director under paragraph (a) of this section and also is required to report under part 803 of this chapter, the manufacturer shall report in accordance with part 803. If a manufacturer is required to report to the Director under paragraph (a) of this section and is not required to report under part 803, the manufacturer shall report in accordance with paragraph (a) of this section. A manufacturer need not file a separate report under this section if an incident involving an accidental radiation occurrence is associated with a defect or noncompliance and is reported pursuant to § 1003.10 of this chapter.

[38 FR 28625, Oct. 15, 1973, as amended at 49 FR 36351, Sept. 14, 1984; 53 FR 11254, Apr. 6, 1988; 60 FR 48386, Sept. 19, 1995]

Subpart D—Manufacturers' Records

§ 1002.30 Records to be maintained by manufacturers.

(a) Manufacturers of products listed under table 1 of § 1002.1 shall establish and maintain the following records with respect to such products:

(1) Description of the quality control procedures with respect to electronic product radiation safety.

(2) Records of the results of tests for electronic product radiation safety, including the control of unnecessary, secondary or leakage electronic product radiation, the methods, devices, and procedures used in such tests, and the basis for selecting such methods, devices, and procedures.

(3) For those products displaying aging effects which may increase electronic product radiation emission, records of the results of tests for durability and stability of the product, and the basis for selecting these tests.

(4) Copies of all written communications between the manufacturer and dealers, distributors, and purchasers concerning radiation safety including complaints, investigations, instructions, or explanations affecting the use, repair, adjustment, maintenance, or testing of the listed product.

(5) Data on production and sales volume levels if available.

(b) In addition to the records required by paragraph (a) of this section, manufacturers of products listed in paragraph (c) of § 1002.61 shall establish and maintain the following records with respect to such products:

(1) A record of the manufacturer's distribution of products in a form which will enable the tracing of specific products or production lots to distributors or to dealers in those instances in which the manufacturer distributes directly to dealers.

(2) Records received from dealers or distributors pursuant to § 1002.41.

[38 FR 28625, Oct. 15, 1973, as amended at 60 FR 48386, Sept. 19, 1995]

§ 1002.31 Preservation and inspection of records.

(a) Every manufacturer required to maintain records pursuant to this part, including records received pursuant to § 1002.41, shall preserve such records for a period of 5 years from the date of the record.

(b) Upon reasonable notice by an officer or employee duly designated by the Department, manufacturers shall permit such officer or employee to inspect appropriate books, records, papers, and documents as are relevant to determining whether the manufacturer has acted or is acting in compliance with Federal standards.

(c) Upon request of the Director, Center for Devices and Radiological Health, a manufacturer of products listed in table 1 of § 1002.1 shall submit to the Director, copies of the records required to be maintained by paragraph (b) of § 1002.30.

[38 FR 28625, Oct. 15, 1973, as amended at 53 FR 11254, Apr. 6, 1988; 60 FR 48386, Sept. 19, 1995]

Subpart E—Dealer and Distributor Records

§ 1002.40 Records to be obtained by dealers and distributors.

(a) Dealers and distributors of electronic products for which there are performance standards and for which the retail price is $50 or more shall obtain such information as is necessary to identify and locate first purchasers if the product is subject to this section by virtue of table 1 of § 1002.1.

(b) Such information shall include:

(1) The name and mailing address of the distributor, dealer, or purchaser to whom the product was transferred.

(2) Identification and brand name of the product.

(3) Model number and serial or other identification number of the product.

(4) Date of sale, award, or lease.

(c) The information obtained pursuant to this section shall be forwarded immediately to the appropriate manufacturer of the electronic product, or preserved as prescribed in § 1002.41.

[38 FR 28625, Oct. 15, 1973, as amended at 42 FR 18063, Apr. 5, 1977; 60 FR 48386, Sept. 19, 1995]

§ 1002.41 Disposition of records obtained by dealers and distributors.

(a) Information obtained by dealers and distributors pursuant to § 1002.40 shall immediately be forwarded to the appropriate manufacturer unless:

(1) The dealer or distributor elects to hold and preserve such information and to immediately furnish it to the manufacturer when advised by the manufacturer or the Director, Center for Devices and Radiological Health, that such information is required for purposes of section 359 of the Act; and

(2) The dealer or distributor, upon making the election under paragraph (a)(1) of this section, promptly notifies the manufacturer of such election; such notification shall be in writing and shall identify the dealer or distributor and the electronic product or products for which the information is being accumulated and preserved.

(b) Every dealer or distributor who elects to hold and preserve information required pursuant to § 1002.40 shall preserve the information for a period of 5 years from the date of the sale, award, or lease of the product, or until the dealer or distributor discontinues dealing in, or distributing the product, whichever is sooner. If the dealer or distributor discontinues dealing in, or distributing the product, such information as obtained pursuant to § 1002.40 shall be furnished at that time, or before, to the manufacturer of the product.

[38 FR 28625, Oct. 15, 1973, as amended at 42 FR 18063, Apr. 5, 1977; 53 FR 11254, Apr. 6, 1988]

§ 1002.42 Confidentiality of records furnished by dealers and distributors.

All information furnished to manufacturers by dealers and distributors pursuant to this part shall be treated by such manufacturers as confidential information which may be used only as necessary to notify persons pursuant to section 359 of the Act.

Subpart F—Exemptions From Records and Reports Requirements

§ 1002.50 Special exemptions.

(a) Manufacturers of electronic products may submit to the Director a request, together with accompanying justification, for exemption from any requirements listed in table 1 of § 1002.1. The request must specify each requirement from which an exemption is requested. In addition to other information that is required, the justification must contain documented evidence showing that the product or product type for which the exemption is requested does not pose a public health risk and meets at least one of the following criteria:

(1) The products cannot emit electronic product radiation in sufficient intensity or of such quality, under any conditions of operation, maintenance, service, or product failure, to be hazardous;

(2) The products are produced in small quantities;

(3) The products are used by trained individuals and are to be used by the same manufacturing corporation or for research, investigation, or training.

(4) The products are custom designed and used by trained individuals knowledgeable of the hazards; or

(5) The products are produced in such a way that the requirements are inappropriate or unnecessary.

(b) The Director may, subject to any conditions that the Director deems necessary to protect the public health, exempt manufacturers from all or part of the record and reporting requirements of this part on the basis of information submitted in accordance with paragraph (a) of this section or such other information which the Director may possess if the Director determines that such exemption is in keeping with the purposes of the Act.

(c) The Director will provide written notification of the reason for any denial. If the exemption is granted, the Director will provide written notification of:

(1) The electronic product or products for which the exemption has been granted;

(2) The requirements from which the product is exempted; and

(3) Such conditions as are deemed necessary to protect the public health and safety. Copies of exemptions shall be available upon request from the Office of Compliance (HFZ–307), Center for Devices and Radiological Health, 2098 Gaither Rd., Rockville, MD 20850.

(d) The Director may, on the Director's own motion, exempt certain classes of products from the reporting requirements listed in table 1 of § 1002.1, provided that the Director finds that such exemption is in keeping with the purposes of the act.

(e) Manufacturers of products for which there is no applicable performance standard under parts 1020 through 1050 of this chapter and for which an investigational device exemption has been approved under § 812.30 of this chapter or for which a premarket approval application has been approved in accordance with § 814.44(d) of this chapter are exempt from submitting all reports listed in table 1 of § 1002.1.

[60 FR 48387, Sept. 19, 1995]

§ 1002.51 Exemptions for manufacturers of products intended for the U.S. Government.

Upon application therefor by the manufacturer, the Director, Center for Devices and Radiological Health, may exempt from the provisions of this part a manufacturer of any electronic product intended for use by departments or agencies of the United States provided such department or agency has prescribed procurement specifications governing emissions of electronic product radiation and provided further that such product is of a type used solely or predominantly by departments or agencies of the United States.

[38 FR 28625, Oct. 15, 1973, as amended at 53 FR 11254, Apr. 6, 1988]

PART 1003—NOTIFICATION OF DEFECTS OR FAILURE TO COMPLY

Subpart A—General Provisions

Sec.
1003.1 Applicability.
1003.2 Defect in an electronic product.
1003.5 Effect of regulations on other laws.

Subpart B—Discovery of Defect or Failure To Comply

1003.10 Discovery of defect or failure of compliance by manufacturer; notice requirements.
1003.11 Determination by Secretary that product fails to comply or has a defect.

Subpart C—Notification

1003.20 Notification by the manufacturer to the Secretary.
1003.21 Notification by the manufacturer to affected persons.
1003.22 Copies of communications sent to purchasers, dealers, or distributors.

Subpart D—Exemptions from Notification Requirements

1003.30 Application for exemption from notification requirements.
1003.31 Granting the exemption.

AUTHORITY: 42 U.S.C. 263b–263n.

SOURCE: 38 FR 28628, Oct. 15, 1973, unless otherwise noted.

Subpart A—General Provisions

§ 1003.1 Applicability.

The provisions of this part are applicable to electronic products which were manufactured after October 18, 1968.

§ 1003.2 Defect in an electronic product.

For the purpose of this part, an electronic product shall be considered to have a defect which relates to the safety of use by reason of the emission of electronic product radiation if:

(a) It is a product which does not utilize the emission of electronic product radiation in order to accomplish its purpose, and from which such emissions are unintended, and as a result of its design, production or assembly;

(1) It emits electronic product radiation which creates a risk of injury, including genetic injury, to any person, or

(2) It fails to conform to its design specifications relating to electronic radiation emissions; or

(b) It is a product which utilizes electronic product radiation to accomplish its primary purpose and from which such emissions are intended, and as a result of its design, production or assembly it;

(1) Fails to conform to its design specifications relating to the emission of electronic product radiation; or

(2) Without regard to the design specifications of the product, emits electronic product radiation unnecessary to the accomplishment of its primary purpose which creates a risk of injury, including genetic injury to any person; or

(3) Fails to accomplish the intended purpose.

§ 1003.5 Effect of regulations on other laws.

The remedies provided for in this subchapter shall be in addition to and not in substitution for any other remedies provided by law and shall not relieve any person from liability at common law or under statutory law.

Subpart B—Discovery of Defect or Failure To Comply

§ 1003.10 Discovery of defect or failure of compliance by manufacturer; notice requirements.

Any manufacturer who discovers that any electronic product produced, assembled, or imported by him, which product has left its place of manufacture, has a defect or fails to comply with an applicable Federal standard shall:

(a) Immediately notify the Secretary in accordance with § 1003.20, and

(b) Except as authorized by § 1003.30, furnish notification with reasonable promptness to the following persons:

(1) The dealers or distributors to whom such product was delivered by the manufacturer; and

(2) The purchaser of such product and any subsequent transferee of such product (where known to the manufacturer or where the manufacturer upon reasonable inquiry to dealers, distributors, or purchasers can identify the present user).

(c) If a manufacturer is required to notify the Secretary under paragraph (a) of this section and also is required to report to the Food and Drug Administration under part 803 of this chapter, the manufacturer shall report in accordance with part 803. If a manufacturer is required to notify the Secretary under paragraph (a) of this section and is not required to report to the Food and Drug Administration under part 803, the manufacturer shall notify the Secretary in accordance with paragraph (a) of this section.

[38 FR 28628, Oct. 15, 1973 and 49 FR 36351, Sept. 14, 1984]

§ 1003.11 Determination by Secretary that product fails to comply or has a defect.

(a) If, the Secretary, through testing, inspection, research, or examination of reports or other data, determines that any electronic product does not comply with an applicable Federal standard issued pursuant to the Act or has a defect, he shall immediately notify the manufacturer of the product in writing specifying:

(1) The defect in the product or the manner in which the product fails to comply with the applicable Federal standard;

(2) The Secretary's findings, with references to the tests, inspections, studies, or reports upon which such findings are based;

(3) A reasonable period of time during which the manufacturer may present his views and evidence to establish that there is no failure of compliance or that the alleged defect does not exist or does not relate to safety of use of the product by reason of the emission of electronic product radiation.

The manufacturer shall have an opportunity for a regulatory hearing before the Food and Drug Administration pursuant to part 16 of this chapter.

(b) Every manufacturer who receives a notice under paragraph (a) of this section shall immediately advise the

Secretary in writing of the total number of such product units produced and the approximate number of such product units which have left the place of manufacture.

(c) If, after the expiration of the period of time specified in the notice, the Secretary determines that the product has a defect or does not comply with an applicable Federal standard and the manufacturer has not applied for an exemption, he shall direct the manufacturer to furnish the notification to the persons specified in § 1003.10(b) in the manner specified in § 1003.21. The manufacturer shall within 14 days from the date of receipt of such directive furnish the required notification.

[38 FR 28628, Oct. 15, 1973, as amended at 41 FR 48269, Nov. 2, 1976; 42 FR 15676, Mar. 22, 1977]

Subpart C—Notification

§ 1003.20 Notification by the manufacturer to the Secretary.

The notification to the Secretary required by § 1003.10(a) shall be confirmed in writing and, in addition to other relevant information which the Secretary may require, shall include the following:

(a) Identification of the product or products involved;

(b) The total number of such product units so produced, and the approximate number of such product units which have left the place of manufacture;

(c) The expected usage for the product if known to the manufacturer;

(d) A description of the defect in the product or the manner in which the product fails to comply with an applicable Federal standard;

(e) An evaluation of the hazards reasonably related to defect or the failure to comply with the Federal standard;

(f) A statement of the measures to be taken to repair such defect or to bring the product into compliance with the Federal standard;

(g) The date and circumstances under which the defect was discovered; and

(h) The identification of any trade secret information which the manufacturer desires kept confidential.

§ 1003.21 Notification by the manufacturer to affected persons.

(a) The notification to the persons specified in § 1003.10(b) shall be in writing and, in addition to other relevant information which the Secretary may require, shall include:

(1) The information prescribed by § 1003.20 (a), (d), and instructions with respect to the use of the product pending the correction of the defect;

(2) A clear evaluation in nontechnical terms of the hazards reasonably related to any defect or failure to comply; and

(3) The following statement:

The manufacturer will, without charge, remedy the defect or bring the product into compliance with each applicable Federal standard in accordance with a plan to be approved by the Secretary of Health and Human Services, the details of which will be included in a subsequent communication to you.

Provided, That if at the time the notification is sent, the Secretary has approved a plan for the repair, replacement or refund of the product, the notification may include the details of the approved plan in lieu of the above statement.

(b) The envelope containing the notice shall not contain advertising or other extraneous material, and such mailings will be made in accordance with this section.

(1) No. 10 white envelopes shall be used, and the name and address of the manufacturer shall appear in the upper left corner of the envelope.

(2) The following statement is to appear in the far left third of the envelope in the type and size indicated and in reverse printing, centered in a red rectangle 3¾ inches wide and 2¼ inches high:

IMPORTANT—ELECTRONIC PRODUCT RADIATION WARNING

The statement shall be in three lines, all capitals, and centered. "Important" shall be in 36-point Gothic Bold type. "Electronic Product" and "Radiation Warning" shall be in 36-point Gothic Condensed type.

(3) Envelopes with markings similar to those prescribed in this section shall not be used by manufacturers for mailings other than those required by this part.

(c) The notification shall be sent:

(1) By certified mail to purchasers of the product and to subsequent transferees.

(2) By certified mail or other more expeditious means to dealers and distributors.

(d) Where products were sold under a name other than that of the manufacturer of the product, the name of the individual or company under whose name the product was sold may be used in the notification required by this section.

§ 1003.22 Copies of communications sent to purchasers, dealers or distributors.

(a) Every manufacturer of electronic products shall furnish to the Secretary a copy of all notices, bulletins, or other communications sent to the dealers or distributors of such manufacturers or to purchasers (or subsequent transferees) of electronic products of such manufacturer regarding any defect in such product or any failure of such product to comply with an applicable Federal standard.

(b) In the event the Secretary deems the content of such notices to be insufficient to protect the public health and safety, the Secretary may require additional notice to such recipients, or may elect to make or cause to be made such notification by whatever means he deems appropriate.

Subpart D—Exemptions From Notification Requirements

§ 1003.30 Application for exemption from notification requirements.

(a) A manufacturer may at the time of giving the written confirmation required by § 1003.20 or within 15 days of the receipt of any notice from the Secretary pursuant to § 1003.11(a), apply for an exemption from the requirement of notice to the persons specified in § 1003.10(b).

(b) The application for exemption shall contain the information required by § 1003.20 and in addition shall set forth in detail the grounds upon which the exemption is sought.

§ 1003.31 Granting the exemption.

(a) If, in the judgment of the Secretary, the application filed pursuant to § 1003.30 states reasonable grounds for an exemption from the requirement of notice, the Secretary shall give the manufacturer written notice specifying a reasonable period of time during which he may present his views and evidence in support of the application.

(b) Such views and evidence shall be confined to matters relevant to whether the defect in the product or its failure to comply with an applicable Federal standard is such as to create a significant risk of injury, including genetic injury, to any person and shall be presented in writing unless the Secretary determines that an oral presentation is desirable. Where such evidence includes nonclinical laboratory studies, the data submitted shall include, with respect to each such study, either a statement that the study was conducted in compliance with the requirements set forth in part 58 of this chapter, or, if the study was not conducted in compliance with such regulations, a brief statement of the reason for the noncompliance. When such evidence includes clinical investigations involving human subjects, the data submitted shall include, with respect to each clinical investigation either a statement that each investigation was conducted in compliance with the requirements set forth in part 56 of this chapter, or a statement that the investigation is not subject to such requirements in accordance with § 56.104 or § 56.105, and a statement that each investigation was conducted in compliance with the requirements set forth in part 50 of this chapter.

(c) If, during the period of time afforded the manufacturer to present his views and evidence, the manufacturer proves to the Secretary's satisfaction that the defect or failure to comply does not create a significant risk of injury, including genetic injury, to any person, the Secretary shall issue an exemption from the requirement of notification to the manufacturer and shall notify the manufacturer in writing specifying:

(1) The electronic product or products for which the exemption has been issued; and

(2) Such conditions as the Secretary deems necessary to protect the public health and safety.

(d) Any person who contests denial of an exemption shall have an opportunity for a regulatory hearing before the Food and Drug Administration pursuant to part 16 of this chapter.

[38 FR 28628, Oct. 15, 1973, as amended at 41 FR 48269, Nov. 2, 1976; 42 FR 15676, Mar. 22, 1977; 50 FR 7518, Feb. 22, 1985]

PART 1004—REPURCHASE, REPAIRS, OR REPLACEMENT OF ELECTRONIC PRODUCTS

Sec.

AUTHORITY: 42 U.S.C. 263b–263n.

SOURCE: 38 FR 28629, Oct. 15, 1973, unless otherwise noted.

§ 1004.1 Manufacturer's obligation to repair, replace, or refund cost of electronic products.

(a) If any electronic product fails to comply with an applicable Federal standard or has a defect and the notification specified in § 1003.10(b) of this chapter is required to be furnished, the manufacturer of such product shall;

(1) Without charge, bring such product into conformity with such standard or remedy such defect and provide reimbursement for any expenses for transportation of such product incurred in connection with having such product brought into conformity or having such defect remedied; or

(2) Replace such product with a like or equivalent product which complies with each applicable Federal standard and which has no defect relating to the safety of its use; or

(3) Make a refund of the cost of the product to the purchaser.

(b) The manufacturer shall take the action required by this section in accordance with a plan approved by the Secretary pursuant to § 1004.6.

§ 1004.2 Plans for the repair of electronic products.

Every plan for bringing an electronic product into conformity with applicable Federal standards or for remedying any defect in such product shall be submitted to the Secretary in writing, and in addition to other relevant information which the Secretary may require, shall include:

(a) Identification of the product involved.

(b) The approximate number of defective product units which have left the place of manufacture.

(c) The specific modifications, alterations, changes, repairs, corrections, or adjustments to be made to bring the product into conformity or remedy any defect.

(d) The manner in which the operations described in paragraph (c) will be accomplished, including the procedure for obtaining access to, or possession of, the products and the location where such operations will be performed.

(e) The technical data, test results or studies demonstrating the effectiveness of the proposed remedial action.

(f) A time limit, reasonable in light of the circumstances, for completion of the operations.

(g) The system by which the manufacturer will provide reimbursement for any transportation expenses incurred in connection with having such product brought into conformity or having any defect remedied.

(h) The text of the statement which the manufacturer will send to the persons specified in § 1003.10(b) of this chapter informing such persons;

(1) That the manufacturer, at his expense, will repair the electronic product involved,

(2) Of the method by which the manufacturer will obtain access to or possession of the product to make such repairs,

(3) That the manufacturer will reimburse such persons for any transportation expenses incurred in connection with making such repairs, and

(4) Of the manner in which such reimbursement will be effected.

(i) An assurance that the manufacturer will provide the Secretary with progress reports on the effectiveness of

208-072 D-20

the plan, including the number of electronic products repaired.

§ 1004.3 Plans for the replacement of electronic products.

Every plan for replacing an electronic product with a like or equivalent product shall be submitted to the Secretary in writing, and in addition to other relevant information which the Secretary may require, shall include:

(a) Identification of the product to be replaced.

(b) A description of the replacement product in sufficient detail to support the manufacturer's contention that the replacement product is like or equivalent to the product being replaced.

(c) The approximate number of defective product units which have left the place of manufacture.

(d) The manner in which the replacement operation will be effected including the procedure for obtaining possession of the product to be replaced.

(e) A time limit, reasonable, in light of the circumstances for completion of the replacement.

(f) The steps which the manufacturer will take to insure that the defective product will not be reintroduced into commerce, until it complies with each applicable Federal standard and has no defect relating to the safety of its use.

(g) The system by which the manufacturer will provide reimbursement for any expenses for transportation of such product incurred in connection with effecting the replacement.

(h) The text of the statement which the manufacturer will send to the persons specified in § 1003.10(b) of this chapter informing such persons;

(1) That the manufacturer, at its expense, will replace the electronic product involved,

(2) Of the method by which the manufacturer will obtain possession of the product and effect the replacement,

(3) That the manufacturer will reimburse such persons for any transportation expenses incurred in connection with effecting such replacement, and

(4) Of the manner in which such reimbursement will be made.

(i) An assurance that the manufacturer will provide the Secretary with progress reports on the effectiveness of the plan, including the number of electronic products replaced.

§ 1004.4 Plans for refunding the cost of electronic products.

Every plan for refunding the cost of an electronic product shall be submitted to the Secretary in writing, and in addition to other relevant information which the Secretary may require, shall include:

(a) Identification of the product involved.

(b) The approximate number of defective product units which have left the place of manufacture.

(c) The manner in which the refund operation will be effected including the procedure for obtaining possession of the product for which the refund is to be made.

(d) The steps which the manufacturer will take to insure that the defective products will not be reintroduced into commerce, until it complies with each applicable Federal standard and has no defect relating to the safety of its use.

(e) A time limit, reasonable in light of the circumstances, for obtaining the product and making the refund.

(f) A statement that the manufacturer will refund the cost of such product together with the information the manufacturer has used to determine the amount of the refund.

(g) The text of the statement which the manufacturer will send to the persons specified in § 1003.10(b) of this chapter informing such persons;

(1) That the manufacturer, at his expense, will refund the cost of the electronic product plus any transportation costs,

(2) Of the amount to be refunded exclusive of transportation costs,

(3) Of the method by which the manufacturer will obtain possession of the product and make the refund.

(h) An assurance that the manufacturer will provide the Secretary with progress reports on the effectiveness of the plan, including the number of refunds made.

§ 1004.6 Approval of plans.

If, after review of any plan submitted pursuant to this subchapter, the Secretary determines that the action to be taken by the manufacturer will expeditiously and effectively fulfill the manufacturer's obligation under §1004.1 in a manner designed to encourage the public to respond to the proposal, the Secretary will send written notice of his approval of such plan to the manufacturer. Such approval may be conditioned upon such additional terms as the Secretary deems necessary to protect the public health and safety. Any person who contests denial of a plan shall have an opportunity for a regulatory hearing before the Food and Drug Administration pursuant to part 16 of this chapter.

[38 FR 28629, Oct. 15, 1973, as amended at 41 FR 48269, Nov. 2, 1976; 42 FR 15676, Mar. 22, 1977]

PART 1005—IMPORTATION OF ELECTRONIC PRODUCTS

Subpart A—General Provisions

Subpart B—Inspection and Testing

Subpart C—Bonding and Compliance Procedures

AUTHORITY: 42 U.S.C. 263d, 263h.

SOURCE: 38 FR 28630, Oct. 15, 1973, unless otherwise noted.

Subpart A—General Provisions

§ 1005.1 Applicability.

(a) The provisions of §§1005.1 through 1005.24 are applicable to electronic products which are subject to the standards prescribed under this subchapter and are offered for importation into the United States.

(b) Section 1005.25 is applicable to every manufacturer of electronic products offering an electronic product for importation into the United States.

[38 FR 28630, Oct. 15, 1973, as amended at 45 FR 81739, Dec. 12, 1980]

§ 1005.2 Definitions.

As used in this part:

The term *owner* or *consignee* means the person who has the rights of a consignee under the provisions of sections 483, 484, and 485 of the Tariff Act of 1930, as amended (19 U.S.C. 1483, 1484, 1485).

§ 1005.3 Importation of noncomplying goods prohibited.

The importation of any electronic product for which standards have been prescribed under section 534 of the Federal Food, Drug, and Cosmetic Act (the act) (21 U.S.C. 360kk) shall be refused admission into the United States unless there is affixed to such product a certification in the form of a label or tag in conformity with section 534(h) of the act (21 U.S.C. 360kk(h)). Merchandise refused admission shall be destroyed or exported under regulations prescribed by the Secretary of the Treasury unless a timely and adequate petition for permission to bring the product into compliance is filed and granted under §§1005.21 and 1005.22.

[69 FR 11314, Mar. 10, 2004]

Subpart B—Inspection and Testing

§ 1005.10 Notice of sampling.

When a sample of a product to be offered for importation has been requested by the Secretary, the District Director of Customs having jurisdiction over the shipment shall, upon the arrival of the shipment, procure the sample and shall give to its owner or consignee prompt notice of the delivery or of the intention to deliver such sample to the Secretary. If the notice so requires, the owner or consignee will hold the shipment of which the sample is typical and not release such shipment until he receives notice of the results of the tests of the sample from the Secretary, stating that the product is in compliance with the requirements of the Act. The District Director of

Customs will be given the results of the tests. If the Secretary notifies the District Director of Customs that the product does not meet the requirements of the Act, the District Director of Customs shall require the exportation or destruction of the shipment in accordance with customs laws.

§ 1005.11 Payment for samples.

The Department of Health and Human Services will pay for all import samples of electronic products rendered unsalable as a result of testing, or will pay the reasonable costs of repackaging such samples for sale, if the samples are found to be in compliance with the requirements of the Radiation Control for Health and Safety Act of 1968. Billing for reimbursement shall be made by the owner or consignee to the Center for Devices and Radiological Health, 5600 Fishers Lane, Rockville, MD 20857. Payment for samples will not be made if the sample is found to be in violation of the Act, even though subsequently brought into compliance pursuant to terms specified in a notice of permission issued under § 1005.22.

[38 FR 28630, Oct. 15, 1973, as amended at 53 FR 11254, Apr. 6, 1988]

Subpart C—Bonding and Compliance Procedures

§ 1005.20 Hearing.

(a) If, from an examination of the sample or otherwise, it appears that the product may be subject to a refusal of admission, the Secretary shall give the owner or consignee a written notice to that effect, stating the reasons therefor. The notice shall specify a place and a period of time during which the owner or consignee shall have an opportunity to introduce testimony unless the owner or consignee indicates his intention to bring the product into compliance. Upon timely request, such time and place may be changed. Such testimony shall be confined to matters relevant to the admissibility of the article and may be introduced orally or in writing.

(b) If the owner or consignee submits or indicates his intention to submit an application for permission to perform such action as is necessary to bring the product into compliance with the Act, such application shall include the information required by § 1005.21.

(c) If the application is not submitted at or prior to the hearing, the Secretary may allow a reasonable time for filing such application.

§ 1005.21 Application for permission to bring product into compliance.

Application for permission to perform such action as is necessary to bring the product into compliance with the Act may be filed only by the owner, consignee, or manufacturer and, in addition to any other information which the Secretary may reasonably require, shall:

(a) Contain a detailed proposal for bringing the product into compliance with the Act;

(b) Specify the time and place where such operations will be effected and the approximate time for their completion; and

(c) Identify the bond required to be filed pursuant to § 1005.23.

§ 1005.22 Granting permission to bring product into compliance.

(a) When permission contemplated by § 1005.21 is granted, the Secretary shall notify the applicant in writing, specifying:

(1) The procedure to be followed;

(2) The disposition of the rejected articles or portions thereof;

(3) That the operations are to be carried out under the supervision of a representative of the Department of Health and Human Services;

(4) A reasonable time limit for completing the operations; and

(5) Such other conditions as he finds necessary to maintain adequate supervision and control over the product.

(b) Upon receipt of a written request for an extension of time to complete the operations necessary to bring the product into compliance, the Secretary may grant such additional time as he deems necessary.

(c) The notice of permission may be amended upon a showing of reasonable grounds thereof and the filing of an amended application for permission with the Secretary.

(d) If ownership of a product included in a notice of permission changes before the operations specified in the notice have been completed, the original owner will remain responsible under its bond, unless the new owner has executed a superseding bond on customs Form 7601 and obtained a new notice.

(e) The Secretary will notify the District Director of Customs having jurisdiction over the shipment involved, of the determination as to whether or not the product has in fact been brought into compliance with the Act.

§ 1005.23 Bonds.

The bond required under section 360(b) of the Act shall be executed by the owner or consignee on the appropriate form of a customs single-entry bond, customs Form 7551 or term bond, customs Form 7553 or 7595, containing a condition for the redelivery of the shipment or any part thereof not complying with the laws and regulations governing its admission into the commerce of the United States upon demand of the District Director of Customs and containing a provision for the performance of any action necessary to bring the product into compliance with all applicable laws and regulations. The bond shall be filed with the District Director of Customs.

§ 1005.24 Costs of bringing product into compliance.

The costs of supervising the operations necessary to bring a product into compliance with the Act shall be paid by the owner or consignee who files an application pursuant to § 1005.21 and executes a bond under section 360(b) of the Act. Such costs shall include:

(a) Travel expenses of the supervising officer;

(b) Per diem in lieu of subsistence of the supervising officer when away from his home station, as provided by law;

(c) *Service fees:* (1) The charge for the services of the supervising officer, which shall include administrative support, shall be computed at a rate per hour equal to 266 percent of the hourly rate of regular pay of a grade GS–11/4 employee, except that such services performed by a customs officer and subject to the provisions of the act of February 13, 1911, as amended (sec. 5, 36 Stat. 901, as amended (19 U.S.C. 267)), shall be calculated as provided in that act.

(2) The charge for the services of the analyst, which shall include administrative and laboratory support, shall be computed at a rate per hour equal to 266 percent of the hourly rate of regular pay of a grade GS–12/4 employee.

(3) The rate per hour equal to 266 percent of the equivalent hourly rate of regular pay of the supervising officer (GS–11/4) and the analyst (GS–12/4) is computed as follows:

	Hours
Gross number of working hours in 52 40-hour weeks	2,080
Less:	
Nine legal public holidays—New Years Day, Washington's Birthday, Memorial Day, Independence Day, Labor Day, Columbus Day, Veterans Day, Thanksgiving Day, and Christmas Day	72
Annual Leave—26 days	208
Sick Leave—13 days	104
Total	384
Net number of working hours	1,696
Gross number of working hours in 52 40-hour weeks	2,080
Working hour equivalent of Government contributions for employee retirement, life insurance, and health benefits computed at 8½% of annual rate of pay of employee	176
Equivalent annual working hours	2,256
Support required to equal to 1 man-year	2,256
Equivalent gross annual working hours charged to Food and Drug appropriation	4,512

NOTE: Ratio of equivalent gross annual number of working hours charged to Food and Drug appropriation to net number of annual working hours (4512/1696)=266 pct.

(d) The minimum charge for services of supervising officers shall be not less than the charge for 1 hour and time after the first hour shall be computed in multiples of 1 hour, disregarding fractional parts less than one-half hour.

[38 FR 28630, Oct. 15, 1973, as amended at 42 FR 55207, Oct. 14, 1977; 42 FR 62130, Dec. 9, 1977]

§ 1005.25 Service of process on manufacturers.

(a) Every manufacturer of electronic products, prior to offering such product for importation into the United States, shall designate a permanent resident of the United States as the manufacturer's agent upon whom service of all processes, notices, orders, decisions, and requirements may be made for and on behalf of the manufacturer as provided in section 360(d) of the Radiation Control for Health and Safety Act of 1968 (42 U.S.C. 263h(d)) and this section. The agent may be an individual, a firm, or a domestic corporation. For purposes of this section, any number of manufacturers may designate the same agent.

(b) A manufacturer designating an agent must address the designation to the Center for Devices and Radiological Health, 9200 Corporate Blvd., Rockville, MD 20850. It must be in writing and dated; all signatures must be in ink. The designation must be made in the legal form required to make it valid and binding on the manufacturer under the laws, corporate bylaws, or other requirements governing the making of the designation by the manufacturer at the place and time where it is made, and the persons or person signing the designation shall certify that it is so made. The designation must disclose the manufacturer's full legal name and the name(s) under which the manufacturer conducts the business, if applicable, the principal place of business, and mailing address. If any of the products of the manufacturer do not bear his legal name, the designation must identify the marks, trade names, or other designations of origin which these products bear. The designation must provide that it will remain in effect until withdrawn or replaced by the manufacturer and shall bear a declaration of acceptance duly signed by the designated agent. The full legal name and mailing address of the agent must be stated. Until rejected by the Secretary, designations are binding on the manufacturer even when not in compliance with all the requirements of this section. The designated agent may not assign performance of his function under the designation to another.

(c) Service of any process, notice, order, requirement, or decision specified in section 360(d) of the Radiation Control for Health and Safety Act of 1968 may be made by registered or certified mail addressed to the agent with return receipt requested, or in any other manner authorized by law. In the absence of such a designation or if for any reason service on the designated agent cannot be effected, service may be made as provided in section 360(d) by posting such process, notice, order, requirement, or decision in the Office of the Director, Center for Devices and Radiological Health and publishing a notice that such service was made in the FEDERAL REGISTER.

[38 FR 28630, Oct. 15, 1973, as amended at 53 FR 11254, Apr. 6, 1988; 65 FR 17137, Mar. 31, 2000]

PART 1010—PERFORMANCE STANDARDS FOR ELECTRONIC PRODUCTS: GENERAL

Subpart A—General Provisions

Subpart B—Alternate Test Procedures

Subpart C—Exportation of Electronic Products

AUTHORITY: 21 U.S.C. 351, 352, 360, 360e–360j, 371, 381; 42 U.S.C. 263b–263n.

SOURCE: 38 FR 28631, Oct. 15, 1973, unless otherwise noted.

Subpart A—General Provisions

§ 1010.1 Scope.

The standards listed in this subchapter are prescribed pursuant to section 358 of the Radiation Control for Health and Safety Act of 1968 (42 U.S.C. 263f) and are applicable to electronic products as specified herein, to control electronic product radiation from such products. Standards so prescribed are subject to amendment or revocation and additional standards may be prescribed as are determined necessary for the protection of the public health and safety.

[40 FR 32257, July 31, 1975]

§ 1010.2 Certification.

(a) Every manufacturer of an electronic product for which an applicable standard is in effect under this subchapter shall furnish to the dealer or distributor, at the time of delivery of such product, the certification that such product conforms to all applicable standards under this subchapter.

(b) The certification shall be in the form of a label or tag permanently affixed to or inscribed on such product so as to be legible and readily accessible to view when the product is fully assembled for use, unless the applicable standard prescribes some other manner of certification. All such labels or tags shall be in the English language.

(c) Such certification shall be based upon a test, in accordance with the standard, of the individual article to which it is attached or upon a testing program which is in accordance with good manufacturing practices. The Director, Center for Devices and Radiological Health may disapprove such a testing program on the grounds that it does not assure the adequacy of safeguards against hazardous electronic product radiation or that it does not assure that electronic products comply with the standards prescribed under this subchapter.

(d) In the case of products for which it is not feasible to certify in accordance with paragraph (b) of this section, upon application by the manufacturer, the Director, Center for Devices and Radiological Health may approve an alternate means by which such certification may be provided.

[38 FR 28631, Oct. 15, 1973, as amended at 40 FR 32257, July 31, 1975; 42 FR 18063, Apr. 5, 1977; 53 FR 11254, Apr. 6, 1988]

§ 1010.3 Identification.

(a) Every manufacturer of an electronic product to which a standard under this subchapter is applicable shall set forth the information specified in paragraphs (a)(1) and (2) of this section. This information shall be provided in the form of a tag or label permanently affixed or inscribed on such product so as to be legible and readily accessible to view when the product is fully assembled for use or in such other manner as may be prescribed in the applicable standard. Except for foreign equivalent abbreviations as authorized in paragraph (a)(1) of this section all such labels or tags shall be in the English language.

(1) The full name and address of the manufacturer of the product; abbreviations such as "Co.," "Inc.," or their foreign equivalents and the first and middle initials of individuals may be used. Where products are sold under a name other than that of the manufacturer of the product, the full name and address of the individual or company under whose name the product was sold may be set forth, provided such individual or company has previously suppled the Director, Center for Devices and Radiological Health with sufficient information to identify the manufacturer of the product.

(2) The place and month and year of manufacture:

(i) The place of manufacture may be expressed in code provided the manufacturer has previously supplied the Director, Center for Devices and Radiological Health with the key to such code.

(ii) The month and year of manufacture shall be provided clearly and legibly, without abbreviation, and with the year shown as a four-digit number as follows:

MANUFACTURED: (INSERT MONTH AND YEAR OF MANUFACTURE.)

(b) In the case of products for which it is not feasible to affix identification labeling in accordance with paragraph (a) of this section, upon application by the manufacturer, the Director, Center for Devices and Radiological Health may approve an alternate means by which such identification may be provided.

(c) Every manufacturer of an electronic product to which a standard under this subchapter is applicable shall provide to the Director, Center for Devices and Radiological Health a list identifying each brand name which is applied to the product together with the full name and address of the individual or company for whom each product so branded is manufactured.

[40 FR 32257, July 31, 1975, as amended at 42 FR 18063, Apr. 5, 1977; 53 FR 11254, Apr. 6, 1988]

§ 1010.4 Variances.

(a) *Criteria for variances.* (1) Upon application by a manufacturer (including an assembler), the Director, Center for Devices and Radiological Health, Food and Drug Administration, may grant a variance from one or more provisions of any performance standard under subchapter J of this chapter for an electronic product subject to such standard when the Director determines that granting such a variance is in keeping with the purposes of the Radiation Control for Health and Safety Act of 1968, and:

(i) The scope of the requested variance is so limited in its applicability as not to justify an amendment to the standard, or

(ii) There is not sufficient time for the promulgation of an amendment to the standard.

(2) The issuance of the variance shall be based upon a determination that:

(i) The product utilizes an alternate means for providing radiation safety or protection equal to or greater than that provided by products meeting all requirements of the applicable standard, or

(ii) The product performs a function or is intended for a purpose which could not be performed or accomplished if required to meet the applicable standards, and suitable means for assuring radiation safety or protection are provided, or

(iii) One or more requirements of the applicable standard are not appropriate, and suitable means for assuring radiation safety or protection are provided.

(b) *Applications for variances.* If you are submitting an application for variances or for amendments or extensions thereof, you must submit an original and two copies to the Division of Dockets Management (HFA–305), Food and Drug Administration, 5630 Fishers Lane, rm. 1061, Rockville, MD 20852.

(1) The application for variance shall include the following information:

(i) A description of the product and its intended use.

(ii) An explanation of how compliance with the applicable standard would restrict or be inappropriate for this intended use.

(iii) A description of the manner in which it is proposed to deviate from the requirements of the applicable standard.

(iv) A description of the advantages to be derived from such deviation.

(v) An explanation of how alternate or suitable means of radiation protection will be provided.

(vi) The period of time it is desired that the variance be in effect, and, if appropriate, the number of units the applicant wishes to manufacture.

(vii) In the case of prototype or experimental equipment, the proposed location of each unit.

(viii) Such other information required by regulation or by the Director, Center for Devices and Radiological Health, to evaluate and act on the application.

(ix) With respect to each nonclinical laboratory study contained in the application, either a statement that the study was conducted in compliance with the good laboratory practice regulations set forth in part 58 of this chapter, or, if the study was not conducted in compliance with such regulations, a brief statement of the reason for the noncompliance.

(x) [Reserved]

(xi) If the electronic product is used in a clinical investigation involving human subjects, is subject to the requirements for institutional review set forth in part 56 of this chapter, and is subject to the requirements for informed consent set forth in part 50 of this chapter, the investigation shall be conducted in compliance with such requirements.

(2) The application for amendment or extension of a variance shall include the following information:

(i) The variance number and expiration date.

(ii) The amendment or extension requested and basis for the amendment or extension.

(iii) A description of the effect of the amendment or extension on protection from radiation produced by the product.

(iv) An explanation of how alternate or suitable means of protection will be provided.

(c) *Ruling on applications.* (1) The Director, Center for Devices and Radiological Health, may approve or deny, in whole or in part, a requested variance or any amendment or extension thereof, and the director shall inform the applicant in writing of this action on a requested variance or amendment or extension. The written notice will state the manner in which the variance differs from the standard, the effective date and the termination date of the variance, a summary of the requirements and conditions attached to the variance, any other information that may be relevant to the application or variance, and, if appropriate, the number of units or other similar limitations for which the variance is approved. Each variance will be assigned an identifying number.

(2) The Director, Center for Devices and Radiological Health, shall amend or withdraw a variance whenever the Director determines that this action is necessary to protect the public health or otherwise is justified by this subchapter. Such action will become effective on the date specified in the written notice of the action sent to the applicant, except that it will become effective immediately upon notification to the applicant when the Director determines that such action is necessary to prevent an imminent health hazard.

(3) All applications for variances and for amendments and extensions thereof and all correspondence (including written notices of approval) on these applications will be available for public disclosure in the office of the Division of Dockets Management, except for information regarded as confidential under section 360A(e) of the act.

(d) *Certification of equipment covered by variance.* The manufacturer of any product for which a variance is granted shall modify the tag, label, or other certification required by §1010.2 to state:

(1) That the product is in conformity with the applicable standard, except with respect to those characteristics covered by the variance;

(2) That the product is in conformity with the provisions of the variance; and

(3) The assigned number and effective date of the variance.

[39 FR 13879, Apr. 18, 1974, as amended at 44 FR 48191, Aug. 17, 1979; 50 FR 7518, Feb. 22, 1985; 50 FR 13565, Apr. 5, 1985; 53 FR 11254, Apr. 6, 1988; 53 FR 52683, Dec. 29, 1988; 59 FR 14365, Mar. 28, 1994; 65 FR 17137, Mar. 31, 2000]

§1010.5 Exemptions for products intended for United States Government use.

(a) *Criteria for exemption.* Upon application by a manufacturer (including assembler) or by a U.S. department or agency, the Director, Center for Devices and Radiological Health, Food and Drug Administration, may grant an exemption from any performance standard under subchapter J of this chapter for an electronic product, or class of products, otherwise subject to such standard when he determines that such electronic product or class is intended for use by departments or agencies of the United States and meets the criteria set forth in paragraph (a) (1) or (2) of this section.

(1) The procuring agency shall prescribe procurement specifications for the product or class of products governing emissions of electronic product radiation, and the product or class shall be of a type used solely or predominantly by a department or agency of the United States.

(2) The product or class of products is intended for research, investigations, studies, demonstration, or training, or for reasons of national security.

(b) *Consultation between the procuring agency and the Food and Drug Administration.* The United States department or agency that intends to procure or manufacture a product or class of products subject to electronic product radiation safety standards contained in this subchapter should consult with the Center for Devices and Radiological Health, Food and Drug Administration, whenever it is anticipated that the specifications for the product or class must deviate from, or be in conflict with, such applicable standards. Such consultation should occur as early as possible during development of such specifications. The department or agency should include in the specifications all requirements of such standards that are not in conflict with, or are not inappropriate for, the special or unique uses for which the product is intended. The procuring agency should indicate to the Center for Devices and Radiological Health if it desires to be notified of the approval, amendment, or withdrawal of the exemption.

(c) *Application for exemption.* If you are submitting an application for exemption, or for amendment or extension thereof, you must submit an original and two copies to the Division of Dockets Management (HFA–305), Food and Drug Administration, 5630 Fishers Lane, rm. 1061, Rockville, MD 20852. For an exemption under the criteria prescribed in paragraph (a)(1) of this section, the application shall include the information prescribed in paragraphs (c)(1) through (c)(13) of this section. For an exemption under the criteria prescribed in paragraph (a)(2) of this section, the application shall include the information prescribed in paragraphs (c)(3) through (c)(13) of this section. An application for exemption, or for amendment or extension thereof, and correspondence relating to such application shall be made available for public disclosure in the Division of Dockets Management, except for confidential or proprietary information submitted in accordance with part 20 of this chapter. Information classified for reasons of national security shall not be included in the application. Except as indicated in this paragraph, the application for exemption shall include the following:

(1) The procurement specifications for the product or class of products that govern emissions of electronic product radiation.

(2) Evidence that the product or class of products is of a type used solely or predominantly by departments or agencies of the United States.

(3) Evidence that such product or class of products is intended for use by a department or agency of the United States.

(4) A description of the product or class of products and its intended use.

(5) An explanation of how compliance with the applicable standard would restrict or be inappropriate for this intended use.

(6) A description of the manner in which it is proposed that the product or class of products shall deviate from the requirements of the applicable standard.

(7) An explanation of the advantages to be derived from such deviation.

(8) An explanation of how means of radiation protection will be provided where the product or class of products deviates from the requirements of the applicable standard.

(9) The period of time it is desired that the exemption be in effect, and, if appropriate, the number of units to be manufactured under the exemption.

(10) The name, address, and telephone number of the manufacturer or his agent.

(11) The name, address, and telephone number of the appropriate office of the United States department or agency purchasing the product or class of products.

(12) Such other information required by regulation or by the Director, Center for Devices and Radiological Health, to evaluate and act on the application. Where such information includes nonclinical laboratory studies, the information shall include, with respect to each nonclinical study, either a statement that each study was conducted in compliance with the requirements set forth in part 58 of this chapter, or, if the study was not conducted in compliance with such regulations, a statement that describes in detail all differences between the practices used in the study and those required in the regulations. When such information includes clinical investigations involving human subjects, the information shall include, with respect to each clinical investigation, either a statement that each investigation was conducted in compliance with the requirements set forth in part 56 of this chapter, or a statement that the investigation is not subject to such requirements in accordance with § 56.104 or § 56.105 and a statement that each investigation was conducted in compliance with the requirements set forth in part 50 of this chapter.

(13) With respect to each nonclinical laboratory study contained in the application, either a statement that the study was conducted in compliance with the requirements set forth in part 58 of this chapter, or, if the study was not conducted in compliance with such regulations, a brief statement of the reason for the noncompliance.

(d) *Amendment or extension of an exemption.* An exemption is granted on the basis of the information contained in the orginal applicaion. Therefore, if changes are needed in the radiation safety specifications for the product, or its use, or related radiation control procedures such that the information in the original application would no longer be correct with respect to radiation safety, the applicant shall submit in advance of such changes a request for an amendment to the exemption. He also shall submit a request for extension of the exemption, if needed, at least 60 days before the expiration date. The application for amendment or extension of an exemption shall include the following information:

(1) The exemption number and expiration date.

(2) The amendment or extension requested and basis for the amendment or extension.

(3) If the radiation safety specifications for the product or class of products or the product's or class of products' use or related radiation control procedures differ from the description provided in the original application, a description of such changes.

(e) *Ruling on an application.* (1) The Director, Center for Devices and Radiological Health, may grant an exemption including in the written notice of exemption such conditions or terms as may be necessary to protect the public health and safety and shall notify the applicant in writing of his action. The conditions or terms of the exemption may include specifications concerning the manufacture, use, control, and disposal of the excess or surplus exempted product of class of products as provided in the Code of Federal Regulations, title 41, subtitle C. Each exemption will be assigned an identifying number.

(2) The Director, Center for Devices and Radiological Health, shall amend or withdraw an exemption whenever he determines that such action is necessary to protect the public health or otherwise is justified by provisions of the act or this subchapter. Such action shall become effective on the date specified in the written notice of the action sent to the applicant, except that it shall become effective immediately when the Director determines that it is necessary to prevent an imminent health hazard.

(f) *Identification of equipment covered by exemption.* The manufacturer of any product for which an exemption is granted shall provide the following identification in the form of a tag or label permanently affixed or inscribed on such product so as to be legible and readily accessible to view when the product is fully assembled for use or in such other manner as may be prescribed in the exemption:

CAUTION

This electronic product has been exempted from Food and Drug Administration radiation safety performance standards prescribed in the Code of Federal Regulations, title 21, chapter I, subchapter J, pursuant to Exemption No. _____, granted on __________

[42 FR 44229, Sept. 2, 1977; 42 FR 61257, Dec. 2, 1977, as amended at 44 FR 17657, Mar. 23, 1979; 46 FR 8460, 8958, Jan. 27, 1981; 50 FR 7518, Feb. 22, 1985; 50 FR 13564, Apr. 5, 1985; 53 FR 11254, Apr. 6, 1988; 59 FR 14365, Mar. 28, 1994; 65 FR 17138, Mar. 31, 2000]

Subpart B—Alternate Test Procedures

§1010.13 Special test procedures.

The Director, Center for Devices and Radiological Health, may, on the basis of a written application by a manufacturer, authorize test programs other than those set forth in the standards under this subchapter for an electronic product if he determines that such products are not susceptible to satisfactory testing by the procedures set forth in the standard and that the alternative test procedures assure compliance with the standard.

[40 FR 32257, July 31, 1975, as amended at 53 FR 11254, Apr. 6, 1988]

Subpart C—Exportation of Electronic Products

§ 1010.20 Electronic products intended for export.

The performance standards prescribed in this subchapter shall not apply to any electronic product which is intended solely for export if:

(a) Such product and the outside of any shipping container used in the export of such product are labeled or tagged to show that such product is intended for export, and

(b) Such product meets all the applicable requirements of the country to which such product is intended for export.

[40 FR 32257, July 31, 1975]

PART 1020—PERFORMANCE STANDARDS FOR IONIZING RADIATION EMITTING PRODUCTS

Sec.

1020.10 Television receivers.
1020.20 Cold-cathode gas discharge tubes.
1020.30 Diagnostic x-ray systems and their major components.
1020.31 Radiographic equipment.
1020.32 Fluoroscopic equipment.
1020.33 Computed tomography (CT) equipment.
1020.40 Cabinet x-ray systems.

AUTHORITY: 21 U.S.C. 351, 352, 360e–360j, 360gg–360ss, 371, 381.

SOURCE: 38 FR 28632, Oct. 15, 1973, unless otherwise noted.

§ 1020.10 Television receivers.

(a) *Applicability.* The provisions of this section are applicable to television receivers manufactured subsequent to January 15, 1970.

(b) *Definitions.* (1) *External surface* means the cabinet or enclosure provided by the manufacturer as part of the receiver. If a cabinet or enclosure is not provided as part of the receiver, the external surface shall be considered to be a hypothetical cabinet, the plane surfaces of which are located at those minimum distances from the chassis sufficient to enclose all components of the receiver except that portion of the neck and socket of the cathode-ray tube which normally extends beyond the plane surfaces of the enclosure.

(2) *Maximum test voltage* means 130 root mean square volts if the receiver is designed to operate from nominal 110 to 120 root mean square volt power sources. If the receiver is designed to operate from a power source having some voltage other than from nominal 110 to 120 root mean square volts, maximum test voltage means 110 percent of the nominal root mean square voltage specified by the manufacturer for the power source.

(3) *Service controls* means all of those controls on a television receiver provided by the manufacturer for purposes of adjustment which, under normal usage, are not accessible to the user.

(4) *Television receiver* means an electronic product designed to receive and display a television picture through broadcast, cable, or closed circuit television.

(5) *Usable picture* means a picture in synchronization and transmitting viewable intelligence.

(6) *User controls* means all of those controls on a television receiver, provided by the manufacturer for purposes of adjustment, which on a fully assembled receiver under normal usage, are accessible to the user.

(c) *Requirements*—(1) *Exposure rate limit.* Radiation exposure rates produced by a television receiver shall not exceed 0.5 milliroentgens per hour at a distance of five (5) centimeters from any point on the external surface of the receiver, as measured in accordance with this section.

(2) *Measurements.* Compliance with the exposure rate limit defined in paragraph (c)(1) of this section shall be determined by measurements made with an instrument, the radiation sensitive volume of which shall have a cross section parallel to the external surface of the receiver with an area of ten (10) square centimeters and no dimension larger than five (5) centimeters. Measurements made with instruments having other areas must be corrected for spatial nonuniformity of the radiation field to obtain the exposure rate average over a ten (10) square centimeter area.

(3) *Test conditions.* All measurements shall be made with the receiver displaying a usable picture and with the power source operated at supply

voltages up to the maximum test voltage of the receiver and, as applicable, under the following specific conditions:

(i) On television receivers manufactured subsequent to January 15, 1970, measurements shall be made with all user controls adjusted so as to produce maximum x-radiation emissions from the receiver.

(ii) On television receivers manufactured subsequent to June 1, 1970, measurements shall be made with all user controls and all service controls adjusted to combinations which result in the production of maximum x-radiation emissions.

(iii) On television receivers manufactured subsequent to June 1, 1971, measurements shall be made under the conditions described in paragraph (c)(3) (ii) of this section, together with conditions identical to those which result from that component or circuit failure which maximizes x-radiation emissions.

(4) *Critical component warning.* The manufacturer shall permanently affix or inscribe a warning label, clearly legible under conditions of service, on all television receivers which could produce radiation exposure rates in excess of the requirements of this section as a result of failure or improper adjustment or improper replacement of a circuit or shield component. The warning label shall include the specification of operating high voltage and an instruction for adjusting the high voltage to the specified value.

§ 1020.20 Cold-cathode gas discharge tubes.

(a) *Applicability.* The provisions of this section are applicable to cold-cathode gas discharge tubes designed to demonstrate the effects of a flow of electrons or the production of x-radiation as specified herein.

(b) *Definitions. Beam blocking device* means a movable or removable portion of any enclosure around a cold-cathode gas discharge tube, which may be opened or closed to permit or prevent the emergence of an exit beam.

Cold-cathode gas discharge tube means an electronic device in which electron flow is produced and sustained by ionization of contained gas atoms and ion bombardment of the cathode.

Exit beam means that portion of the radiation which passes through the aperture resulting from the opening of the beam blocking device.

Exposure means the sum of the electrical charges on all of the ions of one sign produced in air when all electrons liberated by photons in a volume element of air are completely stopped in air divided by the mass of the air in the volume element. The special unit of exposure is the roentgen. One (1) roentgen equals 2.58×10^{-4} coulombs/kilogram.

(c) *Requirements*—(1) *Exposure rate limit.* (i) Radiation exposure rates produced by cold-cathode gas discharge tubes shall not exceed 10 mR./hr. at a distance of thirty (30) centimeters from any point on the external surface of the tube, as measured in accordance with this section.

(ii) The divergence of the exit beam from tubes designed primarily to demonstrate the effects of x radiation, with the beam blocking device in the open position, shall not exceed (Pi) steradians.

(2) *Measurements.* (i) Compliance with the exposure rate limit defined in paragraph (c)(1)(i) of this section shall be determined by measurements averaged over an area of one hundred (100) square centimeters with no linear dimension greater than twenty (20) centimeters.

(ii) Measurements of exposure rates from tubes in enclosures from which the tubes cannot be removed without destroying the function of the tube may be made at a distance of thirty (30) centimeters from any point on the external surface of the enclosure, provided:

(*a*) In the case of enclosures containing tubes designed primarily to demonstrate the production of x radiation, measurements shall be made with any beam blocking device in the beam blocking position, or

(*b*) In the case of enclosures containing tubes designed primarily to demonstrate the effects of a flow of electrons, measurements shall be made with all movable or removable parts of such enclosure in the position which would maximize external exposure levels.

(3) *Test conditions.* (i) Measurements shall be made under the conditions of use specified in instructions provided by the manufacturer.

(ii) Measurements shall be made with the tube operated under forward and reverse polarity.

(4) *Instructions, labels, and warnings.* (i) Manufacturers shall provide, or cause to be provided, with each tube to which this section is applicable, appropriate safety instructions, together with instructions for the use of such tube, including the specification of a power source for use with the tube.

(ii) Each enclosure or tube shall have inscribed on or permanently affixed to it, tags or labels, which identify the intended polarity of the terminals and:

(*a*) In the case of tubes designed primarily to demonstrate the heat effect, fluorescence effect, or magnetic effect, a warning that application of power in excess of that specified may result in the production of x-rays in excess of allowable limits; and (*b*) in the case of tubes designed primarily to demonstrate the production of x-radiation, a warning that this device produces x-rays when energized.

(iii) The tag or label required by this paragraph shall be located on the tube or enclosure so as to be readily visible and legible when the product is fully assembled for use.

§ 1020.30 Diagnostic x-ray systems and their major components.

(a) *Applicability*—(1) The provisions of this section are applicable to:

(i) The following components of diagnostic x-ray systems:

(A) Tube housing assemblies, x-ray controls, x-ray high-voltage generators, x-ray tables, cradles, film changers, vertical cassette holders mounted in a fixed location and cassette holders with front panels, and beam-limiting devices manufactured after August 1, 1974.

(B) Fluoroscopic imaging assemblies manufactured after August 1, 1974, and before April 26, 1977.

(C) Spot-film devices and image intensifiers manufactured after April 26, 1977.

(D) Cephalometric devices manufactured after February 25, 1978.

(E) Image receptor support devices for mammographic x-ray systems manufactured after September 5, 1978.

(ii) Diagnostic x-ray systems, except computed tomography x-ray systems, incorporating one or more of such components; however, such x-ray systems shall be required to comply only with those provisions of this section and §§ 1020.31 and 1020.32 which relate to the components certified in accordance with paragraph (c) of this section and installed into the systems.

(iii) Computed tomography (CT) x-ray systems manufactured before November 29, 1984.

(iv) CT gantries manufactured after September 3, 1985.

(2) The following provisions of this section and § 1020.33 are applicable to CT x-ray systems manufactured or remanufactured on or after November 29, 1984:

(i) Section 1020.30(a);

(ii) Section 1020.30(b) ‘‘Technique factors’’;

(iii) Section 1020.30(b) ‘‘CT,’’ ‘‘Dose,’’ ‘‘Scan,’’ ‘‘Scan time,’’ and ‘‘Tomogram’’;

(iv) Section 1020.30 (h)(3)(vi) through (h)(3)(viii);

(v) Section 1020.30(n);

(vi) Section 1020.33 (a) and (b);

(vii) Section 1020.33(c)(1) as it affects § 1020.33(c)(2); and

(viii) Section 1020.33(c)(2).

(3) The provisions of this section and § 1020.33 in its entirety, including those provisions in paragraph (a)(2) of this section, are applicable to CT x-ray systems manufactured or remanufactured on or after September 3, 1985. The date of manufacture of the CT system is the date of manufacture of the CT gantry.

(b) *Definitions.* As used in this section and §§ 1020.31, 1020.32, and 1020.33, the following definitions apply:

Accessible surface means the external surface of the enclosure or housing provided by the manufacturer.

Accessory component means:

(1) A component used with diagnostic x-ray systems, such as a cradle or film changer, that is not necessary for the compliance of the system with applicable provisions of this subchapter but which requires an initial determination of compatibility with the system; or

(2) A component necessary for compliance of the system with applicable provisions of this subchapter but which may be interchanged with similar compatible components without affecting the system's compliance, such as one of a set of interchangeable beam-limiting devices; or

(3) A component compatible with all x-ray systems with which it may be used and that does not require compatibility or installation instructions, such as a tabletop cassette holder.

Aluminum equivalent means the thickness of aluminum (type 1100 alloy)[1] affording the same attenuation, under specified conditions as the material in question.

Articulated joint means a joint between two separate sections of a tabletop which joint provides the capacity for one of the sections to pivot on the line segment along which the sections join.

Assembler means any person engaged in the business of assembling, replacing, or installing one or more components into a diagnostic x-ray system or subsystem. The term includes the owner of an x-ray system or his or her employee or agent who assembles components into an x-ray system that is subsequently used to provide professional or commercial services.

Attenuation block means a block or stack of type 1100 aluminum alloy or aluminum alloy having equivalent attenuation with dimensions 20 centimeters by 20 centimeters by 3.8 centimeters.

Automatic exposure control means a device which automatically controls one or more technique factors in order to obtain at a preselected location(s) a required quantity of radiation.

Beam axis means a line from the source through the centers of the x-ray fields.

Beam-limiting device means a device which provides a means to restrict the dimensions of the x-ray field.

Cantilevered tabletop means a tabletop designed such that the unsupported portion can be extended at least 100 centimeters beyond the support.

Cassette holder means a device, other than a spot-film device, that supports and/or fixes the position of an x-ray film cassette during an x-ray exposure.

Cephalometric device means a device intended for the radiographic visualization and measurement of the dimensions of the human head.

Coefficient of variation means the ratio of the standard deviation to the mean value of a population of observations. It is estimated using the following equation:

$$C = \frac{s}{\overline{X}} = \frac{1}{\overline{X}} \left[\sum_{i=1}^{n} \frac{(X_i - \overline{X})^2}{n-1} \right]^{1/2}$$

where:

s = Estimated standard deviation of the population.
$\acute{X}$ = Mean value of observations in sample.
X_i = ith observation sampled.
n = Number of observations sampled.

Computed tomography (CT) means the production of a tomogram by the acquisition and computer processing of x-ray transmission data.

Control panel means that part of the x-ray control upon which are mounted the switches, knobs, pushbuttons, and other hardware necessary for manually setting the technique factors.

Cooling curve means the graphical relationship between heat units stored and cooling time.

Cradle means:

(1) A removable device which supports and may restrain a patient above an x-ray table; or

(2) A device;

[1] The nominal chemical composition of type 1100 aluminum alloy is 99.00 percent minimum aluminum, 0.12 percent copper, as given in "Aluminum Standards and Data" (1969). Copies may be obtained from: The Aluminum Association, New York, NY.

(i) Whose patient support structure is interposed between the patient and the image receptor during normal use;

(ii) Which is equipped with means for patient restraint; and

(iii) Which is capable of rotation about its long (longitudinal) axis.

CT gantry means tube housing assemblies, beam-limiting devices, detectors, and the supporting structures, frames, and covers which hold and/or enclose these components.

Diagnostic source assembly means the tube housing assembly with a beam-limiting device attached.

Diagnostic x-ray system means an x-ray system designed for irradiation of any part of the human body for the purpose of diagnosis or visualization.

Dose means the absorbed dose as defined by the International Commission on Radiation Units and Measurements. The absorbed dose, D, is the quotient of de by dm, where de is the mean energy imparted by ionizing radiation to matter of mass dm.

Equipment means x-ray equipment.

Exposure means the quotient of dQ by dm where dQ is the absolute value of the total charge of the ions of one sign produced in air when all the electrons (negatrons and positrons) liberated by photons in a volume element of air having mass dm are completely stopped in air.

Field emission equipment means equipment which uses an x-ray tube in which electron emission from the cathode is due solely to action of an electric field.

Fluoroscopic imaging assembly means a subsystem in which x-ray photons produce a fluoroscopic image. It includes the image receptor(s) such as the image intensifier and spot-film device, electrical interlocks, if any, and structural material providing linkage between the image receptor and diagnostic source assembly.

General purpose radiographic x-ray system means any radiographic x-ray system which, by design, is not limited to radiographic examination of specific anatomical regions.

Half-value layer (HVL) means the thickness of specified material which attenuates the beam of radiation to an extent such that the exposure rate is reduced to one-half of its original value. In this definition the contribution of all scattered radiation, other than any which might be present initially in the beam concerned, is deemed to be excluded.

Image intensifier means a device, installed in its housing, which instantaneously converts an x-ray pattern into a corresponding light image of higher energy density.

Image receptor means any device, such as a fluorescent screen, radiographic film, solid-state detector, or gaseous detector, which transforms incident x-ray photons either into a visible image or into another form which can be made into a visible image by further transformations. In those cases where means are provided to preselect a portion of the image receptor, the term "image receptor" shall mean the preselected portion of the device.

Image receptor support device means, for mammography x-ray systems, that part of the system designed to support the image receptor during a mammographic examination and to provide a primary protective barrier.

Leakage radiation means radiation emanating from the diagnostic source assembly except for:

(1) The useful beam; and

(2) Radiation produced when the exposure switch or timer is not activated.

Leakage technique factors means the technique factors associated with the diagnostic source assembly which are used in measuring leakage radiation. They are defined as follows:

(1) For diagnostic source assemblies intended for capacitor energy storage equipment, the maximum-rated peak tube potential and the maximum-rated number of exposures in an hour for operation at the maximum-rated peak tube potential with the quantity of charge per exposure being 10 millicoulombs (or 10 mAs) or the minimum obtainable from the unit, whichever is larger;

(2) For diagnostic source assemblies intended for field emission equipment rated for pulsed operation, the maximum-rated peak tube potential and the maximum-rated number of x-ray pulses in an hour for operation at the maximum-rated peak tube potential; and

(3) For all other diagnostic source assemblies, the maximum-rated continuous tube current for the maximum-rated continuous tube current for the maximum-rated peak tube potential.

Light field means that area of the intersection of the light beam from the beam-limiting device and one of the set of planes parallel to and including the plane of the image receptor, whose perimeter is the locus of points at which the illuminance is one-fourth of the maximum in the intersection.

Line-voltage regulation means the difference between the no-load and the load line potentials expressed as a percent of the load line potential; that is, Percent line-voltage regulation

$$= \frac{100(V_n - V_i)}{V_i}$$

where:

V_n = No-load line potential and
V_i = Load line potential.

Maximum line current means the root mean square current in the supply line of an x-ray machine operating at its maximum rating.

Movable tabletop means a tabletop which, when assembled for use, is capable of movement with respect to its supporting structure within the plane of the tabletop.

Peak tube potential means the maximum value of the potential difference across the x-ray tube during an exposure.

Primary protective barrier means the material, excluding filters, placed in the useful beam to reduce the radiation exposure for protection purposes.

Pulsed mode means operation of the x-ray system such that the x-ray tube current is pulsed by the x-ray control to produce one or more exposure intervals of duration less than one-half second.

Quick change x-ray tube means an x-ray tube designed for use in its associated tube housing such that:

(1) The tube cannot be inserted in its housing in a manner that would result in noncompliance of the system with the requirements of paragraphs (k) and (m) of this section;

(2) The focal spot position will not cause noncompliance with the provisions of this section or § 1020.31 or § 1020.32;

(3) The shielding within the tube housing cannot be displaced; and

(4) Any removal and subsequent replacement of a beam-limiting device during reloading of the tube in the tube housing will not result in noncompliance of the x-ray system with the applicable field limitation and alignment requirements of §§ 1020.31 and 1020.32.

Radiation therapy simulation system means a radiographic or fluoroscopic x-ray system intended for localizing the volume to be exposed during radiation therapy and confirming the position and size of the therapeutic irradiation field.

Rated line voltage means the range of potentials, in volts, of the supply line specified by the manufacturer at which the x-ray machine is designed to operate.

Rated output current means the maximum allowable load current of the x-ray high-voltage generator.

Rated output voltage means the allowable peak potential, in volts, at the output terminals of the x-ray high-voltage generator.

Rating means the operating limits specified by the manufacturer.

Recording means producing a permanent form of an image resulting from x-ray photons (e.g., film, videotape).

Scan means the complete process of collecting x-ray transmission data for the production of a tomogram. Data may be collected simultaneously during a single scan for the production of one or more tomograms.

Scan time means the period of time between the beginning and end of x-ray transmission data accumulation for a single scan.

Source means the focal spot of the x-ray tube.

Source-image receptor distance (SID) means the distance from the source to the center of the input surface of the image receptor.

Spot-film device means a device intended to transport and/or position a radiographic image receptor between the x-ray source and fluoroscopic image receptor. It includes a device intended to hold a cassette over the input end of an image intensifier for the purpose of a radiograph.

Stationary tabletop means a tabletop which, when assembled for use, is incapable of movement with respect to its supporting structure within the plane of the tabletop.

Technique factors means the following conditions of operation:

(1) For capacitor energy storage equipment, peak tube potential in kilovolts (kV) and quantity of charge in milliamperes-seconds (mAs);

(2) For field emission equipment rated for pulsed operation, peak tube potential in kV and number of x-ray pulses;

(3) For CT equipment designed for pulsed operation, peak tube potential in kV, scan time in seconds, and either tube current in milliamperes (mA), x-ray pulse width in seconds, and the number of x-ray pulses per scan, or the product of the tube current, x-ray pulse width, and the number of x-ray pulses in mAs;

(4) For CT equipment not designed for pulsed operation, peak tube potential in kV, and either tube current in mA and scan time in seconds, or the product of tube current and exposure time in mAs and the scan time when the scan time and exposure time are equivalent; and

(5) For all other equipment, peak tube potential in kV, and either tube current in mA and exposure time in seconds, or the product of tube current and exposure time in mAs.

Tomogram means the depiction of the x-ray attenuation properties of a section through a body.

Tube means an x-ray tube, unless otherwise specified.

Tube housing assembly means the tube housing with tube installed. It includes high-voltage and/or filament transformers and other appropriate elements when they are contained within the tube housing.

Tube rating chart means the set of curves which specify the rated limits of operation of the tube in terms of the technique factors.

Useful beam means the radiation which passes through the tube housing port and the aperture of the beam-limiting device when the exposure switch or timer is activated.

Variable-aperture beam-limiting device means a beam-limiting device which has the capacity for stepless adjustment of the x-ray field size at a given SID.

Visible area means the portion of the input surface of the image receptor over which incident x-ray photons are producing a visible image.

X-ray control means a device which controls input power to the x-ray high-voltage generator and/or the x-ray tube. It includes equipment such as timers, phototimers, automatic brightness stabilizers, and similar devices, which control the technique factors of an x-ray exposure.

X-ray equipment means an x-ray system, subsystem, or component thereof. Types of x-ray equipment are as follows:

(1) *Mobile x-ray equipment* means x-ray equipment mounted on a permanent base with wheels and/or casters for moving while completely assembled;

(2) *Portable x-ray equipment* means x-ray equipment designed to be hand-carried; and

(3) *Stationary x-ray equipment* means x-ray equipment which is installed in a fixed location.

X-ray field means that area of the intersection of the useful beam and any one of the set of planes parallel to and including the plane of the image receptor, whose perimeter is the locus of points at which the exposure rate is one-fourth of the maximum in the intersection.

X-ray high-voltage generator means a device which transforms electrical energy from the potential supplied by the x-ray control to the tube operating potential. The device may also include means for transforming alternating current to direct current, filament transformers for the x-ray tube(s), high-voltage switches, electrical protective devices, and other appropriate elements.

X-ray system means an assemblage of components for the controlled production of x-rays. It includes minimally an x-ray high-voltage generator, an x-ray control, a tube housing assembly, a beam-limiting device, and the necessary supporting structures. Additional components which function with the system are considered integral parts of the system.

X-ray subsystem means any combination of two or more components of an x-ray system for which there are requirements specified in this section and §§ 1020.31 and 1020.32.

X-ray table means a patient support device with its patient support structure (tabletop) interposed between the patient and the image receptor during radiography and/or fluoroscopy. This includes, but is not limited to, any stretcher equipped with a radiolucent panel and any table equipped with a cassette tray (or bucky), cassette tunnel, image intensifier, or spot-film device beneath the tabletop.

X-ray tube means any electron tube which is designed for the conversion of electrical energy into x-ray energy.

(c) *Manufacturers' responsibility*. Manufacturers of products subject to §§ 1020.30 through 1020.33 shall certify that each of their products meet all applicable requirements when installed into a diagnostic x-ray system according to instructions. This certification shall be made under the format specified in § 1010.2 of this chapter. Manufacturers may certify a combination of two or more components if they obtain prior authorization in writing from the Director of the Office of Compliance and Surveillance of the Center for Devices and Radiological Health. Manufacturers shall not be held responsible for noncompliance of their products if that noncompliance is due solely to the improper installation or assembly of that product by another person; however, manufacturers are responsible for providing assembly instructions adequate to assure compliance of their components with the applicable provisions of §§ 1020.30 through 1020.33.

(d) *Assemblers' responsibility*. An assembler who installs one or more components certified as required by paragraph (c) of this section shall install certified components that are of the type required by §§ 1020.31, 1020.32, or 1020.33 and shall assemble, install, adjust, and test the certified components according to the instructions of their respective manufacturers. Assemblers shall not be liable for noncompliance of a certified component if the assembly of that component was according to the component manufacturer's instruction.

(1) *Reports of assembly*. All assemblers who install certified components shall file a report of assembly, except as specified in paragraph (d)(2) of this section. The report will be construed as the assembler's certification and identification under §§ 1010.2 and 1010.3 of this chapter. The assembler shall affirm in the report that the manufacturer's instructions were followed in the assembly or that the certified components as assembled into the system meet all applicable requirements of §§ 1020.30 through 1020.33. All assembler reports must be on a form prescribed by and available from the Director, Center for Devices and Radiological Health, 9200 Corporate Blvd., Rockville, MD 20850. Completed reports must be submitted to the Director, the purchaser, and, where applicable, to the State agency responsible for radiation protection within 15 days following completion of the assembly.

(2) *Exceptions to reporting requirements*. Reports of assembly need not be submitted for any of the following:

(i) Reloaded or replacement tube housing assemblies that are reinstalled in or newly assembled into an existing x-ray system;

(ii) Certified accessory components that have been identified as such to the Center for Devices and Radiological Health in the report required under § 1002.10 of this chapter;

(iii) Repaired components, whether or not removed from the system and reinstalled during the course of repair, provided the original installation into the system was reported; or

(iv) Components installed temporarily in an x-ray system in place of components removed temporarily for repair, provided the temporarily installed component is identified by a tag or label bearing the following information:

Temporarily Installed Component

This certified component has been assembled, installed, adjusted, and tested by me according to the instructions provided by the manufacturer.

Signature
Company Name
Street Address, P.O. Box
City, State, Zip Code
Date of Installation

The replacement of the temporarily installed component by a component other than the component originally removed for repair shall be reported as specified in paragraph (d)(1) of this section.

(e) *Identification of x-ray components.* In addition to the identification requirements specified in § 1010.3 of this chapter, manufacturers of components subject to this section and §§ 1020.31, 1020.32, and 1020.33, except high-voltage generators contained within tube housings and beam-limiting devices that are integral parts of tube housings, shall permanently inscribe or affix thereon the model number and serial number of the product so that they are legible and accessible to view. The word "model" or "type" shall appear as part of the manufacturer's required identification of certified x-ray components. Where the certification of a system or subsystem, consisting of two or more components, has been authorized pursuant to paragraph (c) of this section, a single inscription, tag, or label bearing the model number and serial number may be used to identify the product.

(1) *Tube housing assemblies.* In a similar manner, manufacturers of tube housing assemblies shall also inscribe or affix thereon the name of the manufacturer, model number, and serial number of the x-ray tube which the tube housing assembly incorporates.

(2) *Replacement of tubes.* Except as specified in paragraph (e)(3) of this section, the replacement of an x-ray tube in a previously manufactured tube housing assembly certified pursuant to paragraph (c) of this section constitutes manufacture of a new tube housing assembly, and the manufacturer is subject to the provisions of paragraph (e)(1) of this section. The manufacturer shall remove, cover, or deface any previously affixed inscriptions, tags, or labels, that are no longer applicable.

(3) *Quick-change x-ray tubes.* The requirements of paragraph (e)(2) of this section shall not apply to tube housing assemblies designed and designated by their original manufacturer to contain quick change x-ray tubes. The manufacturer of quick-change x-ray tubes shall include with each replacement tube a label with the tube manufacturer's name, the model, and serial number of the x-ray tube. The manufacturer of the tube shall instruct the assembler who installs the new tube to attach the label to the tube housing assembly and to remove, cover, or deface the previously affixed inscriptions, tags, or labels that are described by the tube manufacturer as no longer applicable.

(f) [Reserved]

(g) *Information to be provided to assemblers.* Manufacturers of components listed in paragraph (a)(1) of this section shall provide to assemblers subject to paragraph (d) of this section and, upon request, to others at a cost not to exceed the cost of publication and distribution, instructions for assembly, installation, adjustment, and testing of such components adequate to assure that the products will comply with applicable provisions of this section and §§ 1020.31, 1020.32, and 1020.33, when assembled, installed, adjusted, and tested as directed. Such instructions shall include specifications of other components compatible with that to be installed when compliance of the system or subsystem depends on their compatibility. Such specifications may describe pertinent physical characteristics of the components and/or may list by manufacturer model number the components which are compatible. For x-ray controls and generators manufactured after May 3, 1994, manufacturers shall provide:

(1) A statement of the rated line voltage and the range of line-voltage regulation for operation at maximum line current;

(2) A statement of the maximum line current of the x-ray system based on the maximum input voltage and current characteristics of the tube housing assembly compatible with rated output voltage and rated output current characteristics of the x-ray control and associated high-voltage generator. If the rated input voltage and current characteristics of the tube housing assembly are not known by the manufacturer of the x-ray control and associated high-voltage generator, he shall provide necessary information to allow the assembler to determine the

maximum line current for the particular tube housing assembly(ies);

(3) A statement of the technique factors that constitute the maximum line current condition described in paragraph (g)(2) of this section.

(h) *Information to be provided to users.* Manufacturers of x-ray equipment shall provide to purchasers and, upon request, to others at a cost not to exceed the cost of publication and distribution, manuals or instruction sheets which shall include the following technical and safety information:

(1) *All x-ray equipment.* For x-ray equipment to which this section and §§ 1020.31, 1020.32, and 1020.33 are applicable, there shall be provided:

(i) Adequate instructions concerning any radiological safety procedures and precautions which may be necessary because of unique features of the equipment; and

(ii) A schedule of the maintenance necessary to keep the equipment in compliance with this section and §§ 1020.31, 1020.32, and 1020.33.

(2) *Tube housing assemblies.* For each tube housing assembly, there shall be provided:

(i) Statements of the leakage technique factors for all combinations of tube housing assemblies and beam-limiting devices for which the tube housing assembly manufacturer states compatibility, the minimum filtration permanently in the useful beam expressed as millimeters of aluminum equivalent, and the peak tube potential at which the aluminum equivalent was obtained;

(ii) Cooling curves for the anode and tube housing; and

(iii) *Tube rating charts.* If the tube is designed to operate from different types of x-ray high-voltage generators (such as single-phase self rectified, single-phase half-wave rectified, single-phase full-wave rectified, 3-phase 6-pulse, 3-phase 12-pulse, constant potential, capacitor energy storage) or under modes of operation such as alternate focal spot sizes or speeds of anode rotation which affect its rating, specific identification of the difference in ratings shall be noted.

(3) *X-ray controls and generators.* For the x-ray control and associated x-ray high-voltage generator, there shall be provided:

(i) A statement of the rated line voltage and the range of line-voltage regulation for operation at maximum line current;

(ii) A statement of the maximum line current of the x-ray system based on the maximum input voltage and output current characteristics of the tube housing assembly compatible with rated output voltage and rated current characteristics of the x-ray control and associated high-voltage generator. If the rated input voltage and current characteristics of the tube housing assembly are not known by the manufacturer of the x-ray control and associated high-voltage generator, the manufacturer shall provide necessary information to allow the purchaser to determine the maximum line current for his particular tube housing assembly(ies);

(iii) A statement of the technique factors that constitute the maximum line current condition described in paragraph (h)(3)(ii) of this section;

(iv) In the case of battery-powered generators, a specification of the minimum state of charge necessary for proper operation;

(v) Generator rating and duty cycle;

(vi) A statement of the maximum deviation from the preindication given by labeled technique factor control settings or indicators during any radiographic or CT exposure where the equipment is connected to a power supply as described in accordance with this paragraph. In the case of fixed technique factors, the maximum deviation from the nominal fixed value of each factor shall be stated;

(vii) A statement of the maximum deviation from the continuous indication of x-ray tube potential and current during any fluoroscopic exposure when the equipment is connected to a power supply as described in accordance with this paragraph; and

(viii) A statement describing the measurement criteria for all technique factors used in paragraphs (h)(3)(iii), (h)(3)(vi), and (h)(3)(vii) of this section; for example, the beginning and endpoints of exposure time measured with respect to a certain percentage of the voltage waveform.

(4) *Beam-limiting device.* For each variable-aperture beam-limiting device, there shall be provided;

(i) Leakage technique factors for all combinations of tube housing assemblies and beam-limiting devices for which the beam-limiting device manufacturer states compatibility; and

(ii) A statement including the minimum aluminum equivalent of that part of the device through which the useful beam passes and including the x-ray tube potential at which the aluminum equivalent was obtained. When two or more filters are provided as part of the device, the statement shall include the aluminum equivalent of each filter.

(i) [Reserved]

(j) *Warning label.* The control panel containing the main power switch shall bear the warning statement, legible and accessible to view:

"Warning: This x-ray unit may be dangerous to patient and operator unless safe exposure factors and operating instructions are observed."

(k) *Leakage radiation from the diagnostic source assembly.* The leakage radiation from the diagnostic source assembly measured at a distance of 1 meter in any direction from the source shall not exceed 2.58×10^{-5} coulombs per kilogram (C/kg) (100 milliroentgens (mR)) in 1 hour when the x-ray tube is operated at the leakage technique factors. If the maximum rated peak tube potential of the tube housing assembly is greater than the maximum rated peak tube potential for the diagnostic source assembly, positive means shall be provided to limit the maximum x-ray tube potential to that of the diagnostic source assembly. Compliance shall be determined by measurements averaged over an area of 100 square centimeters with no linear dimension greater than 20 centimeters.

(l) *Radiation from components other than the diagnostic source assembly.* The radiation emitted by a component other than the diagnostic source assembly shall not exceed 5.16×10^{-7} C/kg (2 mR) in 1 hour at 5 centimeters from any accessible surface of the component when it is operated in an assembled x-ray system under any conditions for which it was designed. Compliance shall be determined by measurements averaged over an area of 100 square centimeters with no linear dimension greater than 20 centimeters.

(m) *Beam quality*—(1) *Half-value layer.* The half-value layer (HVL) of the useful beam for a given x-ray tube potential shall not be less than the appropriate value shown in table I under "Specified dental systems," for any dental x-ray system designed for use with intraoral image receptors and manufactured after December 1, 1980; and under "Other x-ray systems," for all other x-ray systems subject to this section. If it is necessary to determine such HVL at an x-ray tube potential which is not listed in table I, linear interpolation or extrapolation may be made. Positive means[2] shall be provided to insure that at least the minimum filtration needed to achieve the above beam quality requirements is in the useful beam during each exposure.

TABLE I

X-ray tube voltage (kilovolt peak)		Minimum HVL (millimeters of aluminum)	
Designed operating range	Measured operating potential	Specified dental systems	Other X-ray systems
Below 51	30	1.5	0.3
	40	1.5	0.4
	50	1.5	0.5
51 to 70	51	1.5	1.2
	60	1.5	1.3
	70	1.5	1.5
Above 70	71	2.1	2.1
	80	2.3	2.3
	90	2.5	2.5
	100	2.7	2.7
	110	3.0	3.0
	120	3.2	3.2
	130	3.5	3.5
	140	3.8	3.8
	150	4.1	4.1

(2) *Measuring compliance.* For capacitor energy storage equipment, compliance shall be determined with the maximum selectable quantity of charge per exposure.

(n) *Aluminum equivalent of material between patient and image receptor.* Except when used in a CT x-ray system, the aluminum equivalent of each of the items listed in table II, which are used

[2] In the case of a system which is to be operated with more than one thickness of filtration, this requirement can be met by a filter interlock with the kilovoltage selector which will prevent x-ray emission if the minimum required filtration is not in place.

between the patient and image receptor, may not exceed the indicated limits. Compliance shall be determined by x-ray measurements made at a potential of 100 kilovolts peak and with an x-ray beam that has a HVL of 2.7 millimeters of aluminum. This requirement applies to front panel(s) of cassette holders and film changers provided by the manufacturer for patient support or for prevention of foreign object intrusions. It does not apply to screens and their associated mechanical support panels or grids.

TABLE II

Item	Aluminum equivalent (millimeters)
Front panel(s) of cassette holder (total of all)	1.0
Front panel(s) of film changer (total of all)	1.0
Cradle	2.0
Tabletop, stationary, without articulated joint(s)	1.0
Tabletop, movable, without articulated joint(s) (including stationary subtop)	1.5
Tabletop, with radiolucent panel having one articulated joint	1.5
Tabletop, with radiolucent panel having two or more articulated joints	2.0
Tabletop, cantilevered	2.0
Tabletop, radiation therapy simulator	5.0

(o) *Battery charge indicator.* On battery-powered generators, visual means shall be provided on the control panel to indicate whether the battery is in a state of charge adequate for proper operation.

(p) [Reserved]

(q) *Modification of certified diagnostic x-ray components and systems*—(1) Diagnostic x-ray components and systems certified in accordance with § 1010.2 of this chapter shall not be modified such that the component or system fails to comply with any applicable provision of this chapter unless a variance in accordance with § 1010.4 of this chapter or an exemption under sections 358(a)(5) or 360B(b) of the Public Health Service Act has been granted.

(2) The owner of a diagnostic x-ray system who uses the system in a professional or commercial capacity may modify the system, provided the modification does not result in the failure of the system or component to comply with the applicable requirements of this section or of § 1020.31, § 1020.32, or § 1020.33. The owner who causes such modification need not submit the reports required by subpart B of part 1002 of this chapter, provided the owner records the date and the details of the modification, and provided the modification of the x-ray system does not result in a failure to comply with § 1020.31, § 1020.32, or § 1020.33.

[58 FR 26396, May 3, 1993, as amended at 59 FR 26403, May 19, 1994; 64 FR 35927, July 2, 1999; 65 FR 17138, Mar. 31, 2000]

EFFECTIVE DATE NOTE: At 70 FR 34028, June 10, 2005, § 1020.30 was revised, effective June 10, 2006. For the convenience of the user the revised text is set forth as follows:

§ 1020.30 Diagnostic x-ray systems and their major components.

(a) *Applicability.* (1) The provisions of this section are applicable to:

(i) The following components of diagnostic x-ray systems:

(A) Tube housing assemblies, x-ray controls, x-ray high-voltage generators, x-ray tables, cradles, film changers, vertical cassette holders mounted in a fixed location and cassette holders with front panels, and beam-limiting devices manufactured after August 1, 1974.

(B) Fluoroscopic imaging assemblies manufactured after August 1, 1974, and before April 26, 1977, or after June 10, 2006.

(C) Spot-film devices and image intensifiers manufactured after April 26, 1977.

(D) Cephalometric devices manufactured after February 25, 1978.

(E) Image receptor support devices for mammographic x-ray systems manufactured after September 5, 1978.

(F) Image receptors that are electrically powered or connected with the x-ray system manufactured on or after June 10, 2006.

(G) Fluoroscopic air kerma display devices manufactured on or after June 10, 2006.

(ii) Diagnostic x-ray systems, except computed tomography x-ray systems, incorporating one or more of such components; however, such x-ray systems shall be required to comply only with those provisions of this section and §§ 1020.31 and 1020.32, which relate to the components certified in accordance with paragraph (c) of this section and installed into the systems.

(iii) Computed tomography (CT) x-ray systems manufactured before November 29, 1984.

(iv) CT gantries manufactured after September 3, 1985.

(2) The following provisions of this section and § 1020.33 are applicable to CT x-ray systems manufactured or remanufactured on or after November 29, 1984:

(i) Section 1020.30(a);

(ii) Section 1020.30(b) "Technique factors";

(iii) Section 1020.30(b) "CT," "Dose," "Scan," "Scan time," and "Tomogram";

(iv) Section 1020.30(h)(3)(vi) through (h)(3)(viii);

(v) Section 1020.30(n);

(vi) Section 1020.33(a) and (b);

(vii) Section 1020.33(c)(1) as it affects § 1020.33(c)(2); and

(viii) Section 1020.33(c)(2).

(3) The provisions of this section and § 1020.33 in its entirety, including those provisions in paragraph (a)(2) of this section, are applicable to CT x-ray systems manufactured or remanufactured on or after September 3, 1985. The date of manufacture of the CT system is the date of manufacture of the CT gantry.

(b) *Definitions.* As used in this section and §§ 1020.31, 1020.32, and 1020.33, the following definitions apply:

Accessible surface means the external surface of the enclosure or housing provided by the manufacturer.

Accessory component means:

(1) A component used with diagnostic x-ray systems, such as a cradle or film changer, that is not necessary for the compliance of the system with applicable provisions of this subchapter but which requires an initial determination of compatibility with the system; or

(2) A component necessary for compliance of the system with applicable provisions of this subchapter but which may be interchanged with similar compatible components without affecting the system's compliance, such as one of a set of interchangeable beam-limiting devices; or

(3) A component compatible with all x-ray systems with which it may be used and that does not require compatibility or installation instructions, such as a tabletop cassette holder.

Air kerma means kerma in air (see definition of *Kerma*).

Air kerma rate (AKR) means the air kerma per unit time.

Aluminum equivalent means the thickness of aluminum (type 1100 alloy)[1] affording the same attenuation, under specified conditions, as the material in question.

Articulated joint means a joint between two separate sections of a tabletop which joint provides the capacity for one of the sections to pivot on the line segment along which the sections join.

Assembler means any person engaged in the business of assembling, replacing, or installing one or more components into a diagnostic x-ray system or subsystem. The term includes the owner of an x-ray system or his or her employee or agent who assembles components into an x-ray system that is subsequently used to provide professional or commercial services.

Attenuation block means a block or stack of type 1100 aluminum alloy, or aluminum alloy having equivalent attenuation, with dimensions 20 centimeters (cm) or larger by 20 cm or larger by 3.8 cm, that is large enough to intercept the entire x-ray beam.

Automatic exposure control (AEC) means a device which automatically controls one or more technique factors in order to obtain at a preselected location(s) a required quantity of radiation.

Automatic exposure rate control (AERC) means a device which automatically controls one or more technique factors in order to obtain at a preselected location(s) a required quantity of radiation per unit time.

Beam axis means a line from the source through the centers of the x-ray fields.

Beam-limiting device means a device which provides a means to restrict the dimensions of the x-ray field.

C-arm fluoroscope means a fluoroscopic x-ray system in which the image receptor and the x-ray tube housing assembly are connected or coordinated to maintain a spatial relationship. Such a system allows a change in the direction of the beam axis with respect to the patient without moving the patient.

Cantilevered tabletop means a tabletop designed such that the unsupported portion can be extended at least 100 cm beyond the support.

Cassette holder means a device, other than a spot-film device, that supports and/or fixes the position of an x-ray film cassette during an x-ray exposure.

Cephalometric device means a device intended for the radiographic visualization and measurement of the dimensions of the human head.

Coefficient of variation means the ratio of the standard deviation to the mean value of a population of observations. It is estimated using the following equation:

[1] The nominal chemical composition of type 1100 aluminum alloy is 99.00 percent minimum aluminum, 0.12 percent copper, as given in "Aluminum Standards and Data" (1969). Copies may be obtained from The Aluminum Association, New York, NY.

$$C = \frac{s}{\overline{X}} = \frac{1}{\overline{X}}\left[\sum_{i=1}^{n}\frac{\left(X_i - \overline{X}\right)^2}{n-1}\right]^{1/2}$$

where:

s = Estimated standard deviation of the population.
$\bar{X}$ = Mean value of observations in sample.
X_i = ith observation sampled.
n = Number of observations sampled.

Computed tomography (CT) means the production of a tomogram by the acquisition and computer processing of x-ray transmission data.

Control panel means that part of the x-ray control upon which are mounted the switches, knobs, pushbuttons, and other hardware necessary for manually setting the technique factors.

Cooling curve means the graphical relationship between heat units stored and cooling time.

Cradle means:

(1) A removable device which supports and may restrain a patient above an x-ray table; or

(2) A device;

(i) Whose patient support structure is interposed between the patient and the image receptor during normal use;

(ii) Which is equipped with means for patient restraint; and

(iii) Which is capable of rotation about its long (longitudinal) axis.

CT gantry means tube housing assemblies, beam-limiting devices, detectors, and the supporting structures, frames, and covers which hold and/or enclose these components.

Cumulative air kerma means the total air kerma accrued from the beginning of an examination or procedure and includes all contributions from fluoroscopic and radiographic irradiation.

Diagnostic source assembly means the tube housing assembly with a beam-limiting device attached.

Diagnostic x-ray system means an x-ray system designed for irradiation of any part of the human body for the purpose of diagnosis or visualization.

Dose means the absorbed dose as defined by the International Commission on Radiation Units and Measurements. The absorbed dose, D, is the quotient of de by dm, where de is the mean energy imparted to matter of mass dm; thus D=de/dm, in units of J/kg, where the special name for the unit of absorbed dose is gray (Gy).

Equipment means x-ray equipment.

Exposure (X) means the quotient of dQ by dm where dQ is the absolute value of the total charge of the ions of one sign produced in air when all the electrons and positrons liberated or created by photons in air of mass dm are completely stopped in air; thus X=dQ/dm, in units of C/kg. A second meaning of exposure is the process or condition during which the x-ray tube produces x-ray radiation.

Field emission equipment means equipment which uses an x-ray tube in which electron emission from the cathode is due solely to action of an electric field.

Fluoroscopic air kerma display device means a device, subsystem, or component that provides the display of AKR and cumulative air kerma required by § 1020.32(k). It includes radiation detectors, if any, electronic and computer components, associated software, and data displays.

Fluoroscopic imaging assembly means a subsystem in which x-ray photons produce a set of fluoroscopic images or radiographic images recorded from the fluoroscopic image receptor. It includes the image receptor(s), electrical interlocks, if any, and structural material providing linkage between the image receptor and diagnostic source assembly.

Fluoroscopic irradiation time means the cumulative duration during an examination or procedure of operator-applied continuous pressure to the device, enabling x-ray tube activation in any fluoroscopic mode of operation.

Fluoroscopy means a technique for generating x-ray images and presenting them simultaneously and continuously as visible images. This term has the same meaning as the term "radioscopy" in the standards of the International Electrotechnical Commission.

General purpose radiographic x-ray system means any radiographic x-ray system which, by design, is not limited to radiographic examination of specific anatomical regions.

Half-value layer (HVL) means the thickness of specified material which attenuates the beam of radiation to an extent such that the AKR is reduced to one-half of its original value. In this definition the contribution of all scattered radiation, other than any which might be present initially in the beam concerned, is deemed to be excluded.

Image intensifier means a device, installed in its housing, which instantaneously converts an x-ray pattern into a corresponding light image of higher energy density.

Image receptor means any device, such as a fluorescent screen, radiographic film, x-ray image intensifier tube, solid-state detector, or gaseous detector, which transforms incident x-ray photons either into a visible image or into another form which can be made into a visible image by further transformations. In those cases where means are provided to preselect a portion of the image receptor, the term "image receptor" shall mean the preselected portion of the device.

Image receptor support device means, for mammography x-ray systems, that part of the system designed to support the image receptor during a mammographic examination and to provide a primary protective barrier.

Isocenter means the center of the smallest sphere through which the beam axis passes when the equipment moves through a full range of rotations about its common center.

Kerma means the quantity as defined by the International Commission on Radiation Units and Measurements. The kerma, K, is the quotient of dE_{tr} by dm, where dE_{tr} is the sum of the initial kinetic energies of all the charged particles liberated by uncharged particles in a mass dm of material; thus $K=dE_{tr}/dm$, in units of J/kg, where the special name for the unit of kerma is gray (Gy). When the material is air, the quantity is referred to as "air kerma."

Last-image-hold (LIH) radiograph means an image obtained either by retaining one or more fluoroscopic images, which may be temporally integrated, at the end of a fluoroscopic exposure or by initiating a separate and distinct radiographic exposure automatically and immediately in conjunction with termination of the fluoroscopic exposure.

Lateral fluoroscope means the x-ray tube and image receptor combination in a biplane system dedicated to the lateral projection. It consists of the lateral x-ray tube housing assembly and the lateral image receptor that are fixed in position relative to the table with the x-ray beam axis parallel to the plane of the table.

Leakage radiation means radiation emanating from the diagnostic source assembly except for:

(1) The useful beam; and

(2) Radiation produced when the exposure switch or timer is not activated.

Leakage technique factors means the technique factors associated with the diagnostic source assembly which are used in measuring leakage radiation. They are defined as follows:

(1) For diagnostic source assemblies intended for capacitor energy storage equipment, the maximum-rated peak tube potential and the maximum-rated number of exposures in an hour for operation at the maximum-rated peak tube potential with the quantity of charge per exposure being 10 millicoulombs (or 10 mAs) or the minimum obtainable from the unit, whichever is larger;

(2) For diagnostic source assemblies intended for field emission equipment rated for pulsed operation, the maximum-rated peak tube potential and the maximum-rated number of x-ray pulses in an hour for operation at the maximum-rated peak tube potential; and

(3) For all other diagnostic source assemblies, the maximum-rated peak tube potential and the maximum-rated continuous tube current for the maximum-rated peak tube potential.

Light field means that area of the intersection of the light beam from the beam-limiting device and one of the set of planes parallel to and including the plane of the image receptor, whose perimeter is the locus of points at which the illuminance is one-fourth of the maximum in the intersection.

Line-voltage regulation means the difference between the no-load and the load line potentials expressed as a percent of the load line potential; that is,

Percent line-voltage regulation = $100(V_n - V_l)/V_l$

where:

V_n = No-load line potential and
V_l = Load line potential.

Maximum line current means the root mean square current in the supply line of an x-ray machine operating at its maximum rating.

Mode of operation means, for fluoroscopic systems, a distinct method of fluoroscopy or radiography provided by the manufacturer and selected with a set of several technique factors or other control settings uniquely associated with the mode. The set of distinct technique factors and control settings for the mode may be selected by the operation of a single control. Examples of distinct modes of operation include normal fluoroscopy (analog or digital), high-level control fluoroscopy, cineradiography (analog or digital), digital subtraction angiography, electronic radiography using the fluoroscopic image receptor, and photospot recording. In a specific mode of operation, certain system variables affecting air kerma, AKR, or image quality, such as image magnification, x-ray field size, pulse rate, pulse duration, number of pulses, source-image receptor distance (SID), or optical aperture, may be adjustable or may vary; their variation per se does not comprise a mode of operation different from the one that has been selected.

Movable tabletop means a tabletop which, when assembled for use, is capable of movement with respect to its supporting structure within the plane of the tabletop.

Non-image-intensified fluoroscopy means fluoroscopy using only a fluorescent screen.

Peak tube potential means the maximum value of the potential difference across the x-ray tube during an exposure.

Primary protective barrier means the material, excluding filters, placed in the useful beam to reduce the radiation exposure for protection purposes.

Pulsed mode means operation of the x-ray system such that the x-ray tube current is pulsed by the x-ray control to produce one or more exposure intervals of duration less than one-half second.

Quick change x-ray tube means an x-ray tube designed for use in its associated tube housing such that:

(1) The tube cannot be inserted in its housing in a manner that would result in noncompliance of the system with the requirements of paragraphs (k) and (m) of this section;

(2) The focal spot position will not cause noncompliance with the provisions of this section or § 1020.31 or 1020.32;

(3) The shielding within the tube housing cannot be displaced; and

(4) Any removal and subsequent replacement of a beam-limiting device during reloading of the tube in the tube housing will not result in noncompliance of the x-ray system with the applicable field limitation and alignment requirements of §§ 1020.31 and 1020.32.

Radiation therapy simulation system means a radiographic or fluoroscopic x-ray system intended for localizing the volume to be exposed during radiation therapy and confirming the position and size of the therapeutic irradiation field.

Radiography means a technique for generating and recording an x-ray pattern for the purpose of providing the user with an image(s) after termination of the exposure.

Rated line voltage means the range of potentials, in volts, of the supply line specified by the manufacturer at which the x-ray machine is designed to operate.

Rated output current means the maximum allowable load current of the x-ray high-voltage generator.

Rated output voltage means the allowable peak potential, in volts, at the output terminals of the x-ray high-voltage generator.

Rating means the operating limits specified by the manufacturer.

Recording means producing a retrievable form of an image resulting from x-ray photons.

Scan means the complete process of collecting x-ray transmission data for the production of a tomogram. Data may be collected simultaneously during a single scan for the production of one or more tomograms.

Scan time means the period of time between the beginning and end of x-ray transmission data accumulation for a single scan.

Solid state x-ray imaging device means an assembly, typically in a rectangular panel configuration, that intercepts x-ray photons and converts the photon energy into a modulated electronic signal representative of the x-ray intensity over the area of the imaging device. The electronic signal is then used to create an image for display and/or storage.

Source means the focal spot of the x-ray tube.

Source-image receptor distance (SID) means the distance from the source to the center of the input surface of the image receptor.

Source-skin distance (SSD) means the distance from the source to the center of the entrant x-ray field in the plane tangent to the patient skin surface.

Spot-film device means a device intended to transport and/or position a radiographic image receptor between the x-ray source and fluoroscopic image receptor. It includes a device intended to hold a cassette over the input end of the fluoroscopic image receptor for the purpose of producing a radiograph.

Stationary tabletop means a tabletop which, when assembled for use, is incapable of movement with respect to its supporting structure within the plane of the tabletop.

Technique factors means the following conditions of operation:

(1) For capacitor energy storage equipment, peak tube potential in kilovolts (kV) and quantity of charge in milliampere-seconds (mAs);

(2) For field emission equipment rated for pulsed operation, peak tube potential in kV and number of x-ray pulses;

(3) For CT equipment designed for pulsed operation, peak tube potential in kV, scan time in seconds, and either tube current in milliamperes (mA), x-ray pulse width in seconds, and the number of x-ray pulses per scan, or the product of the tube current, x-ray pulse width, and the number of x-ray pulses in mAs;

(4) For CT equipment not designed for pulsed operation, peak tube potential in kV, and either tube current in mA and scan time in seconds, or the product of tube current and exposure time in mAs and the scan time when the scan time and exposure time are equivalent; and

(5) For all other equipment, peak tube potential in kV, and either tube current in mA and exposure time in seconds, or the product of tube current and exposure time in mAs.

Tomogram means the depiction of the x-ray attenuation properties of a section through a body.

Tube means an x-ray tube, unless otherwise specified.

Tube housing assembly means the tube housing with tube installed. It includes high-voltage and/or filament transformers and other appropriate elements when they are contained within the tube housing.

Tube rating chart means the set of curves which specify the rated limits of operation of the tube in terms of the technique factors.

Useful beam means the radiation which passes through the tube housing port and the aperture of the beam-limiting device when the exposure switch or timer is activated.

Variable-aperture beam-limiting device means a beam-limiting device which has the capacity for stepless adjustment of the x-ray field size at a given SID.

Visible area means the portion of the input surface of the image receptor over which incident x-ray photons are producing a visible image.

X-ray control means a device which controls input power to the x-ray high-voltage generator and/or the x-ray tube. It includes equipment such as timers, phototimers, automatic brightness stabilizers, and similar devices, which control the technique factors of an x-ray exposure.

X-ray equipment means an x-ray system, subsystem, or component thereof. Types of x-ray equipment are as follows:

(1) *Mobile x-ray equipment* means x-ray equipment mounted on a permanent base with wheels and/or casters for moving while completely assembled;

(2) *Portable x-ray equipment* means x-ray equipment designed to be hand-carried; and

(3) *Stationary x-ray equipment means* x-ray equipment which is installed in a fixed location.

X-ray field means that area of the intersection of the useful beam and any one of the set of planes parallel to and including the plane of the image receptor, whose perimeter is the locus of points at which the AKR is one-fourth of the maximum in the intersection.

X-ray high-voltage generator means a device which transforms electrical energy from the potential supplied by the x-ray control to the tube operating potential. The device may also include means for transforming alternating current to direct current, filament transformers for the x-ray tube(s), high-voltage switches, electrical protective devices, and other appropriate elements.

X-ray subsystem means any combination of two or more components of an x-ray system for which there are requirements specified in this section and §§ 1020.31 and 1020.32.

X-ray system means an assemblage of components for the controlled production of x-rays. It includes minimally an x-ray high-voltage generator, an x-ray control, a tube housing assembly, a beam-limiting device, and the necessary supporting structures. Additional components which function with the system are considered integral parts of the system.

X-ray table means a patient support device with its patient support structure (tabletop) interposed between the patient and the image receptor during radiography and/or fluoroscopy. This includes, but is not limited to, any stretcher equipped with a radiolucent panel and any table equipped with a cassette tray (or bucky), cassette tunnel, fluoroscopic image receptor, or spot-film device beneath the tabletop.

X-ray tube means any electron tube which is designed for the conversion of electrical energy into x-ray energy.

(c) *Manufacturers' responsibility*. Manufacturers of products subject to §§ 1020.30 through 1020.33 shall certify that each of their products meet all applicable requirements when installed into a diagnostic x-ray system according to instructions. This certification shall be made under the format specified in § 1010.2 of this chapter. Manufacturers may certify a combination of two or more components if they obtain prior authorization in writing from the Director of the Office of Compliance of the Center for Devices and Radiological Health (CDRH). Manufacturers shall not be held responsible for noncompliance of their products if that noncompliance is due solely to the improper installation or assembly of that product by another person; however, manufacturers are responsible for providing assembly instructions adequate to assure compliance of their components with the applicable provisions of §§ 1020.30 through 1020.33.

(d) *Assemblers' responsibility*. An assembler who installs one or more components certified as required by paragraph (c) of this section shall install certified components that are of the type required by § 1020.31, 1020.32, or 1020.33 and shall assemble, install, adjust, and test the certified components according to the instructions of their respective manufacturers. Assemblers shall not be liable for noncompliance of a certified component if the assembly of that component was according to the component manufacturer's instruction.

(1) *Reports of assembly*. All assemblers who install certified components shall file a report of assembly, except as specified in paragraph (d)(2) of this section. The report will be construed as the assembler's certification and identification under §§ 1010.2 and 1010.3 of this chapter. The assembler shall affirm in the report that the manufacturer's instructions were followed in the assembly or that the certified components as assembled into the system meet all applicable requirements of §§ 1020.30 through 1020.33. All assembler reports must be on a form prescribed by the Director, CDRH. Completed reports must be submitted to the Director, the purchaser, and, where applicable, to the State agency responsible for radiation protection within 15 days following completion of the assembly.

(2) *Exceptions to reporting requirements*. Reports of assembly need not be submitted for any of the following:

(i) Reloaded or replacement tube housing assemblies that are reinstalled in or newly assembled into an existing x-ray system;

(ii) Certified accessory components that have been identified as such to CDRH in the report required under § 1002.10 of this chapter;

(iii) Repaired components, whether or not removed from the system and reinstalled during the course of repair, provided the original installation into the system was reported; or

(iv)(A) Components installed temporarily in an x-ray system in place of components removed temporarily for repair, provided the temporarily installed component is identified by a tag or label bearing the following information:

Temporarily Installed Component
This certified component has been assembled, installed, adjusted, and tested by me according to the instructions provided by the manufacturer.
Signature
Company Name
Street Address, P.O. Box
City, State, Zip Code
Date of Installation

(B) The replacement of the temporarily installed component by a component other than the component originally removed for repair shall be reported as specified in paragraph (d)(1) of this section.

(e) *Identification of x-ray components.* In addition to the identification requirements specified in § 1010.3 of this chapter, manufacturers of components subject to this section and §§ 1020.31, 1020.32, and 1020.33, except high-voltage generators contained within tube housings and beam-limiting devices that are integral parts of tube housings, shall permanently inscribe or affix thereon the model number and serial number of the product so that they are legible and accessible to view. The word "model" or "type" shall appear as part of the manufacturer's required identification of certified x-ray components. Where the certification of a system or subsystem, consisting of two or more components, has been authorized under paragraph (c) of this section, a single inscription, tag, or label bearing the model number and serial number may be used to identify the product.

(1) *Tube housing assemblies.* In a similar manner, manufacturers of tube housing assemblies shall also inscribe or affix thereon the name of the manufacturer, model number, and serial number of the x-ray tube which the tube housing assembly incorporates.

(2) *Replacement of tubes.* Except as specified in paragraph (e)(3) of this section, the replacement of an x-ray tube in a previously manufactured tube housing assembly certified under paragraph (c) of this section constitutes manufacture of a new tube housing assembly, and the manufacturer is subject to the provisions of paragraph (e)(1) of this section. The manufacturer shall remove, cover, or deface any previously affixed inscriptions, tags, or labels that are no longer applicable.

(3) *Quick-change x-ray tubes.* The requirements of paragraph (e)(2) of this section shall not apply to tube housing assemblies designed and designated by their original manufacturer to contain quick change x-ray tubes. The manufacturer of quick-change x-ray tubes shall include with each replacement tube a label with the tube manufacturer's name, the model, and serial number of the x-ray tube. The manufacturer of the tube shall instruct the assembler who installs the new tube to attach the label to the tube housing assembly and to remove, cover, or deface the previously affixed inscriptions, tags, or labels that are described by the tube manufacturer as no longer applicable.

(f) [Reserved]

(g) *Information to be provided to assemblers.* Manufacturers of components listed in paragraph (a)(1) of this section shall provide to assemblers subject to paragraph (d) of this section and, upon request, to others at a cost not to exceed the cost of publication and distribution, instructions for assembly, installation, adjustment, and testing of such components adequate to assure that the products will comply with applicable provisions of this section and §§ 1020.31, 1020.32, and 1020.33, when assembled, installed, adjusted, and tested as directed. Such instructions shall include specifications of other components compatible with that to be installed when compliance of the system or subsystem depends on their compatibility. Such specifications may describe pertinent physical characteristics of the components and/or may list by manufacturer model number the components which are compatible. For x-ray controls and generators manufactured after May 3, 1994, manufacturers shall provide:

(1) A statement of the rated line voltage and the range of line-voltage regulation for operation at maximum line current;

(2) A statement of the maximum line current of the x-ray system based on the maximum input voltage and current characteristics of the tube housing assembly compatible with rated output voltage and rated output current characteristics of the x-ray control and associated high-voltage generator. If the rated input voltage and current characteristics of the tube housing assembly are not known by the manufacturer of the x-ray control and associated high-voltage generator, the manufacturer shall provide information necessary to allow the assembler to determine the maximum line current for the particular tube housing assembly(ies);

(3) A statement of the technique factors that constitute the maximum line current condition described in paragraph (g)(2) of this section.

(h) *Information to be provided to users.* Manufacturers of x-ray equipment shall provide to purchasers and, upon request, to others at a cost not to exceed the cost of publication and distribution, manuals or instruction sheets which shall include the following technical and safety information:

(1) *All x-ray equipment.* For x-ray equipment to which this section and §§ 1020.31, 1020.32, and 1020.33 are applicable, there shall be provided:

(i) Adequate instructions concerning any radiological safety procedures and precautions which may be necessary because of unique features of the equipment; and

(ii) A schedule of the maintenance necessary to keep the equipment in compliance with this section and §§ 1020.31, 1020.32, and 1020.33.

(2) *Tube housing assemblies.* For each tube housing assembly, there shall be provided:

(i) Statements of the leakage technique factors for all combinations of tube housing assemblies and beam-limiting devices for which the tube housing assembly manufacturer states compatibility, the minimum filtration permanently in the useful beam expressed as millimeters (mm) of aluminum equivalent, and the peak tube potential at which the aluminum equivalent was obtained;

(ii) Cooling curves for the anode and tube housing; and

(iii) Tube rating charts. If the tube is designed to operate from different types of x-ray high-voltage generators (such as single-phase self rectified, single-phase half-wave rectified, single-phase full-wave rectified, 3-phase 6-pulse, 3-phase 12-pulse, constant potential, capacitor energy storage) or under modes of operation such as alternate focal spot sizes or speeds of anode rotation which affect its rating, specific identification of the difference in ratings shall be noted.

(3) *X-ray controls and generators.* For the x-ray control and associated x-ray high-voltage generator, there shall be provided:

(i) A statement of the rated line voltage and the range of line-voltage regulation for operation at maximum line current;

(ii) A statement of the maximum line current of the x-ray system based on the maximum input voltage and output current characteristics of the tube housing assembly compatible with rated output voltage and rated current characteristics of the x-ray control and associated high-voltage generator. If the rated input voltage and current characteristics of the tube housing assembly are not known by the manufacturer of the x-ray control and associated high-voltage generator, the manufacturer shall provide necessary information to allow the purchaser to determine the maximum line current for his particular tube housing assembly(ies);

(iii) A statement of the technique factors that constitute the maximum line current condition described in paragraph (h)(3)(ii) of this section;

(iv) In the case of battery-powered generators, a specification of the minimum state of charge necessary for proper operation;

(v) Generator rating and duty cycle;

(vi) A statement of the maximum deviation from the preindication given by labeled technique factor control settings or indicators during any radiographic or CT exposure where the equipment is connected to a power supply as described in accordance with this paragraph. In the case of fixed technique factors, the maximum deviation from the nominal fixed value of each factor shall be stated;

(vii) A statement of the maximum deviation from the continuous indication of x-ray tube potential and current during any fluoroscopic exposure when the equipment is connected to a power supply as described in accordance with this paragraph; and

(viii) A statement describing the measurement criteria for all technique factors used in paragraphs (h)(3)(iii), (h)(3)(vi), and (h)(3)(vii) of this section; for example, the beginning and endpoints of exposure time measured with respect to a certain percentage of the voltage waveform.

(4) *Beam-limiting device.* For each variable-aperture beam-limiting device, there shall be provided;

(i) Leakage technique factors for all combinations of tube housing assemblies and beam-limiting devices for which the beam-limiting device manufacturer states compatibility; and

(ii) A statement including the minimum aluminum equivalent of that part of the device through which the useful beam passes and including the x-ray tube potential at which the aluminum equivalent was obtained. When two or more filters are provided as part of the device, the statement shall include the aluminum equivalent of each filter.

(5) *Imaging system information.* For x-ray systems manufactured on or after June 10, 2006, that produce images using the fluoroscopic image receptor, the following information shall be provided in a separate, single section of the user's instruction manual or in a separate manual devoted to this information:

(i) For each mode of operation, a description of the mode and detailed instructions on how the mode is engaged and disengaged. The description of the mode shall identify those technique factors and system controls that are fixed or automatically adjusted by selection of the mode of operation, including the manner in which the automatic adjustment is controlled. This information shall include how the operator can recognize which mode of operation has been selected prior to initiation of x-ray production.

(ii) For each mode of operation, a descriptive example(s) of any specific clinical procedure(s) or imaging task(s) for which the mode is recommended or designed and how each mode should be used. Such recommendations do not preclude other clinical uses.

(6) *Displays of values of AKR and cumulative air kerma.* For fluoroscopic x-ray systems manufactured on or after June 10, 2006, the following shall be provided:

(i) A schedule of maintenance for any system instrumentation associated with the display of air kerma information necessary to maintain the displays of AKR and cumulative air kerma within the limits of allowed uncertainty specified by § 1020.32(k)(6) and, if the capability for user calibration of the display is provided, adequate instructions for such calibration;

(ii) Identification of the distances along the beam axis:

(A) From the focal spot to the isocenter, and

(B) From the focal spot to the reference location to which displayed values of AKR and cumulative air kerma refer according to § 1020.32(k)(4);

(iii) A rationale for specification of a reference irradiation location alternative to 15 cm from the isocenter toward the x-ray source along the beam axis when such alternative specification is made according to § 1020.32(k)(4)(ii).

(i) [Reserved]

(j) *Warning label.* The control panel containing the main power switch shall bear the warning statement, legible and accessible to view:

"Warning: This x-ray unit may be dangerous to patient and operator unless safe exposure factors, operating instructions and maintenance schedules are observed."

(k) *Leakage radiation from the diagnostic source assembly.* The leakage radiation from the diagnostic source assembly measured at a distance of 1 meter in any direction from the source shall not exceed 0.88 milligray (mGy) air kerma (vice 100 milliroentgen (mR) exposure) in 1 hour when the x-ray tube is operated at the leakage technique factors. If the maximum rated peak tube potential of the tube housing assembly is greater than the maximum rated peak tube potential for the diagnostic source assembly, positive means shall be provided to limit the maximum x-ray tube potential to that of the diagnostic source assembly. Compliance shall be determined by measurements averaged over an area of 100 square cm with no linear dimension greater than 20 cm.

(l) *Radiation from components other than the diagnostic source assembly.* The radiation emitted by a component other than the diagnostic source assembly shall not exceed an air kerma of 18 microGy (vice 2 mR exposure) in 1 hour at 5 cm from any accessible surface of the component when it is operated in an assembled x-ray system under any conditions for which it was designed. Compliance shall be determined by measurements averaged over an area of 100 square cm with no linear dimension greater than 20 cm.

(m) *Beam quality*—(1) *Half-value layer (HVL).* The HVL of the useful beam for a given x-ray tube potential shall not be less than the appropriate value shown in table 1 in paragraph (m)(1) of this section under the heading "Specified Dental Systems," for any dental x-ray system designed for use with intraoral image receptors and manufactured after December 1, 1980; under the heading "I—Other X-Ray Systems," for any dental x-ray system designed for use with intraoral image receptors and manufactured before December 1, 1980, and all other x-ray systems subject to this section and manufactured before June 10, 2006; and under the heading "II—Other X-Ray Systems," for all x-ray systems, except dental x-ray systems designed for use with intraoral image receptors, subject to this section and manufactured on or after June 10, 2006. If it is necessary to determine such HVL at an x-ray tube potential which is not listed in table 1 in paragraph (m)(1) of this section, linear interpolation or extrapolation may be made. Positive means[2] shall be provided to ensure that at least the minimum filtration needed to achieve the above beam quality requirements is in the useful beam during each exposure. Table 1 follows:

[2] In the case of a system, which is to be operated with more than one thickness of filtration, this requirement can be met by a filter interlocked with the kilovoltage selector which will prevent x-ray emissions if the minimum required filtration is not in place.

TABLE 1

X-Ray Tube Voltage (kilovolt peak)		Minimum HVL (mm of aluminum)		
Designed Operating Range	Measured Operating Potential	Specified Dental Systems[1]	I—Other X-Ray Systems[2]	II—Other X-Ray Systems[3]
Below 51	30	1.5	0.3	0.3
	40	1.5	0.4	0.4
	50	1.5	0.5	0.5
51 to 70	51	1.5	1.2	1.3
	60	1.5	1.3	1.5
	70	1.5	1.5	1.8
Above 70	71	2.1	2.1	2.5
	80	2.3	2.3	2.9
	90	2.5	2.5	3.2
	100	2.7	2.7	3.6
	110	3.0	3.0	3.9
	120	3.2	3.2	4.3
	130	3.5	3.5	4.7
	140	3.8	3.8	5.0
	150	4.1	4.1	5.4

[1] Dental x-ray systems designed for use with intraoral image receptors and manufactured after December 1, 1980.

[2] Dental x-ray systems designed for use with intraoral image receptors and manufactured before or on December 1, 1980, and all other x-ray systems subject to this section and manufactured before June 10, 2006.

[3] All x-ray systems, except dental x-ray systems designed for use with intraoral image receptors, subject to this section and manufactured on or after June 10, 2006.

(2) *Optional filtration.* Fluoroscopic systems manufactured on or after June 10, 2006, incorporating an x-ray tube(s) with a continuous output of 1 kilowatt or more and an anode heat storage capacity of 1 million heat units or more shall provide the option of adding x-ray filtration to the diagnostic source assembly in addition to the amount needed to meet the HVL provisions of § 1020.30(m)(1). The selection of this additional x-ray filtration shall be either at the option of the user or automatic as part of the selected mode of operation. A means of indicating which combination of additional filtration is in the x-ray beam shall be provided.

(3) *Measuring compliance.* For capacitor energy storage equipment, compliance shall be determined with the maximum selectable quantity of charge per exposure.

(n) *Aluminum equivalent of material between patient and image receptor.* Except when used in a CT x-ray system, the aluminum equivalent of each of the items listed in table 2 in paragraph (n) of this section, which are used between the patient and image receptor, may not exceed the indicated limits. Compliance shall be determined by x-ray measurements made at a potential of 100 kilovolts peak and with an x-ray beam that has an HVL specified in table 1 in paragraph (m)(1) of this section for the potential. This requirement applies to front panel(s) of cassette holders and film changers provided by the manufacturer for patient support or for prevention of foreign object intrusions. It does not apply to screens and their associated mechanical support panels or grids. Table 2 follows:

TABLE 2

Item	Maximum Aluminum Equivalent (millimeters)
1. Front panel(s) of cassette holders (total of all)	1.2
2. Front panel(s) of film changer (total of all)	1.2
3. Cradle	2.3
4. Tabletop, stationary, without articulated joints	1.2
5. Tabletop, movable, without articulated joint(s) (including stationary subtop)	1.7
6. Tabletop, with radiolucent panel having one articulated joint	1.7
7. Tabletop, with radiolucent panel having two or more articulated joints	2.3
8. Tabletop, cantilevered	2.3
9. Tabletop, radiation therapy simulator	5.0

(o) *Battery charge indicator.* On battery-powered generators, visual means shall be provided on the control panel to indicate whether the battery is in a state of charge adequate for proper operation.

(p) [Reserved]

(q) *Modification of certified diagnostic x-ray components and systems.* (1) Diagnostic x-ray components and systems certified in accordance with § 1010.2 of this chapter shall not be modified such that the component or system fails to comply with any applicable provision of this chapter unless a variance in accordance with § 1010.4 of this chapter or an exemption under section 534(a)(5) or 538(b) of the Federal Food, Drug, and Cosmetic Act has been granted.

(2) The owner of a diagnostic x-ray system who uses the system in a professional or commercial capacity may modify the system, provided the modification does not result in the failure of the system or component to comply with the applicable requirements of this section or of § 1020.31, 1020.32, or 1020.33. The owner who causes such modification need not submit the reports required by subpart B of part 1002 of this chapter, provided the owner records the date and the details of the modification in the system records and maintains this information, and provided the modification of the x-ray system does not result in a failure to comply with § 1020.31, 1020.32, or 1020.33.

§ 1020.31 Radiographic equipment.

The provisions of this section apply to equipment for the recording of images, except equipment involving use of an image intensifier or computed tomography x-ray systems manufactured on or after November 28, 1984.

(a) *Control and indication of technique factors*—(1) *Visual indication.* The technique factors to be used during an exposure shall be indicated before the exposure begins, except when automatic exposure controls are used, in which case the technique factors which are set prior to the exposure shall be indicated. On equipment having fixed technique factors, this requirement may be met by permanent markings. Indication of technique factors shall be visible from the operator's position except in the case of spot films made by the fluoroscopist.

(2) *Timers.* Means shall be provided to terminate the exposure at a preset

208-072 D-21

time interval, a preset product of current and time, a preset number of pulses, or a preset radiation exposure to the image receptor.

(i) Except during serial radiography, the operator shall be able to terminate the exposure at any time during an exposure of greater than one-half second. Except during panoramic dental radiography, termination of exposure shall cause automatic resetting of the timer to its initial setting or to zero. It shall not be possible to make an exposure when the timer is set to a zero or off position if either position is provided.

(ii) During serial radiography, the operator shall be able to terminate the x-ray exposure(s) at any time, but means may be provided to permit completion of any single exposure of the series in process.

(3) *Automatic exposure controls.* When an automatic exposure control is provided:

(i) Indication shall be made on the control panel when this mode of operation is selected;

(ii) When the x-ray tube potential is equal to or greater than 51 kilovolts peak (kVp), the minimum exposure time for field emission equipment rated for pulsed operation shall be equal to or less than a time interval equivalent to two pulses and the minimum exposure time for all other equipment shall be equal to or less than 1/60 second or a time interval required to deliver 5 milliamperes-seconds (mAs), whichever is greater;

(iii) Either the product of peak x-ray tube potential, current, and exposure time shall be limited to not more than 60 kilowatt-seconds (kW's) per exposure or the product of x-ray tube current and exposure time shall be limited to not more than 600 mAs per exposure, except when the x-ray tube potential is less than 51 kVp, in which case the product of x-ray tube current and exposure time shall be limited to not more than 2,000 mAs per exposure; and

(iv) A visible signal shall indicate when an exposure has been terminated at the limits described in paragraph (a)(3)(iii) of this section, and manual resetting shall be required before further automatically timed exposures can be made.

(4) *Accuracy.* Deviation of technique factors from indicated values shall not exceed the limits given in the information provided in accordance with § 1020.30(h)(3);

(b) *Reproducibility.* The following requirements shall apply when the equipment is operated on an adequate power supply as specified by the manufacturer in accordance with the requirements of § 1020.30(h)(3);

(1) *Coefficient of variation.* For any specific combination of selected technique factors, the estimated coefficient of variation of radiation exposures shall be no greater than 0.05.

(2) *Measuring compliance.* Determination of compliance shall be based on 10 consecutive measurements taken within a time period of 1 hour. Equipment manufactured after September 5, 1978, shall be subject to the additional requirement that all variable controls for technique factors shall be adjusted to alternate settings and reset to the test setting after each measurement. The percent line-voltage regulation shall be determined for each measurement. All values for percent line-voltage regulation shall be within ±1 of the mean value for all measurements. For equipment having automatic exposure controls, compliance shall be determined with a sufficient thickness of attenuating material in the useful beam such that the technique factors can be adjusted to provide individual exposures of a minimum of 12 pulses on field emission equipment rated for pulsed operation or no less than one-tenth second per exposure on all other equipment.

(c) *Linearity.* The following requirements apply when the equipment is operated on a power supply as specified by the manufacturer in accordance with the requirements of § 1020.30(h)(3) for any fixed x-ray tube potential within the range of 40 percent to 100 percent of the maximum rated.

(1) *Equipment having independent selection of x-ray tube current (mA).* The average ratios of exposure to the indicated milliampere-seconds product (C/kg/mAs (or mR/mAs)) obtained at any two consecutive tube current settings shall not differ by more than 0.10 times their sum. This is: $|X_1 - X_2| \leq 0.10(X_1 + X_2)$; where X_1 and X_2 are the average C/kg/

mAs (or mR/mAs) values obtained at each of two consecutive tube current settings or at two settings differing by no more than a factor of 2 where the tube current selection is continuous.

(2) *Equipment having selection of x-ray tube current-exposure time product (mAs).* For equipment manufactured after May 3, 1994 the average ratios of exposure to the indicated milliampere-seconds product (C/kg/mAs (or mR/mAs)) obtained at any two consecutive mAs selector settings shall not differ by more than 0.10 times their sum. This is: $|X_1 - X_2| \leq 0.10(X_1 + X_2)$; where X_1 and X_2 are the average C/kg/mAs (or mR/mAs) values obtained at each of two consecutive mAs selector settings or at two settings differing by no more than a factor of 2 where the mAs selector provides continuous selection.

(3) *Measuring compliance.* Determination of compliance will be based on 10 exposures, made within ±1 hour, at each of the two settings. These two settings may include any two focal spot sizes except where one is equal to or less than 0.45 millimeters and the other is greater than 0.45 millimeters. For purposes of this requirement, focal spot size is the focal spot size specified by the x-ray tube manufacturer. The percent line-voltage regulation shall be determined for each measurement. All values for percent line-voltage regulation at any one combination of technique factors shall be within #1 of the mean value for all measurements at these technique factors.

(d) *Field limitation and alignment for mobile, portable, and stationary general purpose x-ray systems.* Except when spot-film devices or special attachments for mammography are in service, mobile, portable, and stationary general purpose radiographic x-ray systems shall meet the following requirements:

(1) *Variable x-ray field limitation.* A means for stepless adjustment of the size of the x-ray field shall be provided. Each dimension of the minimum field size at an SID of 100 centimeters shall be equal to or less than 5 centimeters.

(2) *Visual definition.* (i) Means for visually defining the perimeter of the x-ray field shall be provided. The total misalignment of the edges of the visually defined field with the respective edges of the x-ray field along either the length or width of the visually defined field shall not exceed 2 percent of the distance from the source to the center of the visually defined field when the surface upon which it appears is perpendicular to the axis of the x-ray beam.

(ii) When a light localizer is used to define the x-ray field, it shall provide an average illuminance of not less than 160 lux (15 footcandles) at 100 centimeters or at the maximum SID, whichever is less. The average illuminance shall be based upon measurements made in the approximate center of each quadrant of the light field. Radiation therapy simulation systems are exempt from this requirement.

(iii) The edge of the light field at 100 centimeters or at the maximum SID, whichever is less, shall have a contrast ratio, corrected for ambient lighting, of not less than 4 in the case of beam-limiting devices designed for use on stationary equipment, and a contrast ratio of not less than 3 in the case of beam-limiting devices designed for use on mobile and portable equipment. The contrast ratio is defined as I_1 I_2, where I_1 is the illuminance 3 millimeters from the edge of the light field toward the center of the field; and I_2 is the illuminance 3 millimeters from the edge of the light field away from the center of the field. Compliance shall be determined with a measuring aperture of 1 millimeter.

(e) *Field indication and alignment on stationary general purpose x-ray equipment.* Except when spot-film devices or special attachments for mammography are in service, stationary general purpose x-ray systems shall meet the following requirements in addition to those prescribed in paragraph (d) of this section:

(1) Means shall be provided to indicate when the axis of the x-ray beam is perpendicular to the plane of the image receptor, to align the center of the x-ray field with respect to the center of the image receptor to within 2 percent of the SID, and to indicate the SID to within 2 percent;

(2) The beam-limiting device shall numerically indicate the field size in the plane of the image receptor to which it is adjusted;

(3) Indication of field size dimensions and SID's shall be specified in centimeters and/or inches and shall be such that aperture adjustments result in x-ray field dimensions in the plane of the image receptor which correspond to those indicated by the beam-limiting device to within 2 percent of the SID when the beam axis is indicated to be perpendicular to the plane of the image receptor; and

(4) Compliance measurements will be made at discrete SID's and image receptor dimensions in common clinical use (such as SID's of 100, 150, and 200 centimeters and/or 36, 40, 48, and 72 inches and nominal image receptor dimensions of 13, 18, 24, 30, 35, 40, and 43 centimeters and/or 5, 7, 8, 9, 10, 11, 12, 14, and 17 inches) or at any other specific dimensions at which the beam-limiting device or its associated diagnostic x-ray system is uniquely designed to operate.

(f) *Field limitation on radiographic x-ray equipment other than general purpose radiographic systems*—(1) *Equipment for use with intraoral image receptors.* Radiographic equipment designed for use with an intraoral image receptor shall be provided with means to limit the x-ray beam such that:

(i) If the minimum source-to-skin distance (SSD) is 18 centimeters or more, the x-ray field at the minimum SSD shall be containable in a circle having a diameter of no more than 7 centimeters; and

(ii) If the minimum SSD is less than 18 centimeters, the x-ray field at the minimum SSD shall be containable in a circle having a diameter of no more than 6 centimeters.

(2) *X-ray systems designed for one image receptor size.* Radiographic equipment designed for only one image receptor size at a fixed SID shall be provided with means to limit the field at the plane of the image receptor to dimensions no greater than those of the image receptor, and to align the center of the x-ray field with the center of the image receptor to within 2 percent of the SID or shall be provided with means to both size and align the x-ray field such that the x-ray field at the plane of the image receptor does not extend beyond any edge of the image receptor.

(3) *Systems designed for mammography.* (i) Mammographic beam-limiting devices manufactured after September 30, 1999, shall be provided with means to limit the useful beam such that the x-ray field at the plane of the image receptor does not extend beyond any edge of the image receptor by more than 2 percent of the SID. This requirement can be met with a system that performs as prescribed in paragraphs (f)(4)(i), (f)(4)(ii), and (f)(4)(iii) of this section. For systems that allow changes in the SID, the SID indication specified in paragraphs (f)(4)(ii) and (f)(4)(iii) of this section shall be the maximum SID for which the beam-limiting device or aperture is designed.

(ii) Each image receptor support device intended for installation on a system designed for mammography shall have clear and permanent markings to indicate the maximum image receptor size for which it is designed.

(4) *Other x-ray systems.* Radiographic systems not specifically covered in paragraphs (d), (e), (f)(2), (f)(3), and (h) of this section and systems covered in paragraph (f)(1) of this section, which are also designed for use with extraoral image receptors and when used with an extraoral image receptor, shall be provided with means to limit the x-ray field in the plane of the image receptor so that such field does not exceed each dimension of the image receptor by more than 2 percent of the SID, when the axis of the x-ray beam is perpendicular to the plane of the image receptor. In addition, means shall be provided to align the center of the x-ray field with the center of the image receptor to within 2 percent of the SID, or means shall be provided to both size and align the x-ray field such that the x-ray field at the plane of the image receptor does not extend beyond any edge of the image receptor. These requirements may be met with:

(i) A system which performs in accordance with paragraphs (d) and (e) of this section; or when alignment means are also provided, may be met with either;

(ii) An assortment of removable, fixed-aperture, beam-limiting devices sufficient to meet the requirement for each combination of image receptor size and SID for which the unit is designed. Each such device shall have clear and permanent markings to indicate the image receptor size and SID for which it is designed; or

(iii) A beam-limiting device having multiple fixed apertures sufficient to meet the requirement for each combination of image receptor size and SID for which the unit is designed. Permanent, clearly legible markings shall indicate the image receptor size and SID for which each aperture is designed and shall indicate which aperture is in position for use.

(g) *Positive beam limitation (PBL).* The requirements of this paragraph shall apply to radiographic systems which contain PBL.

(1) *Field size.* When a PBL system is provided, it shall prevent x-ray production when:

(i) Either the length or width of the x-ray field in the plane of the image receptor differs from the corresponding image receptor dimension by more than 3 percent of the SID; or

(ii) The sum of the length and width differences as stated in paragraph (g)(1)(i) of this section without regard to sign exceeds 4 percent of the SID.

(iii) The beam limiting device is at an SID for which PBL is not designed for sizing.

(2) *Conditions for PBL.* When provided, the PBL system shall function as described in paragraph (g)(1) of this section whenever all the following conditions are met:

(i) The image receptor is inserted into a permanently mounted cassette holder;

(ii) The image receptor length and width are less than 50 centimeters;

(iii) The x-ray beam axis is within ±3 degrees of vertical and the SID is 90 centimeters to 130 centimeters inclusive; or the x-ray beam axis is within ±3 degrees of horizontal and the SID is 90 centimeters to 205 centimeters inclusive;

(iv) The x-ray beam axis is perpendicular to the plane of the image receptor to within ±3 degrees; and

(v) Neither tomographic nor stereoscopic radiography is being performed.

(3) *Measuring compliance.* Compliance with the requirements of paragraph (g)(1) of this section shall be determined when the equipment indicates that the beam axis is perpendicular to the plane of the image receptor and the provisions of paragraph (g)(2) of this section are met. Compliance shall be determined no sooner than 5 seconds after insertion of the image receptor.

(4) *Operator initiated undersizing.* The PBL system shall be capable of operation such that, at the discretion of the operator, the size of the field may be made smaller than the size of the image receptor through stepless adjustment of the field size. Each dimension of the minimum field size at an SID of 100 centimeters shall be equal to or less than 5 centimeters. Return to PBL function as described in paragraph (g)(1) of this section shall occur automatically upon any change of image receptor size or SID.

(5) *Override of PBL.* A capability may be provided for overriding PBL in case of system failure and for servicing the system. This override may be for all SID's and image receptor sizes. A key shall be required for any override capability that is accessible to the operator. It shall not be possible to remove the key while PBL is overridden. Each such key switch or key shall be clearly and durably labeled as follows:

For X-ray Field Limitation System Failure

The override capability is considered accessible to the operator if it is referenced in the operator's manual or in other material intended for the operator or if its location is such that the operator would consider it part of the operational controls.

(h) *Field limitation and alignment for spot-film devices.* The following requirements shall apply to spot-film devices, except when the spot-film device is provided for use with a radiation therapy simulation system:

(1) Means shall be provided between the source and the patient for adjustment of the x-ray field size in the plane of the image receptor to the size of that portion of the image receptor which has been selected on the spot-film selector. Such adjustment shall be accomplished automatically when the x-ray field size in the plane of the image receptor is greater than the selected portion of the image receptor. If the x-ray field size is less than the size

of the selected portion of the image receptor, the field size shall not open automatically to the size of the selected portion of the image receptor unless the operator has selected that mode of operation.

(2) Neither the length nor the width of the x-ray field in the plane of the image receptor shall differ from the corresponding dimensions of the selected portion of the image receptor by more than 3 percent of the SID when adjusted for full coverage of the selected portion of the image receptor. The sum, without regard to sign, of the length and width differences shall not exceed 4 percent of the SID. On spot-film devices manufactured after February 25, 1978, if the angle between the plane of the image receptor and beam axis is variable, means shall be provided to indicate when the axis of the x-ray beam is perpendicular to the plane of the image receptor, and compliance shall be determined with the beam axis indicated to be perpendicular to the plane of the image receptor.

(3) The center of the x-ray field in the plane of the image receptor shall be aligned with the center of the selected portion of the image receptor to within 2 percent of the SID.

(4) Means shall be provided to reduce the x-ray field size in the plane of the image receptor to a size smaller than the selected portion of the image receptor such that:

(i) For spot-film devices used on fixed-SID fluoroscopic systems which are not required to, and do not provide stepless adjustment of the x-ray field, the minimum field size, at the greatest SID, does not exceed 125 square centimeters; or

(ii) For spot-film devices used on fluoroscopic systems that have a variable SID and/or stepless adjustment of the field size, the minimum field size, at the greatest SID, shall be containable in a square of 5 centimeters by 5 centimeters.

(5) A capability may be provided for overriding the automatic x-ray field size adjustment in case of system failure. If it is so provided, a signal visible at the fluoroscopist's position shall indicate whenever the automatic x-ray field size adjustment override is engaged. Each such system failure override switch shall be clearly labeled as follows:

For X-ray Field Limitation System Failure

(i) *Source-skin distance*—(1) X-ray systems designed for use with an intraoral image receptor shall be provided with means to limit the source-skin distance to not less than:

(i) Eighteen centimeters if operable above 50 kVp; or

(ii) Ten centimeters if not operable above 50 kVp.

(2) Mobile and portable x-ray systems other than dental shall be provided with means to limit the source-skin distance to not less than 30 centimeters.

(j) *Beam-on indicators*. The x-ray control shall provide visual indication whenever x-rays are produced. In addition, a signal audible to the operator shall indicate that the exposure has terminated.

(k) *Multiple tubes*. Where two or more radiographic tubes are controlled by one exposure switch, the tube or tubes which have been selected shall be clearly indicated before initiation of the exposure. This indication shall be both on the x-ray control and at or near the tube housing assembly which has been selected.

(l) *Radiation from capacitor energy storage equipment*. Radiation emitted from the x-ray tube shall not exceed:

(1) 8.6×10^{-9} C/kg (0.03 mR) in 1 minute at 5 centimeters from any accessible surface of the diagnostic source assembly, with the beam-limiting device fully open, the system fully charged, and the exposure switch, timer, or any discharge mechanism not activated. Compliance shall be determined by measurements averaged over an area of 100 square centimeters, with no linear dimension greater than 20 centimeters; and

(2) 2.58×10^{-5} C/kg (100 mR) in 1 hour at 100 centimeters from the x-ray source, with the beam-limiting device fully open, when the system is discharged through the x-ray tube either manually or automatically by use of a discharge switch or deactivation of the

input power. Compliance shall be determined by measurements of the maximum exposure per discharge multiplied by the total number of discharges in 1 hour (duty cycle). The measurements shall be averaged over an area of 100 square centimeters with no linear dimension greater than 20 centimeters.

(m) *Primary protective barrier for mammography x-ray systems.* For mammography x-ray systems manufactured after September 30, 1999:

(1) At any SID where exposures can be made, the image receptor support device shall provide a primary protective barrier that intercepts the cross section of the useful beam along every direction except at the chest wall edge.

(2) The x-ray tube shall not permit exposure unless the appropriate barrier is in place to intercept the useful beam as required in paragraph (m)(1) of this section.

(3) The transmission of the useful beam through the primary protective barrier shall be limited such that the exposure 5 centimeters from any accessible surface beyond the plane of the primary protective barrier does not exceed 2.58×10^{-8} C/kg (0.1 mR) for each activation of the tube.

(4) Compliance for transmission shall be determined with the x-ray system operated at the minimum SID for which it is designed, at the maximum rated peak tube potential, at the maximum rated product of x-ray tube current and exposure time (mAs) for the maximum rated peak tube potential, and by measurements averaged over an area of 100 square centimeters with no linear dimension greater than 20 centimeters. The sensitive volume of the radiation measuring instrument shall not be positioned beyond the edge of the primary protective barrier along the chest wall side.

[58 FR 26401, May 3, 1993; 58 FR 31067, May 28, 1993, as amended at 64 FR 35927, July 2, 1999]

EFFECTIVE DATE NOTE: At 70 FR 34036, June 10, 2005, §1020.31 was revised, effective June 10, 2006. For the convenience of the user, the revised text is set forth as follows:

§ 1020.31 Radiographic equipment.

The provisions of this section apply to equipment for radiography, except equipment for fluoroscopic imaging or for recording images from the fluoroscopic image receptor, or computed tomography x-ray systems manufactured on or after November 29, 1984.

(a) *Control and indication of technique factors*—(1) *Visual indication.* The technique factors to be used during an exposure shall be indicated before the exposure begins, except when automatic exposure controls are used, in which case the technique factors which are set prior to the exposure shall be indicated. On equipment having fixed technique factors, this requirement may be met by permanent markings. Indication of technique factors shall be visible from the operator's position except in the case of spot films made by the fluoroscopist.

(2) *Timers.* Means shall be provided to terminate the exposure at a preset time interval, a preset product of current and time, a preset number of pulses, or a preset radiation exposure to the image receptor.

(i) Except during serial radiography, the operator shall be able to terminate the exposure at any time during an exposure of greater than one-half second. Except during panoramic dental radiography, termination of exposure shall cause automatic resetting of the timer to its initial setting or to zero. It shall not be possible to make an exposure when the timer is set to a zero or off position if either position is provided.

(ii) During serial radiography, the operator shall be able to terminate the x-ray exposure(s) at any time, but means may be provided to permit completion of any single exposure of the series in process.

(3) *Automatic exposure controls.* When an automatic exposure control is provided:

(i) Indication shall be made on the control panel when this mode of operation is selected;

(ii) When the x-ray tube potential is equal to or greater than 51 kilovolts peak (kVp), the minimum exposure time for field emission equipment rated for pulsed operation shall be equal to or less than a time interval equivalent to two pulses and the minimum exposure time for all other equipment shall be equal to or less than 1/60 second or a time interval required to deliver 5 milliampere-seconds (mAs), whichever is greater;

(iii) Either the product of peak x-ray tube potential, current, and exposure time shall be limited to not more than 60 kilowatt-seconds (kWs) per exposure or the product of x-ray tube current and exposure time shall be limited to not more than 600 mAs per exposure, except when the x-ray tube potential is less than 51 kVp, in which case the product of x-ray tube current and exposure time shall be limited to not more than 2,000 mAs per exposure; and

(iv) A visible signal shall indicate when an exposure has been terminated at the limits described in paragraph (a)(3)(iii) of this section, and manual resetting shall be required before further automatically timed exposures can be made.

(4) *Accuracy*. Deviation of technique factors from indicated values shall not exceed the limits given in the information provided in accordance with § 1020.30(h)(3).

(b) *Reproducibility*. The following requirements shall apply when the equipment is operated on an adequate power supply as specified by the manufacturer in accordance with the requirements of § 1020.30(h)(3):

(1) *Coefficient of variation*. For any specific combination of selected technique factors, the estimated coefficient of variation of the air kerma shall be no greater than 0.05.

(2) *Measuring compliance*. Determination of compliance shall be based on 10 consecutive measurements taken within a time period of 1 hour. Equipment manufactured after September 5, 1978, shall be subject to the additional requirement that all variable controls for technique factors shall be adjusted to alternate settings and reset to the test setting after each measurement. The percent line-voltage regulation shall be determined for each measurement. All values for percent line-voltage regulation shall be within ±1 of the mean value for all measurements. For equipment having automatic exposure controls, compliance shall be determined with a sufficient thickness of attenuating material in the useful beam such that the technique factors can be adjusted to provide individual exposures of a minimum of 12 pulses on field emission equipment rated for pulsed operation or no less than one-tenth second per exposure on all other equipment.

(c) *Linearity*. The following requirements apply when the equipment is operated on a power supply as specified by the manufacturer in accordance with the requirements of § 1020.30(h)(3) for any fixed x-ray tube potential within the range of 40 percent to 100 percent of the maximum rated.

(1) *Equipment having independent selection of x-ray tube current (mA)*. The average ratios of air kerma to the indicated milliampere-seconds product (mGy/mAs) obtained at any two consecutive tube current settings shall not differ by more than 0.10 times their sum. This is: $|X_1 - X_2| \leq 0.10(X_1 + X_2)$; where X_1 and X_2 are the average mGy/mAs values obtained at each of two consecutive mAs selector settings or at two settings differing by no more than a factor of 2 where the mAs selector provides continuous selection.

(2) *Equipment having selection of x-ray tube current-exposure time product (mAs)*. For equipment manufactured after May 3, 1994, the average ratios of air kerma to the indicated milliampere-seconds product (mGy/mAs) obtained at any two consecutive mAs selector settings shall not differ by more than 0.10 times their sum. This is: $|X_1 - X_2| \leq 0.10 (X_1 + X_2)$; where X_1 and X_2 are the average mGy/mAs values obtained at each of two consecutive mAs selector settings or at two settings differing by no more than a factor of 2 where the mAs selector provides continuous selection.

(3) *Measuring compliance*. Determination of compliance will be based on 10 exposures, made within 1 hour, at each of the two settings. These two settings may include any two focal spot sizes except where one is equal to or less than 0.45 mm and the other is greater than 0.45 mm. For purposes of this requirement, focal spot size is the focal spot size specified by the x-ray tube manufacturer. The percent line-voltage regulation shall be determined for each measurement. All values for percent line-voltage regulation at any one combination of technique factors shall be within ±1 of the mean value for all measurements at these technique factors.

(d) *Field limitation and alignment for mobile, portable, and stationary general purpose x-ray systems*. Except when spot-film devices are in service, mobile, portable, and stationary general purpose radiographic x-ray systems shall meet the following requirements:

(1) *Variable x-ray field limitation*. A means for stepless adjustment of the size of the x-ray field shall be provided. Each dimension of the minimum field size at an SID of 100 centimeters (cm) shall be equal to or less than 5 cm.

(2) *Visual definition*. (i) Means for visually defining the perimeter of the x-ray field shall be provided. The total misalignment of the edges of the visually defined field with the respective edges of the x-ray field along either the length or width of the visually defined field shall not exceed 2 percent of the distance from the source to the center of the visually defined field when the surface upon which it appears is perpendicular to the axis of the x-ray beam.

(ii) When a light localizer is used to define the x-ray field, it shall provide an average illuminance of not less than 160 lux (15 footcandles) at 100 cm or at the maximum SID, whichever is less. The average illuminance shall be based on measurements made in the approximate center of each quadrant of the light field. Radiation therapy simulation systems are exempt from this requirement.

(iii) The edge of the light field at 100 cm or at the maximum SID, whichever is less, shall have a contrast ratio, corrected for ambient lighting, of not less than 4 in the case of beam-limiting devices designed for use on stationary equipment, and a contrast ratio of not less than 3 in the case of beam-limiting devices designed for use on mobile and portable equipment. The contrast ratio is defined as I_1/I_2, where I_1 is the illuminance 3 mm from the edge of the light field toward the center of the field; and I_2 is the illuminance 3 mm from the edge of the light field away from the center of the field. Compliance shall be determined with a measuring aperture of 1 mm.

(e) *Field indication and alignment on stationary general purpose x-ray equipment.* Except when spot-film devices are in service, stationary general purpose x-ray systems shall meet the following requirements in addition to those prescribed in paragraph (d) of this section:

(1) Means shall be provided to indicate when the axis of the x-ray beam is perpendicular to the plane of the image receptor, to align the center of the x-ray field with respect to the center of the image receptor to within 2 percent of the SID, and to indicate the SID to within 2 percent;

(2) The beam-limiting device shall numerically indicate the field size in the plane of the image receptor to which it is adjusted;

(3) Indication of field size dimensions and SIDs shall be specified in centimeters and/or inches and shall be such that aperture adjustments result in x-ray field dimensions in the plane of the image receptor which correspond to those indicated by the beam-limiting device to within 2 percent of the SID when the beam axis is indicated to be perpendicular to the plane of the image receptor; and

(4) Compliance measurements will be made at discrete SIDs and image receptor dimensions in common clinical use (such as SIDs of 100, 150, and 200 cm and/or 36, 40, 48, and 72 inches and nominal image receptor dimensions of 13, 18, 24, 30, 35, 40, and 43 cm and/or 5, 7, 8, 9, 10, 11, 12, 14, and 17 inches) or at any other specific dimensions at which the beam-limiting device or its associated diagnostic x-ray system is uniquely designed to operate.

(f) *Field limitation on radiographic x-ray equipment other than general purpose radiographic systems*—(1) *Equipment for use with intraoral image receptors.* Radiographic equipment designed for use with an intraoral image receptor shall be provided with means to limit the x-ray beam such that:

(i) If the minimum source-to-skin distance (SSD) is 18 cm or more, the x-ray field at the minimum SSD shall be containable in a circle having a diameter of no more than 7 cm; and

(ii) If the minimum SSD is less than 18 cm, the x-ray field at the minimum SSD shall be containable in a circle having a diameter of no more than 6 cm.

(2) *X-ray systems designed for one image receptor size.* Radiographic equipment designed for only one image receptor size at a fixed SID shall be provided with means to limit the field at the plane of the image receptor to dimensions no greater than those of the image receptor, and to align the center of the x-ray field with the center of the image receptor to within 2 percent of the SID, or shall be provided with means to both size and align the x-ray field such that the x-ray field at the plane of the image receptor does not extend beyond any edge of the image receptor.

(3) *Systems designed for mammography*—(i) Radiographic systems designed only for mammography and general purpose radiography systems, when special attachments for mammography are in service, manufactured on or after November 1, 1977, and before September 30, 1999, shall be provided with means to limit the useful beam such that the x-ray field at the plane of the image receptor does not extend beyond any edge of the image receptor at any designated SID except the edge of the image receptor designed to be adjacent to the chest wall where the x-ray field may not extend beyond this edge by more than 2 percent of the SID. This requirement can be met with a system that performs as prescribed in paragraphs (f)(4)(i), (f)(4)(ii), and (f)(4)(iii) of this section. When the beam-limiting device and image receptor support device are designed to be used to immobilize the breast during a mammographic procedure and the SID may vary, the SID indication specified in paragraphs (f)(4)(ii) and (f)(4)(iii) of this section shall be the maximum SID for which the beam-limiting device or aperture is designed.

(ii) Mammographic beam-limiting devices manufactured on or after September 30, 1999, shall be provided with a means to limit the useful beam such that the x-ray field at the plane of the image receptor does not extend beyond any edge of the image receptor by more than 2 percent of the SID. This requirement can be met with a system that performs as prescribed in paragraphs (f)(4)(i), (f)(4)(ii), and (f)(4)(iii) of this section. For systems that allow changes in the SID, the SID indication specified in paragraphs (f)(4)(ii) and (f)(4)(iii) of this section shall be the maximum SID for which the beam-limiting device or aperture is designed.

(iii) Each image receptor support device manufactured on or after November 1, 1977, intended for installation on a system designed for mammography shall have clear and permanent markings to indicate the maximum image receptor size for which it is designed.

(4) *Other x-ray systems.* Radiographic systems not specifically covered in paragraphs (d), (e), (f)(2), (f)(3), and (h) of this section and systems covered in paragraph (f)(1) of this section, which are also designed for use with extraoral image receptors and when used with an extraoral image receptor, shall be provided with means to limit the x-ray field in the plane of the image receptor so that such field does not exceed each dimension of the image receptor by more than 2 percent of the SID, when the axis of the x-ray beam is perpendicular to the plane of the image receptor. In addition, means shall be provided to align the center of the x-ray field with the center of the image receptor to within 2 percent of the SID, or means shall be provided to both size and align the x-ray field such that the x-ray field at the plane of

the image receptor does not extend beyond any edge of the image receptor. These requirements may be met with:

(i) A system which performs in accordance with paragraphs (d) and (e) of this section; or when alignment means are also provided, may be met with either;

(ii) An assortment of removable, fixed-aperture, beam-limiting devices sufficient to meet the requirement for each combination of image receptor size and SID for which the unit is designed. Each such device shall have clear and permanent markings to indicate the image receptor size and SID for which it is designed; or

(iii) A beam-limiting device having multiple fixed apertures sufficient to meet the requirement for each combination of image receptor size and SID for which the unit is designed. Permanent, clearly legible markings shall indicate the image receptor size and SID for which each aperture is designed and shall indicate which aperture is in position for use.

(g) *Positive beam limitation (PBL).* The requirements of this paragraph shall apply to radiographic systems which contain PBL.

(1) *Field size.* When a PBL system is provided, it shall prevent x-ray production when:

(i) Either the length or width of the x-ray field in the plane of the image receptor differs from the corresponding image receptor dimension by more than 3 percent of the SID; or

(ii) The sum of the length and width differences as stated in paragraph (g)(1)(i) of this section without regard to sign exceeds 4 percent of the SID.

(iii) The beam limiting device is at an SID for which PBL is not designed for sizing.

(2) *Conditions for PBL.* When provided, the PBL system shall function as described in paragraph (g)(1) of this section whenever all the following conditions are met:

(i) The image receptor is inserted into a permanently mounted cassette holder;

(ii) The image receptor length and width are less than 50 cm;

(iii) The x-ray beam axis is within ±3 degrees of vertical and the SID is 90 cm to 130 cm inclusive; or the x-ray beam axis is within ±3 degrees of horizontal and the SID is 90 cm to 205 cm inclusive;

(iv) The x-ray beam axis is perpendicular to the plane of the image receptor to within ±3 degrees; and

(v) Neither tomographic nor stereoscopic radiography is being performed.

(3) *Measuring compliance.* Compliance with the requirements of paragraph (g)(1) of this section shall be determined when the equipment indicates that the beam axis is perpendicular to the plane of the image receptor and the provisions of paragraph (g)(2) of this section are met. Compliance shall be determined no sooner than 5 seconds after insertion of the image receptor.

(4) *Operator initiated undersizing.* The PBL system shall be capable of operation such that, at the discretion of the operator, the size of the field may be made smaller than the size of the image receptor through stepless adjustment of the field size. Each dimension of the minimum field size at an SID of 100 cm shall be equal to or less than 5 cm. Return to PBL function as described in paragraph (g)(1) of this section shall occur automatically upon any change of image receptor size or SID.

(5) *Override of PBL.* A capability may be provided for overriding PBL in case of system failure and for servicing the system. This override may be for all SIDs and image receptor sizes. A key shall be required for any override capability that is accessible to the operator. It shall not be possible to remove the key while PBL is overridden. Each such key switch or key shall be clearly and durably labeled as follows:

For X-ray Field Limitation System Failure

The override capability is considered accessible to the operator if it is referenced in the operator's manual or in other material intended for the operator or if its location is such that the operator would consider it part of the operational controls.

(h) *Field limitation and alignment for spot-film devices.* The following requirements shall apply to spot-film devices, except when the spot-film device is provided for use with a radiation therapy simulation system:

(1) Means shall be provided between the source and the patient for adjustment of the x-ray field size in the plane of the image receptor to the size of that portion of the image receptor which has been selected on the spot-film selector. Such adjustment shall be accomplished automatically when the x-ray field size in the plane of the image receptor is greater than the selected portion of the image receptor. If the x-ray field size is less than the size of the selected portion of the image receptor, the field size shall not open automatically to the size of the selected portion of the image receptor unless the operator has selected that mode of operation.

(2) Neither the length nor the width of the x-ray field in the plane of the image receptor shall differ from the corresponding dimensions of the selected portion of the image receptor by more than 3 percent of the SID when adjusted for full coverage of the selected portion of the image receptor. The sum, without regard to sign, of the length and width differences shall not exceed 4 percent of the SID. On spot-film devices manufactured after February 25, 1978, if the angle between the plane of the image receptor and beam axis is variable, means shall be provided to indicate when the axis of the x-ray

beam is perpendicular to the plane of the image receptor, and compliance shall be determined with the beam axis indicated to be perpendicular to the plane of the image receptor.

(3) The center of the x-ray field in the plane of the image receptor shall be aligned with the center of the selected portion of the image receptor to within 2 percent of the SID.

(4) Means shall be provided to reduce the x-ray field size in the plane of the image receptor to a size smaller than the selected portion of the image receptor such that:

(i) For spot-film devices used on fixed-SID fluoroscopic systems which are not required to, and do not provide stepless adjustment of the x-ray field, the minimum field size, at the greatest SID, does not exceed 125 square cm; or

(ii) For spot-film devices used on fluoroscopic systems that have a variable SID and/or stepless adjustment of the field size, the minimum field size, at the greatest SID, shall be containable in a square of 5 cm by 5 cm.

(5) A capability may be provided for overriding the automatic x-ray field size adjustment in case of system failure. If it is so provided, a signal visible at the fluoroscopist's position shall indicate whenever the automatic x-ray field size adjustment override is engaged. Each such system failure override switch shall be clearly labeled as follows:

For X-ray Field Limitation System Failure

(i) *Source-skin distance*—(1) X-ray systems designed for use with an intraoral image receptor shall be provided with means to limit the source-skin distance to not less than:

(i) Eighteen cm if operable above 50 kVp; or

(ii) Ten cm if not operable above 50 kVp.

(2) Mobile and portable x-ray systems other than dental shall be provided with means to limit the source-skin distance to not less than 30 cm.

(j) *Beam-on indicators*. The x-ray control shall provide visual indication whenever x-rays are produced. In addition, a signal audible to the operator shall indicate that the exposure has terminated.

(k) *Multiple tubes*. Where two or more radiographic tubes are controlled by one exposure switch, the tube or tubes which have been selected shall be clearly indicated before initiation of the exposure. This indication shall be both on the x-ray control and at or near the tube housing assembly which has been selected.

(l) *Radiation from capacitor energy storage equipment*. Radiation emitted from the x-ray tube shall not exceed:

(1) An air kerma of 0.26 microGy (vice 0.03 mR exposure) in 1 minute at 5 cm from any accessible surface of the diagnostic source assembly, with the beam-limiting device fully open, the system fully charged, and the exposure switch, timer, or any discharge mechanism not activated. Compliance shall be determined by measurements averaged over an area of 100 square cm, with no linear dimension greater than 20 cm; and

(2) An air kerma of 0.88 mGy (vice 100 mR exposure) in 1 hour at 100 cm from the x-ray source, with the beam-limiting device fully open, when the system is discharged through the x-ray tube either manually or automatically by use of a discharge switch or deactivation of the input power. Compliance shall be determined by measurements of the maximum air kerma per discharge multiplied by the total number of discharges in 1 hour (duty cycle). The measurements shall be averaged over an area of 100 square cm with no linear dimension greater than 20 cm.

(m) *Primary protective barrier for mammography x-ray systems*—(1) For x-ray systems manufactured after September 5, 1978, and before September 30, 1999, which are designed only for mammography, the transmission of the primary beam through any image receptor support provided with the system shall be limited such that the air kerma 5 cm from any accessible surface beyond the plane of the image receptor supporting device does not exceed 0.88 microGy (vice 0.1 mR exposure) for each activation of the tube.

(2) For mammographic x-ray systems manufactured on or after September 30, 1999:

(i) At any SID where exposures can be made, the image receptor support device shall provide a primary protective barrier that intercepts the cross section of the useful beam along every direction except at the chest wall edge.

(ii) The x-ray system shall not permit exposure unless the appropriate barrier is in place to intercept the useful beam as required in paragraph (m)(2)(i) of this section.

(iii) The transmission of the useful beam through the primary protective barrier shall be limited such that the air kerma 5 cm from any accessible surface beyond the plane of the primary protective barrier does not exceed 0.88 microGy (vice 0.1 mR exposure) for each activation of the tube.

(3) Compliance with the requirements of paragraphs (m)(1) and (m)(2)(iii) of this section for transmission shall be determined with the x-ray system operated at the minimum SID for which it is designed, at the maximum rated peak tube potential, at the maximum rated product of x-ray tube current and exposure time (mAs) for the maximum rated peak tube potential, and by measurements averaged over an area of 100 square cm with no linear dimension greater than 20 cm. The sensitive volume of the radiation measuring instrument shall not be positioned beyond the edge of the primary protective barrier along the chest wall side.

70 FR 34036, June 10, 2005]

§ 1020.32 Fluoroscopic equipment.

The provisions of this section apply to equipment for fluoroscopy and for the recording of images through an image intensifier except computed tomography x-ray systems manufactured on or after November 29, 1984.

(a) *Primary protective barrier*—(1) *Limitation of useful beam.* The fluoroscopic imaging assembly shall be provided with a primary protective barrier which intercepts the entire cross section of the useful beam at any SID. The x-ray tube used for fluoroscopy shall not produce x-rays unless the barrier is in position to intercept the entire useful beam. The exposure rate due to transmission through the barrier with the attenuation block in the useful beam combined with radiation from the image intensifier if provided, shall not exceed 3.34×10^{-3} percent of the entrance exposure rate, at a distance of 10 centimeters from any accessible surface of the fluoroscopic imaging assembly beyond the plane of the image receptor. Radiation therapy simulation systems shall be exempt from this requirement provided the systems are intended only for remote control operation and the manufacturer sets forth instructions for assemblers with respect to control location as part of the information required in § 1020.30(g). Additionally, the manufacturer shall provide to users, pursuant to § 1020.30(h)(1)(i), precautions concerning the importance of remote control operation.

(2) *Measuring compliance.* The entrance exposure rate shall be measured in accordance with paragraph (d) of this section. The exposure rate due to transmission through the primary barrier combined with radiation from the image intensifier shall be determined by measurements averaged over an area of 100 square centimeters with no linear dimension greater than 20 centimeters. If the source is below the tabletop, the measurement shall be made with the input surface of the fluoroscopic imaging assembly positioned 30 centimeters above the tabletop. If the source is above the tabletop and the SID is variable, the measurement shall be made with the end of the beam-limiting device or spacer as close to the tabletop as it can be placed, provided that it shall not be closer than 30 centimeters. Movable grids and compression devices shall be removed from the useful beam during the measurement. For all measurements, the attenuation block shall be positioned in the useful beam 10 centimeters from the point of measurement of entrance exposure rate and between this point and the input surface of the fluoroscopic imaging assembly.

(b) *Field limitation*—(1) *Nonimage-intensified fluoroscopy.* (i) The x-ray field produced by nonimage-intensified fluoroscopic equipment shall not extend beyond the entire visible area of the image receptor. Means shall be provided for stepless adjustment of the field size. The minimum field size, at the greatest SID, shall be containable in a square of 5 centimeters by 5 centimeters.

(ii) For equipment manufactured after February 25, 1978, when the angle between the image receptor and the beam axis of the x-ray beam is variable, means shall be provided to indicate when the axis of the x-ray beam is perpendicular to the plane of the image receptor. Compliance with paragraph (b)(1)(i) of this section shall be determined with the beam axis indicated to be perpendicular to the plane of the image receptor.

(2) *Image-intensified fluoroscopy.* (i) For image-intensified fluoroscopic equipment other than radiation therapy simulation systems, neither the length nor the width of the x-ray field in the plane of the image receptor shall exceed that of the visible area of the image receptor by more than 3 percent of the SID. The sum of the excess length and the excess width shall be no greater than 4 percent of the SID.

(ii) For rectangular x-ray fields used with circular image receptors, the error in alignment shall be determined along the length and width dimensions of the x-ray field which pass through the center of the visible area of the image receptor.

(iii) For equipment manufactured after February 25, 1978, when the angle between the image receptor and beam axis is variable, means shall be provided to indicate when the axis of the x-ray beam is perpendicular to the plane of the image receptor. Compliance with paragraph (b)(2)(i) of this section shall be determined with the beam axis indicated to be perpendicular to the plane of the image receptor.

(iv) Means shall be provided to permit further limitation of the field. Beam-limiting devices manufactured after May 22, 1979, and incorporated in equipment with a variable SID and/or the capability of a visible area of greater than 300 square centimeters shall be provided with means for stepless adjustment of the x-ray field. Equipment with a fixed SID and the capability of a visible area of no greater than 300 square centimeters shall be provided with either stepless adjustment of the x-ray field or with a means to further limit the x-ray field size at the plane of the image receptor to 125 square centimeters or less. Stepless adjustment shall, at the greatest SID, provide continuous field sizes from the maximum obtainable to a field size containable in a square of 5 centimeters by 5 centimeters.

(3) If the fluoroscopic x-ray field size is adjusted automatically as the SID or image receptor size is changed, a capability may be provided for overriding the automatic adjustment in case of system failure. If it is so provided, a signal visible at the fluoroscopist's position shall indicate whenever the automatic field adjustment is overridden. Each such system failure override switch shall be clearly labeled as follows:

For X-ray Field Limitation System Failure

(c) *Activation of tube.* X-ray production in the fluoroscopic mode shall be controlled by a device which requires continuous pressure by the operator for the entire time of any exposure. When recording serial fluoroscopic images, the operator shall be able to terminate the x-ray exposure(s) at any time, but means may be provided to permit completion of any single exposure of the series in process.

(d) *Entrance exposure rates.* For fluoroscopic equipment manufactured before May 19, 1995, the following requirements apply:

(1) *Equipment with automatic exposure rate control (AERC).* Fluoroscopic equipment that is provided with AERC shall not be operable at any combination of tube potential and current that will result in an exposure rate in excess of 2.58×10^{-3} coulomb per kilogram (C/kg) per minute (10 roentgens per minute (10 R/min)) at the point where the center of the useful beam enters the patient, except:

(i) During recording of fluoroscopic images, or

(ii) When an optional high-level control is provided. When so provided, the equipment shall not be operable at any combination of tube potential and current that will result in an exposure rate in excess of 1.29×10^{-3} C/kg per minute (5 R/min) at the point where the center of the useful beam enters the patient, unless the high-level control is activated. Special means of activation of high-level controls shall be required. The high-level control shall be operable only when continuous manual activation is provided by the operator. A continuous signal audible to the fluoroscopist shall indicate that the high-level control is being employed.

(2) *Equipment without AERC (manual mode).* Fluoroscopic equipment that is not provided with AERC shall not be operable at any combination of tube potential and current that will result in an exposure rate in excess of 1.29×10^{-3} C/kg per minute (5 R/min) at the point where the center of the useful beam enters the patient, except:

(i) During recording of fluoroscopic images, or

(ii) When an optional high-level control is activated. Special means of activation of high-level controls shall be required. The high-level control shall be operable only when continuous manual activation is provided by the operator. A continuous signal audible to the fluoroscopist shall indicate that the high-level control is being employed.

(3) *Equipment with both an AERC mode and a manual mode.* Fluoroscopic equipment that is provided with both an AERC mode and a manual mode shall not be operable at any combination of tube potential and current that will result in an exposure rate in excess of

2.58×10^{-3} C/kg per minute (10 R/min) in either mode at the point where the center of the useful beam enters the patient except:

(i) During recording of fluoroscopic images, or

(ii) When the mode or modes have an optional high-level control, in which case that mode or modes shall not be operable at any combination of tube potential and current that will result in an exposure rate in excess of 1.29×10^{-3} C/kg per minute (5 R/min) at the point where the center of the useful beam enters the patient, unless the high-level control is activated. Special means of activation of high-level controls shall be required. The high-level control shall be operable only when continuous manual activation is provided by the operator. A continuous signal audible to the fluoroscopist shall indicate that the high-level is being employed.

(4) *Measuring compliance.* Compliance with paragraph (d) of this section shall be determined as follows:

(i) If the source is below the x-ray table, the exposure rate shall be measured at 1 centimeter above the tabletop or cradle.

(ii) If the source is above the x-ray table, the exposure rate shall be measured at 30 centimeters above the tabletop with the end of the beam-limiting device or spacer positioned as closely as possible to the point of measurement.

(iii) In a C-arm type of fluoroscope, the exposure rate shall be measured at 30 centimeters from the input surface of the fluoroscopic imaging assembly, with the source positioned at any available SID, provided that the end of the beam-limiting device or spacer is no closer than 30 centimeters from the input surface of the imaging assembly.

(iv) In a lateral type of fluoroscope, the exposure rate shall be measured at a point 15 centimeters from the centerline of the x-ray table and in the direction of the x-ray source with the end of the beam-limiting device or spacer positioned as closely as possible to the point of measurement. If the tabletop is movable, it shall be positioned as closely as possible to the lateral x-ray source, with the end of the beam-limiting device or spacer no closer than 15 centimeters to the centerline of the x-ray table.

(5) *Exemptions.* Fluoroscopic radiation therapy simulation systems are exempt from the requirements set forth in paragraph (d) of this section.

(e) *Entrance exposure rate limits.* For fluoroscopic equipment manufactured on and after May 19, 1995, the following requirements apply:

(1) Fluoroscopic equipment operable at any combination of tube potential and current that results in an exposure rate greater than 1.29×10^{-3} C/kg per minute (5 R/min) at the point where the center of the useful beam enters the patient shall be equipped with AERC. Provision for manual selection of technique factors may be provided.

(2) Fluoroscopic equipment shall not be operable at any combination of tube potential and current that will result in an exposure rate in excess of 2.58×10^{-3} C/kg per minute (10 R/min) at the point where the center of the useful beam enters the patient except:

(i) During the recording of images from an x-ray image-intensifier tube using photographic film or a video camera when the x-ray source is operated in a pulsed mode.

(ii) When an optional high-level control is activated. When the high-level control is activated, the equipment shall not be operable at any combination of tube potential and current that will result in an exposure rate in excess of 5.16×10^{-3} C/kg per minute (20 R/min) at the point where the center of the useful beam enters the patient. Special means of activation of high-level controls shall be required. The high-level control shall only be operable when continuous manual activation is provided by the operator. A continuous signal audible to the fluoroscopist shall indicate that the high-level control is being employed.

(3) *Measuring compliance.* Compliance with paragraph (e) of this section shall be determined as follows:

(i) If the source is below the x-ray table, the exposure rate shall be measured at 1 centimeter above the tabletop or cradle.

(ii) If the source is above the x-ray table, the exposure rate shall be measured at 30 centimeters above the tabletop with the end of the beam-limiting device or spacer positioned as closely as possible to the point of measurement.

(iii) In a C-arm type of fluoroscope, the exposure rate shall be measured at 30 centimeters from the input surface of the fluoroscopic imaging assembly, with the source positioned at any available SID, provided that the end of the beam-limiting device or spacer is no closer than 30 centimeters from the input surface of the fluoroscopic imaging assembly.

(iv) In a lateral type of fluoroscope, the exposure rate shall be measured at a point 15 centimeters from the centerline of the x-ray table and in the direction of the x-ray source with the end of the beam-limiting device or spacer positioned as closely as possible to the point of measurement. If the tabletop is movable, it shall be positioned as closely as possible to the lateral x-ray source, with the end of the beam-limiting device or spacer no closer than 15 centimeters to the centerline of the x-ray table.

(4) *Exemptions.* Fluoroscopic radiation therapy simulation systems are exempt from the requirements set forth in paragraph (e) of this section.

(f) *Indication of potential and current.* During fluoroscopy and cinefluorography, x-ray tube potential and current shall be continuously indicated. Deviation of x-ray tube potential and current from the indicated values shall not exceed the maximum deviation as stated by the manufacturer in accordance with § 1020.30(h)(3).

(g) *Source-skin distance.* Means shall be provided to limit the source-skin distance to not less than 38 centimeters on stationary fluoroscopes and to not less than 30 centimeters on mobile and portable fluoroscopes. In addition, for image-intensified fluoroscopes intended for specific surgical application that would be prohibited at the source-skin distances specified in this paragraph, provisions may be made for operation at shorter source-skin distances but in no case less than 20 centimeters. When provided, the manufacturer must set forth precautions with respect to the optional means of spacing, in addition to other information as required in § 1020.30(h).

(h) *Fluoroscopic timer.* Means shall be provided to preset the cumulative on-time of the fluoroscopic tube. The maximum cumulative time of the timing device shall not exceed 5 minutes without resetting. A signal audible to the fluoroscopist shall indicate the completion of any preset cumulative on-time. Such signal shall continue to sound while x-rays are produced until the timing device is reset. As an alternative to the requirements of this paragraph, radiation therapy simulation systems may be provided with a means to indicate the total cumulative exposure time during which x-rays were produced, and which is capable of being reset between x-ray examinations.

(i) *Mobile and portable fluoroscopes.* In addition to the foregoing requirements of this section, mobile and portable fluoroscopes shall provide intensified imaging.

[58 FR 26404, May 3, 1993, as amended at 59 FR 26404, May 19, 1994]

EFFECTIVE DATE NOTE: At 70 FR 34039, June 10, 2005, § 1020.32 was revised, effective June 10, 2006. For the convenience of the user, the revised text is set forth as follows:

§ 1020.32 Fluoroscopic equipment.

The provisions of this section apply to equipment for fluoroscopic imaging or for recording images from the fluoroscopic image receptor, except computed tomography x-ray systems manufactured on or after November 29, 1984.

(a) *Primary protective barrier*—(1) *Limitation of useful beam.* The fluoroscopic imaging assembly shall be provided with a primary protective barrier which intercepts the entire cross section of the useful beam at any SID. The x-ray tube used for fluoroscopy shall not produce x-rays unless the barrier is in position to intercept the entire useful beam. The AKR due to transmission through the barrier with the attenuation block in the useful beam combined with radiation from the fluoroscopic image receptor shall not exceed 3.34×10^{-3} percent of the entrance AKR, at a distance of 10 cm from any accessible surface of the fluoroscopic imaging assembly beyond the plane of the image receptor. Radiation therapy simulation systems shall be exempt from this requirement provided the systems are intended only for remote control operation and the manufacturer sets forth instructions for assemblers with respect to control location as part of the information required in § 1020.30(g). Additionally, the manufacturer shall provide to users, under

§ 1020.30(h)(1)(i), precautions concerning the importance of remote control operation.

(2) *Measuring compliance.* The AKR shall be measured in accordance with paragraph (d) of this section. The AKR due to transmission through the primary barrier combined with radiation from the fluoroscopic image receptor shall be determined by measurements averaged over an area of 100 square cm with no linear dimension greater than 20 cm. If the source is below the tabletop, the measurement shall be made with the input surface of the fluoroscopic imaging assembly positioned 30 cm above the tabletop. If the source is above the tabletop and the SID is variable, the measurement shall be made with the end of the beam-limiting device or spacer as close to the tabletop as it can be placed, provided that it shall not be closer than 30 cm. Movable grids and compression devices shall be removed from the useful beam during the measurement. For all measurements, the attenuation block shall be positioned in the useful beam 10 cm from the point of measurement of entrance AKR and between this point and the input surface of the fluoroscopic imaging assembly.

(b) *Field limitation*—(1) *Angulation.* For fluoroscopic equipment manufactured after February 25, 1978, when the angle between the image receptor and the beam axis of the x-ray beam is variable, means shall be provided to indicate when the axis of the x-ray beam is perpendicular to the plane of the image receptor. Compliance with paragraphs (b)(4) and (b)(5) of this section shall be determined with the beam axis indicated to be perpendicular to the plane of the image receptor.

(2) *Further means for limitation.* Means shall be provided to permit further limitation of the x-ray field to sizes smaller than the limits of paragraphs (b)(4) and (b)(5). Beam-limiting devices manufactured after May 22, 1979, and incorporated in equipment with a variable SID and/or the capability of a visible area of greater than 300 square cm, shall be provided with means for stepless adjustment of the x-ray field. Equipment with a fixed SID and the capability of a visible area of no greater than 300 square cm shall be provided with either stepless adjustment of the x-ray field or with a means to further limit the x-ray field size at the plane of the image receptor to 125 square cm or less. Stepless adjustment shall, at the greatest SID, provide continuous field sizes from the maximum obtainable to a field size containable in a square of 5 cm by 5 cm. This paragraph does not apply to non-image-intensified fluoroscopy.

(3) *Non-image-intensified fluoroscopy.* The x-ray field produced by non-image-intensified fluoroscopic equipment shall not extend beyond the entire visible area of the image receptor. Means shall be provided for stepless adjustment of field size. The minimum field size, at the greatest SID, shall be containable in a square of 5 cm by 5 cm.

(4) *Fluoroscopy and radiography using the fluoroscopic imaging assembly with inherently circular image receptors.* (i) For fluoroscopic equipment manufactured before June 10, 2006, other than radiation therapy simulation systems, the following applies:

(A) Neither the length nor the width of the x-ray field in the plane of the image receptor shall exceed that of the visible area of the image receptor by more than 3 percent of the SID. The sum of the excess length and the excess width shall be no greater than 4 percent of the SID.

(B) For rectangular x-ray fields used with circular image receptors, the error in alignment shall be determined along the length and width dimensions of the x-ray field which pass through the center of the visible area of the image receptor.

(ii) For fluoroscopic equipment manufactured on or after June 10, 2006, other than radiation therapy simulation systems, the maximum area of the x-ray field in the plane of the image receptor shall conform with one of the following requirements:

(A) When any linear dimension of the visible area of the image receptor measured through the center of the visible area is less than or equal to 34 cm in any direction, at least 80 percent of the area of the x-ray field overlaps the visible area of the image receptor, or

(B) When any linear dimension of the visible area of the image receptor measured through the center of the visible area is greater than 34 cm in any direction, the x-ray field measured along the direction of greatest misalignment with the visible area of the image receptor does not extend beyond the edge of the visible area of the image receptor by more than 2 cm.

(5) *Fluoroscopy and radiography using the fluoroscopic imaging assembly with inherently rectangular image receptors.* For x-ray systems manufactured on or after June 10, 2006, the following applies:

(i) Neither the length nor the width of the x-ray field in the plane of the image receptor shall exceed that of the visible area of the image receptor by more than 3 percent of the SID. The sum of the excess length and the excess width shall be no greater than 4 percent of the SID.

(ii) The error in alignment shall be determined along the length and width dimensions of the x-ray field which pass through the center of the visible area of the image receptor.

(6) *Override capability.* If the fluoroscopic x-ray field size is adjusted automatically as the SID or image receptor size is changed, a capability may be provided for overriding the automatic adjustment in case of system failure. If it is so provided, a signal visible at the fluoroscopist's position shall indicate

whenever the automatic field adjustment is overridden. Each such system failure override switch shall be clearly labeled as follows:

For X-ray Field Limitation System Failure

(c) *Activation of tube.* X-ray production in the fluoroscopic mode shall be controlled by a device which requires continuous pressure by the operator for the entire time of any exposure. When recording serial radiographic images from the fluoroscopic image receptor, the operator shall be able to terminate the x-ray exposure(s) at any time, but means may be provided to permit completion of any single exposure of the series in process.

(d) *Air kerma rates.* For fluoroscopic equipment, the following requirements apply:

(1) *Fluoroscopic equipment manufactured before May 19, 1995*—(i) Equipment provided with automatic exposure rate control (AERC) shall not be operable at any combination of tube potential and current that will result in an AKR in excess of 88 mGy per minute (vice 10 R/min exposure rate) at the measurement point specified in § 1020.32(d)(3), except as specified in § 1020.32(d)(1)(v).

(ii) Equipment provided without AERC shall not be operable at any combination of tube potential and current that will result in an AKR in excess of 44 mGy per minute (vice 5 R/min exposure rate) at the measurement point specified in § 1020.32(d)(3), except as specified in § 1020.32(d)(1)(v).

(iii) Equipment provided with both an AERC mode and a manual mode shall not be operable at any combination of tube potential and current that will result in an AKR in excess of 88 mGy per minute (vice 10 R/min exposure rate) in either mode at the measurement point specified in § 1020.32(d)(3), except as specified in § 1020.32(d)(1)(v).

(iv) Equipment may be modified in accordance with § 1020.30(q) to comply with § 1020.32(d)(2). When the equipment is modified, it shall bear a label indicating the date of the modification and the statement:

Modified to comply with 21 CFR 1020.32(h)(2).

(v) Exceptions:

(A) During recording of fluoroscopic images, or

(B) When a mode of operation has an optional high-level control, in which case that mode shall not be operable at any combination of tube potential and current that will result in an AKR in excess of the rates specified in § 1020.32(d)(1)(i), (d)(1)(ii), or (d)(1)(iii) at the measurement point specified in § 1020.32(d)(3), unless the high-level control is activated. Special means of activation of high-level controls shall be required. The high-level control shall be operable only when continuous manual activation is provided by the operator. A continuous signal audible to the fluoroscopist shall indicate that the high-level control is being employed.

(2) *Fluoroscopic equipment manufactured on or after May 19, 1995*—(i) Shall be equipped with AERC if operable at any combination of tube potential and current that results in an AKR greater than 44 mGy per minute (vice 5 R/min exposure rate) at the measurement point specified in § 1020.32(d)(3). Provision for manual selection of technique factors may be provided.

(ii) Shall not be operable at any combination of tube potential and current that will result in an AKR in excess of 88 mGy per minute (vice 10 R/min exposure rate) at the measurement point specified in § 1020.32(d)(3), except as specified in § 1020.32(d)(2)(iii):

(iii) Exceptions:

(A) For equipment manufactured prior to June 10, 2006, during the recording of images from a fluoroscopic image receptor using photographic film or a video camera when the x-ray source is operated in a pulsed mode.

(B) For equipment manufactured on or after June 10, 2006, during the recording of images from the fluoroscopic image receptor for the purpose of providing the user with a recorded image(s) after termination of the exposure. Such recording does not include images resulting from a last-image-hold feature that are not recorded.

(C) When a mode of operation has an optional high-level control and the control is activated, in which case the equipment shall not be operable at any combination of tube potential and current that will result in an AKR in excess of 176 mGy per minute (vice 20 R/min exposure rate) at the measurement point specified in § 1020.32(d)(3). Special means of activation of high-level controls shall be required. The high-level control shall be operable only when continuous manual activation is provided by the operator. A continuous signal audible to the fluoroscopist shall indicate that the high-level control is being employed.

(3) *Measuring compliance.* Compliance with paragraph (d) of this section shall be determined as follows:

(i) If the source is below the x-ray table, the AKR shall be measured at 1 cm above the tabletop or cradle.

(ii) If the source is above the x-ray table, the AKR shall be measured at 30 cm above the tabletop with the end of the beam-limiting device or spacer positioned as closely as possible to the point of measurement.

(iii) In a C-arm type of fluoroscope, the AKR shall be measured at 30 cm from the input surface of the fluoroscopic imaging assembly, with the source positioned at any available SID, provided that the end of the beam-limiting device or spacer is no closer than 30 cm from the input surface of the fluoroscopic imaging assembly.

(iv) In a C-arm type of fluoroscope having an SID less than 45 cm, the AKR shall be measured at the minimum SSD.

(v) In a lateral type of fluoroscope, the air kerma rate shall be measured at a point 15 cm from the centerline of the x-ray table and in the direction of the x-ray source with the end of the beam-limiting device or spacer positioned as closely as possible to the point of measurement. If the tabletop is movable, it shall be positioned as closely as possible to the lateral x-ray source, with the end of the beam-limiting device or spacer no closer than 15 cm to the centerline of the x-ray table.

(4) *Exemptions.* Fluoroscopic radiation therapy simulation systems are exempt from the requirements set forth in paragraph (d) of this section.

(e) [Reserved]

(f) *Indication of potential and current.* During fluoroscopy and cinefluorography, x-ray tube potential and current shall be continuously indicated. Deviation of x-ray tube potential and current from the indicated values shall not exceed the maximum deviation as stated by the manufacturer in accordance with § 1020.30(h)(3).

(g) *Source-skin distance.* (1) Means shall be provided to limit the source-skin distance to not less than 38 cm on stationary fluoroscopes and to not less than 30 cm on mobile and portable fluoroscopes. In addition, for fluoroscopes intended for specific surgical application that would be prohibited at the source-skin distances specified in this paragraph, provisions may be made for operation at shorter source-skin distances but in no case less than 20 cm. When provided, the manufacturer must set forth precautions with respect to the optional means of spacing, in addition to other information as required in § 1020.30(h).

(2) For stationary, mobile, or portable C-arm fluoroscopic systems manufactured on or after June 10, 2006, having a maximum source-image receptor distance of less than 45 cm, means shall be provided to limit the source-skin distance to not less than 19 cm. Such systems shall be labeled for extremity use only. In addition, for those systems intended for specific surgical application that would be prohibited at the source-skin distances specified in this paragraph, provisions may be made for operation at shorter source-skin distances but in no case less than 10 cm. When provided, the manufacturer must set forth precautions with respect to the optional means of spacing, in addition to other information as required in § 1020.30(h).

(h) *Fluoroscopic irradiation time, display, and signal.* (1)(i) Fluoroscopic equipment manufactured before June 10, 2006, shall be provided with means to preset the cumulative irradiation time of the fluoroscopic tube. The maximum cumulative time of the timing device shall not exceed 5 minutes without resetting. A signal audible to the fluoroscopist shall indicate the completion of any preset cumulative irradiation-time. Such signal shall continue to sound while x-rays are produced until the timing device is reset. Fluoroscopic equipment may be modified in accordance with § 1020.30(q) to comply with the requirements of § 1020.32(h)(2). When the equipment is modified, it shall bear a label indicating the statement:

Modified to comply with 21 CFR 1020.32(h)(2).

(ii) As an alternative to the requirements of this paragraph, radiation therapy simulation systems may be provided with a means to indicate the total cumulative exposure time during which x-rays were produced, and which is capable of being reset between x-ray examinations.

(2) For x-ray controls manufactured on or after June 10, 2006, there shall be provided for each fluoroscopic tube:

(i) A display of the fluoroscopic irradiation time at the fluoroscopist's working position. This display shall function independently of the audible signal described in § 1020.32(h)(2)(ii). The following requirements apply:

(A) When the x-ray tube is activated, the fluoroscopic irradiation time in minutes and tenths of minutes shall be continuously displayed and updated at least once every 6 seconds.

(B) The fluoroscopic irradiation time shall also be displayed within 6 seconds of termination of an exposure and remain displayed until reset.

(C) Means shall be provided to reset the display to zero prior to the beginning of a new examination or procedure.

(ii) A signal audible to the fluoroscopist shall sound for each passage of 5 minutes of fluoroscopic irradiation time during an examination or procedure. The signal shall sound until manually reset or, if automatically reset, for at least 2 second.

(i) *Mobile and portable fluoroscopes.* In addition to the other requirements of this section, mobile and portable fluoroscopes shall provide an image receptor incorporating more than a simple fluorescent screen.

(j) *Display of last-image-hold (LIH).* Fluoroscopic equipment manufactured on or after June 10, 2006, shall be equipped with means to display LIH image following termination of the fluoroscopic exposure.

(1) For an LIH image obtained by retaining pretermination fluoroscopic images, if the number of images and method of combining images are selectable by the user, the selection shall be indicated prior to initiation of the fluoroscopic exposure.

(2) For an LIH image obtained by initiating a separate radiographic-like exposure at the termination of fluoroscopic imaging, the techniques factors for the LIH image shall be selectable prior to the fluoroscopic

exposure, and the combination selected shall be indicated prior to initiation of the fluoroscopic exposure.

(3) Means shall be provided to clearly indicate to the user whether a displayed image is the LIH radiograph or fluoroscopy. Display of the LIH radiograph shall be replaced by the fluoroscopic image concurrently with re-initiation of fluoroscopic exposure, unless separate displays are provided for the LIH radiograph and fluoroscopic images.

(4) The predetermined or selectable options for producing the LIH radiograph shall be described in the information required by §1020.30(h). The information shall include a description of any technique factors applicable for the selected option and the impact of the selectable options on image characteristics and the magnitude of radiation emissions.

(k) *Displays of values of AKR and cumulative air kerma.* Fluoroscopic equipment manufactured on or after June 10, 2006, shall display at the fluoroscopist's working position the AKR and cumulative air kerma. The following requirements apply for each x-ray tube used during an examination or procedure:

(1) When the x-ray tube is activated and the number of images produced per unit time is greater than six images per second, the AKR in mGy/min shall be continuously displayed and updated at least once every second.

(2) The cumulative air kerma in units of mGy shall be displayed either within 5 seconds of termination of an exposure or displayed continuously and updated at least once every 5 seconds.

(3) The display of the AKR shall be clearly distinguishable from the display of the cumulative air kerma.

(4) The AKR and cumulative air kerma shall represent the value for conditions of free-in-air irradiation at one of the following reference locations specified according to the type of fluoroscope. The reference location shall be identified and described specifically in the information provided to users according to §1020.30(h)(6)(iii).

(i) For fluoroscopes with x-ray source below the x-ray table, x-ray source above the table, or of lateral type, the reference locations shall be the respective locations specified in §1020.32(d)(3)(i), (d)(3)(ii), or (d)(3)(v) for measuring compliance with air-kerma rate limits.

(ii) For C-arm fluoroscopes, the reference location shall be 15 cm from the isocenter toward the x-ray source along the beam axis. Alternatively, the reference location shall be at a point specified by the manufacturer to represent the location of the intersection of the x-ray beam with the patient's skin.

(5) Means shall be provided to reset to zero the display of cumulative air kerma prior to the commencement of a new examination or procedure.

(6) The displayed AKR and cumulative air kerma shall not deviate from the actual values by more than ±35 percent over the range of 6 mGy/min and 100 mGy to the maximum indication of AKR and cumulative air kerma, respectively. Compliance shall be determined with an irradiation time greater than 3 seconds.

§ 1020.33 Computed tomography (CT) equipment.

(a) *Applicability.* (1) The provisions of this section, except for paragraphs (b), (c)(1), and (c)(2) are applicable as specified herein to CT x-ray systems manufactured or remanufactured on or after September 3, 1985.

(2) The provisions of paragraphs (b), (c)(1), and (c)(2) are applicable to CT x-ray systems manufactured or remanufactured on or after November 29, 1984.

(b) *Definitions.* As used in this section, the following definitions apply:

(1) *Computed tomography dose index (CTDI)* means the integral of the dose profile along a line perpendicular to the tomographic plane divided by the product of the nominal tomographic section thickness and the number of tomograms produced in a single scan; that is:

$$\mathrm{CTDI} = 1/\mathrm{nT} \int_{-7T}^{+7T} D(z)dz$$

where:

z=position along a line perpendicular to the tomographic plane.

D(z)=Dose at position z.

T=Nominal tomographic section thickness.

n=Number of tomograms produced in a single scan.

This definition assumes that the dose profile is centered around z=0 and that, for a multiple tomogram system, the scan increment between adjacent scans is nT.

(2) *Contrast scale* means the change in linear attenuation coefficient per CT number relative to water; that is:

$$\text{Contrast scale} = \frac{\mu_x - \mu_w}{(CT)_x - (CT)_w}$$

where:

μ_w=Linear attenuation coefficient of water.

μ_x=Linear attenuation coefficient of material of interest.
$(CT)_w$=CT number of water.
$(CT)_x$=CT number of material of interest.

(3) *CT conditions of operation* means all selectable parameters governing the operation of a CT x-ray system including nominal tomographic section thickness, filtration, and the technique factors as defined in § 1020.30(b)(36).

(4) *CT number* means the number used to represent the x-ray attenuation associated with each elemental area of the CT image.

(5) [Reserved]

(6) *CT dosimetry phantom* means the phantom used for determination of the dose delivered by a CT x-ray system. The phantom shall be a right circular cylinder of polymethl-methacrylate of density 1.19±0.01 grams per cubic centimeter. The phantom shall be at least 14 centimeters in length and shall have diameters of 32.0 centimeters for testing any CT system designed to image any section of the body (whole body scanners) and 16.0 centimeters for any system designed to image the head (head scanners) or for any whole body scanner operated in the head scanning mode. The phantom shall provide means for the placement of a dosimeter(s) along its axis of rotation and along a line parallel to the axis of rotation 1.0 centimeter from the outer surface and within the phantom. Means for the placement of a dosimeter(s) or alignment device at other locations may be provided for convenience. The means used for placement of a dosimeter(s) (i.e., hole size) and the type of dosimeter(s) used is at the discretion of the manufacturer. Any effect on the doses measured due to the removal of phantom material to accommodate dosimeters shall be accounted for through appropriate corrections to the reported data or included in the statement of maximum deviation for the values obtained using the phantom.

(7) *Dose profile* means the dose as a function of position along a line.

(8) *Modulation transfer function* means the modulus of the Fourier transform of the impulse response of the system.

(9) *Multiple tomogram system* means a CT x-ray system which obtains x-ray transmission data simultaneously during a single scan to produce more than one tomogram.

(10) *Noise* means the standard deviation of the fluctuations in CT number expressed as a percent of the attenuation coefficient of water. Its estimate (S_n) is calculated using the following expression:

$$S_n \frac{100 \times CS \times s}{\mu_w}$$

where:
CS=Contrast scale.
μ_w=Linear attenuation coefficient of water.
s=Estimated standard deviation of the CT numbers of picture elements in a specified area of the CT image.

(11) *Nominal tomographic section thickness* means the full-width at half-maximum of the sensitivity profile taken at the center of the cross-sectional volume over which x-ray transmission data are collected.

(12) *Picture element* means an elemental area of a tomogram.

(13) *Remanufacturing* means modifying a CT system in such a way that the resulting dose and imaging performance become substantially equivalent to any CT x-ray system manufactured by the original manufacturer on or after November 29, 1984. Any reference in this section to "manufacture", "manufacturer", or "manufacturing" includes remanufacture, remanufacturer, or remanufacturing, respectively.

(14) *Scan increment* means the amount of relative displacement of the patient with respect to the CT x-ray system between successive scans measured along the direction of such displacement.

(15) *Scan sequence* means a preselected set of two or more scans performed consecutively under preselected CT conditions of operations.

(16) *Sensitivity profile* means the relative response of the CT x-ray system as a function of position along a line perpendicular to the tomographic plane.

(17) *Single tomogram system* means a CT x-ray system which obtains x-ray transmission data during a scan to produce a single tomogram.

(18) *Tomographic plane* means that geometric plane which the manufacturer identifies as corresponding to the output tomogram.

(19) *Tomographic section* means the volume of an object whose x-ray attenuation properties are imaged in a tomogram.

(c) *Information to be provided for users.* Each manufacturer of a CT x-ray system shall provide the following technical and safety information, in addition to that required under § 1020.30(h), to purchasers and, upon request, to others at a cost not to exceed the cost of publication and distribution of such information. This information shall be identified and provided in a separate section of the user's instruction manual or in a separate manual devoted only to this information.

(1) *Conditions of operation.* A statement of the CT conditions of operation used to provide the information required by paragraph (c) (2) and (3) of this section.

(2) *Dose information.* The following dose information obtained by using the CT dosimetry phantom. For any CT x-ray system designed to image both the head and body, separate dose information shall be provided for each application. All dose measurements shall be performed with the CT dosimetry phantom placed on the patient couch or support device without additional attenuating materials present.

(i) The CTDI at the following locations in the dosimetry phantom:

(*a*) Along the axis of rotation of the phantom.

(*b*) Along a line parallel to the axis of rotation and 1.0 centimeter interior to the surface of the phantom with the phantom positioned so that CTDI is the maximum obtainable at this depth.

(*c*) Along lines parallel to the axis of rotation and 1.0 centimeter interior to the surface of the phantom at positions 90, 180, and 270 degrees from the position in paragraph (c)(2)(i)(*b*) of this section. The CT conditions of operation shall be the typical values suggested by the manufacturer for CT of the head or body. The location of the position where the CTDI is maximum as specified in paragraph (c)(2)(i)(*b*) of this section shall be given by the manufacturer with respect to the housing of the scanning mechanism or other readily identifiable feature of the CT x-ray system in such a manner as to permit placement of the dosimetry phantom in this orientation.

(ii) The CTDI in the center location of the dosimetry phantom for each selectable CT condition of operation that varies either the rate or duration of x-ray exposure. This CTDI shall be presented as a value that is normalized to the CTDI in the center location of the dosimetry phantom from paragraph (c)(2)(i) of this section, with the CTDI of paragraph (c)(2)(i) of this section having a value of one. As each individual CT condition of operation is changed, all other independent CT conditions of operation shall be maintained at the typical values described in paragraph (c)(2)(i) of this section. These data shall encompass the range of each CT condition of operation stated by the manufacturer as appropriate for CT of the head or body. When more than three selections of a CT condition of operation are available, the normalized CTDI shall be provided, at least, for the minimum, maximum, and midrange value of the CT condition of operation.

(iii) The CTDI at the location coincident with the maximum CTDI at 1 centimeter interior to the surface of the dosimetry phantom for each selectable peak tube potential. When more than three selections of peak tube potential are available, the normalized CTDI shall be provided, at least, for the minimum, maximum, and a typical value of peak tube potential. The CTDI shall be presented as a value that is normalized to the maximum CTDI located at 1 centimeter interior to the surface of the dosimetry phantom from paragraph (c)(2)(i) of this section, with the CTDI of paragraph (c)(2)(i) of this section having a value of one.

(iv) The dose profile in the center location of the dosimetry phantom for each selectable nominal tomographic section thickness. When more than three selections of nominal tomographic section thicknesses are available, the information shall be provided, at least, for the minimum, maximum, and midrange value of nominal tomographic section thickness. The dose profile shall be presented on the same graph and to the same scale as the corresponding sensitivity profile required by paragraph (c)(3)(iv) of this section.

(v) A statement of the maximum deviation from the values given in the information provided according to paragraph (c)(2) (i), (ii), (iii), and (iv) of this section. Deviation of actual values may not exceed these limits.

(3) *Imaging performance information.* The following performance data shall be provided for the CT conditions of operation used to provide the information required by paragraph (c)(2)(i) of this section. All other aspects of data collection, including the x-ray attenuation properties of the material in the tomographic section, shall be similar to those used to provide the dose information required by paragraph (c)(2)(i) of this section. For any CT x-ray system designed to image both the head and body, separate imaging performance information shall be provided for each application.

(i) A statement of the noise.

(ii) A graphical presentation of the modulation transfer function for the same image processing and display mode as that used in the statement of the noise.

(iii) A statement of the nominal tomographic section thickness(es).

(iv) A graphical presentation of the sensitivity profile, at the location corresponding to the center location of the dosimetry phantom, for each selectable nominal tomographic section thickness for which the dose profile is given according to paragraph (c)(2)(iv) of this section.

(v) A description of the phantom or device and test protocol or procedure used to determine the specifications and a statement of the maximum deviation from the specifications provided in accordance with paragraphs (c)(3) (i), (ii), (iii), and (iv) of this section. Deviation of actual values may not exceed these limits.

(d) *Quality assurance.* The manufacturer of any CT x-ray system shall provide the following with each system. All information required by this subsection shall be provided in a separate section of the user's instructional manual.

(1) A phantom(s) capable of providing an indication of contrast scale, noise, nominal tomographic section thickness, the spatial resolution capability of the system for low and high contrast objects, and measuring the mean CT number of water or a reference material.

(2) Instructions on the use of the phantom(s) including a schedule of testing appropriate for the system, allowable variations for the indicated parameters, and a method to store as records, quality assurance data.

(3) Representative images obtained with the phantom(s) using the same processing mode and CT conditions of operation as in paragraph (c)(3) of this section for a properly functioning system of the same model. The representative images shall be of two forms as follows:

(i) Photographic copies of the images obtained from the image display device.

(ii) Images stored in digital form on a storage medium compatible with the CT x-ray system. The CT x-ray system shall be provided with the means to display these images on the image display device.

(e) [Reserved]

(f) *Control and indication of conditions of operation*—(1) *Visual indication.* The CT conditions of operation to be used during a scan or a scan sequence shall be indicated prior to initiation of a scan or a scan sequence. On equipment having all or some of these conditions of operation at fixed values, this requirement may be met by permanent markings. Indication of the CT conditions of operation shall be visible from any position from which scan initiation is possible.

(2) *Timers.* (i) Means shall be provided to terminate the x-ray exposure automatically by either deenergizing the x-ray source or shuttering the x-ray beam in the event of equipment failure affecting data collection. Such termination shall occur within an interval that limits the total scan time to no more than 110 percent of its preset value through the use of either a backup timer or devices which monitor equipment function. A visible signal shall indicate when the x-ray exposure has been terminated through these means and manual resetting of the CT

conditions of operation shall be required prior to the initiation of another scan.

(ii) Means shall be provided so that the operator can terminate the x-ray exposure at any time during a scan, or series of scans under x-ray system control, of greater than one-half second duration. Termination of the x-ray exposure shall necessitate resetting of the CT conditions of operation prior to the initiation of another scan.

(g) *Tomographic plane indication and alignment.* (1) For any single tomogram system, means shall be provided to permit visual determination of the tomographic plane or a reference plane offset from the tomographic plane.

(2) For any multiple tomogram system, means shall be provided to permit visual determination of the location of a reference plane. The relationship of the reference plane to the planes of the tomograms shall be provided to the user in addition to other information provided according to § 1020.30(h). This reference plane can be offset from the location of the tomographic planes.

(3) The distance between the indicated location of the tomographic plane or reference plane and its actual location may not exceed 5 millimeters.

(4) For any offset alignment system, the manufacturer shall provide specific instructions with respect to the use of this system for patient positioning, in addition to other information provided according to § 1020.30(h).

(5) If a mechanism using a light source is used to satisfy the requirements of paragraphs (g) (1) and (2) of this section, the light source shall allow visual determination of the location of the tomographic plane or reference plane under ambient light conditions of up to 500 lux.

(h) *Beam-on and shutter status indicators.* (1) Means shall be provided on the control and on or near the housing of the scanning mechanism to provide visual indication when and only when x rays are produced and, if applicable, whether the shutter is open or closed. If the x-ray production period is less than one-half second, the indication of x-ray production shall be actuated for one-half second. Indicators at or near the housing of the scanning mechanism shall be discernible from any point external to the patient opening where insertion of any part of the human body into the primary beam is possible.

(2) For systems that allow high voltage to be applied to the x-ray tube continuously and that control the emission of x rays with a shutter, the radiation emitted may not exceed 100 milliroentgens (2.58×10^{-5} coulomb/kilogram) in 1 hour at any point 5 centimeters outside the external surface of the housing of the scanning mechanism when the shutter is closed. Compliance shall be determined by measurements averaged over an area of 100 square centimeters with no linear dimensions greater than 20 centimeters.

(i) *Scan increment accuracy.* The deviation of indicated scan increment from actual scan increment may not exceed 1 millimeter. Compliance shall be measured as follows: The determination of the deviation of indicated versus actual scan increment shall be based on measurements taken with a mass 100 kilograms or less, on the patient support device. The patient support device shall be incremented from a typical starting position to the maximum incrementation distance or 30 centimeters, whichever is less, and then returned to the starting position. Measurement of actual versus indicated scan increment may be taken anywhere along this travel.

(j) *CT number mean and standard deviation.* (1) A method shall be provided to calculate the mean and standard deviation of CT numbers for an array of picture elements about any location in the image. The number of elements in this array shall be under user control.

(2) The manufacturer shall provide specific instructions concerning the use of the method provided for calculation of CT number mean and standard deviation in addition to other information provided according to § 1020.30(h).

[49 FR 34712, Aug. 31, 1984; 49 37381, Sept. 24, 1984, as amended at 49 FR 47388, Dec. 4, 1984; 56 FR 36098, Aug. 1, 1991; 67 FR 9587, Mar. 4, 2002]

EFFECTIVE DATE NOTE: At 70 FR 34042, June 10, 2005, § 1020.33 was amended by revising paragraph (h)(2), effective June 10, 2006. For the convenience of the user, the revised text is set forth as follows:

§ 1020.33 Computed tomography (CT) equipment.

* * * * *

(h) * * *

(2) For systems that allow high voltage to be applied to the x-ray tube continuously and that control the emission of x-ray with a shutter, the radiation emitted may not exceed 0.88 milligray (vice 100 milliroentgen exposure) in 1 hour at any point 5 cm outside the external surface of the housing of the scanning mechanism when the shutter is closed. Compliance shall be determined by measurements average over an area of 100 square cm with no linear dimension greater than 20 cm.

* * * * *

§ 1020.40 Cabinet x-ray systems.

(a) *Applicability.* The provisions of this section are applicable to cabinet x-ray systems manufactured or assembled on or after April 10, 1975, except that the provisions as applied to x-ray systems designed primarily for the inspection of carry-on baggage are applicable to such systems manufactured or assembled on or after April 25, 1974. The provisions of this section are not applicable to systems which are designed exclusively for microscopic examination of material, e.g., x-ray diffraction, spectroscopic, and electron microscope equipment or to systems for intentional exposure of humans to x-rays.

(b) *Definitions.* As used in this section the following definitions apply:

(1) *Access panel* means any barrier or panel which is designed to be removed or opened for maintenance or service purposes, requires tools to open, and permits access to the interior of the cabinet.

(2) *Aperture* means any opening in the outside surface of the cabinet, other than a port, which remains open during generation of x radiation.

(3) *Cabinet x-ray system* means an x-ray system with the x-ray tube installed in an enclosure (hereinafter termed *cabinet*) which, independently of existing architectural structures except the floor on which it may be placed, is intended to contain at least that portion of a material being irradiated, provide radiation attenuation, and exclude personnel from its interior during generation of x radiation. Included are all x-ray systems designed primarily for the inspection of carry-on baggage at airline, railroad, and bus terminals, and in similar facilities. An x-ray tube used within a shielded part of a building, or x-ray equipment which may temporarily or occasionally incorporate portable shielding is not considered a cabinet x-ray system.

(4) *Door* means any barrier which is designed to be movable or opened for routine operation purposes, does not generally require tools to open, and permits access to the interior of the cabinet. For the purposes of paragraph (c)(4)(i) of this section, inflexible hardware rigidly affixed to the door shall be considered part of the door.

(5) *Exposure* means the quotient of dQ by dm where dQ is the absolute value of the total charge of the ions of one sign produced in air when all the electrons (negatrons and positrons) liberated by photons in a volume element of air having mass dm are completely stopped in air.

(6) *External surface* means the outside surface of the cabinet x-ray system, including the high-voltage generator, doors, access panels, latches, control knobs, and other permanently mounted hardware and including the plane across any aperture or port.

(7) *Floor* means the underside external surface of the cabinet.

(8) *Ground fault* means an accidental electrical grounding of an electrical conductor.

(9) *Port* means any opening in the outside surface of the cabinet which is designed to remain open, during generation of x-rays, for the purpose of conveying material to be irradiated into and out of the cabinet, or for partial insertion for irradiation of an object whose dimensions do not permit complete insertion into the cabinet.

(10) *Primary beam* means the x radiation emitted directly from the from the target and passing through the window of the x-ray tube.

(11) *Safety interlock* means a device which is intended to prevent the generation of x radiation when access by any part of the human body to the interior of the cabinet x-ray system through a door or access panel is possible.

(12) *X-ray system* means an assemblage of components for the controlled generation of x-rays.

(13) *X-ray tube* means any electron tube which is designed for the conversion of electrical energy into x-ray energy.

(c) *Requirements*—(1) *Emission limit.* (i) Radiation emitted from the cabinet x-ray system shall not exceed an exposure of 0.5 milliroentgen in one hour at any point five centimeters outside the external surface.

(ii) Compliance with the exposure limit in paragraph (c)(1)(i) of this section shall be determined by measurements averaged over a cross-sectional area of ten square centimeters with no linear dimension greater than 5 centimeters, with the cabinet x-ray system operated at those combinations of x-ray tube potential, current, beam orientation, and conditions of scatter radiation which produce the maximum x-ray exposure at the external surface, and with the door(s) and access panel(s) fully closed as well as fixed at any other position(s) which will allow the generation of x radiation.

(2) *Floors.* A cabinet x-ray system shall have a permanent floor. Any support surface to which a cabinet x-ray system is permanently affixed may be deemed the floor of the system.

(3) *Ports and apertures.* (i) The insertion of any part of the human body through any port into the primary beam shall not be possible.

(ii) The insertion of any part of the human body through any aperture shall not be possible.

(4) *Safety interlocks.* (i) Each door of a cabinet x-ray system shall have a minimum of two safety interlocks. One, but not both of the required interlocks shall be such that door opening results in physical disconnection of the energy supply circuit to the high-voltage generator, and such disconnection shall not be dependent upon any moving part other than the door.

(ii) Each access panel shall have at least one safety interlock.

(iii) Following interruption of x-ray generation by the functioning of any safety interlock, use of a control provided in accordance with paragraph (c)(6)(ii) of this section shall be necessary for resumption of x-ray generation.

(iv) Failure of any single component of the cabinet x-ray system shall not cause failure of more than one required safety interlock.

(5) *Ground fault.* A ground fault shall not result in the generation of x-rays.

(6) *Controls and indicators for all cabinet x-ray systems.* For all systems to which this section is applicable there shall be provided:

(i) A key-actuated control to insure that x-ray generation is not possible with the key removed.

(ii) A control or controls to initiate and terminate the generation of x-rays other than by functioning of a safety interlock or the main power control.

(iii) Two independent means which indicate when and only when x-rays are being generated, unless the x-ray generation period is less than one-half second, in which case the indicators shall be activated for one-half second, and which are discernible from any point at which initiation of x-ray generation is possible. Failure of a single component of the cabinet x-ray system shall not cause failure of both indicators to perform their intended function. One, but not both, of the indicators required by this subdivision may be a milliammeter labeled to indicate x-ray tube current. All other indicators shall be legibly labeled "X-RAY ON".

(iv) Additional means other than milliammeters which indicate when and only when x-rays are being generated, unless the x-ray generation period is less than one-half second in which case the indicators shall be activated for one-half second, as needed to insure that at least one indicator is visible from each door, access panel, and port, and is legibly labeled "X-RAY ON".

(7) *Additional controls and indicators for cabinet x-ray systems designed to admit humans.* For cabinet x-ray systems designed to admit humans there shall also be provided:

(i) A control within the cabinet for preventing and terminating x-ray generation, which cannot be reset, overridden or bypassed from the outside of the cabinet.

(ii) No means by which x-ray generation can be initiated from within the cabinet.

(iii) Audible and visible warning signals within the cabinet which are actuated for at least 10 seconds immediately prior to the first initiation of x-ray generation after closing any door designed to admit humans. Failure of any single component of the cabinet x-ray system shall not cause failure of both the audible and visible warning signals.

(iv) A visible warning signal within the cabinet which remains actuated when and only when x-rays are being generated, unless the x-ray generation period is less than one-half second in which case the indicators shall be activated for one-half second.

(v) Signs indicating the meaning of the warning signals provided pursuant to paragraphs (c)(7) (iii) and (iv) of this section and containing instructions for the use of the control provided pursuant to paragraph (c)(7)(i) of this section. These signs shall be legible, accessible to view, and illuminated when the main power control is in the "on" position.

(8) *Warning labels.* (i) There shall be permanently affixed or inscribed on the cabinet x-ray system at the location of any controls which can be used to initiate x-ray generation, a clearly legible and visible label bearing the statement:

CAUTION: X-RAYS PRODUCED WHEN ENERGIZED

(ii) There shall be permanently affixed or inscribed on the cabinet x-ray system adjacent to each port a clearly legible and visible label bearing the statement:

CAUTION: DO NOT INSERT ANY PART OF THE BODY WHEN SYSTEM IS ENERGIZED—X-RAY HAZARD

(9) *Instructions.* (i) Manufacturers of cabinet x-ray systems shall provide for purchasers, and to others upon request at a cost not to exceed the cost of preparation and distribution, manuals and instructions which shall include at least the following technical and safety information: Potential, current, and duty cycle ratings of the x-ray generation equipment; adequate instructions concerning any radiological safety procedures and precautions which may be necessary because of unique features of the system; and a schedule of maintenance necessary to keep the system in compliance with this section.

(ii) Manufacturers of cabinet x-ray systems which are intended to be assembled or installed by the purchaser shall provide instructions for assembly, installation, adjustment and testing of the cabinet x-ray system adequate to assure that the system is in compliance with applicable provisions of this section when assembled, installed, adjusted and tested as directed.

(10) *Additional requirements for x-ray baggage inspection systems.* X-ray systems designed primarily for the inspection of carry-on baggage at airline, railroad, and bus terminals, and at similar facilities, shall be provided with means, pursuant to paragraphs (c)(10) (i) and (ii) of this section, to insure operator presence at the control area in a position which permits surveillance of the ports and doors during generation of x-radiation.

(i) During an exposure or preset succession of exposures of one-half second or greater duration, the means provided shall enable the operator to terminate the exposure or preset succession of exposures at any time.

(ii) During an exposure or preset succession of exposures of less than one-half second duration, the means provided may allow completion of the exposure in progress but shall enable the operator to prevent additional exposures.

(d) *Modification of a certified system.* The modification of a cabinet x-ray system, previously certified pursuant to § 1010.2 by any person engaged in the business of manufacturing, assembling or modifying cabinet x-ray systems shall be construed as manufacturing under the act if the modification affects any aspect of the system's performance for which this section has an applicable requirement. The manufacturer who performs such modification shall recertify and reidentify the system in accordance with the provisions of §§ 1010.2 and 1010.3 of this chapter.

[39 FR 12986, Apr. 10, 1974]

PART 1030—PERFORMANCE STANDARDS FOR MICROWAVE AND RADIO FREQUENCY EMITTING PRODUCTS

AUTHORITY: 21 U.S.C. 351, 352, 360, 360e–360j, 371, 381; 42 U.S.C. 263b–263n.

§ 1030.10 Microwave ovens.

(a) *Applicability.* The provisions of this standard are applicable to microwave ovens manufactured after October 6, 1971.

(b) *Definitions.* (1) *Microwave oven* means a device designed to heat, cook, or dry food through the application of electromagnetic energy at frequencies assigned by the Federal Communications Commission in the normal ISM heating bands ranging from 890 megahertz to 6,000 megahertz. As defined in this standard, "microwave ovens" are limited to those manufactured for use in homes, restaurants, food vending, or service establishments, on interstate carriers, and in similar facilities.

(2) *Cavity* means that portion of the microwave oven in which food may be heated, cooked, or dried.

(3) *Door* means the movable barrier which prevents access to the cavity during operation and whose function is to prevent emission of microwave energy from the passage or opening which provides access to the cavity.

(4) *Safety interlock* means a device or system of devices which is intended to prevent generation of microwave energy when access to the cavity is possible.

(5) *Service adjustments or service procedures* means those servicing methods prescribed by the manufacturer for a specific product model.

(6) *Stirrer* means that feature of a microwave oven which is intended to provide uniform heating of the load by constantly changing the standing wave pattern within the cavity or moving the load.

(7) *External surface* means the outside surface of the cabinet or enclosure provided by the manufacturer as part of the microwave oven, including doors, door handles, latches, and control knobs.

(8) *Equivalent plane-wave power density* means the square of the root-mean-square (rms) electric field strength divided by the impedance of free space (377 ohms).

(c) *Requirements*—(1) *Power density limit.* The equivalent plane-wave power density existing in the proximity of the external oven surface shall not exceed 1 milliwatt per square centimeter at any point 5 centimeters or more from the external surface of the oven, measured prior to acquisition by a purchaser, and, thereafter, 5 milliwatts per square centimeter at any such point.

(2) *Safety interlocks.* (i) Microwave ovens shall have a minimum of two operative safety interlocks. At least one operative safety interlock on a fully assembled microwave oven shall not be operable by any part of the human body, or any object with a straight insertable length of 10 centimeters. Such interlock must also be concealed, unless its actuation is prevented when access to the interlock is possible. Any visible actuator or device to prevent actuation of this safety interlock must not be removable without disassembly of the oven or its door. A magnetically operated interlock is considered to be concealed, or its actuation is considered to be prevented, only if a test magnet held in place on the oven by gravity or its own attraction cannot operate the safety interlock. The test magnet shall be capable of lifting vertically at zero air gap at least 4.5 kilograms, and at 1 centimeter air gap at least 450 grams when the face of the magnet, which is toward the interlock when the magnet is in the test position, is pulling against one of the large faces of a mild steel armature having dimensions of 80 millimeters by 50 millimeters by 8 millimeters.

(ii) Failure of any single mechanical or electrical component of the microwave oven shall not cause all safety interlocks to be inoperative.

(iii) Service adjustments or service procedures on the microwave oven shall not cause the safety interlocks to become inoperative or the microwave radiation emission to exceed the power density limits of this section as a result of such service adjustments or procedures.

(iv) Microwave radiation emission in excess of the limits specified in paragraph (c)(1) of this section shall not be caused by insertion of an insulated wire through any opening in the external surfaces of a fully assembled oven into the cavity, waveguide, or other microwave-energy-containing spaces while the door is closed, provided the wire, when inserted, could consist of two straight segments forming an obtuse angle of not less than 170 degrees.

(v) One (the primary) required safety interlock shall prevent microwave radiation emission in excess of the requirement of paragraph (c)(1) of this section; the other (secondary) required safety interlock shall prevent microwave radiation emission in excess of 5 milliwatts per square centimeter at any point 5 centimeters or more from the external surface of the oven. The two required safety interlocks shall be designated as primary or secondary in the service instructions for the oven.

(vi) A means of monitoring one or both of the required safety interlocks shall be provided which shall cause the oven to become inoperable and remain so until repaired if the required safety interlock(s) should fail to perform required functions as specified in this section. Interlock failures shall not disrupt the monitoring function.

(3) *Measurement and test conditions.* (i) Compliance with the power density limit in paragraph (c)(1) of this section shall be determined by measurement of the equivalent plane-wave power density made with an instrument which reaches 90 percent of its steady-state reading within 3 seconds, when the system is subjected to a step-function input signal. Tests for compliance shall account for all measurement errors and uncertainties to ensure that the equivalent plane-wave power density does not exceed the limit prescribed by paragraph (c)(1) of this section.

(ii) Microwave ovens shall be in compliance with the power density limits if the maximum reading obtained at the location of greatest microwave radiation emission, taking into account all measurement errors and uncertainties, does not exceed the limit specified in paragraph (c)(1) of this section, when the emission is measured through at least one stirrer cycle. As provided in § 1010.13 of this chapter, a manufacturer may request alternative test procedures if, as a result of the stirrer characteristics of a microwave oven, such oven is not susceptible to testing by the procedures described in this paragraph.

(iii) Measurements shall be made with the microwave oven operating at its maximum output and containing a load of 275 ±15 milliliters of tap water initially at 20 ±5 °centigrade placed within the cavity at the center of the load-carrying surface provided by the manufacturer. The water container shall be a low form 600-milliliter beaker having an inside diameter of approximately 8.5 centimeters and made of an electrically nonconductive material such as glass or plastic.

(iv) Measurements shall be made with the door fully closed as well as with the door fixed in any other position which allows the oven to operate.

(4) *User instructions.* Manufacturers of microwave ovens to which this section is applicable shall provide, or cause to be provided, with each oven, radiation safety instructions which:

(i) Occupy a separate section and are an integral part of the regularly supplied users' manual and cookbook, if supplied separately, and are located so as to elicit the attention of the reader.

(ii) Are as legible and durable as other instructions with the title emphasized to elicit the attention of the reader by such means as bold-faced type, contrasting color, a heavy-lined border, or by similar means.

(iii) Contain the following wording:

PRECAUTIONS TO AVOID POSSIBLE EXPOSURE TO EXCESSIVE MICROWAVE ENERGY

(*a*) Do not attempt to operate this oven with the door open since open-door operation can result in harmful exposure to microwave energy. It is important not to defeat or tamper with the safety interlocks.

(*b*) Do not place any object between the oven front face and the door or allow soil or cleaner residue to accumulate on sealing surfaces.

(*c*) Do not operate the oven if it is damaged. It is particularly important that the oven door close properly and that there is no damage to the: (*1*) Door (bent), (*2*) hinges and latches (broken or loosened), (*3*) door seals and sealing surfaces.

(*d*) The oven should not be adjusted or repaired by anyone except properly qualified service personnel.

(iv) Include additional radiation safety precautions or instructions which may be necessary for particular oven designs or models, as determined by the Director, Center for Devices and Radiological Health or the manufacturer.

(5) *Service instructions.* Manufacturers of microwave ovens to which this section is applicable shall provide or cause to be provided to servicing dealers and distributors and to others upon request, for each oven model, adequate instructions for service adjustments and service procedures, and, in addition, radiation safety instructions which:

(i) Occupy a separate section and are an integral part of the regularly supplied service manual and are located so as to elicit the attention of the reader.

(ii) Are as legible and durable as other instructions with the title emphasized so as to elicit the attention of the reader by such means as bold-faced type, contrasting color, a heavy-lined border, or by similar means.

(iii) Contain the following wording:

PRECAUTIONS TO BE OBSERVED BEFORE AND DURING SERVICING TO AVOID POSSIBLE EXPOSURE TO EXCESSIVE MICROWAVE ENERGY

(*a*) Do not operate or allow the oven to be operated with the door open.

(*b*) Make the following safety checks on all ovens to be serviced before activating the magnetron or other microwave source, and make repairs as necessary: (*1*) Interlock operation, (*2*) proper door closing, (*3*) seal and sealing surfaces (arcing, wear, and other damage), (*4*) damage to or loosening of hinges and latches, (*5*) evidence of dropping or abuse.

(*c*) Before turning on microwave power for any service test or inspection within the microwave generating compartments, check the magnetron, wave guide or transmission line, and cavity for proper alignment, integrity, and connections.

(*d*) Any defective or misadjusted components in the interlock, monitor, door seal, and microwave generation and transmission systems shall be repaired, replaced, or adjusted by procedures described in this manual before the oven is released to the owner.

(*e*) A Microwave leakage check to verify compliance with the Federal performance standard should be performed on each oven prior to release to the owner.

(iv) Include additional radiation safety precautions or instructions which may be necessary for particular oven designs or models, as determined by the Director, Center for Devices and Radiological Health or the manufacturer.

(6) *Warning labels.* Except as provided in paragraph (c)(6)(iv) of this section, microwave ovens shall have the following warning labels:

(i) A label, permanently attached to or inscribed on the oven, which shall be legible and readily viewable during normal oven use, and which shall have the title emphasized and be so located as to elicit the attention of the user. The label shall bear the following warning statement:

PRECAUTIONS FOR SAFE USE TO AVOID POSSIBLE EXPOSURE TO EXCESSIVE MICROWAVE ENERGY

DO NOT Attempt to Operate This Oven With:

(*a*) Object Caught in Door.

(*b*) Door That Does Not Close Properly.

(*c*) Damaged Door, Hinge, Latch, or Sealing Surface.

(ii) A label, permanently attached to or inscribed on the external surface of the oven, which shall be legible and readily viewable during servicing, and which shall have the word "CAUTION" emphasized and be so located as to elicit the attention of service personnel. The label shall bear the following warning statement:

CAUTION: This Device is to be Serviced Only by Properly Qualified Service Personnel. Consult the Service Manual for Proper Service Procedures to Assure Continued Compliance with the Federal Performance Standard for Microwave Ovens and for Precautions to be Taken to Avoid Possible Exposure to Excessive Microwave Energy.

(iii) The labels provided in accordance with paragraphs (c)(6)(i) and (ii) of this section shall bear only the statements specified in that paragraph, except for additional radiation safety warnings or instructions which may be necessary for particular oven designs or models, as determined by the Director, Center for Devices and Radiological Health or the manufacturer.

(iv) Upon application by a manufacturer, the Director, Center for Devices and Radiological Health, Food and Drug Administration, may grant an exemption from one or more of the statements (radiation safety warnings) specified in paragraph (c)(6)(i) of this section. Such exemption shall be based upon a determination by the Director that the microwave oven model for which the exemption is sought should continue to comply with paragraphs (c) (1), (2), and (3) of this section under the adverse condition of use addressed by such precautionary statement(s). An original and two copies of applications shall be submitted to the Division of Dockets Management (HFA–305), Food and Drug Administration, 5630 Fishers Lane, rm. 1061, Rockville, MD 20852. Copies of the written portion of the application, including supporting data and information, and the Director's action on the application will be maintained by the Branch for public review. The application shall include:

(*a*) The specific microwave oven model(s) for which the exemption is sought.

(*b*) The specific radiation safety warning(s) from which exemption is sought.

(*c*) Data and information which clearly establish that one or more of the radiation safety warnings in paragraph (c)(6)(i) of this section is not necessary for the specified microwave oven model(s).

(*d*) Such other information and a sample of the applicable product if required by regulation or by the Director, Center for Devices and Radiological Health, to evaluate and act on the application.

[38 FR 28640, Oct. 15, 1973, as amended at 40 FR 14752, Apr. 4, 1975; 40 FR 52007, Nov. 7, 1975; 46 FR 8461, Jan. 27, 1981; 48 FR 57482, Dec. 30, 1983; 50 FR 13566, Apr. 5, 1985; 53 FR 11254, Apr. 6, 1988; 59 FR 14365, Mar. 28, 1994]

PART 1040—PERFORMANCE STANDARDS FOR LIGHT-EMITTING PRODUCTS

Sec.

AUTHORITY: 21 U.S.C. 351, 352, 360, 360e–360j, 371, 381; 42 U.S.C. 263b–263n.

§ 1040.10 Laser products.

(a) *Applicability*. The provisions of this section and § 1040.11, as amended, are applicable as specified to all laser products manufactured or assembled after August 1, 1976, except when:

(1) Such a laser product is either sold to a manufacturer of an electronic product for use as a component (or replacement) in such electronic product, or

(2) Sold by or for a manufacturer of an electronic product for use as a component (or replacement) in such electronic product, provided that such laser product:

(i) Is accompanied by a general warning notice that adequate instructions for the safe installation of the laser product are provided in servicing information available from the complete laser product manufacturer under paragraph (h)(2)(ii) of this section, and should be followed,

(ii) Is labeled with a statement that it is designated for use solely as a component of such electronic product and therefore does not comply with the appropriate requirements of this section and § 1040.11 for complete laser products, and

(iii) Is not a removable laser system as described in paragraph (c)(2) of this section; and

(3) The manufacturer of such a laser product, if manufactured after August 20, 1986:

(i) Registers, and provides a listing by type of such laser products manufactured that includes the product name, model number and laser medium or emitted wavelength(s), and the name and address of the manufacturer. The manufacturer must submit the registration and listing to the Director, Office of Compliance (HFZ–300), Center for Devices and Radiological Health, 2094 Gaither Rd., Rockville, MD 20850.

(ii) Maintains and allows access to any sales, shipping, or distribution records that identify the purchaser of such a laser product by name and address, the product by type, the number of units sold, and the date of sale (shipment). These records shall be maintained and made available as specified in § 1002.31.

(b) *Definitions*. As used in this section and § 1040.11, the following definitions apply:

(1) *Accessible emission level* means the magnitude of accessible laser or collateral radiation of a specific wavelength and emission duration at a particular point as measured according to paragraph (e) of this section. Accessible laser or collateral radiation is radiation to which human access is possible, as defined in paragraphs (b) (12), (15), and (22) of this section.

(2) *Accessible emission limit* means the maximum accessible emission level permitted within a particular class as set forth in paragraphs (c), (d), and (e) of this section.

(3) *Aperture* means any opening in the protective housing or other enclosure of a laser product through which laser or collateral radiation is emitted, thereby allowing human access to such radiation.

(4) *Aperture stop* means an opening serving to limit the size and to define the shape of the area over which radiation is measured.

(5) *Class I laser product* means any laser product that does not permit access during the operation to levels of laser radiation in excess of the accessible emission limits contained in table I of paragraph (d) of this section.[1]

(6) *Class IIa laser product* means any laser product that permits human access during operation to levels of visible laser radiation in excess of the accessible emission limits contained in table I, but does not permit human access during operation to levels of laser radiation in excess of the accessible emission limits contained in table II-A of paragraph (d) of this section.[2]

(7) *Class II laser product* means any laser product that permits human access during operation to levels of visible laser radiation in excess of the accessible emission limits contained in table II-A, but does not permit human access during operation to levels of laser radiation in excess of the accessible emission limits contained in table II of paragraph (d) of this section.[3]

(8) *Class IIIa laser product* means any laser product that permits human access during operation to levels of visible laser radiation in excess of the accessible emission limits contained in table II, but does not permit human access during operation to levels of laser radiation in excess of the accessible emission limits contained in table III-A of paragraph (d) of this section.[4]

(9) *Class IIIb laser product* means any laser product that permits human access during operation to levels of laser radiation in excess of the accessible emission limits of table III-A, but does not permit human access during operation to levels of laser radiation in excess of the accessible emission limits contained in table III-B of paragraph (d) of this section.[5]

(10) *Class III laser product* means any Class IIIa or Class IIIb laser product.

(11) *Class IV laser product* means any laser that permits human access during operation to levels of laser radiation in excess of the accessible emission limits contained in table III-B of paragraph (d) of this section.[6]

(12) *Collateral radiation* means any electronic product radiation, except laser radiation, emitted by a laser product as a result of the operation of the laser(s) or any component of the laser product that is physically necessary for the operation of the laser(s).

(13) *Demonstration laser product* means any laser product manufactured, designed, intended, or promoted for purposes of demonstration, entertainment, advertising display, or artistic composition. The term "demonstration laser product" does not apply to laser products which are not manufactured, designed, intended, or promoted for

[1] Class I levels of laser radiation are not considered to be hazardous.

[2] Class IIa levels of laser radiation are not considered to be hazardous if viewed for any period of time less than or equal to 1×10^3 seconds but are considered to be a chronic viewing hazard for any period of time greater than 1×10^3 seconds.

[3] Class II levels of laser radiation are considered to be a chronic viewing hazard.

[4] Class IIIa levels of laser radiation are considered to be, depending upon the irradiance, either an acute intrabeam viewing hazard or chronic viewing hazard, and an acute viewing hazard if viewed directly with optical instruments.

[5] Class IIIb levels of laser radiation are considered to be an acute hazard to the skin and eyes from direct radiation.

[6] Class IV levels of laser radiation are considered to be an acute hazard to the skin and eyes from direct and scattered radiation.

such purposes, even though they may be used for those purposes or are intended to demonstrate other applications.

(14) *Emission duration* means the temporal duration of a pulse, a series of pulses, or continuous operation, expressed in seconds, during which human access to laser or collateral radiation could be permitted as a result of operation, maintenance, or service of a laser product.

(15) *Human access* means the capacity to intercept laser or collateral radiation by any part of the human body. For laser products that contain Class IIIb or IV levels of laser radiation, "human access" also means access to laser radiation that can be reflected directly by any single introduced flat surface from the interior of the product through any opening in the protective housing of the product.

(16) *Integrated radiance* means radiant energy per unit area of a radiating surface per unit solid angle of emission, expressed in joules per square centimeter per steradian ($Jcm^{-2} sr^{-1}$).

(17) *Invisible radiation* means laser or collateral radiation having wavelengths of equal to or greater than 180 nm but less than or equal to 400 nm or greater than 710 nm but less than or equal to 1.0×10^6 nm (1 millimeter).

(18) *Irradiance* means the time-averaged radiant power incident on an element of a surface divided by the area of that element, expressed in watts per square centimeter ($W cm^{-2}$).

(19) *Laser* means any device that can be made to produce or amplify electromagnetic radiation at wavelenghts greater than 250 nm but less than or equal to 13,000 nm or, after August 20, 1986, at wavelengths equal to or greater than 180 nm but less than or equal to 1.0×10^6 nm primarily by the process of controlled stimulated emission.

(20) *Laser energy source* means any device intended for use in conjunction with a laser to supply energy for the operation of the laser. General energy sources such as electrical supply mains or batteries shall not be considered to constitute laser energy sources.

(21) *Laser product* means any manufactured product or assemblage of components which constitutes, incorporates, or is intended to incorporate a laser or laser system. A laser or laser system that is intended for use as a component of an electronic product shall itself be considered a laser product.

(22) *Laser radiation* means all electromagnetic radiation emitted by a laser product within the spectral range specified in paragraph (b)(19) of this section that is produced as a result of controlled stimulated emission or that is detectable with radiation so produced through the appropriate aperture stop and within the appropriate solid angle of acceptance, as specified in paragraph (e) of this section.

(23) *Laser system* means a laser in combination with an appropriate laser energy source with or without additional incorporated components. See paragraph (c)(2) of this section for an explanation of the term "removable laser system."

(24) *Maintenance* means performance of those adjustments or procedures specified in user information provided by the manufacturer with the laser product which are to be performed by the user for the purpose of assuring the intended performance of the product. It does not include operation or service as defined in paragraph (b) (27) and (38) of this section.

(25) *Maximum output* means the maximum radiant power and, where applicable, the maximum radiant energy per pulse of accessible laser radiation emitted by a laser product during operation, as determined under paragraph (e) of this section.

(26) *Medical laser product* means any laser product which is a medical device as defined in 21 U.S.C. 321(h) and is manufactured, designed, intended or promoted for in vivo laser irradiation of any part of the human body for the purpose of: (i) Diagnosis, surgery, or therapy; or (ii) relative positioning of the human body.

(27) *Operation* means the performance of the laser product over the full range of its functions. It does not include maintenance or service as defined in paragraphs (b) (24) and (38) of this section.

(28) *Protective housing* means those portions of a laser product which are designed to prevent human access to laser or collateral radiation in excess

of the prescribed accessible emission limits under conditions specified in this section and in § 1040.11.

(29) *Pulse duration* means the time increment measured between the half-peak-power points at the leading and trailing edges of a pulse.

(30) *Radiance* means time-averaged radiant power per unit area of a radiating surface per unit solid angle of emission, expressed in watts per square centimeter per steradian ($W\ cm^{-2}\ sr^{-1}$).

(31) *Radiant energy* means energy emitted, transferred or received in the form of radiation, expressed in joules (J).

(32) *Radiant exposure* means the radiant energy incident on an element of a surface divided by the area of the element, expressed in joules per square centimeter (Jcm^{-2})

(33) *Radiant power* means time-averaged power emitted, transferred or received in the form of radiation, expressed in watts (W).

(34) *Remote interlock connector* means an electrical connector which permits the connection of external remote interlocks.

(35) *Safety interlock* means a device associated with the protective housing of a laser product to prevent human access to excessive radiation in accordance with paragraph (f)(2) of this section.

(36) *Sampling interval* means the time interval during which the level of accessible laser or collateral radiation is sampled by a measurement process. The magnitude of the sampling interval in units of seconds is represented by the symbol (t).

(37) *Scanned laser radiation* means laser radiation having a time-varying direction, origin or pattern of propagation with respect to a stationary frame of reference.

(38) *Service* means the performance of those procedures or adjustments described in the manufacturer's service instructions which may affect any aspect of the product's performance for which this section and § 1040.11 have applicable requirements. It does not include maintenance or operation as defined in paragraphs (b) (24) and (27) of this section.

(39) *Surveying, leveling, or alignment laser product* means a laser product manufactured, designed, intended or promoted for one or more of the following uses:

(i) Determining and delineating the form, extent, or position of a point, body, or area by taking angular measurement.

(ii) Positioning or adjusting parts in proper relation to one another.

(iii) Defining a plane, level, elevation, or straight line.

(40) *Visible radiation* means laser or collateral radiation having wavelengths of greater than 400 nm but less than or equal to 710 nm.

(41) *Warning logotype* means a logotype as illustrated in either figure 1 or figure 2 of paragraph (g) of this section.

(42) *Wavelength* means the propagation wavelength in air of electromagnetic radiation.

(c) *Classification of laser products*—(1) *All laser products.* Each laser product shall be classified in Class I, IIa, II, IIIa, IIIb, or IV in accordance with definitions set forth in paragraphs (b) (5) through (11) of this section. The product classification shall be based on the highest accessible emission level(s) of laser radiation to which human access is possible during operation in accordance with paragraphs (d), (e), and (f)(1) of this section.

(2) *Removable laser systems.* Any laser system that is incorporated into a laser product subject to the requirements of this section and that is capable, without modification, of producing laser radiation when removed from such laser product, shall itself be considered a laser product and shall be separately subject to the applicable requirements in this subchapter for laser products of its class. It shall be classified on the basis of accessible emission of laser radiation when so removed.

(d) *Accessible emission limits.* Accessible emission limits for laser radiation in each class are specified in tables I, II-A, II, III-A, and III-B of this paragraph. The factors, k_1 and k_2 vary with wavelength and emission duration. These factors are given in table IV of this paragraph, with selected numerical values in table V of this paragraph.

208-072 D-22

Accessible emission limits for collateral radiation are specified in table VI of this paragraph.

NOTES APPLICABLE TO TABLES I, II-A, II, III-A AND III-B: (1) The factors k_1 and k_2 are wavelength-dependent correction factors determined from table IV.

(2) The variable t in the expressions of emission limits is the magnitude of the sampling interval in units of seconds.

TABLE I

CLASS I ACCESSIBLE EMISSION LIMITS FOR LASER RADIATION

Wavelength (nanometers)	Emission duration (seconds)	Class I-Accessible emission limits (value)	(unit)	(quantity)**
≥ 180 but ≤ 400	$\leq 3.0 \times 10^{4}$	$2.4 \times 10^{-5} k_1 k_2$*	Joules(J)*	radiant energy
	$>3.0 \times 10^{4}$	$8.0 \times 10^{-10} k_1 k_2$*	Watts(W)*	radiant power
>400 but ≤ 1400	$>1.0 \times 10^{-9}$ to 2.0×10^{-5}	$2.0 \times 10^{-7} k_1 k_2$	J	radiant energy
	$>2.0 \times 10^{-5}$ to 1.0×10^{1}	$7.0 \times 10^{-4} k_1 k_2 t^{3/4}$	J	radiant energy
	$>1.0 \times 10^{1}$ to 1.0×10^{4}	$3.9 \times 10^{-3} k_1 k_2$	J	radiant energy
	$>1.0 \times 10^{4}$	$3.9 \times 10^{-7} k_1 k_2$	W	radiant power
	and also (See paragraph (d)(4) of this section)			
	$>1.0 \times 10^{-9}$ to 1.0×10^{1}	$10 k_1 k_2 t^{1/3}$	$Jcm^{-2}sr^{-1}$	integrated radiance
	$>1.0 \times 10^{1}$ to 1.0×10^{4}	$20 k_1 k_2$	$Jcm^{-2}sr^{-1}$	integrated radiance
	$>1.0 \times 10^{4}$	$2.0 \times 10^{-3} k_1 k_2$	$Wcm^{-2}sr^{-1}$	radiance
>1400 but ≤ 2500	$>1.0 \times 10^{-9}$ to 1.0×10^{-7}	$7.9 \times 10^{-5} k_1 k_2$	J	radiant energy
	$>1.0 \times 10^{-7}$ to 1.0×10^{1}	$4.4 \times 10^{-3} k_1 k_2 t^{1/4}$	J	radiant energy
	$>1.0 \times 10^{1}$	$7.9 \times 10^{-4} k_1 k_2$	W	radiant power
>2500 but $\leq 1.0 \times 10^{6}$	$>1.0 \times 10^{-9}$ to 1.0×10^{-7}	$1.0 \times 10^{-2} k_1 k_2$	Jcm^{-2}	radiant exposure
	$>1.0 \times 10^{-7}$ to 1.0×10^{1}	$5.6 \times 10^{-1} k_1 k_2 t^{1/4}$	Jcm^{-2}	radiant exposure
	$>1.0 \times 10^{1}$	$1.0 \times 10^{-1} k_1 k_2 t$	Jcm^{-2}	radiant exposure

*Class I accessible emission limits for wavelengths equal to or greater than 180 nm but less than or equal to 400 nm shall not exceed the Class I accessible emission limits for the wavelengths greater than 1400 nm but less than or equal to 1.0×10^{6} nm with a k_1 and k_2 of 1.0 for comparable sampling intervals.

**Measurement parameters and test conditions shall be in accordance with paragraphs (d)(1), (2), (3), and (4), and (e) of this section.

TABLE II-A

CLASS IIa ACCESSIBLE EMISSION LIMITS FOR LASER RADIATION

CLASS IIa ACCESSIBLE EMISSION LIMITS ARE IDENTICAL TO CLASS I ACCESSIBLE EMISSION LIMITS EXCEPT WITHIN THE FOLLOWING RANGE OF WAVELENGTHS AND EMISSION DURATIONS:				
Wavelength	Emission duration	Class IIa-Accessible emission limits		
(nanometers)	(seconds)	(value)	(unit)	(quantity)*
>400 but ≤710	$>1.0 \times 10^{3}$	3.9×10^{-6}	W	radiant power

*Measurement parameters and test conditions shall be in accordance with paragraphs (d)(1), (2), (3), and (4), and (e) of this section.

TABLE II

CLASS II ACCESSIBLE EMISSION LIMITS FOR LASER RADIATION

CLASS II ACCESSIBLE EMISSION LIMITS ARE IDENTICAL TO CLASS I ACCESSIBLE EMISSION LIMITS EXCEPT WITHIN THE FOLLOWING RANGE OF WAVELENGTHS AND EMISSION DURATIONS:				
Wavelength	Emission duration	Class II-Accessible emission limits		
(nanometers)	(seconds)	(value)	(unit)	(quantity)*
>400 but ≤710	$>2.5 \times 10^{-1}$	1.0×10^{-3}	W	radiant power

*Measurement parameters and test conditions shall be in accordance with paragraphs (d)(1), (2), (3), and (4), and (e) of this section.

TABLE III-A

CLASS IIIa ACCESSIBLE EMISSION LIMITS FOR LASER RADIATION

CLASS IIIa ACCESSIBLE EMISSION LIMITS ARE IDENTICAL TO CLASS I ACCESSIBLE EMISSION LIMITS EXCEPT WITHIN THE FOLLOWING RANGE OF WAVELENGTHS AND EMISSION DURATIONS:

Wavelength (nanometers)	Emission duration (seconds)	Class IIIa-Accessible emission limits		
		(value)	(unit)	(quantity)*
>400 but $\leq$710	>3.8 X 10^{-4}	5.0 X 10^{-3}	W	radiant power

*Measurement parameters and test conditions shall be in accordance with paragraphs (d)(1), (2), (3), and (4), and (e) of this section.

TABLE III-B

CLASS IIIb ACCESSIBLE EMISSION LIMITS FOR LASER RADIATION

Wavelength (nanometers)	Emission duration (seconds)	Class IIIb-Accessible emission limits		
		(value)	(unit)	(quantity)*
$\geq$180 but $\leq$400	$\leq$2.5 X 10^{-1}	3.8 X $10^{-4}k_1k_2$	J	radiant energy
	>2.5 X 10^{-1}	1.5 X $10^{-3}k_1k_2$	W	radiant power
>400 but $\leq$1400	>1.0 X 10^{-9} to 2.5 X 10^{-1}	$10k_1k_2t^{1/3}$	Jcm^{-2}	radiant exposure
		to a maximum value of 10	Jcm^{-2}	radiant exposure
	>2.5 X 10^{-1}	5.0 X 10^{-1}	W	radiant power
>1400 but $\leq$1.0 X 10^6	>1.0 X 10^{-9} to 1.0 X 10^{1}	10	Jcm^{-2}	radiant exposure
	>1.0 X 10^{1}	5.0 X 10^{-1}	W	radiant power

*Measurement parameter and test conditions shall be in accordance with paragraphs (d)(1), (2), (3), and (4), and (e) of this section.

TABLE IV

VALUES OF WAVELENGTH DEPENDENT CORRECTION FACTORS k_1 AND k_2

Wavelength (nanometers)	k_1	k_2		
180 to 302.4	1.0	1.0		
> 302.4 to 315	$10^{\left[\frac{\lambda - 302.4}{5}\right]}$	1.0		
> 315 to 400	330.0	1.0		
> 400 to 700	1.0	1.0		
> 700 to 800	$10^{\left[\frac{\lambda - 700}{515}\right]}$	if: $t \leq \frac{10100}{\lambda - 699}$ then: $k_2 = 1.0$	if: $\frac{10100}{\lambda - 699} < t \leq 10^4$ then: $k_2 = \frac{t(\lambda - 699)}{10100}$	if: $t > 10^4$ then: $k_2 = \frac{\lambda - 699}{1.01}$
> 800 to 1060	$10^{\left[\frac{\lambda - 700}{515}\right]}$	if: $t \leq 100$ then: $k_2 = 1.0$	if: $100 < t \leq 10^4$ then: $k_2 = \frac{t}{100}$	if: $t > 10^4$ then: $k_2 = 100$
> 1060 to 1400	5.0			
> 1400 to 1535	1.0	1.0		
> 1535 to 1545	$t \leq 10^{-7}$ $k_1 = 100.0$	1.0		
	$t > 10^{-7}$ $k_1 = 1.0$			
> 1545 to 1.0×10^6	1.0	1.0		

Note: The variables in the expressions are the magnitudes of the sampling interval(t), in units of seconds, and the wavelength (λ), in units of nanometers.

TABLE V

SELECTED NUMERICAL SOLUTIONS FOR k_1 AND k_2

Wavelength (nanometers)	k_1	k_2				
		$t \leq 100$	t=300	t=1000	t=3000	$t \geq 10{,}000$
180	1.0	1.0				
300	1.0					
302	1.0					
303	1.32					
304	2.09					
305	3.31					
306	5.25					
307	8.32					
308	13.2					
309	20.9					
310	33.1					
311	52.5					
312	83.2					
313	132.0					
314	209.0					
315	330.0					
400	330.0					
401	1.0					
500	1.0					
600	1.0					
700	1.0					
710	1.05	1	1	1.1	3.3	11.0
720	1.09	1	1	2.1	6.3	21.0
730	1.14	1	1	3.1	9.3	31.0
740	1.20	1	1.2	4.1	12.0	41.0
750	1.25	1	1.5	5.0	15.0	50.0
760	1.31	1	1.8	6.0	18.0	60.0
770	1.37	1	2.1	7.0	21.0	70.0
780	1.43	1	2.4	8.0	24.0	80.0
790	1.50	1	2.7	9.0	27.0	90.0
800	1.56	1	3.0	10.0	30.0	100.0
850	1.95	1	3.0	10.0	30.0	100.0
900	2.44	1	3.0	10.0	30.0	100.0
950	3.05	1	3.0	10.0	30.0	100.0
1000	3.82	1	3.0	10.0	30.0	100.0
1050	4.78	1	3.0	10.0	30.0	100.0
1060	5.00	1	3.0	10.0	30.0	100.0
1100	5.00	1	3.0	10.0	30.0	100.0
1400	5.00	1	3.0	10.0	30.0	100.0
1500	1.0	1.0				
1540	100.0*					
1600	1.0					
1.0×10^6	1.0					

* The factor k_1 = 100.0 when $t \leq 10^{-7}$, and k_1 = 1.0 when $t > 10^{-7}$

Note: The variable (t) is the magnitude of the sampling interval in units of seconds.

TABLE VI

ACCESSIBLE EMISSION LIMITS FOR COLLATERAL

RADIATION FROM LASER PRODUCTS

1. Accessible emission limits for collateral radiation having wavelengths greater than 180 nanometers but less than or equal to 1.0×10^{6} nanometers are identical to the accessible emission limits of Class I laser radiation, as determined from Tables I and IV in this paragraph.
i. In the wavelength range of less than or equal to 400 nanometers, for all emission durations;
ii. In the wavelength range of greater than 400 nanometers, for all emission durations less than or equal to 1×10^{3} seconds and, when applicable under paragraph (f)(8) of this section, for all emission durations.
2. Accessible emission limit for collateral radiation within the x-ray range of wavelengths is 0.5 milliroentgen in an hour, averaged over a cross-section parallel to the external surface of the product, having an area of 10 square centimeters with no dimension greater than 5 centimeters.

(1) *Beam of a single wavelength.* Laser or collateral radiation of a single wavelength exceeds the accessible emission limits of a class if its accessible emission level is greater than the accessible emission limit of that class within any of the ranges of emission duration specified in tables I, II-A, II, III-A, and III-B of this paragraph.

(2) *Beam of multiple wavelengths in same range.* Laser or collateral radiation having two or more wavelengths within any one of the wavelength ranges specified in tables I, II-A, II, III-A, and III-B of this paragraph exceeds the accessible emission limits of a class if the sum of the ratios of the accessible emission level to the corresponding accessible emission limit at each such wavelength is greater than unity for that combination of emission duration and wavelength distribution which results in the maximum sum.

(3) *Beam with multiple wavelengths in different ranges.* Laser or collateral radiation having wavelengths within two or more of the wavelength ranges specified in tables I, II-A, II, III-A, and III-B of this paragraph exceeds the accessible emission limits of a class if it exceeds the applicable limits within any one of those wavelength ranges. This determination is made for each wavelength range in accordance with paragraph (d) (1) or (2) of this section.

(4) *Class I dual limits.* Laser or collateral radiation in the wavelength range of greater than 400 nm but less than or equal to 1.400 nm exceeds the accessible emission limits of Class I if it exceeds both:

(i) The Class I accessible emission limits for radiant energy within any range of emission duration specified in table I of this paragraph, and

(ii) The Class I accessible emission limits for integrated radiance within any range of emission duration specified in table I of this paragraph.

(e) *Tests for determination of compliance*—(1) *Tests for certification.* Tests on which certification under § 1010.2 is based shall account for all errors and statistical uncertainties in the measurement process. Because compliance with the standard is required for the useful life of a product such tests shall also account for increases in emission and degradation in radiation safety with age.

(2) *Test conditions.* Except as provided in § 1010.13, tests for compliance with each of the applicable requirements of this section and § 1040.11 shall be made during operation, maintenance, or service as appropriate:

(i) Under those conditions and procedures which maximize the accessible emission levels, including start-up, stabilized emission, and shut-down of the laser product; and

(ii) With all controls and adjustments listed in the operation, maintenance, and service instructions adjusted in combination to result in the maximum accessible emission level of radiation; and

(iii) At points in space to which human access is possible in the product configuration which is necessary to determine compliance with each requirement, e.g., if operation may require removal of portions of the protective housing and defeat of safety interlocks, measurements shall be made at points accessible in that product configuration; and

(iv) With the measuring instrument detector so positioned and so oriented with respect to the laser product as to result in the maximum detection of radiation by the instrument; and

(v) For a laser product other than a laser system, with the laser coupled to that type of laser energy source which is specified as compatible by the laser product manufacturer and which produces the maximum emission level of accessible radiation from that product.

(3) *Measurement parameters.* Accessible emission levels of laser and collateral radiation shall be based upon the following measurements as appropriate, or their equivalent:

(i) For laser products intended to be used in a locale where the emitted laser radiation is unlikely to be viewed with optical instruments, the radiant power (W) or radiant energy (J) detectable through a circular aperture stop having a diameter of 7 millimeters and within a circular solid angle of acceptance of 1×10^{-3} steradian with collimating optics of 5 diopters or less. For scanned laser radiation, the direction of the solid angle of acceptance shall change as needed to maximize detectable radiation, with an angular speed of up to 5 radians/second. A 50 millimeter diameter aperture stop with the same collimating optics and acceptance angle stated above shall be used for all other laser products (except that a 7 millimeter diameter aperture stop shall be used in the measurement of scanned laser radiation emitted by laser products manufactured on or before August 20, 1986.

(ii) The irradiance (W cm^{-2}) or radiant exposure (J cm^{-2} equivalent to the radiant power (W) or radiant energy (J) detectable through a circular aperture stop having a diameter of 7 millimeters and, for irradiance, within a circular solid angle of acceptance of 1××10^{-3} steradian with collimating optics of 5 diopters or less, divided by the area of the aperture stop (cm^{-2}).

(iii) The radiance (W cm^{-2} sr^{-1}) or integrated radiance (J cm^{-2} sr^{-1}) equivalent to the radiant power (W) or radiant energy (J) detectable through a circular aperture stop having a diameter of 7 millimeters and within a circular solid angle of acceptance of 1×10^{-5} steradian with collimating optics of 5 diopters or less, divided by that solid angle (sr) and by the area of the aperture stop (cm^{-2}).

(f) *Performance requirements*—(1) *Protective housing.* Each laser product shall have a protective housing that prevents human access during operation to laser and collateral radiation that exceed the limits of Class I and table VI, respectively, wherever and whenever such human access is not necessary for the product to perform its intended function. Wherever and whenever human access to laser radiation levels that exceed the limits of Class I is necessary, these levels shall not exceed the limits of the lowest class necessary to perform the intended function(s) of the product.

(2) *Safety interlocks.* (i) Each laser product, regardless of its class, shall be provided with at least one safety interlock for each portion of the protective housing which is designed to be removed or displaced during operation or maintenance, if removal or displacement of the protective housing could permit, in the absence of such interlock(s), human access to laser or collateral radiation in excess of the accessible emission limit applicable under paragraph (f)(1) of this section.

(ii) Each required safety interlock, unless defeated, shall prevent such human access to laser and collateral radiation upon removal or displacement of such portion of the protective housing

(iii) Either multiple safety interlocks or a means to preclude removal or displacement of the interlocked portion of the protective housing shall be provided, if failure of a single interlock would allow;

(*a*) Human access to a level of laser radiation in excess of the accessible emission limits of Class IIIa; or

(*b*) Laser radiation in excess of the accessible emission limits of Class II to be emitted directly through the opening created by removal or displacement of the interlocked portion of the protective housing.

(iv) Laser products that incorporate safety interlocks designed to allow safety interlock defeat shall incorporate a means of visual or aural indication of interlock defeat. During interlock defeat, such indication shall be visible or audible whenever the laser product is energized, with and without the associated portion of the protective housing removed or displaced.

(v) Replacement of a removed or displaced portion of the protective housing shall not be possible while required safety interlocks are defeated.

(3) *Remote interlock connector.* Each laser system classified as a Class IIIb or IV laser product shall incorporate a readily available remote interlock connector having an electrical potential difference of no greater than 130 root-mean-square volts between terminals. When the terminals of the connector are not electrically joined, human access to all laser and collateral radiation from the laser product in excess of the accessible emission limits of Class I and table VI shall be prevented.

(4) *Key control.* Each laser system classified as a Class IIIb or IV laser product shall incorporate a key-actuated master control. The key shall be removable and the laser shall not be operable when the key is removed.

(5) *Laser radiation emission indicator.* (i) Each laser system classified as a Class II or IIIa laser product shall incorporate an emission indicator that provides a visible or audible signal during emission of accessible laser radiation in excess of the accessible emission limits of Class I.

(ii) Each laser system classified as a Class IIIb or IV laser product shall incorporate an emission indicator which provides a visible or audible signal during emission of accessible laser radiation in excess of the accessible emission limits of Class I, and sufficiently prior to emission of such radiation to allow appropriate action to avoid exposure to the laser radiation.

(iii) For laser systems manufactured on or before August 20, 1986, if the laser and laser energy source are housed separately and can be operated at a separation distance of greater than 2 meters, both laser and laser energy source shall incorporate an emission indicator as required in accordance with paragraph (f)(5) (i) or (ii) of this section. For laser systems manufactured after August 20, 1986, each separately housed laser and operation control of a laser system that regulates the laser or collateral radiation emitted by a product during operation shall incorporate an emission indicator as required in accordance with paragraph (f)(5) (i) or (ii) of this section, if the laser or operation control can be operated at a separation distance greater than 2 meters from any other separately housed portion of the laser product incorporating an emission indicator.

(iv) Any visible signal required by paragraph (f)(5) (i) or (ii) of this section shall be clearly visible through protective eyewear designed specifically for the wavelength(s) of the emitted laser radiation.

(v) Emission indicators required by paragraph (f)(5) (i) or (ii) of this section shall be located so that viewing does not require human exposure to laser or collateral radiation in excess of the accessible emission limits of Class I and table VI.

(6) *Beam attenuator.* (i) Each laser system classified as a Class II, III, or IV laser product shall be provided with one or more permanently attached means, other than laser energy source switch(es), electrical supply main connectors, or the key-actuated master control, capable of preventing access by any part of the human body to all laser and collateral radiation in excess of the accessible emission limits of Class I and table VI.

(ii) If the configuration, design, or function of the laser product would make unnecessary compliance with the requirement in paragraph (f)(6)(i) of this section, the Director, Office of Compliance (HFZ–300), Center for Devices and Radiological Health, may, upon written application by the manufacturer, approve alternate means to accomplish the radiation protection provided by the beam attenuator.

(7) *Location of controls.* Each Class IIa, II, III, or IV laser product shall have operational and adjustment controls located so that human exposure to laser or collateral radiation in excess of the accessible emission limits of Class I and table VI is unnecessary for operation or adjustment of such controls.

(8) *Viewing optics.* All viewing optics, viewports, and display screens incorporated into a laser product, regardless of its class, shall limit the levels of laser and collateral radiation accessible to the human eye by means of such viewing optics, viewports, or display screens during operation or maintenance to less than the accessible emission limits of Class I and table VI. For any shutter or variable attenuator incorporated into such viewing optics, viewports, or display screens, a means shall be provided:

(i) To prevent access by the human eye to laser and collateral radiation in excess of the accessible emission limits of Class I and table VI whenever the shutter is opened or the attenuator varied.

(ii) To preclude, upon failure of such means as required in paragraph (f)(8)(i) of this section, opening the shutter or varying the attenuator when access by the human eye is possible to laser or collateral radiation in excess of the accessible emission limits of Class I and table VI.

(9) *Scanning safeguard.* Laser products that emit accessible scanned laser radiation shall not, as a result of any failure causing a change in either scan velocity or amplitude, permit human access to laser radiation in excess of:

(i) The accessible emission limits of the class of the product, or

(ii) The accessible emission limits of the class of the scanned laser radiation if the product is Class IIIb or IV and the accessible emission limits of Class IIIa would be exceeded solely as result of such failure.

(10) *Manual reset mechanism.* Each laser system manufactured after August 20, 1986, and classified as a Class IV laser product shall be provided with a manual reset to enable resumption of laser radiation emission after interruption of emission caused by the use of a remote interlock or after an interruption of emission in excess of 5 seconds duration due to the unexpected loss of main electrical power.

(g) *Labeling requirements.* In addition to the requirements of §§ 1010.2 and 1010.3, each laser product shall be subject to the applicable labeling requirements of this paragraph.

(1) *Class IIa and II designations and warnings.* (i) Each Class IIa laser product shall have affixed a label bearing the following wording: "Class IIa Laser Product—Avoid Long-Term Viewing of Direct Laser Radiation."

(ii) Each Class II laser product shall have affixed a label bearing the warning logotype A (figure 1 in this paragraph) and including the following wording:

[Position I on the logotype]

"LASER RADIATION—DO NOT STARE INTO BEAM"; and

[Position 3 on the logotype]

"CLASS II LASER PRODUCT".

WARNING LOGOTYPE A

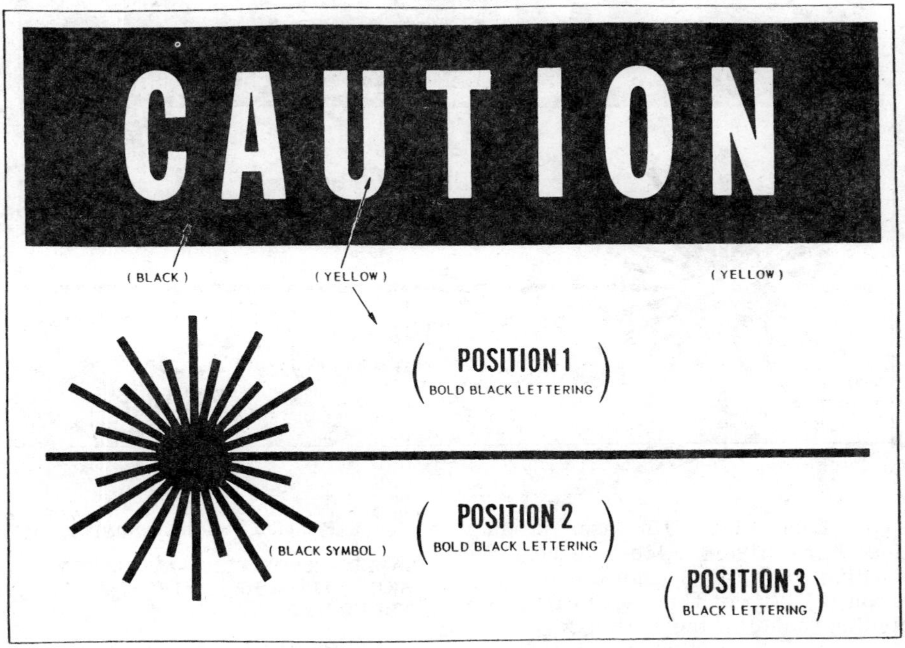

FIGURE 1

(2) *Class IIIa and IIIb designations and warnings.* (i) Each Class IIIa laser product with an irradiance less than or equal to 2.5×10^{-3} W cm^{2-} shall have affixed a label bearing the warning logotype A (figure 1 of paragraph (g)(1)(ii) of this section) and including the following wording:

[Position 1 on the logotype]

"LASER RADIATION—DO NOT STARE INTO BEAM OR VIEW DIRECTLY WITH OPTICAL INSTRUMENTS"; and,

[Position 3 on the logotype]

"CLASS IIIa LASER PRODUCT".

(ii) Each Class IIIa laser product with an irradiance greater than 2.5×10^{-3} W cm^{-2} shall have affixed a label bearing the warning logotype B (figure 2 in this paragraph) and including the following wording:

[Position 1 on the logotype]

"LASER RADIATION—AVOID DIRECT EYE EXPOSURE"; and,

[Position 3 on the logotype]

"CLASS IIIa LASER PRODUCT".

WARNING LOGOTYPE B

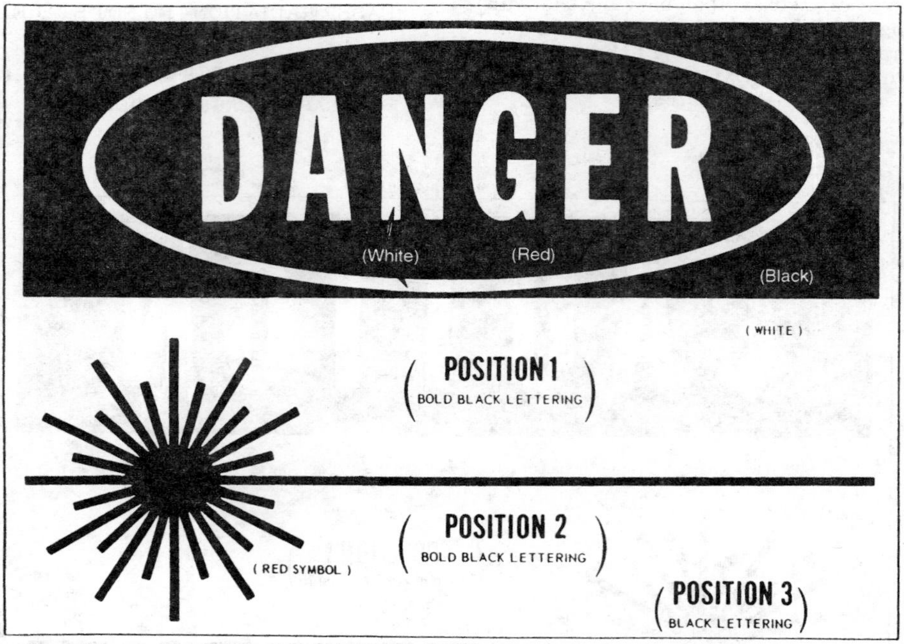

FIGURE 2

(iii) Each Class IIIb laser product shall have affixed a label bearing the warning logotype B (figure 2 of paragraph (g)(2)(ii) of this section) and including the following wording:

[Position 1 on the logotype]

"LASER RADIATION—AVOID DIRECT EXPOSURE TO BEAM"; and,

[Position 3 on the logotype]

"CLASS IIIb LASER PRODUCT".

(3) *Class IV designation and warning.* Each Class IV laser product shall have affixed a label bearing the warning logotype B (figure 2 of paragraph (g)(2)(ii) of this section), and including the following wording:

[Position 1 on the logotype]

"LASER RADIATION—AVOID EYE OR SKIN EXPOSURE TO DIRECT OR SCATTERED RADIATION"; and,

[Position 3 on the logotype]

"CLASS IV LASER PRODUCT".

(4) *Radiation output information on warning logotype.* Each Class II, III, and IV laser product shall state in appropriate units, at position 2 on the required warning logotype, the maximum output of laser radiation, the pulse duration when appropriate, and the laser medium or emitted wavelength(s).

(5) *Aperture label.* Each laser product, except medical laser products and Class IIa laser products, shall have affixed, in close proximity to each aperture through which is emitted accessible laser or collateral radiation in excess of the accessible emission limits of Class I and table VI of paragraph (d) of this section, a label(s) bearing the following wording as applicable.

(i) "AVOID EXPOSURE—Laser radiation is emitted from this aperture," if the radiation emitted through such aperture is laser radiation.

(ii) "AVOID EXPOSURE—Hazardous electromagnetic radiation is emitted from this aperture," if the radiation emitted through such aperture is collateral radiation described in table VI, item 1.

(iii) "AVOID EXPOSURE—Hazardous x-rays are emitted from this aperture," if the radiation emitted through such aperture is collateral radiation described in table VI, item 2.

(6) *Labels for noninterlocked protective housings.* For each laser product, labels shall be provided for each portion of the protective housing which has no safety interlock and which is designed to be displaced or removed during operation, maintenance, or service, and thereby could permit human access to laser or collateral radiation in excess of the limits of Class I and table VI. Such labels shall be visible on the protective housing prior to displacement or removal of such portion of the protective housing and visible on the product in close proximity to the opening created by removal or displacement of such portion of the protective housing, and shall include the wording:

(i) "CAUTION—Laser radiation when open. DO NOT STARE INTO BEAM." for Class II accessible laser radiation.

(ii) "CAUTION—Laser radiation when open. DO NOT STARE INTO BEAM OR VIEW DIRECTLY WITH OPTICAL INSTRUMENTS." for Class IIIa accessible laser radiation with an irradiance less than or equal to 2.5×10^{-3} W cm^{-2}.

(iii) "DANGER—Laser radiation when open. AVOID DIRECT EYE EXPOSURE." for Class IIIa accessible laser radiation with an irradiance greater than 2.5×10^{-3} W cm^{-2}.

(iv) "DANGER—Laser radiation when open. AVOID DIRECT EXPOSURE TO BEAM." for Class IIIb accessible laser radiation.

(v) "DANGER—Laser radiation when open. AVOID EYE OR SKIN EXPOSURE TO DIRECT OR SCATTERED RADIATION." for Class IV accessible laser radiation.

(vi) "CAUTION—Hazardous electromagnetic radiation when open." for collateral radiation in excess of the accessible emission limits in table VI, item 1 of paragraph (d) of this section.

(vii) "CAUTION—Hazardous x-rays when open." for collateral radiation in excess of the accessible emission limits in table VI, item 2 of paragraph (d) of this section.

(7) *Labels for defeatably interlocked protective housings.* For each laser product, labels shall be provided for each defeatably interlocked (as described in paragraph (f)(2)(iv) of this section) portion of the protective housing which is designed to be displaced or removed during operation, maintenance, or service, and which upon interlock defeat could permit human access to laser or collateral radiation in excess of the limits of Class I or table VI. Such labels shall be visible on the product prior to and during interlock defeat and in close proximity to the opening created by the removal or displacement of such portion of the protective housing, and shall include the wording:

(i) "CAUTION—Laser radiation when open and interlock defeated. DO NOT STARE INTO BEAM." for Class II accessible laser radiation.

(ii) "CAUTION—Laser radiation when open and interlock defeated. DO NOT STARE INTO BEAM OR VIEW DIRECTLY WITH OPTICAL INSTRUMENTS." for Class IIIa accessible laser radiation with an irradiance less than or equal to 2.5×10^{-3} W cm^{-2}.

(iii) "DANGER—Laser radiation when open and interlock defeated. AVOID DIRECT EYE EXPOSURE." for Class IIIa accessible laser radiation when an irradiance greater than 2.5×10^{-3} W cm^{-2}.

(iv) "DANGER—Laser radiation when open and interlock defeated. AVOID DIRECT EXPOSURE TO BEAM." for Class IIIb accessible laser radiation.

(v) "DANGER—Laser radiation when open and interlock defeated. AVOID EYE OR SKIN EXPOSURE TO DIRECT OR SCATTERED RADIATION." for Clas IV accessible laser radiation.

(vi) "CAUTION—Hazardous electromagnetic radiation when open and interlock defeated." for collateral radiation in excess of the accessible emission limits in table VI. item 1 of paragraph (d) of this section.

(vii) "CAUTION—Hazardous x-rays when open and interlock defeated." for collateral radiation in excess of the accesible emission limits in table VI. item 2 of paragraph (d) of this section.

(8) *Warning for visible and/or invisible radiation.* On the labels specified in this paragraph, if the laser or collateral radiation referred to is:

(i) Invisible radiation, the word "invisible" shall appropriately precede the word "radiation"; or

(ii) Visible and invisible radiation, the words "visible and invisible" or "visible and/or invisible" shall appropriately precede the word "radiation."

(iii) Visible laser radiation only, the phrase "laser light" may replace the phrase "laser radiation."

(9) *Positioning of labels.* All labels affixed to a laser product shall be positioned so as to make unnecessary, during reading, human exposure to laser radiation in excess of the accessible emission limits of Class I radiation or the limits of collateral radiation established to table VI of paragraph (d) of this section.

(10) *Label specifications.* Labels required by this section and § 1040.11 shall be permanently affixed to, or inscribed on, the laser product, legible, and clearly visible during operation, maintenance, or service, as appropriate. If the size, configuration, design, or function of the laser product would preclude compliance with the requirements for any required label or would render the required wording of such label inappropriate or ineffective, the Director, Office of Compliance (HFZ–300), Center for Devices and Radiological Health, on the Director's own initiative or upon written application by the manufacturer, may approve alternate means of providing such label(s) or alternate wording for such label(s) as applicable.

(h) *Informational requirements*—(1) *User information.* Manufacturers of laser products shall provide as an integral part of any user instruction or operation manual which is regularly supplied with the product, or, if not so supplied, shall cause to be provided with each laser product:

(i) Adequate instructions for assembly, operation, and maintenance, including clear warnings concerning precautions to avoid possible exposure to laser and collateral radiation in excess of the accessible emission limits in tables I, II-A, II, III-A, III-B, and VI of paragraph (d) of this section, and a schedule of maintenance necessary to keep the product in compliance with this section and § 1040.11.

(ii) A statement of the magnitude, in appropriate units, of the pulse durations(s), maximum radiant power and, where applicable, the maximum radiant energy per pulse of the accessible laser radiation detectable in each direction in excess of the accessible emission limits in table I of paragraph (d) of this section determined under paragraph (e) of this section.

(iii) Legible reproductions (color optional) of all labels and hazard warnings required by paragraph (g) of this section and § 1040.11 to be affixed to the laser product or provided with the laser product, including the information required for positions 1, 2, and 3 of the applicable logotype (figure 1 of paragraph (g)(1)(ii) or figure 2 or paragraph (g)(2)(ii) of this section). The corresponding position of each label affixed to the product shall be indicated or, if provided with the product, a statement that such labels could not be affixed to the product but were supplied with the product and a statement of the form and manner in which they were supplied shall be provided.

(iv) A listing of all controls, adjustments, and procedures for operation and maintenance, including the warning "Caution—use of controls or adjustments or performance of procedures other than those specified herein may result in hazardous radiation exposure."

(v) In the case of laser products other than laser systems, a statment of the compatibility requirements for a laser energy source that will assure compliance of the laser product with this section and § 1040.11.

(vi) In the case of laser products classified with a 7 millimeter diameter aperture stop as provided in paragraph (e)(3)(i) of this section, if the use of a 50 millimeter diameter aperture stop would result in a higher clssification of the product, the following warning shall be included in the user information: "CAUTION—The use of optical instruments with this product will increase eye hazard."

(2) *Purchasing and servicing information.* Manufacturers of laser products shall provide or cause to be provided:

(i) In all catalogs, specification sheets, and descriptive brochures pertaining to each laser product, a legible reproduction (color optional) of the class designation and warning required by paragraph (g) of this section to be affixed to that product, including the information required for positions 1, 2, and 3 of the applicable logotype (figure 1 of paragraph (g)(1)(ii) or figure 2 of paragraph (g)(2)(ii) of this section).

(ii) To servicing dealers and distributors and to others upon request at a cost not to exceed the cost of preparation and distribution, adequate instructions for service adjustments and service procedures for each laser product model, including clear warnings and precautions to be taken to avoid possible exposure to laser and collateral radiation in excess of the accessible emission limits in tables I, II-A, II, III-A, III-B, and VI of paragraph (d) of this section, and a schedule of maintenance necessary to keep the product in compliance with this section and §1040.11; and in all such service instructions, a listing of those controls and procedures that could be utilized by persons other than the manufacturers or the manufacturer's agents to increase accessible emission levels of radiation and a clear description of the location of displaceable portions of the protective housing that could allow human access to laser or collateral radiation in excess of the accessible emission limits in tables I, II-A, II, III-A, III-B, and VI of paragraph (d) of this section. The instructions shall include protective procedures for service personnel to avoid exposure to levels of laser and collateral radiation known to be hazardous for each procedure or sequence of procedures to be accomplished, and legible reproductions (color optional) of required labels and hazard warnings.

(i) *Modification of a certified product.* The modification of a laser product, previously certified under §1010.2, by any person engaged in the business of manufacturing, assembling, or modifying laser products shall be construed as manufacturing under the act if the modification affects any aspect of the product's performance or intended function(s) for which this section and §1040.11 have an applicable requirement. The manufacturer who performs such modification shall recertify and reidentify the product in accordance with the provisions of §§1010.2. and 1010.3.

(The information collection requirements contained in paragraph (a)(3)(ii) were approved by the Office of Management and Budget under control number 0910–0176)

[50 FR 33688, Aug. 20, 1985; 50 FR 42156, Oct. 18, 1985; 65 FR 17138, Mar. 31, 2000]

§1040.11 Specific purpose laser products.

(a) *Medical laser products.* Each medical laser product shall comply with all of the applicable requirements of §1040.10 for laser products of its class. In addition, the manufacturer shall:

(1) Incorporate in each Class III or IV medical laser product a means for the measurement of the level of that laser radiation intended for irradiation of the human body. Such means may have an error in measurement of no more than 20 percent when calibrated in accordance with paragraph (a)(2) of this section. Indication of the measurement shall be in International System Units. The requirements of this paragraph do not apply to any laser radiation that is all of the following:

(i) Of a level less than the accessible limits of Class IIIa; and

(ii) Used for relative positioning of the human body; and

(iii) Not used for irradiation of the human eye for ophthalmic purposes.

(2) Supply with each Class III or IV medical laser product instructions specifying a procedure and schedule for calibration of the measurement system required by paragraph (a)(1) of this section.

(3) Affix to each medical laser product, in close proximity to each aperture through which is emitted accessible laser radiation in excess of the accessible emission limits of Class I, a label bearing the wording: "Laser aperture."

(b) *Surveying, leveling, and alignment laser products.* Each surveying, leveling. or alignment laser product shall comply with all of the applicable requirements of § 1040.10 for a Class I, IIa, II or IIIa laser product and shall not permit human access to laser radiation in excess of the accessible emission limits of Class IIIa.

(c) *Demonstration laser products.* Each demonstration laser product shall comply with all of the applicable requirements of § 1040.10 for a Class I, IIa, II, or IIIa laser product and shall not permit human access to laser radiation in excess of the accessible emission limits of Class I and, if applicable, Class IIa, Class II, or Class IIIa.

[50 FR 33702, Aug. 20, 1985]

§ 1040.20 Sunlamp products and ultraviolet lamps intended for use in sunlamp products.

(a) *Applicability.* (1) The provisions of this section, as amended, are applicable as specified herein to the following products manufactured on or after September 8, 1986.

(i) Any sunlamp product.

(ii) Any ultraviolet lamp intended for use in any sunlamp product.

(2) Sunlamp products and ultraviolet lamps manufactured on or after May 7, 1980, but before September 8, 1986, are subject to the provisions of this section as published in the FEDERAL REGISTER of November 9, 1979 (44 FR 65357).

(b) *Definitions.* As used in this section the following definitions apply:

(1) *Exposure position* means any position, distance, orientation, or location relative to the radiating surfaces of the sunlamp product at which the user is intended to be exposed to ultraviolet radiation from the product, as recommended by the manufacturer.

(2) *Intended* means the same as "intended uses" in § 801.4.

(3) *Irradiance* means the radiant power incident on a surface at a specified location and orientation relative to the radiating surface divided by the area of the surface, as the area becomes vanishingly small, expressed in units of watts per square centimeter (W/cm^2).

(4) *Maximum exposure time* means the greatest continuous exposure time interval recommended by the manufacturer of the product.

(5) *Maximum timer interval* means the greatest time interval setting on the timer of a product.

(6) *Protective eyewear* means any device designed to be worn by users of a product to reduce exposure of the eyes to radiation emitted by the product.

(7) *Spectral irradiance* means the irradiance resulting from radiation within a wavelength range divided by the wavelength range as the range becomes vanishingly small, expressed in units of watts per square centimeter per nanometer ($W/(cm^2/nm)$).

(8) *Spectral transmittance* means the spectral irradiance transmitted through protective eyewear divided by the spectral irradiance incident on the protective eyewear.

(9) *Sunlamp product* means any electronic product designed to incorporate one or more ultraviolet lamps and intended for irradiation of any part of the living human body, by ultraviolet radiation with wavelengths in air between 200 and 400 nanometers, to induce skin tanning.

(10) *Timer* means any device incorporated into a product that terminates radiation emission after a preset time interval.

(11) *Ultraviolet lamp* means any lamp that produces ultraviolet radiation in the wavelength interval of 200 to 400 nanometers in air and that is intended for use in any sunlamp product.

(c) *Performance requirements*—(1) *Irradiance ratio limits.* For each sunlamp product and ultraviolet lamp, the ratio of the irradiance within the wavelength range of greater than 200 nanometers through 260 nanometers to the irradiance within the wavelength range of greater than 260 nanometers through 320 nanometers may not exceed 0.003 at any distance and direction from the product or lamp.

(2) *Timer system.* (i) Each sunlamp product shall incorporate a timer system with multiple timer settings adequate for the recommended exposure time intervals for different exposure positions and expected results of the products as specified in the label required by paragraph (d) of this section.

(ii) The maximum timer interval(s) may not exceed the manufacturer's recommended maximum exposure time(s) that is indicated on the label required by paragraph (d)(1)(iv) of this section.

(iii) No timer interval may have an error greater than 10 percent of the maximum timer interval of the product.

(iv) The timer may not automatically reset and cause radiation emission to resume for a period greater than the unused portion of the timer cycle, when emission from the sunlamp product has been terminated.

(v) The timer requirements do not preclude a product from allowing a user to reset the timer before the end of the preset time interval.

(3) *Control for termination of radiation emission.* Each sunlamp product shall incorporate a control on the product to enable the person being exposed to terminate manually radiation emission from the product at any time without disconnecting the electrical plug or removing the ultraviolet lamp.

(4) *Protective eyewear.* (i) Each sunlamp product shall be accompanied by the number of sets of protective eyewear that is equal to the maximum number of persons that the instructions provided under paragraph (e)(1)(ii) of this section recommend to be exposed simultaneously to radiation from such product.

(ii) The spectral transmittance to the eye of the protective eyewear required by paragraph (c)(4)(i) of this section shall not exceed a value of 0.001 over the wavelength range of greater than 200 nanometers 320 nanometers and a value of 0.01 over the wavelength range of greater than 320 nanometers through 400 nanometers, and shall be sufficient over the wavelength greater than 400 nanometers to enable the user to see clearly enough to reset the timer.

(5) *Compatibility of lamps.* An ultraviolet lamp may not be capable of insertion and operation in either the "single-contact medium screw" or the "double-contact medium screw" lampholders described in American National Standard C81.10–1976, Specifications for Electric Lamp Bases and Holders—Screw-Shell Types, which is incorporated by reference. Copies are available from the American National Standards Institute, 1430 Broadway, New York, NY 10018, or available for inspection at the National Archives and Records Administration (NARA). For information on the availability of this material at NARA, call 202–741–6030, or go to: *http://www.archives.gov/federal_register/code_of_federal_regulations/ibr_locations.html.*

(d) *Label requirements.* In addition to the labeling requirements in part 801 and the certification and identification requirements of §§ 1010.2 and 1010.3, each sunlamp product and ultraviolet lamp shall be subject to the labeling requirements prescribed in this paragraph and paragraph (e) of this section.

(1) *Labels for sunlamp products.* Each sunlamp product shall have a label(s) which contains:

(i) A warning statement with the words "DANGER—Ultraviolet radiation. Follow instructions. Avoid overexposure. As with natural sunlight, overexposure can cause eye and skin injury and allergic reactions. Repeated exposure may cause premature aging of the skin and skin cancer. WEAR PROTECTIVE EYEWEAR; FAILURE TO MAY RESULT IN SEVERE BURNS OR LONG-TERM INJURY TO THE EYES. Medications or cosmetics may increase your sensitivity to the ultraviolet radiation. Consult physician before using sunlamp if you are using medications or have a history of skin problems or believe yourself especially sensitive to sunlight. If you do not tan in the sun, you are unlikely to tan from the use of this product."

(ii) Recommended exposure position(s). Any exposure position may be expressed either in terms of a distance specified both in meters and in feet (or in inches) or through the use of markings or other means to indicate clearly the recommended exposure position.

(iii) Directions for achieving the recommended exposure position(s) and a warning that the use of other positions may result in overexposure.

(iv) A recommended exposure schedule including duration and spacing of sequential exposures and maximum exposure time(s) in minutes.

(v) A statement of the time it may take before the expected results appear.

(vi) Designation of the ultraviolet lamp type to be used in the product.

(2) *Labels for ultraviolet lamps.* Each ultraviolet lamp shall have a label which contains:

(i) The words "Sunlamp—DANGER—Ultraviolet radiation. Follow instructions."

(ii) The model identification.

(iii) The words "Use ONLY in fixture equipped with a timer."

(3) *Label specifications.* (i) Any label prescribed in this paragraph for sunlamp products shall be permanently affixed or inscribed on an exterior surface of the product when fully assembled for use so as to be legible and readily accessible to view by the person being exposed immediately before the use of the product.

(ii) Any label prescribed in this paragraph for ultraviolet lamps shall be permanently affixed or inscribed on the product so as to be legible and readily accessible to view.

(iii) If the size, configuration, design, or function of the sunlamp product or ultraviolet lamp would preclude compliance with the requirements for any required label or would render the required wording of such label inappropriate or ineffective, or would render the required label unnecessary, the Director, Office of Compliance (HFZ–300), Center for Devices and Radiological Health, on the Center's own initiative or upon written application by the manufacturer, may approve alternate means of providing such label(s), alernate wording for such label(s), or deletion, as applicable.

(iv) In lieu of permanently affixing or inscribing tags or labels on the ultraviolet lamp as required by §§ 1010.2(b) and 1010.3(a), the manfacturer of the ultraviolet lamp may permanently affix or inscribe such required tags or labels on the lamp packaging uniquely associated with the lamp, if the name of the manufacturer and month and year of manufacture are permanently affixed or inscribed on the exterior surface of the ultraviolet lamp so as to be legible and readily accessible to view. The name of the manufacturer and month and year of manufacture affixed or inscribed on the exterior surface of the lamp may be expressed in code or symbols, if the manufacturer has previously supplied the Director, Office of Compliance (HFZ–300), Center for Devices and Radiological Health, with the key to such code or symbols and the location of the coded information or symbols on the ultraviolet lamp. The label or tag affixed or inscribed on the lamp packaging may provide either the month and year of manufacture without abbreviation, or information to allow the date to be readily decoded.

(v) A label may contain statements or illustrations in addition to those required by this paragraph if the additional statements are not false or misleading in any particular; e.g., if they do not diminish the impact of the required statements; and are not prohibited by this chapter.

(e) *Instructions to be provided to users.* Each manufacturer of a sunlamp product and ultraviolet lamp shall provide or cause to be provided to purchasers and, upon request, to others at a cost not to exceed the cost of publication and distribution, adequate instructions for use to avoid or to minimize potential injury to the user, including the following technical and safety information as applicable:

(1) *Sunlamp products.* The users' instructions for a sunlamp product shall contain:

(i) A reproduction of the label(s) required in paragraph (d)(1) of this section prominently displayed at the beginning of the instructions.

(ii) A statement of the maximum number of people who may be exposed to the product at the same time and a warning that only that number of protective eyewear has been provided.

(iii) Instructions for the proper operation of the product including the function, use, and setting of the timer and other controls, and the use of protective eyewear.

(iv) Instructions for determining the correct exposure time and schedule for persons according to skin type.

(v) Instructions for obtaining repairs and recommended replacement components and accessories which are compatible with the product, including compatible protective eyewear, ultraviolet lamps, timers, reflectors, and filters, and which will, if installed or used as instructed, result in continued compliance with the standard.

(2) *Ultraviolet lamps.* The users' instructions for an ultraviolet lamp not accompanying a sunlamp product shall contain:

(i) A reproduction of the label(s) required in paragraphs (d)(1)(i) and (2) of this section, prominently displayed at the beginning of the instructions.

(ii) A warning that the instructions accompanying the sunlamp product should always be followed to avoid or to minimize potential injury.

(iii) A clear identification by brand and model designation of all lamp models for which replacement lamps are promoted, if applicable.

(f) *Test for determination of compliance.* Tests on which certification pursuant to § 1010.2 is based shall account for all errors and statistical uncertainties in the process and, wherever applicable, for changes in radiation emission or degradation in radiation safety with age of the product. Measurements for certification purposes shall be made under those operational conditions, lamp voltage, current, and position as recommended by the manufacturer. For these measurements, the measuring instrument shall be positioned at the recommended exposure position and so oriented as to result in the maximum detection of the radiation by the instrument.

[50 FR 36550, Sept. 6, 1985, as amended at 67 FR 9587, Mar. 4, 2002; 69 FR 18803, Apr. 9, 2004]

§ 1040.30 High-intensity mercury vapor discharge lamps.

(a) *Applicability.* The provisions of this section apply to any high-intensity mercury vapor discharge lamp that is designed, intended, or promoted for illumination purposes and is manufactured or assembled after March 7, 1980, except as described in paragraph (d)(1)(ii) of this section.

(b) *Definitions.* (1) *High-intensity mercury vapor discharge lamp* means any lamp including any "mercury vapor" and "metal halide" lamp, with the exception of the tungsten filament self-ballasted mercury vapor lamp, incorporating a high-pressure arc discharge tube that has a fill consisting primarily of mercury and that is contained within an outer envelope.

(2) *Advertisement* means any catalog, specification sheet, price list, and any other descriptive or commercial brochure and literature, including videotape and film, pertaining to high-intensity mercury vapor discharge lamps.

(3) *Packaging* means any lamp carton, outer wrapping, or other means of containment that is intended for the storage, shipment, or display of a high-intensity mercury vapor lamp and is intended to identify the contents or recommend its use.

(4) *Outer envelope* means the lamp element, usually glass, surrounding a high-pressure arc discharge tube, that, when intact, attenuates the emission of shortwave ultraviolet radiation.

(5) *Shortwave ultraviolet radiation* means ultraviolet radiation with wavelengths shorter than 320 nanometers.

(6) *Cumulative operating time* means the sum of the times during which electric current passes through the high-pressure arc discharge.

(7) *Self-extinguishing lamp* means a high-intensity mercury vapor discharge lamp that is intended to comply with the requirements of paragraph (d)(1) of this section as applicable.

(8) *Reference ballast* is an inductive reactor designed to have the operating characteristics as listed in Section 7 in the American National Standard Specifications for High-Intensity Discharge Lamp Reference Ballasts (ANSI C82.5–1977)[1] or its equivalent.

(c) *General requirements for all lamps.* (1) Each high-intensity mercury vapor discharge lamp shall:

(i) Meet the requirements of either paragraph (d) or paragraph (e) of this section; and

(ii) Be permanently labeled or marked in such a manner that the name of the manufacturer and the month and year of manufacture of the lamp can be determined on an intact lamp and after the outer envelope of the lamp is broken or removed. The name of the manufacturer and month and year of manufacture may be expressed in code or symbols, provided the manufacturer has previously supplied the Director, Center for Devices and Radiological Health, with the key to the code or symbols and the location

[1]Copies are available from American National Standards Institute, 1430 Broadway, New York, NY 10018.

of the coded information or symbols on the lamp.

(2) In lieu of permanently affixing or inscribing tags or labels on the product as required by §§ 1010.2(b) and 1010.3(a) of this chapter, the manufacturer of any high-intensity mercury vapor discharge lamp may permanently affix or inscribe such required tags or labels on the lamp packaging uniquely associated with the applicable lamp.

(d) *Requirements for self-extinguishing lamps*—(1) *Maximum cumulative operating time.* (i) Each self-extinguishing lamp manufactured after March 7, 1980 shall cease operation within a cumulative operating time not to exceed 15 minutes following complete breakage or removal of the outer envelope (with the exception of fragments extending 50 millimeters or less from the base shell); and

(ii) Each self-extinguishing lamp manufactured after September 7, 1981, shall cease operation within a cumulative operating time not to exceed 15 minutes following breakage or removal of at least 3 square centimeters of contiguous surface of the outer envelope.

(2) *Lamp labeling.* Each self-extinguishing lamp shall be clearly marked with the letter "T" on the outer envelope and on another part of the lamp in such a manner that it is visible after the outer envelope of the lamp is broken or removed.

(3) *Lamp packaging.* Lamp packaging for each self-extinguishing lamp shall clearly and prominently display:

(i) The letter "T"; and

(ii) The words "This lamp should self-extinguish within 15 minutes after the outer envelope is broken or punctured. If such damage occurs, TURN OFF AND REMOVE LAMP to avoid possible injury from hazardous shortwave ultraviolet radiation."

(e) *Requirements for lamps that are not self-extinguishing lamps*—(1) *Lamp labeling.* Any high-intensity mercury vapor discharge lamp that does not comply with paragraph (d)(1) of this section shall be clearly and legibly marked with the letter "R" on the outer envelope and on another part of the lamp in such a manner that it is visible after the outer envelope of the lamp is broken or removed.

(2) *Lamp packaging.* Lamp packaging for each high-intensity mercury vapor discharge lamp that does not comply with paragraph (d)(1) of this section shall clearly and prominently display:

(i) The letter "R"; and

(ii) The words "WARNING: This lamp can cause serious skin burn and eye inflammation from shortwave ultraviolet radiation if outer envelope of the lamp is broken or punctured. Do not use where people will remain for more than a few minutes unless adequate shielding or other safety precautions are used. Lamps that will automatically extinguish when the outer envelope is broken or punctured are commercially available."

(3) *Lamp advertisement.* Advertising for any high-intensity mercury vapor discharge lamp that does not comply with paragraph (d)(1) of this section shall prominently display the following wording: "WARNING: This lamp can cause serious skin burn and eye inflammation from shortwave ultraviolet radiation if outer envelope of the lamp is broken or punctured. Do not use where people will remain for more than a few minutes unless adequate shielding or other safety precautions are used. Lamps that will automatically extinguish when the outer envelope is broken or punctured are commercially available."

(f) *Test conditions.* Any high-intensity mercury vapor discharge lamp under test for compliance with the requirements set forth in paragraph (d)(1) of this section shall be started and operated under the following conditions as applicable:

(1) Lamp voltage, current, and orientation shall be those indicated or recommended by the manufacturer for operation of the intact lamp.

(2) The lamp shall be operated on a reference ballast.

(3) The lamp shall be started in air that has a temperature of 25 ±5 °C. Heating and movement of the air surrounding the lamp shall be that produced by the lamp and ballast alone.

(4) If any test is performed in an enclosure, the enclosure shall be not less than 0.227 cubic meter (8 cubic feet).

(5) Any lamp designed to be operated only in a specific fixture or luminaire that the lamp manufacturer supplies or

specifies shall be tested in that fixture or luminaire. Any other lamp shall be tested with no reflector or other surrounding material.

[44 FR 52195, Sept. 7, 1979, as amended at 53 FR 11254, Apr. 6, 1988]

PART 1050—PERFORMANCE STANDARDS FOR SONIC, INFRASONIC, AND ULTRASONIC RADIATION-EMITTING PRODUCTS

AUTHORITY: 21 U.S.C. 351, 352, 360, 360e–360j, 371, 381; 42 U.S.C. 263b–263n.

§ 1050.10 Ultrasonic therapy products.

(a) *Applicability.* The provisions of this section are applicable as specified herein to any ultrasonic therapy product for use in physical therapy manufactured on or after February 17, 1979.

(b) *Definitions.* The following definitions apply to words and phrases used in this section:

(1) *Amplitude modulated waveform* means a waveform in which the ratio of the temporal-maximum pressure amplitude spatially averaged over the effective radiating surface to the root-mean-square pressure amplitude spatially averaged over the effective radiating surface is greater than 1.05.

(2) *Applicator* means that portion of a fully assembled ultrasonic therapy product that is designed to emit ultrasonic radiation and which includes one or more ultrasonic transducers and any associated housing.

(3) *Beam cross-section* means the surface in any plane consisting of the points at which the intensity is greater than 5 percent of the spatial-maximum intensity in that plane.

(4) *Beam nonuniformity ratio* means the ratio of the temporal-average spatial-maximum intensity to the temporal-average effective intensity.

(5) *Centroid of a surface* means the point whose coordinates are the mean values of the coordinates of the points of the surface.

(6) *Collimating applicator* means an applicator that does not meet the definition of a focusing applicator as specified in paragraph (b)(15) of this section and for which the ratio of the area of at least one beam cross-section, whose centroid is 12 centimeters from the centroid of the effective radiating surface, to the area of the effective radiating surface is less than two.

(7) *Continuous-wave waveform* means a waveform in which the ratio of the temporal-maximum pressure amplitude spatially averaged over the effective radiating surface to the root-mean-square pressure amplitude spatially averaged over the effective radiating surface is less than or equal to 1.05.

(8) *Diverging applicator* means an applicator that does not meet the definition of a collimating applicator or a focusing applicator as specified in paragraphs (b) (6) and (15) of this section.

(9) *Effective intensity* means the ratio of the ultrasonic power to the focal area for a focusing applicator. For all other applicators, the effective intensity is the ratio of the ultrasonic power to the effective radiating area. Effective intensity is expressed in watts per square centimeter ($W\ cm^{-2}$).

(10) *Effective radiating area* means the area consisting of all points of the effective radiating surface at which the intensity is 5 percent or more of the maximum intensity at the effective radiating surface, expressed in square centimeters (cm^2).

(11) *Effective radiating surface* means the surface consisting of all points 5 millimeters from the applicator face.

(12) *Focal area* means the area of the focal surface, expressed in square centimeters (cm^2).

(13) *Focal length* means the distance between the centroids of the effective radiating surface and the focal surface, for a focusing applicator, expressed in centimeters (cm).

(14) *Focal surface* means the beam cross-section with smallest area of a focusing applicator.

(15) *Focusing applicator* means an applicator in which the ratio of the area of the beam cross-section with the smallest area to the effective radiating area is less than one-half.

(16) *Generator* means that portion of a fully assembled ultrasonic therapy product that supplies electrical energy to the applicator. The generator may include, but is not limited to, a power supply, ultrasonic frequency oscillator, service controls, operation controls, and a cabinet to house these components.

(17) *Maximum beam nonuniformity ratio* means the maximum value of the beam nonuniformity ratio characteristic of a model of an ultrasonic therapy product.

(18) *Operation control* means any control used during operation of an ultrasonic therapy product that affects the ultrasonic radiation emitted by the applicator.

(19) *Pressure amplitude* means the instantaneous value of the modulating waveform, and is $p_1(t)$ in the expression for a pressure wave, $p(t)=p_1(t)\ p_2(t)$, where p(t) is the instantaneous pressure, $p_1(t)$ is the modulating envelope, and $p_2(t)$ is the relative amplitude of the carrier wave normalized to a peak height of one. All are periodic functions of time, t, at any point in space. The period of $p_1(t)$ is greater than the period of $p_2(t)$.

(20) *Pulse duration* means a time interval, expressed in seconds, beginning at the first time the pressure amplitude exceeds the minimum pressure amplitude plus 10 percent of the difference between the maximum and minimum pressure amplitudes, and ending at the last time the pressure amplitude returns to this value.

(21) *Pulse repetition rate* means the repetition frequency of the waveform modulating the ultrasonic carrier wave expressed in pulses per second (pps).

(22) *Service control* means any control provided for the purpose of adjustment that is not used during operation and can affect the ultrasonic radiation emitted by the applicator, or can alter the calibration or accuracy of an indicator or operation control.

(23) *Ultrasonic frequency* means the frequency of the ultrasonic radiation carrier wave, expressed in Hertz (Hz), kilohertz (kHz), or megahertz (MHz).

(24) *Ultrasonic power* means the total power emitted in the form of ultrasonic radiation by the applicator averaged over each cycle of the ultrasonic radiation carrier wave, expressed in watts.

(25) *Ultrasonic therapy product* means:

(i) Any device intended to generate and emit ultrasonic radiation for therapeutic purposes at ultrasonic frequencies above 16 kilohertz (kHz); or

(ii) Any generator or applicator designed or specifically designated for use in a device as specified in paragraph (b)(25)(i) of this section.

(26) *Ultrasonic transducer* means a device used to convert electrical energy of ultrasonic frequency into ultrasonic radiation or vice versa.

(c) *Performance requirements.* The requirements of this paragraph are applicable to each ultrasonic therapy product as defined in paragraph (b)(25) of this section when the generator and applicator are designated or intended for use together, or to each generator when the applicator(s) intended for use with the generator does not contain controls that affect the functioning of the generator.

(1) *Ultrasonic power and intensity*—(i) *Continuous-wave waveform operation.* A means shall be incorporated to indicate the magnitudes of the temporal-average ultrasonic power and the temporal-average effective intensity when emission is of continuous-wave waveform. The error in the indication of the temporal-average ultrasonic power shall not exceed ±20 percent for all emissions greater than 10 percent of the maximum emission.

(ii) *Amplitude-modulated waveform operation.* A means shall be incorporated to indicate the magnitudes of the temporal-maximum ultrasonic power and the temporal-maximum effective intensity when the emission is of amplitude-modulated waveform. The sum of the errors in the indications of the temporal-maximum ultrasonic power and the ratio of the temporal-maximum effective intensity to the temporal-average effective intensity specified in paragraph (d)(3)(ii) of this section shall not exceed ±20 percent for all emissions greater than 10 percent of the maximum emission.

(2) *Treatment time.* A means shall be incorporated to enable the duration of emission of ultrasonic radiation for treatment to be preset and such means shall terminate emission at the end of the preset time. Means shall also be incorporated to enable termination of emission at any time. Means shall be incorporated to indicate the magnitude of the duration of emission (expressed in minutes) to within 0.5 minute of the preset duration of emission for setting less than 5 minutes, to within 10 percent of the preset duration of emission

for settings of from 5 minutes to 10 minutes, and to within 1 minute of the preset duration of emission for settings greater than 10 minutes.

(3) *Pulse duration and repetition rate.* A means shall be incorporated for indicating the magnitudes of pulse duration and pulse repetition rate of the emitted ultrasonic radiation, if there are operation controls for varying these quantities.

(4) *Ultrasonic frequency.* A means shall be incorporated for indicating the magnitude of the ultrasonic frequency of the emitted ultrasonic radiation, if there is an operation control for varying this quantity.

(5) *Visual indicator.* A means shall be incorporated to provide a clear, distinct, and readily understood visual indicator when and only when electrical energy of appropriate ultrasonic frequency is being applied to the ultrasonic transducer(s).

(d) *Labeling requirements.* In addition to the labeling requirements in part 801 and the requirements of §§ 1010.2 and 1010.3 of this chapter, each ultrasonic therapy product shall be subject to the applicable labeling requirements of this paragraph.

(1) *Operation controls.* Each operation control shall be clearly labeled identifying the function controlled and, where appropriate, the units of measure of that function. If a separate control and indicator are associated with the same function, then labeling the appropriate units of measure of that function is required for the indicator but not for the control.

(2) *Service controls.* Each service control that is accessible without displacement or removal of any part of the ultrasonic therapy product shall be clearly labeled identifying the function controlled and shall include the phrase "for service adjustment only."

(3) *Generators.* (i) Each generator shall bear a label that states: The brand name, model designation, and unique serial number or other unique identification so that it is individually identifiable; ultrasonic frequency (unless there is an operation control for varying this quantity); and type of waveform (continuous wave or amplitude modulated).

(ii) Generators employing amplitude-modulated waveforms shall also bear a label that provides the following information: Pulse duration and pulse repetition rate (unless there are operation controls for varying these quantities), an illustration of the amplitude-modulated waveform, and the ratio of the temporal-maximum effective intensity to the temporal-average effective intensity. (If this ratio is a function of any operation control setting, then the range of the ratio shall be specified, and the waveform illustration shall be provided for the maximum value of this ratio.)

(4) *Applicators.* Each applicator shall bear a label that provides the following information:

(i) The brand name, model designation, and unique serial number or other unique identification so the applicator is individually identifiable;

(ii) A designation of the generator(s) for which the applicator is intended; and

(iii) The ultrasonic frequency, effective radiating area, maximum beam nonuniformity ratio, type of applicator (focusing, collimating, diverging), and for a focusing applicator the focal length and focal area.

(5) *Label specification.* Labels required by this paragraph shall be permanently affixed to or inscribed on the ultrasonic therapy product; they shall be legible and clearly visible. If the size, configuration, or design of the ultrasonic therapy product would preclude compliance with the requirements of this paragraph, the Director, Center for Devices and Radiological Health, may approve alternate means of providing such labels).

(e) *Tests for determination of compliance*—(1) *Tests for certification.* Tests on which certification pursuant to § 1010.2 of this chapter is based shall account for all measurement errors and uncertainties. Such tests shall also account for increases in emission and degradation in radiation safety that occur with age.

(2) *Test conditions.* Except as provided in § 1010.13 of this chapter, tests for compliance with each of the applicable requirements of this section shall be made:

(i) For all possible combinations of adjustments of the controls listed in the operation instructions.

(ii) With the ultrasonic radiation emitted into the equivalent of an infinite medium of distilled, degassed water at 30 °C for measurements concerning the ultrasonic radiation.

(iii) With line voltage variations in the range of ±10 percent of the rated value specified by the manufacturer.

(3) *Measurement parameters.* Measurements for determination of the spatial distribution of the ultrasonic radiation field shall be made with a detector having dimensions of less than one wavelength in water or an equivalent measurement technique.

(f) *Informational requirements*—(1) *Servicing information.* The manufacturer of an ultrasonic therapy product shall provide or cause to be provided to servicing dealers and distributors, and to others upon request, at a cost not to exceed the cost of preparation and distribution adequate instructions for operations, service, and calibration, including a description of those controls and procedures that could be used to increase radiation emission levels, and a schedule of maintenance necessary to keep equipment in compliance with this section. The instructions shall include adequate safety precautions that may be necessary regarding ultrasonic radiation exposure.

(2) *User information.* The manufacturer of an ultrasonic therapy product shall provide as an integral part of any user instruction or operation manual that is regularly supplied with the product, or, if not so supplied, shall cause to be provided with each ultrasonic therapy product, and to others upon request, at a cost not to exceed the cost of preparation and distribution:

(i) Adequate instructions concerning assembly, operation, safe use, any safety procedures and precautions that may be necessary regarding the use of ultrasonic radiation, and a schedule of maintenance necessary to keep the equipment in compliance with this section. The operation instructions shall include a discussion of all operation controls, and shall describe the effect of each control.

(ii) Adequate description of the spatial distribution of the ultrasonic radiation field and the orientation of the field with respect to the applicator. This will include a textual discussion with diagrams, plots, or photographs representative of the beam pattern. If there is more than one ultrasonic transducer in an applicator and their positions are not fixed relative of each other, then the description must specify the spatial distribution of the ultrasonic radiation field emitted by each ultrasonic transducer and present adequate examples of the combination field of the ultrasonic tranducers with regard to safe use. The description of the ultrasonic radiation field shall state that such description applies under conditions specified in paragraph (e)(2)(ii) of this section.

(iii) Adequate description, as appropriate to the product, of the uncertainties in magnitude expressed in terms of percentage error, of the ultrasonic frequency effective radiating area, and, where applicable, the ratio of the temporal-maximum effective intensity to the temporal-average effective intensity, pulse duration, pulse repetition rate, focal area, and focal length. The errors in indications specified in paragraphs (c)(1) and (c)(2) of this section shall be stated in the instruction manual.

(iv) A listing of controls, adjustments, and procedures for operation and maintenance, including the warning "Caution—use of controls or adjustments or performance of procedures other than those specified herein may result in hazardous exposure to ultrasonic energy."

[43 FR 7166, Feb. 17, 1978, as amended at 45 FR 16483, Mar. 14, 1980; 53 FR 11255, Apr. 6, 1988]

SUBCHAPTER K [RESERVED]

SUBCHAPTER L—REGULATIONS UNDER CERTAIN OTHER ACTS ADMINISTERED BY THE FOOD AND DRUG ADMINISTRATION

PART 1210—REGULATIONS UNDER THE FEDERAL IMPORT MILK ACT

Subpart A—General Provisions

AUTHORITY: 21 U.S.C. 141–149.

SOURCE: 38 FR 32104, Nov. 20, 1973, unless otherwise noted.

CROSS REFERENCES: For Animal and Plant Health Inspection Service regulations concerning tubercular cattle, see 9 CFR parts 50 and 77. For Animal and Plant Health Inspection Service regulations, see 9 CFR chapter I. For customs regulations concerning importation of milk and cream, see 19 CFR 12.7. For regulations of the Agricultural Marketing Service (Marketing Agreements and Orders) covering marketing areas for milk, see 7 CFR chapter X.

Subpart A—General Provisions

§ 1210.1 Authority.

For the purposes of the regulations in this part the act (44 Stat. 1101; 21 U.S.C. 141–149) "To regulate the importation of milk and cream into the United States for the purpose of promoting the dairy industry of the United States and protecting the public health" shall be known and referred to as "the Federal Import Milk Act."

§ 1210.2 Scope of act.

The provisions of the act apply to all milk and cream offered for import into the continental United States.

§ 1210.3 Definitions.

(a) *Secretary*. Secretary means the Secretary of Health and Human Services.

(b) *Commissioner*. Commissioner means the Commissioner of Food and Drugs.

(c) *Milk*. For the purposes of the act and of the regulations in this part:

Milk is the whole, fresh, clean, lacteal secretion obtained by the complete milking of one or more healthy cows, properly fed and kept, excluding that obtained within 15 days before and 5 days after calving, or such longer period as may be necessary to render the milk practically colostrum free.

(d) *Condensed milk*. Condensed milk, as the term is used in section 3, paragraph 2, of the Federal Import Milk Act, includes evaporated milk in the manufacture of which sterilization of the milk and cream is a necessary and usual process; it includes sweetened condensed milk only if it is prepared by a process which insures sterilization of the milk and cream. Condensed milk, as the term is used in section 3, paragraph 3, of the Federal Import Milk Act, means sweetened condensed milk.

(e) *Sweetened condensed milk*. Sweetened condensed milk conforms to the definition and standard of identity for such food as set out in § 131.120 of this chapter.

(f) *Evaporated milk*. Evaporated milk conforms to the definition and standard of identity for such food as set out in § 131.130 of this chapter.

(g) *Cream*. Cream is that portion of the milk, rich in milk fat, which rises to the surface of milk on standing or is separated from it by centrifugal force. (See §§ 131.150 through 131.157 of this chapter).

(h) *Pasteurization*. Pasteurization is the process of heating every particle of milk or cream to at least 143 °F., and holding it at such temperature continuously for at least 30 minutes, or to at least 161 °F., and holding it at such temperature continuously for at least 15 seconds.

(i) *Shipper*. A shipper is anyone, other than a common carrier, who ships, transports, or causes to be shipped or transported into the United States milk or cream owned by him.

[38 FR 32104, Nov. 20, 1973, as amended at 42 FR 14091, Mar. 15, 1977]

Subpart B—Inspection and Testing

§ 1210.10 Availability for examination and inspection.

Dairy farms and plants from which milk or cream is shipped or transported into the United States shall be open at all reasonable times to authorized agents for necessary examinations and inspections. Failure to permit such examinations and inspections may be considered cause for the suspension or revocation of the permit.

§ 1210.11 Sanitary inspection of dairy farms.

The sanitary conditions of any dairy farm producing milk or cream to be shipped or transported into the United States or to a plant from which milk or cream is to be shipped or transported into the United States must score at least 50 points out of 100 points, according to the methods for scoring as provided by the score card for sanitary inspection of dairy farms in the form prescribed by the Secretary.

§ 1210.12 Physical examination of cows.

(a) Physical examination of any and all cows in herds producing milk or cream which is to be shipped or transported into the United States shall be made by an authorized veterinarian of the United States or of any State or municipality thereof or of the country in which such milk or cream is produced to determine whether such cow or cows are in a healthy condition. Such examination shall be made as often as the Secretary may deem necessary and, in any event, shall have been made within one year previous to the time of the importation.

(b) The result of the physical examination shall be set forth in the form prescribed by the Secretary.

§ 1210.13 Tuberculin test.

(a) Except as provided in § 1210.27 any and all animals in herds producing milk or cream which is to be shipped or transported raw into the United States shall be free from tuberculosis, as determined by a tuberculin test applied by an official veterinarian of the United States or of any State or municipality thereof or of the country in which such milk or cream is produced. Such test shall be made as often as the Secretary may deem necessary and, in any event, shall have been made within 1 year previous to the time of the importation. All animals showing positive or suspicious reactions to the tuberculin test must be permanently removed from the herd.

(b) The results of the tuberculin test and all facts concerning the disposal of reacting or suspected animals shall be set forth in the form prescribed by the Secretary.

§ 1210.14 Sanitary inspection of plants.

The sanitary conditions of any plant handling milk or cream any part of which is to be shipped or transported into the United States shall score at least 50 points out of 100 points according to the methods for scoring as provided by the score card for sanitary inspection of such plants in the form prescribed by the Secretary.

§ 1210.15 Pasteurization; equipment and methods.

All dairy farms and plants at which any milk or cream is pasteurized for shipment or transportation into the United States shall employ adequate pasteurization machinery of a type easily cleaned and of sanitary construction capable of holding every portion of the milk or cream at the required temperature for the required time. Such pasteurizing machinery shall be properly equipped with accurate time and temperature recording devices, which shall be kept at all times in good working order. The temperature at the time of heating and holding must invariably be recorded on thermograph charts, initialed, numbered, and dated by the official having jurisdiction over such farms and plants. All thermograph charts shall be held for a period of 2 years unless within that period they have been examined and released by such authorized agents as are designated by the Secretary.

§ 1210.16 Method of bacterial count.

The bacterial count of milk and cream refers to the number of viable bacteria as determined by the standard plate method of the American Public Health Association in use at the time of the examination.

§ 1210.17 Authority to sample and inspect.

Inspectors engaged in the enforcement of the Federal Import Milk Act are empowered to test for temperature, to take samples of milk or cream, and to use such means as may be necessary for these purposes.

§ 1210.18 Scoring.

Scoring of sanitary conditions required by §§ 1210.11, 1210.14 shall be done by an official inspector of the United States or of any State or municipality thereof or of the country in which the dairy farm or plant is located.

Subpart C—Permit Control

§ 1210.20 Application for permit.

Application for a permit to ship or transport milk or cream into the United States shall be made by the actual shipper upon forms prescribed by the Secretary. The request for forms of applications for permits should be addressed to Commissioner of Food and Drugs, Food and Drug Administration, Department of Health and Human Services, 5600 Fishers Lane, Rockville, MD 20857.

§ 1210.21 Permit number.

Each permit issued under the Federal Import Milk Act, including each temporary permit, shall bear an individual number. The right to the use of such number is restricted solely to the permittee.

§ 1210.22 Form of tag.

Each container of milk or cream shipped or transported into the United States by such permittee shall have firmly attached thereto a tag in the following form, bearing the required information in clear and legible type:

Product ______________________

(State whether raw milk, pasteurized milk, raw cream, or pasteurized cream.)

Permit number ______________________

Federal Import Milk Act, Department of Health and Human Services.

Shipper ______________________

Address of shipper ______________________

Provided, That in case of unit shipments consisting of milk only or cream only under one permit number, in lieu of each container being so marked, the vehicle of transportation, if sealed, may be tagged with the above tag, which should, in addition, show the number of containers and quantity of contents of each.

§ 1210.23 Permits granted on certificates.

In the discretion of the Secretary, a permit may be granted on a duly certified statement signed by a duly accredited official of an authorized department of any foreign government or of any State of the United States or any municipality thereof. Such statement shall be in the form of a certificate prescribed by the Secretary, and shall have attached thereto, as a part thereof, signed copies of reports prescribed by §§ 1210.12, 1230.13, and also by §§ 1210.11, 1210.14, as applicable. The necessary inspections and examinations upon which the reports are based shall be made by persons who are acting under the direct supervision of the certifying official.

§ 1210.24 Temporary permits.

A temporary permit will be granted only upon a satisfactory showing that the applicant therefor has been unable to obtain the necessary inspections required by the applicable provisions of section 2 of the Federal Import Milk Act. Temporary permits shall be valid until the Secretary shall provide for inspection to ascertain that clauses 1, 2, and 3 of section 2 of the Federal Import Milk Act have been complied with.

§ 1210.25 Permits for pasteurized milk or cream.

Permits to ship or transport pasteurized milk or cream into the United States will be granted only upon compliance with the requirements of clauses 1 and 3 of section 2 of the Federal Import Milk Act, §§ 1210.11, 1210.12, 1210.14, as applicable.

§ 1210.26 Permits for raw milk or cream.

Except as provided in § 1210.27, permits to ship or transport raw milk or cream into the United States will be granted only when the milk or cream comes from dairy farms or plants where pasteurization is not carried on and then only upon compliance with the requirements of clauses 1, 2, and 3 of section 2 of the Federal Import Milk Act, §§ 1210.11 to 1210.14 as applicable.

§ 1210.27 Permits waiving clauses 2 and 5, section 2 of the Federal Import Milk Act.

A permit to ship or transport raw milk into the United States will contain a waiver of clauses 2 and 5 of section 2 of the Federal Import Milk Act when the shipper is an operator of a creamery or condensery, or is a producer shipping or transporting to a creamery or condensery and the creamery or condensery is located in the United States within a radius of 20 miles of the point of production of such milk, and the milk, prior to its sale, use, or disposal, is pasteurized, condensed, or evaporated.

§ 1210.28 Permits waiving clause 4, section 2 of the Federal Import Milk Act.

The Secretary, in his discretion, will issue to a shipper who is an operator of a condensery a permit waiving the requirements of clause 4, of section 2 of the Federal Import Milk Act and allowing milk and cream containing not to exceed 1,200,000 bacteria per cubic centimeter to be shipped or transported into the United States if the condensery is located within a radius of 15 miles of the point of production of the milk and cream and such milk and cream are to be sterilized in the manufacture of condensed milk.

Subpart D—Hearings

§ 1210.30 Hearing procedure for permit denial, suspension, and revocation.

Any person who contests denial, suspension, or revocation of a permit shall have an opportunity for a regulatory hearing before the Food and Drug Administration pursuant to subpart F of part 16 of this chapter.

[41 FR 48269, Nov. 2, 1976, as amended at 42 FR 15676, Mar. 22, 1977]

§ 1210.31 Hearing before prosecution.

Before violation of the act is referred to the Department of Justice for prosecution under section 5 of the Federal Import Milk Act, an opportunity to be heard will be given to the party against whom prosecution is under consideration. The hearing will be private and confined to questions of fact. The party notified may present evidence, either oral or written, in person or by attorney, to show cause why he should not be prosecuted. After a hearing is held, if it appears that the law has been violated, the facts will be reported to the Department of Justice.

[41 FR 48269, Nov. 2, 1976]

PART 1230—REGULATIONS UNDER THE FEDERAL CAUSTIC POISON ACT

Subpart A—General Provisions

Sec.

AUTHORITY: 15 U.S.C. 1261–1276.

CROSS REFERENCES: For regulations relating to invoices, entry, and assessment of duties, see 19 CFR parts 141, 142, 143, 151, 152. For regulations regarding the examination, classification, and disposition of foods, drugs, devices, cosmetics, insecticides, fungicides, and caustic or corrosive substances, see 19 CFR part 12. For regulations relating to consular invoices, and documentation of merchandise, see 22 CFR parts 91 and 92.

SOURCE: 38 FR 32110, Nov. 20, 1973, unless otherwise noted.

Subpart A—General Provisions

§ 1230.2 Scope of the act.

The provisions of the act apply to any container which has been shipped or delivered for shipment in interstate or foreign commerce, as defined in section 2(c) of the act (44 Stat. 1407; 15 U.S.C. 402) or which has been received from shipment in such commerce for sale or exchange, or which is sold or offered for sale or held for sale or exchange in any Territory or possession or in the District of Columbia.

§ 1230.3 Definitions.

(a) The word *container* as used in the regulations in this part means a retail parcel, package, or container suitable for household use and employed exclusively to hold any dangerous caustic or corrosive substance defined in the act.

(b) The words *suitable for household use* mean and imply adaptability for ready or convenient handling in places where people dwell.

Subpart B—Labeling

§ 1230.10 Placement.

The label or sticker shall be so firmly attached to the container that it will remain thereon while the container is being used, and be so placed as readily to attract attention.

§ 1230.11 Required wording.

(a) The common name of the dangerous caustic or corrosive substance which shall appear on the label or sticker is the name given in section 2(a) of the act (44 Stat. 1406; 15 U.S.C. 402) or any other name commonly employed to designate and identify such substance.

(b) Preparations within the scope of the act bearing trade or fanciful names shall, in addition, be labeled with the common name of the dangerous caustic or corrosive substance contained therein and comply with all the other requirements of the act and of the regulations in this part.

§ 1230.12 Manufacturer; distributor.

If the name on the label or sticker is other than that of the manufacturer, it shall be qualified by such words as "packed for," "packed by," "sold by," or "distributed by," as the case may be, or by other appropriate expression.

§ 1230.13 Labeling of "poison".

The following are styles of uncondensed Gothic capital letters 24-point (type face) size:

POISON

POISON

When letters of not less than 24-point size are required on a label in stating the word "poison" they must not be smaller than those above set forth.

§ 1230.14 Directions for treatment.

Except as provided in § 1230.16, the container shall bear in all cases upon the label or sticker thereof, immediately following the word "Poison," directions for treatment in the case of internal personal injury; in addition, if the substance may cause external injury, directions for appropriate treatment shall be given. The directions shall prescribe such treatments for personal injury as are sanctioned by competent medical authority, and the materials called for by such directions shall be, whenever practicable, such as are usually available in the household.

§ 1230.15 Responsibility for labeling directions for treatment.

A person who receives from a manufacturer or wholesaler any container which under the conditions set forth in section 2(b)(4) of the act and § 1230.16 does not bear at the time of shipment directions for treatment in the case of personal injury must place such directions on the label or sticker if he offers such container for general sale or exchange.

§ 1230.16 Exemption from labeling directions for treatment.

Manufacturers and wholesalers only, at the time of shipment or delivery for shipment, are exempted from placing directions for treatment on the label or sticker of any container for other than household use, but in any event the information required by section 2(b) (1), (2), and (3) of the act (44 Stat. 1407; 15 U.S.C. 402) and the regulations in this part shall be given.

Subpart C—Guaranty

§ 1230.20 General guaranty.

In lieu of a particular guaranty for each lot of dangerous caustic or corrosive substances, a general continuing guaranty may be furnished by the guarantor to actual or prospective purchasers. The following are forms of continuing guaranties:

(a) Substances for both household use and other than household use:

The undersigned guarantees that the retail parcels, packages, or containers of the dangerous caustic or corrosive substance or substances to be sold to ________ are not misbranded within the meaning of the Federal Caustic Poison Act.

(Date)

(Signature and address of guarantor)

(b) Substances for other than household use (this form may be issued only by a manufacturer or wholesaler) (§§ 1230.15, 1230.16):

The dangerous caustic or corrosive substance or substances in retail parcels, packages, or containers suitable for household use to be sold to ________ are for other than household use, and guaranteed not to be misbranded within the meaning of the Federal Caustic Poison Act.

(Date)

(Signature and address of manufacturer or wholesaler)

§ 1230.21 Specific guaranty.

If a guaranty in respect to any specific lot of dangerous caustic or corrosive substances be given, it shall be incorporated in or attached to the bill of sale, invoice, or other schedule bearing the date and the name and quantity of the substance sold, and shall not appear on the label or package. The following are forms of specific guaranties:

(a) Substances for both household use and other than household use:

The undersigned guarantees that the retail parcels, packages, or containers of the dangerous caustic or corrosive substance or substances listed herein (or specifying the substances) are not misbranded within the meaning of the Federal Caustic Poison Act.

(Signature and address of guarantor)

(b) Substances for other than household use (this form may be issued only by a manufacturer or wholesaler (§§ 1230.15, 1230.16):

The dangerous caustic or corrosive substance or substances listed herein (or specifying the substances) in retail parcels, packages, or containers suitable for household use are for other than household use and are guaranteed not to be misbranded within the meaning of the Federal Caustic Poison Act.

(Name and address of manufacturer or wholesaler)

Subpart D—Administrative Procedures

§ 1230.30 Collection of samples.

Samples for examination by or under the direction and supervision of the Food and Drug Administration shall be collected by:

(a) An authorized agent in the employ of the Department of Health and Human Services;

(b) Any officer of any State, Territory, or possession, or of the District of Columbia, authorized by the Secretary of Health and Human Services.

§ 1230.31 Where samples may be collected.

Caustic or corrosive substances within the scope of this act (44 Stat. 1406; 15 U.S.C. 401–411) may be sampled wherever found.

§ 1230.32 Analyzing of samples.

Samples collected by an authorized agent shall be analyzed at the laboratory designated by the Food and Drug Administration. Only such samples as are collected in accordance with §§ 1230.30, 1230.31 may be analyzed by or under the direction and supervision of the Food and Drug Administration. Upon request one subdivision of the sample, if available, shall be delivered to the party or parties interested.

§ 1230.33 Investigations.

Authorized agents in the employ of the Department of Health and Human Services may make investigations, including the inspection of premises where dangerous caustic and corrosive substances subject to the act are manufactured, packed, stored, or held for sale or distribution, and make examinations of freight and other transportation records.

§ 1230.34 Analysis.

(a) The methods of examination or analysis employed shall be those prescribed by the Association of Official Agricultural Chemists, when applicable, provided, however, that any method of analysis or examination satisfactory to the Food and Drug Administration may be employed.

(b) All percentages stated in the definitions in section 2(a) of the Federal Caustic Poison Act shall be determined by weight.

§ 1230.35 Hearings.

Whenever it appears from the inspection, analysis, or test of any container that the provisions of section 3 or 6 of the Federal Caustic Poison Act (44 Stat. 1407, 1409; 15 U.S.C. 403, 406) have been violated and criminal proceedings are contemplated, notice shall be given to the party or parties against whom prosecution is under consideration and to other interested parties, and a date shall be fixed at which such party or parties may be heard. The hearing shall be held at the office of the Food and Drug Administration designated in the notice and shall be private and confined to questions of fact. The parties notified may present evidence, either oral or written, in person or by attorney, to show cause why the matter should not be referred for prosecution as a violation of the Federal Caustic Poison Act.

§ 1230.36 Hearings; when not provided for.

No hearing is provided for when the health, medical, or drug officer or agent of any State, Territory, or possession, or of the District of Columbia, acts under the authority contained in section 8 of the Federal Caustic Poison

Act (44 Stat. 1409; 15 U.S.C. 408) in reporting a violation direct to the United States attorney.

§ 1230.37 Publication.

(a) After judgment of the court in any proceeding under the Federal Caustic Poison Act, notice shall be given by publication. Such notice shall include the findings of the court and may include the findings of the analyst and such explanatory statements of fact as the Secretary of Health and Human Services may deem appropriate.

(b) This publication may be made in the form of a circular, notice, or bulletin, as the Secretary of Health and Human Services may direct.

(c) If an appeal be taken from the judgment of the court before such publication, that fact shall appear.

Subpart E—Imports

§ 1230.40 Required label information.

Containers which are offered for import shall in all cases bear labels or stickers having thereon the information required by section 2(b) (1), (2), and (3) of the Federal Caustic Poison Act and the directions for treatment in the case of personal injury, except such directions need not appear on the label or sticker at the time of shipment by a wholesaler or manufacturer for other than household use.

§ 1230.41 Delivery of containers.

Containers shall not be delivered to the consignee prior to report of examination, unless a bond has been given on the appropriate form for the amount of the full invoice value of such containers, together with the duty thereon, and the refusal of the consignee to return such containers for any cause to the custody of the District Director of Customs when demanded, for the purpose of excluding them from the country or for any other purpose, the consignee shall pay an amount equal to the sum named in the bond, and such part of the duty, if any, as may be payable, as liquidated damages for failure to return to the District Director of Customs on demand all containers covered by the bond.

§ 1230.42 Invoices.

As soon as the importer makes entry, the invoices covering containers and the public stores packages shall be made available, with the least possible delay, for inspection by the representative of the district. When no sample is desired the invoice shall be stamped by the district "No sample desired, Food and Drug Administration, Department of Health and Human Services, per (initials of inspecting officer)."

§ 1230.43 Enforcement.

(a) *Enforcement agency.* The Federal Caustic Poison Act shall be enforced by the Food and Drug Administration, Department of Health and Human Services.

(b) *Enforcement of provisions.* The enforcement of the provisions of the Federal Caustic Poison Act as they relate to imported dangerous caustic or corrosive substances, will, as a general rule, be under the direction of the chief of the local inspection district of the Food and Drug Administration, Department of Health and Human Services, and District Directors of Customs acting as administrative officers in carrying out directions relative to the detention, exportation, and sale, or other disposition of such substances and action under the bond in case of noncompliance with the provisions of the Federal Caustic Poison Act.

(c) *Chief of district as customs officer.* The chief of district shall be deemed a customs officer in enforcing import regulations.

(d) *Nonlaboratory ports.* (1) At the ports of entry where there is no district of the Food and Drug Administration, the District Director of Customs or deputy, on the day when the first notice of expected shipment of containers is received, either by invoice or entry, shall notify the chief of district in whose territory the port is located.

(2) On the day of receipt of such notice the chief of district shall mail to the District Director of Customs appropriate notice, if no sample is desired. This notice serves as an equivalent to stamping the invoices at district ports with the legend "No sample desired,

Food and Drug Administration, Department of Health and Human Services, per (initials of inspecting officer)."

(3) If samples are desired, the Chief of district shall immediately notify the District Director of Customs.

(4) The District Director of Customs at once shall forward samples, accompanied by description of shipment.

(5) When samples are desired from each shipment of containers, the chief of district shall furnish to District Director of Customs and deputies at ports within the district's territory a list of such containers, indicating the size of sample necessary. Samples should then be sent promptly on arrival of containers without awaiting special request.

(6) In all other particulars the procedure shall be the same at nonlaboratory ports as at laboratory ports, except that the time consumed in delivery of notices by mail shall be allowed for.

§ 1230.44 Samples.

On the same day that samples are requested by the district, the District Director of Customs or appraiser shall notify the importer that samples will be taken, that the containers must be held intact pending a notice of the result of inspection and analysis, and that in case the containers do not comply with the requirements of the Federal Caustic Poison Act, they must be returned to the District Director of Customs for disposition. This notification may be given by the District Director of Customs or appraiser through individual notices to the importer or by suitable bulletin notices posted daily in the customhouse.

§ 1230.45 No violation; release.

As soon as examination of the samples is completed, if no violation of the act is detected, the chief of the district shall send a notice of release to the importer and a copy of this notice to the District Director of Customs for his information.

§ 1230.46 Violation.

(a) If a violation of the Federal Caustic Poison Act is disclosed, the chief of the district shall send to the importer due notice of the nature of the violation and of the time and place where evidence may be presented, showing that the containers should not be refused admission. At the same time similar notice regarding detention of the containers shall be sent to the District Director of Customs, requesting him to refuse delivery thereof or to require their return to customs custody if by any chance the containers were released without the bond referred to in § 1230.41. The time allowed the importer for representations regarding the shipment may be extended at his request for a reasonable period to permit him to secure such evidence.

(b) If the importer does not reply to the notice of hearing in person or by letter within the time allowed on the notice, a second notice, marked "second and last notice," shall be sent at once by the chief of the district, advising him that failure to reply will cause definite recommendation to the District Director of Customs that the containers be refused admission and that the containers be exported within 3 months under customs supervision.

§ 1230.47 Rejected containers.

(a) In all cases where the containers are to be refused admission, the chief of the district within 1 day after hearing, or, if the importer does not appear or reply within 3 days after second notice, shall notify the District Director of Customs in duplicate accordingly.

(b) Not later than 1 day after receipt of this notice the District Director of Customs shall sign and transmit to the importer one of the copies, which shall serve as notification to the importer that the containers must be exported under customs supervision within 3 months from such date, as provided by law; the other notice shall be retained as office record and later returned as a report to the chief of the district. In all cases the importer shall return his notice to the District Director of Customs, properly certified as to the information required, as the form provides.

§ 1230.48 Relabeling of containers.

(a) If containers are to be released after relabeling, a notice shall be sent by the chief of district direct to the importer, a carbon copy being sent to the

208-072 D-23

District Director of Customs. This notice must state specifically the conditions to be performed, so as to bring the performance thereof under the provisions of the customs bonds on consumption and warehouse entries, these bonds including provisions requiring compliance with all of the requirements of the Federal Caustic Poison Act and all regulations and instructions issued thereunder. The notice will also state the officer to be notified by the importer when the containers are ready for inspection.

(b) The importer must return the notice to the District Director of Customs or chief of district, as designated, with the certificate thereon filled out, stating that he has complied with the prescribed conditions and that the containers are ready for inspection at the place named.

(c) This notice will be delivered to the inspection officer, who, after inspection, will endorse the result thereof on the back of the notice and return the same to the District Director of Customs or to the chief of district, as the case may be.

(d) When the conditions to be complied with are under the supervision of the chief of district, and these conditions have been fully met, he shall release the containers to the importer, sending a copy of the notice of release to the District Director of Customs for his information. If the containers have not been properly relabeled within the period allowed, the chief of district shall immediately give notice in duplicate to the District Director of Customs of the results of inspection. The District Director of Customs shall sign and immediately transmit one copy of the notice to the importer and proceed in the usual manner.

(e) If the containers are detained subject to relabeling to be performed under the supervision of the District Director of Customs, the District Director of Customs, as soon as relabeling is accomplished, will notify the importer that the containers are released.

(f) If the containers have not been properly relabeled within the period allowed, their sale after labeling as required by the act or other disposition must be effected by the District Director of Customs.

(g) When the final action has been taken on containers which have been refused admission, sold, or otherwise disposed of as provided for by the act or which have been relabeled under the supervision of the District Director of Customs, he shall send to the chief of district a notice of such final action, giving the date and disposition.

(h) When relabeling is allowed the importer must furnish satisfactory evidence as to the identity of the containers before release is given. The relabeling must be done at a stated place and apart from other containers of a similar nature.

(i) When containers are shipped to another port for relabeling or exportation, they must be shipped under customs carrier's manifest, in the same manner as shipments in bond.

(j) District Directors of Customs will perform the inspection service whenever containers are to be exported, sold, or otherwise disposed of, and in other cases when there is no officer of the district available.

(k) District Directors of Customs and representatives of the district will confer and arrange the apportionment of the inspection service according to local conditions. Officers of the district will, whenever feasible, perform the inspection service in connection with relabeling.

§ 1230.49 Penalties.

(a) In case of failure to comply with the instructions or recommendations of the chief of district as to conditions under which containers may be disposed of, the District Director of Customs shall notify the chief of district in all cases coming to his attention within 3 days after inspection or after the expiration of the 3 months allowed by law if no action is taken.

(b) The chief of district, upon receipt of the above-described notice, and in all cases of failure to meet the conditions imposed in order to comply with the provisions of the Federal Caustic Poison Act coming directly under his supervision, shall transmit to the District Director of Customs such evidence as he may have at hand tending to indicate the importer's liability and make a recommendation accordingly.

(c) The District Director of Customs, within 3 days of the receipt of this recommendation, whether favorable or otherwise, shall notify the importer that, the legal period of 3 months for exportation or relabeling having expired, action will be taken within 30 days to enforce the terms of the bond.

PART 1240—CONTROL OF COMMUNICABLE DISEASES

Subpart A—General Provisions

AUTHORITY: 42 U.S.C. 216, 243, 264, 271.

CROSS REFERENCES: For Department of Health and Human Services regulations relating to foreign quarantine, sanitation measures, and control of communicable diseases, see Centers for Disease Control's requirements as set forth in 42 CFR parts 71 and 72.

SOURCE: 40 FR 5620, Feb. 6, 1975, unless otherwise noted.

Subpart A—General Provisions

§ 1240.3 General definitions.

As used in this part, terms shall have the following meaning:

(a) *Bactericidal treatment.* The application of a method or substance for the destruction of pathogens and other organisms as set forth in § 1240.10.

(b) *Communicable diseases.* Illnesses due to infectious agents or their toxic products, which may be transmitted from a reservoir to a susceptible host either directly as from an infected person or animal or indirectly through the agency of an intermediate plant or animal host, vector, or the inanimate environment.

(c) *Communicable period.* The period or periods during which the etiologic agent may be transferred directly or indirectly from the body of the infected person or animal to the body of another.

(d) *Contamination.* The presence of a certain amount of undesirable substance or material, which may contain pathogenic microorganisms.

(e) *Conveyance.* Conveyance means any land or air carrier, or any vessel as defined in paragraph (n) of this section.

(f) *Garbage.* (1) The solid animal and vegetable waste, together with the natural moisture content, resulting from the handling, preparation, or consumption of foods in houses, restaurants, hotels, kitchens, and similar establishments, or (2) any other food waste containing pork.

(g) *Incubation period.* The period between the implanting of disease organisms in a susceptible person and the appearance of clinical manifestation of the disease.

(h) *Interstate traffic.* (1) The movement of any conveyance or the transportation of persons or property, including any portion of such movement or transportation which is entirely within a State or possession,

(i) From a point of origin in any State or possession to a point of destination in any other State or possession, or

(ii) Between a point of origin and a point of destination in the same State or possession but through any other State, possession, or contiguous foreign country.

(2) Interstate traffic does not include the following:

(i) The movement of any conveyance which is solely for the purpose of unloading persons or property transported from a foreign country, or loading persons or property for transportation to a foreign country.

(ii) The movement of any conveyance which is solely for the purpose of effecting its repair, reconstruction, rehabilitation, or storage.

(i) *Milk.* Milk is the product defined in § 131.110 of this chapter.

(j) *Milk products.* Food products made exclusively or principally from the lacteal secretion obtained from one or more healthy milk-producing animals, e.g., cows, goats, sheep, and water buffalo, including, but not limited to, the following: lowfat milk, skim milk, cream, half and half, dry milk, nonfat dry milk, dry cream, condensed or concentrated milk products, cultured or acidified milk or milk products, kefir, eggnog, yogurt, butter, cheese (where not specifically exempted by regulation), whey, condensed or dry whey or whey products, ice cream, ice milk, other frozen dairy desserts and products obtained by modifying the chemical or physical characteristics of milk, cream, or whey by using enzymes, solvents, heat, pressure, cooling, vacuum, genetic engineering, fractionation, or other similar processes, and any such product made by the addition or subtraction of milkfat or the addition of safe and suitable optional ingredients for the protein, vitamin, or mineral fortification of the product.

(k) *Minimum heat treatment.* The causing of all particles in garbage to be heated to a boiling temperature and held at that temperature for a period of not less than 30 minutes.

(l) *Possession.* Any of the possessions of the United States, including Puerto Rico and the Virgin Islands.

(m) *Potable water.* Water which meets the standards prescribed in the Environmental Protection Agency's Primary Drinking Water Regulations as set forth in 40 CFR part 141 and the Food and Drug Administration's sanitation requirements as set forth in this part and part 1250 of this chapter.

(n) *State.* Any State, the District of Columbia, Puerto Rico and the Virgin Islands.

(o) *Utensil.* Includes any kitchenware, tableware, glassware, cutlery, containers, or equipment with which food or drink comes in contact during storage, preparation, or serving.

(p) *Vessel.* Any passenger-carrying, cargo, or towing vessel exclusive of:

(1) Fishing boats including those used for shell-fishing;

(2) Tugs which operate only locally in specific harbors and adjacent waters;

(3) Barges without means of self-propulsion;

(4) Construction-equipment boats and dredges; and

(5) Sand and gravel dredging and handling boats.

(q) *Watering point.* The specific place or water boat from which potable water is loaded on a conveyance.

(r) *Molluscan shellfish.* Any edible species of fresh or frozen oysters, clams, mussels, and scallops or edible portions thereof, except when the product consists entirely of the shucked adductor muscle.

(s) *Certification number* means a unique combination of letters and numbers assigned by a shellfish control authority to a molluscan shellfish processor.

(t) *Shellfish control authority* means a Federal, State, or foreign agency, or sovereign tribal government, legally responsible for the administration of a program that includes activities such as classification of molluscan shellfish growing areas, enforcement of molluscan shellfish harvesting controls, and certification of molluscan shellfish processors.

(u) *Tag* means a record of harvesting information attached to a container of shellstock by the harvester or processor.

[40 FR 5620, Feb. 6, 1975, as amended at 48 FR 11431, Mar. 18, 1983; 57 FR 57344, Dec. 4, 1992; 60 FR 65201, Dec. 18, 1995]

§ 1240.10 Effective bactericidal treatment.

Whenever, under the provisions of this part, bactericidal treatment is required, it shall be accomplished by one or more of the following methods:

(a) By immersion of the utensil or equipment for at least 2 minutes in clean hot water at a temperature of at least 170 °F or for one-half minute in boiling water;

(b) By immersion of the utensil or equipment for at least 2 minutes in a lukewarm chlorine bath containing at least 50 ppm of available chlorine if hypochlorites are used or a concentration of equal bactericidal strength if chloramines are used;

(c) By exposure of the utensil or equipment in a steam cabinet at a temperature of at least 170 °F for at least 15 minutes or at a temperature of 200 °F for at least 5 minutes;

(d) By exposure of the utensil or equipment in an oven or hot air cabinet at a temperature of at least 180 °F for at least 20 minutes;

(e) In the case of utensils or equipment so designed or installed as to make immersion or exposure impractical, the equipment may be treated for the prescribed periods of time either at the temperatures or with chlorine solutions as specified above, (1) with live steam from a hose if the steam can be confined, (2) with boiling rinse water, or (3) by spraying or swabbing with chlorine solution;

(f) Any other method determined by the Commissioner of Food and Drugs, upon application of an owner or operator of a conveyance, to be effective to prevent the spread of communicable disease.

[40 FR 5620, Feb. 6, 1975, as amended at 54 FR 24900, June 12, 1989]

Subpart B—Administrative Procedures

§ 1240.20 Issuance and posting of certificates following inspections.

The Commissioner of Food and Drugs may issue certificates based upon inspections provided for in this part and part 1250. Such certificates shall be prominently posted on conveyances.

[40 FR 5620, Feb. 6, 1975, as amended at 48 FR 11431, Mar. 18, 1983]

§ 1240.30 Measures in the event of inadequate local control.

Whenever the Commissioner of Food and Drugs determines that the measures taken by health authorities of any State or possession (including political subdivisions thereof) are insufficient to prevent the spread of any of the communicable diseases from such State or possession to any other State or possession, he may take such measures to prevent such spread of the diseases as he deems reasonably necessary, including inspection, fumigation, disinfection, sanitation, pest extermination, and destruction of animals or articles believed to be sources of infection.

[40 FR 5620, Feb. 6, 1975, as amended at 48 FR 11431, Mar. 18, 1983]

§ 1240.45 Report of disease.

The master of any vessel or person in charge of any conveyance engaged in interstate traffic, on which a case or suspected case of a communicable disease develops shall, as soon as practicable, notify the local health authority at the next port of call, station, or stop, and shall take such measures to prevent the spread of the disease as the local health authority directs.

Subpart C [Reserved]

Subpart D—Specific Administrative Decisions Regarding Interstate Shipments

§ 1240.60 Molluscan shellfish.

(a) A person shall not offer for transportation, or transport, in interstate traffic any molluscan shellfish handled or stored in such an insanitary manner, or grown in an area so contaminated, as to render such molluscan shellfish likely to become agents in, and their transportation likely to contribute to the spread of communicable disease from one State or possession to another.

(b) All shellstock shall bear a tag that discloses the date and place they were harvested (by State and site), type and quantity of shellfish, and by whom they were harvested (i.e., the identification number assigned to the harvester by the shellfish control authority, where applicable or, if such identification numbers are not assigned, the name of the harvester or the name or registration number of the harvester's vessel). In place of the tag, bulk shellstock shipments may be accompanied by a bill of lading or similar shipping document that contains the same information.

(c) All containers of shucked molluscan shellfish shall bear a label that identifies the name, address, and certification number of the packer or repacker of the molluscan shellfish.

(d) Any molluscan shellfish without such a tag, shipping document, or label, or with a tag, shipping document, or label that does not bear all the information required by paragraphs (b) and (c) of this section, shall be subject to seizure or refusal of entry, and destruction.

[40 FR 5620, Feb. 6, 1975, as amended at 60 FR 65202, Dec. 18, 1995]

§ 1240.61 Mandatory pasteurization for all milk and milk products in final package form intended for direct human consumption.

(a) No person shall cause to be delivered into interstate commerce or shall sell, otherwise distribute, or hold for sale or other distribution after shipment in interstate commerce any milk or milk product in final package form for direct human consumption unless the product has been pasteurized or is made from dairy ingredients (milk or milk products) that have all been pasteurized, except where alternative procedures to pasteurization are provided for by regulation, such as in part 133 of this chapter for curing of certain cheese varieties.

(b) Except as provided in paragraphs (c) and (d) of this section, the terms "pasteurization," "pasteurized," and similar terms shall mean the process of heating every particle of milk and milk product in properly designed and operated equipment to one of the temperatures given in the following table and held continuously at or above that temperature for at least the corresponding specified time:

Temperature	Time
145 °F (63 °C) [1]	30 minutes.
161 °F (72 °C) [1]	15 seconds.
191 °F (89 °C)	1 second.

[1] If the fat content of the milk product is 10 percent or more, or if it contains added sweeteners, the specified temperature shall be increased by 5 °F (3 °C).

Temperature	Time
194 °F (90 °C)	0.5 second.
201 °F (94 °C)	0.1 second.
204 °F (96 °C)	0.05 second.
212 °F (100 °C)	0.01 second.

(c) Eggnog shall be heated to at least the following temperature and time specification:

Temperature	Time
155 °F (69 °C)	30 minutes.
175 °F (80 °C)	25 seconds.
180 °F (83 °C)	15 seconds.

(d) Neither paragraph (b) nor (c) of this section shall be construed as barring any other pasteurization process that has been recognized by the Food and Drug Administration to be equally efficient in the destruction of microbial organisms of public health significance.

[52 FR 29514, Aug. 10, 1987, as amended at 57 FR 57344, Dec. 4, 1992]

§ 1240.62 Turtles intrastate and interstate requirements.

(a) *Definition.* As used in this section the term "turtles" includes all animals commonly known as turtles, tortoises, terrapins, and all other animals of the order Testudinata, class Reptilia, except marine species (families Dermachelidae and Chelonidae).

(b) *Sales; general prohibition.* Except as otherwise provided in this section, viable turtle eggs and live turtles with a carapace length of less than 4 inches shall not be sold, held for sale, or offered for any other type of commercial or public distribution.

(c) *Destruction of turtles or turtle eggs; criminal penalties.* (1) Any viable turtle eggs or live turtles with a carapace length of less than 4 inches which are held for sale or offered for any other type of commercial or public distribution shall be subject to destruction in a humane manner by or under the supervision of an officer or employee of the

Food and Drug Administration in accordance with the following procedures:

(i) Any District Office of the Food and Drug Administration, upon detecting viable turtle eggs or live turtles with a carapace length of less than 4 inches which are held for sale or offered for any other type of commercial or public distribution, shall serve upon the person in whose possession such turtles or turtle eggs are found a written demand that such turtles or turtle eggs be destroyed in a humane manner under the supervision of said District Office, within 10 working days from the date of promulgation of the demand. The demand shall recite with particularity the facts which justify the demand. After service of the demand, the person in possession of the turtles or turtle eggs shall not sell, distribute, or otherwise dispose of any of the turtles or turtle eggs except to destroy them under the supervision of the District Office, unless and until the Director of the Center for Veterinary Medicine withdraws the demand for destruction after an appeal pursuant to paragraph (c)(1)(ii) of this section.

(ii) The person on whom the demand for destruction is served may either comply with the demand or, within 10 working days from the date of its promulgation, appeal the demand for destruction to the Director of the Center for Veterinary Medicine, Food and Drug Administration, 7519 Standish Pl., Rockville, MD 20855. The demand for destruction may also be appealed within the same period of 10 working days by any other person having a pecuniary interest in such turtles or turtle eggs. In the event of such an appeal, the Center Director shall provide an opportunity for hearing by written notice to the appellant(s) specifying a time and place for the hearing, to be held within 14 days from the date of the notice but not within less than 7 days unless by agreement with the appellant(s).

(iii) Appearance by any appellant at the hearing may be by mail or in person, with or without counsel. The hearing shall be conducted by the Center Director or his designee, and a written summary of the proceedings shall be prepared by the person presiding. Any appellant shall have the right to hear and to question the evidence on which the demand for destruction is based, including the right to cross-examine witnesses, and he may present oral or written evidence in response to the demand.

(iv) If, based on the evidence presented at the hearing, the Center Director finds that the turtles or turtle eggs were held for sale or offered for any other type of commercial or public distribution in violation of this section, he shall affirm the demand that they be destroyed under the supervision of an officer or employee of the Food and Drug Administration; otherwise, the Center Director shall issue a written notice that the prior demand by the District Office is withdrawn. If the Center Director affirms the demand for destruction he shall order that the destruction be accomplished in a humane manner within 10 working days from the date of the promulgation of his decision. The Center Director's decision shall be accompanied by a statement of the reasons for the decision. The decision of the Center Director shall constitute final agency action, reviewable in the courts.

(v) If there is no appeal to the Director of the Center for Veterinary Medicine from the demand by the Food and Drug Administration District Office and the person in possession of the turtles or turtle eggs fails to destroy them within 10 working days, or if the demand is affirmed by the Director of the Center for Veterinary Medicine after an appeal and the person in possession of the turtles or turtle eggs fails to destroy them within 10 working days, the District Office shall designate an officer or employee to destroy the turtles or turtle eggs. It shall be unlawful to prevent or to attempt to prevent such destruction of turtles or turtle eggs by the officer or employee designated by the District Office. Such destruction will be stayed if so ordered by a court pursuant to an action for review in the courts as provided in paragraph (c)(1)(iv) of this section.

(2) Any person who violates any provision of this section, including but not limited to any person who sells, offers for sale, or offers for any other type of

commercial or public distribution viable turtle eggs or live turtles with a carapace length of less than 4 inches, or who refuses to comply with a valid final demand for destruction of turtles or turtle eggs (either an unappealed demand by an FDA District Office or a demand which has been affirmed by the Director of the Center for Veterinary Medicine pursuant to appeal), or who fails to comply with the requirement in such a demand that the manner of destruction be humane, shall be subject to a fine of not more than $1,000 or imprisonment for not more than 1 year, or both, for each violation, in accordance with section 368 of the Public Health Service Act (42 U.S.C. 271).

(d) *Exceptions.* The provisions of this section are not applicable to:

(1) The sale, holding for sale, and distribution of live turtles and viable turtle eggs for bona fide scientific, educational, or exhibitional purposes, other than use as pets.

(2) The sale, holding for sale, and distribution of live turtles and viable turtle eggs not in connection with a business.

(3) The sale, holding for sale, and distribution of live turtles and viable turtle eggs intended for export only, provided that the outside of the shipping package is conspicuously labeled "For Export Only."

(4) Marine turtles excluded from this regulation under the provisions of paragraph (a) of this section and eggs of such turtles.

(e) *Petitions.* The Commissioner of Food and Drugs, either on his own initiative or on behalf of any interested person who has submitted a petition, may publish a proposal to amend this regulation. Any such petition shall include an adequate factual basis to support the petition, and will be published for comment if it contains reasonable grounds for the proposed regulation. A petition requesting such a regulation, which would amend this regulation, shall be submitted to the Division of Dockets Management, Food and Drug Administration, 5630 Fishers Lane, rm. 1061, Rockville, MD 20852.

[40 FR 22545, May 23, 1975, as amended at 46 FR 8461, Jan. 27, 1981; 48 FR 11431, Mar. 18, 1983; 54 FR 24900, June 12, 1989; 59 FR 14366, Mar. 28, 1994; 66 FR 56035, Nov. 6, 2001; 70 FR 48073, Aug. 18, 2005]

§ 1240.63 African rodents and other animals that may carry the monkeypox virus.

(a) *What Actions Are Prohibited? What Animals Are Affected?* (1) Except as provided in paragraph (a)(2) of this section,

(i) You must not capture, offer to capture, transport, offer to transport, sell, barter, or exchange, offer to sell, barter, or exchange, distribute, offer to distribute, or release into the environment, any of the following animals, whether dead or alive:

(A) Prairie dogs (*Cynomys* sp.),

(B) African Tree squirrels (*Helioscirurus* sp.),

(C) Rope squirrels (*Funisciurus* sp.),

(D) African Dormice (*Graphiurus* sp.),

(E) Gambian giant pouched rats (*Cricetomys* sp.),

(F) Brush-tailed porcupines (*Atherurus* sp.),

(G) Striped mice (*Hybomys* sp.), or

(H) Any other animal so prohibited by order of the Commissioner of Food and Drugs because of that animal's potential to transmit the monkeypox virus; and

(ii) You must not prevent, or attempt to prevent, the Food and Drug Administration (FDA) from causing an animal to be quarantined or destroyed under a written order for the animal's quarantine or destruction.

(2) The prohibitions in paragraph (a)(1) of this section do not apply if you:

(i) Transport an animal listed in paragraph (a)(1) of this section, or covered by an order by the Commissioner of Food and Drugs, to veterinarians or animal control officials for veterinary care, quarantine, or destruction purposes; or

(ii) Have written permission from FDA to capture, offer to capture, transport, offer to transport, sell, barter, or exchange, offer to sell, barter, or exchange, distribute, offer to distribute, and/or release into the environment an animal listed in paragraph (a)(1) of this section, or covered by an order by the Commissioner of Food and Drugs. You may not seek written permission to sell, barter, or exchange, or offer to sell, barter, or exchange, as a pet, an animal listed in paragraph (a)(1) of this section or covered by an order by the Commissioner of Food and Drugs.

(A) To obtain such written permission from FDA, you must send a written request to the Division of Compliance (HFV–230), Center for Veterinary Medicine, Food and Drug Administration, 7500 Standish Pl., Rockville, MD 20855, Attn: Listed Animal Permit Official. You may also fax your request to the Division of Compliance (using the same address in the previous sentence) at 301–827–1498.

(B) Your request must state the reasons why you need an exemption, describe the animals involved, describe the number of animals involved, describe how the animals will be transported (including carrying containers or cages, precautions for handlers, types of vehicles used, and other procedures to minimize exposure of animals and precautions to prevent animals from escaping into the environment), describe any holding facilities, quarantine procedures, and/or veterinarian evaluation involved in the animals' movement, and explain why an exemption will not result in the spread of monkeypox within the United States.

(C) We (FDA) will respond, in writing, to all requests, and we also may impose conditions in granting an exemption.

(b) *What Actions Can FDA Take?* (1) To prevent the monkeypox virus from spreading and becoming established in the United States, we may, in addition to any other authorities under this part:

(i) Issue an order causing an animal to be placed in quarantine,

(ii) Issue an order causing an animal to be destroyed, or

(iii) Take any other action necessary to prevent the spread of the monkeypox virus.

(2) Any order to cause an animal to be placed in quarantine or to cause an animal to be destroyed will be in writing.

(c) *How Do I Appeal an Order?* (1) If you receive a written order to cause an animal to be placed in quarantine or to cause an animal to be destroyed, you may appeal that order. Your appeal must be in writing and be submitted to the Food and Drug Administration District Director whose office issued the order, and you must submit the appeal within two business days after you receive the order.

(2) As part of your appeal, you may request an informal hearing. Your appeal must include specific facts showing there is a genuine and substantial issue of fact that requires a hearing.

(3) If we grant your request for an informal hearing, we will follow the regulatory hearing requirements at in part 16, except that:

(i) The written order will serve as notice of opportunity for that hearing, for purposes of § 16.22(a) of this chapter;

(ii) The presiding officer will issue a decision rather than a report and a recommended decision. The presiding officer's decision constitutes final agency action.

[68 FR 62368, Nov. 4, 2003]

§ 1240.65 Psittacine birds.

(a) The term psittacine birds shall include all birds commonly known as parrots, Amazons, Mexican double heads, African grays, cocatoos, macaws, parakeets, love birds, lories, lorikeets, and all other birds of the psittacine family.

(b) No person shall transport, or offer for transportation, in interstate traffic any psittacine bird unless the shipment is accompanied by a permit from the State health department of the State of destination where required by such department.

(c) Whenever the Surgeon General finds that psittacine birds or human beings in any area are infected with psittacosis and there is such danger of transmission of psittacosis from such area as to endanger the public health, he may declare it an area of infection. No person shall thereafter transport, or offer for transportation, in interstate traffic any psittacine bird from such area, except shipments authorized by the Surgeon General for purposes of medical research and accompanied by a permit issued by him, until the Surgeon General finds that there is no longer any danger of transmission of psittacosis from such area. As used in this paragraph, the term "area" includes, but is not limited to, specific premises or buildings.

§ 1240.75 Garbage.

(a) A person shall not transport, receive, or cause to be transported or received, garbage in interstate traffic and feed such garbage to swine unless, prior to the feeding, such garbage has received minimum heat treatment.

(b) A person transporting garbage in interstate traffic shall not make, or agree to make, delivery thereof to any person with knowledge of the intent or customary practice of such person to feed to swine garbage which has not been subjected to minimum heat treatment.

Subpart E—Source and Use of Potable Water

§ 1240.80 General requirements for water for drinking and culinary purposes.

Only potable water shall be provided for drinking and culinary purposes by any operator of a conveyance engaged in interstate traffic, except as provided in § 1250.84(b) of this chapter. Such water shall either have been obtained from watering points approved by the Commissioner of Food and Drugs, or, if treated aboard a conveyance, shall have been subjected to treatment approved by the Commissioner of Food and Drugs.

[40 FR 5620, Feb. 6, 1975, as amended at 48 FR 11431, Mar. 18, 1983]

§ 1240.83 Approval of watering points.

(a) The Commissioner of Food and Drugs shall approve any watering point if (1) the water supply thereat meets the standards prescribed in the Environmental Protection Agency's Primary Drinking Water Regulations as set forth in 40 CFR part 141, and (2) the methods of and facilities for delivery of such water to the conveyance and the sanitary conditions surrounding such delivery prevent the introduction, transmission, or spread of communicable diseases.

(b) The Commissioner of Food and Drugs may base his approval or disapproval of a watering point upon investigations made by representatives of State departments of health or of the health authorities of contiguous foreign nations.

(c) If a watering point has not been approved, the Commissioner of Food and Drugs may permit its temporary use under such conditions as, in his judgment, are necessary to prevent the introduction, transmission, or spread of communicable diseases.

(d) Upon request of the Commissioner of Food and Drugs, operators of conveyances shall provide information as to watering points used by them.

[40 FR 5620, Feb. 6, 1975, as amended at 48 FR 11431, Mar. 18, 1983; 48 FR 13978, Apr. 1, 1983]

§ 1240.86 Protection of pier water system.

No vessel engaged in interstate traffic shall make a connection between its nonpotable water system and any pier potable water system unless provisions are made to prevent backflow from the vessel to the pier.

§ 1240.90 Approval of treatment aboard conveyances.

(a) The treatment of water aboard conveyances shall be approved by the Commissioner of Food and Drugs if the apparatus used is of such design and is so operated as to be capable of producing and in fact does produce, potable water.

(b) The Commissioner of Food and Drugs may base his approval or disapproval of the treatment of water upon investigations made by representatives of State departments of health or of the health authorities of contiguous foreign nations.

(c) Overboard water treated on vessels shall be from areas relatively free of contamination and pollution.

[40 FR 5620, Feb. 6, 1975, as amended at 48 FR 11431, Mar. 18, 1983]

§ 1240.95 Sanitation of water boats.

No vessel engaged in interstate traffic shall obtain water for drinking and culinary purposes from any water boat unless the tanks, piping, and other appurtenances used by the water boat in the loading, transportation, and delivery of such drinking and culinary water, have been approved by the Commissioner of Food and Drugs.

[40 FR 5620, Feb. 6, 1975, as amended at 48 FR 11431, Mar. 18, 1983]

PART 1250—INTERSTATE CONVEYANCE SANITATION

Subpart A—General Provisions

Sec.

Subpart B—Food Service Sanitation on Land and Air Conveyances, and Vessels

Subpart C—Equipment and Operation of Land and Air Conveyances

Subpart D—Servicing Areas for Land and Air Conveyances

Subpart E—Sanitation Facilities and Conditions on Vessels

AUTHORITY: 42 U.S.C. 216, 243, 264, 271.

CROSS REFERENCES: For Department of Health and Human Services regulations relating to foreign quarantine and control of communicable diseases, see Centers for Disease Control's requirements as set forth in 42 CFR parts 71 and 72.

SOURCE: 40 FR 5624, Feb. 6, 1975, unless otherwise noted.

Subpart A—General Provisions

§ 1250.3 Definitions.

As used in this part, terms shall have the following meaning:

(a) *Bactericidal treatment*. The application of a method or substance for the destruction of pathogens and other organisms as set forth in § 1240.10 of this chapter.

(b) *Communicable diseases*. Illnesses due to infectious agents or their toxic products, which may be transmitted from a reservoir to a susceptible host either directly as from an infected person or animal or indirectly through the agency of an intermediate plant or animal host, vector, or the inanimate environment.

(c) *Communicable period*. The period or periods during which the etiologic agent may be transferred directly or indirectly from the body of the infected person or animal to the body of another.

(d) *Contamination.* The presence of a certain amount of undesirable substance or material, which may contain pathogenic microorganisms.

(e) *Conveyance.* Conveyance means any land or air carrier, or any vessel as defined in paragraph (m) of this section.

(f) *Existing vessel.* Any vessel the construction of which was started prior to the effective date of the regulations in this part.

(g) *Garbage.* (1) The solid animal and vegetable waste, together with the natural moisture content, resulting from the handling, preparation, or consumption of foods in houses, restaurants, hotels, kitchens, and similar establishments, or (2) any other food waste containing pork.

(h) *Interstate traffic.* (1) The movement of any conveyance or the transportation of persons or property, including any portion of such movement or transportation which is entirely within a State or possession, (i) from a point of origin in any State or possession to a point of destination in any other State or possession, or (ii) between a point of origin and a point of destination in the same State or possession but through any other State, possession, or contiguous foreign country.

(2) Interstate traffic does not include the following:

(i) The movement of any conveyance which is solely for the purpose of unloading persons or property transported from a foreign country, or loading persons or property for transportation to a foreign country.

(ii) The movement of any conveyance which is solely for the purpose of effecting its repair, reconstruction, rehabilitation, or storage.

(i) *Possession.* Any of the possessions of the United States, including Puerto Rico and the Virgin Islands.

(j) *Potable water.* Water which meets the standards prescribed in the Environmental Protection Agency's Primary Drinking Water Regulations as set forth in 40 CFR part 141 and the Food and Drug Administration's sanitation regulations as set forth in this part and part 1240 of this chapter.

(k) *State.* Any State, the District of Columbia, Puerto Rico and the Virgin Islands.

(l) *Utensil.* Includes any kitchenware, tableware, glassware, cutlery, containers, or equipment with which food or drink comes in contact during storage, preparation, or serving.

(m) *Vessel.* Any passenger-carrying, cargo, or towing vessel exclusive of:

(1) Fishing boats including those used for shell-fishing;

(2) Tugs which operate only locally in specific harbors and adjacent waters;

(3) Barges without means of self-propulsion;

(4) Construction-equipment boats and dredges; and

(5) Sand and gravel dredging and handling boats.

(n) *Wash water.* Water suitable for domestic uses other than for drinking and culinary purposes, and medical care purposes excluding hydrotherapy.

(o) *Shellfish.* Any fresh, frozen, or incompletely cooked oysters, clams, or mussels, either shucked or in the shell, and any fresh, frozen, or incompletely cooked edible products thereof.

[40 FR 5624, Feb. 6, 1975, as amended at 48 FR 11432, Mar. 18, 1983]

Subpart B—Food Service Sanitation on Land and Air Conveyances, and Vessels

§ 1250.20 Applicability.

All conveyances engaged in interstate traffic shall comply with the requirements prescribed in this subpart and § 1240.20 of this chapter.

§ 1250.21 Inspection.

The Commissioner of Food and Drugs may inspect such conveyance to determine compliance with the requirements of this subpart and § 1240.20 of this chapter.

[40 FR 5624, Feb. 6, 1975, as amended at 48 FR 11432, Mar. 18, 1983]

§ 1250.22 General requirements.

All food and drink served on conveyances shall be clean, wholesome, and free from spoilage, and shall be prepared, stored, handled, and served in accordance with the requirements prescribed in this subpart and § 1240.20 of this chapter.

§ 1250.25 Source identification and inspection of food and drink.

(a) Operators of conveyances shall identify, when requested by the Commissioner of Food and Drugs, the vendors, distributors or dealers from whom they have acquired or are acquiring their food supply, including milk, fluid milk products, ice cream and other frozen desserts, butter, cheese, bottled water, sandwiches and box lunches.

(b) The Commissioner of Food and Drugs may inspect any source of such food supply in order to determine whether the requirements of the regulations in this subpart and in § 1240.20 of this chapter are being met, and may utilize the results of inspections of such sources made by representatives of State health departments or of the health authorities of contiguous foreign nations.

[40 FR 5624, Feb. 6, 1975, as amended at 48 FR 11432, Mar. 18, 1983]

§ 1250.26 Special food requirements.

Milk, fluid milk products, ice cream and other frozen desserts, butter, cheese, and shellfish served or sold on conveyances shall conform to the following requirements:

(a) Milk and fluid milk products, including cream, buttermilk, skim milk, milk beverages, and reconstituted milk, shall be pasteurized and obtained from a source of supply approved by the Commissioner of Food and Drugs. The Commissioner of Food and Drugs shall approve any source of supply at or from which milk or fluid milk products are produced, processed, and distributed so as to prevent the introduction, transmission, or spread of communicable diseases. If a source of supply of milk or fluid milk products has not been approved, the Commissioner of Food and Drugs may permit its temporary use under such conditions as, in his judgment, are necessary to prevent the introduction, transmission, or spread of communicable diseases. Containers of milk and fluid milk products shall be plainly labeled to show the contents, the word "pasteurized", and the identity of the plant at which the contents were packaged by name and address, provided that a code may be used in lieu of address.

(b) Ice cream, other frozen desserts, and butter shall be manufactured from milk or milk products that have been pasteurized or subjected to equivalent heat treatment.

(c) Cheese shall be (1) pasteurized or subjected to equivalent heat treatment, (2) made from pasteurized milk products or from milk products which have been subjected to equivalent heat treatment, or (3) cured for not less than 60 days at a temperature not less than 35 °F.

(d) Milk, buttermilk, and milk beverages shall be served in or from the original individual containers in which received from the distributor, or from a bulk container equipped with a dispensing device so designed, constructed, installed, and maintained as to prevent the transmission of communicable diseases.

(e) Shellfish purchased for consumption on any conveyance shall originate from a dealer currently listed by the Public Health Service as holding an unexpired and unrevoked certificate issued by a State authority.

(f) Shucked shellfish shall be purchased in the containers in which they are placed at the shucking plant and shall be kept therein until used. The State abbreviation and the certificate number of the packers shall be permanently recorded on the container.

[40 FR 5624, Feb. 6, 1975, as amended at 48 FR 11432, Mar. 18, 1983]

§ 1250.27 Storage of perishables.

All perishable food or drink shall be kept at or below 50 °F, except when being prepared or kept hot for serving.

§ 1250.28 Source and handling of ice.

Ice coming in contact with food or drink and not manufactured on the conveyance shall be obtained from sources approved by competent health authorities. All ice coming in contact with food or drink shall be stored and handled in such manner as to avoid contamination.

§ 1250.30 Construction, maintenance and use of places where food is prepared, served, or stored.

(a) All kitchens, galleys, pantries, and other places where food is prepared, served, or stored shall be adequately lighted and ventilated: *Provided, however,* That ventilation of cold storage rooms shall not be required. All such places where food is prepared, served, or stored shall be so constructed and maintained as to be clean and free from flies, rodents, and other vermin.

(b) Such places shall not be used for sleeping or living quarters.

(c) Water of satisfactory sanitary quality, under head or pressure, and adequate in amount and temperature, shall be easily accessible to all rooms in which food is prepared and utensils are cleaned.

(d) All plumbing shall be so designed, installed, and maintained as to prevent contamination of the water supply, food, and food utensils.

§ 1250.32 Food-handling operations.

(a) All food-handling operations shall be accomplished so as to minimize the possibility of contaminating food, drink, or utensils.

(b) The hands of all persons shall be kept clean while engaged in handling food, drink, utensils, or equipment.

§ 1250.33 Utensils and equipment.

(a) All utensils and working surfaces used in connection with the preparation, storage, and serving of food or beverages, and the cleaning of food utensils, shall be so constructed as to be easily cleaned and self-draining and shall be maintained in good repair. Adequate facilities shall be provided for the cleaning and bactericidal treatment of all multiuse eating and drinking utensils and equipment used in the preparation of food and beverages. An indicating thermometer, suitably located, shall be provided to permit the determination of the hot water temperature when and where hot water is used as the bactericidal agent.

(b) All multiuse eating and drinking utensils shall be thoroughly cleaned in warm water and subjected to an effective bactericidal treatment after each use. All other utensils that come in contact with food and drink shall be similarly treated immediately following the day's operation. All equipment shall be kept clean.

(c) After bactericidal treatment, utensils shall be stored and handled in such manner as to prevent contamination before reuse.

§ 1250.34 Refrigeration equipment.

Each refrigerator shall be equipped with a thermometer located in the warmest portion thereof. Waste water drains from ice boxes, refrigerating equipment, and refrigerated spaces shall be so installed as to prevent backflow of contaminating liquids.

§ 1250.35 Health of persons handling food.

(a) Any person who is known or suspected to be in a communicable period or a carrier of any communicable disease shall not be permitted to engage in the preparation, handling, or serving of water, other beverages, or food.

(b) Any person known or suspected to be suffering from gastrointestinal disturbance or who has on the exposed portion of the body an open lesion or an infected wound shall not be permitted to engage in the preparation, handling, or serving of food or beverages.

§ 1250.38 Toilet and lavatory facilities for use of food-handling employees.

(a) Toilet and lavatory facilities of suitable design and construction shall be provided for use of food-handling employees. Railroad dining car crew lavatory facilities are regulated under § 1250.45.

(b) Signs directing food-handling employees to wash their hands after each use of toilet facilities shall be posted so as to be readily observable by such employees. Hand washing facilities shall include soap, sanitary towels and hot and cold running water or warm running water in lieu of hot and cold running water.

(c) All toilet rooms shall be maintained in a clean condition.

§ 1250.39 Garbage equipment and disposition.

Watertight, readily cleanable nonabsorbent containers with close-fitting covers shall be used to receive and store garbage. Garbage and refuse shall be disposed of as frequently as is necessary and practicable.

Subpart C—Equipment and Operation of Land and Air Conveyances

§ 1250.40 Applicability.

The sanitary equipment and facilities on land and air conveyances engaged in interstate traffic and the use of such equipment and facilities shall comply with the requirements prescribed in this subpart.

§ 1250.41 Submittal of construction plans.

Plans for the construction or major reconstruction of sanitary equipment or facilities for such conveyances shall be submitted to the Commissioner of Food and Drugs for review of the conformity of such plans with the requirements of this subpart, except that submittal of plans shall not be required for any conveyance under reconstruction if the owner or operator thereof has made arrangements satisfactory to the Commissioner of Food and Drugs for inspections of such conveyances while under reconstruction for the purpose of determining conformity with those requirements.

[40 FR 5624, Feb. 6, 1975, as amended at 48 FR 11432, Mar. 18, 1983]

§ 1250.42 Water systems; constant temperature bottles.

(a) The water system, whether of the pressure or gravity type, shall be complete and closed from the filling ends to the discharge taps, except for protected vent openings. The water system shall be protected against backflow.

(b) Filling pipes or connections through which water tanks are supplied shall be provided on both sides of all new railway conveyances and on existing conveyances when they undergo heavy repairs. All filling connections shall be easily cleanable and so located and protected as to minimize the hazard of contamination of the water supply.

(c) On all new or reconstructed conveyances, water coolers shall be an integral part of the closed system.

(d) Water filters if used on dining cars and other conveyances will be permitted only if they are so operated and maintained at all times as to prevent contamination of the water.

(e) Constant temperature bottles and other containers used for storing or dispensing potable water shall be kept clean at all times and shall be subjected to effective bactericidal treatment as often as may be necessary to prevent the contamination of water so stored and dispensed.

§ 1250.43 Ice.

Ice shall not be permitted to come in contact with water in coolers or constant temperature bottles.

§ 1250.44 Drinking utensils and toilet articles.

(a) No cup, glass, or other drinking utensil which may be used by more than one person shall be provided on any conveyance unless such cup, glass, or drinking utensil shall have been thoroughly cleaned and subjected to effective bactericidal treatment after each individual use.

(b) Towels, combs, or brushes for common use shall not be provided.

§ 1250.45 Food handling facilities on railroad conveyances.

(a) Both kitchens and pantries of cars hereafter constructed or reconstructed shall be equipped with double sinks, one of which shall be of sufficient size and depth to permit complete immersion of a basket of dishes during bactericidal treatment; in the pantry a dishwashing machine may be substituted for the double sinks. If chemicals are used for bactericidal treatment, 3-compartment sinks shall be provided.

(b) A sink shall be provided for washing and handling cracked ice used in food or drink and shall be used for no other purpose.

(c) Lavatory facilities for the use of the dining car crew shall be provided on each dining car. Such facilities shall be conveniently located and used for hand and face washing only: *Provided, however*, That where the kitchen and pantry on a dining car hereafter constructed or reconstructed are so partitioned or separated as to impede free passage between them lavatory facilities shall be provided in both the kitchen and the pantry.

(d) Wherever toilet and lavatory facilities required by paragraph (c) of this section are not on the dining car, a lavatory shall be provided on the dining car for the use of employees. The lavatory shall be conveniently located and used only for the purpose for which it is installed.

§ 1250.49 Cleanliness of conveyances.

Conveyances while in transit shall be kept clean and free of flies and mosquitoes. A conveyance which becomes infected with vermin shall be placed out of service until such time as it shall have been effectively treated for the destruction of the vermin.

§ 1250.50 Toilet and lavatory facilities.

Where toilet and lavatory facilities are provided on conveyances they shall be so designed as to permit ready cleaning. On conveyances not equipped with retention facilities, toilet hoppers shall be of such design and so located as to prevent spattering of water filling pipes or hydrants.

§ 1250.51 Railroad conveyances; discharge of wastes.

(a) *New railroad conveyances.* Human wastes, garbage, waste water, or other polluting materials shall not be discharged from any new railroad conveyance except at servicing areas approved by the Commissioner of Food and Drugs. In lieu of retention pending discharge at approved servicing areas, human wastes, garbage, waste water, or other polluting materials that have been suitably treated to prevent the spread of communicable diseases may be discharged from such conveyances, except at stations. For the purposes of this section, "new railroad conveyance" means any such conveyance placed into service for the first time after July 1, 1972, and the terms "waste water or other polluting materials" do not include drainage of drinking water taps or lavatory facilities.

(b) *Nonnew railroad conveyances.* Human wastes, garbage, waste water, or other polluting materials shall not be discharged from any railroad conveyance, other than passenger conveyances for which an extension has been granted pursuant to paragraph (f) of this section, after December 31, 1977, except at servicing areas approved by the Commissioner of Food and Drugs. In lieu of retention pending discharge at approved servicing areas, human wastes, garbage, waste water, or other polluting materials that have been suitably treated to prevent the spread of communicable diseases may be discharged from such conveyances, except at stations. The terms "waste water or other polluting materials" do not include drainage of drinking water taps or lavatory facilities.

(c) *Toilets.* When railroad conveyances, occupied or open to occupancy by travelers, are at a station or servicing area, toilets shall be kept locked unless means are provided to prevent contamination of the area or station.

(d) *Submission of annual report.* Each railroad company shall submit to the Center for Food Safety and Applied Nutrition (HFS–627), Food and Drug Administration, 5100 Paint Branch Pkwy., College Park, MD 20740, an annual report of accomplishments made in modifying conveyances to achieve compliance with paragraph (b) of this section. Annual reports shall be required until a report is submitted showing that 100 percent of the company's conveyances can comply with the requirements of paragraph (b) of this section; annual reports shall be required subsequent to such report if conveyances not capable of complying with the requirements of paragraph (b) of this section are acquired. Every railroad company shall have not less than 10 percent of its nonpassenger conveyances that are in operation capable of complying with the requirements of paragraph (b) of this section by December 31, 1974, not less than 40 percent by December 31, 1975, and not less than 70 percent by December 31, 1976. All conveyances, other than passenger conveyances for which an extension has been granted pursuant to paragraph (f) of this section, in operation after December 31, 1977, shall be capable of complying with paragraph (b) of this section.

(e) *Requirements of annual report.* Annual reports required by paragraph (d) of this section shall be submitted within 60 days of the end of each calendar year. Each report shall contain at least the following information:

(1) Company name and address.

(2) Name, title, and address of the company's chief operating official.

(3) Name, title, address, and telephone number of the person designated by the company to be directly responsible for compliance with this section.

(4) A statement that all new railroad conveyances placed into service after July 1, 1972 meet the requirements of this section.

(5) A complete, factual, narrative statement explaining why retrofitting of noncomplying nonnew conveyances is incomplete, if it is incomplete.

(6) A statement of the percentage of conveyances retrofitted with waste discharge facilities in compliance with this section as of the reporting date and the percentage expected to be completed by December 31st of the following year.

(7) A tabular report with the following vertical columns: equipment type, e.g., locomotive, caboose, passenger car, and any others having toilets; number of toilets per conveyance; number of each equipment type in operation; and number of each to be retrofitted by December 31st of each year until 100 percent compliance with this section is achieved.

(f) *Variances and extensions*—(1) *Variances.* Upon application by a railroad company, the Director, Center for Food Safety and Applied Nutrition, may grant a variance from the compliance schedule prescribed in paragraph (d) of this section for nonpassenger conveyances when the requested variance is required to prevent substantial disruption of the railroad company's operations. Such variance shall not affect the final deadline of compliance established in paragraph (d) of this section.

(2) *Extensions.* Upon application by a railroad company, the Director, Center for Food Safety and Applied Nutrition, may grant an extension of time for compliance with the requirements of paragraph (b) of this section beyond December 31, 1977, for passenger conveyances operated by railroad companies when compliance cannot be achieved without substantial disruption of the railroad company's operations.

(3) *Application for variance or extension.* Application for variances or extensions shall be submitted to the Food and Drug Administration, Center for Food Safety and Applied Nutrition, Manager, Interstate Travel Sanitation Sub-Program, HFF–312, 5100 Paint Branch Pkwy., College Park, MD 20740, and shall include the following information:

(i) A detailed description of the proposed deviation from the requirements of paragraphs (b) or (d) of this section.

(ii) A report, current to the date of the request for a variance or extension, containing the information required by paragraph (e) of this section.

(4) *Administration of variances and extensions.* (i) Written notification of the granting or refusal of a variance or extension will be provided to the applying railroad company by the Director, Center for Food Safety and Applied Nutrition. The notification of a granted variance will state the approved deviation from the compliance schedule provided for in paragraph (d) of this section. The notification of a granted extension will state the final date for compliance with the provisions of paragraph (b) of this section.

(ii) A public file of requested variances and extensions, their disposition, and information relating to pending actions will be maintained in the Division of Dockets Management, 5630 Fishers Lane, rm. 1061, Rockville, MD 20852.

(iii) After notice to the railroad company and opportunity for hearing in accordance with part 16 of this chapter, a variance or extension may be withdrawn prior to its scheduled termination if the Director, Center for Food Safety and Applied Nutrition, determines that such withdrawal is necessary to protect the public health.

CROSS REFERENCE: For statutory exemptions for "intercity rail passenger service," see section 306(i) of 45 U.S.C. 546(i).

[40 FR 5624, Feb. 6, 1975, as amended at 40 FR 30110, July 17, 1975; 46 FR 8461, Jan. 27, 1981; 48 FR 11432, Mar. 18, 1983; 54 FR 24900, June 12, 1989; 59 FR 14366, Mar. 28, 1994; 61 FR 14481, Apr. 2, 1996; 66 FR 56035, Nov. 6, 2001]

EFFECTIVE DATE NOTE: For a document staying the effectiveness of §1250.51 (b) and (d), see 42 FR 57122, Nov. 1, 1977.

§ 1250.52 Discharge of wastes on highway conveyances.

There shall be no discharge of excrement, garbage, or waste water from a highway conveyance except at servicing areas approved by the Commissioner of Food and Drugs.

[40 FR 5624, Feb. 6, 1975, as amended at 48 FR 11432, Mar. 18, 1983]

§ 1250.53 Discharge of wastes on air conveyances.

There shall be no discharge of excrement or garbage from any air conveyance except at servicing areas approved by the Commissioner of Food and Drugs.

[40 FR 5624, Feb. 6, 1975, as amended at 48 FR 11432, Mar. 18, 1983]

Subpart D—Servicing Areas for Land and Air Conveyances

§ 1250.60 Applicability.

Land and air conveyances engaged in interstate traffic shall use only such servicing areas within the United States as have been approved by the Commissioner of Food and Drugs as being in compliance with the requirements prescribed in this subpart.

[40 FR 5624, Feb. 6, 1975, as amended at 48 FR 11432, Mar. 18, 1983]

§ 1250.61 Inspection and approval.

The Commissioner of Food and Drugs may inspect any such areas to determine whether they shall be approved. He may base his approval or disapproval on investigations made by representatives of State departments of health.

[40 FR 5624, Feb. 6, 1975, as amended at 48 FR 11432, Mar. 18, 1983]

§ 1250.62 Submittal of construction plans.

Plans for construction or major reconstruction of sanitation facilities at servicing areas shall be submitted to the Commissioner of Food and Drugs for review of the conformity of the proposed facilities with the requirements of this subpart.

[40 FR 5624, Feb. 6, 1975, as amended at 48 FR 11432, Mar. 18, 1983]

§ 1250.63 General requirements.

Servicing areas shall be provided with all necessary sanitary facilities so operated and maintained as to prevent the spread of communicable diseases.

§ 1250.65 Drainage.

All platforms and other places at which water or food supplies are loaded onto or removed from conveyances shall be adequately drained so as to prevent pooling.

§ 1250.67 Watering equipment.

(a) *General requirements.* All servicing area piping systems, hydrants, taps, faucets, hoses, buckets, and other appurtenances necessary for delivery of drinking and culinary water to a conveyance shall be designed, constructed, maintained and operated in such a manner as to prevent contamination of the water.

(b) *Outlets for nonpotable water.* Outlets for nonpotable water shall be provided with fittings different from those provided for outlets for potable water and each nonpotable water outlet shall be posted with permanent signs warning that the water is unfit for drinking.

(c) *Ice.* If bulk ice is used for the cooling of drinking water or other beverages, or for food preservation purposes, equipment constructed so as not to become a factor in the transmission of communicable diseases shall be provided for the storage, washing, handling, and delivery to conveyances of such bulk ice, and such equipment shall be used for no other purposes.

§ 1250.70 Employee conveniences.

(a) There shall be adequate toilet, washroom, locker, and other essential sanitary facilities readily accessible for use of employees adjacent to places or areas where land and air conveyances are serviced, maintained, and cleaned. These facilities shall be maintained in a clean and sanitary condition at all times.

(b) In the case of diners not in a train but with a crew on board, adequate toilet facilities shall be available to the crew within a reasonable distance but not exceeding 500 feet of such diners.

(c) Drinking fountains and coolers shall be constructed of impervious, nonoxidizing material, and shall be so designed and constructed as to be easily cleaned. The jet of a drinking fountain shall be slanting and the orifice of the jet shall be protected by a guard in such a manner as to prevent contamination thereof by droppings from the mouth. The orifice of such a jet shall be located a sufficient distance above the rim of the basin to prevent backflow.

§ 1250.75 Disposal of human wastes.

(a) At servicing areas and at stations where land and air conveyances are occupied by passengers the operations shall be so conducted as to avoid contamination of such areas and stations by human wastes.

(b) Toilet wastes shall be disposed of through sanitary sewers or by other methods assuring sanitary disposal of such wastes. All soil cans and removable containers shall be thoroughly cleaned before being returned to use. Equipment for cleaning such containers and for flushing nonremovable containers and waste carts shall be so designed as to prevent backflow into the water line, and such equipment shall be used for no purpose connected with the handling of food, water or ice.

(c) All persons who have handled soil cans or other containers which have come in contact with human wastes shall be required to wash their hands thoroughly with soap and warm water and to remove any garments which have become soiled with such wastes before engaging in any work connected with the loading, unloading, transporting or other handling of food, water or ice.

§ 1250.79 Garbage disposal.

(a) Water-tight, readily cleanable, nonabsorbent containers with close-fitting covers shall be used to receive and store garbage.

(b) Can washing and draining facilities shall be provided.

(c) Garbage cans shall be emptied daily and shall be thoroughly washed before being returned for use.

Subpart E—Sanitation Facilities and Conditions on Vessels

§ 1250.80 Applicability.

The sanitation facilities and the sanitary conditions on vessels engaged in interstate traffic shall comply with the requirements prescribed in this subpart, provided that no major structural change will be required on existing vessels.

§ 1250.81 Inspection.

The Commissioner of Food and Drugs may inspect such vessels to determine compliance with the requirements of this subpart.

[40 FR 5624, Feb. 6, 1975, as amended at 48 FR 11432, Mar. 18, 1983]

§ 1250.82 Potable water systems.

The following conditions must be met by vessel water systems used for the storage and distribution of water which has met the requirements of § 1240.80 of this chapter.

(a) The potable water system, including filling hose and lines, pumps, tanks, and distributing pipes, shall be separate and distinct from other water systems and shall be used for no other purposes.

(b) All potable water tanks shall be independent of any tanks holding nonpotable water or other liquid. All potable water tanks shall be independent of the shell of the ship unless (1) the bottom of the tank is at least 2 feet above the maximum load water line, (2) the seams in the shell are continuously welded, and (3) there are no rivets in that part of the shell which forms a side of a tank. A deck may be used as the top of a tank provided there are no access or inspection openings or rivets therein, and the seams are continuously welded. No toilet or urinal shall be installed immediately above that part of the deck which forms the top of a tank. All potable water tanks shall be located at a sufficient height above the bilge to allow for draining and to prevent submergence in bilge water.

(c) Each potable water tank shall be provided with a means of drainage and, if it is equipped with a manhole, overflow, vent, or a device for measuring depth of water, provision shall be made to prevent entrance into the tank of any contaminating substance. No deck or sanitary drain or pipe carrying any nonpotable water or liquid shall be permitted to pass through the tank.

(d) Tanks and piping shall bear clear marks of identification.

(e) There shall be no backflow or cross connection between potable water systems and any other systems. Pipes and fittings conveying potable water to any fixture, apparatus, or equipment shall be installed in such way that backflow will be prevented. Waste pipes from any part of the potable water system, including treatment devices, discharging to a drain, shall be suitably protected against backflow.

(f) Water systems shall be cleaned, disinfected, and flushed whenever the Commissioner of Food and Drugs shall find such treatment necessary to prevent the introduction, transmission, or spread of communicable diseases.

[40 FR 5624, Feb. 6, 1975, as amended at 48 FR 11432, Mar. 18, 1983]

§ 1250.83 Storage of water prior to treatment.

The following requirements with respect to the storage of water on vessels prior to treatment must be met in order to obtain approval of treatment facilities under § 1240.90 of this chapter.

(a) The tank, whether independent or formed by the skin of the ship, deck, tank top, or partitions common with other tanks, shall be free of apparent leakage.

(b) No sanitary drain shall pass through the tank.

(c) The tank shall be adequately protected against both the backflow and discharge into it of bilge or highly contaminated water.

§ 1250.84 Water in galleys and medical care spaces.

(a) Potable water, hot and cold, shall be available in the galley and pantry except that, when potable water storage is inadequate, nonpotable water may be piped to the galley for deck washing and in connection with garbage disposal. Any tap discharging nonpotable water which is installed for deck washing purposes shall not be more than 18 inches above the deck and shall be distinctly marked "For deck washing only".

(b) In the case of existing vessels on which heat treated wash water has been used for the washing of utensils prior to the effective date of the regulations in this part, such water may continue to be so used provided controls are employed to insure the heating of all water to at least 170 °F before discharge from the heater.

(c) Potable water, hot and cold, shall be available in medical care spaces for hand-washing and for medical care purposes excluding hydrotherapy.

§ 1250.85 Drinking fountains and coolers; ice; constant temperature bottles.

(a) Drinking fountains and coolers shall be constructed of impervious, nonoxidizing material, and shall be so designed and constructed as to be easily cleaned. The jet of a drinking fountain shall be slanting and the orifice of the jet shall be protected by a guard in such a manner as to prevent contamination thereof by droppings from the mouth. The orifice of such a jet shall be located a sufficient distance above the rim of the basin to prevent backflow.

(b) Ice shall not be permitted to come in contact with water in coolers or constant temperature bottles.

(c) Constant temperature bottles and other containers used for storing or dispensing potable water shall be kept clean at all times and shall be subjected to effective bactericidal treatment after each occupancy of the space served and at intervals not exceeding one week.

§ 1250.86 Water for making ice.

Only potable water shall be piped into a freezer for making ice for drinking and culinary purposes.

§ 1250.87 Wash water.

Where systems installed on vessels for wash water, as defined in § 1250.3(n), do not comply with the requirements of a potable water system, prescribed in § 1250.82, they shall be constructed so as to minimize the possibility of the water therein being contaminated. The storage tanks shall comply with the requirements of § 1250.83, and the distribution system shall not be cross connected to a system carrying water of a lower sanitary quality. All faucets shall be labeled "Unfit for drinking".

§ 1250.89 Swimming pools.

(a) Fill and draw swimming pools shall not be installed or used.

(b) Swimming pools of the recirculation type shall be equipped so as to provide complete circulation, replacement, and filtration of the water in the pool every six hours or less. Suitable means of chlorination and, if necessary, other treatment of the water shall be provided to maintain the residual chlorine in the pool water at not less than 0.4 part per million and the pH (a measure of the hydrogen ion concentration) not less than 7.0.

(c) Flowing-through types of salt water pools shall be so operated that complete circulation and replacement of the water in the pool will be effected every 6 hours or less. The water delivery pipe to the pool shall be independent of all other pipes and shall originate at a point where maximum flushing of the pump and pipe line is effected after leaving polluted waters.

§ 1250.90 Toilets and lavatories.

Toilet and lavatory equipment and spaces shall be maintained in a clean condition.

§ 1250.93 Discharge of wastes.

Vessels operating on fresh water lakes or rivers shall not discharge sewage, or ballast or bilge water, within such areas adjacent to domestic water intakes as are designated by the Commissioner of Food and Drugs.

CROSS REFERENCE: For Environmental Protection Agency's regulations for vessel sanitary discharges as related to authority under the Federal Water Pollution Control Act, as amended (33 U.S.C. 1314 *et seq.*), see 40 CFR part 140.

[40 FR 5624, Feb. 6, 1975, as amended at 48 FR 11432, Mar. 18, 1983]

§ 1250.95 Insect control.

Vessels shall be maintained free of infestation by flies, mosquitoes, fleas, lice, and other insects known to be vectors in the transmission of communicable diseases, through the use of screening, insecticides, and other generally accepted methods of insect control.

§ 1250.96 Rodent control.

Vessels shall be maintained free of rodent infestation through the use of traps, poisons, and other generally accepted methods of rodent control.

PARTS 1251–1269 [RESERVED]

PART 1270—HUMAN TISSUE INTENDED FOR TRANSPLANTATION

Subpart A—General Provisions

Sec.

Subpart B—Donor Screening and Testing

Subpart C—Procedures and Records

Subpart D—Inspection of Tissue Establishments

AUTHORITY: 42 U.S.C. 216, 243, 264, 271.

SOURCE: 62 FR 40444, July 29, 1997, unless otherwise noted.

Subpart A—General Provisions

§ 1270.1 Scope.

(a) The regulations in this part apply to human tissue and to establishments or persons engaged in the recovery, screening, testing, processing, storage, or distribution of human tissue.

(b) Regulations in this chapter as they apply to drugs, biologics, devices, or other FDA-regulated commodities do not apply to human tissue, except as specified in this part.

(c) Regulations in this chapter do not apply to autologous human tissue.

(d) Regulations in this chapter do not apply to hospitals or other clinical facilities that receive and store human tissue only for transplantation within the same facility.

§ 1270.3 Definitions.

(a) *Act* for the purpose of this part means the Public Health Service Act, section 361 (42 U.S.C. 264).

(b) *Blood component* means any part of a single-donor unit of blood separated by physical or mechanical means.

(c) *Colloid* means a protein or polysaccharide solution that can be used to increase or maintain osmotic (oncotic) pressure in the intravascular compartment such as albumin, dextran, hetastarch; or certain blood components, such as plasma and platelets.

(d) *Contract services* are those functions pertaining to the recovery, screening, testing, processing, storage, or distribution of human tissue that another establishment agrees to perform for a tissue establishment.

(e) *Crystalloid* means a balanced salt and/or glucose solution used for electrolyte replacement or to increase intravascular volume such as saline, Ringer's lactate solution, or 5 percent dextrose in water.

(f) *Distribution* includes any transfer or shipment of human tissue (including importation or exportation), whether or not such transfer or shipment is entirely intrastate and whether or not possession of the tissue is taken.

(g) *Donor* means a human being, living or dead, who is the source of tissue for transplantation.

(h) *Donor medical history interview* means a documented dialogue with an individual or individuals who would be knowledgeable of the donor's relevant medical history and social behavior; such as the donor if living, the next of kin, the nearest available relative, a member of the donor's household, other individual with an affinity relationship, and/or the primary treating physician. The relevant social history includes questions to elicit whether or not the donor met certain descriptions or engaged in certain activities or behaviors considered to place such an individual at increased risk for HIV and hepatitis.

(i) *Establishment* means any facility under one management including all locations, that engages in the recovery, screening, testing, processing, storage, or distribution of human tissue intended for transplantation.

(j) *Human tissue*, for the purpose of this part means any tissue derived from a human body and recovered before May 25, 2005, which:

(1) Is intended for transplantation to another human for the diagnosis, cure, mitigation, treatment, or prevention of any condition or disease;

(2) Is recovered, processed, stored, or distributed by methods that do not change tissue function or characteristics;

(3) Is not currently regulated as a human drug, biological product, or medical device;

(4) Excludes kidney, liver, heart, lung, pancreas, or any other vascularized human organ; and

(5) Excludes semen or other reproductive tissue, human milk, and bone marrow.

(k) *Importer of record* means the person, establishment or their representative responsible for making entry of imported goods in accordance with all laws affecting such importation.

(l) *Legislative consent* means relating to any of the laws of the various States that allow the medical examiner or coroner to procure corneal tissue in the absence of consent of the donor's next-of-kin.

(m) *Person* includes an individual, partnership, corporation, association, or other legal entity.

(n) *Physical assessment* means a limited autopsy or recent antemortem or postmortem physical examination of the donor to assess for any signs of HIV and hepatitis infection or signs suggestive of any risk factor for such infections.

(o) *Plasma dilution* means a decrease in the concentration of the donor's plasma proteins and circulating antigens or antibodies resulting from the transfusion of blood or blood components and/or infusion of fluids.

(p) *Processing* means any activity performed on tissue, other than tissue recovery, including preparation, preservation for storage, and/or removal from storage to assure the quality and/or sterility of human tissue. Processing includes steps to inactivate and remove adventitious agents.

(q) *Quarantine* means the identification of human tissue as not suitable for transplantation, including human tissue that has not yet been characterized as being suitable for transplantation. Quarantine includes the storage of such tissue in an area clearly identified for such use, or other procedures, such as automated designation, for prevention of release of such tissue for transplantation.

(r) *Reconstituted blood* means the extracorporeal resuspension of a blood unit labeled as "Red Blood Cells" by the addition of colloids and/or crystalloids to produce a hematocrit in the normal range.

(s) *Recovery* means the obtaining from a donor of tissue that is intended for use in human transplantation.

(t) *Relevant medical records* means a collection of documents including a donor medical history interview, a physical assessment of the donor, laboratory test results, medical records, existing coroner and autopsy reports, or information obtained from any source or records which may pertain to donor suitability regarding high risk behaviors, clinical signs and symptoms for HIV and hepatitis, and treatments related to medical conditions suggestive of such risk.

(u) *Responsible person* means a person who is authorized to perform designated functions for which he or she is trained and qualified.

(v) *Storage* means holding tissue.

(w) *Summary of records* means a condensed version of the required testing and screening records that contains the identity of the testing laboratory, the listing and interpretation of all required infectious disease tests, and a listing of the documents reviewed as part of the relevant medical records, and the name of the person or establishment determining the suitability of the human tissue for transplantation.

(x) *Vascularized* means containing the original blood vessels which are intended to carry blood after transplantation.

[62 FR 40444, July 29, 1997, as amended at 69 FR 68680, Nov. 24, 2004]

Subpart B—Donor Screening and Testing

§ 1270.21 Determination of donor suitability for human tissue intended for transplantation.

(a) Donor specimens shall be tested for the following communicable viruses, using Food and Drug Administration (FDA) licensed donor screening tests in accordance with manufacturers' instructions:

(1) Human immunodeficiency virus, Type 1 (e.g., FDA licensed screening test for anti-HIV–1);

(2) Human immunodeficiency virus, Type 2 (e.g., FDA licensed screening test for anti-HIV–2);

(3) Hepatitis B (e.g., FDA licensed screening test for HBsAg); and

(4) Hepatitis C (e.g., FDA licensed screening test for anti-HCV).

(b) In the case of a neonate, the mother's specimen is acceptable for testing.

(c) Such infectious disease testing shall be performed by a laboratory certified under the Clinical Laboratories Improvement Amendments of 1988 (CLIA).

(d) Human tissue shall be accompanied by records indicating that the donor's specimen has been tested and found negative using FDA licensed screening tests for HIV–1, HIV–2, hepatitis B, and hepatitis C. FDA licensed screening tests labeled for cadaveric specimens must be used when available.

(e) Human tissue for transplantation shall be accompanied by a summary of records or copies of the original records of the donor's relevant medical records as defined in § 1270.3(t) which documents freedom from risk factors for and clinical evidence of hepatitis B, hepatitis C, or HIV infection. There shall be a responsible person designated and identified in the original record and summary of records as having made the determination that the human tissue is suitable for transplantation.

(f) Determination by the responsible person that a donor of human tissue intended for transplantation is suitable shall include ascertainment of the donor's identity, and accurately recorded relevant medical records (as defined in § 1270.3(t)) which documents freedom from risk factors for and clinical evidence of hepatitis B, hepatitis C, and HIV infection.

(g) For corneal tissue procured under legislative consent where a donor medical history screening interview has not occurred, a physical assessment of the donor is required and other available information shall be reviewed. The corneal tissue shall be accompanied by the summary of records documenting that the corneal tissue was determined to be suitable for transplantation in the absence of the donor medical history interview. Corneal tissue procured under legislative consent shall be documented as such in the summary of records.

(h) Human tissue shall be determined to be not suitable for transplantation if from:

(1) A donor whose specimen has tested repeatedly reactive on a screening test for HIV, hepatitis B, or hepatitis C;

(2) A donor where blood loss is known or suspected to have occurred and transfusion/infusion of more than 2,000 milliliters (mL) of blood (i.e., whole blood, reconstituted blood, or red blood cells), or colloids within 48 hours; or more than 2,000 mL of crystalloids within 1 hour; or any combination thereof prior to the collection of a blood specimen from the tissue donor for testing, unless:

(i) A pretransfusion or preinfusion blood specimen from the tissue donor is available for infectious disease testing; or

(ii) An algorithm is utilized that evaluates the volumes administered in the 48 hours prior to collecting the blood specimen from the tissue donor to ensure that there has not been plasma dilution sufficient to affect test results; or

(3) A donor who is 12 years of age or less and has been transfused or infused at all, unless:

(i) A pretransfusion or preinfusion blood specimen from the tissue donor is available for infectious disease testing; or

(ii) An algorithm is utilized that evaluates the volumes administered in the 48 hours prior to collecting the blood specimen from the tissue donor to ensure that there has not been plasma dilution sufficient to affect test results.

Subpart C—Procedures and Records

§ 1270.31 Written procedures.

(a) There shall be written procedures prepared and followed for all significant steps in the infectious disease testing process under § 1270.21 which shall conform to the manufacturers' instructions for use contained in the package inserts for the required tests. These procedures shall be readily available to the personnel in the area where the procedures are performed unless impractical. Any deviation from the written procedures shall be recorded and justified.

(b) There shall be written procedures prepared and followed for all significant steps for obtaining, reviewing, and assessing the relevant medical records of the donor as provided in § 1270.21. Such procedures shall be readily available to personnel who may perform the procedures. Any deviation from the written procedures shall be recorded and justified.

(c) There shall be written procedures prepared and followed for designating and identifying quarantined tissue.

(d) There shall be written procedures prepared, validated, and followed for prevention of infectious disease contamination or cross-contamination by tissue during processing.

(e) In conformity with this section, any facility may use current standard written procedures such as those in a technical manual prepared by another organization, provided the procedures are consistent with and at least as stringent as the requirements of this part.

§ 1270.33 Records, general requirements.

(a) Records shall be maintained concurrently with the performance of each significant step required in this part in the performance of infectious disease screening and testing of donors of human tissue. All records shall be accurate, indelible, and legible. The records shall identify the person performing the work, the dates of the various entries, and shall be as detailed as necessary to provide a complete history of the work performed and to relate the records to the particular tissue involved.

(b) All human tissue shall be quarantined until the following criteria for donor suitability are satisfied:

(1) All infectious disease testing under § 1270.21 has been completed, reviewed by the responsible person, and found to be negative; and

(2) Donor screening has been completed, reviewed by the responsible person, and determined to assure freedom from risk factors for and clinical evidence of HIV infection, hepatitis B, and hepatitis C.

(c) All human tissue processed or shipped prior to determination of donor suitability must be under quarantine, accompanied by records assuring identification of the donor and indicating that the tissue has not been determined to be suitable for transplantation.

(d) All human tissue determined to be suitable for transplantation must be accompanied by a summary of records, or copies of such original records, documenting that all infectious disease testing and screening under § 1270.21 has been completed, reviewed by the responsible person, and found to be negative, and that the tissue has been determined to be suitable for transplantation.

(e) Human tissue shall be quarantined until the tissue is either determined to be suitable for transplantation or appropriate disposition is accomplished.

(f) All persons or establishments that generate records used in determining the suitability of the donor shall retain such records and make them available for authorized inspection or upon request by FDA. The person(s) or establishment(s) making the determination regarding the suitability of the donor shall retain all records, or true copies of such records required under § 1270.21, including all testing and screening records, and shall make them available for authorized inspection or upon request from FDA. Records that can be retrieved from another location by electronic means meet the requirements of this paragraph.

(g) Records required under this part may be retained electronically, or as original paper records, or as true copies such as photocopies, microfiche, or microfilm, in which case suitable reader and photocopying equipment shall be readily available.

(h) Records shall be retained at least 10 years beyond the date of transplantation if known, distribution, disposition, or expiration, of the tissue, whichever is latest.

[62 FR 40444, July 29, 1997, as amended at 63 FR 16685, Apr. 6, 1998]

§ 1270.35 Specific records.

Records shall be maintained that include, but are not limited to:

(a) Documentation of results and interpretation of all required infectious disease tests;

(b) Information on the identity and relevant medical records of the donor, as required by § 1270.21(e) in English or, if in another language translated to English and accompanied by a statement of authenticity by the translator which specifically identifies the translated document;

(c) Documentation of the receipt and/or distribution of human tissue; and

(d) Documentation of the destruction or other disposition of human tissue.

Subpart D—Inspection of Tissue Establishments

§ 1270.41 Inspections.

(a) An establishment covered by these regulations in this part, including any location performing contract services, shall permit an authorized inspector of the Food and Drug Administration (FDA) to make at any reasonable time and in a reasonable manner such inspection of the establishment, its facilities, equipment, processes, products, and records as may be necessary to determine compliance with the provisions of this part. Such inspections may be made with or without notice and will ordinarily be made during regular business hours.

(b) The frequency of inspection will be at the agency's discretion.

(c) The inspector shall call upon a responsible person of the establishment and may question the personnel of the establishment as the inspector deems necessary.

(d) The inspector may review and copy any records required to be kept pursuant to part 1270.

(e) The public disclosure of records containing the name or other positive identification of donors or recipients of human tissue will be handled in accordance with FDA's procedures on disclosure of information as set forth in 21 CFR part 20 of this chapter.

§ 1270.42 Human tissue offered for import.

(a) When human tissue is offered for entry, the importer of record must notify the director of the district of the Food and Drug Administration having jurisdiction over the port of entry through which the tissue is imported or offered for import, or such officer of the district as the director may designate to act in his or her behalf in administering and enforcing this part.

(b) Human tissue offered for import must be quarantined until the human tissue is released by FDA.

§ 1270.43 Retention, recall, and destruction of human tissue.

(a) Upon a finding that human tissue may be in violation of the regulations in this part, an authorized Food and Drug Administration (FDA) representative may:

(1) Serve upon the person who distributed the tissue a written order that the tissue be recalled and/or destroyed, as appropriate, and upon persons in possession of the tissue that the tissue shall be retained until it is recalled by the distributor, destroyed, or disposed of as agreed by FDA, or the safety of the tissue is confirmed; and/or

(2) Take possession of and/or destroy the violative tissue.

(b) The written order will ordinarily provide that the human tissue be recalled and/or destroyed within 5 working days from the date of receipt of the order and will state with particularity the facts that justify the order.

(c) After receipt of an order under this part, the person in possession of the human tissue shall not distribute or dispose of the tissue in any manner except to recall and/or destroy the tissue consistent with the provisions of the order, under the supervision of an authorized official of FDA.

(d) In lieu of paragraphs (b) and (c) of this section, other arrangements for assuring the proper disposition of the tissue may be agreed upon by the person receiving the written order and an authorized official of FDA. Such arrangements may include providing FDA with records or other written information that adequately assure that the tissue has been recovered, screened, tested, processed, stored, and distributed in conformance with part 1270.

(e) Within 5 working days of receipt of a written order for retention, recall, and/or destruction of tissue (or within 5 working days of the agency's possession of such tissue), the recipient of the written order or prior possessor of such tissue shall request a hearing on the matter in accordance with part 16 of this chapter. The order for destruction will be held in abeyance pending resolution of the hearing request.

PART 1271—HUMAN CELLS, TISSUES, AND CELLULAR AND TISSUE-BASED PRODUCTS

Subpart A—General Provisions

Sec.

Subpart B—Procedures for Registration and Listing

Subpart C—Donor Eligibility

Subpart D—Current Good Tissue Practice

Subpart E—Additional Requirements for Establishments Described in § 1271.10

Subpart F—Inspection and Enforcement of Establishments Described in § 1271.10

AUTHORITY: 42 U.S.C. 216, 243, 263a, 264, 271.

SOURCE: 66 FR 5466, Jan. 19, 2001, unless otherwise noted.

Subpart A—General Provisions

§ 1271.1 What are the purpose and scope of this part?

(a) *Purpose.* The purpose of this part, in conjunction with §§ 207.20(f), 210.1(c), 210.2, 807.20(d), and 820.1(a) of this chapter, is to create a unified registration and listing system for establishments that manufacture human cells, tissues, and cellular and tissue-based products (HCT/P's) and to establish donor-eligibility, current good tissue practice, and other procedures to prevent the introduction, transmission, and spread of communicable diseases by HCT/P's.

(b) *Scope.* (1) If you are an establishment that manufactures HCT/P's that are regulated solely under the authority of section 361 of the Public Health Service Act (the PHS Act), this part requires you to register and list your HCT/P's with the Food and Drug Administration's (FDA's) Center for Biologics Evaluation and Research and to comply with the other requirements contained in this part, whether or not the HCT/P enters into interstate commerce. Those HCT/P's that are regulated solely under the authority of section 361 of the PHS Act are described in § 1271.10.

(2) If you are an establishment that manufactures HCT/P's that are regulated as drugs, devices and/or biological products under section 351 of the PHS Act and/or the Federal Food, Drug, and Cosmetic Act, §§ 207.20(f) and

807.20(d) of this chapter require you to register and list your HCT/P's following the procedures in subpart B of this part. Sections 210.1(c), 210.2, 211.1(b), and 820.1(a) of this chapter require you to comply with the donor-eligibility procedures in subpart C of this part and the current good tissue practice procedures in subpart D of this part, in addition to all other applicable regulations.

[66 FR 5466, Jan. 19, 2001, as amended at 69 FR 29829, May 25, 2004]

§ 1271.3 How does FDA define important terms in this part?

The following definitions apply only to this part:

(a) *Autologous use* means the implantation, transplantation, infusion, or transfer of human cells or tissue back into the individual from whom the cells or tissue were recovered.

(b) *Establishment* means a place of business under one management, at one general physical location, that engages in the manufacture of human cells, tissues, and cellular and tissue-based products. "Establishment" includes:

(1) Any individual, partnership, corporation, association, or other legal entity engaged in the manufacture of human cells, tissues, and cellular and tissue-based products; and

(2) Facilities that engage in contract manufacturing services for a manufacturer of human cells, tissues, and cellular and tissue-based products.

(c) *Homologous use* means the repair, reconstruction, replacement, or supplementation of a recipient's cells or tissues with an HCT/P that performs the same basic function or functions in the recipient as in the donor.

(d) *Human cells, tissues, or cellular or tissue-based products (HCT/Ps)* means articles containing or consisting of human cells or tissues that are intended for implantation, transplantation, infusion, or transfer into a human recipient. Examples of HCT/Ps include, but are not limited to, bone, ligament, skin, dura mater, heart valve, cornea, hematopoietic stem/progenitor cells derived from peripheral and cord blood, manipulated autologous chondrocytes, epithelial cells on a synthetic matrix, and semen or other reproductive tissue. The following articles are not considered HCT/Ps:

(1) Vascularized human organs for transplantation;

(2) Whole blood or blood components or blood derivative products subject to listing under parts 607 and 207 of this chapter, respectively;

(3) Secreted or extracted human products, such as milk, collagen, and cell factors; except that semen is considered an HCT/P;

(4) Minimally manipulated bone marrow for homologous use and not combined with another article (except for water, crystalloids, or a sterilizing, preserving, or storage agent, if the addition of the agent does not raise new clinical safety concerns with respect to the bone marrow);

(5) Ancillary products used in the manufacture of HCT/P;

(6) Cells, tissues, and organs derived from animals other than humans; and

(7) In vitro diagnostic products as defined in § 809.3(a) of this chapter.

(e) *Manufacture means,* but is not limited to, any or all steps in the recovery, processing, storage, labeling, packaging, or distribution of any human cell or tissue, and the screening or testing of the cell or tissue donor.

(f) *Minimal manipulation* means:

(1) For structural tissue, processing that does not alter the original relevant characteristics of the tissue relating to the tissue's utility for reconstruction, repair, or replacement; and

(2) For cells or nonstructural tissues, processing that does not alter the relevant biological characteristics of cells or tissues.

(g) *Transfer* means the placement of human reproductive cells or tissues into a human recipient.

(h) *Biohazard legend* appears on the label as follows and is used to mark HCT/Ps that present a known or suspected relevant communicable disease risk.

BIOHAZARD

(i) *Blood component* means a product containing a part of human blood separated by physical or mechanical means.

(j) *Colloid* means:

(1) A protein or polysaccharide solution, such as albumin, dextran, or hetastarch, that can be used to increase or maintain osmotic (oncotic) pressure in the intravascular compartment; or

(2) Blood components such as plasma and platelets.

(k) *Crystalloid* means an isotonic salt and/or glucose solution used for electrolyte replacement or to increase intravascular volume, such as saline solution, Ringer's lactate solution, or 5 percent dextrose in water.

(l) *Directed reproductive donor* means a donor of reproductive cells or tissue (including semen, oocytes, and embryos to which the donor contributed the spermatozoa or oocyte) to a specific recipient, and who knows and is known by the recipient before donation. The term directed reproductive donor does not include a sexually intimate partner under §1271.90.

(m) *Donor* means a person, living or dead, who is the source of cells or tissue for an HCT/P.

(n) *Donor medical history interview means* a documented dialog about the donor's medical history and relevant social behavior, including activities, behaviors, and descriptions considered to increase the donor's relevant communicable disease risk:

(1) With the donor, if the donor is living and able to participate in the interview, or

(2) If not, with an individual or individuals able to provide the information sought in the interview (e.g., the donor's next-of-kin, the nearest available relative, a member of the donor's household, an individual with an affinity relationship, and/or the primary treating physician).

(o) *Physical assessment of a cadaveric donor* means a limited autopsy or recent antemortem or postmortem physical examination of the donor to assess for signs of a relevant communicable disease and for signs suggestive of any risk factor for a relevant communicable disease.

(p) *Plasma dilution* means a decrease in the concentration of the donor's plasma proteins and circulating antigens or antibodies resulting from the transfusion of blood or blood components and/or infusion of fluids.

(q) *Quarantine* means the storage or identification of an HCT/P, to prevent improper release, in a physically separate area clearly identified for such use, or through use of other procedures, such as automated designation.

(r) *Relevant communicable disease agent or disease* means:

(1)(i) For all human cells and tissues, a communicable disease or disease agent listed as follows:

(A) Human immunodeficiency virus, types 1 and 2;

(B) Hepatitis B virus;

(C) Hepatitis C virus;

(D) Human transmissible spongiform encephalopathy, including Creutzfeldt-Jakob disease; and

(E) *Treponema pallidum.*

(ii) For viable, leukocyte-rich cells and tissues, a cell-associated disease agent or disease listed as follows:

(A) Human T-lymphotropic virus, type I; and

(B) Human T-lymphotropic virus, type II.

(iii) For reproductive cells or tissues, a disease agent or disease of the genitourinary tract listed as follows:

(A) *Chlamydia trachomatis*; and

(B) *Neisseria gonorrhea.*

(2) A disease agent or disease not listed in paragraph (r)(1) of this section:

(i) For which there may be a risk of transmission by an HCT/P, either to the recipient of the HCT/P or to those people who may handle or otherwise come in contact with it, such as medical personnel, because the disease agent or disease:

(A) Is potentially transmissible by an HCT/P and

(B) Either of the following applies:

(*1*) The disease agent or disease has sufficient incidence and/or prevalence to affect the potential donor population, or

(*2*) The disease agent or disease may have been released accidentally or intentionally in a manner that could place potential donors at risk of infection;

(ii) That could be fatal or life-threatening, could result in permanent impairment of a body function or permanent damage to body structure, or could necessitate medical or surgical intervention to preclude permanent impairment of body function or permanent damage to a body structure; and

(iii) For which appropriate screening measures have been developed and/or an appropriate screening test for donor specimens has been licensed, approved, or cleared for such use by FDA and is available.

(s) *Relevant medical records* means a collection of documents that includes a current donor medical history interview; a current report of the physical assessment of a cadaveric donor or the physical examination of a living donor; and, if available, the following:

(1) Laboratory test results (other than results of testing for relevant communicable disease agents required under this subpart);

(2) Medical records;

(3) Coroner and autopsy reports; and

(4) Records or other information received from any source pertaining to risk factors for relevant communicable disease (e.g., social behavior, clinical signs and symptoms of relevant communicable disease, and treatments related to medical conditions suggestive of risk for relevant communicable disease).

(t) *Responsible person* means a person who is authorized to perform designated functions for which he or she is trained and qualified.

(u) *Urgent medical need* means that no comparable HCT/P is available and the recipient is likely to suffer death or serious morbidity without the HCT/P.

(v) *Act* means the Federal Food, Drug, and Cosmetic Act.

(w) *PHS Act* means the Public Health Service Act.

(x) *FDA* means the Food and Drug Administration.

(y) *Adverse reaction* means a noxious and unintended response to any HCT/P for which there is a reasonable possibility that the HCT/P caused the response.

(z) *Available for distribution* means that the HCT/P has been determined to meet all release criteria.

(aa) *Complaint* means any written, oral, or electronic communication about a distributed HCT/P that alleges:

(1) That an HCT/P has transmitted or may have transmitted a communicable disease to the recipient of the HCT/P; or

(2) Any other problem with an HCT/P relating to the potential for transmission of communicable disease, such as the failure to comply with current good tissue practice.

(bb) *Distribution* means any conveyance or shipment (including importation and exportation) of an HCT/P that has been determined to meet all release criteria, whether or not such conveyance or shipment is entirely intrastate. If an entity does not take physical possession of an HCT/P, the entity is not considered a distributor.

(cc) *Establish and maintain* means define, document (in writing or electronically), and implement; then follow, review, and, as needed, revise on an ongoing basis.

(dd) *HCT/P deviation* means an event:

(1) That represents a deviation from applicable regulations in this part or from applicable standards or established specifications that relate to the prevention of communicable disease transmission or HCT/P contamination; or

(2) That is an unexpected or unforeseeable event that may relate to the transmission or potential transmission of a communicable disease or may lead to HCT/P contamination.

(ee) *Importer of record* means the person, establishment, or its representative responsible for making entry of imported goods in accordance with all laws affecting such importation.

(ff) *Processing* means any activity performed on an HCT/P, other than recovery, donor screening, donor testing, storage, labeling, packaging, or distribution, such as testing for microorganisms, preparation, sterilization, steps to inactivate or remove adventitious agents, preservation for storage, and removal from storage.

(gg) *Quality audit* means a documented, independent inspection and review of an establishment's activities related to core CGTP requirements. The purpose of a quality audit is to verify, by examination and evaluation of objective evidence, the degree of compliance with those aspects of the quality program under review.

(hh) *Quality program* means an organization's comprehensive system for manufacturing and tracking HCT/Ps in accordance with this part. A quality program is designed to prevent, detect, and correct deficiencies that may lead to circumstances that increase the risk of introduction, transmission, or spread of communicable diseases.

(ii) *Recovery* means obtaining from a human donor cells or tissues that are intended for use in human implantation, transplantation, infusion, or transfer.

(jj) *Storage* means holding HCT/Ps for future processing and/or distribution.

(kk) *Validation* means confirmation by examination and provision of objective evidence that particular requirements can consistently be fulfilled. Validation of a process, or *process validation*, means establishing by objective evidence that a process consistently produces a result or HCT/P meeting its predetermined specifications.

(ll) *Verification* means confirmation by examination and provision of objective evidence that specified requirements have been fulfilled.

[66 FR 5466, Jan. 19, 2001, as amended at 68 FR 3826, Jan. 27, 2004; 69 FR 29829, May 25, 2004; 69 FR 68680, Nov. 24, 2004]

§1271.10 Are my HCT/P's regulated solely under section 361 of the PHS Act and the regulations in this part, and if so what must I do?

(a) An HCT/P is regulated solely under section 361 of the PHS Act and the regulations in this part if it meets all of the following criteria:

(1) The HCT/P is minimally manipulated;

(2) The HCT/P is intended for homologous use only, as reflected by the labeling, advertising, or other indications of the manufacturer's objective intent;

(3) The manufacture of the HCT/P does not involve the combination of the cells or tissues with another article, except for water, crystalloids, or a sterilizing, preserving, or storage agent, provided that the addition of water, crystalloids, or the sterilizing, preserving, or storage agent does not raise new clinical safety concerns with respect to the HCT/P; and

(4) Either:

(i) The HCT/P does not have a systemic effect and is not dependent upon the metabolic activity of living cells for its primary function; or

(ii) The HCT/P has a systemic effect or is dependent upon the metabolic activity of living cells for its primary function, and:

(*a*) Is for autologous use;

(*b*) Is for allogeneic use in a first-degree or second-degree blood relative; or

(*c*) Is for reproductive use.

(b) If you are a domestic or foreign establishment that manufactures an HCT/P described in paragraph (a) of this section:

(1) You must register with FDA;

(2) You must submit to FDA a list of each HCT/P manufactured; and

(3) You must comply with the other requirements contained in this part.

[66 FR 5466, Jan. 19, 2001, as amended at 69 FR 68681, Nov. 24, 2004]

§1271.15 Are there any exceptions from the requirements of this part?

(a) You are not required to comply with the requirements of this part if you are an establishment that uses HCT/P's solely for nonclinical scientific or educational purposes.

(b) You are not required to comply with the requirements of this part if you are an establishment that removes HCT/P's from an individual and implants such HCT/P's into the same individual during the same surgical procedure.

(c) You are not required to comply with the requirements of this part if you are a carrier who accepts, receives, carries, or delivers HCT/P's in the usual course of business as a carrier.

(d) You are not required to comply with the requirements of this part if you are an establishment that does not recover, screen, test, process, label, package, or distribute, but only receives or stores HCT/P's solely for implantation, transplantation, infusion, or transfer within your facility.

(e) You are not required to comply with the requirements of this part if you are an establishment that only recovers reproductive cells or tissue and immediately transfers them into a sexually intimate partner of the cell or tissue donor.

(f) You are not required to register or list your HCT/P's independently, but you must comply with all other applicable requirements in this part, if you are an individual under contract, agreement, or other arrangement with a registered establishment and engaged solely in recovering cells or tissues and sending the recovered cells or tissues to the registered establishment.

§ 1271.20 If my HCT/P's do not meet the criteria in § 1271.10, and I do not qualify for any of the exceptions in § 1271.15, what regulations apply?

If you are an establishment that manufactures an HCT/P that does not meet the criteria set out in § 1271.10(a), and you do not qualify for any of the exceptions in § 1271.15, your HCT/P will be regulated as a drug, device, and/or biological product under the act and/or section 351 of the PHS Act, and applicable regulations in title 21, chapter I. Applicable regulations include, but are not limited to, §§ 207.20(f), 210.1(c), 210.2, 211.1(b), 807.20(d), and 820.1(a) of this chapter, which require you to follow the procedures in subparts B, C, and D of this part.

Subpart B—Procedures for Registration and Listing

§ 1271.21 When do I register, submit an HCT/P list, and submit updates?

(a) You must register and submit a list of every HCT/P that your establishment manufactures within 5 days after beginning operations or within 30 days of the effective date of this regulation, whichever is later.

(b) You must update your establishment registration annually in December, except as required by § 1271.26. You may accomplish your annual registration in conjunction with updating your HCT/P list under paragraph (c) of this section.

(c)(i) If no change described in § 1271.25(c) has occurred since you previously submitted an HCT/P list, you are not required to update your listing.

(ii) If a change described in § 1271.25(c) has occurred, you must update your HCT/P listing with the new information:

(*a*) At the time of the change, or

(*b*) Each June or December, whichever month occurs first after the change.

[69 FR 68681, Nov. 24, 2004]

§ 1271.25 What information is required for establishment registration and HCT/P listing?

(a) Your establishment registration Form FDA 3356 must include:

(1) The legal name(s) of the establishment;

(2) Each location, including the street address of the establishment and the postal service zip code;

(3) The name, address, and title of the reporting official; and

(4) A dated signature by the reporting official affirming that all information contained in the establishment registration and HCT/P listing form is true and accurate, to the best of his or her knowledge.

(b) Your HCT/P listing must include all HCT/P's (including the established name and the proprietary name) that you recover, process, store, label, package, distribute, or for which you perform donor screening or testing. You must also state whether each HCT/P meets the criteria set out in § 1271.10.

(c) Your HCT/P listing update must include:

(1) A list of each HCT/P that you have begun recovering, processing, storing, labeling, packaging, distributing, or for which you have begun donor screening or testing, that has not been included in any list previously submitted. You must provide all of the information required by § 1271.25(b) for each new HCT/P.

(2) A list of each HCT/P formerly listed in accordance with § 1271.21(a) for which you have discontinued recovery, processing, storage, labeling, packaging, distribution, or donor screening or testing, including for each HCT/P so

listed, the identity by established name and proprietary name, and the date of discontinuance. We request but do not require that you include the reason for discontinuance with this information.

(3) A list of each HCT/P for which a notice of discontinuance was submitted under paragraph (c)(2) of this section and for which you have resumed recovery, processing, storage, labeling, packaging, distribution, or donor screening or testing, including the identity by established name and proprietary name, the date of resumption, and any other information required by § 1271.25(b) not previously submitted.

(4) Any material change in any information previously submitted. Material changes include any change in information submitted on Form FDA 3356, such as whether the HCT/P meets the criteria set out in § 1271.10.

§ 1271.26 When must I amend my establishment registration?

If the ownership or location of your establishment changes, you must submit an amendment to registration within 5 days of the change.

§ 1271.27 Will FDA assign me a registration number?

(a) FDA will assign each location a permanent registration number.

(b) FDA acceptance of an establishment registration and HCT/P listing form does not constitute a determination that an establishment is in compliance with applicable rules and regulations or that the HCT/P is licensed or approved by FDA.

§ 1271.37 Will establishment registrations and HCT/P listings be available for inspection, and how do I request information on registrations and listings?

(a) A copy of the Form FDA 3356 filed by each establishment will be available for public inspection at the Office of Communication, Training, and Manufacturers Assistance (HFM–48), Center for Biologics Evaluation and Research, Food and Drug Administration, 1401 Rockville Pike, suite 200N, Rockville, MD 20852–1448. In addition, there will be available for inspection at each of the Food and Drug Administration district offices the same information for firms within the geographical area of such district office. Upon request and receipt of a self-addressed stamped envelope, verification of a registration number or the location of a registered establishment will be provided. The following information submitted under the HCT/P requirements is illustrative of the type of information that will be available for public disclosure when it is compiled:

(1) A list of all HCT/P's;

(2) A list of all HCT/P's manufactured by each establishment;

(3) A list of all HCT/P's discontinued; and

(4) All data or information that has already become a matter of public record.

(b) You should direct your requests for information regarding HCT/P establishment registrations and HCT/P listings to the Office of Communication, Training and Manufacturers Assistance (HFM–48), Center for Biologics Evaluation and Research, Food and Drug Administration, 1401 Rockville Pike, suite 200N, Rockville, MD 20852–1448.

Subpart C—Donor Eligibility

SOURCE: 69 FR 29830, May 25, 2004, unless otherwise noted.

§ 1271.45 What requirements does this subpart contain?

(a) *General.* This subpart sets out requirements for determining donor eligibility, including donor screening and testing. The requirements contained in this subpart are a component of current good tissue practice (CGTP) requirements. Other CGTP requirements are set out in subpart D of this part.

(b) *Donor-eligibility determination required.* A donor-eligibility determination, based on donor screening and testing for relevant communicable disease agents and diseases, is required for all donors of cells or tissue used in HCT/Ps, except as provided under § 1271.90. In the case of an embryo or of cells derived from an embryo, a donor-eligibility determination is required for both the oocyte donor and the semen donor.

(c) *Prohibition on use.* An HCT/P must not be implanted, transplanted, infused, or transferred until the donor

208-072 D-24

has been determined to be eligible, except as provided under §§ 1271.60(d), 1271.65(b), and 1271.90 of this subpart.

(d) *Applicability of requirements.* If you are an establishment that performs any function described in this subpart, you must comply with the requirements contained in this subpart that are applicable to that function.

[69 FR 29830, May 25, 2004, as amended at 69 FR 68681, Nov. 24, 2004]

§ 1271.47 What procedures must I establish and maintain?

(a) *General.* You must establish and maintain procedures for all steps that you perform in testing, screening, determining donor eligibility, and complying with all other requirements of this subpart. Establish and maintain means define, document (in writing or electronically), and implement; then follow, review, and as needed, revise on an ongoing basis. You must design these procedures to ensure compliance with the requirements of this subpart.

(b) *Review and approval.* Before implementation, a responsible person must review and approve all procedures.

(c) *Availability.* Procedures must be readily available to the personnel in the area where the operations to which they relate are performed, or in a nearby area if such availability is impractical.

(d) *Departures from procedures.* You must record and justify any departure from a procedure relevant to preventing risks of communicable disease transmission at the time of its occurrence. You must not make available for distribution any HCT/P from a donor whose eligibility is determined under such a departure unless a responsible person has determined that the departure does not increase the risks of communicable disease transmission through the use of the HCT/P.

(e) *Standard procedures.* You may adopt current standard procedures, such as those in a technical manual prepared by another organization, provided that you have verified that the procedures are consistent with and at least as stringent as the requirements of this part and appropriate for your operations.

§ 1271.50 How do I determine whether a donor is eligible?

(a) *Determination based on screening and testing.* If you are the establishment responsible for making the donor-eligibility determination, you must determine whether a donor is eligible based upon the results of donor screening in accordance with § 1271.75 and donor testing in accordance with §§ 1271.80 and 1271.85. A responsible person, as defined in § 1271.3(t), must determine and document the eligibility of a cell or tissue donor.

(b) *Eligible donor.* A donor is eligible under these provisions only if:

(1) Donor screening in accordance with § 1271.75 indicates that the donor:

(i) Is free from risk factors for, and clinical evidence of, infection due to relevant communicable disease agents and diseases; and

(ii) Is free from communicable disease risks associated with xenotransplantation; and

(2) The results of donor testing for relevant communicable disease agents in accordance with §§ 1271.80 and 1271.85 are negative or nonreactive, except as provided in § 1271.80(d)(1).

§ 1271.55 What records must accompany an HCT/P after the donor-eligibility determination is complete; and what records must I retain?

(a) *Accompanying records.* Once a donor-eligibility determination has been made, the following must accompany the HCT/P at all times:

(1) A distinct identification code affixed to the HCT/P container, e.g., alphanumeric, that relates the HCT/P to the donor and to all records pertaining to the HCT/P and, except in the case of autologous donations, directed reproductive donations, or donations made by first-degree or second-degree blood relatives, does not include an individual's name, social security number, or medical record number;

(2) A statement whether, based on the results of screening and testing, the donor has been determined to be eligible or ineligible; and

(3) A summary of the records used to make the donor-eligibility determination.

(b) *Summary of records.* The summary of records required by paragraph (a)(3) of this section must contain the following information:

(1) A statement that the communicable disease testing was performed by a laboratory:

(i) Certified to perform such testing on human specimens under the Clinical Laboratory Improvement Amendments of 1988 (42 U.S.C. 263a) and 42 CFR part 493; or

(ii) That has met equivalent requirements as determined by the Centers for Medicare and Medicaid Services in accordance with those provisions;

(2) A listing and interpretation of the results of all communicable disease tests performed;

(3) The name and address of the establishment that made the donor-eligibility determination; and

(4) In the case of an HCT/P from a donor who is ineligible based on screening and released under paragraph (b) of §1271.65, a statement noting the reason(s) for the determination of ineligibility.

(c) *Deletion of personal information.* The accompanying records required by this section must not contain the donor's name or other personal information that might identify the donor.

(d) *Record retention requirements.* (1) You must maintain documentation of:

(i) Results and interpretation of all testing for relevant communicable disease agents in compliance with §§1271.80 and 1271.85, as well as the name and address of the testing laboratory or laboratories;

(ii) Results and interpretation of all donor screening for communicable diseases in compliance with §1271.75; and

(iii) The donor-eligibility determination, including the name of the responsible person who made the determination and the date of the determination.

(2) All records must be accurate, indelible, and legible. Information on the identity and relevant medical records of the donor, as defined in §1271.3(s), must be in English or, if in another language, must be retained and translated to English and accompanied by a statement of authenticity by the translator that specifically identifies the translated document.

(3) You must retain required records and make them available for authorized inspection by or upon request from FDA. Records that can be readily retrieved from another location by electronic means are considered "retained."

(4) You must retain the records pertaining to a particular HCT/P at least 10 years after the date of its administration, or if the date of administration is not known, then at least 10 years after the date of the HCT/P's distribution, disposition, or expiration, whichever is latest.

[69 FR 29830, May 25, 2004, as amended at 70 FR 29952, May 25, 2005]

§1271.60 What quarantine and other requirements apply before the donor-eligibility determination is complete?

(a) *Quarantine.* You must keep an HCT/P in quarantine, as defined in §1271.3(q), until completion of the donor-eligibility determination required by §1271.50. You must quarantine semen from anonymous donors until the retesting required under §1271.85(d) is complete.

(b) *Identification of HCT/Ps in quarantine.* You must clearly identify as quarantined an HCT/P that is in quarantine pending completion of a donor-eligibility determination. The quarantined HCT/P must be easily distinguishable from HCT/Ps that are available for release and distribution.

(c) *Shipping of HCT/Ps in quarantine.* If you ship an HCT/P before completion of the donor-eligibility determination, you must keep it in quarantine during shipment. The HCT/P must be accompanied by records:

(1) Identifying the donor (e.g., by a distinct identification code affixed to the HCT/P container);

(2) Stating that the donor-eligibility determination has not been completed; and

(3) Stating that the product must not be implanted, transplanted, infused, or transferred until completion of the donor-eligibility determination, except under the terms of paragraph (d) of this section.

(d) *Use in cases of urgent medical need.* (1) This subpart C does not prohibit the implantation, transplantation, infusion, or transfer of an HCT/P from a donor for whom the donor-eligibility determination is not complete if there is a documented urgent medical need for the HCT/P, as defined in § 1271.3(u).

(2) If you make an HCT/P available for use under the provisions of paragraph (d)(1) of this section, you must prominently label it "NOT EVALUATED FOR INFECTIOUS SUBSTANCES," and " WARNING: Advise patient of communicable disease risks." The following information must accompany the HCT/P:

(i) The results of any donor screening required under § 1271.75 that has been completed;

(ii) The results of any testing required under § 1271.80 or 1271.85 that has been completed; and

(iii) A list of any screening or testing required under § 1271.75, 1271.80 or 1271.85 that has not yet been completed.

(3) If you are the establishment that manufactured an HCT/P used under the provisions of paragraph (d)(1) of this section, you must document that you notified the physician using the HCT/P that the testing and screening were not complete.

(4) In the case of an HCT/P used for an urgent medical need under the provisions of paragraph (d)(1) of this section, you must complete the donor-eligibility determination during or after the use of the HCT/P, and you must inform the physician of the results of the determination.

§ 1271.65 How do I store an HCT/P from a donor determined to be ineligible, and what uses of the HCT/P are not prohibited?

(a) *Storage*. If you are the establishment that stores the HCT/P, you must store or identify HCT/Ps from donors who have been determined to be ineligible in a physically separate area clearly identified for such use, or follow other procedures, such as automated designation, that are adequate to prevent improper release until destruction or other disposition of the HCT/P in accordance with paragraph (b) or (c) of this section.

(b) *Limited uses of HCT/P from ineligible donor*. (1) An HCT/P from a donor who has been determined to be ineligible, based on the results of required testing and/or screening, is not prohibited by subpart C of this part from use for implantation, transplantation, infusion, or transfer under the following circumstances:

(i) The HCT/P is for allogeneic use in a first-degree or second-degree blood relative;

(ii) The HCT/P consists of reproductive cells or tissue from a directed reproductive donor, as defined in § 1271.3(l); or

(iii) There is a documented urgent medical need as defined in § 1271.3(u).

(2) You must prominently label an HCT/P made available for use under the provisions of paragraph (b)(1) of this section with the Biohazard legend shown in § 1271.3(h) with the statement "WARNING: Advise patient of communicable disease risks," and, in the case of reactive test results, "WARNING: Reactive test results for (name of disease agent or disease)." The HCT/P must be accompanied by the records required under § 1271.55.

(3) If you are the establishment that manufactured an HCT/P used under the provisions of paragraph (b)(1) of this section, you must document that you notified the physician using the HCT/P of the results of testing and screening.

(c) *Nonclinical use*. You may make available for nonclinical purposes an HCT/P from a donor who has been determined to be ineligible, based on the results of required testing and/or screening, provided that it is labeled:

(1) "For Nonclinical Use Only" and

(2) With the Biohazard legend shown in § 1271.3(h).

§ 1271.75 How do I screen a donor?

(a) *All donors*. Except as provided under § 1271.90, if you are the establishment that performs donor screening, you must screen a donor of cells or tissue by reviewing the donor's relevant medical records for:

(1) Risk factors for, and clinical evidence of, relevant communicable disease agents and diseases, including:

(i) Human immunodeficiency virus;

(ii) Hepatitis B virus;

(iii) Hepatitis C virus;

(iv) Human transmissible spongiform encephalopathy, including Creutzfeldt-Jakob disease;

(v) *Treponema pallidum*; and

(2) Communicable disease risks associated with xenotransplantation.

(b) *Donors of viable, leukocyte-rich cells or tissue.* In addition to the relevant communicable disease agents and diseases for which screening is required under paragraph (a) of this section, and except as provided under § 1271.90, you must screen the donor of viable, leukocyte-rich cells or tissue by reviewing the donor's relevant medical records for risk factors for and clinical evidence of relevant cell-associated communicable disease agents and diseases, including Human T-lymphotropic virus.

(c) *Donors of reproductive cells or tissue.* In addition to the relevant communicable disease agents and diseases for which screening is required under paragraphs (a) and (b) of this section, as applicable, and except as provided under § 1271.90, you must screen the donor of reproductive cells or tissue by reviewing the donor's relevant medical records for risk factors for and clinical evidence of infection due to relevant communicable diseases of the genitourinary tract. Such screening must include screening for the communicable disease agents listed in paragraphs (c)(1) and (c)(2) of this section. However, if the reproductive cells or tissues are recovered by a method that ensures freedom from contamination of the cells or tissue by infectious disease organisms that may be present in the genitourinary tract, then screening for the communicable disease agents listed in paragraphs (c)(1) and (c)(2) of this section is not required. Communicable disease agents of the genitourinary tract for which you must screen include:

(1) *Chlamydia trachomatis*; and

(2) *Neisseria gonorrhea.*

(d) *Ineligible donors.* You must determine ineligible a donor who is identified as having either of the following:

(1) A risk factor for or clinical evidence of any of the relevant communicable disease agents or diseases for which screening is required under paragraphs (a)(1), (b), or (c) of this section; or

(2) Any communicable disease risk associated with xenotransplantation.

(e) *Abbreviated procedure for repeat donors.* If you have performed a complete donor screening procedure on a living donor within the previous 6 months, you may use an abbreviated donor screening procedure on repeat donations. The abbreviated procedure must determine and document any changes in the donor's medical history since the previous donation that would make the donor ineligible, including relevant social behavior.

[66 FR 5466, Jan. 19, 2001, as amended at 71 FR 14798, Mar. 24, 2006]

§ 1271.80 What are the general requirements for donor testing?

(a) *Testing for relevant communicable diseases is required.* To adequately and appropriately reduce the risk of transmission of relevant communicable diseases, and except as provided under § 1271.90, if you are the establishment that performs donor testing, you must test a donor specimen for evidence of infection due to communicable disease agents in accordance with paragraph (c) of this section. You must test for those communicable disease agents specified in § 1271.85. In the case of a donor 1 month of age or younger, you must test a specimen from the birth mother instead of a specimen from the donor.

(b) *Timing of specimen collection.* You must collect the donor specimen for testing at the time of recovery of cells or tissue from the donor; or up to 7 days before or after recovery, except:

(1) For donors of peripheral blood stem/progenitor cells, bone marrow (if not excepted under § 1271.3(d)(4)), or oocytes, you may collect the donor specimen for testing up to 30 days before recovery; or

(2) In the case of a repeat semen donor from whom a specimen has already been collected and tested, and for whom retesting is required under § 1271.85(d), you are not required to collect a donor specimen at the time of each donation.

(c) *Tests.* You must test using appropriate FDA-licensed, approved, or cleared donor screening tests, in accordance with the manufacturer's instructions, to adequately and appropriately reduce the risk of transmission of relevant communicable disease agents or diseases; however, until such time as appropriate FDA-licensed, approved, or cleared donor screening tests for *Chlamydia trachomatis* and for

Neisseria gonorrhea are available, you must use FDA-licensed, approved, or cleared tests labeled for the detection of those organisms in an asymptomatic, low-prevalence population. You must use a test specifically labeled for cadaveric specimens instead of a more generally labeled test when applicable and when available. Required testing under this section must be performed by a laboratory that either is certified to perform such testing on human specimens under the Clinical Laboratory Improvement Amendments of 1988 (42 U.S.C. 263a) and 42 CFR part 493, or has met equivalent requirements as determined by the Centers for Medicare and Medicaid Services.

(d) *Ineligible donors.* You must determine the following donors to be ineligible:

(1) A donor whose specimen tests reactive on a screening test for a communicable disease agent in accordance with § 1271.85, except for a donor whose specimen tests reactive on a nontreponemal screening test for syphilis and negative on a specific treponemal confirmatory test;

(2)(i) A donor in whom plasma dilution sufficient to affect the results of communicable disease testing is suspected, unless:

(A) You test a specimen taken from the donor before transfusion or infusion and up to 7 days before recovery of cells or tissue; or

(B) You use an appropriate algorithm designed to evaluate volumes administered in the 48 hours before specimen collection, and the algorithm shows that plasma dilution sufficient to affect the results of communicable disease testing has not occurred.

(ii) Clinical situations in which you must suspect plasma dilution sufficient to affect the results of communicable disease testing include but are not limited to the following:

(A) Blood loss is known or suspected in a donor over 12 years of age, and the donor has received a transfusion or infusion of any of the following, alone or in combination:

(*1*) More than 2,000 milliliters (mL) of blood (e.g., whole blood, red blood cells) or colloids within 48 hours before death or specimen collection, whichever occurred earlier, or

(*2*) More than 2,000 mL of crystalloids within 1 hour before death or specimen collection, whichever occurred earlier.

(B) Regardless of the presence or absence of blood loss, the donor is 12 years of age or younger and has received a transfusion or infusion of any amount of any of the following, alone or in combination:

(*1*) Blood (e.g., whole blood, red blood cells) or colloids within 48 hours before death or specimen collection, whichever occurred earlier, or

(*2*) Crystalloids within 1 hour before death or specimen collection, whichever occurred earlier.

[69 FR 29830, May 25, 2004, as amended at 70 FR 29952, May 25, 2005]

§ 1271.85 What donor testing is required for different types of cells and tissues?

(a) *All donors.* To adequately and appropriately reduce the risk of transmission of relevant communicable diseases, and except as provided under § 1271.90, you must test a specimen from the donor of cells or tissue, whether viable or nonviable, for evidence of infection due to relevant communicable disease agents, including:

(1) Human immunodeficiency virus, type 1;

(2) Human immunodeficiency virus, type 2;

(3) Hepatitis B virus;

(4) Hepatitis C virus; and

(5) *Treponema pallidum.*

(b) *Donors of viable, leukocyte-rich cells or tissue.* In addition to the relevant communicable disease agents for which testing is required under paragraph (a) of this section, and except as provided under § 1271.90,

(1) You must test a specimen from the donor of viable, leukocyte-rich cells or tissue to adequately and appropriately reduce the risk of transmission of relevant cell-associated communicable diseases, including:

(i) Human T-lymphotropic virus, type I; and

(ii) Human T-lymphotropic virus, type II.

(2) You must test a specimen from the donor of viable, leukocyte-rich cells or tissue for evidence of infection

due to cytomegalovirus (CMV), to adequately and appropriately reduce the risk of transmission. You must establish and maintain a standard operating procedure governing the release of an HCT/P from a donor whose specimen tests reactive for CMV.

(c) *Donors of reproductive cells or tissue.* In addition to the communicable disease agents for which testing is required under paragraphs (a) and (b) of this section, as applicable, and except as provided under § 1271.90, you must test a specimen from the donor of reproductive cells or tissue to adequately and appropriately reduce the risk of transmission of relevant communicable disease agents of the genitourinary tract. Such testing must include testing for the communicable disease agents listed in paragraphs (c)(1) and (c)(2) of this section. However, if the reproductive cells or tissues are recovered by a method that ensures freedom from contamination of the cells or tissue by infectious disease organisms that may be present in the genitourinary tract, then testing for the communicable disease agents listed in paragraphs (c)(1) and (c)(2) of this section is not required. Communicable disease agents of the genitourinary tract for which you must test include:

(1) *Chlamydia trachomatis*; and

(2) *Neisseria gonorrhea.*

(d) *Retesting anonymous semen donors.* Except as provided under § 1271.90 and except for directed reproductive donors as defined in § 1271.3(l), at least 6 months after the date of donation of semen from anonymous donors, you must collect a new specimen from the donor and test it for evidence of infection due to the communicable disease agents for which testing is required under paragraphs (a), (b), and (c) of this section.

(e) *Dura mater.* For donors of dura mater, you must perform an adequate assessment designed to detect evidence of transmissible spongiform encephalopathy.

§ 1271.90 Are there exceptions from the requirement of determining donor eligibility, and what labeling requirements apply?

(a) *Donor-eligibility determination not required.* You are not required to make a donor-eligibility determination under § 1271.50 or to perform donor screening or testing under §§ 1271.75, 1271.80 and 1271.85 for:

(1) Cells and tissues for autologous use; or

(2) Reproductive cells or tissue donated by a sexually intimate partner of the recipient for reproductive use; or

(3) Cryopreserved cells or tissue for reproductive use, other than embryos, originally exempt under paragraphs (a)(1) or (a)(2) of this section at the time of donation, that are subsequently intended for directed donation, provided that

(i) Additional donations are unavailable, for example, due to the infertility or health of a donor of the cryopreserved reproductive cells or tissue; and

(ii) Appropriate measures are taken to screen and test the donor(s) before transfer to the recipient.

(4) A cryopreserved embryo, originally exempt under paragraph (a)(2) of this section at the time of cryopreservation, that is subsequently intended for directed or anonymous donation. When possible, appropriate measures should be taken to screen and test the semen and oocyte donors before transfer of the embryo to the recipient.

(b) *Required labeling.* As applicable, you must prominently label an HCT/P described in paragraph (a) of this section as follows:

(1) "FOR AUTOLOGOUS USE ONLY," if it is stored for autologous use.

(2) "NOT EVALUATED FOR INFECTIOUS SUBSTANCES," unless you have performed all otherwise applicable screening and testing under §§ 1271.75, 1271.80, and 1271.85. This paragraph does not apply to reproductive cells or tissue labeled in accordance with paragraph (b)(6) of this section.

(3) Unless the HCT/P is for autologous use only, "WARNING: Advise recipient of communicable disease risks,"

(i) When the donor-eligibility determination under § 1271.50(a) is not performed or is not completed; or

(ii) If the results of any screening or testing performed indicate:

(A) The presence of relevant communicable disease agents and/or

(B) Risk factors for or clinical evidence of relevant communicable disease agents or diseases.

(4) With the Biohazard legend shown in § 1271.3(h), if the results of any screening or testing performed indicate:

(i) The presence of relevant communicable disease agents and/or

(ii) Risk factors for or clinical evidence of relevant communicable disease agents or diseases.

(5) "WARNING: Reactive test results for (name of disease agent or disease)," in the case of reactive test results.

(6) "Advise recipient that screening and testing of the donor(s) were not performed at the time of cryopreservation of the reproductive cells or tissue, but have been performed subsequently," for paragraphs (a)(3) or (a)(4) of this section.

[69 FR 29830, May 25, 2004, as amended at 70 FR 29952, May 25, 2005]

Subpart D—Current Good Tissue Practice

SOURCE: 69 FR 68681, Nov. 24, 2004, unless otherwise noted.

§ 1271.145 Prevention of the introduction, transmission, or spread of communicable diseases.

You must recover, process, store, label, package, and distribute HCT/Ps, and screen and test cell and tissue donors, in a way that prevents the introduction, transmission, or spread of communicable diseases.

§ 1271.150 Current good tissue practice requirements.

(a) *General.* This subpart D and subpart C of this part set forth current good tissue practice (CGTP) requirements. You must follow CGTP requirements to prevent the introduction, transmission, or spread of communicable diseases by HCT/Ps (e.g., by ensuring that the HCT/Ps do not contain communicable disease agents, that they are not contaminated, and that they do not become contaminated during manufacturing). Communicable diseases include, but are not limited to, those transmitted by viruses, bacteria, fungi, parasites, and transmissible spongiform encephalopathy agents. CGTP requirements govern the methods used in, and the facilities and controls used for, the manufacture of HCT/Ps, including but not limited to all steps in recovery, donor screening, donor testing, processing, storage, labeling, packaging, and distribution. The CGTP provisions specifically governing determinations of donor eligibility, including donor screening and testing, are set out separately in subpart C of this part.

(b) *Core CGTP requirements.* The following are core CGTP requirements:

(1) Requirements relating to facilities in § 1271.190(a) and (b);

(2) Requirements relating to environmental control in § 1271.195(a);

(3) Requirements relating to equipment in § 1271.200(a);

(4) Requirements relating to supplies and reagents in § 1271.210(a) and (b);

(5) Requirements relating to recovery in § 1271.215;

(6) Requirements relating to processing and process controls in § 1271.220;

(7) Requirements relating to labeling controls in § 1271.250(a) and (b);

(8) Requirements relating to storage in § 1271.260 (a) through (d);

(9) Requirements relating to receipt, predistribution shipment, and distribution of an HCT/P in § 1271.265(a) through (d); and

(10) Requirements relating to donor eligibility determinations, donor screening, and donor testing in §§ 1271.50, 1271.75, 1271.80, and 1271.85.

(c) *Compliance with applicable requirements*—(1) *Manufacturing arrangements* (i) If you are an establishment that engages in only some operations subject to the regulations in this subpart and subpart C of this part, and not others, then you need only comply with those requirements applicable to the operations that you perform.

(ii) If you engage another establishment (e.g., a laboratory to perform communicable disease testing, or an irradiation facility to perform terminal sterilization), under a contract, agreement, or other arrangement, to perform any step in manufacture for you, that establishment is responsible for complying with requirements applicable to that manufacturing step.

(iii) Before entering into a contract, agreement, or other arrangement with another establishment to perform any step in manufacture for you, you must ensure that the establishment complies with applicable CGTP requirements. If, during the course of this contract, agreement, or other arrangement, you become aware of information suggesting that the establishment may no longer be in compliance with such requirements, you must take reasonable steps to ensure the establishment complies with those requirements. If you determine that the establishment is not in compliance with those requirements, you must terminate your contract, agreement, or other arrangement with the establishment.

(2) If you are the establishment that determines that an HCT/P meets all release criteria and makes the HCT/P available for distribution, whether or not you are the actual distributor, you are responsible for reviewing manufacturing and tracking records to determine that the HCT/P has been manufactured and tracked in compliance with the requirements of this subpart and subpart C of this part and any other applicable requirements.

(3) With the exception of §§ 1271.150(c) and 1271.155 of this subpart, the regulations in this subpart are not being implemented for reproductive HCT/Ps described in § 1271.10 and regulated solely under section 361 of the Public Health Service Act and the regulations in this part, or for the establishments that manufacture them.

(d) *Compliance with parts 210, 211, and 820 of this chapter.* With respect to HCT/Ps that are drugs (subject to review under an application submitted under section 505 of the Federal Food, Drug, and Cosmetic Act or under a biological product license application under section 351 of the Public Health Service Act) or that are devices (subject to premarket review or notification under the device provisions of the act or under a biological product license application under section 351 of the Public Health Service Act), the procedures contained in this subpart and in subpart C of this part and the current good manufacturing practice regulations in parts 210 and 211 of this chapter and the quality system regulations in part 820 of this chapter supplement, and do not supersede, each other unless the regulations explicitly provide otherwise. In the event that a regulation in part 1271 of this chapter is in conflict with a requirement in parts 210, 211, or 820 of this chapter, the regulations more specifically applicable to the product in question will supersede the more general.

(e) *Where appropriate.* When a requirement is qualified by "where appropriate," it is deemed to be "appropriate" unless you can document justification otherwise. A requirement is "appropriate" if nonimplementation of the requirement could reasonably be expected to result in the HCT/P not meeting its specified requirements related to prevention of introduction, transmission, or spread of communicable diseases, or in your inability to carry out any necessary corrective action.

§ 1271.155 Exemptions and alternatives.

(a) *General.* You may request an exemption from or alternative to any requirement in subpart C or D of this part.

(b) *Request for exemption or alternative.* Submit your request under this section to the Director of the appropriate Center (the Director), e.g., the Center for Biologics Evaluation and Research or the Center for Devices and Radiological Health. The request must be accompanied by supporting documentation, including all relevant valid scientific data, and must contain either:

(1) Information justifying the requested exemption from the requirement, or

(2) A description of a proposed alternative method of meeting the requirement.

(c) *Criteria for granting an exemption or alternative.* The Director may grant an exemption or alternative if he or she finds that such action is consistent with the goals of protecting the public health and/or preventing the introduction, transmission, or spread of communicable diseases and that:

(1) The information submitted justifies an exemption; or

(2) The proposed alternative satisfies the purpose of the requirement.

(d) *Form of request.* You must ordinarily make your request for an exemption or alternative in writing (hard copy or electronically). However, if circumstances make it difficult (e.g., there is inadequate time) to submit your request in writing, you may make the request orally, and the Director may orally grant an exemption or alternative. You must follow your oral request with an immediate written request, to which the Director will respond in writing.

(e) *Operation under exemption or alternative.* You must not begin operating under the terms of a requested exemption or alternative until the exemption or alternative has been granted. You may apply for an extension of an exemption or alternative beyond its expiration date, if any.

(f) *Documentation.* If you operate under the terms of an exemption or alternative, you must maintain documentation of:

(1) FDA's grant of the exemption or alternative, and

(2) The date on which you began operating under the terms of the exemption or alternative.

(g) *Issuance of an exemption or alternative by the Director.* In a public health emergency, the Director may issue an exemption from, or alternative to, any requirement in part 1271. The Director may issue an exemption or alternative under this section if the exemption or alternative is necessary to assure that certain HCT/Ps will be available in a specified location to respond to an unanticipated immediate need for those HCT/Ps.

§ 1271.160 Establishment and maintenance of a quality program.

(a) *General.* If you are an establishment that performs any step in the manufacture of HCT/Ps, you must establish and maintain a quality program intended to prevent the introduction, transmission, or spread of communicable diseases through the manufacture and use of HCT/Ps. The quality program must be appropriate for the specific HCT/Ps manufactured and the manufacturing steps performed. The quality program must address all core CGTP requirements listed in § 1271.150(b).

(b) *Functions.* Functions of the quality program must include:

(1) Establishing and maintaining appropriate procedures relating to core CGTP requirements, and ensuring compliance with the requirements of § 1271.180 with respect to such procedures, including review, approval, and revision;

(2) Ensuring that procedures exist for receiving, investigating, evaluating, and documenting information relating to core CGTP requirements, including complaints, and for sharing any information pertaining to the possible contamination of the HCT/P or the potential for transmission of a communicable disease by the HCT/P with the following:

(i) Other establishments that are known to have recovered HCT/Ps from the same donor;

(ii) Other establishments that are known to have performed manufacturing steps with respect to the same HCT/P; and

(iii) Relating to consignees, in the case of such information received after the HCT/P is made available for distribution, shipped to the consignee, or administered to the recipient, procedures must include provisions for assessing risk and appropriate followup, and evaluating the effect this information has on the HCT/P and for the notification of all entities to whom the affected HCT/P was distributed, the quarantine and recall of the HCT/P, and/or reporting to FDA, as necessary.

(3) Ensuring that appropriate corrective actions relating to core CGTP requirements, including reaudits of deficiencies, are taken and documented, as necessary. You must verify corrective actions to ensure that such actions are effective and are in compliance with CGTP. Where appropriate, corrective actions must include both short-term action to address the immediate problem and long-term action to prevent the problem's recurrence. Documentation of corrective actions must include, where appropriate:

(i) Identification of the HCT/P affected and a description of its disposition;

(ii) The nature of the problem requiring corrective action;

(iii) A description of the corrective action taken; and

(iv) The date(s) of the corrective action.

(4) Ensuring the proper training and education of personnel involved in activities related to core CGTP requirements;

(5) Establishing and maintaining appropriate monitoring systems as necessary to comply with the requirements of this subpart (e.g., environmental monitoring);

(6) Investigating and documenting HCT/P deviations and trends of HCT/P deviations relating to core CGTP requirements and making reports if required under § 1271.350(b) or other applicable regulations. Each investigation must include a review and evaluation of the HCT/P deviation, the efforts made to determine the cause, and the implementation of corrective action(s) to address the HCT/P deviation and prevent recurrence.

(c) *Audits.* You must periodically perform for management review a quality audit, as defined in § 1271.3(gg), of activities related to core CGTP requirements.

(d) *Computers.* You must validate the performance of computer software for the intended use, and the performance of any changes to that software for the intended use, if you rely upon the software to comply with core CGTP requirements and if the software either is custom software or is commercially available software that has been customized or programmed (including software programmed to perform a user defined calculation or table) to perform a function related to core CGTP requirements. You must verify the performance of all other software for the intended use if you rely upon it to comply with core CGTP requirements. You must approve and document these activities and results before implementation.

§ 1271.170 Personnel.

(a) *General.* You must have personnel sufficient to ensure compliance with the requirements of this part.

(b) *Competent performance of functions.* You must have personnel with the necessary education, experience, and training to ensure competent performance of their assigned functions. Personnel must perform only those activities for which they are qualified and authorized.

(c) *Training.* You must train all personnel, and retrain as necessary, to perform their assigned responsibilities adequately.

§ 1271.180 Procedures.

(a) *General.* You must establish and maintain procedures appropriate to meet core CGTP requirements for all steps that you perform in the manufacture of HCT/Ps. You must design these procedures to prevent circumstances that increase the risk of the introduction, transmission, or spread of communicable diseases through the use of HCT/Ps.

(b) *Review and approval.* Before implementation, a responsible person must review and approve these procedures.

(c) *Availability.* These procedures must be readily available to the personnel in the area where the operations to which they relate are performed, or in a nearby area if such availability is impractical.

(d) *Standard procedures.* If you adopt current standard procedures from another organization, you must verify that the procedures meet the requirements of this part and are appropriate for your operations.

§ 1271.190 Facilities.

(a) *General.* Any facility used in the manufacture of HCT/Ps must be of suitable size, construction, and location to prevent contamination of HCT/Ps with communicable disease agents and to ensure orderly handling of HCT/Ps without mix-ups. You must maintain the facility in a good state of repair. You must provide lighting, ventilation, plumbing, drainage, and access to sinks and toilets that are adequate to prevent the introduction, transmission, or spread of communicable disease.

(b) *Facility cleaning and sanitation.* (1) You must maintain any facility used in the manufacture of HCT/Ps in a clean, sanitary, and orderly manner, to prevent the introduction, transmission, or spread of communicable disease.

(2) You must dispose of sewage, trash, and other refuse in a timely, safe, and sanitary manner.

(c) *Operations.* You must divide a facility used in the manufacture of HCT/Ps into separate or defined areas of adequate size for each operation that takes place in the facility, or you must establish and maintain other control systems to prevent improper labeling, mix-ups, contamination, cross-contamination, and accidental exposure of HCT/Ps to communicable disease agents.

(d) *Procedures and records.* (1) You must establish and maintain procedures for facility cleaning and sanitation for the purpose of preventing the introduction, transmission, or spread of communicable disease. These procedures must assign responsibility for sanitation and must describe in sufficient detail the cleaning methods to be used and the schedule for cleaning the facility.

(2) You must document, and maintain records of, all cleaning and sanitation activities performed to prevent contamination of HCT/Ps. You must retain such records 3 years after their creation.

§ 1271.195 Environmental control and monitoring.

(a) *Environmental control.* Where environmental conditions could reasonably be expected to cause contamination or cross-contamination of HCT/Ps or equipment, or accidental exposure of HCT/Ps to communicable disease agents, you must adequately control environmental conditions and provide proper conditions for operations. Where appropriate, you must provide for the following control activities or systems:

(1) Temperature and humidity controls;

(2) Ventilation and air filtration;

(3) Cleaning and disinfecting of rooms and equipment to ensure aseptic processing operations; and

(4) Maintenance of equipment used to control conditions necessary for aseptic processing operations.

(b) *Inspections.* You must inspect each environmental control system periodically to verify that the system, including necessary equipment, is adequate and functioning properly. You must take appropriate corrective action as necessary.

(c) *Environmental monitoring.* You must monitor environmental conditions where environmental conditions could reasonably be expected to cause contamination or cross-contamination of HCT/Ps or equipment, or accidental exposure of HCT/Ps to communicable disease agents. Where appropriate, you must provide environmental monitoring for microorganisms.

(d) *Records.* You must document, and maintain records of, environmental control and monitoring activities.

§ 1271.200 Equipment.

(a) *General.* To prevent the introduction, transmission, or spread of communicable diseases, equipment used in the manufacture of HCT/Ps must be of appropriate design for its use and must be suitably located and installed to facilitate operations, including cleaning and maintenance. Any automated, mechanical, electronic, or other equipment used for inspection, measuring, or testing in accordance with this part must be capable of producing valid results. You must clean, sanitize, and maintain equipment according to established schedules.

(b) *Procedures and schedules.* You must establish and maintain procedures for cleaning, sanitizing, and maintaining equipment to prevent malfunctions, contamination or cross-contamination, accidental exposure of HCT/Ps to communicable disease agents, and other events that could reasonably be expected to result in the introduction, transmission, or spread of communicable diseases.

(c) *Calibration of equipment.* Where appropriate, you must routinely calibrate according to established procedures and schedules all automated, mechanical, electronic, or other equipment used for inspection, measuring, and testing in accordance with this part.

(d) *Inspections.* You must routinely inspect equipment for cleanliness, sanitation, and calibration, and to ensure adherence to applicable equipment maintenance schedules.

(e) *Records.* You must document and maintain records of all equipment maintenance, cleaning, sanitizing, calibration, and other activities performed in accordance with this section. You must display records of recent maintenance, cleaning, sanitizing, calibration, and other activities on or near each piece of equipment, or make the records readily available to the individuals responsible for performing these activities and to the personnel using the equipment. You must maintain records of the use of each piece of equipment, including the identification of each HCT/P manufactured with that equipment.

§ 1271.210 Supplies and reagents.

(a) *Verification.* You must not use supplies and reagents until they have been verified to meet specifications designed to prevent circumstances that increase the risk of the introduction, transmission, or spread of communicable diseases. Verification may be accomplished by the establishment that uses the supply or reagent, or by the vendor of the supply or reagent.

(b) *Reagents.* Reagents used in processing and preservation of HCT/Ps must be sterile, where appropriate.

(c) *In-house reagents.* You must validate and/or verify the processes used for production of in-house reagents.

(d) *Records.* You must maintain the following records pertaining to supplies and reagents:

(1) Records of the receipt of each supply or reagent, including the type, quantity, manufacturer, lot number, date of receipt, and expiration date;

(2) Records of the verification of each supply or reagent, including test results or, in the case of vendor verification, a certificate of analysis from the vendor; and

(3) Records of the lot of supply or reagent used in the manufacture of each HCT/P.

§ 1271.215 Recovery.

If you are an establishment that recovers HCT/Ps, you must recover each HCT/P in a way that does not cause contamination or cross-contamination during recovery, or otherwise increase the risk of the introduction, transmission, or spread of communicable disease through the use of the HCT/P.

§ 1271.220 Processing and process controls.

(a) *General.* If you are an establishment that processes HCT/Ps, you must process each HCT/P in a way that does not cause contamination or cross-contamination during processing, and that prevents the introduction, transmission, or spread of communicable disease through the use of the HCT/P.

(b) *Pooling.* Human cells or tissue from two or more donors must not be pooled (placed in physical contact or mixed in a single receptacle) during manufacturing.

(c) *In-process control and testing.* You must ensure that specified requirements, consistent with paragraph (a) of this section, for in-process controls are met, and that each in-process HCT/P is controlled until the required inspection and tests or other verification activities have been completed, or necessary approvals are received and documented. Sampling of in-process HCT/Ps must be representative of the material to be evaluated.

(d) *Dura mater.* (1) When there is a published validated process that reduces the risk of transmissible spongiform encephalopathy, you must use this process for dura mater (or an equivalent process that you have validated), unless following this process adversely affects the clinical utility of the dura mater.

(2) When you use a published validated process, you must verify such a process in your establishment.

§ 1271.225 Process changes.

Any change to a process must be verified or validated in accordance with § 1271.230, to ensure that the change does not create an adverse impact elsewhere in the operation, and must be approved before implementation by a responsible person with appropriate knowledge and background. You must communicate approved changes to the appropriate personnel in a timely manner.

§ 1271.230 Process validation.

(a) *General.* Where the results of processing described in § 1271.220 cannot be fully verified by subsequent inspection and tests, you must validate and approve the process according to established procedures. The validation activities and results must be documented, including the date and signature of the individual(s) approving the validation.

(b) *Written representation.* Any written representation that your processing methods reduce the risk of transmission of communicable disease by an HCT/P, including but not limited to, a representation of sterility or pathogen inactivation of an HCT/P, must be based on a fully verified or validated process.

(c) *Changes.* When changes to a validated process subject to paragraph (a) of this section occur, you must review and evaluate the process and perform revalidation where appropriate. You must document these activities.

§ 1271.250 Labeling controls.

(a) *General.* You must establish and maintain procedures to control the labeling of HCT/Ps. You must design these procedures to ensure proper HCT/P identification and to prevent mix-ups.

(b) *Verification.* Procedures must include verification of label accuracy, legibility, and integrity.

(c) *Labeling requirements.* Procedures must ensure that each HCT/P is labeled in accordance with all applicable labeling requirements, including those in §§ 1271.55, 1271.60, 1271.65, 1271.90, 1271.290, and 1271.370, and that each HCT/P made available for distribution is accompanied by documentation of the donor eligibility determination as required under § 1271.55.

§ 1271.260 Storage.

(a) *Control of storage areas.* You must control your storage areas and stock rooms to prevent:

(1) Mix-ups, contamination, and cross-contamination of HCT/Ps, supplies, and reagents, and

(2) An HCT/P from being improperly made available for distribution.

(b) *Temperature.* You must store HCT/Ps at an appropriate temperature.

(c) *Expiration date.* Where appropriate, you must assign an expiration date to each HCT/P based on the following factors:

(1) HCT/P type;

(2) Processing, including the method of preservation;

(3) Storage conditions; and

(4) Packaging.

(d) *Corrective action.* You must take and document corrective action whenever proper storage conditions are not met.

(e) *Acceptable temperature limits.* You must establish acceptable temperature limits for storage of HCT/Ps at each step of the manufacturing process to inhibit the growth of infectious agents. You must maintain and record storage temperatures for HCT/Ps. You must periodically review recorded temperatures to ensure that temperatures have been within acceptable limits.

§ 1271.265 Receipt, predistribution shipment, and distribution of an HCT/P.

(a) *Receipt.* You must evaluate each incoming HCT/P for the presence and significance of microorganisms and inspect for damage and contamination. You must determine whether to accept, reject, or place in quarantine each incoming HCT/P, based upon pre-established criteria designed to prevent communicable disease transmission.

(b) *Predistribution shipment.* If you ship an HCT/P within your establishment or between establishments (e.g., procurer to processor) and the HCT/P is not available for distribution as described in paragraph (c) of this section, you must first determine and document whether pre-established criteria designed to prevent communicable disease transmission have been met, and you must ship the HCT/P in quarantine.

(c) *Availability for distribution.* (1) Before making an HCT/P available for distribution, you must review manufacturing and tracking records pertaining to the HCT/P, and, on the basis of that record review, you must verify and document that the release criteria have been met. A responsible person must document and date the determination that an HCT/P is available for distribution.

(2) You must not make available for distribution an HCT/P that is in quarantine, is contaminated, is recovered from a donor who has been determined

to be ineligible or for whom a donor-eligibility determination has not been completed (except as provided under §§ 1271.60, 1271.65, and 1271.90), or that otherwise does not meet release criteria designed to prevent communicable disease transmission.

(3) You must not make available for distribution any HCT/P manufactured under a departure from a procedure relevant to preventing risks of communicable disease transmission, unless a responsible person has determined that the departure does not increase the risk of communicable disease through the use of the HCT/P. You must record and justify any departure from a procedure at the time of its occurrence.

(d) *Packaging and shipping.* Packaging and shipping containers must be designed and constructed to protect the HCT/P from contamination. For each type of HCT/P, you must establish appropriate shipping conditions to be maintained during transit.

(e) *Procedures.* You must establish and maintain procedures, including release criteria, for the activities in paragraphs (a) through (d) of this section. You must document these activities. Documentation must include:

(1) Identification of the HCT/P and the establishment that supplied the HCT/P;

(2) Activities performed and the results of each activity;

(3) Date(s) of activity;

(4) Quantity of HCT/P subject to the activity; and

(5) Disposition of the HCT/P (e.g., identity of consignee).

(f) *Return to inventory.* You must establish and maintain procedures to determine if an HCT/P that is returned to your establishment is suitable to be returned to inventory.

§ 1271.270 Records.

(a) *General.* You must maintain records concurrently with the performance of each step required in this subpart and subpart C of this part. Any requirement in this part that an action be documented involves the creation of a record, which is subject to the requirements of this section. All records must be accurate, indelible, and legible. The records must identify the person performing the work and the dates of the various entries, and must be as detailed as necessary to provide a complete history of the work performed and to relate the records to the particular HCT/P involved.

(b) *Records management system.* You must establish and maintain a records management system relating to core CGTP requirements. Under this system, records pertaining to a particular HCT/P must be maintained in such a way as to facilitate review of the HCT/Ps history before making it available for distribution and, if necessary, subsequent to the HCT/Ps release as part of a followup evaluation or investigation. Records pertinent to the manufacture of HCT/Ps (e.g., labeling and packaging procedures, and equipment logs) must also be maintained and organized under the records management system. If records are maintained in more than one location, then the records management system must be designed to ensure prompt identification, location, and retrieval of all records.

(c) *Methods of retention.* You may maintain records required under this subpart electronically, as original paper records, or as true copies such as photocopies, microfiche, or microfilm. Equipment that is necessary to make the records available and legible, such as computer and reader equipment, must be readily available. Records stored in electronic systems must be backed up.

(d) *Length of retention.* You must retain all records for 10 years after their creation, unless stated otherwise in this part. However, you must retain the records pertaining to a particular HCT/P at least 10 years after the date of its administration, or if the date of administration is not known, then at least 10 years after the date of the HCT/Ps distribution, disposition, or expiration, whichever is latest. You must retain records for archived specimens of dura mater for 10 years after the appropriate disposition of the specimens.

(e) *Contracts and agreements.* You must maintain the name and address and a list of the responsibilities of any establishment that performs a manufacturing step for you. This information must be available during an inspection conducted under § 1271.400.

§ 1271.290 Tracking.

(a) *General.* If you perform any step in the manufacture of an HCT/P in which you handle the HCT/P, you must track each such HCT/P in accordance with this section, to facilitate the investigation of actual or suspected transmission of communicable disease and take appropriate and timely corrective action.

(b) *System of HCT/P tracking.* (1) You must establish and maintain a system of HCT/P tracking that enables the tracking of all HCT/Ps from:

(i) The donor to the consignee or final disposition; and

(ii) The consignee or final disposition to the donor.

(2) Alternatively, if you are an establishment that performs some but not all of the steps in the manufacture of an HCT/P in which you handle the HCT/P, you may participate in a system of HCT/P tracking established and maintained by another establishment responsible for other steps in the manufacture of the same HCT/P, provided that the tracking system complies with all the requirements of this section.

(c) *Distinct identification code.* As part of your tracking system, you must ensure: That each HCT/P that you manufacture is assigned and labeled with a distinct identification code, e.g., alphanumeric, that relates the HCT/P to the donor and to all records pertaining to the HCT/P; and that labeling includes information designed to facilitate effective tracking, using the distinct identification code, from the donor to the recipient and from the recipient to the donor. Except as described in § 1271.55(a)(1), you must create such a code specifically for tracking, and it may not include an individual's name, social security number, or medical record number. You may adopt a distinct identification code assigned by another establishment engaged in the manufacturing process, or you may assign a new code. If you assign a new code to an HCT/P, you must establish and maintain procedures for relating the new code to the old code.

(d) *Tracking from consignee to donor.* As part of your tracking system, you must establish and maintain a method for recording the distinct identification code and type of each HCT/P distributed to a consignee to enable tracking from the consignee to the donor.

(e) *Tracking from donor to consignee or final disposition.* As part of your tracking system, you must establish and maintain a method for documenting the disposition of each of your HCT/Ps, to enable tracking from the donor to the consignee or final disposition. The information you maintain must permit the prompt identification of the consignee of the HCT/P, if any.

(f) *Consignees.* At or before the time of distribution of an HCT/P to a consignee, you must inform the consignee in writing of the requirements in this section and of the tracking system that you have established and are maintaining to comply with these requirements.

(g) *Requirements specific to dura mater donors.* You must archive appropriate specimens from each donor of dura mater, under appropriate storage conditions, and for the appropriate duration, to enable testing of the archived material for evidence of transmissible spongiform encephalopathy, and to enable appropriate disposition of any affected nonadministered dura mater tissue, if necessary.

[69 FR 68681, Nov. 24, 2004, as amended at 70 FR 29952, May 25, 2005]

§ 1271.320 Complaint file.

(a) *Procedures.* You must establish and maintain procedures for the review, evaluation, and documentation of complaints as defined in §1271.3(aa), relating to core current good tissue practice (CGTP) requirements, and the investigation of complaints as appropriate.

(b) *Complaint file.* You must maintain a record of complaints that you receive in a file designated for complaints. The complaint file must contain sufficient information about each complaint for proper review and evaluation of the complaint (including the distinct identification code of the HCT/P that is the subject of the complaint) and for determining whether the complaint is an isolated event or represents a trend. You must make the complaint file available for review and copying upon request from FDA.

(c) *Review and evaluation of complaints.* You must review and evaluate each complaint relating to core CGTP requirements to determine if the complaint is related to an HCT/P deviation or to an adverse reaction, and to determine if a report under § 1271.350 or another applicable regulation is required. As soon as practical, you must review, evaluate, and investigate each complaint that represents an event required to be reported to FDA, as described in § 1271.350. You must review and evaluate a complaint relating to core CGTP requirements that does not represent an event required to be reported to determine whether an investigation is necessary; an investigation may include referring a copy of the complaint to another establishment that performed manufacturing steps pertinent to the complaint. When no investigation is made, you must maintain a record that includes the reason no investigation was made, and the name of the individual(s) responsible for the decision not to investigate.

Subpart E—Additional Requirements for Establishments Described in § 1271.10

SOURCE: 69 FR 68686, Nov. 24, 2004, unless otherwise noted.

§ 1271.330 Applicability.

The provisions set forth in this subpart are being implemented for non-reproductive HCT/Ps described in § 1271.10 and regulated solely under section 361 of the Public Health Service Act and the regulations in this part, and for the establishments that manufacture those HCT/Ps. HCT/Ps that are drugs or devices regulated under the act, or are biological products regulated under section 351 of the Public Health Service Act, are not subject to the regulations set forth in this subpart.

§ 1271.350 Reporting.

(a) *Adverse reaction reports.* (1) You must investigate any adverse reaction involving a communicable disease related to an HCT/P that you made available for distribution. You must report to FDA an adverse reaction involving a communicable disease if it:

(i) Is fatal;

(ii) Is life-threatening;

(iii) Results in permanent impairment of a body function or permanent damage to body structure; or

(iv) Necessitates medical or surgical intervention, including hospitalization.

(2) You must submit each report on a Form FDA–3500A to the address in paragraph (a)(5) of this section within 15 calendar days of initial receipt of the information.

(3) You must, as soon as practical, investigate all adverse reactions that are the subject of these 15-day reports and must submit followup reports within 15 calendar days of the receipt of new information or as requested by FDA. If additional information is not obtainable, a followup report may be required that describes briefly the steps taken to seek additional information and the reasons why it could not be obtained.

(4) You may obtain copies of the reporting form (FDA–3500A) from the Center for Biologics Evaluation and Research (see address in paragraph (a)(5) of this section). Electronic Form FDA–3500A may be obtained at *http://www.fda.gov/medwatch* or at *http://www.hhs.gov/forms.*

(5) You must submit two copies of each report described in this paragraph to the Center for Biologics Evaluation and Research (HFM–210), Food and Drug Administration, 1401 Rockville Pike, suite 200N, Rockville, MD 20852–1448. FDA may waive the requirement for the second copy in appropriate circumstances.

(b) *Reports of HCT/P deviations.* (1) You must investigate all HCT/P deviations related to a distributed HCT/P for which you performed a manufacturing step.

(2) You must report any such HCT/P deviation relating to the core CGTP requirements, if the HCT/P deviation occurred in your facility or in a facility that performed a manufacturing step for you under contract, agreement, or other arrangement. Each report must contain a description of the HCT/P deviation, information relevant to the event and the manufacture of the HCT/P involved, and information on all follow-up actions that have been or will

be taken in response to the HCT/P deviation (e.g., recalls).

(3) You must report each such HCT/P deviation that relates to a core CGTP requirement on Form FDA–3486 available at *http://www.fda.gov/cber/biodev/bpdrform.pdf*, within 45 days of the discovery of the event either electronically at *http://www.fda.gov/cber/biodev/biodevsub.htm* or by mail to the Director, Office of Compliance and Biologics Quality, Center for Biologics Evaluation and Research (HFM–600), 1401 Rockville Pike, suite 200N, Rockville, MD 20852–1448.

§ 1271.370 Labeling.

The following requirements apply in addition to §§ 1271.55, 1271.60, 1271.65, and 1271.90:

(a) You must label each HCT/P made available for distribution clearly and accurately.

(b) The following information must appear on the HCT/P label:

(1) Distinct identification code affixed to the HCT/P container, and assigned in accordance with § 1271.290(c);

(2) Description of the type of HCT/P;

(3) Expiration date, if any; and

(4) Warnings required under § 1271.60(d)(2), § 1271.65(b)(2), or § 1271.90(b), if applicable and physically possible. If it is not physically possible to include these warnings on the label, the warnings must, instead, accompany the HCT/P.

(c) The following information must either appear on the HCT/P label or accompany the HCT/P:

(1) Name and address of the establishment that determines that the HCT/P meets release criteria and makes the HCT/P available for distribution;

(2) Storage temperature;

(3) Other warnings, where appropriate; and

(4) Instructions for use when related to the prevention of the introduction, transmission, or spread of communicable diseases.

[69 FR 68686, Nov. 24, 2004, as amended at 70 FR 29952, May 25, 2005]

Subpart F—Inspection and Enforcement of Establishments Described in § 1271.10

SOURCE: 69 FR 68687, Nov. 24, 2004, unless otherwise noted.

§ 1271.390 Applicability.

The provisions set forth in this subpart are applicable only to HCT/Ps described in § 1271.10 and regulated solely under section 361 of the Public Health Service Act and the regulations in this part, and to the establishments that manufacture those HCT/Ps. HCT/Ps that are drugs or devices regulated under the act, or are biological products regulated under section 351 of the Public Health Service Act, are not subject to the regulations set forth in this subpart.

§ 1271.400 Inspections.

(a) If you are an establishment that manufactures HCT/Ps described in § 1271.10, whether or not under contract, you must permit the Food and Drug Administration (FDA) to inspect any manufacturing location at any reasonable time and in a reasonable manner to determine compliance with applicable provisions of this part. The inspection will be conducted as necessary in the judgment of the FDA and may include your establishment, facilities, equipment, finished and unfinished materials, containers, processes, HCT/Ps, procedures, labeling, records, files, papers, and controls required to be maintained under the part. The inspection may be made with or without prior notification and will ordinarily be made during regular business hours.

(b) The frequency of inspection will be at the agency's discretion.

(c) FDA will call upon the most responsible person available at the time of the inspection of the establishment and may question the personnel of the establishment as necessary to determine compliance with the provisions of this part.

(d) FDA's representatives may take samples, may review and copy any records required to be kept under this part, and may use other appropriate means to record evidence of observations during inspections conducted under this subpart.

(e) The public disclosure of records containing the name or other positive identification of donors or recipients of HCT/Ps will be handled in accordance with FDA's procedures on disclosure of information as set forth in parts 20 and 21 of this chapter.

§ 1271.420 HCT/Ps offered for import.

(a) Except as provided in paragraphs (c) and (d) of this section, when an HCT/P is offered for import, the importer of record must notify, either before or at the time of importation, the director of the district of the Food and Drug Administration (FDA) having jurisdiction over the port of entry through which the HCT/P is imported or offered for import, or such officer of the district as the director may designate to act in his or her behalf in administering and enforcing this part, and must provide sufficient information for FDA to make an admissibility decision.

(b) Except as provided in paragraphs (c) and (d) of this section, an HCT/P offered for import must be held intact by the importer or consignee, under conditions necessary to prevent transmission of communicable disease, until an admissibility decision is made by FDA. The HCT/P may be transported under quarantine to the consignee, while the FDA district reviews the documentation accompanying the HCT/P. When FDA makes a decision regarding the admissibility of the HCT/P, FDA will notify the importer of record.

(c) This section does not apply to reproductive HCT/Ps regulated solely under section 361 of the Public Health Service Act and the regulations in this part, and donated by a sexually intimate partner of the recipient for reproductive use.

(d) This section does not apply to peripheral blood stem/progenitor cells regulated solely under section 361 of the Public Health Service Act and the regulations in this part, except that paragraphs (a) and (b) of this section apply when circumstances occur under which such imported peripheral blood stem/progenitor cells may present an unreasonable risk of communicable disease transmission which indicates the need to review the information referenced in paragraph (a) of this section.

§ 1271.440 Orders of retention, recall, destruction, and cessation of manufacturing.

(a) Upon an agency finding that there are reasonable grounds to believe that an HCT/P is a violative HCT/P because it was manufactured in violation of the regulations in this part and, therefore, the conditions of manufacture of the HCT/P do not provide adequate protections against risks of communicable disease transmission; or the HCT/P is infected or contaminated so as to be a source of dangerous infection to humans; or an establishment is in violation of the regulations in this part and, therefore, does not provide adequate protections against the risks of communicable disease transmission, the Food and Drug Administration (FDA) may take one or more of the following actions:

(1) Serve upon the person who distributed the HCT/P a written order that the HCT/P be recalled and/or destroyed, as appropriate, and upon persons in possession of the HCT/P that the HCT/P must be retained until it is recalled by the distributor, destroyed, or disposed of as agreed by FDA, or the safety of the HCT/P is confirmed;

(2) Take possession of and/or destroy the violative HCT/P; or

(3) Serve upon the establishment an order to cease manufacturing until compliance with the regulations of this part has been achieved. When FDA determines there are reasonable grounds to believe there is a danger to health, such order will be effective immediately. In other situations, such order will be effective after one of the following events, whichever is later:

(i) Passage of 5 working days from the establishment's receipt of the order; or

(ii) If the establishment requests a hearing in accordance with paragraph (e) of this section and part 16 of this chapter, a decision in, and in accordance with, those proceedings.

(b) A written order issued under paragraph (a) of this section will state with particularity the facts that justify the order.

(c)(1) A written order issued under paragraph (a)(1) of this section will ordinarily provide that the HCT/P be recalled and/or destroyed within 5 working days from the date of receipt of the order. After receipt of an order issued under paragraph (a)(1) of this section, the establishment in possession of the HCT/P must not distribute or dispose of the HCT/P in any manner except to recall and/or destroy the HCT/P consistent with the provisions of the order, under the supervision of FDA.

(2) In lieu of paragraph (c)(1) of this section, other arrangements for assuring the proper disposition of the HCT/P may be agreed upon by the person receiving the written order and FDA. Such arrangements may include, among others, providing FDA with records or other written information that adequately ensure that the HCT/P has been recovered, processed, stored, and distributed in conformance with this part, and that, except as provided under §§1271.60, 1271.65, and 1271.90, the donor of the cells or tissue for the HCT/P has been determined to be eligible.

(d) A written order issued under paragraph (a)(3) of this section will specify the regulations with which you must achieve compliance and will ordinarily specify the particular operations covered by the order. After receipt of an order that is in effect and issued under paragraph (a)(3) of this section, you must not resume operations without prior written authorization of FDA.

(e) The recipient of an order issued under this section may request a hearing in accordance with part 16 of this chapter. To request a hearing, the recipient of the written order or prior possessor of such HCT/P must make the request within 5 working days of receipt of a written order for retention, recall, destruction, and/or cessation (or within 5 working days of the agency's possession of an HCT/P under paragraph (a)(2) of this section), in accordance with part 16 of this chapter. An order of destruction will be held in abeyance pending resolution of the hearing request. Upon request under part 16 of this chapter, FDA will provide an opportunity for an expedited hearing for an order of cessation that is not stayed by the Commissioner of Food and Drugs.

(f) FDA will not issue an order for the destruction of reproductive tissue under paragraph (a)(1) of this section, nor will it carry out such destruction itself under paragraph (a)(2) of this section.

PARTS 1272–1299 [RESERVED]

FINDING AIDS

A list of CFR titles, subtitles, chapters, subchapters and parts and an alphabetical list of agencies publishing in the CFR are included in the CFR Index and Finding Aids volume to the Code of Federal Regulations which is published separately and revised annually.

Material Approved for Incorporation by Reference

(Revised as of April 1, 2006)

The Director of the Federal Register has approved under 5 U.S.C. 552(a) and 1 CFR Part 51 the incorporation by reference of the following publications. This list contains only those incorporations by reference effective as of the revision date of this volume. Incorporations by reference found within a regulation are effective upon the effective date of that regulation. For more information on incorporation by reference, see the preliminary pages of this volume.

21 CFR (PARTS 800 TO 1299)
FOOD AND DRUG ADMINISTRATION, DEPARTMENT OF HEALTH AND HUMAN SERVICES

	21 CFR
American College of Radiology, Committee on Quality Assurance in Mammography, Mammography Accreditation Program	
1891 Preston White Dr., Reston, VA 22091–5431	
Mammography Quality Control: Radiologist's Manual, Radiologic Technologists's Manual, Medical Physicist's Manual, 1992 edition.	900.12(d)
Mammography Quality Control Manual, 1994 edition	900.12(d)
American Society for Testing and Materials	
100 Barr Harbor Drive, West Conshohocken, PA 19428-2959, Telephone (610) 832-9585, FAX (610) 832-9555	
ASTM D 412–68 Standard Method of Tension Testing of Vulcanized Rubber.	801.410
ASTM D 412–97 Standard Test Methods for Vulcanized Rubber and Thermoplastic Rubbers and Thermoplastic Elastomers—Tension.	801.410(d)(2)
ASTM D 1415–68 Test for International Hardness of Vulcanized Rubber.	801.410
ASTM D 1415–88 Standard Test Method for Rubber Property—International Hardness.	801.410(d)(2)
ASTM D 3492–83 Standard Specification for Rubber Contraceptives (Condoms).	801.430(f)(2)
ASTM D 3492–96 Standard Specification for Rubber Contraceptives (Male Condoms).	801.430(f)(2)
American National Standards Institute	
11 West 42nd Street, New York, NY 10036 Telephone: (212) 642–4900	
ANSI C81.10–1976 Specifications for Electric Lamp Bases and Holders-Screw Shall Types.	1040.20(c)
ANSI S3.22–1987 Specification of Hearing Aid Characteristics (ASA 70–1987).	801.420(c)(4)
Available from: Standards Secretariat, Accoustical Society of America, 120 Wall Street, 32nd Floor, New York, NY 10005–3993	
ANSI S3.22–1996, Specification of Hearing Aid Characteristics (ASA 70–1996).	801.420(c)(4)

Table of CFR Titles and Chapters

(Revised as of April 1, 2006)

Title 1—General Provisions

Title 2—Grants and Agreements

Title 3—The President

Title 4—Accounts

Title 5—Administrative Personnel

Title 7—Agriculture—Continued

Chap.	
XVIII	Rural Housing Service, Rural Business-Cooperative Service, Rural Utilities Service, and Farm Service Agency, Department of Agriculture (Parts 1800—2099)
XX	Local Television Loan Guarantee Board (Parts 2200—2299)
XXVI	Office of Inspector General, Department of Agriculture (Parts 2600—2699)
XXVII	Office of Information Resources Management, Department of Agriculture (Parts 2700—2799)
XXVIII	Office of Operations, Department of Agriculture (Parts 2800—2899)
XXIX	Office of Energy Policy and New Uses, Department of Agriculture (Parts 2900—2999)
XXX	Office of the Chief Financial Officer, Department of Agriculture (Parts 3000—3099)
XXXI	Office of Environmental Quality, Department of Agriculture (Parts 3100—3199)
XXXII	Office of Procurement and Property Management, Department of Agriculture (Parts 3200—3299)
XXXIII	Office of Transportation, Department of Agriculture (Parts 3300—3399)
XXXIV	Cooperative State Research, Education, and Extension Service, Department of Agriculture (Parts 3400—3499)
XXXV	Rural Housing Service, Department of Agriculture (Parts 3500—3599)
XXXVI	National Agricultural Statistics Service, Department of Agriculture (Parts 3600—3699)
XXXVII	Economic Research Service, Department of Agriculture (Parts 3700—3799)
XXXVIII	World Agricultural Outlook Board, Department of Agriculture (Parts 3800—3899)
XLI	[Reserved]
XLII	Rural Business-Cooperative Service and Rural Utilities Service, Department of Agriculture (Parts 4200—4299)

Title 8—Aliens and Nationality

I	Department of Homeland Security (Immigration and Naturalization) (Parts 1—499)
V	Executive Office for Immigration Review, Department of Justice (Parts 1000—1399)

Title 9—Animals and Animal Products

I	Animal and Plant Health Inspection Service, Department of Agriculture (Parts 1—199)
II	Grain Inspection, Packers and Stockyards Administration (Packers and Stockyards Programs), Department of Agriculture (Parts 200—299)
III	Food Safety and Inspection Service, Department of Agriculture (Parts 300—599)

Title 10—Energy

Chap.	
I	Nuclear Regulatory Commission (Parts 0—199)
II	Department of Energy (Parts 200—699)
III	Department of Energy (Parts 700—999)
X	Department of Energy (General Provisions) (Parts 1000—1099)
XIII	Nuclear Waste Technical Review Board (Parts 1303—1399)
XVII	Defense Nuclear Facilities Safety Board (Parts 1700—1799)
XVIII	Northeast Interstate Low-Level Radioactive Waste Commission (Part 1800)

Title 11—Federal Elections

I	Federal Election Commission (Parts 1—9099)

Title 12—Banks and Banking

I	Comptroller of the Currency, Department of the Treasury (Parts 1—199)
II	Federal Reserve System (Parts 200—299)
III	Federal Deposit Insurance Corporation (Parts 300—399)
IV	Export-Import Bank of the United States (Parts 400—499)
V	Office of Thrift Supervision, Department of the Treasury (Parts 500—599)
VI	Farm Credit Administration (Parts 600—699)
VII	National Credit Union Administration (Parts 700—799)
VIII	Federal Financing Bank (Parts 800—899)
IX	Federal Housing Finance Board (Parts 900—999)
XI	Federal Financial Institutions Examination Council (Parts 1100—1199)
XIV	Farm Credit System Insurance Corporation (Parts 1400—1499)
XV	Department of the Treasury (Parts 1500—1599)
XVII	Office of Federal Housing Enterprise Oversight, Department of Housing and Urban Development (Parts 1700—1799)
XVIII	Community Development Financial Institutions Fund, Department of the Treasury (Parts 1800—1899)

Title 13—Business Credit and Assistance

I	Small Business Administration (Parts 1—199)
III	Economic Development Administration, Department of Commerce (Parts 300—399)
IV	Emergency Steel Guarantee Loan Board, Department of Commerce (Parts 400—499)
V	Emergency Oil and Gas Guaranteed Loan Board, Department of Commerce (Parts 500—599)

Title 14—Aeronautics and Space

Chap.

I Federal Aviation Administration, Department of Transportation (Parts 1—199)

II Office of the Secretary, Department of Transportation (Aviation Proceedings) (Parts 200—399)

III Commercial Space Transportation, Federal Aviation Administration, Department of Transportation (Parts 400—499)

V National Aeronautics and Space Administration (Parts 1200—1299)

VI Air Transportation System Stabilization (Parts 1300—1399)

Title 15—Commerce and Foreign Trade

SUBTITLE A—OFFICE OF THE SECRETARY OF COMMERCE (PARTS 0—29)

SUBTITLE B—REGULATIONS RELATING TO COMMERCE AND FOREIGN TRADE

I Bureau of the Census, Department of Commerce (Parts 30—199)

II National Institute of Standards and Technology, Department of Commerce (Parts 200—299)

III International Trade Administration, Department of Commerce (Parts 300—399)

IV Foreign-Trade Zones Board, Department of Commerce (Parts 400—499)

VII Bureau of Industry and Security, Department of Commerce (Parts 700—799)

VIII Bureau of Economic Analysis, Department of Commerce (Parts 800—899)

IX National Oceanic and Atmospheric Administration, Department of Commerce (Parts 900—999)

XI Technology Administration, Department of Commerce (Parts 1100—1199)

XIII East-West Foreign Trade Board (Parts 1300—1399)

XIV Minority Business Development Agency (Parts 1400—1499)

SUBTITLE C—REGULATIONS RELATING TO FOREIGN TRADE AGREEMENTS

XX Office of the United States Trade Representative (Parts 2000—2099)

SUBTITLE D—REGULATIONS RELATING TO TELECOMMUNICATIONS AND INFORMATION

XXIII National Telecommunications and Information Administration, Department of Commerce (Parts 2300—2399)

Title 16—Commercial Practices

I Federal Trade Commission (Parts 0—999)

II Consumer Product Safety Commission (Parts 1000—1799)

Title 17—Commodity and Securities Exchanges

Title 18—Conservation of Power and Water Resources

Title 19—Customs Duties

Title 20—Employees' Benefits

Title 21—Food and Drugs

Title 22—Foreign Relations

Title 23—Highways

Title 24—Housing and Urban Development

Title 25—Indians

Title 26—Internal Revenue

Title 27—Alcohol, Tobacco Products and Firearms

Title 28—Judicial Administration

Title 29—Labor

Title 30—Mineral Resources

Chap.	
I	Mine Safety and Health Administration, Department of Labor (Parts 1—199)
II	Minerals Management Service, Department of the Interior (Parts 200—299)
III	Board of Surface Mining and Reclamation Appeals, Department of the Interior (Parts 300—399)
IV	Geological Survey, Department of the Interior (Parts 400—499)
VII	Office of Surface Mining Reclamation and Enforcement, Department of the Interior (Parts 700—999)

Title 31—Money and Finance: Treasury

	SUBTITLE A—OFFICE OF THE SECRETARY OF THE TREASURY (PARTS 0—50)
	SUBTITLE B—REGULATIONS RELATING TO MONEY AND FINANCE
I	Monetary Offices, Department of the Treasury (Parts 51—199)
II	Fiscal Service, Department of the Treasury (Parts 200—399)
IV	Secret Service, Department of the Treasury (Parts 400—499)
V	Office of Foreign Assets Control, Department of the Treasury (Parts 500—599)
VI	Bureau of Engraving and Printing, Department of the Treasury (Parts 600—699)
VII	Federal Law Enforcement Training Center, Department of the Treasury (Parts 700—799)
VIII	Office of International Investment, Department of the Treasury (Parts 800—899)
IX	Federal Claims Collection Standards (Department of the Treasury—Department of Justice) (Parts 900—999)

Title 32—National Defense

	SUBTITLE A—DEPARTMENT OF DEFENSE
I	Office of the Secretary of Defense (Parts 1—399)
V	Department of the Army (Parts 400—699)
VI	Department of the Navy (Parts 700—799)
VII	Department of the Air Force (Parts 800—1099)
	SUBTITLE B—OTHER REGULATIONS RELATING TO NATIONAL DEFENSE
XII	Defense Logistics Agency (Parts 1200—1299)
XVI	Selective Service System (Parts 1600—1699)
XVIII	National Counterintelligence Center (Parts 1800—1899)
XIX	Central Intelligence Agency (Parts 1900—1999)
XX	Information Security Oversight Office, National Archives and Records Administration (Parts 2000—2099)
XXI	National Security Council (Parts 2100—2199)
XXIV	Office of Science and Technology Policy (Parts 2400—2499)
XXVII	Office for Micronesian Status Negotiations (Parts 2700—2799)

Title 32—National Defense—Continued

Chap.

XXVIII Office of the Vice President of the United States (Parts 2800—2899)

Title 33—Navigation and Navigable Waters

I Coast Guard, Department of Homeland Security (Parts 1—199)

II Corps of Engineers, Department of the Army (Parts 200—399)

IV Saint Lawrence Seaway Development Corporation, Department of Transportation (Parts 400—499)

Title 34—Education

SUBTITLE A—OFFICE OF THE SECRETARY, DEPARTMENT OF EDUCATION (PARTS 1—99)

SUBTITLE B—REGULATIONS OF THE OFFICES OF THE DEPARTMENT OF EDUCATION

I Office for Civil Rights, Department of Education (Parts 100—199)

II Office of Elementary and Secondary Education, Department of Education (Parts 200—299)

III Office of Special Education and Rehabilitative Services, Department of Education (Parts 300—399)

IV Office of Vocational and Adult Education, Department of Education (Parts 400—499)

V Office of Bilingual Education and Minority Languages Affairs, Department of Education (Parts 500—599)

VI Office of Postsecondary Education, Department of Education (Parts 600—699)

XI National Institute for Literacy (Parts 1100—1199)

SUBTITLE C—REGULATIONS RELATING TO EDUCATION

XII National Council on Disability (Parts 1200—1299)

Title 35 [Reserved]

Title 36—Parks, Forests, and Public Property

I National Park Service, Department of the Interior (Parts 1—199)

II Forest Service, Department of Agriculture (Parts 200—299)

III Corps of Engineers, Department of the Army (Parts 300—399)

IV American Battle Monuments Commission (Parts 400—499)

V Smithsonian Institution (Parts 500—599)

VII Library of Congress (Parts 700—799)

VIII Advisory Council on Historic Preservation (Parts 800—899)

IX Pennsylvania Avenue Development Corporation (Parts 900—999)

X Presidio Trust (Parts 1000—1099)

XI Architectural and Transportation Barriers Compliance Board (Parts 1100—1199)

XII National Archives and Records Administration (Parts 1200—1299)

XV Oklahoma City National Memorial Trust (Part 1501)

Title 37—Patents, Trademarks, and Copyrights

Title 38—Pensions, Bonuses, and Veterans' Relief

Title 39—Postal Service

Title 40—Protection of Environment

Title 41—Public Contracts and Property Management

Title 42—Public Health

Title 43—Public Lands: Interior

Title 44—Emergency Management and Assistance

Title 45—Public Welfare

Title 46—Shipping

Title 47—Telecommunication

Title 48—Federal Acquisition Regulations System

Title 48—Federal Acquisition Regulations System—Continued

Chap.

52 Department of the Navy Acquisition Regulations (Parts 5200—5299)

53 Department of the Air Force Federal Acquisition Regulation Supplement (Parts 5300—5399)

54 Defense Logistics Agency, Department of Defense (Parts 5400—5499)

57 African Development Foundation (Parts 5700—5799)

61 General Services Administration Board of Contract Appeals (Parts 6100—6199)

63 Department of Transportation Board of Contract Appeals (Parts 6300—6399)

99 Cost Accounting Standards Board, Office of Federal Procurement Policy, Office of Management and Budget (Parts 9900—9999)

Title 49—Transportation

SUBTITLE A—OFFICE OF THE SECRETARY OF TRANSPORTATION (PARTS 1—99)

SUBTITLE B—OTHER REGULATIONS RELATING TO TRANSPORTATION

I Pipeline and Hazardous Materials Safety Administration, Department of Transportation (Parts 100—199)

II Federal Railroad Administration, Department of Transportation (Parts 200—299)

III Federal Motor Carrier Safety Administration, Department of Transportation (Parts 300—399)

IV Coast Guard, Department of Homeland Security (Parts 400—499)

V National Highway Traffic Safety Administration, Department of Transportation (Parts 500—599)

VI Federal Transit Administration, Department of Transportation (Parts 600—699)

VII National Railroad Passenger Corporation (AMTRAK) (Parts 700—799)

VIII National Transportation Safety Board (Parts 800—999)

X Surface Transportation Board, Department of Transportation (Parts 1000—1399)

XI Research and Innovative Technology Administration, Department of Transportation (Parts 1400—1499)

XII Transportation Security Administration, Department of Homeland Security (Parts 1500—1699)

Title 50—Wildlife and Fisheries

I United States Fish and Wildlife Service, Department of the Interior (Parts 1—199)

II National Marine Fisheries Service, National Oceanic and Atmospheric Administration, Department of Commerce (Parts 200—299)

III International Fishing and Related Activities (Parts 300—399)

CFR Index and Finding Aids

Alphabetical List of Agencies Appearing in the CFR

(Revised as of April 1, 2006)

Agency	CFR Title, Subtitle or Chapter
Administrative Committee of the Federal Register	1, I
Advanced Research Projects Agency	32, I
Advisory Council on Historic Preservation	36, VIII
African Development Foundation	22, XV
Federal Acquisition Regulation	48, 57
Agency for International Development, United States	22, II
Federal Acquisition Regulation	48, 7
Agricultural Marketing Service	7, I, IX, X, XI
Agricultural Research Service	7, V
Agriculture Department	5, LXXIII
Agricultural Marketing Service	7, I, IX, X, XI
Agricultural Research Service	7, V
Animal and Plant Health Inspection Service	7, III; 9, I
Chief Financial Officer, Office of	7, XXX
Commodity Credit Corporation	7, XIV
Cooperative State Research, Education, and Extension Service	7, XXXIV
Economic Research Service	7, XXXVII
Energy, Office of	7, XXIX
Environmental Quality, Office of	7, XXXI
Farm Service Agency	7, VII, XVIII
Federal Acquisition Regulation	48, 4
Federal Crop Insurance Corporation	7, IV
Food and Nutrition Service	7, II
Food Safety and Inspection Service	9, III
Foreign Agricultural Service	7, XV
Forest Service	36, II
Grain Inspection, Packers and Stockyards Administration	7, VIII; 9, II
Information Resources Management, Office of	7, XXVII
Inspector General, Office of	7, XXVI
National Agricultural Library	7, XLI
National Agricultural Statistics Service	7, XXXVI
Natural Resources Conservation Service	7, VI
Operations, Office of	7, XXVIII
Procurement and Property Management, Office of	7, XXXII
Rural Business-Cooperative Service	7, XVIII, XLII
Rural Development Administration	7, XLII
Rural Housing Service	7, XVIII, XXXV
Rural Telephone Bank	7, XVI
Rural Utilities Service	7, XVII, XVIII, XLII
Secretary of Agriculture, Office of	7, Subtitle A
Transportation, Office of	7, XXXIII
World Agricultural Outlook Board	7, XXXVIII
Air Force Department	32, VII
Federal Acquisition Regulation Supplement	48, 53
Air Transportation Stabilization Board	14, VI
Alcohol and Tobacco Tax and Trade Bureau	27, I
Alcohol, Tobacco, Firearms, and Explosives, Bureau of	27, II
AMTRAK	49, VII
American Battle Monuments Commission	36, IV
American Indians, Office of the Special Trustee	25, VII
Animal and Plant Health Inspection Service	7, III; 9, I
Appalachian Regional Commission	5, IX

Agency	CFR Title, Subtitle or Chapter
Architectural and Transportation Barriers Compliance Board	36, XI
Arctic Research Commission	45, XXIII
Armed Forces Retirement Home	5, XI
Army Department	32, V
Engineers, Corps of	33, II; 36, III
Federal Acquisition Regulation	48, 51
Benefits Review Board	20, VII
Bilingual Education and Minority Languages Affairs, Office of	34, V
Blind or Severely Disabled, Committee for Purchase From People Who Are	41, 51
Broadcasting Board of Governors	22, V
Federal Acquisition Regulation	48, 19
Census Bureau	15, I
Centers for Medicare & Medicaid Services	42, IV
Central Intelligence Agency	32, XIX
Chief Financial Officer, Office of	7, XXX
Child Support Enforcement, Office of	45, III
Children and Families, Administration for	45, II, III, IV, X
Civil Rights, Commission on	45, VII
Civil Rights, Office for	34, I
Coast Guard	33, I; 46, I; 49, IV
Coast Guard (Great Lakes Pilotage)	46, III
Commerce Department	44, IV
Census Bureau	15, I
Economic Affairs, Under Secretary	37, V
Economic Analysis, Bureau of	15, VIII
Economic Development Administration	13, III
Emergency Management and Assistance	44, IV
Federal Acquisition Regulation	48, 13
Fishery Conservation and Management	50, VI
Foreign-Trade Zones Board	15, IV
Industry and Security, Bureau of	15, VII
International Trade Administration	15, III; 19, III
National Institute of Standards and Technology	15, II
National Marine Fisheries Service	50, II, IV, VI
National Oceanic and Atmospheric Administration	15, IX; 50, II, III, IV, VI
National Telecommunications and Information Administration	15, XXIII; 47, III
National Weather Service	15, IX
Patent and Trademark Office, United States	37, I
Productivity, Technology and Innovation, Assistant Secretary for	37, IV
Secretary of Commerce, Office of	15, Subtitle A
Technology, Under Secretary for	37, V
Technology Administration	15, XI
Technology Policy, Assistant Secretary for	37, IV
Commercial Space Transportation	14, III
Commodity Credit Corporation	7, XIV
Commodity Futures Trading Commission	5, XLI; 17, I
Community Planning and Development, Office of Assistant Secretary for	24, V, VI
Community Services, Office of	45, X
Comptroller of the Currency	12, I
Construction Industry Collective Bargaining Commission	29, IX
Consumer Product Safety Commission	5, LXXI; 16, II
Cooperative State Research, Education, and Extension Service	7, XXXIV
Copyright Office	37, II
Copyright Royalty Board	37, III
Corporation for National and Community Service	45, XII, XXV
Cost Accounting Standards Board	48, 99
Council on Environmental Quality	40, V
Court Services and Offender Supervision Agency for the District of Columbia	28, VIII
Customs and Border Protection Bureau	19, I
Defense Contract Audit Agency	32, I
Defense Department	5, XXVI; 32, Subtitle A; 40, VII

Agency	CFR Title, Subtitle or Chapter
Advanced Research Projects Agency	32, I
Air Force Department	32, VII
Army Department	32, V; 33, II; 36, III, 48, 51
Defense Acquisition Regulations System	48, II
Defense Intelligence Agency	32, I
Defense Logistics Agency	32, I, XII; 48, 54
Engineers, Corps of	33, II; 36, III
National Imagery and Mapping Agency	32, I
Navy Department	32, VI; 48, 52
Secretary of Defense, Office of	32, I
Defense Contract Audit Agency	32, I
Defense Intelligence Agency	32, I
Defense Logistics Agency	32, XII; 48, 54
Defense Nuclear Facilities Safety Board	10, XVII
Delaware River Basin Commission	18, III
District of Columbia, Court Services and Offender Supervision Agency for the	28, VIII
Drug Enforcement Administration	21, II
East-West Foreign Trade Board	15, XIII
Economic Affairs, Under Secretary	37, V
Economic Analysis, Bureau of	15, VIII
Economic Development Administration	13, III
Economic Research Service	7, XXXVII
Education, Department of	5, LIII
Bilingual Education and Minority Languages Affairs, Office of	34, V
Civil Rights, Office for	34, I
Educational Research and Improvement, Office of	34, VII
Elementary and Secondary Education, Office of	34, II
Federal Acquisition Regulation	48, 34
Postsecondary Education, Office of	34, VI
Secretary of Education, Office of	34, Subtitle A
Special Education and Rehabilitative Services, Office of	34, III
Vocational and Adult Education, Office of	34, IV
Educational Research and Improvement, Office of	34, VII
Elementary and Secondary Education, Office of	34, II
Emergency Oil and Gas Guaranteed Loan Board	13, V
Emergency Steel Guarantee Loan Board	13, IV
Employee Benefits Security Administration	29, XXV
Employees' Compensation Appeals Board	20, IV
Employees Loyalty Board	5, V
Employment and Training Administration	20, V
Employment Standards Administration	20, VI
Endangered Species Committee	50, IV
Energy, Department of	5, XXIII; 10, II, III, X
Federal Acquisition Regulation	48, 9
Federal Energy Regulatory Commission	5, XXIV; 18, I
Property Management Regulations	41, 109
Energy, Office of	7, XXIX
Engineers, Corps of	33, II; 36, III
Engraving and Printing, Bureau of	31, VI
Environmental Protection Agency	5, LIV; 40, I, IV, VII
Federal Acquisition Regulation	48, 15
Property Management Regulations	41, 115
Environmental Quality, Office of	7, XXXI
Equal Employment Opportunity Commission	5, LXII; 29, XIV
Equal Opportunity, Office of Assistant Secretary for	24, I
Executive Office of the President	3, I
Administration, Office of	5, XV
Environmental Quality, Council on	40, V
Management and Budget, Office of	5, III, LXXVII; 14, VI; 48, 99
National Drug Control Policy, Office of	21, III
National Security Council	32, XXI; 47, 2
Presidential Documents	3
Science and Technology Policy, Office of	32, XXIV; 47, II

Agency	CFR Title, Subtitle or Chapter
Trade Representative, Office of the United States	15, XX
Export-Import Bank of the United States	5, LII; 12, IV
Family Assistance, Office of	45, II
Farm Credit Administration	5, XXXI; 12, VI
Farm Credit System Insurance Corporation	5, XXX; 12, XIV
Farm Service Agency	7, VII, XVIII
Federal Acquisition Regulation	48, 1
Federal Aviation Administration	14, I
Commercial Space Transportation	14, III
Federal Claims Collection Standards	31, IX
Federal Communications Commission	5, XXIX; 47, I
Federal Contract Compliance Programs, Office of	41, 60
Federal Crop Insurance Corporation	7, IV
Federal Deposit Insurance Corporation	5, XXII; 12, III
Federal Election Commission	11, I
Federal Emergency Management Agency	44, I
Federal Acquisition Regulation	48, 44
Federal Employees Group Life Insurance Federal Acquisition Regulation	48, 21
Federal Employees Health Benefits Acquisition Regulation	48, 16
Federal Energy Regulatory Commission	5, XXIV; 18, I
Federal Financial Institutions Examination Council	12, XI
Federal Financing Bank	12, VIII
Federal Highway Administration	23, I, II
Federal Home Loan Mortgage Corporation	1, IV
Federal Housing Enterprise Oversight Office	12, XVII
Federal Housing Finance Board	12, IX
Federal Labor Relations Authority, and General Counsel of the Federal Labor Relations Authority	5, XIV; 22, XIV
Federal Law Enforcement Training Center	31, VII
Federal Management Regulation	41, 102
Federal Maritime Commission	46, IV
Federal Mediation and Conciliation Service	29, XII
Federal Mine Safety and Health Review Commission	5, LXXIV; 29, XXVII
Federal Motor Carrier Safety Administration	49, III
Federal Prison Industries, Inc.	28, III
Federal Procurement Policy Office	48, 99
Federal Property Management Regulations	41, 101
Federal Railroad Administration	49, II
Federal Register, Administrative Committee of	1, I
Federal Register, Office of	1, II
Federal Reserve System	12, II
Board of Governors	5, LVIII
Federal Retirement Thrift Investment Board	5, VI, LXXVI
Federal Service Impasses Panel	5, XIV
Federal Trade Commission	5, XLVII; 16, I
Federal Transit Administration	49, VI
Federal Travel Regulation System	41, Subtitle F
Fine Arts, Commission on	45, XXI
Fiscal Service	31, II
Fish and Wildlife Service, United States	50, I, IV
Fishery Conservation and Management	50, VI
Food and Drug Administration	21, I
Food and Nutrition Service	7, II
Food Safety and Inspection Service	9, III
Foreign Agricultural Service	7, XV
Foreign Assets Control, Office of	31, V
Foreign Claims Settlement Commission of the United States	45, V
Foreign Service Grievance Board	22, IX
Foreign Service Impasse Disputes Panel	22, XIV
Foreign Service Labor Relations Board	22, XIV
Foreign-Trade Zones Board	15, IV
Forest Service	36, II
General Services Administration	5, LVII; 41, 105
Contract Appeals, Board of	48, 61
Federal Acquisition Regulation	48, 5
Federal Management Regulation	41, 102

Agency	CFR Title, Subtitle or Chapter
Inspector General	
Agriculture Department	7, XXVI
Health and Human Services Department	42, V
Housing and Urban Development Department	24, XII
Institute of Peace, United States	22, XVII
Inter-American Foundation	5, LXIII; 22, X
Interior Department	
American Indians, Office of the Special Trustee	25, VII
Endangered Species Committee	50, IV
Federal Acquisition Regulation	48, 14
Federal Property Management Regulations System	41, 114
Fish and Wildlife Service, United States	50, I, IV
Geological Survey	30, IV
Indian Affairs, Bureau of	25, I, V
Indian Affairs, Office of the Assistant Secretary	25, VI
Indian Arts and Crafts Board	25, II
Land Management, Bureau of	43, II
Minerals Management Service	30, II
National Indian Gaming Commission	25, III
National Park Service	36, I
Reclamation, Bureau of	43, I
Secretary of the Interior, Office of	43, Subtitle A
Surface Mining and Reclamation Appeals, Board of	30, III
Surface Mining Reclamation and Enforcement, Office of	30, VII
Internal Revenue Service	26, I
International Boundary and Water Commission, United States and Mexico, United States Section	22, XI
International Development, United States Agency for	22, II
Federal Acquisition Regulation	48, 7
International Development Cooperation Agency, United States	22, XII
International Fishing and Related Activities	50, III
International Investment, Office of	31, VIII
International Joint Commission, United States and Canada	22, IV
International Organizations Employees Loyalty Board	5, V
International Trade Administration	15, III; 19, III
International Trade Commission, United States	19, II
Interstate Commerce Commission	5, XL
James Madison Memorial Fellowship Foundation	45, XXIV
Japan–United States Friendship Commission	22, XVI
Joint Board for the Enrollment of Actuaries	20, VIII
Justice Department	5, XXVIII; 28, I, XI; 40, IV
Alcohol, Tobacco, Firearms, and Explosives, Bureau of	27, II
Drug Enforcement Administration	21, II
Federal Acquisition Regulation	48, 28
Federal Claims Collection Standards	31, IX
Federal Prison Industries, Inc.	28, III
Foreign Claims Settlement Commission of the United States	45, V
Immigration Review, Executive Office for	8, V
Offices of Independent Counsel	28, VI
Prisons, Bureau of	28, V
Property Management Regulations	41, 128
Labor Department	5, XLII
Benefits Review Board	20, VII
Employee Benefits Security Administration	29, XXV
Employees' Compensation Appeals Board	20, IV
Employment and Training Administration	20, V
Employment Standards Administration	20, VI
Federal Acquisition Regulation	48, 29
Federal Contract Compliance Programs, Office of	41, 60
Federal Procurement Regulations System	41, 50
Labor-Management Standards, Office of	29, II, IV
Mine Safety and Health Administration	30, I
Occupational Safety and Health Administration	29, XVII
Public Contracts	41, 50

Agency	CFR Title, Subtitle or Chapter
Secretary of Labor, Office of	29, Subtitle A
Veterans' Employment and Training Service, Office of the Assistant Secretary for	41, 61; 20, IX
Wage and Hour Division	29, V
Workers' Compensation Programs, Office of	20, I
Labor-Management Standards, Office of	29, II, IV
Land Management, Bureau of	43, II
Legal Services Corporation	45, XVI
Library of Congress	36, VII
Copyright Office	37, II
Copyright Royalty Board	37, III
Local Television Loan Guarantee Board	7, XX
Management and Budget, Office of	5, III, LXXVII; 14, VI; 48, 99
Marine Mammal Commission	50, V
Maritime Administration	46, II
Merit Systems Protection Board	5, II
Micronesian Status Negotiations, Office for	32, XXVII
Mine Safety and Health Administration	30, I
Minerals Management Service	30, II
Minority Business Development Agency	15, XIV
Miscellaneous Agencies	1, IV
Monetary Offices	31, I
Morris K. Udall Scholarship and Excellence in National Environmental Policy Foundation	36, XVI
National Aeronautics and Space Administration	5, LIX; 14, V
Federal Acquisition Regulation	48, 18
National Agricultural Library	7, XLI
National Agricultural Statistics Service	7, XXXVI
National and Community Service, Corporation for	45, XII, XXV
National Archives and Records Administration	5, LXVI; 36, XII
Information Security Oversight Office	32, XX
National Bureau of Standards	15, II
National Capital Planning Commission	1, IV
National Commission for Employment Policy	1, IV
National Commission on Libraries and Information Science	45, XVII
National Council on Disability	34, XII
National Counterintelligence Center	32, XVIII
National Credit Union Administration	12, VII
National Crime Prevention and Privacy Compact Council	28, IX
National Drug Control Policy, Office of	21, III
National Foundation on the Arts and the Humanities	45, XI
National Highway Traffic Safety Administration	23, II, III; 49, V
National Imagery and Mapping Agency	32, I
National Indian Gaming Commission	25, III
National Institute for Literacy	34, XI
National Institute of Standards and Technology	15, II
National Labor Relations Board	5, LXI; 29, I
National Marine Fisheries Service	50, II, IV, VI
National Mediation Board	29, X
National Oceanic and Atmospheric Administration	15, IX; 50, II, III, IV, VI
National Park Service	36, I
National Railroad Adjustment Board	29, III
National Railroad Passenger Corporation (AMTRAK)	49, VII
National Science Foundation	5, XLIII; 45, VI
Federal Acquisition Regulation	48, 25
National Security Council	32, XXI
National Security Council and Office of Science and Technology Policy	47, II
National Telecommunications and Information Administration	15, XXIII; 47, III
National Transportation Safety Board	49, VIII
National Weather Service	15, IX
Natural Resources Conservation Service	7, VI
Navajo and Hopi Indian Relocation, Office of	25, IV
Navy Department	32, VI
Federal Acquisition Regulation	48, 52

Agency	CFR Title, Subtitle or Chapter
Neighborhood Reinvestment Corporation	24, XXV
Northeast Interstate Low-Level Radioactive Waste Commission	10, XVIII
Nuclear Regulatory Commission	5, XLVIII; 10, I
Federal Acquisition Regulation	48, 20
Occupational Safety and Health Administration	29, XVII
Occupational Safety and Health Review Commission	29, XX
Offices of Independent Counsel	28, VI
Oklahoma City National Memorial Trust	36, XV
Operations Office	7, XXVIII
Overseas Private Investment Corporation	5, XXXIII; 22, VII
Patent and Trademark Office, United States	37, I
Payment From a Non-Federal Source for Travel Expenses	41, 304
Payment of Expenses Connected With the Death of Certain Employees	41, 303
Peace Corps	22, III
Pennsylvania Avenue Development Corporation	36, IX
Pension Benefit Guaranty Corporation	29, XL
Personnel Management, Office of	5, I, XXXV; 45, VIII
Federal Acquisition Regulation	48, 17
Federal Employees Group Life Insurance Federal Acquisition Regulation	48, 21
Federal Employees Health Benefits Acquisition Regulation	48, 16
Pipeline and Hazardous Materials Safety Administration	49, I
Postal Rate Commission	5, XLVI; 39, III
Postal Service, United States	5, LX; 39, I
Postsecondary Education, Office of	34, VI
President's Commission on White House Fellowships	1, IV
Presidential Documents	3
Presidio Trust	36, X
Prisons, Bureau of	28, V
Procurement and Property Management, Office of	7, XXXII
Productivity, Technology and Innovation, Assistant Secretary	37, IV
Public Contracts, Department of Labor	41, 50
Public and Indian Housing, Office of Assistant Secretary for	24, IX
Public Health Service	42, I
Railroad Retirement Board	20, II
Reclamation, Bureau of	43, I
Refugee Resettlement, Office of	45, IV
Regional Action Planning Commissions	13, V
Relocation Allowances	41, 302
Research and Innovative Technology Administration	49, XI
Rural Business-Cooperative Service	7, XVIII, XLII
Rural Development Administration	7, XLII
Rural Housing Service	7, XVIII, XXXV
Rural Telephone Bank	7, XVI
Rural Utilities Service	7, XVII, XVIII, XLII
Saint Lawrence Seaway Development Corporation	33, IV
Science and Technology Policy, Office of	32, XXIV
Science and Technology Policy, Office of, and National Security Council	47, II
Secret Service	31, IV
Securities and Exchange Commission	17, II
Selective Service System	32, XVI
Small Business Administration	13, I
Smithsonian Institution	36, V
Social Security Administration	20, III; 48, 23
Soldiers' and Airmen's Home, United States	5, XI
Special Counsel, Office of	5, VIII
Special Education and Rehabilitative Services, Office of	34, III
State Department	22, I; 28, XI
Federal Acquisition Regulation	48, 6
Surface Mining and Reclamation Appeals, Board of	30, III
Surface Mining Reclamation and Enforcement, Office of	30, VII
Surface Transportation Board	49, X
Susquehanna River Basin Commission	18, VIII

List of CFR Sections Affected

All changes in this volume of the Code of Federal Regulations that were made by documents published in the FEDERAL REGISTER since January 1, 2001, are enumerated in the following list. Entries indicate the nature of the changes effected. Page numbers refer to FEDERAL REGISTER pages. The user should consult the entries for chapters and parts as well as sections for revisions.

For the period before January 1, 2001, see the "List of CFR Sections Affected", 1949–1963, 1964–1972, 1973–1985, and 1986—2000 published in 11 separate volumes.

2001

21 CFR

66 FR Page

Chapter I

803.3 (f) and (r)(2)(ii) revised; second (ee) redesignated as (ff) 23156
803.10 (b) revised; (c)(5) removed 23157
803.17 (b)(3) revised 23157
803.19 (a)(2) revised 23157
803.20 (a) introductory text, (2) and (c)(1) revised 23157
803.50 (b)(1)(i) and (2) revised 23157
803.52 (d)(1), (f)(11)(i) and (ii) revised 23157
803.58 (b)(3) removed; (b)(4), (5) and (6) redesignated as (b)(3), (4) and (5) 23157
807 Authority citation revised 5466, 59159
807.3 (b) and (r) revised 59159
807.20—807.39 (Subpart B) Heading revised 59160
807.20 Heading revised, eff. 4–4–01; (d) added, eff. 1–21–03 5466
(a) revised 59160
807.25 (a) amended 59160
807.40 Revised 59160
807.65 Introductory text revised 59160
809.10 Regulation at 65 FR 18234 eff. date delayed to 6–8–01 17359
809.20 (b) revised 31165
809.40 Regulation at 65 FR 18234 eff. date delayed to 6–8–01 17359
862.1117 Added 12734
862.1190 (b) revised 38787
862.1210 (b) revised 38787
862.1255 (b) revised 38787

21 CFR—Continued

66 FR Page

Chapter I—Continued

862.1290 (b) revised 38787
862.1305 (b) revised 38787
862.1320 (b) revised 38787
862.1365 (b) revised 38787
862.1380 (b) revised 38787
862.1420 (b) revised 38788
862.1470 (b) revised 38788
862.1490 (b) revised 38788
862.1515 (b) revised 38788
862.1565 (b) revised 38788
862.1575 (b) revised 38788
862.1640 (b) revised 38788
862.1670 (b) revised 38788
862.1720 (b) revised 38788
862.1815 (b) revised 38788
862.2050 (b) revised 38788
862.2100 (b) revised 38788
862.2230 (b) revised 38788
862.2310 (b) revised 38788
862.2320 (b) revised 38788
862.2485 (b) revised 38788
862.2720 (b) revised 38788
862.2800 (b) revised 38788
862.2920 (b) revised 38788
864.1850 (b) revised 38789
864.2220 (a) revised 27024
(b) revised 38789
864.2240 (b) revised 38789
864.2260 (b) revised 38789
864.2360 (b) revised 38789
864.2800 (b) revised 38789
864.2875 (b) revised 38789
864.3010 (b) revised 38789
864.3250 Regulation at 65 FR 18234 eff. date delayed to 6–8–01 17359
864.3260 Regulation at 65 FR 18234 eff. date delayed to 6–8–01 17359

21 CFR—Continued

21 CFR—Continued

21 CFR—Continued

66 FR Page

Chapter I—Continued

21 CFR—Continued

66 FR Page

Chapter I—Continued

21 CFR—Continued

21 CFR—Continued

21 CFR—Continued

Chapter I—Continued

Section	Action	66 FR Page
882.4600	(b) revised	38808
882.4900	(b) revised	38808
884.1550	(b) revised	38808
884.1640	(b)(2) revised	38808
884.1690	(b)(2) revised	38808
884.1700	(b)(2) revised	38808
884.1720	(b)(2) revised	38808
884.1730	(b)(2) revised	38809
884.2730	Added; eff. 4–9–01	14076
884.2900	(b) revised	38809
884.2980	(a)(2) revised	38809, 46953
884.2982	(a)(2) revised	38809, 46953
884.4520	(b) revised	38809
884.4530	(b)(2) revised	38809
884.5150	(b) revised	38809
884.5425	(b)(1) revised	38809
884.5900	(b)(2) revised	38809
884.5920	(b) revised	38809
884.6190	(b) revised	38809
886.1040	(b) revised	38809
886.1050	(b) revised	38809
886.1070	(b) revised	38810
886.1090	(b) revised	38810
886.1140	(b) revised	38810
886.1150	(b) revised	38810
886.1160	(b) revised	38810
886.1170	(b) revised	38810
886.1190	(b) revised	38810
886.1200	(b) revised	38810
886.1250	(b) revised	38810
886.1270	(b) revised	38810
886.1290	(b) revised	38810
886.1320	(b) revised	38810
886.1330	(b) revised	38810
886.1340	(b) revised	38810
886.1375	(b) revised	38810
886.1380	(b) revised	38810
886.1390	(b) revised	38811
886.1395	(b) revised	38811
886.1400	(b) revised	38811
886.1405	(b) revised	38811
886.1410	(b) revised	38811
886.1415	(b) revised	38811
886.1420	(b) revised	38811
886.1425	(b) revised	38811
886.1430	(b) revised	38811
886.1435	(b) revised	38811
886.1450	(b) revised	38811
886.1460	(b) revised	38811
886.1500	(b) revised	38811
886.1605	(b) revised	38811
886.1650	(b) revised	38812
886.1655	(b) revised	38812
886.1660	(b) revised	38812
886.1665	(b) revised	38812
886.1680	(b) revised	38812

21 CFR—Continued

Chapter I—Continued

Section	Action	66 FR Page
886.1690	(b) revised	38812
886.1700	(b) revised	38812
886.1750	(b) revised	38812
886.1760	(b) revised	38812
886.1770	(b) revised	38812
886.1790	(b) revised	38812
886.1800	(b) revised	38812
886.1810	(b) revised	38812
886.1840	(b) revised	38812
886.1860	(b) revised	38812
886.1870	(b) revised	38813
886.1880	(b) revised	38813
886.1905	(b) revised	38813
886.1910	(b) revised	38813
886.1945	(b) revised	38813
886.3200	(b) revised	38813
886.3920	(a) amended	18542
886.4230	(b) revised	38813
886.4250	(b) revised	38813
886.4335	(b) revised	38813
886.4350	(b) revised	38813
886.4360	(b) revised	38813
886.4445	(b) revised	38813
886.4570	(b) revised	38813
886.4770	(b) revised	38813
886.4855	(b) revised	38814
886.5120	(b) revised	38814
886.5420	(b) revised	38814
886.5540	(b) revised	38814
886.5600	(b) revised	38814
886.5800	(b) revised	38814
886.5810	(b) revised	38814
886.5820	(b) revised	38814
886.5840	(b) revised	38814
886.5842	(b) revised	38814
886.5844	(b) revised	38814
886.5870	(b) revised	38814
886.5900	(b) revised	38814
886.5910	(b) revised	38814
886.5915	(b) revised	38815
888.1100	(b)(2) revised	38815
888.1520	(b) revised	38815
888.3000	(b) revised	38815
888.3070	Revised	28053
888.3670	Added	12737
888.4150	(b) revised	38815
888.4200	(b) revised	38815
888.4210	(b) revised	38815
888.4220	(b) revised	38815
888.4230	(b) revised	38815
888.4300	(b) revised	38815
888.4540	(b) revised	38815
888.4600	(b) revised	38815
888.4800	(b) revised	38815
888.5850	(b) revised	38815
888.5890	(b) revised	38815

21 CFR—Continued

66 FR Page

Chapter I—Continued

888.5940 (b) revised 38815
888.5960 (b) revised 38816
888.5980 (b) revised 38816
890.1575 (b) revised 38816
890.1600 (b) revised 38816
890.1615 (b) revised 38816
890.3025 (b) revised 38816
890.3075 (b) revised 38816
890.3100 (b) revised 38816
890.3150 (b) revised 38816
890.3175 (b) revised 38816
890.3410 (b) revised 38816
890.3420 (b) revised 38816
890.3475 (b) revised 38816
890.3490 (b) revised 38816
890.3520 (b) revised 38817
890.3640 (b) revised 38817
890.3665 (b) revised 38817
890.3675 (b) revised 38817
890.3700 (b) revised 38817
890.3750 (b) revised 38817
890.3790 (b) revised 38817
890.3825 (b) revised 38817
890.3910 (b) revised 38817
890.3920 (b) revised 38817
890.3940 (b) revised 38817
890.5050 (b) revised 38817
890.5125 (b) revised 38818
890.5350 (b) revised 38818
890.5370 (b) revised 38818
890.5380 (b) revised 38818
890.5410 (b) revised 38818
890.5660 (b) revised 38818
890.5730 (b) revised 38818
890.5765 (b) revised 38818
890.5925 (b) revised 38818
890.5940 (b) revised 38818
890.5950 (b) revised 38818
890.5975 (b) revised 38818
892.1100 (b) revised 38818, 46953
892.1110 (b) revised 38818, 46953
892.1130 (b) revised 38818
892.1300 (b) revised 38818
892.1370 (b) revised 38818
892.1380 (b) revised 38819
892.1400 (b) revised 38819
892.1420 (b) revised 38819
892.1640 (b) revised 38819
892.1650 (b) revised 57369
892.1700 (b) revised 38819
892.1760 (b) revised 38819
892.1770 (b) revised 38819
892.1830 (b) revised 38819
892.1840 (b) revised 38819
892.1880 (b) revised 38819
892.1920 (b) revised 38819
892.1940 (b) revised 38819
892.1950 (b) revised 38819
892.5740 (b) revised 38819
892.5780 (b) revised 38819
1240 Nomenclature change 56035
1250 Nomenclature change 56035
1271 Added, eff. in part 4–4–01 5466
1271.3 (d)(2) added, eff. 1–21–03 5467

2002

21 CFR

67 FR Page

Chapter I

814.20 OMB number 9587
814.39 OMB number 9587
814.84 OMB number 9587
821.1 (a) and (b) revised; (c) removed; (d) and (e) redesignated as (c) and (d) (OMB number pending); eff. 5–9–02 5951
821.2 (d) removed (OMB number pending); eff. 5–9–02 5951
821.3 (b) and (f) revised (OMB number pending); eff. 5–9–02 5951
821.20 Revised (OMB number pending); eff. 5–9–02 5951
821.25 (a)(2) introductory text, (iii), (3) introductory text and (iv) revised (OMB number pending); eff. 5–9–02 5951
821.30 (b)(3) and (c)(1)(ii) amended (OMB number pending); eff. 5–9–02 5952
821.55 (a) and (b) redesignated as (b) and (c); new (a) added (OMB number pending); eff. 5–9–02 5952
822 Added 38887
862.1 (d) added 58329
862.1235 Added 58329
862.1678 Added 58329
864.5220 Revised 1607
868.1 (e) added 76681
868.2375 (a) revised 46852
868.2377 Added 46852
868.2480 (b) revised 76681
868.2500 Added 76681
872.5570 Added 68512
874.1 (e) added 67790
874.3850 Revised 20894
874.3950 Added 67790
876.1300 Added 3433
878.1 (e) added 77676
878.4840 Added 77676
880.6991 Added 69121
880.6992 Added 69121
884.5320 (c) revised 40849

21 CFR—Continued

67 FR Page

Chapter I—Continued

888.3027 Revised 46855
888.3310 Revised 21173
900.2 Introductory text and (i) revised; (zz), (aaa) and (bbb) added; eff. 5–7–02 5467
900.20—900.25 (Subpart C) Added; eff. 5–7–02 5467
1020.33 OMB number 9587
1040.20 OMB number 9587

2003

21 CFR

68 FR Page

Chapter I

807.20 Regulation at 66 FR 5466 eff. date delayed to 1–21–04 2690
808 Nomenclature change 24879
812 Nomenclature change 24879
814 Nomenclature change 24879
862.3080 Added 40127
864.9245 (b) and (c) redesignated as (c) and (d); new (b) added; new (c) and (d) revised 9532
866.1 (e) added; eff. 5–6–03 5827
866.1645 Added; eff. 5–6–03 5827
866.3610 Added 62008
866.3940 Added 61745
870.1 (e) added 61344
870.1025 Revised 61344
870.5310 Added 61344
872.1 (e) added 19737
872.2050 Added 67367
872.2060 Added 67367
872.3660 (b) revised 19738
872.3661 Added 19738
878.4493 (b) revised 32984
878.4495 (b) revised 32984
878.4830 (b) revised 32984
878.5000 (b) revised 32984
878.5010 (b) revised 32984
878.5020 (b) revised 32985
878.5030 (b) revised 32985
878.5035 (b) revised 32985
882.1 (e) added 70436
882.5975 Added 70436
884.1 (e) added 44415
884.2990 Added 44415
886.1500 (b) correctly added; CFR correction 49351
888.1 (e) added 14137
888.3045 Added 32636
888.3535 Added 14137
888.3565 Added 14137
1030 Nomenclature change 24879
1240 Nomenclature change 24879
1240.63 Added; interim 62368

21 CFR—Continued

68 FR Page

Chapter I—Continued

1250 Nomenclature change 24879
1271.3 Regulation at 66 FR 5466 eff. date delayed to 1–21–04 2690

2004

21 CFR

69 FR Page

Chapter I

800.55 (g)(4) revised 17292
801.430 (e)(1) table revised; eff. 2-27-06 52171
803.18 (e) revised 11311
806.10 (f) revised 11311
807.22 (b) revised 11311
(a) correctly revised 18473
(a) corrected 25489
807.25 (a), (f)(1), (2), (6) and (7) revised 11312
807.26 Revised 11312
807.30 Revised 11312
807.35 Revised 11312
807.37 Revised 11313
808 Nomenclature change 13717
812 Nomenclature change 13717
814 Nomenclature change 13717
814.39 (e) and (f) revised 11313
820 Authority citation revised 29829
820.1 (a)(1) amended; (b) revised; eff. 5–25–05 29829
820.198 (d) revised 11313
820.200 (c) revised 11313
860 Nomenclature change 13717
861 Nomenclature change 13717
862.1055 Added 68255
862.3840 Added 58259
864.1 (d) added 12273
864.7280 Added 12273
866.3050 Added 56936
866.6020 Added 26038
868.1120 (c) revised 34920
870.3300 Revised 77899
870.4320 (c) revised 34920
870.5310 Corrected 10615
872.3060 Revised 51766
872.3630 Added 26304
872.3640 Revised 26304
872.3710 Revised 51766
876.1 (e) added 77623
876.5020 Added 77623
878.4025 Added 48148
880.1 (e) added 71704
880.6300 Added 71704
882.1790 (c) revised 34920
882.4700 (a) heading and section revised 10332
882.5950 Revised 77900

21 CFR—Continued

69 FR Page

Chapter I—Continued

884.6200 Added 77624
888.3410 Revised 59134
895 Nomenclature change 13717
900 Nomenclature change 13717
1002.3 Revised 17292
1005.3 Revised 11314
1010 Nomenclature change 13717
1030 Nomenclature change 13717
1240 Nomenclature change 13717
1250 Nomenclature change 13717
1270.3 (j) introductory text revised; eff. 5-25-05 68680
1271 Technical correction 5272
Authority citation revised 29829
1271.1 Amended; eff. 5-25-05 29829
1271.3 (d)(2) introductory text amended; (d)(2)(vi) and (vii) revised; (d)(2)(viii) added; interim 3826
(h) through (x) added 29829
(c) and (d) revised; (y) through (ll) added; eff. 5-25-05 68680
1271.10 (a)(3) revised; eff. 5-25-05 68681
1271.22 Revised; eff. 5-25-05 68681
1271.45—1271.90 (Subpart C) Added; eff. 5-25-05 29830
1271.45 (a) amended; eff. 5-25-05 68681
1271.145—1271.320 (Subpart D) Added; eff. 5-25-05 68681
1271.330—1271.370 (Subpart E) Added; eff. 5-25-05 68686
1271.390—1271.440 (Subpart F) Added; eff. 5-25-05 68687

2005

21 CFR

70 FR Page

Chapter I

803 Revised; eff. 7-13-05 9519
Effective date confirmation at 70 FR 9519 confirmed 34652
803.55 (b)(9) and (10) stayed 9519
803.58 Stayed 9519
807.90 (a)(2) amended 14986

21 CFR—Continued

70 FR Page

Chapter I—Continued

822.8 Amended 14986
862.2570 Added; eff. 4-22-05 11868
866.3210 Redesignated from 866.3610 53069
862.3360 Added; eff. 4-22-05 11867
866.3610 Redesignated as 866.3210 53069
866.4070 Added 49863
866.4700 Added; eff. 4-22-05 14534
866.5900 Added 61738
866.6030 Added 57749
872.3930 Revised 21949
872.5580 Added 55028
878.4410 Added 67355
1020.30 Revised; eff. 6-10-06 34028
1020.31 Revised; eff. 6-10-06 34036
1020.32 Revised; eff. 6-10-06 34039
1020.33 (h)(2) revised; eff. 6-10-06 34042
1240.62 (c)(1)(i), (ii), (v) and (2) amended 48073
1271.55 (a)(1) revised; interim 29952
1271.80 (b) revised; interim 29952
1271.90 (a)(3) introductory text and (b) revised; (a)(4) added; interim 29952
1271.290 (c) amended; interim 29952
1271.370 (b)(4) revised; interim 29952

2006

(Regulations published from January 1, 2006, through April 1, 2006)

21 CFR

71 FR Page

Chapter I

803.12 (c) revised 1488
814.126 (b)(1)(iv) amended 16228
820.198 (a)(3) amended 16228
866 Technical correction 10433
866.3310 Added 6679
866.3332 Added; eff. 4-21-06 14379
870.2855 Added 7871
1271.75 (d)(1) correctly amended 14798

○